PARRY'S

VALUATION

AND

INVESTMENT TABLES

by

A.W. DAVIDSON

BSc FRICS

*Sometime: Visiting Professor, University of Westminster,
Head of Valuation Section, University of Reading and
Head of Valuation Department, College of Estate
Management.*

THE ESTATES GAZETTE
1 PROCTER STREET, LONDON, WC1V 6EU

Twelfth Edition 2002

Reprinted 2003

Reprinted 2004

ISBN 0 7282 0368 5

© College of Estate Management, 2002

Printed and bound in Great Britain by
The Lavenham Press Ltd, Lavenham, Suffolk

FOREWORD

I doubt very much whether, in 1913, Richard Parry would have anticipated his book of valuation tables still to be in print nearly 90 years later and to be in its 12th edition. That it is so is a remarkable tribute to Richard Parry and his vision for the need for such a publication. Richard Parry became the first Principal of The College of Estate Management (to whom he granted the copyright) in 1919. This 12th edition is also a remarkable tribute to its publishers, Estates Gazette.

Estates Gazette has had a close relationship with the College and its forebears for more than 100 years. We have always worked together well on the publication of professional texts and look forward to that relationship continuing.

In an era when calculators and computers might have been thought to have superseded a book of valuation tables, it is pleasing to report that the profession has indicated its wish to have a new edition. This 12th edition deals with not only updated life tables but also quarterly in advance valuation tables.

The College is indebted to Alick Davidson for his meticulous care and attention to detail which has enabled this latest edition to be published. Alick Davidson was head of the Valuation Department at The College of Estate Management and produced the 9th, 10th and 11th editions. The property profession is most fortunate in having such a distinguished academic, who also has considerable professional experience, prepared to devote so much time to this task.

I commend this publication to the profession.

Peter Goodacre RD MSc FRICS FCIOB
Principal of The College of Estate Management

PREFACE

TO THE TWELFTH EDITION

———

THE VALUATION TABLES were prepared for the first time in 1913 by the late Richard Parry. They were introduced at a time when considerable controversy existed over the use of dual rate tables of Years' Purchase in preference to single rate tables for the valuation of terminable incomes. The decision to include both tables in the first edition and to print them on the same page for ease of comparison was one of the main reasons for the general acceptance of the dual rate principle in valuation theory.

The purpose of this book is to provide a comprehensive set of tables available in one volume to meet the requirements of current practice. Certain tables have been added and others omitted over the years and in the eighth edition new tables were added dealing with the various effects of taxation.

In this edition, Years' Purchase tables based on the assumption that income is received quarterly in advance have been extended and given more prominence to facilitate their use. The traditional approach to Years' Purchase, where income is deemed to be received annually in arrear is retained. In this way, a more comprehensive approach is achieved where both sets of tables are available within one volume.

Two new tables providing for the conversion of Nominal Yields to Effective Yields (and vice versa) have been introduced. In addition, there has been a demand for a much wider range of tax rates from surveyors in countries where different tax regimes exist and lower rates of tax have been included.

It has been a tradition in past editions to provide tables in response to the demand for them and not to advocate which tables should be used in a particular set of circumstances and I have endeavoured to follow this principle in the present edition. In practice today, numerous calculations are required for a variety of purposes which often justify more than one approach. With this in mind, the Internal Rates of Return (IRR) tables which may be used for property investment analysis have been retained in a modified form. Using the IRR tables, both growth and non-growth scenarios can be analysed to provide a more detailed appraisal and form the basis for more in-depth investment advice. With a comprehensive set of tables available, users of this book are free to decide which table to use according to the circumstances prevailing and the assumptions to be made.

In a book of tables, manual typesetting is always a possible source of error and to avoid this the new methods which were introduced for most of the tables in the ninth edition have been extended. These methods involved the use of different computers and filmsetters for which a considerable number of programs were written for computation, tabulation and pagination. For example, ICL 4100 and Elliot 903 computers were used by Elliot Automation Ltd to punch nearly three miles of paper tape to drive a Photon 713 filmsetter for the production of many tables in the ninth edition.

Subsequently, an IBM 360/40 computer and Fototronic CRT filmsetter (capable of transmitting data at speeds varying up to 60,000 characters per second) and APS5 phototypesetter were used by C.R. Barber & Partners to produce the calculations and new tables for the two previous editions. Keyword Typesetting Services used Microsoft Excel 2000 for the new tables and extensions included in this edition.

All the tables in this edition have been typeset by these methods with the exception of the Internal Rates of Return without growth, the English Life Tables and decimals of £1 which were all computed elsewhere and typeset in the conventional way.

Metric Conversion Tables are included in this volume in one of the sections coloured green for ease of reference. The present life tables are now based on the English Life Tables No. 15 and the years' purchase for life on a dual rate principle is considered in the section covering the construction and use of the tables.

I am indebted to the late Richard Parry and to those who followed him for their work in earlier editions. I am particularly grateful to Peter Sayers, MSc FIA, for his advice on many aspects of this book. His contribution in the section dealing with the Life Tables has been invaluable. My warmest thanks are due to Colin Greasby for all the help and advice he has given on publishing the book, to Stephen Johnson for writing the programs required for this edition and Lin Blagrove for overseeing many production tasks. I should also like to thank Anthony Banfield for his interest and for our helpful discussions concerning the content of this book.

In a volume of tables, checking the page proofs is a difficult and tedious task and I am grateful to Patricia Brown for her diligent work in this respect. My wife undertook many unforeseen tasks and her encouragement during the months of preparation was an immense help.

The tables produced on the 900 Series, the 4100 Series, the 360/40 and the 4341 Series computers were calculated to a degree of accuracy of 0.50 of the last decimal retained where the printed multipliers do not exceed seven figures. It will be realised that the degree of accuracy will depend on the numbers of digits held by the computer. For example, in all the calculations on the IBM 360/40 computer, including the higher rates used in the Amount of £1 and the Amount of £1 per annum tables, a "word-length" equivalent to 16 decimal digits was used. The new tables in this edition were produced on a desktop PC running Windows 98SE and Microsoft Excel 2000.

Although accuracy is a matter of degree and in a book of this nature can never be absolutely guaranteed, it is believed that the system of computer typesetting and the extreme care exercised in the subsequent checking will enable the user to have considerable confidence in the level of accuracy achieved.

The English Life Tables on pages 306 and 307 were reproduced from the English Life Tables No. 15. Permission was granted for the reproduction of these tables and the source: 'English Life Tables', National Statistics, © Grown Copyright 2000, is acknowledged.

Finally I should like to record my sincere thanks to Peter Goodacre for his help and interest in the preparation of this book.

ALICK DAVIDSON

October 2001

The majority of these tables were calculated on the 900 series, the 4100 series, the 360/40 series and the 4341 series computers to a degree of accuracy of 0.50 of the last decimal retained where the printed multipliers do not exceed seven figures. It will be realised that the degree of accuracy will depend on the number of digits held by the computer. For example, in all the calculations on the IBM 360/40 computer, including the higher rates used in the Amount of £1 and the Amount of £1 per annum tables, a "word-length" equivalent to 16 decimal digits was used.

All the tables calculated on the computers mentioned above were printed using a computer-linked Photon typesetter, a Fototronic filmsetter or an APS5 phototypesetter. The new tables in this edition were produced on a desktop PC running Windows 98SE and Microsoft Excel 2000. The above approach should ensure that a high level of accuracy has been achieved. Further information is given in the preface to this edition. It will be appreciated that absolute accuracy can never be guaranteed in a book of this nature.

TABLE OF CONTENTS

PAGE
COLOUR   PAGE

The Construction and Use of the Tables   ix

TABLES WITHOUT ALLOWANCE FOR TAX

Green	Years' Purchase (Single Rate) Quarterly in Advance	1
Green	Years' Purchase (Dual Rate) Quarterly in Advance	15
White	Yield Conversion Tables	
	(Nominal and Effective Yields)	29
White	Years' Purchase in Perpetuity	35
White	Years' Purchase (Single Rate)	39
White	Years' Purchase (Dual Rate)	53
White	Years' Purchase of a Reversion to a Perpetuity	73
White	Present Value of £1	89
White	Annual Sinking Fund	111
White	Amount of £1	123
White	Amount of £1 per annum	141
White	Annuity £1 will purchase (Single Rate)	157
White	Internal Rates of Return (no growth)	179
White	Internal Rates of Return (growth)	195

CONVERSION TABLES

Green	Decimals of £1	240
Green	Feet–Yards/Metres	242
Green	Square Feet–Square Yards/Square Metres	248
Green	Acres/Hectares	254
Green	Cubic Feet/Cubic Metres	258
Green	Price per Square Foot to Price per Square Metre	260
Green	Price per Cubic Foot to Price per Cubic Metre	262
Green	Price per Acre to Price per Hectare	264

Life Tables

White	Years' Purchase for a Single Life	267
White	Years' Purchase for the Joint of Continuation of 2 Lives	277
White	Rule for Years' Purchase for the Longer of 2 Lives	295
White	Annual Premium to secure £1 at death	299
White	English Life Tables No. 15	305
White	Tax Adjustment Factors	309

Tables With Allowance for Tax

Years' Purchase Quarterly in Advance Tables

Green	Years' Purchase (Dual Rate) — tax at 10%	313
Green	Years' Purchase (Dual Rate) — tax at 20%	327
Green	Years' Purchase (Dual Rate) — tax at 25%	341
Green	Years' Purchase (Dual Rate) — tax at 30%	355
Green	Years' Purchase (Dual Rate) — tax at 35%	369
Green	Years' Purchase (Dual Rate) — tax at 40%	383
Green	Years' Purchase (Dual Rate) — tax at 50%	397

Years' Purchase Annually in Arrear Tables

Yellow	Years' Purchase (Dual Rate) — tax at 10%	411
Yellow	Years' Purchase (Dual Rate) — tax at 20%	431
Yellow	Years' Purchase (Dual Rate) — tax at 25%	451
Yellow	Years' Purchase (Dual Rate) — tax at 30%	471
Yellow	Years' Purchase (Dual Rate) — tax at 35%	491
Yellow	Years' Purchase (Dual Rate) — tax at 40%	511
Yellow	Years' Purchase (Dual Rate) — tax at 50%	531

THE CONSTRUCTION AND USE
of the
TABLES

INTRODUCTION

In valuation practice and other studies in land use where some aspect of financial analysis is involved, compound interest calculations of a tedious nature are frequently required. Valuation and financial tables may be used to reduce the time-consuming element involved in these calculations.

Due to the various assumptions that may be made in valuation theory, financial analysis and other studies, it is desirable that a wide range of tables should be available. With this approach, a free choice may be made and valuers can select the appropriate table to be used for a particular problem having regard to the circumstances involved.

A number of significant changes have occurred since the first valuation tables were produced; new tables have been introduced and others discarded as the demand for them changes. In this way the book has kept up with the requirements of modern practice.

In this edition a number of changes have been introduced.

1. The Years' Purchase Tables based on the assumption that income is received quarterly in advance have been extended and given more prominence to facilitate their use in response to the increased interest in this concept and recent discussions by investment surveyors concerning their use. However, as many valuers still use the more traditional approach, years' purchase figures based on incomes received annually in arrear still form an important section in this volume. This means that a more comprehensive coverage is available by including both bases. To avoid confusion, the quarterly in advance single and dual rate years' purchase tables have all been printed in separate sections on green pages, so that these tables can be easily identified.

However, the much shorter Years' Purchase in perpetuity tables for both quarterly in advance and annually in arrear have been printed on the same page in the white section for ease of comparison.

In a volume of this nature, the objective is to provide the necessary tables to enable a wide choice to be made. It is not considered appropriate to suggest which table should be used for a particular valuation where alternative approaches are available. This should be decided by reference to text books and relevant articles in professional journals. Where a choice exists, this must be determined by the user having regard to experience, methods of analysis and the circumstances involved.

2. The concept of valuing incomes on a quarterly in advance basis introduces the need to quantify the difference between nominal yields where incomes are considered annually in arrear and effective yields where incomes are on a quarterly in advance basis. Two tables have now been introduced to provide for the conversion of Nominal Yields to Effective Yields and vice versa.

3. There has been a demand for a much wider range of tax rates from surveyors in countries where different tax regimes exist. Lower rates of tax have now been included for both sets of Years' Purchase dual rate tax adjusted tables.

4. The Tables of Internal Rates of Return (IRR) which are included in this volume, enable freehold investment transactions to be analysed where certain variables are known (see pages xx to xxiii).

The IRR (without growth) tables will indicate the internal rate of return that has been achieved for a completed or a prospective transaction. The years to review or reversion have now been restricted to 10 years to match those available in the 'growth table' and to make room for other changes.

The second IRR Table (with growth) is included to allow further analysis to be undertaken showing the internal rates of return which will be obtained if anticipated growth of either 3%, 5% or 7% is achieved over the period to the next rent review or reversion. In this way, a matrix of returns can be considered to show the yield which may flow from the investment under different assumptions. By considering which assumption is most likely, more analytical information is available for investors. Furthermore, a range of returns can be presented according to the likelihood or probability of more optimistic or pessimistic estimates of growth being achieved.

5. Other tables have been extended or updated and some have been omitted. Far more rates of interest are now available in the PV of £1 and the Single Rate Annuity £1 will Purchase tables.

The Mortgage instalment table and the Dual Rate Annuity tables have not been included. Mortgage instalments and dual rate annuity figures can be obtained directly from the single rate annuity and dual rate years' purchase tables respectively (see pages xii to xiv).

6. The Present Value of £1 adjusted for tax table was introduced in the 1960s mainly to deal with the concept of tax-free capital appreciation which existed in the pre-capital gains tax area. This table has now been omitted. Tax adjusted Present Value figures can be obtained if required by adjusting the yield (using the appropriate Net Tax Adjustment Factor on page 311) and using the adjusted yield in the PV of £1 formula given below.

7. The Life tables have been recalculated to reflect the updated mortality statistics and these are now based on the English Life Tables No. 15. These are considered in more detail later in this chapter on pages (xxiv to xxvi) and an alternative dual rate approach is also discussed in this section.

CONSTRUCTION OF THE TABLES

A brief description of most of the tables included in this volume is given on the introductory page at the start of each table.

A. The formulae for the four basic compound interest tables are given below. Proof of these formulae may be found in most financial and valuation text books.

(a) Amount of £1 table (pages 123 to 140)

$$A = (1 + i)^n$$

(b) Present Value of £1 table (pages 89 to 110)

$$PV = (1 + i)^{-n}$$

(c) Amount of £1 per annum table (pages 141 to 156)

$$Apa = \frac{(1 + i)^n - 1}{i}$$

(d) Annual Sinking Fund table (pages 111 to 122)

$$Asf = \frac{i}{(1 + i)^n - 1}$$

In the above tables it is assumed that interest is added at the end of each year.

$n =$ the number of years

$i =$ the interest receivable on each £1. Therefore if the rate % $=$ R, then $i = R/100$

B. THE ANNUITY £1 WILL PURCHASE TABLE

This table shows the annuity or income that may be obtained over a given number of years at various rates of interest for each pound of capital outlay. The Annuity is receivable at the end of each year.

The table may be used to find the annual equivalent of a capital payment or receipt. Two separate concepts exist so that either the dual rate or single rate basis may be considered.

Construction of the table

Two elements are involved in the formula:

(i) the interest received on the capital outlay, i.e. the remunerative rate expected,

(ii) the replacement of capital over the given period, i.e. the annual sinking fund to replace the capital invested over this period.

Where the remunerative rate on each £1 $= i$
and the a.s.f. to replace £1 over the term $= s$
The formula for the Annuity £1 will purchase is:

$$i + s$$

(a) Single Rate Annuity table (pages 157 to 178)

With this table the sinking fund is deemed to accumulate at the same rate of interest as that required on the capital outlay. For this reason, only one rate of interest, the remunerative rate, is used in the single rate approach.

Example

Find the annual equivalent of a sum of £3,000 invested in a capital asset with an expected life of 12 years on an 8% basis.

Capital Outlay	£3,000
Annuity £1 will purchase for 12 yrs. at 8% (see page 168)	0·132695
Annuity £3,000 will purchase for 12 yrs. at 8%	= £398·085
Annual Equivalent of £3,000 over 12 yrs. at 8%	= £398·09

The construction of the single rate annuity table is illustrated by the following calculation:

As the required rate of return is 8%	$i = 0·08$
SF to replace £1 in 12 yrs. at 8% (see page 121)	$= 0·052695$
	$i + s = 0·132695$
The Annuity £1 will purchase for 12 yrs. at 8% (as above).	$= 0·132695$

(b) The Dual Rate Annuity

This concept is based on the same formula $(i + s)$ but two separate rates of interest are involved. The sinking fund is allowed to accumulate over the required period at a low rate, say 3% but the rate of interest allowed on capital outlay is the appropriate remunerative rate.

If the previous example is considered on a dual rate basis at 8% and 3%, the annual equivalent calculation would be:

Capital Outlay	£3,000
Annuity £1 will purchase for 12 yrs. at 8% and 3%	0·150462
	£451·386

There is no separate table for the dual rate annuity figures. However, the relevant annuity can be found from the Years' Purchase dual rate table (without tax). E.g. YP for 12 years at 8% and 3% (no tax) (see page 62) = 6·6462

Then Annuity £1 will Purchase dual rate

$$= \frac{1}{\text{YP dual rate}} = \frac{1}{6\cdot6462} = 0\cdot1504619$$

The basis for this approach will be apparent if the above formula $(i + s)$ is compared with the YP dual rate (without tax) formula on pages xv and xviii.

C. Monthly Mortgage Instalments

The monthly Mortgage Instalment is the monthly instalment to redeem each £1 of borrowed capital over a period of years at a given rate of interest. There is no separate table for these instalments, but the figures can be computed from the Single Rate Annuity £1 will Purchase table.

The monthly Mortgage Instalment formula is:

$$\frac{(i + s)}{12}$$

Example

Calculate the monthly instalment to redeem £60,000 borrowed on mortgage at 7%, if the loan is to be repaid over 20 years.

Annuity £1 will purchase for 20 years at 7% (page 166)   = 0·094393

Annuity £60,000 will purchase $= £60,000 \times 0.094393 = \underline{\underline{£5663.58}}$

Note, this is the *Annual* mortgage instalment to redeem £60,000
in 20 years at 7% $= £5663.58$

Monthly instalment $= \dfrac{5663.58}{12} = \underline{\underline{£471.97}}$

It will be seen from this calculation that compounding is on an annual basis. The annual instalment is converted to a monthly instalment by most Building Societies without allowing for compound interest each month.

D. TABLE OF TAX ADJUSTMENT FACTORS

It is sometimes necessary to calculate the true net income or return that will be obtained from an investment after tax has been deducted from the income received.

Alternatively, a true net income free of tax may be required from an investment; in this case, it will often be necessary to calculate the equivalent grossed up income which must be received before tax is deducted.

The factors in the table given on page 311 may be used in these calculations.

Net Adjustment Factor—T_N

If the net income from an investment is multiplied by the Net Adjustment Factor the product will be the true net income that will remain after deduction of tax at the appropriate rate.

Example

A net income of £1,250 p.a. is received. Calculate the true net income that will be available after deduction of income tax at 30%.

Net Income received	£1,250 p.a.
$T_N(30)$ (page 311)	0.7000
Net Income after tax	$\underline{\underline{£875}}$

Gross Adjustment Factor—T_G

If the true net income required from an investment after tax is multiplied by the Gross Adjustment Factor, the product will be the equivalent grossed up income that must be received before deduction of tax.

Example

A true net income of £1,400 p.a. which is free of tax is required from an investment. If the appropriate rate of tax is 45%, calculate the income that must be received before the deduction for tax is made in order to satisfy this requirement.

True net income required after tax	£1,400 p.a.
$T_G(45)$ (page 311)	1·81818
Net income required before tax	£2,545·45

The formulae for these factors are given below:

$$T_N = \frac{100 - Rate\ of\ Tax}{100}$$

$$T_G = \frac{100}{100 - Rate\ of\ Tax}$$

Further examples using these factors are given in subsequent sections.

E. YEARS' PURCHASE TABLES (Present Value of £1 per annum)

These are provided in separate groups of tables. Multipliers are now available within the same volume based on two different fundamental assumptions as follows:

(i) Tables using the more traditional approach where income is deemed to be received annually in arrear, and

(ii) Tables based on the assumption that income is received quarterly in advance.

Apart from the Years' Purchase in perpetuity figures, the Years' Purchase quarterly in advance tables are included in separate sections coloured green, to enable these tables to be clearly identified.

(i) YEARS' PURCHASE TABLES (annually in arrear)

(a) Years' Purchase in perpetuity (pages 35 to 37)

$$YP\ in\ perpetuity = \frac{1}{i} \quad or \quad \frac{100}{Rate\ \%}$$

(b) Years' Purchase for a given term: Single Rate (pages 39 to 51)

$$YP\ for\ n\ years = \frac{1 - (1 + i)^{-n}}{i}$$

(c) Years' Purchase for a given term: Dual Rate (pages 53 to 71)

$$YP\ for\ n\ years = \frac{1}{i + s}$$

where $s =$ annual sinking fund to replace £1 in n years at an appropriate accumulative rate of interest

and i is the return (yield on each £1) required by the investor.

Using the sinking fund formula, this may be extended as follows (where the sinking fund rate of accumulation $= a$):

$$YP \text{ (dual rate)} = \frac{1}{i + \dfrac{a}{(1 + a)^n - 1}}$$

(ii) YEARS' PURCHASE TABLES (quarterly in advance)

The Single Rate and Dual Rate tables considered in (b) and (c) below are included in the Years' Purchase Green Sections of this volume.

(a) Years' Purchase in perpetuity (Quarterly in Advance) pages 35 to 37

$$YP \text{ in perpetuity} = \frac{1}{4\left[1 - (1 + r)^{-1/4}\right]}$$

(b) Years' Purchase for a given term: Single Rate (Quarterly in Advance) pages 1 to 13

$$YP \text{ for } n \text{ years} = \frac{1 - (1 + r)^{-n}}{4\left[1 - (1 + r)^{-1/4}\right]}$$

(c) Years' Purchase for a given term: Dual Rate (Quarterly in Advance) pages 15 to 27

$$YP \text{ for } n \text{ years} = \frac{1}{4\left[1 - (1 + r)^{-1/4}\right] + \dfrac{4\left[1 - (1 + a)^{-1/4}\right]}{(1 + a)^n - 1}}$$

where $r =$ effective yield (see below)

$a =$ sinking fund rate of accumulation

Nominal and Effective Rates of Interest (i and r)

In all the tables other than the quarterly in advance tables, the yield at the top of each column is the nominal yield which represents the total interest (income) received during the year and assuming no reinvestment of any periodic payments received during that year.

In the quarterly in advance tables the yield at the top of each column is the effective yield. This is the income received for that year if periodic payments (quarterly in this case) received during the year are, or are deemed to be, re-invested at the time of their receipt at the same annual rate of interest.

In general, the relationship between nominal and effective yields will depend on the timing of the periodic payments and whether they are received in advance or in arrear.

Where income is received periodically in arrear, the effective yield (*r*) is:

$$r \text{ (effective yield)} = \left(1 + \frac{i}{p}\right)^{p} - 1$$

where *i* is the nominal yield

p is the number of times the income is received and compounded during the year.

The formulae where income is received quarterly in advance are considered in the next section.

F. Conversion Tables:
Nominal Yields – Effective Yields (Quarterly in Advance)

The tables on pages 29 to 33 provide for the direct conversion of nominal yields (where income is received annually in arrear) to effective yields (where the income is received quarterly in advance) and vice versa.

When using these tables it is essential to understand the difference between nominal and effective yields (sometimes termed True Equivalent Yields) as explained above.

The tables are based on the following formulae:

$$i = 4\left[1 - (1 + r)^{-1/4}\right]$$

$$r = \left[\frac{1}{\left(1 - \frac{i}{4}\right)^{4}} - 1\right]$$

where *i* is the nominal yield, and

r is the effective yield.

G. Years' Purchase Tables with Allowance for Taxation

The Years' Purchase figures given in these tables are calculated on a dual-rate basis with allowance for the effect of taxation on that part of the income which must be invested in a sinking fund to replace the capital outlay at the end of the term.

The Years' Purchase tax adjusted figures are now available in separate groups of tables, similar to the unadjusted tax tables, on either (1) the traditional approach where income is deemed to be received annually in arrear, or (2) based on the assumption that income is received quarterly in advance.

1. YEARS' PURCHASE: Dual Rate (Annually in Arrear) with adjustment for tax

This table is provided in the Yellow Section of this volume on pages 411 to 549.

The use of this table is shown in the following valuation.

Example

Value the leasehold interest in a property producing a net income (profit rent) of £65,000 p.a. The property is held on lease with 20 years to run and a return of 7% is required on the capital invested. The sinking fund will accumulate at 3% and the appropriate rate of tax is 40%.

Net Income (Profit Rent)	£65,000 p.a.
YP for 20 yrs. at 7% and 3% (allowing for tax at 40% see page 520)	7·5743
Capital Value	£492,330

Construction of the tables

(i) YEARS' PURCHASE: Dual Rate (Annually in Arrear) Tax Adjusted Table

The Years' Purchase formula on a dual rate basis (annually in arrear) with no allowance for tax is:

$$\frac{1}{i+s}$$

where i = rate of interested expected on each £1
s = sinking fund instalment.

To allow for the effect of taxation on the sinking fund instalment, it is necessary to calculate the gross sinking fund required before tax which will provide the net sinking fund "s" after tax has been deducted.

To allow of the effect of tax the Gross Adjustment Factors discussed in Section D may be used.

Gross SF = Net SF × Gross Adjustment Factor
∴ Gross SF = $s.T_G$

The formula then becomes:

$$\text{YP dual rate} = \frac{1}{i + s.T_G}$$

$$T_G = \frac{100}{100 - \text{Rate of Tax}}$$

Example

Calculate the Years' Purchase for 22 years at 8% and 2·5% adjusted for income tax at 35%

$$i = 0·08$$

$$s = 0·0346466 \text{ (see page 114)}$$

$$T_G(35) = 1·53846 \text{ (see page 311)}$$

$$\text{YP} = \frac{1}{0·08 + (0·0346466 \times 1·53846)}$$

$$= \frac{1}{0·1333024} = 7·5017404$$

Compare the appropriate Years' Purchase figure (page 494) $= 7·5017$

Net sinking fund rates

It is the usual practice for valuers to use net sinking fund rates of interest when using dual rate tables of Years' Purchase. If a net rate of 2·5% is required, the sinking fund investment must yield 2·5% after tax on the gross interest has been paid.

For example, if tax is charged at 40% on a sinking fund investment and a net rate of accumulation of 2·5% is required, the gross rate of interest to be received before tax must be between 4·1% and 4·2% (a more precise figure is given in (a) below).

The Gross Adjustment and Net Adjustment factors T_G and T_N which were discussed in Section D may be used to find equivalent gross or net rates of interest as follows:

(a) A net rate of interest of 2·5% is required. If tax is considered at 40%, calculate the equivalent grossed up rate which must be available before tax

2·5% $\times T_G(40) = 2·5 \times 1·66667 = 4·167\%$

(b) If the gross rate of interest received before tax is 5%, calculate the equivalent net rate of interest after tax at 30%

5% $\times T_N(30) = 5 \times 0·7 = 3·5\%$

In this edition, years' purchase tables with an allowance for tax on the sinking fund instalment have been calculated at net sinking fund rates of 2·5%, 3% and 4%.

(2) YEARS' PURCHASE: Dual Rate (Quarterly in Advance) Tax Adjusted Table, in the green pages 313 to 409.

The Years' Purchase formula on a dual rate basis (quarterly in advance) with allowance for taxation is:

$$\cfrac{1}{4\left[1-(1+r)^{-1/4}\right]+\left\{\cfrac{4\left[1-(1+a)^{-1/4}\right]}{(1+a)^n-1}\right\}T_G}$$

Where T_G is the Gross Tax Adjustment Factor (see Section D)

r = effective yield (see page xvii)

a = sinking fund rate of accumulation

H. INTERNAL RATES OF RETURN (IRR)

These tables are included for investment analysis where a transaction in freehold property has taken place or is being considered. Two tables are available which will enable the analysis to be carried out, either reflecting adjustment for anticipated growth, or without adjustment for growth.

(i) INTERNAL RATES OF RETURN (IRR) without reflecting rental growth, pages 179 to 193.

Certain input data are required. The following example identifies this and illustrates how the analysis may be carried out.

Example

The freehold interest in a property has recently been purchased for £160,000. The property is let on a lease which now has five years to run and produces a net rent of £8,000 per annum. The net full rental value is £20,000 per annum. Calculate the true return (IRR) received.

Preliminary Calculations

$$Initial\ Yield = \frac{Net\ Income}{Price} \times 100\% = \frac{8,000}{160,000} \times 100\% = 5\%$$

$$Rental\ Factor = \frac{Full\ Rental\ Value}{Net\ Income} = \frac{20,000}{8,000} = 2\cdot5$$

Reference to the IRR (without growth) Table, on page 188 gives the figure for the Internal Rate of Return

$$\underline{IRR = 9 \cdot 72\%}$$

This table has been computed based on the following formula with some adjustment to accommodate the table headings.

$$V = I\left[\frac{1 - (1 + e)^{-n}}{e}\right] + R\left[\frac{1}{e}(1 + e)^{-n}\right]$$

Where I = current income $\quad R$ = full rental value

$\quad V$ = price paid

$\quad e$ = IRR $\qquad n$ = period to reversion or next review

When applying the above equation to the example, the variables V, I, R and n are known. The value of e (the internal rate of return) can then be found. Its value will be such that when used in the formula, the left-hand side of the equation (V) just equals the right hand side.

The solution for e can be found by using calculus (the Newton-Raphson Method) or by trial and error. The figures in the tables were computed using an iterative approach to find the values for e. The solution is demonstrated as follows, setting out the calculation in the normal net present value format.

Year	Cash flow	PV of £1 at 9.72%	Present value
1	8,000	0.9114108	7,291·29
2	8,000	0.8306697	6,645·36
3	8,000	0.7570814	6,056·65
4	8,000	0.6900122	5,520·10
5	8,000	0.6288846	5,031·08
Value of future cash flow	205,761*	0.6288846	129,399·92
	Present Value of Income Flow		159,944·40
	Less Purchase Value (V)		160,000
	Net Present Value		−£55·60

* See calculation below.

* The calculation is made on the assumption that the income receivable on reversion, or at the next review, is taken as the current rental value today. The figure of £205,761 therefore represents the value in 5 years' time of the future cash flow in terms of the current rental value capitalised at the inherent rate of interest (i.e. the computed figure for IRR):

	FRV	£20,000	
YP in perp at 9·72%		10·288066	£205,761

The future cash flow is therefore dealt with as a cash flow in perpetuity deferred for the period to the next review or reversion. This follows conventional valuation practice.

The true IRR will occur when the Present Value of the Income Flow just equates to the Value (Purchase Price), i.e. the Net Present Value = 0. If, as a check, the above calculation is repeated at a lower rate of 9.71%, the present value will be +£144.61. This indicates that the true IRR falls between 9.72% and 9.71% and that the IRR correct to two decimal places is 9.72%.

(ii) INTERNAL RATES OF RETURN (IRR) reflecting anticipated growth to the reversion or next review, pages 195–237.

If the above example is considered further and the investor or analyst wishes to calculate the true rate of return that will be obtained if future rental growth is considered, reference to the IRR table with growth will give the required Internal Rate of Return.

Example: As above

The preliminary calculations in the above example are the same. In addition, if it is now assumed that a rate of growth of 7% per annum is considered appropriate, the Table on page 232 shows the new Internal Rate of Return figure is:

$$\text{IRR (7\%)} = 12 \cdot 08\%$$

This can be demonstrated in the same way as in the first example except that the Present Value of the future cash flow in five years will be different.

The full rental value in 5 years with 7% growth will be:

$$£20,000 \times (1 \cdot 07)^5 = £28,051$$

The present value of this future income flow is therefore:

£28,051 × YP in perp at 12·08% = £28,051 × 8·2781457

= £232,210 (instead of £205,761 in the above example).

If the calculation is completed using PV of £1 figures at 12·08%, this gives a Net Present Value of +£73·75. However, as a check, if a rate of 12·09% is used, the Net Present Value is −£100·42. This shows that the true IRR correct to two decimal places is 12·08%.

The appropriate formula for this table is:

$$V = I\left[\frac{1-(1+e)^{-n}}{e}\right] + R(1+g)^n\left[\frac{1}{e}(1+e)^{-n}\right]$$

Where the variables are the same as for the previous table and in addition:

g = rate of anticipated growth (expressed as the rate of growth on each £1)

The value of e (IRR with growth) is found in the same way as for the previous table.

It should be noted that in this table the growth is only considered for the period to the reversion or to the next rent review. If growth were built in for further periods, alternative rent review patterns would need to be covered. This would require far more space to be devoted to this table, perhaps at the expense of other tables in order to contain the size of this edition.

It would also pose a further problem. Whilst it may be realistic to anticipate future annual growth for a relatively short period, it may not be realistic to project the same average rate of growth too far distant into the future. If this type of approach is contemplated, more complex models more suited to computer analysis should be used.

Note for both IRR Tables:

(i) The tables provide for a range of Rental Factors, Initial Yields, Periods to review or reversion and, in the second table, for Rates of Growth. In some cases this may result in a combination which is unlikely to occur in practice; for example, a combination where there is a high rental factor, a high initial yield and where the reversion is imminent. This has led to relatively high internal rates of return occurring in some cases, which are included to complete the tables. However, returns in excess of 70% have not been printed, and asterisks indicate that the return is in excess of 70%.

(ii) In some cases the exact initial yield or rental factor will not be included in the set of tables. For example, the initial yield may be 4·3%. In these cases the IRR for 4% and 4·5% can be found and these will provide a range of IRR figures for consideration, within which the true figure will lie. Alternatively, a close approximation of the actual IRR can be found by interpolation between the two figures obtained.

I. The Metric Conversion Tables

These are included in the green coloured section on pages 239 to 265.

In some conversion tables exact equivalent figures are given, whereas in other tables the figures have been rounded to the number of decimal places printed. The basis of computation is printed below each table and where this is an exact conversion factor this is mentioned, e.g.:

(i) 1ft. = 0·3048 m. (exactly)—see page 243.

The tabulated figures given in this table are exact conversion figures.

(ii) 1m. = 1·093613298 yds. (see page 247).

The omission of the word (exactly) signifies that the basis is a rounded figure and the tabulated figures are rounded to the number of decimal places printed.

e.g.,   10 metres = 10·9361 yards
1,000 metres = 1093·6133 yards

(iii) 1 sq. ft. = 0·09290304 sq. m. (exactly) (see page 249).

In this case, although the basis is an exact conversion factor, the tabulated figures have less decimal places than the actual basis so that rounded figures are given in the tables. This will be apparent on inspection for 10 sq. ft. is given as 0·92903 sq. m.

Supplementary tables have been given in most cases. In this way it has been possible to extend the range of the tables without increasing the size of this section.

J. The Life Tables

Following the precedent set in previous editions, the Life Tables used in this edition are based on the most recent tables available from the Office of National Statistics. These are English Life Tables No. 15, prepared by the Government Actuary. They were constructed on the mortality experience of the population of England and Wales during the three years 1990, 1991 and 1992.

The inclusion of English Life Tables No. 15 does not mean that they are necessarily recommended for all valuation problems involving life interests. The main purpose of the English Life Tables is historical; to show the general level of mortality in England and Wales at the time of the census. If, as in the past, mortality rates continue to improve in the future, the future expectations of life will be greater than those indicated in the Tables. The effect of improvement in mortality rates in the past can be seen from a comparison of expectation of life for males and females at age 40 taken from five of the English Life Tables, No. 8 (1910–12); No. 10 (1930–32); No. 12 (1960–62); No. 14 (1980–82) and the current tables.

	E.L.T.8	E.L.T.10	E.L.T.12	E.L.T.14	E.L.T.15
Males	27·74	29·62	31·62	33·34	35·35
Females	30·30	32·55	36·69	38·67	40·24

It will be seen from the above figures that mortality rates have continually improved. However, there is currently more uncertainty about the future trend. Although actuaries are concerned at the possible future level of deaths of person contracting the AIDS virus, there is also a more positive view due to the considerable progress in medical research in recent years and the consequent improvement in the treatment of previous terminal conditions.

In valuation problems involving a life interest it should be remembered that location of residence, status and occupation as well as general state of health, are all factors that affect an individual's life span to varying degrees. Where the problem is particularly complex or important, actuarial advice will no doubt be sought.

Tables of Years' Purchase for Single Lives (Males and Females), pages 267 to 275, and the Joint Continuation of Two Lives (pages 277 to 293) are included in this Edition.

A range of rates of interest for annual premiums to secure a sum assured of £1 payable at the end of the year of death is provided. In previous editions, specimen life assurance office company rates for ordinary non-profit business were also included. With the significant changes in mortality being experienced, any quoted 'specimen rates' may relatively quickly become out of date. For this reason they have been omitted from this volume.

The Years' Purchase Table for a Single Life at Dual Rates has been excluded from this and the previous edition. In the traditional Dual Rate Years' Purchase Tables in the early parts of the book, it is assumed that the Sinking Fund is invested from the income which becomes available annually in arrear. Where, in the case of a Life Interest, it is desired to replace the capital by a life policy payable on death, such an approach is not appropriate since the investment must be protected by effecting the policy and paying the first annual premium at the outset. A valuation can be carried out allowing for this as in the following example.

Example

What is the value of a life interest in a property yielding a net income of £10,000 p.a. held by a male life aged 50 on an 8% basis? The life can be assured for an annual premium at a rate of 3% and it is to be assumed that the income arises annually in arrear from the valuation date.

$$\text{Capital Value} = \text{Net Income} \times \left[\frac{1 - P(1 + i)}{i + P(1 + i)}\right]$$

where i = the interest rate of the valuation

and P = the annual premium for the life policy.

$$\text{Capital Value} = 10,000\left[\frac{1 - 0.0266\,(1.08)}{0.08 + 0.0266\,(1.08)}\right]$$

$$= 10,000 \times \frac{0.971272}{0.108728}$$

$$= 10,000 \times 8.93304 = £89,330$$

Capital Value say £89,350

It will be noted that the life assurance premium has been taken from the table of annual premiums on page 300 based on the English Life Tables No. 15.

Tables of Years' Purchase for the Longer of Two Lives has not been included in this Edition since the appropriate value can easily be calculated from the Years' Purchase of the two single lives involved and the Years' Purchase for the Joint continuation of the two lives. The method and example are given on page 297.

Permission for the reproduction of the English Life Tables No. 15 was granted and the source 'English Life Tables', National Statistics © Crown Copyright 2000 is acknowledged.

Construction of the Life Tables

The Tables of Years' Purchase for Single Lives and the Joint Continuation of Two Lives were calculated by computer based on formulae using the tabulated values of "lx", the number of lives attaining a given age. The values of this function tabulated for each age are whole numbers and are shown on pages 306 and 307.

The Tables of Annual Premiums were calculated from the value of Years' Purchase for Single Lives produced by the computer.

YEARS' PURCHASE

(SINGLE RATE % PRINCIPLE)

OR

PRESENT VALUE OF
ONE POUND PER ANNUM

receivable quarterly in advance, allow-
ing for a sinking fund to accumulate at
the same rate of interest as that which
is required on the invested capital and
ignoring the effect of income tax on that
part of the income used to provide the
annual sinking fund instalment.

AT RATES OF INTEREST*
FROM
2% to 28%

---

* *Note:*—In computing the quarterly in advance figures in these tables, all the rates
of interest quoted are effective rates, see the Introductory Section, pages
xvi-xvii.

				Rate per Cent*					
Yrs.	2	2.25	2.5	2.75	3	3.25	3.5	3.75	Yrs.
1	0.9926	0.9917	0.9908	0.9899	0.9890	0.9881	0.9872	0.9863	1
2	1.9658	1.9616	1.9574	1.9533	1.9492	1.9451	1.9411	1.9370	2
3	2.9198	2.9101	2.9005	2.8909	2.8815	2.8720	2.8627	2.8534	3
4	3.8552	3.8378	3.8206	3.8035	3.7865	3.7697	3.7531	3.7366	4
5	4.7722	4.7451	4.7182	4.6916	4.6653	4.6392	4.6134	4.5879	5
6	5.6713	5.6324	5.5939	5.5559	5.5184	5.4813	5.4446	5.4084	6
7	6.5527	6.5001	6.4483	6.3971	6.3467	6.2969	6.2477	6.1992	7
8	7.4168	7.3488	7.2818	7.2158	7.1508	7.0868	7.0237	6.9615	8
9	8.2640	8.1788	8.0950	8.0126	7.9316	7.8518	7.7734	7.6962	9
10	9.0946	8.9906	8.8884	8.7881	8.6896	8.5928	8.4977	8.4044	10
11	9.9089	9.7844	9.6624	9.5428	9.4255	9.3104	9.1976	9.0870	11
12	10.7072	10.5608	10.4175	10.2773	10.1400	10.0055	9.8738	9.7449	12
13	11.4899	11.3201	11.1543	10.9921	10.8336	10.6787	10.5271	10.3790	13
14	12.2572	12.0628	11.8730	11.6878	11.5071	11.3306	11.1584	10.9902	14
15	13.0095	12.7890	12.5742	12.3649	12.1609	11.9621	11.7683	11.5793	15
16	13.7470	13.4993	13.2584	13.0239	12.7958	12.5737	12.3575	12.1471	16
17	14.4701	14.1940	13.9258	13.6652	13.4121	13.1660	12.9269	12.6944	17
18	15.1790	14.8734	14.5769	14.2894	14.0104	13.7397	13.4770	13.2219	18
19	15.8740	15.5378	15.2122	14.8969	14.5914	14.2954	14.0085	13.7303	19
20	16.5553	16.1876	15.8320	15.4881	15.1554	14.8335	14.5220	14.2204	20
21	17.2233	16.8231	16.4367	16.0635	15.7030	15.3547	15.0181	14.6927	21
22	17.8782	17.4446	17.0266	16.6235	16.2346	15.8595	15.4975	15.1480	22
23	18.5203	18.0524	17.6021	17.1684	16.7508	16.3484	15.9606	15.5869	23
24	19.1498	18.6469	18.1636	17.6989	17.2519	16.8219	16.4081	16.0098	24
25	19.7669	19.2283	18.7114	18.2151	17.7384	17.2805	16.8405	16.4175	25
26	20.3719	19.7969	19.2458	18.7175	18.2108	17.7247	17.2582	16.8104	26
27	20.9651	20.3530	19.7672	19.2064	18.6694	18.1549	17.6619	17.1892	27
28	21.5466	20.8968	20.2759	19.6823	19.1146	18.5716	18.0518	17.5542	28
29	22.1168	21.4287	20.7722	20.1454	19.5469	18.9751	18.4286	17.9061	29
30	22.6757	21.9489	21.2563	20.5962	19.9666	19.3659	18.7926	18.2452	30
31	23.2237	22.4576	21.7287	21.0348	20.3741	19.7445	19.1444	18.5721	31
32	23.7610	22.9551	22.1895	21.4618	20.7696	20.1111	19.4842	18.8871	32
33	24.2877	23.4417	22.6391	21.8773	21.1537	20.4662	19.8125	19.1908	33
34	24.8041	23.9176	23.0777	22.2816	21.5266	20.8101	20.1298	19.4835	34
35	25.3103	24.3830	23.5057	22.6752	21.8886	21.1432	20.4363	19.7656	35
36	25.8067	24.8382	23.9232	23.0582	22.2401	21.4658	20.7324	20.0375	36
37	26.2933	25.2833	24.3305	23.4310	22.5813	21.7782	21.0186	20.2996	37
38	26.7703	25.7187	24.7279	23.7938	22.9126	22.0808	21.2950	20.5523	38
39	27.2380	26.1444	25.1156	24.1469	23.2343	22.3739	21.5621	20.7957	39
40	27.6966	26.5608	25.4938	24.4905	23.5466	22.6577	21.8202	21.0304	40
41	28.1461	26.9681	25.8628	24.8250	23.8498	22.9327	22.0696	21.2566	41
42	28.5869	27.3664	26.2228	25.1505	24.1441	23.1989	22.3105	21.4747	42
43	29.0189	27.7559	26.5740	25.4673	24.4299	23.4568	22.5432	21.6848	43
44	29.4426	28.1368	26.9167	25.7756	24.7074	23.7066	22.7681	21.8874	44
45	29.8579	28.5094	27.2510	26.0756	24.9767	23.9485	22.9854	22.0826	45
46	30.2650	28.8738	27.5771	26.3676	25.2383	24.1828	23.1954	22.2708	46
47	30.6642	29.2301	27.8953	26.6518	25.4922	24.4097	23.3982	22.4522	47
48	31.0556	29.5786	28.2058	26.9284	25.7387	24.6295	23.5942	22.6270	48
49	31.4393	29.9194	28.5086	27.1976	25.9780	24.8423	23.7835	22.7955	49
50	31.8154	30.2528	28.8041	27.4596	26.2104	25.0485	23.9665	22.9579	50

* *Note:*—In computing the quarterly in advance figures in these tables, all the rates % quoted above are effective rates, see the Introductory Section, pages xvi-xvii.

2

YEARS' PURCHASE*

Yrs.	2	2.25	2.5	2.75	3	3.25	3.5	3.75	Yrs.
				Rate per Cent*					
51	32.1842	30.5788	29.0924	27.7146	26.4360	25.2481	24.1433	23.1144	51
52	32.5458	30.8976	29.3736	27.9627	26.6550	25.4415	24.3141	23.2653	52
53	32.9002	31.2094	29.6480	28.2043	26.8677	25.6288	24.4791	23.4107	53
54	33.2477	31.5144	29.9157	28.4393	27.0742	25.8102	24.6385	23.5509	54
55	33.5884	31.8126	30.1768	28.6681	27.2746	25.9859	24.7925	23.6860	55
56	33.9225	32.1043	30.4316	28.8907	27.4692	26.1561	24.9414	23.8162	56
57	34.2499	32.3895	30.6802	29.1074	27.6581	26.3209	25.0852	23.9417	57
58	34.5710	32.6685	30.9227	29.3182	27.8416	26.4805	25.2241	24.0627	58
59	34.8857	32.9414	31.1593	29.5235	28.0197	26.6351	25.3584	24.1793	59
60	35.1943	33.2082	31.3901	29.7232	28.1926	26.7848	25.4881	24.2917	60
61	35.4969	33.4692	31.6153	29.9176	28.3604	26.9298	25.6134	24.4000	61
62	35.7935	33.7244	31.8350	30.1068	28.5234	27.0703	25.7344	24.5045	62
63	36.0842	33.9740	32.0494	30.2909	28.6816	27.2063	25.8514	24.6051	63
64	36.3693	34.2181	32.2585	30.4701	28.8353	27.3380	25.9644	24.7021	64
65	36.6488	34.4569	32.4625	30.6445	28.9844	27.4656	26.0737	24.7956	65
66	36.9228	34.6904	32.6615	30.8143	29.1292	27.5892	26.1792	24.8857	66
67	37.1915	34.9187	32.8557	30.9795	29.2698	27.7089	26.2811	24.9726	67
68	37.4548	35.1420	33.0452	31.1402	29.4063	27.8248	26.3796	25.0563	68
69	37.7130	35.3604	33.2300	31.2967	29.5388	27.9371	26.4748	25.1370	69
70	37.9662	35.5741	33.4103	31.4490	29.6675	28.0458	26.5667	25.2148	70
71	38.2144	35.7830	33.5862	31.5972	29.7924	28.1512	26.6555	25.2897	71
72	38.4577	35.9873	33.7579	31.7414	29.9136	28.2532	26.7414	25.3620	72
73	38.6962	36.1871	33.9253	31.8818	30.0314	28.3520	26.8243	25.4316	73
74	38.9301	36.3825	34.0887	32.0184	30.1457	28.4476	26.9044	25.4987	74
75	39.1594	36.5736	34.2480	32.1514	30.2567	28.5403	26.9819	25.5634	75
76	39.3842	36.7605	34.4035	32.2808	30.3644	28.6301	27.0566	25.6258	76
77	39.6045	36.9433	34.5552	32.4068	30.4690	28.7170	27.1289	25.6859	77
78	39.8206	37.1221	34.7032	32.5293	30.5706	28.8012	27.1987	25.7438	78
79	40.0324	37.2969	34.8476	32.6486	30.6692	28.8827	27.2662	25.7997	79
80	40.2401	37.4679	34.9885	32.7647	30.7649	28.9617	27.3314	25.8535	80
81	40.4437	37.6352	35.1259	32.8777	30.8579	29.0382	27.3944	25.9054	81
82	40.6433	37.7987	35.2600	32.9877	30.9481	29.1123	27.4552	25.9554	82
83	40.8390	37.9587	35.3908	33.0947	31.0357	29.1840	27.5140	26.0036	83
84	41.0308	38.1151	35.5184	33.1989	31.1208	29.2535	27.5708	26.0500	84
85	41.2189	38.2681	35.6429	33.3002	31.2034	29.3208	27.6257	26.0948	85
86	41.4033	38.4177	35.7644	33.3989	31.2835	29.3860	27.6787	26.1380	86
87	41.5841	38.5641	35.8829	33.4949	31.3614	29.4491	27.7299	26.1796	87
88	41.7614	38.7072	35.9985	33.5884	31.4369	29.5103	27.7794	26.2197	88
89	41.9351	38.8471	36.1113	33.6793	31.5103	29.5695	27.8273	26.2583	89
90	42.1055	38.9840	36.2213	33.7678	31.5816	29.6269	27.8735	26.2955	90
91	42.2725	39.1179	36.3287	33.8540	31.6507	29.6824	27.9181	26.3314	91
92	42.4362	39.2488	36.4334	33.9378	31.7179	29.7362	27.9613	26.3661	92
93	42.5968	39.3769	36.5356	34.0194	31.7830	29.7883	28.0029	26.3994	93
94	42.7542	39.5021	36.6353	34.0988	31.8463	29.8388	28.0432	26.4316	94
95	42.9085	39.6246	36.7326	34.1761	31.9078	29.8877	28.0821	26.4625	95
96	43.0597	39.7443	36.8275	34.2513	31.9674	29.9350	28.1197	26.4924	96
97	43.2080	39.8615	36.9200	34.3245	32.0254	29.9809	28.1560	26.5212	97
98	43.3534	39.9760	37.0104	34.3958	32.0816	30.0253	28.1911	26.5489	98
99	43.4960	40.0881	37.0985	34.4651	32.1362	30.0683	28.2250	26.5757	99
100	43.6357	40.1977	37.1844	34.5326	32.1892	30.1100	28.2578	26.6014	100
PERP	50.6235	45.0677	40.6231	36.9865	33.9560	31.3917	29.1937	27.2888	PERP

* *Note:*—In computing the quarterly in advance figures in these tables, all the rates % quoted above are effective rates, see the Introductory Section, pages xvi-xvii.

3

Rate per Cent*

Yrs.	4	4.25	4.5	4.75	5	5.25	5.5	5.75	Yrs.
1	0.9855	0.9846	0.9837	0.9828	0.9820	0.9811	0.9802	0.9794	1
2	1.9330	1.9290	1.9250	1.9211	1.9172	1.9133	1.9094	1.9055	2
3	2.8441	2.8350	2.8259	2.8168	2.8078	2.7989	2.7901	2.7813	3
4	3.7202	3.7040	3.6879	3.6719	3.6561	3.6404	3.6248	3.6094	4
5	4.5626	4.5375	4.5128	4.4882	4.4639	4.4399	4.4161	4.3925	5
6	5.3725	5.3371	5.3021	5.2675	5.2333	5.1995	5.1661	5.1331	6
7	6.1514	6.1041	6.0575	6.0115	5.9661	5.9213	5.8770	5.8333	7
8	6.9002	6.8399	6.7804	6.7217	6.6640	6.6070	6.5509	6.4955	8
9	7.6203	7.5456	7.4721	7.3998	7.3286	7.2585	7.1896	7.1217	9
10	8.3127	8.2226	8.1340	8.0470	7.9616	7.8776	7.7950	7.7138	10
11	8.9784	8.8719	8.7675	8.6650	8.5644	8.4657	8.3689	8.2738	11
12	9.6186	9.4948	9.3736	9.2549	9.1385	9.0245	8.9128	8.8033	12
13	10.2341	10.0923	9.9537	9.8180	9.6853	9.5555	9.4284	9.3040	13
14	10.8259	10.6655	10.5088	10.3557	10.2061	10.0599	9.9171	9.7775	14
15	11.3950	11.2152	11.0399	10.8689	10.7020	10.5392	10.3803	10.2252	15
16	11.9422	11.7426	11.5482	11.3589	11.1744	10.9946	10.8194	10.6486	16
17	12.4683	12.2485	12.0346	11.8266	11.6242	11.4273	11.2356	11.0490	17
18	12.9742	12.7337	12.5001	12.2732	12.0527	11.8384	11.6301	11.4276	18
19	13.4607	13.1992	12.9455	12.6995	12.4607	12.2289	12.0040	11.7856	19
20	13.9284	13.6457	13.3718	13.1064	12.8493	12.6000	12.3584	12.1241	20
21	14.3782	14.0739	13.7797	13.4949	13.2194	12.9526	12.6944	12.4443	21
22	14.8106	14.4848	14.1700	13.8658	13.5718	13.2876	13.0128	12.7470	22
23	15.2264	14.8788	14.5435	14.2199	13.9075	13.6059	13.3147	13.0333	23
24	15.6263	15.2568	14.9009	14.5579	14.2272	13.9083	13.6008	13.3040	24
25	16.0107	15.6194	15.2430	14.8806	14.5317	14.1957	13.8720	13.5600	25
26	16.3804	15.9673	15.5703	15.1886	14.8217	14.4687	14.1290	13.8020	26
27	16.7358	16.3009	15.8835	15.4827	15.0978	14.7281	14.3726	14.0310	27
28	17.0776	16.6209	16.1832	15.7635	15.3609	14.9745	14.6036	14.2474	28
29	17.4062	16.9279	16.4700	16.0315	15.6113	15.2086	14.8225	14.4521	29
30	17.7222	17.2224	16.7445	16.2874	15.8499	15.4311	15.0300	14.6457	30
31	18.0261	17.5049	17.0071	16.5316	16.0771	15.6425	15.2267	14.8287	31
32	18.3182	17.7758	17.2585	16.7648	16.2935	15.8433	15.4131	15.0018	32
33	18.5991	18.0357	17.4990	16.9874	16.4996	16.0341	15.5898	15.1655	33
34	18.8692	18.2850	17.7292	17.1999	16.6958	16.2154	15.7573	15.3202	34
35	19.1289	18.5242	17.9494	17.4028	16.8828	16.3877	15.9161	15.4666	35
36	19.3787	18.7536	18.1602	17.5965	17.0608	16.5513	16.0665	15.6050	36
37	19.6188	18.9736	18.3619	17.7814	17.2303	16.7068	16.2092	15.7359	37
38	19.8497	19.1847	18.5549	17.9579	17.3918	16.8546	16.3444	15.8596	38
39	20.0717	19.3872	18.7395	18.1264	17.5456	16.9949	16.4725	15.9766	39
40	20.2852	19.5814	18.9163	18.2873	17.6920	17.1283	16.5940	16.0873	40
41	20.4904	19.7677	19.0854	18.4409	17.8315	17.2550	16.7092	16.1920	41
42	20.6878	19.9464	19.2473	18.5875	17.9644	17.3754	16.8183	16.2909	42
43	20.8776	20.1178	19.4021	18.7274	18.0909	17.4898	16.9217	16.3845	43
44	21.0600	20.2822	19.5503	18.8611	18.2114	17.5985	17.0198	16.4730	44
45	21.2355	20.4400	19.6922	18.9886	18.3261	17.7017	17.1127	16.5567	45
46	21.4042	20.5912	19.8279	19.1104	18.4354	17.7998	17.2008	16.6358	46
47	21.5664	20.7364	19.9577	19.2266	18.5395	17.8931	17.2844	16.7106	47
48	21.7224	20.8756	20.0820	19.3376	18.6386	17.9816	17.3635	16.7814	48
49	21.8724	21.0091	20.2010	19.4436	18.7330	18.0658	17.4385	16.8483	49
50	22.0166	21.1372	20.3148	19.5447	18.8229	18.1457	17.5096	16.9116	50

* *Note:*—In computing the quarterly in advance figures in these tables, all the rates % quoted above are effective rates, see the Introductory Section, pages xvi-xvii.

YEARS' PURCHASE*

Yrs.	4	4.25	4.5	4.75	5	5.25	5.5	5.75	Yrs.
					Rate per Cent*				
51	22.1553	21.2601	20.4237	19.6413	18.9086	18.2217	17.5771	16.9714	51
52	22.2886	21.3780	20.5279	19.7334	18.9901	18.2939	17.6409	17.0280	52
53	22.4168	21.4910	20.6276	19.8214	19.0678	18.3624	17.7015	17.0815	53
54	22.5401	21.5995	20.7230	19.9054	19.1418	18.4276	17.7589	17.1321	54
55	22.6586	21.7035	20.8144	19.9856	19.2122	18.4895	17.8133	17.1799	55
56	22.7726	21.8033	20.9018	20.0622	19.2793	18.5483	17.8649	17.2251	56
57	22.8822	21.8990	20.9854	20.1353	19.3432	18.6042	17.9138	17.2679	57
58	22.9876	21.9908	21.0654	20.2051	19.4041	18.6573	17.9601	17.3084	58
59	23.0889	22.0789	21.1420	20.2717	19.4620	18.7077	18.0040	17.3466	59
60	23.1863	22.1634	21.2153	20.3353	19.5172	18.7557	18.0457	17.3828	60
61	23.2800	22.2444	21.2854	20.3960	19.5698	18.8012	18.0851	17.4170	61
62	23.3701	22.3221	21.3525	20.4539	19.6199	18.8445	18.1225	17.4494	62
63	23.4567	22.3967	21.4167	20.5092	19.6675	18.8856	18.1580	17.4800	63
64	23.5399	22.4682	21.4782	20.5621	19.7130	18.9246	18.1916	17.5089	64
65	23.6200	22.5368	21.5370	20.6125	19.7562	18.9618	18.2235	17.5362	65
66	23.6970	22.6026	21.5933	20.6606	19.7974	18.9970	18.2537	17.5621	66
67	23.7711	22.6658	21.6471	20.7066	19.8366	19.0305	18.2823	17.5866	67
68	23.8422	22.7263	21.6986	20.7504	19.8740	19.0623	18.3094	17.6097	68
69	23.9107	22.7844	21.7480	20.7923	19.9096	19.0926	18.3351	17.6316	69
70	23.9765	22.8401	21.7951	20.8323	19.9434	19.1213	18.3595	17.6522	70
71	24.0398	22.8936	21.8403	20.8705	19.9757	19.1486	18.3826	17.6718	71
72	24.1006	22.9448	21.8835	20.9069	20.0065	19.1746	18.4045	17.6903	72
73	24.1592	22.9940	21.9249	20.9417	20.0357	19.1992	18.4252	17.7078	73
74	24.2154	23.0412	21.9644	20.9749	20.0636	19.2226	18.4449	17.7243	74
75	24.2695	23.0864	22.0023	21.0066	20.0902	19.2449	18.4636	17.7400	75
76	24.3215	23.1299	22.0385	21.0369	20.1154	19.2660	18.4812	17.7547	76
77	24.3715	23.1715	22.0732	21.0658	20.1395	19.2861	18.4980	17.7687	77
78	24.4196	23.2114	22.1064	21.0933	20.1625	19.3052	18.5139	17.7820	78
79	24.4659	23.2497	22.1381	21.1197	20.1843	19.3233	18.5289	17.7945	79
80	24.5103	23.2865	22.1685	21.1448	20.2051	19.3405	18.5432	17.8063	80
81	24.5531	23.3217	22.1976	21.1688	20.2249	19.3569	18.5567	17.8175	81
82	24.5942	23.3555	22.2254	21.1917	20.2438	19.3724	18.5695	17.8280	82
83	24.6337	23.3880	22.2521	21.2136	20.2618	19.3872	18.5817	17.8380	83
84	24.6717	23.4191	22.2775	21.2345	20.2789	19.4012	18.5932	17.8475	84
85	24.7083	23.4489	22.3019	21.2544	20.2952	19.4146	18.6041	17.8564	85
86	24.7434	23.4776	22.3253	21.2734	20.3107	19.4273	18.6145	17.8649	86
87	24.7772	23.5050	22.3476	21.2916	20.3255	19.4393	18.6243	17.8729	87
88	24.8097	23.5314	22.3690	21.3089	20.3396	19.4507	18.6336	17.8804	88
89	24.8409	23.5566	22.3894	21.3255	20.3530	19.4616	18.6424	17.8876	89
90	24.8710	23.5809	22.4090	21.3413	20.3658	19.4719	18.6508	17.8944	90
91	24.8999	23.6041	22.4277	21.3564	20.3779	19.4817	18.6587	17.9008	91
92	24.9276	23.6264	22.4456	21.3708	20.3895	19.4911	18.6662	17.9068	92
93	24.9543	23.6478	22.4628	21.3845	20.4005	19.4999	18.6733	17.9125	93
94	24.9800	23.6683	22.4792	21.3977	20.4110	19.5083	18.6800	17.9179	94
95	25.0047	23.6880	22.4949	21.4102	20.4211	19.5163	18.6864	17.9230	95
96	25.0284	23.7069	22.5099	21.4222	20.4306	19.5239	18.6925	17.9279	96
97	25.0513	23.7250	22.5243	21.4336	20.4397	19.5311	18.6982	17.9324	97
98	25.0732	23.7424	22.5380	21.4445	20.4483	19.5380	18.7037	17.9368	98
99	25.0943	23.7591	22.5512	21.4549	20.4565	19.5445	18.7088	17.9408	99
100	25.1146	23.7750	22.5638	21.4648	20.4644	19.5507	18.7137	17.9447	100
PERP	25.6219	24.1512	22.8438	21.6740	20.6212	19.6686	18.8026	18.0119	PERP

*Note:—In computing the quarterly in advance figures in these tables, all the rates % quoted above are effective rates, see the Introductory Section, pages xvi-xvii.

				Rate per Cent*					
Yrs.	6	6.25	6.5	6.75	7	7.25	7.5	7.75	Yrs.
1	0.9785	0.9777	0.9768	0.9760	0.9751	0.9743	0.9734	0.9726	1
2	1.9016	1.8978	1.8940	1.8902	1.8865	1.8827	1.8790	1.8753	2
3	2.7725	2.7638	2.7552	2.7467	2.7382	2.7297	2.7213	2.7130	3
4	3.5941	3.5789	3.5639	3.5490	3.5341	3.5195	3.5049	3.4905	4
5	4.3692	4.3461	4.3232	4.3005	4.2781	4.2558	4.2338	4.2120	5
6	5.1004	5.0681	5.0361	5.0045	4.9733	4.9424	4.9119	4.8817	6
7	5.7902	5.7476	5.7056	5.6641	5.6231	5.5826	5.5426	5.5032	7
8	6.4410	6.3872	6.3342	6.2819	6.2303	6.1795	6.1294	6.0800	8
9	7.0549	6.9891	6.9244	6.8606	6.7979	6.7361	6.6752	6.6153	9
10	7.6341	7.5557	7.4786	7.4028	7.3283	7.2550	7.1829	7.1121	10
11	8.1805	8.0889	7.9990	7.9107	7.8240	7.7388	7.6552	7.5731	11
12	8.6959	8.5907	8.4876	8.3864	8.2872	8.1900	8.0946	8.0010	12
13	9.1822	9.0631	8.9464	8.8321	8.7202	8.6106	8.5033	8.3982	13
14	9.6410	9.5076	9.3771	9.2496	9.1248	9.0028	8.8835	8.7667	14
15	10.0738	9.9260	9.7816	9.6407	9.5030	9.3685	9.2372	9.1088	15
16	10.4821	10.3198	10.1615	10.0071	9.8564	9.7095	9.5661	9.4262	16
17	10.8673	10.6904	10.5181	10.3503	10.1868	10.0274	9.8722	9.7209	17
18	11.2307	11.0392	10.8529	10.6718	10.4955	10.3239	10.1569	9.9943	18
19	11.5735	11.3675	11.1674	10.9729	10.7840	10.6003	10.4217	10.2480	19
20	11.8969	11.6765	11.4626	11.2551	11.0536	10.8580	10.6680	10.4836	20
21	12.2020	11.9673	11.7398	11.5193	11.3056	11.0983	10.8972	10.7021	21
22	12.4899	12.2410	12.0001	11.7669	11.5411	11.3223	11.1104	10.9050	22
23	12.7614	12.4986	12.2445	11.9988	11.7612	11.5312	11.3087	11.0932	23
24	13.0176	12.7411	12.4740	12.2161	11.9669	11.7260	11.4931	11.2680	24
25	13.2592	12.9692	12.6895	12.4196	12.1591	11.9076	11.6647	11.4301	25
26	13.4872	13.1840	12.8918	12.6103	12.3388	12.0770	11.8244	11.5806	26
27	13.7023	13.3861	13.0818	12.7889	12.5067	12.2348	11.9728	11.7203	27
28	13.9052	13.5764	13.2602	12.9562	12.6636	12.3821	12.1110	11.8499	28
29	14.0967	13.7554	13.4277	13.1129	12.8103	12.5193	12.2395	11.9702	29
30	14.2773	13.9240	13.5850	13.2597	12.9473	12.6473	12.3590	12.0818	30
31	14.4476	14.0826	13.7327	13.3972	13.0754	12.7666	12.4702	12.1854	31
32	14.6083	14.2318	13.8714	13.5261	13.1952	12.8779	12.5736	12.2816	32
33	14.7600	14.3723	14.0016	13.6467	13.3070	12.9817	12.6698	12.3708	33
34	14.9030	14.5046	14.1238	13.7598	13.4116	13.0784	12.7593	12.4537	34
35	15.0380	14.6290	14.2386	13.8657	13.5093	13.1686	12.8426	12.5305	35
36	15.1653	14.7461	14.3464	13.9649	13.6007	13.2527	12.9200	12.6019	36
37	15.2854	14.8564	14.4476	14.0579	13.6860	13.3311	12.9921	12.6681	37
38	15.3987	14.9601	14.5426	14.1449	13.7658	13.4042	13.0591	12.7295	38
39	15.5056	15.0578	14.6319	14.2265	13.8404	13.4724	13.1214	12.7866	39
40	15.6064	15.1497	14.7157	14.3029	13.9100	13.5359	13.1794	12.8395	40
41	15.7016	15.2362	14.7943	14.3744	13.9752	13.5952	13.2334	12.8886	41
42	15.7913	15.3176	14.8682	14.4415	14.0360	13.6505	13.2836	12.9342	42
43	15.8760	15.3943	14.9376	14.5043	14.0929	13.7020	13.3302	12.9765	43
44	15.9558	15.4664	15.0027	14.5631	14.1461	13.7500	13.3737	13.0158	44
45	16.0312	15.5342	15.0639	14.6182	14.1957	13.7948	13.4141	13.0522	45
46	16.1023	15.5981	15.1213	14.6699	14.2422	13.8366	13.4516	13.0860	46
47	16.1694	15.6583	15.1752	14.7182	14.2855	13.8755	13.4866	13.1174	47
48	16.2326	15.7148	15.2258	14.7635	14.3261	13.9118	13.5191	13.1465	48
49	16.2923	15.7681	15.2734	14.8060	14.3640	13.9457	13.5494	13.1736	49
50	16.3486	15.8182	15.3180	14.8457	14.3994	13.9772	13.5775	13.1987	50

* *Note:*—In computing the quarterly in advance figures in these tables, all the rates % quoted above are effective rates, see the Introductory Section, pages xvi-xvii.

Yrs.	6	6.25	6.5	6.75	7	7.25	7.5	7.75	Yrs.
					Rate per Cent*				
51	16.4017	15.8654	15.3599	14.8830	14.4325	14.0067	13.6037	13.2219	51
52	16.4519	15.9098	15.3993	14.9179	14.4635	14.0341	13.6280	13.2436	52
53	16.4991	15.9516	15.4362	14.9505	14.4924	14.0597	13.6507	13.2636	53
54	16.5437	15.9909	15.4709	14.9812	14.5194	14.0836	13.6718	13.2822	54
55	16.5858	16.0280	15.5035	15.0098	14.5447	14.1058	13.6914	13.2995	55
56	16.6255	16.0628	15.5341	15.0367	14.5683	14.1266	13.7096	13.3155	56
57	16.6630	16.0956	15.5628	15.0619	14.5903	14.1459	13.7265	13.3304	57
58	16.6983	16.1265	15.5898	15.0854	14.6109	14.1639	13.7423	13.3442	58
59	16.7316	16.1555	15.6151	15.1075	14.6302	14.1807	13.7570	13.3570	59
60	16.7631	16.1828	15.6389	15.1282	14.6482	14.1964	13.7707	13.3689	60
61	16.7927	16.2086	15.6612	15.1476	14.6650	14.2110	13.7834	13.3800	61
62	16.8207	16.2328	15.6822	15.1658	14.6808	14.2247	13.7952	13.3902	62
63	16.8471	16.2556	15.7018	15.1828	14.6955	14.2374	13.8062	13.3997	63
64	16.8720	16.2770	15.7203	15.1987	14.7092	14.2492	13.8164	13.4085	64
65	16.8955	16.2972	15.7377	15.2136	14.7220	14.2603	13.8259	13.4167	65
66	16.9177	16.3162	15.7540	15.2276	14.7340	14.2706	13.8347	13.4243	66
67	16.9386	16.3341	15.7693	15.2407	14.7452	14.2802	13.8430	13.4314	67
68	16.9583	16.3509	15.7837	15.2530	14.7557	14.2891	13.8506	13.4379	68
69	16.9769	16.3668	15.7971	15.2645	14.7655	14.2975	13.8577	13.4440	69
70	16.9945	16.3817	15.8098	15.2752	14.7747	14.3053	13.8644	13.4496	70
71	17.0111	16.3957	15.8217	15.2853	14.7832	14.3125	13.8705	13.4549	71
72	17.0267	16.4089	15.8329	15.2947	14.7912	14.3193	13.8763	13.4597	72
73	17.0414	16.4214	15.8434	15.3036	14.7987	14.3256	13.8816	13.4642	73
74	17.0553	16.4331	15.8532	15.3119	14.8057	14.3315	13.8865	13.4684	74
75	17.0684	16.4441	15.8625	15.3197	14.8122	14.3370	13.8912	13.4723	75
76	17.0808	16.4544	15.8711	15.3269	14.8183	14.3421	13.8955	13.4759	76
77	17.0925	16.4642	15.8793	15.3337	14.8240	14.3468	13.8994	13.4793	77
78	17.1035	16.4734	15.8869	15.3401	14.8293	14.3513	13.9032	13.4824	78
79	17.1139	16.4820	15.8941	15.3461	14.8343	14.3554	13.9066	13.4852	79
80	17.1237	16.4902	15.9009	15.3517	14.8390	14.3593	13.9098	13.4879	80
81	17.1330	16.4978	15.9072	15.3570	14.8433	14.3629	13.9128	13.4904	81
82	17.1417	16.5050	15.9132	15.3619	14.8474	14.3663	13.9156	13.4927	82
83	17.1499	16.5118	15.9187	15.3665	14.8512	14.3694	13.9182	13.4948	83
84	17.1577	16.5182	15.9240	15.3708	14.8547	14.3723	13.9206	13.4968	84
85	17.1650	16.5242	15.9289	15.3748	14.8580	14.3751	13.9228	13.4986	85
86	17.1719	16.5298	15.9335	15.3786	14.8611	14.3776	13.9249	13.5004	86
87	17.1784	16.5351	15.9379	15.3822	14.8640	14.3800	13.9269	13.5019	87
88	17.1846	16.5402	15.9420	15.3855	14.8667	14.3822	13.9287	13.5034	88
89	17.1904	16.5449	15.9458	15.3886	14.8693	14.3842	13.9303	13.5048	89
90	17.1959	16.5493	15.9494	15.3915	14.8716	14.3862	13.9319	13.5060	90
91	17.2010	16.5535	15.9528	15.3943	14.8738	14.3879	13.9333	13.5072	91
92	17.2059	16.5574	15.9559	15.3968	14.8759	14.3896	13.9347	13.5083	92
93	17.2105	16.5611	15.9589	15.3992	14.8778	14.3912	13.9359	13.5093	93
94	17.2148	16.5646	15.9617	15.4015	14.8796	14.3926	13.9371	13.5103	94
95	17.2189	16.5679	15.9643	15.4036	14.8813	14.3940	13.9382	13.5111	95
96	17.2228	16.5709	15.9668	15.4055	14.8829	14.3952	13.9392	13.5119	96
97	17.2264	16.5738	15.9691	15.4074	14.8844	14.3964	13.9402	13.5127	97
98	17.2299	16.5766	15.9713	15.4091	14.8858	14.3975	13.9410	13.5134	98
99	17.2331	16.5791	15.9733	15.4107	14.8870	14.3985	13.9418	13.5140	99
100	17.2362	16.5816	15.9752	15.4122	14.8883	14.3995	13.9426	13.5146	100
PERP	17.2871	16.6203	16.0047	15.4347	14.9054	14.4126	13.9527	13.5224	PERP

* *Note:*—In computing the quarterly in advance figures in these tables, all the rates % quoted above are effective rates, see the Introductory Section, pages xvi-xvii.

				Rate per Cent*					
Yrs.	8	8.25	8.5	8.75	9	9.5	10	11	Yrs.
1	0.9718	0.9709	0.9701	0.9693	0.9685	0.9669	0.9652	0.9620	1
2	1.8716	1.8679	1.8642	1.8606	1.8570	1.8498	1.8427	1.8287	2
3	2.7047	2.6965	2.6883	2.6802	2.6721	2.6562	2.6404	2.6095	3
4	3.4761	3.4619	3.4478	3.4339	3.4200	3.3926	3.3656	3.3130	4
5	4.1904	4.1690	4.1479	4.1269	4.1061	4.0651	4.0249	3.9467	5
6	4.8518	4.8223	4.7930	4.7641	4.7355	4.6793	4.6242	4.5176	6
7	5.4642	5.4257	5.3877	5.3501	5.3130	5.2402	5.1691	5.0319	7
8	6.0312	5.9831	5.9357	5.8889	5.8428	5.7524	5.6644	5.4953	8
9	6.5562	6.4981	6.4408	6.3844	6.3288	6.2202	6.1147	5.9128	9
10	7.0424	6.9738	6.9064	6.8400	6.7748	6.6474	6.5240	6.2888	10
11	7.4925	7.4133	7.3354	7.2590	7.1839	7.0375	6.8962	6.6276	11
12	7.9093	7.8192	7.7309	7.6442	7.5592	7.3938	7.2345	· 6.9329	12
13	8.2952	8.1943	8.0954	7.9985	7.9035	7.7192	7.5420	7.2079	13
14	8.6525	8.5407	8.4313	8.3242	8.2194	8.0163	7.8216	7.4556	14
15	8.9833	8.8607	8.7409	8.6238	8.5092	8.2877	8.0758	7.6788	15
16	9.2897	9.1564	9.0263	8.8992	8.7751	8.5355	8.3069	7.8799	16
17	9.5733	9.4295	9.2892	9.1525	9.0190	8.7619	8.5169	8.0610	17
18	9.8360	9.6818	9.5316	9.3854	9.2428	8.9685	8.7079	8.2242	18
19	10.0792	9.9149	9.7550	9.5995	9.4481	9.1573	8.8815	8.3712	19
20	10.3043	10.1302	9.9610	9.7964	9.6365	9.3297	9.0393	8.5037	20
21	10.5128	10.3291	10.1507	9.9775	9.8093	9.4871	9.1828	8.6230	21
22	10.7059	10.5128	10.3256	10.1440	9.9678	9.6309	9.3132	8.7305	22
23	10.8846	10.6826	10.4868	10.2971	10.1133	9.7622	9.4318	8.8273	23
24	11.0501	10.8394	10.6354	10.4379	10.2467	9.8821	9.5396	8.9146	24
25	11.2034	10.9842	10.7723	10.5674	10.3691	9.9916	9.6376	8.9932	25
26	11.3453	11.1180	10.8985	10.6865	10.4815	10.0916	9.7267	9.0640	26
27	11.4767	11.2417	11.0149	10.7959	10.5845	10.1829	9.8077	9.1278	27
28	11.5983	11.3559	11.1221	10.8966	10.6790	10.2663	9.8813	9.1853	28
29	11.7110	11.4613	11.2209	10.9892	10.7658	10.3425	9.9482	9.2370	29
30	11.8153	11.5588	11.3119	11.0743	10.8453	10.4120	10.0091	9.2837	30
31	11.9118	11.6488	11.3959	11.1525	10.9183	10.4756	10.0644	9.3257	31
32	12.0012	11.7320	11.4732	11.2245	10.9853	10.5336	10.1147	9.3636	32
33	12.0840	11.8088	11.5445	11.2907	11.0467	10.5865	10.1604	9.3977	33
34	12.1607	11.8798	11.6102	11.3515	11.1031	10.6349	10.2020	9.4284	34
35	12.2317	11.9453	11.6708	11.4075	11.1548	10.6791	10.2397	9.4561	35
36	12.2974	12.0059	11.7266	11.4590	11.2022	10.7195	10.2741	9.4810	36
37	12.3583	12.0619	11.7781	11.5063	11.2458	10.7563	10.3053	9.5035	37
38	12.4146	12.1135	11.8255	11.5498	11.2857	10.7900	10.3337	9.5237	38
39	12.4668	12.1613	11.8692	11.5898	11.3223	10.8207	10.3595	9.5420	39
40	12.5151	12.2054	11.9095	11.6266	11.3559	10.8488	10.3830	9.5584	40
41	12.5598	12.2461	11.9466	11.6604	11.3868	10.8744	10.4043	9.5732	41
42	12.6013	12.2838	11.9808	11.6915	11.4151	10.8978	10.4237	9.5865	42
43	12.6396	12.3185	12.0123	11.7201	11.4410	10.9192	10.4413	9.5986	43
44	12.6751	12.3507	12.0414	11.7464	11.4648	10.9387	10.4573	9.6094	44
45	12.7080	12.3803	12.0682	11.7706	11.4867	10.9565	10.4719	9.6191	45
46	12.7384	12.4078	12.0929	11.7928	11.5067	10.9728	10.4851	9.6279	46
47	12.7666	12.4331	12.1156	11.8133	11.5251	10.9877	10.4972	9.6358	47
48	12.7927	12.4565	12.1366	11.8321	11.5420	11.0013	10.5081	9.6430	48
49	12.8169	12.4781	12.1559	11.8494	11.5574	11.0137	10.5181	9.6494	49
50	12.8393	12.4980	12.1738	11.8653	11.5716	11.0250	10.5271	9.6552	50

* *Note:*—In computing the quarterly in advance figures in these tables, all the rates % quoted above are effective rates, see the Introductory Section, pages xvi-xvii.

Rate per Cent*

Yrs.	8	8.25	8.5	8.75	9	9.5	10	11	Yrs.
51	12.8600	12.5165	12.1902	11.8799	11.5847	11.0354	10.5353	9.6604	51
52	12.8792	12.5335	12.2053	11.8934	11.5966	11.0448	10.5428	9.6651	52
53	12.8969	12.5493	12.2193	11.9057	11.6076	11.0534	10.5496	9.6693	53
54	12.9134	12.5638	12.2321	11.9171	11.6176	11.0613	10.5558	9.6731	54
55	12.9286	12.5772	12.2440	11.9275	11.6269	11.0685	10.5614	9.6765	55
56	12.9427	12.5896	12.2549	11.9372	11.6353	11.0751	10.5665	9.6796	56
57	12.9558	12.6011	12.2649	11.9460	11.6431	11.0811	10.5711	9.6824	57
58	12.9679	12.6117	12.2742	11.9541	11.6502	11.0865	10.5754	9.6849	58
59	12.9791	12.6215	12.2828	11.9616	11.6568	11.0916	10.5792	9.6872	59
60	12.9894	12.6305	12.2906	11.9685	11.6627	11.0961	10.5827	9.6892	60
61	12.9990	12.6388	12.2979	11.9748	11.6682	11.1003	10.5859	9.6911	61
62	13.0079	12.6466	12.3046	11.9806	11.6733	11.1041	10.5887	9.6927	62
63	13.0161	12.6537	12.3108	11.9859	11.6779	11.1076	10.5914	9.6942	63
64	13.0238	12.6603	12.3164	11.9909	11.6822	11.1108	10.5937	9.6956	64
65	13.0308	12.6663	12.3217	11.9954	11.6861	11.1137	10.5959	9.6968	65
66	13.0373	12.6720	12.3265	11.9995	11.6896	11.1163	10.5979	9.6979	66
67	13.0434	12.6771	12.3310	12.0034	11.6929	11.1187	10.5997	9.6988	67
68	13.0490	12.6819	12.3351	12.0069	11.6959	11.1210	10.6013	9.6997	68
69	13.0542	12.6864	12.3389	12.0101	11.6987	11.1230	10.6028	9.7005	69
70	13.0590	12.6904	12.3423	12.0131	11.7012	11.1248	10.6041	9.7012	70
71	13.0634	12.6942	12.3455	12.0158	11.7036	11.1265	10.6053	9.7019	71
72	13.0675	12.6977	12.3485	12.0183	11.7057	11.1280	10.6064	9.7025	72
73	13.0713	12.7009	12.3512	12.0206	11.7076	11.1294	10.6075	9.7030	73
74	13.0749	12.7039	12.3538	12.0227	11.7094	11.1307	10.6084	9.7035	74
75	13.0781	12.7067	12.3561	12.0247	11.7111	11.1319	10.6092	9.7039	75
76	13.0812	12.7092	12.3582	12.0265	11.7126	11.1330	10.6100	9.7043	76
77	13.0840	12.7116	12.3602	12.0281	11.7140	11.1339	10.6107	9.7046	77
78	13.0866	12.7137	12.3620	12.0297	11.7153	11.1348	10.6113	9.7049	78
79	13.0890	12.7157	12.3637	12.0311	11.7164	11.1356	10.6119	9.7052	79
80	13.0912	12.7176	12.3652	12.0323	11.7175	11.1364	10.6124	9.7055	80
81	13.0932	12.7193	12.3666	12.0335	11.7185	11.1371	10.6128	9.7057	81
82	13.0952	12.7209	12.3679	12.0346	11.7194	11.1377	10.6133	9.7059	82
83	13.0969	12.7223	12.3691	12.0356	11.7202	11.1383	10.6137	9.7061	83
84	13.0986	12.7237	12.3702	12.0365	11.7210	11.1388	10.6140	9.7062	84
85	13.1001	12.7249	12.3713	12.0374	11.7217	11.1393	10.6143	9.7064	85
86	13.1015	12.7261	12.3722	12.0381	11.7223	11.1397	10.6146	9.7065	86
87	13.1028	12.7271	12.3731	12.0389	11.7229	11.1401	10.6149	9.7067	87
88	13.1040	12.7281	12.3739	12.0395	11.7234	11.1404	10.6151	9.7068	88
89	13.1051	12.7290	12.3746	12.0401	11.7239	11.1408	10.6154	9.7069	89
90	13.1061	12.7299	12.3753	12.0407	11.7244	11.1411	10.6156	9.7069	90
91	13.1071	12.7306	12.3759	12.0412	11.7248	11.1413	10.6157	9.7070	91
92	13.1079	12.7314	12.3765	12.0417	11.7252	11.1416	10.6159	9.7071	92
93	13.1088	12.7320	12.3771	12.0421	11.7255	11.1418	10.6161	9.7072	93
94	13.1095	12.7326	12.3775	12.0425	11.7258	11.1420	10.6162	9.7072	94
95	13.1102	12.7332	12.3780	12.0428	11.7261	11.1422	10.6163	9.7073	95
96	13.1109	12.7337	12.3784	12.0432	11.7264	11.1424	10.6164	9.7073	96
97	13.1115	12.7342	12.3788	12.0435	11.7266	11.1426	10.6165	9.7074	97
98	13.1120	12.7346	12.3792	12.0438	11.7269	11.1427	10.6166	9.7074	98
99	13.1125	12.7350	12.3795	12.0440	11.7271	11.1428	10.6167	9.7074	99
100	13.1130	12.7354	12.3798	12.0443	11.7273	11.1430	10.6168	9.7075	100
PERP	13.1190	12.7400	12.3833	12.0470	11.7294	11.1442	10.6176	9.7078	PERP

* *Note:*—In computing the quarterly in advance figures in these tables, all the rates % quoted above are effective rates, see the Introductory Section, pages xvi-xvii.

YEARS' PURCHASE*

Rate per Cent*

Yrs.	12	13	14	15	16	17	18	19	Yrs.
1	0.9589	0.9558	0.9527	0.9497	0.9467	0.9437	0.9408	0.9380	1
2	1.8150	1.8016	1.7884	1.7755	1.7628	1.7503	1.7381	1.7262	2
3	2.5794	2.5501	2.5214	2.4935	2.4663	2.4397	2.4138	2.3885	3
4	3.2619	3.2125	3.1645	3.1180	3.0728	3.0290	2.9864	2.9451	4
5	3.8713	3.7986	3.7286	3.6609	3.5957	3.5326	3.4717	3.4128	5
6	4.4154	4.3174	4.2234	4.1331	4.0464	3.9631	3.8829	3.8059	6
7	4.9012	4.7765	4.6574	4.5436	4.4349	4.3310	4.2315	4.1362	7
8	5.3349	5.1827	5.0381	4.9007	4.7699	4.6454	4.5268	4.4137	8
9	5.7222	5.5422	5.3721	5.2111	5.0587	4.9142	4.7771	4.6470	9
10	6.0680	5.8604	5.6651	5.4811	5.3076	5.1439	4.9892	4.8430	10
11	6.3767	6.1419	5.9220	5.7158	5.5222	5.3402	5.1690	5.0077	11
12	6.6524	6.3911	6.1475	5.9199	5.7072	5.5080	5.3213	5.1461	12
13	6.8985	6.6116	6.3452	6.0974	5.8667	5.6514	5.4504	5.2624	13
14	7.1182	6.8067	6.5187	6.2518	6.0041	5.7740	5.5598	5.3602	14
15	7.3144	6.9794	6.6708	6.3860	6.1227	5.8788	5.6525	5.4423	15
16	7.4896	7.1322	6.8043	6.5027	6.2248	5.9683	5.7311	5.5113	16
17	7.6460	7.2675	6.9214	6.6042	6.3129	6.0449	5.7977	5.5693	17
18	7.7857	7.3872	7.0241	6.6924	6.3888	6.1103	5.8541	5.6180	18
19	7.9104	7.4931	7.1141	6.7692	6.4543	6.1662	5.9020	5.6590	19
20	8.0217	7.5868	7.1932	6.8359	6.5107	6.2140	5.9425	5.6934	20
21	8.1211	7.6697	7.2625	6.8939	6.5594	6.2548	5.9768	5.7223	21
22	8.2099	7.7431	7.3233	6.9444	6.6013	6.2897	6.0059	5.7467	22
23	8.2891	7.8081	7.3766	6.9883	6.6375	6.3196	6.0306	5.7671	23
24	8.3599	7.8656	7.4234	7.0264	6.6686	6.3451	6.0515	5.7842	24
25	8.4230	7.9165	7.4645	7.0596	6.6955	6.3669	6.0692	5.7987	25
26	8.4794	7.9615	7.5005	7.0884	6.7187	6.3855	6.0842	5.8108	26
27	8.5298	8.0013	7.5320	7.1135	6.7386	6.4014	6.0969	5.8210	27
28	8.5748	8.0366	7.5597	7.1353	6.7558	6.4150	6.1077	5.8295	28
29	8.6149	8.0678	7.5840	7.1543	6.7707	6.4267	6.1169	5.8367	29
30	8.6508	8.0954	7.6054	7.1708	6.7835	6.4366	6.1246	5.8428	30
31	8.6828	8.1198	7.6241	7.1851	6.7945	6.4451	6.1312	5.8478	31
32	8.7113	8.1414	7.6405	7.1976	6.8040	6.4524	6.1367	5.8521	32
33	8.7369	8.1606	7.6549	7.2085	6.8122	6.4586	6.1414	5.8557	33
34	8.7596	8.1775	7.6675	7.2179	6.8193	6.4639	6.1454	5.8587	34
35	8.7800	8.1925	7.6785	7.2261	6.8253	6.4684	6.1488	5.8612	35
36	8.7981	8.2057	7.6883	7.2332	6.8306	6.4723	6.1517	5.8634	36
37	8.8144	8.2175	7.6968	7.2394	6.8351	6.4756	6.1541	5.8652	37
38	8.8288	8.2279	7.7042	7.2448	6.8390	6.4784	6.1562	5.8667	38
39	8.8418	8.2371	7.7108	7.2495	6.8424	6.4809	6.1579	5.8679	39
40	8.8533	8.2452	7.7166	7.2536	6.8453	6.4829	6.1594	5.8690	40
41	8.8636	8.2524	7.7216	7.2571	6.8478	6.4847	6.1607	5.8699	41
42	8.8728	8.2588	7.7260	7.2602	6.8499	6.4862	6.1617	5.8706	42
43	8.8810	8.2644	7.7299	7.2629	6.8518	6.4875	6.1626	5.8713	43
44	8.8884	8.2694	7.7333	7.2652	6.8534	6.4886	6.1634	5.8718	44
45	8.8949	8.2738	7.7363	7.2672	6.8548	6.4895	6.1640	5.8722	45
46	8.9008	8.2777	7.7389	7.2690	6.8560	6.4903	6.1646	5.8726	46
47	8.9060	8.2812	7.7412	7.2705	6.8570	6.4910	6.1651	5.8729	47
48	8.9106	8.2842	7.7432	7.2719	6.8579	6.4916	6.1654	5.8732	48
49	8.9148	8.2869	7.7450	7.2730	6.8586	6.4921	6.1658	5.8734	49
50	8.9185	8.2893	7.7465	7.2740	6.8593	6.4926	6.1661	5.8736	50

* *Note:*—In computing the quarterly in advance figures in these tables, all the rates % quoted above are effective rates, see the Introductory Section, pages xvi-xvii.

				Rate per Cent*					
Yrs.	12	13	14	15	16	17	18	19	Yrs.
51	8.9218	8.2915	7.7479	7.2749	6.8599	6.4929	6.1663	5.8738	51
52	8.9248	8.2933	7.7491	7.2757	6.8604	6.4932	6.1665	5.8739	52
53	8.9274	8.2950	7.7501	7.2763	6.8608	6.4935	6.1667	5.8740	53
54	8.9298	8.2965	7.7511	7.2769	6.8611	6.4937	6.1668	5.8741	54
55	8.9319	8.2978	7.7519	7.2774	6.8615	6.4939	6.1669	5.8742	55
56	8.9338	8.2989	7.7526	7.2778	6.8617	6.4941	6.1670	5.8742	56
57	8.9355	8.2999	7.7532	7.2782	6.8620	6.4942	6.1671	5.8743	57
58	8.9370	8.3008	7.7537	7.2786	6.8622	6.4944	6.1672	5.8743	58
59	8.9383	8.3016	7.7542	7.2788	6.8623	6.4945	6.1673	5.8744	59
60	8.9395	8.3023	7.7546	7.2791	6.8625	6.4946	6.1673	5.8744	60
61	8.9406	8.3030	7.7550	7.2793	6.8626	6.4946	6.1674	5.8744	61
62	8.9415	8.3035	7.7553	7.2795	6.8627	6.4947	6.1674	5.8745	62
63	8.9424	8.3040	7.7556	7.2797	6.8628	6.4948	6.1674	5.8745	63
64	8.9431	8.3044	7.7559	7.2798	6.8629	6.4948	6.1675	5.8745	64
65	8.9438	8.3048	7.7561	7.2799	6.8630	6.4949	6.1675	5.8745	65
66	8.9444	8.3052	7.7563	7.2800	6.8630	6.4949	6.1675	5.8745	66
67	8.9450	8.3055	7.7564	7.2801	6.8631	6.4949	6.1675	5.8745	67
68	8.9455	8.3057	7.7566	7.2802	6.8631	6.4949	6.1676	5.8745	68
69	8.9459	8.3060	7.7567	7.2803	6.8632	6.4950	6.1676	5.8745	69
70	8.9463	8.3062	7.7568	7.2803	6.8632	6.4950	6.1676	5.8745	70
71	8.9466	8.3063	7.7569	7.2804	6.8632	6.4950	6.1676	5.8745	71
72	8.9469	8.3065	7.7570	7.2804	6.8633	6.4950	6.1676	5.8746	72
73	8.9472	8.3067	7.7571	7.2805	6.8633	6.4950	6.1676	5.8746	73
74	8.9474	8.3068	7.7571	7.2805	6.8633	6.4950	6.1676	5.8746	74
75	8.9477	8.3069	7.7572	7.2805	6.8633	6.4950	6.1676	5.8746	75
76	8.9479	8.3070	7.7573	7.2806	6.8633	6.4950	6.1676	5.8746	76
77	8.9480	8.3071	7.7573	7.2806	6.8633	6.4951	6.1676	5.8746	77
78	8.9482	8.3072	7.7573	7.2806	6.8633	6.4951	6.1676	5.8746	78
79	8.9483	8.3072	7.7574	7.2806	6.8634	6.4951	6.1676	5.8746	79
80	8.9484	8.3073	7.7574	7.2807	6.8634	6.4951	6.1676	5.8746	80
81	8.9486	8.3073	7.7574	7.2807	6.8634	6.4951	6.1676	5.8746	81
82	8.9487	8.3074	7.7575	7.2807	6.8634	6.4951	6.1676	5.8746	82
83	8.9487	8.3074	7.7575	7.2807	6.8634	6.4951	6.1676	5.8746	83
84	8.9488	8.3075	7.7575	7.2807	6.8634	6.4951	6.1676	5.8746	84
85	8.9489	8.3075	7.7575	7.2807	6.8634	6.4951	6.1676	5.8746	85
86	8.9490	8.3075	7.7575	7.2807	6.8634	6.4951	6.1676	5.8746	86
87	8.9490	8.3076	7.7575	7.2807	6.8634	6.4951	6.1676	5.8746	87
88	8.9491	8.3076	7.7575	7.2807	6.8634	6.4951	6.1676	5.8746	88
89	8.9491	8.3076	7.7576	7.2807	6.8634	6.4951	6.1676	5.8746	89
90	8.9491	8.3076	7.7576	7.2807	6.8634	6.4951	6.1676	5.8746	90
91	8.9492	8.3076	7.7576	7.2807	6.8634	6.4951	6.1676	5.8746	91
92	8.9492	8.3077	7.7576	7.2807	6.8634	6.4951	6.1676	5.8746	92
93	8.9492	8.3077	7.7576	7.2807	6.8634	6.4951	6.1676	5.8746	93
94	8.9493	8.3077	7.7576	7.2807	6.8634	6.4951	6.1676	5.8746	94
95	8.9493	8.3077	7.7576	7.2807	6.8634	6.4951	6.1676	5.8746	95
96	8.9493	8.3077	7.7576	7.2807	6.8634	6.4951	6.1676	5.8746	96
97	8.9493	8.3077	7.7576	7.2807	6.8634	6.4951	6.1676	5.8746	97
98	8.9493	8.3077	7.7576	7.2807	6.8634	6.4951	6.1676	5.8746	98
99	8.9494	8.3077	7.7576	7.2807	6.8634	6.4951	6.1676	5.8746	99
100	8.9494	8.3077	7.7576	7.2807	6.8634	6.4951	6.1676	5.8746	100
PERP	8.9495	8.3078	7.7576	7.2808	6.8634	6.4951	6.1676	5.8746	PERP

* *Note:*—In computing the quarterly in advance figures in these tables, all the rates % quoted above are effective rates,
see the Introductory Section, pages xvi-xvii.

Rate per Cent*

Yrs.	20	21	22	23	24	25	26	28	Yrs.
1	0.9351	0.9323	0.9296	0.9269	0.9242	0.9215	0.9189	0.9138	1
2	1.7144	1.7029	1.6915	1.6804	1.6695	1.6587	1.6482	1.6276	2
3	2.3638	2.3397	2.3161	2.2930	2.2705	2.2485	2.2270	2.1853	3
4	2.9050	2.8659	2.8280	2.7911	2.7552	2.7203	2.6863	2.6210	4
5	3.3559	3.3009	3.2476	3.1961	3.1461	3.0978	3.0509	2.9614	5
6	3.7317	3.6603	3.5916	3.5253	3.4614	3.3997	3.3403	3.2274	6
7	4.0449	3.9574	3.8735	3.7929	3.7156	3.6413	3.5699	3.4352	7
8	4.3059	4.2029	4.1046	4.0105	3.9206	3.8346	3.7521	3.5975	8
9	4.5234	4.4058	4.2940	4.1875	4.0860	3.9892	3.8968	3.7243	9
10	4.7046	4.5735	4.4492	4.3313	4.2193	4.1128	4.0116	3.8233	10
11	4.8556	4.7121	4.5765	4.4482	4.3268	4.2118	4.1027	3.9007	11
12	4.9815	4.8266	4.6808	4.5433	4.4135	4.2910	4.1750	3.9612	12
13	5.0864	4.9213	4.7663	4.6206	4.4835	4.3543	4.2324	4.0084	13
14	5.1738	4.9995	4.8364	4.6834	4.5399	4.4049	4.2779	4.0454	14
15	5.2466	5.0642	4.8938	4.7345	4.5854	4.4455	4.3141	4.0742	15
16	5.3073	5.1176	4.9409	4.7761	4.6220	4.4779	4.3428	4.0967	16
17	5.3579	5.1617	4.9795	4.8098	4.6516	4.5038	4.3655	4.1143	17
18	5.4000	5.1982	5.0111	4.8373	4.6755	4.5246	4.3836	4.1281	18
19	5.4351	5.2284	5.0371	4.8596	4.6947	4.5412	4.3980	4.1388	19
20	5.4644	5.2533	5.0583	4.8778	4.7102	4.5545	4.4093	4.1472	20
21	5.4888	5.2739	5.0757	4.8925	4.7227	4.5651	4.4184	4.1537	21
22	5.5091	5.2910	5.0900	4.9045	4.7328	4.5736	4.4255	4.1589	22
23	5.5261	5.3050	5.1017	4.9143	4.7410	4.5804	4.4312	4.1629	23
24	5.5402	5.3167	5.1113	4.9222	4.7475	4.5858	4.4357	4.1660	24
25	5.5519	5.3263	5.1192	4.9286	4.7528	4.5902	4.4393	4.1684	25
26	5.5618	5.3342	5.1256	4.9339	4.7571	4.5937	4.4422	4.1703	26
27	5.5699	5.3408	5.1309	4.9381	4.7605	4.5964	4.4444	4.1718	27
28	5.5767	5.3462	5.1352	4.9416	4.7633	4.5987	4.4462	4.1730	28
29	5.5824	5.3507	5.1388	4.9444	4.7655	4.6005	4.4476	4.1739	29
30	5.5871	5.3544	5.1417	4.9467	4.7674	4.6019	4.4488	4.1746	30
31	5.5911	5.3574	5.1441	4.9486	4.7688	4.6030	4.4497	4.1752	31
32	5.5944	5.3600	5.1460	4.9501	4.7700	4.6039	4.4504	4.1756	32
33	5.5971	5.3621	5.1476	4.9513	4.7709	4.6047	4.4509	4.1760	33
34	5.5994	5.3638	5.1490	4.9523	4.7717	4.6052	4.4514	4.1762	34
35	5.6013	5.3652	5.1500	4.9531	4.7723	4.6057	4.4517	4.1764	35
36	5.6028	5.3664	5.1509	4.9538	4.7728	4.6061	4.4520	4.1766	36
37	5.6042	5.3674	5.1516	4.9543	4.7732	4.6064	4.4523	4.1767	37
38	5.6053	5.3682	5.1522	4.9548	4.7735	4.6066	4.4524	4.1768	38
39	5.6062	5.3688	5.1527	4.9551	4.7738	4.6068	4.4526	4.1769	39
40	5.6069	5.3694	5.1531	4.9554	4.7740	4.6070	4.4527	4.1769	40
41	5.6076	5.3699	5.1534	4.9556	4.7742	4.6071	4.4528	4.1770	41
42	5.6081	5.3702	5.1537	4.9558	4.7743	4.6072	4.4528	4.1770	42
43	5.6086	5.3705	5.1539	4.9560	4.7744	4.6073	4.4529	4.1771	43
44	5.6089	5.3708	5.1541	4.9561	4.7745	4.6073	4.4529	4.1771	44
45	5.6092	5.3710	5.1543	4.9562	4.7746	4.6074	4.4530	4.1771	45
46	5.6095	5.3712	5.1544	4.9563	4.7746	4.6074	4.4530	4.1771	46
47	5.6097	5.3713	5.1545	4.9564	4.7747	4.6075	4.4530	4.1771	47
48	5.6099	5.3715	5.1546	4.9564	4.7747	4.6075	4.4530	4.1771	48
49	5.6100	5.3716	5.1546	4.9565	4.7747	4.6075	4.4531	4.1771	49
50	5.6101	5.3716	5.1547	4.9565	4.7748	4.6075	4.4531	4.1771	50

* *Note:*—In computing the quarterly in advance figures in these tables, all the rates % quoted above are effective rates, see the Introductory Section, pages xvi-xvii.

12

				Rate per Cent*					
Yrs.	20	21	22	23	24	25	26	28	Yrs.
51	5.6103	5.3717	5.1547	4.9565	4.7748	4.6075	4.4531	4.1771	51
52	5.6103	5.3718	5.1548	4.9566	4.7748	4.6075	4.4531	4.1772	52
53	5.6104	5.3718	5.1548	4.9566	4.7748	4.6075	4.4531	4.1772	53
54	5.6105	5.3718	5.1548	4.9566	4.7748	4.6076	4.4531	4.1772	54
55	5.6105	5.3719	5.1548	4.9566	4.7748	4.6076	4.4531	4.1772	55
56	5.6106	5.3719	5.1549	4.9566	4.7748	4.6076	4.4531	4.1772	56
57	5.6106	5.3719	5.1549	4.9566	4.7749	4.6076	4.4531	4.1772	57
58	5.6106	5.3719	5.1549	4.9566	4.7749	4.6076	4.4531	4.1772	58
59	5.6106	5.3720	5.1549	4.9566	4.7749	4.6076	4.4531	4.1772	59
60	5.6107	5.3720	5.1549	4.9566	4.7749	4.6076	4.4531	4.1772	60
61	5.6107	5.3720	5.1549	4.9566	4.7749	4.6076	4.4531	4.1772	61
62	5.6107	5.3720	5.1549	4.9567	4.7749	4.6076	4.4531	4.1772	62
63	5.6107	5.3720	5.1549	4.9567	4.7749	4.6076	4.4531	4.1772	63
64	5.6107	5.3720	5.1549	4.9567	4.7749	4.6076	4.4531	4.1772	64
65	5.6107	5.3720	5.1549	4.9567	4.7749	4.6076	4.4531	4.1772	65
66	5.6107	5.3720	5.1549	4.9567	4.7749	4.6076	4.4531	4.1772	66
67	5.6107	5.3720	5.1549	4.9567	4.7749	4.6076	4.4531	4.1772	67
68	5.6107	5.3720	5.1549	4.9567	4.7749	4.6076	4.4531	4.1772	68
69	5.6107	5.3720	5.1549	4.9567	4.7749	4.6076	4.4531	4.1772	69
70	5.6107	5.3720	5.1549	4.9567	4.7749	4.6076	4.4531	4.1772	70
71	5.6108	5.3720	5.1549	4.9567	4.7749	4.6076	4.4531	4.1772	71
72	5.6108	5.3720	5.1549	4.9567	4.7749	4.6076	4.4531	4.1772	72
73	5.6108	5.3720	5.1549	4.9567	4.7749	4.6076	4.4531	4.1772	73
74	5.6108	5.3720	5.1549	4.9567	4.7749	4.6076	4.4531	4.1772	74
75	5.6108	5.3720	5.1549	4.9567	4.7749	4.6076	4.4531	4.1772	75
76	5.6108	5.3720	5.1549	4.9567	4.7749	4.6076	4.4531	4.1772	76
77	5.6108	5.3720	5.1549	4.9567	4.7749	4.6076	4.4531	4.1772	77
78	5.6108	5.3720	5.1549	4.9567	4.7749	4.6076	4.4531	4.1772	78
79	5.6108	5.3720	5.1549	4.9567	4.7749	4.6076	4.4531	4.1772	79
80	5.6108	5.3720	5.1549	4.9567	4.7749	4.6076	4.4531	4.1772	80
81	5.6108	5.3720	5.1549	4.9567	4.7749	4.6076	4.4531	4.1772	81
82	5.6108	5.3720	5.1549	4.9567	4.7749	4.6076	4.4531	4.1772	82
83	5.6108	5.3720	5.1549	4.9567	4.7749	4.6076	4.4531	4.1772	83
84	5.6108	5.3720	5.1549	4.9567	4.7749	4.6076	4.4531	4.1772	84
85	5.6108	5.3720	5.1549	4.9567	4.7749	4.6076	4.4531	4.1772	85
86	5.6108	5.3720	5.1549	4.9567	4.7749	4.6076	4.4531	4.1772	86
87	5.6108	5.3720	5.1549	4.9567	4.7749	4.6076	4.4531	4.1772	87
88	5.6108	5.3720	5.1549	4.9567	4.7749	4.6076	4.4531	4.1772	88
89	5.6108	5.3720	5.1549	4.9567	4.7749	4.6076	4.4531	4.1772	89
90	5.6108	5.3720	5.1549	4.9567	4.7749	4.6076	4.4531	4.1772	90
91	5.6108	5.3720	5.1549	4.9567	4.7749	4.6076	4.4531	4.1772	91
92	5.6108	5.3720	5.1549	4.9567	4.7749	4.6076	4.4531	4.1772	92
93	5.6108	5.3720	5.1549	4.9567	4.7749	4.6076	4.4531	4.1772	93
94	5.6108	5.3720	5.1549	4.9567	4.7749	4.6076	4.4531	4.1772	94
95	5.6108	5.3720	5.1549	4.9567	4.7749	4.6076	4.4531	4.1772	95
96	5.6108	5.3720	5.1549	4.9567	4.7749	4.6076	4.4531	4.1772	96
97	5.6108	5.3720	5.1549	4.9567	4.7749	4.6076	4.4531	4.1772	97
98	5.6108	5.3720	5.1549	4.9567	4.7749	4.6076	4.4531	4.1772	98
99	5.6108	5.3720	5.1549	4.9567	4.7749	4.6076	4.4531	4.1772	99
100	5.6108	5.3720	5.1549	4.9567	4.7749	4.6076	4.4531	4.1772	100
PERP	5.6108	5.3720	5.1549	4.9567	4.7749	4.6076	4.4531	4.1772	PERP

* *Note:*—In computing the quarterly in advance figures in these tables, all the rates % quoted above are effective rates, see the Introductory Section, pages xvi-xvii.

YEARS' PURCHASE

(DUAL RATE % PRINCIPLE)

OR

PRESENT VALUE OF
ONE POUND PER ANNUM

*receivable quarterly in advance, allow-
ing for a sinking fund at a given rate
to replace the invested capital and ignoring the effect of income tax on that part
of the income used to provide the annual
sinking fund instalment.*

AT RATES OF INTEREST*

FROM

4% to 20%

AND

ALLOWING FOR THE INVESTMENT

OF SINKING FUNDS AT

3% and 4%

				Rate per Cent*					
Yrs.	4	4.5	5	5.5	6	6.25	6.5	6.75	Yrs.
1	0.9797	0.9752	0.9707	0.9663	0.9620	0.9599	0.9577	0.9556	1
2	1.9135	1.8963	1.8794	1.8630	1.8470	1.8391	1.8313	1.8236	2
3	2.8041	2.7672	2.7316	2.6970	2.6635	2.6471	2.6310	2.6152	3
4	3.6540	3.5917	3.5319	3.4743	3.4189	3.3920	3.3656	3.3396	4
5	4.4657	4.3730	4.2846	4.2002	4.1195	4.0805	4.0423	4.0050	5
6	5.2413	5.1141	4.9936	4.8793	4.7708	4.7185	4.6676	4.6178	6
7	5.9829	5.8177	5.6623	5.5158	5.3775	5.3112	5.2467	5.1840	7
8	6.6924	6.4864	6.2937	6.1133	5.9439	5.8630	5.7845	5.7083	8
9	7.3715	7.1223	6.8907	6.6750	6.4735	6.3777	6.2849	6.1951	9
10	8.0218	7.7276	7.4557	7.2038	6.9697	6.8588	6.7516	6.6481	10
11	8.6449	8.3042	7.9911	7.7024	7.4354	7.3092	7.1877	7.0704	11
12	9.2422	8.8538	8.4988	8.1730	7.8730	7.7317	7.5958	7.4650	12
13	9.8150	9.3781	8.9808	8.6178	8.2849	8.1286	7.9785	7.8343	13
14	10.3646	9.8786	9.4387	9.0386	8.6731	8.5019	8.3379	8.1805	14
15	10.8921	10.3567	9.8742	9.4371	9.0394	8.8536	8.6759	8.5056	15
16	11.3986	10.8135	10.2886	9.8150	9.3855	9.1854	8.9942	8.8113	16
17	11.8851	11.2504	10.6833	10.1736	9.7128	9.4987	9.2944	9.0993	17
18	12.3526	11.6685	11.0596	10.5142	10.0228	9.7950	9.5779	9.3708	18
19	12.8020	12.0686	11.4184	10.8380	10.3167	10.0754	9.8459	9.6272	19
20	13.2341	12.4519	11.7610	11.1461	10.5955	10.3412	10.0995	9.8695	20
21	13.6498	12.8192	12.0881	11.4395	10.8603	10.5932	10.3398	10.0988	21
22	14.0497	13.1713	12.4007	11.7191	11.1119	10.8325	10.5676	10.3161	22
23	14.4346	13.5091	12.6996	11.9857	11.3513	11.0599	10.7839	10.5221	23
24	14.8052	13.8331	12.9856	12.2401	11.5793	11.2762	10.9894	10.7177	24
25	15.1621	14.1442	13.2593	12.4830	11.7964	11.4821	11.1849	10.9035	25
26	15.5059	14.4429	13.5215	12.7151	12.0035	11.6781	11.3708	11.0801	26
27	15.8371	14.7299	13.7727	12.9370	12.2010	11.8650	11.5480	11.2482	27
28	16.1564	15.0056	14.0135	13.1492	12.3896	12.0433	11.7168	11.4083	28
29	16.4641	15.2707	14.2444	13.3524	12.5698	12.2135	11.8778	11.5609	29
30	16.7608	15.5257	14.4660	13.5469	12.7420	12.3760	12.0315	11.7065	30
31	17.0470	15.7709	14.6787	13.7332	12.9068	12.5314	12.1782	11.8454	31
32	17.3231	16.0069	14.8829	13.9118	13.0644	12.6799	12.3185	11.9780	32
33	17.5895	16.2341	15.0791	14.0831	13.2153	12.8220	12.4526	12.1047	33
34	17.8465	16.4528	15.2676	14.2474	13.3599	12.9581	12.5808	12.2259	34
35	18.0946	16.6635	15.4488	14.4051	13.4985	13.0884	12.7036	12.3419	35
36	18.3341	16.8664	15.6231	14.5565	13.6313	13.2133	12.8212	12.4528	36
37	18.5654	17.0619	15.7907	14.7019	13.7587	13.3330	12.9339	12.5591	37
38	18.7888	17.2504	15.9520	14.8416	13.8810	13.4478	13.0419	12.6609	38
39	19.0045	17.4321	16.1073	14.9759	13.9984	13.5579	13.1455	12.7585	39
40	19.2130	17.6073	16.2567	15.1050	14.1112	13.6637	13.2449	12.8521	40
41	19.4143	17.7762	16.4007	15.2292	14.2195	13.7652	13.3403	12.9419	41
42	19.6089	17.9393	16.5393	15.3487	14.3236	13.8628	13.4319	13.0281	42
43	19.7970	18.0965	16.6729	15.4637	14.4237	13.9565	13.5199	13.1109	43
44	19.9788	18.2483	16.8017	15.5744	14.5200	14.0466	13.6044	13.1904	44
45	20.1546	18.3949	16.9259	15.6810	14.6126	14.1333	13.6857	13.2667	45
46	20.3246	18.5363	17.0456	15.7837	14.7018	14.2167	13.7638	13.3402	46
47	20.4889	18.6729	17.1610	15.8826	14.7876	14.2969	13.8390	13.4108	47
48	20.6479	18.8049	17.2724	15.9780	14.8702	14.3741	13.9113	13.4787	48
49	20.8016	18.9323	17.3798	16.0699	14.9498	14.4484	13.9810	13.5440	49
50	20.9503	19.0554	17.4835	16.1585	15.0264	14.5200	14.0480	13.6069	50

* *Note:*—In computing the quarterly in advance figures in these tables, all the rates % quoted above are effective rates, see the Introductory Section, page xvi.

16

Yrs.	4	4.5	5	5.5	6	6.25	6.5	6.75	Yrs.
51	21.0942	19.1744	17.5836	16.2440	15.1003	14.5890	14.1125	13.6675	51
52	21.2334	19.2893	17.6802	16.3264	15.1715	14.6554	14.1747	13.7258	52
53	21.3682	19.4005	17.7736	16.4059	15.2402	14.7195	14.2346	13.7820	53
54	21.4985	19.5079	17.8637	16.4827	15.3064	14.7813	14.2924	13.8361	54
55	21.6247	19.6117	17.9507	16.5567	15.3702	14.8408	14.3480	13.8882	55
56	21.7469	19.7121	18.0348	16.6283	15.4318	14.8982	14.4017	13.9385	56
57	21.8651	19.8093	18.1160	16.6973	15.4913	14.9536	14.4535	13.9870	57
58	21.9796	19.9032	18.1946	16.7640	15.5487	15.0071	14.5034	14.0338	58
59	22.0905	19.9940	18.2705	16.8284	15.6041	15.0587	14.5516	14.0789	59
60	22.1978	20.0819	18.3438	16.8906	15.6575	15.1085	14.5981	14.1224	60
61	22.3017	20.1669	18.4147	16.9507	15.7092	15.1566	14.6430	14.1644	61
62	22.4024	20.2492	18.4833	17.0088	15.7590	15.2030	14.6863	14.2049	62
63	22.4999	20.3288	18.5496	17.0649	15.8072	15.2478	14.7281	14.2441	63
64	22.5943	20.4059	18.6138	17.1192	15.8538	15.2911	14.7685	14.2819	64
65	22.6858	20.4805	18.6758	17.1717	15.8988	15.3330	14.8076	14.3183	65
66	22.7744	20.5527	18.7358	17.2224	15.9422	15.3734	14.8453	14.3536	66
67	22.8602	20.6225	18.7939	17.2714	15.9842	15.4125	14.8817	14.3876	67
68	22.9434	20.6902	18.8500	17.3189	16.0249	15.4502	14.9169	14.4205	68
69	23.0240	20.7557	18.9044	17.3647	16.0641	15.4867	14.9509	14.4523	69
70	23.1021	20.8192	18.9570	17.4091	16.1021	15.5220	14.9838	14.4831	70
71	23.1777	20.8806	19.0079	17.4521	16.1388	15.5561	15.0156	14.5128	71
72	23.2511	20.9401	19.0572	17.4936	16.1744	15.5891	15.0463	14.5415	72
73	23.3221	20.9977	19.1050	17.5338	16.2087	15.6211	15.0761	14.5693	73
74	23.3910	21.0535	19.1512	17.5727	16.2420	15.6519	15.1048	14.5961	74
75	23.4578	21.1076	19.1959	17.6104	16.2741	15.6818	15.1326	14.6221	75
76	23.5225	21.1600	19.2392	17.6468	16.3052	15.7107	15.1595	14.6472	76
77	23.5852	21.2107	19.2811	17.6821	16.3354	15.7386	15.1856	14.6715	77
78	23.6460	21.2599	19.3218	17.7163	16.3645	15.7657	15.2108	14.6950	78
79	23.7050	21.3075	19.3611	17.7493	16.3927	15.7919	15.2351	14.7177	79
80	23.7621	21.3537	19.3992	17.7813	16.4200	15.8172	15.2587	14.7398	80
81	23.8176	21.3984	19.4361	17.8124	16.4465	15.8418	15.2815	14.7611	81
82	23.8713	21.4418	19.4719	17.8424	16.4721	15.8655	15.3036	14.7817	82
83	23.9234	21.4838	19.5065	17.8715	16.4969	15.8885	15.3250	14.8016	83
84	23.9739	21.5245	19.5401	17.8996	16.5209	15.9108	15.3457	14.8210	84
85	24.0229	21.5640	19.5726	17.9269	16.5441	15.9323	15.3658	14.8397	85
86	24.0703	21.6023	19.6041	17.9534	16.5666	15.9532	15.3852	14.8578	86
87	24.1164	21.6394	19.6347	17.9790	16.5884	15.9734	15.4040	14.8753	87
88	24.1611	21.6753	19.6643	18.0038	16.6095	15.9930	15.4222	14.8923	88
89	24.2044	21.7102	19.6930	18.0278	16.6300	16.0120	15.4399	14.9087	89
90	24.2464	21.7440	19.7208	18.0511	16.6498	16.0303	15.4569	14.9247	90
91	24.2871	21.7767	19.7477	18.0737	16.6690	16.0481	15.4735	14.9401	91
92	24.3266	21.8085	19.7738	18.0956	16.6876	16.0654	15.4895	14.9550	92
93	24.3650	21.8393	19.7991	18.1168	16.7056	16.0821	15.5050	14.9695	93
94	24.4022	21.8692	19.8237	18.1373	16.7231	16.0983	15.5201	14.9835	94
95	24.4382	21.8981	19.8475	18.1572	16.7400	16.1140	15.5347	14.9971	95
96	24.4732	21.9262	19.8705	18.1765	16.7564	16.1292	15.5488	15.0103	96
97	24.5071	21.9534	19.8929	18.1952	16.7723	16.1439	15.5625	15.0230	97
98	24.5400	21.9798	19.9146	18.2134	16.7878	16.1582	15.5758	15.0354	98
99	24.5720	22.0054	19.9356	18.2310	16.8027	16.1720	15.5886	15.0474	99
100	24.6030	22.0303	19.9560	18.2480	16.8172	16.1854	15.6011	15.0590	100

Rate per Cent*

* *Note:*—In computing the quarterly in advance figures in these tables, all the rates % quoted above are effective rates,
    see the Introductory Section, page xvi.

Rate per Cent*

Yrs.	7	7.25	7.5	8	8.5	9	9.5	10	Yrs.
1	0.9535	0.9514	0.9494	0.9453	0.9413	0.9373	0.9334	0.9295	1
2	1.8160	1.8084	1.8010	1.7863	1.7720	1.7580	1.7443	1.7308	2
3	2.5995	2.5841	2.5689	2.5392	2.5103	2.4823	2.4550	2.4285	3
4	3.3142	3.2892	3.2646	3.2168	3.1706	3.1260	3.0828	3.0411	4
5	3.9684	3.9326	3.8976	3.8296	3.7643	3.7016	3.6412	3.5831	5
6	4.5693	4.5219	4.4756	4.3862	4.3008	4.2191	4.1409	4.0659	6
7	5.1229	5.0634	5.0054	4.8938	4.7877	4.6867	4.5904	4.4985	7
8	5.6343	5.5624	5.4925	5.3585	5.2316	5.1112	4.9968	4.8881	8
9	6.1080	6.0236	5.9418	5.7852	5.6375	5.4980	5.3659	5.2407	9
10	6.5479	6.4510	6.3572	6.1783	6.0102	5.8518	5.7024	5.5613	10
11	6.9572	6.8479	6.7423	6.5415	6.3533	6.1766	6.0104	5.8538	11
12	7.3389	7.2174	7.1002	6.8778	6.6701	6.4756	6.2932	6.1217	12
13	7.6956	7.5621	7.4335	7.1901	6.9633	6.7517	6.5536	6.3678	13
14	8.0294	7.8842	7.7445	7.4806	7.2355	7.0073	6.7941	6.5947	14
15	8.3424	8.1857	8.0353	7.7516	7.4887	7.2445	7.0169	6.8044	15
16	8.6363	8.4685	8.3076	8.0047	7.7247	7.4651	7.2237	6.9986	16
17	8.9127	8.7341	8.5631	8.2416	7.9451	7.6707	7.4161	7.1791	17
18	9.1730	8.9840	8.8031	8.4638	8.1513	7.8628	7.5954	7.3470	18
19	9.4186	9.2194	9.0290	8.6723	8.3446	8.0425	7.7630	7.5037	19
20	9.6504	9.4414	9.2418	8.8685	8.5261	8.2109	7.9198	7.6501	20
21	9.8695	9.6510	9.4426	9.0532	8.6967	8.3690	8.0668	7.7872	21
22	10.0769	9.8493	9.6323	9.2275	8.8574	8.5177	8.2048	7.9157	22
23	10.2734	10.0369	9.8117	9.3919	9.0088	8.6576	8.3346	8.0365	23
24	10.4598	10.2147	9.9815	9.5474	9.1518	8.7896	8.4569	8.1501	24
25	10.6367	10.3833	10.1424	9.6946	9.2869	8.9142	8.5721	8.2571	25
26	10.8047	10.5434	10.2951	9.8340	9.4147	9.0319	8.6809	8.3580	26
27	10.9645	10.6955	10.4401	9.9662	9.5358	9.1433	8.7838	8.4533	27
28	11.1166	10.8402	10.5779	10.0917	9.6507	9.2488	8.8811	8.5434	28
29	11.2614	10.9778	10.7089	10.2109	9.7596	9.3488	8.9733	8.6287	29
30	11.3995	11.1090	10.8337	10.3243	9.8631	9.4438	9.0607	8.7095	30
31	11.5311	11.2340	10.9525	10.4321	9.9616	9.5340	9.1437	8.7861	31
32	11.6568	11.3532	11.0658	10.5349	10.0552	9.6197	9.2225	8.8589	32
33	11.7768	11.4670	11.1739	10.6328	10.1444	9.7013	9.2975	8.9280	33
34	11.8915	11.5757	11.2771	10.7262	10.2293	9.7790	9.3688	8.9938	34
35	12.0011	11.6796	11.3757	10.8153	10.3104	9.8530	9.4368	9.0564	35
36	12.1060	11.7789	11.4699	10.9004	10.3877	9.9236	9.5015	9.1160	36
37	12.2064	11.8739	11.5600	10.9818	10.4615	9.9910	9.5632	9.1728	37
38	12.3026	11.9649	11.6462	11.0596	10.5321	10.0553	9.6222	9.2270	38
39	12.3947	12.0520	11.7287	11.1340	10.5995	10.1167	9.6784	9.2787	39
40	12.4830	12.1355	11.8078	11.2052	10.6641	10.1755	9.7322	9.3281	40
41	12.5677	12.2156	11.8835	11.2734	10.7258	10.2317	9.7836	9.3753	41
42	12.6490	12.2923	11.9562	11.3387	10.7849	10.2855	9.8328	9.4205	42
43	12.7270	12.3660	12.0258	11.4013	10.8416	10.3370	9.8798	9.4637	43
44	12.8019	12.4367	12.0927	11.4614	10.8959	10.3864	9.9249	9.5050	44
45	12.8738	12.5045	12.1568	11.5190	10.9480	10.4337	9.9681	9.5446	45
46	12.9429	12.5698	12.2185	11.5744	10.9979	10.4790	10.0095	9.5826	46
47	13.0094	12.6324	12.2777	11.6275	11.0459	10.5226	10.0492	9.6189	47
48	13.0733	12.6927	12.3346	11.6785	11.0919	10.5643	10.0873	9.6538	48
49	13.1348	12.7506	12.3893	11.7275	11.1361	10.6044	10.1238	9.6873	49
50	13.1939	12.8063	12.4419	11.7746	11.1786	10.6429	10.1589	9.7194	50

* *Note:*—In computing the quarterly in advance figures in these tables, all the rates % quoted above are effective rates, see the Introductory Section, page xvi.

Rate per Cent*

Yrs.	7	7.25	7.5	8	8.5	9	9.5	10	Yrs.
51	13.2508	12.8599	12.4925	11.8200	11.2194	10.6800	10.1927	9.7503	51
52	13.3056	12.9115	12.5412	11.8635	11.2587	10.7155	10.2250	9.7799	52
53	13.3584	12.9612	12.5881	11.9055	11.2965	10.7497	10.2562	9.8084	53
54	13.4092	13.0091	12.6332	11.9458	11.3328	10.7826	10.2861	9.8358	54
55	13.4582	13.0552	12.6767	11.9847	11.3678	10.8143	10.3149	9.8621	55
56	13.5054	13.0996	12.7185	12.0221	11.4014	10.8447	10.3426	9.8875	56
57	13.5510	13.1424	12.7589	12.0582	11.4339	10.8741	10.3693	9.9118	57
58	13.5948	13.1837	12.7978	12.0929	11.4651	10.9023	10.3950	9.9353	58
59	13.6372	13.2235	12.8353	12.1264	11.4952	10.9295	10.4197	9.9579	59
60	13.6780	13.2619	12.8715	12.1587	11.5242	10.9557	10.4435	9.9796	60
61	13.7174	13.2989	12.9063	12.1898	11.5521	10.9810	10.4665	10.0006	61
62	13.7554	13.3346	12.9400	12.2198	11.5791	11.0053	10.4886	10.0208	62
63	13.7921	13.3691	12.9724	12.2487	11.6051	11.0288	10.5099	10.0402	63
64	13.8275	13.4024	13.0038	12.2767	11.6301	11.0514	10.5305	10.0590	64
65	13.8617	13.4345	13.0340	12.3036	11.6543	11.0733	10.5503	10.0771	65
66	13.8948	13.4656	13.0632	12.3296	11.6777	11.0944	10.5694	10.0945	66
67	13.9267	13.4955	13.0914	12.3548	11.7002	11.1147	10.5879	10.1114	67
68	13.9575	13.5245	13.1187	12.3790	11.7219	11.1343	10.6057	10.1276	68
69	13.9873	13.5524	13.1450	12.4024	11.7429	11.1533	10.6229	10.1433	69
70	14.0161	13.5795	13.1704	12.4251	11.7632	11.1715	10.6395	10.1584	70
71	14.0439	13.6056	13.1949	12.4469	11.7828	11.1892	10.6555	10.1730	71
72	14.0708	13.6308	13.2187	12.4680	11.8017	11.2063	10.6710	10.1871	72
73	14.0968	13.6552	13.2416	12.4884	11.8200	11.2228	10.6859	10.2007	73
74	14.1219	13.6788	13.2638	12.5082	11.8377	11.2387	10.7003	10.2139	74
75	14.1462	13.7016	13.2852	12.5272	11.8547	11.2541	10.7143	10.2266	75
76	14.1697	13.7236	13.3060	12.5457	11.8713	11.2689	10.7278	10.2389	76
77	14.1924	13.7450	13.3260	12.5635	11.8872	11.2833	10.7408	10.2507	77
78	14.2144	13.7656	13.3454	12.5807	11.9026	11.2972	10.7534	10.2622	78
79	14.2357	13.7856	13.3642	12.5974	11.9176	11.3107	10.7656	10.2733	79
80	14.2563	13.8049	13.3823	12.6135	11.9320	11.3237	10.7773	10.2840	80
81	14.2762	13.8235	13.3999	12.6291	11.9459	11.3362	10.7887	10.2944	81
82	14.2955	13.8416	13.4169	12.6442	11.9594	11.3484	10.7997	10.3044	82
83	14.3142	13.8591	13.4333	12.6588	11.9725	11.3601	10.8104	10.3141	83
84	14.3323	13.8761	13.4492	12.6729	11.9851	11.3715	10.8207	10.3235	84
85	14.3498	13.8925	13.4646	12.6866	11.9974	11.3825	10.8307	10.3325	85
86	14.3667	13.9083	13.4795	12.6998	12.0092	11.3932	10.8403	10.3413	86
87	14.3831	13.9237	13.4939	12.7126	12.0206	11.4035	10.8496	10.3498	87
88	14.3990	13.9386	13.5079	12.7250	12.0317	11.4135	10.8587	10.3580	88
89	14.4143	13.9530	13.5214	12.7370	12.0425	11.4231	10.8674	10.3660	89
90	14.4292	13.9669	13.5345	12.7487	12.0529	11.4325	10.8759	10.3737	90
91	14.4436	13.9804	13.5472	12.7599	12.0629	11.4415	10.8840	10.3811	91
92	14.4576	13.9935	13.5595	12.7708	12.0727	11.4503	10.8920	10.3883	92
93	14.4711	14.0062	13.5714	12.7814	12.0821	11.4588	10.8997	10.3953	93
94	14.4842	14.0185	13.5829	12.7916	12.0912	11.4670	10.9071	10.4021	94
95	14.4969	14.0304	13.5941	12.8015	12.1001	11.4749	10.9143	10.4086	95
96	14.5092	14.0419	13.6049	12.8111	12.1086	11.4826	10.9213	10.4150	96
97	14.5212	14.0530	13.6154	12.8204	12.1169	11.4901	10.9280	10.4211	97
98	14.5327	14.0639	13.6256	12.8294	12.1250	11.4973	10.9346	10.4271	98
99	14.5439	14.0743	13.6354	12.8381	12.1328	11.5043	10.9409	10.4328	99
100	14.5547	14.0845	13.6449	12.8466	12.1403	11.5111	10.9470	10.4384	100

* *Note:*—In computing the quarterly in advance figures in these tables, all the rates % quoted above are effective rates, see the Introductory Section, page xvi.

				Rate per Cent*					
Yrs.	11	12	13	14	15	16	18	20	Yrs.
1	0.9219	0.9146	0.9074	0.9004	0.8936	0.8870	0.8743	0.8622	1
2	1.7048	1.6798	1.6558	1.6327	1.6105	1.5891	1.5487	1.5110	2
3	2.3775	2.3292	2.2833	2.2396	2.1981	2.1584	2.0845	2.0168	3
4	2.9616	2.8870	2.8168	2.7507	2.6882	2.6292	2.5203	2.4221	4
5	3.4733	3.3711	3.2758	3.1867	3.1032	3.0248	2.8815	2.7538	5
6	3.9251	3.7951	3.6747	3.5629	3.4589	3.3618	3.1857	3.0304	6
7	4.3267	4.1692	4.0244	3.8908	3.7670	3.6521	3.4453	3.2643	7
8	4.6859	4.5018	4.3334	4.1789	4.0364	3.9048	3.6693	3.4647	8
9	5.0090	4.7992	4.6083	4.4339	4.2739	4.1266	3.8645	3.6382	9
10	5.3011	5.0666	4.8544	4.6612	4.4847	4.3228	4.0360	3.7899	10
11	5.5662	5.3083	5.0758	4.8650	4.6730	4.4975	4.1879	3.9235	11
12	5.8079	5.5277	5.2760	5.0486	4.8422	4.6540	4.3233	4.0421	12
13	6.0290	5.7276	5.4578	5.2148	4.9949	4.7949	4.4446	4.1479	13
14	6.2319	5.9105	5.6236	5.3660	5.1334	4.9224	4.5539	4.2430	14
15	6.4189	6.0783	5.7753	5.5040	5.2596	5.0383	4.6530	4.3288	15
16	6.5915	6.2329	5.9147	5.6304	5.3749	5.1440	4.7430	4.4066	16
17	6.7513	6.3756	6.0431	5.7466	5.4807	5.2408	4.8252	4.4775	17
18	6.8996	6.5077	6.1616	5.8537	5.5781	5.3298	4.9005	4.5423	18
19	7.0376	6.6303	6.2714	5.9528	5.6679	5.4117	4.9697	4.6017	19
20	7.1662	6.7444	6.3734	6.0445	5.7510	5.4875	5.0335	4.6563	20
21	7.2864	6.8507	6.4682	6.1298	5.8282	5.5576	5.0925	4.7067	21
22	7.3988	6.9500	6.5567	6.2092	5.8999	5.6228	5.1471	4.7534	22
23	7.5042	7.0429	6.6393	6.2832	5.9667	5.6835	5.1979	4.7967	23
24	7.6031	7.1300	6.7166	6.3524	6.0291	5.7400	5.2452	4.8369	24
25	7.6961	7.2117	6.7891	6.4172	6.0874	5.7929	5.2893	4.8744	25
26	7.7837	7.2886	6.8572	6.4780	6.1421	5.8424	5.3305	4.9094	26
27	7.8663	7.3609	6.9212	6.5351	6.1934	5.8888	5.3691	4.9421	27
28	7.9443	7.4292	6.9815	6.5889	6.2416	5.9324	5.4053	4.9728	28
29	8.0180	7.4936	7.0384	6.6395	6.2870	5.9734	5.4393	5.0016	29
30	8.0877	7.5545	7.0920	6.6872	6.3298	6.0120	5.4713	5.0286	30
31	8.1538	7.6121	7.1428	6.7323	6.3702	6.0484	5.5015	5.0541	31
32	8.2164	7.6666	7.1908	6.7749	6.4084	6.0828	5.5299	5.0780	32
33	8.2758	7.7183	7.2363	6.8153	6.4445	6.1153	5.5568	5.1007	33
34	8.3323	7.7674	7.2794	6.8536	6.4787	6.1461	5.5822	5.1221	34
35	8.3860	7.8141	7.3204	6.8898	6.5111	6.1753	5.6062	5.1423	35
36	8.4371	7.8584	7.3593	6.9243	6.5418	6.2029	5.6290	5.1615	36
37	8.4857	7.9006	7.3962	6.9570	6.5710	6.2292	5.6506	5.1796	37
38	8.5321	7.9408	7.4314	6.9881	6.5988	6.2541	5.6712	5.1969	38
39	8.5763	7.9790	7.4650	7.0178	6.6252	6.2779	5.6907	5.2133	39
40	8.6185	8.0156	7.4969	7.0460	6.6504	6.3004	5.7092	5.2288	40
41	8.6588	8.0504	7.5274	7.0729	6.6743	6.3219	5.7268	5.2436	41
42	8.6973	8.0837	7.5564	7.0986	6.6972	6.3424	5.7437	5.2577	42
43	8.7341	8.1154	7.5842	7.1231	6.7190	6.3620	5.7597	5.2711	43
44	8.7693	8.1458	7.6107	7.1465	6.7398	6.3806	5.7750	5.2839	44
45	8.8030	8.1749	7.6361	7.1688	6.7597	6.3985	5.7896	5.2962	45
46	8.8353	8.2027	7.6604	7.1902	6.7787	6.4155	5.8035	5.3078	46
47	8.8662	8.2294	7.6836	7.2107	6.7969	6.4318	5.8168	5.3190	47
48	8.8958	8.2549	7.7059	7.2303	6.8143	6.4474	5.8296	5.3296	48
49	8.9242	8.2793	7.7272	7.2490	6.8309	6.4623	5.8418	5.3398	49
50	8.9515	8.3028	7.7476	7.2670	6.8469	6.4766	5.8534	5.3495	50

* *Note:*—In computing the quarterly in advance figures in these tables, all the rates % quoted above are effective rates, see the Introductory Section, page xvi.

Yrs.	11	12	13	14	15	16	18	20	Yrs.
				Rate per Cent*					
51	8.9776	8.3253	7.7672	7.2842	6.8622	6.4902	5.8646	5.3589	51
52	9.0028	8.3469	7.7860	7.3008	6.8769	6.5034	5.8753	5.3678	52
53	9.0269	8.3676	7.8040	7.3166	6.8909	6.5159	5.8856	5.3764	53
54	9.0501	8.3876	7.8214	7.3319	6.9044	6.5280	5.8954	5.3846	54
55	9.0724	8.4067	7.8380	7.3465	6.9174	6.5396	5.9049	5.3925	55
56	9.0938	8.4251	7.8540	7.3605	6.9299	6.5507	5.9140	5.4001	56
57	9.1144	8.4428	7.8694	7.3740	6.9418	6.5614	5.9227	5.4073	57
58	9.1343	8.4598	7.8841	7.3870	6.9533	6.5717	5.9310	5.4143	58
59	9.1533	8.4762	7.8983	7.3995	6.9644	6.5816	5.9391	5.4210	59
60	9.1717	8.4919	7.9120	7.4115	6.9750	6.5911	5.9468	5.4274	60
61	9.1894	8.5071	7.9252	7.4230	6.9852	6.6002	5.9542	5.4336	61
62	9.2065	8.5217	7.9379	7.4341	6.9951	6.6090	5.9614	5.4396	62
63	9.2229	8.5358	7.9501	7.4448	7.0046	6.6174	5.9683	5.4453	63
64	9.2387	8.5493	7.9618	7.4552	7.0137	6.6256	5.9749	5.4508	64
65	9.2540	8.5624	7.9732	7.4651	7.0225	6.6334	5.9813	5.4561	65
66	9.2687	8.5750	7.9841	7.4747	7.0309	6.6410	5.9874	5.4612	66
67	9.2829	8.5871	7.9946	7.4839	7.0391	6.6483	5.9933	5.4662	67
68	9.2965	8.5988	8.0047	7.4928	7.0470	6.6553	5.9990	5.4709	68
69	9.3097	8.6101	8.0145	7.5013	7.0546	6.6620	6.0045	5.4755	69
70	9.3225	8.6210	8.0240	7.5096	7.0619	6.6686	6.0098	5.4799	70
71	9.3348	8.6315	8.0331	7.5176	7.0689	6.6749	6.0149	5.4841	71
72	9.3467	8.6417	8.0419	7.5253	7.0757	6.6809	6.0199	5.4882	72
73	9.3581	8.6515	8.0504	7.5327	7.0823	6.6868	6.0246	5.4922	73
74	9.3692	8.6610	8.0586	7.5399	7.0886	6.6924	6.0292	5.4960	74
75	9.3799	8.6701	8.0665	7.5468	7.0948	6.6979	6.0336	5.4997	75
76	9.3902	8.6789	8.0741	7.5535	7.1007	6.7031	6.0379	5.5032	76
77	9.4002	8.6874	8.0815	7.5600	7.1064	6.7082	6.0420	5.5066	77
78	9.4098	8.6957	8.0886	7.5662	7.1119	6.7131	6.0460	5.5099	78
79	9.4192	8.7036	8.0955	7.5722	7.1172	6.7179	6.0499	5.5131	79
80	9.4282	8.7113	8.1021	7.5780	7.1223	6.7225	6.0536	5.5162	80
81	9.4369	8.7188	8.1086	7.5837	7.1273	6.7269	6.0572	5.5192	81
82	9.4453	8.7260	8.1148	7.5891	7.1321	6.7312	6.0606	5.5221	82
83	9.4535	8.7329	8.1208	7.5944	7.1368	6.7353	6.0640	5.5249	83
84	9.4613	8.7396	8.1266	7.5995	7.1413	6.7393	6.0672	5.5276	84
85	9.4689	8.7461	8.1322	7.6044	7.1456	6.7432	6.0704	5.5302	85
86	9.4763	8.7524	8.1377	7.6091	7.1498	6.7469	6.0734	5.5327	86
87	9.4834	8.7585	8.1429	7.6137	7.1538	6.7505	6.0763	5.5351	87
88	9.4903	8.7644	8.1480	7.6182	7.1578	6.7540	6.0792	5.5374	88
89	9.4970	8.7701	8.1529	7.6225	7.1616	6.7574	6.0819	5.5397	89
90	9.5035	8.7756	8.1577	7.6266	7.1652	6.7607	6.0845	5.5419	90
91	9.5097	8.7809	8.1623	7.6306	7.1688	6.7638	6.0871	5.5440	91
92	9.5158	8.7861	8.1668	7.6345	7.1722	6.7669	6.0896	5.5461	92
93	9.5216	8.7911	8.1711	7.6383	7.1756	6.7699	6.0920	5.5481	93
94	9.5273	8.7959	8.1753	7.6420	7.1788	6.7727	6.0943	5.5500	94
95	9.5328	8.8006	8.1793	7.6455	7.1819	6.7755	6.0965	5.5519	95
96	9.5381	8.8051	8.1832	7.6489	7.1849	6.7782	6.0987	5.5537	96
97	9.5433	8.8095	8.1870	7.6522	7.1878	6.7808	6.1008	5.5554	97
98	9.5483	8.8138	8.1907	7.6554	7.1907	6.7833	6.1029	5.5571	98
99	9.5531	8.8179	8.1942	7.6585	7.1934	6.7857	6.1048	5.5588	99
100	9.5578	8.8219	8.1977	7.6615	7.1961	6.7881	6.1067	5.5603	100

* Note:—In computing the quarterly in advance figures in these tables, all the rates % quoted above are effective rates, see the Introductory Section, page xvi.

21

YEARS' PURCHASE*

Rate per Cent*

Yrs.	4	4.5	5	5.5	6	6.25	6.5	6.75	Yrs.
1	0.9855	0.9809	0.9764	0.9719	0.9675	0.9653	0.9632	0.9611	1
2	1.9330	1.9154	1.8983	1.8815	1.8652	1.8571	1.8492	1.8413	2
3	2.8441	2.8062	2.7696	2.7341	2.6996	2.6828	2.6663	2.6500	3
4	3.7202	3.6556	3.5937	3.5341	3.4768	3.4490	3.4217	3.3949	4
5	4.5626	4.4659	4.3737	4.2858	4.2018	4.1612	4.1215	4.0827	5
6	5.3725	5.2389	5.1126	4.9929	4.8793	4.8246	4.7714	4.7194	6
7	6.1514	5.9769	5.8129	5.6587	5.5132	5.4435	5.3758	5.3100	7
8	6.9002	6.6814	6.4772	6.2862	6.1072	6.0219	5.9391	5.8588	8
9	7.6203	7.3543	7.1077	6.8784	6.6646	6.5631	6.4649	6.3699	9
10	8.3127	7.9971	7.7064	7.4375	7.1883	7.0703	6.9565	6.8466	10
11	8.9784	8.6114	8.2752	7.9660	7.6807	7.5462	7.4167	7.2919	11
12	9.6186	9.1986	8.8160	8.4659	8.1444	7.9933	7.8482	7.7086	12
13	10.2341	9.7600	9.3303	8.9391	8.5815	8.4139	8.2532	8.0989	13
14	10.8259	10.2968	9.8197	9.3874	8.9937	8.8098	8.6338	8.4652	14
15	11.3950	10.8103	10.2857	9.8123	9.3830	9.1830	8.9920	8.8092	15
16	11.9422	11.3016	10.7294	10.2154	9.7509	9.5351	9.3293	9.1327	16
17	12.4683	11.7717	11.1522	10.5979	10.0989	9.8676	9.6473	9.4372	17
18	12.9742	12.2216	11.5553	10.9612	10.4282	10.1818	9.9474	9.7242	18
19	13.4607	12.6523	11.9396	11.3064	10.7402	10.4790	10.2309	9.9950	19
20	13.9284	13.0647	12.3061	11.6346	11.0359	10.7603	10.4989	10.2506	20
21	14.3782	13.4596	12.6559	11.9467	11.3164	11.0268	10.7524	10.4921	21
22	14.8106	13.8378	12.9897	12.2438	11.5826	11.2794	10.9924	10.7205	22
23	15.2264	14.2002	13.3085	12.5266	11.8353	11.5189	11.2198	10.9367	23
24	15.6263	14.5473	13.6129	12.7959	12.0755	11.7463	11.4354	11.1415	24
25	16.0107	14.8799	13.9038	13.0526	12.3038	11.9622	11.6400	11.3355	25
26	16.3804	15.1987	14.1817	13.2972	12.5210	12.1674	11.8341	11.5196	26
27	16.7358	15.5042	14.4474	13.5305	12.7276	12.3624	12.0186	11.6943	27
28	17.0776	15.7971	14.7013	13.7530	12.9243	12.5479	12.1938	11.8601	28
29	17.4062	16.0779	14.9442	13.9654	13.1116	12.7244	12.3604	12.0177	29
30	17.7222	16.3471	15.1766	14.1681	13.2901	12.8924	12.5189	12.1675	30
31	18.0261	16.6053	15.3988	14.3616	13.4603	13.0525	12.6698	12.3099	31
32	18.3182	16.8529	15.6115	14.5464	13.6225	13.2050	12.8134	12.4455	32
33	18.5991	17.0904	15.8151	14.7230	13.7772	13.3503	12.9502	12.5745	33
34	18.8692	17.3182	16.0100	14.8917	13.9249	13.4889	13.0806	12.6974	34
35	19.1289	17.5367	16.1965	15.0530	14.0658	13.6211	13.2049	12.8145	35
36	19.3787	17.7463	16.3752	15.2073	14.2004	13.7473	13.3234	12.9261	36
37	19.6188	17.9475	16.5464	15.3547	14.3289	13.8677	13.4365	13.0325	37
38	19.8497	18.1405	16.7103	15.4958	14.4517	13.9827	13.5444	13.1339	38
39	20.0717	18.3258	16.8673	15.6308	14.5690	14.0925	13.6474	13.2308	39
40	20.2852	18.5036	17.0178	15.7599	14.6811	14.1974	13.7458	13.3232	40
41	20.4904	18.6742	17.1621	15.8835	14.7884	14.2976	13.8397	13.4114	41
42	20.6878	18.8380	17.3003	16.0019	14.8909	14.3934	13.9295	13.4957	42
43	20.8776	18.9952	17.4328	16.1152	14.9890	14.4850	14.0152	13.5762	43
44	21.0600	19.1462	17.5599	16.2237	15.0828	14.5726	14.0972	13.6531	44
45	21.2355	19.2911	17.6817	16.3276	15.1726	14.6564	14.1756	13.7267	45
46	21.4042	19.4302	17.7985	16.4272	15.2585	14.7366	14.2506	13.7970	46
47	21.5664	19.5638	17.9105	16.5226	15.3408	14.8133	14.3223	13.8642	47
48	21.7224	19.6920	18.0180	16.6140	15.4195	14.8867	14.3910	13.9285	48
49	21.8724	19.8152	18.1210	16.7015	15.4949	14.9570	14.4566	13.9900	49
50	22.0166	19.9335	18.2199	16.7855	15.5672	15.0243	14.5195	14.0488	50

* *Note:*—In computing the quarterly in advance figures in these tables, all the rates % quoted above are effective rates, see the Introductory Section, page xvi.

22

Rate per Cent*

Yrs.	4	4.5	5	5.5	6	6.25	6.5	6.75	Yrs.
51	22.1553	20.0471	18.3148	16.8660	15.6364	15.0888	14.5797	14.1052	51
52	22.2886	20.1562	18.4058	16.9431	15.7027	15.1505	14.6373	14.1591	52
53	22.4168	20.2610	18.4931	17.0171	15.7662	15.2096	14.6925	14.2107	53
54	22.5401	20.3617	18.5769	17.0881	15.8271	15.2663	14.7453	14.2602	54
55	22.6586	20.4583	18.6574	17.1561	15.8854	15.3206	14.7960	14.3075	55
56	22.7726	20.5512	18.7346	17.2214	15.9414	15.3726	14.8445	14.3529	56
57	22.8822	20.6404	18.8087	17.2840	15.9950	15.4224	14.8910	14.3963	57
58	22.9876	20.7261	18.8798	17.3440	16.0464	15.4702	14.9355	14.4380	58
59	23.0889	20.8084	18.9481	17.4016	16.0957	15.5161	14.9782	14.4779	59
60	23.1863	20.8875	19.0137	17.4569	16.1430	15.5600	15.0192	14.5161	60
61	23.2800	20.9635	19.0766	17.5100	16.1883	15.6021	15.0584	14.5528	61
62	23.3701	21.0365	19.1371	17.5609	16.2318	15.6425	15.0961	14.5879	62
63	23.4567	21.1067	19.1951	17.6097	16.2736	15.6813	15.1322	14.6216	63
64	23.5399	21.1741	19.2509	17.6566	16.3136	15.7185	15.1668	14.6540	64
65	23.6200	21.2389	19.3044	17.7016	16.3520	15.7541	15.2000	14.6849	65
66	23.6970	21.3011	19.3558	17.7449	16.3889	15.7883	15.2318	14.7147	66
67	23.7711	21.3609	19.4051	17.7863	16.4243	15.8212	15.2624	14.7432	67
68	23.8422	21.4184	19.4526	17.8262	16.4582	15.8527	15.2917	14.7705	68
69	23.9107	21.4736	19.4981	17.8644	16.4908	15.8829	15.3198	14.7968	69
70	23.9765	21.5267	19.5418	17.9011	16.5221	15.9119	15.3468	14.8220	70
71	24.0398	21.5777	19.5839	17.9364	16.5521	15.9398	15.3727	14.8461	71
72	24.1006	21.6267	19.6242	17.9702	16.5810	15.9665	15.3976	14.8693	72
73	24.1592	21.6738	19.6630	18.0027	16.6086	15.9922	15.4214	14.8916	73
74	24.2154	21.7190	19.7003	18.0339	16.6352	16.0168	15.4443	14.9129	74
75	24.2695	21.7626	19.7360	18.0639	16.6607	16.0404	15.4663	14.9334	75
76	24.3215	21.8044	19.7704	18.0927	16.6852	16.0632	15.4874	14.9531	76
77	24.3715	21.8446	19.8035	18.1204	16.7087	16.0850	15.5077	14.9720	77
78	24.4196	21.8832	19.8352	18.1470	16.7313	16.1059	15.5272	14.9901	78
79	24.4659	21.9203	19.8657	18.1725	16.7530	16.1260	15.5458	15.0075	79
80	24.5103	21.9560	19.8950	18.1970	16.7739	16.1453	15.5638	15.0242	80
81	24.5531	21.9903	19.9232	18.2206	16.7939	16.1638	15.5810	15.0403	81
82	24.5942	22.0233	19.9502	18.2432	16.8131	16.1816	15.5976	15.0557	82
83	24.6337	22.0550	19.9762	18.2649	16.8316	16.1987	15.6135	15.0705	83
84	24.6717	22.0854	20.0012	18.2858	16.8493	16.2152	15.6287	15.0847	84
85	24.7083	22.1147	20.0252	18.3059	16.8663	16.2309	15.6434	15.0984	85
86	24.7434	22.1429	20.0483	18.3252	16.8827	16.2461	15.6574	15.1115	86
87	24.7772	22.1699	20.0705	18.3437	16.8984	16.2607	15.6710	15.1241	87
88	24.8097	22.1959	20.0918	18.3615	16.9135	16.2746	15.6840	15.1362	88
89	24.8409	22.2209	20.1123	18.3786	16.9280	16.2881	15.6964	15.1478	89
90	24.8710	22.2450	20.1320	18.3950	16.9420	16.3010	15.7084	15.1590	90
91	24.8999	22.2681	20.1509	18.4108	16.9554	16.3134	15.7199	15.1697	91
92	24.9276	22.2903	20.1691	18.4260	16.9682	16.3253	15.7310	15.1800	92
93	24.9543	22.3116	20.1865	18.4406	16.9806	16.3368	15.7416	15.1899	93
94	24.9800	22.3321	20.2033	18.4546	16.9925	16.3478	15.7519	15.1994	94
95	25.0047	22.3519	20.2195	18.4681	17.0039	16.3583	15.7617	15.2086	95
96	25.0284	22.3708	20.2350	18.4810	17.0149	16.3685	15.7711	15.2173	96
97	25.0513	22.3891	20.2499	18.4935	17.0254	16.3782	15.7802	15.2258	97
98	25.0732	22.4066	20.2643	18.5054	17.0356	16.3876	15.7889	15.2339	98
99	25.0943	22.4234	20.2781	18.5169	17.0453	16.3966	15.7972	15.2417	99
100	25.1146	22.4397	20.2913	18.5280	17.0547	16.4053	15.8053	15.2492	100

Note:—In computing the quarterly in advance figures in these tables, all the rates % quoted above are effective rates, see the Introductory Section, page xvi.

Rate per Cent*

Yrs.	7	7.25	7.5	8	8.5	9	9.5	10	Yrs.
1	0.9589	0.9568	0.9547	0.9506	0.9465	0.9425	0.9386	0.9347	1
2	1.8336	1.8259	1.8183	1.8034	1.7887	1.7745	1.7605	1.7468	2
3	2.6339	2.6181	2.6025	2.5720	2.5424	2.5136	2.4857	2.4585	3
4	3.3686	3.3427	3.3174	3.2680	3.2203	3.1743	3.1298	3.0868	4
5	4.0447	4.0075	3.9711	3.9006	3.8329	3.7679	3.7054	3.6453	5
6	4.6687	4.6192	4.5709	4.4777	4.3887	4.3037	4.2224	4.1445	6
7	5.2459	5.1835	5.1228	5.0060	4.8950	4.7894	4.6889	4.5931	7
8	5.7809	5.7053	5.6318	5.4909	5.3577	5.2315	5.1118	4.9981	8
9	6.2779	6.1888	6.1024	5.9374	5.7819	5.6352	5.4966	5.3653	9
10	6.7404	6.6378	6.5385	6.3494	6.1720	6.0051	5.8479	5.6995	10
11	7.1716	7.0555	6.9435	6.7306	6.5316	6.3450	6.1697	6.0048	11
12	7.5742	7.4449	7.3202	7.0841	6.8639	6.6581	6.4654	6.2846	12
13	7.9508	7.8084	7.6714	7.4124	7.1717	6.9473	6.7378	6.5416	13
14	8.3035	8.1483	7.9992	7.7180	7.4574	7.2151	6.9894	6.7785	14
15	8.6342	8.4665	8.3057	8.0029	7.7230	7.4635	7.2222	6.9973	15
16	8.9447	8.7649	8.5926	8.2690	7.9706	7.6945	7.4382	7.1999	16
17	9.2367	9.0450	8.8617	8.5179	8.2016	7.9095	7.6390	7.3878	17
18	9.5114	9.3084	9.1143	8.7510	8.4175	8.1101	7.8260	7.5626	18
19	9.7703	9.5561	9.3517	8.9697	8.6196	8.2976	8.0004	7.7253	19
20	10.0144	9.7895	9.5751	9.1750	8.8090	8.4730	8.1633	7.8771	20
21	10.2448	10.0096	9.7855	9.3680	8.9868	8.6373	8.3158	8.0190	21
22	10.4625	10.2173	9.9839	9.5497	9.1538	8.7915	8.4586	8.1517	22
23	10.6683	10.4134	10.1712	9.7209	9.3110	8.9364	8.5926	8.2761	23
24	10.8630	10.5989	10.3481	9.8823	9.4590	9.0726	8.7185	8.3928	24
25	11.0474	10.7744	10.5153	10.0347	9.5985	9.2009	8.8369	8.5025	25
26	11.2222	10.9405	10.6735	10.1786	9.7302	9.3218	8.9484	8.6056	26
27	11.3879	11.0980	10.8232	10.3148	9.8545	9.4358	9.0534	8.7027	27
28	11.5451	11.2472	10.9652	10.4436	9.9720	9.5435	9.1525	8.7942	28
29	11.6944	11.3888	11.0997	10.5656	10.0832	9.6453	9.2461	8.8806	29
30	11.8361	11.5233	11.2274	10.6812	10.1884	9.7415	9.3345	8.9621	30
31	11.9709	11.6510	11.3485	10.7908	10.2881	9.8326	9.4181	9.0392	31
32	12.0991	11.7723	11.4636	10.8948	10.3826	9.9189	9.4972	9.1120	32
33	12.2210	11.8877	11.5730	10.9936	10.4722	10.0007	9.5722	9.1810	33
34	12.3370	11.9975	11.6770	11.0874	10.5573	10.0783	9.6432	9.2463	34
35	12.4475	12.1019	11.7760	11.1765	10.6381	10.1519	9.7106	9.3083	35
36	12.5528	12.2014	11.8702	11.2613	10.7149	10.2218	9.7745	9.3670	36
37	12.6531	12.2962	11.9598	11.3420	10.7879	10.2882	9.8353	9.4228	37
38	12.7487	12.3865	12.0452	11.4188	10.8574	10.3514	9.8929	9.4757	38
39	12.8399	12.4726	12.1266	11.4919	10.9235	10.4114	9.9478	9.5260	39
40	12.9270	12.5547	12.2042	11.5616	10.9864	10.4686	9.9999	9.5738	40
41	13.0100	12.6330	12.2782	11.6280	11.0463	10.5230	10.0496	9.6193	41
42	13.0893	12.7077	12.3488	11.6913	11.1034	10.5748	10.0968	9.6626	42
43	13.1650	12.7791	12.4162	11.7516	11.1579	10.6241	10.1418	9.7038	43
44	13.2373	12.8472	12.4805	11.8092	11.2098	10.6712	10.1847	9.7430	44
45	13.3064	12.9123	12.5419	11.8642	11.2593	10.7161	10.2255	9.7804	45
46	13.3725	12.9745	12.6006	11.9167	11.3065	10.7589	10.2645	9.8160	46
47	13.4356	13.0339	12.6566	11.9668	11.3516	10.7997	10.3016	9.8500	47
48	13.4960	13.0907	12.7102	12.0146	11.3947	10.8387	10.3371	9.8824	48
49	13.5537	13.1451	12.7614	12.0604	11.4358	10.8759	10.3709	9.9133	49
50	13.6090	13.1970	12.8103	12.1041	11.4751	10.9114	10.4033	9.9428	50

* *Note:*—In computing the quarterly in advance figures in these tables, all the rates % quoted above are effective rates, see the Introductory Section, page xvi.

24

Rate per Cent*

Yrs.	7	7.25	7.5	8	8.5	9	9.5	10	Yrs.
51	13.6618	13.2467	12.8572	12.1459	11.5127	10.9454	10.4341	9.9710	51
52	13.7124	13.2943	12.9019	12.1859	11.5486	10.9778	10.4636	9.9979	52
53	13.7608	13.3398	12.9448	12.2241	11.5829	11.0088	10.4918	10.0237	53
54	13.8072	13.3833	12.9858	12.2607	11.6157	11.0385	10.5187	10.0482	54
55	13.8516	13.4250	13.0251	12.2956	11.6471	11.0668	10.5444	10.0717	55
56	13.8941	13.4649	13.0626	12.3291	11.6772	11.0939	10.5690	10.0942	56
57	13.9348	13.5032	13.0986	12.3612	11.7059	11.1199	10.5926	10.1157	57
58	13.9738	13.5398	13.1331	12.3919	11.7334	11.1447	10.6151	10.1362	58
59	14.0112	13.5749	13.1661	12.4212	11.7598	11.1685	10.6367	10.1558	59
60	14.0470	13.6085	13.1977	12.4494	11.7850	11.1912	10.6573	10.1746	60
61	14.0813	13.6407	13.2280	12.4763	11.8092	11.2130	10.6770	10.1926	61
62	14.1142	13.6716	13.2571	12.5022	11.8323	11.2338	10.6960	10.2099	62
63	14.1458	13.7012	13.2849	12.5269	11.8545	11.2538	10.7141	10.2264	63
64	14.1760	13.7296	13.3116	12.5506	11.8757	11.2730	10.7314	10.2422	64
65	14.2050	13.7568	13.3371	12.5734	11.8960	11.2913	10.7480	10.2573	65
66	14.2329	13.7829	13.3616	12.5951	11.9155	11.3088	10.7639	10.2718	66
67	14.2595	13.8079	13.3851	12.6160	11.9342	11.3257	10.7792	10.2857	67
68	14.2851	13.8319	13.4077	12.6360	11.9521	11.3418	10.7938	10.2990	68
69	14.3097	13.8549	13.4293	12.6552	11.9693	11.3573	10.8078	10.3117	69
70	14.3332	13.8769	13.4500	12.6737	11.9858	11.3721	10.8212	10.3240	70
71	14.3558	13.8981	13.4699	12.6913	12.0016	11.3863	10.8341	10.3357	71
72	14.3775	13.9184	13.4890	12.7083	12.0167	11.4000	10.8464	10.3469	72
73	14.3983	13.9379	13.5073	12.7245	12.0313	11.4130	10.8583	10.3577	73
74	14.4182	13.9566	13.5249	12.7401	12.0452	11.4256	10.8696	10.3680	74
75	14.4374	13.9746	13.5418	12.7551	12.0586	11.4376	10.8805	10.3779	75
76	14.4558	13.9918	13.5579	12.7694	12.0714	11.4491	10.8910	10.3874	76
77	14.4734	14.0084	13.5735	12.7832	12.0837	11.4602	10.9010	10.3965	77
78	14.4904	14.0242	13.5884	12.7964	12.0955	11.4708	10.9106	10.4053	78
79	14.5067	14.0395	13.6027	12.8091	12.1068	11.4810	10.9198	10.4136	79
80	14.5223	14.0541	13.6164	12.8213	12.1177	11.4908	10.9287	10.4217	80
81	14.5373	14.0681	13.6296	12.8330	12.1282	11.5002	10.9371	10.4294	81
82	14.5517	14.0816	13.6422	12.8442	12.1382	11.5092	10.9453	10.4368	82
83	14.5655	14.0946	13.6544	12.8549	12.1478	11.5179	10.9531	10.4439	83
84	14.5788	14.1070	13.6661	12.8653	12.1570	11.5262	10.9606	10.4508	84
85	14.5915	14.1190	13.6773	12.8752	12.1659	11.5341	10.9678	10.4573	85
86	14.6038	14.1304	13.6880	12.8848	12.1744	11.5418	10.9747	10.4636	86
87	14.6156	14.1414	13.6984	12.8939	12.1826	11.5491	10.9814	10.4696	87
88	14.6269	14.1520	13.7083	12.9027	12.1904	11.5562	10.9878	10.4754	88
89	14.6377	14.1622	13.7178	12.9111	12.1980	11.5630	10.9939	10.4810	89
90	14.6481	14.1719	13.7270	12.9193	12.2052	11.5695	10.9998	10.4863	90
91	14.6581	14.1813	13.7358	12.9270	12.2122	11.5757	11.0054	10.4915	91
92	14.6678	14.1903	13.7442	12.9345	12.2189	11.5817	11.0108	10.4964	92
93	14.6770	14.1990	13.7523	12.9417	12.2253	11.5875	11.0160	10.5011	93
94	14.6859	14.2073	13.7601	12.9486	12.2314	11.5930	11.0210	10.5057	94
95	14.6944	14.2153	13.7676	12.9552	12.2373	11.5983	11.0258	10.5100	95
96	14.7026	14.2229	13.7748	12.9616	12.2430	11.6034	11.0305	10.5142	96
97	14.7105	14.2303	13.7817	12.9677	12.2485	11.6083	11.0349	10.5183	97
98	14.7181	14.2374	13.7884	12.9736	12.2537	11.6130	11.0391	10.5221	98
99	14.7253	14.2442	13.7947	12.9793	12.2588	11.6176	11.0432	10.5258	99
100	14.7323	14.2507	13.8009	12.9847	12.2636	11.6219	11.0472	10.5294	100

* *Note:*—In computing the quarterly in advance figures in these tables, all the rates % quoted above are effective rates, see the Introductory Section, page xvi.

Rate per Cent*

Yrs.	11	12	13	14	15	16	18	20	Yrs.
1	0.9270	0.9196	0.9123	0.9053	0.8984	0.8917	0.8788	0.8666	1
2	1.7203	1.6948	1.6704	1.6469	1.6243	1.6026	1.5614	1.5232	2
3	2.4063	2.3568	2.3098	2.2651	2.2226	2.1821	2.1066	2.0375	3
4	3.0049	2.9282	2.8560	2.7880	2.7239	2.6633	2.5516	2.4510	4
5	3.5316	3.4260	3.3276	3.2357	3.1497	3.0689	2.9216	2.7904	5
6	3.9982	3.8634	3.7387	3.6231	3.5155	3.4153	3.2337	3.0738	6
7	4.4141	4.2504	4.0999	3.9613	3.8331	3.7142	3.5005	3.3138	7
8	4.7869	4.5949	4.4196	4.2590	4.1111	3.9747	3.7309	3.5196	8
9	5.1227	4.9035	4.7044	4.5227	4.3564	4.2035	3.9318	3.6978	9
10	5.4265	5.1811	4.9594	4.7579	4.5742	4.4059	4.1084	3.8536	10
11	5.7026	5.4322	5.1889	4.9688	4.7688	4.5861	4.2646	3.9908	11
12	5.9543	5.6601	5.3965	5.1588	4.9435	4.7475	4.4039	4.1124	12
13	6.1845	5.8678	5.5849	5.3308	5.1012	4.8928	4.5286	4.2210	13
14	6.3958	6.0577	5.7567	5.4870	5.2441	5.0241	4.6408	4.3183	14
15	6.5903	6.2318	5.9137	5.6295	5.3741	5.1433	4.7424	4.4061	15
16	6.7696	6.3920	6.0578	5.7599	5.4928	5.2519	4.8345	4.4856	16
17	6.9356	6.5397	6.1903	5.8796	5.6015	5.3512	4.9186	4.5578	17
18	7.0893	6.6762	6.3125	5.9897	5.7014	5.4423	4.9954	4.6237	18
19	7.2321	6.8027	6.4255	6.0914	5.7934	5.5260	5.0659	4.6841	19
20	7.3650	6.9202	6.5301	6.1854	5.8784	5.6033	5.1307	4.7394	20
21	7.4889	7.0294	6.6273	6.2725	5.9570	5.6747	5.1906	4.7904	21
22	7.6045	7.1312	6.7177	6.3534	6.0300	5.7408	5.2459	4.8375	22
23	7.7127	7.2262	6.8020	6.4287	6.0978	5.8023	5.2971	4.8810	23
24	7.8140	7.3151	6.8806	6.4989	6.1609	5.8594	5.3447	4.9214	24
25	7.9089	7.3982	6.9542	6.5645	6.2198	5.9126	5.3889	4.9589	25
26	7.9981	7.4762	7.0230	6.6258	6.2748	5.9623	5.4302	4.9938	26
27	8.0819	7.5494	7.0876	6.6832	6.3263	6.0088	5.4687	5.0263	27
28	8.1608	7.6181	7.1481	6.7371	6.3745	6.0523	5.5047	5.0567	28
29	8.2351	7.6829	7.2051	6.7876	6.4197	6.0930	5.5384	5.0852	29
30	8.3051	7.7438	7.2587	6.8351	6.4622	6.1313	5.5700	5.1118	30
31	8.3712	7.8013	7.3091	6.8799	6.5022	6.1673	5.5996	5.1368	31
32	8.4337	7.8555	7.3567	6.9220	6.5398	6.2011	5.6275	5.1602	32
33	8.4928	7.9067	7.4016	6.9617	6.5753	6.2330	5.6538	5.1823	33
34	8.5486	7.9551	7.4440	6.9992	6.6087	6.2630	5.6785	5.2030	34
35	8.6015	8.0009	7.4841	7.0347	6.6403	6.2914	5.7018	5.2226	35
36	8.6517	8.0443	7.5220	7.0682	6.6701	6.3181	5.7237	5.2410	36
37	8.6992	8.0853	7.5579	7.0999	6.6983	6.3435	5.7445	5.2584	37
38	8.7443	8.1243	7.5919	7.1299	6.7250	6.3674	5.7641	5.2749	38
39	8.7871	8.1612	7.6242	7.1583	6.7503	6.3901	5.7827	5.2904	39
40	8.8278	8.1963	7.6548	7.1853	6.7743	6.4116	5.8003	5.3051	40
41	8.8665	8.2296	7.6838	7.2109	6.7970	6.4319	5.8170	5.3191	41
42	8.9032	8.2613	7.7114	7.2352	6.8186	6.4512	5.8328	5.3323	42
43	8.9382	8.2914	7.7376	7.2582	6.8391	6.4696	5.8477	5.3448	43
44	8.9715	8.3200	7.7626	7.2802	6.8586	6.4870	5.8620	5.3567	44
45	9.0031	8.3472	7.7863	7.3010	6.8771	6.5036	5.8755	5.3680	45
46	9.0333	8.3732	7.8088	7.3208	6.8947	6.5193	5.8883	5.3787	46
47	9.0621	8.3979	7.8303	7.3397	6.9114	6.5343	5.9005	5.3889	47
48	9.0895	8.4214	7.8508	7.3577	6.9274	6.5485	5.9122	5.3985	48
49	9.1157	8.4439	7.8703	7.3748	6.9426	6.5621	5.9232	5.4078	49
50	9.1406	8.4653	7.8889	7.3912	6.9570	6.5750	5.9337	5.4165	50

* *Note:*—In computing the quarterly in advance figures in these tables, all the rates % quoted above are effective rates, see the Introductory Section, page xvi.

				Rate per Cent*					
Yrs.	11	12	13	14	15	16	18	20	Yrs.
51	9.1644	8.4857	7.9066	7.4067	6.9708	6.5873	5.9438	5.4249	51
52	9.1872	8.5052	7.9235	7.4216	6.9840	6.5990	5.9533	5.4328	52
53	9.2089	8.5238	7.9397	7.4357	6.9965	6.6102	5.9624	5.4404	53
54	9.2296	8.5416	7.9551	7.4492	7.0085	6.6209	5.9711	5.4477	54
55	9.2494	8.5585	7.9698	7.4621	7.0199	6.6311	5.9794	5.4546	55
56	9.2684	8.5747	7.9839	7.4745	7.0308	6.6408	5.9873	5.4611	56
57	9.2865	8.5902	7.9973	7.4862	7.0412	6.6501	5.9948	5.4674	57
58	9.3038	8.6050	8.0101	7.4975	7.0511	6.6590	6.0021	5.4734	58
59	9.3203	8.6192	8.0224	7.5082	7.0606	6.6675	6.0089	5.4791	59
60	9.3362	8.6327	8.0341	7.5185	7.0697	6.6756	6.0155	5.4846	60
61	9.3513	8.6457	8.0453	7.5283	7.0784	6.6833	6.0218	5.4898	61
62	9.3658	8.6581	8.0561	7.5377	7.0867	6.6907	6.0278	5.4948	62
63	9.3797	8.6699	8.0663	7.5467	7.0947	6.6978	6.0336	5.4996	63
64	9.3930	8.6813	8.0762	7.5553	7.1023	6.7046	6.0391	5.5042	64
65	9.4057	8.6922	8.0856	7.5635	7.1095	6.7110	6.0443	5.5085	65
66	9.4179	8.7026	8.0946	7.5714	7.1165	6.7172	6.0493	5.5127	66
67	9.4296	8.7125	8.1032	7.5790	7.1231	6.7232	6.0542	5.5167	67
68	9.4408	8.7221	8.1114	7.5862	7.1295	6.7289	6.0588	5.5205	68
69	9.4515	8.7312	8.1193	7.5931	7.1356	6.7343	6.0632	5.5242	69
70	9.4617	8.7400	8.1269	7.5997	7.1415	6.7395	6.0674	5.5277	70
71	9.4716	8.7484	8.1342	7.6061	7.1471	6.7445	6.0714	5.5311	71
72	9.4810	8.7564	8.1411	7.6121	7.1525	6.7493	6.0753	5.5343	72
73	9.4901	8.7641	8.1478	7.6180	7.1576	6.7539	6.0790	5.5373	73
74	9.4987	8.7715	8.1542	7.6236	7.1625	6.7583	6.0826	5.5403	74
75	9.5070	8.7786	8.1603	7.6289	7.1673	6.7625	6.0860	5.5431	75
76	9.5150	8.7854	8.1662	7.6340	7.1718	6.7665	6.0893	5.5458	76
77	9.5226	8.7919	8.1718	7.6390	7.1761	6.7704	6.0924	5.5484	77
78	9.5300	8.7982	8.1772	7.6437	7.1803	6.7741	6.0954	5.5509	78
79	9.5370	8.8042	8.1824	7.6482	7.1843	6.7776	6.0983	5.5533	79
80	9.5438	8.8099	8.1874	7.6525	7.1881	6.7810	6.1010	5.5556	80
81	9.5502	8.8154	8.1921	7.6567	7.1918	6.7843	6.1037	5.5578	81
82	9.5565	8.8207	8.1967	7.6607	7.1953	6.7874	6.1062	5.5599	82
83	9.5624	8.8258	8.2011	7.6645	7.1987	6.7904	6.1086	5.5619	83
84	9.5681	8.8307	8.2053	7.6682	7.2019	6.7933	6.1110	5.5638	84
85	9.5736	8.8354	8.2093	7.6717	7.2050	6.7961	6.1132	5.5657	85
86	9.5789	8.8399	8.2132	7.6751	7.2080	6.7987	6.1154	5.5675	86
87	9.5840	8.8442	8.2169	7.6784	7.2109	6.8013	6.1174	5.5692	87
88	9.5888	8.8483	8.2205	7.6815	7.2136	6.8037	6.1194	5.5708	88
89	9.5935	8.8523	8.2239	7.6845	7.2163	6.8061	6.1213	5.5724	89
90	9.5980	8.8561	8.2272	7.6873	7.2188	6.8083	6.1231	5.5739	90
91	9.6023	8.8597	8.2304	7.6901	7.2212	6.8105	6.1249	5.5754	91
92	9.6064	8.8633	8.2334	7.6928	7.2236	6.8126	6.1266	5.5768	92
93	9.6103	8.8666	8.2363	7.6953	7.2258	6.8146	6.1282	5.5781	93
94	9.6142	8.8699	8.2391	7.6977	7.2280	6.8165	6.1297	5.5794	94
95	9.6178	8.8730	8.2418	7.7001	7.2300	6.8183	6.1312	5.5806	95
96	9.6213	8.8760	8.2444	7.7023	7.2320	6.8201	6.1326	5.5818	96
97	9.6247	8.8788	8.2468	7.7045	7.2339	6.8218	6.1340	5.5829	97
98	9.6279	8.8816	8.2492	7.7066	7.2358	6.8234	6.1353	5.5840	98
99	9.6310	8.8842	8.2515	7.7086	7.2375	6.8250	6.1366	5.5851	99
100	9.6340	8.8868	8.2537	7.7105	7.2392	6.8265	6.1378	5.5861	100

* *Note:*—In computing the quarterly in advance figures in these tables, all the rates % quoted above are effective rates,
see the Introductory Section, page xvi.

YIELD CONVERSION TABLES

TABLE 1: **Nominal Yields to Effective Yields**

TABLE 2: **Effective Yields to Nominal Yields**

Note:—Nominal Yields (Annually in Arrear)
Effective Yields (Quarterly in Advance)
Effective Yields are also known as True Equivalent Yields—see Introduction
page xvii.

YIELD CONVERSION TABLE 1.

Nominal Yields (AIA)* - Effective Yields (QIA)†

Nominal Yield%*	Effective Yield%†	Nominal Yield%*	Effective Yield%†	Nominal Yield%*	Effective Yield%†
1.00	1.0063	4.00	4.1020	7.00	7.3173
1.10	1.1076	4.10	4.2073	7.10	7.4266
1.20	1.2091	4.20	4.3126	7.20	7.5360
1.25	1.2598	4.25	4.3653	7.25	7.5908
1.30	1.3106	4.30	4.4181	7.30	7.6456
1.40	1.4123	4.40	4.5237	7.40	7.7553
1.50	1.5142	4.50	4.6295	7.50	7.8652
1.60	1.6161	4.60	4.7354	7.60	7.9752
1.70	1.7182	4.70	4.8414	7.70	8.0853
1.75	1.7693	4.75	4.8944	7.75	8.1404
1.80	1.8204	4.80	4.9475	7.80	8.1956
1.90	1.9228	4.90	5.0538	7.90	8.3060
2.00	2.0253	5.00	5.1602	8.00	8.4166
2.10	2.1279	5.10	5.2668	8.10	8.5273
2.20	2.2306	5.20	5.3735	8.20	8.6381
2.25	2.2820	5.25	5.4269	8.25	8.6936
2.30	2.3334	5.30	5.4803	8.30	8.7491
2.40	2.4364	5.40	5.5873	8.40	8.8602
2.50	2.5396	5.50	5.6944	8.50	8.9715
2.60	2.6428	5.60	5.8016	8.60	9.0829
2.70	2.7462	5.70	5.9090	8.70	9.1945
2.75	2.7979	5.75	5.9627	8.75	9.2503
2.80	2.8497	5.80	6.0165	8.80	9.3061
2.90	2.9533	5.90	6.1242	8.90	9.4180
3.00	3.0571	6.00	6.2319	9.00	9.5300
3.10	3.1610	6.10	6.3398	9.10	9.6421
3.20	3.2650	6.20	6.4479	9.20	9.7544
3.25	3.3171	6.25	6.5020	9.25	9.8105
3.30	3.3692	6.30	6.5561	9.30	9.8668
3.40	3.4735	6.40	6.6644	9.40	9.9793
3.50	3.5779	6.50	6.7729	9.50	10.0920
3.60	3.6825	6.60	6.8815	9.60	10.2049
3.70	3.7872	6.70	6.9902	9.70	10.3178
3.75	3.8396	6.75	7.0447	9.75	10.3744
3.80	3.8920	6.80	7.0991	9.80	10.4310
3.90	3.9969	6.90	7.2081	9.90	10.5443

* Nominal Yield (Annually in Arrear)
†Effective Yield (Quarterly in Advance)
†Effective Yields are also known as `True Equivalent Yields' - see Introduction page xvii

YIELD CONVERSION TABLE 1.

Nominal Yields (AIA)* - Effective Yields (QIA)†

Nominal Yield%*	Effective Yield%†	Nominal Yield%*	Effective Yield%†	Nominal Yield%*	Effective Yield%†
10.00	10.6577	13.00	14.1290	16.00	17.7376
10.10	10.7712	13.10	14.2471	16.10	17.8603
10.20	10.8850	13.20	14.3653	16.20	17.9832
10.25	10.9419	13.25	14.4244	16.25	18.0447
10.30	10.9988	13.30	14.4836	16.30	18.1062
10.40	11.1128	13.40	14.6021	16.40	18.2294
10.50	11.2270	13.50	14.7207	16.50	18.3528
10.60	11.3413	13.60	14.8396	16.60	18.4763
10.70	11.4557	13.70	14.9585	16.70	18.6000
10.75	11.5130	13.75	15.0180	16.75	18.6619
10.80	11.5703	13.80	15.0776	16.80	18.7238
10.90	11.6850	13.90	15.1969	16.90	18.8479
11.00	11.7999	14.00	15.3163	17.00	18.9720
11.10	11.9150	14.10	15.4359	17.10	19.0964
11.20	12.0302	14.20	15.5556	17.20	19.2209
11.25	12.0878	14.25	15.6155	17.25	19.2832
11.30	12.1455	14.30	15.6755	17.30	19.3455
11.40	12.2610	14.40	15.7955	17.40	19.4703
11.50	12.3766	14.50	15.9157	17.50	19.5953
11.60	12.4924	14.60	16.0361	17.60	19.7205
11.70	12.6083	14.70	16.1566	17.70	19.8458
11.75	12.6663	14.75	16.2169	17.75	19.9085
11.80	12.7244	14.80	16.2773	17.80	19.9713
11.90	12.8406	14.90	16.3981	17.90	20.0969
12.00	12.9570	15.00	16.5191	18.00	20.2227
12.10	13.0735	15.10	16.6402	18.10	20.3487
12.20	13.1902	15.20	16.7615	18.20	20.4748
12.25	13.2486	15.25	16.8222	18.25	20.5379
12.30	13.3070	15.30	16.8830	18.30	20.6011
12.40	13.4240	15.40	17.0046	18.40	20.7276
12.50	13.5411	15.50	17.1263	18.50	20.8542
12.60	13.6584	15.60	17.2483	18.60	20.9810
12.70	13.7758	15.70	17.3704	18.70	21.1080
12.75	13.8346	15.75	17.4315	18.75	21.1715
12.80	13.8934	15.80	17.4926	18.80	21.2351
12.90	14.0111	15.90	17.6150	18.90	21.3624
				19.00	21.4899

* Nominal Yield (Annually in Arrear)
†Effective Yield (Quarterly in Advance)
†Effective Yields are also known as `True Equivalent Yields' - see Introduction page xvii

YIELD CONVERSION TABLE 2.

Effective Yields (QIA)† - Nominal Yields (AIA)*

Effective Yield%†	Nominal Yield%*	Effective Yield%†	Nominal Yield%*	Effective Yield%†	Nominal Yield%*
1.00	0.9938	4.00	3.9029	7.00	6.7090
1.10	1.0925	4.10	3.9981	7.10	6.8008
1.20	1.1911	4.20	4.0931	7.20	6.8925
1.25	1.2403	4.25	4.1406	7.25	6.9384
1.30	1.2895	4.30	4.1880	7.30	6.9842
1.40	1.3879	4.40	4.2829	7.40	7.0757
1.50	1.4861	4.50	4.3776	7.50	7.1671
1.60	1.5842	4.60	4.4721	7.60	7.2584
1.70	1.6822	4.70	4.5666	7.70	7.3496
1.75	1.7311	4.75	4.6138	7.75	7.3951
1.80	1.7800	4.80	4.6610	7.80	7.4407
1.90	1.8778	4.90	4.7552	7.90	7.5317
2.00	1.9754	5.00	4.8494	8.00	7.6225
2.10	2.0729	5.10	4.9434	8.10	7.7133
2.20	2.1702	5.20	5.0373	8.20	7.8040
2.25	2.2189	5.25	5.0842	8.25	7.8493
2.30	2.2675	5.30	5.1311	8.30	7.8946
2.40	2.3646	5.40	5.2248	8.40	7.9850
2.50	2.4617	5.50	5.3184	8.50	8.0754
2.60	2.5586	5.60	5.4119	8.60	8.1656
2.70	2.6553	5.70	5.5052	8.70	8.2558
2.75	2.7037	5.75	5.5519	8.75	8.3008
2.80	2.7520	5.80	5.5985	8.80	8.3458
2.90	2.8486	5.90	5.6916	8.90	8.4358
3.00	2.9450	6.00	5.7847	9.00	8.5256
3.10	3.0413	6.10	5.8776	9.10	8.6153
3.20	3.1375	6.20	5.9704	9.20	8.7050
3.25	3.1856	6.25	6.0168	9.25	8.7497
3.30	3.2336	6.30	6.0631	9.30	8.7945
3.40	3.3295	6.40	6.1557	9.40	8.8839
3.50	3.4254	6.50	6.2482	9.50	8.9733
3.60	3.5211	6.60	6.3405	9.60	9.0625
3.70	3.6167	6.70	6.4328	9.70	9.1516
3.75	3.6645	6.75	6.4789	9.75	9.1961
3.80	3.7122	6.80	6.5250	9.80	9.2406
3.90	3.8076	6.90	6.6170	9.90	9.3295

* Nominal Yield (Annually in Arrear)
†Effective Yield (Quarterly in Advance)
†Effective Yields are also known as 'True Equivalent Yields' - see Introduction page xvii

YIELD CONVERSION TABLE 2.

Effective Yields (QIA)† - Nominal Yields (AIA)*

Effective Yield%†	Nominal Yield%*	Effective Yield%†	Nominal Yield%*	Effective Yield%†	Nominal Yield%*
10.00	9.4184	13.00	12.0369	16.00	14.5700
10.10	9.5071	13.10	12.1227	16.10	14.6530
10.20	9.5957	13.20	12.2084	16.20	14.7360
10.25	9.6400	13.25	12.2512	16.25	14.7774
10.30	9.6842	13.30	12.2940	16.30	14.8188
10.40	9.7726	13.40	12.3795	16.40	14.9016
10.50	9.8610	13.50	12.4649	16.50	14.9842
10.60	9.9492	13.60	12.5502	16.60	15.0668
10.70	10.0373	13.70	12.6354	16.70	15.1493
10.75	10.0813	13.75	12.6780	16.75	15.1905
10.80	10.1253	13.80	12.7206	16.80	15.2317
10.90	10.2132	13.90	12.8056	16.90	15.3140
11.00	10.3010	14.00	12.8905	17.00	15.3962
11.10	10.3888	14.10	12.9754	17.10	15.4784
11.20	10.4764	14.20	13.0601	17.20	15.5604
11.25	10.5202	14.25	13.1025	17.25	15.6014
11.30	10.5639	14.30	13.1448	17.30	15.6424
11.40	10.6513	14.40	13.2294	17.40	15.7243
11.50	10.7387	14.50	13.3138	17.50	15.8060
11.60	10.8259	14.60	13.3982	17.60	15.8877
11.70	10.9130	14.70	13.4825	17.70	15.9694
11.75	10.9565	14.75	13.5246	17.75	16.0101
11.80	11.0001	14.80	13.5667	17.80	16.0509
11.90	11.0870	14.90	13.6508	17.90	16.1323
12.00	11.1738	15.00	13.7348	18.00	16.2137
12.10	11.2606	15.10	13.8188	18.10	16.2949
12.20	11.3472	15.20	13.9026	18.20	16.3761
12.25	11.3905	15.25	13.9445	18.25	16.4167
12.30	11.4338	15.30	13.9863	18.30	16.4572
12.40	11.5202	15.40	14.0700	18.40	16.5382
12.50	11.6066	15.50	14.1536	18.50	16.6192
12.60	11.6928	15.60	14.2370	18.60	16.7000
12.70	11.7790	15.70	14.3204	18.70	16.7808
12.75	11.8221	15.75	14.3621	18.75	16.8211
12.80	11.8651	15.80	14.4037	18.80	16.8614
12.90	11.9511	15.90	14.4869	18.90	16.9420
				19.00	17.0225

* Nominal Yield (Annually in Arrear)
†Effective Yield (Quarterly in Advance)
†Effective Yields are also known as `True Equivalent Yields' - see Introduction page xvii

YEARS' PURCHASE

(IN PERPETUITY)

OR

PRESENT VALUE OF
ONE POUND PER ANNUM

(IN PERPETUITY)

Two separate sets of Years' Purchase figures based on the assumptions that:

- (i) income is received annually in arrear, and
- (ii) income is received quarterly in advance.

Rates of Interest from

1% to 100%*

---

* See note in Introductory Section on page xvi regarding different assumptions in respect of Rates % used for each column.

YEAR'S PURCHASE IN PERPETUITY

Rate %*	ANNUALLY IN ARREAR	QUARTERLY IN ADVANCE	Rate %*	ANNUALLY IN ARREAR	QUARTERLY IN ADVANCE
1	100.0000	100.6242	15	6.6667	7.2808
1.25	80.0000	80.6240	15.25	6.5574	7.1713
1.5	66.6667	67.2905	15.5	6.4516	7.0654
1.75	57.1429	57.7665	15.75	6.3492	6.9628
2	50.0000	50.6235	16	6.2500	6.8634
2.25	44.4444	45.0677	16.25	6.1538	6.7671
2.5	40.0000	40.6231	16.5	6.0606	6.6737
2.75	36.3636	36.9865	16.75	5.9701	6.5831
3	33.3333	33.9560	17	5.8824	6.4951
3.25	30.7692	31.3917	17.25	5.7971	6.4097
3.5	28.5714	29.1937	17.5	5.7143	6.3267
3.75	26.6667	27.2888	17.75	5.6338	6.2460
4	25.0000	25.6219	18	5.5556	6.1676
4.25	23.5294	24.1512	18.25	5.4795	6.0914
4.5	22.2222	22.8438	18.5	5.4054	6.0172
4.75	21.0526	21.6740	18.75	5.3333	5.9449
5	20.0000	20.6212	19	5.2632	5.8746
5.25	19.0476	19.6686	19.25	5.1948	5.8061
5.5	18.1818	18.8026	19.5	5.1282	5.7393
5.75	17.3913	18.0119	19.75	5.0633	5.6742
6	16.6667	17.2871	20	5.0000	5.6108
6.25	16.0000	16.6203	20.25	4.9383	5.5489
6.5	15.3846	16.0047	20.5	4.8780	5.4885
6.75	14.8148	15.4347	20.75	4.8193	5.4296
7	14.2857	14.9054	21	4.7619	5.3720
7.25	13.7931	14.4126	21.25	4.7059	5.3158
7.5	13.3333	13.9527	21.5	4.6512	5.2610
7.75	12.9032	13.5224	21.75	4.5977	5.2073
8	12.5000	13.1190	22	4.5455	5.1549
8.25	12.1212	12.7400	22.25	4.4944	5.1037
8.5	11.7647	12.3833	22.5	4.4444	5.0536
8.75	11.4286	12.0470	22.75	4.3956	5.0046
9	11.1111	11.7294	23	4.3478	4.9567
9.25	10.8108	11.4289	23.25	4.3011	4.9098
9.5	10.5263	11.1442	23.5	4.2553	4.8638
9.75	10.2564	10.8741	23.75	4.2105	4.8189
10	10.0000	10.6176	24	4.1667	4.7749
10.25	9.7561	10.3735	24.25	4.1237	4.7318
10.5	9.5238	10.1410	24.5	4.0816	4.6895
10.75	9.3023	9.9194	24.75	4.0404	4.6481
11	9.0909	9.7078	25	4.0000	4.6076
11.25	8.8889	9.5056	25.25	3.9604	4.5678
11.5	8.6957	9.3121	25.5	3.9216	4.5288
11.75	8.5106	9.1270	25.75	3.8835	4.4906
12	8.3333	8.9495	26	3.8462	4.4531
12.25	8.1633	8.7792	26.25	3.8095	4.4163
12.5	8.0000	8.6158	26.5	3.7736	4.3802
12.75	7.8431	8.4588	26.75	3.7383	4.3448
13	7.6923	8.3078	27	3.7037	4.3100
13.25	7.5472	8.1625	27.25	3.6697	4.2759
13.5	7.4074	8.0225	27.5	3.6364	4.2424
13.75	7.2727	7.8877	27.75	3.6036	4.2095
14	7.1429	7.7576	28	3.5714	4.1772
14.25	7.0175	7.6321	28.25	3.5398	4.1454
14.5	6.8966	7.5110	28.5	3.5088	4.1142
14.75	6.7797	7.3939	28.75	3.4783	4.0835

* See note in Introductory Section on page xvi on different assumptions in respect of Rates % used for each column.

YEAR'S PURCHASE IN PERPETUITY

Rate %*	ANNUALLY IN ARREAR	QUARTERLY IN ADVANCE	Rate %*	ANNUALLY IN ARREAR	QUARTERLY IN ADVANCE
29	3.4483	4.0534	43	2.3256	2.9227
29.25	3.4188	4.0238	43.5	2.2989	2.8957
29.5	3.3898	3.9947	44	2.2727	2.8693
29.75	3.3613	3.9660	44.5	2.2472	2.8435
30	3.3333	3.9379	45	2.2222	2.8183
30.25	3.3058	3.9102	45.5	2.1978	2.7936
30.5	3.2787	3.8829	46	2.1739	2.7694
30.75	3.2520	3.8561	46.5	2.1505	2.7458
31	3.2258	3.8297	47	2.1277	2.7226
31.25	3.2000	3.8038	47.5	2.1053	2.7000
31.5	3.1746	3.7782	48	2.0833	2.6778
31.75	3.1496	3.7531	48.5	2.0619	2.6560
32	3.1250	3.7283	49	2.0408	2.6347
32.25	3.1008	3.7040	49.5	2.0202	2.6139
32.5	3.0769	3.6800	50	2.0000	2.5934
32.75	3.0534	3.6563	50.5	1.9802	2.5734
33	3.0303	3.6331	51	1.9608	2.5537
33.25	3.0075	3.6101	52	1.9231	2.5155
33.5	2.9851	3.5875	53	1.8868	2.4787
33.75	2.9630	3.5653	54	1.8519	2.4432
34	2.9412	3.5433	55	1.8182	2.4091
34.25	2.9197	3.5217	56	1.7857	2.3761
34.5	2.8986	3.5004	57	1.7544	2.3443
34.75	2.8777	3.4794	58	1.7241	2.3135
35	2.8571	3.4587	59	1.6949	2.2838
35.25	2.8369	3.4383	60	1.6667	2.2551
35.5	2.8169	3.4182	61	1.6393	2.2273
35.75	2.7972	3.3984	62	1.6129	2.2004
36	2.7778	3.3788	63	1.5873	2.1743
36.25	2.7586	3.3595	64	1.5625	2.1490
36.5	2.7397	3.3405	65	1.5385	2.1245
36.75	2.7211	3.3217	66	1.5152	2.1007
37	2.7027	3.3032	67	1.4925	2.0777
37.25	2.6846	3.2849	68	1.4706	2.0552
37.5	2.6667	3.2668	69	1.4493	2.0335
37.75	2.6490	3.2490	70	1.4286	2.0123
38	2.6316	3.2315	71	1.4085	1.9917
38.25	2.6144	3.2141	72	1.3889	1.9717
38.5	2.5974	3.1970	73	1.3699	1.9523
38.75	2.5806	3.1801	74	1.3514	1.9333
39	2.5641	3.1634	75	1.3333	1.9149
39.25	2.5478	3.1470	76	1.3158	1.8969
39.5	2.5316	3.1307	77	1.2987	1.8793
39.75	2.5157	3.1146	78	1.2821	1.8623
40	2.5000	3.0988	79	1.2658	1.8456
40.25	2.4845	3.0831	80	1.2500	1.8294
40.5	2.4691	3.0676	81	1.2346	1.8135
40.75	2.4540	3.0523	82	1.2195	1.7980
41	2.4390	3.0372	83	1.2048	1.7829
41.25	2.4242	3.0223	84	1.1905	1.7681
41.5	2.4096	3.0076	85	1.1765	1.7537
41.75	2.3952	2.9930	86	1.1628	1.7396
42	2.3810	2.9786	88	1.1364	1.7124
42.25	2.3669	2.9644	90	1.1111	1.6863
42.5	2.3529	2.9503	95	1.0526	1.6259
42.75	2.3392	2.9364	100	1.0000	1.5713

* See note in Introductory Section on page xvi on different assumptions in respect of Rates % used for each column.

YEARS' PURCHASE

(SINGLE RATE % PRINCIPLE)

OR

PRESENT VALUE OF
ONE POUND PER ANNUM

receivable at the end of each year after allowing for a sinking fund to
accumulate at the same rate of interest as that which is required on
the invested capital and ignoring the effect of income tax on that
part of the income used to provide the annual sinking fund instalment.

At Rates of Interest from
2% to 28%

YEARS' PURCHASE

Rate Per Cent

Yrs.	2	2.25	2.5	2.75	3	3.25	3.5	3.75	Yrs.
1	0.9804	0.9780	0.9756	0.9732	0.9709	0.9685	0.9662	0.9639	1
2	1.9416	1.9345	1.9274	1.9204	1.9135	1.9066	1.8997	1.8929	2
3	2.8839	2.8699	2.8560	2.8423	2.8286	2.8151	2.8016	2.7883	3
4	3.8077	3.7847	3.7620	3.7394	3.7171	3.6950	3.6731	3.6514	4
5	4.7135	4.6795	4.6458	4.6126	4.5797	4.5472	4.5151	4.4833	5
6	5.6014	5.5545	5.5081	5.4624	5.4172	5.3726	5.3286	5.2851	6
7	6.4720	6.4102	6.3494	6.2894	6.2303	6.1720	6.1145	6.0579	7
8	7.3255	7.2472	7.1701	7.0943	7.0197	6.9462	6.8740	6.8028	8
9	8.1622	8.0657	7.9709	7.8777	7.7861	7.6961	7.6077	7.5208	9
10	8.9826	8.8662	8.7521	8.6401	8.5302	8.4224	8.3166	8.2128	10
11	9.7868	9.6491	9.5142	9.3821	9.2526	9.1258	9.0016	8.8798	11
12	10.5753	10.4148	10.2578	10.1042	9.9540	9.8071	9.6633	9.5227	12
13	11.3484	11.1636	10.9832	10.8070	10.6350	10.4669	10.3027	10.1424	13
14	12.1062	11.8959	11.6909	11.4910	11.2961	11.1060	10.9205	10.7396	14
15	12.8493	12.6122	12.3814	12.1567	11.9379	11.7249	11.5174	11.3153	15
16	13.5777	13.3126	13.0550	12.8046	12.5611	12.3244	12.0941	11.8702	16
17	14.2919	13.9977	13.7122	13.4351	13.1661	12.9049	12.6513	12.4050	17
18	14.9920	14.6677	14.3534	14.0488	13.7535	13.4673	13.1897	12.9205	18
19	15.6785	15.3229	14.9789	14.6460	14.3238	14.0119	13.7098	13.4173	19
20	16.3514	15.9637	15.5892	15.2273	14.8775	14.5393	14.2124	13.8962	20
21	17.0112	16.5904	16.1845	15.7929	15.4150	15.0502	14.6980	14.3578	21
22	17.6580	17.2034	16.7654	16.3435	15.9369	15.5450	15.1671	14.8027	22
23	18.2922	17.8028	17.3321	16.8793	16.4436	16.0242	15.6204	15.2315	23
24	18.9139	18.3890	17.8850	17.4008	16.9355	16.4883	16.0584	15.6448	24
25	19.5235	18.9624	18.4244	17.9083	17.4131	16.9379	16.4815	16.0432	25
26	20.1210	19.5231	18.9506	18.4023	17.8768	17.3732	16.8904	16.4272	26
27	20.7069	20.0715	19.4640	18.8830	18.3270	17.7949	17.2854	16.7973	27
28	21.2813	20.6078	19.9649	19.3508	18.7641	18.2033	17.6670	17.1540	28
29	21.8444	21.1323	20.4535	19.8062	19.1885	18.5988	18.0358	17.4978	29
30	22.3965	21.6453	20.9303	20.2493	19.6004	18.9819	18.3920	17.8292	30
31	22.9377	22.1470	21.3954	20.6806	20.0004	19.3529	18.7363	18.1487	31
32	23.4683	22.6377	21.8492	21.1003	20.3888	19.7123	19.0689	18.4565	32
33	23.9886	23.1175	22.2919	21.5088	20.7658	20.0603	19.3902	18.7533	33
34	24.4986	23.5868	22.7238	21.9064	21.1318	20.3974	19.7007	19.0393	34
35	24.9986	24.0458	23.1452	22.2933	21.4872	20.7239	20.0007	19.3150	35
36	25.4888	24.4947	23.5563	22.6699	21.8323	21.0401	20.2905	19.5807	36
37	25.9695	24.9337	23.9573	23.0364	22.1672	21.3463	20.5705	19.8369	37
38	26.4406	25.3630	24.3486	23.3931	22.4925	21.6429	20.8411	20.0837	38
39	26.9026	25.7829	24.7303	23.7402	22.8082	21.9302	21.1025	20.3217	39
40	27.3555	26.1935	25.1028	24.0781	23.1148	22.2084	21.3551	20.5510	40
41	27.7995	26.5951	25.4661	24.4069	23.4124	22.4779	21.5991	20.7720	41
42	28.2348	26.9879	25.8206	24.7269	23.7014	22.7389	21.8349	20.9851	42
43	28.6616	27.3720	26.1664	25.0384	23.9819	22.9917	22.0627	21.1905	43
44	29.0800	27.7477	26.5038	25.3415	24.2543	23.2365	22.2828	21.3884	44
45	29.4902	28.1151	26.8330	25.6365	24.5187	23.4736	22.4955	21.5792	45
46	29.8923	28.4744	27.1542	25.9236	24.7754	23.7032	22.7009	21.7631	46
47	30.2866	28.8259	27.4675	26.2030	25.0247	23.9256	22.8994	21.9403	47
48	30.6731	29.1695	27.7732	26.4749	25.2667	24.1411	23.0912	22.1111	48
49	31.0521	29.5057	28.0714	26.7396	25.5017	24.3497	23.2766	22.2758	49
50	31.4236	29.8344	28.3623	26.9972	25.7298	24.5518	23.4556	22.4345	50

YEARS' PURCHASE

				Rate Per Cent					
Yrs.	2	2.25	2.5	2.75	3	3.25	3.5	3.75	Yrs.
51	31.7878	30.1559	28.6462	27.2479	25.9512	24.7475	23.6286	22.5875	51
52	32.1449	30.4703	28.9231	27.4918	26.1662	24.9370	23.7958	22.7349	52
53	32.4950	30.7778	29.1932	27.7293	26.3750	25.1206	23.9573	22.8770	53
54	32.8383	31.0785	29.4568	27.9604	26.5777	25.2984	24.1133	23.0140	54
55	33.1748	31.3727	29.7140	28.1853	26.7744	25.4706	24.2641	23.1460	55
56	33.5047	31.6603	29.9649	28.4042	26.9655	25.6374	24.4097	23.2733	56
57	33.8281	31.9416	30.2096	28.6172	27.1509	25.7989	24.5504	23.3959	57
58	34.1452	32.2167	30.4484	28.8245	27.3310	25.9554	24.6864	23.5141	58
59	34.4561	32.4858	30.6814	29.0263	27.5058	26.1069	24.8178	23.6281	59
60	34.7609	32.7490	30.9087	29.2227	27.6756	26.2537	24.9447	23.7379	60
61	35.0597	33.0063	31.1304	29.4138	27.8404	26.3958	25.0674	23.8438	61
62	35.3526	33.2580	31.3467	29.5998	28.0003	26.5335	25.1859	23.9458	62
63	35.6398	33.5042	31.5578	29.7808	28.1557	26.6668	25.3004	24.0442	63
64	35.9214	33.7449	31.7637	29.9570	28.3065	26.7959	25.4110	24.1389	64
65	36.1975	33.9803	31.9646	30.1285	28.4529	26.9210	25.5178	24.2303	65
66	36.4681	34.2106	32.1606	30.2953	28.5950	27.0421	25.6211	24.3184	66
67	36.7334	34.4358	32.3518	30.4578	28.7330	27.1594	25.7209	24.4032	67
68	36.9936	34.6560	32.5383	30.6158	28.8670	27.2731	25.8173	24.4851	68
69	37.2486	34.8714	32.7203	30.7697	28.9971	27.3831	25.9104	24.5639	69
70	37.4986	35.0821	32.8979	30.9194	29.1234	27.4897	26.0004	24.6399	70
71	37.7437	35.2881	33.0711	31.0651	29.2460	27.5929	26.0873	24.7132	71
72	37.9841	35.4896	33.2401	31.2069	29.3651	27.6929	26.1713	24.7838	72
73	38.2197	35.6866	33.4050	31.3449	29.4807	27.7897	26.2525	24.8518	73
74	38.4507	35.8794	33.5658	31.4792	29.5929	27.8835	26.3309	24.9174	74
75	38.6771	36.0678	33.7227	31.6100	29.7018	27.9744	26.4067	24.9807	75
76	38.8991	36.2522	33.8758	31.7372	29.8076	28.0623	26.4799	25.0416	76
77	39.1168	36.4324	34.0252	31.8610	29.9103	28.1475	26.5506	25.1003	77
78	39.3302	36.6087	34.1709	31.9815	30.0100	28.2301	26.6190	25.1569	78
79	39.5394	36.7812	34.3131	32.0988	30.1068	28.3100	26.6850	25.2115	79
80	39.7445	36.9498	34.4518	32.2129	30.2008	28.3874	26.7488	25.2641	80
81	39.9456	37.1147	34.5871	32.3240	30.2920	28.4624	26.8104	25.3148	81
82	40.1427	37.2760	34.7192	32.4321	30.3806	28.5350	26.8700	25.3637	82
83	40.3360	37.4337	34.8480	32.5374	30.4666	28.6053	26.9275	25.4108	83
84	40.5255	37.5880	34.9736	32.6398	30.5501	28.6734	26.9831	25.4562	84
85	40.7113	37.7389	35.0962	32.7394	30.6312	28.7394	27.0368	25.4999	85
86	40.8934	37.8864	35.2158	32.8364	30.7099	28.8033	27.0887	25.5421	86
87	41.0720	38.0307	35.3325	32.9308	30.7863	28.8652	27.1388	25.5827	87
88	41.2470	38.1719	35.4463	33.0227	30.8605	28.9251	27.1873	25.6219	88
89	41.4187	38.3099	35.5574	33.1121	30.9325	28.9831	27.2341	25.6597	89
90	41.5869	38.4449	35.6658	33.1992	31.0024	29.0394	27.2793	25.6961	90
91	41.7519	38.5769	35.7715	33.2838	31.0703	29.0938	27.3230	25.7312	91
92	41.9136	38.7060	35.8746	33.3663	31.1362	29.1466	27.3652	25.7650	92
93	42.0722	38.8323	35.9752	33.4465	31.2002	29.1976	27.4060	25.7976	93
94	42.2276	38.9558	36.0734	33.5246	31.2623	29.2471	27.4454	25.8290	94
95	42.3800	39.0766	36.1692	33.6006	31.3227	29.2950	27.4835	25.8592	95
96	42.5294	39.1947	36.2626	33.6745	31.3812	29.3414	27.5203	25.8884	96
97	42.6759	39.3102	36.3538	33.7465	31.4381	29.3864	27.5558	25.9166	97
98	42.8195	39.4232	36.4427	33.8165	31.4933	29.4299	27.5902	25.9437	98
99	42.9603	39.5337	36.5295	33.8847	31.5469	29.4720	27.6234	25.9698	99
100	43.0984	39.6417	36.6141	33.9510	31.5989	29.5129	27.6554	25.9950	100
PERP	50.0000	44.4444	40.0000	36.3636	33.3333	30.7692	28.5714	26.6667	PERP

YEARS' PURCHASE

				Rate Per Cent					
Yrs.	**4**	**4.25**	**4.5**	**4.75**	**5**	**5.25**	**5.5**	**5.75**	Yrs.
1	0.9615	0.9592	0.9569	0.9547	0.9524	0.9501	0.9479	0.9456	1
2	1.8861	1.8794	1.8727	1.8660	1.8594	1.8528	1.8463	1.8398	2
3	2.7751	2.7620	2.7490	2.7361	2.7232	2.7105	2.6979	2.6854	3
4	3.6299	3.6086	3.5875	3.5666	3.5460	3.5255	3.5052	3.4850	4
5	4.4518	4.4207	4.3900	4.3596	4.3295	4.2997	4.2703	4.2412	5
6	5.2421	5.1997	5.1579	5.1165	5.0757	5.0354	4.9955	4.9562	6
7	6.0021	5.9470	5.8927	5.8392	5.7864	5.7343	5.6830	5.6323	7
8	6.7327	6.6638	6.5959	6.5290	6.4632	6.3984	6.3346	6.2717	8
9	7.4353	7.3513	7.2688	7.1876	7.1078	7.0294	6.9522	6.8763	9
10	8.1109	8.0109	7.9127	7.8163	7.7217	7.6288	7.5376	7.4481	10
11	8.7605	8.6435	8.5289	8.4166	8.3064	8.1984	8.0925	7.9887	11
12	9.3851	9.2504	9.1186	8.9896	8.8633	8.7396	8.6185	8.5000	12
13	9.9856	9.8325	9.6829	9.5366	9.3936	9.2538	9.1171	8.9834	13
14	10.5631	10.3909	10.2228	10.0588	9.8986	9.7423	9.5896	9.4406	14
15	11.1184	10.9265	10.7395	10.5573	10.3797	10.2065	10.0376	9.8729	15
16	11.6523	11.4403	11.2340	11.0332	10.8378	10.6475	10.4622	10.2817	16
17	12.1657	11.9332	11.7072	11.4876	11.2741	11.0665	10.8646	10.6683	17
18	12.6593	12.4059	12.1600	11.9213	11.6896	11.4646	11.2461	11.0338	18
19	13.1339	12.8594	12.5933	12.3354	12.0853	11.8428	11.6077	11.3795	19
20	13.5903	13.2944	13.0079	12.7307	12.4622	12.2022	11.9504	11.7064	20
21	14.0292	13.7116	13.4047	13.1080	12.8212	12.5437	12.2752	12.0155	21
22	14.4511	14.1119	13.7844	13.4683	13.1630	12.8681	12.5832	12.3078	22
23	14.8568	14.4958	14.1478	13.8122	13.4886	13.1763	12.8750	12.5842	23
24	15.2470	14.8641	14.4955	14.1405	13.7986	13.4692	13.1517	12.8456	24
25	15.6221	15.2173	14.8282	14.4540	14.0939	13.7475	13.4139	13.0927	25
26	15.9828	15.5562	15.1466	14.7532	14.3752	14.0118	13.6625	13.3265	26
27	16.3296	15.8812	15.4513	15.0389	14.6430	14.2630	13.8981	13.5475	27
28	16.6631	16.1930	15.7429	15.3116	14.8981	14.5017	14.1214	13.7565	28
29	16.9837	16.4921	16.0219	15.5719	15.1411	14.7285	14.3331	13.9541	29
30	17.2920	16.7790	16.2889	15.8204	15.3725	14.9439	14.5337	14.1410	30
31	17.5885	17.0542	16.5444	16.0577	15.5928	15.1486	14.7239	14.3178	31
32	17.8736	17.3182	16.7889	16.2842	15.8027	15.3431	14.9042	14.4849	32
33	18.1476	17.5714	17.0229	16.5004	16.0025	15.5279	15.0751	14.6429	33
34	18.4112	17.8143	17.2468	16.7068	16.1929	15.7034	15.2370	14.7923	34
35	18.6646	18.0473	17.4610	16.9039	16.3742	15.8703	15.3906	14.9337	35
36	18.9083	18.2708	17.6660	17.0920	16.5469	16.0287	15.5361	15.0673	36
37	19.1426	18.4852	17.8622	17.2716	16.7113	16.1793	15.6740	15.1937	37
38	19.3679	18.6908	18.0500	17.4431	16.8679	16.3224	15.8047	15.3131	38
39	19.5845	18.8881	18.2297	17.6068	17.0170	16.4583	15.9287	15.4261	39
40	19.7928	19.0773	18.4016	17.7630	17.1591	16.5875	16.0461	15.5330	40
41	19.9931	19.2588	18.5661	17.9122	17.2944	16.7102	16.1575	15.6340	41
42	20.1856	19.4329	18.7235	18.0546	17.4232	16.8268	16.2630	15.7296	42
43	20.3708	19.5999	18.8742	18.1905	17.5459	16.9376	16.3630	15.8199	43
44	20.5488	19.7601	19.0184	18.3203	17.6628	17.0428	16.4579	15.9054	44
45	20.7200	19.9137	19.1563	18.4442	17.7741	17.1428	16.5477	15.9862	45
46	20.8847	20.0611	19.2884	18.5625	17.8801	17.2378	16.6329	16.0626	46
47	21.0429	20.2025	19.4147	18.6754	17.9810	17.3281	16.7137	16.1348	47
48	21.1951	20.3382	19.5356	18.7832	18.0772	17.4139	16.7902	16.2031	48
49	21.3415	20.4683	19.6513	18.8861	18.1687	17.4954	16.8628	16.2678	49
50	21.4822	20.5931	19.7620	18.9844	18.2559	17.5728	16.9315	16.3288	50

YEARS' PURCHASE

Rate Per Cent

Yrs.	4	4.25	4.5	4.75	5	5.25	5.5	5.75	Yrs.
51	21.6175	20.7128	19.8679	19.0782	18.3390	17.6464	16.9967	16.3866	51
52	21.7476	20.8276	19.9693	19.1677	18.4181	17.7163	17.0585	16.4412	52
53	21.8727	20.9377	20.0663	19.2532	18.4934	17.7827	17.1170	16.4929	53
54	21.9930	21.0434	20.1592	19.3348	18.5651	17.8458	17.1726	16.5417	54
55	22.1086	21.1447	20.2480	19.4127	18.6335	17.9057	17.2252	16.5879	55
56	22.2198	21.2420	20.3330	19.4870	18.6985	17.9627	17.2750	16.6316	56
57	22.3267	21.3352	20.4144	19.5580	18.7605	18.0168	17.3223	16.6729	57
58	22.4296	21.4247	20.4922	19.6258	18.8195	18.0682	17.3671	16.7120	58
59	22.5284	21.5105	20.5667	19.6905	18.8758	18.1171	17.4096	16.7489	59
60	22.6235	21.5928	20.6380	19.7523	18.9293	18.1635	17.4499	16.7839	60
61	22.7149	21.6717	20.7062	19.8112	18.9803	18.2076	17.4880	16.8169	61
62	22.8028	21.7475	20.7715	19.8675	19.0288	18.2495	17.5242	16.8481	62
63	22.8873	21.8201	20.8340	19.9213	19.0751	18.2893	17.5585	16.8777	63
64	22.9685	21.8898	20.8938	19.9726	19.1191	18.3271	17.5910	16.9056	64
65	23.0467	21.9566	20.9510	20.0215	19.1611	18.3631	17.6218	16.9320	65
66	23.1218	22.0208	21.0057	20.0683	19.2010	18.3972	17.6510	16.9570	66
67	23.1940	22.0823	21.0581	20.1129	19.2391	18.4297	17.6786	16.9806	67
68	23.2635	22.1413	21.1082	20.1555	19.2753	18.4605	17.7049	17.0029	68
69	23.3303	22.1978	21.1562	20.1962	19.3098	18.4898	17.7297	17.0240	69
70	23.3945	22.2521	21.2021	20.2351	19.3427	18.5176	17.7533	17.0440	70
71	23.4563	22.3042	21.2460	20.2721	19.3740	18.5440	17.7756	17.0629	71
72	23.5156	22.3542	21.2881	20.3075	19.4038	18.5692	17.7968	17.0807	72
73	23.5727	22.4021	21.3283	20.3413	19.4322	18.5930	17.8169	17.0976	73
74	23.6276	22.4480	21.3668	20.3736	19.4592	18.6157	17.8359	17.1136	74
75	23.6804	22.4921	21.4036	20.4044	19.4850	18.6372	17.8539	17.1287	75
76	23.7312	22.5344	21.4389	20.4338	19.5095	18.6577	17.8710	17.1430	76
77	23.7800	22.5750	21.4726	20.4618	19.5329	18.6772	17.8872	17.1565	77
78	23.8269	22.6139	21.5049	20.4886	19.5551	18.6956	17.9026	17.1692	78
79	23.8720	22.6512	21.5358	20.5142	19.5763	18.7132	17.9172	17.1813	79
80	23.9154	22.6870	21.5653	20.5386	19.5965	18.7299	17.9310	17.1927	80
81	23.9571	22.7213	21.5936	20.5619	19.6157	18.7457	17.9440	17.2035	81
82	23.9972	22.7543	21.6207	20.5842	19.6340	18.7608	17.9564	17.2137	82
83	24.0358	22.7859	21.6466	20.6054	19.6514	18.7751	17.9682	17.2234	83
84	24.0729	22.8162	21.6714	20.6257	19.6680	18.7887	17.9793	17.2325	84
85	24.1085	22.8453	21.6951	20.6451	19.6838	18.8016	17.9899	17.2412	85
86	24.1428	22.8732	21.7178	20.6635	19.6989	18.8139	17.9999	17.2493	86
87	24.1758	22.8999	21.7395	20.6812	19.7132	18.8255	18.0094	17.2570	87
88	24.2075	22.9256	21.7603	20.6980	19.7269	18.8366	18.0184	17.2643	88
89	24.2380	22.9502	21.7802	20.7141	19.7399	18.8471	18.0269	17.2712	89
90	24.2673	22.9738	21.7992	20.7295	19.7523	18.8571	18.0350	17.2778	90
91	24.2955	22.9965	21.8175	20.7441	19.7641	18.8666	18.0426	17.2840	91
92	24.3226	23.0182	21.8349	20.7581	19.7753	18.8757	18.0499	17.2898	92
93	24.3486	23.0390	21.8516	20.7715	19.7860	18.8842	18.0567	17.2953	93
94	24.3737	23.0590	21.8675	20.7842	19.7962	18.8924	18.0633	17.3005	94
95	24.3978	23.0782	21.8828	20.7964	19.8059	18.9001	18.0694	17.3055	95
96	24.4209	23.0966	21.8974	20.8080	19.8151	18.9075	18.0753	17.3101	96
97	24.4432	23.1142	21.9114	20.8191	19.8239	18.9145	18.0809	17.3145	97
98	24.4646	23.1312	21.9248	20.8297	19.8323	18.9211	18.0861	17.3187	98
99	24.4852	23.1474	21.9376	20.8398	19.8403	18.9274	18.0911	17.3227	99
100	24.5050	23.1630	21.9499	20.8494	19.8479	18.9334	18.0958	17.3264	100
PERP	25.0000	23.5294	22.2222	21.0526	20.0000	19.0476	18.1818	17.3913	PERP

YEARS' PURCHASE

					Rate Per Cent				
Yrs.	6	6.25	6.5	6.75	7	7.25	7.5	7.75	Yrs.
1	0.9434	0.9412	0.9390	0.9368	0.9346	0.9324	0.9302	0.9281	1
2	1.8334	1.8270	1.8206	1.8143	1.8080	1.8018	1.7956	1.7894	2
3	2.6730	2.6607	2.6485	2.6363	2.6243	2.6124	2.6005	2.5888	3
4	3.4651	3.4454	3.4258	3.4064	3.3872	3.3682	3.3493	3.3306	4
5	4.2124	4.1839	4.1557	4.1278	4.1002	4.0729	4.0459	4.0192	5
6	4.9173	4.8789	4.8410	4.8036	4.7665	4.7300	4.6938	4.6582	6
7	5.5824	5.5331	5.4845	5.4366	5.3893	5.3426	5.2966	5.2512	7
8	6.2098	6.1488	6.0888	6.0296	5.9713	5.9139	5.8573	5.8016	8
9	6.8017	6.7283	6.6561	6.5851	6.5152	6.4465	6.3789	6.3124	9
10	7.3601	7.2737	7.1888	7.1055	7.0236	6.9431	6.8641	6.7864	10
11	7.8869	7.7870	7.6890	7.5929	7.4987	7.4062	7.3154	7.2264	11
12	8.3838	8.2701	8.1587	8.0496	7.9427	7.8379	7.7353	7.6347	12
13	8.8527	8.7248	8.5997	8.4774	8.3577	8.2405	8.1258	8.0136	13
14	9.2950	9.1528	9.0138	8.8781	8.7455	8.6158	8.4892	8.3653	14
15	9.7122	9.5555	9.4027	9.2535	9.1079	8.9658	8.8271	8.6917	15
16	10.1059	9.9346	9.7678	9.6051	9.4466	9.2921	9.1415	8.9946	16
17	10.4773	10.2914	10.1106	9.9346	9.7632	9.5964	9.4340	9.2757	17
18	10.8276	10.6272	10.4325	10.2432	10.0591	9.8801	9.7060	9.5367	18
19	11.1581	10.9433	10.7347	10.5322	10.3356	10.1446	9.9591	9.7788	19
20	11.4699	11.2407	11.0185	10.8030	10.5940	10.3912	10.1945	10.0035	20
21	11.7641	11.5207	11.2850	11.0567	10.8355	10.6212	10.4135	10.2121	21
22	12.0416	11.7842	11.5352	11.2943	11.0612	10.8356	10.6172	10.4057	22
23	12.3034	12.0322	11.7701	11.5169	11.2722	11.0355	10.8067	10.5853	23
24	12.5504	12.2656	11.9907	11.7255	11.4693	11.2220	10.9830	10.7520	24
25	12.7834	12.4852	12.1979	11.9208	11.6536	11.3958	11.1469	10.9067	25
26	13.0032	12.6920	12.3924	12.1038	11.8258	11.5578	11.2995	11.0503	26
27	13.2105	12.8866	12.5750	12.2752	11.9867	11.7089	11.4414	11.1836	27
28	13.4062	13.0697	12.7465	12.4358	12.1371	11.8498	11.5734	11.3073	28
29	13.5907	13.2421	12.9075	12.5862	12.2777	11.9812	11.6962	11.4221	29
30	13.7648	13.4043	13.0587	12.7272	12.4090	12.1037	11.8104	11.5286	30
31	13.9291	13.5570	13.2006	12.8592	12.5318	12.2179	11.9166	11.6275	31
32	14.0840	13.7007	13.3339	12.9828	12.6466	12.3244	12.0155	11.7192	32
33	14.2302	13.8360	13.4591	13.0987	12.7538	12.4236	12.1074	11.8044	33
34	14.3681	13.9633	13.5766	13.2072	12.8540	12.5162	12.1929	11.8834	34
35	14.4982	14.0831	13.6870	13.3088	12.9477	12.6025	12.2725	11.9568'	35
36	14.6210	14.1958	13.7906	13.4041	13.0352	12.6830	12.3465	12.0249	36
37	14.7368	14.3020	13.8879	13.4933	13.1170	12.7581	12.4154	12.0880	37
38	14.8460	14.4018	13.9792	13.5768	13.1935	12.8280	12.4794	12.1467	38
39	14.9491	14.4958	14.0650	13.6551	13.2649	12.8933	12.5390	12.2011	39
40	15.0463	14.5843	14.1455	13.7284	13.3317	12.9541	12.5944	12.2516	40
41	15.1380	14.6676	14.2212	13.7971	13.3941	13.0108	12.6460	12.2985	41
42	15.2245	14.7460	14.2922	13.8615	13.4524	13.0637	12.6939	12.3420	42
43	15.3062	14.8197	14.3588	13.9218	13.5070	13.1130	12.7385	12.3823	43
44	15.3832	14.8892	14.4214	13.9782	13.5579	13.1590	12.7800	12.4198	44
45	15.4558	14.9545	14.4802	14.0311	13.6055	13.2018	12.8186	12.4546	45
46	15.5244	15.0160	14.5354	14.0807	13.6500	13.2418	12.8545	12.4868	46
47	15.5890	15.0739	14.5873	14.1271	13.6916	13.2791	12.8879	12.5168	47
48	15.6500	15.1284	14.6359	14.1706	13.7305	13.3138	12.9190	12.5446	48
49	15.7076	15.1796	14.6816	14.2113	13.7668	13.3462	12.9479	12.5704	49
50	15.7619	15.2279	14.7245	14.2495	13.8007	13.3764	12.9748	12.5943	50

YEARS' PURCHASE

				Rate Per Cent					
Yrs.	6	6.25	6.5	6.75	7	7.25	7.5	7.75	Yrs.
51	15.8131	15.2733	14.7648	14.2852	13.8325	13.4046	12.9998	12.6165	51
52	15.8614	15.3161	14.8026	14.3187	13.8621	13.4309	13.0231	12.6372	52
53	15.9070	15.3563	14.8382	14.3501	13.8898	13.4553	13.0447	12.6563	53
54	15.9500	15.3942	14.8715	14.3795	13.9157	13.4782	13.0649	12.6741	54
55	15.9905	15.4298	14.9028	14.4070	13.9399	13.4995	13.0836	12.6905	55
56	16.0288	15.4633	14.9322	14.4328	13.9626	13.5193	13.1010	12.7058	56
57	16.0649	15.4949	14.9598	14.4569	13.9837	13.5378	13.1172	12.7200	57
58	16.0990	15.5246	14.9858	14.4796	14.0035	13.5551	13.1323	12.7332	58
59	16.1311	15.5526	15.0101	14.5008	14.0219	13.5712	13.1463	12.7454	59
60	16.1614	15.5789	15.0330	14.5206	14.0392	13.5862	13.1594	12.7568	60
61	16.1900	15.6037	15.0544	14.5392	14.0553	13.6002	13.1715	12.7673	61
62	16.2170	15.6270	15.0746	14.5567	14.0704	13.6132	13.1828	12.7771	62
63	16.2425	15.6489	15.0935	14.5730	14.0845	13.6254	13.1933	12.7862	63
64	16.2665	15.6696	15.1113	14.5883	14.0976	13.6367	13.2031	12.7946	64
65	16.2891	15.6890	15.1280	14.6026	14.1099	13.6473	13.2122	12.8024	65
66	16.3105	15.7073	15.1436	14.6160	14.1214	13.6571	13.2206	12.8097	66
67	16.3307	15.7245	15.1583	14.6286	14.1322	13.6663	13.2285	12.8164	67
68	16.3497	15.7407	15.1721	14.6404	14.1422	13.6749	13.2358	12.8226	68
69	16.3676	15.7560	15.1851	14.6514	14.1516	13.6829	13.2426	12.8284	69
70	16.3845	15.7703	15.1973	14.6617	14.1604	13.6903	13.2489	12.8338	70
71	16.4005	15.7838	15.2087	14.6714	14.1686	13.6973	13.2548	12.8388	71
72	16.4156	15.7966	15.2195	14.6805	14.1763	13.7038	13.2603	12.8434	72
73	16.4298	15.8085	15.2295	14.6890	14.1834	13.7098	13.2654	12.8477	73
74	16.4432	15.8198	15.2390	14.6969	14.1901	13.7154	13.2701	12.8517	74
75	16.4558	15.8304	15.2479	14.7044	14.1964	13.7207	13.2745	12.8554	75
76	16.4678	15.8404	15.2562	14.7114	14.2022	13.7256	13.2786	12.8589	76
77	16.4790	15.8498	15.2641	14.7179	14.2077	13.7301	13.2825	12.8621	77
78	16.4897	15.8586	15.2714	14.7240	14.2128	13.7344	13.2860	12.8650	78
79	16.4997	15.8669	15.2783	14.7298	14.2175	13.7384	13.2893	12.8678	79
80	16.5091	15.8747	15.2848	14.7352	14.2220	13.7421	13.2924	12.8703	80
81	16.5180	15.8821	15.2909	14.7402	14.2262	13.7455	13.2952	12.8727	81
82	16.5265	15.8890	15.2966	14.7449	14.2301	13.7487	13.2979	12.8749	82
83	16.5344	15.8956	15.3020	14.7493	14.2337	13.7517	13.3004	12.8769	83
84	16.5419	15.9017	15.3070	14.7535	14.2371	13.7545	13.3027	12.8788	84
85	16.5489	15.9075	15.3118	14.7573	14.2403	13.7571	13.3048	12.8806	85
86	16.5556	15.9129	15.3162	14.7610	14.2433	13.7596	13.3068	12.8822	86
87	16.5619	15.9181	15.3204	14.7644	14.2460	13.7618	13.3087	12.8837	87
88	16.5678	15.9229	15.3243	14.7676	14.2486	13.7639	13.3104	12.8851	88
89	16.5734	15.9274	15.3280	14.7706	14.2511	13.7659	13.3120	12.8864	89
90	16.5787	15.9317	15.3315	14.7734	14.2533	13.7678	13.3135	12.8876	90
91	16.5837	15.9357	15.3347	14.7760	14.2554	13.7695	13.3149	12.8887	91
92	16.5884	15.9395	15.3377	14.7784	14.2574	13.7711	13.3161	12.8898	92
93	16.5928	15.9430	15.3406	14.7807	14.2593	13.7726	13.3173	12.8908	93
94	16.5970	15.9464	15.3433	14.7829	14.2610	13.7739	13.3185	12.8917	94
95	16.6009	15.9495	15.3458	14.7849	14.2626	13.7752	13.3195	12.8925	95
96	16.6047	15.9525	15.3482	14.7868	14.2641	13.7764	13.3205	12.8933	96
97	16.6082	15.9553	15.3504	14.7886	14.2655	13.7776	13.3214	12.8940	97
98	16.6115	15.9579	15.3525	14.7902	14.2669	13.7786	13.3222	12.8946	98
99	16.6146	15.9604	15.3545	14.7918	14.2681	13.7796	13.3230	12.8953	99
100	16.6175	15.9627	15.3563	14.7932	14.2693	13.7805	13.3237	12.8958	100
PERP	16.6667	16.0000	15.3846	14.8148	14.2857	13.7931	13.3333	12.9032	PERP

YEARS' PURCHASE

					Rate Per Cent				
Yrs.	8	8.25	8.5	8.75	9	9.5	10	11	Yrs.
1	0.9259	0.9238	0.9217	0.9195	0.9174	0.9132	0.9091	0.9009	1
2	1.7833	1.7772	1.7711	1.7651	1.7591	1.7473	1.7355	1.7125	2
3	2.5771	2.5655	2.5540	2.5426	2.5313	2.5089	2.4869	2.4437	3
4	3.3121	3.2938	3.2756	3.2576	3.2397	3.2045	3.1699	3.1024	4
5	3.9927	3.9665	3.9406	3.9150	3.8897	3.8397	3.7908	3.6959	5
6	4.6229	4.5880	4.5536	4.5196	4.4859	4.4198	4.3553	4.2305	6
7	5.2064	5.1621	5.1185	5.0755	5.0330	4.9496	4.8684	4.7122	7
8	5.7466	5.6925	5.6392	5.5866	5.5348	5.4334	5.3349	5.1461	8
9	6.2469	6.1825	6.1191	6.0567	5.9952	5.8753	5.7590	5.5370	9
10	6.7101	6.6351	6.5613	6.4889	6.4177	6.2788	6.1446	5.8892	10
11	7.1390	7.0532	6.9690	6.8863	6.8052	6.6473	6.4951	6.2065	11
12	7.5361	7.4394	7.3447	7.2518	7.1607	6.9838	6.8137	6.4924	12
13	7.9038	7.7962	7.6910	7.5879	7.4869	7.2912	7.1034	6.7499	13
14	8.2442	8.1259	8.0101	7.8969	7.7862	7.5719	7.3667	6.9819	14
15	8.5595	8.4304	8.3042	8.1810	8.0607	7.8282	7.6061	7.1909	15
16	8.8514	8.7116	8.5753	8.4423	8.3126	8.0623	7.8237	7.3792	16
17	9.1216	8.9715	8.8252	8.6826	8.5436	8.2760	8.0216	7.5488	17
18	9.3719	9.2115	9.0555	8.9035	8.7556	8.4713	8.2014	7.7016	18
19	9.6036	9.4333	9.2677	9.1067	8.9501	8.6496	8.3649	7.8393	19
20	9.8181	9.6381	9.4633	9.2935	9.1285	8.8124	8.5136	7.9633	20
21	10.0168	9.8274	9.6436	9.4653	9.2922	8.9611	8.6487	8.0751	21
22	10.2007	10.0022	9.8098	9.6233	9.4424	9.0969	8.7715	8.1757	22
23	10.3711	10.1637	9.9629	9.7685	9.5802	9.2209	8.8832	8.2664	23
24	10.5288	10.3129	10.1041	9.9021	9.7066	9.3341	8.9847	8.3481	24
25	10.6748	10.4507	10.2342	10.0249	9.8226	9.4376	9.0770	8.4217	25
26	10.8100	10.5780	10.3541	10.1379	9.9290	9.5320	9.1609	8.4881	26
27	10.9352	10.6956	10.4646	10.2417	10.0266	9.6183	9.2372	8.5478	27
28	11.0511	10.8043	10.5665	10.3372	10.1161	9.6971	9.3066	8.6016	28
29	11.1584	10.9046	10.6603	10.4250	10.1983	9.7690	9.3696	8.6501	29
30	11.2578	10.9974	10.7468	10.5058	10.2737	9.8347	9.4269	8.6938	30
31	11.3498	11.0830	10.8266	10.5800	10.3428	9.8947	9.4790	8.7331	31
32	11.4350	11.1621	10.9001	10.6483	10.4062	9.9495	9.5264	8.7686	32
33	11.5139	11.2352	10.9678	10.7111	10.4644	9.9996	9.5694	8.8005	33
34	11.5869	11.3028	11.0302	10.7688	10.5178	10.0453	9.6086	8.8293	34
35	11.6546	11.3651	11.0878	10.8219	10.5668	10.0870	9.6442	8.8552	35
36	11.7172	11.4228	11.1408	10.8707	10.6118	10.1251	9.6765	8.8786	36
37	11.7752	11.4760	11.1897	10.9156	10.6530	10.1599	9.7059	8.8996	37
38	11.8289	11.5252	11.2347	10.9569	10.6908	10.1917	9.7327	8.9186	38
39	11.8786	11.5706	11.2763	10.9948	10.7255	10.2207	9.7570	8.9357	39
40	11.9246	11.6125	11.3145	11.0297	10.7574	10.2472	9.7791	8.9511	40
41	11.9672	11.6513	11.3498	11.0618	10.7866	10.2715	9.7991	8.9649	41
42	12.0067	11.6871	11.3823	11.0913	10.8134	10.2936	9.8174	8.9774	42
43	12.0432	11.7202	11.4123	11.1184	10.8380	10.3138	9.8340	8.9886	43
44	12.0771	11.7508	11.4399	11.1434	10.8605	10.3322	9.8491	8.9988	44
45	12.1084	11.7790	11.4653	11.1663	10.8812	10.3490	9.8628	9.0079	45
46	12.1374	11.8051	11.4888	11.1874	10.9002	10.3644	9.8753	9.0161	46
47	12.1643	11.8292	11.5104	11.2068	10.9176	10.3785	9.8866	9.0235	47
48	12.1891	11.8514	11.5303	11.2247	10.9336	10.3913	9.8969	9.0302	48
49	12.2122	11.8720	11.5487	11.2411	10.9482	10.4030	9.9063	9.0362	49
50	12.2335	11.8910	11.5656	11.2562	10.9617	10.4137	9.9148	9.0417	50

YEARS' PURCHASE

				Rate Per Cent					
Yrs.	8	8.25	8.5	8.75	9	9.5	10	11	Yrs.
51	12.2532	11.9085	11.5812	11.2700	10.9740	10.4235	9.9226	9.0465	51
52	12.2715	11.9247	11.5956	11.2828	10.9853	10.4324	9.9296	9.0509	52
53	12.2884	11.9397	11.6088	11.2945	10.9957	10.4405	9.9360	9.0549	53
54	12.3041	11.9535	11.6210	11.3053	11.0053	10.4480	9.9418	9.0585	54
55	12.3186	11.9663	11.6323	11.3152	11.0140	10.4548	9.9471	9.0617	55
56	12.3321	11.9781	11.6427	11.3244	11.0220	10.4610	9.9519	9.0646	56
57	12.3445	11.9890	11.6522	11.3327	11.0294	10.4667	9.9563	9.0672	57
58	12.3560	11.9991	11.6610	11.3404	11.0361	10.4718	9.9603	9.0695	58
59	12.3667	12.0084	11.6692	11.3475	11.0423	10.4766	9.9639	9.0717	59
60	12.3766	12.0170	11.6766	11.3541	11.0480	10.4809	9.9672	9.0736	60
61	12.3857	12.0250	11.6835	11.3601	11.0532	10.4848	9.9701	9.0753	61
62	12.3942	12.0323	11.6899	11.3656	11.0580	10.4884	9.9729	9.0768	62
63	12.4020	12.0391	11.6958	11.3706	11.0624	10.4917	9.9753	9.0782	63
64	12.4093	12.0453	11.7012	11.3753	11.0664	10.4947	9.9776	9.0795	64
65	12.4160	12.0511	11.7061	11.3796	11.0701	10.4975	9.9796	9.0806	65
66	12.4222	12.0565	11.7107	11.3835	11.0735	10.5000	9.9815	9.0816	66
67	12.4280	12.0614	11.7150	11.3871	11.0766	10.5022	9.9831	9.0826	67
68	12.4333	12.0659	11.7189	11.3905	11.0794	10.5043	9.9847	9.0834	68
69	12.4382	12.0702	11.7224	11.3935	11.0820	10.5062	9.9861	9.0841	69
70	12.4428	12.0741	11.7258	11.3964	11.0844	10.5080	9.9873	9.0848	70
71	12.4471	12.0776	11.7288	11.3990	11.0867	10.5096	9.9885	9.0854	71
72	12.4510	12.0810	11.7316	11.4013	11.0887	10.5110	9.9895	9.0860	72
73	12.4546	12.0840	11.7342	11.4035	11.0905	10.5124	9.9905	9.0864	73
74	12.4580	12.0869	11.7366	11.4055	11.0922	10.5136	9.9914	9.0869	74
75	12.4611	12.0895	11.7388	11.4074	11.0938	10.5147	9.9921	9.0873	75
76	12.4640	12.0919	11.7408	11.4091	11.0952	10.5157	9.9929	9.0876	76
77	12.4666	12.0941	11.7427	11.4107	11.0965	10.5166	9.9935	9.0880	77
78	12.4691	12.0962	11.7444	11.4121	11.0977	10.5174	9.9941	9.0883	78
79	12.4714	12.0981	11.7460	11.4134	11.0988	10.5182	9.9946	9.0885	79
80	12.4735	12.0999	11.7475	11.4147	11.0998	10.5189	9.9951	9.0888	80
81	12.4755	12.1015	11.7488	11.4158	11.1008	10.5196	9.9956	9.0890	81
82	12.4773	12.1030	11.7501	11.4168	11.1016	10.5201	9.9960	9.0892	82
83	12.4790	12.1044	11.7512	11.4177	11.1024	10.5207	9.9963	9.0893	83
84	12.4805	12.1057	11.7523	11.4186	11.1031	10.5212	9.9967	9.0895	84
85	12.4820	12.1069	11.7532	11.4194	11.1038	10.5216	9.9970	9.0896	85
86	12.4833	12.1079	11.7541	11.4202	11.1044	10.5220	9.9972	9.0898	86
87	12.4845	12.1090	11.7550	11.4208	11.1050	10.5224	9.9975	9.0899	87
88	12.4857	12.1099	11.7557	11.4215	11.1055	10.5227	9.9977	9.0900	88
89	12.4868	12.1108	11.7564	11.4220	11.1059	10.5230	9.9979	9.0901	89
90	12.4877	12.1116	11.7571	11.4226	11.1064	10.5233	9.9981	9.0902	90
91	12.4886	12.1123	11.7577	11.4230	11.1067	10.5236	9.9983	9.0902	91
92	12.4895	12.1130	11.7582	11.4235	11.1071	10.5238	9.9984	9.0903	92
93	12.4903	12.1136	11.7587	11.4239	11.1074	10.5240	9.9986	9.0904	93
94	12.4910	12.1142	11.7592	11.4243	11.1077	10.5242	9.9987	9.0904	94
95	12.4917	12.1147	11.7596	11.4246	11.1080	10.5244	9.9988	9.0905	95
96	12.4923	12.1152	11.7600	11.4249	11.1083	10.5246	9.9989	9.0905	96
97	12.4928	12.1157	11.7604	11.4252	11.1085	10.5247	9.9990	9.0905	97
98	12.4934	12.1161	11.7607	11.4255	11.1087	10.5249	9.9991	9.0906	98
99	12.4939	12.1165	11.7610	11.4257	11.1089	10.5250	9.9992	9.0906	99
100	12.4943	12.1168	11.7613	11.4260	11.1091	10.5251	9.9993	9.0906	100
PERP	12.5000	12.1212	11.7647	11.4286	11.1111	10.5263	10.0000	9.0909	PERP

YEARS' PURCHASE

				Rate Per Cent					
Yrs.	12	13	14	15	16	17	18	19	Yrs.
1	0.8929	0.8850	0.8772	0.8696	0.8621	0.8547	0.8475	0.8403	1
2	1.6901	1.6681	1.6467	1.6257	1.6052	1.5852	1.5656	1.5465	2
3	2.4018	2.3612	2.3216	2.2832	2.2459	2.2096	2.1743	2.1399	3
4	3.0373	2.9745	2.9137	2.8550	2.7982	2.7432	2.6901	2.6386	4
5	3.6048	3.5172	3.4331	3.3522	3.2743	3.1993	3.1272	3.0576	5
6	4.1114	3.9975	3.8887	3.7845	3.6847	3.5892	3.4976	3.4098	6
7	4.5638	4.4226	4.2883	4.1604	4.0386	3.9224	3.8115	3.7057	7
8	4.9676	4.7988	4.6389	4.4873	4.3436	4.2072	4.0776	3.9544	8
9	5.3282	5.1317	4.9464	4.7716	4.6065	4.4506	4.3030	4.1633	9
10	5.6502	5.4262	5.2161	5.0188	4.8332	4.6586	4.4941	4.3389	10
11	5.9377	5.6869	5.4527	5.2337	5.0286	4.8364	4.6560	4.4865	11
12	6.1944	5.9176	5.6603	5.4206	5.1971	4.9884	4.7932	4.6105	12
13	6.4235	6.1218	5.8424	5.5831	5.3423	5.1183	4.9095	4.7147	13
14	6.6282	6.3025	6.0021	5.7245	5.4675	5.2293	5.0081	4.8023	14
15	6.8109	6.4624	6.1422	5.8474	5.5755	5.3242	5.0916	4.8759	15
16	6.9740	6.6039	6.2651	5.9542	5.6685	5.4053	5.1624	4.9377	16
17	7.1196	6.7291	6.3729	6.0472	5.7487	5.4746	5.2223	4.9897	17
18	7.2497	6.8399	6.4674	6.1280	5.8178	5.5339	5.2732	5.0333	18
19	7.3658	6.9380	6.5504	6.1982	5.8775	5.5845	5.3162	5.0700	19
20	7.4694	7.0248	6.6231	6.2593	5.9288	5.6278	5.3527	5.1009	20
21	7.5620	7.1016	6.6870	6.3125	5.9731	5.6648	5.3837	5.1268	21
22	7.6446	7.1695	6.7429	6.3587	6.0113	5.6964	5.4099	5.1486	22
23	7.7184	7.2297	6.7921	6.3988	6.0442	5.7234	5.4321	5.1668	23
24	7.7843	7.2829	6.8351	6.4338	6.0726	5.7465	5.4509	5.1822	24
25	7.8431	7.3300	6.8729	6.4641	6.0971	5.7662	5.4669	5.1951	25
26	7.8957	7.3717	6.9061	6.4906	6.1182	5.7831	5.4804	5.2060	26
27	7.9426	7.4086	6.9352	6.5135	6.1364	5.7975	5.4919	5.2151	27
28	7.9844	7.4412	6.9607	6.5335	6.1520	5.8099	5.5016	5.2228	28
29	8.0218	7.4701	6.9830	6.5509	6.1656	5.8204	5.5098	5.2292	29
30	8.0552	7.4957	7.0027	6.5660	6.1772	5.8294	5.5168	5.2347	30
31	8.0850	7.5183	7.0199	6.5791	6.1872	5.8371	5.5227	5.2392	31
32	8.1116	7.5383	7.0350	6.5905	6.1959	5.8437	5.5277	5.2430	32
33	8.1354	7.5560	7.0482	6.6005	6.2034	5.8493	5.5320	5.2462	33
34	8.1566	7.5717	7.0599	6.6091	6.2098	5.8541	5.5356	5.2489	34
35	8.1755	7.5856	7.0700	6.6166	6.2153	5.8582	5.5386	5.2512	35
36	8.1924	7.5979	7.0790	6.6231	6.2201	5.8617	5.5412	5.2531	36
37	8.2075	7.6087	7.0868	6.6288	6.2242	5.8647	5.5434	5.2547	37
38	8.2210	7.6183	7.0937	6.6338	6.2278	5.8673	5.5452	5.2561	38
39	8.2330	7.6268	7.0997	6.6380	6.2309	5.8695	5.5468	5.2572	39
40	8.2438	7.6344	7.1050	6.6418	6.2335	5.8713	5.5482	5.2582	40
41	8.2534	7.6410	7.1097	6.6450	6.2358	5.8729	5.5493	5.2590	41
42	8.2619	7.6469	7.1138	6.6478	6.2377	5.8743	5.5502	5.2596	42
43	8.2696	7.6522	7.1173	6.6503	6.2394	5.8755	5.5510	5.2602	43
44	8.2764	7.6568	7.1205	6.6524	6.2409	5.8765	5.5517	5.2607	44
45	8.2825	7.6609	7.1232	6.6543	6.2421	5.8773	5.5523	5.2611	45
46	8.2880	7.6645	7.1256	6.6559	6.2432	5.8781	5.5528	5.2614	46
47	8.2928	7.6677	7.1277	6.6573	6.2442	5.8787	5.5532	5.2617	47
48	8.2972	7.6705	7.1296	6.6585	6.2450	5.8792	5.5536	5.2619	48
49	8.3010	7.6730	7.1312	6.6596	6.2457	5.8797	5.5539	5.2621	49
50	8.3045	7.6752	7.1327	6.6605	6.2463	5.8801	5.5541	5.2623	50

YEARS' PURCHASE

				Rate Per Cent					
Yrs.	12	13	14	15	16	17	18	19	Yrs.
51	8.3076	7.6772	7.1339	6.6613	6.2468	5.8804	5.5544	5.2624	51
52	8.3103	7.6789	7.1350	6.6620	6.2472	5.8807	5.5545	5.2625	52
53	8.3128	7.6805	7.1360	6.6626	6.2476	5.8809	5.5547	5.2626	53
54	8.3150	7.6818	7.1368	6.6631	6.2479	5.8811	5.5548	5.2627	54
55	8.3170	7.6830	7.1376	6.6636	6.2482	5.8813	5.5549	5.2628	55
56	8.3187	7.6841	7.1382	6.6640	6.2485	5.8815	5.5550	5.2628	56
57	8.3203	7.6851	7.1388	6.6644	6.2487	5.8816	5.5551	5.2629	57
58	8.3217	7.6859	7.1393	6.6647	6.2489	5.8817	5.5552	5.2629	58
59	8.3229	7.6866	7.1397	6.6649	6.2490	5.8818	5.5552	5.2630	59
60	8.3240	7.6873	7.1401	6.6651	6.2492	5.8819	5.5553	5.2630	60
61	8.3250	7.6879	7.1404	6.6653	6.2493	5.8819	5.5553	5.2630	61
62	8.3259	7.6884	7.1407	6.6655	6.2494	5.8820	5.5554	5.2630	62
63	8.3267	7.6888	7.1410	6.6657	6.2495	5.8821	5.5554	5.2631	63
64	8.3274	7.6892	7.1412	6.6658	6.2495	5.8821	5.5554	5.2631	64
65	8.3281	7.6896	7.1414	6.6659	6.2496	5.8821	5.5554	5.2631	65
66	8.3286	7.6899	7.1416	6.6660	6.2497	5.8822	5.5555	5.2631	66
67	8.3291	7.6902	7.1418	6.6661	6.2497	5.8822	5.5555	5.2631	67
68	8.3296	7.6904	7.1419	6.6662	6.2497	5.8822	5.5555	5.2631	68
69	8.3300	7.6906	7.1420	6.6662	6.2498	5.8822	5.5555	5.2631	69
70	8.3303	7.6908	7.1421	6.6663	6.2498	5.8823	5.5555	5.2631	70
71	8.3307	7.6910	7.1422	6.6663	6.2498	5.8823	5.5555	5.2631	71
72	8.3310	7.6911	7.1423	6.6664	6.2499	5.8823	5.5555	5.2631	72
73	8.3312	7.6913	7.1424	6.6664	6.2499	5.8823	5.5555	5.2631	73
74	8.3314	7.6914	7.1424	6.6665	6.2499	5.8823	5.5555	5.2631	74
75	8.3316	7.6915	7.1425	6.6665	6.2499	5.8823	5.5555	5.2631	75
76	8.3318	7.6916	7.1425	6.6665	6.2499	5.8823	5.5555	5.2631	76
77	8.3320	7.6917	7.1426	6.6665	6.2499	5.8823	5.5555	5.2631	77
78	8.3321	7.6918	7.1426	6.6665	6.2499	5.8823	5.5555	5.2632	78
79	8.3323	7.6918	7.1426	6.6666	6.2499	5.8823	5.5555	5.2632	79
80	8.3324	7.6919	7.1427	6.6666	6.2500	5.8823	5.5555	5.2632	80
81	8.3325	7.6919	7.1427	6.6666	6.2500	5.8823	5.5555	5.2632	81
82	8.3326	7.6920	7.1427	6.6666	6.2500	5.8823	5.5555	5.2632	82
83	8.3326	7.6920	7.1427	6.6666	6.2500	5.8823	5.5555	5.2632	83
84	8.3327	7.6920	7.1427	6.6666	6.2500	5.8823	5.5556	5.2632	84
85	8.3328	7.6921	7.1428	6.6666	6.2500	5.8823	5.5556	5.2632	85
86	8.3328	7.6921	7.1428	6.6666	6.2500	5.8823	5.5556	5.2632	86
87	8.3329	7.6921	7.1428	6.6666	6.2500	5.8823	5.5556	5.2632	87
88	8.3329	7.6921	7.1428	6.6666	6.2500	5.8823	5.5556	5.2632	88
89	8.3330	7.6922	7.1428	6.6666	6.2500	5.8823	5.5556	5.2632	89
90	8.3330	7.6922	7.1428	6.6666	6.2500	5.8823	5.5556	5.2632	90
91	8.3331	7.6922	7.1428	6.6666	6.2500	5.8823	5.5556	5.2632	91
92	8.3331	7.6922	7.1428	6.6666	6.2500	5.8823	5.5556	5.2632	92
93	8.3331	7.6922	7.1428	6.6667	6.2500	5.8824	5.5556	5.2632	93
94	8.3331	7.6922	7.1428	6.6667	6.2500	5.8824	5.5556	5.2632	94
95	8.3332	7.6922	7.1428	6.6667	6.2500	5.8824	5.5556	5.2632	95
96	8.3332	7.6922	7.1428	6.6667	6.2500	5.8824	5.5556	5.2632	96
97	8.3332	7.6923	7.1428	6.6667	6.2500	5.8824	5.5556	5.2632	97
98	8.3332	7.6923	7.1428	6.6667	6.2500	5.8824	5.5556	5.2632	98
99	8.3332	7.6923	7.1428	6.6667	6.2500	5.8824	5.5556	5.2632	99
100	8.3332	7.6923	7.1428	6.6667	6.2500	5.8824	5.5556	5.2632	100
PERP	8.3333	7.6923	7.1429	6.6667	6.2500	5.8824	5.5556	5.2632	PERP

YEARS' PURCHASE

				Rate Per Cent					
Yrs.	20	21	22	23	24	25	26	28	Yrs.
1	0.8333	0.8264	0.8197	0.8130	0.8065	0.8000	0.7937	0.7812	1
2	1.5278	1.5095	1.4915	1.4740	1.4568	1.4400	1.4235	1.3916	2
3	2.1065	2.0739	2.0422	2.0114	1.9813	1.9520	1.9234	1.8684	3
4	2.5887	2.5404	2.4936	2.4483	2.4043	2.3616	2.3202	2.2410	4
5	2.9906	2.9260	2.8636	2.8035	2.7454	2.6893	2.6351	2.5320	5
6	3.3255	3.2446	3.1669	3.0923	3.0205	2.9514	2.8850	2.7594	6
7	3.6046	3.5079	3.4155	3.3270	3.2423	3.1611	3.0833	2.9370	7
8	3.8372	3.7256	3.6193	3.5179	3.4212	3.3289	3.2407	3.0758	8
9	4.0310	3.9054	3.7863	3.6731	3.5655	3.4631	3.3657	3.1842	9
10	4.1925	4.0541	3.9232	3.7993	3.6819	3.5705	3.4648	3.2689	10
11	4.3271	4.1769	4.0354	3.9018	3.7757	3.6564	3.5435	3.3351	11
12	4.4392	4.2784	4.1274	3.9852	3.8514	3.7251	3.6059	3.3868	12
13	4.5327	4.3624	4.2028	4.0530	3.9124	3.7801	3.6555	3.4272	13
14	4.6106	4.4317	4.2646	4.1082	3.9616	3.8241	3.6949	3.4587	14
15	4.6755	4.4890	4.3152	4.1530	4.0013	3.8593	3.7261	3.4834	15
16	4.7296	4.5364	4.3567	4.1894	4.0333	3.8874	3.7509	3.5026	16
17	4.7746	4.5755	4.3908	4.2190	4.0591	3.9099	3.7705	3.5177	17
18	4.8122	4.6079	4.4187	4.2431	4.0799	3.9279	3.7861	3.5294	18
19	4.8435	4.6346	4.4415	4.2627	4.0967	3.9424	3.7985	3.5386	19
20	4.8696	4.6567	4.4603	4.2786	4.1103	3.9539	3.8083	3.5458	20
21	4.8913	4.6750	4.4756	4.2916	4.1212	3.9631	3.8161	3.5514	21
22	4.9094	4.6900	4.4882	4.3021	4.1300	3.9705	3.8223	3.5558	22
23	4.9245	4.7025	4.4985	4.3106	4.1371	3.9764	3.8273	3.5592	23
24	4.9371	4.7128	4.5070	4.3176	4.1428	3.9811	3.8312	3.5619	24
25	4.9476	4.7213	4.5139	4.3232	4.1474	3.9849	3.8342	3.5640	25
26	4.9563	4.7284	4.5196	4.3278	4.1511	3.9879	3.8367	3.5656	26
27	4.9636	4.7342	4.5243	4.3316	4.1542	3.9903	3.8387	3.5669	27
28	4.9697	4.7390	4.5281	4.3346	4.1566	3.9923	3.8402	3.5679	28
29	4.9747	4.7430	4.5312	4.3371	4.1585	3.9938	3.8414	3.5687	29
30	4.9789	4.7463	4.5338	4.3391	4.1601	3.9950	3.8424	3.5693	30
31	4.9824	4.7490	4.5359	4.3407	4.1614	3.9960	3.8432	3.5697	31
32	4.9854	4.7512	4.5376	4.3421	4.1624	3.9968	3.8438	3.5701	32
33	4.9878	4.7531	4.5390	4.3431	4.1632	3.9975	3.8443	3.5704	33
34	4.9898	4.7546	4.5402	4.3440	4.1639	3.9980	3.8447	3.5706	34
35	4.9915	4.7559	4.5411	4.3447	4.1644	3.9984	3.8450	3.5708	35
36	4.9929	4.7569	4.5419	4.3453	4.1649	3.9987	3.8452	3.5709	36
37	4.9941	4.7578	4.5426	4.3458	4.1652	3.9990	3.8454	3.5710	37
38	4.9951	4.7585	4.5431	4.3462	4.1655	3.9992	3.8456	3.5711	38
39	4.9959	4.7591	4.5435	4.3465	4.1657	3.9993	3.8457	3.5712	39
40	4.9966	4.7596	4.5439	4.3467	4.1659	3.9995	3.8458	3.5712	40
41	4.9972	4.7600	4.5441	4.3469	4.1661	3.9996	3.8459	3.5713	41
42	4.9976	4.7603	4.5444	4.3471	4.1662	3.9997	3.8459	3.5713	42
43	4.9980	4.7606	4.5446	4.3472	4.1663	3.9997	3.8460	3.5713	43
44	4.9984	4.7608	4.5447	4.3473	4.1663	3.9998	3.8460	3.5714	44
45	4.9986	4.7610	4.5449	4.3474	4.1664	3.9998	3.8460	3.5714	45
46	4.9989	4.7612	4.5450	4.3475	4.1665	3.9999	3.8461	3.5714	46
47	4.9991	4.7613	4.5451	4.3476	4.1665	3.9999	3.8461	3.5714	47
48	4.9992	4.7614	4.5451	4.3476	4.1665	3.9999	3.8461	3.5714	48
49	4.9993	4.7615	4.5452	4.3477	4.1666	3.9999	3.8461	3.5714	49
50	4.9995	4.7616	4.5452	4.3477	4.1666	3.9999	3.8461	3.5714	50

YEARS' PURCHASE

				Rate Per Cent					
Yrs.	20	21	22	23	24	25	26	28	Yrs.
51	4.9995	4.7616	4.5453	4.3477	4.1666	4.0000	3.8461	3.5714	51
52	4.9996	4.7617	4.5453	4.3477	4.1666	4.0000	3.8461	3.5714	52
53	4.9997	4.7617	4.5453	4.3478	4.1666	4.0000	3.8461	3.5714	53
54	4.9997	4.7617	4.5454	4.3478	4.1666	4.0000	3.8461	3.5714	54
55	4.9998	4.7618	4.5454	4.3478	4.1666	4.0000	3.8461	3.5714	55
56	4.9998	4.7618	4.5454	4.3478	4.1666	4.0000	3.8461	3.5714	56
57	4.9998	4.7618	4.5454	4.3478	4.1666	4.0000	3.8461	3.5714	57
58	4.9999	4.7618	4.5454	4.3478	4.1667	4.0000	3.8461	3.5714	58
59	4.9999	4.7618	4.5454	4.3478	4.1667	4.0000	3.8461	3.5714	59
60	4.9999	4.7619	4.5454	4.3478	4.1667	4.0000	3.8462	3.5714	60
61	4.9999	4.7619	4.5454	4.3478	4.1667	4.0000	3.8462	3.5714	61
62	4.9999	4.7619	4.5454	4.3478	4.1667	4.0000	3.8462	3.5714	62
63	4.9999	4.7619	4.5454	4.3478	4.1667	4.0000	3.8462	3.5714	63
64	5.0000	4.7619	4.5454	4.3478	4.1667	4.0000	3.8462	3.5714	64
65	5.0000	4.7619	4.5454	4.3478	4.1667	4.0000	3.8462	3.5714	65
66	5.0000	4.7619	4.5454	4.3478	4.1667	4.0000	3.8462	3.5714	66
67	5.0000	4.7619	4.5454	4.3478	4.1667	4.0000	3.8462	3.5714	67
68	5.0000	4.7619	4.5454	4.3478	4.1667	4.0000	3.8462	3.5714	68
69	5.0000	4.7619	4.5454	4.3478	4.1667	4.0000	3.8462	3.5714	69
70	5.0000	4.7619	4.5455	4.3478	4.1667	4.0000	3.8462	3.5714	70
71	5.0000	4.7619	4.5455	4.3478	4.1667	4.0000	3.8462	3.5714	71
72	5.0000	4.7619	4.5455	4.3478	4.1667	4.0000	3.8462	3.5714	72
73	5.0000	4.7619	4.5455	4.3478	4.1667	4.0000	3.8462	3.5714	73
74	5.0000	4.7619	4.5455	4.3478	4.1667	4.0000	3.8462	3.5714	74
75	5.0000	4.7619	4.5455	4.3478	4.1667	4.0000	3.8462	3.5714	75
76	5.0000	4.7619	4.5455	4.3478	4.1667	4.0000	3.8462	3.5714	76
77	5.0000	4.7619	4.5455	4.3478	4.1667	4.0000	3.8462	3.5714	77
78	5.0000	4.7619	4.5455	4.3478	4.1667	4.0000	3.8462	3.5714	78
79	5.0000	4.7619	4.5455	4.3478	4.1667	4.0000	3.8462	3.5714	79
80	5.0000	4.7619	4.5455	4.3478	4.1667	4.0000	3.8462	3.5714	80
81	5.0000	4.7619	4.5455	4.3478	4.1667	4.0000	3.8462	3.5714	81
82	5.0000	4.7619	4.5455	4.3478	4.1667	4.0000	3.8462	3.5714	82
83	5.0000	4.7619	4.5455	4.3478	4.1667	4.0000	3.8462	3.5714	83
84	5.0000	4.7619	4.5455	4.3478	4.1667	4.0000	3.8462	3.5714	84
85	5.0000	4.7619	4.5455	4.3478	4.1667	4.0000	3.8462	3.5714	85
86	5.0000	4.7619	4.5455	4.3478	4.1667	4.0000	3.8462	3.5714	86
87	5.0000	4.7619	4.5455	4.3478	4.1667	4.0000	3.8462	3.5714	87
88	5.0000	4.7619	4.5455	4.3478	4.1667	4.0000	3.8462	3.5714	88
89	5.0000	4.7619	4.5455	4.3478	4.1667	4.0000	3.8462	3.5714	89
90	5.0000	4.7619	4.5455	4.3478	4.1667	4.0000	3.8462	3.5714	90
91	5.0000	4.7619	4.5455	4.3478	4.1667	4.0000	3.8462	3.5714	91
92	5.0000	4.7619	4.5455	4.3478	4.1667	4.0000	3.8462	3.5714	92
93	5.0000	4.7619	4.5455	4.3478	4.1667	4.0000	3.8462	3.5714	93
94	5.0000	4.7619	4.5455	4.3478	4.1667	4.0000	3.8462	3.5714	94
95	5.0000	4.7619	4.5455	4.3478	4.1667	4.0000	3.8462	3.5714	95
96	5.0000	4.7619	4.5455	4.3478	4.1667	4.0000	3.8462	3.5714	96
97	5.0000	4.7619	4.5455	4.3478	4.1667	4.0000	3.8462	3.5714	97
98	5.0000	4.7619	4.5455	4.3478	4.1667	4.0000	3.8462	3.5714	98
99	5.0000	4.7619	4.5455	4.3478	4.1667	4.0000	3.8462	3.5714	99
100	5.0000	4.7619	4.5455	4.3478	4.1667	4.0000	3.8462	3.5714	100
PERP	5.0000	4.7619	4.5455	4.3478	4.1667	4.0000	3.8462	3.5714	PERP

YEARS' PURCHASE

(DUAL RATE % PRINCIPLE)

OR

PRESENT VALUE OF
ONE POUND PER ANNUM

*receivable at the end of each year after
allowing for a sinking fund at a given
rate to replace the invested capital and
ignoring the effect of income tax on
that part of the income used to provide
the annual sinking fund instalment.*

AT RATES OF INTEREST FROM

4% to 20%

AND
ALLOWING FOR THE INVESTMENT
OF SINKING FUNDS AT
2·5%, 3% and 4%

Note:—The Tables on the following pages are first grouped by rates of sinking fund
interest and then placed in order of remunerative rates

YEARS' PURCHASE

				Rate Per Cent					
Yrs.	4	4.5	5	5.5	6	6.25	6.5	6.75	Yrs.
1	0.9615	0.9569	0.9524	0.9479	0.9434	0.9412	0.9390	0.9368	1
2	1.8733	1.8559	1.8388	1.8221	1.8056	1.7975	1.7895	1.7815	2
3	2.7387	2.7017	2.6657	2.6306	2.5965	2.5797	2.5632	2.5469	3
4	3.5610	3.4987	3.4386	3.3805	3.3243	3.2969	3.2699	3.2434	4
5	4.3432	4.2509	4.1624	4.0775	3.9961	3.9565	3.9178	3.8798	5
6	5.0878	4.9615	4.8414	4.7270	4.6179	4.5652	4.5137	4.4633	6
7	5.7973	5.6339	5.4796	5.3335	5.1949	5.1283	5.0634	5.0001	7
8	6.4739	6.2709	6.0802	5.9008	5.7317	5.6508	5.5720	5.4955	8
9	7.1196	6.8749	6.6464	6.4327	6.2322	6.1366	6.0439	5.9539	9
10	7.7364	7.4483	7.1809	6.9320	6.6998	6.5894	6.4826	6.3792	10
11	8.3260	7.9932	7.6860	7.4016	7.1375	7.0123	6.8915	6.7748	11
12	8.8899	8.5116	8.1641	7.8439	7.5479	7.4081	7.2734	7.1435	12
13	9.4297	9.0051	8.6171	8.2612	7.9335	7.7792	7.6308	7.4879	13
14	9.9466	9.4754	9.0468	8.6553	8.2962	8.1277	7.9658	7.8103	14
15	10.4421	9.9239	9.4548	9.0280	8.6381	8.4555	8.2804	8.1125	15
16	10.9172	10.3521	9.8426	9.3810	8.9607	8.7643	8.5764	8.3964	16
17	11.3730	10.7610	10.2116	9.7156	9.2655	9.0557	8.8552	8.6634	17
18	11.8105	11.1520	10.5630	10.0331	9.5538	9.3310	9.1183	8.9150	18
19	12.2308	11.5260	10.8979	10.3348	9.8270	9.5913	9.3668	9.1524	19
20	12.6347	11.8839	11.2174	10.6217	10.0860	9.8380	9.6018	9.3767	20
21	13.0230	12.2268	11.5224	10.8947	10.3319	10.0718	9.8244	9.5889	21
22	13.3965	12.5555	11.8138	11.1549	10.5656	10.2937	10.0355	9.7899	22
23	13.7558	12.8706	12.0924	11.4030	10.7879	10.5046	10.2358	9.9804	23
24	14.1018	13.1730	12.3590	11.6397	10.9995	10.7052	10.4261	10.1613	24
25	14.4350	13.4633	12.6142	11.8658	11.2012	10.8961	10.6072	10.3331	25
26	14.7561	13.7422	12.8586	12.0819	11.3936	11.0780	10.7795	10.4966	26
27	15.0655	14.0101	13.0930	12.2885	11.5772	11.2515	10.9437	10.6523	27
28	15.3638	14.2678	13.3177	12.4863	11.7525	11.4171	11.1003	10.8005	28
29	15.6516	14.5156	13.5334	12.6757	11.9202	11.5752	11.2497	10.9420	29
30	15.9292	14.7541	13.7405	12.8572	12.0806	11.7264	11.3924	11.0769	30
31	16.1972	14.9837	13.9394	13.0312	12.2341	11.8710	11.5288	11.2059	31
32	16.4559	15.2049	14.1306	13.1981	12.3811	12.0094	11.6593	11.3291	32
33	16.7058	15.4180	14.3145	13.3584	12.5220	12.1419	11.7842	11.4470	33
34	16.9472	15.6234	14.4913	13.5123	12.6571	12.2689	11.9038	11.5598	34
35	17.1805	15.8214	14.6616	13.6602	12.7868	12.3907	12.0184	11.6679	35
36	17.4060	16.0124	14.8255	13.8023	12.9113	12.5076	12.1283	11.7714	36
37	17.6240	16.1967	14.9833	13.9391	13.0309	12.6197	12.2338	11.8707	37
38	17.8348	16.3746	15.1354	14.0706	13.1458	12.7275	12.3350	11.9660	38
39	18.0388	16.5464	15.2821	14.1973	13.2562	12.8310	12.4322	12.0575	39
40	18.2361	16.7123	15.4235	14.3192	13.3625	12.9305	12.5256	12.1453	40
41	18.4271	16.8726	15.5599	14.4367	13.4648	13.0263	12.6155	12.2297	41
42	18.6120	17.0274	15.6915	14.5500	13.5632	13.1184	12.7018	12.3109	42
43	18.7910	17.1771	15.8186	14.6591	13.6581	13.2071	12.7850	12.3890	43
44	18.9644	17.3219	15.9412	14.7644	13.7494	13.2925	12.8650	12.4641	44
45	19.1323	17.4619	16.0597	14.8660	13.8375	13.3748	12.9420	12.5364	45
46	19.2951	17.5973	16.1742	14.9641	13.9224	13.4541	13.0163	12.6061	46
47	19.4527	17.7284	16.2849	15.0587	14.0043	13.5306	13.0879	12.6732	47
48	19.6055	17.8552	16.3918	15.1501	14.0833	13.6043	13.1569	12.7379	48
49	19.7537	17.9780	16.4953	15.2384	14.1596	13.6755	13.2234	12.8002	49
50	19.8973	18.0969	16.5953	15.3238	14.2332	13.7442	13.2876	12.8604	50

YEARS' PURCHASE

Yrs.					Rate Per Cent				Yrs.
	4	4.5	5	5.5	6	6.25	6.5	6.75	
51	20.0366	18.2121	16.6921	15.4063	14.3044	13.8105	13.3496	12.9184	51
52	20.1717	18.3236	16.7857	15.4860	14.3731	13.8745	13.4094	12.9745	52
53	20.3027	18.4317	16.8764	15.5631	14.4395	13.9364	13.4672	13.0286	53
54	20.4298	18.5364	16.9641	15.6377	14.5037	13.9962	13.5230	13.0808	54
55	20.5532	18.6379	17.0491	15.7099	14.5658	14.0540	13.5770	13.1313	55
56	20.6729	18.7363	17.1314	15.7797	14.6258	14.1099	13.6291	13.1800	56
57	20.7891	18.8317	17.2111	15.8473	14.6838	14.1639	13.6795	13.2272	57
58	20.9019	18.9242	17.2883	15.9128	14.7400	14.2162	13.7283	13.2727	58
59	21.0115	19.0139	17.3632	15.9762	14.7944	14.2667	13.7754	13.3168	59
60	21.1178	19.1010	17.4358	16.0376	14.8471	14.3157	13.8211	13.3594	60
61	21.2211	19.1854	17.5061	16.0971	14.8980	14.3631	13.8652	13.4007	61
62	21.3214	19.2674	17.5743	16.1548	14.9474	14.4090	13.9080	13.4406	62
63	21.4188	19.3469	17.6404	16.2106	14.9952	14.4534	13.9494	13.4793	63
64	21.5135	19.4241	17.7046	16.2648	15.0416	14.4964	13.9894	13.5167	64
65	21.6054	19.4990	17.7668	16.3173	15.0865	14.5381	14.0283	13.5530	65
66	21.6948	19.5718	17.8272	16.3682	15.1300	14.5785	14.0659	13.5881	66
67	21.7816	19.6424	17.8858	16.4176	15.1722	14.6177	14.1023	13.6221	67
68	21.8660	19.7110	17.9427	16.4655	15.2131	14.6557	14.1377	13.6550	68
69	21.9481	19.7777	17.9979	16.5120	15.2527	14.6925	14.1719	13.6870	69
70	22.0278	19.8424	18.0515	16.5571	15.2912	14.7282	14.2051	13.7180	70
71	22.1054	19.9053	18.1035	16.6008	15.3285	14.7628	14.2373	13.7480	71
72	22.1807	19.9664	18.1540	16.6433	15.3647	14.7964	14.2686	13.7771	72
73	22.2540	20.0258	18.2031	16.6846	15.3999	14.8289	14.2989	13.8054	73
74	22.3253	20.0835	18.2508	16.7246	15.4340	14.8606	14.3282	13.8327	74
75	22.3946	20.1395	18.2971	16.7635	15.4670	14.8912	14.3568	13.8593	75
76	22.4620	20.1940	18.3420	16.8012	15.4992	14.9210	14.3844	13.8851	76
77	22.5276	20.2470	18.3857	16.8379	15.5304	14.9499	14.4113	13.9101	77
78	22.5914	20.2985	18.4282	16.8735	15.5607	14.9780	14.4374	13.9344	78
79	22.6535	20.3486	18.4695	16.9081	15.5901	15.0052	14.4627	13.9580	79
80	22.7138	20.3973	18.5096	16.9417	15.6186	15.0317	14.4873	13.9809	80
81	22.7726	20.4447	18.5486	16.9743	15.6464	15.0574	14.5112	14.0032	81
82	22.8297	20.4907	18.5865	17.0061	15.6734	15.0824	14.5343	14.0247	82
83	22.8853	20.5355	18.6233	17.0369	15.6996	15.1066	14.5569	14.0457	83
84	22.9395	20.5791	18.6592	17.0669	15.7250	15.1302	14.5788	14.0661	84
85	22.9922	20.6215	18.6940	17.0960	15.7497	15.1531	14.6000	14.0859	85
86	23.0434	20.6627	18.7279	17.1244	15.7738	15.1754	14.6207	14.1051	86
87	23.0933	20.7028	18.7608	17.1519	15.7972	15.1970	14.6407	14.1238	87
88	23.1419	20.7419	18.7929	17.1787	15.8199	15.2180	14.6603	14.1419	88
89	23.1892	20.7799	18.8240	17.2047	15.8420	15.2384	14.6792	14.1596	89
90	23.2352	20.8168	18.8544	17.2301	15.8634	15.2583	14.6976	14.1767	90
91	23.2801	20.8528	18.8839	17.2547	15.8843	15.2776	14.7156	14.1934	91
92	23.3237	20.8878	18.9126	17.2787	15.9046	15.2964	14.7330	14.2096	92
93	23.3662	20.9219	18.9405	17.3020	15.9244	15.3147	14.7499	14.2254	93
94	23.4076	20.9550	18.9677	17.3246	15.9436	15.3324	14.7664	14.2407	94
95	23.4478	20.9873	18.9941	17.3467	15.9622	15.3497	14.7824	14.2556	95
96	23.4871	21.0187	19.0199	17.3682	15.9804	15.3665	14.7980	14.2701	96
97	23.5253	21.0493	19.0449	17.3890	15.9981	15.3828	14.8132	14.2842	97
98	23.5625	21.0791	19.0693	17.4094	16.0153	15.3987	14.8279	14.2979	98
99	23.5987	21.1081	19.0930	17.4291	16.0320	15.4142	14.8423	14.3112	99
100	23.6340	21.1363	19.1161	17.4484	16.0483	15.4293	14.8562	14.3242	100

YEARS' PURCHASE

				Rate Per Cent					
Yrs.	7	7.25	7.5	8	8.5	9	9.5	10	Yrs.
1	0.9346	0.9324	0.9302	0.9259	0.9217	0.9174	0.9132	0.9091	1
2	1.7736	1.7658	1.7580	1.7427	1.7276	1.7128	1.6983	1.6840	2
3	2.5308	2.5149	2.4991	2.4683	2.4382	2.4088	2.3802	2.3522	3
4	3.2173	3.1916	3.1664	3.1170	3.0692	3.0228	2.9778	2.9341	4
5	3.8425	3.8059	3.7701	3.7003	3.6331	3.5683	3.5057	3.4453	5
6	4.4140	4.3659	4.3187	4.2274	4.1399	4.0560	3.9754	3.8979	6
7	4.9384	4.8782	4.8194	4.7060	4.5978	4.4945	4.3957	4.3012	7
8	5.4210	5.3485	5.2780	5.1423	5.0134	4.8908	4.7740	4.6627	8
9	5.8666	5.7818	5.6994	5.5415	5.3921	5.2505	5.1162	4.9886	9
10	6.2791	6.1820	6.0880	5.9081	5.7386	5.5785	5.4272	5.2838	10
11	6.6620	6.5528	6.4472	6.2459	6.0567	5.8787	5.7108	5.5523	11
12	7.0182	6.8972	6.7803	6.5579	6.3497	6.1543	5.9706	5.7975	12
13	7.3503	7.2177	7.0898	6.8471	6.6204	6.4083	6.2093	6.0223	13
14	7.6607	7.5167	7.3781	7.1156	6.8711	6.6429	6.4294	6.2291	14
15	7.9512	7.7963	7.6472	7.3656	7.1040	6.8603	6.6328	6.4199	15
16	8.2238	8.0581	7.8990	7.5988	7.3207	7.0622	6.8213	6.5964	16
17	8.4798	8.3037	8.1349	7.8169	7.5229	7.2502	6.9965	6.7601	17
18	8.7207	8.5346	8.3563	8.0212	7.7119	7.4256	7.1597	6.9123	18
19	8.9477	8.7519	8.5645	8.2128	7.8889	7.5895	7.3120	7.0541	19
20	9.1619	8.9568	8.7606	8.3930	8.0550	7.7431	7.4545	7.1866	20
21	9.3644	9.1502	8.9456	8.5626	8.2110	7.8872	7.5880	7.3106	21
22	9.5560	9.3330	9.1202	8.7225	8.3579	8.0227	7.7133	7.4268	22
23	9.7374	9.5060	9.2854	8.8734	8.4964	8.1502	7.8311	7.5360	23
24	9.9095	9.6700	9.4417	9.0161	8.6272	8.2704	7.9420	7.6387	24
25	10.0729	9.8255	9.5899	9.1511	8.7507	8.3839	8.0466	7.7354	25
26	10.2282	9.9732	9.7306	9.2791	8.8677	8.4912	8.1454	7.8266	26
27	10.3759	10.1136	9.8642	9.4005	8.9785	8.5928	8.2388	7.9128	27
28	10.5166	10.2472	9.9912	9.5158	9.0836	8.6890	8.3272	7.9944	28
29	10.6506	10.3744	10.1121	9.6254	9.1835	8.7803	8.4110	8.0716	29
30	10.7785	10.4956	10.2273	9.7297	9.2784	8.8670	8.4906	8.1448	30
31	10.9005	10.6113	10.3371	9.8291	9.3686	8.9494	8.5661	8.2143	31
32	11.0171	10.7218	10.4419	9.9238	9.4546	9.0279	8.6379	8.2803	32
33	11.1285	10.8273	10.5419	10.0141	9.5366	9.1025	8.7063	8.3431	33
34	11.2351	10.9282	10.6375	10.1003	9.6148	9.1737	8.7714	8.4029	34
35	11.3372	11.0247	10.7290	10.1827	9.6894	9.2417	8.8335	8.4598	35
36	11.4349	11.1171	10.8165	10.2615	9.7607	9.3065	8.8927	8.5141	36
37	11.5286	11.2056	10.9003	10.3369	9.8289	9.3685	8.9493	8.5660	37
38	11.6184	11.2905	10.9805	10.4091	9.8941	9.4277	9.0033	8.6155	38
39	11.7046	11.3719	11.0575	10.4782	9.9566	9.4844	9.0550	8.6628	39
40	11.7874	11.4500	11.1314	10.5445	10.0164	9.5387	9.1045	8.7081	40
41	11.8669	11.5250	11.2022	10.6081	10.0738	9.5907	9.1518	8.7514	41
42	11.9433	11.5971	11.2703	10.6691	10.1288	9.6405	9.1972	8.7929	42
43	12.0168	11.6663	11.3357	10.7277	10.1815	9.6883	9.2407	8.8326	43
44	12.0875	11.7329	11.3986	10.7840	10.2322	9.7342	9.2824	8.8707	44
45	12.1555	11.7970	11.4590	10.8381	10.2809	9.7783	9.3225	8.9073	45
46	12.2209	11.8586	11.5172	10.8901	10.3277	9.8206	9.3610	8.9424	46
47	12.2840	11.9180	11.5732	10.9401	10.3727	9.8613	9.3979	8.9761	47
48	12.3448	11.9752	11.6271	10.9883	10.4160	9.9004	9.4334	9.0085	48
49	12.4033	12.0303	11.6790	11.0347	10.4577	9.9380	9.4676	9.0397	49
50	12.4598	12.0834	11.7291	11.0793	10.4978	9.9743	9.5005	9.0696	50

YEARS' PURCHASE

| | | | | Rate Per Cent | | | | | |
Yrs.	7	7.25	7.5	8	8.5	9	9.5	10	Yrs.
51	12.5143	12.1346	11.7774	11.1224	10.5364	10.0091	9.5321	9.0985	51
52	12.5668	12.1841	11.8239	11.1639	10.5737	10.0427	9.5626	9.1262	52
53	12.6176	12.2317	11.8688	11.2039	10.6096	10.0751	9.5919	9.1529	53
54	12.6666	12.2778	11.9121	11.2425	10.6442	10.1063	9.6202	9.1787	54
55	12.7139	12.3222	11.9540	11.2798	10.6776	10.1364	9.6475	9.2035	55
56	12.7596	12.3652	11.9944	11.3157	10.7098	10.1655	9.6738	9.2274	56
57	12.8038	12.4066	12.0334	11.3505	10.7409	10.1935	9.6991	9.2505	57
58	12.8465	12.4467	12.0711	11.3840	10.7709	10.2205	9.7236	9.2728	58
59	12.8877	12.4855	12.1076	11.4164	10.7999	10.2466	9.7472	9.2943	59
60	12.9277	12.5229	12.1428	11.4478	10.8280	10.2719	9.7701	9.3150	60
61	12.9663	12.5592	12.1769	11.4780	10.8551	10.2962	9.7921	9.3351	61
62	13.0037	12.5943	12.2098	11.5073	10.8812	10.3198	9.8134	9.3544	62
63	13.0399	12.6282	12.2417	11.5356	10.9066	10.3426	9.8340	9.3731	63
64	13.0749	12.6610	12.2726	11.5630	10.9311	10.3646	9.8539	9.3912	64
65	13.1088	12.6928	12.3024	11.5895	10.9547	10.3859	9.8732	9.4087	65
66	13.1416	12.7236	12.3314	11.6152	10.9777	10.4065	9.8918	9.4256	66
67	13.1735	12.7534	12.3594	11.6401	10.9999	10.4264	9.9098	9.4420	67
68	13.2043	12.7823	12.3865	11.6641	11.0213	10.4457	9.9272	9.4578	68
69	13.2342	12.8103	12.4128	11.6874	11.0422	10.4644	9.9441	9.4731	69
70	13.2631	12.8374	12.4383	11.7100	11.0623	10.4825	9.9604	9.4879	70
71	13.2912	12.8637	12.4629	11.7319	11.0818	10.5000	9.9763	9.5023	71
72	13.3184	12.8892	12.4869	11.7531	11.1007	10.5170	9.9916	9.5162	72
73	13.3448	12.9139	12.5101	11.7736	11.1191	10.5334	10.0064	9.5296	73
74	13.3704	12.9379	12.5325	11.7935	11.1368	10.5494	10.0208	9.5427	74
75	13.3952	12.9612	12.5544	11.8128	11.1540	10.5648	10.0348	9.5553	75
76	13.4193	12.9837	12.5755	11.8316	11.1707	10.5798	10.0483	9.5676	76
77	13.4427	13.0056	12.5960	11.8497	11.1869	10.5943	10.0614	9.5795	77
78	13.4654	13.0268	12.6160	11.8674	11.2026	10.6084	10.0741	9.5910	78
79	13.4874	13.0474	12.6353	11.8845	11.2179	10.6221	10.0864	9.6021	79
80	13.5088	13.0674	12.6541	11.9011	11.2327	10.6354	10.0984	9.6130	80
81	13.5295	13.0869	12.6723	11.9172	11.2470	10.6482	10.1099	9.6235	81
82	13.5497	13.1057	12.6899	11.9328	11.2609	10.6607	10.1212	9.6337	82
83	13.5692	13.1240	12.7071	11.9480	11.2745	10.6728	10.1321	9.6436	83
84	13.5883	13.1418	12.7238	11.9627	11.2876	10.6846	10.1427	9.6532	84
85	13.6067	13.1591	12.7400	11.9770	11.3003	10.6960	10.1530	9.6625	85
86	13.6247	13.1759	12.7557	11.9909	11.3127	10.7071	10.1630	9.6715	86
87	13.6421	13.1922	12.7710	12.0044	11.3247	10.7178	10.1727	9.6803	87
88	13.6590	13.2080	12.7858	12.0175	11.3364	10.7283	10.1821	9.6888	88
89	13.6755	13.2234	12.8002	12.0303	11.3477	10.7384	10.1912	9.6971	89
90	13.6915	13.2384	12.8143	12.0427	11.3587	10.7483	10.2001	9.7051	90
91	13.7070	13.2529	12.8279	12.0547	11.3694	10.7579	10.2087	9.7130	91
92	13.7222	13.2670	12.8411	12.0664	11.3798	10.7672	10.2171	9.7205	92
93	13.7368	13.2808	12.8540	12.0777	11.3899	10.7762	10.2253	9.7279	93
94	13.7511	13.2941	12.8665	12.0888	11.3997	10.7850	10.2332	9.7351	94
95	13.7650	13.3071	12.8787	12.0995	11.4093	10.7936	10.2409	9.7420	95
96	13.7785	13.3197	12.8905	12.1100	11.4186	10.8019	10.2484	9.7488	96
97	13.7917	13.3320	12.9020	12.1201	11.4276	10.8099	10.2556	9.7554	97
98	13.8045	13.3439	12.9132	12.1300	11.4364	10.8178	10.2627	9.7618	98
99	13.8169	13.3556	12.9240	12.1396	11.4449	10.8254	10.2696	9.7680	99
100	13.8290	13.3669	12.9346	12.1489	11.4532	10.8328	10.2762	9.7740	100

YEARS' PURCHASE

				Rate Per Cent					
Yrs.	**11**	**12**	**13**	**14**	**15**	**16**	**18**	**20**	Yrs.
1	0.9009	0.8929	0.8850	0.8772	0.8696	0.8621	0.8475	0.8333	1
2	1.6561	1.6291	1.6030	1.5777	1.5532	1.5295	1.4841	1.4413	2
3	2.2981	2.2465	2.1971	2.1499	2.1047	2.0613	1.9797	1.9043	3
4	2.8505	2.7715	2.6967	2.6259	2.5587	2.4949	2.3763	2.2685	4
5	3.3306	3.2232	3.1226	3.0280	2.9390	2.8551	2.7009	2.5625	5
6	3.7516	3.6160	3.4898	3.3721	3.2621	3.1591	2.9713	2.8047	6
7	4.1238	3.9605	3.8096	3.6698	3.5399	3.4189	3.2000	3.0076	7
8	4.4550	4.2650	4.0905	3.9298	3.7812	3.6434	3.3960	3.1800	8
9	4.7516	4.5360	4.3392	4.1587	3.9927	3.8394	3.5656	3.3283	9
10	5.0186	4.7788	4.5608	4.3619	4.1796	4.0119	3.7139	3.4571	10
11	5.2602	4.9974	4.7595	4.5433	4.3458	4.1648	3.8446	3.5701	11
12	5.4798	5.1952	4.9386	4.7062	4.4946	4.3013	3.9606	3.6699	12
13	5.6803	5.3749	5.1008	4.8532	4.6286	4.4238	4.0642	3.7587	13
14	5.8638	5.5390	5.2483	4.9866	4.7498	4.5344	4.1574	3.8382	14
15	6.0326	5.6894	5.3831	5.1081	4.8599	4.6346	4.2415	3.9098	15
16	6.1882	5.8275	5.5066	5.2192	4.9603	4.7259	4.3178	3.9746	16
17	6.3320	5.9549	5.6203	5.3212	5.0523	4.8094	4.3874	4.0334	17
18	6.4654	6.0727	5.7251	5.4151	5.1369	4.8859	4.4510	4.0871	18
19	6.5893	6.1820	5.8221	5.5017	5.2148	4.9564	4.5094	4.1363	19
20	6.7048	6.2835	5.9120	5.5820	5.2869	5.0214	4.5631	4.1815	20
21	6.8126	6.3781	5.9957	5.6565	5.3537	5.0816	4.6128	4.2232	21
22	6.9134	6.4664	6.0736	5.7258	5.4158	5.1375	4.6588	4.2617	22
23	7.0079	6.5489	6.1464	5.7905	5.4736	5.1895	4.7015	4.2974	23
24	7.0966	6.6263	6.2145	5.8509	5.5275	5.2380	4.7413	4.3306	24
25	7.1800	6.6990	6.2784	5.9075	5.5780	5.2833	4.7784	4.3616	25
26	7.2585	6.7673	6.3384	5.9606	5.6253	5.3257	4.8130	4.3904	26
27	7.3326	6.8317	6.3948	6.0105	5.6697	5.3655	4.8455	4.4174	27
28	7.4026	6.8924	6.4480	6.0574	5.7114	5.4028	4.8760	4.4427	28
29	7.4687	6.9497	6.4981	6.1016	5.7507	5.4380	4.9046	4.4665	29
30	7.5314	7.0039	6.5455	6.1433	5.7878	5.4711	4.9315	4.4888	30
31	7.5908	7.0552	6.5903	6.1828	5.8228	5.5024	4.9569	4.5098	31
32	7.6471	7.1039	6.6327	6.2201	5.8559	5.5319	4.9809	4.5296	32
33	7.7006	7.1500	6.6729	6.2555	5.8872	5.5599	5.0035	4.5484	33
34	7.7515	7.1939	6.7111	6.2890	5.9169	5.5864	5.0250	4.5661	34
35	7.8000	7.2356	6.7474	6.3209	5.9451	5.6115	5.0453	4.5828	35
36	7.8461	7.2753	6.7819	6.3512	5.9719	5.6353	5.0645	4.5987	36
37	7.8901	7.3131	6.8147	6.3800	5.9973	5.6580	5.0828	4.6138	37
38	7.9321	7.3492	6.8460	6.4074	6.0216	5.6796	5.1002	4.6281	38
39	7.9722	7.3836	6.8759	6.4335	6.0446	5.7001	5.1168	4.6417	39
40	8.0105	7.4164	6.9043	6.4584	6.0666	5.7196	5.1325	4.6547	40
41	8.0471	7.4478	6.9316	6.4822	6.0876	5.7383	5.1475	4.6671	41
42	8.0822	7.4778	6.9575	6.5050	6.1077	5.7561	5.1619	4.6788	42
43	8.1158	7.5066	6.9824	6.5267	6.1268	5.7731	5.1755	4.6901	43
44	8.1479	7.5341	7.0062	6.5475	6.1451	5.7894	5.1886	4.7008	44
45	8.1788	7.5604	7.0290	6.5674	6.1627	5.8049	5.2011	4.7110	45
46	8.2084	7.5857	7.0509	6.5865	6.1794	5.8198	5.2130	4.7208	46
47	8.2368	7.6100	7.0718	6.6047	6.1955	5.8341	5.2245	4.7302	47
48	8.2641	7.6332	7.0919	6.6223	6.2109	5.8477	5.2354	4.7392	48
49	8.2903	7.6556	7.1112	6.6391	6.2257	5.8609	5.2459	4.7478	49
50	8.3155	7.6771	7.1297	6.6552	6.2399	5.8734	5.2560	4.7561	50

YEARS' PURCHASE

				Rate Per Cent					
Yrs.	11	12	13	14	15	16	18	20	Yrs.
51	8.3397	7.6977	7.1475	6.6707	6.2536	5.8855	5.2657	4.7640	51
52	8.3630	7.7176	7.1646	6.6856	6.2667	5.8971	5.2750	4.7716	52
53	8.3854	7.7367	7.1811	6.7000	6.2793	5.9083	5.2839	4.7789	53
54	8.4070	7.7551	7.1969	6.7138	6.2914	5.9190	5.2925	4.7859	54
55	8.4279	7.7728	7.2122	6.7270	6.3030	5.9293	5.3007	4.7926	55
56	8.4479	7.7898	7.2269	6.7398	6.3142	5.9392	5.3086	4.7991	56
57	8.4673	7.8063	7.2410	6.7521	6.3250	5.9488	5.3163	4.8053	57
58	8.4859	7.8221	7.2547	6.7640	6.3354	5.9580	5.3236	4.8113	58
59	8.5039	7.8374	7.2678	6.7754	6.3455	5.9668	5.3307	4.8171	59
60	8.5213	7.8522	7.2805	6.7864	6.3551	5.9754	5.3375	4.8227	60
61	8.5380	7.8664	7.2927	6.7970	6.3644	5.9836	5.3441	4.8281	61
62	8.5542	7.8801	7.3045	6.8073	6.3734	5.9916	5.3504	4.8332	62
63	8.5699	7.8934	7.3159	6.8172	6.3821	5.9992	5.3565	4.8382	63
64	8.5850	7.9062	7.3269	6.8268	6.3905	6.0066	5.3624	4.8430	64
65	8.5996	7.9186	7.3376	6.8360	6.3986	6.0138	5.3681	4.8477	65
66	8.6137	7.9306	7.3479	6.8449	6.4064	6.0207	5.3736	4.8522	66
67	8.6274	7.9422	7.3578	6.8535	6.4139	6.0274	5.3789	4.8565	67
68	8.6406	7.9534	7.3674	6.8619	6.4212	6.0338	5.3841	4.8607	68
69	8.6534	7.9642	7.3767	6.8699	6.4283	6.0400	5.3890	4.8647	69
70	8.6657	7.9747	7.3857	6.8777	6.4351	6.0461	5.3938	4.8686	70
71	8.6777	7.9848	7.3944	6.8853	6.4417	6.0519	5.3985	4.8724	71
72	8.6893	7.9946	7.4028	6.8926	6.4481	6.0575	5.4029	4.8760	72
73	8.7005	8.0041	7.4109	6.8996	6.4543	6.0630	5.4073	4.8796	73
74	8.7114	8.0133	7.4188	6.9064	6.4603	6.0682	5.4115	4.8830	74
75	8.7219	8.0222	7.4265	6.9131	6.4661	6.0734	5.4155	4.8863	75
76	8.7321	8.0309	7.4339	6.9195	6.4717	6.0783	5.4195	4.8895	76
77	8.7420	8.0392	7.4410	6.9257	6.4771	6.0831	5.4233	4.8926	77
78	8.7516	8.0473	7.4480	6.9317	6.4824	6.0877	5.4270	4.8956	78
79	8.7609	8.0552	7.4547	6.9375	6.4875	6.0922	5.4306	4.8985	79
80	8.7699	8.0628	7.4612	6.9432	6.4924	6.0966	5.4340	4.9013	80
81	8.7787	8.0702	7.4676	6.9487	6.4972	6.1008	5.4374	4.9041	81
82	8.7872	8.0774	7.4737	6.9540	6.5018	6.1049	5.4406	4.9067	82
83	8.7954	8.0843	7.4797	6.9591	6.5063	6.1089	5.4438	4.9093	83
84	8.8034	8.0911	7.4854	6.9641	6.5107	6.1127	5.4468	4.9118	84
85	8.8111	8.0976	7.4910	6.9690	6.5150	6.1165	5.4498	4.9142	85
86	8.8186	8.1040	7.4965	6.9737	6.5191	6.1201	5.4527	4.9165	86
87	8.8259	8.1101	7.5017	6.9782	6.5230	6.1236	5.4555	4.9188	87
88	8.8330	8.1161	7.5069	6.9827	6.5269	6.1270	5.4582	4.9210	88
89	8.8399	8.1219	7.5118	6.9870	6.5307	6.1303	5.4608	4.9231	89
90	8.8466	8.1276	7.5166	6.9911	6.5343	6.1335	5.4633	4.9252	90
91	8.8531	8.1330	7.5213	6.9952	6.5379	6.1367	5.4658	4.9272	91
92	8.8594	8.1384	7.5259	6.9991	6.5413	6.1397	5.4682	4.9291	92
93	8.8655	8.1435	7.5303	7.0030	6.5446	6.1426	5.4706	4.9310	93
94	8.8714	8.1485	7.5346	7.0067	6.5479	6.1455	5.4728	4.9329	94
95	8.8772	8.1534	7.5388	7.0103	6.5510	6.1483	5.4750	4.9347	95
96	8.8828	8.1582	7.5428	7.0138	6.5541	6.1509	5.4772	4.9364	96
97	8.8883	8.1628	7.5467	7.0172	6.5571	6.1536	5.4792	4.9381	97
98	8.8936	8.1672	7.5506	7.0205	6.5599	6.1561	5.4812	4.9397	98
99	8.8988	8.1716	7.5543	7.0237	6.5627	6.1586	5.4832	4.9413	99
100	8.9038	8.1758	7.5579	7.0268	6.5655	6.1610	5.4851	4.9429	100

YEARS' PURCHASE

				Rate Per Cent					
Yrs.	4	4.5	5	5.5	6	6.25	6.5	6.75	Yrs.
1	0.9615	0.9569	0.9524	0.9479	0.9434	0.9412	0.9390	0.9368	1
2	1.8775	1.8601	1.8429	1.8261	1.8096	1.8014	1.7934	1.7854	2
3	2.7508	2.7135	2.6772	2.6418	2.6074	2.5905	2.5738	2.5573	3
4	3.5839	3.5208	3.4599	3.4010	3.3442	3.3165	3.2892	3.2624	4
5	4.3792	4.2853	4.1954	4.1092	4.0265	3.9864	3.9470	3.9085	5
6	5.1388	5.0101	4.8876	4.7710	4.6599	4.6062	4.5538	4.5025	6
7	5.8649	5.6978	5.5400	5.3907	5.2492	5.1812	5.1149	5.0503	7
8	6.5593	6.3510	6.1555	5.9717	5.7986	5.7157	5.6352	5.5569	8
9	7.2237	6.9719	6.7370	6.5175	6.3118	6.2137	6.1187	6.0265	9
10	7.8598	7.5626	7.2870	7.0308	6.7921	6.6787	6.5690	6.4628	10
11	8.4690	8.1250	7.8078	7.5144	7.2423	7.1135	6.9892	6.8692	11
12	9.0529	8.6609	8.3014	7.9705	7.6651	7.5209	7.3821	7.2484	12
13	9.6127	9.1718	8.7697	8.4013	8.0626	7.9033	7.7502	7.6029	13
14	10.1496	9.6594	9.2144	8.8085	8.4369	8.2627	8.0954	7.9348	14
15	10.6648	10.1249	9.6370	9.1940	8.7899	8.6009	8.4199	8.2463	15
16	11.1594	10.5696	10.0391	9.5592	9.1232	8.9197	8.7252	8.5389	16
17	11.6343	10.9947	10.4218	9.9056	9.4382	9.2206	9.0129	8.8143	17
18	12.0906	11.4014	10.7865	10.2345	9.7363	9.5049	9.2843	9.0737	18
19	12.5291	11.7905	11.1341	10.5470	10.0186	9.7738	9.5407	9.3185	19
20	12.9507	12.1631	11.4658	10.8441	10.2864	10.0285	9.7832	9.5497	20
21	13.3562	12.5201	11.7825	11.1270	10.5405	10.2699	10.0128	9.7683	21
22	13.7462	12.8622	12.0850	11.3963	10.7820	10.4990	10.2305	9.9753	22
23	14.1215	13.1902	12.3741	11.6531	11.0115	10.7165	10.4369	10.1715	23
24	14.4828	13.5049	12.6506	11.8980	11.2300	10.9233	10.6329	10.3576	24
25	14.8307	13.8068	12.9152	12.1318	11.4380	11.1200	10.8192	10.5343	25
26	15.1657	14.0968	13.1686	12.3551	11.6363	11.3073	10.9965	10.7023	26
27	15.4885	14.3752	13.4113	12.5685	11.8253	11.4858	11.1652	10.8620	27
28	15.7995	14.6427	13.6438	12.7725	12.0058	11.6559	11.3259	11.0140	28
29	16.0993	14.8999	13.8668	12.9677	12.1781	11.8183	11.4791	11.1589	29
30	16.3883	15.1471	14.0807	13.1546	12.3427	11.9733	11.6253	11.2970	30
31	16.6670	15.3849	14.2859	13.3335	12.5002	12.1214	11.7649	11.4287	31
32	16.9358	15.6136	14.4830	13.5050	12.6508	12.2629	11.8982	11.5545	32
33	17.1951	15.8338	14.6722	13.6694	12.7949	12.3983	12.0256	11.6746	33
34	17.4453	16.0457	14.8540	13.8271	12.9329	12.5279	12.1474	11.7894	34
35	17.6868	16.2498	15.0287	13.9783	13.0652	12.6519	12.2640	11.8992	35
36	17.9199	16.4463	15.1967	14.1235	13.1920	12.7708	12.3757	12.0043	36
37	18.1450	16.6357	15.3582	14.2630	13.3135	12.8847	12.4826	12.1048	37
38	18.3623	16.8182	15.5137	14.3969	13.4301	12.9939	12.5851	12.2012	38
39	18.5722	16.9941	15.6632	14.5256	13.5421	13.0986	12.6833	12.2935	39
40	18.7750	17.1637	15.8072	14.6494	13.6496	13.1992	12.7775	12.3820	40
41	18.9709	17.3273	15.9458	14.7683	13.7528	13.2957	12.8680	12.4669	41
42	19.1601	17.4851	16.0793	14.8828	13.8520	13.3884	12.9548	12.5484	42
43	19.3431	17.6373	16.2080	14.9929	13.9474	13.4774	13.0381	12.6266	43
44	19.5199	17.7841	16.3319	15.0989	14.0391	13.5630	13.1182	12.7017	44
45	19.6908	17.9259	16.4514	15.2010	14.1273	13.6453	13.1952	12.7738	45
46	19.8560	18.0628	16.5666	15.2993	14.2121	13.7245	13.2692	12.8431	46
47	20.0158	18.1949	16.6776	15.3940	14.2938	13.8006	13.3404	12.9098	47
48	20.1703	18.3225	16.7848	15.4852	14.3724	13.8739	13.4088	12.9739	48
49	20.3198	18.4457	16.8881	15.5731	14.4481	13.9444	13.4747	13.0356	49
50	20.4643	18.5648	16.9879	15.6579	14.5211	14.0124	13.5381	13.0949	50

YEARS' PURCHASE

				Rate Per Cent					
Yrs.	4	4.5	5	5.5	6	6.25	6.5	6.75	Yrs.
51	20.6042	18.6798	17.0841	15.7396	14.5913	14.0778	13.5992	13.1520	51
52	20.7395	18.7909	17.1771	15.8185	14.6591	14.1408	13.6580	13.2070	52
53	20.8704	18.8983	17.2668	15.8945	14.7244	14.2016	13.7147	13.2600	53
54	20.9971	19.0022	17.3534	15.9679	14.7873	14.2601	13.7693	13.3110	54
55	21.1197	19.1025	17.4371	16.0387	14.8480	14.3166	13.8219	13.3602	55
56	21.2384	19.1996	17.5179	16.1071	14.9066	14.3710	13.8726	13.4076	56
57	21.3533	19.2934	17.5960	16.1731	14.9631	14.4235	13.9215	13.4533	57
58	21.4645	19.3842	17.6714	16.2368	15.0176	14.4742	13.9687	13.4974	58
59	21.5722	19.4719	17.7444	16.2983	15.0702	14.5231	14.0143	13.5399	59
60	21.6765	19.5569	17.8148	16.3578	15.1211	14.5703	14.0582	13.5809	60
61	21.7774	19.6390	17.8830	16.4152	15.1701	14.6158	14.1006	13.6204	61
62	21.8752	19.7185	17.9489	16.4707	15.2175	14.6598	14.1415	13.6586	62
63	21.9699	19.7954	18.0126	16.5243	15.2633	14.7022	14.1810	13.6955	63
64	22.0616	19.8698	18.0742	16.5762	15.3075	14.7433	14.2192	13.7311	64
65	22.1504	19.9418	18.1337	16.6263	15.3502	14.7829	14.2560	13.7654	65
66	22.2365	20.0116	18.1914	16.6747	15.3915	14.8212	14.2916	13.7986	66
67	22.3199	20.0791	18.2471	16.7215	15.4314	14.8582	14.3260	13.8307	67
68	22.4006	20.1444	18.3011	16.7668	15.4699	14.8939	14.3592	13.8616	68
69	22.4789	20.2077	18.3533	16.8106	15.5072	14.9285	14.3914	13.8916	69
70	22.5547	20.2689	18.4038	16.8530	15.5432	14.9619	14.4224	13.9205	70
71	22.6282	20.3282	18.4527	16.8940	15.5781	14.9942	14.4524	13.9484	71
72	22.6994	20.3857	18.5000	16.9336	15.6118	15.0254	14.4814	13.9755	72
73	22.7684	20.4413	18.5458	16.9720	15.6444	15.0556	14.5095	14.0016	73
74	22.8353	20.4952	18.5902	17.0091	15.6760	15.0848	14.5366	14.0268	74
75	22.9001	20.5474	18.6331	17.0451	15.7065	15.1131	14.5628	14.0513	75
76	22.9629	20.5980	18.6747	17.0799	15.7360	15.1404	14.5882	14.0749	76
77	23.0238	20.6469	18.7149	17.1135	15.7646	15.1668	14.6128	14.0977	77
78	23.0828	20.6944	18.7539	17.1461	15.7922	15.1924	14.6365	14.1199	78
79	23.1401	20.7404	18.7917	17.1777	15.8190	15.2172	14.6595	14.1413	79
80	23.1955	20.7849	18.8282	17.2082	15.8449	15.2412	14.6818	14.1619	80
81	23.2493	20.8281	18.8636	17.2378	15.8700	15.2644	14.7033	14.1820	81
82	23.3015	20.8700	18.8980	17.2665	15.8943	15.2868	14.7241	14.2014	82
83	23.3520	20.9105	18.9312	17.2942	15.9178	15.3086	14.7443	14.2201	83
84	23.4010	20.9498	18.9634	17.3211	15.9405	15.3296	14.7638	14.2383	84
85	23.4486	20.9879	18.9946	17.3471	15.9626	15.3500	14.7827	14.2559	85
86	23.4947	21.0248	19.0248	17.3723	15.9839	15.3698	14.8010	14.2729	86
87	23.5394	21.0606	19.0541	17.3967	16.0046	15.3889	14.8188	14.2894	87
88	23.5827	21.0953	19.0825	17.4204	16.0246	15.4074	14.8359	14.3053	88
89	23.6248	21.1289	19.1101	17.4433	16.0440	15.4253	14.8526	14.3208	89
90	23.6655	21.1615	19.1367	17.4656	16.0628	15.4427	14.8687	14.3358	90
91	23.7051	21.1931	19.1626	17.4871	16.0810	15.4595	14.8843	14.3503	91
92	23.7434	21.2238	19.1876	17.5079	16.0987	15.4758	14.8994	14.3643	92
93	23.7806	21.2535	19.2119	17.5282	16.1158	15.4916	14.9140	14.3779	93
94	23.8167	21.2823	19.2354	17.5477	16.1323	15.5069	14.9282	14.3911	94
95	23.8517	21.3103	19.2583	17.5667	16.1484	15.5217	14.9419	14.4039	95
96	23.8856	21.3373	19.2804	17.5851	16.1639	15.5361	14.9552	14.4162	96
97	23.9185	21.3636	19.3018	17.6030	16.1790	15.5500	14.9681	14.4282	97
98	23.9505	21.3891	19.3226	17.6203	16.1936	15.5635	14.9806	14.4398	98
99	23.9815	21.4138	19.3428	17.6370	16.2078	15.5766	14.9928	14.4511	99
100	24.0115	21.4378	19.3623	17.6533	16.2215	15.5893	15.0045	14.4620	100

YEARS' PURCHASE

Rate Per Cent

Yrs.	7	7.25	7.5	8	8.5	9	9.5	10	Yrs.
1	0.9346	0.9324	0.9302	0.9259	0.9217	0.9174	0.9132	0.9091	1
2	1.7774	1.7696	1.7618	1.7464	1.7313	1.7164	1.7018	1.6874	2
3	2.5411	2.5251	2.5092	2.4781	2.4478	2.4182	2.3893	2.3611	3
4	3.2360	3.2100	3.1844	3.1345	3.0862	3.0393	2.9938	2.9496	4
5	3.8706	3.8336	3.7972	3.7264	3.6583	3.5925	3.5291	3.4680	5
6	4.4524	4.4034	4.3554	4.2626	4.1737	4.0883	4.0065	3.9278	6
7	4.9874	4.9260	4.8660	4.7505	4.6402	4.5350	4.4345	4.3383	7
8	5.4808	5.4067	5.3346	5.1960	5.0644	4.9393	4.8203	4.7068	8
9	5.9370	5.8502	5.7659	5.6043	5.4516	5.3069	5.1697	5.0395	9
10	6.3601	6.2605	6.1641	5.9798	5.8062	5.6424	5.4876	5.3410	10
11	6.7532	6.6411	6.5326	6.3260	6.1321	5.9496	5.7778	5.6155	11
12	7.1194	6.9949	6.8746	6.6462	6.4324	6.2320	6.0437	5.8664	12
13	7.4610	7.3244	7.1927	6.9430	6.7101	6.4923	6.2881	6.0965	13
14	7.7805	7.6321	7.4892	7.2188	6.9674	6.7328	6.5135	6.3081	14
15	8.0797	7.9198	7.7660	7.4757	7.2063	6.9557	6.7219	6.5034	15
16	8.3604	8.1893	8.0250	7.7154	7.4288	7.1628	6.9151	6.6840	16
17	8.6242	8.4422	8.2677	7.9395	7.6364	7.3555	7.0946	6.8515	17
18	8.8724	8.6799	8.4955	8.1494	7.8303	7.5353	7.2617	7.0073	18
19	9.1063	8.9036	8.7097	8.3463	8.0119	7.7033	7.4176	7.1524	19
20	9.3270	9.1145	8.9114	8.5313	8.1823	7.8607	7.5634	7.2878	20
21	9.5355	9.3134	9.1015	8.7054	8.3422	8.0082	7.6999	7.4144	21
22	9.7326	9.5014	9.2810	8.8694	8.4928	8.1468	7.8279	7.5331	22
23	9.9193	9.6792	9.4506	9.0241	8.6345	8.2772	7.9482	7.6444	23
24	10.0962	9.8476	9.6110	9.1703	8.7683	8.4000	8.0614	7.7491	24
25	10.2640	10.0072	9.7630	9.3086	8.8946	8.5159	8.1681	7.8476	25
26	10.4234	10.1586	9.9070	9.4395	9.0140	8.6253	8.2687	7.9404	26
27	10.5748	10.3025	10.0438	9.5635	9.1271	8.7287	8.3637	8.0280	27
28	10.7189	10.4392	10.1736	9.6812	9.2342	8.8267	8.4536	8.1107	28
29	10.8560	10.5692	10.2971	9.7929	9.3358	8.9194	8.5386	8.1890	29
30	10.9867	10.6930	10.4146	9.8991	9.4322	9.0074	8.6193	8.2631	30
31	11.1112	10.8109	10.5264	10.0001	9.5239	9.0910	8.6957	8.3334	31
32	11.2301	10.9234	10.6330	10.0963	9.6111	9.1704	8.7683	8.4001	32
33	11.3435	11.0307	10.7347	10.1879	9.6940	9.2459	8.8373	8.4634	33
34	11.4519	11.1331	10.8317	10.2752	9.7731	9.3178	8.9030	8.5236	34
35	11.5554	11.2310	10.9243	10.3585	9.8484	9.3862	8.9655	8.5808	35
36	11.6545	11.3245	11.0128	10.4380	9.9203	9.4515	9.0250	8.6353	36
37	11.7493	11.4140	11.0973	10.5140	9.9888	9.5137	9.0817	8.6872	37
38	11.8400	11.4996	11.1783	10.5866	10.0544	9.5731	9.1358	8.7367	38
39	11.9269	11.5816	11.2557	10.6560	10.1170	9.6298	9.1875	8.7840	39
40	12.0102	11.6601	11.3299	10.7224	10.1768	9.6841	9.2368	8.8291	40
41	12.0901	11.7354	11.4009	10.7860	10.2341	9.7359	9.2840	8.8721	41
42	12.1667	11.8075	11.4690	10.8470	10.2889	9.7855	9.3291	8.9133	42
43	12.2402	11.8768	11.5343	10.9054	10.3415	9.8330	9.3722	8.9527	43
44	12.3107	11.9432	11.5969	10.9613	10.3918	9.8785	9.4136	8.9904	44
45	12.3785	12.0069	11.6570	11.0150	10.4400	9.9221	9.4531	9.0265	45
46	12.4436	12.0682	11.7147	11.0665	10.4863	9.9639	9.4910	9.0610	46
47	12.5062	12.1270	11.7702	11.1160	10.5307	10.0040	9.5274	9.0942	47
48	12.5663	12.1836	11.8234	11.1635	10.5733	10.0424	9.5623	9.1259	48
49	12.6242	12.2379	11.8746	11.2091	10.6142	10.0793	9.5957	9.1564	49
50	12.6798	12.2902	11.9239	11.2530	10.6535	10.1148	9.6278	9.1856	50

YEARS' PURCHASE

				Rate Per Cent					
Yrs.	7	7.25	7.5	8	8.5	9	9.5	10	Yrs.
51	12.7334	12.3405	11.9712	11.2951	10.6913	10.1488	9.6587	9.2137	51
52	12.7849	12.3889	12.0167	11.3357	10.7276	10.1815	9.6883	9.2407	52
53	12.8345	12.4355	12.0606	11.3747	10.7626	10.2130	9.7168	9.2666	53
54	12.8824	12.4804	12.1028	11.4122	10.7962	10.2432	9.7442	9.2915	54
55	12.9284	12.5236	12.1434	11.4483	10.8285	10.2723	9.7705	9.3154	55
56	12.9728	12.5653	12.1826	11.4831	10.8596	10.3003	9.7958	9.3384	56
57	13.0156	12.6054	12.2203	11.5166	10.8896	10.3273	9.8202	9.3606	57
58	13.0568	12.6441	12.2566	11.5489	10.9184	10.3532	9.8436	9.3819	58
59	13.0966	12.6814	12.2917	11.5800	10.9462	10.3782	9.8662	9.4024	59
60	13.1349	12.7173	12.3254	11.6100	10.9730	10.4023	9.8880	9.4221	60
61	13.1719	12.7520	12.3580	11.6389	10.9988	10.4255	9.9089	9.4412	61
62	13.2076	12.7855	12.3894	11.6667	11.0237	10.4478	9.9291	9.4595	62
63	13.2421	12.8177	12.4198	11.6936	11.0477	10.4694	9.9486	9.4772	63
64	13.2753	12.8489	12.4490	11.7195	11.0708	10.4901	9.9673	9.4942	64
65	13.3075	12.8790	12.4773	11.7446	11.0931	10.5102	9.9854	9.5106	65
66	13.3385	12.9080	12.5045	11.7687	11.1147	10.5295	10.0029	9.5264	66
67	13.3684	12.9361	12.5308	11.7920	11.1355	10.5482	10.0197	9.5417	67
68	13.3974	12.9632	12.5563	11.8145	11.1555	10.5662	10.0360	9.5564	68
69	13.4253	12.9893	12.5808	11.8363	11.1749	10.5836	10.0516	9.5706	69
70	13.4523	13.0146	12.6045	11.8572	11.1936	10.6003	10.0668	9.5844	70
71	13.4784	13.0391	12.6274	11.8775	11.2117	10.6165	10.0814	9.5976	71
72	13.5037	13.0627	12.6496	11.8971	11.2291	10.6322	10.0955	9.6104	72
73	13.5280	13.0855	12.6710	11.9160	11.2460	10.6473	10.1091	9.6227	73
74	13.5516	13.1075	12.6917	11.9343	11.2623	10.6619	10.1223	9.6347	74
75	13.5744	13.1289	12.7117	11.9520	11.2780	10.6760	10.1350	9.6462	75
76	13.5965	13.1495	12.7310	11.9691	11.2932	10.6896	10.1473	9.6573	76
77	13.6178	13.1695	12.7497	11.9856	11.3080	10.7028	10.1592	9.6681	77
78	13.6384	13.1887	12.7678	12.0016	11.3222	10.7156	10.1706	9.6785	78
79	13.6584	13.2074	12.7853	12.0170	11.3359	10.7279	10.1817	9.6885	79
80	13.6777	13.2255	12.8022	12.0320	11.3492	10.7398	10.1925	9.6982	80
81	13.6964	13.2429	12.8185	12.0464	11.3621	10.7513	10.2028	9.7076	81
82	13.7145	13.2598	12.8344	12.0604	11.3745	10.7624	10.2129	9.7167	82
83	13.7320	13.2762	12.8497	12.0740	11.3866	10.7732	10.2226	9.7255	83
84	13.7489	13.2920	12.8645	12.0871	11.3982	10.7836	10.2319	9.7340	84
85	13.7653	13.3073	12.8789	12.0997	11.4095	10.7937	10.2410	9.7422	85
86	13.7812	13.3222	12.8928	12.1120	11.4204	10.8035	10.2498	9.7501	86
87	13.7965	13.3365	12.9062	12.1239	11.4309	10.8129	10.2583	9.7578	87
88	13.8114	13.3504	12.9192	12.1353	11.4411	10.8221	10.2665	9.7653	88
89	13.8258	13.3639	12.9318	12.1465	11.4510	10.8309	10.2745	9.7725	89
90	13.8398	13.3769	12.9441	12.1572	11.4606	10.8395	10.2822	9.7794	90
91	13.8533	13.3896	12.9559	12.1677	11.4699	10.8477	10.2896	9.7862	91
92	13.8664	13.4018	12.9673	12.1778	11.4788	10.8558	10.2969	9.7927	92
93	13.8790	13.4136	12.9784	12.1875	11.4875	10.8635	10.3039	9.7990	93
94	13.8913	13.4251	12.9891	12.1970	11.4959	10.8711	10.3106	9.8051	94
95	13.9032	13.4362	12.9995	12.2062	11.5041	10.8783	10.3172	9.8111	95
96	13.9147	13.4470	13.0096	12.2151	11.5120	10.8854	10.3235	9.8168	96
97	13.9259	13.4574	13.0194	12.2237	11.5196	10.8922	10.3297	9.8224	97
98	13.9367	13.4675	13.0288	12.2320	11.5270	10.8988	10.3356	9.8277	98
99	13.9472	13.4773	13.0380	12.2401	11.5342	10.9053	10.3414	9.8330	99
100	13.9574	13.4868	13.0469	12.2479	11.5411	10.9110	10.3470	9.8380	100

YEARS' PURCHASE

				Rate Per Cent					
Yrs.	11	12	13	14	15	16	18	20	Yrs.
1	0.9009	0.8929	0.8850	0.8772	0.8696	0.8621	0.8475	0.8333	1
2	1.6594	1.6324	1.6061	1.5808	1.5562	1.5323	1.4867	1.4438	2
3	2.3066	2.2546	2.2049	2.1574	2.1118	2.0681	1.9860	1.9101	3
4	2.8651	2.7853	2.7098	2.6383	2.5705	2.5061	2.3865	2.2778	4
5	3.3517	3.2430	3.1412	3.0455	2.9555	2.8706	2.7148	2.5750	5
6	3.7793	3.6417	3.5137	3.3945	3.2830	3.1787	2.9887	2.8201	6
7	4.1579	3.9919	3.8387	3.6968	3.5650	3.4423	3.2205	3.0257	7
8	4.4953	4.3019	4.1245	3.9611	3.8102	3.6703	3.4193	3.2004	8
9	4.7977	4.5780	4.3776	4.1940	4.0252	3.8695	3.5915	3.3508	9
10	5.0702	4.8255	4.6034	4.4008	4.2153	4.0448	3.7421	3.4815	10
11	5.3170	5.0485	4.8059	4.5855	4.3845	4.2003	3.8748	3.5961	11
12	5.5413	5.2504	4.9885	4.7514	4.5359	4.3391	3.9926	3.6974	12
13	5.7462	5.4339	5.1539	4.9013	4.6723	4.4637	4.0979	3.7875	13
14	5.9338	5.6014	5.3043	5.0371	4.7956	4.5761	4.1924	3.8681	14
15	6.1063	5.7548	5.4417	5.1608	4.9076	4.6780	4.2778	3.9406	15
16	6.2652	5.8958	5.5676	5.2740	5.0097	4.7707	4.3552	4.0062	16
17	6.4122	6.0258	5.6834	5.3777	5.1033	4.8555	4.4257	4.0658	17
18	6.5484	6.1460	5.7901	5.4732	5.1892	4.9332	4.4902	4.1202	18
19	6.6749	6.2573	5.8888	5.5613	5.2683	5.0047	4.5493	4.1699	19
20	6.7928	6.3607	5.9803	5.6428	5.3414	5.0706	4.6037	4.2156	20
21	6.9027	6.4570	6.0653	5.7185	5.4092	5.1316	4.6539	4.2576	21
22	7.0054	6.5468	6.1445	5.7888	5.4720	5.1881	4.7004	4.2965	22
23	7.1016	6.6307	6.2184	5.8543	5.5305	5.2407	4.7435	4.3325	23
24	7.1918	6.7093	6.2874	5.9155	5.5851	5.2897	4.7836	4.3659	24
25	7.2765	6.7830	6.3521	5.9727	5.6361	5.3354	4.8210	4.3970	25
26	7.3563	6.8522	6.4128	6.0263	5.6838	5.3781	4.8558	4.4260	26
27	7.4314	6.9173	6.4698	6.0767	5.7286	5.4182	4.8884	4.4531	27
28	7.5023	6.9787	6.5234	6.1240	5.7706	5.4557	4.9190	4.4784	28
29	7.5692	7.0366	6.5740	6.1685	5.8101	5.4910	4.9477	4.5022	29
30	7.6325	7.0912	6.6217	6.2104	5.8473	5.5243	4.9746	4.5245	30
31	7.6924	7.1429	6.6667	6.2500	5.8824	5.5556	5.0000	4.5455	31
32	7.7491	7.1918	6.7093	6.2875	5.9155	5.5851	5.0239	4.5652	32
33	7.8030	7.2382	6.7496	6.3229	5.9469	5.6131	5.0465	4.5839	33
34	7.8541	7.2822	6.7879	6.3564	5.9765	5.6395	5.0679	4.6015	34
35	7.9027	7.3239	6.8241	6.3882	6.0046	5.6645	5.0880	4.6181	35
36	7.9489	7.3636	6.8585	6.4183	6.0312	5.6882	5.1072	4.6338	36
37	7.9929	7.4013	6.8912	6.4470	6.0565	5.7106	5.1253	4.6487	37
38	8.0348	7.4372	6.9224	6.4742	6.0805	5.7320	5.1425	4.6629	38
39	8.0747	7.4714	6.9520	6.5001	6.1034	5.7523	5.1588	4.6763	39
40	8.1128	7.5040	6.9802	6.5248	6.1251	5.7716	5.1743	4.6891	40
41	8.1491	7.5351	7.0071	6.5483	6.1458	5.7900	5.1891	4.7012	41
42	8.1839	7.5648	7.0328	6.5707	6.1655	5.8075	5.2031	4.7127	42
43	8.2171	7.5931	7.0573	6.5920	6.1844	5.8242	5.2165	4.7237	43
44	8.2488	7.6202	7.0807	6.6125	6.2023	5.8401	5.2293	4.7342	44
45	8.2792	7.6461	7.1030	6.6320	6.2195	5.8553	5.2415	4.7442	45
46	8.3082	7.6709	7.1244	6.6506	6.2359	5.8698	5.2531	4.7537	46
47	8.3361	7.6946	7.1449	6.6684	6.2515	5.8837	5.2643	4.7628	47
48	8.3628	7.7174	7.1645	6.6855	6.2665	5.8970	5.2749	4.7715	48
49	8.3883	7.7392	7.1832	6.7018	6.2809	5.9097	5.2850	4.7798	49
50	8.4129	7.7600	7.2012	6.7175	6.2946	5.9219	5.2948	4.7878	50

YEARS' PURCHASE

Yrs.	11	12	13	14	15	16	18	20	Yrs.
				Rate Per Cent					
51	8.4364	7.7801	7.2185	6.7325	6.3078	5.9335	5.3041	4.7954	51
52	8.4590	7.7993	7.2350	6.7469	6.3204	5.9447	5.3130	4.8027	52
53	8.4807	7.8177	7.2509	6.7607	6.3325	5.9554	5.3216	4.8097	53
54	8.5016	7.8354	7.2661	6.7739	6.3441	5.9657	5.3298	4.8164	54
55	8.5216	7.8524	7.2807	6.7866	6.3553	5.9755	5.3376	4.8228	55
56	8.5408	7.8688	7.2948	6.7988	6.3660	5.9850	5.3452	4.8289	56
57	8.5594	7.8845	7.3083	6.8105	6.3763	5.9941	5.3524	4.8349	57
58	8.5772	7.8996	7.3213	6.8218	6.3862	6.0028	5.3594	4.8405	58
59	8.5943	7.9142	7.3337	6.8327	6.3957	6.0112	5.3661	4.8460	59
60	8.6108	7.9281	7.3458	6.8431	6.4048	6.0193	5.3725	4.8512	60
61	8.6267	7.9416	7.3573	6.8531	6.4136	6.0270	5.3787	4.8563	61
62	8.6420	7.9546	7.3684	6.8628	6.4220	6.0345	5.3846	4.8611	62
63	8.6567	7.9671	7.3792	6.8721	6.4302	6.0417	5.3903	4.8658	63
64	8.6710	7.9791	7.3895	6.8810	6.4380	6.0486	5.3959	4.8703	64
65	8.6846	7.9907	7.3994	6.8896	6.4455	6.0553	5.4011	4.8746	65
66	8.6978	8.0019	7.4090	6.8979	6.4528	6.0617	5.4062	4.8787	66
67	8.7106	8.0126	7.4182	6.9059	6.4598	6.0678	5.4112	4.8827	67
68	8.7228	8.0230	7.4271	6.9136	6.4666	6.0738	5.4159	4.8866	68
69	8.7347	8.0330	7.4357	6.9211	6.4731	6.0795	5.4205	4.8903	69
70	8.7461	8.0427	7.4440	6.9282	6.4793	6.0851	5.4249	4.8939	70
71	8.7571	8.0520	7.4520	6.9352	6.4854	6.0904	5.4291	4.8973	71
72	8.7678	8.0610	7.4597	6.9418	6.4912	6.0956	5.4332	4.9007	72
73	8.7780	8.0697	7.4671	6.9483	6.4969	6.1005	5.4371	4.9039	73
74	8.7880	8.0781	7.4743	6.9545	6.5023	6.1053	5.4409	4.9070	74
75	8.7976	8.0862	7.4812	6.9605	6.5075	6.1099	5.4446	4.9100	75
76	8.8068	8.0940	7.4879	6.9663	6.5126	6.1144	5.4482	4.9128	76
77	8.8158	8.1015	7.4944	6.9719	6.5175	6.1187	5.4516	4.9156	77
78	8.8244	8.1088	7.5006	6.9773	6.5222	6.1229	5.4549	4.9183	78
79	8.8327	8.1159	7.5067	6.9825	6.5268	6.1269	5.4581	4.9209	79
80	8.8408	8.1227	7.5125	6.9875	6.5312	6.1308	5.4611	4.9234	80
81	8.8486	8.1293	7.5181	6.9924	6.5354	6.1345	5.4641	4.9258	81
82	8.8562	8.1357	7.5236	6.9971	6.5395	6.1381	5.4670	4.9282	82
83	8.8635	8.1418	7.5288	7.0017	6.5435	6.1416	5.4698	4.9304	83
84	8.8705	8.1478	7.5339	7.0061	6.5474	6.1450	5.4725	4.9326	84
85	8.8773	8.1535	7.5388	7.0103	6.5511	6.1483	5.4751	4.9347	85
86	8.8839	8.1591	7.5436	7.0145	6.5547	6.1515	5.4776	4.9367	86
87	8.8903	8.1645	7.5482	7.0184	6.5582	6.1545	5.4800	4.9387	87
88	8.8965	8.1697	7.5526	7.0223	6.5615	6.1575	5.4823	4.9406	88
89	8.9025	8.1747	7.5570	7.0260	6.5648	6.1603	5.4846	4.9425	89
90	8.9082	8.1796	7.5611	7.0296	6.5679	6.1631	5.4868	4.9442	90
91	8.9138	8.1843	7.5652	7.0331	6.5709	6.1658	5.4889	4.9460	91
92	8.9193	8.1889	7.5691	7.0365	6.5739	6.1684	5.4910	4.9476	92
93	8.9245	8.1933	7.5728	7.0397	6.5767	6.1709	5.4930	4.9492	93
94	8.9296	8.1976	7.5765	7.0429	6.5795	6.1733	5.4949	4.9508	94
95	8.9345	8.2017	7.5800	7.0459	6.5822	6.1757	5.4967	4.9523	95
96	8.9393	8.2057	7.5834	7.0489	6.5847	6.1779	5.4985	4.9538	96
97	8.9439	8.2096	7.5868	7.0518	6.5872	6.1801	5.5003	4.9552	97
98	8.9483	8.2134	7.5900	7.0545	6.5897	6.1823	5.5020	4.9566	98
99	8.9526	8.2170	7.5931	7.0572	6.5920	6.1843	5.5036	4.9579	99
100	8.9568	8.2205	7.5961	7.0598	6.5943	6.1863	5.5052	4.9592	100

YEARS' PURCHASE

					Rate Per Cent				
Yrs.	4	4.5	5	5.5	6	6.25	6.5	6.75	Yrs.
1	0.9615	0.9569	0.9524	0.9479	0.9434	0.9412	0.9390	0.9368	1
2	1.8861	1.8685	1.8512	1.8342	1.8175	1.8093	1.8012	1.7931	2
3	2.7751	2.7371	2.7002	2.6642	2.6292	2.6120	2.5951	2.5783	3
4	3.6299	3.5652	3.5027	3.4425	3.3842	3.3558	3.3279	3.3004	4
5	4.4518	4.3549	4.2621	4.1732	4.0879	4.0465	4.0060	3.9663	5
6	5.2421	5.1082	4.9810	4.8600	4.7447	4.6891	4.6347	4.5817	6
7	6.0021	5.8272	5.6622	5.5063	5.3588	5.2879	5.2189	5.1517	7
8	6.7327	6.5135	6.3080	6.1152	5.9337	5.8470	5.7628	5.6809	8
9	7.4353	7.1688	6.9208	6.6893	6.4728	6.3697	6.2699	6.1731	9
10	8.1109	7.7948	7.5024	7.2311	6.9788	6.8591	6.7435	6.6317	10
11	8.7605	8.3928	8.0548	7.7430	7.4544	7.3180	7.1865	7.0597	11
12	9.3851	8.9644	8.5798	8.2269	7.9019	7.7488	7.6015	7.4598	12
13	9.9856	9.5108	9.0790	8.6848	8.3234	8.1537	7.9908	7.8343	13
14	10.5631	10.0332	9.5539	9.1183	8.7208	8.5347	8.3564	8.1854	14
15	11.1184	10.5328	10.0059	9.5292	9.0958	8.8935	8.7001	8.5149	15
16	11.6523	11.0108	10.4362	9.9187	9.4500	9.2319	9.0236	8.8246	16
17	12.1657	11.4681	10.8462	10.2882	9.7849	9.5512	9.3285	9.1159	17
18	12.6593	11.9057	11.2368	10.6391	10.1017	9.8529	9.6160	9.3903	18
19	13.1339	12.3246	11.6092	10.9723	10.4016	10.1380	9.8874	9.6489	19
20	13.5903	12.7256	11.9643	11.2890	10.6858	10.4078	10.1439	9.8930	20
21	14.0292	13.1096	12.3031	11.5902	10.9553	10.6632	10.3864	10.1235	21
22	14.4511	13.4773	12.6265	11.8767	11.2109	10.9053	10.6158	10.3414	22
23	14.8568	13.8295	12.9351	12.1493	11.4536	11.1347	10.8332	10.5475	23
24	15.2470	14.1669	13.2298	12.4090	11.6840	11.3524	11.0391	10.7427	24
25	15.6221	14.4902	13.5113	12.6563	11.9031	11.5591	11.2344	10.9275	25
26	15.9828	14.8000	13.7803	12.8920	12.1113	11.7554	11.4198	11.1028	26
27	16.3296	15.0970	14.0373	13.1167	12.3094	11.9419	11.5957	11.2691	27
28	16.6631	15.3815	14.2831	13.3310	12.4980	12.1193	11.7629	11.4269	28
29	16.9837	15.6544	14.5180	13.5355	12.6775	12.2880	11.9218	11.5768	29
30	17.2920	15.9159	14.7427	13.7306	12.8485	12.4486	12.0729	11.7192	30
31	17.5885	16.1667	14.9577	13.9168	13.0115	12.6015	12.2167	11.8546	31
32	17.8736	16.4073	15.1633	14.0947	13.1668	12.7472	12.3535	11.9834	32
33	18.1476	16.6379	15.3601	14.2646	13.3149	12.8860	12.4838	12.1060	33
34	18.4112	16.8592	15.5485	14.4269	13.4563	13.0183	12.6080	12.2227	34
35	18.6646	17.0715	15.7289	14.5821	13.5911	13.1445	12.7263	12.3339	35
36	18.9083	17.2751	15.9016	14.7304	13.7199	13.2649	12.8391	12.4398	36
37	19.1426	17.4704	16.0670	14.8722	13.8428	13.3798	12.9467	12.5408	37
38	19.3679	17.6579	16.2254	15.0078	13.9603	13.4895	13.0494	12.6371	38
39	19.5845	17.8378	16.3771	15.1376	14.0725	13.5942	13.1474	12.7290	39
40	19.7928	18.0104	16.5225	15.2617	14.1797	13.6942	13.2409	12.8167	40
41	19.9931	18.1761	16.6618	15.3805	14.2822	13.7898	13.3302	12.9003	41
42	20.1856	18.3351	16.7954	15.4942	14.3802	13.8811	13.4156	12.9802	42
43	20.3708	18.4877	16.9234	15.6031	14.4739	13.9685	13.4971	13.0566	43
44	20.5488	18.6343	17.0461	15.7073	14.5636	14.0519	13.5751	13.1295	44
45	20.7200	18.7750	17.1637	15.8072	14.6493	14.1318	13.6496	13.1991	45
46	20.8847	18.9100	17.2765	15.9028	14.7314	14.2082	13.7208	13.2658	46
47	21.0429	19.0397	17.3847	15.9944	14.8100	14.2812	13.7889	13.3294	47
48	21.1951	19.1642	17.4884	16.0822	14.8852	14.3512	13.8541	13.3903	48
49	21.3415	19.2838	17.5879	16.1663	14.9573	14.4181	13.9165	13.4486	49
50	21.4822	19.3986	17.6834	16.2469	15.0263	14.4822	13.9762	13.5044	50

YEARS' PURCHASE

					Rate Per Cent				
Yrs.	4	4.5	5	5.5	6	6.25	6.5	6.75	Yrs.
51	21.6175	19.5088	17.7750	16.3242	15.0923	14.5436	14.0333	13.5577	51
52	21.7476	19.6147	17.8628	16.3982	15.1556	14.6024	14.0881	13.6088	52
53	21.8727	19.7164	17.9472	16.4693	15.2163	14.6586	14.1404	13.6576	53
54	21.9930	19.8141	18.0281	16.5374	15.2744	14.7126	14.1906	13.7044	54
55	22.1086	19.9079	18.1057	16.6027	15.3301	14.7642	14.2387	13.7492	55
56	22.2198	19.9981	18.1802	16.6653	15.3835	14.8137	14.2847	13.7922	56
57	22.3267	20.0846	18.2517	16.7254	15.4346	14.8612	14.3288	13.8333	57
58	22.4296	20.1678	18.3204	16.7830	15.4837	14.9067	14.3711	13.8727	58
59	22.5284	20.2477	18.3863	16.8383	15.5308	14.9503	14.4116	13.9105	59
60	22.6235	20.3244	18.4496	16.8914	15.5759	14.9921	14.4505	13.9466	60
61	22.7149	20.3982	18.5103	16.9423	15.6191	15.0322	14.4877	13.9813	61
62	22.8028	20.4690	18.5686	16.9911	15.6607	15.0706	14.5234	14.0146	62
63	22.8873	20.5371	18.6246	17.0380	15.7005	15.1075	14.5577	14.0464	63
64	22.9685	20.6025	18.6784	17.0830	15.7387	15.1428	14.5905	14.0770	64
65	23.0467	20.6653	18.7300	17.1262	15.7753	15.1768	14.6220	14.1063	65
66	23.1218	20.7257	18.7796	17.1676	15.8105	15.2093	14.6522	14.1344	66
67	23.1940	20.7838	18.8272	17.2074	15.8442	15.2405	14.6812	14.1614	67
68	23.2635	20.8395	18.8730	17.2456	15.8766	15.2705	14.7090	14.1873	68
69	23.3303	20.8931	18.9169	17.2823	15.9077	15.2992	14.7356	14.2121	69
70	23.3945	20.9446	18.9591	17.3175	15.9375	15.3268	14.7612	14.2359	70
71	23.4563	20.9941	18.9997	17.3513	15.9661	15.3533	14.7858	14.2587	71
72	23.5156	21.0416	19.0386	17.3838	15.9936	15.3787	14.8094	14.2806	72
73	23.5727	21.0873	19.0760	17.4150	16.0200	15.4031	14.8320	14.3017	73
74	23.6276	21.1312	19.1119	17.4449	16.0454	15.4265	14.8537	14.3219	74
75	23.6804	21.1734	19.1465	17.4737	16.0697	15.4490	14.8745	14.3412	75
76	23.7312	21.2140	19.1796	17.5013	16.0930	15.4706	14.8945	14.3598	76
77	23.7800	21.2530	19.2115	17.5278	16.1155	15.4913	14.9138	14.3777	77
78	23.8269	21.2905	19.2421	17.5533	16.1370	15.5112	14.9322	14.3948	78
79	23.8720	21.3265	19.2715	17.5778	16.1577	15.5303	14.9499	14.4113	79
80	23.9154	21.3611	19.2998	17.6013	16.1775	15.5487	14.9669	14.4271	80
81	23.9571	21.3944	19.3269	17.6239	16.1966	15.5663	14.9832	14.4423	81
82	23.9972	21.4264	19.3530	17.6456	16.2149	15.5832	14.9989	14.4568	82
83	24.0358	21.4571	19.3781	17.6664	16.2325	15.5995	15.0140	14.4708	83
84	24.0729	21.4866	19.4022	17.6864	16.2495	15.6151	15.0284	14.4842	84
85	24.1085	21.5150	19.4254	17.7057	16.2657	15.6301	15.0423	14.4971	85
86	24.1428	21.5424	19.4476	17.7242	16.2813	15.6445	15.0557	14.5095	86
87	24.1758	21.5686	19.4690	17.7419	16.2963	15.6583	15.0685	14.5214	87
88	24.2075	21.5938	19.4896	17.7590	16.3107	15.6716	15.0808	14.5329	88
89	24.2380	21.6181	19.5093	17.7754	16.3245	15.6844	15.0926	14.5439	89
90	24.2673	21.6414	19.5283	17.7911	16.3378	15.6967	15.1040	14.5544	90
91	24.2955	21.6638	19.5465	17.8063	16.3506	15.7085	15.1149	14.5645	91
92	24.3226	21.6853	19.5641	17.8208	16.3628	15.7198	15.1254	14.5743	92
93	24.3486	21.7061	19.5809	17.8348	16.3746	15.7307	15.1354	14.5836	93
94	24.3737	21.7260	19.5971	17.8483	16.3860	15.7411	15.1451	14.5926	94
95	24.3978	21.7451	19.6127	17.8612	16.3968	15.7512	15.1544	14.6012	95
96	24.4209	21.7635	19.6277	17.8736	16.4073	15.7608	15.1633	14.6095	96
97	24.4432	21.7812	19.6420	17.8855	16.4173	15.7701	15.1719	14.6175	97
98	24.4646	21.7982	19.6559	17.8970	16.4270	15.7790	15.1802	14.6251	98
99	24.4852	21.8145	19.6692	17.9080	16.4363	15.7876	15.1881	14.6325	99
100	24.5050	21.8302	19.6819	17.9186	16.4452	15.7958	15.1957	14.6396	100

YEARS' PURCHASE

					Rate Per Cent				
Yrs.	7	7.25	7.5	8	8.5	9	9.5	10	Yrs.
1	0.9346	0.9324	0.9302	0.9259	0.9217	0.9174	0.9132	0.9091	1
2	1.7851	1.7772	1.7693	1.7538	1.7385	1.7236	1.7088	1.6944	2
3	2.5618	2.5455	2.5294	2.4978	2.4670	2.4370	2.4076	2.3790	3
4	3.2734	3.2469	3.2207	3.1697	3.1202	3.0723	3.0258	2.9807	4
5	3.9273	3.8891	3.8517	3.7789	3.7088	3.6413	3.5762	3.5134	5
6	4.5298	4.4790	4.4294	4.3335	4.2416	4.1535	4.0690	3.9878	6
7	5.0862	5.0224	4.9601	4.8400	4.7257	4.6166	4.5124	4.4129	7
8	5.6014	5.5240	5.4488	5.3043	5.1672	5.0371	4.9133	4.7955	8
9	6.0793	5.9883	5.8999	5.7309	5.5712	5.4203	5.2772	5.1416	9
10	6.5235	6.4189	6.3175	6.1240	5.9421	5.7706	5.6088	5.4558	10
11	6.9373	6.8190	6.7047	6.4872	6.2834	6.0920	5.9119	5.7422	11
12	7.3232	7.1915	7.0645	6.8235	6.5984	6.3876	6.1900	6.0041	12
13	7.6838	7.5390	7.3995	7.1355	6.8897	6.6603	6.4456	6.2444	13
14	8.0212	7.8636	7.7119	7.4256	7.1598	6.9123	6.6814	6.4654	14
15	8.3374	8.1672	8.0038	7.6958	7.4106	7.1459	6.8994	6.6693	15
16	8.6341	8.4517	8.2768	7.9479	7.6441	7.3627	7.1013	6.8578	16
17	8.9128	8.7185	8.5325	8.1834	7.8617	7.5644	7.2887	7.0324	17
18	9.1749	8.9691	8.7724	8.4038	8.0649	7.7523	7.4631	7.1946	18
19	9.4216	9.2048	8.9978	8.6104	8.2550	7.9278	7.6255	7.3455	19
20	9.6542	9.4267	9.2097	8.8042	8.4330	8.0918	7.7772	7.4861	20
21	9.8736	9.6358	9.4091	8.9863	8.5999	8.2454	7.9189	7.6173	21
22	10.0808	9.8330	9.5970	9.1576	8.7567	8.3893	8.0516	7.7400	22
23	10.2765	10.0191	9.7743	9.3189	8.9040	8.5245	8.1760	7.8549	23
24	10.4617	10.1950	9.9417	9.4709	9.0427	8.6515	8.2928	7.9626	24
25	10.6369	10.3614	10.0998	9.6143	9.1733	8.7710	8.4025	8.0637	25
26	10.8029	10.5189	10.2493	9.7497	9.2965	8.8836	8.5058	8.1588	26
27	10.9603	10.6680	10.3908	9.8777	9.4128	8.9897	8.6030	8.2482	27
28	11.1095	10.8093	10.5249	9.9987	9.5226	9.0898	8.6947	8.3324	28
29	11.2511	10.9433	10.6519	10.1133	9.6265	9.1844	8.7812	8.4118	29
30	11.3856	11.0705	10.7724	10.2218	9.7248	9.2738	8.8629	8.4868	30
31	11.5134	11.1913	10.8867	10.3247	9.8178	9.3584	8.9401	8.5576	31
32	11.6349	11.3060	10.9952	10.4222	9.9060	9.4385	9.0132	8.6245	32
33	11.7504	11.4151	11.0983	10.5149	9.9897	9.5144	9.0824	8.6878	33
34	11.8603	11.5188	11.1964	10.6028	10.0690	9.5864	9.1479	8.7478	34
35	11.9650	11.6175	11.2896	10.6863	10.1443	9.6546	9.2100	8.8046	35
36	12.0646	11.7114	11.3783	10.7658	10.2159	9.7194	9.2690	8.8584	36
37	12.1596	11.8009	11.4627	10.8413	10.2839	9.7809	9.3249	8.9095	37
38	12.2501	11.8861	11.5431	10.9132	10.3485	9.8394	9.3781	8.9580	38
39	12.3364	11.9673	11.6197	10.9817	10.4101	9.8950	9.4286	9.0041	39
40	12.4187	12.0448	11.6927	11.0469	10.4686	9.9479	9.4766	9.0479	40
41	12.4973	12.1187	11.7623	11.1090	10.5244	9.9983	9.5222	9.0895	41
42	12.5723	12.1891	11.8287	11.1682	10.5775	10.0462	9.5657	9.1291	42
43	12.6438	12.2564	11.8920	11.2246	10.6281	10.0918	9.6071	9.1668	43
44	12.7122	12.3207	11.9525	11.2785	10.6764	10.1354	9.6465	9.2026	44
45	12.7775	12.3820	12.0102	11.3298	10.7224	10.1768	9.6841	9.2368	45
46	12.8399	12.4406	12.0653	11.3789	10.7663	10.2164	9.7199	9.2694	46
47	12.8996	12.4966	12.1180	11.4257	10.8083	10.2541	9.7540	9.3004	47
48	12.9566	12.5501	12.1683	11.4704	10.8483	10.2901	9.7866	9.3300	48
49	13.0112	12.6013	12.2164	11.5132	10.8865	10.3245	9.8177	9.3583	49
50	13.0633	12.6502	12.2624	11.5540	10.9230	10.3573	9.8473	9.3852	50

YEARS' PURCHASE

Rate Per Cent

Yrs.	7	7.25	7.5	8	8.5	9	9.5	10	Yrs.
51	13.1132	12.6970	12.3064	11.5930	10.9578	10.3887	9.8757	9.4110	51
52	13.1610	12.7418	12.3484	11.6303	10.9912	10.4186	9.9027	9.4356	52
53	13.2067	12.7846	12.3886	11.6660	11.0230	10.4472	9.9286	9.4590	53
54	13.2505	12.8256	12.4271	11.7001	11.0535	10.4746	9.9533	9.4814	54
55	13.2923	12.8648	12.4640	11.7328	11.0826	10.5008	9.9769	9.5029	55
56	13.3325	12.9024	12.4992	11.7640	11.1105	10.5258	9.9995	9.5234	56
57	13.3709	12.9384	12.5330	11.7939	11.1372	10.5497	10.0211	9.5430	57
58	13.4077	12.9729	12.5653	11.8226	11.1627	10.5726	10.0418	9.5617	58
59	13.4430	13.0059	12.5963	11.8500	11.1871	10.5945	10.0615	9.5796	59
60	13.4768	13.0375	12.6260	11.8762	11.2105	10.6155	10.0805	9.5968	60
61	13.5091	13.0678	12.6544	11.9014	11.2329	10.6356	10.0986	9.6132	61
62	13.5402	13.0968	12.6816	11.9254	11.2544	10.6548	10.1159	9.6289	62
63	13.5699	13.1247	12.7077	11.9485	11.2749	10.6732	10.1325	9.6439	63
64	13.5984	13.1514	12.7327	11.9706	11.2946	10.6909	10.1484	9.6583	64
65	13.6258	13.1769	12.7567	11.9918	11.3135	10.7078	10.1636	9.6721	65
66	13.6520	13.2015	12.7797	12.0121	11.3315	10.7240	10.1782	9.6853	66
67	13.6772	13.2250	12.8017	12.0316	11.3489	10.7395	10.1922	9.6980	67
68	13.7013	13.2475	12.8229	12.0503	11.3655	10.7543	10.2056	9.7101	68
69	13.7244	13.2692	12.8431	12.0682	11.3814	10.7686	10.2184	9.7217	69
70	13.7466	13.2899	12.8626	12.0853	11.3967	10.7822	10.2307	9.7328	70
71	13.7679	13.3098	12.8812	12.1018	11.4113	10.7953	10.2425	9.7435	71
72	13.7884	13.3289	12.8991	12.1176	11.4253	10.8079	10.2538	9.7537	72
73	13.8080	13.3472	12.9162	12.1327	11.4388	10.8199	10.2646	9.7635	73
74	13.8268	13.3648	12.9327	12.1472	11.4517	10.8315	10.2750	9.7729	74
75	13.8449	13.3817	12.9485	12.1612	11.4641	10.8426	10.2850	9.7820	75
76	13.8622	13.3979	12.9637	12.1745	11.4760	10.8532	10.2946	9.7906	76
77	13.8788	13.4134	12.9782	12.1874	11.4874	10.8634	10.3037	9.7989	77
78	13.8948	13.4283	12.9922	12.1997	11.4983	10.8732	10.3125	9.8069	78
79	13.9101	13.4427	13.0056	12.2115	11.5088	10.8826	10.3210	9.8145	79
80	13.9248	13.4564	13.0184	12.2228	11.5189	10.8916	10.3291	9.8218	80
81	13.9390	13.4696	13.0308	12.2337	11.5285	10.9002	10.3369	9.8289	81
82	13.9525	13.4823	13.0427	12.2442	11.5378	10.9085	10.3443	9.8356	82
83	13.9656	13.4944	13.0540	12.2542	11.5467	10.9165	10.3515	9.8421	83
84	13.9781	13.5061	13.0650	12.2638	11.5553	10.9241	10.3583	9.8483	84
85	13.9901	13.5173	13.0755	12.2731	11.5635	10.9315	10.3649	9.8542	85
86	14.0016	13.5281	13.0855	12.2820	11.5714	10.9385	10.3713	9.8600	86
87	14.0127	13.5384	13.0952	12.2905	11.5789	10.9453	10.3774	9.8655	87
88	14.0234	13.5484	13.1045	12.2987	11.5862	10.9518	10.3832	9.8707	88
89	14.0336	13.5579	13.1134	12.3065	11.5932	10.9580	10.3888	9.8758	89
90	14.0434	13.5671	13.1220	12.3141	11.5999	10.9640	10.3942	9.8807	90
91	14.0528	13.5759	13.1303	12.3213	11.6063	10.9697	10.3993	9.8853	91
92	14.0619	13.5844	13.1382	12.3283	11.6125	10.9752	10.4043	9.8898	92
93	14.0706	13.5925	13.1458	12.3350	11.6184	10.9806	10.4091	9.8941	93
94	14.0790	13.6003	13.1531	12.3414	11.6241	10.9856	10.4136	9.8983	94
95	14.0870	13.6078	13.1601	12.3476	11.6296	10.9905	10.4180	9.9022	95
96	14.0947	13.6150	13.1668	12.3535	11.6349	10.9952	10.4223	9.9060	96
97	14.1021	13.6219	13.1733	12.3592	11.6399	10.9997	10.4263	9.9097	97
98	14.1093	13.6286	13.1795	12.3647	11.6448	11.0041	10.4302	9.9132	98
99	14.1161	13.6349	13.1855	12.3700	11.6494	11.0082	10.4339	9.9166	99
100	14.1227	13.6411	13.1912	12.3750	11.6539	11.0122	10.4375	9.9198	100

YEARS' PURCHASE

				Rate Per Cent					
Yrs.	11	12	13	14	15	16	18	20	Yrs.
1	0.9009	0.8929	0.8850	0.8772	0.8696	0.8621	0.8475	0.8333	1
2	1.6661	1.6388	1.6124	1.5868	1.5620	1.5380	1.4921	1.4489	2
3	2.3237	2.2709	2.2205	2.1723	2.1261	2.0818	1.9986	1.9218	3
4	2.8944	2.8130	2.7361	2.6632	2.5941	2.5285	2.4068	2.2963	4
5	3.3941	3.2827	3.1784	3.0805	2.9884	2.9017	2.7425	2.5999	5
6	3.8349	3.6933	3.5617	3.4392	3.3249	3.2179	3.0233	2.8509	6
7	4.2264	4.0550	3.8970	3.7508	3.6152	3.4891	3.2615	3.0618	7
8	4.5761	4.3758	4.1924	4.0237	3.8681	3.7240	3.4659	3.2412	8
9	4.8901	4.6622	4.4545	4.2645	4.0901	3.9294	3.6431	3.3957	9
10	5.1735	4.9191	4.6884	4.4785	4.2865	4.1103	3.7981	3.5299	10
11	5.4304	5.1507	4.8984	4.6696	4.4613	4.2708	3.9347 ·	3.6477	11
12	5.6640	5.3604	5.0877	4.8414	4.6178	4.4140	4.0559	3.7516	12
13	5.8774	5.5511	5.2592	4.9964	4.7586	4.5425	4.1642	3.8440	13
14	6.0728	5.7251	5.4151	5.1369	4.8859	4.6583	4.2613	3.9267	14
15	6.2523	5.8844	5.5574	5.2648	5.0015	4.7632	4.3489	4.0009	15
16	6.4177	6.0306	5.6876	5.3816	5.1067	4.8586	4.4283	4.0680	16
17	6.5704	6.1653	5.8073	5.4885	5.2030	4.9456	4.5005	4.1288	17
18	6.7117	6.2896	5.9174	5.5868	5.2912	5.0253	4.5663	4.1842	18
19	6.8428	6.4046	6.0191	5.6773	5.3723	5.0984	4.6267	4.2348	19
20	6.9647	6.5112	6.1132	5.7610	5.4472	5.1658	4.6820	4.2812	20
21	7.0781	6.6103	6.2004	5.8384	5.5163	5.2279	4.7331	4.3238	21
22	7.1840	6.7025	6.2815	5.9102	5.5804	5.2854	4.7801	4.3630	22
23	7.2828	6.7884	6.3569	5.9770	5.6399	5.3388	4.8237	4.3993	23
24	7.3753	6.8688	6.4273	6.0391	5.6952	5.3883	4.8641	4.4329	24
25	7.4620	6.9439	6.4930	6.0971	5.7467	5.4344	4.9017	4.4640	25
26	7.5433	7.0142	6.5545	6.1513	5.7948	5.4774	4.9366	4.4930	26
27	7.6197	7.0802	6.6121	6.2020	5.8398	5.5176	4.9692	4.5200	27
28	7.6915	7.1422	6.6661	6.2495	5.8819	5.5552	4.9997	4.5452	28
29	7.7592	7.2005	6.7168	6.2941	5.9214	5.5903	5.0282	4.5687	29
30	7.8229	7.2553	6.7645	6.3359	5.9584	5.6233	5.0548	4.5907	30
31	7.8830	7.3070	6.8094	6.3753	5.9932	5.6543	5.0799	4.6114	31
32	7.9397	7.3557	6.8517	6.4124	6.0260	5.6835	5.1034	4.6307	32
33	7.9934	7.4017	6.8916	6.4473	6.0568	5.7109	5.1255	4.6489	33
34	8.0441	7.4452	6.9293	6.4803	6.0859	5.7367	5.1463	4.6660	34
35	8.0921	7.4863	6.9649	6.5114	6.1133	5.7611	5.1659	4.6821	35
36	8.1376	7.5252	6.9985	6.5408	6.1392	5.7841	5.1844	4.6973	36
37	8.1807	7.5620	7.0304	6.5686	6.1637	5.8059	5.2018	4.7117	37
38	8.2215	7.5969	7.0606	6.5949	6.1869	5.8264	5.2183	4.7252	38
39	8.2603	7.6300	7.0891	6.6198	6.2088	5.8459	5.2339	4.7380	39
40	8.2971	7.6615	7.1162	6.6435	6.2296	5.8643	5.2487	4.7501	40
41	8.3321	7.6913	7.1420	6.6659	6.2493	5.8818	5.2627	4.7615	41
42	8.3654	7.7196	7.1664	6.6872	6.2680	5.8983	5.2759	4.7724	42
43	8.3970	7.7465	7.1896	6.7074	6.2858	5.9140	5.2885	4.7826	43
44	8.4271	7.7721	7.2116	6.7266	6.3026	5.9289	5.3004	4.7924	44
45	8.4558	7.7965	7.2326	6.7448	6.3186	5.9431	5.3117	4.8016	45
46	8.4831	7.8197	7.2526	6.7621	6.3338	5.9566	5.3225	4.8104	46
47	8.5091	7.8418	7.2716	6.7787	6.3483	5.9694	5.3327	4.8188	47
48	8.5338	7.8628	7.2897	6.7944	6.3621	5.9816	5.3424	4.8267	48
49	8.5575	7.8829	7.3069	6.8093	6.3752	5.9932	5.3517	4.8343	49
50	8.5800	7.9020	7.3233	6.8236	6.3877	6.0042	5.3605	4.8414	50

YEARS' PURCHASE

Yrs.	11	12	13	14	15	16	18	20	Yrs.
51	8.6015	7.9202	7.3390	6.8372	6.3996	6.0147	5.3689	4.8483	51
52	8.6220	7.9376	7.3539	6.8502	6.4110	6.0247	5.3769	4.8548	52
53	8.6416	7.9542	7.3682	6.8625	6.4218	6.0343	5.3845	4.8610	53
54	8.6603	7.9701	7.3818	6.8743	6.4321	6.0434	5.3917	4.8669	54
55	8.6782	7.9852	7.3947	6.8856	6.4420	6.0521	5.3987	4.8726	55
56	8.6953	7.9997	7.4071	6.8963	6.4514	6.0604	5.4053	4.8779	56
57	8.7116	8.0135	7.4190	6.9066	6.4604	6.0684	5.4116	4.8831	57
58	8.7272	8.0267	7.4303	6.9164	6.4690	6.0759	5.4176	4.8880	58
59	8.7421	8.0393	7.4411	6.9258	6.4772	6.0832	5.4233	4.8926	59
60	8.7564	8.0514	7.4515	6.9347	6.4850	6.0901	5.4288	4.8971	60
61	8.7701	8.0630	7.4614	6.9433	6.4925	6.0967	5.4341	4.9014	61
62	8.7832	8.0740	7.4708	6.9515	6.4997	6.1030	5.4391	4.9055	62
63	8.7957	8.0846	7.4799	6.9593	6.5065	6.1090	5.4439	4.9094	63
64	8.8076	8.0947	7.4885	6.9668	6.5131	6.1148	5.4485	4.9131	64
65	8.8191	8.1044	7.4968	6.9740	6.5193	6.1203	5.4529	4.9167	65
66	8.8301	8.1136	7.5047	6.9808	6.5253	6.1256	5.4570	4.9201	66
67	8.8406	8.1225	7.5123	6.9874	6.5311	6.1307	5.4611	4.9233	67
68	8:8507	8.1310	7.5196	6.9937	6.5366	6.1355	5.4649	4.9265	68
69	8.8603	8.1392	7.5266	6.9997	6.5418	6.1401	5.4686	4.9294	69
70	8.8696	8.1470	7.5332	7.0055	6.5469	6.1446	5.4721	4.9323	70
71	8.8784	8.1544	7.5396	7.0110	6.5517	6.1488	5.4755	4.9350	71
72	8.8869	8.1616	7.5458	7.0163	6.5563	6.1529	5.4787	4.9377	72
73	8.8951	8.1685	7.5516	7.0214	6.5607	6.1568	5.4818	4.9402	73
74	8.9029	8.1751	7.5572	7.0263	6.5650	6.1605	5.4848	4.9426	74
75	8.9104	8.1814	7.5626	7.0309	6.5691	6.1641	5.4876	4.9449	75
76	8.9175	8.1874	7.5678	7.0354	6.5730	6.1676	5.4903	4.9471	76
77	8.9244	8.1932	7.5728	7.0397	6.5767	6.1709	5.4929	4.9492	77
78	8.9310	8.1988	7.5775	7.0438	6.5803	6.1740	5.4954	4.9512	78
79	8.9373	8.2041	7.5821	7.0477	6.5837	6.1770	5.4978	4.9532	79
80	8.9434	8.2092	7.5864	7.0515	6.5870	6.1799	5.5001	4.9551	80
81	8.9492	8.2141	7.5906	7.0551	6.5902	6.1827	5.5023	4.9568	81
82	8.9548	8.2189	7.5947	7.0586	6.5932	6.1854	5.5044	4.9586	82
83	8.9602	8.2234	7.5985	7.0619	6.5961	6.1879	5.5065	4.9602	83
84	8.9654	8.2277	7.6022	7.0651	6.5989	6.1904	5.5084	4.9618	84
85	8.9703	8.2319	7.6058	7.0682	6.6016	6.1928	5.5103	4.9633	85
86	8.9750	8.2359	7.6092	7.0711	6.6041	6.1950	5.5121	4.9647	86
87	8.9796	8.2397	7.6125	7.0740	6.6066	6.1972	5.5138	4.9661	87
88	8.9840	8.2434	7.6156	7.0767	6.6090	6.1993	5.5154	4.9675	88
89	8.9882	8.2469	7.6186	7.0793	6.6112	6.2013	5.5170	4.9688	89
90	8.9922	8.2503	7.6215	7.0818	6.6134	6.2032	5.5185	4.9700	90
91	8.9960	8.2536	7.6243	7.0842	6.6155	6.2050	5.5200	4.9712	91
92	8.9998	8.2567	7.6269	7.0865	6.6175	6.2068	5.5214	4.9723	92
93	9.0033	8.2597	7.6295	7.0887	6.6194	6.2085	5.5227	4.9734	93
94	9.0067	8.2626	7.6320	7.0908	6.6213	6.2101	5.5240	4.9744	94
95	9.0100	8.2653	7.6343	7.0928	6.6231	6.2117	5.5252	4.9754	95
96	9.0132	8.2680	7.6366	7.0948	6.6248	6.2132	5.5264	4.9764	96
97	9.0162	8.2705	7.6388	7.0967	6.6264	6.2146	5.5276	4.9773	97
98	9.0191	8.2730	7.6409	7.0985	6.6280	6.2160	5.5287	4.9782	98
99	9.0219	8.2753	7.6429	7.1002	6.6295	6.2173	5.5297	4.9791	99
100	9.0246	8.2776	7.6448	7.1019	6.6309	6.2186	5.5307	4.9799	100

YEARS' PURCHASE

OF A REVERSION

TO A PERPETUITY

(*i.e.*, the Present Value of £1 per annum receivable after the
expiration of a given number of years)
At Rates of Interest from
2% to 16%

Note:—No allowance has been made for the effect of income tax on interest
accumulations.

YEARS' PURCHASE OF A REVERSION TO A PERPETUITY

AFTER A GIVEN NUMBER OF YEARS

After Yrs.	Rate Per Cent					After Yrs.
	2	2.25	2.5	2.75	3	
0	50.00000	44.44444	40.00000	36.36364	33.33333	0
1	49.01961	43.46645	39.02439	35.39040	32.36246	1
2	48.05844	42.50997	38.07258	34.44321	31.41986	2
3	47.11612	41.57455	37.14398	33.52137	30.50472	3
4	46.19227	40.65970	36.23803	32.62421	29.61623	4
5	45.28654	39.76499	35.35417	31.75105	28.75363	5
6	44.39857	38.88997	34.49187	30.90127	27.91614	6
7	43.52801	38.03420	33.65061	30.07423	27.10305	7
8	42.67452	37.19726	32.82986	29.26932	26.31364	8
9	41.83776	36.37874	32.02913	28.48596	25.54722	9
10	41.01741	35.57823	31.24794	27.72356	24.80313	10
11	40.21315	34.79533	30.48579	26.98157	24.08071	11
12	39.42466	34.02967	29.74224	26.25943	23.37933	12
13	38.65163	33.28085	29.01682	25.55663	22.69838	13
14	37.89375	32.54851	28.30909	24.87263	22.03726	14
15	37.15074	31.83228	27.61862	24.20694	21.39540	15
16	36.42229	31.13181	26.94500	23.55906	20.77223	16
17	35.70813	30.44676	26.28780	22.92853	20.16721	17
18	35.00797	29.77678	25.64664	22.31487	19.57982	18
19	34.32154	29.12155	25.02111	21.71763	19.00953	19
20	33.64857	28.48073	24.41084	21.13638	18.45586	20
21	32.98879	27.85402	23.81545	20.57069	17.91831	21
22	32.34195	27.24199	23.23459	20.02014	17.39642	22
23	31.70780	26.64165	22.66789	19.48432	16.88972	23
24	31.08607	26.05541	22.11501	18.96284	16.39779	24
25	30.47654	25.48206	21.57562	18.45532	15.92019	25
26	29.87896	24.92133	21.04939	17.96138	15.45649	26
27	29.29310	24.37294	20.53599	17.48066	15.00630	27
28	28.71873	23.83662	20.03511	17.01281	14.56923	28
29	28.15562	23.31209	19.54645	16.55748	14.14488	29
30	27.60354	22.79911	19.06971	16.11434	13.73289	30
31	27.06230	22.29742	18.60459	15.68305	13.33290	31
32	26.53167	21.80677	18.15082	15.26331	12.94457	32
33	26.01144	21.32691	17.70812	14.85480	12.56754	33
34	25.50141	20.85762	17.27621	14.45723	12.20150	34
35	25.00138	20.39865	16.85484	14.07030	11.84611	35
36	24.51116	19.94978	16.44375	13.69372	11.50108	36
37	24.03055	19.51079	16.04268	13.32722	11.16610	37
38	23.55936	19.08145	15.65140	12.97053	10.84087	38
39	23.09741	18.66157	15.26966	12.62339	10.52512	39
40	22.64452	18.25092	14.89722	12.28554	10.21856	40
41	22.20051	17.84931	14.53388	11.95673	9.92093	41
42	21.76521	17.45654	14.17939	11.63672	9.63197	42
43	21.33844	17.07241	13.83355	11.32527	9.35143	43
44	20.92004	16.69673	13.49615	11.02216	9.07906	44
45	20.50984	16.32932	13.16698	10.72716	8.81462	45
46	20.10769	15.97000	12.84583	10.44006	8.55788	46
47	19.71342	15.61858	12.53252	10.16064	8.30863	47
48	19.32688	15.27490	12.22685	9.88871	8.06663	48
49	18.94792	14.93877	11.92863	9.62404	7.83168	49

YEARS' PURCHASE OF A REVERSION TO A PERPETUITY
AFTER A GIVEN NUMBER OF YEARS

After Yrs.	Rate Per Cent					After Yrs.
	2	**2.25**	**2.5**	**2.75**	**3**	
50	18.57639	14.61005	11.63769	9.36647	7.60357	50
51	18.21215	14.28856	11.35384	9.11578	7.38211	51
52	17.85505	13.97414	11.07692	8.87181	7.16709	52
53	17.50495	13.66664	10.80675	8.63436	6.95834	53
54	17.16172	13.36591	10.54317	8.40327	6.75567	54
55	16.82521	13.07179	10.28602	8.17837	6.55891	55
56	16.49531	12.78415	10.03514	7.95948	6.36787	56
57	16.17187	12.50283	9.79038	7.74645	6.18240	57
58	15.85477	12.22771	9.55159	7.53913	6.00233	58
59	15.54390	11.95864	9.31863	7.33735	5.82750	59
60	15.23911	11.69549	9.09134	7.14097	5.65777	60
61	14.94031	11.43813	8.86960	6.94985	5.49298	61
62	14.64736	11.18644	8.65327	6.76385	5.33299	62
63	14.36016	10.94028	8.44222	6.58282	5.17766	63
64	14.07859	10.69954	8.23631	6.40664	5.02686	64
65	13.80253	10.46410	8.03542	6.23517	4.88044	65
66	13.53190	10.23384	7.83944	6.06829	4.73829	66
67	13.26657	10.00864	7.64823	5.90588	4.60028	67
68	13.00644	9.78841	7.46169	5.74782	4.46630	68
69	12.75141	9.57301	7.27970	5.59398	4.33621	69
70	12.50138	9.36236	7.10214	5.44426	4.20991	70
71	12.25626	9.15634	6.92892	5.29855	4.08729	71
72	12.01594	8.95486	6.75992	5.15674	3.96825	72
73	11.78033	8.75781	6.59505	5.01873	3.85267	73
74	11.54934	8.56509	6.43419	4.88441	3.74045	74
75	11.32289	8.37662	6.27726	4.75368	3.63151	75
76	11.10087	8.19229	6.12416	4.62645	3.52574	76
77	10.88320	8.01202	5.97479	4.50263	3.42304	77
78	10.66981	7.83572	5.82906	4.38212	3.32334	78
79	10.46060	7.66329	5.68689	4.26484	3.22655	79
80	10.25549	7.49466	5.54818	4.15070	3.13257	80
81	10.05440	7.32974	5.41286	4.03961	3.04133	81
82	9.85725	7.16845	5.28084	3.93149	2.95275	82
83	9.66397	7.01071	5.15204	3.82627	2.86675	83
84	9.47448	6.85644	5.02638	3.72386	2.78325	84
85	9.28871	6.70557	4.90379	3.62420	2.70218	85
86	9.10658	6.55801	4.78418	3.52720	2.62348	86
87	8.92802	6.41370	4.66749	3.43280	2.54707	87
88	8.75296	6.27257	4.55365	3.34092	2.47288	88
89	8.58133	6.13454	4.44259	3.25150	2.40085	89
90	8.41307	5.99955	4.33423	3.16448	2.33093	90
91	8.24811	5.86753	4.22852	3.07979	2.26304	91
92	8.08638	5.73842	4.12538	2.99736	2.19712	92
93	7.92782	5.61215	4.02476	2.91714	2.13313	93
94	7.77238	5.48865	3.92660	2.83906	2.07100	94
95	7.61998	5.36788	3.83083	2.76308	2.01068	95
96	7.47057	5.24976	3.73739	2.68913	1.95211	96
97	7.32408	5.13424	3.64624	2.61716	1.89526	97
98	7.18047	5.02126	3.55731	2.54711	1.84005	98
99	7.03968	4.91076	3.47054	2.47894	1.78646	99

YEARS' PURCHASE OF A REVERSION TO A PERPETUITY
AFTER A GIVEN NUMBER OF YEARS

After Yrs.	Rate Per Cent					After Yrs.
	3.25	**3.5**	**3.75**	**4**	**4.25**	
0	30.76923	28.57143	26.66667	25.00000	23.52941	0
1	29.80071	27.60524	25.70281	24.03846	22.57018	1
2	28.86267	26.67173	24.77379	23.11391	21.65005	2
3	27.95416	25.76979	23.87836	22.22491	20.76744	3
4	27.07425	24.89835	23.01528	21.37010	19.92080	4
5	26.22203	24.05638	22.18340	20.54818	19.10868	5
6	25.39664	23.24288	21.38160	19.75786	18.32967	6
7	24.59723	22.45688	20.60877	18.99795	17.58242	7
8	23.82298	21.69747	19.86387	18.26726	16.86563	8
9	23.07311	20.96374	19.14590	17.56467	16.17806	9
10	22.34684	20.25482	18.45388	16.88910	15.51852	10
11	21.64342	19.56988	17.78687	16.23952	14.88588	11
12	20.96215	18.90809	17.14397	15.61493	14.27902	12
13	20.30233	18.26869	16.52431	15.01435	13.69690	13
14	19.66327	17.65091	15.92705	14.43688	13.13851	14
15	19.04433	17.05402	15.35137	13.88161	12.60289	15
16	18.44487	16.47731	14.79650	13.34770	12.08910	16
17	17.86428	15.92011	14.26169	12.83433	11.59626	17
18	17.30197	15.38175	13.74621	12.34070	11.12351	18
19	16.75736	14.86159	13.24935	11.86606	10.67004	19
20	16.22988	14.35903	12.77046	11.40967	10.23505	20
21	15.71902	13.87345	12.30888	10.97084	9.81779	21
22	15.22423	13.40430	11.86398	10.54888	9.41754	22
23	14.74502	12.95102	11.43516	10.14316	9.03362	23
24	14.28089	12.51306	11.02184	9.75304	8.66534	24
25	13.83137	12.08991	10.62346	9.37792	8.31208	25
26	13.39600	11.68108	10.23948	9.01723	7.97321	26
27	12.97433	11.28606	9.86938	8.67041	7.64817	27
28	12.56594	10.90441	9.51266	8.33694	7.33637	28
29	12.17040	10.53566	9.16882	8.01629	7.03729	29
30	11.78731	10.17938	8.83742	7.70797	6.75039	30
31	11.41628	9.83515	8.51800	7.41151	6.47520	31
32	11.05693	9.50256	8.21012	7.12645	6.21122	32
33	10.70889	9.18122	7.91337	6.85235	5.95801	33
34	10.37181	8.87074	7.62734	6.58880	5.71511	34
35	10.04534	8.57077	7.35165	6.33539	5.48212	35
36	9.72914	8.28093	7.08593	6.09172	5.25863	36
37	9.42290	8.00090	6.82981	5.85742	5.04425	37
38	9.12629	7.73034	6.58295	5.63214	4.83861	38
39	8.83902	7.46893	6.34501	5.41552	4.64135	39
40	8.56080	7.21636	6.11568	5.20723	4.45214	40
41	8.29133	6.97232	5.89463	5.00695	4.27064	41
42	8.03034	6.73655	5.68157	4.81437	4.09653	42
43	7.77757	6.50874	5.47621	4.62921	3.92953	43
44	7.53276	6.28864	5.27828	4.45116	3.76933	44
45	7.29565	6.07598	5.08749	4.27996	3.61567	45
46	7.06600	5.87051	4.90361	4.11535	3.46826	46
47	6.84359	5.67199	4.72637	3.95706	3.32687	47
48	6.62817	5.48018	4.55554	3.80487	3.19124	48
49	6.41954	5.29486	4.39088	3.65853	3.06115	49

YEARS' PURCHASE OF A REVERSION TO A PERPETUITY

AFTER A GIVEN NUMBER OF YEARS

After Yrs.	Rate Per Cent					After Yrs.
	3.25	**3.5**	**3.75**	**4**	**4.25**	
50	6.21747	5.11581	4.23217	3.51782	2.93635	50
51	6.02176	4.94281	4.07920	3.38251	2.81664	51
52	5.83221	4.77566	3.93176	3.25242	2.70182	52
53	5.64863	4.61417	3.78965	3.12733	2.59167	53
54	5.47083	4.45813	3.65268	3.00704	2.48601	54
55	5.29863	4.30738	3.52065	2.89139	2.38467	55
56	5.13184	4.16172	3.39340	2.78018	2.28745	56
57	4.97031	4.02098	3.27075	2.67325	2.19420	57
58	4.81386	3.88501	3.15253	2.57043	2.10474	58
59	4.66233	3.75363	3.03858	2.47157	2.01894	59
60	4.51557	3.62669	2.92875	2.37651	1.93663	60
61	4.37344	3.50405	2.82289	2.28511	1.85768	61
62	4.23578	3.38556	2.72086	2.19722	1.78195	62
63	4.10245	3.27107	2.62252	2.11271	1.70930	63
64	3.97331	3.16045	2.52773	2.03145	1.63962	64
65	3.84825	3.05358	2.43636	1.95332	1.57278	65
66	3.72711	2.95032	2.34830	1.87819	1.50866	66
67	3.60980	2.85055	2.26342	1.80595	1.44715	67
68	3.49617	2.75415	2.18161	1.73649	1.38816	68
69	3.38612	2.66102	2.10276	1.66970	1.33157	69
70	3.27954	2.57103	2.02676	1.60549	1.27728	70
71	3.17631	2.48409	1.95350	1.54374	1.22521	71
72	3.07633	2.40009	1.88289	1.48436	1.17526	72
73	2.97949	2.31892	1.81483	1.42727	1.12735	73
74	2.88571	2.24051	1.74924	1.37238	1.08139	74
75	2.79487	2.16474	1.68601	1.31959	1.03730	75
76	2.70690	2.09154	1.62507	1.26884	0.99502	76
77	2.62169	2.02081	1.56633	1.22004	0.95445	77
78	2.53917	1.95247	1.50972	1.17311	0.91554	78
79	2.45925	1.88645	1.45515	1.12799	0.87822	79
80	2.38184	1.82265	1.40256	1.08461	0.84241	80
81	2.30686	1.76102	1.35186	1.04289	0.80807	81
82	2.23425	1.70147	1.30300	1.00278	0.77513	82
83	2.16392	1.64393	1.25590	0.96421	0.74353	83
84	2.09581	1.58834	1.21051	0.92713	0.71322	84
85	2.02984	1.53462	1.16676	0.89147	0.68414	85
86	1.96595	1.48273	1.12458	0.85718	0.65625	86
87	1.90406	1.43259	1.08394	0.82421	0.62950	87
88	1.84413	1.38414	1.04476	0.79251	0.60383	88
89	1.78608	1.33734	1.00700	0.76203	0.57922	89
90	1.72986	1.29211	0.97060	0.73272	0.55560	90
91	1.67541	1.24842	0.93552	0.70454	0.53295	91
92	1.62267	1.20620	0.90170	0.67744	0.51123	92
93	1.57160	1.16541	0.86911	0.65139	0.49038	93
94	1.52213	1.12600	0.83770	0.62633	0.47039	94
95	1.47422	1.08792	0.80742	0.60224	0.45122	95
96	1.42781	1.05113	0.77823	0.57908	0.43282	96
97	1.38287	1.01559	0.75011	0.55681	0.41518	97
98	1.33934	0.98125	0.72299	0.53539	0.39825	98
99	1.29718	0.94806	0.69686	0.51480	0.38202	99

YEARS' PURCHASE OF A REVERSION TO A PERPETUITY

AFTER A GIVEN NUMBER OF YEARS

After Yrs.	Rate Per Cent					After Yrs.
	4.5	**4.75**	**5**	**5.25**	**5.5**	
0	22.22222	21.05263	20.00000	19.04762	18.18182	0
1	21.26528	20.09798	19.04762	18.09750	17.23395	1
2	20.34955	19.18661	18.14059	17.19477	16.33550	2
3	19.47326	18.31658	17.27675	16.33708	15.48388	3
4	18.63470	17.48599	16.45405	15.52216	14.67667	4
5	17.83225	16.69307	15.67052	14.74790	13.91153	5
6	17.06435	15.93611	14.92431	14.01226	13.18629	6
7	16.32952	15.21347	14.21363	13.31331	12.49885	7
8	15.62634	14.52360	13.53679	12.64922	11.84725	8
9	14.95343	13.86501	12.89218	12.01826	11.22962	9
10	14.30950	13.23628	12.27827	11.41878	10.64419	10
11	13.69331	12.63607	11.69359	10.84920	10.08928	11
12	13.10364	12.06307	11.13675	10.30802	9.56330	12
13	12.53937	11.51606	10.60643	9.79385	9.06474	13
14	11.99940	10.99385	10.10136	9.30532	8.59217	14
15	11.48268	10.49533	9.62034	8.84116	8.14424	15
16	10.98821	10.01940	9.16223	8.40015	7.71966	16
17	10.51503	9.56506	8.72593	7.98114	7.31721	17
18	10.06223	9.13133	8.31041	7.58303	6.93574	18
19	9.62893	8.71726	7.91468	7.20478	6.57416	19
20	9.21429	8.32196	7.53779	6.84540	6.23144	20
21	8.81750	7.94459	7.17885	6.50394	5.90657	21
22	8.43780	7.58434	6.83700	6.17952	5.59865	22
23	8.07445	7.24042	6.51143	5.87127	5.30678	23
24	7.72674	6.91209	6.20136	5.57841	5.03012	24
25	7.39401	6.59866	5.90606	5.30015	4.76789	25
26	7.07561	6.29943	5.62481	5.03577	4.51932	26
27	6.77092	6.01378	5.35697	4.78458	4.28372	27
28	6.47935	5.74108	5.10187	4.54592	4.06040	28
29	6.20033	5.48074	4.85893	4.31916	3.84872	29
30	5.93333	5.23221	4.62755	4.10372	3.64807	30
31	5.67783	4.99495	4.40719	3.89902	3.45789	31
32	5.43333	4.76845	4.19732	3.70453	3.27762	32
33	5.19936	4.55222	3.99745	3.51975	3.10675	33
34	4.97546	4.34580	3.80710	3.34418	2.94479	34
35	4.76121	4.14873	3.62581	3.17736	2.79127	35
36	4.55618	3.96060	3.45315	3.01887	2.64575	36
37	4.35998	3.78100	3.28871	2.86829	2.50782	37
38	4.17223	3.60955	3.13211	2.72521	2.37708	38
39	3.99257	3.44587	2.98296	2.58928	2.25316	39
40	3.82064	3.28962	2.84091	2.46012	2.13569	40
41	3.65611	3.14044	2.70563	2.33741	2.02435	41
42	3.49867	2.99804	2.57679	2.22081	1.91882	42
43	3.34801	2.86209	2.45409	2.11004	1.81879	43
44	3.20384	2.73230	2.33723	2.00479	1.72397	44
45	3.06587	2.60840	2.22593	1.90479	1.63409	45
46	2.93385	2.49012	2.11993	1.80977	1.54890	46
47	2.80751	2.37721	2.01898	1.71950	1.46815	47
48	2.68662	2.26941	1.92284	1.63373	1.39162	48
49	2.57092	2.16650	1.83128	1.55224	1.31907	49

YEARS' PURCHASE OF A REVERSION TO A PERPETUITY

AFTER A GIVEN NUMBER OF YEARS

After Yrs.	Rate Per Cent					After Yrs.
	4.5	**4.75**	**5**	**5.25**	**5.5**	
50	2.46021	2.06826	1.74407	1.47481	1.25030	50
51	2.35427	1.97447	1.66102	1.40124	1.18512	51
52	2.25289	1.88494	1.58193	1.33135	1.12334	52
53	2.15588	1.79946	1.50660	1.26494	1.06477	53
54	2.06304	1.71786	1.43485	1.20184	1.00926	54
55	1.97420	1.63997	1.36653	1.14189	0.95665	55
56	1.88919	1.56560	1.30146	1.08493	0.90678	56
57	1.80784	1.49461	1.23948	1.03082	0.85950	57
58	1.72999	1.42683	1.18046	0.97940	0.81469	58
59	1.65549	1.36213	1.12425	0.93054	0.77222	59
60	1.58420	1.30036	1.07071	0.88413	0.73196	60
61	1.51598	1.24140	1.01972	0.84003	0.69380	61
62	1.45070	1.18510	0.97117	0.79812	0.65763	62
63	1.38823	1.13136	0.92492	0.75831	0.62335	63
64	1.32845	1.08006	0.88088	0.72049	0.59085	64
65	1.27124	1.03108	0.83893	0.68455	0.56005	65
66	1.21650	0.98433	0.79898	0.65040	0.53085	66
67	1.16412	0.93969	0.76093	0.61796	0.50318	67
68	1.11399	0.89708	0.72470	0.58713	0.47695	68
69	1.06602	0.85640	0.69019	0.55785	0.45208	69
70	1.02011	0.81757	0.65732	0.53002	0.42851	70
71	0.97618	0.78050	0.62602	0.50358	0.40617	71
72	0.93415	0.74510	0.59621	0.47846	0.38500	72
73	0.89392	0.71132	0.56782	0.45460	0.36493	73
74	0.85543	0.67906	0.54078	0.43192	0.34590	74
75	0.81859	0.64827	0.51503	0.41038	0.32787	75
76	0.78334	0.61887	0.49050	0.38991	0.31078	76
77	0.74961	0.59081	0.46715	0.37046	0.29458	77
78	0.71733	0.56402	0.44490	0.35198	0.27922	78
79	0.68644	0.53844	0.42372	0.33442	0.26466	79
80	0.65688	0.51402	0.40354	0.31774	0.25087	80
81	0.62859	0.49072	0.38432	0.30189	0.23779	81
82	0.60152	0.46846	0.36602	0.28683	0.22539	82
83	0.57562	0.44722	0.34859	0.27252	0.21364	83
84	0.55083	0.42694	0.33199	0.25893	0.20250	84
85	0.52711	0.40758	0.31618	0.24602	0.19195	85
86	0.50441	0.38910	0.30113	0.23374	0.18194	86
87	0.48269	0.37145	0.28679	0.22208	0.17245	87
88	0.46191	0.35461	0.27313	0.21101	0.16346	88
89	0.44202	0.33853	0.26013	0.20048	0.15494	89
90	0.42298	0.32318	0.24774	0.19048	0.14686	90
91	0.40477	0.30852	0.23594	0.18098	0.13921	91
92	0.38734	0.29453	0.22471	0.17195	0.13195	92
93	0.37066	0.28118	0.21401	0.16337	0.12507	93
94	0.35470	0.26843	0.20381	0.15523	0.11855	94
95	0.33942	0.25626	0.19411	0.14748	0.11237	95
96	0.32481	0.24464	0.18487	0.14013	0.10651	96
97	0.31082	0.23354	0.17606	0.13314	0.10096	97
98	0.29743	0.22295	0.16768	0.12650	0.09570	98
99	0.28463	0.21284	0.15969	0.12019	0.09071	99

YEARS' PURCHASE OF A REVERSION TO A PERPETUITY

AFTER A GIVEN NUMBER OF YEARS

After Yrs.	Rate Per Cent					After Yrs.
	5.75	**6**	**6.25**	**6.5**	**6.75**	
0	17.39130	16.66667	16.00000	15.38462	14.81481	0
1	16.44568	15.72327	15.05882	14.44565	13.87805	1
2	15.55147	14.83327	14.17301	13.56399	13.00051	2
3	14.70588	13.99365	13.33930	12.73614	12.17847	3
4	13.90627	13.20156	12.55464	11.95882	11.40840	4
5	13.15014	12.45430	11.81613	11.22894	10.68702	5
6	12.43512	11.74934	11.12106	10.54360	10.01126	6
7	11.75898	11.08429	10.46688	9.90010	9.37823	7
8	11.11960	10.45687	9.85118	9.29586	8.78523	8
9	10.51499	9.86497	9.27170	8.72851	8.22972	9
10	9.94325	9.30658	8.72631	8.19579	7.70934	10
11	9.40260	8.77979	8.21300	7.69557	7.22187	11
12	8.89135	8.28282	7.72988	7.22589	6.76522	12
13	8.40789	7.81398	7.27518	6.78487	6.33744	13
14	7.95073	7.37168	6.84723	6.37077	5.93671	14
15	7.51842	6.95442	6.44445	5.98195	5.56132	15
16	7.10962	6.56077	6.06537	5.61685	5.20967	16
17	6.72304	6.18941	5.70858	5.27404	4.88025	17
18	6.35749	5.83906	5.37278	4.95215	4.57166	18
19	6.01181	5.50855	5.05673	4.64991	4.28259	19
20	5.68492	5.19675	4.75928	4.36611	4.01179	20
21	5.37581	4.90259	4.47932	4.09963	3.75812	21
22	5.08351	4.62508	4.21583	3.84942	3.52049	22
23	4.80710	4.36329	3.96784	3.61448	3.29788	23
24	4.54572	4.11631	3.73444	3.39388	3.08935	24
25	4.29856	3.88331	3.51477	3.18674	2.89400	25
26	4.06483	3.66350	3.30802	2.99224	2.71101	26
27	3.84381	3.45613	3.11343	2.80962	2.53959	27
28	3.63481	3.26050	2.93028	2.63814	2.37901	28
29	3.43717	3.07595	2.75791	2.47713	2.22858	29
30	3.25028	2.90184	2.59568	2.32594	2.08766	30
31	3.07355	2.73758	2.44300	2.18398	1.95565	31
32	2.90643	2.58262	2.29929	2.05069	1.83199	32
33	2.74840	2.43644	2.16404	1.92553	1.71615	33
34	2.59896	2.29853	2.03674	1.80801	1.60764	34
35	2.45764	2.16842	1.91693	1.69766	1.50598	35
36	2.32401	2.04568	1.80417	1.59405	1.41076	36
37	2.19765	1.92989	1.69805	1.49676	1.32155	37
38	2.07815	1.82065	1.59816	1.40541	1.23799	38
39	1.96516	1.71759	1.50415	1.31963	1.15971	39
40	1.85831	1.62037	1.41567	1.23909	1.08638	40
41	1.75726	1.52865	1.33240	1.16346	1.01768	41
42	1.66171	1.44212	1.25402	1.09245	0.95333	42
43	1.57136	1.36049	1.18025	1.02578	0.89305	43
44	1.48592	1.28348	1.11083	0.96317	0.83658	44
45	1.40513	1.21083	1.04548	0.90439	0.78368	45
46	1.32872	1.14230	0.98399	0.84919	0.73413	46
47	1.25648	1.07764	0.92610	0.79736	0.68771	47
48	1.18816	1.01664	0.87163	0.74870	0.64423	48
49	1.12355	0.95909	0.82036	0.70300	0.60349	49

YEARS' PURCHASE OF A REVERSION TO A PERPETUITY

AFTER A GIVEN NUMBER OF YEARS

After Yrs.	Rate Per Cent					After Yrs.
	5.75	**6**	**6.25**	**6.5**	**6.75**	
50	1.06246	0.90481	0.77210	0.66009	0.56533	50
51	1.00469	0.85359	0.72668	0.61981	0.52958	51
52	0.95006	0.80527	0.68394	0.58198	0.49610	52
53	0.89841	0.75969	0.64370	0.54646	0.46473	53
54	0.84956	0.71669	0.60584	0.51311	0.43534	54
55	0.80336	0.67612	0.57020	0.48179	0.40781	55
56	0.75968	0.63785	0.53666	0.45239	0.38203	56
57	0.71837	0.60175	0.50509	0.42478	0.35787	57
58	0.67931	0.56769	0.47538	0.39885	0.33524	58
59	0.64238	0.53555	0.44742	0.37451	0.31404	59
60	0.60745	0.50524	0.42110	0.35165	0.29419	60
61	0.57442	0.47664	0.39633	0.33019	0.27558	61
62	0.54319	0.44966	0.37301	0.31004	0.25816	62
63	0.51365	0.42421	0.35107	0.29111	0.24184	63
64	0.48572	0.40020	0.33042	0.27335	0.22654	64
65	0.45931	0.37754	0.31098	0.25666	0.21222	65
66	0.43434	0.35617	0.29269	0.24100	0.19880	66
67	0.41072	0.33601	0.27547	0.22629	0.18623	67
68	0.38839	0.31699	0.25927	0.21248	0.17445	68
69	0.36727	0.29905	0.24402	0.19951	0.16342	69
70	0.34730	0.28212	0.22966	0.18733	0.15309	70
71	0.32842	0.26615	0.21616	0.17590	0.14341	71
72	0.31056	0.25109	0.20344	0.16516	0.13434	72
73	0.29367	0.23688	0.19147	0.15508	0.12585	73
74	0.27771	0.22347	0.18021	0.14562	0.11789	74
75	0.26261	0.21082	0.16961	0.13673	0.11043	75
76	0.24833	0.19889	0.15963	0.12839	0.10345	76
77	0.23482	0.18763	0.15024	0.12055	0.09691	77
78	0.22206	0.17701	0.14140	0.11319	0.09078	78
79	0.20998	0.16699	0.13309	0.10628	0.08504	79
80	0.19856	0.15754	0.12526	0.09980	0.07966	80
81	0.18777	0.14862	0.11789	0.09371	0.07463	81
82	0.17756	0.14021	0.11096	0.08799	0.06991	82
83	0.16790	0.13227	0.10443	0.08262	0.06549	83
84	0.15877	0.12478	0.09829	0.07757	0.06135	84
85	0.15014	0.11772	0.09250	0.07284	0.05747	85
86	0.14198	0.11106	0.08706	0.06839	0.05383	86
87	0.13426	0.10477	0.08194	0.06422	0.05043	87
88	0.12696	0.09884	0.07712	0.06030	0.04724	88
89	0.12005	0.09325	0.07258	0.05662	0.04425	89
90	0.11353	0.08797	0.06831	0.05316	0.04146	90
91	0.10735	0.08299	0.06430	0.04992	0.03883	91
92	0.10152	0.07829	0.06051	0.04687	0.03638	92
93	0.09600	0.07386	0.05695	0.04401	0.03408	93
94	0.09078	0.06968	0.05360	0.04133	0.03192	94
95	0.08584	0.06573	0.05045	0.03880	0.02991	95
96	0.08117	0.06201	0.04748	0.03644	0.02801	96
97	0.07676	0.05850	0.04469	0.03421	0.02624	97
98	0.07259	0.05519	0.04206	0.03212	0.02458	98
99	0.06864	0.05207	0.03959	0.03016	0.02303	99

YEARS' PURCHASE OF A REVERSION TO A PERPETUITY

AFTER A GIVEN NUMBER OF YEARS

After Yrs.	7	7.25	7.5	7.75	8	After Yrs.
			Rate Per Cent			
0	14.28571	13.79310	13.33333	12.90323	12.50000	0
1	13.35113	12.86070	12.40310	11.97515	11.57407	1
2	12.47770	11.99133	11.53777	11.11383	10.71674	2
3	11.66140	11.18073	10.73281	10.31446	9.92290	3
4	10.89850	10.42492	9.98401	9.57258	9.18787	4
5	10.18552	9.72021	9.28745	8.88407	8.50729	5
6	9.51917	9.06313	8.63949	8.24508	7.87712	6
7	8.89642	8.45047	8.03673	7.65204	7.29363	7
8	8.31442	7.87923	7.47603	7.10166	6.75336	8
9	7.77048	7.34660	6.95445	6.59087	6.25311	9
10	7.26213	6.84997	6.46925	6.11682	5.78992	10
11	6.78704	6.38692	6.01791	5.67686	5.36104	11
12	6.34303	5.95517	5.59806	5.26855	4.96392	12
13	5.92806	5.55261	5.20749	4.88960	4.59622	13
14	5.54025	5.17726	4.84418	4.53792	4.25576	14
15	5.17780	4.82728	4.50621	4.21152	3.94052	15
16	4.83907	4.50096	4.19183	3.90861	3.64863	16
17	4.52249	4.19670	3.89937	3.62748	3.37836	17
18	4.22663	3.91301	3.62732	3.36657	3.12811	18
19	3.95012	3.64849	3.37426	3.12442	2.89640	19
20	3.69170	3.40186	3.13884	2.89970	2.68185	20
21	3.45019	3.17189	2.91985	2.69113	2.48320	21
22	3.22447	2.95748	2.71614	2.49757	2.29926	22
23	3.01353	2.75755	2.52664	2.31793	2.12894	23
24	2.81638	2.57115	2.35037	2.15121	1.97124	24
25	2.63213	2.39734	2.18639	1.99649	1.82522	25
26	2.45994	2.23528	2.03385	1.85289	1.69002	26
27	2.29901	2.08418	1.89195	1.71962	1.56484	27
28	2.14860	1.94329	1.75996	1.59593	1.44892	28
29	2.00804	1.81193	1.63717	1.48114	1.34159	29
30	1.87667	1.68944	1.52295	1.37461	1.24222	30
31	1.75390	1.57524	1.41669	1.27574	1.15020	31
32	1.63916	1.46875	1.31786	1.18398	1.06500	32
33	1.53192	1.36947	1.22591	1.09882	0.98611	33
34	1.43170	1.27689	1.14038	1.01979	0.91307	34
35	1.33804	1.19057	1.06082	0.94644	0.84543	35
36	1.25051	1.11009	0.98681	0.87837	0.78281	36
37	1.16870	1.03505	0.91796	0.81519	0.72482	37
38	1.09224	0.96508	0.85392	0.75656	0.67113	38
39	1.02079	0.89984	0.79434	0.70214	0.62142	39
40	0.95401	0.83902	0.73892	0.65164	0.57539	40
41	0.89159	0.78230	0.68737	0.60477	0.53277	41
42	0.83327	0.72942	0.63942	0.56127	0.49330	42
43	0.77875	0.68011	0.59481	0.52090	0.45676	43
44	0.72781	0.63413	0.55331	0.48343	0.42293	44
45	0.68019	0.59127	0.51470	0.44866	0.39160	45
46	0.63569	0.55130	0.47879	0.41639	0.36259	46
47	0.59411	0.51403	0.44539	0.38644	0.33573	47
48	0.55524	0.47928	0.41432	0.35865	0.31086	48
49	0.51892	0.44688	0.38541	0.33285	0.28784	49

YEARS' PURCHASE OF A REVERSION TO A PERPETUITY

AFTER A GIVEN NUMBER OF YEARS

After Yrs.	7	7.25	7.5	7.75	8	After Yrs.
50	0.48497	0.41667	0.35852	0.30891	0.26652	50
51	0.45324	0.38851	0.33351	0.28669	0.24677	51
52	0.42359	0.36224	0.31024	0.26607	0.22849	52
53	0.39588	0.33776	0.28860	0.24693	0.21157	53
54	0.36998	0.31493	0.26846	0.22917	0.19590	54
55	0.34578	0.29364	0.24973	0.21269	0.18139	55
56	0.32315	0.27379	0.23231	0.19739	0.16795	56
57	0.30201	0.25528	0.21610	0.18319	0.15551	57
58	0.28226	0.23802	0.20102	0.17002	0.14399	58
59	0.26379	0.22193	0.18700	0.15779	0.13332	59
60	0.24653	0.20693	0.17395	0.14644	0.12345	60
61	0.23040	0.19294	0.16182	0.13591	0.11430	61
62	0.21533	0.17990	0.15053	0.12613	0.10584	62
63	0.20124	0.16774	0.14002	0.11706	0.09800	63
64	0.18808	0.15640	0.13026	0.10864	0.09074	64
65	0.17577	0.14583	0.12117	0.10083	0.08402	65
66	0.16428	0.13597	0.11271	0.09357	0.07779	66
67	0.15353	0.12678	0.10485	0.08684	0.07203	67
68	0.14348	0.11821	0.09754	0.08060	0.06670	68
69	0.13410	0.11022	0.09073	0.07480	0.06175	69
70	0.12532	0.10277	0.08440	0.06942	0.05718	70
71	0.11713	0.09582	0.07851	0.06443	0.05294	71
72	0.10946	0.08934	0.07303	0.05979	0.04902	72
73	0.10230	0.08330	0.06794	0.05549	0.04539	73
74	0.09561	0.07767	0.06320	0.05150	0.04203	74
75	0.08935	0.07242	0.05879	0.04780	0.03892	75
76	0.08351	0.06753	0.05469	0.04436	0.03603	76
77	0.07805	0.06296	0.05087	0.04117	0.03336	77
78	0.07294	0.05870	0.04732	0.03821	0.03089	78
79	0.06817	0.05474	0.04402	0.03546	0.02860	79
80	0.06371	0.05104	0.04095	0.03291	0.02649	80
81	0.05954	0.04759	0.03809	0.03054	0.02452	81
82	0.05565	0.04437	0.03544	0.02835	0.02271	82
83	0.05201	0.04137	0.03296	0.02631	0.02103	83
84	0.04860	0.03857	0.03066	0.02441	0.01947	84
85	0.04542	0.03597	0.02852	0.02266	0.01803	85
86	0.04245	0.03353	0.02653	0.02103	0.01669	86
87	0.03967	0.03127	0.02468	0.01952	0.01545	87
88	0.03708	0.02915	0.02296	0.01811	0.01431	88
89	0.03465	0.02718	0.02136	0.01681	0.01325	89
90	0.03239	0.02535	0.01987	0.01560	0.01227	90
91	0.03027	0.02363	0.01848	0.01448	0.01136	91
92	0.02829	0.02203	0.01719	0.01344	0.01052	92
93	0.02644	0.02055	0.01599	0.01247	0.00974	93
94	0.02471	0.01916	0.01488	0.01157	0.00902	94
95	0.02309	0.01786	0.01384	0.01074	0.00835	95
96	0.02158	0.01665	0.01287	0.00997	0.00773	96
97	0.02017	0.01553	0.01198	0.00925	0.00716	97
98	0.01885	0.01448	0.01114	0.00859	0.00663	98
99	0.01762	0.01350	0.01036	0.00797	0.00614	99

YEARS' PURCHASE OF A REVERSION TO A PERPETUITY

AFTER A GIVEN NUMBER OF YEARS

After Yrs.	Rate Per Cent					After Yrs.
	8.5	9	9.5	10	11	
0	11.76471	11.11111	10.52632	10.00000	9.09091	0
1	10.84305	10.19368	9.61307	9.09091	8.19001	1
2	9.99359	9.35200	8.77906	8.26446	7.37839	2
3	9.21068	8.57982	8.01741	7.51315	6.64719	3
4	8.48911	7.87139	7.32183	6.83013	5.98846	4
5	7.82406	7.22146	6.68661	6.20921	5.39501	5
6	7.21112	6.62519	6.10649	5.64474	4.86037	6
7	6.64619	6.07816	5.57670	5.13158	4.37871	7
8	6.12552	5.57629	5.09288	4.66507	3.94479	8
9	5.64564	5.11586	4.65103	4.24098	3.55386	9
10	5.20336	4.69345	4.24752	3.85543	3.20168	10
11	4.79572	4.30592	3.87901	3.50494	2.88439	11
12	4.42002	3.95039	3.54248	3.18631	2.59855	12
13	4.07375	3.62421	3.23514	2.89664	2.34104	13
14	3.75461	3.32496	2.95446	2.63331	2.10904	14
15	3.46047	3.05042	2.69814	2.39392	1.90004	15
16	3.18937	2.79855	2.46406	2.17629	1.71175	16
17	2.93951	2.56748	2.25028	1.97845	1.54211	17
18	2.70923	2.35549	2.05505	1.79859	1.38929	18
19	2.49699	2.16100	1.87676	1.63508	1.25161	19
20	2.30137	1.98257	1.71393	1.48644	1.12758	20
21	2.12108	1.81887	1.56524	1.35131	1.01584	21
22	1.95491	1.66869	1.42944	1.22846	0.91517	22
23	1.80176	1.53090	1.30542	1.11678	0.82448	23
24	1.66061	1.40450	1.19217	1.01526	0.74277	24
25	1.53052	1.28853	1.08874	0.92296	0.66916	25
26	1.41061	1.18214	0.99428	0.83905	0.60285	26
27	1.30010	1.08453	0.90802	0.76278	0.54311	27
28	1.19825	0.99498	0.82924	0.69343	0.48929	28
29	1.10438	0.91283	0.75730	0.63039	0.44080	29
30	1.01786	0.83746	0.69160	0.57309	0.39712	30
31	0.93812	0.76831	0.63160	0.52099	0.35776	31
32	0.86463	0.70487	0.57680	0.47362	0.32231	32
33	0.79689	0.64667	0.52676	0.43057	0.29037	33
34	0.73446	0.59328	0.48106	0.39143	0.26159	34
35	0.67692	0.54429	0.43932	0.35584	0.23567	35
36	0.62389	0.49935	0.40121	0.32349	0.21231	36
37	0.57502	0.45812	0.36640	0.29408	0.19127	37
38	0.52997	0.42029	0.33461	0.26735	0.17232	38
39	0.48845	0.38559	0.30558	0.24304	0.15524	39
40	0.45019	0.35375	0.27907	0.22095	0.13986	40
41	0.41492	0.32454	0.25486	0.20086	0.12600	41
42	0.38241	0.29775	0.23275	0.18260	0.11351	42
43	0.35245	0.27316	0.21255	0.16600	0.10226	43
44	0.32484	0.25061	0.19411	0.15091	0.09213	44
45	0.29939	0.22991	0.17727	0.13719	0.08300	45
46	0.27594	0.21093	0.16189	0.12472	0.07477	46
47	0.25432	0.19351	0.14785	0.11338	0.06736	47
48	0.23440	0.17754	0.13502	0.10307	0.06069	48
49	0.21603	0.16288	0.12331	0.09370	0.05467	49

YEARS' PURCHASE OF A REVERSION TO A PERPETUITY

AFTER A GIVEN NUMBER OF YEARS

After Yrs.	8.5	9	9.5	10	11	After Yrs.
			Rate Per Cent			
50	0.19911	0.14943	0.11261	0.08519	0.04926	50
51	0.18351	0.13709	0.10284	0.07744	0.04437	51
52	0.16914	0.12577	0.09392	0.07040	0.03998	52
53	0.15589	0.11539	0.08577	0.06400	0.03602	53
54	0.14367	0.10586	0.07833	0.05818	0.03245	54
55	0.13242	0.09712	0.07153	0.05289	0.02923	55
56	0.12204	0.08910	0.06533	0.04808	0.02633	56
57	0.11248	0.08174	0.05966	0.04371	0.02372	57
58	0.10367	0.07499	0.05448	0.03974	0.02137	58
59	0.09555	0.06880	0.04976	0.03613	0.01926	59
60	0.08806	0.06312	0.04544	0.03284	0.01735	60
61	0.08116	0.05791	0.04150	0.02986	0.01563	61
62	0.07481	0.05313	0.03790	0.02714	0.01408	62
63	0.06895	0.04874	0.03461	0.02468	0.01268	63
64	0.06354	0.04472	0.03161	0.02243	0.01143	64
65	0.05857	0.04102	0.02886	0.02039	0.01029	65
66	0.05398	0.03764	0.02636	0.01854	0.00927	66
67	0.04975	0.03453	0.02407	0.01685	0.00836	67
68	0.04585	0.03168	0.02198	0.01532	0.00753	68
69	0.04226	0.02906	0.02008	0.01393	0.00678	69
70	0.03895	0.02666	0.01834	0.01266	0.00611	70
71	0.03590	0.02446	0.01674	0.01151	0.00550	71
72	0.03309	0.02244	0.01529	0.01046	0.00496	72
73	0.03049	0.02059	0.01397	0.00951	0.00447	73
74	0.02810	0.01889	0.01275	0.00865	0.00402	74
75	0.02590	0.01733	0.01165	0.00786	0.00363	75
76	0.02387	0.01590	0.01064	0.00715	0.00327	76
77	0.02200	0.01459	0.00971	0.00650	0.00294	77
78	0.02028	0.01338	0.00887	0.00591	0.00265	78
79	0.01869	0.01228	0.00810	0.00537	0.00239	79
80	0.01723	0.01126	0.00740	0.00488	0.00215	80
81	0.01588	0.01033	0.00676	0.00444	0.00194	81
82	0.01463	0.00948	0.00617	0.00403	0.00175	82
83	0.01349	0.00870	0.00564	0.00367	0.00157	83
84	0.01243	0.00798	0.00515	0.00333	0.00142	84
85	0.01146	0.00732	0.00470	0.00303	0.00128	85
86	0.01056	0.00672	0.00429	0.00276	0.00115	86
87	0.00973	0.00616	0.00392	0.00251	0.00104	87
88	0.00897	0.00565	0.00358	0.00228	0.00093	88
89	0.00827	0.00519	0.00327	0.00207	0.00084	89
90	0.00762	0.00476	0.00299	0.00188	0.00076	90
91	0.00702	0.00436	0.00273	0.00171	0.00068	91
92	0.00647	0.00400	0.00249	0.00156	0.00062	92
93	0.00597	0.00367	0.00227	0.00141	0.00055	93
94	0.00550	0.00337	0.00208	0.00129	0.00050	94
95	0.00507	0.00309	0.00190	0.00117	0.00045	95
96	0.00467	0.00284	0.00173	0.00106	0.00041	96
97	0.00430	0.00260	0.00158	0.00097	0.00036	97
98	0.00397	0.00239	0.00144	0.00088	0.00033	98
99	0.00366	0.00219	0.00132	0.00080	0.00030	99

YEARS' PURCHASE OF A REVERSION TO A PERPETUITY

AFTER A GIVEN NUMBER OF YEARS

After Yrs.	Rate Per Cent					After Yrs.
	12	**13**	**14**	**15**	**16**	
0	8.33333	7.69231	7.14286	6.66667	6.25000	0
1	7.44048	6.80735	6.26566	5.79710	5.38793	1
2	6.64328	6.02421	5.49620	5.04096	4.64477	2
3	5.93150	5.33116	4.82123	4.38344	4.00411	3
4	5.29598	4.71784	4.22914	3.81169	3.45182	4
5	4.72856	4.17508	3.70978	3.31451	2.97571	5
6	4.22193	3.69476	3.25419	2.88218	2.56526	6
7	3.76958	3.26970	2.85455	2.50625	2.21143	7
8	3.36569	2.89354	2.50399	2.17935	1.90641	8
9	3.00508	2.56065	2.19649	1.89508	1.64346	9
10	2.68311	2.26606	1.92674	1.64790	1.41677	10
11	2.39563	2.00537	1.69012	1.43295	1.22136	11
12	2.13896	1.77466	1.48257	1.24605	1.05289	12
13	1.90978	1.57050	1.30050	1.08352	0.90767	13
14	1.70517	1.38982	1.14079	0.94219	0.78247	14
15	1.52247	1.22993	1.00069	0.81930	0.67454	15
16	1.35935	1.08843	0.87780	0.71243	0.58150	16
17	1.21370	0.96321	0.77000	0.61951	0.50130	17
18	1.08366	0.85240	0.67544	0.53870	0.43215	18
19	0.96756	0.75434	0.59249	0.46844	0.37254	19
20	0.86389	0.66756	0.51973	0.40734	0.32116	20
21	0.77133	0.59076	0.45590	0.35420	0.27686	21
22	0.68869	0.52279	0.39991	0.30800	0.23867	22
23	0.61490	0.46265	0.35080	0.26783	0.20575	23
24	0.54902	0.40942	0.30772	0.23290	0.17737	24
25	0.49019	0.36232	0.26993	0.20252	0.15291	25
26	0.43767	0.32064	0.23678	0.17610	0.13182	26
27	0.39078	0.28375	0.20770	0.15313	0.11364	27
28	0.34891	0.25111	0.18219	0.13316	0.09796	28
29	0.31153	0.22222	0.15982	0.11579	0.08445	29
30	0.27815	0.19665	0.14019	0.10069	0.07280	30
31	0.24835	0.17403	0.12298	0.08755	0.06276	31
32	0.22174	0.15401	0.10787	0.07613	0.05410	32
33	0.19798	0.13629	0.09463	0.06620	0.04664	33
34	0.17677	0.12061	0.08301	0.05757	0.04021	34
35	0.15783	0.10674	0.07281	0.05006	0.03466	35
36	0.14092	0.09446	0.06387	0.04353	0.02988	36
37	0.12582	0.08359	0.05603	0.03785	0.02576	37
38	0.11234	0.07397	0.04915	0.03291	0.02221	38
39	0.10030	0.06546	0.04311	0.02862	0.01914	39
40	0.08956	0.05793	0.03782	0.02489	0.01650	40
41	0.07996	0.05127	0.03317	0.02164	0.01423	41
42	0.07139	0.04537	0.02910	0.01882	0.01226	42
43	0.06374	0.04015	0.02552	0.01636	0.01057	43
44	0.05691	0.03553	0.02239	0.01423	0.00911	44
45	0.05082	0.03144	0.01964	0.01237	0.00786	45
46	0.04537	0.02783	0.01723	0.01076	0.00677	46
47	0.04051	0.02462	0.01511	0.00936	0.00584	47
48	0.03617	0.02179	0.01326	0.00814	0.00503	48
49	0.03230	0.01928	0.01163	0.00707	0.00434	49

YEARS' PURCHASE OF A REVERSION TO A PERPETUITY

AFTER A GIVEN NUMBER OF YEARS

After Yrs.	Rate Per Cent					After Yrs.
	12	**13**	**14**	**15**	**16**	
50	0.02883	0.01707	0.01020	0.00615	0.00374	50
51	0.02575	0.01510	0.00895	0.00535	0.00322	51
52	0.02299	0.01337	0.00785	0.00465	0.00278	52
53	0.02052	0.01183	0.00689	0.00405	0.00240	53
54	0.01833	0.01047	0.00604	0.00352	0.00207	54
55	0.01636	0.00926	0.00530	0.00306	0.00178	55
56	0.01461	0.00820	0.00465	0.00266	0.00154	56
57	0.01304	0.00725	0.00408	0.00231	0.00132	57
58	0.01165	0.00642	0.00358	0.00201	0.00114	58
59	0.01040	0.00568	0.00314	0.00175	0.00098	59
60	0.00928	0.00503	0.00275	0.00152	0.00085	60
61	0.00829	0.00445	0.00241	0.00132	0.00073	61
62	0.00740	0.00394	0.00212	0.00115	0.00063	62
63	0.00661	0.00348	0.00186	0.00100	0.00054	63
64	0.00590	0.00308	0.00163	0.00087	0.00047	64
65	0.00527	0.00273	0.00143	0.00076	0.00040	65
66	0.00470	0.00241	0.00125	0.00066	0.00035	66
67	0.00420	0.00214	0.00110	0.00057	0.00030	67
68	0.00375	0.00189	0.00096	0.00050	0.00026	68
69	0.00335	0.00167	0.00085	0.00043	0.00022	69
70	0.00299	0.00148	0.00074	0.00038	0.00019	70
71	0.00267	0.00131	0.00065	0.00033	0.00017	71
72	0.00238	0.00116	0.00057	0.00028	0.00014	72
73	0.00213	0.00103	0.00050	0.00025	0.00012	73
74	0.00190	0.00091	0.00044	0.00021	0.00011	74
75	0.00170	0.00080	0.00039	0.00019	0.00009	75
76	0.00151	0.00071	0.00034	0.00016	0.00008	76
77	0.00135	0.00063	0.00030	0.00014	0.00007	77
78	0.00121	0.00056	0.00026	0.00012	0.00006	78
79	0.00108	0.00049	0.00023	0.00011	0.00005	79
80	0.00096	0.00044	0.00020	0.00009	0.00004	80
81	0.00086	0.00039	0.00018	0.00008	0.00004	81
82	0.00077	0.00034	0.00015	0.00007	0.00003	82
83	0.00069	0.00030	0.00014	0.00006	0.00003	83
84	0.00061	0.00027	0.00012	0.00005	0.00002	84
85	0.00055	0.00024	0.00010	0.00005	0.00002	85
86	0.00049	0.00021	0.00009	0.00004	0.00002	86
87	0.00044	0.00019	0.00008	0.00003	0.00002	87
88	0.00039	0.00016	0.00007	0.00003	0.00001	88
89	0.00035	0.00015	0.00006	0.00003	0.00001	89
90	0.00031	0.00013	0.00005	0.00002	0.00001	90
91	0.00028	0.00011	0.00005	0.00002	0.00001	91
92	0.00025	0.00010	0.00004	0.00002	0.00001	92
93	0.00022	0.00009	0.00004	0.00002	0.00001	93
94	0.00020	0.00008	0.00003	0.00001	0.00001	94
95	0.00018	0.00007	0.00003	0.00001	0.00000	95
96	0.00016	0.00006	0.00002	0.00001	0.00000	96
97	0.00014	0.00005	0.00002	0.00001	0.00000	97
98	0.00013	0.00005	0.00002	0.00001	0.00000	98
99	0.00011	0.00004	0.00002	0.00001	0.00000	99

Rules for finding the

YEARS' PURCHASE OF A REVERSION TO A PERPETUITY

(1) Multiply the Years' Purchase for Perpetuity by the Present Value of £1 receivable at the time the income commences to accrue.

Example—
   What is the Present Value of a Reversion to a Perpetuity of £1 per annum at 10% after 43 years?

Answer—
   Y.P. for Perp. at 10% (see page 36)   ...   .   = 10·0000
   P.V. of £1 at 10% receivable in 43 years (see page 102)   ...   =   0·0166002

   10 × 0·0166002 (see page 84)   = 0·166002

OR

(2) Subtract the Years' Purchase (according to the Single Rate % Principle) for the period intervening before the income commences from the Years' Purchase for Perpetuity.

Example. (Same as in Rule 1.)
   Y.P. for Perp. at 10%   ...   . (see page 36)   =   10·0000
   Y.P. for 43 yrs. at 10%   ...   . (see page 46)   =   9·8340

   Result (see page 84)   0·1660

*Note:—*Using the figures given in the tables the first approach will give an answer correct to more decimal places.

PRESENT VALUE OF ONE POUND

Receivable at the expiration of a given
Number of Years at Rates of Interest ranging
from
2% to 30%
(*i.e.*, the amount which must be invested now in order to accumulate
to £1 at Compound Interest)

Note:—No allowance has been made for the effect of income tax on interest
accumulations.

PRESENT VALUE OF £1

			Rate Per Cent			
Yrs.	2	2.25	2.5	2.75	3	Yrs.
1	.9803922	.9779951	.9756098	.9732360	.9708738	1
2	.9611688	.9564744	.9518144	.9471883	.9425959	2
3	.9423223	.9354273	.9285994	.9218378	.9151417	3
4	.9238454	.9148433	.9059506	.8971657	.8884870	4
5	.9057308	.8947123	.8838543	.8731540	.8626088	5
6	.8879714	.8750243	.8622969	.8497849	.8374843	6
7	.8705602	.8557695	.8412652	.8270413	.8130915	7
8	.8534904	.8369383	.8207466	.8049064	.7894092	8
9	.8367553	.8185216	.8007284	.7833638	.7664167	9
10	.8203483	.8005101	.7811984	.7623979	.7440939	10
11	.8042630	.7828950	.7621448	.7419931	.7224213	11
12	.7884932	.7656675	.7435559	.7221344	.7013799	12
13	.7730325	.7488190	.7254204	.7028072	.6809513	13
14	.7578750	.7323414	.7077272	.6839973	.6611178	14
15	.7430147	.7162263	.6904656	.6656908	.6418619	15
16	.7284458	.7004658	.6736249	.6478742	.6231669	16
17	.7141626	.6850521	.6571951	.6305345	.6050164	17
18	.7001594	.6699776	.6411659	.6136589	.5873946	18
19	.6864308	.6552348	.6255277	.5972350	.5702860	19
20	.6729713	.6408165	.6102709	.5812506	.5536758	20
21	.6597758	.6267154	.5953863	.5656940	.5375493	21
22	.6468390	.6129246	.5808647	.5505538	.5218925	22
23	.6341559	.5994372	.5666972	.5358187	.5066917	23
24	.6217215	.5862467	.5528754	.5214781	.4919337	24
25	.6095309	.5733464	.5393906	.5075213	.4776056	25
26	.5975793	.5607300	.5262347	.4939380	.4636947	26
27	.5858620	.5483912	.5133997	.4807182	.4501891	27
28	.5743746	.5363239	.5008778	.4678523	.4370768	28
29	.5631123	.5245221	.4886613	.4553307	.4243464	29
30	.5520709	.5129801	.4767427	.4431442	.4119868	30
31	.5412460	.5016920	.4651148	.4312839	.3999871	31
32	.5306333	.4906523	.4537706	.4197410	.3883370	32
33	.5202287	.4798556	.4427030	.4085071	.3770262	33
34	.5100282	.4692964	.4319053	.3975738	.3660449	34
35	.5000276	.4589696	.4213711	.3869331	.3553834	35
36	.4902232	.4488700	.4110937	.3765773	.3450324	36
37	.4806109	.4389927	.4010670	.3664986	.3349829	37
38	.4711872	.4293327	.3912849	.3566896	.3252262	38
39	.4619482	.4198853	.3817414	.3471432	.3157535	39
40	.4528904	.4106458	.3724306	.3378522	.3065568	40
41	.4440102	.4016095	.3633469	.3288099	.2976280	41
42	.4353041	.3927722	.3544848	.3200097	.2889592	42
43	.4267688	.3841293	.3458389	.3114449	.2805429	43
44	.4184007	.3756765	.3374038	.3031094	.2723718	44
45	.4101968	.3674098	.3291744	.2949970	.2644386	45
46	.4021537	.3593250	.3211458	.2871017	.2567365	46
47	.3942684	.3514181	.3133129	.2794177	.2492588	47
48	.3865376	.3436852	.3056712	.2719394	.2419988	48
49	.3789584	.3361224	.2982158	.2646612	.2349503	49
50	.3715279	.3287261	.2909422	.2575778	.2281071	50

PRESENT VALUE OF £1

			Rate Per Cent			
Yrs.	2	2.25	2.5	2.75	3	Yrs.
51	.3642430	.3214925	.2838461	.2506840	.2214632	51
52	.3571010	.3144181	.2769230	.2439747	.2150128	52
53	.3500990	.3074994	.2701688	.2374450	.2087503	53
54	.3432343	.3007329	.2635793	.2310900	.2026702	54
55	.3365042	.2941153	.2571505	.2249051	.1967672	55
56	.3299061	.2876433	.2508786	.2188858	.1910361	56
57	.3234374	.2813137	.2447596	.2130275	.1854719	57
58	.3170955	.2751235	.2387898	.2073260	.1800698	58
59	.3108779	.2690694	.2329657	.2017772	.1748251	59
60	.3047823	.2631486	.2272836	.1963768	.1697331	60
61	.2988061	.2573580	.2217401	.1911210	.1647894	61
62	.2929472	.2516949	.2163318	.1860058	.1599897	62
63	.2872031	.2461564	.2110554	.1810276	.1553298	63
64	.2815717	.2407397	.2059077	.1761825	.1508057	64
65	.2760507	.2354423	.2008856	.1714672	.1464133	65
66	.2706379	.2302614	.1959859	.1668780	.1421488	66
67	.2653313	.2251945	.1912058	.1624117	.1380085	67
68	.2601287	.2202391	.1865422	.1580649	.1339889	68
69	.2550282	.2153928	.1819924	.1538345	.1300863	69
70	.2500276	.2106531	.1775536	.1497173	.1262974	70
71	.2451251	.2060177	.1732230	.1457102	.1226188	71
72	.2403187	.2014843	.1689980	.1418104	.1190474	72
73	.2356066	.1970507	.1648761	.1380150	.1155800	73
74	.2309869	.1927146	.1608548	.1343212	.1122136	74
75	.2264577	.1884739	.1569315	.1307262	.1089452	75
76	.2220174	.1843266	.1531039	.1272275	.1057721	76
77	.2176641	.1802705	.1493697	.1238224	.1026913	77
78	.2133962	.1763036	.1457265	.1205084	.0997003	78
79	.2092119	.1724241	.1421722	.1172831	.0967964	79
80	.2051097	.1686299	.1387046	.1141441	.0939771	80
81	.2010880	.1649192	.1353215	.1110892	.0912399	81
82	.1971451	.1612902	.1320210	.1081160	.0885824	82
83	.1932795	.1577410	.1288010	.1052224	.0860024	83
84	.1894897	.1542700	.1256595	.1024062	.0834974	84
85	.1857742	.1508753	.1225946	.0996654	.0810655	85
86	.1821316	.1475553	.1196045	.0969980	.0787043	86
87	.1785604	.1443083	.1166873	.0944019	.0764120	87
88	.1750592	.1411329	.1138413	.0918753	.0741864	88
89	.1716266	.1380272	.1110647	.0894164	.0720256	89
90	.1682614	.1349900	.1083558	.0870232	.0699278	90
91	.1649622	.1320195	.1057130	.0846942	.0678911	91
92	.1617276	.1291145	.1031346	.0824274	.0659136	92
93	.1585565	.1262733	.1006191	.0802213	.0639938	93
94	.1554475	.1234947	.0981650	.0780743	.0621299	94
95	.1523995	.1207772	.0957707	.0759847	.0603203	95
96	.1494113	.1181195	.0934349	.0739510	.0585634	96
97	.1464817	.1155203	.0911560	.0719718	.0568577	97
98	.1436095	.1129783	.0889326	.0700456	.0552016	98
99	.1407936	.1104922	.0867636	.0681709	.0535938	99
100	.1380330	.1080608	.0846474	.0663463	.0520328	100

PRESENT VALUE OF £1

			Rate Per Cent			
Yrs.	3.25	3.5	3.75	4	4.25	Yrs.
1	.9685230	.9661836	.9638554	.9615385	.9592326	1
2	.9380368	.9335107	.9290173	.9245562	.9201272	2
3	.9085102	.9019427	.8954383	.8889964	.8826160	3
4	.8799130	.8714422	.8630731	.8548042	.8466341	4
5	.8522160	.8419732	.8318777	.8219271	.8121190	5
6	.8253908	.8135006	.8018098	.7903145	.7790111	6
7	.7994100	.7859910	.7728287	.7599178	.7472528	7
8	.7742470	.7594116	.7448952	.7306902	.7167893	8
9	.7498760	.7337310	.7179712	.7025867	.6875676	9
10	.7262722	.7089188	.6920205	.6755642	.6595373	10
11	.7034113	.6849457	.6670077	.6495809	.6326497	11
12	.6812700	.6617833	.6428990	.6245970	.6068582	12
13	.6598257	.6394042	.6196617	.6005741	.5821182	13
14	.6390564	.6177818	.5972643	.5774751	.5583868	14
15	.6189408	.5968906	.5756764	.5552645	.5356228	15
16	.5994584	.5767059	.5548688	.5339082	.5137868	16
17	.5805892	.5572038	.5348133	.5133732	.4928411	17
18	.5623140	.5383611	.5154827	.4936281	.4727493	18
19	.5446141	.5201557	.4968508	.4746424	.4534765	19
20	.5274713	.5025659	.4788923	.4563869	.4349895	20
21	.5108680	.4855709	.4615830	.4388336	.4172561	21
22	.4947874	.4691506	.4448993	.4219554	.4002456	22
23	.4792130	.4532856	.4288186	.4057263	.3839287	23
24	.4641288	.4379571	.4133191	.3901215	.3682769	24
25	.4495195	.4231470	.3983799	.3751168	.3532632	25
26	.4353699	.4088377	.3839806	.3606892	.3388616	26
27	.4216658	.3950122	.3701018	.3468166	.3250471	27
28	.4083930	.3816543	.3567246	.3334775	.3117958	28
29	.3955380	.3687482	.3438309	.3206514	.2990847	29
30	.3830877	.3562784	.3314033	.3083187	.2868918	30
31	.3710292	.3442303	.3194249	.2964603	.2751959	31
32	.3593503	.3325897	.3078794	.2850579	.2639769	32
33	.3480391	.3213427	.2967512	.2740942	.2532153	33
34	.3370839	.3104761	.2860253	.2635521	.2428923	34
35	.3264735	.2999769	.2756870	.2534155	.2329903	35
36	.3161971	.2898327	.2657224	.2436687	.2234919	36
37	.3062441	.2800316	.2561180	.2342968	.2143807	37
38	.2966045	.2705619	.2468607	.2252854	.2056409	38
39	.2872683	.2614125	.2379380	.2166206	.1972575	39
40	.2782259	.2525725	.2293379	.2082890	.1892158	40
41	.2694682	.2440314	.2210486	.2002779	.1815020	41
42	.2609862	.2357791	.2130588	.1925749	.1741026	42
43	.2527711	.2278059	.2053579	.1851682	.1670049	43
44	.2448146	.2201023	.1979353	.1780463	.1601966	44
45	.2371086	.2126592	.1907811	.1711984	.1536658	45
46	.2296451	.2054679	.1838854	.1646139	.1474012	46
47	.2224166	.1985197	.1772389	.1582826	.1413921	47
48	.2154156	.1918065	.1708327	.1521948	.1356279	48
49	.2086349	.1853202	.1646580	.1463411	.1300987	49
50	.2020677	.1790534	.1587065	.1407126	.1247949	50

PRESENT VALUE OF £1

			Rate Per Cent			
Yrs.	**3.25**	**3.5**	**3.75**	**4**	**4.25**	Yrs.
51	.1957073	.1729984	.1529701	.1353006	.1197073	**51**
52	.1895470	.1671482	.1474411	.1300967	.1148272	**52**
53	.1835806	.1614959	.1421119	.1250930	.1101460	**53**
54	.1778020	.1560347	.1369753	.1202817	.1056556	**54**
55	.1722054	.1507581	.1320244	.1156555	.1013483	**55**
56	.1667849	.1456600	.1272524	.1112072	.0972166	**56**
57	.1615350	.1407343	.1226529	.1069300	.0932533	**57**
58	.1564503	.1359752	.1182197	.1028173	.0894516	**58**
59	.1515257	.1313770	.1139467	.0988628	.0858049	**59**
60	.1467562	.1269343	.1098282	.0950604	.0823069	**60**
61	.1421367	.1226418	.1058585	.0914042	.0789515	**61**
62	.1376627	.1184945	.1020323	.0878887	.0757328	**62**
63	.1333295	.1144875	.0983443	.0845084	.0726454	**63**
64	.1291327	.1106159	.0947897	.0812580	.0696838	**64**
65	.1250680	.1068753	.0913636	.0781327	.0668430	**65**
66	.1211312	.1032611	.0880613	.0751276	.0641180	**66**
67	.1173184	.0997692	.0848784	.0722381	.0615041	**67**
68	.1136255	.0963954	.0818105	.0694597	.0589967	**68**
69	.1100489	.0931356	.0788535	.0667882	.0565916	**69**
70	.1065849	.0899861	.0760033	.0642194	.0542845	**70**
71	.1032299	.0869431	.0732562	.0617494	.0520714	**71**
72	.0999806	.0840030	.0706084	.0593744	.0499486	**72**
73	.0968335	.0811623	.0680563	.0570908	.0479123	**73**
74	.0937855	.0784177	.0655964	.0548950	.0459591	**74**
75	.0908334	.0757659	.0632255	.0527837	.0440854	**75**
76	.0879742	.0732038	.0609402	.0507535	.0422882	**76**
77	.0852051	.0707283	.0587376	.0488015	.0405642	**77**
78	.0825231	.0683365	.0566145	.0469245	.0389105	**78**
79	.0799255	.0660256	.0545682	.0451197	.0373242	**79**
80	.0774097	.0637929	.0525959	.0433843	.0358026	**80**
81	.0749730	.0616356	.0506948	.0417157	.0343430	**81**
82	.0726131	.0595513	.0488625	.0401112	.0329430	**82**
83	.0703275	.0575375	.0470964	.0385685	.0316000	**83**
84	.0681138	.0555918	.0453941	.0370851	.0303117	**84**
85	.0659698	.0537119	.0437533	.0356588	.0290760	**85**
86	.0638932	.0518955	.0421719	.0342873	.0278906	**86**
87	.0618821	.0501406	.0406476	.0329685	.0267536	**87**
88	.0599342	.0484450	.0391784	.0317005	.0256629	**88**
89	.0580476	.0468068	.0377623	.0304813	.0246167	**89**
90	.0562205	.0452240	.0363974	.0293089	.0236132	**90**
91	.0544508	.0436946	.0350818	.0281816	.0226505	**91**
92	.0527369	.0422170	.0338138	.0270977	.0217271	**92**
93	.0510769	.0407894	.0325916	.0260555	.0208414	**93**
94	.0494691	.0394101	.0314136	.0250534	.0199917	**94**
95	.0479120	.0380774	.0302782	.0240898	.0191767	**95**
96	.0464039	.0367897	.0291838	.0231632	.0183949	**96**
97	.0449432	.0355456	.0281290	.0222724	.0176450	**97**
98	.0435285	.0343436	.0271123	.0214157	.0169257	**98**
99	.0421584	.0331822	.0261323	.0205920	.0162357	**99**
100	.0408314	.0320601	.0251878	.0198000	.0155738	**100**

PRESENT VALUE OF £1

	Rate Per Cent					
Yrs.	4.5	4.75	5	5.25	5.5	Yrs.
1	.9569378	.9546539	.9523810	.9501188	.9478673	1
2	.9157300	.9113641	.9070295	.9027257	.8984524	2
3	.8762966	.8700374	.8638376	.8576966	.8516137	3
4	.8385613	.8305846	.8227025	.8149136	.8072167	4
5	.8024510	.7929209	.7835262	.7742647	.7651344	5
6	.7678957	.7569650	.7462154	.7356435	.7252458	6
7	.7348285	.7226396	.7106813	.6989486	.6874368	7
8	.7031851	.6898708	.6768394	.6640842	.6515989	8
9	.6729044	.6585879	.6446089	.6309589	.6176293	9
10	.6439277	.6287235	.6139133	.5994859	.5854306	10
11	.6161987	.6002134	.5846793	.5695828	.5549105	11
12	.5896639	.5729960	.5568374	.5411713	.5259815	12
13	.5642716	.5470129	.5303214	.5141770	.4985607	13
14	.5399729	.5222080	.5050680	.4885292	.4725694	14
15	.5167204	.4985280	.4810171	.4641608	.4479330	15
16	.4944693	.4759217	.4581115	.4410079	.4245811	16
17	.4731764	.4543405	.4362967	.4190098	.4024465	17
18	.4528004	.4337380	.4155207	.3981091	.3814659	18
19	.4333018	.4140696	.3957340	.3782509	.3615791	19
20	.4146429	.3952932	.3768895	.3593833	.3427290	20
21	.3967874	.3773682	.3589424	.3414568	.3248616	21
22	.3797009	.3602561	.3418499	.3244245	.3079257	22
23	.3633501	.3439199	.3255713	.3082418	.2918727	23
24	.3477035	.3283245	.3100679	.2928664	.2766566	24
25	.3327306	.3134362	.2953028	.2782578	.2622337	25
26	.3184025	.2992231	.2812407	.2643780	.2485628	26
27	.3046914	.2856545	.2678483	.2511905	.2356045	27
28	.2915707	.2727012	.2550936	.2386608	.2233218	28
29	.2790150	.2603353	.2429463	.2267561	.2116794	29
30	.2670000	.2485301	.2313774	.2154452	.2006440	30
31	.2555024	.2372603	.2203595	.2046985	.1901839	31
32	.2444999	.2265015	.2098662	.1944879	.1802691	32
33	.2339712	.2162305	.1998725	.1847866	.1708712	33
34	.2238959	.2064253	.1903548	.1755692	.1619632	34
35	.2142544	.1970647	.1812903	.1668116	.1535196	35
36	.2050282	.1881286	.1726574	.1584909	.1455162	36
37	.1961992	.1795977	.1644356	.1505851	.1379301	37
38	.1877504	.1714537	.1566054	.1430738	.1307394	38
39	.1796655	.1636789	.1491480	.1359371	.1239236	39
40	.1719287	.1562567	.1420457	.1291564	.1174631	40
41	.1645251	.1491711	.1352816	.1227139	.1113395	41
42	.1574403	.1424068	.1288396	.1165928	.1055350	42
43	.1506605	.1359492	.1227044	.1107770	.1000332	43
44	.1441728	.1297844	.1168613	.1052513	.0948182	44
45	.1379644	.1238992	.1112965	.1000012	.0898751	45
46	.1320233	.1182809	.1059967	.0950130	.0851897	46
47	.1263381	.1129173	.1009492	.0902737	.0807485	47
48	.1208977	.1077970	.0961421	.0857707	.0765389	48
49	.1156916	.1029088	.0915639	.0814924	.0725487	49
50	.1107096	.0982423	.0872037	.0774274	.0687665	50

PRESENT VALUE OF £1

			Rate Per Cent			
Yrs.	4.5	4.75	5	5.25	5.5	Yrs.
51	.1059422	.0937874	.0830512	.0735652	.0651815	51
52	.1013801	.0895345	.0790964	.0698957	.0617834	52
53	.0970145	.0854745	.0753299	.0664092	.0585625	53
54	.0928368	.0815985	.0717427	.0630967	.0555095	54
55	.0888391	.0778984	.0683264	.0599493	.0526156	55
56	.0850135	.0743660	.0650728	.0569590	.0498726	56
57	.0813526	.0709938	.0619741	.0541178	.0472726	57
58	.0778494	.0677745	.0590229	.0514183	.0448082	58
59	.0744970	.0647012	.0562123	.0488535	.0424722	59
60	.0712890	.0617672	.0535355	.0464166	.0402580	60
61	.0682191	.0589663	.0509862	.0441013	.0381593	61
62	.0652815	.0562924	.0485583	.0419015	.0361699	62
63	.0624703	.0537398	.0462460	.0398114	.0342843	63
64	.0597802	.0513029	.0440438	.0378256	.0324969	64
65	.0572059	.0489765	.0419465	.0359388	.0308028	65
66	.0547425	.0467556	.0399490	.0341461	.0291970	66
67	.0523852	.0446354	.0380467	.0324428	.0276748	67
68	.0501294	.0426114	.0362349	.0308246	.0262321	68
69	.0479707	.0406791	.0345095	.0292870	.0248645	69
70	.0459050	.0388345	.0328662	.0278261	.0235683	70
71	.0439282	.0370735	.0313011	.0264381	.0223396	71
72	.0420366	.0353924	.0298106	.0251194	.0211750	72
73	.0402264	.0337875	.0283910	.0238664	.0200711	73
74	.0384941	.0322553	.0270391	.0226759	.0190247	74
75	.0368365	.0307927	.0257515	.0215448	.0180329	75
76	.0352502	.0293964	.0245252	.0204701	.0170928	76
77	.0337323	.0280634	.0233574	.0194490	.0162017	77
78	.0322797	.0267908	.0222451	.0184789	.0153571	78
79	.0308897	.0255759	.0211858	.0175571	.0145565	79
80	.0295595	.0244162	.0201770	.0166814	.0137976	80
81	.0282866	.0233090	.0192162	.0158493	.0130783	81
82	.0270685	.0222520	.0183011	.0150587	.0123965	82
83	.0259029	.0212430	.0174296	.0143076	.0117502	83
84	.0247874	.0202797	.0165996	.0135939	.0111376	84
85	.0237200	.0193601	.0158092	.0129158	.0105570	85
86	.0226986	.0184822	.0150564	.0122715	.0100066	86
87	.0217211	.0176441	.0143394	.0116594	.0094850	87
88	.0207858	.0168440	.0136566	.0110778	.0089905	88
89	.0198907	.0160802	.0130063	.0105253	.0085218	89
90	.0190342	.0153510	.0123869	.0100002	.0080775	90
91	.0182145	.0146549	.0117971	.0095014	.0076564	91
92	.0174302	.0139904	.0112353	.0090275	.0072573	92
93	.0166796	.0133560	.0107003	.0085772	.0068789	93
94	.0159613	.0127503	.0101907	.0081493	.0065203	94
95	.0152740	.0121721	.0097055	.0077428	.0061804	95
96	.0146163	.0116202	.0092433	.0073566	.0058582	96
97	.0139868	.0110933	.0088031	.0069897	.0055528	97
98	.0133845	.0105902	.0083840	.0066410	.0052633	98
99	.0128082	.0101100	.0079847	.0063097	.0049889	99
100	.0122566	.0096515	.0076045	.0059950	.0047288	100

PRESENT VALUE OF £1

			Rate Per Cent			
Yrs.	5.75	6	6.25	6.5	6.75	Yrs.
1	.9456265	.9433962	.9411765	.9389671	.9367681	1
2	.8942094	.8899964	.8858131	.8816593	.8775346	2
3	.8455881	.8396193	.8337065	.8278491	.8220464	3
4	.7996105	.7920937	.7846649	.7773231	.7700669	4
5	.7561329	.7472582	.7385082	.7298808	.7213742	5
6	.7150193	.7049605	.6950665	.6853341	.6757603	6
7	.6761411	.6650571	.6541803	.6435062	.6330308	7
8	.6393770	.6274124	.6156991	.6042312	.5930031	8
9	.6046118	.5918985	.5794815	.5673532	.5555064	9
10	.5717369	.5583948	.5453943	.5327260	.5203807	10
11	.5406496	.5267875	.5133123	.5002122	.4874760	11
12	.5112526	.4969694	.4831175	.4696829	.4566520	12
13	.4834539	.4688390	.4546988	.4410168	.4277771	13
14	.4571669	.4423010	.4279518	.4141002	.4007279	14
15	.4323091	.4172651	.4027782	.3888265	.3753892	15
16	.4088029	.3936463	.3790853	.3650953	.3516526	16
17	.3865749	.3713644	.3567862	.3428125	.3294170	17
18	.3655554	.3503438	.3357988	.3218897	.3085873	18
19	.3456789	.3305130	.3160459	.3022438	.2890748	19
20	.3268831	.3118047	.2974550	.2837970	.2707961	20
21	.3091093	.2941554	.2799576	.2664761	.2536731	21
22	.2923020	.2775051	.2634895	.2502123	.2376329	22
23	.2764085	.2617973	.2479901	.2349411	.2226069	23
24	.2613792	.2469785	.2334025	.2206020	.2085311	24
25	.2471671	.2329986	.2196729	.2071380	.1953453	25
26	.2337277	.2198100	.2067510	.1944958	.1829932	26
27	.2210191	.2073680	.1945892	.1826252	.1714222	27
28	.2090015	.1956301	.1831427	.1714790	.1605829	28
29	.1976374	.1845567	.1723696	.1610132	.1504289	29
30	.1868911	.1741101	.1622303	.1511861	.1409170	30
31	.1767292	.1642548	.1526873	.1419587	.1320066	31
32	.1671198	.1549574	.1437057	.1332946	.1236596	32
33	.1580329	.1461862	.1352524	.1251592	.1158403	33
34	.1494401	.1379115	.1272964	.1175204	.1085155	34
35	.1413145	.1301052	.1198084	.1103478	.1016539	35
36	.1336308	.1227408	.1127608	.1036130	.0952261	36
37	.1263648	.1157932	.1061278	.0972892	.0892048	37
38	.1194939	.1092389	.0998850	.0913513	.0835642	38
39	.1129966	.1030555	.0940094	.0857759	.0782803	39
40	.1068526	.0972222	.0884795	.0805408	.0733305	40
41	.1010426	.0917190	.0832748	.0756251	.0686937	41
42	.0955486	.0865274	.0783763	.0710095	.0643500	42
43	.0903533	.0816296	.0737659	.0666756	.0602811	43
44	.0854404	.0770091	.0694267	.0626062	.0564694	44
45	.0807947	.0726501	.0653428	.0587852	.0528987	45
46	.0764016	.0685378	.0614991	.0551973	.0495538	46
47	.0722474	.0646583	.0578815	.0518285	.0464205	47
48	.0683191	.0609984	.0544767	.0486652	.0434852	48
49	.0646043	.0575457	.0512722	.0456951	.0407356	49
50	.0610916	.0542884	.0482562	.0429062	.0381598	50

PRESENT VALUE OF £1

			Rate Per Cent			
Yrs.	5.75	6	6.25	6.5	6.75	Yrs.
51	.0577698	.0512154	.0454176	.0402875	.0357469	51
52	.0546286	.0483164	.0427460	.0378286	.0334865	52
53	.0516583	.0455816	.0402315	.0355198	.0313691	53
54	.0488495	.0430015	.0378649	.0333519	.0293856	54
55	.0461933	.0405674	.0356376	.0313164	.0275275	55
56	.0436816	.0382712	.0335413	.0294051	.0257869	56
57	.0413065	.0361049	.0315682	.0276104	.0241563	57
58	.0390605	.0340612	.0297113	.0259252	.0226289	58
59	.0369367	.0321332	.0279636	.0243429	.0211980	59
60	.0349283	.0303143	.0263187	.0228572	.0198576	60
61	.0330291	.0285984	.0247705	.0214622	.0186020	61
62	.0312332	.0269797	.0233134	.0201523	.0174257	62
63	.0295350	.0254525	.0219420	.0189223	.0163239	63
64	.0279290	.0240118	.0206513	.0177675	.0152917	64
65	.0264104	.0226526	.0194365	.0166831	.0143248	65
66	.0249744	.0213704	.0182932	.0156648	.0134190	66
67	.0236165	.0201608	.0172171	.0147088	.0125705	67
68	.0223324	.0190196	.0162044	.0138110	.0117756	68
69	.0211181	.0179430	.0152512	.0129681	.0110310	69
70	.0199698	.0169274	.0143540	.0121766	.0103335	70
71	.0188840	.0159692	.0135097	.0114335	.0096801	71
72	.0178572	.0150653	.0127150	.0107356	.0090680	72
73	.0168862	.0142125	.0119671	.0100804	.0084946	73
74	.0159681	.0134081	.0112631	.0094652	.0079575	74
75	.0150998	.0126491	.0106006	.0088875	.0074543	75
76	.0142788	.0119331	.0099770	.0083451	.0069830	76
77	.0135024	.0112577	.0093901	.0078357	.0065414	77
78	.0127682	.0106204	.0088378	.0073575	.0061278	78
79	.0120740	.0100193	.0083179	.0069085	.0057403	79
80	.0114175	.0094522	.0078286	.0064868	.0053774	80
81	.0107967	.0089171	.0073681	.0060909	.0050373	81
82	.0102096	.0084124	.0069347	.0057192	.0047188	82
83	.0096545	.0079362	.0065268	.0053701	.0044204	83
84	.0091295	.0074870	.0061428	.0050423	.0041409	84
85	.0086331	.0070632	.0057815	.0047346	.0038791	85
86	.0081637	.0066634	.0054414	.0044456	.0036338	86
87	.0077198	.0062862	.0051213	.0041743	.0034040	87
88	.0073001	.0059304	.0048201	.0039195	.0031888	88
89	.0069031	.0055947	.0045365	.0036803	.0029872	89
90	.0065278	.0052780	.0042697	.0034557	.0027983	90
91	.0061729	.0049793	.0040185	.0032448	.0026213	91
92	.0058372	.0046974	.0037821	.0030467	.0024556	92
93	.0055198	.0044315	.0035597	.0028608	.0023003	93
94	.0052197	.0041807	.0033503	.0026862	.0021549	94
95	.0049359	.0039441	.0031532	.0025222	.0020186	95
96	.0046675	.0037208	.0029677	.0023683	.0018910	96
97	.0044137	.0035102	.0027931	.0022238	.0017714	97
98	.0041737	.0033115	.0026288	.0020880	.0016594	98
99	.0039468	.0031241	.0024742	.0019606	.0015545	99
100	.0037322	.0029472	.0023287	.0018409	.0014562	100

PRESENT VALUE OF £1

			Rate Per Cent			
Yrs.	7	7.25	7.5	7.75	8	Yrs.
1	.9345794	.9324009	.9302326	.9280742	.9259259	1
2	.8734387	.8693715	.8653326	.8613218	.8573388	2
3	.8162979	.8106028	.8049606	.7993706	.7938322	3
4	.7628952	.7558068	.7488005	.7418753	.7350299	4
5	.7129862	.7047150	.6965586	.6885153	.6805832	5
6	.6663422	.6570769	.6479615	.6389933	.6301696	6
7	.6227497	.6126591	.6027549	.5930333	.5834904	7
8	.5820091	.5712439	.5607022	.5503789	.5402689	8
9	.5439337	.5326284	.5215835	.5107925	.5002490	9
10	.5083493	.4966232	.4851939	.4740533	.4631935	10
11	.4750928	.4630519	.4513432	.4399567	.4288829	11
12	.4440120	.4317500	.4198541	.4083125	.3971138	12
13	.4149644	.4025641	.3905620	.3789443	.3676979	13
14	.3878172	.3753512	.3633135	.3516884	.3404610	14
15	.3624460	.3499778	.3379660	.3263930	.3152417	15
16	.3387346	.3263196	.3143870	.3029169	.2918905	16
17	.3165744	.3042607	.2924530	.2811294	.2702690	17
18	.2958639	.2836930	.2720493	.2609090	.2502490	18
19	.2765083	.2645156	.2530691	.2421429	.2317121	19
20	.2584190	.2466346	.2354131	.2247266	.2145482	20
21	.2415131	.2299623	.2189890	.2085629	.1986557	21
22	.2257132	.2144171	.2037107	.1935619	.1839405	22
23	.2109469	.1999227	.1894983	.1796398	.1703153	23
24	.1971466	.1864081	.1762775	.1667191	.1576993	24
25	.1842492	.1738071	.1639791	.1547277	.1460179	25
26	.1721955	.1620579	.1525387	.1435988	.1352018	26
27	.1609304	.1511029	.1418964	.1332703	.1251868	27
28	.1504022	.1408885	.1319967	.1236848	.1159137	28
29	.1405628	.1313646	.1227876	.1147886	.1073275	29
30	.1313671	.1224845	.1142210	.1065324	.0993773	30
31	.1227730	.1142046	.1062521	.0988700	.0920160	31
32	.1147411	.1064845	.0988392	.0917587	.0852000	32
33	.1072347	.0992862	.0919434	.0851589	.0788889	33
34	.1002193	.0925746	.0855288	.0790337	.0730453	34
35	.0936629	.0863166	.0795616	.0733492	.0676345	35
36	.0875355	.0804817	.0740108	.0680735	.0626246	36
37	.0818088	.0750412	.0688473	.0631772	.0579857	37
38	.0764569	.0699685	.0640440	.0586332	.0536905	38
39	.0714550	.0652387	.0595758	.0544159	.0497134	39
40	.0667804	.0608286	.0554194	.0505020	.0460309	40
41	.0624116	.0567167	.0515529	.0468696	.0426212	41
42	.0583286	.0528827	.0479562	.0434985	.0394641	42
43	.0545127	.0493078	.0446104	.0403698	.0365408	43
44	.0509464	.0459747	.0414980	.0374662	.0338341	44
45	.0476135	.0428668	.0386028	.0347714	.0313279	45
46	.0444986	.0399691	.0359096	.0322705	.0290073	46
47	.0415875	.0372672	.0334043	.0299494	.0268586	47
48	.0388668	.0347480	.0310738	.0277953	.0248691	48
49	.0363241	.0323990	.0289058	.0257961	.0230269	49
50	.0339478	.0302089	.0268891	.0239407	.0213212	50

PRESENT VALUE OF £1

			Rate Per Cent			
Yrs.	7	7.25	7.5	7.75	8	Yrs.
51	.0317269	.0281668	.0250131	.0222187	.0197419	51
52	.0296513	.0262628	.0232680	.0206206	.0182795	52
53	.0277115	.0244874	.0216447	.0191375	.0169255	53
54	.0258986	.0228321	.0201346	.0177610	.0156717	54
55	.0242043	.0212887	.0187299	.0164835	.0145109	55
56	.0226208	.0198496	.0174231	.0152979	.0134360	56
57	.0211410	.0185078	.0162076	.0141976	.0124407	57
58	.0197579	.0172567	.0150768	.0131764	.0115192	58
59	.0184653	.0160901	.0140249	.0122287	.0106659	59
60	.0172573	.0150024	.0130464	.0113491	.0098759	60
61	.0161283	.0139883	.0121362	.0105329	.0091443	61
62	.0150732	.0130427	.0112895	.0097753	.0084670	62
63	.0140871	.0121610	.0105019	.0090722	.0078398	63
64	.0131655	.0113389	.0097692	.0084197	.0072590	64
65	.0123042	.0105724	.0090876	.0078141	.0067213	65
66	.0114993	.0098578	.0084536	.0072520	.0062235	66
67	.0107470	.0091914	.0078638	.0067304	.0057625	67
68	.0100439	.0085701	.0073152	.0062463	.0053356	68
69	.0093868	.0079907	.0068048	.0057971	.0049404	69
70	.0087727	.0074506	.0063301	.0053801	.0045744	70
71	.0081988	.0069469	.0058884	.0049931	.0042356	71
72	.0076625	.0064773	.0054776	.0046340	.0039218	72
73	.0071612	.0060394	.0050954	.0043007	.0036313	73
74	.0066927	.0056312	.0047399	.0039914	.0033623	74
75	.0062548	.0052505	.0044093	.0037043	.0031133	75
76	.0058457	.0048956	.0041016	.0034378	.0028827	76
77	.0054632	.0045647	.0038155	.0031906	.0026691	77
78	.0051058	.0042561	.0035493	.0029611	.0024714	78
79	.0047718	.0039684	.0033017	.0027481	.0022884	79
80	.0044596	.0037001	.0030713	.0025505	.0021188	80
81	.0041679	.0034500	.0028570	.0023670	.0019619	81
82	.0038952	.0032168	.0026577	.0021968	.0018166	82
83	.0036404	.0029993	.0024723	.0020388	.0016820	83
84	.0034022	.0027966	.0022998	.0018921	.0015574	84
85	.0031796	.0026075	.0021393	.0017560	.0014421	85
86	.0029716	.0024313	.0019901	.0016297	.0013352	86
87	.0027772	.0022669	.0018512	.0015125	.0012363	87
88	.0025955	.0021137	.0017221	.0014037	.0011447	88
89	.0024257	.0019708	.0016019	.0013028	.0010600	89
90	.0022670	.0018376	.0014902	.0012091	.0009814	90
91	.0021187	.0017133	.0013862	.0011221	.0009087	91
92	.0019801	.0015975	.0012895	.0010414	.0008414	92
93	.0018506	.0014895	.0011995	.0009665	.0007791	93
94	.0017295	.0013888	.0011158	.0008970	.0007214	94
95	.0016164	.0012950	.0010380	.0008325	.0006679	95
96	.0015106	.0012074	.0009656	.0007726	.0006185	96
97	.0014118	.0011258	.0008982	.0007170	.0005727	97
98	.0013194	.0010497	.0008355	.0006654	.0005302	98
99	.0012331	.0009787	.0007773	.0006176	.0004910	99
100	.0011525	.0009126	.0007230	.0005732	.0004546	100

PRESENT VALUE OF £1

			Rate Per Cent			
Yrs.	8.25	8.5	8.75	9	9.25	Yrs.
1	.9237875	.9216590	.9195402	.9174312	.9153318	1
2	.8533834	.8494553	.8455542	.8416800	.8378323	2
3	.7883449	.7829081	.7775211	.7721835	.7668946	3
4	.7282632	.7215743	.7149620	.7084252	.7019630	4
5	.6727605	.6650454	.6574363	.6499314	.6425291	5
6	.6214877	.6129451	.6045391	.5962673	.5881273	6
7	.5741226	.5649264	.5558980	.5470342	.5383316	7
8	.5303673	.5206694	.5111706	.5018663	.4927520	8
9	.4899467	.4798797	.4700419	.4604278	.4510316	9
10	.4526067	.4422854	.4322225	.4224108	.4128436	10
11	.4181124	.4076363	.3974460	.3875329	.3778889	11
12	.3862470	.3757017	.3654675	.3555347	.3458937	12
13	.3568102	.3462688	.3360621	.3261786	.3166075	13
14	.3296168	.3191418	.3090226	.2992465	.2898009	14
15	.3044959	.2941399	.2841587	.2745380	.2652640	15
16	.2812895	.2710967	.2612954	.2518698	.2428046	16
17	.2598517	.2498587	.2402716	.2310732	.2222468	17
18	.2400478	.2302845	.2209394	.2119937	.2034295	18
19	.2217532	.2122438	.2031627	.1944897	.1862055	19
20	.2048528	.1956164	.1868163	.1784309	.1704398	20
21	.1892405	.1802916	.1717851	.1636981	.1560090	21
22	.1748180	.1661674	.1579633	.1501817	.1428000	22
23	.1614947	.1531497	.1452536	.1377814	.1307094	23
24	.1491868	.1411518	.1335665	.1264049	.1196425	24
25	.1378169	.1300938	.1228198	.1159678	.1095125	25
26	.1273135	.1199021	.1129377	.1063925	.1002403	26
27	.1176106	.1105089	.1038508	.0976078	.0917531	27
28	.1086472	.1018515	.0954950	.0895484	.0839846	28
29	.1003670	.0938723	.0878115	.0821545	.0768738	29
30	.0927177	.0865183	.0807462	.0753711	.0703650	30
31	.0856515	.0797403	.0742494	.0691478	.0644073	31
32	.0791238	.0734934	.0682753	.0634384	.0589541	32
33	.0730936	.0677359	.0627819	.0582003	.0539625	33
34	.0675229	.0624294	.0577305	.0533948	.0493936	34
35	.0623768	.0575386	.0530855	.0489861	.0452116	35
36	.0576229	.0530310	.0488142	.0449413	.0413836	36
37	.0532314	.0488765	.0448866	.0412306	.0378797	37
38	.0491745	.0450474	.0412751	.0378262	.0346725	38
39	.0454268	.0415184	.0379541	.0347030	.0317368	39
40	.0419647	.0382658	.0349003	.0318376	.0290497	40
41	.0387664	.0352680	.0320922	.0292088	.0265901	41
42	.0358120	.0325051	.0295101	.0267971	.0243388	42
43	.0330826	.0299586	.0271357	.0245845	.0222781	43
44	.0305613	.0276116	.0249524	.0225545	.0203918	44
45	.0282322	.0254485	.0229447	.0206922	.0186653	45
46	.0260805	.0234548	.0210986	.0189837	.0170849	46
47	.0240929	.0216173	.0194010	.0174162	.0156384	47
48	.0222567	.0199238	.0178400	.0159782	.0143143	48
49	.0205605	.0183630	.0164046	.0146589	.0131023	49
50	.0189935	.0169244	.0150847	.0134485	.0119930	50

PRESENT VALUE OF £1

			Rate Per Cent			
Yrs.	8.25	8.5	8.75	9	9.25	Yrs.
51	.0175459	.0155985	.0138710	.0123381	.0109776	51
52	.0162087	.0143765	.0127549	.0113194	.0100481	52
53	.0149734	.0132502	.0117287	.0103847	.0091974	53
54	.0138323	.0122122	.0107850	.0095273	.0084186	54
55	.0127781	.0112555	.0099172	.0087406	.0077058	55
56	.0118042	.0103737	.0091193	.0080189	.0070534	56
57	.0109046	.0095610	.0083856	.0073568	.0064562	57
58	.0100735	.0088120	.0077109	.0067494	.0059096	58
59	.0093058	.0081217	.0070904	.0061921	.0054092	59
60	.0085966	.0074854	.0065199	.0056808	.0049512	60
61	.0079414	.0068990	.0059954	.0052118	.0045320	61
62	.0073362	.0063585	.0055130	.0047814	.0041483	62
63	.0067771	.0058604	.0050694	.0043866	.0037971	63
64	.0062606	.0054013	.0046615	.0040244	.0034756	64
65	.0057834	.0049781	.0042864	.0036921	.0031813	65
66	.0053427	.0045881	.0039416	.0033873	.0029120	66
67	.0049355	.0042287	.0036244	.0031076	.0026654	67
68	.0045593	.0038974	.0033328	.0028510	.0024397	68
69	.0042119	.0035921	.0030646	.0026156	.0022332	69
70	.0038909	.0033107	.0028181	.0023996	.0020441	70
71	.0035943	.0030513	.0025913	.0022015	.0018710	71
72	.0033204	.0028123	.0023828	.0020197	.0017126	72
73	.0030673	.0025920	.0021911	.0018530	.0015676	73
74	.0028336	.0023889	.0020148	.0017000	.0014349	74
75	.0026176	.0022018	.0018527	.0015596	.0013134	75
76	.0024181	.0020293	.0017036	.0014308	.0012022	76
77	.0022338	.0018703	.0015666	.0013127	.0011004	77
78	.0020636	.0017238	.0014405	.0012043	.0010072	78
79	.0019063	.0015887	.0013246	.0011049	.0009219	79
80	.0017610	.0014643	.0012180	.0010136	.0008439	80
81	.0016268	.0013496	.0011200	.0009299	.0007724	81
82	.0015028	.0012438	.0010299	.0008532	.0007070	82
83	.0013883	.0011464	.0009470	.0007827	.0006472	83
84	.0012825	.0010566	.0008708	.0007181	.0005924	84
85	.0011848	.0009738	.0008008	.0006588	.0005422	85
86	.0010945	.0008975	.0007363	.0006044	.0004963	86
87	.0010110	.0008272	.0006771	.0005545	.0004543	87
88	.0009340	.0007624	.0006226	.0005087	.0004158	88
89	.0008628	.0007027	.0005725	.0004667	.0003806	89
90	.0007971	.0006476	.0005265	.0004282	.0003484	90
91	.0007363	.0005969	.0004841	.0003928	.0003189	91
92	.0006802	.0005501	.0004452	.0003604	.0002919	92
93	.0006284	.0005070	.0004093	.0003306	.0002672	93
94	.0005805	.0004673	.0003764	.0003033	.0002446	94
95	.0005362	.0004307	.0003461	.0002783	.0002239	95
96	.0004954	.0003970	.0003183	.0002553	.0002049	96
97	.0004576	.0003659	.0002927	.0002342	.0001876	97
98	.0004227	.0003372	.0002691	.0002149	.0001717	98
99	.0003905	.0003108	.0002475	.0001971	.0001571	99
100	.0003608	.0002864	.0002275	.0001809	.0001438	100

PRESENT VALUE OF £1

			Rate Per Cent			
Yrs.	9.5	10	10.5	11	11.5	Yrs.
1	0.9132420	0.9090909	0.9049774	0.9009009	0.8968610	1
2	0.8340110	0.8264463	0.8189841	0.8116224	0.8043596	2
3	0.7616539	0.7513148	0.7411620	0.7311914	0.7213988	3
4	0.6955743	0.6830135	0.6707349	0.6587310	0.6469944	4
5	0.6352277	0.6209213	0.6069999	0.5934513	0.5802640	5
6	0.5801166	0.5644739	0.5493212	0.5346408	0.5204162	6
7	0.5297868	0.5131581	0.4971232	0.4816584	0.4667410	7
8	0.4838236	0.4665074	0.4498853	0.4339265	0.4186018	8
9	0.4418480	0.4240976	0.4071360	0.3909248	0.3754276	9
10	0.4035142	0.3855433	0.3684489	0.3521845	0.3367064	10
11	0.3685061	0.3504939	0.3334379	0.3172833	0.3019788	11
12	0.3365353	0.3186308	0.3017537	0.2858408	0.2708330	12
13	0.3073381	0.2896644	0.2730803	0.2575143	0.2428996	13
14	0.2806741	0.2633313	0.2471315	0.2319948	0.2178471	14
15	0.2563234	0.2393920	0.2236484	0.2090043	0.1953786	15
16	0.2340853	0.2176291	0.2023968	0.1882922	0.1752274	16
17	0.2137765	0.1978447	0.1831645	0.1696326	0.1571547	17
18	0.1952297	0.1798588	0.1657597	0.1528222	0.1409459	18
19	0.1782920	0.1635080	0.1500088	0.1376776	0.1264089	19
20	0.1628237	0.1486436	0.1357546	0.1240339	0.1133712	20
21	0.1486974	0.1351306	0.1228548	0.1117423	0.1016782	21
22	0.1357968	0.1228460	0.1111808	0.1006687	0.0911912	22
23	0.1240153	0.1116782	0.1006161	0.0906925	0.0817858	23
24	0.1132560	0.1015256	0.0910553	0.0817050	0.0733505	24
25	0.1034301	0.0922960	0.0824030	0.0736081	0.0657852	25
26	0.0944567	0.0839055	0.0745729	0.0663136	0.0590002	26
27	0.0862619	0.0762777	0.0674867	0.0597420	0.0529150	27
28	0.0787779	0.0693433	0.0610740	0.0538216	0.0474574	28
29	0.0719433	0.0630394	0.0552706	0.0484879	0.0425627	29
30	0.0657017	0.0573086	0.0500186	0.0436828	0.0381728	30
31	0.0600015	0.0520987	0.0452657	0.0393539	0.0342357	31
32	0.0547959	0.0473624	0.0409644	0.0354540	0.0307047	32
33	0.0500419	0.0430568	0.0370719	0.0319405	0.0275378	33
34	0.0457004	0.0391425	0.0335492	0.0287752	0.0246976	34
35	0.0417355	0.0355841	0.0303613	0.0259236	0.0221503	35
36	0.0381146	0.0323492	0.0274763	0.0233546	0.0198657	36
37	0.0348079	0.0294083	0.0248654	0.0210402	0.0178168	37
38	0.0317880	0.0267349	0.0225026	0.0189551	0.0159792	38
39	0.0290302	0.0243044	0.0203644	0.0170767	0.0143311	39
40	0.0265116	0.0220949	0.0184293	0.0153844	0.0128530	40
41	0.0242115	0.0200863	0.0166781	0.0138598	0.0115274	41
42	0.0221109	0.0182603	0.0150933	0.0124863	0.0103385	42
43	0.0201926	0.0166002	0.0136591	0.0112489	0.0092722	43
44	0.0184408	0.0150911	0.0123612	0.0101342	0.0083158	44
45	0.0168409	0.0137192	0.0111866	0.0091299	0.0074581	45
46	0.0153798	0.0124720	0.0101236	0.0082251	0.0066889	46
47	0.0140455	0.0113382	0.0091616	0.0074100	0.0059990	47
48	0.0128269	0.0103074	0.0082911	0.0066757	0.0053803	48
49	0.0117141	0.0093704	0.0075032	0.0060141	0.0048254	49
50	0.0106978	0.0085186	0.0067903	0.0054182	0.0043277	50

PRESENT VALUE OF £1

			Rate Per Cent			
Yrs.	**9.5**	**10**	**10.5**	**11**	**11.5**	Yrs.
51	0.0097697	0.0077441	0.0061450	0.0048812	0.0038813	**51**
52	0.0089221	0.0070401	0.0055611	0.0043975	0.0034810	**52**
53	0.0081480	0.0064001	0.0050327	0.0039617	0.0031220	**53**
54	0.0074411	0.0058183	0.0045545	0.0035691	0.0028000	**54**
55	0.0067955	0.0052894	0.0041217	0.0032154	0.0025112	**55**
56	0.0062060	0.0048085	0.0037300	0.0028968	0.0022522	**56**
57	0.0056675	0.0043714	0.0033756	0.0026097	0.0020199	**57**
58	0.0051758	0.0039740	0.0030548	0.0023511	0.0018116	**58**
59	0.0047268	0.0036127	0.0027646	0.0021181	0.0016247	**59**
60	0.0043167	0.0032843	0.0025019	0.0019082	0.0014572	**60**
61	0.0039422	0.0029857	0.0022641	0.0017191	0.0013069	**61**
62	0.0036002	0.0027143	0.0020490	0.0015487	0.0011721	**62**
63	0.0032878	0.0024675	0.0018543	0.0013953	0.0010512	**63**
64	0.0030026	0.0022432	0.0016781	0.0012570	0.0009428	**64**
65	0.0027421	0.0020393	0.0015186	0.0011324	0.0008455	**65**
66	0.0025042	0.0018539	0.0013743	0.0010202	0.0007583	**66**
67	0.0022869	0.0016853	0.0012437	0.0009191	0.0006801	**67**
68	0.0020885	0.0015321	0.0011256	0.0008280	0.0006100	**68**
69	0.0019073	0.0013929	0.0010186	0.0007460	0.0005471	**69**
70	0.0017419	0.0012662	0.0009218	0.0006720	0.0004906	**70**
71	0.0015907	0.0011511	0.0008342	0.0006054	0.0004400	**71**
72	0.0014527	0.0010465	0.0007549	0.0005454	0.0003946	**72**
73	0.0013267	0.0009513	0.0006832	0.0004914	0.0003539	**73**
74	0.0012116	0.0008649	0.0006183	0.0004427	0.0003174	**74**
75	0.0011065	0.0007862	0.0005595	0.0003988	0.0002847	**75**
76	0.0010105	0.0007148	0.0005064	0.0003593	0.0002553	**76**
77	0.0009228	0.0006498	0.0004583	0.0003237	0.0002290	**77**
78	0.0008427	0.0005907	0.0004147	0.0002916	0.0002054	**78**
79	0.0007696	0.0005370	0.0003753	0.0002627	0.0001842	**79**
80	0.0007029	0.0004882	0.0003396	0.0002367	0.0001652	**80**
81	0.0006419	0.0004438	0.0003074	0.0002132	0.0001482	**81**
82	0.0005862	0.0004035	0.0002782	0.0001921	0.0001329	**82**
83	0.0005353	0.0003668	0.0002517	0.0001731	0.0001192	**83**
84	0.0004889	0.0003334	0.0002278	0.0001559	0.0001069	**84**
85	0.0004465	0.0003031	0.0002062	0.0001405	0.0000959	**85**
86	0.0004077	0.0002756	0.0001866	0.0001265	0.0000860	**86**
87	0.0003724	0.0002505	0.0001688	0.0001140	0.0000771	**87**
88	0.0003401	0.0002277	0.0001528	0.0001027	0.0000692	**88**
89	0.0003106	0.0002070	0.0001383	0.0000925	0.0000620	**89**
90	0.0002836	0.0001882	0.0001251	0.0000834	0.0000556	**90**
91	0.0002590	0.0001711	0.0001132	0.0000751	0.0000499	**91**
92	0.0002365	0.0001556	0.0001025	0.0000677	0.0000447	**92**
93	0.0002160	0.0001414	0.0000927	0.0000609	0.0000401	**93**
94	0.0001973	0.0001286	0.0000839	0.0000549	0.0000360	**94**
95	0.0001802	0.0001169	0.0000760	0.0000495	0.0000323	**95**
96	0.0001645	0.0001062	0.0000687	0.0000446	0.0000289	**96**
97	0.0001503	0.0000966	0.0000622	0.0000401	0.0000260	**97**
98	0.0001372	0.0000878	0.0000563	0.0000362	0.0000233	**98**
99	0.0001253	0.0000798	0.0000509	0.0000326	0.0000209	**99**
100	0.0001144	0.0000726	0.0000461	0.0000294	0.0000187	**100**

PRESENT VALUE OF £1

			Rate Per Cent			
Yrs.	**12**	**12.5**	**13**	**13.5**	**14**	Yrs.
1	0.8928571	0.8888889	0.8849558	0.8810573	0.8771930	1
2	0.7971939	0.7901235	0.7831467	0.7762619	0.7694675	2
3	0.7117802	0.7023320	0.6930502	0.6839312	0.6749715	3
4	0.6355181	0.6242951	0.6133187	0.6025826	0.5920803	4
5	0.5674269	0.5549290	0.5427599	0.5309097	0.5193687	5
6	0.5066311	0.4932702	0.4803185	0.4677619	0.4555865	6
7	0.4523492	0.4384624	0.4250606	0.4121250	0.3996373	7
8	0.4038832	0.3897443	0.3761599	0.3631057	0.3505591	8
9	0.3606100	0.3464394	0.3328848	0.3199169	0.3075079	9
10	0.3219732	0.3079461	0.2945883	0.2818652	0.2697438	10
11	0.2874761	0.2737299	0.2606977	0.2483393	0.2366174	11
12	0.2566751	0.2433155	0.2307059	0.2188012	0.2075591	12
13	0.2291742	0.2162804	0.2041645	0.1927764	0.1820694	13
14	0.2046198	0.1922493	0.1806766	0.1698470	0.1597100	14
15	0.1826963	0.1708882	0.1598908	0.1496450	0.1400965	15
16	0.1631217	0.1519007	0.1414962	0.1318458	0.1228917	16
17	0.1456443	0.1350228	0.1252179	0.1161637	0.1077997	17
18	0.1300396	0.1200203	0.1108123	0.1023469	0.0945611	18
19	0.1161068	0.1066847	0.0980640	0.0901734	0.0829484	19
20	0.1036668	0.0948308	0.0867823	0.0794480	0.0727617	20
21	0.0925596	0.0842941	0.0767985	0.0699982	0.0638261	21
22	0.0826425	0.0749281	0.0679633	0.0616724	0.0559878	22
23	0.0737880	0.0666027	0.0601445	0.0543369	0.0491121	23
24	0.0658821	0.0592024	0.0532252	0.0478740	0.0430808	24
25	0.0588233	0.0526244	0.0471020	0.0421797	0.0377902	25
26	0.0525208	0.0467772	0.0416831	0.0371627	0.0331493	26
27	0.0468936	0.0415798	0.0368877	0.0327425	0.0290783	27
28	0.0418693	0.0369598	0.0326440	0.0288480	0.0255073	28
29	0.0373833	0.0328531	0.0288885	0.0254167	0.0223748	29
30	0.0333779	0.0292028	0.0255651	0.0223936	0.0196270	30
31	0.0298017	0.0259580	0.0226239	0.0197301	0.0172167	31
32	0.0266087	0.0230738	0.0200212	0.0173833	0.0151024	32
33	0.0237577	0.0205101	0.0177179	0.0153157	0.0132477	33
34	0.0212123	0.0182312	0.0156795	0.0134940	0.0116208	34
35	0.0189395	0.0162055	0.0138757	0.0118890	0.0101937	35
36	0.0169103	0.0144049	0.0122794	0.0104749	0.0089418	36
37	0.0150985	0.0128043	0.0108667	0.0092290	0.0078437	37
38	0.0134808	0.0113816	0.0096165	0.0081312	0.0068804	38
39	0.0120364	0.0101170	0.0085102	0.0071641	0.0060355	39
40	0.0107468	0.0089929	0.0075312	0.0063120	0.0052943	40
41	0.0095954	0.0079937	0.0066647	0.0055612	0.0046441	41
42	0.0085673	0.0071055	0.0058980	0.0048997	0.0040738	42
43	0.0076494	0.0063160	0.0052195	0.0043170	0.0035735	43
44	0.0068298	0.0056142	0.0046190	0.0038035	0.0031346	44
45	0.0060980	0.0049904	0.0040876	0.0033511	0.0027497	45
46	0.0054447	0.0044359	0.0036174	0.0029525	0.0024120	46
47	0.0048613	0.0039430	0.0032012	0.0026013	0.0021158	47
48	0.0043405	0.0035049	0.0028329	0.0022919	0.0018560	48
49	0.0038754	0.0031155	0.0025070	0.0020193	0.0016280	49
50	0.0034602	0.0027693	0.0022186	0.0017791	0.0014281	50

PRESENT VALUE OF £1

			Rate Per Cent			
Yrs.	12	12.5	13	13.5	14	Yrs.
51	0.0030894	0.0024616	0.0019634	0.0015675	0.0012527	51
52	0.0027584	0.0021881	0.0017375	0.0013811	0.0010989	52
53	0.0024629	0.0019450	0.0015376	0.0012168	0.0009639	53
54	0.0021990	0.0017289	0.0013607	0.0010721	0.0008455	54
55	0.0019634	0.0015368	0.0012042	0.0009446	0.0007417	55
56	0.0017530	0.0013660	0.0010656	0.0008322	0.0006506	56
57	0.0015652	0.0012142	0.0009430	0.0007332	0.0005707	57
58	0.0013975	0.0010793	0.0008345	0.0006460	0.0005006	58
59	0.0012478	0.0009594	0.0007385	0.0005692	0.0004392	59
60	0.0011141	0.0008528	0.0006536	0.0005015	0.0003852	60
61	0.0009947	0.0007580	0.0005784	0.0004418	0.0003379	61
62	0.0008881	0.0006738	0.0005118	0.0003893	0.0002964	62
63	0.0007930	0.0005990	0.0004530	0.0003430	0.0002600	63
64	0.0007080	0.0005324	0.0004008	0.0003022	0.0002281	64
65	0.0006322	0.0004732	0.0003547	0.0002662	0.0002001	65
66	0.0005644	0.0004207	0.0003139	0.0002346	0.0001755	66
67	0.0005040	0.0003739	0.0002778	0.0002067	0.0001539	67
68	0.0004500	0.0003324	0.0002458	0.0001821	0.0001350	68
69	0.0004018	0.0002954	0.0002176	0.0001604	0.0001185	69
70	0.0003587	0.0002626	0.0001925	0.0001413	0.0001039	70
71	0.0003203	0.0002334	0.0001704	0.0001245	0.0000911	71
72	0.0002860	0.0002075	0.0001508	0.0001097	0.0000800	72
73	0.0002553	0.0001844	0.0001334	0.0000967	0.0000701	73
74	0.0002280	0.0001640	0.0001181	0.0000852	0.0000615	74
75	0.0002035	0.0001457	0.0001045	0.0000750	0.0000540	75
76	0.0001817	0.0001295	0.0000925	0.0000661	0.0000473	76
77	0.0001623	0.0001151	0.0000818	0.0000583	0.0000415	77
78	0.0001449	0.0001024	0.0000724	0.0000513	0.0000364	78
79	0.0001294	0.0000910	0.0000641	0.0000452	0.0000320	79
80	0.0001155	0.0000809	0.0000567	0.0000398	0.0000280	80
81	0.0001031	0.0000719	0.0000502	0.0000351	0.0000246	81
82	0.0000921	0.0000639	0.0000444	0.0000309	0.0000216	82
83	0.0000822	0.0000568	0.0000393	0.0000272	0.0000189	83
84	0.0000734	0.0000505	0.0000348	0.0000240	0.0000166	84
85	0.0000655	0.0000449	0.0000308	0.0000212	0.0000146	85
86	0.0000585	0.0000399	0.0000272	0.0000186	0.0000128	86
87	0.0000522	0.0000355	0.0000241	0.0000164	0.0000112	87
88	0.0000466	0.0000315	0.0000213	0.0000145	0.0000098	88
89	0.0000416	0.0000280	0.0000189	0.0000127	0.0000086	89
90	0.0000372	0.0000249	0.0000167	0.0000112	0.0000076	90
91	0.0000332	0.0000221	0.0000148	0.0000099	0.0000066	91
92	0.0000296	0.0000197	0.0000131	0.0000087	0.0000058	92
93	0.0000265	0.0000175	0.0000116	0.0000077	0.0000051	93
94	0.0000236	0.0000155	0.0000102	0.0000068	0.0000045	94
95	0.0000211	0.0000138	0.0000091	0.0000060	0.0000039	95
96	0.0000188	0.0000123	0.0000080	0.0000053	0.0000034	96
97	0.0000168	0.0000109	0.0000071	0.0000046	0.0000030	97
98	0.0000150	0.0000097	0.0000063	0.0000041	0.0000027	98
99	0.0000134	0.0000086	0.0000056	0.0000036	0.0000023	99
100	0.0000120	0.0000077	0.0000049	0.0000032	0.0000020	100

PRESENT VALUE OF £1

			Rate Per Cent			
Yrs.	**14.5**	**15**	**16**	**17**	**18**	Yrs.
1	0.8733624	0.8695652	0.8620690	0.8547009	0.8474576	1
2	0.7627620	0.7561437	0.7431629	0.7305136	0.7181844	2
3	0.6661677	0.6575162	0.6406577	0.6243706	0.6086309	3
4	0.5818058	0.5717532	0.5522911	0.5336500	0.5157889	4
5	0.5081273	0.4971767	0.4761130	0.4561112	0.4371092	5
6	0.4437793	0.4323276	0.4104423	0.3898386	0.3704315	6
7	0.3875802	0.3759370	0.3538295	0.3331954	0.3139250	7
8	0.3384980	0.3269018	0.3050255	0.2847824	0.2660382	8
9	0.2956314	0.2842624	0.2629530	0.2434037	0.2254561	9
10	0.2581934	0.2471847	0.2266836	0.2080374	0.1910645	10
11	0.2254964	0.2149432	0.1954169	0.1778097	0.1619190	11
12	0.1969401	0.1869072	0.1684628	0.1519741	0.1372195	12
13	0.1720001	0.1625280	0.1452266	0.1298924	0.1162877	13
14	0.1502184	0.1413287	0.1251953	0.1110192	0.0985489	14
15	0.1311951	0.1228945	0.1079270	0.0948882	0.0835160	15
16	0.1145809	0.1068648	0.0930405	0.0811010	0.0707763	16
17	0.1000707	0.0929259	0.0802074	0.0693171	0.0599799	17
18	0.0873979	0.0808051	0.0691443	0.0592454	0.0508304	18
19	0.0763301	0.0702653	0.0596071	0.0506371	0.0430766	19
20	0.0666638	0.0611003	0.0513855	0.0432796	0.0365056	20
21	0.0582217	0.0531307	0.0442978	0.0369911	0.0309370	21
22	0.0508486	0.0462006	0.0381878	0.0316163	0.0262178	22
23	0.0444093	0.0401744	0.0329205	0.0270225	0.0222185	23
24	0.0387854	0.0349343	0.0283797	0.0230961	0.0188292	24
25	0.0338737	0.0303776	0.0244653	0.0197403	0.0159569	25
26	0.0295840	0.0264153	0.0210908	0.0168720	0.0135228	26
27	0.0258376	0.0229699	0.0181817	0.0144205	0.0114600	27
28	0.0225656	0.0199738	0.0156739	0.0123253	0.0097119	28
29	0.0197079	0.0173685	0.0135120	0.0105344	0.0082304	29
30	0.0172122	0.0151031	0.0116482	0.0090038	0.0069749	30
31	0.0150325	0.0131331	0.0100416	0.0076955	0.0059110	31
32	0.0131288	0.0114201	0.0086565	0.0065774	0.0050093	32
33	0.0114662	0.0099305	0.0074625	0.0056217	0.0042452	33
34	0.0100141	0.0086352	0.0064332	0.0048049	0.0035976	34
35	0.0087460	0.0075089	0.0055459	0.0041067	0.0030488	35
36	0.0076384	0.0065295	0.0047809	0.0035100	0.0025837	36
37	0.0066711	0.0056778	0.0041215	0.0030000	0.0021896	37
38	0.0058263	0.0049372	0.0035530	0.0025641	0.0018556	38
39	0.0050885	0.0042932	0.0030629	0.0021916	0.0015725	39
40	0.0044441	0.0037332	0.0026405	0.0018731	0.0013327	40
41	0.0038813	0.0032463	0.0022763	0.0016010	0.0011294	41
42	0.0033898	0.0028229	0.0019623	0.0013683	0.0009571	42
43	0.0029605	0.0024547	0.0016916	0.0011695	0.0008111	43
44	0.0025856	0.0021345	0.0014583	0.0009996	0.0006874	44
45	0.0022582	0.0018561	0.0012572	0.0008544	0.0005825	45
46	0.0019722	0.0016140	0.0010838	0.0007302	0.0004937	46
47	0.0017224	0.0014035	0.0009343	0.0006241	0.0004184	47
48	0.0015043	0.0012204	0.0008054	0.0005334	0.0003545	48
49	0.0013138	0.0010612	0.0006943	0.0004559	0.0003005	49
50	0.0011474	0.0009228	0.0005986	0.0003897	0.0002546	50

PRESENT VALUE OF £1

			Rate Per Cent			
Yrs.	14.5	15	16	17	18	Yrs.
51	0.0010021	0.0008024	0.0005160	0.0003331	0.0002158	51
52	0.0008752	0.0006978	0.0004448	0.0002847	0.0001829	52
53	0.0007644	0.0006068	0.0003835	0.0002433	0.0001550	53
54	0.0006676	0.0005276	0.0003306	0.0002080	0.0001313	54
55	0.0005830	0.0004588	0.0002850	0.0001777	0.0001113	55
56	0.0005092	0.0003990	0.0002457	0.0001519	0.0000943	56
57	0.0004447	0.0003469	0.0002118	0.0001298	0.0000799	57
58	0.0003884	0.0003017	0.0001826	0.0001110	0.0000677	58
59	0.0003392	0.0002623	0.0001574	0.0000948	0.0000574	59
60	0.0002963	0.0002281	0.0001357	0.0000811	0.0000486	60
61	0.0002587	0.0001983	0.0001170	0.0000693	0.0000412	61
62	0.0002260	0.0001725	0.0001008	0.0000592	0.0000349	62
63	0.0001974	0.0001500	0.0000869	0.0000506	0.0000296	63
64	0.0001724	0.0001304	0.0000749	0.0000433	0.0000251	64
65	0.0001505	0.0001134	0.0000646	0.0000370	0.0000213	65
66	0.0001315	0.0000986	0.0000557	0.0000316	0.0000180	66
67	0.0001148	0.0000858	0.0000480	0.0000270	0.0000153	67
68	0.0001003	0.0000746	0.0000414	0.0000231	0.0000129	68
69	0.0000876	0.0000648	0.0000357	0.0000197	0.0000110	69
70	0.0000765	0.0000564	0.0000308	0.0000169	0.0000093	70
71	0.0000668	0.0000490	0.0000265	0.0000144	0.0000079	71
72	0.0000583	0.0000426	0.0000229	0.0000123	0.0000067	72
73	0.0000510	0.0000371	0.0000197	0.0000105	0.0000057	73
74	0.0000445	0.0000322	0.0000170	0.0000090	0.0000048	74
75	0.0000389	0.0000280	0.0000146	0.0000077	0.0000041	75
76	0.0000339	0.0000244	0.0000126	0.0000066	0.0000034	76
77	0.0000296	0.0000212	0.0000109	0.0000056	0.0000029	77
78	0.0000259	0.0000184	0.0000094	0.0000048	0.0000025	78
79	0.0000226	0.0000160	0.0000081	0.0000041	0.0000021	79
80	0.0000197	0.0000139	0.0000070	0.0000035	0.0000018	80
81	0.0000172	0.0000121	0.0000060	0.0000030	0.0000015	81
82	0.0000151	0.0000105	0.0000052	0.0000026	0.0000013	82
83	0.0000132	0.0000092	0.0000045	0.0000022	0.0000011	83
84	0.0000115	0.0000080	0.0000039	0.0000019	0.0000009	84
85	0.0000100	0.0000069	0.0000033	0.0000016	0.0000008	85
86	0.0000088	0.0000060	0.0000029	0.0000014	0.0000007	86
87	0.0000077	0.0000052	0.0000025	0.0000012	0.0000006	87
88	0.0000067	0.0000046	0.0000021	0.0000010	0.0000005	88
89	0.0000058	0.0000040	0.0000018	0.0000009	0.0000004	89
90	0.0000051	0.0000034	0.0000016	0.0000007	0.0000003	90
91	0.0000045	0.0000030	0.0000014	0.0000006	0.0000003	91
92	0.0000039	0.0000026	0.0000012	0.0000005	0.0000002	92
93	0.0000034	0.0000023	0.0000010	0.0000005	0.0000002	93
94	0.0000030	0.0000020	0.0000009	0.0000004	0.0000002	94
95	0.0000026	0.0000017	0.0000008	0.0000003	0.0000001	95
96	0.0000023	0.0000015	0.0000006	0.0000003	0.0000001	96
97	0.0000020	0.0000013	0.0000006	0.0000002	0.0000001	97
98	0.0000017	0.0000011	0.0000005	0.0000002	0.0000001	98
99	0.0000015	0.0000010	0.0000004	0.0000002	0.0000001	99
100	0.0000013	0.0000009	0.0000004	0.0000002	0.0000001	100

PRESENT VALUE OF £1

			Rate Per Cent			
Yrs.	**19**	**20**	**21**	**22**	**23**	Yrs.
1	0.8403361	0.8333333	0.8264463	0.8196721	0.8130081	1
2	0.7061648	0.6944444	0.6830135	0.6718624	0.6609822	2
3	0.5934158	0.5787037	0.5644739	0.5507069	0.5373839	3
4	0.4986688	0.4822531	0.4665074	0.4513991	0.4368975	4
5	0.4190494	0.4018776	0.3855433	0.3699993	0.3552012	5
6	0.3521423	0.3348980	0.3186308	0.3032781	0.2887815	6
7	0.2959179	0.2790816	0.2633313	0.2485886	0.2347817	7
8	0.2486705	0.2325680	0.2176291	0.2037611	0.1908794	8
9	0.2089668	0.1938067	0.1798588	0.1670173	0.1551865	9
10	0.1756024	0.1615056	0.1486436	0.1368994	0.1261679	10
11	0.1475650	0.1345880	0.1228460	0.1122127	0.1025755	11
12	0.1240042	0.1121567	0.1015256	0.0919776	0.0833947	12
13	0.1042052	0.0934639	0.0839055	0.0753915	0.0678006	13
14	0.0875674	0.0778866	0.0693433	0.0617963	0.0551224	14
15	0.0735861	0.0649055	0.0573086	0.0506527	0.0448150	15
16	0.0618370	0.0540879	0.0473624	0.0415186	0.0364350	16
17	0.0519639	0.0450732	0.0391425	0.0340316	0.0296219	17
18	0.0436671	0.0375610	0.0323492	0.0278948	0.0240829	18
19	0.0366951	0.0313009	0.0267349	0.0228646	0.0195796	19
20	0.0308362	0.0260841	0.0220949	0.0187415	0.0159183	20
21	0.0259128	0.0217367	0.0182603	0.0153619	0.0129417	21
22	0.0217754	0.0181139	0.0150911	0.0125917	0.0105217	22
23	0.0182987	0.0150949	0.0124720	0.0103211	0.0085543	23
24	0.0153770	0.0125791	0.0103074	0.0084599	0.0069547	24
25	0.0129219	0.0104826	0.0085186	0.0069343	0.0056542	25
26	0.0108587	0.0087355	0.0070401	0.0056839	0.0045969	26
27	0.0091250	0.0072796	0.0058183	0.0046589	0.0037373	27
28	0.0076681	0.0060663	0.0048085	0.0038188	0.0030385	28
29	0.0064437	0.0050553	0.0039740	0.0031301	0.0024703	29
30	0.0054149	0.0042127	0.0032843	0.0025657	0.0020084	30
31	0.0045503	0.0035106	0.0027143	0.0021030	0.0016328	31
32	0.0038238	0.0029255	0.0022432	0.0017238	0.0013275	32
33	0.0032133	0.0024379	0.0018539	0.0014129	0.0010793	33
34	0.0027002	0.0020316	0.0015321	0.0011582	0.0008775	34
35	0.0022691	0.0016930	0.0012662	0.0009493	0.0007134	35
36	0.0019068	0.0014108	0.0010465	0.0007781	0.0005800	36
37	0.0016024	0.0011757	0.0008649	0.0006378	0.0004715	37
38	0.0013465	0.0009797	0.0007148	0.0005228	0.0003834	38
39	0.0011315	0.0008165	0.0005907	0.0004285	0.0003117	39
40	0.0009509	0.0006804	0.0004882	0.0003512	0.0002534	40
45	0.0003985	0.0002734	0.0001882	0.0001300	0.0000900	45
50	0.0001670	0.0001099	0.0000726	0.0000481	0.0000320	50
55	0.0000700	0.0000442	0.0000280	0.0000178	0.0000114	55
60	0.0000293	0.0000177	0.0000108	0.0000066	0.0000040	60
65	0.0000123	0.0000071	0.0000042	0.0000024	0.0000014	65
70	0.0000051	0.0000029	0.0000016	0.0000009	0.0000005	70
75	0.0000022	0.0000012	0.0000006	0.0000003	0.0000002	75
80	0.0000009	0.0000005	0.0000002	0.0000001	0.0000001	80
90	0.0000002	0.0000001	0.0000000	0.0000000	0.0000000	90
100	0.0000000	0.0000000	0.0000000	0.0000000	0.0000000	100

PRESENT VALUE OF £1

			Rate Per Cent			
Yrs.	**24**	**25**	**26**	**28**	**30**	Yrs.
1	0.8064516	0.8000000	0.7936508	0.7812500	0.7692308	1
2	0.6503642	0.6400000	0.6298816	0.6103516	0.5917160	2
3	0.5244873	0.5120000	0.4999060	0.4768372	0.4551661	3
4	0.4229736	0.4096000	0.3967508	0.3725290	0.3501278	4
5	0.3411077	0.3276800	0.3148816	0.2910383	0.2693291	5
6	0.2750869	0.2621440	0.2499060	0.2273737	0.2071762	6
7	0.2218443	0.2097152	0.1983381	0.1776357	0.1593663	7
8	0.1789067	0.1677722	0.1574112	0.1387779	0.1225895	8
9	0.1442796	0.1342177	0.1249295	0.1084202	0.0942996	9
10	0.1163545	0.1073742	0.0991504	0.0847033	0.0725382	10
11	0.0938343	0.0858993	0.0786908	0.0661744	0.0557986	11
12	0.0756728	0.0687195	0.0624530	0.0516988	0.0429220	12
13	0.0610264	0.0549756	0.0495659	0.0403897	0.0330169	13
14	0.0492149	0.0439805	0.0393380	0.0315544	0.0253976	14
15	0.0396894	0.0351844	0.0312206	0.0246519	0.0195366	15
16	0.0320076	0.0281475	0.0247783	0.0192593	0.0150282	16
17	0.0258126	0.0225180	0.0196653	0.0150463	0.0115601	17
18	0.0208166	0.0180144	0.0156074	0.0117549	0.0088924	18
19	0.0167876	0.0144115	0.0123868	0.0091835	0.0068403	19
20	0.0135384	0.0115292	0.0098308	0.0071746	0.0052618	20
21	0.0109180	0.0092234	0.0078022	0.0056052	0.0040475	21
22	0.0088049	0.0073787	0.0061922	0.0043791	0.0031135	22
23	0.0071007	0.0059030	0.0049145	0.0034211	0.0023950	23
24	0.0057264	0.0047224	0.0039004	0.0026728	0.0018423	24
25	0.0046180	0.0037779	0.0030955	0.0020881	0.0014172	25
26	0.0037242	0.0030223	0.0024568	0.0016313	0.0010901	26
27	0.0030034	0.0024179	0.0019498	0.0012745	0.0008386	27
28	0.0024221	0.0019343	0.0015475	0.0009957	0.0006450	28
29	0.0019533	0.0015474	0.0012282	0.0007779	0.0004962	29
30	0.0015752	0.0012379	0.0009747	0.0006077	0.0003817	30
31	0.0012704	0.0009904	0.0007736	0.0004748	0.0002936	31
32	0.0010245	0.0007923	0.0006140	0.0003709	0.0002258	32
33	0.0008262	0.0006338	0.0004873	0.0002898	0.0001737	33
34	0.0006663	0.0005071	0.0003867	0.0002264	0.0001336	34
35	0.0005373	0.0004056	0.0003069	0.0001769	0.0001028	35
36	0.0004333	0.0003245	0.0002436	0.0001382	0.0000791	36
37	0.0003495	0.0002596	0.0001933	0.0001080	0.0000608	37
38	0.0002818	0.0002077	0.0001534	0.0000843	0.0000468	38
39	0.0002273	0.0001662	0.0001218	0.0000659	0.0000360	39
40	0.0001833	0.0001329	0.0000966	0.0000515	0.0000277	40
45	0.0000625	0.0000436	0.0000304	0.0000150	0.0000075	45
50	0.0000213	0.0000143	0.0000096	0.0000044	0.0000020	50
55	0.0000073	0.0000047	0.0000030	0.0000013	0.0000005	55
60	0.0000025	0.0000015	0.0000010	0.0000004	0.0000001	60
65	0.0000008	0.0000005	0.0000003	0.0000001	0.0000000	65
70	0.0000003	0.0000002	0.0000001	0.0000000	0.0000000	70
75	0.0000001	0.0000001	0.0000000	0.0000000	0.0000000	75
80	0.0000000	0.0000000	0.0000000	0.0000000	0.0000000	80
90	0.0000000	0.0000000	0.0000000	0.0000000	0.0000000	90
100	0.0000000	0.0000000	0.0000000	0.0000000	0.0000000	100

Rule for extending the Table of
PRESENT VALUE OF ONE POUND

Rule—

P.V. of £1 receivable in N years × P.V. of £1 receivable in M years =
P.V. of £1 receivable in N + M years.

Example—

What is the present value of £1 receivable in 120 years at 4 per cent?

Answer—

P.V. of £1 receivable in 20 years at 4% (see page 92)   =   ·4563869
P.V. of £1 receivable in 100 years at 4% (see page 93)   =   ·0198000
P.V. of £1 receivable in 120 years at 4%
                = (·4563869 × ·0198000)   =   ·009036 *Ans.*

ANNUAL SINKING FUND

being the annual sum required to be invested
to amount to £1 in a given number of years.
(*i.e.*, the sum which, if invested at the end of each year, will
accumulate at Compound Interest to £1)

At Rates of Interest from
1% to 15%

ANNUAL SINKING FUND

FOR THE REDEMPTION OF £1 CAPITAL INVESTED

			Rate Per Cent			
Yrs.	1	1.25	1.5	1.75	2	Yrs.
1	1.0000000	1.0000000	1.0000000	1.0000000	1.0000000	1
2	.4975124	.4968944	.4962779	.4956629	.4950495	2
3	.3300221	.3292012	.3283830	.3275675	.3267547	3
4	.2462811	.2453610	.2444448	.2435324	.2426238	4
5	.1960398	.1950621	.1940893	.1931214	.1921584	5
6	.1625484	.1615338	.1605252	.1595226	.1585258	6
7	.1386283	.1375887	.1365562	.1355306	.1345120	7
8	.1206903	.1196331	.1185840	.1175429	.1165098	8
9	.1067404	.1056706	.1046098	.1035581	.1025154	9
10	.0955821	.0945031	.0934342	.0923753	.0913265	10
11	.0864541	.0853684	.0842938	.0832304	.0821779	11
12	.0788488	.0777583	.0766800	.0756138	.0745596	12
13	.0724148	.0713210	.0702404	.0691728	.0681184	13
14	.0669012	.0658051	.0647233	.0636556	.0626020	14
15	.0621238	.0610265	.0599444	.0588774	.0578255	15
16	.0579446	.0568467	.0557651	.0546996	.0536501	16
17	.0542581	.0531602	.0520797	.0510162	.0499698	17
18	.0509820	.0498848	.0488058	.0477449	.0467021	18
19	.0480518	.0469555	.0458785	.0448206	.0437818	19
20	.0454153	.0443204	.0432457	.0421912	.0411567	20
21	.0430308	.0419375	.0408655	.0398146	.0387848	21
22	.0408637	.0397724	.0387033	.0376564	.0366314	22
23	.0388858	.0377967	.0367308	.0356880	.0346681	23
24	.0370735	.0359866	.0349241	.0338857	.0328711	24
25	.0354068	.0343225	.0332635	.0322295	.0312204	25
26	.0338689	.0327873	.0317320	.0307027	.0296992	26
27	.0324455	.0313668	.0303153	.0292908	.0282931	27
28	.0311244	.0300486	.0290011	.0279815	.0269897	28
29	.0298950	.0288223	.0277788	.0267642	.0257784	29
30	.0287481	.0276785	.0266392	.0256298	.0246499	30
31	.0276757	.0266094	.0255743	.0245701	.0235963	31
32	.0266709	.0256079	.0245771	.0235781	.0226106	32
33	.0257274	.0246679	.0236414	.0226478	.0216865	33
34	.0248400	.0237839	.0227619	.0217736	.0208187	34
35	.0240037	.0229511	.0219336	.0209508	.0200022	35
36	.0232143	.0221653	.0211524	.0201751	.0192329	36
37	.0224680	.0214227	.0204144	.0194426	.0185068	37
38	.0217615	.0207198	.0197161	.0187499	.0178206	38
39	.0210916	.0200537	.0190546	.0180940	.0171711	39
40	.0204556	.0194214	.0184271	.0174721	.0165557	40
41	.0198510	.0188206	.0178311	.0168817	.0159719	41
42	.0192756	.0182491	.0172643	.0163206	.0154173	42
43	.0187274	.0177047	.0167246	.0157867	.0148899	43
44	.0182044	.0171856	.0162104	.0152781	.0143879	44
45	.0177050	.0166901	.0157198	.0147932	.0139096	45
46	.0172277	.0162167	.0152512	.0143304	.0134534	46
47	.0167711	.0157641	.0148034	.0138884	.0130179	47
48	.0163338	.0153307	.0143750	.0134657	.0126018	48
49	.0159147	.0149156	.0139648	.0130612	.0122040	49
50	.0155127	.0145176	.0135717	.0126739	.0118232	50

ANNUAL SINKING FUND

FOR THE REDEMPTION OF £1 CAPITAL INVESTED

			Rate Per Cent			
Yrs.	1	1.25	1.5	1.75	2	Yrs.
51	.0151268	.0141357	.0131947	.0123027	.0114586	51
52	.0147560	.0137690	.0128329	.0119467	.0111091	52
53	.0143996	.0134165	.0124854	.0116049	.0107739	53
54	.0140566	.0130776	.0121514	.0112767	.0104523	54
55	.0137264	.0127514	.0118302	.0109613	.0101434	55
56	.0134082	.0124374	.0115211	.0106579	.0098466	56
57	.0131016	.0121348	.0112234	.0103661	.0095612	57
58	.0128057	.0118430	.0109366	.0100850	.0092867	58
59	.0125202	.0115616	.0106601	.0098143	.0090224	59
60	.0122444	.0112899	.0103934	.0095534	.0087680	60
61	.0119780	.0110276	.0101360	.0093017	.0085228	61
62	.0117204	.0107741	.0098875	.0090589	.0082864	62
63	.0114713	.0105290	.0096474	.0088246	.0080585	63
64	.0112301	.0102920	.0094153	.0085982	.0078385	64
65	.0109967	.0100627	.0091909	.0083795	.0076262	65
66	.0107705	.0098406	.0089739	.0081681	.0074212	66
67	.0105514	.0096256	.0087638	.0079637	.0072232	67
68	.0103389	.0094172	.0085603	.0077660	.0070317	68
69	.0101328	.0092153	.0083633	.0075746	.0068467	69
70	.0099328	.0090194	.0081724	.0073893	.0066676	70
71	.0097387	.0088294	.0079873	.0072099	.0064945	71
72	.0095502	.0086450	.0078078	.0070360	.0063268	72
73	.0093671	.0084660	.0076337	.0068675	.0061645	73
74	.0091891	.0082921	.0074647	.0067041	.0060074	74
75	.0090161	.0081232	.0073007	.0065457	.0058551	75
76	.0088478	.0079591	.0071415	.0063920	.0057075	76
77	.0086842	.0077995	.0069868	.0062428	.0055645	77
78	.0085249	.0076444	.0068365	.0060981	.0054258	78
79	.0083698	.0074934	.0066904	.0059575	.0052912	79
80	.0082189	.0073465	.0065483	.0058209	.0051607	80
81	.0080718	.0072036	.0064102	.0056883	.0050340	81
82	.0079285	.0070644	.0062758	.0055594	.0049111	82
83	.0077889	.0069288	.0061451	.0054341	.0047917	83
84	.0076527	.0067968	.0060178	.0053122	.0046758	84
85	.0075200	.0066681	.0058940	.0051937	.0045632	85
86	.0073905	.0065427	.0057733	.0050785	.0044538	86
87	.0072642	.0064204	.0056558	.0049664	.0043475	87
88	.0071409	.0063012	.0055414	.0048572	.0042442	88
89	.0070206	.0061849	.0054298	.0047510	.0041437	89
90	.0069031	.0060715	.0053211	.0046476	.0040460	90
91	.0067883	.0059608	.0052152	.0045469	.0039510	91
92	.0066762	.0058527	.0051118	.0044488	.0038586	92
93	.0065667	.0057472	.0050110	.0043533	.0037687	93
94	.0064597	.0056442	.0049127	.0042602	.0036812	94
95	.0063551	.0055437	.0048168	.0041694	.0035960	95
96	.0062528	.0054454	.0047232	.0040810	.0035131	96
97	.0061528	.0053494	.0046319	.0039948	.0034324	97
98	.0060550	.0052556	.0045427	.0039107	.0033538	98
99	.0059594	.0051639	.0044556	.0038288	.0032773	99
100	.0058657	.0050743	.0043706	.0037488	.0032027	100

ANNUAL SINKING FUND

FOR THE REDEMPTION OF £1 CAPITAL INVESTED

			Rate Per Cent			
Yrs.	**2.25**	**2.5**	**2.75**	**3**	**3.25**	Yrs.
1	1.0000000	1.0000000	1.0000000	1.0000000	1.0000000	1
2	.4944376	.4938272	.4932182	.4926108	.4920049	2
3	.3259446	.3251372	.3243324	.3235304	.3227309	3
4	.2417189	.2408179	.2399206	.2390270	.2381372	4
5	.1912002	.1902469	.1892983	.1883546	.1874156	5
6	.1575350	.1565500	.1555708	.1545975	.1536300	6
7	.1335002	.1324954	.1314975	.1305064	.1295220	7
8	.1154846	.1144673	.1134579	.1124564	.1114626	8
9	.1014817	.1004569	.0994410	.0984339	.0974356	9
10	.0902877	.0892588	.0882397	.0872305	.0862311	10
11	.0811365	.0801060	.0790863	.0780774	.0770794	11
12	.0735174	.0724871	.0714687	.0704621	.0694672	12
13	.0670769	.0660483	.0650325	.0640295	.0630393	13
14	.0615623	.0605365	.0595246	.0585263	.0575418	14
15	.0567885	.0557665	.0547592	.0537666	.0527886	15
16	.0526166	.0515990	.0505971	.0496108	.0486401	16
17	.0489404	.0479278	.0469319	.0459525	.0449897	17
18	.0456772	.0446701	.0436806	.0427087	.0417541	18
19	.0427618	.0417606	.0407780	.0398139	.0388680	19
20	.0401421	.0391471	.0381717	.0372157	.0362789	20
21	.0377757	.0367873	.0358194	.0348718	.0339442	21
22	.0356282	.0346466	.0336864	.0327474	.0318294	22
23	.0336710	.0326964	.0317441	.0308139	.0299056	23
24	.0318802	.0309128	.0299686	.0290474	.0281489	24
25	.0302360	.0292759	.0283400	.0274279	.0265393	25
26	.0287213	.0277687	.0268412	.0259383	.0250598	26
27	.0273219	.0263769	.0254578	.0245642	.0236959	27
28	.0260253	.0250879	.0241774	.0232932	.0224351	28
29	.0248208	.0238913	.0229894	.0221147	.0212668	29
30	.0236993	.0227776	.0218844	.0210193	.0201817	30
31	.0226528	.0217390	.0208545	.0199989	.0191717	31
32	.0216741	.0207683	.0198926	.0190466	.0182298	32
33	.0207572	.0198594	.0189925	.0181561	.0173496	33
34	.0198965	.0190068	.0181487	.0173220	.0165258	34
35	.0190873	.0182056	.0173564	.0165393	.0157535	35
36	.0183252	.0174516	.0166113	.0158038	.0150283	36
37	.0176064	.0167409	.0159095	.0151116	.0143465	37
38	.0169275	.0160701	.0152476	.0144593	.0137044	38
39	.0162854	.0154362	.0146226	.0138439	.0130992	39
40	.0156774	.0148362	.0140315	.0132624	.0125279	40
41	.0151009	.0142679	.0134720	.0127124	.0119881	41
42	.0145536	.0137288	.0129418	.0121917	.0114775	42
43	.0140336	.0132169	.0124387	.0116981	.0109940	43
44	.0135390	.0127304	.0119610	.0112298	.0105358	44
45	.0130681	.0122675	.0115069	.0107852	.0101011	45
46	.0126192	.0118268	.0110749	.0103625	.0096883	46
47	.0121911	.0114067	.0106636	.0099605	.0092962	47
48	.0117823	.0110060	.0102716	.0095778	.0089232	48
49	.0113918	.0106235	.0098977	.0092131	.0085683	49
50	.0110184	.0102581	.0095409	.0088655	.0082303	50

ANNUAL SINKING FUND

FOR THE REDEMPTION OF £1 CAPITAL INVESTED

			Rate Per Cent			
Yrs.	2.25	2.5	2.75	3	3.25	Yrs.
51	.0106610	.0099087	.0092001	.0085338	.0079082	51
52	.0103188	.0095745	.0088744	.0082172	.0076010	52
53	.0099909	.0092545	.0085630	.0079147	.0073080	53
54	.0096765	.0089480	.0082649	.0076256	.0070282	54
55	.0093749	.0086542	.0079795	.0073491	.0067609	55
56	.0090853	.0083724	.0077061	.0070845	.0065055	56
57	.0088071	.0081020	.0074440	.0068311	.0062613	57
58	.0085398	.0078424	.0071927	.0065885	.0060277	58
59	.0082827	.0075931	.0069515	.0063559	.0058040	59
60	.0080353	.0073534	.0067200	.0061330	.0055899	60
61	.0077972	.0071229	.0064977	.0059191	.0053848	61
62	.0075679	.0069013	.0062840	.0057138	.0051883	62
63	.0073470	.0066879	.0060787	.0055168	.0049998	63
64	.0071341	.0064825	.0058812	.0053276	.0048191	64
65	.0069288	.0062846	.0056912	.0051458	.0046457	65
66	.0067307	.0060940	.0055084	.0049711	.0044794	66
67	.0065395	.0059102	.0053324	.0048031	.0043196	67
68	.0063550	.0057330	.0051629	.0046416	.0041662	68
69	.0061768	.0055621	.0049996	.0044862	.0040189	69
70	.0060046	.0053971	.0048422	.0043366	.0038773	70
71	.0058382	.0052379	.0046905	.0041927	.0037412	71
72	.0056773	.0050842	.0045442	.0040540	.0036103	72
73	.0055217	.0049357	.0044031	.0039205	.0034845	73
74	.0053712	.0047922	.0042670	.0037919	.0033635	74
75	.0052255	.0046536	.0041356	.0036680	.0032470	75
76	.0050846	.0045196	.0040088	.0035485	.0031350	76
77	.0049481	.0043900	.0038863	.0034333	.0030271	77
78	.0048159	.0042646	.0037681	.0033222	.0029232	78
79	.0046878	.0041434	.0036538	.0032151	.0028232	79
80	.0045638	.0040260	.0035434	.0031117	.0027269	80
81	.0044435	.0039125	.0034367	.0030120	.0026341	81
82	.0043269	.0038025	.0033336	.0029158	.0025447	82
83	.0042139	.0036961	.0032339	.0028228	.0024585	83
84	.0041042	.0035930	.0031375	.0027331	.0023755	84
85	.0039979	.0034931	.0030442	.0026465	.0022954	85
86	.0038947	.0033963	.0029540	.0025628	.0022183	86
87	.0037945	.0033025	.0028667	.0024820	.0021438	87
88	.0036973	.0032117	.0027822	.0024039	.0020720	88
89	.0036029	.0031235	.0027004	.0023285	.0020028	89
90	.0035113	.0030381	.0026212	.0022556	.0019360	90
91	.0034222	.0029552	.0025446	.0021851	.0018716	91
92	.0033358	.0028749	.0024704	.0021169	.0018094	92
93	.0032518	.0027969	.0023985	.0020511	.0017494	93
94	.0031701	.0027213	.0023289	.0019874	.0016914	94
95	.0030908	.0026479	.0022614	.0019258	.0016355	95
96	.0030137	.0025766	.0021961	.0018662	.0015815	96
97	.0029387	.0025075	.0021327	.0018086	.0015294	97
98	.0028658	.0024403	.0020713	.0017528	.0014791	98
99	.0027949	.0023752	.0020118	.0016989	.0014305	99
100	.0027259	.0023119	.0019542	.0016467	.0013835	100

ANNUAL SINKING FUND
FOR THE REDEMPTION OF £1 CAPITAL INVESTED

Rate Per Cent

Yrs.	3.5	3.75	4	4.25	4.5	Yrs.
1	1.0000000	1.0000000	1.0000000	1.0000000	1.0000000	1
2	.4914005	.4907975	.4901961	.4895961	.4889976	2
3	.3219342	.3211400	.3203485	.3195596	.3187734	3
4	.2372511	.2363687	.2354900	.2346150	.2337436	4
5	.1864814	.1855519	.1846271	.1837070	.1827916	5
6	.1526682	.1517122	.1507619	.1498173	.1488784	6
7	.1285445	.1275737	.1266096	.1256522	.1247015	7
8	.1104766	.1094984	.1085278	.1075649	.1066097	8
9	.0964460	.0954652	.0944930	.0935294	.0925745	9
10	.0852414	.0842613	.0832909	.0823301	.0813788	10
11	.0760920	.0751152	.0741490	.0731934	.0722482	11
12	.0684839	.0675123	.0665522	.0656035	.0646662	12
13	.0620616	.0610964	.0601437	.0592034	.0582754	13
14	.0565707	.0556132	.0546690	.0537381	.0528203	14
15	.0518251	.0508759	.0499411	.0490204	.0481138	15
16	.0476848	.0467448	.0458200	.0449102	.0440154	16
17	.0440431	.0431128	.0421985	.0413002	.0404176	17
18	.0408168	.0398966	.0389933	.0381068	.0372369	18
19	.0379403	.0370306	.0361386	.0352643	.0344073	19
20	.0353611	.0344621	.0335818	.0327198	.0318761	20
21	.0330366	.0321486	.0312801	.0304308	.0296006	21
22	.0309321	.0300553	.0291988	.0283623	.0275456	22
23	.0290188	.0281534	.0273091	.0264855	.0256825	23
24	.0272728	.0264189	.0255868	.0247763	.0239870	24
25	.0256740	.0248317	.0240120	.0232145	.0224390	25
26	.0242054	.0233747	.0225674	.0217831	.0210214	26
27	.0228524	.0220334	.0212385	.0204674	.0197195	27
28	.0216026	.0207954	.0200130	.0192549	.0185208	28
29	.0204454	.0196499	.0188799	.0181350	.0174146	29
30	.0193713	.0185876	.0178301	.0170982	.0163915	30
31	.0183724	.0176005	.0168554	.0161365	.0154434	31
32	.0174415	.0166813	.0159486	.0152428	.0145632	32
33	.0165724	.0158239	.0151036	.0144106	.0137445	33
34	.0157597	.0150229	.0143148	.0136347	.0129819	34
35	.0149983	.0142732	.0135773	.0129100	.0122704	35
36	.0142842	.0135706	.0128869	.0122322	.0116058	36
37	.0136132	.0129112	.0122396	.0115974	.0109840	37
38	.0129821	.0122916	.0116319	.0110023	.0104017	38
39	.0123878	.0117086	.0110608	.0104435	.0098557	39
40	.0118273	.0111595	.0105235	.0099184	.0093431	40
41	.0112982	.0106416	.0100174	.0094244	.0088616	41
42	.0107983	.0101529	.0095402	.0089592	.0084087	42
43	.0103254	.0096911	.0090899	.0085207	.0079823	43
44	.0098777	.0092543	.0086645	.0081071	.0075807	44
45	.0094534	.0088410	.0082625	.0077166	.0072020	45
46	.0090511	.0084494	.0078820	.0073476	.0068447	46
47	.0086692	.0080782	.0075219	.0069987	.0065073	47
48	.0083065	.0077261	.0071806	.0066686	.0061886	48
49	.0079617	.0073918	.0068571	.0063561	.0058872	49
50	.0076337	.0070742	.0065502	.0060600	.0056021	50

ANNUAL SINKING FUND

FOR THE REDEMPTION OF £1 CAPITAL INVESTED

			Rate Per Cent			
Yrs.	3.5	3.75	4	4.25	4.5	Yrs.
51	.0073216	.0067723	.0062588	.0057794	.0053323	51
52	.0070243	.0064852	.0059821	.0055132	.0050768	52
53	.0067410	.0062120	.0057191	.0052606	.0048347	53
54	.0064709	.0059518	.0054691	.0050208	.0046052	54
55	.0062132	.0057040	.0052312	.0047931	.0043875	55
56	.0059673	.0054678	.0050049	.0045766	.0041811	56
57	.0057325	.0052425	.0047893	.0043709	.0039851	57
58	.0055081	.0050276	.0045840	.0041752	.0037990	58
59	.0052937	.0048225	.0043884	.0039890	.0036222	59
60	.0050886	.0046267	.0042018	.0038118	.0034543	60
61	.0048925	.0044397	.0040240	.0036431	.0032946	61
62	.0047048	.0042610	.0038543	.0034824	.0031428	62
63	.0045251	.0040902	.0036924	.0033293	.0029985	63
64	.0043531	.0039268	.0035378	.0031834	.0028611	64
65	.0041883	.0037706	.0033902	.0030443	.0027305	65
66	.0040303	.0036212	.0032492	.0029117	.0026061	66
67	.0038789	.0034782	.0031145	.0027852	.0024876	67
68	.0037338	.0033412	.0029858	.0026646	.0023749	68
69	.0035945	.0032101	.0028627	.0025494	.0022675	69
70	.0034610	.0030846	.0027451	.0024395	.0021651	70
71	.0033328	.0029643	.0026325	.0023346	.0020676	71
72	.0032097	.0028490	.0025249	.0022344	.0019747	72
73	.0030916	.0027385	.0024219	.0021387	.0018861	73
74	.0029782	.0026326	.0023233	.0020474	.0018016	74
75	.0028692	.0025310	.0022290	.0019600	.0017210	75
76	.0027645	.0024336	.0021387	.0018766	.0016442	76
77	.0026639	.0023401	.0020522	.0017969	.0015709	77
78	.0025672	.0022505	.0019694	.0017206	.0015010	78
79	.0024743	.0021644	.0018901	.0016478	.0014343	79
80	.0023849	.0020818	.0018141	.0015781	.0013707	80
81	.0022989	.0020026	.0017413	.0015115	.0013100	81
82	.0022163	.0019265	.0016715	.0014478	.0012520	82
83	.0021368	.0018534	.0016046	.0013868	.0011966	83
84	.0020602	.0017832	.0015405	.0013285	.0011438	84
85	.0019866	.0017158	.0014791	.0012727	.0010933	85
86	.0019158	.0016511	.0014202	.0012194	.0010452	86
87	.0018476	.0015889	.0013637	.0011683	.0009992	87
88	.0017819	.0015291	.0013095	.0011194	.0009552	88
89	.0017187	.0014717	.0012576	.0010726	.0009132	89
90	.0016578	.0014165	.0012078	.0010278	.0008732	90
91	.0015992	.0013634	.0011600	.0009850	.0008349	91
92	.0015427	.0013124	.0011141	.0009439	.0007983	92
93	.0014883	.0012634	.0010701	.0009046	.0007633	93
94	.0014359	.0012162	.0010279	.0008670	.0007299	94
95	.0013855	.0011709	.0009874	.0008309	.0006980	95
96	.0013368	.0011273	.0009485	.0007964	.0006675	96
97	.0012899	.0010854	.0009112	.0007634	.0006383	97
98	.0012448	.0010450	.0008754	.0007317	.0006105	98
99	.0012012	.0010063	.0008410	.0007014	.0005838	99
100	.0011593	.0009689	.0008080	.0006724	.0005584	100

ANNUAL SINKING FUND

FOR THE REDEMPTION OF £1 CAPITAL INVESTED

Yrs.	4.75	5	5.25	5.5	5.75	Yrs.
			Rate Per Cent			
1	1.0000000	1.0000000	1.0000000	1.0000000	1.0000000	1
2	.4884005	.4878049	.4872107	.4866180	.4860267	2
3	.3179897	.3172086	.3164300	.3156541	.3148807	3
4	.2328759	.2320118	.2311514	.2302945	.2294412	4
5	.1818809	.1809748	.1800733	.1791764	.1782841	5
6	.1479451	.1470175	.1460954	.1451789	.1442680	6
7	.1237573	.1228198	.1218889	.1209644	.1200465	7
8	.1056620	.1047218	.1037892	.1028640	.1019463	8
9	.0916280	.0906901	.0897606	.0888395	.0879267	9
10	.0804370	.0795046	.0785815	.0776678	.0767633	10
11	.0713134	.0703889	.0694747	.0685707	.0676768	11
12	.0637402	.0628254	.0619218	.0610292	.0601477	12
13	.0573595	.0564558	.0555640	.0546843	.0538163	13
14	.0519157	.0510240	.0501452	.0492791	.0484257	14
15	.0472211	.0463423	.0454771	.0446256	.0437875	15
16	.0431353	.0422699	.0414190	.0405825	.0397603	16
17	.0395506	.0386991	.0378630	.0370420	.0362360	17
18	.0363834	.0355462	.0347251	.0339199	.0331305	18
19	.0335677	.0327450	.0319392	.0311501	.0303773	19
20	.0310505	.0302426	.0294523	.0286793	.0279235	20
21	.0287891	.0279961	.0272214	.0264648	.0257259	21
22	.0267485	.0259705	.0252115	.0244712	.0237493	22
23	.0248997	.0241368	.0233936	.0226696	.0219647	23
24	.0232187	.0224709	.0217434	.0210358	.0203478	24
25	.0216851	.0209525	.0202407	.0195494	.0188782	25
26	.0202819	.0195643	.0188682	.0181931	.0175386	26
27	.0189944	.0182919	.0176113	.0169523	.0163144	27
28	.0178102	.0171225	.0164574	.0158144	.0151929	28
29	.0167183	.0160455	.0153958	.0147686	.0141634	29
30	.0157095	.0150514	.0144169	.0138054	.0132162	30
31	.0147755	.0141321	.0135127	.0129167	.0123434	31
32	.0139093	.0132804	.0126759	.0120952	.0115375	32
33	.0131046	.0124900	.0119003	.0113347	.0107925	33
34	.0123557	.0117554	.0111803	.0106296	.0101025	34
35	.0116579	.0110717	.0105110	.0099749	.0094628	35
36	.0110068	.0104345	.0098879	.0093663	.0088689	36
37	.0103984	.0098398	.0093073	.0087999	.0083169	37
38	.0098293	.0092842	.0087655	.0082722	.0078034	38
39	.0092964	.0087646	.0082595	.0077799	.0073250	39
40	.0087967	.0082782	.0077864	.0073203	.0068791	40
41	.0083279	.0078223	.0073436	.0068909	.0064630	41
42	.0078876	.0073947	.0069290	.0064893	.0060744	42
43	.0074736	.0069933	.0065403	.0061134	.0057114	43
44	.0070842	.0066163	.0061757	.0057613	.0053718	44
45	.0067175	.0062617	.0058334	.0054313	.0050540	45
46	.0063720	.0059282	.0055119	.0051218	.0047565	46
47	.0060463	.0056142	.0052097	.0048313	.0044777	47
48	.0057390	.0053184	.0049254	.0045585	.0042164	48
49	.0054489	.0050396	.0046579	.0043023	.0039713	49
50	.0051749	.0047767	.0044061	.0040615	.0037413	50

ANNUAL SINKING FUND

FOR THE REDEMPTION OF £1 CAPITAL INVESTED

			Rate Per Cent			
Yrs.	4.75	5	5.25	5.5	5.75	Yrs.
51	.0049160	.0045287	.0041689	.0038350	.0035254	51
52	.0046711	.0042945	.0039453	.0036219	.0033227	52
53	.0044395	.0040733	.0037345	.0034213	.0031322	53
54	.0042203	.0038644	.0035357	.0032325	.0029531	54
55	.0040128	.0036669	.0033481	.0030546	.0027848	55
56	.0038162	.0034801	.0031710	.0028870	.0026264	56
57	.0036299	.0033034	.0030037	.0027290	.0024775	57
58	.0034533	.0031363	.0028458	.0025801	.0023373	58
59	.0032859	.0029780	.0026965	.0024396	.0022053	59
60	.0031271	.0028282	.0025555	.0023071	.0020811	60
61	.0029764	.0026863	.0024221	.0021820	.0019640	61
62	.0028334	.0025518	.0022960	.0020640	.0018538	62
63	.0026976	.0024244	.0021768	.0019526	.0017499	63
64	.0025687	.0023037	.0020639	.0018474	.0016521	64
65	.0024462	.0021892	.0019571	.0017480	.0015598	65
66	.0023298	.0020806	.0018560	.0016541	.0014728	66
67	.0022192	.0019776	.0017604	.0015654	.0013908	67
68	.0021141	.0018799	.0016698	.0014816	.0013134	68
69	.0020142	.0017871	.0015840	.0014024	.0012405	69
70	.0019192	.0016992	.0015027	.0013275	.0011717	70
71	.0018288	.0016156	.0014257	.0012568	.0011067	71
72	.0017428	.0015363	.0013527	.0011898	.0010455	72
73	.0016610	.0014610	.0012836	.0011265	.0009876	73
74	.0015832	.0013895	.0012181	.0010667	.0009331	74
75	.0015091	.0013216	.0011560	.0010100	.0008816	75
76	.0014386	.0012571	.0010971	.0009565	.0008329	76
77	.0013715	.0011958	.0010413	.0009058	.0007870	77
78	.0013076	.0011376	.0009884	.0008578	.0007437	78
79	.0012467	.0010822	.0009382	.0008124	.0007027	79
80	.0011888	.0010296	.0008906	.0007695	.0006641	80
81	.0011336	.0009796	.0008455	.0007288	.0006276	81
82	.0010810	.0009321	.0008027	.0006904	.0005931	82
83	.0010309	.0008869	.0007620	.0006539	.0005605	83
84	.0009832	.0008440	.0007235	.0006195	.0005298	84
85	.0009378	.0008032	.0006870	.0005868	.0005007	85
86	.0008944	.0007643	.0006523	.0005559	.0004733	86
87	.0008531	.0007274	.0006193	.0005267	.0004473	87
88	.0008138	.0006923	.0005881	.0004990	.0004228	88
89	.0007763	.0006589	.0005585	.0004727	.0003997	89
90	.0007405	.0006271	.0005303	.0004479	.0003778	90
91	.0007065	.0005969	.0005036	.0004244	.0003571	91
92	.0006740	.0005681	.0004783	.0004021	.0003376	92
93	.0006430	.0005408	.0004542	.0003810	.0003192	93
94	.0006135	.0005148	.0004314	.0003610	.0003017	94
95	.0005853	.0004900	.0004097	.0003420	.0002852	95
96	.0005584	.0004665	.0003891	.0003241	.0002696	96
97	.0005328	.0004441	.0003695	.0003071	.0002549	97
98	.0005084	.0004227	.0003510	.0002910	.0002410	98
99	.0004851	.0004024	.0003334	.0002758	.0002278	99
100	.0004629	.0003831	.0003166	.0002613	.0002154	100

ANNUAL SINKING FUND

FOR THE REDEMPTION OF £1 CAPITAL INVESTED

Rate Per Cent

Yrs.	6	6.25	6.5	6.75	7	Yrs.
1	1.0000000	1.0000000	1.0000000	1.0000000	1.0000000	1
2	.4854369	.4848485	.4842615	.4836759	.4830918	2
3	.3141098	.3133415	.3125757	.3118124	.3110517	3
4	.2285915	.2277453	.2269027	.2260637	.2252281	4
5	.1773964	.1765132	.1756345	.1747604	.1738907	5
6	.1433626	.1424627	.1415683	.1406793	.1397958	6
7	.1191350	.1182300	.1173314	.1164391	.1155532	7
8	.1010359	.1001330	.0992373	.0983489	.0974678	8
9	.0870222	.0861260	.0852390	.0843582	.0834865	9
10	.0758680	.0749818	.0741047	.0732366	.0723775	10
11	.0667929	.0659191	.0650552	.0642012	.0633569	11
12	.0592770	.0584172	.0575682	.0567298	.0559020	12
13	.0529601	.0521156	.0512826	.0504610	.0496508	13
14	.0475849	.0467565	.0459405	.0451367	.0443449	14
15	.0429628	.0421512	.0413528	.0405673	.0397946	15
16	.0389521	.0381580	.0373776	.0366109	.0358576	16
17	.0354448	.0346683	.0339063	.0331587	.0324252	17
18	.0323565	.0315980	.0308546	.0301262	.0294126	18
19	.0296209	.0288804	.0281558	.0274467	.0267530	19
20	.0271846	.0264623	.0257564	.0250667	.0243929	20
21	.0250045	.0243004	.0236133	.0229429	.0222890	21
22	.0230456	.0223596	.0216912	.0210400	.0204058	22
23	.0212785	.0206106	.0199608	.0193287	.0187139	23
24	.0196790	.0190291	.0183977	.0177845	.0171890	24
25	.0182267	.0175946	.0169815	.0163869	.0158105	25
26	.0169043	.0162899	.0156948	.0151187	.0145610	26
27	.0156972	.0151001	.0145229	.0139649	.0134257	27
28	.0145926	.0140128	.0134531	.0129129	.0123919	28
29	.0135796	.0130168	.0124744	.0119519	.0114487	29
30	.0126489	.0121028	.0115774	.0110722	.0105864	30
31	.0117922	.0112626	.0107539	.0102656	.0097969	31
32	.0110023	.0104889	.0099966	.0095249	.0090729	32
33	.0102729	.0097754	.0092992	.0088437	.0084081	33
34	.0095984	.0091165	.0086561	.0082164	.0077967	34
35	.0089739	.0085073	.0080623	.0076381	.0072340	35
36	.0083948	.0079432	.0075133	.0071043	.0067153	36
37	.0078574	.0074205	.0070053	.0066111	.0062368	37
38	.0073581	.0069356	.0065348	.0061549	.0057951	38
39	.0068938	.0064853	.0060985	.0057327	.0053868	39
40	.0064615	.0060667	.0056937	.0053415	.0050091	40
41	.0060589	.0056775	.0053178	.0049788	.0046596	41
42	.0056834	.0053151	.0049684	.0046424	.0043359	42
43	.0053331	.0049775	.0046435	.0043300	.0040359	43
44	.0050061	.0046629	.0043412	.0040398	.0037577	44
45	.0047005	.0043694	.0040597	.0037701	.0034996	45
46	.0044149	.0040956	.0037974	.0035193	.0032600	46
47	.0041477	.0038399	.0035530	.0032859	.0030374	47
48	.0038977	.0036010	.0033251	.0030687	.0028307	48
49	.0036636	.0033777	.0031124	.0028664	.0026385	49
50	.0034443	.0031689	.0029139	.0026780	.0024598	50

ANNUAL SINKING FUND
FOR THE REDEMPTION OF £1 CAPITAL INVESTED

			Rate Per Cent			
Yrs.	7.25	7.5	8	8.5	9	Yrs.
1	1.0000000	1.0000000	1.0000000	1.0000000	1.0000000	1
2	.4825090	.4819277	.4807692	.4796163	.4784689	2
3	.3102934	.3095376	.3080335	.3065392	.3050548	3
4	.2243961	.2235675	.2219208	.2202879	.2186687	4
5	.1730255	.1721647	.1704565	.1687658	.1670925	5
6	.1389177	.1380449	.1363154	.1346071	.1329198	6
7	.1146736	.1138003	.1120724	.1103692	.1086905	7
8	.0965938	.0957270	.0940148	.0923307	.0906744	8
9	.0826228	.0817672	.0800797	.0784237	.0767988	9
10	.0715273	.0706859	.0690295	.0674077	.0658201	10
11	.0625224	.0616975	.0600763	.0584929	.0569467	11
12	.0550847	.0542778	.0526950	.0511529	.0496507	12
13	.0488519	.0480642	.0465218	.0450229	.0435666	13
14	.0435652	.0427974	.0412969	.0398424	.0384332	14
15	.0390347	.0382872	.0368295	.0354205	.0340589	15
16	.0351178	.0343912	.0329769	.0316135	.0302999	16
17	.0317057	.0310000	.0296294	.0283120	.0270462	17
18	.0287136	.0280290	.0267021	.0254304	.0242123	18
19	.0260745	.0254109	.0241276	.0229014	.0217304	19
20	.0237348	.0230922	.0218522	.0206710	.0195465	20
21	.0216512	.0210294	.0198323	.0186954	.0176166	21
22	.0197882	.0191869	.0180321	.0169389	.0159050	22
23	.0181162	.0175353	.0164222	.0153719	.0143819	23
24	.0166110	.0160501	.0149780	.0139698	.0130226	24
25	.0152519	.0147107	.0136788	.0127117	.0118063	25
26	.0140215	.0134996	.0125071	.0115802	.0107154	26
27	.0129049	.0124020	.0114481	.0105603	.0097349	27
28	.0118895	.0114052	.0104889	.0096391	.0088520	28
29	.0109642	.0104981	.0096185	.0088058	.0080557	29
30	.0101196	.0096712	.0088274	.0080506	.0073364	30
31	.0093473	.0089163	.0081073	.0073652	.0066856	31
32	.0086402	.0082260	.0074508	.0067425	.0060962	32
33	.0079917	.0075940	.0068516	.0061759	.0055617	33
34	.0073964	.0070146	.0063041	.0056598	.0050766	34
35	.0068492	.0064829	.0058033	.0051894	.0046358	35
36	.0063456	.0059945	.0053447	.0047601	.0042350	36
37	.0058819	.0055453	.0049244	.0043680	.0038703	37
38	.0054543	.0051320	.0045389	.0040097	.0035382	38
39	.0050599	.0047512	.0041851	.0036819	.0032356	39
40	.0046957	.0044003	.0038602	.0033820	.0029596	40
41	.0043592	.0040766	.0035615	.0031074	.0027079	41
42	.0040481	.0037779	.0032868	.0028558	.0024781	42
43	.0037602	.0035020	.0030341	.0026251	.0022684	43
44	.0034938	.0032471	.0028015	.0024136	.0020767	44
45	.0032470	.0030115	.0025873	.0022196	.0019017	45
46	.0030184	.0027935	.0023899	.0020415	.0017416	46
47	.0028065	.0025919	.0022080	.0018781	.0015952	47
48	.0026099	.0024053	.0020403	.0017280	.0014614	48
49	.0024276	.0022325	.0018856	.0015901	.0013389	49
50	.0022584	.0020724	.0017429	.0014633	.0012269	50

ANNUAL SINKING FUND
FOR THE REDEMPTION OF £1 CAPITAL INVESTED

			Rate Per Cent			
Yrs.	10	11	12	13	15	Yrs.
1	1.0000000	1.0000000	1.0000000	1.0000000	1.0000000	1
2	.4761905	.4739336	.4716981	.4694836	.4651163	2
3	.3021148	.2992131	.2963490	.2935220	.2879770	3
4	.2154708	.2123264	.2092344	.2061942	.2002654	4
5	.1637975	.1605703	.1574097	.1543145	.1483156	5
6	.1296074	.1263766	.1232257	.1201532	.1142369	6
7	.1054055	.1022153	.0991177	.0961108	.0903604	7
8	.0874440	.0843211	.0813028	.0783867	.0728501	8
9	.0736405	.0706017	.0676789	.0648689	.0595740	9
10	.0627454	.0598014	.0569842	.0542896	.0492521	10
11	.0539631	.0511210	.0484154	.0458415	.0410690	11
12	.0467633	.0440273	.0414368	.0389861	.0344808	12
13	.0407785	.0381510	.0356772	.0333503	.0291105	13
14	.0357462	.0332282	.0308712	.0286675	.0246885	14
15	.0314738	.0290652	.0268242	.0247418	.0210171	15
16	.0278166	.0255167	.0233900	.0214262	.0179477	16
17	.0246641	.0224715	.0204567	.0186084	.0153669	17
18	.0219302	.0198429	.0179373	.0162009	.0131863	18
19	.0195469	.0175625	.0157630	.0141344	.0113364	19
20	.0174596	.0155756	.0138788	.0123538	.0097615	20
21	.0156244	.0138379	.0122401	.0108143	.0084168	21
22	.0140051	.0123131	.0108105	.0094795	.0072658	22
23	.0125718	.0109712	.0095600	.0083191	.0062784	23
24	.0112998	.0097872	.0084634	.0073083	.0054298	24
25	.0101681	.0087402	.0075000	.0064259	.0046994	25
26	.0091590	.0078126	.0066519	.0056545	.0040698	26
27	.0082576	.0069892	.0059041	.0049791	.0035265	27
28	.0074510	.0062571	.0052439	.0043869	.0030571	28
29	.0067281	.0056055	.0046602	.0038672	.0026513	29
30	.0060792	.0050246	.0041437	.0034107	.0023002	30
31	.0054962	.0045063	.0036861	.0030092	.0019962	31
32	.0049717	.0040433	.0032803	.0026559	.0017328	32
33	.0044994	.0036294	.0029203	.0023449	.0015045	33
34	.0040737	.0032591	.0026006	.0020708	.0013066	34
35	.0036897	.0029275	.0023166	.0018292	.0011349	35
36	.0033431	.0026304	.0020641	.0016162	.0009859	36
37	.0030299	.0023642	.0018396	.0014282	.0008565	37
38	.0027469	.0021254	.0016398	.0012623	.0007443	38
39	.0024910	.0019111	.0014620	.0011158	.0006468	39
40	.0022594	.0017187	.0013036	.0009865	.0005621	40
41	.0020498	.0015460	.0011626	.0008722	.0004885	41
42	.0018600	.0013909	.0010370	.0007713	.0004246	42
43	.0016880	.0012515	.0009250	.0006821	.0003691	43
44	.0015322	.0011262	.0008252	.0006033	.0003209	44
45	.0013910	.0010135	.0007363	.0005336	.0002789	45
46	.0012630	.0009123	.0006569	.0004720	.0002425	46
47	.0011468	.0008212	.0005862	.0004175	.0002108	47
48	.0010415	.0007393	.0005231	.0003693	.0001833	48
49	.0009459	.0006656	.0004669	.0003267	.0001594	49
50	.0008592	.0005992	.0004167	.0002891	.0001385	50

AMOUNT OF ONE POUND

in a given Number of Years at Rates of
Interest ranging from
1% to 28%

(*i.e.*, the amount to which £1 invested now will accumulate at
Compound Interest)

Note:–No allowance has been made for the effect of income tax on interest
accumulations.

AMOUNT OF £1

			Rate Per Cent			
Yrs.	1	1.25	1.5	1.75	2	Yrs.
1	1.0100	1.0125	1.0150	1.0175	1.0200	1
2	1.0201	1.0252	1.0302	1.0353	1.0404	2
3	1.0303	1.0380	1.0457	1.0534	1.0612	3
4	1.0406	1.0509	1.0614	1.0719	1.0824	4
5	1.0510	1.0641	1.0773	1.0906	1.1041	5
6	1.0615	1.0774	1.0934	1.1097	1.1262	6
7	1.0721	1.0909	1.1098	1.1291	1.1487	7
8	1.0829	1.1045	1.1265	1.1489	1.1717	8
9	1.0937	1.1183	1.1434	1.1690	1.1951	9
10	1.1046	1.1323	1.1605	1.1894	1.2190	10
11	1.1157	1.1464	1.1779	1.2103	1.2434	11
12	1.1268	1.1608	1.1956	1.2314	1.2682	12
13	1.1381	1.1753	1.2136	1.2530	1.2936	13
14	1.1495	1.1900	1.2318	1.2749	1.3195	14
15	1.1610	1.2048	1.2502	1.2972	1.3459	15
16	1.1726	1.2199	1.2690	1.3199	1.3728	16
17	1.1843	1.2351	1.2880	1.3430	1.4002	17
18	1.1961	1.2506	1.3073	1.3665	1.4282	18
19	1.2081	1.2662	1.3270	1.3904	1.4568	19
20	1.2202	1.2820	1.3469	1.4148	1.4859	20
21	1.2324	1.2981	1.3671	1.4395	1.5157	21
22	1.2447	1.3143	1.3876	1.4647	1.5460	22
23	1.2572	1.3307	1.4084	1.4904	1.5769	23
24	1.2697	1.3474	1.4295	1.5164	1.6084	24
25	1.2824	1.3642	1.4509	1.5430	1.6406	25
26	1.2953	1.3812	1.4727	1.5700	1.6734	26
27	1.3082	1.3985	1.4948	1.5975	1.7069	27
28	1.3213	1.4160	1.5172	1.6254	1.7410	28
29	1.3345	1.4337	1.5400	1.6539	1.7758	29
30	1.3478	1.4516	1.5631	1.6828	1.8114	30
31	1.3613	1.4698	1.5865	1.7122	1.8476	31
32	1.3749	1.4881	1.6103	1.7422	1.8845	32
33	1.3887	1.5067	1.6345	1.7727	1.9222	33
34	1.4026	1.5256	1.6590	1.8037	1.9607	34
35	1.4166	1.5446	1.6839	1.8353	1.9999	35
36	1.4308	1.5639	1.7091	1.8674	2.0399	36
37	1.4451	1.5835	1.7348	1.9001	2.0807	37
38	1.4595	1.6033	1.7608	1.9333	2.1223	38
39	1.4741	1.6233	1.7872	1.9672	2.1647	39
40	1.4889	1.6436	1.8140	2.0016	2.2080	40
41	1.5038	1.6642	1.8412	2.0366	2.2522	41
42	1.5188	1.6850	1.8688	2.0723	2.2972	42
43	1.5340	1.7060	1.8969	2.1085	2.3432	43
44	1.5493	1.7274	1.9253	2.1454	2.3901	44
45	1.5648	1.7489	1.9542	2.1830	2.4379	45
46	1.5805	1.7708	1.9835	2.2212	2.4866	46
47	1.5963	1.7929	2.0133	2.2600	2.5363	47
48	1.6122	1.8154	2.0435	2.2996	2.5871	48
49	1.6283	1.8380	2.0741	2.3398	2.6388	49
50	1.6446	1.8610	2.1052	2.3808	2.6916	50

AMOUNT OF £1

			Rate Per Cent			
Yrs.	1	1.25	1.5	1.75	2	Yrs.
51	1.6611	1.8843	2.1368	2.4225	2.7454	51
52	1.6777	1.9078	2.1689	2.4648	2.8003	52
53	1.6945	1.9317	2.2014	2.5080	2.8563	53
54	1.7114	1.9558	2.2344	2.5519	2.9135	54
55	1.7285	1.9803	2.2679	2.5965	2.9717	55
56	1.7458	2.0050	2.3020	2.6420	3.0312	56
57	1.7633	2.0301	2.3365	2.6882	3.0918	57
58	1.7809	2.0555	2.3715	2.7352	3.1536	58
59	1.7987	2.0812	2.4071	2.7831	3.2167	59
60	1.8167	2.1072	2.4432	2.8318	3.2810	60
61	1.8349	2.1335	2.4799	2.8814	3.3467	61
62	1.8532	2.1602	2.5171	2.9318	3.4136	62
63	1.8717	2.1872	2.5548	2.9831	3.4819	63
64	1.8905	2.2145	2.5931	3.0353	3.5515	64
65	1.9094	2.2422	2.6320	3.0884	3.6225	65
66	1.9285	2.2702	2.6715	3.1425	3.6950	66
67	1.9477	2.2986	2.7116	3.1975	3.7689	67
68	1.9672	2.3274	2.7523	3.2534	3.8443	68
69	1.9869	2.3564	2.7936	3.3104	3.9211	69
70	2.0068	2.3859	2.8355	3.3683	3.9996	70
71	2.0268	2.4157	2.8780	3.4272	4.0795	71
72	2.0471	2.4459	2.9212	3.4872	4.1611	72
73	2.0676	2.4765	2.9650	3.5482	4.2444	73
74	2.0882	2.5075	3.0094	3.6103	4.3293	74
75	2.1091	2.5388	3.0546	3.6735	4.4158	75
76	2.1302	2.5705	3.1004	3.7378	4.5042	76
77	2.1515	2.6027	3.1469	3.8032	4.5942	77
78	2.1730	2.6352	3.1941	3.8698	4.6861	78
79	2.1948	2.6681	3.2420	3.9375	4.7798	79
80	2.2167	2.7015	3.2907	4.0064	4.8754	80
81	2.2389	2.7353	3.3400	4.0765	4.9729	81
82	2.2613	2.7694	3.3901	4.1478	5.0724	82
83	2.2839	2.8041	3.4410	4.2204	5.1739	83
84	2.3067	2.8391	3.4926	4.2943	5.2773	84
85	2.3298	2.8746	3.5450	4.3694	5.3829	85
86	2.3531	2.9105	3.5982	4.4459	5.4905	86
87	2.3766	2.9469	3.6521	4.5237	5.6003	87
88	2.4004	2.9838	3.7069	4.6029	5.7124	88
89	2.4244	3.0210	3.7625	4.6834	5.8266	89
90	2.4486	3.0588	3.8189	4.7654	5.9431	90
91	2.4731	3.0970	3.8762	4.8488	6.0620	91
92	2.4979	3.1358	3.9344	4.9336	6.1832	92
93	2.5228	3.1750	3.9934	5.0200	6.3069	93
94	2.5481	3.2146	4.0533	5.1078	6.4330	94
95	2.5735	3.2548	4.1141	5.1972	6.5617	95
96	2.5993	3.2955	4.1758	5.2882	6.6929	96
97	2.6253	3.3367	4.2384	5.3807	6.8268	97
98	2.6515	3.3784	4.3020	5.4749	6.9633	98
99	2.6780	3.4206	4.3665	5.5707	7.1026	99
100	2.7048	3.4634	4.4320	5.6682	7.2446	100

AMOUNT OF £1

			Rate Per Cent			
Yrs.	**2.25**	**2.5**	**2.75**	**3**	**3.25**	Yrs.
1	1.0225	1.0250	1.0275	1.0300	1.0325	1
2	1.0455	1.0506	1.0558	1.0609	1.0661	2
3	1.0690	1.0769	1.0848	1.0927	1.1007	3
4	1.0931	1.1038	1.1146	1.1255	1.1365	4
5	1.1177	1.1314	1.1453	1.1593	1.1734	5
6	1.1428	1.1597	1.1768	1.1941	1.2115	6
7	1.1685	1.1887	1.2091	1.2299	1.2509	7
8	1.1948	1.2184	1.2424	1.2668	1.2916	8
9	1.2217	1.2489	1.2765	1.3048	1.3336	9
10	1.2492	1.2801	1.3117	1.3439	1.3769	10
11	1.2773	1.3121	1.3477	1.3842	1.4216	11
12	1.3060	1.3449	1.3848	1.4258	1.4678	12
13	1.3354	1.3785	1.4229	1.4685	1.5156	13
14	1.3655	1.4130	1.4620	1.5126	1.5648	14
15	1.3962	1.4483	1.5022	1.5580	1.6157	15
16	1.4276	1.4845	1.5435	1.6047	1.6682	16
17	1.4597	1.5216	1.5860	1.6528	1.7224	17
18	1.4926	1.5597	1.6296	1.7024	1.7784	18
19	1.5262	1.5987	1.6744	1.7535	1.8362	19
20	1.5605	1.6386	1.7204	1.8061	1.8958	20
21	1.5956	1.6796	1.7677	1.8603	1.9575	21
22	1.6315	1.7216	1.8164	1.9161	2.0211	22
23	1.6682	1.7646	1.8663	1.9736	2.0868	23
24	1.7058	1.8087	1.9176	2.0328	2.1546	24
25	1.7441	1.8539	1.9704	2.0938	2.2246	25
26	1.7834	1.9003	2.0245	2.1566	2.2969	26
27	1.8235	1.9478	2.0802	2.2213	2.3715	27
28	1.8645	1.9965	2.1374	2.2879	2.4486	28
29	1.9065	2.0464	2.1962	2.3566	2.5282	29
30	1.9494	2.0976	2.2566	2.4273	2.6104	30
31	1.9933	2.1500	2.3187	2.5001	2.6952	31
32	2.0381	2.2038	2.3824	2.5751	2.7828	32
33	2.0840	2.2589	2.4479	2.6523	2.8732	33
34	2.1308	2.3153	2.5153	2.7319	2.9666	34
35	2.1788	2.3732	2.5844	2.8139	3.0630	35
36	2.2278	2.4325	2.6555	2.8983	3.1626	36
37	2.2779	2.4933	2.7285	2.9852	3.2654	37
38	2.3292	2.5557	2.8036	3.0748	3.3715	38
39	2.3816	2.6196	2.8807	3.1670	3.4811	39
40	2.4352	2.6851	2.9599	3.2620	3.5942	40
41	2.4900	2.7522	3.0413	3.3599	3.7110	41
42	2.5460	2.8210	3.1249	3.4607	3.8316	42
43	2.6033	2.8915	3.2108	3.5645	3.9561	43
44	2.6619	2.9638	3.2991	3.6715	4.0847	44
45	2.7218	3.0379	3.3899	3.7816	4.2175	45
46	2.7830	3.1139	3.4831	3.8950	4.3545	46
47	2.8456	3.1917	3.5789	4.0119	4.4961	47
48	2.9096	3.2715	3.6773	4.1323	4.6422	48
49	2.9751	3.3533	3.7784	4.2562	4.7931	49
50	3.0420	3.4371	3.8823	4.3839	4.9488	50

AMOUNT OF £1

			Rate Per Cent			
Yrs.	2.25	2.5	2.75	3	3.25	Yrs.
51	3.1105	3.5230	3.9891	4.5154	5.1097	51
52	3.1805	3.6111	4.0988	4.6509	5.2757	52
53	3.2520	3.7014	4.2115	4.7904	5.4472	53
54	3.3252	3.7939	4.3273	4.9341	5.6242	54
55	3.4000	3.8888	4.4463	5.0821	5.8070	55
56	3.4765	3.9860	4.5686	5.2346	5.9957	56
57	3.5547	4.0856	4.6942	5.3917	6.1906	57
58	3.6347	4.1878	4.8233	5.5534	6.3918	58
59	3.7165	4.2925	4.9560	5.7200	6.5995	59
60	3.8001	4.3998	5.0923	5.8916	6.8140	60
61	3.8856	4.5098	5.2323	6.0684	7.0355	61
62	3.9731	4.6225	5.3762	6.2504	7.2641	62
63	4.0625	4.7381	5.5240	6.4379	7.5002	63
64	4.1539	4.8565	5.6759	6.6311	7.7440	64
65	4.2473	4.9780	5.8320	6.8300	7.9957	65
66	4.3429	5.1024	5.9924	7.0349	8.2555	66
67	4.4406	5.2300	6.1572	7.2459	8.5238	67
68	4.5405	5.3607	6.3265	7.4633	8.8008	68
69	4.6427	5.4947	6.5005	7.6872	9.0869	69
70	4.7471	5.6321	6.6793	7.9178	9.3822	70
71	4.8540	5.7729	6.8629	8.1554	9.6871	71
72	4.9632	5.9172	7.0517	8.4000	10.0019	72
73	5.0748	6.0652	7.2456	8.6520	10.3270	73
74	5.1890	6.2168	7.4448	8.9116	10.6626	74
75	5.3058	6.3722	7.6496	9.1789	11.0092	75
76	5.4252	6.5315	7.8599	9.4543	11.3670	76
77	5.5472	6.6948	8.0761	9.7379	11.7364	77
78	5.6720	6.8622	8.2982	10.0301	12.1178	78
79	5.7997	7.0337	8.5264	10.3310	12.5117	79
80	5.9301	7.2096	8.7609	10.6409	12.9183	80
81	6.0636	7.3898	9.0018	10.9601	13.3381	81
82	6.2000	7.5746	9.2493	11.2889	13.7716	82
83	6.3395	7.7639	9.5037	11.6276	14.2192	83
84	6.4821	7.9580	9.7650	11.9764	14.6813	84
85	6.6280	8.1570	10.0336	12.3357	15.1585	85
86	6.7771	8.3609	10.3095	12.7058	15.6511	86
87	6.9296	8.5699	10.5930	13.0870	16.1598	87
88	7.0855	8.7842	10.8843	13.4796	16.6850	88
89	7.2449	9.0038	11.1836	13.8839	17.2272	89
90	7.4080	9.2289	11.4912	14.3005	17.7871	90
91	7.5746	9.4596	11.8072	14.7295	18.3652	91
92	7.7451	9.6961	12.1319	15.1714	18.9621	92
93	7.9193	9.9385	12.4655	15.6265	19.5783	93
94	8.0975	10.1869	12.8083	16.0953	20.2146	94
95	8.2797	10.4416	13.1605	16.5782	20.8716	95
96	8.4660	10.7026	13.5225	17.0755	21.5499	96
97	8.6565	10.9702	13.8943	17.5878	22.2503	97
98	8.8513	11.2445	14.2764	18.1154	22.9734	98
99	9.0504	11.5256	14.6690	18.6589	23.7201	99
100	9.2540	11.8137	15.0724	19.2186	24.4910	100

AMOUNT OF £1

			Rate Per Cent			
Yrs.	**3.5**	**3.75**	**4**	**4.25**	**4.5**	Yrs.
1	1.0350	1.0375	1.0400	1.0425	1.0450	1
2	1.0712	1.0764	1.0816	1.0868	1.0920	2
3	1.1087	1.1168	1.1249	1.1330	1.1412	3
4	1.1475	1.1587	1.1699	1.1811	1.1925	4
5	1.1877	1.2021	1.2167	1.2313	1.2462	5
6	1.2293	1.2472	1.2653	1.2837	1.3023	6
7	1.2723	1.2939	1.3159	1.3382	1.3609	7
8	1.3168	1.3425	1.3686	1.3951	1.4221	8
9	1.3629	1.3928	1.4233	1.4544	1.4861	9
10	1.4106	1.4450	1.4802	1.5162	1.5530	10
11	1.4600	1.4992	1.5395	1.5807	1.6229	11
12	1.5111	1.5555	1.6010	1.6478	1.6959	12
13	1.5640	1.6138	1.6651	1.7179	1.7722	13
14	1.6187	1.6743	1.7317	1.7909	1.8519	14
15	1.6753	1.7371	1.8009	1.8670	1.9353	15
16	1.7340	1.8022	1.8730	1.9463	2.0224	16
17	1.7947	1.8698	1.9479	2.0291	2.1134	17
18	1.8575	1.9399	2.0258	2.1153	2.2085	18
19	1.9225	2.0127	2.1068	2.2052	2.3079	19
20	1.9898	2.0882	2.1911	2.2989	2.4117	20
21	2.0594	2.1665	2.2788	2.3966	2.5202	21
22	2.1315	2.2477	2.3699	2.4985	2.6337	22
23	2.2061	2.3320	2.4647	2.6047	2.7522	23
24	2.2833	2.4194	2.5633	2.7153	2.8760	24
25	2.3632	2.5102	2.6658	2.8308	3.0054	25
26	2.4460	2.6043	2.7725	2.9511	3.1407	26
27	2.5316	2.7020	2.8834	3.0765	3.2820	27
28	2.6202	2.8033	2.9987	3.2072	3.4297	28
29	2.7119	2.9084	3.1187	3.3435	3.5840	29
30	2.8068	3.0175	3.2434	3.4856	3.7453	30
31	2.9050	3.1306	3.3731	3.6338	3.9139	31
32	3.0067	3.2480	3.5081	3.7882	4.0900	32
33	3.1119	3.3698	3.6484	3.9492	4.2740	33
34	3.2209	3.4962	3.7943	4.1171	4.4664	34
35	3.3336	3.6273	3.9461	4.2920	4.6673	35
36	3.4503	3.7633	4.1039	4.4744	4.8774	36
37	3.5710	3.9045	4.2681	4.6646	5.0969	37
38	3.6960	4.0509	4.4388	4.8628	5.3262	38
39	3.8254	4.2028	4.6164	5.0695	5.5659	39
40	3.9593	4.3604	4.8010	5.2850	5.8164	40
41	4.0978	4.5239	4.9931	5.5096	6.0781	41
42	4.2413	4.6935	5.1928	5.7437	6.3516	42
43	4.3897	4.8695	5.4005	5.9878	6.6374	43
44	4.5433	5.0522	5.6165	6.2423	6.9361	44
45	4.7024	5.2416	5.8412	6.5076	7.2482	45
46	4.8669	5.4382	6.0748	6.7842	7.5744	46
47	5.0373	5.6421	6.3178	7.0725	7.9153	47
48	5.2136	5.8537	6.5705	7.3731	8.2715	48
49	5.3961	6.0732	6.8333	7.6865	8.6437	49
50	5.5849	6.3009	7.1067	8.0131	9.0326	50

AMOUNT OF £1

			Rate Per Cent			
Yrs.	3.5	3.75	4	4.25	4.5	Yrs.
51	5.7804	6.5372	7.3910	8.3537	9.4391	51
52	5.9827	6.7824	7.6866	8.7087	9.8639	52
53	6.1921	7.0367	7.9941	9.0789	10.3077	53
54	6.4088	7.3006	8.3138	9.4647	10.7716	54
55	6.6331	7.5744	8.6464	9.8670	11.2563	55
56	6.8653	7.8584	8.9922	10.2863	11.7628	56
57	7.1056	8.1531	9.3519	10.7235	12.2922	57
58	7.3543	8.4588	9.7260	11.1792	12.8453	58
59	7.6117	8.7760	10.1150	11.6543	13.4234	59
60	7.8781	9.1051	10.5196	12.1497	14.0274	60
61	8.1538	9.4466	10.9404	12.6660	14.6586	61
62	8.4392	9.8008	11.3780	13.2043	15.3183	62
63	8.7346	10.1684	11.8332	13.7655	16.0076	63
64	9.0403	10.5497	12.3065	14.3505	16.7279	64
65	9.3567	10.9453	12.7987	14.9604	17.4807	65
66	9.6842	11.3557	13.3107	15.5963	18.2673	66
67	10.0231	11.7816	13.8431	16.2591	19.0894	67
68	10.3739	12.2234	14.3968	16.9501	19.9484	68
69	10.7370	12.6818	14.9727	17.6705	20.8461	69
70	11.1128	13.1573	15.5716	18.4215	21.7841	70
71	11.5018	13.6507	16.1945	19.2044	22.7644	71
72	11.9043	14.1626	16.8423	20.0206	23.7888	72
73	12.3210	14.6937	17.5160	20.8715	24.8593	73
74	12.7522	15.2447	18.2166	21.7585	25.9780	74
75	13.1986	15.8164	18.9453	22.6832	27.1470	75
76	13.6605	16.4095	19.7031	23.6473	28.3686	76
77	14.1386	17.0249	20.4912	24.6523	29.6452	77
78	14.6335	17.6633	21.3108	25.7000	30.9792	78
79	15.1456	18.3257	22.1633	26.7922	32.3733	79
80	15.6757	19.0129	23.0498	27.9309	33.8301	80
81	16.2244	19.7259	23.9718	29.1180	35.3525	81
82	16.7922	20.4656	24.9307	30.3555	36.9433	82
83	17.3800	21.2331	25.9279	31.6456	38.6058	83
84	17.9883	22.0293	26.9650	32.9905	40.3430	84
85	18.6179	22.8554	28.0436	34.3926	42.1585	85
86	19.2695	23.7125	29.1653	35.8543	44.0556	86
87	19.9439	24.6017	30.3320	37.3781	46.0381	87
88	20.6420	25.5243	31.5452	38.9667	48.1098	88
89	21.3644	26.4814	32.8071	40.6228	50.2747	89
90	22.1122	27.4745	34.1193	42.3493	52.5371	90
91	22.8861	28.5048	35.4841	44.1491	54.9013	91
92	23.6871	29.5737	36.9035	46.0254	57.3718	92
93	24.5162	30.6827	38.3796	47.9815	59.9536	93
94	25.3742	31.8333	39.9148	50.0207	62.6515	94
95	26.2623	33.0271	41.5114	52.1466	65.4708	95
96	27.1815	34.2656	43.1718	54.3628	68.4170	96
97	28.1329	35.5505	44.8987	56.6733	71.4957	97
98	29.1175	36.8837	46.6947	59.0819	74.7130	98
99	30.1366	38.2668	48.5625	61.5929	78.0751	99
100	31.1914	39.7018	50.5049	64.2105	81.5885	100

AMOUNT OF £1

			Rate Per Cent			
Yrs.	**4.75**	**5**	**5.25**	**5.5**	**5.75**	Yrs.
1	1.0475	1.0500	1.0525	1.0550	1.0575	1
2	1.0973	1.1025	1.1078	1.1130	1.1183	2
3	1.1494	1.1576	1.1659	1.1742	1.1826	3
4	1.2040	1.2155	1.2271	1.2388	1.2506	4
5	1.2612	1.2763	1.2915	1.3070	1.3225	5
6	1.3211	1.3401	1.3594	1.3788	1.3986	6
7	1.3838	1.4071	1.4307	1.4547	1.4790	7
8	1.4495	1.4775	1.5058	1.5347	1.5640	8
9	1.5184	1.5513	1.5849	1.6191	1.6540	9
10	1.5905	1.6289	1.6681	1.7081	1.7491	10
11	1.6661	1.7103	1.7557	1.8021	1.8496	11
12	1.7452	1.7959	1.8478	1.9012	1.9560	12
13	1.8281	1.8856	1.9449	2.0058	2.0684	13
14	1.9149	1.9799	2.0470	2.1161	2.1874	14
15	2.0059	2.0789	2.1544	2.2325	2.3132	15
16	2.1012	2.1829	2.2675	2.3553	2.4462	16
17	2.2010	2.2920	2.3866	2.4848	2.5868	17
18	2.3055	2.4066	2.5119	2.6215	2.7356	18
19	2.4151	2.5270	2.6437	2.7656	2.8929	19
20	2.5298	2.6533	2.7825	2.9178	3.0592	20
21	2.6499	2.7860	2.9286	3.0782	3.2351	21
22	2.7758	2.9253	3.0824	3.2475	3.4211	22
23	2.9077	3.0715	3.2442	3.4262	3.6178	23
24	3.0458	3.2251	3.4145	3.6146	3.8259	24
25	3.1904	3.3864	3.5938	3.8134	4.0458	25
26	3.3420	3.5557	3.7825	4.0231	4.2785	26
27	3.5007	3.7335	3.9810	4.2444	4.5245	27
28	3.6670	3.9201	4.1900	4.4778	4.7847	28
29	3.8412	4.1161	4.4100	4.7241	5.0598	29
30	4.0237	4.3219	4.6416	4.9840	5.3507	30
31	4.2148	4.5380	4.8852	5.2581	5.6584	31
32	4.4150	4.7649	5.1417	5.5473	5.9837	32
33	4.6247	5.0032	5.4116	5.8524	6.3278	33
34	4.8444	5.2533	5.6958	6.1742	6.6916	34
35	5.0745	5.5160	5.9948	6.5138	7.0764	35
36	5.3155	5.7918	6.3095	6.8721	7.4833	36
37	5.5680	6.0814	6.6408	7.2501	7.9136	37
38	5.8325	6.3855	6.9894	7.6488	8.3686	38
39	6.1095	6.7048	7.3563	8.0695	8.8498	39
40	6.3997	7.0400	7.7426	8.5133	9.3587	40
41	6.7037	7.3920	8.1490	8.9815	9.8968	41
42	7.0221	7.7616	8.5769	9.4755	10.4659	42
43	7.3557	8.1497	9.0271	9.9967	11.0677	43
44	7.7051	8.5572	9.5011	10.5465	11.7041	44
45	8.0711	8.9850	9.9999	11.1266	12.3770	45
46	8.4545	9.4343	10.5249	11.7385	13.0887	46
47	8.8560	9.9060	11.0774	12.3841	13.8413	47
48	9.2767	10.4013	11.6590	13.0653	14.6372	48
49	9.7173	10.9213	12.2711	13.7838	15.4788	49
50	10.1789	11.4674	12.9153	14.5420	16.3689	50

AMOUNT OF £1

			Rate Per Cent			
Yrs.	**4.75**	**5**	**5.25**	**5.5**	**5.75**	Yrs.
51	10.6624	12.0408	13.5934	15.3418	17.3101	51
52	11.1689	12.6428	14.3070	16.1856	18.3054	52
53	11.6994	13.2749	15.0581	17.0758	19.3580	53
54	12.2551	13.9387	15.8487	18.0149	20.4711	54
55	12.8372	14.6356	16.6808	19.0058	21.6481	55
56	13.4470	15.3674	17.5565	20.0511	22.8929	56
57	14.0857	16.1358	18.4782	21.1539	24.2093	57
58	14.7548	16.9426	19.4483	22.3174	25.6013	58
59	15.4557	17.7897	20.4694	23.5448	27.0734	59
60	16.1898	18.6792	21.5440	24.8398	28.6301	60
61	16.9588	19.6131	22.6751	26.2060	30.2763	61
62	17.7644	20.5938	23.8655	27.6473	32.0172	62
63	18.6082	21.6235	25.1184	29.1679	33.8582	63
64	19.4921	22.7047	26.4372	30.7721	35.8050	64
65	20.4179	23.8399	27.8251	32.4646	37.8638	65
66	21.3878	25.0319	29.2859	34.2501	40.0410	66
67	22.4037	26.2835	30.8234	36.1339	42.3433	67
68	23.4679	27.5977	32.4417	38.1213	44.7781	68
69	24.5826	28.9775	34.1449	40.2179	47.3528	69
70	25.7503	30.4264	35.9375	42.4299	50.0756	70
71	26.9734	31.9477	37.8242	44.7636	52.9550	71
72	28.2547	33.5451	39.8099	47.2256	55.9999	72
73	29.5968	35.2224	41.9000	49.8230	59.2199	73
74	31.0026	36.9835	44.0997	52.5632	62.6250	74
75	32.4752	38.8327	46.4149	55.4542	66.2260	75
76	34.0178	40.7743	48.8517	58.5042	70.0339	76
77	35.6337	42.8130	51.4164	61.7219	74.0609	77
78	37.3263	44.9537	54.1158	65.1166	78.3194	78
79	39.0993	47.2014	56.9569	68.6980	82.8228	79
80	40.9565	49.5614	59.9471	72.4764	87.5851	80
81	42.9019	52.0395	63.0943	76.4626	92.6212	81
82	44.9397	54.6415	66.4068	80.6681	97.9469	82
83	47.0744	57.3736	69.8932	85.1048	103.5789	83
84	49.3104	60.2422	73.5626	89.7856	109.5347	84
85	51.6527	63.2544	77.4246	94.7238	115.8329	85
86	54.1062	66.4171	81.4894	99.9336	122.4933	86
87	56.6762	69.7379	85.7676	105.4299	129.5367	87
88	59.3683	73.2248	90.2704	111.2286	136.9850	88
89	62.1883	76.8861	95.0096	117.3462	144.8617	89
90	65.1423	80.7304	99.9976	123.8002	153.1912	90
91	68.2365	84.7669	105.2474	130.6092	161.9997	91
92	71.4778	89.0052	110.7729	137.7927	171.3147	92
93	74.8729	93.4555	116.5885	145.3713	181.1653	93
94	78.4294	98.1283	122.7094	153.3667	191.5823	94
95	82.1548	103.0347	129.1516	161.8019	202.5983	95
96	86.0572	108.1864	135.9321	170.7010	214.2477	96
97	90.1449	113.5957	143.0685	180.0896	226.5669	97
98	94.4268	119.2755	150.5796	189.9945	239.5945	98
99	98.9120	125.2393	158.4851	200.4442	253.3712	99
100	103.6104	131.5013	166.8055	211.4686	267.9400	100

AMOUNT OF £1

			Rate Per Cent			
Yrs.	6	6.25	6.5	7	7.5	Yrs.
1	1.0600	1.0625	1.0650	1.0700	1.0750	1
2	1.1236	1.1289	1.1342	1.1449	1.1556	2
3	1.1910	1.1995	1.2079	1.2250	1.2423	3
4	1.2625	1.2744	1.2865	1.3108	1.3355	4
5	1.3382	1.3541	1.3701	1.4026	1.4356	5
6	1.4185	1.4387	1.4591	1.5007	1.5433	6
7	1.5036	1.5286	1.5540	1.6058	1.6590	7
8	1.5938	1.6242	1.6550	1.7182	1.7835	8
9	1.6895	1.7257	1.7626	1.8385	1.9172	9
10	1.7908	1.8335	1.8771	1.9672	2.0610	10
11	1.8983	1.9481	1.9992	2.1049	2.2156	11
12	2.0122	2.0699	2.1291	2.2522	2.3818	12
13	2.1329	2.1993	2.2675	2.4098	2.5604	13
14	2.2609	2.3367	2.4149	2.5785	2.7524	14
15	2.3966	2.4828	2.5718	2.7590	2.9589	15
16	2.5404	2.6379	2.7390	2.9522	3.1808	16
17	2.6928	2.8028	2.9170	3.1588	3.4194	17
18	2.8543	2.9780	3.1067	3.3799	3.6758	18
19	3.0256	3.1641	3.3086	3.6165	3.9515	19
20	3.2071	3.3619	3.5236	3.8697	4.2479	20
21	3.3996	3.5720	3.7527	4.1406	4.5664	21
22	3.6035	3.7952	3.9966	4.4304	4.9089	22
23	3.8197	4.0324	4.2564	4.7405	5.2771	23
24	4.0489	4.2844	4.5331	5.0724	5.6729	24
25	4.2919	4.5522	4.8277	5.4274	6.0983	25
26	4.5494	4.8367	5.1415	5.8074	6.5557	26
27	4.8223	5.1390	5.4757	6.2139	7.0474	27
28	5.1117	5.4602	5.8316	6.6488	7.5759	28
29	5.4184	5.8015	6.2107	7.1143	8.1441	29
30	5.7435	6.1641	6.6144	7.6123	8.7550	30
31	6.0881	6.5493	7.0443	8.1451	9.4116	31
32	6.4534	6.9587	7.5022	8.7153	10.1174	32
33	6.8406	7.3936	7.9898	9.3253	10.8763	33
34	7.2510	7.8557	8.5092	9.9781	11.6920	34
35	7.6861	8.3467	9.0623	10.6766	12.5689	35
36	8.1473	8.8683	9.6513	11.4239	13.5115	36
37	8.6361	9.4226	10.2786	12.2236	14.5249	37
38	9.1543	10.0115	10.9467	13.0793	15.6143	38
39	9.7035	10.6372	11.6583	13.9948	16.7853	39
40	10.2857	11.3021	12.4161	14.9745	18.0442	40
41	10.9029	12.0084	13.2231	16.0227	19.3976	41
42	11.5570	12.7590	14.0826	17.1443	20.8524	42
43	12.2505	13.5564	14.9980	18.3444	22.4163	43
44	12.9855	14.4037	15.9729	19.6285	24.0975	44
45	13.7646	15.3039	17.0111	21.0025	25.9048	45
46	14.5905	16.2604	18.1168	22.4726	27.8477	46
47	15.4659	17.2767	19.2944	24.0457	29.9363	47
48	16.3939	18.3565	20.5485	25.7289	32.1815	48
49	17.3775	19.5037	21.8842	27.5299	34.5951	49
50	18.4202	20.7227	23.3067	29.4570	37.1897	50

AMOUNT OF £1

			Rate Per Cent			
Yrs.	6	6.25	6.5	7	7.5	Yrs.
51	19.5254	22.0179	24.8216	31.5190	39.9790	51
52	20.6969	23.3940	26.4350	33.7253	42.9774	52
53	21.9387	24.8561	28.1533	36.0861	46.2007	53
54	23.2550	26.4097	29.9833	38.6122	49.6658	54
55	24.6503	28.0603	31.9322	41.3150	53.3907	55
56	26.1293	29.8140	34.0078	44.2071	57.3950	56
57	27.6971	31.6774	36.2183	47.3015	61.6996	57
58	29.3589	33.6572	38.5725	50.6127	66.3271	58
59	31.1205	35.7608	41.0797	54.1555	71.3016	59
60	32.9877	37.9959	43.7498	57.9464	76.6492	60
61	34.9670	40.3706	46.5936	62.0027	82.3979	61
62	37.0650	42.8938	49.6222	66.3429	88.5778	62
63	39.2889	45.5746	52.8476	70.9869	95.2211	63
64	41.6462	48.4230	56.2827	75.9559	102.3627	64
65	44.1450	51.4495	59.9411	81.2729	110.0399	65
66	46.7937	54.6651	63.8372	86.9620	118.2929	66
67	49.6013	58.0816	67.9867	93.0493	127.1649	67
68	52.5774	61.7117	72.4058	99.5627	136.7022	68
69	55.7320	65.5687	77.1122	106.5321	146.9549	69
70	59.0759	69.6668	82.1245	113.9894	157.9765	70
71	62.6205	74.0209	87.4626	121.9686	169.8247	71
72	66.3777	78.6473	93.1476	130.5065	182.5616	72
73	70.3604	83.5627	99.2022	139.6419	196.2537	73
74	74.5820	88.7854	105.6504	149.4168	210.9727	74
75	79.0569	94.3345	112.5176	159.8760	226.7957	75
76	83.8003	100.2304	119.8313	171.0673	243.8054	76
77	88.8284	106.4948	127.6203	183.0421	262.0908	77
78	94.1581	113.1507	135.9156	195.8550	281.7476	78
79	99.8075	120.2226	144.7501	209.5648	302.8787	79
80	105.7960	127.7365	154.1589	224.2344	325.5946	80
81	112.1438	135.7201	164.1792	239.9308	350.0142	81
82	118.8724	144.2026	174.8509	256.7260	376.2652	82
83	126.0047	153.2152	186.2162	274.6968	404.4851	83
84	133.5650	162.7912	198.3202	293.9255	434.8215	84
85	141.5789	172.9656	211.2111	314.5003	467.4331	85
86	150.0736	183.7760	224.9398	336.5154	502.4906	86
87	159.0781	195.2620	239.5609	360.0714	540.1774	87
88	168.6227	207.4658	255.1323	385.2764	580.6907	88
89	178.7401	220.4325	271.7159	412.2458	624.2425	89
90	189.4645	234.2095	289.3775	441.1030	671.0607	90
91	200.8324	248.8476	308.1870	471.9802	721.3902	91
92	212.8823	264.4006	328.2191	505.0188	775.4945	92
93	225.6553	280.9256	349.5534	540.3701	833.6566	93
94	239.1946	298.4834	372.2744	578.1960	896.1808	94
95	253.5463	317.1387	396.4722	618.6697	963.3944	95
96	268.7590	336.9598	422.2429	661.9766	1035.6489	96
97	284.8846	358.0198	449.6887	708.3150	1113.3226	97
98	301.9776	380.3960	478.9184	757.8970	1196.8218	98
99	320.0963	404.1708	510.0481	810.9498	1286.5835	99
100	339.3021	429.4315	543.2013	867.7163	1383.0772	100

AMOUNT OF £1

			Rate Per Cent			
Yrs.	8	9	10	11	12	Yrs.
1	1.0800	1.0900	1.1000	1.1100	1.1200	1
2	1.1664	1.1881	1.2100	1.2321	1.2544	2
3	1.2597	1.2950	1.3310	1.3676	1.4049	3
4	1.3605	1.4116	1.4641	1.5181	1.5735	4
5	1.4693	1.5386	1.6105	1.6851	1.7623	5
6	1.5869	1.6771	1.7716	1.8704	1.9738	6
7	1.7138	1.8280	1.9487	2.0762	2.2107	7
8	1.8509	1.9926	2.1436	2.3045	2.4760	8
9	1.9990	2.1719	2.3579	2.5580	2.7731	9
10	2.1589	2.3674	2.5937	2.8394	3.1058	10
11	2.3316	2.5804	2.8531	3.1518	3.4785	11
12	2.5182	2.8127	3.1384	3.4985	3.8960	12
13	2.7196	3.0658	3.4523	3.8833	4.3635	13
14	2.9372	3.3417	3.7975	4.3104	4.8871	14
15	3.1722	3.6425	4.1772	4.7846	5.4736	15
16	3.4259	3.9703	4.5950	5.3109	6.1304	16
17	3.7000	4.3276	5.0545	5.8951	6.8660	17
18	3.9960	4.7171	5.5599	6.5436	7.6900	18
19	4.3157	5.1417	6.1159	7.2633	8.6128	19
20	4.6610	5.6044	6.7275	8.0623	9.6463	20
21	5.0338	6.1088	7.4002	8.9492	10.8038	21
22	5.4365	6.6586	8.1403	9.9336	12.1003	22
23	5.8715	7.2579	8.9543	11.0263	13.5523	23
24	6.3412	7.9111	9.8497	12.2392	15.1786	24
25	6.8485	8.6231	10.8347	13.5855	17.0001	25
26	7.3964	9.3992	11.9182	15.0799	19.0401	26
27	7.9881	10.2451	13.1100	16.7386	21.3249	27
28	8.6271	11.1671	14.4210	18.5799	23.8839	28
29	9.3173	12.1722	15.8631	20.6237	26.7499	29
30	10.0627	13.2677	17.4494	22.8923	29.9599	30
31	10.8677	14.4618	19.1943	25.4104	33.5551	31
32	11.7371	15.7633	21.1138	28.2056	37.5817	32
33	12.6760	17.1820	23.2252	31.3082	42.0915	33
34	13.6901	18.7284	25.5477	34.7521	47.1425	34
35	14.7853	20.4140	28.1024	38.5749	52.7996	35
36	15.9682	22.2512	30.9127	42.8181	59.1356	36
37	17.2456	24.2538	34.0039	47.5281	66.2318	37
38	18.6253	26.4367	37.4043	52.7562	74.1797	38
39	20.1153	28.8160	41.1448	58.5593	83.0812	39
40	21.7245	31.4094	45.2593	65.0009	93.0510	40
41	23.4625	34.2363	49.7852	72.1510	104.2171	41
42	25.3395	37.3175	54.7637	80.0876	116.7231	42
43	27.3666	40.6761	60.2401	88.8972	130.7299	43
44	29.5560	44.3370	66.2641	98.6759	146.4175	44
45	31.9204	48.3273	72.8905	109.5302	163.9876	45
46	34.4741	52.6767	80.1795	121.5786	183.6661	46
47	37.2320	57.4176	88.1975	134.9522	205.7061	47
48	40.2106	62.5852	97.0172	149.7970	230.3908	48
49	43.4274	68.2179	106.7190	166.2746	258.0377	49
50	46.9016	74.3575	117.3909	184.5648	289.0022	50

AMOUNT OF £1

Yrs.	8	9	10	11	12	Yrs.
			Rate Per Cent			
51	50.6537	81.0497	129.1299	204.8670	323.6825	51
52	54.7060	88.3442	142.0429	227.4023	362.5243	52
53	59.0825	96.2951	156.2472	252.4166	406.0273	53
54	63.8091	104.9617	171.8719	280.1824	454.7505	54
55	68.9139	114.4083	189.0591	311.0025	509.3206	55
56	74.4270	124.7050	207.9651	345.2127	570.4391	56
57	80.3811	135.9285	228.7616	383.1861	638.8918	57
58	86.8116	148.1620	251.6377	425.3366	715.5588	58
59	93.7565	161.4966	276.8015	472.1236	801.4258	59
60	101.2571	176.0313	304.4816	524.0572	897.5969	60
61	109.3576	191.8741	334.9298	581.7035	1005.3086	61
62	118.1062	209.1428	368.4228	645.6909	1125.9456	62
63	127.5547	227.9656	405.2651	716.7169	1261.0591	63
64	137.7591	248.4825	445.7916	795.5558	1412.3862	64
65	148.7798	270.8460	490.3707	883.0669	1581.8725	65
66	160.6822	295.2221	539.4078	980.2043	1771.6972	66
67	173.5368	321.7921	593.3486	1088.0268	1984.3009	67
68	187.4198	350.7534	652.6834	1207.7097	2222.4170	68
69	202.4133	382.3212	717.9518	1340.5578	2489.1070	69
70	218.6064	416.7301	789.7470	1488.0191	2787.7998	70
71	236.0949	454.2358	868.7217	1651.7012	3122.3358	71
72	254.9825	495.1170	955.5938	1833.3884	3497.0161	72
73	275.3811	539.6775	1051.1532	2035.0611	3916.6580	73
74	297.4116	588.2485	1156.2685	2258.9178	4386.6570	74
75	321.2045	641.1909	1271.8954	2507.3988	4913.0558	75
76	346.9009	698.8981	1399.0849	2783.2126	5502.6225	76
77	374.6530	761.7989	1538.9934	3089.3660	6162.9372	77
78	404.6252	830.3608	1692.8927	3429.1963	6902.4897	78
79	436.9952	905.0933	1862.1820	3806.4079	7730.7885	79
80	471.9548	986.5517	2048.4002	4225.1128	8658.4831	80
81	509.7112	1075.3413	2253.2402	4689.8752	9697.5011	81
82	550.4881	1172.1220	2478.5643	5205.7614	10861.2012	82
83	594.5272	1277.6130	2726.4207	5778.3952	12164.5453	83
84	642.0893	1392.5982	2999.0628	6414.0186	13624.2908	84
85	693.4565	1517.9320	3298.9690	7119.5607	15259.2057	85
86	748.9330	1654.5459	3628.8659	7902.7124	17090.3104	86
87	808.8476	1803.4550	3991.7525	8772.0107	19141.1476	87
88	873.5555	1965.7660	4390.9278	9736.9319	21438.0853	88
89	943.4399	2142.6849	4830.0206	10807.9944	24010.6556	89
90	1018.9151	2335.5266	5313.0226	11996.8738	26891.9342	90
91	1100.4283	2545.7240	5844.3249	13316.5299	30118.9663	91
92	1188.4626	2774.8391	6428.7574	14781.3482	33733.2423	92
93	1283.5396	3024.5747	7071.6331	16407.2965	37781.2314	93
94	1386.2227	3296.7864	7778.7964	18212.0991	42314.9791	94
95	1497.1205	3593.4971	8556.6760	20215.4301	47392.7766	95
96	1616.8902	3916.9119	9412.3437	22439.1274	53079.9098	96
97	1746.2414	4269.4340	10353.5780	24907.4314	59449.4990	97
98	1885.9407	4653.6830	11388.9358	27647.2488	66583.4389	98
99	2036.8160	5072.5145	12527.8294	30688.4462	74573.4515	99
100	2199.7613	5529.0408	13780.6123	34064.1753	83522.2657	100

AMOUNT OF £1

		Rate Per Cent			
Yrs.	**13**	**14**	**15**	**16**	Yrs.
1	1.1300	1.1400	1.1500	1.1600	1
2	1.2769	1.2996	1.3225	1.3456	2
3	1.4429	1.4815	1.5209	1.5609	3
4	1.6305	1.6890	1.7490	1.8106	4
5	1.8424	1.9254	2.0114	2.1003	5
6	2.0820	2.1950	2.3131	2.4364	6
7	2.3526	2.5023	2.6600	2.8262	7
8	2.6584	2.8526	3.0590	3.2784	8
9	3.0040	3.2519	3.5179	3.8030	9
10	3.3946	3.7072	4.0456	4.4114	10
11	3.8359	4.2262	4.6524	5.1173	11
12	4.3345	4.8179	5.3503	5.9360	12
13	4.8980	5.4924	6.1528	6.8858	13
14	5.5348	6.2613	7.0757	7.9875	14
15	6.2543	7.1379	8.1371	9.2655	15
16	7.0673	8.1372	9.3576	10.7480	16
17	7.9861	9.2765	10.7613	12.4677	17
18	9.0243	10.5752	12.3755	14.4625	18
19	10.1974	12.0557	14.2318	16.7765	19
20	11.5231	13.7435	16.3665	19.4608	20
21	13.0211	15.6676	18.8215	22.5745	21
22	14.7138	17.8610	21.6447	26.1864	22
23	16.6266	20.3616	24.8915	30.3762	23
24	18.7881	23.2122	28.6252	35.2364	24
25	21.2305	26.4619	32.9190	40.8742	25
26	23.9905	30.1666	37.8568	47.4141	26
27	27.1093	34.3899	43.5353	55.0004	27
28	30.6335	39.2045	50.0656	63.8004	28
29	34.6158	44.6931	57.5755	74.0085	29
30	39.1159	50.9502	66.2118	85.8499	30
31	44.2010	58.0832	76.1435	99.5859	31
32	49.9471	66.2148	87.5651	115.5196	32
33	56.4402	75.4849	100.6998	134.0027	33
34	63.7774	86.0528	115.8048	155.4432	34
35	72.0685	98.1002	133.1755	180.3141	35
36	81.4374	111.8342	153.1519	209.1643	36
37	92.0243	127.4910	176.1246	242.6306	37
38	103.9874	145.3397	202.5433	281.4515	38
39	117.5058	165.6873	232.9248	326.4838	39
40	132.7816	188.8835	267.8635	378.7212	40
45	244.6414	363.6791	538.7693	795.4438	45
50	450.7359	700.2330	1083.6574	1670.7038	50
55	830.4517	1348.2388	2179.6222	3509.0488	55
60	1530.0535	2595.9187	4383.9987	7370.2014	60
65	2819.0243	4998.2196	8817.7874	15479.9410	65
70	5193.8696	9623.6450	17735.7200	32513.1648	70
75	9569.3681	18529.5064	35672.8680	68288.7545	75
80	17630.9405	35676.9818	71750.8794	143429.7159	80
90	59849.4155	132262.4674	290272.3252	632730.8800	90
100	203162.8742	490326.2381	1174313.4507	2791251.1994	100

AMOUNT OF £1

		Rate Per Cent			
Yrs.	17	18	19	20	Yrs.
1	1.1700	1.1800	1.1900	1.2000	1
2	1.3689	1.3924	1.4161	1.4400	2
3	1.6016	1.6430	1.6852	1.7280	3
4	1.8739	1.9388	2.0053	2.0736	4
5	2.1924	2.2878	2.3864	2.4883	5
6	2.5652	2.6996	2.8398	2.9860	6
7	3.0012	3.1855	3.3793	3.5832	7
8	3.5115	3.7589	4.0214	4.2998	8
9	4.1084	4.4355	4.7854	5.1598	9
10	4.8068	5.2338	5.6947	6.1917	10
11	5.6240	6.1759	6.7767	7.4301	11
12	6.5801	7.2876	8.0642	8.9161	12
13	7.6987	8.5994	9.5964	10.6993	13
14	9.0075	10.1472	11.4198	12.8392	14
15	10.5387	11.9737	13.5895	15.4070	15
16	12.3303	14.1290	16.1715	18.4884	16
17	14.4265	16.6722	19.2441	22.1861	17
18	16.8790	19.6733	22.9005	26.6233	18
19	19.7484	23.2144	27.2516	31.9480	19
20	23.1056	27.3930	32.4294	38.3376	20
21	27.0336	32.3238	38.5910	46.0051	21
22	31.6293	38.1421	45.9233	55.2061	22
23	37.0062	45.0076	54.6487	66.2474	23
24	43.2973	53.1090	65.0320	79.4968	24
25	50.6578	62.6686	77.3881	95.3962	25
26	59.2697	73.9490	92.0918	114.4755	26
27	69.3455	87.2598	109.5893	137.3706	27
28	81.1342	102.9666	130.4112	164.8447	28
29	94.9271	121.5005	155.1893	197.8136	29
30	111.0647	143.3706	184.6753	237.3763	30
31	129.9456	169.1774	219.7636	284.8516	31
32	152.0364	199.6293	261.5187	341.8219	32
33	177.8826	235.5625	311.2073	410.1863	33
34	208.1226	277.9638	370.3366	492.2235	34
35	243.5035	327.9973	440.7006	590.6682	35
36	284.8991	387.0368	524.4337	708.8019	36
37	333.3319	456.7034	624.0761	850.5622	37
38	389.9983	538.9100	742.6506	1020.6747	38
39	456.2980	635.9139	883.7542	1224.8096	39
40	533.8687	750.3783	1051.6675	1469.7716	40
45	1170.4794	1716.6839	2509.6506	3657.2620	45
50	2566.2153	3927.3569	5988.9139	9100.4381	50
55	5626.2937	8984.8411	14291.6666	22644.8023	55
60	12335.3565	20555.1400	34104.9709	56347.5144	60
65	27044.6281	47025.1809	81386.5222	140210.6469	65
70	59293.9417	107582.2224	194217.0251	348888.9569	70
75	129998.8861	246122.0637	463470.5086	868147.3693	75
80	285015.8024	563067.6604	1106004.5444	2160228.4620	80
90	1370022.0504	2947003.5401	6298346.1505	13375565.249	90
100	6585460.8860	15424131.906	35867089.729	82817974.524	100

AMOUNT OF £1

		Rate Per Cent			
Yrs.	21	22	23	24	Yrs.
1	1.2100	1.2200	1.2300	1.2400	1
2	1.4641	1.4884	1.5129	1.5376	2
3	1.7716	1.8158	1.8609	1.9066	3
4	2.1436	2.2153	2.2889	2.3642	4
5	2.5937	2.7027	2.8153	2.9316	5
6	3.1384	3.2973	3.4628	3.6352	6
7	3.7975	4.0227	4.2593	4.5077	7
8	4.5950	4.9077	5.2389	5.5895	8
9	5.5599	5.9874	6.4439	6.9310	9
10	6.7275	7.3046	7.9259	8.5944	10
11	8.1403	8.9117	9.7489	10.6571	11
12	9.8497	10.8722	11.9912	13.2148	12
13	11.9182	13.2641	14.7491	16.3863	13
14	14.4210	16.1822	18.1414	20.3191	14
15	17.4494	19.7423	22.3140	25.1956	15
16	21.1138	24.0856	27.4462	31.2426	16
17	25.5477	29.3844	33.7588	38.7408	17
18	30.9127	35.8490	41.5233	48.0386	18
19	37.4043	43.7358	51.0737	59.5679	19
20	45.2593	53.3576	62.8206	73.8641	20
21	54.7637	65.0963	77.2694	91.5915	21
22	66.2641	79.4175	95.0413	113.5735	22
23	80.1795	96.8894	116.9008	140.8312	23
24	97.0172	118.2050	143.7880	174.6306	24
25	117.3909	144.2101	176.8593	216.5420	25
26	142.0429	175.9364	217.5369	268.5121	26
27	171.8719	214.6424	267.5704	332.9550	27
28	207.9651	261.8637	329.1115	412.8642	28
29	251.6377	319.4737	404.8072	511.9516	29
30	304.4816	389.7579	497.9129	634.8199	30
31	368.4228	475.5046	612.4328	787.1767	31
32	445.7916	580.1156	753.2924	976.0991	32
33	539.4078	707.7411	926.5496	1210.3629	33
34	652.6834	863.4441	1139.6560	1500.8500	34
35	789.7470	1053.4018	1401.7769	1861.0540	35
36	955.5938	1285.1502	1724.1856	2307.7070	36
37	1156.2685	1567.8833	2120.7483	2861.5567	37
38	1399.0849	1912.8176	2608.5204	3548.3303	38
39	1692.8927	2333.6375	3208.4801	4399.9295	39
40	2048.4002	2847.0378	3946.4305	5455.9126	40
45	5313.0226	7694.7122	11110.4082	15994.6902	45
50	13780.6123	20796.5615	31279.1953	46890.4346	50
55	35743.3594	56207.0364	88060.4964	137465.1733	55
60	92709.0688	151911.2161	247917.2160	402996.3473	60
65	240463.4482	410571.6839	697962.7475	1181434.192	65
70	623700.2558	1109655.442	1964978.490	3463522.086	70
75	1617717.836	2999074.820	5532015.114	10153748.15	75
80	4195943.439	8105623.999	15574313.60	29766982.56	80
90	28228209.27	59208595.71	123441170.0	255830114.1	90
100	189905276.5	432496968.3	978388059.8	2198712858.	100

No Income Tax

AMOUNT OF £1

		Rate Per Cent			
Yrs.	25	26	27	28	Yrs.
1	1.2500	1.2600	1.2700	1.2800	1
2	1.5625	1.5876	1.6129	1.6384	2
3	1.9531	2.0004	2.0484	2.0972	3
4	2.4414	2.5205	2.6014	2.6844	4
5	3.0518	3.1758	3.3038	3.4360	5
6	3.8147	4.0015	4.1959	4.3980	6
7	4.7684	5.0419	5.3288	5.6295	7
8	5.9605	6.3528	6.7675	7.2058	8
9	7.4506	8.0045	8.5948	9.2234	9
10	9.3132	10.0857	10.9153	11.8059	10
11	11.6415	12.7080	13.8625	15.1116	11
12	14.5519	16.0120	17.6053	19.3428	12
13	18.1899	20.1752	22.3588	24.7588	13
14	22.7374	25.4207	28.3957	31.6913	14
15	28.4217	32.0301	36.0625	40.5648	15
16	35.5271	40.3579	45.7994	51.9230	16
17	44.4089	50.8510	58.1652	66.4614	17
18	55.5112	64.0722	73.8698	85.0706	18
19	69.3889	80.7310	93.8147	108.8904	19
20	86.7362	101.7211	119.1446	139.3797	20
21	108.4202	128.1685	151.3137	178.4060	21
22	135.5253	161.4924	192.1683	228.3596	22
23	169.4066	203.4804	244.0538	292.3003	23
24	211.7582	256.3853	309.9483	374.1444	24
25	264.6978	323.0454	393.6344	478.9049	25
26	330.8722	407.0373	499.9157	612.9982	26
27	413.5903	512.8670	634.8929	784.6377	27
28	516.9879	646.2124	806.3140	1004.3363	28
29	646.2349	814.2276	1024.0187	1285.5504	29
30	807.7936	1025.9267	1300.5038	1645.5046	30
31	1009.7420	1292.6677	1651.6398	2106.2458	31
32	1262.1774	1628.7613	2097.5826	2695.9947	32
33	1577.7218	2052.2392	2663.9299	3450.8732	33
34	1972.1523	2585.8215	3383.1910	4417.1177	34
35	2465.1903	3258.1350	4296.6525	5653.9106	35
36	3081.4879	4105.2501	5456.7487	7237.0056	36
37	3851.8599	5172.6152	6930.0709	9263.3671	37
38	4814.8249	6517.4951	8801.1900	11857.1099	38
39	6018.5311	8212.0438	11177.5113	15177.1007	39
40	7523.1638	10347.1752	14195.4393	19426.6889	40
45	22958.8740	32860.5275	46899.4169	66749.5949	45
50	70064.9232	104358.3625	154948.0260	229349.8616	50
55	213821.1768	331420.9680	511923.0122	788040.1239	55
60	652530.4468	1052525.6953	1691310.1584	2707685.2482	60
65	1991364.8889	3342607.8798	5587812.9796	9303535.6710	65
70	6077163.3573	10615443.868	18461222.940	31966705.155	70
75	18546030.753	33712494.128	60992870.319	109836762.56	75
80	56597994.243	107064035.61	201510498.08	377396242.48	80
90	527109897.16	1079814265.3	2199555304.0	4455508415.6	90
100	4909093465.3	10890667822.	24008890761.	52601359015.	100

Rule for extending the Table of
AMOUNT OF ONE POUND

Rule—
Amount of £1 in N years × Amount of £1 in M years = Amount of £1 in N + M years.

Example—
To what sum will £1 accumulate in 160 years at 3·5 per cent, Compound Interest?

Answer—
Amount of £1 in 60 years at 3·5% (see page 129) = 7·8781
Amount of £1 in 100 years at 3·5% (see page 129) = 31·1914

Amount of £1 in 160 years at 3.5% ⎫
       (7·8781 × 31·1914) ⎭   ...   = 245·729 *Ans.*

AMOUNT OF
ONE POUND PER ANNUM

in a given Number of Years at Rates of
Interest ranging from

1% to 16%

i.e., the amount to which £1 per annum invested at the end of
each year will accumulate at Compound Interest)

Note:—No allowance has been made for the effect of income tax on interest
accumulations.

AMOUNT OF £1 PER ANNUM

			Rate Per Cent			
Yrs.	1	1.25	1.5	1.75	2	Yrs.
1	1.0000	1.0000	1.0000	1.0000	1.0000	1
2	2.0100	2.0125	2.0150	2.0175	2.0200	2
3	3.0301	3.0377	3.0452	3.0528	3.0604	3
4	4.0604	4.0756	4.0909	4.1062	4.1216	4
5	5.1010	5.1266	5.1523	5.1781	5.2040	5
6	6.1520	6.1907	6.2296	6.2687	6.3081	6
7	7.2135	7.2680	7.3230	7.3784	7.4343	7
8	8.2857	8.3589	8.4328	8.5075	8.5830	8
9	9.3685	9.4634	9.5593	9.6564	9.7546	9
10	10.4622	10.5817	10.7027	10.8254	10.9497	10
11	11.5668	11.7139	11.8633	12.0148	12.1687	11
12	12.6825	12.8604	13.0412	13.2251	13.4121	12
13	13.8093	14.0211	14.2368	14.4565	14.6803	13
14	14.9474	15.1964	15.4504	15.7095	15.9739	14
15	16.0969	16.3863	16.6821	16.9844	17.2934	15
16	17.2579	17.5912	17.9324	18.2817	18.6393	16
17	18.4304	18.8111	19.2014	19.6016	20.0121	17
18	19.6147	20.0462	20.4894	20.9446	21.4123	18
19	20.8109	21.2968	21.7967	22.3112	22.8406	19
20	22.0190	22.5630	23.1237	23.7016	24.2974	20
21	23.2392	23.8450	24.4705	25.1164	25.7833	21
22	24.4716	25.1431	25.8376	26.5559	27.2990	22
23	25.7163	26.4574	27.2251	28.0207	28.8450	23
24	26.9735	27.7881	28.6335	29.5110	30.4219	24
25	28.2432	29.1354	30.0630	31.0275	32.0303	25
26	29.5256	30.4996	31.5140	32.5704	33.6709	26
27	30.8209	31.8809	32.9867	34.1404	35.3443	27
28	32.1291	33.2794	34.4815	35.7379	37.0512	28
29	33.4504	34.6954	35.9987	37.3633	38.7922	29
30	34.7849	36.1291	37.5387	39.0172	40.5681	30
31	36.1327	37.5807	39.1018	40.7000	42.3794	31
32	37.4941	39.0504	40.6883	42.4122	44.2270	32
33	38.8690	40.5386	42.2986	44.1544	46.1116	33
34	40.2577	42.0453	43.9331	45.9271	48.0338	34
35	41.6603	43.5709	45.5921	47.7308	49.9945	35
36	43.0769	45.1155	47.2760	49.5661	51.9944	36
37	44.5076	46.6794	48.9851	51.4335	54.0343	37
38	45.9527	48.2629	50.7199	53.3336	56.1149	38
39	47.4123	49.8662	52.4807	55.2670	58.2372	39
40	48.8864	51.4896	54.2679	57.2341	60.4020	40
41	50.3752	53.1332	56.0819	59.2357	62.6100	41
42	51.8790	54.7973	57.9231	61.2724	64.8622	42
43	53.3978	56.4823	59.7920	63.3446	67.1595	43
44	54.9318	58.1883	61.6889	65.4532	69.5027	44
45	56.4811	59.9157	63.6142	67.5986	71.8927	45
46	58.0459	61.6646	65.5684	69.7816	74.3306	46
47	59.6263	63.4354	67.5519	72.0027	76.8172	47
48	61.2226	65.2284	69.5652	74.2628	79.3535	48
49	62.8348	67.0437	71.6087	76.5624	81.9406	49
50	64.4632	68.8818	73.6828	78.9022	84.5794	50

AMOUNT OF £1 PER ANNUM

			Rate Per Cent			
Yrs.	1	1.25	1.5	1.75	2	Yrs.
51	66.1078	70.7428	75.7881	81.2830	87.2710	51
52	67.7689	72.6271	77.9249	83.7055	90.0164	52
53	69.4466	74.5349	80.0938	86.1703	92.8167	53
54	71.1410	76.4666	82.2952	88.6783	95.6731	54
55	72.8525	78.4225	84.5296	91.2302	98.5865	55
56	74.5810	80.4027	86.7975	93.8267	101.5583	56
57	76.3268	82.4078	89.0995	96.4687	104.5894	57
58	78.0901	84.4379	91.4360	99.1569	107.6812	58
59	79.8710	86.4933	93.8075	101.8921	110.8348	59
60	81.6697	88.5745	96.2147	104.6752	114.0515	60
61	83.4864	90.6817	98.6579	107.5070	117.3326	61
62	85.3212	92.8152	101.1377	110.3884	120.6792	62
63	87.1744	94.9754	103.6548	113.3202	124.0928	63
64	89.0462	97.1626	106.2096	116.3033	127.5747	64
65	90.9366	99.3771	108.8028	119.3386	131.1262	65
66	92.8460	101.6193	111.4348	122.4270	134.7487	66
67	94.7745	103.8896	114.1063	125.5695	138.4437	67
68	96.7222	106.1882	116.8179	128.7670	142.2125	68
69	98.6894	108.5156	119.5702	132.0204	146.0568	69
70	100.6763	110.8720	122.3638	135.3308	149.9779	70
71	102.6831	113.2579	125.1992	138.6990	153.9775	71
72	104.7099	115.6736	128.0772	142.1263	158.0570	72
73	106.7570	118.1195	130.9984	145.6135	162.2182	73
74	108.8246	120.5960	133.9633	149.1617	166.4625	74
75	110.9128	123.1035	136.9728	152.7721	170.7918	75
76	113.0220	125.6423	140.0274	156.4456	175.2076	76
77	115.1522	128.2128	143.1278	160.1834	179.7118	77
78	117.3037	130.8155	146.2747	163.9866	184.3060	78
79	119.4768	133.4507	149.4688	167.8563	188.9921	79
80	121.6715	136.1188	152.7109	171.7938	193.7720	80
81	123.8882	138.8203	156.0015	175.8002	198.6474	81
82	126.1271	141.5555	159.3415	179.8767	203.6203	82
83	128.3884	144.3250	162.7317	184.0246	208.6928	83
84	130.6723	147.1290	166.1726	188.2450	213.8666	84
85	132.9790	149.9682	169.6652	192.5393	219.1439	85
86	135.3088	152.8428	173.2102	196.9087	224.5268	86
87	137.6619	155.7533	176.8084	201.3546	230.0174	87
88	140.0385	158.7002	180.4605	205.8783	235.6177	88
89	142.4389	161.6840	184.1674	210.4812	241.3301	89
90	144.8633	164.7050	187.9299	215.1646	247.1567	90
91	147.3119	167.7638	191.7488	219.9300	253.0998	91
92	149.7850	170.8609	195.6251	224.7788	259.1618	92
93	152.2829	173.9966	199.5595	229.7124	265.3450	93
94	154.8057	177.1716	203.5528	234.7324	271.6519	94
95	157.3538	180.3862	207.6061	239.8402	278.0850	95
96	159.9273	183.6411	211.7202	245.0374	284.6467	96
97	162.5266	186.9366	215.8960	250.3255	291.3396	97
98	165.1518	190.2733	220.1345	255.7062	298.1664	98
99	167.8033	193.6517	224.4365	261.1811	305.1297	99
100	170.4814	197.0723	228.8030	266.7518	312.2323	100

AMOUNT OF £1 PER ANNUM

		Rate Per Cent				
Yrs.	2.25	2.5	2.75	3	3.25	Yrs.
1	1.0000	1.0000	1.0000	1.0000	1.0000	1
2	2.0225	2.0250	2.0275	2.0300	2.0325	2
3	3.0680	3.0756	3.0833	3.0909	3.0986	3
4	4.1370	4.1525	4.1680	4.1836	4.1993	4
5	5.2301	5.2563	5.2827	5.3091	5.3357	5
6	6.3478	6.3877	6.4279	6.4684	6.5091	6
7	7.4906	7.5474	7.6047	7.6625	7.7207	7
8	8.6592	8.7361	8.8138	8.8923	8.9716	8
9	9.8540	9.9545	10.0562	10.1591	10.2632	9
10	11.0757	11.2034	11.3328	11.4639	11.5967	10
11	12.3249	12.4835	12.6444	12.8078	12.9736	11
12	13.6022	13.7956	13.9921	14.1920	14.3953	12
13	14.9083	15.1404	15.3769	15.6178	15.8631	13
14	16.2437	16.5190	16.7998	17.0863	17.3787	14
15	17.6092	17.9319	18.2618	18.5989	18.9435	15
16	19.0054	19.3802	19.7640	20.1569	20.5592	16
17	20.4330	20.8647	21.3075	21.7616	22.2273	17
18	21.8928	22.3863	22.8934	23.4144	23.9497	18
19	23.3853	23.9460	24.5230	25.1169	25.7281	19
20	24.9115	25.5447	26.1974	26.8704	27.5642	20
21	26.4720	27.1833	27.9178	28.6765	29.4601	21
22	28.0676	28.8629	29.6856	30.5368	31.4175	22
23	29.6992	30.5844	31.5019	32.4529	33.4386	23
24	31.3674	32.3490	33.3682	34.4265	35.5254	24
25	33.0732	34.1578	35.2858	36.4593	37.6799	25
26	34.8173	36.0117	37.2562	38.5530	39.9045	26
27	36.6007	37.9120	39.2808	40.7096	42.2014	27
28	38.4242	39.8598	41.3610	42.9309	44.5730	28
29	40.2888	41.8563	43.4984	45.2189	47.0216	29
30	42.1953	43.9027	45.6946	47.5754	49.5498	30
31	44.1447	46.0003	47.9512	50.0027	52.1602	31
32	46.1379	48.1503	50.2699	52.5028	54.8554	32
33	48.1760	50.3540	52.6523	55.0778	57.6382	33
34	50.2600	52.6129	55.1002	57.7302	60.5114	34
35	52.3908	54.9282	57.6155	60.4621	63.4780	35
36	54.5696	57.3014	60.1999	63.2759	66.5411	36
37	56.7974	59.7339	62.8554	66.1742	69.7037	37
38	59.0754	62.2273	65.5839	69.1594	72.9690	38
39	61.4046	64.7830	68.3875	72.2342	76.3405	39
40	63.7862	67.4026	71.2681	75.4013	79.8216	40
41	66.2214	70.0876	74.2280	78.6633	83.4158	41
42	68.7113	72.8398	77.2693	82.0232	87.1268	42
43	71.2574	75.6608	80.3942	85.4839	90.9584	43
44	73.8606	78.5523	83.6050	89.0484	94.9146	44
45	76.5225	81.5161	86.9042	92.7199	98.9993	45
46	79.2443	84.5540	90.2940	96.5015	103.2168	46
47	82.0273	87.6679	93.7771	100.3965	107.5713	47
48	84.8729	90.8596	97.3560	104.4084	112.0674	48
49	87.7825	94.1311	101.0333	108.5406	116.7096	49
50	90.7576	97.4843	104.8117	112.7969	121.5026	50

AMOUNT OF £1 PER ANNUM

			Rate Per Cent			
Yrs.	2.25	2.5	2.75	3	3.25	Yrs.
51	93.7997	100.9215	108.6940	117.1808	126.4515	51
52	96.9102	104.4445	112.6831	121.6962	131.5611	52
53	100.0906	108.0556	116.7819	126.3471	136.8369	53
54	103.3427	111.7570	120.9934	131.1375	142.2841	54
55	106.6679	115.5509	125.3207	136.0716	147.9083	55
56	110.0679	119.4397	129.7670	141.1538	153.7153	56
57	113.5444	123.4257	134.3356	146.3884	159.7111	57
58	117.0992	127.5113	139.0299	151.7800	165.9017	58
59	120.7339	131.6991	143.8532	157.3334	172.2935	59
60	124.4504	135.9916	148.8091	163.0534	178.8930	60
61	128.2506	140.3914	153.9014	168.9450	185.7071	61
62	132.1362	144.9012	159.1337	175.0134	192.7425	62
63	136.1093	149.5237	164.5099	181.2638	200.0067	63
64	140.1717	154.2618	170.0339	187.7017	207.5069	64
65	144.3256	159.1183	175.7098	194.3328	215.2509	65
66	148.5729	164.0963	181.5418	201.1627	223.2465	66
67	152.9158	169.1987	187.5342	208.1976	231.5020	67
68	157.3564	174.4287	193.6914	215.4436	240.0258	68
69	161.8969	179.7894	200.0179	222.9069	248.8267	69
70	166.5396	185.2841	206.5184	230.5941	257.9135	70
71	171.2868	190.9162	213.1977	238.5119	267.2957	71
72	176.1407	196.6891	220.0606	246.6672	276.9828	72
73	181.1039	202.6064	227.1123	255.0673	286.9848	73
74	186.1787	208.6715	234.3579	263.7193	297.3118	74
75	191.3677	214.8883	241.8027	272.6309	307.9744	75
76	196.6735	221.2605	249.4523	281.8098	318.9836	76
77	202.0987	227.7920	257.3122	291.2641	330.3506	77
78	207.6459	234.4868	265.3883	301.0020	342.0869	78
79	213.3179	241.3490	273.6865	311.0321	354.2048	79
80	219.1176	248.3827	282.2129	321.3630	366.7164	80
81	225.0477	255.5923	290.9737	332.0039	379.6347	81
82	231.1113	262.9821	299.9755	342.9640	392.9728	82
83	237.3113	270.5566	309.2248	354.2529	406.7445	83
84	243.6508	278.3206	318.7285	365.8805	420.9637	84
85	250.1329	286.2786	328.4935	377.8570	435.6450	85
86	256.7609	294.4355	338.5271	390.1927	450.8034	86
87	263.5381	302.7964	348.8366	402.8984	466.4545	87
88	270.4677	311.3663	359.4296	415.9854	482.6143	88
89	277.5532	320.1505	370.3139	429.4650	499.2993	89
90	284.7981	329.1543	381.4976	443.3489	516.5265	90
91	292.2061	338.3831	392.9888	457.6494	534.3136	91
92	299.7807	347.8427	404.7959	472.3789	552.6788	92
93	307.5258	357.5388	416.9278	487.5502	571.6409	93
94	315.4451	367.4772	429.3933	503.1767	591.2192	94
95	323.5426	377.6642	442.2017	519.2720	611.4338	95
96	331.8223	388.1058	455.3622	535.8502	632.3054	96
97	340.2883	398.8084	468.8847	552.9257	653.8554	97
98	348.9448	409.7786	482.7790	570.5135	676.1057	98
99	357.7961	421.0231	497.0554	588.6289	699.0791	99
100	366.8465	432.5487	511.7244	607.2877	722.7992	100

AMOUNT OF £1 PER ANNUM

			Rate Per Cent			
Yrs.	3.5	3.75	4	4.25	4.5	Yrs.
1	1.0000	1.0000	1.0000	1.0000	1.0000	1
2	2.0350	2.0375	2.0400	2.0425	2.0450	2
3	3.1062	3.1139	3.1216	3.1293	3.1370	3
4	4.2149	4.2307	4.2465	4.2623	4.2782	4
5	5.3625	5.3893	5.4163	5.4434	5.4707	5
6	6.5502	6.5914	6.6330	6.6748	6.7169	6
7	7.7794	7.8386	7.8983	7.9585	8.0192	7
8	9.0517	9.1326	9.2142	9.2967	9.3800	8
9	10.3685	10.4750	10.5828	10.6918	10.8021	9
10	11.7314	11.8678	12.0061	12.1462	12.2882	10
11	13.1420	13.3129	13.4864	13.6624	13.8412	11
12	14.6020	14.8121	15.0258	15.2431	15.4640	12
13	16.1130	16.3676	16.6268	16.8909	17.1599	13
14	17.6770	17.9814	18.2919	18.6088	18.9321	14
15	19.2957	19.6557	20.0236	20.3997	20.7841	15
16	20.9710	21.3927	21.8245	22.2666	22.7193	16
17	22.7050	23.1950	23.6975	24.2130	24.7417	17
18	24.4997	25.0648	25.6454	26.2420	26.8551	18
19	26.3572	27.0047	27.6712	28.3573	29.0636	19
20	28.2797	29.0174	29.7781	30.5625	31.3714	20
21	30.2695	31.1055	31.9692	32.8614	33.7831	21
22	32.3289	33.2720	34.2480	35.2580	36.3034	22
23	34.4604	35.5197	36.6179	37.7565	38.9370	23
24	36.6665	37.8517	39.0826	40.3611	41.6892	24
25	38.9499	40.2711	41.6459	43.0765	44.5652	25
26	41.3131	42.7813	44.3117	45.9072	47.5706	26
27	43.7591	45.3856	47.0842	48.8583	50.7113	27
28	46.2906	48.0875	49.9676	51.9348	53.9933	28
29	48.9108	50.8908	52.9663	55.1420	57.4230	29
30	51.6227	53.7992	56.0849	58.4855	61.0071	30
31	54.4295	56.8167	59.3283	61.9712	64.7524	31
32	57.3345	59.9473	62.7015	65.6049	68.6662	32
33	60.3412	63.1954	66.2095	69.3931	72.7562	33
34	63.4532	66.5652	69.8579	73.3424	77.0303	34
35	66.6740	70.0614	73.6522	77.4594	81.4966	35
36	70.0076	73.6887	77.5983	81.7514	86.1640	36
37	73.4579	77.4520	81.7022	86.2259	91.0413	37
38	77.0289	81.3565	85.9703	90.8905	96.1382	38
39	80.7249	85.4073	90.4091	95.7533	101.4644	39
40	84.5503	89.6101	95.0255	100.8228	107.0303	40
41	88.5095	93.9705	99.8265	106.1078	112.8467	41
42	92.6074	98.4944	104.8196	111.6174	118.9248	42
43	96.8486	103.1879	110.0124	117.3611	125.2764	43
44	101.2383	108.0575	115.4129	123.3490	131.9138	44
45	105.7817	113.1096	121.0294	129.5913	138.8500	45
46	110.4840	118.3512	126.8706	136.0989	146.0982	46
47	115.3510	123.7894	132.9454	142.8831	153.6726	47
48	120.3883	129.4315	139.2632	149.9557	161.5879	48
49	125.6018	135.2852	145.8337	157.3288	169.8594	49
50	130.9979	141.3584	152.6671	165.0153	178.5030	50

AMOUNT OF £1 PER ANNUM

			Rate Per Cent			
Yrs.	3.5	3.75	4	4.25	4.5	Yrs.
51	136.5828	147.6593	159.7738	173.0284	187.5357	51
52	142.3632	154.1965	167.1647	181.3821	196.9748	52
53	148.3459	160.9789	174.8513	190.0909	206.8386	53
54	154.5381	168.0156	182.8454	199.1697	217.1464	54
55	160.9469	175.3162	191.1592	208.6344	227.9180	55
56	167.5800	182.8906	199.8055	218.5014	239.1743	56
57	174.4453	190.7490	208.7978	228.7877	250.9371	57
58	181.5509	198.9020	218.1497	239.5112	263.2293	58
59	188.9052	207.3609	227.8757	250.6904	276.0746	59
60	196.5169	216.1369	237.9907	262.3447	289.4980	60
61	204.3950	225.2420	248.5103	274.4944	303.5254	61
62	212.5488	234.6886	259.4507	287.1604	318.1840	62
63	220.9880	244.4894	270.8288	300.3647	333.5023	63
64	229.7226	254.6578	282.6619	314.1302	349.5099	64
65	238.7629	265.2074	294.9684	328.4808	366.2378	65
66	248.1196	276.1527	307.7671	343.4412	383.7185	66
67	257.8038	287.5085	321.0778	359.0374	401.9859	67
68	267.8269	299.2900	334.9209	375.2965	421.0752	68
69	278.2008	311.5134	349.3177	392.2466	441.0236	69
70	288.9379	324.1952	364.2905	409.9171	461.8697	70
71	300.0507	337.3525	379.8621	428.3386	483.6538	71
72	311.5525	351.0032	396.0566	447.5430	506.4182	72
73	323.4568	365.1658	412.8988	467.5636	530.2071	73
74	335.7778	379.8595	430.4148	488.4350	555.0664	74
75	348.5300	395.1043	448.6314	510.1935	581.0444	75
76	361.7286	410.9207	467.5766	532.8767	608.1914	76
77	375.3891	427.3302	487.2797	556.5240	636.5600	77
78	389.5277	444.3551	507.7709	581.1762	666.2052	78
79	404.1611	462.0184	529.0817	606.8762	697.1844	79
80	419.3068	480.3441	551.2450	633.6685	729.5577	80
81	434.9825	499.3570	574.2948	661.5994	763.3878	81
82	451.2069	519.0829	598.2666	690.7174	798.7402	82
83	467.9992	539.5485	623.1972	721.0729	835.6836	83
84	485.3791	560.7815	649.1251	752.7184	874.2893	84
85	503.3674	582.8109	676.0901	785.7090	914.6323	85
86	521.9853	605.6663	704.1337	820.1016	956.7908	86
87	541.2547	629.3787	733.2991	855.9559	1000.8464	87
88	561.1987	653.9804	763.6310	893.3341	1046.8845	88
89	581.8406	679.5047	795.1763	932.3008	1094.9943	89
90	603.2050	705.9861	827.9833	972.9235	1145.2690	90
91	625.3172	733.4606	862.1027	1015.2728	1197.8061	91
92	648.2033	761.9654	897.5868	1059.4219	1252.7074	92
93	671.8904	791.5391	934.4902	1105.4473	1310.0792	93
94	696.4066	822.2218	972.8699	1153.4288	1370.0328	94
95	721.7808	854.0551	1012.7846	1203.4496	1432.6843	95
96	748.0431	887.0822	1054.2960	1255.5962	1498.1551	96
97	775.2247	921.3478	1097.4679	1309.9590	1566.5720	97
98	803.3575	956.8983	1142.3666	1366.6322	1638.0678	98
99	832.4750	993.7820	1189.0613	1425.7141	1712.7808	99
100	862.6117	1032.0488	1237.6237	1487.3070	1790.8560	100

AMOUNT OF £1 PER ANNUM

Rate Per Cent

Yrs.	4.75	5	5.25	5.5	5.75	Yrs.
1	1.0000	1.0000	1.0000	1.0000	1.0000	1
2	2.0475	2.0500	2.0525	2.0550	2.0575	2
3	3.1448	3.1525	3.1603	3.1680	3.1758	3
4	4.2941	4.3101	4.3262	4.3423	4.3584	4
5	5.4981	5.5256	5.5533	5.5811	5.6090	5
6	6.7593	6.8019	6.8448	6.8881	6.9315	6
7	8.0803	8.1420	8.2042	8.2669	8.3301	7
8	9.4641	9.5491	9.6349	9.7216	9.8091	8
9	10.9137	11.0266	11.1407	11.2563	11.3731	9
10	12.4321	12.5779	12.7256	12.8754	13.0271	10
11	14.0226	14.2068	14.3937	14.5835	14.7761	11
12	15.6887	15.9171	16.1494	16.3856	16.6257	12
13	17.4339	17.7130	17.9972	18.2868	18.5817	13
14	19.2620	19.5986	19.9421	20.2926	20.6502	14
15	21.1770	21.5786	21.9891	22.4087	22.8376	15
16	23.1829	23.6575	24.1435	24.6411	25.1507	16
17	25.2840	25.8404	26.4110	26.9964	27.5969	17
18	27.4850	28.1324	28.7976	29.4812	30.1837	18
19	29.7906	30.5390	31.3095	32.1027	32.9193	19
20	32.2056	33.0660	33.9532	34.8683	35.8121	20
21	34.7354	35.7193	36.7358	37.7861	38.8713	21
22	37.3853	38.5052	39.6644	40.8643	42.1064	22
23	40.1611	41.4305	42.7468	44.1118	45.5275	23
24	43.0688	44.5020	45.9910	47.5380	49.1454	24
25	46.1146	47.7271	49.4055	51.1526	52.9712	25
26	49.3050	51.1135	52.9993	54.9660	57.0171	26
27	52.6470	54.6691	56.7818	58.9891	61.2956	27
28	56.1477	58.4026	60.7628	63.2335	65.8201	28
29	59.8147	62.3227	64.9529	67.7114	70.6047	29
30	63.6559	66.4388	69.3629	72.4355	75.6645	30
31	67.6796	70.7608	74.0044	77.4194	81.0152	31
32	71.8944	75.2988	78.8897	82.6775	86.6736	32
33	76.3094	80.0638	84.0314	88.2248	92.6573	33
34	80.9341	85.0670	89.4430	94.0771	98.9851	34
35	85.7784	90.3203	95.1388	100.2514	105.6767	35
36	90.8529	95.8363	101.1336	106.7652	112.7532	36
37	96.1684	101.6281	107.4431	113.6373	120.2365	37
38	101.7364	107.7095	114.0838	120.8873	128.1501	38
39	107.5689	114.0950	121.0732	128.5361	136.5187	39
40	113.6784	120.7998	128.4296	136.6056	145.3685	40
41	120.0781	127.8398	136.1721	145.1189	154.7272	41
42	126.7818	135.2318	144.3212	154.1005	164.6240	42
43	133.8040	142.9933	152.8980	163.5760	175.0899	43
44	141.1597	151.1430	161.9252	173.5727	186.1576	44
45	148.8648	159.7002	171.4262	184.1192	197.8616	45
46	156.9358	168.6852	181.4261	195.2457	210.2387	46
47	165.3903	178.1194	191.9510	206.9842	223.3274	47
48	174.2463	188.0254	203.0284	219.3684	237.1687	48
49	183.5230	198.4267	214.6874	232.4336	251.8059	49
50	193.2404	209.3480	226.9585	246.2175	267.2848	50

AMOUNT OF £1 PER ANNUM

			Rate Per Cent			
Yrs.	**4.75**	**5**	**5.25**	**5.5**	**5.75**	Yrs.
51	203.4193	220.8154	239.8738	260.7594	283.6536	51
52	214.0817	232.8562	253.4672	276.1012	300.9637	52
53	225.2506	245.4990	267.7742	292.2868	319.2691	53
54	236.9500	258.7739	282.8324	309.3625	338.6271	54
55	249.2051	272.7126	298.6811	327.3775	359.0982	55
56	262.0423	287.3482	315.3618	346.3832	380.7463	56
57	275.4894	302.7157	332.9183	366.4343	403.6392	57
58	289.5751	318.8514	351.3965	387.5882	427.8485	58
59	304.3299	335.7940	370.8449	409.9056	453.4498	59
60	319.7856	353.5837	391.3142	433.4504	480.5231	60
61	335.9754	372.2629	412.8582	458.2901	509.1532	61
62	352.9342	391.8760	435.5333	484.4961	539.4295	62
63	370.6986	412.4699	459.3988	512.1434	571.4467	63
64	389.3068	434.0933	484.5172	541.3113	605.3049	64
65	408.7989	456.7980	510.9544	572.0834	641.1099	65
66	429.2168	480.6379	538.7795	604.5480	678.9738	66
67	450.6046	505.6698	568.0654	638.7981	719.0148	67
68	473.0083	531.9533	598.8888	674.9320	761.3581	68
69	496.4762	559.5510	631.3305	713.0533	806.1362	69
70	521.0588	588.5285	665.4753	753.2712	853.4890	70
71	546.8091	618.9549	701.4128	795.7011	903.5646	71
72	573.7826	650.9027	739.2370	840.4647	956.5196	72
73	602.0373	684.4478	779.0469	887.6902	1012.5195	73
74	631.6340	719.6702	820.9469	937.5132	1071.7394	74
75	662.6366	756.6537	865.0466	990.0764	1134.3644	75
76	695.1119	795.4864	911.4615	1045.5306	1200.5903	76
77	729.1297	836.2607	960.3132	1104.0348	1270.6243	77
78	764.7634	879.0738	1011.7297	1165.7567	1344.6852	78
79	802.0896	924.0274	1065.8455	1230.8734	1423.0046	79
80	841.1889	971.2288	1122.8024	1299.5714	1505.8273	80
81	882.1453	1020.7903	1182.7495	1372.0478	1593.4124	81
82	925.0472	1072.8298	1245.8439	1448.5104	1686.0336	82
83	969.9870	1127.4713	1312.2507	1529.1785	1783.9805	83
84	1017.0614	1184.8448	1382.1438	1614.2833	1887.5594	84
85	1066.3718	1245.0871	1455.7064	1704.0689	1997.0941	85
86	1118.0244	1308.3414	1533.1310	1798.7927	2112.9270	86
87	1172.1306	1374.7585	1614.6203	1898.7263	2235.4203	87
88	1228.8068	1444.4964	1700.3879	2004.1563	2364.9570	88
89	1288.1751	1517.7212	1790.6583	2115.3848	2501.9420	89
90	1350.3635	1594.6073	1885.6678	2232.7310	2642.8036	90
91	1415.5057	1675.3377	1985.6654	2356.5312	2799.9948	91
92	1483.7422	1760.1045	2090.9128	2487.1404	2961.994€	92
93	1555.2200	1849.1098	2201.6857	2624.9332	3133.3092	93
94	1630.0929	1942.5653	2318.2742	2770.3045	3314.4745	94
95	1708.5224	2040.6935	2440.9836	2923.6712	3506.0568	95
96	1790.6772	2143.7282	2570.1353	3085.4732	3708.6551	96
97	1876.7343	2251.9146	2706.0674	3256.1742	3922.9027	97
98	1966.8792	2365.5103	2849.1359	3436.2638	4149.4696	98
99	2061.3060	2484.7859	2999.7156	3626.2583	4389.0641	99
100	2160.2180	2610.0252	3158.2006	3826.7025	4642.4353	100

AMOUNT OF £1 PER ANNUM

			Rate Per Cent			
Yrs.	6	6.5	7	7.5	8	Yrs.
1	1.0000	1.0000	1.0000	1.0000	1.0000	1
2	2.0600	2.0650	2.0700	2.0750	2.0800	2
3	3.1836	3.1992	3.2149	3.2306	3.2464	3
4	4.3746	4.4072	4.4399	4.4729	4.5061	4
5	5.6371	5.6936	5.7507	5.8084	5.8666	5
6	6.9753	7.0637	7.1533	7.2440	7.3359	6
7	8.3938	8.5229	8.6540	8.7873	8.9228	7
8	9.8975	10.0769	10.2598	10.4464	10.6366	8
9	11.4913	11.7319	11.9780	12.2298	12.4876	9
10	13.1808	13.4944	13.8164	14.1471	14.4866	10
11	14.9716	15.3716	15.7836	16.2081	16.6455	11
12	16.8699	17.3707	17.8885	18.4237	18.9771	12
13	18.8821	19.4998	20.1406	20.8055	21.4953	13
14	21.0151	21.7673	22.5505	23.3659	24.2149	14
15	23.2760	24.1822	25.1290	26.1184	27.1521	15
16	25.6725	26.7540	27.8881	29.0772	30.3243	16
17	28.2129	29.4930	30.8402	32.2580	33.7502	17
18	30.9057	32.4101	33.9990	35.6774	37.4502	18
19	33.7600	35.5167	37.3790	39.3532	41.4463	19
20	36.7856	38.8253	40.9955	43.3047	45.7620	20
21	39.9927	42.3490	44.8652	47.5525	50.4229	21
22	43.3923	46.1016	49.0057	52.1190	55.4568	22
23	46.9958	50.0982	53.4361	57.0279	60.8933	23
24	50.8156	54.3546	58.1767	62.3050	66.7648	24
25	54.8645	58.8877	63.2490	67.9779	73.1059	25
26	59.1564	63.7154	68.6765	74.0762	79.9544	26
27	63.7058	68.8569	74.4838	80.6319	87.3508	27
28	68.5281	74.3326	80.6977	87.6793	95.3388	28
29	73.6398	80.1642	87.3465	95.2553	103.9659	29
30	79.0582	86.3749	94.4608	103.3994	113.2832	30
31	84.8017	92.9892	102.0730	112.1544	123.3459	31
32	90.8898	100.0335	110.2182	121.5659	134.2135	32
33	97.3432	107.5357	118.9334	131.6834	145.9506	33
34	104.1838	115.5255	128.2588	142.5596	158.6267	34
35	111.4348	124.0347	138.2369	154.2516	172.3168	35
36	119.1209	133.0969	148.9135	166.8205	187.1021	36
37	127.2681	142.7482	160.3374	180.3320	203.0703	37
38	135.9042	153.0269	172.5610	194.8569	220.3159	38
39	145.0585	163.9736	185.6403	210.4712	238.9412	39
40	154.7620	175.6319	199.6351	227.2565	259.0565	40
41	165.0477	188.0480	214.6096	245.3008	280.7810	41
42	175.9505	201.2711	230.6322	264.6983	304.2435	42
43	187.5076	215.3537	247.7765	285.5507	329.5830	43
44	199.7580	230.3517	266.1209	307.9670	356.9496	44
45	212.7435	246.3246	285.7493	332.0645	386.5056	45
46	226.5081	263.3357	306.7518	357.9694	418.4261	46
47	241.0986	281.4525	329.2244	385.8171	452.9002	47
48	256.5645	300.7469	353.2701	415.7533	490.1322	48
49	272.9584	321.2955	378.9990	447.9348	530.3427	49
50	290.3359	343.1797	406.5289	482.5299	573.7702	50

AMOUNT OF £1 PER ANNUM

	Rate Per Cent					
Yrs.	6	6.5	7	7.5	8	Yrs.
51	308.7561	366.4864	435.9860	519.7197	620.6718	51
52	328.2814	391.3080	467.5050	559.6987	671.3255	52
53	348.9783	417.7430	501.2303	602.6761	726.0316	53
54	370.9170	445.8963	537.3164	648.8768	785.1141	54
55	394.1720	475.8795	575.9286	698.5425	848.9232	55
56	418.8223	507.8117	617.2436	751.9332	917.8371	56
57	444.9517	541.8195	661.4506	809.3282	992.2640	57
58	472.6488	578.0377	708.7522	871.0278	1072.6451	58
59	502.0077	616.6102	759.3648	937.3549	1159.4568	59
60	533.1282	657.6898	813.5204	1008.6565	1253.2133	60
61	566.1159	701.4397	871.4668	1085.3058	1354.4704	61
62	601.0828	748.0333	933.4695	1167.7037	1463.8280	62
63	638.1478	797.6554	999.8124	1256.2815	1581.9342	63
64	677.4367	850.5030	1070.7992	1351.5026	1709.4890	64
65	719.0829	906.7857	1146.7552	1453.8653	1847.2481	65
66	763.2278	966.7268	1228.0280	1563.9052	1996.0279	66
67	810.0215	1030.5640	1314.9900	1682.1981	2156.7102	67
68	859.6228	1098.5507	1408.0393	1809.3629	2330.2470	68
69	912.2002	1170.9565	1507.6020	1946.0652	2517.6667	69
70	967.9322	1248.0687	1614.1342	2093.0200	2720.0801	70
71	1027.0081	1330.1931	1728.1236	2250.9966	2938.6865	71
72	1089.6286	1417.6557	1850.0922	2420.8213	3174.7814	72
73	1156.0063	1510.8033	1980.5987	2603.3829	3429.7639	73
74	1226.3667	1610.0055	2120.2406	2799.6366	3705.1450	74
75	1300.9487	1715.6559	2269.6574	3010.6094	4002.5566	75
76	1380.0056	1828.1735	2429.5334	3237.4051	4323.7612	76
77	1463.8059	1948.0048	2600.6008	3481.2104	4670.6620	77
78	1552.6343	2075.6251	2783.6428	3743.3012	5045.3150	78
79	1646.7924	2211.5407	2979.4978	4025.0488	5449.9402	79
80	1746.5999	2356.2909	3189.0627	4327.9275	5886.9354	80
81	1852.3959	2510.4498	3413.2971	4653.5220	6358.8903	81
82	1964.5396	2674.6290	3653.2279	5003.5362	6868.6015	82
83	2083.4120	2849.4799	3909.9538	5379.8014	7419.0896	83
84	2209.4167	3035.6961	4184.6506	5784.2865	8013.6168	84
85	2342.9817	3234.0163	4478.5761	6219.1080	8655.7061	85
86	2484.5606	3445.2274	4793.0764	6686.5411	9349.1626	86
87	2634.6343	3670.1672	5129.5918	7189.0317	10098.0956	87
88	2793.7123	3909.7281	5489.6632	7729.2090	10906.9433	88
89	2962.3351	4164.8604	5874.9397	8309.8997	11780.4987	89
90	3141.0752	4436.5763	6287.1854	8934.1422	12723.9386	90
91	3330.5397	4725.9538	6728.2884	9605.2029	13742.8537	91
92	3531.3721	5034.1408	7200.2686	10326.5931	14843.2820	92
93	3744.2544	5362.3599	7705.2874	11102.0876	16031.7446	93
94	3969.9097	5711.9133	8245.6575	11935.7441	17315.2841	94
95	4209.1042	6084.1877	8823.8535	12831.9249	18701.5069	95
96	4462.6505	6480.6599	9442.5233	13795.3193	20198.6274	96
97	4731.4095	6902.9028	10104.4999	14830.9682	21815.5176	97
98	5016.2941	7352.5914	10812.8149	15944.2909	23561.7590	98
99	5318.2718	7831.5099	11570.7120	17141.1127	25447.6997	99
100	5638.3681	8341.5580	12381.6618	18427.6961	27484.5157	100

AMOUNT OF £1 PER ANNUM

			Rate Per Cent			
Yrs.	8.5	9	10	11	12	Yrs.
1	1.0000	1.0000	1.0000	1.0000	1.0000	1
2	2.0850	2.0900	2.1000	2.1100	2.1200	2
3	3.2622	3.2781	3.3100	3.3421	3.3744	3
4	4.5395	4.5731	4.6410	4.7097	4.7793	4
5	5.9254	5.9847	6.1051	6.2278	6.3528	5
6	7.4290	7.5233	7.7156	7.9129	8.1152	6
7	9.0605	9.2004	9.4872	9.7833	10.0890	7
8	10.8306	11.0285	11.4359	11.8594	12.2997	8
9	12.7512	13.0210	13.5795	14.1640	14.7757	9
10	14.8351	15.1929	15.9374	16.7220	17.5487	10
11	17.0961	17.5603	18.5312	19.5614	20.6546	11
12	19.5492	20.1407	21.3843	22.7132	24.1331	12
13	22.2109	22.9534	24.5227	26.2116	28.0291	13
14	25.0989	26.0192	27.9750	30.0949	32.3926	14
15	28.2323	29.3609	31.7725	34.4054	37.2797	15
16	31.6320	33.0034	35.9497	39.1899	42.7533	16
17	35.3207	36.9737	40.5447	44.5008	48.8837	17
18	39.3230	41.3013	45.5992	50.3959	55.7497	18
19	43.6654	46.0185	51.1591	56.9395	63.4397	19
20	48.3770	51.1601	57.2750	64.2028	72.0524	20
21	53.4891	56.7645	64.0025	72.2651	81.6987	21
22	59.0356	62.8733	71.4027	81.2143	92.5026	22
23	65.0537	69.5319	79.5430	91.1479	104.6029	23
24	71.5832	76.7898	88.4973	102.1742	118.1552	24
25	78.6678	84.7009	98.3471	114.4133	133.3339	25
26	86.3546	93.3240	109.1818	127.9988	150.3339	26
27	94.6947	102.7231	121.0999	143.0786	169.3740	27
28	103.7437	112.9682	134.2099	159.8173	190.6989	28
29	113.5620	124.1354	148.6309	178.3972	214.5828	29
30	124.2147	136.3075	164.4940	199.0209	241.3327	30
31	135.7730	149.5752	181.9434	221.9132	271.2926	31
32	148.3137	164.0370	201.1378	247.3236	304.8477	32
33	161.9203	179.8003	222.2515	275.5292	342.4294	33
34	176.6836	196.9823	245.4767	306.8374	384.5210	34
35	192.7017	215.7108	271.0244	341.5896	431.6635	35
36	210.0813	236.1247	299.1268	380.1644	484.4631	36
37	228.9382	258.3759	330.0395	422.9825	543.5987	37
38	249.3980	282.6298	364.0434	470.5106	609.8305	38
39	271.5968	309.0665	401.4478	523.2667	684.0102	39
40	295.6825	337.8824	442.5926	581.8261	767.0914	40
41	321.8156	369.2919	487.8518	646.8269	860.1424	41
42	350.1699	403.5281	537.6370	718.9779	964.3595	42
43	380.9343	440.8457	592.4007	799.0655	1081.0826	43
44	414.3137	481.5218	652.6408	887.9627	1211.8125	44
45	450.5304	525.8587	718.9048	986.6386	1358.2300	45
46	489.8255	574.1860	791.7953	1096.1688	1522.2176	46
47	532.4606	626.8628	871.9749	1217.7474	1705.8838	47
48	578.7198	684.2804	960.1723	1352.6996	1911.5898	48
49	628.9110	746.8656	1057.1896	1502.4965	2141.9806	49
50	683.3684	815.0836	1163.9085	1668.7712	2400.0182	50

AMOUNT OF £1 PER ANNUM

			Rate Per Cent			
Yrs.	8.5	9	10	11	12	Yrs.
51	742.4547	889.4411	1281.2994	1853.3360	2689.0204	51
52	806.5634	970.4908	1410.4293	2058.2029	3012.7029	52
53	876.1213	1058.8349	1552.4723	2285.6053	3375.2272	53
54	951.5916	1155.1301	1708.7195	2538.0218	3781.2545	54
55	1033.4769	1260.0918	1880.5914	2818.2042	4236.0050	55
56	1122.3224	1374.5001	2069.6506	3129.2067	4745.3257	56
57	1218.7198	1499.2051	2277.6156	3474.4194	5315.7647	57
58	1323.3110	1635.1335	2506.3772	3857.6056	5954.6565	58
59	1436.7924	1783.2955	2758.0149	4282.9422	6670.2153	59
60	1559.9198	1944.7921	3034.8164	4755.0658	7471.6411	60
61	1693.5130	2120.8234	3339.2980	5279.1231	8369.2380	61
62	1838.4616	2312.6975	3674.2278	5860.8266	9374.5466	62
63	1995.7308	2521.8403	4042.6506	6506.5175	10500.4922	63
64	2166.3679	2749.8059	4447.9157	7223.2345	11761.5513	64
65	2351.5092	2998.2885	4893.7073	8018.7903	13173.9374	65
66	2552.3875	3269.1344	5384.0780	8901.8572	14755.8099	66
67	2770.3404	3564.3565	5923.4858	9882.0615	16527.5071	67
68	3006.8193	3886.1486	6516.8344	10970.0883	18511.8080	68
69	3263.3990	4236.9020	7169.5178	12177.7980	20734.2249	69
70	3541.7879	4619.2232	7887.4696	13518.3557	23223.3319	70
71	3843.8399	5035.9533	8677.2165	15006.3749	26011.1317	71
72	4171.5662	5490.1891	9545.9382	16658.0761	29133.4675	72
73	4527.1494	5985.3061	10501.5320	18491.4645	32630.4836	73
74	4912.9571	6524.9836	11552.6852	20526.5256	36547.1417	74
75	5331.5584	7113.2321	12708.9537	22785.4434	40933.7987	75
76	5785.7409	7754.4230	13980.8491	25292.8422	45846.8545	76
77	6278.5289	8453.3211	15379.9340	28076.0548	51349.4771	77
78	6813.2038	9215.1200	16918.9274	31165.4208	57512.4143	78
79	7393.3261	10045.4808	18611.8201	34594.6171	64414.9040	79
80	8022.7589	10950.5741	20474.0021	38401.0250	72145.6925	80
81	8705.6934	11937.1258	22522.4024	42626.1378	80804.1756	81
82	9446.6773	13012.4671	24775.6426	47316.0129	90501.6767	82
83	10250.6449	14184.5891	27254.2069	52521.7743	101362.8779	83
84	11122.9497	15462.2021	29980.6275	58300.1695	113527.4232	84
85	12069.4004	16854.8003	32979.6903	64714.1881	127151.7140	85
86	13096.2994	18372.7324	36278.6593	71833.7488	142410.9197	86
87	14210.4849	20027.2783	39907.5253	79736.4612	159501.2300	87
88	15419.3761	21830.7333	43899.2778	88508.4720	178642.3777	88
89	16731.0231	23796.4993	48290.2056	98245.4039	200080.4630	89
90	18154.1600	25939.1842	53120.2261	109053.3983	224091.1185	90
91	19698.2637	28274.7108	58433.2487	121050.2721	250983.0528	91
92	21373.6161	30820.4348	64277.5736	134366.8020	281102.0191	92
93	23191.3734	33595.2739	70706.3310	149148.1503	314835.2614	93
94	25163.6402	36619.8486	77777.9641	165555.4468	352616.4927	94
95	27303.5496	39916.6350	85556.7605	183767.5459	394931.4719	95
96	29625.3513	43510.1321	94113.4365	203982.9760	442324.2485	96
97	32144.5062	47427.0440	103525.7802	226422.1033	495404.1583	97
98	34877.7892	51696.4780	113879.3582	251329.5347	554853.6573	98
99	37843.4013	56350.1610	125268.2940	278976.7835	621437.0962	99
100	41061.0904	61422.6755	137796.1234	309665.2297	696010.5477	100

AMOUNT OF £1 PER ANNUM

Yrs.	13	14	15	16	Yrs.
	Rate Per Cent				
1	1.0000	1.0000	1.0000	1.0000	1
2	2.1300	2.1400	2.1500	2.1600	2
3	3.4069	3.4396	3.4725	3.5056	3
4	4.8498	4.9211	4.9934	5.0665	4
5	6.4803	6.6101	6.7424	6.8771	5
6	8.3227	8.5355	8.7537	8.9775	6
7	10.4047	10.7305	11.0668	11.4139	7
8	12.7573	13.2328	13.7268	14.2401	8
9	15.4157	16.0853	16.7858	17.5185	9
10	18.4197	19.3373	20.3037	21.3215	10
11	21.8143	23.0445	24.3493	25.7329	11
12	25.6502	27.2707	29.0017	30.8502	12
13	29.9847	32.0887	34.3519	36.7862	13
14	34.8827	37.5811	40.5047	43.6720	14
15	40.4175	43.8424	47.5804	51.6595	15
16	46.6717	50.9804	55.7175	60.9250	16
17	53.7391	59.1176	65.0751	71.6730	17
18	61.7251	68.3941	75.8364	84.1407	18
19	70.7494	78.9692	88.2118	98.6032	19
20	80.9468	91.0249	102.4436	115.3797	20
21	92.4699	104.7684	118.8101	134.8405	21
22	105.4910	120.4360	137.6316	157.4150	22
23	120.2048	138.2970	159.2764	183.6014	23
24	136.8315	158.6586	184.1678	213.9776	24
25	155.6196	181.8708	212.7930	249.2140	25
26	176.8501	208.3327	245.7120	290.0883	26
27	200.8406	238.4993	283.5688	337.5024	27
28	227.9499	272.8892	327.1041	392.5028	28
29	258.5834	312.0937	377.1697	456.3032	29
30	293.1992	356.7868	434.7451	530.3117	30
31	332.3151	407.7370	500.9569	616.1616	31
32	376.5161	465.8202	577.1005	715.7475	32
33	426.4632	532.0350	664.6655	831.2671	33
34	482.9034	607.5199	765.3654	965.2698	34
35	546.6808	693.5727	881.1702	1120.7130	35
36	618.7493	791.6729	1014.3457	1301.0270	36
37	700.1867	903.5071	1167.4975	1510.1914	37
38	792.2110	1030.9981	1343.6222	1752.8220	38
39	896.1984	1176.3378	1546.1655	2034.2735	39
40	1013.7042	1342.0251	1779.0903	2360.7572	40
41	1146.4858	1530.9086	2046.9539	2739.4784	41
42	1296.5289	1746.2358	2354.9969	3178.7949	42
43	1466.0777	1991.7088	2709.2465	3688.4021	43
44	1657.6678	2271.5481	3116.6334	4279.5465	44
45	1874.1646	2590.5648	3585.1285	4965.2739	45
46	2118.8060	2954.2439	4123.8977	5760.7177	46
47	2395.2508	3368.8380	4743.4824	6683.4326	47
48	2707.6334	3841.4753	5456.0047	7753.7818	48
49	3060.6258	4380.2819	6275.4055	8995.3869	49
50	3459.5071	4994.5213	7217.7163	10435.6488	50

AMOUNT OF £1 PER ANNUM

		Rate Per Cent			
Yrs.	13	14	15	16	Yrs.
51	3910.2430	5694.7543	8301.3737	12106.3526	51
52	4419.5746	6493.0199	9547.5798	14044.3690	52
53	4995.1193	7403.0427	10980.7167	16292.4680	53
54	5645.4849	8440.4687	12628.8243	18900.2629	54
55	6380.3979	9623.1343	14524.1479	21925.3050	55
56	7210.8496	10971.3731	16703.7701	25434.3538	56
57	8149.2601	12508.3654	19210.3356	29504.8504	57
58	9209.6639	14260.5365	22092.8859	34226.6264	58
59	10407.9202	16258.0117	25407.8188	39703.8867	59
60	11761.9498	18535.1333	29219.9916	46057.5085	60
61	13292.0033	21131.0519	33603.9904	53427.7099	61
62	15020.9637	24090.3992	38645.5889	61977.1435	62
63	16974.6890	27464.0551	44443.4273	71894.4864	63
64	19182.3985	31310.0228	51110.9414	83398.6043	64
65	21677.1103	35694.4260	58778.5826	96743.3810	65
66	24496.1347	40692.6457	67596.3700	112223.3219	66
67	27681.6322	46390.6160	77736.8255	130180.0534	67
68	31281.2444	52886.3023	89398.3493	151009.8620	68
69	35348.8062	60291.3846	102809.102	175172.4399	69
70	39945.1510	68733.1785	118231.467	203201.0302	70
71	45139.0206	78356.8234	135967.187	235714.1951	71
72	51008.0933	89327.7787	156363.265	273429.4663	72
73	57640.1454	101834.668	179818.755	317179.1809	73
74	65134.3643	116092.521	206792.568	367928.8499	74
75	73602.8316	132346.474	237812.453	426798.4658	75
76	83172.1997	150875.981	273485.321	495087.2204	76
77	93985.5857	171999.618	314509.119	574302.1756	77
78	106204.712	196080.564	361686.487	666191.5237	78
79	120012.324	223532.843	415940.460	772783.1675	79
80	135614.927	254828.441	478332.529	896429.4743	80
81	153245.867	290505.423	550083.409	1039859.190	81
82	173168.830	331177.183	632596.920	1206237.661	82
83	195681.778	377542.988	727487.458	1399236.686	83
84	221121.409	430400.006	836611.577	1623115.556	84
85	249868.192	490657.007	962104.313	1882815.045	85
86	282352.057	559349.988	1106420.960	2184066.452	86
87	319058.824	637659.987	1272385.104	2533518.085	87
88	360537.471	726933.385	1463243.870	2938881.978	88
89	407408.343	828705.059	1682731.450	3409104.095	89
90	460372.427	944724.767	1935142.168	3954561.750	90
91	520221.843	1076987.234	2225414.493	4587292.630	91
92	587851.682	1227766.447	2559227.667	5321260.451	92
93	664273.401	1399654.750	2943112.817	6172663.123	93
94	750629.943	1595607.415	3384580.740	7160290.223	94
95	848212.835	1818993.453	3892268.851	8305937.658	95
96	958481.504	2073653.536	4476110.179	9634888.684	96
97	1083085.100	2363966.031	5147527.705	11176471.873	97
98	1223887.163	2694922.276	5919657.861	12964708.373	98
99	1382993.494	3072212.394	6807607.540	15039062.712	99
100	1562783.648	3502323.129	7828749.671	17445313.746	100

Rule for extending the Table of
AMOUNT OF ONE POUND PER ANNUM

Rule—
Amount of £1 p.a. in N years × Amount of £1 in M years + Amount of £1 p.a. in M years = Amount of £1 p.a. in N + M years.

Example—
What is the amount of £1 per annum in 150 years at 4%?

Answer—

Amount of £1 p.a. in 50 years at 4% (see page 146) ... =		152·6671
Amount of £1 in 100 years at 4% (see page 129) ... =		50·5049
(152·6671 × 50·5049) =		7710·4366
Amount of £1 p.a. in 100 years at 4% (see page 147) =		1237·6237
Amount of £1 p.a. in 150 years at 4% ⎱ = (7710·4366 + 1237·6237) ⎰		8948·0603

ANNUITY ONE POUND WILL PURCHASE
(SINGLE RATE % PRINCIPLE)

Allowing for a sinking fund to accumulate at the same rate as that which is required on the invested capital and ignoring the effect of income tax on that part of the income used to provide the annual sinking fund instalment. The annuity is receivable at the end of each year.

AT RATES OF INTEREST FROM

3% to 20%

ANNUITY £1 WILL PURCHASE (SINGLE RATE)

			Rate Per Cent				
Yrs.	3	3.1	3.2	3.25	3.3	3.4	Yrs.
1	1.030000	1.031000	1.032000	1.032500	1.033000	1.034000	1
2	0.522611	0.523368	0.524126	0.524505	0.524884	0.525642	2
3	0.353530	0.354210	0.354891	0.355231	0.355571	0.356253	3
4	0.269027	0.269671	0.270315	0.270637	0.270960	0.271605	4
5	0.218355	0.218978	0.219603	0.219916	0.220228	0.220854	5
6	0.184598	0.185210	0.185823	0.186130	0.186437	0.187052	6
7	0.160506	0.161112	0.161718	0.162022	0.162326	0.162935	7
8	0.142456	0.143058	0.143661	0.143963	0.144265	0.144870	8
9	0.128434	0.129033	0.129635	0.129936	0.130237	0.130841	9
10	0.117231	0.117830	0.118430	0.118731	0.119032	0.119636	10
11	0.108077	0.108677	0.109278	0.109579	0.109881	0.110486	11
12	0.100462	0.101063	0.101665	0.101967	0.102270	0.102876	12
13	0.094030	0.094632	0.095236	0.095539	0.095843	0.096451	13
14	0.088526	0.089131	0.089738	0.090042	0.090346	0.090958	14
15	0.083767	0.084374	0.084983	0.085289	0.085595	0.086209	15
16	0.079611	0.080221	0.080833	0.081140	0.081448	0.082065	16
17	0.075953	0.076565	0.077181	0.077490	0.077799	0.078420	17
18	0.072709	0.073325	0.073944	0.074254	0.074565	0.075190	18
19	0.069814	0.070433	0.071056	0.071368	0.071681	0.072309	19
20	0.067216	0.067839	0.068465	0.068779	0.069094	0.069726	20
21	0.064872	0.065498	0.066128	0.066444	0.066761	0.067397	21
22	0.062747	0.063378	0.064011	0.064329	0.064648	0.065289	22
23	0.060814	0.061448	0.062085	0.062406	0.062726	0.063371	23
24	0.059047	0.059685	0.060327	0.060649	0.060972	0.061621	24
25	0.057428	0.058070	0.058715	0.059039	0.059364	0.060017	25
26	0.055938	0.056584	0.057234	0.057560	0.057887	0.058544	26
27	0.054564	0.055214	0.055868	0.056196	0.056525	0.057187	27
28	0.053293	0.053947	0.054605	0.054935	0.055267	0.055933	28
29	0.052115	0.052772	0.053434	0.053767	0.054100	0.054771	29
30	0.051019	0.051681	0.052347	0.052682	0.053017	0.053692	30
31	0.049999	0.050665	0.051335	0.051672	0.052010	0.052689	31
32	0.049047	0.049716	0.050391	0.050730	0.051070	0.051753	32
33	0.048156	0.048830	0.049509	0.049850	0.050192	0.050880	33
34	0.047322	0.048000	0.048683	0.049026	0.049370	0.050063	34
35	0.046539	0.047221	0.047908	0.048253	0.048600	0.049297	35
36	0.045804	0.046490	0.047181	0.047528	0.047877	0.048578	36
37	0.045112	0.045802	0.046497	0.046846	0.047197	0.047903	37
38	0.044459	0.045153	0.045853	0.046204	0.046557	0.047267	38
39	0.043844	0.044542	0.045245	0.045599	0.045954	0.046668	39
40	0.043262	0.043964	0.044672	0.045028	0.045385	0.046104	40
41	0.042712	0.043419	0.044130	0.044488	0.044847	0.045570	41
42	0.042192	0.042902	0.043618	0.043978	0.044339	0.045066	42
43	0.041698	0.042412	0.043132	0.043494	0.043858	0.044589	43
44	0.041230	0.041948	0.042672	0.043036	0.043401	0.044137	44
45	0.040785	0.041507	0.042235	0.042601	0.042969	0.043708	45
46	0.040363	0.041088	0.041820	0.042188	0.042558	0.043302	46
47	0.039961	0.040690	0.041426	0.041796	0.042168	0.042916	47
48	0.039578	0.040311	0.041051	0.041423	0.041797	0.042549	48
49	0.039213	0.039951	0.040694	0.041068	0.041444	0.042200	49
50	0.038865	0.039607	0.040354	0.040730	0.041108	0.041868	50

ANNUITY £1 WILL PURCHASE (SINGLE RATE)

			Rate Per Cent				
Yrs.	3.5	3.6	3.7	3.75	3.8	3.9	Yrs.
1	1.035000	1.036000	1.037000	1.037500	1.038000	1.039000	1
2	0.526400	0.527159	0.527918	0.528298	0.528677	0.529436	2
3	0.356934	0.357616	0.358299	0.358640	0.358982	0.359665	3
4	0.272251	0.272898	0.273545	0.273869	0.274193	0.274841	4
5	0.221481	0.222109	0.222737	0.223052	0.223367	0.223996	5
6	0.187668	0.188285	0.188903	0.189212	0.189522	0.190141	6
7	0.163544	0.164155	0.164767	0.165074	0.165380	0.165994	7
8	0.145477	0.146084	0.146693	0.146998	0.147304	0.147915	8
9	0.131446	0.132053	0.132661	0.132965	0.133270	0.133881	9
10	0.120241	0.120848	0.121457	0.121761	0.122066	0.122678	10
11	0.111092	0.111700	0.112310	0.112615	0.112921	0.113534	11
12	0.103484	0.104094	0.104706	0.105012	0.105319	0.105935	12
13	0.097062	0.097674	0.098288	0.098596	0.098905	0.099523	13
14	0.091571	0.092186	0.092804	0.093113	0.093423	0.094045	14
15	0.086825	0.087444	0.088065	0.088376	0.088688	0.089313	15
16	0.082685	0.083307	0.083932	0.084245	0.084559	0.085188	16
17	0.079043	0.079669	0.080298	0.080613	0.080929	0.081562	17
18	0.075817	0.076447	0.077079	0.077397	0.077715	0.078353	18
19	0.072940	0.073574	0.074211	0.074531	0.074851	0.075493	19
20	0.070361	0.070999	0.071640	0.071962	0.072285	0.072932	20
21	0.068037	0.068679	0.069325	0.069649	0.069973	0.070625	21
22	0.065932	0.066579	0.067229	0.067555	0.067882	0.068539	22
23	0.064019	0.064670	0.065325	0.065653	0.065983	0.066644	23
24	0.062273	0.062929	0.063588	0.063919	0.064251	0.064917	24
25	0.060674	0.061334	0.061998	0.062332	0.062666	0.063337	25
26	0.059205	0.059870	0.060539	0.060875	0.061211	0.061888	26
27	0.057852	0.058522	0.059195	0.059533	0.059873	0.060554	27
28	0.056603	0.057277	0.057955	0.058295	0.058637	0.059323	28
29	0.055445	0.056124	0.056807	0.057150	0.057494	0.058185	29
30	0.054371	0.055055	0.055742	0.056088	0.056434	0.057130	30
31	0.053372	0.054060	0.054753	0.055100	0.055449	0.056150	31
32	0.052442	0.053134	0.053831	0.054181	0.054533	0.055238	32
33	0.051572	0.052270	0.052971	0.053324	0.053678	0.054388	33
34	0.050760	0.051461	0.052168	0.052523	0.052879	0.053595	34
35	0.049998	0.050705	0.051416	0.051773	0.052132	0.052852	35
36	0.049284	0.049995	0.050711	0.051071	0.051431	0.052157	36
37	0.048613	0.049329	0.050049	0.050411	0.050774	0.051505	37
38	0.047982	0.048702	0.049427	0.049792	0.050157	0.050892	38
39	0.047388	0.048112	0.048842	0.049209	0.049577	0.050316	39
40	0.046827	0.047556	0.048290	0.048659	0.049030	0.049774	40
41	0.046298	0.047032	0.047770	0.048142	0.048514	0.049263	41
42	0.045798	0.046536	0.047279	0.047653	0.048028	0.048781	42
43	0.045325	0.046068	0.046815	0.047191	0.047568	0.048326	43
44	0.044878	0.045624	0.046376	0.046754	0.047134	0.047897	44
45	0.044453	0.045204	0.045961	0.046341	0.046723	0.047490	45
46	0.044051	0.044806	0.045567	0.045949	0.046333	0.047105	46
47	0.043669	0.044429	0.045194	0.045578	0.045964	0.046740	47
48	0.043306	0.044070	0.044839	0.045226	0.045614	0.046395	48
49	0.042962	0.043729	0.044503	0.044892	0.045282	0.046067	49
50	0.042634	0.043406	0.044183	0.044574	0.044967	0.045756	50

ANNUITY £1 WILL PURCHASE (SINGLE RATE)

Rate Per Cent

Yrs.	4	4.1	4.2	4.25	4.3	4.4	Yrs.
1	1.040000	1.041000	1.042000	1.042500	1.043000	1.044000	1
2	0.530196	0.530956	0.531716	0.532096	0.532476	0.533237	2
3	0.360349	0.361033	0.361717	0.362060	0.362402	0.363088	3
4	0.275490	0.276140	0.276790	0.277115	0.277440	0.278092	4
5	0.224627	0.225259	0.225891	0.226207	0.226524	0.227157	5
6	0.190762	0.191383	0.192006	0.192317	0.192629	0.193253	6
7	0.166610	0.167226	0.167843	0.168152	0.168462	0.169081	7
8	0.148528	0.149142	0.149757	0.150065	0.150373	0.150991	8
9	0.134493	0.135107	0.135721	0.136029	0.136338	0.136955	9
10	0.123291	0.123905	0.124522	0.124830	0.125139	0.125758	10
11	0.114149	0.114766	0.115384	0.115693	0.116004	0.116625	11
12	0.106552	0.107171	0.107792	0.108103	0.108415	0.109040	12
13	0.100144	0.100766	0.101390	0.101703	0.102017	0.102645	13
14	0.094669	0.095295	0.095923	0.096238	0.096553	0.097186	14
15	0.089941	0.090571	0.091203	0.091520	0.091838	0.092475	15
16	0.085820	0.086454	0.087091	0.087410	0.087730	0.088372	16
17	0.082199	0.082837	0.083479	0.083800	0.084122	0.084769	17
18	0.078993	0.079637	0.080283	0.080607	0.080932	0.081583	18
19	0.076139	0.076787	0.077438	0.077764	0.078091	0.078748	19
20	0.073582	0.074235	0.074891	0.075220	0.075550	0.076211	20
21	0.071280	0.071938	0.072599	0.072931	0.073263	0.073930	21
22	0.069199	0.069862	0.070528	0.070862	0.071197	0.071870	22
23	0.067309	0.067977	0.068649	0.068986	0.069323	0.070001	23
24	0.065587	0.066260	0.066937	0.067276	0.067617	0.068300	24
25	0.064012	0.064690	0.065372	0.065715	0.066058	0.066747	25
26	0.062567	0.063251	0.063938	0.064283	0.064629	0.065323	26
27	0.061239	0.061927	0.062620	0.062967	0.063316	0.064016	27
28	0.060013	0.060707	0.061405	0.061755	0.062106	0.062812	28
29	0.058880	0.059579	0.060282	0.060635	0.060989	0.061700	29
30	0.057830	0.058534	0.059243	0.059598	0.059955	0.060671	30
31	0.056855	0.057565	0.058278	0.058637	0.058996	0.059718	31
32	0.055949	0.056663	0.057382	0.057743	0.058105	0.058832	32
33	0.055104	0.055823	0.056547	0.056911	0.057275	0.058008	33
34	0.054315	0.055039	0.055768	0.056135	0.056502	0.057240	34
35	0.053577	0.054307	0.055041	0.055410	0.055780	0.056523	35
36	0.052887	0.053622	0.054361	0.054732	0.055105	0.055853	36
37	0.052240	0.052979	0.053724	0.054097	0.054472	0.055226	37
38	0.051632	0.052377	0.053126	0.053502	0.053880	0.054638	38
39	0.051061	0.051810	0.052565	0.052944	0.053324	0.054087	39
40	0.050523	0.051278	0.052037	0.052418	0.052801	0.053570	40
41	0.050017	0.050776	0.051541	0.051924	0.052309	0.053083	41
42	0.049540	0.050304	0.051073	0.051459	0.051847	0.052625	42
43	0.049090	0.049858	0.050632	0.051021	0.051411	0.052194	43
44	0.048665	0.049438	0.050216	0.050607	0.050999	0.051788	44
45	0.048262	0.049040	0.049823	0.050217	0.050611	0.051404	45
46	0.047882	0.048664	0.049452	0.049848	0.050245	0.051042	46
47	0.047522	0.048309	0.049101	0.049499	0.049898	0.050700	47
48	0.047181	0.047972	0.048768	0.049169	0.049570	0.050377	48
49	0.046857	0.047653	0.048454	0.048856	0.049260	0.050071	49
50	0.046550	0.047350	0.048155	0.048560	0.048966	0.049782	50

ANNUITY £1 WILL PURCHASE (SINGLE RATE)

				Rate Per Cent			
Yrs.	**4.5**	**4.6**	**4.7**	**4.75**	**4.8**	**4.9**	Yrs.
1	1.045000	1.046000	1.047000	1.047500	1.048000	1.049000	1
2	0.533998	0.534759	0.535520	0.535900	0.536281	0.537043	2
3	0.363773	0.364460	0.365146	0.365490	0.365833	0.366521	3
4	0.278744	0.279396	0.280049	0.280376	0.280703	0.281357	4
5	0.227792	0.228427	0.229063	0.229381	0.229699	0.230337	5
6	0.193878	0.194504	0.195131	0.195445	0.195759	0.196388	6
7	0.169701	0.170323	0.170946	0.171257	0.171569	0.172194	7
8	0.151610	0.152230	0.152851	0.153162	0.153473	0.154097	8
9	0.137574	0.138195	0.138817	0.139128	0.139440	0.140064	9
10	0.126379	0.127001	0.127625	0.127937	0.128250	0.128876	10
11	0.117248	0.117873	0.118500	0.118813	0.119128	0.119757	11
12	0.109666	0.110294	0.110924	0.111240	0.111556	0.112190	12
13	0.103275	0.103908	0.104542	0.104860	0.105178	0.105816	13
14	0.097820	0.098457	0.099096	0.099416	0.099736	0.100379	14
15	0.093114	0.093755	0.094399	0.094721	0.095044	0.095692	15
16	0.089015	0.089662	0.090310	0.090635	0.090961	0.091614	16
17	0.085418	0.086069	0.086723	0.087051	0.087379	0.088038	17
18	0.082237	0.082894	0.083553	0.083883	0.084215	0.084879	18
19	0.079407	0.080069	0.080734	0.081068	0.081402	0.082072	19
20	0.076876	0.077544	0.078214	0.078550	0.078887	0.079564	20
21	0.074601	0.075274	0.075950	0.076289	0.076629	0.077311	21
22	0.072546	0.073224	0.073906	0.074248	0.074591	0.075279	22
23	0.070682	0.071367	0.072055	0.072400	0.072746	0.073440	23
24	0.068987	0.069677	0.070371	0.070719	0.071067	0.071768	24
25	0.067439	0.068135	0.068834	0.069185	0.069537	0.070243	25
26	0.066021	0.066723	0.067428	0.067782	0.068137	0.068849	26
27	0.064719	0.065427	0.066138	0.066494	0.066852	0.067570	27
28	0.063521	0.064234	0.064950	0.065310	0.065671	0.066395	28
29	0.062415	0.063133	0.063856	0.064218	0.064582	0.065312	29
30	0.061392	0.062116	0.062844	0.063209	0.063576	0.064312	30
31	0.060443	0.061173	0.061907	0.062276	0.062645	0.063387	31
32	0.059563	0.060299	0.061038	0.061409	0.061782	0.062529	32
33	0.058745	0.059485	0.060230	0.060605	0.060980	0.061733	33
34	0.057982	0.058728	0.059479	0.059856	0.060234	0.060993	34
35	0.057270	0.058022	0.058778	0.059158	0.059539	0.060303	35
36	0.056606	0.057363	0.058124	0.058507	0.058890	0.059660	36
37	0.055984	0.056746	0.057513	0.057898	0.058285	0.059060	37
38	0.055402	0.056169	0.056942	0.057329	0.057718	0.058499	38
39	0.054856	0.055629	0.056406	0.056796	0.057188	0.057974	39
40	0.054343	0.055121	0.055904	0.056297	0.056691	0.057482	40
41	0.053862	0.054645	0.055432	0.055828	0.056225	0.057021	41
42	0.053409	0.054197	0.054989	0.055388	0.055787	0.056589	42
43	0.052982	0.053775	0.054573	0.054974	0.055375	0.056182	43
44	0.052581	0.053379	0.054181	0.054584	0.054988	0.055800	44
45	0.052202	0.053005	0.053812	0.054218	0.054624	0.055441	45
46	0.051845	0.052652	0.053464	0.053872	0.054281	0.055102	46
47	0.051507	0.052319	0.053136	0.053546	0.053958	0.054784	47
48	0.051189	0.052005	0.052827	0.053239	0.053653	0.054483	48
49	0.050887	0.051708	0.052534	0.052949	0.053365	0.054200	49
50	0.050602	0.051428	0.052258	0.052675	0.053093	0.053933	50

ANNUITY £1 WILL PURCHASE (SINGLE RATE)

			Rate Per Cent				
Yrs.	5	5.1	5.2	5.25	5.3	5.4	Yrs.
1	1.050000	1.051000	1.052000	1.052500	1.053000	1.054000	1
2	0.537805	0.538567	0.539329	0.539711	0.540092	0.540855	2
3	0.367209	0.367897	0.368586	0.368930	0.369275	0.369964	3
4	0.282012	0.282667	0.283323	0.283651	0.283980	0.284637	4
5	0.230975	0.231614	0.232253	0.232573	0.232894	0.233535	5
6	0.197017	0.197648	0.198279	0.198595	0.198912	0.199545	6
7	0.172820	0.173447	0.174075	0.174389	0.174703	0.175333	7
8	0.154722	0.155348	0.155975	0.156289	0.156604	0.157233	8
9	0.140690	0.141317	0.141946	0.142261	0.142576	0.143207	9
10	0.129505	0.130134	0.130765	0.131082	0.131398	0.132032	10
11	0.120389	0.121022	0.121657	0.121975	0.122293	0.122931	11
12	0.112825	0.113463	0.114102	0.114422	0.114742	0.115385	12
13	0.106456	0.107098	0.107741	0.108064	0.108387	0.109035	13
14	0.101024	0.101671	0.102320	0.102645	0.102971	0.103624	14
15	0.096342	0.096995	0.097649	0.097977	0.098306	0.098965	15
16	0.092270	0.092928	0.093588	0.093919	0.094251	0.094915	16
17	0.088699	0.089363	0.090029	0.090363	0.090698	0.091369	17
18	0.085546	0.086216	0.086888	0.087225	0.087563	0.088240	18
19	0.082745	0.083421	0.084099	0.084439	0.084780	0.085464	19
20	0.080243	0.080924	0.081609	0.081952	0.082296	0.082986	20
21	0.077996	0.078684	0.079375	0.079721	0.080069	0.080765	21
22	0.075971	0.076665	0.077362	0.077712	0.078062	0.078765	22
23	0.074137	0.074837	0.075541	0.075894	0.076247	0.076957	23
24	0.072471	0.073177	0.073887	0.074243	0.074600	0.075316	24
25	0.070952	0.071665	0.072381	0.072741	0.073101	0.073823	25
26	0.069564	0.070283	0.071006	0.071368	0.071731	0.072461	26
27	0.068292	0.069017	0.069746	0.070111	0.070478	0.071213	27
28	0.067123	0.067854	0.068589	0.068957	0.069327	0.070069	28
29	0.066046	0.066783	0.067524	0.067896	0.068269	0.069017	29
30	0.065051	0.065795	0.066542	0.066917	0.067293	0.068047	30
31	0.064132	0.064882	0.065635	0.066013	0.066392	0.067152	31
32	0.063280	0.064036	0.064795	0.065176	0.065558	0.066325	32
33	0.062490	0.063251	0.064016	0.064400	0.064785	0.065558	33
34	0.061755	0.062522	0.063293	0.063680	0.064068	0.064847	34
35	0.061072	0.061844	0.062621	0.063011	0.063402	0.064186	35
36	0.060434	0.061213	0.061995	0.062388	0.062782	0.063572	36
37	0.059840	0.060624	0.061412	0.061807	0.062204	0.063000	37
38	0.059284	0.060074	0.060867	0.061265	0.061665	0.062467	38
39	0.058765	0.059559	0.060358	0.060759	0.061162	0.061969	39
40	0.058278	0.059078	0.059883	0.060286	0.060691	0.061504	40
41	0.057822	0.058628	0.059437	0.059844	0.060251	0.061069	41
42	0.057395	0.058205	0.059020	0.059429	0.059839	0.060662	42
43	0.056993	0.057809	0.058629	0.059040	0.059453	0.060281	43
44	0.056616	0.057437	0.058262	0.058676	0.059091	0.059924	44
45	0.056262	0.057087	0.057917	0.058333	0.058751	0.059589	45
46	0.055928	0.056758	0.057593	0.058012	0.058432	0.059275	46
47	0.055614	0.056449	0.057288	0.057710	0.058132	0.058980	47
48	0.055318	0.056158	0.057002	0.057425	0.057850	0.058702	48
49	0.055040	0.055884	0.056732	0.057158	0.057585	0.058442	49
50	0.054777	0.055625	0.056478	0.056906	0.057335	0.058196	50

ANNUITY £1 WILL PURCHASE (SINGLE RATE)

			Rate Per Cent				
Yrs.	**5.5**	**5.6**	**5.7**	**5.75**	**5.8**	**5.9**	Yrs.
1	1.055000	1.056000	1.057000	1.057500	1.058000	1.059000	1
2	0.541618	0.542381	0.543145	0.543527	0.543909	0.544673	2
3	0.370654	0.371344	0.372035	0.372381	0.372726	0.373418	3
4	0.285294	0.285953	0.286612	0.286941	0.287271	0.287931	4
5	0.234176	0.234819	0.235462	0.235784	0.236106	0.236751	5
6	0.200179	0.200814	0.201450	0.201768	0.202087	0.202724	6
7	0.175964	0.176596	0.177230	0.177546	0.177864	0.178499	7
8	0.157864	0.158496	0.159129	0.159446	0.159764	0.160399	8
9	0.143839	0.144473	0.145109	0.145427	0.145745	0.146383	9
10	0.132668	0.133305	0.133943	0.134263	0.134583	0.135225	10
11	0.123571	0.124212	0.124855	0.125177	0.125499	0.126145	11
12	0.116029	0.116675	0.117323	0.117648	0.117973	0.118624	12
13	0.109684	0.110336	0.110989	0.111316	0.111644	0.112301	13
14	0.104279	0.104936	0.105595	0.105926	0.106257	0.106920	14
15	0.099626	0.100289	0.100954	0.101288	0.101621	0.102291	15
16	0.095583	0.096252	0.096924	0.097260	0.097598	0.098274	16
17	0.092042	0.092718	0.093396	0.093736	0.094077	0.094760	17
18	0.088920	0.089602	0.090287	0.090630	0.090974	0.091664	18
19	0.086150	0.086839	0.087531	0.087877	0.088225	0.088922	19
20	0.083679	0.084375	0.085073	0.085423	0.085774	0.086478	20
21	0.081465	0.082167	0.082872	0.083226	0.083580	0.084291	21
22	0.079471	0.080180	0.080892	0.081249	0.081607	0.082325	22
23	0.077670	0.078385	0.079104	0.079465	0.079826	0.080551	23
24	0.076036	0.076758	0.077484	0.077848	0.078213	0.078944	24
25	0.074549	0.075278	0.076011	0.076378	0.076746	0.077485	25
26	0.073193	0.073929	0.074668	0.075039	0.075410	0.076156	26
27	0.071952	0.072695	0.073440	0.073814	0.074189	0.074942	27
28	0.070814	0.071563	0.072316	0.072693	0.073071	0.073830	28
29	0.069769	0.070524	0.071283	0.071663	0.072045	0.072811	29
30	0.068805	0.069567	0.070332	0.070716	0.071101	0.071873	30
31	0.067917	0.068685	0.069456	0.069843	0.070231	0.071010	31
32	0.067095	0.067869	0.068647	0.069038	0.069429	0.070214	32
33	0.066335	0.067115	0.067899	0.068292	0.068687	0.069478	33
34	0.065630	0.066416	0.067206	0.067603	0.068000	0.068797	34
35	0.064975	0.065767	0.066563	0.066963	0.067363	0.068167	35
36	0.064366	0.065165	0.065967	0.066369	0.066772	0.067582	36
37	0.063800	0.064604	0.065412	0.065817	0.066223	0.067038	37
38	0.063272	0.064082	0.064895	0.065303	0.065712	0.066533	38
39	0.062780	0.063595	0.064414	0.064825	0.065237	0.066063	39
40	0.062320	0.063141	0.063965	0.064379	0.064794	0.065626	40
41	0.061891	0.062717	0.063547	0.063963	0.064380	0.065218	41
42	0.061489	0.062320	0.063155	0.063574	0.063994	0.064837	42
43	0.061113	0.061950	0.062790	0.063211	0.063634	0.064482	43
44	0.060761	0.061603	0.062448	0.062872	0.063297	0.064150	44
45	0.060431	0.061277	0.062128	0.062554	0.062981	0.063839	45
46	0.060122	0.060973	0.061828	0.062256	0.062686	0.063549	46
47	0.059831	0.060687	0.061547	0.061978	0.062410	0.063277	47
48	0.059559	0.060419	0.061283	0.061716	0.062151	0.063022	48
49	0.059302	0.060167	0.061036	0.061471	0.061908	0.062784	49
50	0.059061	0.059931	0.060803	0.061241	0.061680	0.062560	50

ANNUITY £1 WILL PURCHASE (SINGLE RATE)

			Rate Per Cent				
Yrs.	6	6.1	6.2	6.25	6.3	6.4	Yrs.
1	1.060000	1.061000	1.062000	1.062500	1.063000	1.064000	1
2	0.545437	0.546201	0.546966	0.547348	0.547731	0.548496	2
3	0.374110	0.374802	0.375495	0.375841	0.376188	0.376882	3
4	0.288591	0.289253	0.289914	0.290245	0.290577	0.291239	4
5	0.237396	0.238043	0.238689	0.239013	0.239337	0.239985	5
6	0.203363	0.204002	0.204642	0.204963	0.205283	0.205925	6
7	0.179135	0.179772	0.180410	0.180730	0.181050	0.181690	7
8	0.161036	0.161674	0.162313	0.162633	0.162953	0.163595	8
9	0.147022	0.147663	0.148305	0.148626	0.148948	0.149592	9
10	0.135868	0.136512	0.137158	0.137482	0.137806	0.138454	10
11	0.126793	0.127442	0.128093	0.128419	0.128746	0.129400	11
12	0.119277	0.119932	0.120588	0.120917	0.121247	0.121907	12
13	0.112960	0.113621	0.114284	0.114616	0.114948	0.115614	13
14	0.107585	0.108252	0.108921	0.109257	0.109592	0.110265	14
15	0.102963	0.103637	0.104312	0.104651	0.104990	0.105671	15
16	0.098952	0.099633	0.100316	0.100658	0.101001	0.101688	16
17	0.095445	0.096132	0.096822	0.097168	0.097515	0.098209	17
18	0.092357	0.093051	0.093748	0.094098	0.094448	0.095150	18
19	0.089621	0.090323	0.091027	0.091380	0.091734	0.092444	19
20	0.087185	0.087894	0.088605	0.088962	0.089320	0.090037	20
21	0.085005	0.085721	0.086440	0.086800	0.087162	0.087886	21
22	0.083046	0.083769	0.084495	0.084860	0.085225	0.085956	22
23	0.081278	0.082009	0.082743	0.083111	0.083479	0.084219	23
24	0.079679	0.080417	0.081158	0.081529	0.081901	0.082648	24
25	0.078227	0.078972	0.079720	0.080095	0.080470	0.081224	25
26	0.076904	0.077656	0.078411	0.078790	0.079169	0.079931	26
27	0.075697	0.076456	0.077218	0.077600	0.077983	0.078751	27
28	0.074593	0.075358	0.076127	0.076513	0.076899	0.077675	28
29	0.073580	0.074352	0.075128	0.075517	0.075907	0.076689	29
30	0.072649	0.073428	0.074210	0.074603	0.074996	0.075785	30
31	0.071792	0.072578	0.073367	0.073763	0.074159	0.074955	31
32	0.071002	0.071794	0.072590	0.072989	0.073389	0.074191	32
33	0.070273	0.071071	0.071873	0.072275	0.072679	0.073487	33
34	0.069598	0.070403	0.071211	0.071617	0.072023	0.072838	34
35	0.068974	0.069785	0.070599	0.071007	0.071417	0.072238	35
36	0.068395	0.069212	0.070032	0.070443	0.070856	0.071683	36
37	0.067857	0.068680	0.069506	0.069921	0.070336	0.071169	37
38	0.067358	0.068186	0.069018	0.069436	0.069854	0.070693	38
39	0.066894	0.067728	0.068565	0.068985	0.069406	0.070251	39
40	0.066462	0.067301	0.068144	0.068567	0.068990	0.069840	40
41	0.066059	0.066904	0.067752	0.068177	0.068604	0.069459	41
42	0.065683	0.066533	0.067387	0.067815	0.068244	0.069105	42
43	0.065333	0.066188	0.067047	0.067478	0.067909	0.068775	43
44	0.065006	0.065866	0.066730	0.067163	0.067597	0.068467	44
45	0.064700	0.065565	0.066434	0.066869	0.067306	0.068181	45
46	0.064415	0.065285	0.066158	0.066596	0.067034	0.067914	46
47	0.064148	0.065022	0.065900	0.066340	0.066781	0.067665	47
48	0.063898	0.064776	0.065659	0.066101	0.066544	0.067433	48
49	0.063664	0.064547	0.065433	0.065878	0.066323	0.067216	49
50	0.063444	0.064332	0.065222	0.065669	0.066116	0.067014	50

ANNUITY £1 WILL PURCHASE (SINGLE RATE)

			Rate Per Cent				
Yrs.	**6.5**	**6.6**	**6.7**	**6.75**	**6.8**	**6.9**	Yrs.
1	1.065000	1.066000	1.067000	1.067500	1.068000	1.069000	1
2	0.549262	0.550027	0.550793	0.551176	0.551559	0.552325	2
3	0.377576	0.378270	0.378965	0.379312	0.379660	0.380356	3
4	0.291903	0.292567	0.293231	0.293564	0.293896	0.294562	4
5	0.240635	0.241284	0.241935	0.242260	0.242586	0.243238	5
6	0.206568	0.207212	0.207857	0.208179	0.208502	0.209149	6
7	0.182331	0.182974	0.183617	0.183939	0.184261	0.184907	7
8	0.164237	0.164881	0.165526	0.165849	0.166172	0.166819	8
9	0.150238	0.150885	0.151534	0.151858	0.152183	0.152834	9
10	0.139105	0.139756	0.140410	0.140737	0.141064	0.141720	10
11	0.130055	0.130712	0.131371	0.131701	0.132032	0.132693	11
12	0.122568	0.123232	0.123897	0.124230	0.124563	0.125232	12
13	0.116283	0.116953	0.117624	0.117961	0.118298	0.118974	13
14	0.110940	0.111617	0.112296	0.112637	0.112977	0.113660	14
15	0.106353	0.107037	0.107723	0.108067	0.108412	0.109102	15
16	0.102378	0.103069	0.103763	0.104111	0.104459	0.105157	16
17	0.098906	0.099606	0.100307	0.100659	0.101011	0.101717	17
18	0.095855	0.096561	0.097271	0.097626	0.097982	0.098696	18
19	0.093156	0.093870	0.094587	0.094947	0.095307	0.096029	19
20	0.090756	0.091479	0.092203	0.092567	0.092931	0.093661	20
21	0.088613	0.089343	0.090076	0.090443	0.090811	0.091549	21
22	0.086691	0.087429	0.088169	0.088540	0.088912	0.089657	22
23	0.084961	0.085706	0.086454	0.086829	0.087204	0.087958	23
24	0.083398	0.084150	0.084906	0.085284	0.085664	0.086425	24
25	0.081981	0.082741	0.083504	0.083887	0.084270	0.085039	25
26	0.080695	0.081462	0.082232	0.082619	0.083006	0.083782	26
27	0.079523	0.080297	0.081075	0.081465	0.081856	0.082639	27
28	0.078453	0.079235	0.080019	0.080413	0.080807	0.081598	28
29	0.077474	0.078263	0.079055	0.079452	0.079850	0.080648	29
30	0.076577	0.077373	0.078172	0.078572	0.078973	0.079778	30
31	0.075754	0.076556	0.077362	0.077766	0.078170	0.078982	31
32	0.074997	0.075806	0.076618	0.077025	0.077433	0.078251	32
33	0.074299	0.075115	0.075933	0.076344	0.076755	0.077580	33
34	0.073656	0.074478	0.075303	0.075716	0.076131	0.076962	34
35	0.073062	0.073890	0.074721	0.075138	0.075556	0.076393	35
36	0.072513	0.073347	0.074184	0.074604	0.075025	0.075869	36
37	0.072005	0.072845	0.073688	0.074111	0.074535	0.075384	37
38	0.071535	0.072380	0.073229	0.073655	0.074081	0.074937	38
39	0.071099	0.071950	0.072804	0.073233	0.073662	0.074523	39
40	0.070694	0.071550	0.072410	0.072842	0.073273	0.074140	40
41	0.070318	0.071180	0.072045	0.072479	0.072913	0.073785	41
42	0.069968	0.070836	0.071706	0.072142	0.072580	0.073456	42
43	0.069644	0.070516	0.071391	0.071830	0.072270	0.073151	43
44	0.069341	0.070218	0.071099	0.071540	0.071982	0.072868	44
45	0.069060	0.069941	0.070826	0.071270	0.071715	0.072606	45
46	0.068797	0.069684	0.070573	0.071019	0.071466	0.072362	46
47	0.068553	0.069444	0.070338	0.070786	0.071235	0.072135	47
48	0.068325	0.069220	0.070118	0.070569	0.071020	0.071924	48
49	0.068112	0.069012	0.069914	0.070366	0.070819	0.071728	49
50	0.067914	0.068817	0.069724	0.070178	0.070633	0.071545	50

ANNUITY £1 WILL PURCHASE (SINGLE RATE)

Rate Per Cent

Yrs.	7	7.1	7.2	7.25	7.3	7.4	Yrs.
1	1.070000	1.071000	1.072000	1.072500	1.073000	1.074000	1
2	0.553092	0.553859	0.554625	0.555009	0.555393	0.556160	2
3	0.381052	0.381748	0.382445	0.382793	0.383142	0.383840	3
4	0.295228	0.295895	0.296562	0.296896	0.297230	0.297899	4
5	0.243891	0.244544	0.245198	0.245525	0.245853	0.246508	5
6	0.209796	0.210444	0.211093	0.211418	0.211743	0.212393	6
7	0.185553	0.186201	0.186849	0.187174	0.187498	0.188149	7
8	0.167468	0.168117	0.168768	0.169094	0.169420	0.170073	8
9	0.153486	0.154140	0.154795	0.155123	0.155451	0.156108	9
10	0.142378	0.143036	0.143697	0.144027	0.144358	0.145021	10
11	0.133357	0.134022	0.134688	0.135022	0.135357	0.136026	11
12	0.125902	0.126574	0.127247	0.127585	0.127922	0.128599	12
13	0.119651	0.120330	0.121011	0.121352	0.121694	0.122378	13
14	0.114345	0.115032	0.115720	0.116065	0.116411	0.117103	14
15	0.109795	0.110489	0.111186	0.111535	0.111884	0.112585	15
16	0.105858	0.106560	0.107265	0.107618	0.107971	0.108680	16
17	0.102425	0.103136	0.103848	0.104206	0.104563	0.105281	17
18	0.099413	0.100131	0.100852	0.101214	0.101576	0.102301	18
19	0.096753	0.097480	0.098209	0.098574	0.098941	0.099675	19
20	0.094393	0.095128	0.095865	0.096235	0.096605	0.097347	20
21	0.092289	0.093032	0.093778	0.094151	0.094526	0.095276	21
22	0.090406	0.091157	0.091910	0.092288	0.092667	0.093425	22
23	0.088714	0.089473	0.090234	0.090616	0.090999	0.091766	23
24	0.087189	0.087956	0.088725	0.089111	0.089497	0.090272	24
25	0.085811	0.086585	0.087362	0.087752	0.088142	0.088925	25
26	0.084561	0.085343	0.086128	0.086521	0.086916	0.087706	26
27	0.083426	0.084215	0.085008	0.085405	0.085803	0.086601	27
28	0.082392	0.083189	0.083989	0.084390	0.084791	0.085597	28
29	0.081449	0.082253	0.083060	0.083464	0.083870	0.084682	29
30	0.080586	0.081397	0.082211	0.082620	0.083028	0.083848	30
31	0.079797	0.080615	0.081436	0.081847	0.082260	0.083087	31
32	0.079073	0.079898	0.080725	0.081140	0.081556	0.082389	32
33	0.078408	0.079239	0.080073	0.080492	0.080911	0.081751	33
34	0.077797	0.078634	0.079475	0.079896	0.080319	0.081165	34
35	0.077234	0.078078	0.078925	0.079349	0.079774	0.080627	35
36	0.076715	0.077565	0.078418	0.078846	0.079274	0.080133	36
37	0.076237	0.077093	0.077951	0.078382	0.078813	0.079678	37
38	0.075795	0.076657	0.077521	0.077954	0.078388	0.079259	38
39	0.075387	0.076254	0.077124	0.077560	0.077997	0.078873	39
40	0.075009	0.075882	0.076757	0.077196	0.077635	0.078516	40
41	0.074660	0.075537	0.076418	0.076859	0.077301	0.078188	41
42	0.074336	0.075219	0.076104	0.076548	0.076993	0.077884	42
43	0.074036	0.074923	0.075814	0.076260	0.076707	0.077603	43
44	0.073758	0.074650	0.075545	0.075994	0.076443	0.077344	44
45	0.073500	0.074396	0.075296	0.075747	0.076199	0.077104	45
46	0.073260	0.074161	0.075065	0.075518	0.075972	0.076882	46
47	0.073037	0.073943	0.074851	0.075306	0.075762	0.076676	47
48	0.072831	0.073740	0.074653	0.075110	0.075568	0.076485	48
49	0.072639	0.073552	0.074468	0.074928	0.075387	0.076309	49
50	0.072460	0.073377	0.074297	0.074758	0.075220	0.076145	50

ANNUITY £1 WILL PURCHASE (SINGLE RATE)

				Rate Per Cent			
Yrs.	7.5	7.6	7.7	7.75	7.8	7.9	Yrs.
1	1.075000	1.076000	1.077000	1.077500	1.078000	1.079000	1
2	0.556928	0.557696	0.558464	0.558848	0.559232	0.560000	2
3	0.384538	0.385236	0.385935	0.386284	0.386634	0.387334	3
4	0.298568	0.299237	0.299907	0.300242	0.300578	0.301249	4
5	0.247165	0.247822	0.248479	0.248808	0.249138	0.249797	5
6	0.213045	0.213697	0.214351	0.214677	0.215005	0.215660	6
7	0.188800	0.189453	0.190106	0.190433	0.190761	0.191416	7
8	0.170727	0.171382	0.172039	0.172367	0.172696	0.173355	8
9	0.156767	0.157427	0.158088	0.158419	0.158751	0.159415	9
10	0.145686	0.146352	0.147019	0.147353	0.147688	0.148358	10
11	0.136697	0.137370	0.138044	0.138382	0.138720	0.139398	11
12	0.129278	0.129958	0.130640	0.130981	0.131323	0.132008	12
13	0.123064	0.123752	0.124442	0.124788	0.125134	0.125827	13
14	0.117797	0.118494	0.119192	0.119541	0.119891	0.120593	14
15	0.113287	0.113992	0.114698	0.115052	0.115407	0.116117	15
16	0.109391	0.110104	0.110819	0.111178	0.111536	0.112256	16
17	0.106000	0.106722	0.107445	0.107808	0.108171	0.108899	17
18	0.103029	0.103759	0.104491	0.104859	0.105226	0.105963	18
19	0.100411	0.101150	0.101891	0.102262	0.102634	0.103380	19
20	0.098092	0.098839	0.099589	0.099965	0.100341	0.101095	20
21	0.096029	0.096785	0.097543	0.097923	0.098304	0.099067	21
22	0.094187	0.094951	0.095717	0.096102	0.096486	0.097258	22
23	0.092535	0.093307	0.094082	0.094471	0.094860	0.095640	23
24	0.091050	0.091830	0.092613	0.093006	0.093399	0.094187	24
25	0.089711	0.090499	0.091290	0.091686	0.092084	0.092880	25
26	0.088500	0.089296	0.090095	0.090495	0.090896	0.091700	26
27	0.087402	0.088206	0.089012	0.089417	0.089822	0.090633	27
28	0.086405	0.087216	0.088030	0.088438	0.088847	0.089667	28
29	0.085498	0.086317	0.087138	0.087550	0.087962	0.088789	29
30	0.084671	0.085497	0.086325	0.086741	0.087157	0.087991	30
31	0.083916	0.084749	0.085584	0.086003	0.086423	0.087264	31
32	0.083226	0.084065	0.084908	0.085330	0.085753	0.086600	32
33	0.082594	0.083440	0.084289	0.084714	0.085140	0.085995	33
34	0.082015	0.082867	0.083722	0.084151	0.084580	0.085441	34
35	0.081483	0.082341	0.083203	0.083635	0.084067	0.084934	35
36	0.080994	0.081859	0.082726	0.083161	0.083596	0.084469	36
37	0.080545	0.081416	0.082289	0.082726	0.083165	0.084043	37
38	0.080132	0.081008	0.081887	0.082327	0.082768	0.083652	38
39	0.079751	0.080633	0.081517	0.081960	0.082404	0.083293	39
40	0.079400	0.080287	0.081176	0.081622	0.082068	0.082963	40
41	0.079077	0.079968	0.080863	0.081311	0.081760	0.082659	41
42	0.078778	0.079675	0.080574	0.081024	0.081476	0.082380	42
43	0.078502	0.079403	0.080307	0.080760	0.081214	0.082123	43
44	0.078247	0.079153	0.080061	0.080517	0.080972	0.081886	44
45	0.078011	0.078922	0.079835	0.080292	0.080750	0.081667	45
46	0.077794	0.078708	0.079625	0.080084	0.080544	0.081466	46
47	0.077592	0.078510	0.079431	0.079893	0.080355	0.081280	47
48	0.077405	0.078328	0.079252	0.079716	0.080180	0.081109	48
49	0.077232	0.078159	0.079087	0.079552	0.080018	0.080951	49
50	0.077072	0.078002	0.078934	0.079401	0.079868	0.080805	50

ANNUITY £1 WILL PURCHASE (SINGLE RATE)

			Rate Per Cent				
Yrs.	**8**	**8.1**	**8.2**	**8.25**	**8.3**	**8.4**	Yrs.
1	1.080000	1.081000	1.082000	1.082500	1.083000	1.084000	1
2	0.560769	0.561538	0.562307	0.562692	0.563077	0.563846	2
3	0.388034	0.388734	0.389435	0.389785	0.390136	0.390837	3
4	0.301921	0.302593	0.303266	0.303603	0.303939	0.304613	4
5	0.250456	0.251117	0.251778	0.252109	0.252440	0.253102	5
6	0.216315	0.216972	0.217630	0.217959	0.218288	0.218947	6
7	0.192072	0.192730	0.193388	0.193718	0.194048	0.194708	7
8	0.174015	0.174676	0.175338	0.175669	0.176001	0.176665	8
9	0.160080	0.160746	0.161414	0.161748	0.162082	0.162752	9
10	0.149029	0.149702	0.150377	0.150714	0.151052	0.151729	10
11	0.140076	0.140757	0.141438	0.141780	0.142122	0.142807	11
12	0.132695	0.133383	0.134073	0.134419	0.134765	0.135458	12
13	0.126522	0.127219	0.127917	0.128267	0.128617	0.129319	13
14	0.121297	0.122002	0.122710	0.123064	0.123419	0.124130	14
15	0.116830	0.117544	0.118260	0.118619	0.118978	0.119698	15
16	0.112977	0.113700	0.114425	0.114789	0.115153	0.115882	16
17	0.109629	0.110362	0.111096	0.111464	0.111833	0.112571	17
18	0.106702	0.107443	0.108187	0.108559	0.108933	0.109680	18
19	0.104128	0.104878	0.105630	0.106007	0.106385	0.107142	19
20	0.101852	0.102611	0.103373	0.103754	0.104137	0.104903	20
21	0.099832	0.100600	0.101370	0.101756	0.102143	0.102918	21
22	0.098032	0.098809	0.099588	0.099978	0.100369	0.101153	22
23	0.096422	0.097207	0.097995	0.098389	0.098785	0.099577	23
24	0.094978	0.095771	0.096567	0.096966	0.097366	0.098166	24
25	0.093679	0.094480	0.095284	0.095687	0.096091	0.096900	25
26	0.092507	0.093317	0.094129	0.094536	0.094943	0.095760	26
27	0.091448	0.092265	0.093085	0.093496	0.093908	0.094733	27
28	0.090489	0.091314	0.092141	0.092556	0.092971	0.093804	28
29	0.089619	0.090451	0.091286	0.091704	0.092123	0.092963	29
30	0.088827	0.089667	0.090509	0.090931	0.091354	0.092201	30
31	0.088107	0.088954	0.089803	0.090228	0.090654	0.091509	31
32	0.087451	0.088304	0.089160	0.089589	0.090018	0.090879	32
33	0.086852	0.087711	0.088574	0.089006	0.089439	0.090306	33
34	0.086304	0.087170	0.088039	0.088474	0.088910	0.089784	34
35	0.085803	0.086675	0.087550	0.087988	0.088427	0.089307	35
36	0.085345	0.086223	0.087103	0.087545	0.087986	0.088872	36
37	0.084924	0.085808	0.086694	0.087139	0.087583	0.088474	37
38	0.084539	0.085428	0.086320	0.086767	0.087214	0.088111	38
39	0.084185	0.085080	0.085977	0.086426	0.086876	0.087778	39
40	0.083860	0.084760	0.085662	0.086114	0.086566	0.087473	40
41	0.083561	0.084466	0.085373	0.085827	0.086282	0.087194	41
42	0.083287	0.084196	0.085108	0.085564	0.086021	0.086937	42
43	0.083034	0.083948	0.084864	0.085323	0.085782	0.086703	43
44	0.082802	0.083720	0.084640	0.085101	0.085562	0.086487	44
45	0.082587	0.083509	0.084434	0.084897	0.085360	0.086289	45
46	0.082390	0.083316	0.084244	0.084709	0.085175	0.086107	46
47	0.082208	0.083138	0.084070	0.084537	0.085004	0.085940	47
48	0.082040	0.082974	0.083909	0.084378	0.084847	0.085787	48
49	0.081886	0.082823	0.083762	0.084232	0.084703	0.085645	49
50	0.081743	0.082683	0.083625	0.084097	0.084570	0.085516	50

ANNUITY £1 WILL PURCHASE (SINGLE RATE)

				Rate Per Cent			
Yrs.	**8.5**	**8.6**	**8.7**	**8.75**	**8.8**	**8.9**	Yrs.
1	1.085000	1.086000	1.087000	1.087500	1.088000	1.089000	1
2	0.564616	0.565386	0.566157	0.566542	0.566927	0.567698	2
3	0.391539	0.392242	0.392944	0.393296	0.393647	0.394351	3
4	0.305288	0.305963	0.306639	0.306977	0.307315	0.307991	4
5	0.253766	0.254430	0.255094	0.255427	0.255760	0.256426	5
6	0.219607	0.220268	0.220930	0.221261	0.221592	0.222256	6
7	0.195369	0.196032	0.196695	0.197027	0.197359	0.198024	7
8	0.177331	0.177997	0.178665	0.178999	0.179334	0.180003	8
9	0.163424	0.164096	0.164770	0.165107	0.165445	0.166121	9
10	0.152408	0.153087	0.153769	0.154110	0.154451	0.155135	10
11	0.143493	0.144181	0.144870	0.145215	0.145561	0.146253	11
12	0.136153	0.136849	0.137547	0.137897	0.138247	0.138948	12
13	0.130023	0.130728	0.131435	0.131789	0.132144	0.132854	13
14	0.124842	0.125557	0.126273	0.126632	0.126992	0.127711	14
15	0.120420	0.121144	0.121870	0.122234	0.122598	0.123327	15
16	0.116614	0.117347	0.118082	0.118451	0.118819	0.119559	16
17	0.113312	0.114055	0.114800	0.115173	0.115546	0.116295	17
18	0.110430	0.111183	0.111937	0.112315	0.112693	0.113452	18
19	0.107901	0.108663	0.109426	0.109809	0.110192	0.110960	19
20	0.105671	0.106442	0.107214	0.107602	0.107990	0.108767	20
21	0.103695	0.104475	0.105257	0.105649	0.106041	0.106828	21
22	0.101939	0.102727	0.103518	0.103915	0.104312	0.105107	22
23	0.100372	0.101169	0.101969	0.102370	0.102771	0.103575	23
24	0.098970	0.099776	0.100584	0.100989	0.101394	0.102207	24
25	0.097712	0.098526	0.099342	0.099751	0.100161	0.100983	25
26	0.096580	0.097402	0.098227	0.098640	0.099054	0.099884	26
27	0.095560	0.096390	0.097223	0.097640	0.098058	0.098895	27
28	0.094639	0.095477	0.096317	0.096738	0.097160	0.098005	28
29	0.093806	0.094651	0.095498	0.095923	0.096348	0.097201	29
30	0.093051	0.093903	0.094758	0.095186	0.095615	0.096474	30
31	0.092365	0.093224	0.094086	0.094518	0.094950	0.095817	31
32	0.091742	0.092608	0.093477	0.093912	0.094348	0.095221	32
33	0.091176	0.092048	0.092923	0.093361	0.093800	0.094680	33
34	0.090660	0.091538	0.092419	0.092861	0.093303	0.094189	34
35	0.090189	0.091074	0.091961	0.092405	0.092850	0.093742	35
36	0.089760	0.090650	0.091543	0.091990	0.092438	0.093336	36
37	0.089368	0.090264	0.091162	0.091612	0.092063	0.092965	37
38	0.089010	0.089911	0.090814	0.091267	0.091720	0.092628	38
39	0.088682	0.089588	0.090497	0.090952	0.091408	0.092321	39
40	0.088382	0.089293	0.090207	0.090664	0.091122	0.092040	40
41	0.088107	0.089023	0.089941	0.090401	0.090862	0.091784	41
42	0.087856	0.088776	0.089699	0.090161	0.090623	0.091550	42
43	0.087625	0.088550	0.089477	0.089941	0.090405	0.091336	43
44	0.087414	0.088342	0.089273	0.089739	0.090206	0.091140	44
45	0.087220	0.088152	0.089087	0.089555	0.090023	0.090962	45
46	0.087042	0.087978	0.088916	0.089386	0.089856	0.090798	46
47	0.086878	0.087818	0.088760	0.089231	0.089703	0.090648	47
48	0.086728	0.087671	0.088616	0.089089	0.089563	0.090511	48
49	0.086590	0.087536	0.088485	0.088959	0.089434	0.090386	49
50	0.086463	0.087413	0.088364	0.088840	0.089317	0.090271	50

ANNUITY £1 WILL PURCHASE (SINGLE RATE)

			Rate Per Cent				
Yrs.	9	9.1	9.2	9.25	9.3	9.4	Yrs.
1	1.090000	1.091000	1.092000	1.092500	1.093000	1.094000	1
2	0.568469	0.569240	0.570011	0.570397	0.570783	0.571555	2
3	0.395055	0.395759	0.396464	0.396816	0.397169	0.397874	3
4	0.308669	0.309346	0.310025	0.310364	0.310704	0.311383	4
5	0.257092	0.257760	0.258428	0.258762	0.259097	0.259766	5
6	0.222920	0.223585	0.224251	0.224584	0.224917	0.225585	6
7	0.198691	0.199358	0.200026	0.200360	0.200695	0.201365	7
8	0.180674	0.181346	0.182020	0.182357	0.182694	0.183369	8
9	0.166799	0.167478	0.168157	0.168498	0.168839	0.169521	9
10	0.155820	0.156507	0.157195	0.157539	0.157884	0.158574	10
11	0.146947	0.147642	0.148338	0.148687	0.149036	0.149736	11
12	0.139651	0.140355	0.141061	0.141414	0.141768	0.142477	12
13	0.133567	0.134280	0.134996	0.135354	0.135713	0.136432	13
14	0.128433	0.129157	0.129882	0.130245	0.130609	0.131338	14
15	0.124059	0.124792	0.125527	0.125896	0.126264	0.127003	15
16	0.120300	0.121043	0.121788	0.122161	0.122535	0.123284	16
17	0.117046	0.117799	0.118554	0.118932	0.119311	0.120070	17
18	0.114212	0.114975	0.115740	0.116123	0.116506	0.117275	18
19	0.111730	0.112503	0.113277	0.113665	0.114054	0.114832	19
20	0.109546	0.110328	0.111112	0.111505	0.111898	0.112686	20
21	0.107617	0.108408	0.109201	0.109598	0.109996	0.110794	21
22	0.105905	0.106705	0.107507	0.107909	0.108312	0.109119	22
23	0.104382	0.105191	0.106002	0.106409	0.106816	0.107631	23
24	0.103023	0.103840	0.104660	0.105071	0.105482	0.106307	24
25	0.101806	0.102632	0.103461	0.103876	0.104291	0.105124	25
26	0.100715	0.101550	0.102386	0.102805	0.103225	0.104066	26
27	0.099735	0.100577	0.101421	0.101845	0.102268	0.103117	27
28	0.098852	0.099702	0.100554	0.100981	0.101408	0.102265	28
29	0.098056	0.098913	0.099772	0.100203	0.100634	0.101498	29
30	0.097336	0.098201	0.099067	0.099501	0.099936	0.100807	30
31	0.096686	0.097557	0.098430	0.098868	0.099306	0.100184	31
32	0.096096	0.096974	0.097854	0.098295	0.098736	0.099621	32
33	0.095562	0.096446	0.097332	0.097776	0.098221	0.099112	33
34	0.095077	0.095967	0.096859	0.097306	0.097754	0.098651	34
35	0.094636	0.095532	0.096430	0.096880	0.097331	0.098233	35
36	0.094235	0.095137	0.096041	0.096493	0.096946	0.097854	36
37	0.093870	0.094777	0.095686	0.096142	0.096598	0.097511	37
38	0.093538	0.094450	0.095365	0.095822	0.096281	0.097199	38
39	0.093236	0.094153	0.095072	0.095532	0.095993	0.096916	39
40	0.092960	0.093881	0.094805	0.095267	0.095731	0.096658	40
41	0.092708	0.093634	0.094562	0.095027	0.095492	0.096424	41
42	0.092478	0.093409	0.094341	0.094808	0.095275	0.096211	42
43	0.092268	0.093203	0.094139	0.094608	0.095077	0.096017	43
44	0.092077	0.093015	0.093955	0.094426	0.094897	0.095840	44
45	0.091902	0.092843	0.093787	0.094259	0.094732	0.095679	45
46	0.091742	0.092687	0.093634	0.094108	0.094582	0.095532	46
47	0.091595	0.092544	0.093494	0.093970	0.094446	0.095399	47
48	0.091461	0.092413	0.093366	0.093843	0.094321	0.095277	48
49	0.091339	0.092293	0.093250	0.093728	0.094207	0.095166	49
50	0.091227	0.092184	0.093143	0.093623	0.094103	0.095065	50

ANNUITY £1 WILL PURCHASE (SINGLE RATE)

			Rate Per Cent				
Yrs.	9.5	9.6	9.7	9.75	9.8	9.9	Yrs.
1	1.095000	1.096000	1.097000	1.097500	1.098000	1.099000	1
2	0.572327	0.573099	0.573872	0.574258	0.574644	0.575417	2
3	0.398580	0.399286	0.399993	0.400346	0.400700	0.401407	3
4	0.312063	0.312743	0.313425	0.313765	0.314106	0.314788	4
5	0.260436	0.261107	0.261779	0.262115	0.262451	0.263124	5
6	0.226253	0.226922	0.227592	0.227928	0.228263	0.228935	6
7	0.202036	0.202708	0.203381	0.203718	0.204055	0.204730	7
8	0.184046	0.184723	0.185402	0.185741	0.186081	0.186762	8
9	0.170205	0.170889	0.171575	0.171919	0.172263	0.172951	9
10	0.159266	0.159959	0.160654	0.161002	0.161350	0.162047	10
11	0.150437	0.151139	0.151843	0.152196	0.152548	0.153255	11
12	0.143188	0.143900	0.144613	0.144971	0.145328	0.146045	12
13	0.137152	0.137874	0.138598	0.138960	0.139323	0.140050	13
14	0.132068	0.132800	0.133534	0.133902	0.134270	0.135007	14
15	0.127744	0.128486	0.129230	0.129603	0.129976	0.130724	15
16	0.124035	0.124787	0.125542	0.125920	0.126298	0.127057	16
17	0.120831	0.121594	0.122358	0.122741	0.123125	0.123894	17
18	0.118046	0.118819	0.119594	0.119982	0.120371	0.121149	18
19	0.115613	0.116396	0.117180	0.117574	0.117967	0.118756	19
20	0.113477	0.114269	0.115064	0.115462	0.115860	0.116659	20
21	0.111594	0.112396	0.113200	0.113602	0.114006	0.114814	21
22	0.109928	0.110739	0.111552	0.111960	0.112368	0.113185	22
23	0.108449	0.109270	0.110092	0.110504	0.110916	0.111743	23
24	0.107134	0.107962	0.108794	0.109210	0.109627	0.110462	24
25	0.105959	0.106797	0.107636	0.108057	0.108478	0.109322	25
26	0.104909	0.105755	0.106603	0.107027	0.107453	0.108305	26
27	0.103969	0.104822	0.105678	0.106106	0.106536	0.107396	27
28	0.103124	0.103985	0.104848	0.105281	0.105714	0.106581	28
29	0.102364	0.103233	0.104104	0.104540	0.104976	0.105851	29
30	0.101681	0.102556	0.103434	0.103873	0.104314	0.105195	30
31	0.101064	0.101946	0.102831	0.103274	0.103717	0.104606	31
32	0.100507	0.101396	0.102287	0.102733	0.103180	0.104075	32
33	0.100004	0.100899	0.101796	0.102246	0.102695	0.103596	33
34	0.099549	0.100450	0.101353	0.101805	0.102258	0.103165	34
35	0.099138	0.100044	0.100953	0.101408	0.101863	0.102775	35
36	0.098764	0.099676	0.100590	0.101048	0.101506	0.102424	36
37	0.098426	0.099343	0.100262	0.100722	0.101183	0.102106	37
38	0.098119	0.099041	0.099965	0.100427	0.100890	0.101818	38
39	0.097840	0.098767	0.099695	0.100160	0.100626	0.101557	39
40	0.097587	0.098518	0.099451	0.099918	0.100385	0.101322	40
41	0.097357	0.098292	0.099229	0.099698	0.100168	0.101108	41
42	0.097148	0.098087	0.099028	0.099499	0.099970	0.100914	42
43	0.096958	0.097901	0.098845	0.099318	0.099791	0.100739	43
44	0.096785	0.097731	0.098679	0.099154	0.099629	0.100580	44
45	0.096627	0.097577	0.098528	0.099005	0.099481	0.100435	45
46	0.096484	0.097437	0.098391	0.098869	0.099347	0.100304	46
47	0.096353	0.097309	0.098267	0.098746	0.099225	0.100186	47
48	0.096234	0.097193	0.098153	0.098634	0.099115	0.100078	48
49	0.096126	0.097088	0.098050	0.098532	0.099014	0.099980	49
50	0.096027	0.096991	0.097957	0.098440	0.098923	0.099891	50

ANNUITY £1 WILL PURCHASE (SINGLE RATE)

			Rate Per Cent				
Yrs.	**10**	**10.1**	**10.2**	**10.25**	**10.3**	**10.4**	Yrs.
1	1.100000	1.101000	1.102000	1.102500	1.103000	1.104000	1
2	0.576190	0.576964	0.577737	0.578124	0.578511	0.579285	2
3	0.402115	0.402823	0.403531	0.403886	0.404240	0.404950	3
4	0.315471	0.316154	0.316838	0.317180	0.317522	0.318207	4
5	0.263797	0.264472	0.265147	0.265484	0.265822	0.266499	5
6	0.229607	0.230281	0.230955	0.231292	0.231630	0.232305	6
7	0.205405	0.206082	0.206760	0.207099	0.207439	0.208118	7
8	0.187444	0.188127	0.188811	0.189153	0.189496	0.190182	8
9	0.173641	0.174331	0.175023	0.175370	0.175716	0.176411	9
10	0.162745	0.163445	0.164146	0.164497	0.164849	0.165552	10
11	0.153963	0.154673	0.155384	0.155740	0.156096	0.156810	11
12	0.146763	0.147483	0.148204	0.148565	0.148927	0.149651	12
13	0.140779	0.141509	0.142240	0.142607	0.142974	0.143709	13
14	0.135746	0.136487	0.137229	0.137601	0.137973	0.138719	14
15	0.131474	0.132225	0.132978	0.133355	0.133733	0.134490	15
16	0.127817	0.128579	0.129342	0.129725	0.130108	0.130875	16
17	0.124664	0.125437	0.126211	0.126599	0.126987	0.127765	17
18	0.121930	0.122713	0.123498	0.123891	0.124284	0.125073	18
19	0.119547	0.120340	0.121135	0.121533	0.121931	0.122730	19
20	0.117460	0.118262	0.119067	0.119470	0.119874	0.120683	20
21	0.115624	0.116437	0.117251	0.117659	0.118068	0.118886	21
22	0.114005	0.114827	0.115651	0.116063	0.116476	0.117304	22
23	0.112572	0.113403	0.114236	0.114653	0.115071	0.115908	23
24	0.111300	0.112139	0.112981	0.113403	0.113825	0.114671	24
25	0.110168	0.111016	0.111866	0.112292	0.112719	0.113573	25
26	0.109159	0.110015	0.110874	0.111304	0.111734	0.112597	26
27	0.108258	0.109122	0.109988	0.110422	0.110856	0.111727	27
28	0.107451	0.108323	0.109197	0.109634	0.110072	0.110950	28
29	0.106728	0.107607	0.108488	0.108929	0.109371	0.110256	29
30	0.106079	0.106965	0.107853	0.108298	0.108743	0.109635	30
31	0.105496	0.106389	0.107283	0.107731	0.108180	0.109078	31
32	0.104972	0.105871	0.106771	0.107222	0.107674	0.108579	32
33	0.104499	0.105404	0.106311	0.106765	0.107220	0.108130	33
34	0.104074	0.104984	0.105897	0.106354	0.106811	0.107727	34
35	0.103690	0.104606	0.105524	0.105983	0.106443	0.107365	35
36	0.103343	0.104264	0.105187	0.105649	0.106112	0.107038	36
37	0.103030	0.103956	0.104884	0.105349	0.105814	0.106745	37
38	0.102747	0.103678	0.104610	0.105077	0.105544	0.106480	38
39	0.102491	0.103426	0.104363	0.104832	0.105301	0.106241	39
40	0.102259	0.103199	0.104140	0.104611	0.105082	0.106026	40
41	0.102050	0.102993	0.103938	0.104411	0.104884	0.105832	41
42	0.101860	0.102807	0.103755	0.104230	0.104705	0.105657	42
43	0.101688	0.102639	0.103590	0.104067	0.104544	0.105498	43
44	0.101532	0.102486	0.103441	0.103919	0.104398	0.105355	44
45	0.101391	0.102348	0.103306	0.103786	0.104265	0.105226	45
46	0.101263	0.102223	0.103184	0.103665	0.104146	0.105109	46
47	0.101147	0.102109	0.103073	0.103555	0.104038	0.105004	47
48	0.101041	0.102007	0.102973	0.103456	0.103940	0.104908	48
49	0.100946	0.101913	0.102882	0.103367	0.103852	0.104822	49
50	0.100859	0.101829	0.102800	0.103285	0.103771	0.104744	50

ANNUITY £1 WILL PURCHASE (SINGLE RATE)

			Rate Per Cent				
Yrs.	10.5	10.6	10.7	10.75	10.8	10.9	Yrs.
1	1.105000	1.106000	1.107000	1.107500	1.108000	1.109000	1
2	0.580059	0.580834	0.581608	0.581996	0.582383	0.583158	2
3	0.405659	0.406369	0.407080	0.407435	0.407790	0.408502	3
4	0.318892	0.319578	0.320264	0.320608	0.320951	0.321638	4
5	0.267175	0.267853	0.268531	0.268871	0.269210	0.269890	5
6	0.232982	0.233659	0.234337	0.234677	0.235016	0.235696	6
7	0.208799	0.209480	0.210163	0.210504	0.210846	0.211530	7
8	0.190869	0.191558	0.192247	0.192592	0.192937	0.193629	8
9	0.177106	0.177803	0.178501	0.178850	0.179200	0.179900	9
10	0.166257	0.166964	0.167671	0.168025	0.168380	0.169090	10
11	0.157525	0.158241	0.158959	0.159319	0.159678	0.160399	11
12	0.150377	0.151104	0.151833	0.152197	0.152563	0.153294	12
13	0.144445	0.145183	0.145923	0.146293	0.146664	0.147407	13
14	0.139467	0.140216	0.140966	0.141342	0.141719	0.142473	14
15	0.135248	0.136008	0.136770	0.137151	0.137533	0.138298	15
16	0.131644	0.132415	0.133188	0.133575	0.133963	0.134739	16
17	0.128545	0.129327	0.130110	0.130503	0.130895	0.131683	17
18	0.125863	0.126655	0.127449	0.127847	0.128245	0.129043	18
19	0.123531	0.124333	0.125138	0.125541	0.125944	0.126752	19
20	0.121493	0.122306	0.123120	0.123528	0.123937	0.124755	20
21	0.119707	0.120529	0.121353	0.121766	0.122180	0.123008	21
22	0.118134	0.118966	0.119800	0.120218	0.120636	0.121473	22
23	0.116747	0.117588	0.118431	0.118853	0.119276	0.120122	23
24	0.115519	0.116368	0.117220	0.117647	0.118074	0.118930	24
25	0.114429	0.115288	0.116148	0.116579	0.117010	0.117874	25
26	0.113461	0.114328	0.115196	0.115631	0.116066	0.116938	26
27	0.112599	0.113473	0.114349	0.114788	0.115227	0.116107	27
28	0.111830	0.112712	0.113595	0.114038	0.114481	0.115368	28
29	0.111143	0.112032	0.112922	0.113368	0.113815	0.114709	29
30	0.110528	0.111424	0.112322	0.112771	0.113221	0.114122	30
31	0.109978	0.110882	0.111784	0.112237	0.112690	0.113597	31
32	0.109485	0.110393	0.111303	0.111759	0.112215	0.113128	32
33	0.109042	0.109956	0.110872	0.111331	0.111790	0.112709	33
34	0.108645	0.109564	0.110486	0.110947	0.111409	0.112333	34
35	0.108288	0.109212	0.110139	0.110603	0.111067	0.111996	35
36	0.107967	0.108896	0.109827	0.110294	0.110760	0.111695	36
37	0.107677	0.108612	0.109548	0.110016	0.110485	0.111424	37
38	0.107417	0.108356	0.109296	0.109767	0.110238	0.111181	38
39	0.107183	0.108126	0.109070	0.109543	0.110016	0.110963	39
40	0.106971	0.107918	0.108866	0.109341	0.109816	0.110767	40
41	0.106781	0.107731	0.108683	0.109159	0.109636	0.110590	41
42	0.106609	0.107563	0.108518	0.108996	0.109474	0.110432	42
43	0.106454	0.107411	0.108369	0.108849	0.109329	0.110290	43
44	0.106314	0.107274	0.108236	0.108717	0.109198	0.110162	44
45	0.106188	0.107151	0.108115	0.108597	0.109080	0.110046	45
46	0.106074	0.107039	0.108006	0.108490	0.108974	0.109943	46
47	0.105971	0.106939	0.107908	0.108393	0.108878	0.109849	47
48	0.105878	0.106848	0.107820	0.108306	0.108792	0.109765	48
49	0.105794	0.106766	0.107740	0.108227	0.108714	0.109689	49
50	0.105718	0.106692	0.107668	0.108156	0.108644	0.109621	50

ANNUITY £1 WILL PURCHASE (SINGLE RATE)

				Rate Per Cent			
Yrs.	**11**	**11.25**	**11.5**	**11.75**	**12**	**12.25**	Yrs.
1	1.110000	1.112500	1.115000	1.117500	1.120000	1.122500	1
2	0.583934	0.585873	0.587813	0.589755	0.591698	0.593643	2
3	0.409213	0.410994	0.412776	0.414562	0.416349	0.418139	3
4	0.322326	0.324048	0.325774	0.327503	0.329234	0.330970	4
5	0.270570	0.272274	0.273982	0.275694	0.277410	0.279130	5
6	0.236377	0.238081	0.239791	0.241506	0.243226	0.244950	6
7	0.212215	0.213932	0.215655	0.217384	0.219118	0.220858	7
8	0.194321	0.196057	0.197799	0.199548	0.201303	0.203064	8
9	0.180602	0.182360	0.184126	0.185899	0.187679	0.189466	9
10	0.169801	0.171585	0.173377	0.175177	0.176984	0.178799	10
11	0.161121	0.162932	0.164751	0.166579	0.168415	0.170260	11
12	0.154027	0.155866	0.157714	0.159571	0.161437	0.163311	12
13	0.148151	0.150018	0.151895	0.153782	0.155677	0.157582	13
14	0.143228	0.145124	0.147030	0.148946	0.150871	0.152806	14
15	0.139065	0.140990	0.142924	0.144869	0.146824	0.148789	15
16	0.135517	0.137469	0.139432	0.141406	0.143390	0.145384	16
17	0.132471	0.134452	0.136443	0.138444	0.140457	0.142479	17
18	0.129843	0.131850	0.133868	0.135897	0.137937	0.139988	18
19	0.127563	0.129596	0.131641	0.133696	0.135763	0.137840	19
20	0.125576	0.127634	0.129705	0.131786	0.133879	0.135982	20
21	0.123838	0.125921	0.128016	0.130123	0.132240	0.134368	21
22	0.122313	0.124420	0.126539	0.128669	0.130811	0.132962	22
23	0.120971	0.123101	0.125243	0.127396	0.129560	0.131735	23
24	0.119787	0.121939	0.124103	0.126278	0.128463	0.130660	24
25	0.118740	0.120913	0.123098	0.125294	0.127500	0.129717	25
26	0.117813	0.120006	0.122210	0.124426	0.126652	0.128888	26
27	0.116989	0.119202	0.121425	0.123659	0.125904	0.128159	27
28	0.116257	0.118488	0.120730	0.122982	0.125244	0.127516	28
29	0.115605	0.117854	0.120112	0.122381	0.124660	0.126949	29
30	0.115025	0.117289	0.119564	0.121849	0.124144	0.126447	30
31	0.114506	0.116786	0.119077	0.121377	0.123686	0.126004	31
32	0.114043	0.116338	0.118643	0.120957	0.123280	0.125612	32
33	0.113629	0.115938	0.118257	0.120584	0.122920	0.125265	33
34	0.113259	0.115581	0.117912	0.120252	0.122601	0.124957	34
35	0.112927	0.115262	0.117605	0.119957	0.122317	0.124684	35
36	0.112630	0.114976	0.117331	0.119694	0.122064	0.124442	36
37	0.112364	0.114721	0.117086	0.119459	0.121840	0.124227	37
38	0.112125	0.114492	0.116867	0.119250	0.121640	0.124036	38
39	0.111911	0.114288	0.116672	0.119064	0.121462	0.123867	39
40	0.111719	0.114104	0.116497	0.118897	0.121304	0.123716	40
41	0.111546	0.113940	0.116341	0.118749	0.121163	0.123582	41
42	0.111391	0.113793	0.116201	0.118616	0.121037	0.123463	42
43	0.111251	0.113661	0.116076	0.118498	0.120925	0.123357	43
44	0.111126	0.113542	0.115964	0.118392	0.120825	0.123263	44
45	0.111014	0.113436	0.115864	0.118298	0.120736	0.123179	45
46	0.110912	0.113341	0.115774	0.118213	0.120657	0.123105	46
47	0.110821	0.113255	0.115694	0.118138	0.120586	0.123039	47
48	0.110739	0.113178	0.115622	0.118071	0.120523	0.122980	48
49	0.110666	0.113109	0.115558	0.118010	0.120467	0.122927	49
50	0.110599	0.113047	0.115500	0.117956	0.120417	0.122880	50

ANNUITY £1 WILL PURCHASE (SINGLE RATE)

Yrs.	12.5	12.75	13	13.25	13.5	13.75	Yrs.
			Rate Per Cent				
1	1.125000	1.127500	1.130000	1.132500	1.135000	1.137500	1
2	0.595588	0.597535	0.599484	0.601433	0.603384	0.605336	2
3	0.419931	0.421725	0.423522	0.425321	0.427122	0.428926	3
4	0.332708	0.334449	0.336194	0.337942	0.339693	0.341447	4
5	0.280854	0.282582	0.284315	0.286051	0.287791	0.289535	5
6	0.246680	0.248414	0.250153	0.251897	0.253646	0.255399	6
7	0.222603	0.224354	0.226111	0.227873	0.229641	0.231414	7
8	0.204832	0.206606	0.208387	0.210173	0.211966	0.213765	8
9	0.191260	0.193061	0.194869	0.196684	0.198505	0.200333	9
10	0.180622	0.182452	0.184290	0.186135	0.187987	0.189847	10
11	0.172112	0.173973	0.175841	0.177718	0.179602	0.181494	11
12	0.165194	0.167086	0.168986	0.170895	0.172811	0.174736	12
13	0.159496	0.161419	0.163350	0.165291	0.167240	0.169198	13
14	0.154751	0.156704	0.158667	0.160640	0.162621	0.164611	14
15	0.150764	0.152748	0.154742	0.156745	0.158757	0.160779	15
16	0.147388	0.149402	0.151426	0.153460	0.155502	0.157554	16
17	0.144512	0.146556	0.148608	0.150671	0.152743	0.154825	17
18	0.142049	0.144120	0.146201	0.148292	0.150392	0.152502	18
19	0.139928	0.142026	0.144134	0.146252	0.148380	0.150517	19
20	0.138096	0.140220	0.142354	0.144498	0.146651	0.148814	20
21	0.136507	0.138656	0.140814	0.142983	0.145161	0.147348	21
22	0.135125	0.137297	0.139479	0.141672	0.143873	0.146084	22
23	0.133919	0.136114	0.138319	0.140533	0.142757	0.144990	23
24	0.132866	0.135082	0.137308	0.139544	0.141788	0.144041	24
25	0.131943	0.134180	0.136426	0.138681	0.140945	0.143218	25
26	0.131134	0.133390	0.135655	0.137928	0.140211	0.142501	26
27	0.130423	0.132697	0.134979	0.137270	0.139570	0.141877	27
28	0.129797	0.132088	0.134387	0.136695	0.139010	0.141334	28
29	0.129246	0.131552	0.133867	0.136190	0.138521	0.140859	29
30	0.128760	0.131081	0.133411	0.135748	0.138092	0.140444	30
31	0.128331	0.130666	0.133009	0.135360	0.137717	0.140082	31
32	0.127952	0.130300	0.132656	0.135019	0.137388	0.139764	32
33	0.127617	0.129978	0.132345	0.134719	0.137100	0.139487	33
34	0.127321	0.129693	0.132071	0.134456	0.136847	0.139244	34
35	0.127059	0.129441	0.131829	0.134224	0.136624	0.139030	35
36	0.126827	0.129218	0.131616	0.134020	0.136429	0.138844	36
37	0.126621	0.129022	0.131428	0.133840	0.136258	0.138680	37
38	0.126439	0.128848	0.131262	0.133682	0.136107	0.138536	38
39	0.126278	0.128694	0.131116	0.133543	0.135974	0.138410	39
40	0.126134	0.128558	0.130986	0.133420	0.135858	0.138299	40
41	0.126007	0.128437	0.130872	0.133312	0.135755	0.138202	41
42	0.125895	0.128331	0.130771	0.133216	0.135665	0.138117	42
43	0.125795	0.128236	0.130682	0.133132	0.135585	0.138042	43
44	0.125706	0.128153	0.130603	0.133058	0.135515	0.137976	44
45	0.125627	0.128078	0.130534	0.132992	0.135454	0.137919	45
46	0.125557	0.128013	0.130472	0.132934	0.135400	0.137868	46
47	0.125495	0.127955	0.130417	0.132883	0.135352	0.137823	47
48	0.125440	0.127903	0.130369	0.132838	0.135310	0.137784	48
49	0.125391	0.127857	0.130327	0.132799	0.135273	0.137750	49
50	0.125347	0.127817	0.130289	0.132764	0.135241	0.137719	50

ANNUITY £1 WILL PURCHASE (SINGLE RATE)

				Rate Per Cent			
Yrs.	14	14.25	14.5	14.75	15	15.25	Yrs.
1	1.140000	1.142500	1.145000	1.147500	1.150000	1.152500	1
2	0.607290	0.609244	0.611200	0.613158	0.615116	0.617076	2
3	0.430731	0.432539	0.434350	0.436162	0.437977	0.439794	3
4	0.343205	0.344965	0.346729	0.348496	0.350265	0.352038	4
5	0.291284	0.293036	0.294792	0.296552	0.298316	0.300083	5
6	0.257157	0.258920	0.260688	0.262460	0.264237	0.266018	6
7	0.233192	0.234976	0.236766	0.238560	0.240360	0.242166	7
8	0.215570	0.217381	0.219198	0.221021	0.222850	0.224685	8
9	0.202168	0.204010	0.205858	0.207713	0.209574	0.211442	9
10	0.191714	0.193588	0.195469	0.197357	0.199252	0.201154	10
11	0.183394	0.185302	0.187217	0.189139	0.191069	0.193006	11
12	0.176669	0.178610	0.180559	0.182516	0.184481	0.186453	12
13	0.171164	0.173138	0.175121	0.177112	0.179110	0.181117	13
14	0.166609	0.168616	0.170632	0.172656	0.174688	0.176729	14
15	0.162809	0.164848	0.166896	0.168952	0.171017	0.173090	15
16	0.159615	0.161685	0.163764	0.165852	0.167948	0.170052	16
17	0.156915	0.159015	0.161124	0.163241	0.165367	0.167501	17
18	0.154621	0.156749	0.158886	0.161032	0.163186	0.165349	18
19	0.152663	0.154818	0.156982	0.159155	0.161336	0.163526	19
20	0.150986	0.153167	0.155357	0.157555	0.159761	0.161976	20
21	0.149545	0.151750	0.153964	0.156186	0.158417	0.160655	21
22	0.148303	0.150531	0.152768	0.155013	0.157266	0.159526	22
23	0.147231	0.149481	0.151739	0.154005	0.156278	0.158560	23
24	0.146303	0.148573	0.150851	0.153137	0.155430	0.157730	24
25	0.145498	0.147787	0.150084	0.152388	0.154699	0.157018	25
26	0.144800	0.147106	0.149420	0.151742	0.154070	0.156405	26
27	0.144193	0.146516	0.148846	0.151183	0.153526	0.155877	27
28	0.143664	0.146003	0.148348	0.150699	0.153057	0.155421	28
29	0.143204	0.145556	0.147915	0.150280	0.152651	0.155028	29
30	0.142803	0.145168	0.147539	0.149917	0.152300	0.154689	30
31	0.142453	0.144830	0.147213	0.149602	0.151996	0.154396	31
32	0.142147	0.144535	0.146929	0.149328	0.151733	0.154142	32
33	0.141880	0.144278	0.146682	0.149091	0.151505	0.153923	33
34	0.141646	0.144054	0.146467	0.148884	0.151307	0.153733	34
35	0.141442	0.143858	0.146279	0.148705	0.151135	0.153569	35
36	0.141263	0.143687	0.146116	0.148549	0.150986	0.153426	36
37	0.141107	0.143538	0.145974	0.148413	0.150857	0.153303	37
38	0.140970	0.143408	0.145850	0.148295	0.150744	0.153196	38
39	0.140850	0.143294	0.145742	0.148193	0.150647	0.153104	39
40	0.140745	0.143195	0.145647	0.148103	0.150562	0.153024	40
41	0.140653	0.143108	0.145565	0.148025	0.150489	0.152954	41
42	0.140573	0.143031	0.145493	0.147958	0.150425	0.152894	42
43	0.140502	0.142965	0.145431	0.147899	0.150369	0.152842	43
44	0.140440	0.142907	0.145376	0.147847	0.150321	0.152796	44
45	0.140386	0.142856	0.145328	0.147803	0.150279	0.152757	45
46	0.140338	0.142811	0.145287	0.147764	0.150242	0.152723	46
47	0.140297	0.142773	0.145250	0.147730	0.150211	0.152694	47
48	0.140260	0.142738	0.145218	0.147700	0.150183	0.152668	48
49	0.140228	0.142709	0.145191	0.147674	0.150159	0.152646	49
50	0.140200	0.142683	0.145167	0.147652	0.150139	0.152626	50

ANNUITY £1 WILL PURCHASE (SINGLE RATE)

			Rate Per Cent				
Yrs.	15.5	15.75	16	16.25	16.5	16.75	Yrs.
1	1.155000	1.157500	1.160000	1.162500	1.165000	1.167500	1
2	0.619037	0.620999	0.622963	0.624928	0.626894	0.628861	2
3	0.441613	0.443434	0.445258	0.447084	0.448911	0.450741	3
4	0.353814	0.355593	0.357375	0.359160	0.360948	0.362739	4
5	0.301855	0.303630	0.305409	0.307192	0.308979	0.310770	5
6	0.267804	0.269595	0.271390	0.273189	0.274993	0.276802	6
7	0.243976	0.245792	0.247613	0.249439	0.251270	0.253106	7
8	0.226526	0.228372	0.230224	0.232082	0.233946	0.235815	8
9	0.213316	0.215196	0.217082	0.218975	0.220874	0.222779	9
10	0.203063	0.204979	0.206901	0.208830	0.210766	0.212708	10
11	0.194951	0.196902	0.198861	0.200826	0.202799	0.204779	11
12	0.188433	0.190420	0.192415	0.194417	0.196426	0.198442	12
13	0.183132	0.185154	0.187184	0.189222	0.191266	0.193319	13
14	0.178777	0.180834	0.182898	0.184970	0.187049	0.189136	14
15	0.175171	0.177260	0.179358	0.181462	0.183575	0.185695	15
16	0.172165	0.174285	0.176414	0.178550	0.180694	0.182845	16
17	0.169643	0.171794	0.173952	0.176118	0.178292	0.180473	17
18	0.167520	0.169698	0.171885	0.174079	0.176281	0.178490	18
19	0.165723	0.167929	0.170142	0.172362	0.174590	0.176825	19
20	0.164199	0.166429	0.168667	0.170912	0.173165	0.175424	20
21	0.162901	0.165155	0.167416	0.169684	0.171960	0.174242	21
22	0.161795	0.164070	0.166353	0.168642	0.170938	0.173241	22
23	0.160848	0.163144	0.165447	0.167756	0.170071	0.172393	23
24	0.160038	0.162352	0.164673	0.167001	0.169334	0.171674	24
25	0.159343	0.161675	0.164013	0.166357	0.168707	0.171062	25
26	0.158746	0.161094	0.163447	0.165807	0.168172	0.170542	26
27	0.158233	0.160595	0.162963	0.165336	0.167715	0.170099	27
28	0.157791	0.160167	0.162548	0.164934	0.167325	0.169721	28
29	0.157411	0.159799	0.162192	0.164589	0.166992	0.169399	29
30	0.157083	0.159482	0.161886	0.164294	0.166707	0.169124	30
31	0.156800	0.159209	0.161623	0.164041	0.166463	0.168889	31
32	0.156556	0.158975	0.161397	0.163824	0.166254	0.168688	32
33	0.156345	0.158772	0.161203	0.163637	0.166075	0.168517	33
34	0.156164	0.158598	0.161036	0.163477	0.165922	0.168370	34
35	0.156006	0.158448	0.160892	0.163340	0.165791	0.168245	35
36	0.155871	0.158318	0.160769	0.163222	0.165678	0.168137	36
37	0.155753	0.158206	0.160662	0.163121	0.165582	0.168046	37
38	0.155652	0.158110	0.160571	0.163034	0.165499	0.167967	38
39	0.155564	0.158027	0.160492	0.162959	0.165428	0.167900	39
40	0.155488	0.157955	0.160424	0.162895	0.165368	0.167842	40
41	0.155422	0.157893	0.160365	0.162839	0.165315	0.167793	41
42	0.155366	0.157839	0.160315	0.162792	0.165271	0.167751	42
43	0.155316	0.157793	0.160271	0.162751	0.165232	0.167715	43
44	0.155274	0.157753	0.160234	0.162716	0.165199	0.167684	44
45	0.155237	0.157718	0.160201	0.162686	0.165171	0.167658	45
46	0.155205	0.157689	0.160174	0.162660	0.165147	0.167635	46
47	0.155178	0.157663	0.160150	0.162637	0.165126	0.167616	47
48	0.155154	0.157641	0.160129	0.162618	0.165108	0.167599	48
49	0.155133	0.157622	0.160111	0.162602	0.165093	0.167585	49
50	0.155115	0.157605	0.160096	0.162587	0.165080	0.167573	50

ANNUITY £1 WILL PURCHASE (SINGLE RATE)

				Rate Per Cent			
Yrs.	**17**	**17.5**	**18**	**18.5**	**19**	**20**	Yrs
1	1.170000	1.175000	1.180000	1.185000	1.190000	1.200000	1
2	0.630829	0.634770	0.638716	0.642666	0.646621	0.654545	2
3	0.452574	0.456245	0.459924	0.463612	0.467308	0.474725	3
4	0.364533	0.368130	0.371739	0.375359	0.378991	0.386289	4
5	0.312564	0.316163	0.319778	0.323407	0.327050	0.334380	5
6	0.278615	0.282254	0.285910	0.289584	0.293274	0.300706	6
7	0.254947	0.258645	0.262362	0.266099	0.269855	0.277424	7
8	0.237690	0.241456	0.245244	0.249054	0.252885	0.260609	8
9	0.224691	0.228531	0.232395	0.236282	0.240192	0.248079	9
10	0.214657	0.218573	0.222515	0.226481	0.230471	0.238523	10
11	0.206765	0.210757	0.214776	0.218821	0.222891	0.231104	11
12	0.200466	0.204533	0.208628	0.212749	0.216896	0.225265	12
13	0.195378	0.199518	0.203686	0.207881	0.212102	0.220620	13
14	0.191230	0.195440	0.199678	0.203943	0.208235	0.216893	14
15	0.187822	0.192098	0.196403	0.200734	0.205092	0.213882	15
16	0.185004	0.189343	0.193710	0.198104	0.202523	0.211436	16
17	0.182662	0.187060	0.191485	0.195937	0.200414	0.209440	17
18	0.180706	0.185159	0.189639	0.194145	0.198676	0.207805	18
19	0.179067	0.183572	0.188103	0.192658	0.197238	0.206462	19
20	0.177690	0.182243	0.186820	0.191421	0.196045	0.205357	20
21	0.176530	0.181126	0.185746	0.190390	0.195054	0.204444	21
22	0.175550	0.180187	0.184846	0.189528	0.194229	0.203690	22
23	0.174721	0.179395	0.184090	0.188806	0.193542	0.203065	23
24	0.174019	0.178726	0.183454	0.188202	0.192967	0.202548	24
25	0.173423	0.178161	0.182919	0.187695	0.192487	0.202119	25
26	0.172917	0.177683	0.182467	0.187269	0.192086	0.201762	26
27	0.172487	0.177278	0.182087	0.186911	0.191750	0.201467	27
28	0.172121	0.176935	0.181765	0.186610	0.191468	0.201221	28
29	0.171810	0.176644	0.181494	0.186357	0.191232	0.201016	29
30	0.171545	0.176398	0.181264	0.186144	0.191034	0.200846	30
31	0.171318	0.176188	0.181070	0.185964	0.190869	0.200705	31
32	0.171126	0.176010	0.180906	0.185813	0.190729	0.200587	32
33	0.170961	0.175859	0.180767	0.185686	0.190612	0.200489	33
34	0.170821	0.175730	0.180650	0.185578	0.190514	0.200407	34
35	0.170701	0.175621	0.180550	0.185488	0.190432	0.200339	35
36	0.170599	0.175528	0.180466	0.185411	0.190363	0.200283	36
37	0.170512	0.175450	0.180395	0.185347	0.190305	0.200235	37
38	0.170437	0.175382	0.180335	0.185293	0.190256	0.200196	38
39	0.170373	0.175325	0.180284	0.185247	0.190215	0.200163	39
40	0.170319	0.175277	0.180240	0.185208	0.190181	0.200136	40
41	0.170273	0.175236	0.180204	0.185176	0.190152	0.200113	41
42	0.170233	0.175200	0.180172	0.185148	0.190128	0.200095	42
43	0.170199	0.175171	0.180146	0.185125	0.190107	0.200079	43
44	0.170170	0.175145	0.180124	0.185106	0.190090	0.200066	44
45	0.170145	0.175123	0.180105	0.185089	0.190076	0.200055	45
46	0.170124	0.175105	0.180089	0.185075	0.190064	0.200046	46
47	0.170106	0.175089	0.180075	0.185063	0.190053	0.200038	47
48	0.170091	0.175076	0.180064	0.185054	0.190045	0.200032	48
49	0.170078	0.175065	0.180054	0.185045	0.190038	0.200026	49
50	0.170066	0.175055	0.180046	0.185038	0.190032	0.200022	50

INTERNAL RATES OF RETURN[†]
(IRR)

with **No** Projected Rental Growth.
In respect of Freehold Investments with the next
rent review or reversion in

1 to 10 YEARS

with
INITIAL YIELDS

1% to 10%

† For an explanation of the headings and the significance of asterisks and high returns
which appear on some pages, see Introductory Section, pages xx to xxiii.

INTERNAL RATE OF RETURN (IRR)†
(with NO projected Rental Growth)

RENTAL FACTOR†	YEARS TO REVIEW/REVERSION									
	1	2	3	4	5	6	7	8	9	10
1.1	1.10	1.10	1.10	1.10	1.09	1.09	1.09	1.09	1.09	1.09
1.2	1.20	1.20	1.19	1.19	1.19	1.19	1.18	1.18	1.18	1.18
1.3	1.30	1.29	1.29	1.29	1.28	1.28	1.27	1.27	1.27	1.26
1.4	1.39	1.39	1.38	1.38	1.37	1.37	1.36	1.36	1.35	1.35
1.5	1.49	1.49	1.48	1.47	1.46	1.46	1.45	1.45	1.44	1.43
1.6	1.59	1.58	1.57	1.56	1.56	1.55	1.54	1.53	1.52	1.52
1.7	1.69	1.68	1.67	1.66	1.65	1.64	1.63	1.62	1.61	1.60
1.8	1.79	1.77	1.76	1.75	1.73	1.72	1.71	1.70	1.69	1.68
1.9	1.88	1.87	1.85	1.84	1.82	1.81	1.79	1.78	1.77	1.76
2.0	1.98	1.96	1.94	1.93	1.91	1.89	1.88	1.86	1.85	1.83
2.2	2.17	2.15	2.13	2.10	2.08	2.06	2.04	2.02	2.00	1.99
2.4	2.37	2.34	2.31	2.28	2.25	2.23	2.20	2.18	2.16	2.13
2.5	2.46	2.43	2.40	2.37	2.34	2.31	2.28	2.25	2.23	2.21
2.6	2.56	2.52	2.49	2.45	2.42	2.39	2.36	2.33	2.30	2.28
2.8	2.75	2.71	2.66	2.62	2.58	2.55	2.51	2.48	2.45	2.42
3.0	2.94	2.89	2.84	2.79	2.75	2.70	2.66	2.63	2.59	2.55
3.2	3.13	3.07	3.01	2.96	2.91	2.86	2.81	2.77	2.73	2.69
3.4	3.32	3.25	3.18	3.12	3.06	3.01	2.96	2.91	2.86	2.82
3.5	3.42	3.34	3.27	3.20	3.14	3.08	3.03	2.98	2.93	2.88
3.6	3.51	3.43	3.35	3.28	3.22	3.16	3.10	3.05	2.99	2.95
3.8	3.70	3.61	3.52	3.45	3.37	3.30	3.24	3.18	3.12	3.07
4.0	3.89	3.79	3.69	3.60	3.52	3.45	3.38	3.31	3.25	3.19
4.2	4.07	3.96	3.86	3.76	3.67	3.59	3.51	3.44	3.37	3.31
4.4	4.26	4.14	4.02	3.92	3.82	3.73	3.65	3.57	3.50	3.43
4.5	4.35	4.22	4.10	3.99	3.89	3.80	3.71	3.63	3.56	3.48
4.6	4.45	4.31	4.18	4.07	3.96	3.87	3.78	3.69	3.62	3.54
4.8	4.63	4.48	4.34	4.22	4.11	4.00	3.91	3.82	3.73	3.65
5.0	4.82	4.65	4.50	4.37	4.25	4.14	4.03	3.94	3.85	3.76
5.2	5.00	4.82	4.66	4.52	4.39	4.27	4.16	4.06	3.96	3.87
5.4	5.18	4.99	4.82	4.67	4.53	4.40	4.28	4.17	4.07	3.98
5.5	5.27	5.08	4.90	4.74	4.59	4.46	4.34	4.23	4.13	4.03
5.6	5.37	5.16	4.98	4.81	4.66	4.53	4.40	4.29	4.18	4.08
5.8	5.55	5.33	5.13	4.96	4.80	4.65	4.52	4.40	4.29	4.19
6.0	5.73	5.49	5.28	5.10	4.93	4.78	4.64	4.51	4.40	4.29
6.5	6.18	5.90	5.66	5.45	5.26	5.08	4.93	4.78	4.65	4.53
7.0	6.63	6.31	6.03	5.79	5.57	5.38	5.21	5.05	4.90	4.77
7.5	7.07	6.71	6.40	6.12	5.88	5.67	5.48	5.30	5.14	4.99
8.0	7.51	7.10	6.75	6.45	6.19	5.95	5.74	5.55	5.37	5.21
8.5	7.95	7.49	7.10	6.77	6.48	6.22	5.99	5.78	5.59	5.42
9.0	8.38	7.87	7.45	7.08	6.77	6.49	6.24	6.01	5.81	5.63
9.5	8.81	8.25	7.79	7.39	7.05	6.75	6.48	6.24	6.02	5.83
10.0	9.24	8.63	8.12	7.69	7.32	7.00	6.71	6.46	6.23	6.02
11.0	10.08	9.36	8.77	8.28	7.85	7.49	7.16	6.87	6.62	6.39
12.0	10.92	10.08	9.40	8.84	8.36	7.95	7.59	7.27	6.99	6.73
13.0	11.74	10.78	10.01	9.38	8.85	8.40	8.00	7.65	7.34	7.06
14.0	12.55	11.46	10.61	9.91	9.32	8.83	8.39	8.02	7.68	7.38
15.0	13.35	12.13	11.19	10.42	9.78	9.24	8.77	8.36	8.00	7.68
16.0	14.14	12.79	11.75	10.91	10.22	9.64	9.13	8.70	8.31	7.97
18.0	15.69	14.07	12.83	11.86	11.06	10.39	9.82	9.33	8.90	8.51
20.0	17.21	15.29	13.87	12.75	11.85	11.10	10.47	9.92	9.44	9.01

†For an explanation of the headings and the significance of asterisks and high returns which appear on some pages, see Introductory Section, pages xx–xxiiii.

INTERNAL RATE OF RETURN (IRR)[†]
(with NO projected Rental Growth)

RENTAL FACTOR[†]	YEARS TO REVIEW/REVERSION									
	1	2	3	4	5	6	7	8	9	10
1.1	1.65	1.65	1.64	1.64	1.64	1.64	1.63	1.63	1.63	1.63
1.2	1.79	1.79	1.78	1.78	1.77	1.77	1.77	1.76	1.76	1.75
1.3	1.94	1.93	1.92	1.92	1.91	1.90	1.89	1.89	1.88	1.87
1.4	2.09	2.08	2.06	2.05	2.04	2.03	2.02	2.01	2.00	1.99
1.5	2.23	2.22	2.20	2.19	2.17	2.16	2.15	2.13	2.12	2.11
1.6	2.38	2.36	2.34	2.32	2.30	2.29	2.27	2.25	2.24	2.22
1.7	2.52	2.50	2.48	2.45	2.43	2.41	2.39	2.37	2.35	2.33
1.8	2.67	2.64	2.61	2.58	2.56	2.53	2.51	2.49	2.46	2.44
1.9	2.81	2.78	2.74	2.71	2.68	2.65	2.63	2.60	2.57	2.55
2.0	2.96	2.92	2.88	2.84	2.81	2.77	2.74	2.71	2.68	2.65
2.2	3.24	3.19	3.14	3.09	3.05	3.01	2.97	2.93	2.89	2.86
2.4	3.53	3.46	3.40	3.34	3.29	3.23	3.19	3.14	3.10	3.05
2.5	3.67	3.60	3.53	3.46	3.40	3.35	3.29	3.24	3.20	3.15
2.6	3.81	3.73	3.65	3.58	3.52	3.46	3.40	3.34	3.29	3.24
2.8	4.09	4.00	3.91	3.82	3.75	3.67	3.61	3.54	3.48	3.43
3.0	4.37	4.26	4.16	4.06	3.97	3.89	3.81	3.74	3.67	3.61
3.2	4.65	4.52	4.40	4.29	4.19	4.09	4.01	3.93	3.85	3.78
3.4	4.93	4.78	4.64	4.52	4.40	4.30	4.20	4.11	4.02	3.94
3.5	5.07	4.91	4.76	4.63	4.51	4.40	4.29	4.20	4.11	4.03
3.6	5.21	5.04	4.88	4.74	4.61	4.50	4.39	4.29	4.19	4.11
3.8	5.48	5.29	5.12	4.96	4.82	4.69	4.57	4.46	4.36	4.27
4.0	5.76	5.54	5.35	5.18	5.02	4.88	4.75	4.63	4.52	4.42
4.2	6.03	5.79	5.58	5.39	5.22	5.07	4.93	4.80	4.68	4.57
4.4	6.30	6.04	5.81	5.60	5.42	5.25	5.10	4.96	4.83	4.72
4.5	6.43	6.16	5.92	5.71	5.51	5.34	5.19	5.04	4.91	4.79
4.6	6.57	6.28	6.03	5.81	5.61	5.43	5.27	5.12	4.99	4.86
4.8	6.84	6.52	6.25	6.01	5.80	5.61	5.44	5.28	5.13	5.00
5.0	7.10	6.76	6.47	6.21	5.99	5.78	5.60	5.43	5.28	5.14
5.2	7.37	7.00	6.69	6.41	6.17	5.95	5.76	5.58	5.42	5.27
5.4	7.63	7.24	6.90	6.61	6.35	6.12	5.91	5.73	5.56	5.40
5.5	7.76	7.36	7.01	6.71	6.44	6.20	5.99	5.80	5.62	5.46
5.6	7.90	7.47	7.11	6.80	6.53	6.29	6.07	5.87	5.69	5.53
5.8	8.16	7.71	7.32	6.99	6.70	6.45	6.22	6.01	5.83	5.65
6.0	8.42	7.94	7.53	7.18	6.88	6.61	6.37	6.15	5.96	5.78
6.5	9.06	8.51	8.04	7.64	7.30	7.00	6.73	6.49	6.27	6.07
7.0	9.70	9.07	8.54	8.09	7.71	7.37	7.08	6.81	6.57	6.36
7.5	10.34	9.61	9.02	8.53	8.10	7.74	7.41	7.12	6.86	6.63
8.0	10.96	10.15	9.50	8.95	8.49	8.09	7.73	7.42	7.14	6.89
8.5	11.58	10.68	9.96	9.36	8.86	8.42	8.05	7.71	7.41	7.14
9.0	12.20	11.20	10.41	9.77	9.22	8.75	8.35	7.99	7.67	7.38
9.5	12.80	11.72	10.86	10.16	9.57	9.07	8.64	8.26	7.92	7.62
10.0	13.40	12.22	11.29	10.54	9.91	9.38	8.92	8.52	8.16	7.84
11.0	14.59	13.20	12.14	11.28	10.57	9.98	9.46	9.02	8.62	8.27
12.0	15.75	14.16	12.95	11.99	11.20	10.54	9.98	9.49	9.06	8.68
13.0	16.90	15.09	13.73	12.67	11.80	11.08	10.47	9.94	9.47	9.06
14.0	18.02	15.99	14.49	13.32	12.38	11.60	10.93	10.36	9.86	9.42
15.0	19.13	16.87	15.23	13.95	12.93	12.09	11.38	10.77	10.24	9.77
16.0	20.22	17.73	15.94	14.56	13.46	12.56	11.80	11.15	10.59	10.10
18.0	22.34	19.39	17.30	15.72	14.47	13.46	12.61	11.88	11.26	10.71
20.0	24.41	20.97	18.59	16.81	15.42	14.29	13.35	12.56	11.88	11.28

[†]For an explanation of the headings and the significance of asterisks and high returns which appear on some pages, see Introductory Section, pages xx–xxiiii.

INTERNAL RATE OF RETURN (IRR)†
(with NO projected Rental Growth)

RENTAL FACTOR†	YEARS TO REVIEW/REVERSION									
	1	2	3	4	5	6	7	8	9	10
1.1	2.20	2.19	2.19	2.18	2.18	2.18	2.17	2.17	2.16	2.16
1.2	2.39	2.38	2.37	2.36	2.36	2.35	2.34	2.33	2.33	2.32
1.3	2.58	2.57	2.56	2.54	2.53	2.52	2.50	2.49	2.48	2.47
1.4	2.78	2.76	2.74	2.72	2.70	2.68	2.67	2.65	2.63	2.62
1.5	2.97	2.94	2.92	2.89	2.87	2.85	2.82	2.80	2.78	2.76
1.6	3.16	3.13	3.10	3.06	3.03	3.00	2.98	2.95	2.93	2.90
1.7	3.35	3.31	3.27	3.23	3.20	3.16	3.13	3.10	3.07	3.04
1.8	3.55	3.49	3.45	3.40	3.36	3.32	3.28	3.24	3.20	3.17
1.9	3.74	3.67	3.62	3.56	3.51	3.47	3.42	3.38	3.34	3.30
2.0	3.92	3.85	3.79	3.73	3.67	3.62	3.57	3.52	3.47	3.43
2.2	4.30	4.21	4.13	4.05	3.97	3.91	3.84	3.78	3.73	3.67
2.4	4.67	4.56	4.46	4.36	4.27	4.19	4.11	4.04	3.97	3.91
2.5	4.86	4.73	4.62	4.51	4.42	4.33	4.24	4.16	4.09	4.02
2.6	5.05	4.91	4.78	4.67	4.56	4.46	4.37	4.29	4.21	4.13
2.8	5.42	5.25	5.10	4.97	4.84	4.73	4.62	4.53	4.44	4.35
3.0	5.78	5.59	5.41	5.26	5.12	4.99	4.87	4.76	4.66	4.56
3.2	6.15	5.92	5.72	5.55	5.39	5.24	5.11	4.98	4.87	4.76
3.4	6.51	6.25	6.03	5.83	5.65	5.48	5.34	5.20	5.07	4.96
3.5	6.69	6.42	6.18	5.97	5.78	5.60	5.45	5.31	5.18	5.05
3.6	6.87	6.58	6.33	6.10	5.90	5.72	5.56	5.41	5.27	5.15
3.8	7.22	6.90	6.62	6.37	6.15	5.96	5.78	5.62	5.47	5.33
4.0	7.58	7.22	6.91	6.64	6.40	6.19	5.99	5.82	5.66	5.51
4.2	7.93	7.53	7.20	6.90	6.64	6.41	6.20	6.01	5.84	5.68
4.4	8.28	7.85	7.48	7.16	6.88	6.63	6.40	6.20	6.02	5.85
4.5	8.45	8.00	7.62	7.28	6.99	6.73	6.50	6.30	6.11	5.93
4.6	8.63	8.16	7.75	7.41	7.11	6.84	6.60	6.39	6.19	6.01
4.8	8.97	8.46	8.03	7.66	7.33	7.05	6.80	6.57	6.36	6.17
5.0	9.32	8.76	8.30	7.90	7.56	7.26	6.99	6.75	6.53	6.33
5.2	9.66	9.06	8.56	8.14	7.78	7.46	7.17	6.92	6.69	6.48
5.4	10.00	9.36	8.83	8.38	7.99	7.65	7.35	7.09	6.85	6.63
5.5	10.17	9.51	8.96	8.50	8.10	7.75	7.44	7.17	6.93	6.70
5.6	10.34	9.65	9.09	8.61	8.20	7.85	7.53	7.25	7.00	6.78
5.8	10.67	9.94	9.34	8.84	8.41	8.04	7.71	7.42	7.15	6.92
6.0	11.01	10.23	9.60	9.07	8.62	8.22	7.88	7.58	7.30	7.06
6.5	11.84	10.94	10.22	9.62	9.11	8.68	8.30	7.96	7.66	7.39
7.0	12.65	11.63	10.82	10.15	9.59	9.11	8.69	8.33	8.00	7.71
7.5	13.46	12.31	11.40	10.67	10.05	9.53	9.08	8.68	8.33	8.01
8.0	14.25	12.97	11.97	11.17	10.50	9.93	9.44	9.02	8.64	8.30
8.5	15.04	13.62	12.53	11.65	10.93	10.32	9.80	9.34	8.94	8.58
9.0	15.82	14.26	13.07	12.12	11.35	10.70	10.14	9.65	9.23	8.85
9.5	16.58	14.88	13.60	12.58	11.75	11.06	10.47	9.96	9.51	9.11
10.0	17.34	15.49	14.11	13.03	12.15	11.41	10.79	10.25	9.78	9.36
11.0	18.83	16.69	15.11	13.89	12.90	12.09	11.40	10.80	10.29	9.83
12.0	20.29	17.84	16.07	14.71	13.62	12.72	11.97	11.33	10.77	10.27
13.0	21.72	18.96	16.99	15.49	14.30	13.33	12.51	11.82	11.22	10.69
14.0	23.12	20.04	17.87	16.24	14.95	13.90	13.03	12.29	11.65	11.09
15.0	24.49	21.09	18.73	16.96	15.58	14.46	13.52	12.73	12.05	11.46
16.0	25.84	22.12	19.56	17.66	16.18	14.98	13.99	13.16	12.44	11.82
18.0	28.47	24.08	21.13	18.97	17.31	15.97	14.88	13.96	13.17	12.48
20.0	31.01	25.95	22.61	20.20	18.36	16.89	15.70	14.69	13.84	13.10

†For an explanation of the headings and the significance of asterisks and high returns which appear on some pages, see Introductory Section, pages xx–xxiiii.

INTERNAL RATE OF RETURN (IRR)†
(with NO projected Rental Growth)

RENTAL FACTOR†	YEARS TO REVIEW/REVERSION									
	1	2	3	4	5	6	7	8	9	10
1.1	2.74	2.74	2.73	2.72	2.72	2.71	2.71	2.70	2.70	2.69
1.2	2.99	2.97	2.96	2.95	2.93	2.92	2.91	2.90	2.89	2.88
1.3	3.23	3.20	3.18	3.16	3.14	3.12	3.11	3.09	3.07	3.06
1.4	3.47	3.43	3.40	3.38	3.35	3.32	3.30	3.27	3.25	3.23
1.5	3.71	3.66	3.62	3.59	3.55	3.52	3.48	3.45	3.42	3.40
1.6	3.94	3.89	3.84	3.79	3.75	3.71	3.67	3.63	3.59	3.56
1.7	4.18	4.11	4.05	4.00	3.94	3.89	3.84	3.80	3.76	3.72
1.8	4.42	4.34	4.26	4.20	4.13	4.07	4.02	3.97	3.92	3.87
1.9	4.65	4.56	4.47	4.39	4.32	4.25	4.19	4.13	4.07	4.02
2.0	4.88	4.78	4.68	4.59	4.51	4.43	4.36	4.29	4.22	4.16
2.2	5.35	5.21	5.09	4.97	4.87	4.77	4.68	4.59	4.52	4.44
2.4	5.81	5.64	5.48	5.34	5.21	5.10	4.99	4.89	4.80	4.71
2.5	6.04	5.85	5.68	5.52	5.38	5.26	5.14	5.03	4.93	4.84
2.6	6.26	6.06	5.87	5.70	5.55	5.42	5.29	5.17	5.06	4.96
2.8	6.72	6.47	6.25	6.06	5.88	5.72	5.58	5.44	5.32	5.21
3.0	7.17	6.88	6.62	6.40	6.20	6.02	5.86	5.71	5.57	5.44
3.2	7.61	7.28	6.99	6.74	6.51	6.31	6.13	5.96	5.81	5.67
3.4	8.05	7.68	7.35	7.07	6.82	6.59	6.39	6.21	6.04	5.89
3.5	8.27	7.87	7.53	7.23	6.96	6.73	6.52	6.33	6.15	5.99
3.6	8.49	8.07	7.70	7.39	7.11	6.86	6.64	6.44	6.26	6.10
3.8	8.93	8.45	8.05	7.70	7.40	7.13	6.89	6.67	6.48	6.30
4.0	9.36	8.83	8.39	8.01	7.68	7.39	7.13	6.90	6.69	6.50
4.2	9.79	9.21	8.72	8.31	7.96	7.64	7.36	7.12	6.89	6.69
4.4	10.21	9.58	9.05	8.61	8.23	7.89	7.59	7.33	7.09	6.87
4.5	10.42	9.76	9.22	8.75	8.36	8.01	7.70	7.43	7.19	6.96
4.6	10.63	9.95	9.38	8.90	8.49	8.13	7.81	7.53	7.28	7.05
4.8	11.05	10.31	9.70	9.18	8.75	8.37	8.03	7.73	7.47	7.23
5.0	11.47	10.67	10.01	9.46	9.00	8.60	8.24	7.93	7.65	7.40
5.2	11.88	11.02	10.32	9.74	9.25	8.82	8.45	8.12	7.83	7.56
5.4	12.30	11.37	10.63	10.01	9.49	9.04	8.65	8.31	8.00	7.73
5.5	12.50	11.54	10.78	10.14	9.61	9.15	8.75	8.40	8.09	7.81
5.6	12.70	11.71	10.93	10.28	9.73	9.26	8.85	8.49	8.17	7.88
5.8	13.11	12.06	11.22	10.54	9.96	9.47	9.05	8.67	8.34	8.04
6.0	13.51	12.40	11.51	10.80	10.19	9.68	9.24	8.85	8.50	8.19
6.5	14.51	13.23	12.23	11.42	10.75	10.18	9.69	9.27	8.89	8.55
7.0	15.49	14.03	12.92	12.02	11.29	10.67	10.13	9.67	9.26	8.90
7.5	16.45	14.82	13.59	12.61	11.80	11.13	10.55	10.05	9.61	9.22
8.0	17.41	15.60	14.24	13.17	12.30	11.57	10.95	10.42	9.95	9.54
8.5	18.34	16.35	14.87	13.71	12.78	12.00	11.34	10.77	10.28	9.84
9.0	19.27	17.09	15.49	14.24	13.24	12.41	11.71	11.11	10.59	10.12
9.5	20.18	17.81	16.08	14.75	13.69	12.81	12.07	11.44	10.89	10.40
10.0	21.08	18.52	16.67	15.25	14.12	13.20	12.42	11.75	11.17	10.67
11.0	22.85	19.89	17.80	16.21	14.95	13.93	13.08	12.35	11.72	11.17
12.0	24.58	21.22	18.87	17.12	15.74	14.62	13.70	12.91	12.23	11.64
13.0	26.26	22.49	19.90	17.98	16.49	15.28	14.28	13.44	12.72	12.09
14.0	27.91	23.73	20.89	18.81	17.20	15.91	14.84	13.94	13.17	12.50
15.0	29.52	24.93	21.85	19.60	17.88	16.50	15.37	14.42	13.60	12.90
16.0	31.10	26.09	22.77	20.37	18.53	17.07	15.87	14.87	14.02	13.28
18.0	34.18	28.31	24.52	21.81	19.76	18.14	16.82	15.72	14.79	13.98
20.0	37.14	30.42	26.16	23.15	20.89	19.12	17.69	16.50	15.49	14.63

†For an explanation of the headings and the significance of asterisks and high returns which appear on some pages, see Introductory Section, pages xx–xxiiii.

INTERNAL RATE OF RETURN (IRR)[†]
(with NO projected Rental Growth)

RENTAL FACTOR[†]	YEARS TO REVIEW/REVERSION									
	1	2	3	4	5	6	7	8	9	10
1.1	3.29	3.28	3.27	3.26	3.26	3.25	3.24	3.23	3.23	3.22
1.2	3.58	3.56	3.54	3.52	3.51	3.49	3.47	3.46	3.44	3.43
1.3	3.87	3.83	3.80	3.78	3.75	3.72	3.70	3.67	3.65	3.63
1.4	4.15	4.11	4.06	4.02	3.99	3.95	3.92	3.88	3.85	3.82
1.5	4.44	4.38	4.32	4.27	4.22	4.17	4.13	4.09	4.05	4.01
1.6	4.72	4.64	4.57	4.51	4.45	4.39	4.34	4.29	4.24	4.19
1.7	5.00	4.91	4.82	4.74	4.67	4.60	4.54	4.48	4.42	4.37
1.8	5.28	5.17	5.07	4.98	4.89	4.81	4.74	4.67	4.60	4.54
1.9	5.56	5.43	5.31	5.20	5.10	5.01	4.93	4.85	4.77	4.70
2.0	5.83	5.69	5.55	5.43	5.32	5.21	5.12	5.03	4.94	4.87
2.2	6.38	6.19	6.02	5.87	5.73	5.60	5.48	5.37	5.27	5.17
2.4	6.93	6.69	6.48	6.29	6.12	5.97	5.83	5.70	5.58	5.47
2.5	7.20	6.94	6.70	6.50	6.31	6.15	5.99	5.85	5.73	5.61
2.6	7.47	7.18	6.93	6.70	6.50	6.32	6.16	6.01	5.87	5.75
2.8	8.00	7.66	7.36	7.10	6.87	6.67	6.48	6.31	6.15	6.01
3.0	8.53	8.13	7.79	7.49	7.23	7.00	6.79	6.60	6.43	6.27
3.2	9.05	8.60	8.21	7.87	7.58	7.32	7.09	6.88	6.69	6.51
3.4	9.57	9.05	8.62	8.24	7.92	7.63	7.38	7.15	6.94	6.75
3.5	9.83	9.28	8.82	8.43	8.08	7.78	7.52	7.28	7.06	6.86
3.6	10.09	9.50	9.02	8.61	8.25	7.93	7.65	7.40	7.18	6.97
3.8	10.60	9.95	9.41	8.96	8.57	8.23	7.93	7.66	7.41	7.19
4.0	11.10	10.39	9.80	9.30	8.88	8.51	8.19	7.90	7.64	7.41
4.2	11.60	10.82	10.18	9.64	9.19	8.79	8.44	8.14	7.86	7.61
4.4	12.10	11.24	10.55	9.97	9.48	9.06	8.69	8.36	8.07	7.81
4.5	12.35	11.45	10.73	10.14	9.63	9.19	8.81	8.48	8.18	7.91
4.6	12.59	11.66	10.92	10.30	9.77	9.33	8.93	8.59	8.28	8.00
4.8	13.08	12.08	11.27	10.61	10.06	9.58	9.17	8.80	8.48	8.19
5.0	13.57	12.48	11.63	10.93	10.34	9.83	9.40	9.02	8.68	8.37
5.2	14.05	12.89	11.97	11.23	10.61	10.08	9.62	9.22	8.87	8.55
5.4	14.53	13.29	12.32	11.53	10.88	10.32	9.84	9.42	9.05	8.72
5.5	14.76	13.48	12.49	11.68	11.01	10.44	9.95	9.52	9.14	8.81
5.6	15.00	13.68	12.65	11.83	11.14	10.56	10.06	9.62	9.23	8.89
5.8	15.47	14.07	12.98	12.11	11.40	10.79	10.27	9.81	9.41	9.05
6.0	15.94	14.45	13.31	12.40	11.65	11.01	10.47	10.00	9.58	9.21
6.5	17.09	15.39	14.11	13.09	12.26	11.56	10.96	10.45	10.00	9.60
7.0	18.23	16.31	14.87	13.75	12.84	12.08	11.44	10.88	10.39	9.96
7.5	19.34	17.20	15.62	14.39	13.40	12.58	11.88	11.29	10.77	10.31
8.0	20.44	18.07	16.34	15.00	13.94	13.06	12.31	11.68	11.13	10.64
8.5	21.52	18.91	17.04	15.60	14.46	13.52	12.73	12.05	11.47	10.96
9.0	22.58	19.74	17.71	16.18	14.96	13.96	13.12	12.41	11.80	11.26
9.5	23.63	20.55	18.37	16.73	15.44	14.38	13.51	12.76	12.11	11.55
10.0	24.66	21.34	19.02	17.27	15.91	14.80	13.87	13.09	12.42	11.83
11.0	26.68	22.87	20.25	18.31	16.80	15.58	14.57	13.72	12.99	12.36
12.0	28.65	24.34	21.43	19.29	17.64	16.32	15.23	14.32	13.53	12.85
13.0	30.57	25.76	22.56	20.23	18.44	17.02	15.85	14.87	14.04	13.31
14.0	32.45	27.13	23.64	21.12	19.20	17.68	16.44	15.40	14.52	13.75
15.0	34.28	28.45	24.67	21.97	19.93	18.31	17.00	15.90	14.97	14.17
16.0	36.07	29.74	25.67	22.79	20.62	18.91	17.53	16.38	15.40	14.56
18.0	39.55	32.19	27.57	24.34	21.93	20.04	18.52	17.26	16.20	15.29
20.0	42.89	34.51	29.34	25.78	23.14	21.09	19.44	18.08	16.94	15.96

[†]For an explanation of the headings and the significance of asterisks and high returns which appear on some pages, see Introductory Section, pages xx–xxiiii.

INTERNAL RATE OF RETURN (IRR)[†]
(with NO projected Rental Growth)

RENTAL FACTOR[†]	YEARS TO REVIEW/REVERSION									
	1	2	3	4	5	6	7	8	9	10
1.1	3.84	3.82	3.81	3.80	3.79	3.78	3.77	3.76	3.75	3.74
1.2	4.17	4.15	4.12	4.10	4.07	4.05	4.03	4.01	3.99	3.97
1.3	4.50	4.46	4.42	4.38	4.35	4.31	4.28	4.25	4.22	4.20
1.4	4.84	4.78	4.72	4.67	4.62	4.57	4.53	4.49	4.45	4.41
1.5	5.16	5.08	5.01	4.94	4.88	4.82	4.76	4.71	4.66	4.61
1.6	5.49	5.39	5.30	5.21	5.13	5.06	4.99	4.93	4.87	4.81
1.7	5.82	5.69	5.58	5.48	5.38	5.30	5.22	5.14	5.07	5.00
1.8	6.14	5.99	5.86	5.74	5.63	5.53	5.43	5.35	5.26	5.19
1.9	6.46	6.29	6.13	6.00	5.87	5.75	5.64	5.55	5.45	5.37
2.0	6.78	6.58	6.41	6.25	6.10	5.97	5.85	5.74	5.64	5.54
2.2	7.41	7.16	6.93	6.74	6.56	6.40	6.25	6.11	5.99	5.87
2.4	8.04	7.72	7.45	7.21	6.99	6.80	6.63	6.47	6.32	6.19
2.5	8.35	8.00	7.70	7.44	7.21	7.00	6.81	6.64	6.48	6.34
2.6	8.65	8.28	7.95	7.67	7.42	7.19	6.99	6.81	6.64	6.49
2.8	9.27	8.82	8.44	8.11	7.82	7.57	7.34	7.13	6.94	6.77
3.0	9.87	9.35	8.92	8.54	8.22	7.93	7.67	7.44	7.23	7.04
3.2	10.47	9.88	9.38	8.96	8.60	8.28	7.99	7.74	7.51	7.30
3.4	11.06	10.39	9.84	9.37	8.97	8.62	8.31	8.03	7.78	7.55
3.5	11.36	10.65	10.06	9.57	9.15	8.78	8.46	8.17	7.91	7.68
3.6	11.65	10.90	10.28	9.77	9.33	8.94	8.61	8.31	8.04	7.80
3.8	12.23	11.40	10.72	10.16	9.68	9.26	8.90	8.57	8.29	8.03
4.0	12.81	11.89	11.15	10.53	10.02	9.57	9.18	8.83	8.53	8.25
4.2	13.38	12.37	11.57	10.90	10.35	9.87	9.45	9.09	8.76	8.47
4.4	13.94	12.85	11.98	11.26	10.67	10.16	9.72	9.33	8.99	8.68
4.5	14.22	13.08	12.18	11.44	10.83	10.30	9.85	9.45	9.10	8.78
4.6	14.50	13.31	12.38	11.62	10.98	10.44	9.98	9.57	9.20	8.88
4.8	15.06	13.77	12.77	11.96	11.29	10.72	10.23	9.80	9.42	9.08
5.0	15.61	14.23	13.16	12.30	11.59	10.99	10.47	10.02	9.62	9.27
5.2	16.16	14.68	13.54	12.63	11.88	11.25	10.71	10.24	9.82	9.46
5.4	16.70	15.12	13.92	12.96	12.17	11.51	10.94	10.45	10.02	9.64
5.5	16.97	15.34	14.10	13.12	12.31	11.64	11.06	10.56	10.12	9.73
5.6	17.23	15.56	14.29	13.28	12.45	11.76	11.17	10.66	10.21	9.81
5.8	17.77	15.99	14.65	13.59	12.73	12.01	11.39	10.86	10.40	9.99
6.0	18.29	16.41	15.01	13.90	13.00	12.25	11.61	11.06	10.58	10.15
6.5	19.60	17.45	15.87	14.64	13.65	12.83	12.13	11.54	11.02	10.56
7.0	20.87	18.46	16.71	15.36	14.28	13.38	12.63	11.99	11.43	10.94
7.5	22.13	19.45	17.52	16.05	14.87	13.91	13.11	12.42	11.82	11.30
8.0	23.36	20.40	18.30	16.71	15.45	14.42	13.56	12.83	12.20	11.64
8.5	24.57	21.33	19.06	17.34	16.00	14.91	13.99	13.22	12.55	11.97
9.0	25.76	22.24	19.79	17.96	16.53	15.37	14.41	13.60	12.90	12.29
9.5	26.94	23.12	20.50	18.56	17.04	15.82	14.81	13.96	13.23	12.59
10.0	28.09	23.99	21.20	19.14	17.54	16.26	15.20	14.31	13.54	12.88
11.0	30.35	25.66	22.53	20.24	18.49	17.09	15.93	14.97	14.14	13.43
12.0	32.55	27.27	23.79	21.29	19.38	17.86	16.62	15.59	14.70	13.94
13.0	34.68	28.81	25.00	22.28	20.22	18.60	17.27	16.17	15.23	14.42
14.0	36.77	30.30	26.16	23.23	21.02	19.29	17.88	16.71	15.72	14.87
15.0	38.80	31.74	27.27	24.14	21.79	19.95	18.46	17.23	16.19	15.30
16.0	40.79	33.12	28.34	25.00	22.52	20.58	19.02	17.73	16.64	15.71
18.0	44.64	35.78	30.36	26.64	23.89	21.76	20.05	18.65	17.47	16.46
20.0	48.33	38.28	32.25	28.15	25.16	22.85	21.01	19.50	18.23	17.15

[†]For an explanation of the headings and the significance of asterisks and high returns which appear on some pages, see Introductory Section, pages xx–xxiiii.

INTERNAL RATE OF RETURN (IRR)[†]
(with NO projected Rental Growth)

RENTAL FACTOR[†]	YEARS TO REVIEW/REVERSION									
	1	2	3	4	5	6	7	8	9	10
1.1	4.38	4.37	4.35	4.34	4.32	4.31	4.30	4.29	4.27	4.26
1.2	4.76	4.73	4.70	4.67	4.64	4.61	4.58	4.56	4.54	4.51
1.3	5.14	5.09	5.04	4.99	4.94	4.90	4.86	4.82	4.79	4.75
1.4	5.52	5.44	5.37	5.30	5.24	5.18	5.13	5.08	5.03	4.98
1.5	5.89	5.79	5.69	5.61	5.53	5.45	5.39	5.32	5.26	5.20
1.6	6.26	6.13	6.01	5.91	5.81	5.72	5.64	5.56	5.48	5.42
1.7	6.63	6.47	6.33	6.20	6.08	5.98	5.88	5.79	5.70	5.62
1.8	6.99	6.81	6.64	6.49	6.35	6.23	6.11	6.01	5.91	5.82
1.9	7.35	7.14	6.94	6.77	6.61	6.47	6.34	6.22	6.11	6.01
2.0	7.71	7.46	7.24	7.05	6.87	6.71	6.56	6.43	6.31	6.19
2.2	8.43	8.11	7.83	7.58	7.36	7.17	6.99	6.83	6.68	6.55
2.4	9.13	8.74	8.40	8.10	7.84	7.61	7.40	7.21	7.04	6.88
2.5	9.48	9.05	8.67	8.35	8.07	7.82	7.59	7.39	7.21	7.04
2.6	9.83	9.35	8.95	8.60	8.30	8.03	7.79	7.57	7.37	7.19
2.8	10.51	9.96	9.49	9.08	8.74	8.43	8.16	7.91	7.69	7.50
3.0	11.19	10.55	10.01	9.55	9.16	8.82	8.52	8.24	8.00	7.78
3.2	11.87	11.13	10.52	10.01	9.57	9.19	8.86	8.56	8.30	8.06
3.4	12.53	11.69	11.02	10.45	9.97	9.55	9.19	8.87	8.58	8.32
3.5	12.86	11.98	11.26	10.67	10.16	9.73	9.35	9.01	8.71	8.45
3.6	13.19	12.25	11.50	10.88	10.35	9.90	9.51	9.16	8.85	8.57
3.8	13.84	12.80	11.98	11.30	10.73	10.24	9.82	9.44	9.11	8.81
4.0	14.48	13.34	12.44	11.71	11.09	10.57	10.11	9.72	9.36	9.05
4.2	15.12	13.87	12.90	12.10	11.45	10.89	10.40	9.98	9.61	9.27
4.4	15.75	14.39	13.34	12.49	11.79	11.19	10.68	10.24	9.84	9.49
4.5	16.06	14.65	13.56	12.68	11.96	11.35	10.82	10.36	9.96	9.60
4.6	16.37	14.91	13.78	12.87	12.13	11.50	10.96	10.49	10.07	9.70
4.8	16.99	15.41	14.20	13.24	12.45	11.79	11.22	10.73	10.29	9.91
5.0	17.60	15.91	14.62	13.61	12.77	12.07	11.48	10.96	10.51	10.11
5.2	18.21	16.40	15.04	13.96	13.08	12.35	11.73	11.19	10.72	10.30
5.4	18.81	16.88	15.44	14.31	13.39	12.62	11.97	11.41	10.92	10.49
5.5	19.11	17.12	15.64	14.48	13.54	12.76	12.09	11.52	11.02	10.58
5.6	19.41	17.36	15.84	14.65	13.69	12.89	12.21	11.63	11.12	10.67
5.8	20.00	17.83	16.23	14.98	13.98	13.15	12.45	11.84	11.32	10.85
6.0	20.59	18.29	16.61	15.31	14.27	13.40	12.67	12.05	11.51	11.03
6.5	22.03	19.43	17.55	16.11	14.96	14.01	13.22	12.55	11.96	11.44
7.0	23.44	20.52	18.44	16.87	15.62	14.60	13.74	13.02	12.39	11.84
7.5	24.83	21.59	19.31	17.60	16.25	15.15	14.24	13.46	12.80	12.21
8.0	26.19	22.62	20.15	18.30	16.85	15.68	14.71	13.89	13.18	12.57
8.5	27.52	23.63	20.95	18.97	17.43	16.19	15.17	14.30	13.56	12.91
9.0	28.84	24.61	21.74	19.63	17.99	16.68	15.60	14.69	13.91	13.23
9.5	30.13	25.56	22.50	20.26	18.53	17.15	16.02	15.06	14.25	13.55
10.0	31.40	26.50	23.24	20.87	19.05	17.61	16.42	15.43	14.58	13.84
11.0	33.88	28.30	24.65	22.04	20.04	18.47	17.18	16.11	15.20	14.41
12.0	36.29	30.03	26.00	23.14	20.98	19.28	17.90	16.75	15.77	14.94
13.0	38.63	31.68	27.28	24.18	21.86	20.04	18.57	17.35	16.32	15.43
14.0	40.90	33.28	28.50	25.18	22.70	20.76	19.20	17.91	16.83	15.89
15.0	43.13	34.81	29.68	26.13	23.50	21.45	19.81	18.45	17.31	16.33
16.0	45.30	36.30	30.81	27.04	24.26	22.10	20.38	18.96	17.77	16.75
18.0	49.49	39.13	32.94	28.75	25.68	23.33	21.45	19.91	18.62	17.53
20.0	53.51	41.80	34.93	30.34	27.00	24.45	22.43	20.78	19.40	18.23

[†]For an explanation of the headings and the significance of asterisks and high returns which appear on some pages, see Introductory Section, pages xx–xxiii.

INTERNAL RATE OF RETURN (IRR)[†]
(with NO projected Rental Growth)

RENTAL FACTOR[†]	YEARS TO REVIEW/REVERSION									
	1	2	3	4	5	6	7	8	9	10
1.1	4.93	4.91	4.89	4.87	4.86	4.84	4.82	4.81	4.80	4.78
1.2	5.35	5.31	5.27	5.23	5.20	5.17	5.13	5.10	5.08	5.05
1.3	5.78	5.71	5.64	5.59	5.53	5.48	5.43	5.39	5.34	5.31
1.4	6.19	6.10	6.01	5.93	5.85	5.78	5.72	5.66	5.60	5.55
1.5	6.61	6.48	6.37	6.26	6.17	6.08	6.00	5.92	5.85	5.78
1.6	7.02	6.86	6.72	6.59	6.47	6.36	6.26	6.17	6.09	6.01
1.7	7.43	7.24	7.07	6.91	6.77	6.64	6.52	6.42	6.32	6.22
1.8	7.84	7.61	7.41	7.22	7.06	6.91	6.78	6.65	6.54	6.43
1.9	8.24	7.97	7.74	7.53	7.34	7.17	7.02	6.88	6.75	6.63
2.0	8.64	8.33	8.07	7.83	7.62	7.43	7.26	7.10	6.96	6.83
2.2	9.43	9.04	8.70	8.41	8.15	7.92	7.71	7.52	7.35	7.20
2.4	10.22	9.73	9.32	8.97	8.66	8.39	8.14	7.92	7.72	7.54
2.5	10.60	10.07	9.62	9.24	8.91	8.61	8.35	8.12	7.90	7.71
2.6	10.99	10.41	9.92	9.51	9.15	8.83	8.55	8.30	8.08	7.87
2.8	11.75	11.07	10.50	10.03	9.62	9.26	8.95	8.67	8.41	8.19
3.0	12.50	11.71	11.07	10.53	10.07	9.67	9.32	9.01	8.74	8.49
3.2	13.24	12.34	11.62	11.02	10.51	10.07	9.68	9.34	9.04	8.77
3.4	13.98	12.96	12.16	11.49	10.93	10.45	10.03	9.66	9.34	9.04
3.5	14.34	13.27	12.42	11.72	11.14	10.63	10.20	9.82	9.48	9.18
3.6	14.70	13.57	12.68	11.95	11.34	10.82	10.37	9.97	9.62	9.31
3.8	15.42	14.17	13.19	12.40	11.73	11.17	10.69	10.27	9.89	9.56
4.0	16.13	14.75	13.69	12.83	12.12	11.52	11.00	10.55	10.15	9.80
4.2	16.83	15.33	14.17	13.25	12.49	11.85	11.30	10.83	10.41	10.03
4.4	17.52	15.89	14.65	13.67	12.86	12.18	11.60	11.09	10.65	10.26
4.5	17.86	16.17	14.89	13.87	13.04	12.34	11.74	11.22	10.77	10.37
4.6	18.21	16.45	15.12	14.07	13.21	12.49	11.88	11.35	10.89	10.48
4.8	18.88	16.99	15.58	14.46	13.56	12.80	12.16	11.60	11.12	10.69
5.0	19.56	17.53	16.02	14.85	13.89	13.10	12.43	11.85	11.34	10.90
5.2	20.22	18.06	16.46	15.22	14.22	13.39	12.69	12.09	11.56	11.10
5.4	20.88	18.58	16.90	15.59	14.54	13.68	12.94	12.32	11.77	11.29
5.5	21.21	18.84	17.11	15.77	14.70	13.82	13.07	12.43	11.88	11.39
5.6	21.53	19.09	17.32	15.95	14.86	13.95	13.19	12.54	11.98	11.48
5.8	22.18	19.60	17.74	16.30	15.16	14.22	13.44	12.76	12.18	11.67
6.0	22.82	20.10	18.14	16.65	15.46	14.49	13.67	12.98	12.37	11.85
6.5	24.40	21.32	19.14	17.49	16.19	15.13	14.24	13.49	12.84	12.28
7.0	25.94	22.49	20.09	18.29	16.88	15.74	14.78	13.98	13.29	12.68
7.5	27.45	23.64	21.01	19.06	17.54	16.31	15.30	14.44	13.71	13.07
8.0	28.93	24.74	21.89	19.80	18.17	16.87	15.79	14.88	14.11	13.43
8.5	30.38	25.82	22.75	20.50	18.78	17.39	16.26	15.30	14.49	13.78
9.0	31.81	26.87	23.58	21.19	19.36	17.90	16.71	15.71	14.85	14.11
9.5	33.21	27.89	24.38	21.85	19.92	18.39	17.14	16.09	15.20	14.43
10.0	34.59	28.88	25.16	22.49	20.46	18.86	17.55	16.46	15.54	14.74
11.0	37.28	30.80	26.65	23.71	21.50	19.76	18.34	17.17	16.17	15.32
12.0	39.89	32.64	28.07	24.86	22.47	20.59	19.08	17.83	16.77	15.86
13.0	42.42	34.40	29.41	25.96	23.38	21.38	19.77	18.44	17.32	16.36
14.0	44.88	36.09	30.70	26.99	24.25	22.13	20.43	19.02	17.85	16.84
15.0	47.28	37.72	31.93	27.98	25.08	22.84	21.05	19.57	18.34	17.29
16.0	49.62	39.29	33.12	28.93	25.87	23.51	21.64	20.10	18.81	17.71
18.0	54.13	42.29	35.35	30.71	27.34	24.77	22.73	21.07	19.68	18.51
20.0	58.46	45.11	37.44	32.36	28.71	25.93	23.74	21.96	20.48	19.23

[†]For an explanation of the headings and the significance of asterisks and high returns which appear on some pages, see introductory Section, pages xx–xxiiii.

INTERNAL RATE OF RETURN (IRR)[†]
(with NO projected Rental Growth)

RENTAL FACTOR[†]	YEARS TO REVIEW/REVERSION									
	1	**2**	**3**	**4**	**5**	**6**	**7**	**8**	**9**	**10**
1.1	5.47	5.45	5.43	5.41	5.38	5.37	5.35	5.33	5.31	5.30
1.2	5.94	5.89	5.84	5.80	5.76	5.72	5.68	5.64	5.61	5.58
1.3	6.41	6.33	6.25	6.18	6.11	6.05	6.00	5.95	5.90	5.85
1.4	6.87	6.75	6.65	6.55	6.46	6.38	6.30	6.23	6.17	6.11
1.5	7.33	7.18	7.04	6.91	6.80	6.69	6.60	6.51	6.43	6.35
1.6	7.78	7.59	7.42	7.27	7.13	7.00	6.88	6.78	6.68	6.59
1.7	8.23	8.00	7.79	7.61	7.44	7.29	7.16	7.03	6.92	6.81
1.8	8.68	8.40	8.16	7.95	7.75	7.58	7.42	7.28	7.15	7.03
1.9	9.12	8.80	8.52	8.27	8.05	7.86	7.68	7.52	7.37	7.24
2.0	9.56	9.19	8.87	8.60	8.35	8.13	7.93	7.75	7.59	7.44
2.2	10.43	9.96	9.56	9.22	8.91	8.65	8.41	8.20	8.00	7.82
2.4	11.29	10.71	10.23	9.81	9.46	9.14	8.86	8.61	8.39	8.19
2.5	11.71	11.08	10.55	10.10	9.72	9.38	9.08	8.82	8.58	8.36
2.6	12.13	11.44	10.87	10.39	9.97	9.61	9.29	9.01	8.76	8.53
2.8	12.97	12.15	11.49	10.94	10.47	10.06	9.71	9.39	9.11	8.85
3.0	13.79	12.85	12.10	11.48	10.95	10.49	10.10	9.75	9.44	9.16
3.2	14.60	13.53	12.69	11.99	11.41	10.91	10.48	10.10	9.76	9.46
3.4	15.40	14.20	13.26	12.49	11.85	11.31	10.84	10.43	10.06	9.74
3.5	15.79	14.53	13.54	12.74	12.07	11.50	11.02	10.59	10.21	9.87
3.6	16.19	14.85	13.82	12.98	12.28	11.69	11.19	10.75	10.36	10.01
3.8	16.97	15.50	14.36	13.45	12.70	12.07	11.52	11.05	10.64	10.27
4.0	17.74	16.12	14.89	13.91	13.10	12.43	11.85	11.35	10.91	10.52
4.2	18.50	16.74	15.41	14.36	13.50	12.78	12.16	11.63	11.17	10.76
4.4	19.26	17.35	15.92	14.79	13.88	13.12	12.47	11.91	11.42	10.99
4.5	19.63	17.64	16.16	15.00	14.06	13.28	12.62	12.05	11.55	11.11
4.6	20.00	17.94	16.41	15.21	14.25	13.44	12.76	12.18	11.67	11.22
4.8	20.74	18.52	16.90	15.63	14.61	13.76	13.05	12.44	11.90	11.43
5.0	21.47	19.10	17.37	16.03	14.96	14.08	13.33	12.69	12.13	11.65
5.2	22.19	19.67	17.84	16.43	15.30	14.38	13.60	12.94	12.36	11.85
5.4	22.90	20.22	18.29	16.81	15.64	14.67	13.87	13.17	12.58	12.05
5.5	23.25	20.50	18.52	17.01	15.80	14.82	13.99	13.29	12.68	12.15
5.6	23.61	20.77	18.74	17.19	15.97	14.96	14.12	13.41	12.79	12.25
5.8	24.31	21.31	19.18	17.56	16.29	15.24	14.37	13.63	12.99	12.43
6.0	25.00	21.84	19.61	17.93	16.60	15.52	14.62	13.85	13.19	12.62
6.5	26.70	23.14	20.66	18.80	17.35	16.18	15.21	14.38	13.68	13.06
7.0	28.37	24.39	21.66	19.64	18.07	16.81	15.77	14.89	14.13	13.47
7.5	30.00	25.60	22.63	20.44	18.76	17.41	16.30	15.36	14.56	13.87
8.0	31.60	26.78	23.56	21.21	19.41	17.98	16.80	15.81	14.97	14.24
8.5	33.16	27.92	24.45	21.95	20.04	18.53	17.28	16.25	15.36	14.60
9.0	34.70	29.03	25.32	22.67	20.65	19.05	17.75	16.66	15.74	14.94
9.5	36.20	30.11	26.16	23.36	21.23	19.55	18.19	17.06	16.09	15.27
10.0	37.68	31.16	26.98	24.02	21.79	20.04	18.62	17.44	16.44	15.58
11.0	40.57	33.19	28.54	25.29	22.86	20.96	19.43	18.16	17.09	16.17
12.0	43.36	35.12	30.02	26.49	23.86	21.82	20.18	18.83	17.69	16.72
13.0	46.07	36.98	31.43	27.62	24.81	22.64	20.90	19.46	18.26	17.23
14.0	48.71	38.76	32.77	28.70	25.71	23.40	21.57	20.06	18.80	17.72
15.0	51.27	40.47	34.06	29.72	26.56	24.13	22.20	20.62	19.30	18.18
16.0	53.77	42.13	35.29	30.70	27.37	24.83	22.81	21.16	19.78	18.61
18.0	58.60	45.28	37.62	32.54	28.89	26.12	23.93	22.15	20.67	19.42
20.0	63.21	48.23	39.78	34.25	30.30	27.31	24.96	23.06	21.48	20.15

[†]For an explanation of the headings and the significance of asterisks and high returns which appear on some pages, see Introductory Section, pages xx–xxiiii.

INTERNAL RATE OF RETURN (IRR)[†]
(with NO projected Rental Growth)

RENTAL FACTOR[†]	YEARS TO REVIEW/REVERSION									
	1	2	3	4	5	6	7	8	9	10
1.1	6.56	6.53	6.50	6.47	6.44	6.41	6.39	6.37	6.34	6.32
1.2	7.12	7.05	6.98	6.92	6.86	6.81	6.76	6.71	6.67	6.63
1.3	7.67	7.56	7.45	7.36	7.27	7.19	7.11	7.04	6.98	6.92
1.4	8.22	8.06	7.91	7.78	7.66	7.55	7.45	7.36	7.28	7.20
1.5	8.76	8.55	8.36	8.19	8.04	7.90	7.78	7.66	7.56	7.46
1.6	9.29	9.03	8.80	8.59	8.40	8.24	8.09	7.95	7.83	7.71
1.7	9.82	9.50	9.22	8.98	8.76	8.57	8.39	8.23	8.09	7.95
1.8	10.35	9.97	9.64	9.36	9.10	8.88	8.68	8.50	8.34	8.19
1.9	10.87	10.43	10.05	9.73	9.44	9.19	8.96	8.76	8.58	8.41
2.0	11.39	10.88	10.45	10.09	9.77	9.48	9.23	9.01	8.81	8.62
2.2	12.41	11.76	11.23	10.78	10.39	10.05	9.75	9.49	9.25	9.03
2.4	13.41	12.62	11.98	11.45	10.99	10.59	10.24	9.94	9.66	9.42
2.5	13.90	13.04	12.35	11.77	11.28	10.85	10.48	10.15	9.86	9.60
2.6	14.39	13.46	12.71	12.08	11.56	11.10	10.71	10.36	10.05	9.78
2.8	15.36	14.27	13.41	12.70	12.10	11.59	11.15	10.77	10.42	10.12
3.0	16.32	15.06	14.08	13.29	12.62	12.06	11.57	11.15	10.78	10.44
3.2	17.26	15.84	14.74	13.86	13.12	12.51	11.98	11.52	11.11	10.75
3.4	18.18	16.59	15.38	14.41	13.61	12.94	12.37	11.87	11.43	11.05
3.5	18.64	16.96	15.69	14.67	13.84	13.15	12.55	12.04	11.59	11.19
3.6	19.10	17.33	16.00	14.94	14.08	13.35	12.74	12.21	11.74	11.33
3.8	20.00	18.05	16.60	15.45	14.53	13.75	13.10	12.53	12.04	11.60
4.0	20.89	18.76	17.19	15.96	14.96	14.14	13.44	12.85	12.32	11.87
4.2	21.77	19.46	17.76	16.44	15.39	14.51	13.78	13.15	12.60	12.12
4.4	22.63	20.13	18.32	16.92	15.80	14.88	14.10	13.44	12.86	12.36
4.5	23.06	20.47	18.59	17.15	16.00	15.05	14.26	13.58	12.99	12.48
4.6	23.49	20.80	18.86	17.38	16.20	15.23	14.42	13.72	13.12	12.60
4.8	24.34	21.46	19.40	17.83	16.59	15.57	14.72	14.00	13.37	12.82
5.0	25.17	22.10	19.92	18.27	16.96	15.90	15.01	14.26	13.61	13.04
5.2	26.00	22.73	20.43	18.70	17.33	16.22	15.30	14.52	13.84	13.26
5.4	26.82	23.35	20.93	19.11	17.69	16.54	15.58	14.77	14.07	13.46
5.5	27.22	23.66	21.17	19.32	17.87	16.69	15.72	14.89	14.18	13.57
5.6	27.63	23.96	21.42	19.52	18.04	16.85	15.85	15.01	14.29	13.67
5.8	28.43	24.56	21.90	19.92	18.39	17.14	16.12	15.25	14.51	13.86
6.0	29.22	25.15	22.37	20.32	18.72	17.44	16.38	15.48	14.72	14.05
6.5	31.16	26.59	23.51	21.26	19.53	18.14	17.00	16.04	15.22	14.51
7.0	33.06	27.98	24.61	22.16	20.29	18.80	17.58	16.56	15.69	14.94
7.5	34.91	29.32	25.66	23.03	21.02	19.44	18.14	17.06	16.14	15.35
8.0	36.72	30.62	26.67	23.85	21.72	20.04	18.67	17.53	16.57	15.74
8.5	38.49	31.88	27.64	24.64	22.39	20.62	19.18	17.98	16.97	16.11
9.0	40.23	33.10	28.58	25.41	23.03	21.17	19.66	18.42	17.36	16.46
9.5	41.93	34.28	29.49	26.14	23.65	21.70	20.13	18.83	17.73	16.80
10.0	43.60	35.44	30.37	26.85	24.24	22.21	20.57	19.23	18.09	17.12
11.0	46.86	37.66	32.05	28.21	25.37	23.18	21.42	19.98	18.76	17.73
12.0	50.00	39.78	33.65	29.48	26.43	24.08	22.21	20.68	19.39	18.30
13.0	53.04	41.81	35.16	30.68	27.43	24.93	22.95	21.33	19.98	18.83
14.0	56.00	43.75	36.60	31.83	28.37	25.74	23.65	21.95	20.53	19.33
15.0	58.87	45.62	37.98	32.91	29.27	26.50	24.31	22.53	21.05	19.80
16.0	61.67	47.42	39.30	33.95	30.12	27.22	24.94	23.08	21.55	20.24
18.0	67.06	50.83	41.79	35.90	31.72	28.58	26.11	24.11	22.46	21.07
20.0	****	54.04	44.10	37.70	33.19	29.82	27.18	25.06	23.30	21.83

[†]For an explanation of the headings and the significance of asterisks and high returns which appear on some pages, see Introductory Section, pages xx–xxiiii.

INTERNAL RATE OF RETURN (IRR)[†]
(with NO projected Rental Growth)

RENTAL FACTOR[†]	YEARS TO REVIEW/REVERSION									
	1	2	3	4	5	6	7	8	9	10
1.1	7.65	7.60	7.56	7.52	7.49	7.45	7.42	7.40	7.37	7.34
1.2	8.29	8.20	8.11	8.03	7.95	7.89	7.83	7.77	7.72	7.67
1.3	8.93	8.77	8.64	8.51	8.40	8.30	8.21	8.12	8.05	7.97
1.4	9.56	9.34	9.15	8.98	8.83	8.70	8.57	8.46	8.36	8.27
1.5	10.18	9.90	9.65	9.44	9.25	9.08	8.92	8.78	8.66	8.54
1.6	10.79	10.44	10.14	9.88	9.65	9.44	9.26	9.09	8.94	8.81
1.7	11.40	10.98	10.62	10.31	10.04	9.80	9.58	9.39	9.22	9.06
1.8	12.00	11.50	11.09	10.73	10.41	10.14	9.89	9.68	9.48	9.30
1.9	12.60	12.02	11.54	11.13	10.78	10.47	10.19	9.95	9.73	9.53
2.0	13.18	12.53	11.98	11.52	11.13	10.79	10.48	10.21	9.97	9.76
2.2	14.35	13.52	12.85	12.28	11.81	11.40	11.04	10.72	10.44	10.18
2.4	15.49	14.48	13.67	13.01	12.45	11.97	11.56	11.19	10.87	10.58
2.5	16.05	14.95	14.07	13.36	12.76	12.25	11.81	11.42	11.08	10.77
2.6	16.61	15.41	14.47	13.70	13.06	12.52	12.05	11.64	11.28	10.96
2.8	17.70	16.31	15.23	14.37	13.65	13.04	12.52	12.07	11.67	11.31
3.0	18.79	17.19	15.98	15.00	14.21	13.54	12.96	12.47	12.03	11.65
3.2	19.85	18.05	16.69	15.62	14.74	14.01	13.39	12.85	12.38	11.97
3.4	20.90	18.89	17.39	16.21	15.26	14.47	13.80	13.22	12.72	12.28
3.5	21.41	19.30	17.73	16.50	15.51	14.69	14.00	13.40	12.88	12.43
3.6	21.93	19.70	18.06	16.78	15.76	14.91	14.19	13.57	13.04	12.57
3.8	22.94	20.50	18.71	17.34	16.24	15.33	14.57	13.91	13.35	12.85
4.0	23.94	21.28	19.35	17.88	16.70	15.74	14.93	14.24	13.64	13.12
4.2	24.93	22.04	19.97	18.40	17.15	16.13	15.28	14.55	13.93	13.38
4.4	25.90	22.79	20.58	18.91	17.59	16.51	15.62	14.86	14.20	13.63
4.5	26.39	23.15	20.87	19.15	17.80	16.70	15.78	15.01	14.34	13.75
4.6	26.86	23.52	21.17	19.40	18.01	16.88	15.95	15.15	14.47	13.87
4.8	27.81	24.23	21.74	19.88	18.42	17.24	16.26	15.44	14.73	14.11
5.0	28.75	24.94	22.30	20.35	18.82	17.59	16.57	15.71	14.97	14.33
5.2	29.67	25.63	22.86	20.80	19.21	17.93	16.87	15.98	15.22	14.55
5.4	30.59	26.31	23.39	21.25	19.59	18.26	17.16	16.24	15.45	14.77
5.5	31.04	26.64	23.66	21.47	19.78	18.42	17.31	16.37	15.57	14.87
5.6	31.49	26.97	23.92	21.69	19.96	18.58	17.45	16.49	15.68	14.98
5.8	32.38	27.63	24.44	22.11	20.32	18.89	17.72	16.74	15.90	15.18
6.0	33.26	28.27	24.94	22.53	20.68	19.20	17.99	16.98	16.12	15.37
6.5	35.43	29.84	26.17	23.53	21.53	19.94	18.64	17.56	16.64	15.84
7.0	37.54	31.35	27.34	24.49	22.33	20.63	19.25	18.10	17.13	16.29
7.5	39.59	32.80	28.46	25.40	23.10	21.29	19.83	18.61	17.59	16.71
8.0	41.60	34.21	29.54	26.27	23.83	21.92	20.38	19.10	18.03	17.10
8.5	43.57	35.57	30.58	27.11	24.53	22.52	20.90	19.57	18.44	17.48
9.0	45.49	36.89	31.58	27.92	25.20	23.10	21.41	20.01	18.84	17.84
9.5	47.37	38.17	32.55	28.69	25.85	23.65	21.89	20.44	19.22	18.19
10.0	49.22	39.41	33.49	29.44	26.47	24.18	22.35	20.85	19.59	18.52
11.0	52.81	41.81	35.28	30.87	27.65	25.19	23.23	21.62	20.28	19.14
12.0	56.27	44.09	36.97	32.21	28.76	26.13	24.04	22.34	20.93	19.72
13.0	59.62	46.26	38.57	33.47	29.80	27.01	24.81	23.02	21.53	20.27
14.0	62.87	48.35	40.10	34.67	30.78	27.84	25.53	23.65	22.09	20.78
15.0	66.03	50.35	41.55	35.81	31.72	28.63	26.21	24.25	22.63	21.26
16.0	69.10	52.28	42.95	36.90	32.61	29.38	26.86	24.82	23.13	21.71
18.0	****	55.94	45.57	38.94	34.27	30.78	28.06	25.88	24.08	22.56
20.0	****	59.37	48.01	40.82	35.80	32.07	29.17	26.85	24.93	23.33

[†]For an explanation of the headings and the significance of asterisks and high returns which appear on some pages, see Introductory Section, pages xx–xxiiii.

INTERNAL RATE OF RETURN (IRR)[†]
(with NO projected Rental Growth)

RENTAL FACTOR[†]	YEARS TO REVIEW/REVERSION									
	1	**2**	**3**	**4**	**5**	**6**	**7**	**8**	**9**	**10**
1.1	8.74	8.68	8.62	8.58	8.53	8.49	8.45	8.42	8.39	8.36
1.2	9.46	9.34	9.23	9.13	9.04	8.96	8.88	8.81	8.75	8.70
1.3	10.18	9.98	9.81	9.66	9.52	9.40	9.29	9.19	9.10	9.01
1.4	10.89	10.62	10.38	10.17	9.99	9.82	9.68	9.54	9.42	9.31
1.5	11.58	11.23	10.93	10.67	10.44	10.23	10.05	9.88	9.73	9.60
1.6	12.28	11.84	11.47	11.15	10.87	10.62	10.40	10.21	10.03	9.87
1.7	12.96	12.43	11.99	11.61	11.28	10.99	10.74	10.52	10.31	10.13
1.8	13.63	13.01	12.50	12.06	11.68	11.36	11.07	10.81	10.59	10.38
1.9	14.30	13.58	12.99	12.50	12.07	11.71	11.38	11.10	10.85	10.62
2.0	14.96	14.14	13.48	12.92	12.45	12.04	11.69	11.38	11.10	10.85
2.2	16.26	15.23	14.41	13.74	13.17	12.69	12.27	11.90	11.58	11.29
2.4	17.53	16.28	15.31	14.51	13.85	13.30	12.82	12.40	12.03	11.70
2.5	18.16	16.80	15.74	14.89	14.18	13.59	13.08	12.63	12.24	11.90
2.6	18.78	17.30	16.17	15.25	14.50	13.87	13.33	12.86	12.45	12.09
2.8	20.00	18.29	16.99	15.96	15.12	14.42	13.82	13.30	12.85	12.45
3.0	21.20	19.25	17.79	16.64	15.71	14.94	14.28	13.72	13.23	12.80
3.2	22.38	20.18	18.56	17.30	16.28	15.44	14.73	14.12	13.59	13.13
3.4	23.54	21.09	19.31	17.93	16.82	15.92	15.15	14.50	13.93	13.44
3.5	24.11	21.54	19.67	18.23	17.09	16.15	15.36	14.68	14.10	13.59
3.6	24.68	21.98	20.03	18.54	17.35	16.37	15.56	14.86	14.26	13.74
3.8	25.81	22.84	20.73	19.12	17.85	16.82	15.95	15.21	14.58	14.03
4.0	26.91	23.69	21.41	19.69	18.34	17.24	16.33	15.55	14.88	14.30
4.2	28.00	24.51	22.07	20.25	18.81	17.65	16.69	15.88	15.18	14.57
4.4	29.07	25.32	22.72	20.78	19.27	18.05	17.04	16.19	15.46	14.83
4.5	29.60	25.72	23.03	21.04	19.49	18.24	17.21	16.34	15.60	14.95
4.6	30.13	26.11	23.35	21.30	19.71	18.44	17.38	16.49	15.73	15.07
4.8	31.18	26.88	23.96	21.81	20.14	18.81	17.71	16.79	16.00	15.31
5.0	32.20	27.64	24.56	22.30	20.56	19.17	18.03	17.07	16.25	15.54
5.2	33.22	28.38	25.14	22.78	20.97	19.52	18.34	17.35	16.50	15.77
5.4	34.22	29.11	25.72	23.25	21.37	19.87	18.64	17.61	16.74	15.99
5.5	34.72	29.47	26.00	23.48	21.56	20.04	18.79	17.74	16.86	16.09
5.6	35.22	29.83	26.28	23.71	21.75	20.20	18.93	17.87	16.97	16.20
5.8	36.19	30.54	26.82	24.16	22.13	20.53	19.22	18.13	17.20	16.41
6.0	37.16	31.23	27.36	24.60	22.50	20.84	19.50	18.37	17.42	16.61
6.5	39.53	32.91	28.66	25.65	23.39	21.61	20.16	18.97	17.95	17.09
7.0	41.84	34.52	29.90	26.65	24.23	22.33	20.79	19.52	18.45	17.54
7.5	44.09	36.08	31.09	27.61	25.02	23.01	21.39	20.05	18.93	17.96
8.0	46.28	37.58	32.22	28.52	25.78	23.66	21.96	20.55	19.38	18.37
8.5	48.42	39.04	33.32	29.40	26.51	24.28	22.50	21.03	19.80	18.76
9.0	50.52	40.45	34.38	30.24	27.21	24.88	23.01	21.49	20.21	19.12
9.5	52.57	41.81	35.40	31.05	27.88	25.45	23.51	21.92	20.60	19.47
10.0	54.58	43.14	36.38	31.83	28.53	26.00	23.99	22.34	20.97	19.81
11.0	58.48	45.69	38.27	33.32	29.75	27.03	24.89	23.14	21.68	20.45
12.0	62.24	48.11	40.04	34.72	30.90	28.00	25.72	23.87	22.34	21.04
13.0	65.87	50.43	41.72	36.03	31.98	28.91	26.51	24.56	22.95	21.59
14.0	69.39	52.64	43.32	37.28	33.00	29.77	27.25	25.21	23.53	22.11
15.0	****	54.76	44.85	38.47	33.96	30.59	27.95	25.83	24.07	22.60
16.0	****	56.80	46.31	39.60	34.88	31.36	28.61	26.41	24.59	23.06
18.0	****	60.68	49.06	41.72	36.60	32.80	29.85	27.49	25.55	23.92
20.0	****	64.30	51.61	43.67	38.18	34.12	30.98	28.48	26.42	24.71

[†]For an explanation of the headings and the significance of asterisks and high returns which appear on some pages, see Introductory Section, pages xx–xxiiii.

INTERNAL RATE OF RETURN (IRR)[†]
(with NO projected Rental Growth)

RENTAL FACTOR[†]	YEARS TO REVIEW/REVERSION									
	1	**2**	**3**	**4**	**5**	**6**	**7**	**8**	**9**	**10**
1.1	9.82	9.75	9.68	9.62	9.57	9.52	9.48	9.44	9.40	9.37
1.2	10.63	10.47	10.34	10.22	10.11	10.02	9.93	9.85	9.78	9.71
1.3	11.42	11.18	10.98	10.79	10.63	10.48	10.35	10.24	10.13	10.04
1.4	12.21	11.88	11.59	11.34	11.12	10.93	10.76	10.61	10.47	10.35
1.5	12.98	12.55	12.19	11.87	11.60	11.36	11.15	10.96	10.79	10.64
1.6	13.75	13.21	12.77	12.39	12.06	11.77	11.52	11.29	11.09	10.92
1.7	14.50	13.86	13.33	12.88	12.50	12.16	11.87	11.62	11.39	11.18
1.8	15.25	14.49	13.88	13.36	12.92	12.54	12.21	11.92	11.67	11.44
1.9	15.98	15.11	14.41	13.83	13.33	12.91	12.54	12.22	11.94	11.68
2.0	16.71	15.72	14.93	14.28	13.73	13.26	12.86	12.51	12.20	11.92
2.2	18.14	16.90	15.93	15.14	14.49	13.94	13.46	13.05	12.69	12.37
2.4	19.54	18.04	16.89	15.97	15.21	14.57	14.03	13.56	13.15	12.78
2.5	20.23	18.60	17.35	16.36	15.55	14.87	14.30	13.80	13.37	12.98
2.6	20.91	19.14	17.81	16.75	15.89	15.17	14.56	14.04	13.58	13.18
2.8	22.25	20.21	18.69	17.50	16.54	15.74	15.07	14.49	13.99	13.55
3.0	23.57	21.24	19.54	18.22	17.16	16.28	15.55	14.92	14.37	13.90
3.2	24.86	22.25	20.36	18.91	17.75	16.80	16.00	15.33	14.74	14.23
3.4	26.13	23.23	21.15	19.57	18.32	17.29	16.44	15.72	15.09	14.55
3.5	26.75	23.70	21.53	19.89	18.59	17.53	16.65	15.91	15.26	14.71
3.6	27.37	24.18	21.91	20.21	18.86	17.77	16.86	16.09	15.43	14.86
3.8	28.60	25.10	22.66	20.82	19.39	18.23	17.26	16.45	15.75	15.15
4.0	29.80	26.01	23.38	21.42	19.90	18.67	17.65	16.80	16.06	15.43
4.2	30.99	26.89	24.08	22.00	20.39	19.09	18.03	17.13	16.36	15.70
4.4	32.15	27.75	24.76	22.56	20.86	19.51	18.39	17.45	16.65	15.96
4.5	32.73	28.17	25.09	22.84	21.10	19.71	18.56	17.61	16.79	16.09
4.6	33.31	28.59	25.42	23.11	21.33	19.90	18.74	17.76	16.93	16.21
4.8	34.44	29.42	26.07	23.64	21.77	20.29	19.08	18.06	17.20	16.45
5.0	35.56	30.23	26.70	24.15	22.21	20.66	19.40	18.35	17.46	16.69
5.2	36.66	31.02	27.32	24.65	22.63	21.03	19.72	18.63	17.71	16.92
5.4	37.75	31.80	27.92	25.15	23.04	21.38	20.03	18.91	17.96	17.14
5.5	38.29	32.18	28.21	25.39	23.24	21.56	20.18	19.04	18.08	17.25
5.6	38.82	32.56	28.51	25.62	23.44	21.73	20.33	19.17	18.20	17.36
5.8	39.88	33.31	29.08	26.09	23.83	22.06	20.63	19.43	18.43	17.56
6.0	40.93	34.04	29.65	26.55	24.22	22.39	20.91	19.69	18.65	17.77
6.5	43.50	35.83	31.01	27.65	25.13	23.17	21.59	20.29	19.19	18.26
7.0	45.99	37.54	32.31	28.69	26.00	23.92	22.24	20.86	19.70	18.71
7.5	48.42	39.19	33.56	29.68	26.83	24.62	22.85	21.40	20.18	19.15
8.0	50.78	40.79	34.75	30.63	27.61	25.29	23.43	21.91	20.64	19.56
8.5	53.09	42.32	35.90	31.54	28.37	25.93	23.99	22.40	21.07	19.95
9.0	55.35	43.81	37.00	32.42	29.09	26.54	24.52	22.86	21.49	20.32
9.5	57.55	45.26	38.07	33.26	29.78	27.12	25.02	23.31	21.89	20.68
10.0	59.72	46.66	39.10	34.07	30.45	27.69	25.51	23.74	22.27	21.02
11.0	63.91	49.35	41.06	35.61	31.71	28.75	26.43	24.55	22.98	21.66
12.0	67.95	51.90	42.92	37.06	32.89	29.75	27.29	25.30	23.65	22.26
13.0	****	54.34	44.67	38.42	34.00	30.68	28.09	26.00	24.27	22.82
14.0	****	56.67	46.34	39.71	35.05	31.56	28.85	26.66	24.86	23.35
15.0	****	58.90	47.93	40.94	36.04	32.40	29.56	27.29	25.41	23.85
16.0	****	61.05	49.45	42.11	36.99	33.19	30.24	27.88	25.94	24.31
18.0	****	65.12	52.31	44.29	38.75	34.66	31.50	28.98	26.91	25.19
20.0	****	68.93	54.96	46.31	40.37	36.01	32.66	29.98	27.80	25.98

[†]For an explanation of the headings and the significance of asterisks and high returns which appear on some pages, see Introductory Section, pages xx–xxiiii.

INTERNAL RATE OF RETURN (IRR)†
(with NO projected Rental Growth)

RENTAL FACTOR†	YEARS TO REVIEW/REVERSION									
	1	2	3	4	5	6	7	8	9	10
1.1	10.90	10.81	10.74	10.67	10.60	10.55	10.50	10.45	10.41	10.37
1.2	11.79	11.61	11.44	11.30	11.18	11.07	10.97	10.88	10.79	10.72
1.3	12.66	12.38	12.13	11.91	11.72	11.56	11.41	11.28	11.16	11.05
1.4	13.52	13.13	12.79	12.50	12.25	12.02	11.83	11.66	11.50	11.36
1.5	14.37	13.86	13.43	13.06	12.74	12.47	12.23	12.02	11.83	11.66
1.6	15.21	14.57	14.05	13.60	13.22	12.90	12.61	12.36	12.14	11.94
1.7	16.03	15.27	14.65	14.13	13.69	13.31	12.98	12.69	12.44	12.21
1.8	16.85	15.95	15.23	14.63	14.13	13.70	13.33	13.01	12.72	12.47
1.9	17.65	16.62	15.80	15.12	14.56	14.08	13.67	13.31	13.00	12.72
2.0	18.44	17.27	16.35	15.60	14.98	14.45	14.00	13.60	13.26	12.96
2.2	20.00	18.54	17.41	16.51	15.77	15.15	14.62	14.16	13.76	13.41
2.4	21.52	19.76	18.43	17.38	16.52	15.80	15.20	14.68	14.23	13.83
2.5	22.27	20.36	18.92	17.79	16.88	16.12	15.48	14.93	14.45	14.03
2.6	23.01	20.94	19.40	18.20	17.23	16.42	15.75	15.17	14.67	14.23
2.8	24.46	22.08	20.33	18.98	17.90	17.01	16.27	15.63	15.08	14.61
3.0	25.89	23.18	21.23	19.73	18.54	17.57	16.76	16.07	15.48	14.96
3.2	27.28	24.25	22.09	20.45	19.16	18.11	17.23	16.49	15.85	15.30
3.4	28.65	25.29	22.92	21.14	19.75	18.62	17.68	16.89	16.21	15.62
3.5	29.33	25.80	23.33	21.48	20.03	18.86	17.90	17.08	16.38	15.78
3.6	30.00	26.30	23.73	21.81	20.31	19.11	18.11	17.27	16.55	15.93
3.8	31.32	27.28	24.51	22.45	20.86	19.58	18.52	17.64	16.88	16.23
4.0	32.62	28.24	25.26	23.08	21.38	20.03	18.92	17.99	17.19	16.51
4.2	33.90	29.18	26.00	23.68	21.89	20.47	19.30	18.33	17.50	16.78
4.4	35.16	30.09	26.71	24.26	22.38	20.89	19.67	18.65	17.79	17.05
4.5	35.78	30.54	27.06	24.55	22.62	21.10	19.85	18.81	17.93	17.17
4.6	36.39	30.98	27.41	24.83	22.86	21.30	20.03	18.97	18.07	17.30
4.8	37.61	31.86	28.08	25.38	23.32	21.70	20.38	19.28	18.35	17.55
5.0	38.82	32.71	28.74	25.91	23.77	22.08	20.71	19.57	18.61	17.78
5.2	40.00	33.55	29.39	26.44	24.21	22.46	21.04	19.86	18.87	18.01
5.4	41.17	34.37	30.02	26.94	24.63	22.82	21.35	20.14	19.11	18.24
5.5	41.75	34.77	30.33	27.19	24.84	23.00	21.51	20.28	19.24	18.35
5.6	42.32	35.17	30.63	27.44	25.05	23.17	21.66	20.41	19.36	18.46
5.8	43.46	35.96	31.24	27.92	25.45	23.52	21.96	20.67	19.59	18.67
6.0	44.58	36.74	31.83	28.40	25.84	23.85	22.25	20.93	19.82	18.87
6.5	47.33	38.62	33.25	29.54	26.79	24.66	22.95	21.55	20.37	19.37
7.0	50.00	40.43	34.60	30.61	27.68	25.42	23.61	22.13	20.88	19.83
7.5	52.60	42.16	35.90	31.64	28.53	26.14	24.23	22.67	21.37	20.27
8.0	55.12	43.84	37.14	32.63	29.34	26.82	24.82	23.19	21.83	20.68
8.5	57.59	45.45	38.33	33.57	30.11	27.48	25.39	23.69	22.28	21.08
9.0	60.00	47.01	39.48	34.47	30.85	28.10	25.93	24.16	22.70	21.45
9.5	62.35	48.53	40.59	35.34	31.56	28.70	26.45	24.62	23.10	21.81
10.0	64.66	50.00	41.66	36.17	32.25	29.28	26.94	25.05	23.48	22.16
11.0	69.13	52.82	43.70	37.76	33.54	30.37	27.88	25.87	24.21	22.81
12.0	****	55.49	45.62	39.25	34.76	31.39	28.75	26.63	24.89	23.42
13.0	****	58.04	47.44	40.66	35.89	32.34	29.57	27.35	25.52	23.98
14.0	****	60.48	49.17	41.99	36.97	33.24	30.34	28.02	26.11	24.51
15.0	****	62.81	50.81	43.25	37.99	34.09	31.07	28.65	26.67	25.01
16.0	****	65.06	52.39	44.45	38.95	34.90	31.76	29.25	27.20	25.49
18.0	****	69.31	55.35	46.70	40.76	36.40	33.04	30.37	28.19	26.37
20.0	****	****	58.09	48.78	42.42	37.78	34.22	31.39	29.09	27.17

†For an explanation of the headings and the significance of asterisks and high returns which appear on some pages, see Introductory Section, pages xx–xxiiii.

INTERNAL RATES OF RETURN[†]
(IRR)

In respect of Freehold Investments.
Reflecting Projected **Rental Growth**
to next rent review or reversion in

1 to 10 YEARS

with
INITIAL YIELDS

1% to 10%

(GROWTH RATES, 3%, 5% and 7%**)**

† For an explanation of the headings and the significance of asterisks and high returns which appear on some pages, see Introductory Section, pages xx to xxiii.

INTERNAL RATE OF RETURN (IRR) †
(Reflecting projected growth in Full Rental Value to next Review/Reversion)

RENTAL FACTOR †	YEARS TO REVIEW/REVERSION								
	1	2	3	4	5	6	7	8	10
1.1	1.13	1.16	1.19	1.23	1.26	1.29	1.32	1.35	1.42
1.2	1.23	1.27	1.30	1.33	1.37	1.40	1.43	1.46	1.53
1.3	1.33	1.37	1.40	1.44	1.47	1.50	1.54	1.57	1.64
1.4	1.44	1.47	1.51	1.54	1.58	1.61	1.64	1.68	1.74
1.5	1.54	1.57	1.61	1.64	1.68	1.71	1.75	1.78	1.85
1.6	1.64	1.67	1.71	1.75	1.78	1.82	1.85	1.88	1.95
1.7	1.74	1.78	1.81	1.85	1.88	1.92	1.95	1.99	2.05
1.8	1.84	1.88	1.91	1.95	1.98	2.02	2.05	2.09	2.15
1.9	1.94	1.98	2.01	2.05	2.08	2.12	2.15	2.18	2.24
2.0	2.04	2.08	2.11	2.15	2.18	2.22	2.25	2.28	2.34
2.2	2.24	2.28	2.31	2.35	2.38	2.41	2.44	2.47	2.52
2.4	2.44	2.47	2.51	2.54	2.57	2.60	2.63	2.65	2.70
2.5	2.54	2.57	2.60	2.63	2.66	2.69	2.72	2.74	2.79
2.6	2.63	2.67	2.70	2.73	2.76	2.78	2.81	2.83	2.88
2.8	2.83	2.86	2.89	2.92	2.94	2.97	2.99	3.01	3.05
3.0	3.03	3.06	3.08	3.10	3.12	3.14	3.16	3.18	3.21
3.2	3.22	3.25	3.27	3.29	3.30	3.32	3.33	3.35	3.37
3.4	3.42	3.44	3.45	3.47	3.48	3.49	3.50	3.51	3.52
3.5	3.52	3.53	3.54	3.56	3.57	3.58	3.58	3.59	3.60
3.6	3.61	3.63	3.64	3.64	3.65	3.66	3.66	3.67	3.68
3.8	3.81	3.81	3.82	3.82	3.82	3.82	3.82	3.82	3.82
4.0	4.00	4.00	4.00	3.99	3.99	3.99	3.98	3.98	3.97
4.2	4.19	4.18	4.17	4.17	4.16	4.15	4.14	4.13	4.11
4.4	4.38	4.37	4.35	4.34	4.32	4.30	4.29	4.27	4.24
4.5	4.48	4.46	4.44	4.42	4.40	4.38	4.36	4.34	4.31
4.6	4.57	4.55	4.53	4.50	4.48	4.46	4.44	4.42	4.38
4.8	4.76	4.73	4.70	4.67	4.64	4.61	4.58	4.56	4.51
5.0	4.95	4.91	4.87	4.83	4.79	4.76	4.73	4.70	4.64
5.2	5.14	5.09	5.04	4.99	4.95	4.91	4.87	4.83	4.76
5.4	5.33	5.27	5.21	5.15	5.10	5.05	5.01	4.96	4.88
5.5	5.42	5.36	5.29	5.23	5.18	5.12	5.08	5.03	4.94
5.6	5.52	5.44	5.38	5.31	5.25	5.20	5.14	5.09	5.00
5.8	5.71	5.62	5.54	5.47	5.40	5.34	5.28	5.22	5.12
6.0	5.89	5.79	5.70	5.62	5.55	5.48	5.41	5.35	5.24
6.5	6.35	6.23	6.11	6.00	5.91	5.82	5.73	5.66	5.52
7.0	6.81	6.65	6.50	6.37	6.25	6.14	6.05	5.95	5.79
7.5	7.27	7.07	6.89	6.73	6.59	6.46	6.35	6.24	6.05
8.0	7.72	7.48	7.27	7.09	6.92	6.77	6.64	6.51	6.30
8.5	8.17	7.89	7.64	7.43	7.24	7.07	6.92	6.78	6.53
9.0	8.61	8.29	8.01	7.77	7.55	7.36	7.19	7.04	6.77
9.5	9.06	8.69	8.37	8.10	7.86	7.65	7.46	7.29	6.99
10.0	9.49	9.08	8.72	8.42	8.16	7.92	7.72	7.53	7.20
11.0	10.36	9.84	9.41	9.05	8.73	8.46	8.21	7.99	7.62
12.0	11.21	10.59	10.08	9.65	9.28	8.96	8.68	8.43	8.00
13.0	12.06	11.32	10.73	10.23	9.81	9.45	9.13	8.85	8.37
14.0	12.89	12.04	11.36	10.79	10.32	9.91	9.56	9.25	8.72
15.0	13.71	12.73	11.97	11.34	10.81	10.36	9.97	9.63	9.05
16.0	14.52	13.42	12.56	11.86	11.28	10.79	10.37	9.99	9.37
18.0	16.11	14.74	13.70	12.87	12.18	11.60	11.11	10.68	9.97
20.0	17.66	16.02	14.79	13.82	13.03	12.37	11.81	11.32	10.52

† For an explanation of the headings and the significance of asterisks and high returns which appear on some pages, see Introductory Section, pages xx-xxiii.

INTERNAL RATE OF RETURN (IRR) †
(Reflecting projected growth in Full Rental Value to next Review/Reversion)

RENTAL FACTOR †	YEARS TO REVIEW/REVERSION								
	1	2	3	4	5	6	7	8	10
1.1	1.70	1.74	1.79	1.83	1.88	1.92	1.96	2.00	2.08
1.2	1.85	1.89	1.94	1.99	2.03	2.07	2.12	2.16	2.24
1.3	2.00	2.05	2.09	2.14	2.18	2.23	2.27	2.31	2.39
1.4	2.15	2.20	2.24	2.29	2.33	2.38	2.42	2.46	2.53
1.5	2.30	2.35	2.39	2.44	2.48	2.52	2.56	2.60	2.67
1.6	2.45	2.50	2.54	2.58	2.63	2.67	2.70	2.74	2.81
1.7	2.60	2.64	2.69	2.73	2.77	2.81	2.84	2.88	2.94
1.8	2.75	2.79	2.83	2.87	2.91	2.95	2.98	3.01	3.07
1.9	2.90	2.94	2.98	3.02	3.05	3.09	3.12	3.15	3.20
2.0	3.04	3.08	3.12	3.16	3.19	3.22	3.25	3.28	3.33
2.2	3.34	3.37	3.40	3.43	3.46	3.49	3.51	3.53	3.57
2.4	3.63	3.66	3.68	3.71	3.73	3.74	3.76	3.78	3.80
2.5	3.78	3.80	3.82	3.84	3.86	3.87	3.88	3.89	3.91
2.6	3.92	3.94	3.96	3.97	3.98	4.00	4.00	4.01	4.02
2.8	4.21	4.22	4.23	4.23	4.24	4.24	4.24	4.24	4.24
3.0	4.50	4.50	4.49	4.49	4.48	4.48	4.47	4.46	4.44
3.2	4.79	4.77	4.76	4.74	4.73	4.71	4.69	4.68	4.64
3.4	5.07	5.04	5.02	4.99	4.96	4.94	4.91	4.89	4.84
3.5	5.21	5.18	5.14	5.11	5.08	5.05	5.02	4.99	4.93
3.6	5.36	5.31	5.27	5.23	5.20	5.16	5.12	5.09	5.03
3.8	5.64	5.58	5.52	5.47	5.42	5.38	5.33	5.29	5.21
4.0	5.92	5.84	5.77	5.71	5.65	5.59	5.53	5.48	5.38
4.2	6.20	6.10	6.02	5.94	5.86	5.80	5.73	5.67	5.56
4.4	6.48	6.36	6.26	6.17	6.08	6.00	5.92	5.85	5.72
4.5	6.61	6.49	6.38	6.28	6.19	6.10	6.02	5.94	5.81
4.6	6.75	6.62	6.50	6.39	6.29	6.20	6.11	6.03	5.89
4.8	7.03	6.87	6.74	6.61	6.50	6.39	6.30	6.21	6.05
5.0	7.30	7.13	6.97	6.83	6.70	6.59	6.48	6.38	6.20
5.2	7.57	7.38	7.20	7.04	6.90	6.77	6.66	6.55	6.35
5.4	7.85	7.62	7.43	7.26	7.10	6.96	6.83	6.71	6.50
5.5	7.98	7.75	7.54	7.36	7.20	7.05	6.91	6.79	6.57
5.6	8.12	7.87	7.65	7.46	7.29	7.14	7.00	6.87	6.64
5.8	8.38	8.11	7.88	7.67	7.48	7.32	7.17	7.03	6.79
6.0	8.65	8.35	8.10	7.87	7.67	7.49	7.33	7.18	6.92
6.5	9.31	8.95	8.64	8.37	8.13	7.92	7.73	7.56	7.26
7.0	9.97	9.53	9.17	8.85	8.57	8.33	8.11	7.92	7.58
7.5	10.62	10.11	9.68	9.32	9.00	8.72	8.48	8.26	7.88
8.0	11.26	10.67	10.18	9.77	9.41	9.11	8.83	8.59	8.17
8.5	11.90	11.22	10.67	10.21	9.82	9.47	9.17	8.90	8.45
9.0	12.52	11.76	11.15	10.64	10.20	9.83	9.50	9.21	8.72
9.5	13.15	12.30	11.62	11.06	10.58	10.17	9.82	9.51	8.97
10.0	13.76	12.82	12.08	11.46	10.95	10.51	10.13	9.79	9.22
11.0	14.98	13.85	12.97	12.25	11.66	11.15	10.72	10.33	9.70
12.0	16.17	14.84	13.82	13.00	12.33	11.76	11.27	10.85	10.14
13.0	17.34	15.81	14.65	13.72	12.97	12.34	11.80	11.33	10.56
14.0	18.49	16.75	15.44	14.42	13.58	12.89	12.30	11.79	10.95
15.0	19.62	17.66	16.21	15.08	14.17	13.42	12.78	12.23	11.33
16.0	20.73	18.55	16.96	15.72	14.74	13.92	13.24	12.65	11.68
18.0	22.91	20.27	18.38	16.95	15.81	14.88	14.10	13.43	12.35
20.0	25.02	21.91	19.73	18.09	16.81	15.76	14.89	14.16	12.97

† For an explanation of the headings and the significance of asterisks and high returns which appear on some pages, see Introductory Section, pages xx-xxiii.

INTERNAL RATE OF RETURN (IRR) †

(Reflecting projected growth in Full Rental Value to next Review/Reversion)

RENTAL FACTOR †	YEARS TO REVIEW/REVERSION								
	1	2	3	4	5	6	7	8	10
1.1	2.26	2.32	2.38	2.43	2.49	2.54	2.59	2.64	2.73
1.2	2.46	2.52	2.58	2.63	2.69	2.74	2.79	2.83	2.92
1.3	2.66	2.72	2.77	2.83	2.88	2.93	2.98	3.02	3.10
1.4	2.86	2.92	2.97	3.02	3.07	3.12	3.16	3.20	3.28
1.5	3.06	3 11	3.16	3.21	3.26	3.30	3.34	3.38	3.45
1.6	3.26	3.31	3.36	3.40	3.44	3.48	3.52	3.55	3.61
1.7	3.45	3 50	3.55	3.59	3.62	3.66	3.69	3.72	3.77
1.8	3.65	3.69	3.73	3.77	3.80	3.83	3.86	3.89	3.93
1.9	3.84	3.88	3.92	3.95	3.98	4.00	4.03	4.05	4.08
2.0	4.04	4.07	4.10	4.13	4.15	4.17	4.19	4.21	4.23
2.2	4.42	4.45	4.46	4.48	4.49	4.50	4.51	.4.51	4.52
2.4	4.81	4.81	4.82	4.82	4.82	4.81	4.81	4.80	4.79
2.5	5.00	5.00	4.99	4.99	4.98	4.97	4.96	4.95	4.92
2.6	5.19	5.18	5.17	5.15	5.14	5.12	5.10	5.08	5.05
2.8	5.57	5.54	5.51	5.48	5.45	5.42	5.39	5.36	5.30
3.0	5.95	5.89	5.84	5.79	5.75	5.70	5.66	5.62	5.54
3.2	6.32	6.24	6.17	6.11	6.04	5.98	5.92	5.87	5.77
3.4	6.69	6.59	6.50	6.41	6.33	6.25	6.18	6.11	5.99
3.5	6.87	6.76	6.66	6.56	6.47	6.39	6.31	6.23	6.10
3.6	7.06	6.93	6.81	6.71	6.61	6.52	6.43	6.35	6.20
3.8	7.43	7.27	7.13	7.00	6.88	6.77	6.67	6.58	6.41
4.0	7.79	7.60	7.44	7.29	7.15	7.03	6.91	6.80	6.61
4.2	8.15	7.93	7.74	7.57	7.41	7.27	7.14	7.02	6.81
4.4	8.51	8.26	8.04	7.84	7.67	7.51	7.37	7.23	7.00
4.5	8.69	8.42	8.19	7.98	7.79	7.63	7.48	7.34	7.09
4.6	8.87	8.58	8.33	8.11	7.92	7.74	7.58	7.44	7.18
4.8	9.22	8.90	8.62	8.38	8.17	7.97	7.80	7.64	7.36
5.0	9.57	9.22	8.91	8.64	8.41	8.20	8.01	7.83	7.53
5.2	9.93	9.53	9.19	8.90	8.64	8.42	8.21	8.03	7.70
5.4	10.27	9.84	9.47	9.15	8.88	8.63	8.41	8.21	7.87
5.5	10.45	9.99	9.61	9.28	8.99	8.74	8.51	8.30	7.95
5.6	10.62	10.15	9.75	9.40	9.10	8.84	8.61	8.40	8.03
5.8	10.96	10.45	10.02	9.65	9.33	9.05	8.80	8.57	8.19
6.0	11.31	10.75	10.28	9.89	9.55	9.25	8.99	8.75	8.34
6.5	12.16	11.49	10.94	10.48	10.08	9.74	9.44	9.17	8.71
7.0	12.99	12.21	11.57	11.05	10.60	10.21	9.87	9.57	9.06
7.5	13.82	12.91	12.19	11.60	11.09	10.66	10.29	9.96	9.40
8.0	14.63	13.60	12.79	12.13	11.57	11.10	10.69	10.32	9.72
8.5	15.44	14.28	13.37	12.64	12.03	11.52	11.07	10.68	10.02
9.0	16.23	14.94	13.94	13.14	12.48	11.92	11.44	11.02	10.31
9.5	17.02	15.59	14.50	13.63	12.91	12.31	11.79	11.34	10.60
10.0	17.79	16.23	15.04	14.10	13.33	12.69	12.14	11.66	10.87
11.0	19.32	17.47	16.09	15.01	14.14	13.41	12.79	12.26	11.38
12.0	20.81	18.66	17.09	15.87	14.90	14.09	13.41	12.82	11.86
13.0	22.27	19.82	18.05	16.70	15.62	14.73	13.99	13.35	12.31
14.0	23.70	20.94	18.98	17.49	16.31	15.35	14.54	13.85	12.74
15.0	25.10	22.03	19.87	18.25	16.97	15.93	15.07	14.33	13.14
16.0	26.48	23.09	20.73	18.98	17.60	16.49	15.57	14.79	13.53
18.0	29.16	25.12	22.37	20.36	18.79	17.54	16.51	15.64	14.25
20.0	31.75	27.05	23.92	21.65	19.90	18.51	17.38	16.42	14.90

† For an explanation of the headings and the significance of asterisks and high returns which appear on some pages, see Introductory Section, pages xx-xxiii.

INTERNAL RATE OF RETURN (IRR) †
(Reflecting projected growth in Full Rental Value to next Review/Reversion)

RENTAL FACTOR †	YEARS TO REVIEW/REVERSION								
	1	2	3	4	5	6	7	8	10
1.1	2.82	2.89	2.96	3.03	3.09	3.15	3.21	3.26	3.36
1.2	3.07	3.14	3.21	3.27	3.33	3.39	3.44	3.49	3.58
1.3	3.32	3.39	3.45	3.51	3.56	3.62	3.66	3.71	3.79
1.4	3.57	3.63	3.69	3.74	3.79	3.84	3.88	3.92	3.99
1.5	3.81	3.87	3.92	3.97	4.02	4.06	4.09	4.13	4.19
1.6	4.06	4.11	4.16	4.20	4.24	4.27	4.30	4.33	4.37
1.7	4.30	4.35	4.39	4.42	4.45	4.48	4.50	4.52	4.56
1.8	4.54	4.58	4.61	4.64	4.66	4.68	4.70	4.71	4.73
1.9	4.78	4.81	4.84	4.85	4.87	4.88	4.89	4.90	4.91
2.0	5.02	5.04	5.06	5.07	5.07	5.08	5.08	5.08	5.07
2.2	5.50	5.50	5.49	5.48	5.47	5.46	5.44	5.43	5.39
2.4	5.97	5.94	5.91	5.88	5.85	5.82	5.79	5.76	5.70
2.5	6.21	6.16	6.12	6.08	6.04	6.00	5.96	5.92	5.84
2.6	6.44	6.38	6.33	6.28	6.22	6.17	6.12	6.08	5.99
2.8	6.91	6.82	6.73	6.66	6.58	6.51	6.45	6.38	6.26
3.0	7.37	7.24	7.13	7.03	6.93	6.84	6.76	6.67	6.53
3.2	7.82	7.67	7.52	7.39	7.27	7.16	7.05	6.96	6.78
3.4	8.28	8.08	7.90	7.74	7.60	7.47	7.34	7.23	7.03
3.5	8.50	8.28	8.09	7.92	7.76	7.62	7.48	7.36	7.14
3.6	8.73	8.49	8.28	8.09	7.92	7.77	7.62	7.49	7.26
3.8	9.17	8.89	8.65	8.43	8.23	8.06	7.90	7.75	7.49
4.0	9.62	9.29	9.01	8.76	8.54	8.34	8.16	8.00	7.71
4.2	10.06	9.68	9.36	9.08	8.83	8.61	8.42	8.23	7.92
4.4	10.49	10.07	9.71	9.40	9.13	8.88	8.66	8.47	8.12
4.5	10.71	10.26	9.88	9.55	9.27	9.01	8.79	8.58	8.22
4.6	10.92	10.45	10.05	9.71	9.41	9.14	8.91	8.69	8.32
4.8	11.35	10.83	10.39	10.01	9.69	9.40	9.14	8.91	8.52
5.0	11.78	11.20	10.72	10.31	9.96	9.65	9.37	9.13	8.71
5.2	12.21	11.57	11.05	10.61	10.23	9.89	9.60	9.34	8.89
5.4	12.63	11.94	11.37	10.89	10.49	10.13	9.82	9.54	9.07
5.5	12.84	12.12	11.53	11.04	10.62	10.25	9.93	9.64	9.15
5.6	13.04	12.30	11.69	11.18	10.74	10.37	10.04	9.74	9.24
5.8	13.46	12.65	12.00	11.46	10.99	10.60	10.25	9.94	9.41
6.0	13.87	13.00	12.31	11.73	11.24	10.82	10.45	10.13	9.58
6.5	14.89	13.87	13.06	12.39	11.84	11.36	10.95	10.59	9.97
7.0	15.90	14.71	13.78	13.03	12.41	11.88	11.42	11.02	10.35
7.5	16.88	15.53	14.49	13.65	12.96	12.38	11.87	11.44	10.71
8.0	17.86	16.33	15.17	14.25	13.49	12.85	12.31	11.83	11.05
8.5	18.82	17.11	15.83	14.82	14.00	13.31	12.72	12.21	11.37
9.0	19.76	17.88	16.48	15.38	14.49	13.75	13.12	12.58	11.69
9.5	20.70	18.63	17.10	15.92	14.96	14.17	13.50	12.93	11.98
10.0	21.62	19.36	17.72	16.44	15.43	14.58	13.88	13.27	12.27
11.0	23.42	20.78	18.89	17.45	16.31	15.37	14.58	13.91	12.82
12.0	25.19	22.15	20.02	18.41	17.14	16.10	15.24	14.51	13.33
13.0	26.91	23.48	21.09	19.32	17.92	16.80	15.87	15.08	13.80
14.0	28.59	24.75	22.12	20.18	18.67	17.46	16.46	15.61	14.25
15.0	30.24	25.99	23.12	21.01	19.39	18.09	17.02	16.12	14.68
16.0	31.85	27.19	24.07	21.81	20.08	18.69	17.55	16.60	15.08
18.0	34.99	29.48	25.89	23.32	21.36	19.82	18.55	17.50	15.83
20.0	38.01	31.66	27.60	24.72	22.56	20.86	19.48	18.33	16.52

† For an explanation of the headings and the significance of asterisks and high returns which appear on some pages, see Introductory Section, pages xx-xxiii.

INTERNAL RATE OF RETURN (IRR) †
(Reflecting projected growth in Full Rental Value to next Review/Reversion)

RENTAL FACTOR †	YEARS TO REVIEW/REVERSION								
	1	2	3	4	5	6	7	8	10
1.1	3.39	3.47	3.55	3.62	3.69	3.75	3.81	3.87	3.97
1.2	3.68	3.76	3.83	3.90	3.97	4.02	4.08	4.13	4.22
1.3	3.98	4.05	4.12	4.18	4.24	4.29	4.33	4.38	4.45
1.4	4.27	4.34	4.40	4.45	4.50	4.54	4.58	4.62	4.67
1.5	4.56	4.62	4.67	4.72	4.76	4.79	4.82	4.85	4.89
1.6	4.85	4.90	4.94	4.98	5.01	5.03	5.06	5.07	5.10
1.7	5.14	5.18	5.21	5.23	5.25	5.27	5.28	5.29	5.30
1.8	5.43	5.45	5.47	5.49	5.49	5.50	5.50	5.50	5.49
1.9	5.72	5.73	5.73	5.73	5.73	5.73	5.72	5.71	5.68
2.0	6.00	6.00	5.99	5.98	5.96	5.94	5.93	5.91	5.86
2.2	6.56	6.53	6.49	6.45	6.41	6.37	6.33	6.29	6.21
2.4	7.12	7.05	6.98	6.91	6.84	6.78	6.71	6.66	6.54
2.5	7.40	7.30	7.22	7.13	7.05	6.97	6.90	6.83	6.70
2.6	7.68	7.56	7.45	7.35	7.26	7.17	7.08	7.00	6.86
2.8	8.22	8.06	7.92	7.78	7.66	7.54	7.44	7.34	7.15
3.0	8.76	8.56	8.37	8.20	8.05	7.91	7.78	7.66	7.44
3.2	9.30	9.04	8.81	8.61	8.42	8.26	8.10	7.96	7.71
3.4	9.83	9.52	9.25	9.01	8.79	8.60	8.42	8.26	7.97
3.5	10.10	9.76	9.46	9.20	8.97	8.76	8.57	8.40	8.10
3.6	10.36	9.99	9.67	9.39	9.15	8.92	8.73	8.54	8.22
3.8	10.88	10.45	10.09	9.77	9.49	9.24	9.02	8.82	8.47
4.0	11.40	10.91	10.50	10.14	9.83	9.55	9.31	9.09	8.70
4.2	11.92	11.36	10.90	10.50	10.16	9.85	9.58	9.34	8.93
4.4	12.42	11.80	11.29	10.85	10.48	10.15	9.85	9.59	9.14
4.5	12.68	12.02	11.48	11.03	10.63	10.29	9.99	9.72	9.25
4.6	12.93	12.24	11.67	11.20	10.79	10.43	10.12	9.84	9.36
4.8	13.43	12.67	12.05	11.53	11.09	10.71	10.37	10.07	9.56
5.0	13.93	13.10	12.42	11.87	11.39	10.98	10.62	10.30	9.76
5.2	14.42	13.52	12.79	12.19	11.68	11.25	10.86	10.53	9.95
5.4	14.91	13.93	13.15	12.51	11.97	11.50	11.10	10.74	10.14
5.5	15.15	14.13	13.33	12.66	12.11	11.63	11.22	10.85	10.24
5.6	15.40	14.34	13.50	12.82	12.25	11.76	11.33	10.96	10.33
5.8	15.88	14.74	13.85	13.13	12.52	12.00	11.56	11.17	10.51
6.0	16.36	15.14	14.19	13.43	12.79	12.25	11.78	11.37	10.68
6.5	17.54	16.12	15.03	14.16	13.44	12.83	12.31	11.86	11.10
7.0	18.70	17.07	15.83	14.86	14.06	13.39	12.81	12.32	11.50
7.5	19.84	17.99	16.61	15.53	14.65	13.92	13.30	12.76	11.87
8.0	20.96	18.89	17.37	16.18	15.22	14.43	13.76	13.18	12.23
8.5	22.06	19.77	18.10	16.81	15.77	14.92	14.20	13.58	12.57
9.0	23.15	20.62	18.81	17.41	16.30	15.39	14.62	13.97	12.90
9.5	24.22	21.46	19.49	18.00	16.81	15.84	15.03	14.34	13.21
10.0	25.27	22.28	20.16	18.57	17.31	16.28	15.42	14.69	13.51
11.0	27.34	23.86	21.45	19.66	18.25	17.11	16.17	15.37	14.08
12.0	29.35	25.39	22.68	20.69	19.14	17.89	16.86	16.00	14.61
13.0	31.31	26.85	23.85	21.66	19.98	18.63	17.52	16.59	15.10
14.0	33.22	28.26	24.98	22.60	20.78	19.33	18.14	17.15	15.57
15.0	35.09	29.63	26.05	23.49	21.54	19.99	18.73	17.68	16.01
16.0	36.92	30.95	27.09	24.34	22.27	20.63	19.30	18.19	16.43
18.0	40.46	33.47	29.06	25.96	23.63	21.82	20.35	19.13	17.21
20.0	43.87	35.86	30.90	27.45	24.90	22.91	21.31	19.99	17.93

† For an explanation of the headings and the significance of asterisks and high returns which appear on some pages, see Introductory Section, pages xx-xxiii.

INTERNAL RATE OF RETURN (IRR) †

(Reflecting projected growth in Full Rental Value to next Review/Reversion)

RENTAL FACTOR †	YEARS TO REVIEW/REVERSION								
	1	2	3	4	5	6	7	8	10
1.1	3.95	4.04	4.13	4.21	4.28	4.35	4.41	4.47	4.57
1.2	4.29	4.38	4.46	4.53	4.59	4.65	4.71	4.76	4.84
1.3	4.63	4.71	4.78	4.84	4.90	4.95	4.99	5.03	5.09
1.4	4.97	5.04	5.10	5.15	5.19	5.23	5.26	5.29	5.33
1.5	5.31	5.36	5.41	5.45	5.48	5.51	5.53	5.55	5.57
1.6	5.65	5.69	5.72	5.74	5.76	5.78	5.78	5.79	5.79
1.7	5.98	6.00	6.02	6.03	6.03	6.04	6.03	6.03	6.01
1.8	6.31	6.32	6.32	6.31	6.30	6.29	6.27	6.26	6.22
1.9	6.64	6.63	6.61	6.59	6.56	6.54	6.51	6.48	6.42
2.0	6.97	6.93	6.90	6.86	6.82	6.78	6.74	6.70	6.61
2.2	7.62	7.54	7.46	7.39	7.31	7.24	7.18	7.11	6.99
2.4	8.26	8.13	8.01	7.89	7.79	7.69	7.59	7.50	7.34
2.5	8.58	8.42	8.28	8.14	8.02	7.90	7.79	7.69	7.51
2.6	8.89	8.71	8.54	8.39	8.24	8.11	7.99	7.88	7.67
2.8	9.52	9.28	9.06	8.86	8.68	8.52	8.37	8.23	7.99
3.0	10.14	9.83	9.56	9.32	9.11	8.91	8.74	8.57	8.29
3.2	10.76	10.38	10.06	9.77	9.52	9.29	9.09	8.90	8.57
3.4	11.36	10.92	10.54	10.21	9.92	9.66	9.43	9.22	8.85
3.5	11.67	11.18	10.77	10.42	10.11	9.84	9.59	9.37	8.98
3.6	11.97	11.44	11.01	10.63	10.30	10.01	9.75	9.52	9.11
3.8	12.56	11.96	11.47	11.04	10.68	10.35	10.07	9.81	9.37
4.0	13.15	12.47	11.92	11.45	11.04	10.69	10.37	10.09	9.62
4.2	13.74	12.98	12.36	11.84	11.39	11.01	10.67	10.37	9.85
4.4	14.31	13.47	12.79	12.22	11.74	11.32	10.96	10.63	10.08
4.5	14.60	13.72	13.00	12.41	11.91	11.48	11.10	10.76	10.19
4.6	14.89	13.96	13.21	12.60	12.08	11.63	11.24	10.89	10.30
4.8	15.46	14.44	13.63	12.96	12.40	11.92	11.51	11.14	10.52
5.0	16.02	14.91	14.04	13.32	12.72	12.21	11.77	11.38	10.73
5.2	16.58	15.38	14.44	13.67	13.04	12.50	12.03	11.62	10.93
5.4	17.13	15.84	14.83	14.02	13.34	12.77	12.28	11.85	11.13
5.5	17.41	16.06	15.02	14.19	13.49	12.91	12.40	11.96	11.22
5.6	17.68	16.29	15.22	14.35	13.64	13.04	12.52	12.07	11.32
5.8	18.23	16.74	15.60	14.68	13.93	13.30	12.76	12.29	11.50
6.0	18.77	17.18	15.97	15.01	14.22	13.56	12.99	12.50	11.69
6.5	20.10	18.26	16.88	15.80	14.91	14.18	13.55	13.01	12.12
7.0	21.40	19.30	17.75	16.55	15.58	14.77	14.09	13.50	12.53
7.5	22.69	20.32	18.60	17.27	16.21	15.33	14.59	13.96	12.92
8.0	23.94	21.31	19.41	17.97	16.81	15.87	15.08	14.40	13.30
8.5	25.18	22.27	20.20	18.64	17.40	16.38	15.54	14.82	13.65
9.0	26.40	23.21	20.97	19.28	17.96	16.88	15.98	15.22	13.99
9.5	27.60	24.12	21.71	19.91	18.50	17.36	16.41	15.61	14.31
10.0	28.78	25.02	22.43	20.52	19.02	17.82	16.82	15.98	14.62
11.0	31.08	26.75	23.82	21.67	20.02	18.69	17.60	16.68	15.21
12.0	33.32	28.40	25.14	22.77	20.95	19.51	18.33	17.33	15.76
13.0	35.50	29.99	26.39	23.81	21.84	20.28	19.01	17.95	16.27
14.0	37.63	31.53	27.59	24.80	22.68	21.01	19.66	18.53	16.75
15.0	39.70	33.01	28.74	25.74	23.48	21.71	20.27	19.08	17.21
16.0	41.73	34.44	29.85	26.64	24.24	22.37	20.86	19.60	17.64
18.0	45.65	37.16	31.95	28.34	25.68	23.61	21.95	20.58	18.44
20.0	49.41	39.74	33.90	29.92	27.00	24.75	22.95	21.47	19.17

† For an explanation of the headings and the significance of asterisks and high returns which appear on some pages, see Introductory Section, pages xx-xxiii.

INTERNAL RATE OF RETURN (IRR) †
(Reflecting projected growth in Full Rental Value to next Review/Reversion)

RENTAL FACTOR †	YEARS TO REVIEW/REVERSION								
	1	2	3	4	5	6	7	8	10
1.1	4.51	4.61	4.70	4.79	4.87	4.94	5.00	5.06	5.16
1.2	4.90	4.99	5.07	5.15	5.21	5.27	5.32	5.37	5.44
1.3	5.29	5.37	5.44	5.50	5.55	5.59	5.63	5.66	5.71
1.4	5.67	5.74	5.79	5.84	5.87	5.90	5.93	5.95	5.97
1.5	6.06	6.10	6.14	6.17	6.19	6.21	6.22	6.22	6.22
1.6	6.44	6.46	6.48	6.49	6.50	6.50	6.49	6.48	6.46
1.7	6.81	6.82	6.81	6.81	6.80	6.78	6.76	6.74	6.69
1.8	7.19	7.17	7.14	7.12	7.09	7.05	7.02	6.98	6.91
1.9	7.56	7.51	7.47	7.42	7.37	7.32	7.27	7.22	7.12
2.0	7.93	7.86	7.79	7.72	7.65	7.58	7.52	7.45	7.33
2.2	8.66	8.53	8.41	8.29	8.18	8.08	7.98	7.89	7.72
2.4	9.38	9.19	9.01	8.85	8.70	8.56	8.43	8.31	8.09
2.5	9.74	9.51	9.30	9.12	8.95	8.79	8.64	8.51	8.27
2.6	10.10	9.83	9.59	9.38	9.19	9.02	8.85	8.71	8.44
2.8	10.80	10.46	10.16	9.90	9.66	9.45	9.26	9.08	8.77
3.0	11.50	11.08	10.71	10.40	10.12	9.87	9.65	9.44	9.08
3.2	12.19	11.68	11.25	10.88	10.56	10.27	10.02	9.79	9.38
3.4	12.87	12.27	11.78	11.35	10.99	10.66	10.38	10.12	9.67
3.5	13.20	12.56	12.03	11.58	11.19	10.85	10.55	10.28	9.81
3.6	13.54	12.85	12.29	11.81	11.40	11.04	10.72	10.44	9.95
3.8	14.21	13.43	12.79	12.25	11.80	11.40	11.05	10.74	10.21
4.0	14.86	13.99	13.28	12.69	12.19	11.75	11.38	11.04	10.47
4.2	15.52	14.54	13.75	13.11	12.56	12.10	11.69	11.33	10.71
4.4	16.16	15.08	14.22	13.52	12.93	12.43	11.99	11.60	10.95
4.5	16.48	15.35	14.45	13.72	13.11	12.59	12.14	11.74	11.07
4.6	16.80	15.61	14.68	13.92	13.29	12.75	12.28	11.87	11.18
4.8	17.43	16.14	15.13	14.31	13.64	13.06	12.57	12.13	11.40
5.0	18.06	16.65	15.57	14.70	13.98	13.37	12.84	12.38	11.62
5.2	18.68	17.16	16.00	15.07	14.31	13.66	13.11	12.63	11.83
5.4	19.30	17.66	16.42	15.44	14.63	13.95	13.37	12.87	12.03
5.5	19.60	17.91	16.63	15.62	14.79	14.09	13.50	12.99	12.13
5.6	19.91	18.16	16.84	15.80	14.95	14.24	13.63	13.10	12.23
5.8	20.51	18.64	17.25	16.15	15.26	14.51	13.88	13.33	12.43
6.0	21.11	19.12	17.65	16.49	15.56	14.78	14.12	13.55	12.61
6.5	22.58	20.30	18.62	17.33	16.29	15.43	14.71	14.08	13.06
7.0	24.03	21.43	19.56	18.13	16.99	16.05	15.26	14.59	13.49
7.5	25.44	22.53	20.46	18.90	17.65	16.64	15.79	15.07	13.89
8.0	26.83	23.60	21.34	19.63	18.29	17.20	16.29	15.52	14.27
8.5	28.20	24.64	22.18	20.34	18.90	17.74	16.77	15.95	14.64
9.0	29.54	25.66	22.99	21.02	19.49	18.26	17.23	16.37	14.98
9.5	30.85	26.64	23.78	21.68	20.06	18.75	17.68	16.77	15.32
10.0	32.15	27.60	24.55	22.32	20.61	19.23	18.10	17.15	15.64
11.0	34.68	29.46	26.02	23.54	21.65	20.14	18.91	17.88	16.24
12.0	37.14	31.24	27.42	24.69	22.63	20.99	19.66	18.56	16.80
13.0	39.52	32.95	28.75	25.78	23.55	21.80	20.37	19.19	17.33
14.0	41.84	34.59	30.02	26.82	24.43	22.55	21.04	19.79	17.82
15.0	44.11	36.17	31.24	27.81	25.26	23.27	21.68	20.36	18.29
16.0	46.32	37.70	32.41	28.75	26.05	23.96	22.28	20.89	18.73
18.0	50.59	40.61	34.61	30.53	27.54	25.24	23.40	21.89	19.55
20.0	54.68	43.35	36.67	32.18	28.92	26.42	24.43	22.81	20.30

† For an explanation of the headings and the significance of asterisks and high returns which appear on some pages, see Introductory Section, pages xx-xxiii.

INTERNAL RATE OF RETURN (IRR) †

(Reflecting projected growth in Full Rental Value to next Review/Reversion)

RENTAL FACTOR †	YEARS TO REVIEW/REVERSION								
	1	2	3	4	5	6	7	8	10
1.1	5.07	5.18	5.28	5.37	5.45	5.52	5.59	5.64	5.73
1.2	5.51	5.60	5.69	5.76	5.83	5.88	5.93	5.97	6.03
1.3	5.94	6.02	6.09	6.14	6.19	6.23	6.26	6.29	6.32
1.4	6.37	6.43	6.48	6.51	6.54	6.56	6.58	6.59	6.59
1.5	6.80	6.83	6.86	6.87	6.88	6.89	6.89	6.88	6.86
1.6	7.22	7.23	7.23	7.23	7.22	7.20	7.18	7.16	7.11
1.7	7.64	7.62	7.60	7.57	7.54	7.50	7.47	7.43	7.35
1.8	8.06	8.01	7.96	7.91	7.85	7.80	7.74	7.69	7.58
1.9	8.47	8.39	8.31	8.23	8.16	8.08	8.01	7.94	7.80
2.0	8.88	8.77	8.66	8.55	8.45	8.36	8.27	8.18	8.01
2.2	9.69	9.51	9.33	9.18	9.03	8.89	8.76	8.64	8.42
2.4	10.49	10.23	9.99	9.77	9.58	9.40	9.23	9.08	8.81
2.5	10.89	10.58	10.31	10.06	9.84	9.64	9.46	9.29	8.99
2.6	11.29	10.93	10.62	10.35	10.10	9.88	9.68	9.50	9.17
2.8	12.07	11.62	11.23	10.90	10.61	10.34	10.11	9.89	9.51
3.0	12.84	12.29	11.83	11.44	11.09	10.79	10.51	10.27	9.84
3.2	13.60	12.95	12.41	11.95	11.56	11.21	10.90	10.63	10.15
3.4	14.35	13.59	12.97	12.45	12.01	11.62	11.28	10.97	10.45
3.5	14.72	13.91	13.25	12.70	12.23	11.82	11.46	11.14	10.59
3.6	15.09	14.22	13.52	12.94	12.44	12.01	11.64	11.30	10.73
3.8	15.82	14.84	14.06	13.41	12.87	12.40	11.98	11.62	11.01
4.0	16.55	15.45	14.58	13.87	13.28	12.76	12.32	11.93	11.27
4.2	17.26	16.05	15.09	14.32	13.67	13.12	12.65	12.23	11.52
4.4	17.97	16.63	15.59	14.76	14.06	13.47	12.96	12.51	11.77
4.5	18.32	16.92	15.84	14.97	14.25	13.64	13.11	12.65	11.89
4.6	18.67	17.21	16.08	15.18	14.44	13.81	13.26	12.79	12.00
4.8	19.37	17.78	16.56	15.60	14.80	14.13	13.56	13.06	12.23
5.0	20.06	18.33	17.03	16.00	15.16	14.45	13.85	13.32	12.46
5.2	20.74	18.88	17.49	16.40	15.51	14.76	14.13	13.58	12.67
5.4	21.41	19.42	17.94	16.78	15.85	15.06	14.40	13.83	12.88
5.5	21.74	19.69	18.16	16.97	16.01	15.21	14.53	13.95	12.98
5.6	22.08	19.95	18.38	17.16	16.18	15.36	14.67	14.07	13.09
5.8	22.74	20.48	18.82	17.53	16.50	15.65	14.92	14.30	13.28
6.0	23.39	20.99	19.25	17.90	16.82	15.93	15.18	14.53	13.48
6.5	25.00	22.25	20.28	18.78	17.58	16.60	15.78	15.08	13.94
7.0	26.58	23.47	21.27	19.62	18.31	17.25	16.36	15.60	14.38
7.5	28.12	24.65	22.23	20.42	19.01	17.86	16.90	16.09	14.79
8.0	29.63	25.79	23.15	21.19	19.67	18.44	17.42	16.56	15.18
8.5	31.12	26.90	24.04	21.94	20.31	19.00	17.92	17.01	15.55
9.0	32.57	27.98	24.90	22.65	20.92	19.53	18.40	17.44	15.91
9.5	34.00	29.04	25.74	23.34	21.51	20.05	18.85	17.85	16.25
10.0	35.41	30.06	26.54	24.01	22.08	20.55	19.29	18.24	16.58
11.0	38.15	32.04	28.09	25.29	23.16	21.49	20.12	18.99	17.19
12.0	40.81	33.93	29.56	26.49	24.18	22.37	20.90	19.68	17.77
13.0	43.38	35.74	30.96	27.62	25.14	23.19	21.63	20.33	18.30
14.0	45.89	37.48	32.29	28.70	26.04	23.98	22.32	20.95	18.81
15.0	48.34	39.16	33.57	29.73	26.91	24.72	22.97	21.53	19.28
16.0	50.72	40.77	34.79	30.72	27.73	25.43	23.59	22.08	19.73
18.0	55.32	43.85	37.10	32.56	29.27	26.74	24.74	23.10	20.57
20.0	59.72	46.75	39.25	34.28	30.69	27.96	25.80	24.04	21.34

† For an explanation of the headings and the significance of asterisks and high returns which appear on some pages, see Introductory Section, pages xx-xxiii.

INTERNAL RATE OF RETURN (IRR) †
(Reflecting projected growth in Full Rental Value to next Review/Reversion)

RENTAL FACTOR †	YEARS TO REVIEW/REVERSION								
	1	2	3	4	5	6	7	8	10
1.1	5.63	5.75	5.85	5.94	6.03	6.10	6.16	6.21	6.30
1.2	6.11	6.21	6.30	6.37	6.43	6.48	6.53	6.56	6.61
1.3	6.59	6.67	6.73	6.78	6.82	6.86	6.88	6.90	6.91
1.4	7.06	7.11	7.15	7.18	7.20	7.21	7.22	7.21	7.20
1.5	7.53	7.56	7.57	7.57	7.57	7.56	7.54	7.52	7.47
1.6	8.00	7.99	7.97	7.95	7.92	7.89	7.85	7.81	7.73
1.7	8.46	8.42	8.37	8.32	8.26	8.21	8.15	8.09	7.98
1.8	8.92	8.84	8.76	8.68	8.60	8.52	8.44	8.37	8.22
1.9	9.37	9.25	9.14	9.03	8.92	8.82	8.72	8.63	8.45
2.0	9.83	9.66	9.51	9.37	9.24	9.11	8.99	8.88	8.67
2.2	10.72	10.47	10.24	10.03	9.85	9.67	9.51	9.37	9.10
2.4	11.60	11.25	10.94	10.67	10.43	10.21	10.01	9.82	9.49
2.5	12.03	11.63	11.28	10.98	10.71	10.46	10.24	10.04	9.68
2.6	12.46	12.01	11.62	11.28	10.98	10.71	10.47	10.25	9.87
2.8	13.31	12.75	12.28	11.87	11.51	11.20	10.92	10.66	10.22
3.0	14.15	13.48	12.91	12.44	12.02	11.66	11.34	11.05	10.56
3.2	14.98	14.18	13.53	12.98	12.51	12.11	11.75	11.43	10.88
3.4	15.80	14.88	14.13	13.51	12.99	12.53	12.14	11.78	11.18
3.5	16.21	15.22	14.43	13.77	13.22	12.74	12.32	11.96	11.33
3.6	16.61	15.56	14.72	14.03	13.44	12.95	12.51	12.13	11.48
3.8	17.41	16.22	15.29	14.52	13.89	13.34	12.87	12.46	11.76
4.0	18.20	16.87	15.84	15.01	14.32	13.73	13.22	12.77	12.03
4.2	18.98	17.51	16.38	15.48	14.73	14.10	13.55	13.08	12.29
4.4	19.75	18.14	16.91	15.94	15.13	14.46	13.88	13.38	12.54
4.5	20.13	18.45	17.17	16.16	15.33	14.64	14.04	13.52	12.66
4.6	20.51	18.76	17.43	16.38	15.53	14.81	14.20	13.66	12.78
4.8	21.26	19.36	17.94	16.82	15.91	15.15	14.50	13.94	13.02
5.0	22.01	19.96	18.43	17.24	16.28	15.48	14.80	14.21	13.24
5.2	22.74	20.54	18.92	17.66	16.64	15.80	15.09	14.47	13.47
5.4	23.47	21.12	19.40	18.07	17.00	16.11	15.37	14.73	13.68
5.5	23.84	21.40	19.63	18.27	17.17	16.27	15.51	14.85	13.78
5.6	24.20	21.68	19.86	18.46	17.34	16.42	15.64	14.98	13.89
5.8	24.91	22.24	20.32	18.85	17.68	16.72	15.91	15.22	14.09
6.0	25.62	22.79	20.77	19.23	18.01	17.01	16.17	15.45	14.29
6.5	27.36	24.13	21.86	20.15	18.81	17.71	16.80	16.02	14.76
7.0	29.06	25.43	22.91	21.03	19.56	18.37	17.39	16.55	15.21
7.5	30.72	26.68	23.91	21.87	20.28	19.00	17.95	17.06	15.63
8.0	32.35	27.89	24.88	22.67	20.97	19.61	18.48	17.54	16.03
8.5	33.95	29.07	25.81	23.45	21.63	20.18	18.99	18.00	16.41
9.0	35.51	30.21	26.71	24.19	22.26	20.73	19.48	18.43	16.77
9.5	37.05	31.32	27.58	24.91	22.87	21.26	19.95	18.85	17.12
10.0	38.56	32.41	28.43	25.60	23.46	21.77	20.40	19.26	17.45
11.0	41.50	34.49	30.05	26.93	24.58	22.74	21.25	20.02	18.08
12.0	44.35	36.49	31.58	28.17	25.63	23.65	22.05	20.73	18.66
13.0	47.11	38.39	33.04	29.35	26.62	24.50	22.80	21.40	19.21
14.0	49.79	40.22	34.43	30.47	27.55	25.30	23.50	22.02	19.72
15.0	52.41	41.99	35.76	31.53	28.44	26.07	24.17	22.62	20.21
16.0	54.95	43.69	37.03	32.55	29.29	26.79	24.80	23.18	20.67
18.0	59.86	46.92	39.43	34.46	30.87	28.14	25.98	24.22	21.52
20.0	64.55	49.96	41.67	36.23	32.33	29.39	27.06	25.18	22.29

† For an explanation of the headings and the significance of asterisks and high returns which appear on some pages, see Introductory Section, pages xx-xxiii.

INTERNAL RATE OF RETURN (IRR) †

(Reflecting projected growth in Full Rental Value to next Review/Reversion)

RENTAL FACTOR †	YEARS TO REVIEW/REVERSION								
	1	2	3	4	5	6	7	8	10
1.1	6.75	6.88	6.99	7.09	7.17	7.24	7.29	7.34	7.40
1.2	7.32	7.42	7.50	7.57	7.63	7.67	7.70	7.72	7.74
1.3	7.89	7.95	8.00	8.04	8.06	8.08	8.08	8.08	8.06
1.4	8.45	8.47	8.49	8.49	8.49	8.47	8.45	8.43	8.37
1.5	9.00	8.99	8.96	8.93	8.90	8.85	8.81	8.76	8.66
1.6	9.55	9.49	9.43	9.36	9.29	9.22	9.15	9.08	8.93
1.7	10.09	9.99	9.88	9.77	9.67	9.57	9.47	9.38	9.20
1.8	10.63	10.47	10.32	10.18	10.04	9.91	9.79	9.67	9.45
1.9	11.17	10.95	10.75	10.57	10.40	10.24	10.09	9.95	9.69
2.0	11.69	11.42	11.18	10.95	10.75	10.56	10.39	10.22	9.93
2.2	12.74	12.34	12.00	11.69	11.42	11.17	10.95	10.74	10.37
2.4	13.76	13.24	12.79	12.40	12.05	11.75	11.47	11.23	10.79
2.5	14.27	13.67	13.17	12.74	12.36	12.03	11.73	11.46	10.99
2.6	14.77	14.10	13.55	13.07	12.66	12.30	11.97	11.68	11.18
2.8	15.76	14.95	14.28	13.72	13.24	12.82	12.45	12.12	11.55
3.0	16.74	15.77	14.99	14.34	13.79	13.32	12.90	12.53	11.91
3.2	17.70	16.57	15.68	14.94	14.32	13.79	13.33	12.93	12.24
3.4	18.65	17.36	16.34	15.52	14.84	14.25	13.75	13.30	12.56
3.5	19.12	17.74	16.67	15.81	15.09	14.48	13.95	13.49	12.71
3.6	19.59	18.12	16.99	16.08	15.33	14.69	14.15	13.67	12.87
3.8	20.51	18.87	17.62	16.63	15.81	15.12	14.53	14.01	13.16
4.0	21.42	19.60	18.24	17.15	16.27	15.53	14.90	14.35	13.44
4.2	22.32	20.32	18.83	17.67	16.72	15.93	15.25	14.67	13.71
4.4	23.20	21.03	19.42	18.16	17.15	16.31	15.60	14.98	13.97
4.5	23.64	21.37	19.70	18.41	17.36	16.50	15.76	15.13	14.10
4.6	24.08	21.71	19.99	18.65	17.57	16.68	15.93	15.28	14.22
4.8	24.94	22.39	20.54	19.12	17.98	17.04	16.25	15.58	14.47
5.0	25.79	23.06	21.09	19.58	18.38	17.39	16.57	15.86	14.70
5.2	26.64	23.71	21.62	20.03	18.77	17.73	16.87	16.13	14.93
5.4	27.47	24.35	22.14	20.47	19.15	18.07	17.17	16.40	15.16
5.5	27.89	24.67	22.39	20.68	19.33	18.23	17.31	16.53	15.26
5.6	28.30	24.98	22.65	20.89	19.51	18.39	17.45	16.66	15.37
5.8	29.11	25.60	23.15	21.31	19.87	18.71	17.74	16.91	15.58
6.0	29.92	26.21	23.64	21.72	20.23	19.01	18.01	17.16	15.79
6.5	31.91	27.69	24.83	22.71	21.07	19.75	18.67	17.75	16.27
7.0	33.84	29.13	25.96	23.65	21.88	20.45	19.29	18.30	16.74
7.5	35.73	30.51	27.05	24.55	22.64	21.12	19.87	18.83	17.17
8.0	37.58	31.84	28.10	25.41	23.37	21.75	20.43	19.33	17.58
8.5	39.38	33.14	29.11	26.24	24.07	22.36	20.97	19.81	17.97
9.0	41.15	34.40	30.08	27.03	24.74	22.94	21.48	20.26	18.35
9.5	42.89	35.62	31.02	27.80	25.39	23.50	21.97	20.70	18.71
10.0	44.59	36.81	31.94	28.54	26.01	24.03	22.44	21.12	19.05
11.0	47.91	39.09	33.68	29.95	27.19	25.04	23.33	21.91	19.70
12.0	51.11	41.27	35.33	31.27	28.29	25.99	24.15	22.64	20.30
13.0	54.21	43.35	36.89	32.52	29.33	26.88	24.93	23.33	20.86
14.0	57.22	45.34	38.38	33.71	30.32	27.72	25.66	23.98	21.39
15.0	60.14	47.26	39.80	34.83	31.25	28.52	26.36	24.60	21.89
16.0	62.99	49.11	41.16	35.91	32.14	29.27	27.01	25.18	22.36
18.0	68.47	52.62	43.73	37.93	33.80	30.68	28.24	26.26	23.23
20.0	****	55.91	46.11	39.79	35.33	31.98	29.36	27.25	24.03

† For an explanation of the headings and the significance of asterisks and high returns which appear on some pages, see Introductory Section, pages xx-xxiii.

INTERNAL RATE OF RETURN (IRR) †

(Reflecting projected growth in Full Rental Value to next Review/Reversion)

RENTAL FACTOR †	YEARS TO REVIEW/REVERSION								
	1	2	3	4	5	6	7	8	10
1.1	7.86	8.00	8.12	8.22	8.29	8.36	8.40	8.44	8.48
1.2	8.52	8.62	8.70	8.75	8.80	8.82	8.84	8.85	8.84
1.3	9.17	9.22	9.26	9.27	9.28	9.27	9.26	9.23	9.17
1.4	9.82	9.82	9.80	9.78	9.74	9.70	9.65	9.60	9.49
1.5	10.45	10.40	10.33	10.26	10.18	10.11	10.03	9.95	9.79
1.6	11.08	10.96	10.85	10.73	10.61	10.50	10.39	10.28	10.08
1.7	11.71	11.52	11.35	11.18	11.03	10.88	10.74	10.61	10.36
1.8	12.32	12.07	11.84	11.63	11.43	11.24	11.07	10.91	10.62
1.9	12.93	12.61	12.32	12.05	11.82	11.60	11.40	11.21	10.87
2.0	13.54	13.14	12.78	12.47	12.19	11.94	11.71	11.50	11.12
2.2	14.72	14.16	13.69	13.28	12.91	12.59	12.30	12.04	11.58
2.4	15.89	15.16	14.56	14.04	13.60	13.21	12.86	12.55	12.01
2.5	16.47	15.65	14.98	14.41	13.92	13.50	13.13	12.79	12.22
2.6	17.04	16.13	15.39	14.77	14.24	13.79	13.39	13.03	12.42
2.8	18.16	17.07	16.19	15.47	14.86	14.34	13.88	13.48	12.80
3.0	19.27	17.98	16.97	16.14	15.45	14.87	14.36	13.91	13.16
3.2	20.35	18.87	17.71	16.79	16.02	15.37	14.81	14.32	13.51
3.4	21.42	19.73	18.44	17.41	16.57	15.86	15.25	14.72	13.84
3.5	21.95	20.15	18.79	17.72	16.83	16.09	15.46	14.91	14.00
3.6	22.48	20.57	19.14	18.01	17.09	16.32	15.66	15.09	14.15
3.8	23.51	21.40	19.83	18.60	17.60	16.77	16.06	15.46	14.45
4.0	24.54	22.20	20.49	19.16	18.09	17.20	16.45	15.80	14.74
4.2	25.54	22.99	21.14	19.71	18.56	17.62	16.82	16.14	15.02
4.4	26.54	23.76	21.76	20.24	19.02	18.02	17.18	16.46	15.29
4.5	27.03	24.14	22.07	20.50	19.24	18.22	17.35	16.62	15.42
4.6	27.52	24.52	22.38	20.75	19.46	18.41	17.53	16.77	15.55
4.8	28.49	25.26	22.98	21.26	19.90	18.79	17.86	17.08	15.80
5.0	29.44	25.98	23.56	21.74	20.32	19.16	18.19	17.37	16.04
5.2	30.39	26.70	24.13	22.22	20.73	19.51	18.51	17.65	16.28
5.4	31.32	27.40	24.69	22.69	21.12	19.86	18.81	17.93	16.51
5.5	31.78	27.74	24.97	22.92	21.32	20.03	18.96	18.06	16.62
5.6	32.24	28.08	25.24	23.14	21.51	20.20	19.11	18.20	16.73
5.8	33.15	28.76	25.78	23.59	21.89	20.53	19.41	18.46	16.94
6.0	34.05	29.42	26.30	24.02	22.26	20.85	19.69	18.71	17.15
6.5	36.26	31.04	27.57	25.07	23.15	21.62	20.37	19.32	17.65
7.0	38.41	32.59	28.79	26.06	23.99	22.35	21.01	19.90	18.13
7.5	40.50	34.09	29.95	27.01	24.79	23.05	21.62	20.44	18.57
8.0	42.55	35.53	31.07	27.92	25.56	23.71	22.20	20.96	18.99
8.5	44.55	36.93	32.14	28.79	26.29	24.34	22.76	21.45	19.39
9.0	46.51	38.29	33.18	29.63	26.99	24.94	23.28	21.92	19.78
9.5	48.43	39.61	34.18	30.44	27.67	25.52	23.79	22.37	20.14
10.0	50.31	40.89	35.15	31.22	28.32	26.07	24.28	22.80	20.50
11.0	53.97	43.35	37.00	32.69	29.55	27.13	25.19	23.61	21.16
12.0	57.49	45.69	38.75	34.08	30.70	28.11	26.05	24.37	21.77
13.0	60.90	47.92	40.40	35.39	31.78	29.03	26.85	25.08	22.34
14.0	64.21	50.06	41.97	36.64	32.81	29.90	27.61	25.74	22.88
15.0	67.42	52.12	43.48	37.82	33.78	30.72	28.32	26.37	23.39
16.0	****	54.09	44.91	38.94	34.70	31.50	29.00	26.97	23.87
18.0	****	57.84	47.62	41.06	36.43	32.96	30.26	28.08	24.76
20.0	****	61.36	50.14	43.00	38.01	34.30	31.41	29.09	25.58

† For an explanation of the headings and the significance of asterisks and high returns which appear on some pages, see Introductory Section, pages xx-xxiii.

INTERNAL RATE OF RETURN (IRR) †

(Reflecting projected growth in Full Rental Value to next Review/Reversion)

RENTAL FACTOR †	YEARS TO REVIEW/REVERSION								
	1	2	3	4	5	6	7	8	10
1.1	8.98	9.12	9.24	9.33	9.40	9.46	9.50	9.52	9.54
1.2	9.72	9.81	9.88	9.92	9.95	9.96	9.96	9.95	9.91
1.3	10.46	10.48	10.49	10.49	10.47	10.44	10.40	10.35	10.25
1.4	11.18	11.14	11.09	11.03	10.96	10.89	10.81	10.74	10.58
1.5	11.90	11.79	11.67	11.56	11.44	11.32	11.21	11.10	10.89
1.6	12.60	12.42	12.23	12.06	11.90	11.74	11.59	11.45	11.19
1.7	13.30	13.03	12.78	12.55	12.34	12.14	11.96	11.78	11.47
1.8	13.99	13.64	13.32	13.03	12.77	12.53	12.31	12.11	11.74
1.9	14.68	14.23	13.84	13.49	13.18	12.90	12.65	12.41	12.00
2.0	15.35	14.81	14.34	13.94	13.58	13.26	12.97	12.71	12.25
2.2	16.68	15.94	15.32	14.80	14.35	13.95	13.59	13.27	12.72
2.4	17.98	17.03	16.26	15.62	15.07	14.59	14.17	13.80	13.17
2.5	18.62	17.56	16.71	16.01	15.42	14.90	14.45	14.05	13.38
2.6	19.26	18.09	17.16	16.40	15.75	15.20	14.72	14.30	13.58
2.8	20.51	19.11	18.02	17.14	16.41	15.78	15.24	14.77	13.98
3.0	21.73	20.10	18.85	17.85	17.03	16.33	15.74	15.21	14.35
3.2	22.94	21.07	19.66	18.54	17.63	16.86	16.21	15.64	14.70
3.4	24.13	22.01	20.43	19.20	18.20	17.37	16.66	16.04	15.04
3.5	24.71	22.47	20.81	19.52	18.48	17.61	16.87	16.24	15.20
3.6	25.29	22.93	21.19	19.84	18.75	17.85	17.09	16.43	15.36
3.8	26.44	23.82	21.92	20.45	19.28	18.32	17.50	16.81	15.67
4.0	27.57	24.69	22.63	21.05	19.79	18.77	17.90	17.16	15.96
4.2	28.68	25.54	23.31	21.63	20.29	19.20	18.29	17.51	16.25
4.4	29.77	26.37	23.98	22.19	20.77	19.62	18.66	17.84	16.52
4.5	30.32	26.78	24.31	22.46	21.00	19.82	18.84	18.00	16.65
4.6	30.85	27.19	24.64	22.73	21.23	20.02	19.02	18.16	16.78
4.8	31.92	27.99	25.27	23.26	21.69	20.42	19.36	18.47	17.04
5.0	32.97	28.77	25.90	23.77	22.12	20.80	19.70	18.77	17.29
5.2	34.00	29.53	26.50	24.28	22.55	21.17	20.03	19.07	17.53
5.4	35.03	30.29	27.10	24.76	22.97	21.53	20.34	19.35	17.76
5.5	35.54	30.66	27.39	25.00	23.17	21.71	20.50	19.49	17.87
5.6	36.04	31.03	27.68	25.24	23.37	21.88	20.65	19.63	17.99
5.8	37.04	31.75	28.25	25.71	23.77	22.22	20.95	19.89	18.21
6.0	38.02	32.46	28.80	26.16	24.15	22.55	21.25	20.15	18.42
6.5	40.44	34.19	30.15	27.26	25.08	23.35	21.95	20.78	18.93
7.0	42.79	35.85	31.43	28.31	25.95	24.11	22.61	21.37	19.41
7.5	45.08	37.46	32.66	29.30	26.79	24.82	23.24	21.93	19.86
8.0	47.32	39.00	33.84	30.25	27.58	25.51	23.83	22.45	20.29
8.5	49.50	40.49	34.97	31.16	28.34	26.16	24.40	22.96	20.70
9.0	51.63	41.94	36.06	32.03	29.07	26.78	24.94	23.44	21.09
9.5	53.72	43.35	37.11	32.87	29.77	27.38	25.46	23.90	21.47
10.0	55.76	44.71	38.13	33.69	30.44	27.95	25.96	24.34	21.82
11.0	59.74	47.33	40.08	35.23	31.72	29.03	26.91	25.17	22.50
12.0	63.56	49.81	41.91	36.67	32.91	30.05	27.78	25.95	23.12
13.0	67.26	52.18	43.64	38.04	34.03	30.99	28.61	26.67	23.70
14.0	****	54.45	45.29	39.33	35.09	31.89	29.38	27.35	24.25
15.0	****	56.63	46.87	40.56	36.09	32.74	30.11	27.99	24.77
16.0	****	58.73	48.37	41.73	37.04	33.54	30.81	28.60	25.26
18.0	****	62.69	51.20	43.92	38.82	35.04	32.10	29.73	26.16
20.0	****	66.41	53.83	45.94	40.46	36.41	33.28	30.77	26.99

† For an explanation of the headings and the significance of asterisks and high returns which appear on some pages, see Introductory Section, pages xx-xxiii.

INTERNAL RATE OF RETURN (IRR) †
(Reflecting projected growth in Full Rental Value to next Review/Reversion)

RENTAL FACTOR †	YEARS TO REVIEW/REVERSION								
	1	2	3	4	5	6	7	8	10
1.1	10.09	10.24	10.35	10.44	10.50	10.55	10.57	10.58	10.58
1.2	10.91	10.99	11.05	11.07	11.08	11.07	11.06	11.03	10.95
1.3	11.73	11.73	11.71	11.68	11.63	11.58	11.51	11.45	11.30
1.4	12.53	12.45	12.36	12.26	12.16	12.05	11.95	11.84	11.64
1.5	13.33	13.16	12.99	12.82	12.66	12.51	12.36	12.22	11.96
1.6	14.11	13.84	13.59	13.36	13.15	12.95	12.76	12.58	12.26
1.7	14.88	14.51	14.18	13.89	13.62	13.37	13.14	12.93	12.55
1.8	15.65	15.17	14.76	14.39	14.07	13.77	13.50	13.26	12.82
1.9	16.40	15.82	15.32	14.88	14.50	14.16	13.85	13.57	13.09
2.0	17.14	16.45	15.86	15.36	14.92	14.53	14.19	13.88	13.34
2.2	18.61	17.67	16.91	16.27	15.72	15.25	14.83	14.46	13.82
2.4	20.04	18.85	17.91	17.13	16.48	15.92	15.43	15.00	14.27
2.5	20.74	19.43	18.39	17.55	16.84	16.24	15.72	15.26	14.49
2.6	21.44	19.99	18.87	17.96	17.20	16.55	16.00	15.51	14.70
2.8	22.81	21.09	19.79	18.74	17.88	17.16	16.54	15.99	15.10
3.0	24.15	22.16	20.67	19.49	18.53	17.73	17.04	16.45	15.47
3.2	25.47	23.20	21.52	20.21	19.15	18.27	17.53	16.89	15.83
3.4	26.76	24.21	22.34	20.91	19.75	18.80	17.99	17.30	16.17
3.5	27.40	24.70	22.75	21.24	20.04	19.05	18.22	17.50	16.34
3.6	28.04	25.19	23.14	21.57	20.32	19.30	18.44	17.70	16.50
3.8	29.29	26.15	23.91	22.22	20.88	19.78	18.86	18.08	16.81
4.0	30.51	27.08	24.66	22.84	21.41	20.24	19.27	18.45	17.11
4.2	31.72	27.99	25.39	23.45	21.92	20.69	19.67	18.80	17.40
4.4	32.92	28.88	26.09	24.03	22.42	21.12	20.05	19.14	17.68
4.5	33.50	29.31	26.44	24.32	22.67	21.33	20.23	19.31	17.82
4.6	34.09	29.74	26.78	24.60	22.90	21.54	20.42	19.47	17.95
4.8	35.25	30.60	27.45	25.15	23.37	21.95	20.77	19.79	18.21
5.0	36.39	31.43	28.11	25.69	23.83	22.34	21.12	20.09	18.46
5.2	37.51	32.24	28.75	26.21	24.27	22.72	21.46	20.39	18.70
5.4	38.62	33.04	29.37	26.72	24.70	23.09	21.78	20.68	18.94
5.5	39.17	33.44	29.68	26.97	24.91	23.28	21.94	20.82	19.05
5.6	39.71	33.83	29.98	27.22	25.12	23.46	22.10	20.96	19.17
5.8	40.80	34.60	30.58	27.71	25.53	23.81	22.41	21.24	19.39
6.0	41.86	35.36	31.16	28.18	25.93	24.15	22.71	21.50	19.61
6.5	44.48	37.19	32.57	29.32	26.88	24.97	23.43	22.14	20.13
7.0	47.02	38.95	33.92	30.41	27.79	25.75	24.10	22.74	20.61
7.5	49.49	40.65	35.20	31.44	28.65	26.49	24.75	23.31	21.07
8.0	51.90	42.28	36.44	32.42	29.47	27.19	25.36	23.85	21.51
8.5	54.25	43.86	37.62	33.37	30.25	27.85	25.94	24.36	21.92
9.0	56.55	45.39	38.76	34.28	31.00	28.49	26.49	24.85	22.32
9.5	58.79	46.88	39.86	35.15	31.72	29.10	27.02	25.32	22.70
10.0	60.99	48.31	40.92	35.99	32.42	29.69	27.53	25.77	23.06
11.0	65.26	51.07	42.95	37.58	33.73	30.80	28.50	26.62	23.74
12.0	69.37	53.69	44.86	39.08	34.96	31.84	29.39	27.41	24.37
13.0	****	56.19	46.67	40.49	36.11	32.81	30.23	28.15	24.96
14.0	****	58.58	48.39	41.83	37.20	33.73	31.02	28.84	25.52
15.0	****	60.87	50.02	43.09	38.23	34.60	31.77	29.49	26.04
16.0	****	63.07	51.59	44.30	39.21	35.42	32.48	30.11	26.54
18.0	****	67.23	54.53	46.56	41.04	36.95	33.79	31.26	27.45
20.0	****	****	57.26	48.65	42.72	38.36	34.99	32.31	28.29

† For an explanation of the headings and the significance of asterisks and high returns which appear on some page, see Introductory Section, pages xx-xxiii.

INTERNAL RATE OF RETURN (IRR) †
(Reflecting projected growth in Full Rental Value to next Review/Reversion)

RENTAL FACTOR †	YEARS TO REVIEW/REVERSION								
	1	2	3	4	5	6	7	8	10
1.1	11.20	11.35	11.46	11.54	11.59	11.62	11.63	11.63	11.60
1.2	12.11	12.17	12.20	12.21	12.20	12.17	12.13	12.09	11.98
1.3	13.00	12.97	12.92	12.86	12.78	12.70	12.61	12.52	12.33
1.4	13.88	13.75	13.61	13.47	13.33	13.19	13.06	12.93	12.67
1.5	14.75	14.51	14.28	14.07	13.86	13.67	13.48	13.31	12.99
1.6	15.61	15.25	14.93	14.64	14.37	14.12	13.89	13.68	13.30
1.7	16.45	15.97	15.56	15.19	14.86	14.56	14.28	14.03	13.59
1.8	17.28	16.68	16.17	15.72	15.33	14.98	14.66	14.37	13.87
1.9	18.10	17.37	16.76	16.24	15.78	15.38	15.02	14.70	14.14
2.0	18.91	18.05	17.34	16.74	16.22	15.77	15.37	15.01	14.40
2.2	20.51	19.36	18.45	17.69	17.05	16.51	16.03	15.60	14.89
2.4	22.06	20.63	19.51	18.60	17.84	17.20	16.64	16.16	15.34
2.5	22.82	21.24	20.02	19.03	18.22	17.53	16.94	16.42	15.56
2.6	23.58	21.84	20.52	19.46	18.59	17.85	17.22	16.68	15.77
2.8	25.06	23.02	21.49	20.28	19.30	18.47	17.77	17.17	16.17
3.0	26.52	24.16	22.42	21.06	19.97	19.06	18.30	17.64	16.55
3.2	27.95	25.26	23.31	21.82	20.61	19.63	18.79	18.08	16.92
3.4	29.34	26.33	24.18	22.54	21.23	20.16	19.27	18.50	17.26
3.5	30.03	26.86	24.60	22.89	21.53	20.42	19.50	18.71	17.43
3.6	30.72	27.38	25.02	23.23	21.83	20.68	19.72	18.91	17.59
3.8	32.06	28.39	25.82	23.90	22.40	21.17	20.16	19.30	17.91
4.0	33.39	29.38	26.61	24.55	22.95	21.65	20.58	19.67	18.21
4.2	34.69	30.34	27.37	25.18	23.48	22.11	20.98	20.03	18.50
4.4	35.98	31.28	28.11	25.79	23.99	22.55	21.37	20.37	18.79
4.5	36.61	31.74	28.47	26.08	24.24	22.77	21.56	20.54	18.92
4.6	37.24	32.20	28.83	26.38	24.49	22.98	21.75	20.71	19.06
4.8	38.48	33.10	29.53	26.95	24.97	23.40	22.11	21.03	19.32
5.0	39.71	33.98	30.22	27.51	25.44	23.80	22.46	21.35	19.57
5.2	40.91	34.84	30.88	28.05	25.90	24.20	22.81	21.65	19.82
5.4	42.10	35.69	31.53	28.58	26.34	24.58	23.14	21.94	20.06
5.5	42.69	36.10	31.85	28.84	26.56	24.76	23.30	22.09	20.17
5.6	43.28	36.51	32.17	29.09	26.77	24.95	23.46	22.23	20.29
5.8	44.44	37.33	32.79	29.60	27.19	25.31	23.78	22.51	20.51
6.0	45.58	38.12	33.40	30.09	27.60	25.66	24.08	22.78	20.73
6.5	48.38	40.06	34.87	31.27	28.59	26.50	24.82	23.43	21.26
7.0	51.10	41.91	36.27	32.39	29.52	27.29	25.51	24.04	21.75
7.5	53.74	43.69	37.61	33.46	30.40	28.05	26.16	24.62	22.21
8.0	56.32	45.41	38.89	34.48	31.25	28.76	26.79	25.16	22.65
8.5	58.83	47.07	40.12	35.45	32.05	29.45	27.38	25.69	23.07
9.0	61.28	48.67	41.31	36.39	32.82	30.10	27.94	26.18	23.47
9.5	63.67	50.23	42.45	37.29	33.56	30.73	28.48	26.66	23.85
10.0	66.02	51.74	43.56	38.15	34.27	31.33	29.00	27.12	24.22
11.0	****	54.63	45.66	39.80	35.62	32.46	29.98	27.98	24.91
12.0	****	57.37	47.64	41.34	36.88	33.52	30.90	28.78	25.55
13.0	****	59.98	49.51	42.79	38.06	34.51	31.75	29.53	26.14
14.0	****	62.48	51.29	44.16	39.17	35.45	32.55	30.23	26.70
15.0	****	64.87	52.99	45.47	40.23	36.33	33.31	30.89	27.23
16.0	****	67.17	54.60	46.71	41.23	37.17	34.03	31.52	27.73
18.0	****	****	57.65	49.04	43.10	38.74	35.37	32.69	28.66
20.0	****	****	60.47	51.18	44.82	40.17	36.60	33.76	29.50

† For an explanation of the headings and the significance of asterisks and high returns which appear on some pages, see Introductory Section, pages xx-xxiii.

INTERNAL RATE OF RETURN (IRR) †
(Reflecting projected growth in Full Rental Value to next Review/Reversion)

RENTAL FACTOR †	YEARS TO REVIEW/REVERSION								
	1	2	3	4	5	6	7	8	10
1.1	1.15	1.21	1.26	1.32	1.38	1.44	1.49	1.55	1.67
1.2	1.26	1.31	1.37	1.43	1.49	1.55	1.62	1.68	1.80
1.3	1.36	1.42	1.48	1.55	1.61	1.67	1.74	1.80	1.92
1.4	1.46	1.53	1.59	1.66	1.72	1.79	1.85	1.92	2.05
1.5	1.57	1.63	1.70	1.77	1.83	1.90	1.97	2.04	2.17
1.6	1.67	1.74	1.81	1.88	1.95	2.02	2.08	2.15	2.28
1.7	1.77	1.84	1.91	1.99	2.06	2.13	2.20	2.26	2.40
1.8	1.87	1.95	2.02	2.09	2.17	2.24	2.31	2.38	2.51
1.9	1.98	2.05	2.13	2.20	2.27	2.35	2.42	2.48	2.62
2.0	2.08	2.15	2.23	2.31	2.38	2.45	2.52	2.59	2.73
2.2	2.28	2.36	2.44	2.52	2.59	2.66	2.73	2.80	2.93
2.4	2.48	2.56	2.64	2.72	2.80	2.87	2.94	3.01	3.14
2.5	2.58	2.67	2.75	2.82	2.90	2.97	3.04	3.11	3.23
2.6	2.68	2.77	2.85	2.93	3.00	3.07	3.14	3.21	3.33
2.8	2.89	2.97	3.05	3.13	3.20	3.27	3.34	3.40	3.52
3.0	3.09	3.17	3.25	3.32	3.39	3.46	3.53	3.59	3.70
3.2	3.28	3.37	3.44	3.52	3.59	3.65	3.71	3.77	3.88
3.4	3.48	3.56	3.64	3.71	3.77	3.84	3.90	3.95	4.05
3.5	3.58	3.66	3.73	3.80	3.87	3.93	3.99	4.04	4.13
3.6	3.68	3.76	3.83	3.90	3.96	4.02	4.07	4.13	4.22
3.8	3.88	3.95	4.02	4.08	4.14	4.20	4.25	4.30	4.38
4.0	4.07	4.14	4.21	4.27	4.32	4.37	4.42	4.46	4.54
4.2	4.27	4.34	4.39	4.45	4.50	4.54	4.59	4.63	4.69
4.4	4.47	4.52	4.58	4.63	4.67	4.71	4.75	4.78	4.84
4.5	4.56	4.62	4.67	4.72	4.76	4.80	4.83	4.86	4.92
4.6	4.66	4.71	4.76	4.81	4.84	4.88	4.91	4.94	4.99
4.8	4.85	4.90	4.94	4.98	5.01	5.04	5.07	5.09	5.13
5.0	5.05	5.09	5.12	5.15	5.18	5.20	5.23	5.24	5.27
5.2	5.24	5.27	5.30	5.32	5.34	5.36	5.38	5.39	5.41
5.4	5.43	5.45	5.48	5.49	5.51	5.52	5.53	5.53	5.54
5.5	5.52	5.55	5.56	5.58	5.59	5.60	5.60	5.61	5.61
5.6	5.62	5.64	5.65	5.66	5.67	5.67	5.67	5.68	5.68
5.8	5.81	5.82	5.82	5.82	5.82	5.82	5.82	5.82	5.80
6.0	6.00	6.00	5.99	5.99	5.98	5.97	5.96	5.95	5.93
6.5	6.47	6.44	6.41	6.39	6.36	6.33	6.31	6.28	6.24
7.0	6.94	6.88	6.83	6.78	6.73	6.68	6.64	6.60	6.53
7.5	7.40	7.31	7.23	7.16	7.09	7.02	6.96	6.91	6.81
8.0	7.86	7.74	7.63	7.53	7.43	7.35	7.27	7.20	7.07
8.5	8.32	8.16	8.01	7.89	7.77	7.67	7.57	7.49	7.33
9.0	8.77	8.57	8.40	8.24	8.10	7.98	7.87	7.76	7.58
9.5	9.22	8.98	8.77	8.59	8.42	8.28	8.15	8.03	7.82
10.0	9.66	9.38	9.14	8.92	8.74	8.57	8.42	8.29	8.05
11.0	10.54	10.17	9.85	9.58	9.34	9.13	8.95	8.78	8.49
12.0	11.41	10.94	10.54	10.21	9.92	9.67	9.45	9.25	8.90
13.0	12.27	11.69	11.21	10.82	10.47	10.18	9.92	9.69	9.30
14.0	13.11	12.42	11.86	11.40	11.01	10.67	10.37	10.11	9.67
15.0	13.94	13.14	12.49	11.97	11.52	11.14	10.80	10.51	10.02
16.0	14.77	13.84	13.11	12.51	12.01	11.59	11.22	10.90	10.36
18.0	16.38	15.20	14.29	13.56	12.95	12.44	12.00	11.62	10.99
20.0	17.96	16.51	15.41	14.54	13.83	13.24	12.73	12.29	11.57

† For an explanation of the headings and the significance of asterisks and high returns which appear on some pages, see Introductory Section, pages xx-xxiii.

INTERNAL RATE OF RETURN (IRR) †
(Reflecting projected growth in Full Rental Value to next Review/Reversion)

RENTAL FACTOR †	YEARS TO REVIEW/REVERSION								
	1	2	3	4	5	6	7	8	10
1.1	1.73	1.81	1.89	1.97	2.05	2.13	2.21	2.28	2.43
1.2	1.88	1.97	2.05	2.13	2.21	2.30	2.38	2.45	2.61
1.3	2.04	2.12	2.21	2.29	2.38	2.46	2.54	2.62	2.77
1.4	2.19	2.28	2.37	2.46	2.54	2.62	2.71	2.79	2.94
1.5	2.34	2.43	2.53	2.61	2.70	2.78	2.87	2.95	3.10
1.6	2.50	2.59	2.68	2.77	2.86	2.94	3.02	3.10	3.25
1.7	2.65	2.74	2.84	2.93	3.01	3.10	3.18	3.25	3.40
1.8	2.80	2.89	2.99	3.08	3.17	3.25	3.33	3.40	3.55
1.9	2.95	3.05	3.14	3.23	3.32	3.40	3.48	3.55	3.69
2.0	3.10	3.20	3.29	3.38	3.46	3.54	3.62	3.69	3.83
2.2	3.40	3.50	3.59	3.67	3.76	3.83	3.90	3.97	4.09
2.4	3.70	3.79	3.88	3.96	4.04	4.11	4.18	4.24	4.35
2.5	3.85	3.94	4.02	4.10	4.18	4.25	4.31	4.37	4.47
2.6	4.00	4.08	4.17	4.24	4.32	4.38	4.44	4.50	4.60
2.8	4.29	4.37	4.45	4.52	4.59	4.64	4.70	4.75	4.83
3.0	4.58	4.66	4.73	4.79	4.85	4.90	4.95	4.99	5.06
3.2	4.88	4.94	5.00	5.06	5.11	5.15	5.19	5.22	5.28
3.4	5.17	5.22	5.27	5.32	5.36	5.39	5.42	5.45	5.49
3.5	5.31	5.36	5.41	5.45	5.48	5.51	5.54	5.56	5.59
3.6	5.45	5.50	5.54	5.58	5.61	5.63	5.65	5.67	5.69
3.8	5.74	5.78	5.80	5.83	5.85	5.86	5.87	5.88	5.89
4.0	6.03	6.05	6.06	6.08	6.08	6.09	6.09	6.09	6.08
4.2	6.31	6.32	6.32	6.32	6.32	6.31	6.30	6.29	6.27
4.4	6.59	6.58	6.57	6.56	6.54	6.53	6.51	6.49	6.45
4.5	6.73	6.72	6.70	6.68	6.66	6.63	6.61	6.59	6.54
4.6	6.88	6.85	6.82	6.79	6.77	6.74	6.71	6.68	6.63
4.8	7.16	7.11	7.07	7.03	6.99	6.95	6.91	6.87	6.80
5.0	7.43	7.37	7.31	7.26	7.20	7.15	7.10	7.05	6.97
5.2	7.71	7.63	7.55	7.48	7.41	7.35	7.29	7.23	7.13
5.4	7.99	7.88	7.79	7.70	7.62	7.55	7.48	7.41	7.29
5.5	8.12	8.01	7.91	7.81	7.72	7.64	7.57	7.50	7.37
5.6	8.26	8.14	8.02	7.92	7.83	7.74	7.66	7.58	7.44
5.8	8.53	8.39	8.26	8.14	8.03	7.93	7.84	7.75	7.59
6.0	8.81	8.64	8.49	8.35	8.23	8.11	8.01	7.91	7.74
6.5	9.48	9.25	9.05	8.87	8.71	8.56	8.43	8.31	8.10
7.0	10.15	9.85	9.59	9.37	9.17	9.00	8.84	8.69	8.44
7.5	10.81	10.44	10.13	9.86	9.62	9.41	9.23	9.06	8.76
8.0	11.46	11.02	10.65	10.33	10.06	9.81	9.60	9.41	9.07
8.5	12.10	11.58	11.15	10.79	10.48	10.20	9.96	9.74	9.37
9.0	12.74	12.14	11.65	11.24	10.88	10.58	10.31	10.06	9.65
9.5	13.37	12.69	12.14	11.67	11.28	10.94	10.64	10.38	9.93
10.0	14.00	13.23	12.61	12.10	11.66	11.29	10.96	10.68	10.19
11.0	15.23	14.28	13.53	12.91	12.40	11.96	11.58	11.25	10.69
12.0	16.44	15.30	14.41	13.70	13.10	12.60	12.17	11.79	11.16
13.0	17.63	16.29	15.26	14.44	13.77	13.20	12.72	12.30	11.60
14.0	18.80	17.25	16.08	15.16	14.41	13.78	13.24	12.78	12.02
15.0	19.95	18.19	16.88	15.85	15.02	14.33	13.74	13.24	12.41
16.0	21.07	19.10	17.64	16.51	15.61	14.86	14.22	13.68	12.79
18.0	23.28	20.85	19.11	17.78	16.72	15.85	15.12	14.50	13.49
20.0	25.42	22.53	20.49	18.96	17.75	16.77	15.95	15.25	14.13

† For an explanation of the headings and the significance of asterisks and high returns which appear on some pages, see Introductory Section, pages xx-xxiii.

INTERNAL RATE OF RETURN (IRR) †
(Reflecting projected growth in Full Rental Value to next Review/Reversion)

RENTAL FACTOR †	YEARS TO REVIEW/REVERSION								
	1	2	3	4	5	6	7	8	10
1.1	2.30	2.41	2.51	2.61	2.71	2.80	2.90	2.99	3.16
1.2	2.51	2.61	2.72	2.82	2.92	3.02	3.11	3.20	3.37
1.3	2.71	2.82	2.93	3.03	3.13	3.23	3.32	3.41	3.57
1.4	2.91	3.02	3.13	3.24	3.34	3.43	3.52	3.61	3.77
1.5	3.12	3.23	3.33	3.44	3.54	3.63	3.72	3.80	3.96
1.6	3.32	3.43	3.54	3.64	3.73	3.83	3.91	3.99	4.14
1.7	3.52	3.63	3.73	3.83	3.93	4.02	4.10	4.18	4.32
1.8	3.72	3.83	3.93	4.03	4.12	4.21	4.29	4.36	4.49
1.9	3.92	4.02	4.13	4.22	4.31	4.39	4.47	4.53	4.66
2.0	4.11	4.22	4.32	4.41	4.49	4.57	4.64	4.71	4.82
2.2	4.51	4.61	4.70	4.78	4.85	4.92	4.98	5.04	5.13
2.4	4.90	4.99	5.07	5.14	5.20	5.26	5.31	5.35	5.43
2.5	5.09	5.18	5.25	5.31	5.37	5.42	5.47	5.51	5.57
2.6	5.29	5.36	5.43	5.49	5.54	5.59	5.62	5.66	5.71
2.8	5.67	5.73	5.79	5.83	5.87	5.90	5.93	5.95	5.98
3.0	6.05	6.10	6.14	6.17	6.19	6.21	6.22	6.23	6.24
3.2	6.43	6.46	6.48	6.49	6.50	6.51	6.51	6.50	6.49
3.4	6.81	6.82	6.82	6.81	6.81	6.79	6.78	6.77	6.73
3.5	7.00	6.99	6.98	6.97	6.95	6.94	6.92	6.89	6.85
3.6	7.19	7.17	7.15	7.13	7.10	7.08	7.05	7.02	6.96
3.8	7.56	7.52	7.48	7.43	7.39	7.35	7.31	7.27	7.19
4.0	7.93	7.86	7.80	7.73	7.67	7.61	7.56	7.50	7.40
4.2	8.30	8.20	8.11	8.03	7.95	7.87	7.80	7.74	7.61
4.4	8.66	8.54	8.42	8.32	8.22	8.13	8.04	7.96	7.81
4.5	8.84	8.70	8.58	8.46	8.35	8.25	8.16	8.07	7.91
4.6	9.03	8.87	8.73	8.60	8.48	8.37	8.27	8.18	8.01
4.8	9.39	9.20	9.03	8.88	8.74	8.62	8.50	8.39	8.20
5.0	9.75	9.52	9.33	9.15	9.00	8.85	8.72	8.60	8.39
5.2	10.10	9.85	9.62	9.42	9.25	9.08	8.94	8.80	8.57
5.4	10.46	10.16	9.91	9.69	9.49	9.31	9.15	9.00	8.74
5.5	10.63	10.32	10.05	9.82	9.61	9.42	9.25	9.10	8.83
5.6	10.81	10.48	10.19	9.95	9.73	9.53	9.36	9.20	8.92
5.8	11.16	10.79	10.48	10.20	9.96	9.75	9.56	9.39	9.08
6.0	11.51	11.10	10.75	10.46	10.19	9.96	9.76	9.57	9.25
6.5	12.37	11.86	11.43	11.07	10.76	10.48	10.24	10.02	9.64
7.0	13.22	12.60	12.09	11.66	11.29	10.97	10.69	10.44	10.01
7.5	14.06	13.32	12.73	12.23	11.81	11.45	11.13	10.85	10.37
8.0	14.88	14.03	13.35	12.78	12.31	11.90	11.55	11.23	10.70
8.5	15.70	14.72	13.95	13.32	12.79	12.34	11.95	11.60	11.03
9.0	16.51	15.40	14.54	13.84	13.26	12.76	12.33	11.96	11.34
9.5	17.30	16.07	15.11	14.34	13.71	13.17	12.71	12.30	11.63
10.0	18.09	16.72	15.67	14.83	14.14	13.56	13.07	12.63	11.92
11.0	19.64	17.99	16.75	15.77	14.98	14.31	13.75	13.26	12.46
12.0	21.15	19.21	17.78	16.67	15.77	15.02	14.39	13.85	12.96
13.0	22.63	20.40	18.77	17.52	16.52	15.69	15.00	14.41	13.44
14.0	24.08	21.54	19.72	18.34	17.23	16.33	15.58	14.93	13.88
15.0	25.51	22.66	20.64	19.12	17.92	16.94	16.12	15.43	14.31
16.0	26.90	23.74	21.53	19.87	18.57	17.52	16.64	15.90	14.71
18.0	29.62	25.81	23.21	21.29	19.81	18.61	17.62	16.79	15.46
20.0	32.25	27.78	24.80	22.62	20.95	19.62	18.52	17.60	16.14

† For an explanation of the headings and the significance of asterisks and high returns which appear on some pages see Introductory Section, pages xx–xxiii.

INTERNAL RATE OF RETURN (IRR) †
(Reflecting projected growth in Full Rental Value to next Review/Reversion)

RENTAL FACTOR †	YEARS TO REVIEW/REVERSION								
	1	2	3	4	5	6	7	8	10
1.1	2.88	3.00	3.12	3.24	3.36	3.47	3.57	3.67	3.86
1.2	3.13	3.26	3.38	3.50	3.61	3.72	3.82	3.92	4.10
1.3	3.38	3.51	3.63	3.75	3.86	3.97	4.07	4.16	4.33
1.4	3.63	3.76	3.88	4.00	4.11	4.21	4.31	4.39	4.55
1.5	3.88	4.01	4.13	4.24	4.35	4.45	4.54	4.62	4.77
1.6	4.13	4.26	4.37	4.48	4.58	4.67	4.76	4.84	4.97
1.7	4.38	4.50	4.61	4.72	4.81	4.90	4.98	5.05	5.17
1.8	4.63	4.74	4.85	4.95	5.04	5.12	5.19	5.25	5.36
1.9	4.87	4.98	5.08	5.18	5.26	5.33	5.40	5.45	5.55
2.0	5.12	5.22	5.31	5.40	5.47	5.54	5.60	5.65	5.73
2.2	5.60	5.69	5.77	5.84	5.89	5.94	5.99	6.02	6.08
2.4	6.08	6.15	6.21	6.26	6.30	6.33	6.36	6.38	6.41
2.5	6.32	6.38	6.43	6.47	6.50	6.52	6.54	6.55	6.57
2.6	6.56	6.61	6.64	6.67	6.69	6.71	6.72	6.72	6.72
2.8	7.03	7.05	7.07	7.07	7.07	7.07	7.06	7.05	7.02
3.0	7.50	7.49	7.48	7.46	7.44	7.42	7.39	7.36	7.30
3.2	7.96	7.93	7.88	7.84	7.80	7.75	7.71	7.66	7.57
3.4	8.43	8.35	8.28	8.21	8.14	8.08	8.01	7.95	7.84
3.5	8.65	8.56	8.48	8.39	8.31	8.24	8.16	8.09	7.96
3.6	8.88	8.77	8.67	8.57	8.48	8.39	8.31	8.23	8.09
3.8	9.34	9.19	9.05	8.93	8.81	8.70	8.60	8.50	8.33
4.0	9.79	9.60	9.43	9.27	9.13	9.00	8.88	8.77	8.56
4.2	10.23	10.00	9.79	9.61	9.44	9.29	9.15	9.02	8.79
4.4	10.68	10.40	10.16	9.94	9.75	9.57	9.41	9.27	9.01
4.5	10.90	10.60	10.33	10.10	9.90	9.71	9.54	9.39	9.11
4.6	11.12	10.79	10.51	10.26	10.05	9.85	9.67	9.51	9.22
4.8	11.55	11.18	10.86	10.58	10.34	10.12	9.92	9.74	9.43
5.0	11.99	11.56	11.20	10.89	10.62	10.38	10.16	9.97	9.63
5.2	12.42	11.94	11.54	11.20	10.90	10.64	10.40	10.19	9.82
5.4	12.85	12.32	11.88	11.50	11.17	10.89	10.63	10.40	10.01
5.5	13.06	12.50	12.04	11.65	11.31	11.01	10.75	10.51	10.10
5.6	13.27	12.69	12.20	11.79	11.44	11.13	10.86	10.61	10.19
5.8	13.69	13.05	12.53	12.08	11.70	11.37	11.08	10.82	10.37
6.0	14.11	13.41	12.84	12.37	11.96	11.61	11.30	11.02	10.55
6.5	15.15	14.30	13.62	13.06	12.58	12.18	11.82	11.50	10.97
7.0	16.17	15.16	14.37	13.72	13.18	12.72	12.31	11.96	11.36
7.5	17.17	16.00	15.10	14.36	13.75	13.23	12.79	12.40	11.74
8.0	18.16	16.82	15.80	14.98	14.30	13.73	13.24	12.81	12.10
8.5	19.13	17.63	16.48	15.58	14.83	14.21	13.67	13.21	12.44
9.0	20.09	18.41	17.15	16.15	15.34	14.67	14.09	13.59	12.77
9.5	21.04	19.18	17.79	16.71	15.84	15.11	14.49	13.96	13.08
10.0	21.97	19.93	18.42	17.25	16.31	15.54	14.88	14.31	13.38
11.0	23.80	21.38	19.63	18.29	17.23	16.35	15.61	14.98	13.95
12.0	25.59	22.78	20.79	19.28	18.09	17.11	16.30	15.61	14.49
13.0	27.34	24.13	21.89	20.22	18.90	17.84	16.95	16.20	14.98
14.0	29.04	25.44	22.95	21.11	19.68	18.52	17.56	16.75	15.45
15.0	30.71	26.70	23.97	21.97	20.42	19.17	18.14	17.28	15.90
16.0	32.35	27.92	24.95	22.79	21.12	19.79	18.70	17.78	16.32
18.0	35.52	30.26	26.82	24.34	22.45	20.96	19.74	18.72	17.10
20.0	38.58	32.48	28.56	25.78	23.68	22.03	20.69	19.58	17.82

† For an explanation of the headings and the significance of asterisks and high returns which appear on some pages, see Introductory Section, pages xx-xxiii.

INTERNAL RATE OF RETURN (IRR) †
(Reflecting projected growth in Full Rental Value to next Review/Reversion)

RENTAL FACTOR †	YEARS TO REVIEW/REVERSION								
	1	2	3	4	5	6	7	8	10
1.1	3.45	3.59	3.73	3.87	4.00	4.12	4.23	4.34	4.53
1.2	3.75	3.90	4.04	4.17	4.29	4.41	4.52	4.62	4.79
1.3	4.05	4.20	4.33	4.46	4.58	4.69	4.79	4.89	5.05
1.4	4.35	4.49	4.63	4.75	4.86	4.97	5.06	5.15	5.29
1.5	4.65	4.79	4.91	5.03	5.14	5.23	5.32	5.40	5.53
1.6	4.94	5.08	5.20	5.30	5.40	5.49	5.57	5.64	5.75
1.7	5.24	5.36	5.47	5.58	5.66	5.74	5.81	5.87	5.97
1.8	5.53	5.65	5.75	5.84	5.92	5.99	6.05	6.10	6.18
1.9	5.82	5.93	6.02	6.10	6.17	6.23	6.28	6.32	6.38
2.0	6.11	6.20	6.29	6.36	6.41	6.46	6.50	6.53	6.58
2.2	6.68	6.75	6.81	6.85	6.89	6.91	6.93	6.95	6.96
2.4	7.25	7.29	7.32	7.33	7.34	7.35	7.34	7.34	7.31
2.5	7.53	7.55	7.57	7.57	7.56	7.55	7.54	7.52	7.48
2.6	7.81	7.82	7.81	7.80	7.78	7.76	7.73	7.71	7.65
2.8	8.37	8.33	8.29	8.25	8.21	8.16	8.11	8.06	7.96
3.0	8.92	8.84	8.77	8.69	8.61	8.54	8.47	8.40	8.27
3.2	9.47	9.34	9.23	9.12	9.01	8.91	8.82	8.73	8.56
3.4	10.01	9.83	9.68	9.53	9.39	9.27	9.15	9.04	8.84
3.5	10.28	10.08	9.90	9.73	9.58	9.44	9.31	9.19	8.97
3.6	10.54	10.32	10.12	9.93	9.77	9.61	9.47	9.34	9.10
3.8	11.08	10.79	10.55	10.33	10.13	9.95	9.78	9.63	9.36
4.0	11.60	11.26	10.97	10.71	10.48	10.27	10.09	9.91	9.61
4.2	12.12	11.73	11.38	11.09	10.82	10.59	10.38	10.19	9.85
4.4	12.64	12.18	11.79	11.45	11.16	10.90	10.66	10.45	10.08
4.5	12.90	12.41	11.99	11.63	11.32	11.05	10.80	10.58	10.19
4.6	13.15	12.63	12.19	11.81	11.48	11.20	10.94	10.71	10.30
4.8	13.66	13.07	12.58	12.16	11.80	11.49	11.21	10.96	10.52
5.0	14.17	13.51	12.96	12.51	12.11	11.77	11.47	11.20	10.73
5.2	14.67	13.94	13.34	12.84	12.42	12.05	11.72	11.43	10.94
5.4	15.17	14.36	13.71	13.17	12.72	12.32	11.97	11.66	11.14
5.5	15.41	14.57	13.90	13.34	12.86	12.45	12.09	11.77	11.23
5.6	15.66	14.78	14.08	13.50	13.01	12.58	12.21	11.89	11.33
5.8	16.15	15.20	14.44	13.82	13.29	12.84	12.45	12.10	11.52
6.0	16.63	15.60	14.79	14.13	13.57	13.09	12.68	12.32	11.70
6.5	17.83	16.61	15.65	14.88	14.25	13.70	13.24	12.83	12.14
7.0	19.01	17.58	16.48	15.61	14.89	14.28	13.76	13.31	12.56
7.5	20.16	18.52	17.28	16.31	15.51	14.84	14.27	13.77	12.95
8.0	21.30	19.44	18.06	16.98	16.10	15.37	14.75	14.21	13.33
8.5	22.42	20.34	18.81	17.62	16.67	15.88	15.21	14.63	13.69
9.0	23.52	21.22	19.54	18.25	17.22	16.37	15.65	15.03	14.03
9.5	24.61	22.07	20.25	18.86	17.75	16.84	16.07	15.42	14.35
10.0	25.68	22.91	20.94	19.44	18.26	17.29	16.48	15.79	14.67
11.0	27.77	24.53	22.26	20.56	19.23	18.15	17.25	16.49	15.26
12.0	29.81	26.08	23.52	21.63	20.15	18.96	17.98	17.15	15.82
13.0	31.79	27.58	24.72	22.63	21.02	19.73	18.66	17.76	16.33
14.0	33.73	29.02	25.87	23.59	21.84	20.45	19.31	18.35	16.82
15.0	35.63	30.41	26.98	24.51	22.63	21.14	19.92	18.90	17.28
16.0	37.48	31.76	28.04	25.39	23.38	21.79	20.50	19.42	17.72
18.0	41.07	34.33	30.05	27.04	24.79	23.02	21.58	20.39	18.53
20.0	44.52	36.76	31.94	28.58	26.09	24.14	22.58	21.29	19.27

† For an explanation of the headings and the significance of asterisks and high returns which appear on some pages, see Introductory Section, pages xx-xxiii.

GROWTH 5%

INTERNAL RATE OF RETURN (IRR) †
(Reflecting projected growth in Full Rental Value to next Review/Reversion)

RENTAL FACTOR †	YEARS TO REVIEW/REVERSION								
	1	2	3	4	5	6	7	8	10
1.1	4.02	4.19	4.34	4.49	4.63	4.76	4.87	4.98	5.17
1.2	4.37	4.53	4.69	4.83	4.96	5.08	5.19	5.29	5.46
1.3	4.72	4.88	5.03	5.16	5.28	5.40	5.50	5.59	5.74
1.4	5.07	5.22	5.36	5.48	5.60	5.70	5.79	5.87	6.00
1.5	5.41	5.55	5.68	5.80	5.90	5.99	6.07	6.14	6.25
1.6	5.75	5.89	6.00	6.11	6.20	6.28	6.35	6.41	6.50
1.7	6.09	6.21	6.32	6.41	6.49	6.56	6.61	6.66	6.73
1.8	6.43	6.54	6.63	6.71	6.77	6.83	6.87	6.90	6.95
1.9	6.76	6.86	6.93	7.00	7.05	7.09	7.12	7.14	7.17
2.0	7.09	7.17	7.23	7.28	7.32	7.34	7.36	7.37	7.38
2.2	7.76	7.79	7.82	7.83	7.84	7.84	7.83	7.81	7.78
2.4	8.41	8.40	8.39	8.37	8.34	8.31	8.27	8.23	8.15
2.5	8.73	8.70	8.67	8.63	8.58	8.53	8.48	8.43	8.33
2.6	9.05	9.00	8.94	8.88	8.82	8.75	8.69	8.63	8.51
2.8	9.69	9.58	9.48	9.38	9.28	9.19	9.09	9.01	8.84
3.0	10.32	10.16	10.00	9.86	9.73	9.60	9.48	9.37	9.16
3.2	10.95	10.72	10.51	10.33	10.16	10.00	9.85	9.71	9.47
3.4	11.56	11.27	11.01	10.78	10.57	10.38	10.21	10.05	9.76
3.5	11.87	11.54	11.26	11.00	10.77	10.57	10.38	10.21	9.90
3.6	12.17	11.81	11.50	11.22	10.97	10.75	10.55	10.37	10.04
3.8	12.78	12.34	11.97	11.65	11.37	11.11	10.88	10.67	10.31
4.0	13.38	12.87	12.44	12.07	11.75	11.46	11.20	10.97	10.57
4.2	13.97	13.38	12.89	12.48	12.12	11.80	11.51	11.26	10.82
4.4	14.56	13.89	13.34	12.88	12.47	12.12	11.81	11.54	11.06
4.5	14.85	14.14	13.56	13.07	12.65	12.29	11.96	11.67	11.18
4.6	15.14	14.39	13.78	13.26	12.83	12.44	12.11	11.81	11.29
4.8	15.72	14.88	14.21	13.64	13.17	12.75	12.39	12.07	11.52
5.0	16.29	15.37	14.63	14.02	13.50	13.06	12.67	12.32	11.74
5.2	16.86	15.84	15.04	14.38	13.83	13.35	12.94	12.57	11.95
5.4	17.42	16.31	15.45	14.74	14.14	13.64	13.20	12.81	12.16
5.5	17.70	16.55	15.65	14.91	14.30	13.78	13.33	12.93	12.26
5.6	17.98	16.78	15.84	15.09	14.45	13.92	13.45	13.05	12.36
5.8	18.53	17.24	16.24	15.43	14.76	14.19	13.70	13.27	12.56
6.0	19.08	17.69	16.62	15.76	15.06	14.46	13.95	13.50	12.75
6.5	20.43	18.79	17.56	16.58	15.78	15.10	14.53	14.03	13.21
7.0	21.75	19.86	18.46	17.36	16.46	15.72	15.08	14.54	13.64
7.5	23.05	20.90	19.33	18.10	17.12	16.30	15.61	15.02	14.05
8.0	24.33	21.91	20.16	18.82	17.74	16.86	16.11	15.48	14.43
8.5	25.59	22.90	20.98	19.51	18.35	17.39	16.60	15.91	14.80
9.0	26.82	23.86	21.76	20.18	18.93	17.91	17.06	16.33	15.16
9.5	28.03	24.79	22.52	20.82	19.49	18.40	17.50	16.73	15.50
10.0	29.23	25.70	23.26	21.45	20.03	18.88	17.93	17.12	15.82
11.0	31.57	27.47	24.69	22.64	21.06	19.78	18.73	17.85	16.43
12.0	33.84	29.16	26.04	23.77	22.02	20.63	19.49	18.53	17.00
13.0	36.04	30.78	27.32	24.83	22.93	21.43	20.20	19.17	17.54
14.0	38.20	32.34	28.55	25.85	23.80	22.18	20.87	19.77	18.04
15.0	40.30	33.85	29.73	26.82	24.62	22.90	21.50	20.34	18.51
16.0	42.35	35.31	30.86	27.75	25.41	23.58	22.11	20.88	18.96
18.0	46.32	38.09	33.01	29.49	26.88	24.86	23.23	21.89	19.79
20.0	50.13	40.71	35.01	31.11	28.24	26.03	24.26	22.81	20.55

† For an explanation of the headings and the significance of asterisks and high returns which appear on some pages, see Introductory Section, pages xx-xxiii.

INTERNAL RATE OF RETURN (IRR) †
(Reflecting projected growth in Full Rental Value to next Review/Reversion)

RENTAL FACTOR †	YEARS TO REVIEW/REVERSION								
	1	2	3	4	5	6	7	8	10
1.1	4.59	4.78	4.95	5.10	5.25	5.38	5.51	5.62	5.8C
1.2	4.99	5.17	5.33	5.48	5.62	5.74	5.85	5.95	6.1
1.3	5.39	5.56	5.71	5.85	5.97	6.08	6.18	6.26	6.4C
1.4	5.78	5.94	6.08	6.21	6.32	6.41	6.50	6.57	6.68
1.5	6.17	6.31	6.44	6.55	6.65	6.73	6.80	6.86	6.9
1.6	6.55	6.69	6.80	6.89	6.98	7.04	7.10	7.14	7.2C
1.7	6.94	7.05	7.15	7.23	7.29	7.34	7.38	7.41	7.4
1.8	7.32	7.41	7.49	7.55	7.60	7.63	7.66	7.67	7.69
1.9	7.70	7.77	7.83	7.87	7.90	7.92	7.92	7.93	7.91
2.0	8.07	8.12	8.16	8.18	8.19	8.19	8.18	8.17	8.13
2.2	8.82	8.82	8.80	8.78	8.75	8.72	8.68	8.64	8.55
2.4	9.55	9.49	9.43	9.36	9.29	9.22	9.15	9.08	8.94
2.5	9.91	9.82	9.73	9.64	9.55	9.46	9.38	9.29	9.13
2.6	10.28	10.15	10.03	9.92	9.81	9.70	9.60	9.50	9.31
2.8	10.99	10.80	10.62	10.46	10.30	10.16	10.02	9.90	9.66
3.0	11.70	11.43	11.19	10.98	10.78	10.60	10.43	10.28	10.00
3.2	12.40	12.05	11.75	11.48	11.24	11.02	10.82	10.64	10.31
3.4	13.09	12.66	12.29	11.97	11.69	11.43	11.20	10.99	10.62
3.5	13.43	12.96	12.56	12.21	11.90	11.63	11.38	11.16	10.76
3.6	13.77	13.26	12.82	12.45	12.12	11.82	11.56	11.32	10.91
3.8	14.45	13.84	13.34	12.91	12.53	12.20	11.91	11.65	11.19
4.0	15.12	14.42	13.84	13.36	12.94	12.57	12.25	11.96	11.46
4.2	15.78	14.98	14.34	13.79	13.33	12.93	12.57	12.26	11.72
4.4	16.44	15.54	14.82	14.22	13.71	13.27	12.89	12.55	11.97
4.5	16.76	15.81	15.05	14.43	13.90	13.44	13.04	12.69	12.09
4.6	17.08	16.09	15.29	14.63	14.08	13.61	13.19	12.83	12.21
4.8	17.73	16.62	15.75	15.04	14.44	13.93	13.49	13.10	12.44
5.0	18.36	17.15	16.21	15.44	14.80	14.25	13.78	13.37	12.67
5.2	18.99	17.67	16.65	15.83	15.14	14.56	14.06	13.62	12.89
5.4	19.62	18.19	17.09	16.20	15.48	14.86	14.33	13.87	13.10
5.5	19.93	18.44	17.30	16.39	15.64	15.01	14.47	14.00	13.21
5.6	20.23	18.69	17.51	16.58	15.81	15.16	14.60	14.12	13.31
5.8	20.85	19.19	17.93	16.94	16.13	15.44	14.86	14.35	13.51
6.0	21.46	19.68	18.35	17.30	16.44	15.72	15.11	14.59	13.71
6.5	22.95	20.88	19.35	18.16	17.20	16.40	15.72	15.14	14.18
7.0	24.42	22.04	20.31	18.98	17.92	17.04	16.30	15.67	14.63
7.5	25.85	23.17	21.24	19.77	18.61	17.65	16.85	16.16	15.05
8.0	27.26	24.26	22.14	20.53	19.27	18.23	17.37	16.64	15.45
8.5	28.64	25.32	23.00	21.26	19.90	18.79	17.87	17.09	15.82
9.0	30.00	26.35	23.84	21.97	20.51	19.33	18.35	17.52	16.19
9.5	31.33	27.36	24.65	22.65	21.09	19.84	18.81	17.93	16.53
10.0	32.65	28.34	25.44	23.30	21.66	20.34	19.25	18.33	16.87
11.0	35.21	30.24	26.94	24.56	22.73	21.28	20.08	19.08	17.50
12.0	37.70	32.05	28.37	25.74	23.74	22.16	20.86	19.79	18.08
13.0	40.11	33.79	29.74	26.86	24.69	22.98	21.60	20.44	18.63
14.0	42.47	35.46	31.04	27.92	25.59	23.77	22.29	21.06	19.14
15.0	44.76	37.08	32.28	28.94	26.45	24.51	22.94	21.65	19.62
16.0	46.99	38.63	33.47	29.91	27.27	25.21	23.57	22.21	20.08
18.0	51.32	41.60	35.73	31.73	28.80	26.53	24.72	23.24	20.93
20.0	55.46	44.39	37.84	33.42	30.21	27.75	25.79	24.19	21.71

† For an explanation of the headings and the significance of asterisks and high returns which appear on some page see Introductory Section, pages xx-xxiii.

INTERNAL RATE OF RETURN (IRR) †
(Reflecting projected growth in Full Rental Value to next Review/Reversion)

RENTAL FACTOR †	YEARS TO REVIEW/REVERSION								
	1	2	3	4	5	6	7	8	10
1.1	5.16	5.36	5.55	5.71	5.87	6.00	6.13	6.23	6.41
1.2	5.61	5.80	5.97	6.13	6.27	6.39	6.49	6.59	6.74
1.3	6.05	6.23	6.39	6.53	6.65	6.76	6.85	6.92	7.05
1.4	6.49	6.65	6.79	6.92	7.02	7.11	7.19	7.25	7.34
1.5	6.92	7.07	7.19	7.30	7.38	7.45	7.51	7.56	7.62
1.6	7.35	7.48	7.58	7.66	7.73	7.78	7.82	7.85	7.88
1.7	7.78	7.88	7.96	8.02	8.07	8.10	8.13	8.14	8.14
1.8	8.20	8.28	8.34	8.38	8.40	8.41	8.42	8.41	8.39
1.9	8.62	8.67	8.70	8.72	8.72	8.71	8.70	8.68	8.62
2.0	9.04	9.06	9.06	9.05	9.03	9.01	8.97	8.94	8.85
2.2	9.87	9.82	9.76	9.70	9.64	9.57	9.50	9.43	9.28
2.4	10.68	10.56	10.44	10.32	10.21	10.10	9.99	9.89	9.69
2.5	11.08	10.92	10.77	10.63	10.49	10.36	10.23	10.11	9.89
2.6	11.48	11.28	11.10	10.92	10.76	10.61	10.46	10.33	10.07
2.8	12.28	11.99	11.73	11.50	11.29	11.09	10.91	10.74	10.44
3.0	13.06	12.68	12.35	12.05	11.79	11.55	11.34	11.14	10.78
3.2	13.83	13.35	12.95	12.59	12.28	12.00	11.74	11.51	11.11
3.4	14.59	14.01	13.53	13.11	12.75	12.43	12.14	11.88	11.42
3.5	14.97	14.34	13.81	13.37	12.98	12.63	12.33	12.05	11.58
3.6	15.35	14.66	14.10	13.62	13.20	12.84	12.51	12.22	11.72
3.8	16.09	15.30	14.65	14.11	13.64	13.24	12.88	12.56	12.01
4.0	16.83	15.92	15.19	14.58	14.07	13.62	13.23	12.88	12.29
4.2	17.55	16.53	15.72	15.05	14.48	13.99	13.57	13.19	12.56
4.4	18.27	17.13	16.23	15.50	14.88	14.35	13.90	13.49	12.81
4.5	18.63	17.43	16.48	15.72	15.08	14.53	14.06	13.64	12.94
4.6	18.99	17.72	16.73	15.94	15.27	14.70	14.21	13.78	13.06
4.8	19.69	18.30	17.23	16.37	15.65	15.04	14.52	14.07	13.30
5.0	20.39	18.87	17.71	16.78	16.02	15.38	14.82	14.34	13.53
5.2	21.08	19.43	18.18	17.19	16.38	15.70	15.11	14.61	13.76
5.4	21.76	19.98	18.65	17.59	16.73	16.01	15.40	14.86	13.98
5.5	22.10	20.26	18.88	17.79	16.91	16.17	15.54	14.99	14.09
5.6	22.44	20.53	19.10	17.98	17.08	16.32	15.67	15.11	14.19
5.8	23.11	21.06	19.55	18.37	17.41	16.62	15.94	15.36	14.40
6.0	23.77	21.59	19.99	18.74	17.74	16.91	16.20	15.60	14.60
6.5	25.40	22.88	21.05	19.65	18.53	17.61	16.84	16.17	15.09
7.0	27.00	24.12	22.07	20.52	19.28	18.28	17.43	16.71	15.54
7.5	28.56	25.33	23.05	21.35	20.00	18.91	18.00	17.22	15.97
8.0	30.10	26.49	24.00	22.14	20.69	19.51	18.54	17.71	16.38
8.5	31.60	27.63	24.91	22.90	21.34	20.09	19.05	18.17	16.77
9.0	33.07	28.73	25.79	23.64	21.98	20.64	19.54	18.62	17.14
9.5	34.52	29.80	26.65	24.35	22.58	21.18	20.02	19.04	17.49
10.0	35.95	30.85	27.48	25.04	23.17	21.69	20.47	19.45	17.83
11.0	38.72	32.87	29.06	26.35	24.29	22.66	21.33	20.22	18.47
12.0	41.41	34.79	30.56	27.58	25.33	23.57	22.13	20.94	19.07
13.0	44.02	36.64	31.99	28.74	26.32	24.42	22.89	21.62	19.63
14.0	46.56	38.41	33.35	29.85	27.25	25.23	23.60	22.25	20.15
15.0	49.04	40.11	34.66	30.91	28.14	25.99	24.27	22.85	20.64
16.0	51.45	41.76	35.91	31.91	28.98	26.72	24.91	23.42	21.11
18.0	56.10	44.89	38.27	33.81	30.56	28.07	26.10	24.48	21.98
20.0	60.56	47.84	40.47	35.56	32.02	29.32	27.19	25.45	22.77

† For an explanation of the headings and the significance of asterisks and high returns which appear on some pages, see Introductory Section, pages xx–xxiii.

INTERNAL RATE OF RETURN (IRR) †
(Reflecting projected growth in Full Rental Value to next Review/Reversion)

RENTAL FACTOR †	YEARS TO REVIEW/REVERSION								
	1	2	3	4	5	6	7	8	10
1.1	5.73	5.95	6.14	6.32	6.48	6.61	6.74	6.84	7.0
1.2	6.22	6.43	6.61	6.76	6.90	7.02	7.13	7.21	7.3
1.3	6.71	6.90	7.06	7.20	7.32	7.42	7.50	7.57	7.6
1.4	7.19	7.36	7.50	7.62	7.71	7.79	7.86	7.91	7.9
1.5	7.67	7.81	7.93	8.02	8.10	8.16	8.20	8.23	8.2
1.6	8.14	8.26	8.35	8.42	8.47	8.51	8.53	8.54	8.5
1.7	8.61	8.70	8.76	8.80	8.83	8.84	8.85	8.84	8.8
1.8	9.08	9.13	9.17	9.18	9.18	9.17	9.15	9.13	9.0
1.9	9.54	9.56	9.56	9.55	9.52	9.49	9.45	9.40	9.3
2.0	10.00	9.98	9.95	9.90	9.85	9.80	9.73	9.67	9.5
2.2	10.91	10.81	10.70	10.60	10.49	10.38	10.28	10.18	9.99
2.4	11.80	11.61	11.43	11.26	11.10	10.94	10.80	10.66	10.4
2.5	12.24	12.00	11.78	11.58	11.39	11.21	11.05	10.89	10.6
2.6	12.68	12.39	12.13	11.89	11.67	11.47	11.29	11.11	10.8
2.8	13.54	13.15	12.81	12.50	12.23	11.98	11.75	11.54	11.1
3.0	14.40	13.89	13.46	13.09	12.76	12.46	12.20	11.95	11.53
3.2	15.24	14.62	14.10	13.66	13.27	12.93	12.62	12.35	11.86
3.4	16.07	15.33	14.72	14.21	13.76	13.37	13.03	12.72	12.19
3.5	16.48	15.68	15.03	14.48	14.00	13.59	13.23	12.90	12.34
3.6	16.89	16.03	15.33	14.74	14.24	13.80	13.42	13.08	12.49
3.8	17.70	16.71	15.91	15.25	14.70	14.22	13.80	13.42	12.79
4.0	18.50	17.38	16.48	15.76	15.14	14.62	14.16	13.76	13.07
4.2	19.29	18.03	17.04	16.24	15.57	15.00	14.51	14.08	13.35
4.4	20.07	18.67	17.59	16.72	15.99	15.38	14.85	14.39	13.6
4.5	20.46	18.99	17.86	16.95	16.20	15.56	15.01	14.54	13.74
4.6	20.85	19.30	18.12	17.18	16.40	15.74	15.18	14.68	13.86
4.8	21.61	19.92	18.64	17.63	16.79	16.09	15.50	14.97	14.1
5.0	22.37	20.53	19.15	18.07	17.18	16.44	15.80	15.26	14.35
5.2	23.11	21.13	19.65	18.49	17.55	16.77	16.11	15.53	14.58
5.4	23.85	21.72	20.14	18.91	17.92	17.10	16.40	15.79	14.80
5.5	24.22	22.01	20.38	19.12	18.10	17.26	16.54	15.92	14.91
5.6	24.59	22.30	20.62	19.32	18.28	17.41	16.68	16.05	15.02
5.8	25.31	22.87	21.09	19.72	18.63	17.72	16.96	16.30	15.23
6.0	26.03	23.43	21.55	20.12	18.97	18.02	17.23	16.55	15.44
6.5	27.79	24.80	22.67	21.06	19.79	18.75	17.88	17.14	15.93
7.0	29.51	26.12	23.74	21.97	20.57	19.44	18.49	17.69	16.40
7.5	31.20	27.39	24.77	22.83	21.31	20.09	19.07	18.22	16.83
8.0	32.85	28.63	25.76	23.66	22.02	20.71	19.63	18.71	17.25
8.5	34.47	29.83	26.72	24.45	22.70	21.31	20.16	19.19	17.65
9.0	36.05	31.00	27.64	25.22	23.36	21.88	20.66	19.64	18.02
9.5	37.61	32.13	28.54	25.95	23.99	22.42	21.15	20.08	18.38
10.0	39.14	33.24	29.40	26.67	24.59	22.95	21.61	20.50	18.73
11.0	42.12	35.36	31.06	28.02	25.74	23.95	22.49	21.29	19.38
12.0	45.00	37.39	32.63	29.30	26.82	24.88	23.32	22.02	19.99
13.0	47.80	39.34	34.12	30.51	27.84	25.76	24.09	22.71	20.56
14.0	50.51	41.20	35.54	31.66	28.80	26.59	24.81	23.36	21.09
15.0	53.15	42.99	36.89	32.75	29.71	27.37	25.50	23.97	21.59
16.0	55.73	44.72	38.20	33.79	30.58	28.11	26.15	24.55	22.06
18.0	60.70	48.01	40.65	35.75	32.20	29.51	27.37	25.63	22.95
20.0	65.44	51.10	42.93	37.55	33.70	30.78	28.48	26.61	23.75

† For an explanation of the headings and the significance of asterisks and high returns which appear on some pages see Introductory Section, pages xx-xxiii.

INTERNAL RATE OF RETURN (IRR) †
(Reflecting projected growth in Full Rental Value to next Review/Reversion)

RENTAL FACTOR †	YEARS TO REVIEW/REVERSION								
	1	**2**	**3**	**4**	**5**	**6**	**7**	**8**	**10**
1.1	6.87	7.11	7.33	7.51	7.67	7.81	7.93	8.02	8.17
1.2	7.45	7.67	7.86	8.02	8.16	8.27	8.36	8.43	8.53
1.3	8.03	8.22	8.38	8.51	8.62	8.70	8.76	8.81	8.87
1.4	8.60	8.76	8.88	8.98	9.06	9.11	9.15	9.18	9.19
1.5	9.16	9.28	9.38	9.44	9.49	9.51	9.52	9.52	9.50
1.6	9.72	9.80	9.86	9.89	9.90	9.90	9.88	9.86	9.79
1.7	10.27	10.31	10.33	10.32	10.30	10.27	10.23	10.18	10.07
1.8	10.82	10.81	10.78	10.74	10.69	10.62	10.56	10.48	10.34
1.9	11.36	11.30	11.23	11.15	11.06	10.97	10.87	10.78	10.59
2.0	11.90	11.79	11.67	11.55	11.42	11.30	11.18	11.07	10.84
2.2	12.96	12.73	12.52	12.31	12.12	11.94	11.77	11.61	11.31
2.4	14.00	13.65	13.33	13.04	12.78	12.54	12.32	12.12	11.75
2.5	14.51	14.09	13.73	13.40	13.10	12.83	12.59	12.36	11.96
2.6	15.02	14.54	14.11	13.74	13.41	13.12	12.85	12.60	12.16
2.8	16.03	15.40	14.87	14.41	14.01	13.66	13.34	13.05	12.55
3.0	17.02	16.25	15.60	15.06	14.59	14.18	13.81	13.49	12.92
3.2	18.00	17.07	16.31	15.68	15.14	14.68	14.26	13.90	13.27
3.4	18.96	17.87	17.00	16.28	15.67	15.15	14.70	14.29	13.60
3.5	19.44	18.26	17.33	16.57	15.93	15.38	14.90	14.48	13.77
3.6	19.91	18.65	17.67	16.86	16.19	15.61	15.11	14.67	13.92
3.8	20.85	19.42	18.31	17.42	16.68	16.05	15.51	15.03	14.23
4.0	21.77	20.17	18.94	17.97	17.16	16.48	15.89	15.38	14.53
4.2	22.68	20.90	19.56	18.49	17.62	16.89	16.26	15.72	14.81
4.4	23.58	21.62	20.16	19.01	18.07	17.29	16.62	16.04	15.08
4.5	24.02	21.98	20.45	19.26	18.29	17.48	16.79	16.20	15.21
4.6	24.46	22.33	20.74	19.51	18.50	17.67	16.96	16.35	15.35
4.8	25.34	23.02	21.31	19.99	18.93	18.05	17.30	16.66	15.60
5.0	26.21	23.70	21.87	20.47	19.34	18.41	17.62	16.95	15.85
5.2	27.06	24.36	22.42	20.93	19.74	18.76	17.94	17.24	16.09
5.4	27.91	25.02	22.95	21.38	20.13	19.11	18.25	17.51	16.32
5.5	28.33	25.34	23.21	21.60	20.32	19.27	18.40	17.65	16.43
5.6	28.74	25.66	23.47	21.82	20.51	19.44	18.55	17.78	16.54
5.8	29.57	26.29	23.99	22.25	20.88	19.77	18.84	18.05	16.76
6.0	30.39	26.92	24.49	22.67	21.25	20.09	19.12	18.30	16.97
6.5	32.40	28.43	25.71	23.69	22.12	20.85	19.80	18.91	17.48
7.0	34.36	29.89	26.87	24.66	22.95	21.57	20.44	19.49	17.96
7.5	36.27	31.30	27.99	25.58	23.73	22.26	21.05	20.03	18.42
8.0	38.14	32.66	29.06	26.46	24.49	22.91	21.63	20.55	18.84
8.5	39.97	33.98	30.09	27.31	25.20	23.54	22.18	21.05	19.25
9.0	41.76	35.26	31.09	28.13	25.89	24.14	22.71	21.52	19.64
9.5	43.52	36.51	32.05	28.91	26.56	24.71	23.21	21.97	20.01
10.0	45.24	37.72	32.98	29.67	27.20	25.26	23.70	22.40	20.37
11.0	48.60	40.04	34.76	31.11	28.41	26.31	24.62	23.22	21.04
12.0	51.84	42.26	36.45	32.47	29.54	27.28	25.47	23.98	21.66
13.0	54.98	44.38	38.04	33.75	30.61	28.20	26.27	24.69	22.24
14.0	58.02	46.41	39.56	34.96	31.62	29.06	27.03	25.36	22.79
15.0	60.98	48.36	41.02	36.12	32.58	29.88	27.74	25.99	23.31
16.0	63.86	50.23	42.40	37.22	33.49	30.66	28.42	26.59	23.79
18.0	69.40	53.80	45.02	39.28	35.19	32.10	29.67	27.71	24.70
20.0	****	57.14	47.46	41.19	36.76	33.43	30.83	28.72	25.52

† For an explanation of the headings and the significance of asterisks and high returns which appear on some pages, see Introductory Section, pages xx-xxiii.

INTERNAL RATE OF RETURN (IRR) †
(Reflecting projected growth in Full Rental Value to next Review/Reversion)

RENTAL FACTOR †	YEARS TO REVIEW/REVERSION								
	1	2	3	4	5	6	7	8	10
1.1	8.00	8.27	8.50	8.69	8.85	8.98	9.09	9.17	9.28
1.2	8.67	8.91	9.10	9.25	9.38	9.47	9.55	9.60	9.66
1.3	9.34	9.53	9.68	9.79	9.88	9.94	9.98	10.01	10.01
1.4	9.99	10.14	10.24	10.32	10.36	10.39	10.40	10.39	10.35
1.5	10.64	10.73	10.79	10.82	10.83	10.82	10.79	10.76	10.67
1.6	11.28	11.32	11.32	11.31	11.28	11.23	11.17	11.11	10.97
1.7	11.91	11.89	11.84	11.78	11.71	11.63	11.54	11.45	11.26
1.8	12.54	12.45	12.35	12.24	12.12	12.01	11.89	11.77	11.54
1.9	13.16	13.00	12.84	12.68	12.53	12.37	12.23	12.08	11.80
2.0	13.77	13.54	13.33	13.12	12.92	12.73	12.55	12.38	12.06
2.2	14.98	14.60	14.26	13.95	13.67	13.41	13.17	12.95	12.55
2.4	16.16	15.62	15.15	14.74	14.38	14.05	13.75	13.48	13.00
2.5	16.74	16.12	15.59	15.12	14.72	14.36	14.03	13.73	13.22
2.6	17.32	16.61	16.01	15.50	15.05	14.65	14.30	13.98	13.43
2.8	18.46	17.57	16.84	16.22	15.69	15.23	14.82	14.46	13.83
3.0	19.59	18.50	17.63	16.91	16.30	15.78	15.32	14.91	14.21
3.2	20.69	19.41	18.40	17.58	16.89	16.30	15.79	15.33	14.57
3.4	21.77	20.30	19.15	18.22	17.46	16.80	16.24	15.74	14.91
3.5	22.31	20.73	19.51	18.54	17.73	17.05	16.46	15.94	15.08
3.6	22.84	21.16	19.87	18.85	18.00	17.28	16.67	16.14	15.24
3.8	23.89	22.00	20.57	19.45	18.52	17.75	17.09	16.51	15.56
4.0	24.93	22.82	21.25	20.03	19.03	18.20	17.49	16.87	15.86
4.2	25.95	23.63	21.92	20.59	19.52	18.63	17.87	17.22	16.15
4.4	26.96	24.42	22.56	21.14	19.99	19.04	18.24	17.56	16.43
4.5	27.46	24.80	22.88	21.40	20.22	19.25	18.43	17.72	16.57
4.6	27.95	25.19	23.19	21.67	20.45	19.45	18.60	17.88	16.70
4.8	28.93	25.94	23.81	22.18	20.89	19.84	18.95	18.20	16.96
5.0	29.90	26.68	24.41	22.69	21.33	20.22	19.29	18.50	17.21
5.2	30.86	27.41	24.99	23.18	21.75	20.59	19.62	18.79	17.46
5.4	31.80	28.12	25.57	23.66	22.16	20.95	19.94	19.08	17.70
5.5	32.27	28.47	25.85	23.89	22.36	21.12	20.09	19.22	17.81
5.6	32.74	28.82	26.13	24.12	22.56	21.30	20.25	19.36	17.93
5.8	33.66	29.51	26.68	24.58	22.95	21.64	20.55	19.63	18.15
6.0	34.57	30.19	27.22	25.03	23.33	21.97	20.84	19.89	18.37
6.5	36.81	31.83	28.52	26.10	24.25	22.77	21.55	20.53	18.89
7.0	38.98	33.42	29.76	27.12	25.12	23.52	22.21	21.12	19.38
7.5	41.11	34.94	30.95	28.10	25.94	24.23	22.84	21.68	19.85
8.0	43.18	36.42	32.09	29.03	26.73	24.91	23.44	22.22	20.28
8.5	45.20	37.84	33.19	29.92	27.48	25.56	24.01	22.72	20.70
9.0	47.19	39.22	34.25	30.78	28.20	26.18	24.56	23.21	21.10
9.5	49.13	40.56	35.27	31.61	28.89	26.78	25.08	23.67	21.48
10.0	51.03	41.87	36.26	32.41	29.56	27.35	25.58	24.12	21.84
11.0	54.73	44.37	38.15	33.92	30.82	28.43	26.52	24.96	22.53
12.0	58.30	46.75	39.93	35.34	32.00	29.44	27.41	25.74	23.16
13.0	61.75	49.02	41.62	36.68	33.11	30.39	28.23	26.47	23.76
14.0	65.09	51.20	43.23	37.96	34.16	31.28	29.01	27.16	24.31
15.0	68.34	53.29	44.76	39.16	35.16	32.13	29.74	27.81	24.84
16.0	****	55.30	46.23	40.31	36.11	32.93	30.44	28.42	25.34
18.0	****	59.11	48.99	42.47	37.87	34.43	31.73	29.56	26.26
20.0	****	62.68	51.55	44.46	39.50	35.80	32.92	30.61	27.10

† For an explanation of the headings and the significance of asterisks and high returns which appear on some pages see Introductory Section, pages xx-xxiii.

INTERNAL RATE OF RETURN (IRR) †

(Reflecting projected growth in Full Rental Value to next Review/Reversion)

RENTAL FACTOR †	YEARS TO REVIEW/REVERSION								
	1	2	3	4	5	6	7	8	10
1.1	9.14	9.42	9.66	9.85	10.01	10.13	10.22	10.29	10.36
1.2	9.89	10.13	10.32	10.46	10.57	10.65	10.70	10.74	10.75
1.3	10.64	10.82	10.96	11.05	11.11	11.15	11.16	11.16	11.12
1.4	11.38	11.50	11.57	11.62	11.63	11.62	11.60	11.56	11.46
1.5	12.10	12.16	12.17	12.16	12.13	12.08	12.02	11.95	11.79
1.6	12.82	12.80	12.76	12.69	12.60	12.51	12.41	12.31	12.10
1.7	13.53	13.44	13.32	13.20	13.06	12.93	12.79	12.66	12.40
1.8	14.23	14.05	13.87	13.69	13.51	13.33	13.16	13.00	12.68
1.9	14.93	14.66	14.41	14.17	13.94	13.72	13.51	13.32	12.96
2.0	15.61	15.26	14.93	14.63	14.35	14.09	13.85	13.63	13.22
2.2	16.96	16.41	15.94	15.52	15.15	14.81	14.50	14.22	13.72
2.4	18.28	17.53	16.90	16.37	15.89	15.48	15.10	14.77	14.18
2.5	18.93	18.08	17.37	16.77	16.25	15.80	15.39	15.03	14.40
2.6	19.57	18.61	17.83	17.17	16.60	16.11	15.67	15.29	14.62
2.8	20.84	19.66	18.72	17.94	17.28	16.71	16.21	15.77	15.03
3.0	22.09	20.68	19.57	18.67	17.92	17.28	16.73	16.24	15.41
3.2	23.31	21.66	20.40	19.38	18.54	17.83	17.22	16.68	15.78
3.4	24.51	22.62	21.19	20.06	19.13	18.35	17.68	17.10	16.13
3.5	25.11	23.09	21.58	20.39	19.42	18.61	17.91	17.31	16.30
3.6	25.69	23.56	21.97	20.72	19.70	18.85	18.13	17.51	16.47
3.8	26.86	24.47	22.71	21.35	20.25	19.34	18.56	17.89	16.79
4.0	28.00	25.36	23.44	21.96	20.78	19.80	18.97	18.26	17.10
4.2	29.13	26.23	24.15	22.56	21.29	20.25	19.37	18.62	17.40
4.4	30.24	27.08	24.83	23.13	21.78	20.68	19.76	18.97	17.68
4.5	30.79	27.50	25.17	23.41	22.03	20.89	19.94	19.14	17.82
4.6	31.33	27.91	25.50	23.69	22.26	21.10	20.13	19.30	17.96
4.8	32.41	28.72	26.16	24.24	22.73	21.51	20.49	19.62	18.23
5.0	33.47	29.52	26.79	24.76	23.18	21.90	20.84	19.94	18.48
5.2	34.52	30.30	27.41	25.28	23.62	22.28	21.17	20.24	18.73
5.4	35.56	31.07	28.02	25.78	24.05	22.65	21.50	20.53	18.98
5.5	36.07	31.45	28.32	26.03	24.26	22.84	21.66	20.68	19.09
5.6	36.58	31.82	28.62	26.27	24.46	23.02	21.82	20.82	19.21
5.8	37.59	32.56	29.20	26.75	24.87	23.37	22.13	21.10	19.44
6.0	38.59	33.29	29.77	27.22	25.27	23.71	22.44	21.37	19.66
6.5	41.04	35.05	31.14	28.35	26.22	24.54	23.16	22.01	20.19
7.0	43.42	36.74	32.45	29.41	27.12	25.31	23.84	22.62	20.69
7.5	45.74	38.37	33.71	30.43	27.98	26.05	24.49	23.20	21.16
8.0	48.00	39.94	34.91	31.41	28.79	26.75	25.11	23.74	21.61
8.5	50.21	41.46	36.07	32.34	29.57	27.42	25.69	24.26	22.03
9.0	52.37	42.94	37.18	33.23	30.32	28.06	26.25	24.76	22.44
9.5	54.48	44.37	38.26	34.10	31.04	28.68	26.79	25.24	22.82
10.0	56.55	45.75	39.30	34.93	31.73	29.26	27.30	25.69	23.19
11.0	60.56	48.41	41.28	36.50	33.03	30.38	28.27	26.55	23.89
12.0	64.44	50.94	43.15	37.98	34.26	31.42	29.18	27.35	24.54
13.0	68.18	53.35	44.92	39.38	35.40	32.39	30.02	28.09	25.14
14.0	****	55.66	46.61	40.70	36.49	33.31	30.82	28.80	25.71
15.0	****	57.87	48.21	41.95	37.52	34.18	31.57	29.46	26.24
16.0	****	60.00	49.75	43.15	38.49	35.01	32.28	30.09	26.74
18.0	****	64.03	52.63	45.39	40.32	36.54	33.61	31.25	27.68
20.0	****	67.80	55.31	47.45	41.99	37.95	34.82	32.31	28.53

† For an explanation of the headings and the significance of asterisks and high returns which appear on some pages, see Introductory Section, pages xx-xxiii.

INTERNAL RATE OF RETURN (IRR) †
(Reflecting projected growth in Full Rental Value to next Review/Reversion)

RENTAL FACTOR †	YEARS TO REVIEW/REVERSION								
	1	2	3	4	5	6	7	8	10
1.1	10.27	10.57	10.81	11.00	11.14	11.25	11.33	11.38	11.42
1.2	11.11	11.34	11.52	11.66	11.75	11.80	11.83	11.84	11.81
1.3	11.93	12.10	12.22	12.28	12.32	12.33	12.31	12.28	12.19
1.4	12.75	12.84	12.88	12.89	12.87	12.82	12.77	12.70	12.54
1.5	13.56	13.56	13.53	13.47	13.39	13.30	13.20	13.09	12.87
1.6	14.35	14.27	14.16	14.03	13.89	13.75	13.61	13.47	13.19
1.7	15.14	14.95	14.76	14.57	14.38	14.19	14.01	13.83	13.49
1.8	15.91	15.63	15.35	15.09	14.84	14.61	14.38	14.17	13.78
1.9	16.68	16.29	15.93	15.60	15.29	15.01	14.75	14.50	14.06
2.0	17.43	16.93	16.49	16.09	15.73	15.40	15.10	14.82	14.33
2.2	18.91	18.19	17.57	17.03	16.56	16.14	15.77	15.43	14.83
2.4	20.37	19.39	18.60	17.92	17.35	16.84	16.39	15.99	15.30
2.5	21.08	19.98	19.09	18.35	17.72	17.17	16.69	16.26	15.53
2.6	21.78	20.56	19.58	18.77	18.09	17.50	16.98	16.52	15.75
2.8	23.17	21.69	20.52	19.58	18.79	18.12	17.54	17.03	16.16
3.0	24.54	22.78	21.43	20.35	19.46	18.71	18.07	17.50	16.56
3.2	25.87	23.84	22.30	21.09	20.11	19.28	18.57	17.95	16.93
3.4	27.19	24.87	23.15	21.81	20.72	19.82	19.05	18.39	17.29
3.5	27.83	25.37	23.56	22.15	21.02	20.08	19.28	18.59	17.46
3.6	28.47	25.87	23.96	22.49	21.31	20.34	19.51	18.80	17.63
3.8	29.74	26.84	24.75	23.16	21.88	20.83	19.95	19.19	17.96
4.0	30.99	27.79	25.52	23.80	22.43	21.31	20.37	19.57	18.27
4.2	32.21	28.72	26.27	24.42	22.96	21.77	20.78	19.94	18.57
4.4	33.42	29.63	26.99	25.02	23.47	22.22	21.18	20.29	18.86
4.5	34.02	30.07	27.34	25.31	23.72	22.44	21.37	20.46	19.00
4.6	34.61	30.51	27.69	25.60	23.97	22.65	21.56	20.63	19.14
4.8	35.78	31.38	28.38	26.17	24.45	23.07	21.93	20.96	19.41
5.0	36.93	32.23	29.05	26.72	24.92	23.48	22.28	21.28	19.67
5.2	38.07	33.06	29.70	27.26	25.38	23.87	22.63	21.59	19.93
5.4	39.20	33.87	30.34	27.78	25.82	24.25	22.97	21.89	20.17
5.5	39.75	34.28	30.66	28.04	26.04	24.44	23.13	22.04	20.29
5.6	40.30	34.67	30.97	28.29	26.25	24.63	23.30	22.18	20.41
5.8	41.40	35.46	31.58	28.79	26.67	24.99	23.61	22.46	20.64
6.0	42.48	36.23	32.17	29.28	27.08	25.34	23.92	22.74	20.87
6.5	45.12	38.10	33.62	30.45	28.06	26.19	24.67	23.40	21.40
7.0	47.69	39.89	34.99	31.56	28.99	26.99	25.37	24.02	21.91
7.5	50.20	41.62	36.30	32.62	29.88	27.74	26.03	24.61	22.39
8.0	52.63	43.28	37.56	33.63	30.72	28.46	26.66	25.17	22.84
8.5	55.01	44.89	38.77	34.59	31.52	29.15	27.25	25.70	23.27
9.0	57.34	46.44	39.94	35.52	32.29	29.81	27.83	26.20	23.68
9.5	59.61	47.95	41.06	36.41	33.03	30.44	28.37	26.68	24.07
10.0	61.83	49.41	42.14	37.27	33.74	31.04	28.90	27.15	24.45
11.0	66.15	52.22	44.21	38.91	35.09	32.18	29.89	28.02	25.15
12.0	****	54.88	46.16	40.44	36.34	33.25	30.81	28.84	25.81
13.0	****	57.42	48.00	41.88	37.53	34.25	31.68	29.60	26.42
14.0	****	59.85	49.75	43.24	38.64	35.19	32.49	30.31	26.99
15.0	****	62.17	51.42	44.54	39.70	36.08	33.25	30.98	27.53
16.0	****	64.41	53.02	45.77	40.70	36.92	33.98	31.62	28.04
18.0	****	68.64	56.01	48.08	42.57	38.49	35.33	32.81	28.99
20.0	****	****	58.79	50.21	44.29	39.93	36.57	33.89	29.85

† For an explanation of the headings and the significance of asterisks and high returns which appear on some pages, see Introductory Section, pages xx-xxiii.

INTERNAL RATE OF RETURN (IRR) †
(Reflecting projected growth in Full Rental Value to next Review/Reversion)

RENTAL FACTOR †	YEARS TO REVIEW/REVERSION								
	1	2	3	4	5	6	7	8	10
1.1	11.39	11.70	11.95	12.13	12.26	12.36	12.41	12.45	12.45
1.2	12.31	12.55	12.72	12.83	12.90	12.93	12.94	12.92	12.85
1.3	13.22	13.37	13.46	13.50	13.50	13.48	13.43	13.37	13.23
1.4	14.12	14.17	14.17	14.14	14.07	13.99	13.90	13.80	13.58
1.5	15.00	14.95	14.86	14.75	14.62	14.49	14.35	14.20	13.92
1.6	15.87	15.71	15.53	15.34	15.15	14.96	14.77	14.59	14.24
1.7	16.73	16.45	16.17	15.91	15.65	15.41	15.18	14.96	14.55
1.8	17.57	17.17	16.80	16.46	16.14	15.84	15.57	15.31	14.84
1.9	18.40	17.88	17.41	16.99	16.61	16.26	15.94	15.65	15.12
2.0	19.23	18.57	18.00	17.51	17.06	16.66	16.30	15.97	15.39
2.2	20.84	19.91	19.15	18.49	17.93	17.43	16.99	16.59	15.90
2.4	22.42	21.20	20.23	19.43	18.74	18.15	17.63	17.17	16.38
2.5	23.19	21.83	20.76	19.87	19.13	18.49	17.93	17.44	16.61
2.6	23.96	22.45	21.27	20.31	19.51	18.83	18.23	17.71	16.83
2.8	25.46	23.65	22.26	21.15	20.24	19.47	18.80	18.22	17.25
3.0	26.94	24.81	23.22	21.96	20.94	20.08	19.34	18.71	17.65
3.2	28.38	25.94	24.14	22.73	21.60	20.66	19.86	19.17	18.03
3.4	29.80	27.03	25.02	23.48	22.24	21.21	20.35	19.61	18.39
3.5	30.50	27.57	25.46	23.84	22.55	21.48	20.59	19.82	18.56
3.6	31.19	28.09	25.88	24.19	22.85	21.74	20.82	20.03	18.73
3.8	32.56	29.13	26.71	24.88	23.44	22.26	21.27	20.43	19.06
4.0	33.90	30.14	27.51	25.55	24.00	22.75	21.70	20.82	19.38
4.2	35.22	31.12	28.29	26.19	24.55	23.22	22.12	21.19	19.69
4.4	36.52	32.08	29.05	26.81	25.08	23.68	22.52	21.55	19.98
4.5	37.16	32.55	29.42	27.12	25.34	23.90	22.72	21.72	20.12
4.6	37.79	33.01	29.78	27.42	25.59	24.12	22.91	21.89	20.26
4.8	39.05	33.93	30.50	28.01	26.09	24.55	23.29	22.23	20.53
5.0	40.29	34.82	31.20	28.58	26.57	24.97	23.65	22.55	20.80
5.2	41.52	35.70	31.88	29.13	27.04	25.37	24.01	22.87	21.05
5.4	42.72	36.56	32.55	29.68	27.49	25.76	24.35	23.17	21.30
5.5	43.32	36.99	32.88	29.94	27.72	25.95	24.52	23.32	21.42
5.6	43.91	37.40	33.20	30.20	27.94	26.14	24.69	23.47	21.54
5.8	45.08	38.23	33.84	30.72	28.37	26.52	25.01	23.76	21.78
6.0	46.24	39.04	34.46	31.22	28.79	26.88	25.33	24.04	22.01
6.5	49.07	41.01	35.96	32.43	29.80	27.74	26.08	24.71	22.55
7.0	51.82	42.90	37.39	33.58	30.76	28.56	26.80	25.34	23.06
7.5	54.50	44.71	38.76	34.67	31.66	29.34	27.47	25.93	23.54
8.0	57.10	46.46	40.06	35.72	32.53	30.07	28.11	26.50	24.00
8.5	59.64	48.14	41.32	36.71	33.35	30.77	28.72	27.04	24.43
9.0	62.12	49.77	42.53	37.67	34.14	31.44	29.30	27.55	24.85
9.5	64.54	51.36	43.69	38.59	34.90	32.09	29.86	28.04	25.24
10.0	66.92	52.89	44.82	39.48	35.63	32.71	30.39	28.51	25.62
11.0	****	55.83	46.97	41.16	37.01	33.87	31.40	29.40	26.33
12.0	****	58.61	48.98	42.73	38.30	34.96	32.34	30.23	27.00
13.0	****	61.27	50.89	44.22	39.51	35.98	33.22	31.00	27.61
14.0	****	63.80	52.70	45.62	40.65	36.94	34.05	31.72	28.19
15.0	****	66.23	54.43	46.95	41.73	37.84	34.83	32.40	28.74
16.0	****	68.56	56.08	48.22	42.76	38.71	35.57	33.05	29.26
18.0	****	****	59.18	50.59	44.67	40.31	36.94	34.25	30.21
20.0	****	****	62.05	52.78	46.43	41.77	38.20	35.35	31.08

† For an explanation of the headings and the significance of asterisks and high returns which appear on some pages, see Introductory Section, pages xx-xxiii.

INTERNAL RATE OF RETURN (IRR) †

(Reflecting projected growth in Full Rental Value to next Review/Reversion)

RENTAL FACTOR †	YEARS TO REVIEW/REVERSION								
	1	2	3	4	5	6	7	8	10
1.1	1.17	1.25	1.33	1.42	1.50	1.59	1.68	1.77	1.96
1.2	1.28	1.36	1.45	1.54	1.63	1.72	1.82	1.91	2.10
1.3	1.39	1.47	1.57	1.66	1.75	1.85	1.95	2.05	2.25
1.4	1.49	1.58	1.68	1.78	1.88	1.98	2.08	2.18	2.39
1.5	1.60	1.69	1.79	1.90	2.00	2.10	2.21	2.31	2.52
1.6	1.70	1.80	1.91	2.01	2.12	2.23	2.34	2.44	2.65
1.7	1.80	1.91	2.02	2.13	2.24	2.35	2.46	2.57	2.78
1.8	1.91	2.02	2.13	2.24	2.36	2.47	2.58	2.69	2.91
1.9	2.01	2.13	2.24	2.36	2.47	2.59	2.70	2.81	3.03
2.0	2.12	2.23	2.35	2.47	2.59	2.71	2.82	2.93	3.15
2.2	2.32	2.45	2.57	2.69	2.82	2.93	3.05	3.17	3.39
2.4	2.53	2.66	2.79	2.91	3.04	3.16	3.28	3.39	3.61
2.5	2.63	2.76	2.89	3.02	3.15	3.27	3.39	3.50	3.72
2.6	2.73	2.87	3.00	3.13	3.25	3.38	3.50	3.61	3.83
2.8	2.94	3.08	3.21	3.34	3.47	3.59	3.71	3.82	4.04
3.0	3.14	3.28	3.42	3.55	3.68	3.80	3.92	4.03	4.24
3.2	3.35	3.49	3.62	3.76	3.88	4.00	4.12	4.23	4.43
3.4	3.55	3.69	3.83	3.96	4.08	4.20	4.32	4.42	4.62
3.5	3.65	3.79	3.93	4.06	4.18	4.30	4.41	4.52	4.71
3.6	3.75	3.89	4.03	4.16	4.28	4.40	4.51	4.61	4.80
3.8	3.95	4.09	4.23	4.36	4.48	4.59	4.70	4.80	4.98
4.0	4.15	4.29	4.43	4.55	4.67	4.78	4.88	4.98	5.15
4.2	4.35	4.49	4.62	4.74	4.86	4.97	5.06	5.16	5.32
4.4	4.55	4.68	4.81	4.93	5.04	5.15	5.24	5.33	5.49
4.5	4.65	4.78	4.91	5.03	5.14	5.24	5.33	5.41	5.57
4.6	4.74	4.88	5.00	5.12	5.23	5.32	5.42	5.50	5.65
4.8	4.94	5.07	5.19	5.30	5.41	5.50	5.59	5.66	5.80
5.0	5.14	5.26	5.38	5.49	5.58	5.67	5.75	5.83	5.95
5.2	5.33	5.45	5.56	5.67	5.76	5.84	5.92	5.98	6.10
5.4	5.53	5.64	5.75	5.84	5.93	6.01	6.08	6.14	6.25
5.5	5.62	5.74	5.84	5.93	6.01	6.09	6.16	6.22	6.32
5.6	5.72	5.83	5.93	6.02	6.10	6.17	6.23	6.29	6.39
5.8	5.92	6.02	6.11	6.19	6.27	6.33	6.39	6.44	6.53
6.0	6.11	6.20	6.29	6.36	6.43	6.49	6.54	6.59	6.67
6.5	6.59	6.66	6.73	6.78	6.83	6.87	6.91	6.94	6.99
7.0	7.06	7.11	7.16	7.19	7.22	7.25	7.27	7.28	7.31
7.5	7.53	7.56	7.58	7.59	7.60	7.61	7.61	7.61	7.61
8.0	8.00	8.00	7.99	7.98	7.97	7.95	7.94	7.92	7.89
8.5	8.46	8.43	8.39	8.36	8.32	8.29	8.26	8.23	8.17
9.0	8.92	8.85	8.79	8.73	8.67	8.62	8.57	8.52	8.43
9.5	9.38	9.27	9.17	9.09	9.01	8.93	8.87	8.80	8.69
10.0	9.83	9.69	9.56	9.44	9.34	9.24	9.16	9.08	8.93
11.0	10.73	10.50	10.30	10.12	9.97	9.83	9.71	9.60	9.40
12.0	11.61	11.29	11.01	10.78	10.58	10.40	10.24	10.09	9.84
13.0	12.48	12.06	11.71	11.41	11.16	10.93	10.73	10.56	10.26
14.0	13.34	12.81	12.38	12.02	11.71	11.44	11.21	11.00	10.65
15.0	14.18	13.54	13.03	12.61	12.25	11.94	11.66	11.43	11.02
16.0	15.02	14.26	13.67	13.17	12.76	12.41	12.10	11.83	11.38
18.0	16.65	15.66	14.88	14.26	13.74	13.30	12.92	12.59	12.04
20.0	18.25	17.00	16.04	15.28	14.65	14.13	13.68	13.29	12.65

† For an explanation of the headings and the significance of asterisks and high returns which appear on some pages, see Introductory Section, pages xx-xxiii.

INTERNAL RATE OF RETURN (IRR) †

(Reflecting projected growth in Full Rental Value to next Review/Reversion)

RENTAL FACTOR †	YEARS TO REVIEW/REVERSION								
	1	2	3	4	5	6	7	8	10
1.1	1.76	1.87	1.99	2.11	2.23	2.35	2.47	2.59	2.82
1.2	1.92	2.04	2.16	2.29	2.41	2.53	2.66	2.78	3.02
1.3	2.07	2.20	2.33	2.46	2.59	2.71	2.84	2.96	3.20
1.4	2.23	2.36	2.50	2.63	2.76	2.89	3.02	3.15	3.39
1.5	2.39	2.52	2.66	2.80	2.93	3.07	3.20	3.32	3.56
1.6	2.54	2.68	2.82	2.96	3.10	3.24	3.37	3.49	3.73
1.7	2.70	2.84	2.99	3.13	3.27	3.40	3.53	3.66	3.90
1.8	2.85	3.00	3.15	3.29	3.43	3.57	3.70	3.82	4.06
1.9	3.00	3.16	3.31	3.45	3.59	3.73	3.86	3.98	4.22
2.0	3.16	3.31	3.46	3.61	3.75	3.89	4.02	4.14	4.37
2.2	3.46	3.62	3.78	3.92	4.06	4.20	4.32	4.44	4.66
2.4	3.77	3.93	4.08	4.23	4.37	4.50	4.62	4.74	4.94
2.5	3.92	4.08	4.23	4.38	4.51	4.64	4.76	4.88	5.08
2.6	4.07	4.23	4.38	4.53	4.66	4.79	4.91	5.02	5.21
2.8	4.37	4.53	4.68	4.82	4.95	5.07	5.18	5.29	5.47
3.0	4.67	4.82	4.97	5.10	5.23	5.34	5.45	5.55	5.72
3.2	4.96	5.12	5.26	5.38	5.50	5.61	5.71	5.80	5.95
3.4	5.26	5.41	5.54	5.66	5.77	5.87	5.96	6.04	6.18
3.5	5.41	5.55	5.68	5.80	5.90	6.00	6.08	6.16	6.29
3.6	5.55	5.69	5.82	5.93	6.03	6.12	6.21	6.28	6.40
3.8	5.85	5.98	6.09	6.20	6.29	6.37	6.44	6.51	6.62
4.0	6.14	6.26	6.36	6.46	6.54	6.61	6.67	6.73	6.82
4.2	6.42	6.53	6.63	6.71	6.78	6.85	6.90	6.95	7.02
4.4	6.71	6.81	6.89	6.96	7.02	7.08	7.12	7.16	7.22
4.5	6.86	6.95	7.02	7.09	7.14	7.19	7.23	7.26	7.31
4.6	7.00	7.08	7.15	7.21	7.26	7.30	7.34	7.37	7.41
4.8	7.28	7.35	7.41	7.45	7.49	7.52	7.55	7.57	7.59
5.0	7.57	7.62	7.66	7.69	7.72	7.74	7.75	7.76	7.77
5.2	7.85	7.88	7.91	7.93	7.94	7.95	7.95	7.95	7.95
5.4	8.13	8.15	8.16	8.16	8.16	8.16	8.15	8.14	8.11
5.5	8.27	8.28	8.28	8.28	8.27	8.26	8.25	8.23	8.20
5.6	8.41	8.41	8.40	8.39	8.38	8.36	8.34	8.32	8.28
5.8	8.68	8.67	8.64	8.62	8.59	8.56	8.53	8.50	8.44
6.0	8.96	8.92	8.88	8.84	8.80	8.76	8.72	8.68	8.60
6.5	9.65	9.55	9.46	9.38	9.30	9.23	9.16	9.10	8.98
7.0	10.32	10.17	10.03	9.91	9.79	9.69	9.59	9.50	9.34
7.5	10.99	10.77	10.58	10.41	10.26	10.12	10.00	9.89	9.69
8.0	11.66	11.37	11.12	10.91	10.72	10.55	10.39	10.26	10.01
8.5	12.31	11.95	11.65	11.38	11.15	10.95	10.77	10.61	10.33
9.0	12.96	12.52	12.16	11.85	11.58	11.34	11.14	10.95	10.63
9.5	13.60	13.08	12.66	12.30	11.99	11.72	11.49	11.28	10.92
10.0	14.24	13.64	13.15	12.74	12.39	12.09	11.83	11.59	11.19
11.0	15.49	14.72	14.10	13.59	13.16	12.80	12.48	12.19	11.72
12.0	16.72	15.76	15.01	14.40	13.89	13.46	13.09	12.76	12.21
13.0	17.92	16.77	15.89	15.17	14.59	14.09	13.66	13.29	12.68
14.0	19.11	17.76	16.73	15.92	15.25	14.69	14.21	13.80	13.11
15.0	20.27	18.71	17.55	16.63	15.88	15.26	14.73	14.28	13.53
16.0	21.42	19.65	18.34	17.32	16.49	15.81	15.23	14.73	13.92
18.0	23.65	21.44	19.84	18.62	17.64	16.84	16.16	15.59	14.65
20.0	25.82	23.16	21.27	19.84	18.71	17.79	17.03	16.38	15.32

† For an explanation of the headings and the significance of asterisks and high returns which appear on some pages, see Introductory Section, pages xx–xxiii.

INTERNAL RATE OF RETURN (IRR) †
(Reflecting projected growth in Full Rental Value to next Review/Reversion)

RENTAL FACTOR †	YEARS TO REVIEW/REVERSION								
	1	2	3	4	5	6	7	8	10
1.1	2.35	2.49	2.64	2.79	2.94	3.08	3.23	3.37	3.63
1.2	2.55	2.71	2.86	3.02	3.17	3.32	3.46	3.60	3.86
1.3	2.76	2.92	3.08	3.24	3.39	3.54	3.69	3.83	4.09
1.4	2.97	3.13	3.30	3.46	3.61	3.76	3.91	4.05	4.30
1.5	3.17	3.34	3.51	3.67	3.83	3.98	4.12	4.26	4.51
1.6	3.38	3.55	3.72	3.88	4.04	4.19	4.33	4.47	4.71
1.7	3.58	3.76	3.93	4.09	4.25	4.40	4.54	4.67	4.90
1.8	3.78	3.96	4.13	4.30	4.45	4.60	4.73	4.86	5.09
1.9	3.99	4.17	4.34	4.50	4.65	4.80	4.93	5.05	5.27
2.0	4.19	4.37	4.54	4.70	4.85	4.99	5.12	5.24	5.45
2.2	4.59	4.77	4.93	5.09	5.23	5.36	5.49	5.60	5.79
2.4	4.99	5.16	5.32	5.47	5.60	5.73	5.84	5.94	6.11
2.5	5.18	5.36	5.51	5.65	5.78	5.90	6.01	6.10	6.27
2.6	5.38	5.55	5.70	5.84	5.96	6.07	6.17	6.26	6.42
2.8	5.77	5.93	6.07	6.20	6.31	6.41	6.50	6.58	6.71
3.0	6.16	6.31	6.44	6.55	6.65	6.74	6.81	6.88	6.99
3.2	6.55	6.68	6.79	6.89	6.98	7.05	7.12	7.17	7.26
3.4	6.93	7.05	7.15	7.23	7.30	7.36	7.41	7.45	7.51
3.5	7.12	7.23	7.32	7.39	7.46	7.51	7.55	7.59	7.64
3.6	7.32	7.41	7.49	7.56	7.61	7.66	7.69	7.72	7.76
3.8	7.69	7.77	7.83	7.88	7.92	7.95	7.97	7.98	8.00
4.0	8.07	8.12	8.16	8.19	8.21	8.23	8.23	8.24	8.23
4.2	8.44	8.47	8.49	8.50	8.50	8.50	8.49	8.48	8.45
4.4	8.82	8.82	8.81	8.80	8.79	8.77	8.75	8.72	8.67
4.5	9.00	8.99	8.97	8.95	8.93	8.90	8.87	8.84	8.77
4.6	9.18	9.16	9.13	9.10	9.07	9.03	8.99	8.95	8.88
4.8	9.55	9.50	9.45	9.39	9.34	9.28	9.23	9.18	9.08
5.0	9.92	9.83	9.75	9.68	9.60	9.53	9.46	9.40	9.28
5.2	10.28	10.16	10.06	9.96	9.86	9.78	9.69	9.61	9.47
5.4	10.64	10.49	10.36	10.23	10.12	10.01	9.92	9.82	9.66
5.5	10.82	10.65	10.50	10.37	10.25	10.13	10.03	9.93	9.75
5.6	10.99	10.81	10.65	10.50	10.37	10.25	10.13	10.03	9.84
5.8	11.35	11.13	10.94	10.77	10.62	10.48	10.35	10.23	10.02
6.0	11.70	11.45	11.23	11.03	10.86	10.70	10.56	10.42	10.19
6.5	12.58	12.23	11.93	11.67	11.44	11.24	11.06	10.89	10.60
7.0	13.44	12.99	12.61	12.29	12.00	11.76	11.54	11.34	11.00
7.5	14.29	13.73	13.27	12.88	12.54	12.25	11.99	11.76	11.37
8.0	15.13	14.46	13.91	13.45	13.06	12.73	12.43	12.17	11.73
8.5	15.96	15.17	14.53	14.01	13.56	13.18	12.85	12.56	12.06
9.0	16.78	15.86	15.14	14.54	14.05	13.62	13.26	12.93	12.39
9.5	17.59	16.54	15.73	15.07	14.52	14.05	13.65	13.29	12.70
10.0	18.39	17.21	16.30	15.57	14.97	14.46	14.02	13.64	13.00
11.0	19.96	18.51	17.41	16.55	15.84	15.24	14.73	14.29	13.57
12.0	21.49	19.76	18.48	17.47	16.66	15.98	15.40	14.91	14.09
13.0	22.99	20.97	19.49	18.35	17.43	16.67	16.04	15.49	14.59
14.0	24.46	22.15	20.47	19.19	18.17	17.34	16.63	16.03	15.06
15.0	25.91	23.28	21.42	20.00	18.88	17.97	17.20	16.55	15.50
16.0	27.32	24.39	22.32	20.77	19.56	18.57	17.74	17.04	15.92
18.0	30.08	26.50	24.05	22.24	20.83	19.69	18.75	17.96	16.70
20.0	32.74	28.52	25.68	23.61	22.01	20.73	19.69	18.81	17.41

† For an explanation of the headings and the significance of asterisks and high returns which appear on some pages see Introductory Section, pages xx-xxiii.

INTERNAL RATE OF RETURN (IRR) †
(Reflecting projected growth in Full Rental Value to next Review/Reversion)

RENTAL FACTOR †	YEARS TO REVIEW/REVERSION								
	1	2	3	4	5	6	7	8	10
1.1	2.93	3.11	3.29	3.46	3.64	3.80	3.96	4.11	4.39
1.2	3.19	3.37	3.56	3.74	3.91	4.08	4.23	4.38	4.66
1.3	3.44	3.64	3.82	4.00	4.18	4.34	4.50	4.64	4.91
1.4	3.70	3.90	4.09	4.27	4.44	4.60	4.75	4.90	5.15
1.5	3.95	4.15	4.34	4.52	4.69	4.85	5.00	5.14	5.39
1.6	4.21	4.41	4.60	4.78	4.94	5.10	5.24	5.38	5.61
1.7	4.46	4.66	4.85	5.02	5.19	5.34	5.48	5.60	5.83
1.8	4.71	4.91	5.10	5.27	5.43	5.57	5.70	5.83	6.04
1.9	4.96	5.16	5.34	5.51	5.66	5.80	5.93	6.04	6.24
2.0	5.21	5.40	5.58	5.74	5.89	6.02	6.14	6.25	6.43
2.2	5.70	5.89	6.05	6.20	6.34	6.45	6.56	6.65	6.81
2.4	6.19	6.36	6.51	6.65	6.76	6.87	6.96	7.03	7.16
2.5	6.43	6.60	6.74	6.86	6.97	7.07	7.15	7.22	7.33
2.6	6.68	6.83	6.96	7.08	7.18	7.26	7.34	7.40	7.49
2.8	7.16	7.29	7.40	7.50	7.58	7.65	7.70	7.75	7.81
3.0	7.63	7.74	7.83	7.91	7.97	8.01	8.05	8.08	8.12
3.2	8.11	8.19	8.25	8.30	8.34	8.37	8.39	8.40	8.41
3.4	8.57	8.63	8.67	8.69	8.71	8.71	8.71	8.71	8.68
3.5	8.81	8.85	8.87	8.88	8.88	8.88	8.87	8.86	8.82
3.6	9.04	9.06	9.07	9.07	9.06	9.05	9.03	9.00	8.95
3.8	9.50	9.49	9.47	9.44	9.41	9.37	9.33	9.29	9.21
4.0	9.96	9.91	9.85	9.80	9.74	9.68	9.63	9.57	9.46
4.2	10.41	10.32	10.24	10.15	10.07	9.99	9.91	9.84	9.70
4.4	10.86	10.73	10.61	10.50	10.39	10.29	10.19	10.10	9.93
4.5	11.09	10.93	10.79	10.67	10.54	10.43	10.32	10.22	10.04
4.6	11.31	11.14	10.98	10.83	10.70	10.57	10.46	10.35	10.15
4.8	11.75	11.53	11.34	11.16	11.00	10.86	10.72	10.60	10.37
5.0	12.19	11.93	11.69	11.49	11.30	11.13	10.98	10.83	10.58
5.2	12.63	12.32	12.04	11.81	11.59	11.40	11.23	11.07	10.79
5.4	13.06	12.70	12.39	12.12	11.88	11.66	11.47	11.29	10.98
5.5	13.28	12.89	12.56	12.27	12.02	11.79	11.59	11.40	11.08
5.6	13.50	13.08	12.73	12.42	12.16	11.92	11.71	11.51	11.18
5.8	13.92	13.45	13.06	12.72	12.43	12.17	11.94	11.73	11.37
6.0	14.35	13.83	13.39	13.02	12.70	12.42	12.16	11.94	11.55
6.5	15.40	14.73	14.19	13.74	13.35	13.01	12.71	12.45	11.99
7.0	16.43	15.62	14.96	14.42	13.97	13.57	13.23	12.93	12.41
7.5	17.45	16.48	15.71	15.09	14.56	14.11	13.72	13.38	12.81
8.0	18.46	17.32	16.44	15.72	15.13	14.63	14.19	13.82	13.18
8.5	19.44	18.14	17.14	16.34	15.68	15.12	14.65	14.23	13.54
9.0	20.42	18.94	17.82	16.94	16.21	15.60	15.08	14.63	13.88
9.5	21.38	19.73	18.49	17.51	16.72	16.06	15.50	15.01	14.21
10.0	22.32	20.49	19.13	18.07	17.22	16.50	15.90	15.38	14.53
11.0	24.18	21.98	20.38	19.15	18.16	17.35	16.66	16.08	15.12
12.0	25.99	23.41	21.57	20.16	19.05	18.14	17.38	16.73	15.68
13.0	27.76	24.79	22.70	21.13	19.89	18.89	18.05	17.34	16.19
14.0	29.49	26.12	23.79	22.05	20.69	19.60	18.69	17.92	16.68
15.0	31.18	27.41	24.83	22.93	21.45	20.27	19.29	18.47	17.14
16.0	32.84	28.66	25.84	23.77	22.18	20.91	19.86	18.99	17.58
18.0	36.05	31.05	27.75	25.37	23.55	22.11	20.94	19.95	18.40
20.0	39.15	33.31	29.53	26.85	24.82	23.22	21.92	20.84	19.14

† For an explanation of the headings and the significance of asterisks and high returns which appear on some pages, see Introductory Section, pages xx-xxiii.

INTERNAL RATE OF RETURN (IRR) †

(Reflecting projected growth in Full Rental Value to next Review/Reversion)

RENTAL FACTOR †	YEARS TO REVIEW/REVERSION								
	1	2	3	4	5	6	7	8	10
1.1	3.51	3.72	3.93	4.13	4.32	4.50	4.67	4.83	5.12
1.2	3.82	4.04	4.24	4.44	4.63	4.81	4.98	5.13	5.41
1.3	4.13	4.35	4.56	4.75	4.94	5.11	5.28	5.43	5.69
1.4	4.43	4.65	4.86	5.06	5.24	5.41	5.56	5.71	5.95
1.5	4.73	4.95	5.16	5.35	5.53	5.69	5.84	5.97	6.21
1.6	5.03	5.25	5.46	5.64	5.81	5.97	6.11	6.23	6.45
1.7	5.33	5.55	5.75	5.93	6.09	6.24	6.37	6.49	6.68
1.8	5.63	5.84	6.03	6.21	6.36	6.50	6.62	6.73	6.91
1.9	5.93	6.13	6.31	6.48	6.62	6.75	6.87	6.96	7.13
2.0	6.22	6.42	6.59	6.75	6.88	7.00	7.10	7.19	7.34
2.2	6.80	6.98	7.14	7.27	7.38	7.48	7.56	7.63	7.74
2.4	7.38	7.53	7.66	7.77	7.86	7.94	8.00	8.05	8.12
2.5	7.67	7.81	7.92	8.02	8.09	8.16	8.21	8.25	8.30
2.6	7.95	8.08	8.18	8.26	8.32	8.37	8.41	8.44	8.47
2.8	8.52	8.61	8.68	8.73	8.77	8.79	8.81	8.82	8.81
3.0	9.08	9.13	9.17	9.19	9.20	9.20	9.19	9.18	9.13
3.2	9.63	9.65	9.65	9.63	9.61	9.59	9.55	9.52	9.44
3.4	10.18	10.15	10.11	10.07	10.02	9.96	9.91	9.85	9.74
3.5	10.46	10.40	10.34	10.28	10.21	10.14	10.08	10.01	9.88
3.6	10.73	10.65	10.57	10.49	10.41	10.32	10.25	10.17	10.02
3.8	11.27	11.14	11.01	10.90	10.78	10.68	10.57	10.48	10.29
4.0	11.80	11.62	11.45	11.30	11.15	11.02	10.89	10.77	10.55
4.2	12.33	12.09	11.88	11.69	11.51	11.35	11.20	11.06	10.81
4.4	12.86	12.56	12.30	12.07	11.86	11.67	11.50	11.34	11.05
4.5	13.12	12.79	12.51	12.25	12.03	11.83	11.64	11.47	11.17
4.6	13.38	13.02	12.71	12.44	12.20	11.98	11.78	11.60	11.29
4.8	13.89	13.47	13.12	12.80	12.53	12.29	12.07	11.87	11.52
5.0	14.41	13.92	13.51	13.16	12.85	12.58	12.34	12.12	11.74
5.2	14.91	14.36	13.90	13.51	13.17	12.87	12.61	12.37	11.95
5.4	15.42	14.80	14.29	13.85	13.48	13.15	12.86	12.61	12.16
5.5	15.67	15.01	14.47	14.02	13.63	13.29	12.99	12.72	12.26
5.6	15.92	15.23	14.66	14.19	13.78	13.43	13.12	12.84	12.36
5.8	16.42	15.65	15.03	14.52	14.08	13.70	13.36	13.07	12.56
6.0	16.91	16.07	15.40	14.84	14.37	13.96	13.61	13.29	12.76
6.5	18.12	17.09	16.28	15.62	15.07	14.60	14.19	13.83	13.22
7.0	19.32	18.09	17.14	16.37	15.74	15.20	14.74	14.33	13.65
7.5	20.49	19.06	17.96	17.09	16.38	15.78	15.26	14.81	14.07
8.0	21.64	20.00	18.76	17.79	16.99	16.33	15.76	15.27	14.46
8.5	22.78	20.92	19.53	18.45	17.58	16.85	16.24	15.71	14.83
9.0	23.90	21.81	20.28	19.10	18.15	17.36	16.69	16.12	15.19
9.5	25.00	22.69	21.01	19.72	18.69	17.85	17.13	16.52	15.53
10.0	26.08	23.54	21.72	20.33	19.22	18.32	17.56	16.91	15.86
11.0	28.20	25.19	23.08	21.48	20.23	19.21	18.36	17.64	16.47
12.0	30.27	26.78	24.37	22.58	21.18	20.05	19.11	18.32	17.05
13.0	32.28	28.30	25.60	23.61	22.07	20.84	19.82	18.96	17.59
14.0	34.24	29.77	26.78	24.60	22.92	21.58	20.48	19.56	18.09
15.0	36.16	31.19	27.91	25.54	23.73	22.29	21.12	20.13	18.57
16.0	38.03	32.56	29.00	26.44	24.50	22.97	21.72	20.67	19.02
18.0	41.67	35.19	31.06	28.14	25.95	24.23	22.84	21.68	19.86
20.0	45.16	37.66	32.98	29.72	27.29	25.39	23.87	22.60	20.63

† For an explanation of the headings and the significance of asterisks and high returns which appear on some pages, see Introductory Section, pages xx-xxiii.

INTERNAL RATE OF RETURN (IRR) †
(Reflecting projected growth in Full Rental Value to next Review/Reversion)

RENTAL FACTOR †	YEARS TO REVIEW/REVERSION								
	1	2	3	4	5	6	7	8	10
1.1	4.10	4.33	4.56	4.78	4.99	5.18	5.36	5.53	5.81
1.2	4.45	4.69	4.92	5.14	5.34	5.53	5.70	5.86	6.13
1.3	4.81	5.05	5.28	5.49	5.69	5.86	6.03	6.17	6.42
1.4	5.16	5.40	5.62	5.83	6.02	6.19	6.34	6.48	6.71
1.5	5.51	5.75	5.96	6.16	6.34	6.50	6.64	6.77	6.98
1.6	5.85	6.09	6.30	6.49	6.66	6.80	6.93	7.05	7.24
1.7	6.20	6.42	6.63	6.80	6.96	7.10	7.22	7.32	7.49
1.8	6.54	6.76	6.95	7.11	7.26	7.38	7.49	7.58	7.73
1.9	6.88	7.09	7.26	7.42	7.55	7.66	7.76	7.83	7.96
2.0	7.22	7.41	7.58	7.72	7.83	7.93	8.01	8.08	8.18
2.2	7.89	8.05	8.19	8.29	8.38	8.45	8.51	8.55	8.60
2.4	8.56	8.68	8.78	8.85	8.91	8.95	8.97	8.99	9.00
2.5	8.88	8.99	9.06	9.12	9.16	9.18	9.20	9.20	9.19
2.6	9.21	9.29	9.35	9.39	9.41	9.42	9.42	9.41	9.38
2.8	9.86	9.89	9.91	9.91	9.89	9.87	9.84	9.81	9.73
3.0	10.50	10.48	10.45	10.41	10.36	10.31	10.25	10.19	10.07
3.2	11.13	11.06	10.98	10.89	10.81	10.72	10.64	10.55	10.39
3.4	11.76	11.63	11.49	11.37	11.24	11.13	11.01	10.90	10.70
3.5	12.07	11.90	11.75	11.60	11.45	11.32	11.19	11.07	10.85
3.6	12.38	12.18	12.00	11.82	11.66	11.51	11.37	11.24	11.00
3.8	13.00	12.73	12.49	12.27	12.07	11.89	11.72	11.56	11.28
4.0	13.61	13.27	12.97	12.70	12.47	12.25	12.06	11.88	11.55
4.2	14.21	13.79	13.44	13.13	12.85	12.60	12.38	12.18	11.82
4.4	14.80	14.31	13.90	13.54	13.23	12.95	12.70	12.47	12.07
4.5	15.10	14.57	14.13	13.74	13.41	13.11	12.85	12.61	12.20
4.6	15.40	14.83	14.35	13.94	13.59	13.28	13.00	12.75	12.32
4.8	15.98	15.33	14.79	14.34	13.95	13.60	13.30	13.02	12.55
5.0	16.56	15.83	15.23	14.72	14.29	13.92	13.59	13.29	12.79
5.2	17.14	16.31	15.65	15.10	14.63	14.22	13.87	13.55	13.01
5.4	17.71	16.80	16.07	15.47	14.96	14.52	14.14	13.80	13.23
5.5	17.99	17.04	16.27	15.65	15.12	14.67	14.27	13.92	13.33
5.6	18.27	17.27	16.48	15.83	15.28	14.81	14.41	14.05	13.44
5.8	18.83	17.74	16.88	16.18	15.60	15.10	14.66	14.28	13.64
6.0	19.39	18.20	17.28	16.53	15.91	15.38	14.92	14.52	13.84
6.5	20.76	19.33	18.24	17.37	16.65	16.05	15.53	15.07	14.32
7.0	22.10	20.43	19.17	18.17	17.36	16.68	16.10	15.60	14.77
7.5	23.42	21.49	20.06	18.94	18.04	17.29	16.65	16.10	15.20
8.0	24.72	22.52	20.92	19.68	18.69	17.87	17.17	16.58	15.60
8.5	25.99	23.53	21.75	20.39	19.31	18.42	17.67	17.03	15.99
9.0	27.24	24.51	22.56	21.08	19.91	18.95	18.15	17.47	16.35
9.5	28.47	25.46	23.34	21.75	20.49	19.46	18.61	17.88	16.71
10.0	29.68	26.39	24.10	22.39	21.04	19.96	19.05	18.28	17.04
11.0	32.05	28.19	25.56	23.61	22.10	20.89	19.89	19.04	17.68
12.0	34.35	29.92	26.94	24.77	23.10	21.76	20.67	19.74	18.27
13.0	36.58	31.57	28.26	25.87	24.04	22.59	21.40	20.41	18.83
14.0	38.76	33.16	29.52	26.91	24.93	23.37	22.09	21.03	19.35
15.0	40.89	34.70	30.72	27.90	25.78	24.11	22.75	21.62	19.84
16.0	42.96	36.18	31.88	28.86	26.59	24.81	23.37	22.18	20.30
18.0	46.98	39.01	34.07	30.65	28.10	26.12	24.53	23.22	21.16
20.0	50.84	41.68	36.12	32.30	29.50	27.33	25.59	24.17	21.95

† For an explanation of the headings and the significance of asterisks and high returns which appear on some pages, see Introductory Section, pages xx-xxiii.

INTERNAL RATE OF RETURN (IRR) †

(Reflecting projected growth in Full Rental Value to next Review/Reversion)

RENTAL FACTOR †	YEARS TO REVIEW/REVERSION								
	1	2	3	4	5	6	7	8	10
1.1	4.68	4.94	5.19	5.43	5.65	5.85	6.03	6.20	6.48
1.2	5.08	5.35	5.60	5.83	6.04	6.23	6.40	6.56	6.81
1.3	5.48	5.75	5.99	6.21	6.41	6.59	6.75	6.89	7.13
1.4	5.88	6.14	6.38	6.59	6.78	6.94	7.09	7.22	7.43
1.5	6.28	6.53	6.75	6.95	7.13	7.28	7.42	7.53	7.71
1.6	6.67	6.91	7.12	7.31	7.47	7.61	7.73	7.83	7.98
1.7	7.06	7.29	7.49	7.66	7.80	7.93	8.03	8.12	8.25
1.8	7.45	7.66	7.84	8.00	8.13	8.23	8.32	8.39	8.50
1.9	7.83	8.03	8.19	8.33	8.44	8.53	8.60	8.66	8.74
2.0	8.21	8.39	8.54	8.65	8.75	8.82	8.88	8.92	8.97
2.2	8.97	9.10	9.21	9.28	9.34	9.38	9.40	9.41	9.41
2.4	9.72	9.80	9.85	9.89	9.90	9.91	9.90	9.88	9.83
2.5	10.09	10.14	10.17	10.18	10.18	10.16	10.13	10.10	10.03
2.6	10.45	10.48	10.48	10.47	10.44	10.41	10.37	10.32	10.22
2.8	11.18	11.14	11.09	11.03	10.96	10.89	10.81	10.74	10.59
3.0	11.90	11.79	11.68	11.57	11.46	11.35	11.24	11.14	10.94
3.2	12.61	12.43	12.26	12.09	11.94	11.79	11.65	11.52	11.28
3.4	13.31	13.05	12.82	12.60	12.40	12.22	12.05	11.89	11.60
3.5	13.66	13.36	13.09	12.85	12.63	12.42	12.24	12.06	11.75
3.6	14.01	13.66	13.36	13.09	12.85	12.63	12.42	12.24	11.90
3.8	14.69	14.26	13.90	13.57	13.28	13.02	12.79	12.57	12.20
4.0	15.37	14.85	14.42	14.04	13.70	13.41	13.14	12.90	12.48
4.2	16.04	15.43	14.92	14.49	14.11	13.78	13.48	13.21	12.75
4.4	16.71	16.00	15.42	14.93	14.51	14.14	13.81	13.52	13.01
4.5	17.04	16.28	15.67	15.15	14.70	14.31	13.97	13.66	13.14
4.6	17.37	16.56	15.91	15.36	14.89	14.49	14.13	13.81	13.26
4.8	18.02	17.11	16.38	15.78	15.27	14.83	14.44	14.09	13.51
5.0	18.66	17.65	16.85	16.19	15.63	15.16	14.74	14.37	13.75
5.2	19.30	18.19	17.31	16.59	15.99	15.48	15.03	14.64	13.98
5.4	19.94	18.71	17.76	16.98	16.34	15.79	15.32	14.90	14.20
5.5	20.25	18.97	17.98	17.18	16.51	15.94	15.46	15.03	14.31
5.6	20.56	19.23	18.20	17.37	16.68	16.10	15.59	15.15	14.42
5.8	21.18	19.74	18.63	17.74	17.01	16.39	15.86	15.40	14.63
6.0	21.80	20.24	19.05	18.11	17.34	16.69	16.13	15.64	14.84
6.5	23.32	21.46	20.08	19.00	18.12	17.39	16.76	16.22	15.33
7.0	24.80	22.65	21.07	19.85	18.86	18.05	17.36	16.77	15.79
7.5	26.26	23.80	22.03	20.66	19.58	18.68	17.93	17.28	16.23
8.0	27.68	24.92	22.94	21.44	20.26	19.28	18.47	17.77	16.64
8.5	29.08	26.00	23.83	22.20	20.91	19.86	18.99	18.24	17.04
9.0	30.46	27.05	24.69	22.92	21.53	20.41	19.48	18.69	17.42
9.5	31.81	28.08	25.52	23.62	22.14	20.94	19.95	19.12	17.78
10.0	33.14	29.08	26.32	24.29	22.72	21.46	20.41	19.53	18.12
11.0	35.74	31.02	27.87	25.58	23.83	22.42	21.28	20.31	18.77
12.0	38.26	32.86	29.33	26.80	24.86	23.33	22.08	21.04	19.38
13.0	40.70	34.64	30.72	27.94	25.84	24.19	22.84	21.72	19.95
14.0	43.08	36.34	32.05	29.04	26.77	24.99	23.55	22.36	20.48
15.0	45.40	37.98	33.33	30.08	27.65	25.76	24.23	22.96	20.98
16.0	47.67	39.56	34.55	31.07	28.49	26.48	24.87	23.54	21.45
18.0	52.04	42.58	36.85	32.94	30.06	27.84	26.06	24.60	22.33
20.0	56.23	45.42	39.00	34.67	31.51	29.09	27.16	25.58	23.14

† For an explanation of the headings and the significance of asterisks and high returns which appear on some page see Introductory Section, pages xx-xxiii.

INTERNAL RATE OF RETURN (IRR) †

(Reflecting projected growth in Full Rental Value to next Review/Reversion)

RENTAL FACTOR †	YEARS TO REVIEW/REVERSION								
	1	2	3	4	5	6	7	8	10
1.1	5.26	5.55	5.82	6.07	6.30	6.51	6.69	6.86	7.13
1.2	5.71	6.00	6.26	6.50	6.72	6.91	7.08	7.23	7.48
1.3	6.16	6.44	6.70	6.92	7.13	7.30	7.46	7.59	7.81
1.4	6.60	6.88	7.12	7.33	7.52	7.68	7.82	7.93	8.12
1.5	7.04	7.30	7.53	7.73	7.90	8.04	8.16	8.26	8.41
1.6	7.48	7.73	7.94	8.11	8.26	8.39	8.49	8.58	8.70
1.7	7.92	8.14	8.33	8.49	8.62	8.73	8.81	8.88	8.97
1.8	8.35	8.55	8.72	8.86	8.97	9.05	9.12	9.17	9.23
1.9	8.77	8.96	9.10	9.21	9.30	9.37	9.42	9.45	9.48
2.0	9.20	9.35	9.47	9.56	9.63	9.68	9.70	9.72	9.72
2.2	10.04	10.13	10.20	10.24	10.26	10.26	10.25	10.24	10.18
2.4	10.86	10.90	10.90	10.89	10.86	10.82	10.77	10.72	10.61
2.5	11.27	11.27	11.24	11.20	11.15	11.09	11.02	10.96	10.81
2.6	11.68	11.64	11.58	11.51	11.43	11.35	11.27	11.18	11.01
2.8	12.49	12.36	12.23	12.11	11.98	11.86	11.74	11.62	11.40
3.0	13.28	13.07	12.87	12.68	12.51	12.34	12.18	12.03	11.76
3.2	14.06	13.76	13.49	13.24	13.01	12.80	12.61	12.43	12.10
3.4	14.84	14.44	14.09	13.78	13.50	13.25	13.02	12.81	12.43
3.5	15.22	14.77	14.39	14.04	13.74	13.47	13.22	12.99	12.59
3.6	15.60	15.10	14.68	14.30	13.97	13.68	13.41	13.17	12.75
3.8	16.36	15.75	15.25	14.81	14.43	14.09	13.79	13.52	13.05
4.0	17.10	16.39	15.80	15.30	14.87	14.49	14.16	13.86	13.34
4.2	17.84	17.02	16.34	15.78	15.30	14.88	14.51	14.18	13.62
4.4	18.57	17.63	16.87	16.25	15.72	15.26	14.85	14.50	13.89
4.5	18.93	17.93	17.14	16.48	15.92	15.44	15.02	14.65	14.02
4.6	19.29	18.23	17.39	16.70	16.12	15.62	15.18	14.80	14.15
4.8	20.01	18.83	17.90	17.15	16.51	15.97	15.50	15.09	14.40
5.0	20.72	19.41	18.40	17.58	16.90	16.32	15.82	15.38	14.64
5.2	21.42	19.98	18.88	18.00	17.27	16.65	16.12	15.65	14.88
5.4	22.11	20.55	19.36	18.41	17.63	16.98	16.41	15.92	15.11
5.5	22.45	20.83	19.59	18.62	17.81	17.14	16.56	16.05	15.22
5.6	22.79	21.10	19.83	18.82	17.99	17.29	16.70	16.18	15.33
5.8	23.47	21.65	20.29	19.21	18.34	17.60	16.98	16.44	15.54
6.0	24.15	22.19	20.74	19.60	18.68	17.91	17.25	16.69	15.76
6.5	25.80	23.50	21.83	20.53	19.49	18.63	17.91	17.28	16.26
7.0	27.42	24.77	22.87	21.42	20.27	19.32	18.52	17.84	16.73
7.5	29.00	26.00	23.88	22.28	21.01	19.97	19.11	18.37	17.18
8.0	30.56	27.19	24.85	23.09	21.72	20.60	19.67	18.88	17.60
8.5	32.08	28.35	25.78	23.88	22.39	21.19	20.20	19.36	18.01
9.0	33.57	29.48	26.69	24.64	23.04	21.77	20.71	19.82	18.39
9.5	35.04	30.57	27.56	25.36	23.67	22.31	21.20	20.26	18.76
10.0	36.48	31.64	28.41	26.07	24.27	22.84	21.67	20.68	19.11
11.0	39.29	33.69	30.03	27.41	25.42	23.84	22.56	21.48	19.78
12.0	42.02	35.65	31.57	28.68	26.50	24.78	23.38	22.23	20.40
13.0	44.66	37.53	33.03	29.87	27.51	25.66	24.16	22.92	20.97
14.0	47.23	39.33	34.42	31.01	28.47	26.49	24.89	23.58	21.51
15.0	49.73	41.07	35.75	32.09	29.38	27.27	25.59	24.19	22.03
16.0	52.17	42.75	37.03	33.12	30.24	28.02	26.24	24.78	22.51
18.0	56.88	45.93	39.44	35.06	31.87	29.42	27.47	25.87	23.40
20.0	61.38	48.93	41.68	36.85	33.36	30.70	28.59	26.87	24.22

† For an explanation of the headings and the significance of asterisks and high returns which appear on some pages, see Introductory Section, pages xx-xxiii.

INTERNAL RATE OF RETURN (IRR) †
(Reflecting projected growth in Full Rental Value to next Review/Reversion)

RENTAL FACTOR †	YEARS TO REVIEW/REVERSION								
	1	2	3	4	5	6	7	8	10
1.1	5.84	6.15	6.44	6.70	6.94	7.15	7.33	7.50	7.76
1.2	6.34	6.64	6.92	7.17	7.39	7.58	7.75	7.89	8.12
1.3	6.83	7.13	7.39	7.62	7.82	8.00	8.14	8.27	8.46
1.4	7.32	7.60	7.85	8.06	8.24	8.39	8.52	8.63	8.78
1.5	7.81	8.07	8.30	8.49	8.65	8.78	8.88	8.97	9.09
1.6	8.29	8.53	8.73	8.90	9.04	9.14	9.23	9.30	9.38
1.7	8.76	8.98	9.16	9.30	9.41	9.50	9.56	9.61	9.66
1.8	9.24	9.43	9.58	9.69	9.78	9.84	9.89	9.91	9.93
1.9	9.71	9.87	9.99	10.08	10.14	10.18	10.20	10.20	10.19
2.0	10.17	10.30	10.39	10.45	10.48	10.50	10.50	10.49	10.44
2.2	11.09	11.15	11.17	11.17	11.15	11.11	11.07	11.02	10.91
2.4	12.00	11.97	11.92	11.85	11.78	11.70	11.61	11.53	11.35
2.5	12.45	12.37	12.28	12.19	12.08	11.98	11.87	11.77	11.56
2.6	12.89	12.77	12.64	12.51	12.38	12.25	12.13	12.00	11.76
2.8	13.77	13.55	13.34	13.15	12.96	12.78	12.61	12.45	12.16
3.0	14.64	14.32	14.02	13.76	13.51	13.28	13.08	12.88	12.53
3.2	15.49	15.06	14.68	14.34	14.04	13.77	13.52	13.29	12.88
3.4	16.34	15.79	15.32	14.91	14.55	14.23	13.94	13.68	13.22
3.5	16.76	16.15	15.63	15.19	14.80	14.46	14.15	13.87	13.38
3.6	17.17	16.50	15.94	15.46	15.05	14.68	14.35	14.05	13.54
3.8	17.99	17.20	16.55	16.00	15.52	15.11	14.74	14.41	13.85
4.0	18.80	17.88	17.13	16.51	15.98	15.52	15.12	14.76	14.15
4.2	19.61	18.55	17.71	17.02	16.43	15.93	15.48	15.09	14.43
4.4	20.40	19.21	18.27	17.50	16.86	16.31	15.84	15.42	14.71
4.5	20.79	19.53	18.54	17.74	17.07	16.50	16.01	15.57	14.84
4.6	21.18	19.85	18.82	17.98	17.28	16.69	16.18	15.73	14.97
4.8	21.96	20.48	19.35	18.44	17.69	17.06	16.51	16.03	15.23
5.0	22.72	21.11	19.88	18.90	18.09	17.41	16.83	16.32	15.48
5.2	23.48	21.72	20.39	19.34	18.48	17.76	17.14	16.61	15.72
5.4	24.23	22.32	20.89	19.77	18.86	18.09	17.44	16.88	15.95
5.5	24.60	22.62	21.14	19.98	19.04	18.26	17.59	17.02	16.07
5.6	24.97	22.91	21.38	20.19	19.23	18.42	17.74	17.15	16.18
5.8	25.71	23.49	21.87	20.60	19.59	18.74	18.03	17.41	16.40
6.0	26.43	24.07	22.34	21.01	19.94	19.05	18.31	17.67	16.61
6.5	28.22	25.46	23.49	21.98	20.79	19.80	18.98	18.28	17.13
7.0	29.97	26.81	24.59	22.91	21.59	20.51	19.61	18.85	17.61
7.5	31.68	28.11	25.64	23.80	22.36	21.19	20.22	19.39	18.07
8.0	33.35	29.37	26.66	24.65	23.09	21.83	20.79	19.91	18.50
8.5	34.98	30.60	27.64	25.46	23.79	22.44	21.34	20.40	18.91
9.0	36.59	31.79	28.58	26.25	24.46	23.03	21.86	20.87	19.30
9.5	38.17	32.94	29.49	27.01	25.11	23.60	22.36	21.32	19.68
10.0	39.71	34.07	30.38	27.74	25.73	24.14	22.84	21.75	20.03
11.0	42.73	36.23	32.07	29.13	26.91	25.17	23.75	22.57	20.71
12.0	45.65	38.30	33.68	30.44	28.02	26.13	24.60	23.33	21.34
13.0	48.48	40.28	35.20	31.68	29.06	27.03	25.39	24.04	21.92
14.0	51.22	42.17	36.65	32.85	30.05	27.88	26.14	24.71	22.47
15.0	53.90	44.00	38.03	33.97	30.98	28.68	26.85	25.34	22.99
16.0	56.50	45.76	39.36	35.03	31.88	29.45	27.52	25.94	23.48
18.0	61.52	49.10	41.86	37.04	33.54	30.88	28.77	27.05	24.39
20.0	66.33	52.24	44.19	38.89	35.08	32.19	29.91	28.06	25.22

† For an explanation of the headings and the significance of asterisks and high returns which appear on some pages, see Introductory Section, pages xx-xxiii.

INTERNAL RATE OF RETURN (IRR) †

(Reflecting projected growth in Full Rental Value to next Review/Reversion)

RENTAL FACTOR †	YEARS TO REVIEW/REVERSION								
	1	2	3	4	5	6	7	8	10
1.1	6.99	7.35	7.67	7.95	8.20	8.41	8.58	8.73	8.96
1.2	7.58	7.93	8.22	8.48	8.70	8.88	9.04	9.16	9.34
1.3	8.17	8.49	8.76	8.99	9.18	9.34	9.46	9.56	9.70
1.4	8.75	9.04	9.29	9.49	9.65	9.78	9.87	9.95	10.04
1.5	9.32	9.58	9.80	9.96	10.09	10.19	10.26	10.32	10.37
1.6	9.89	10.12	10.29	10.43	10.53	10.59	10.64	10.32	10.37
1.7	10.45	10.64	10.78	10.88	10.94	10.98	11.00	11.00	10.67
1.8	11.01	11.15	11.25	11.31	11.35	11.35	11.35	11.32	11.25
1.9	11.56	11.66	11.71	11.74	11.74	11.71	11.68	11.63	11.52
2.0	12.10	12.15	12.17	12.15	12.11	12.06	12.00	11.93	11.78
2.2	13.18	13.12	13.04	12.95	12.84	12.73	12.62	12.50	12.27
2.4	14.24	14.06	13.88	13.70	13.53	13.36	13.19	13.03	12.73
2.5	14.76	14.52	14.29	14.07	13.86	13.66	13.47	13.29	12.95
2.6	15.28	14.97	14.69	14.43	14.18	13.95	13.74	13.54	13.17
2.8	16.30	15.86	15.47	15.12	14.81	14.52	14.25	14.01	13.57
3.0	17.30	16.72	16.22	15.79	15.40	15.06	14.75	14.46	13.96
3.2	18.29	17.56	16.95	16.43	15.98	15.57	15.21	14.89	14.33
3.4	19.27	18.38	17.66	17.05	16.53	16.07	15.66	15.30	14.68
3.5	19.75	18.79	18.00	17.35	16.79	16.31	15.88	15.50	14.85
3.6	20.23	19.19	18.34	17.65	17.06	16.54	16.09	15.69	15.01
3.8	21.18	19.97	19.01	18.23	17.57	17.00	16.51	16.07	15.33
4.0	22.12	20.73	19.66	18.79	18.06	17.44	16.90	16.43	15.64
4.2	23.04	21.48	20.29	19.33	18.54	17.87	17.29	16.78	15.93
4.4	23.95	22.22	20.90	19.86	19.00	18.28	17.66	17.12	16.22
4.5	24.40	22.58	21.21	20.12	19.23	18.48	17.84	17.28	16.36
4.6	24.85	22.94	21.50	20.37	19.45	18.68	18.02	17.44	16.49
4.8	25.74	23.64	22.09	20.87	19.89	19.06	18.36	17.76	16.76
5.0	26.61	24.34	22.66	21.36	20.31	19.44	18.70	18.06	17.01
5.2	27.48	25.02	23.22	21.84	20.72	19.81	19.03	18.36	17.26
5.4	28.34	25.68	23.77	22.30	21.13	20.16	19.35	18.65	17.50
5.5	28.76	26.01	24.04	22.53	21.32	20.33	19.50	18.79	17.62
5.6	29.19	26.34	24.31	22.75	21.52	20.51	19.66	18.93	17.74
5.8	30.02	26.99	24.83	23.20	21.90	20.84	19.96	19.20	17.97
6.0	30.85	27.62	25.35	23.63	22.28	21.17	20.25	19.46	18.19
6.5	32.89	29.17	26.59	24.68	23.18	21.96	20.95	20.10	18.72
7.0	34.87	30.65	27.78	25.67	24.03	22.71	21.62	20.69	19.21
7.5	36.81	32.09	28.92	26.62	24.84	23.41	22.24	21.26	19.68
8.0	38.70	33.48	30.02	27.52	25.61	24.09	22.84	21.79	20.13
8.5	40.56	34.82	31.08	28.39	26.35	24.73	23.41	22.30	20.55
9.0	42.37	36.13	32.10	29.23	27.06	25.35	23.95	22.79	20.95
9.5	44.15	37.39	33.08	30.03	27.74	25.94	24.47	23.26	21.34
10.0	45.89	38.62	34.03	30.81	28.40	26.50	24.97	23.70	21.71
11.0	49.29	40.99	35.85	32.29	29.64	27.58	25.92	24.55	22.40
12.0	52.56	43.25	37.57	33.68	30.80	28.58	26.80	25.33	23.05
13.0	55.74	45.40	39.20	34.99	31.90	29.52	27.62	26.07	23.65
14.0	58.82	47.47	40.75	36.23	32.93	30.41	28.40	26.76	24.21
15.0	61.81	49.45	42.23	37.41	33.92	31.25	29.13	27.41	24.74
16.0	64.72	51.36	43.65	38.53	34.85	32.05	29.83	28.02	25.25
18.0	****	54.98	46.32	40.65	36.59	33.53	31.12	29.17	26.18
20.0	****	58.38	48.80	42.59	38.20	34.89	32.31	30.22	27.03

† For an explanation of the headings and the significance of asterisks and high returns which appear on some pages, see Introductory Section, pages xx-xxiii.

INTERNAL RATE OF RETURN (IRR) †

(Reflecting projected growth in Full Rental Value to next Review/Reversion)

RENTAL FACTOR †	YEARS TO REVIEW/REVERSION								
	1	2	3	4	5	6	7	8	10
1.1	8.15	8.54	8.88	9.18	9.42	9.62	9.79	9.92	10.11
1.2	8.83	9.19	9.51	9.76	9.97	10.14	10.27	10.37	10.51
1.3	9.50	9.83	10.11	10.33	10.50	10.63	10.73	10.80	10.88
1.4	10.16	10.46	10.69	10.87	11.00	11.10	11.16	11.21	11.23
1.5	10.82	11.07	11.26	11.39	11.49	11.55	11.58	11.59	11.57
1.6	11.47	11.67	11.81	11.90	11.95	11.98	11.98	11.96	11.89
1.7	12.11	12.26	12.34	12.39	12.40	12.39	12.36	12.31	12.19
1.8	12.75	12.83	12.87	12.86	12.83	12.79	12.72	12.65	12.48
1.9	13.38	13.40	13.38	13.33	13.25	13.17	13.07	12.97	12.76
2.0	14.00	13.95	13.87	13.77	13.66	13.54	13.41	13.29	13.03
2.2	15.23	15.03	14.84	14.64	14.44	14.25	14.06	13.88	13.54
2.4	16.43	16.08	15.76	15.45	15.17	14.91	14.67	14.44	14.02
2.5	17.02	16.59	16.20	15.85	15.53	15.23	14.95	14.70	14.24
2.6	17.61	17.09	16.64	16.23	15.87	15.54	15.24	14.96	14.46
2.8	18.76	18.07	17.49	16.98	16.53	16.14	15.78	15.45	14.88
3.0	19.90	19.03	18.31	17.70	17.17	16.70	16.29	15.92	15.28
3.2	21.02	19.96	19.10	18.38	17.78	17.25	16.78	16.37	15.66
3.4	22.12	20.86	19.87	19.05	18.36	17.77	17.25	16.79	16.01
3.5	22.66	21.31	20.24	19.37	18.64	18.02	17.47	17.00	16.19
3.6	23.20	21.74	20.61	19.69	18.92	18.26	17.70	17.20	16.36
3.8	24.27	22.60	21.33	20.30	19.46	18.74	18.13	17.59	16.69
4.0	25.32	23.44	22.03	20.90	19.98	19.21	18.54	17.96	17.00
4.2	26.36	24.26	22.71	21.48	20.48	19.65	18.94	18.32	17.30
4.4	27.38	25.07	23.37	22.04	20.97	20.08	19.33	18.67	17.60
4.5	27.88	25.46	23.69	22.32	21.21	20.29	19.51	18.84	17.74
4.6	28.38	25.86	24.01	22.59	21.45	20.50	19.70	19.01	17.88
4.8	29.38	26.63	24.64	23.12	21.90	20.90	20.06	19.33	18.15
5.0	30.36	27.38	25.26	23.64	22.35	21.30	20.41	19.65	18.41
5.2	31.33	28.12	25.86	24.14	22.79	21.68	20.75	19.96	18.67
5.4	32.28	28.85	26.44	24.63	23.21	22.05	21.08	20.25	18.91
5.5	32.76	29.21	26.73	24.87	23.42	22.23	21.24	20.40	19.03
5.6	33.23	29.57	27.02	25.11	23.62	22.41	21.40	20.54	19.15
5.8	34.16	30.27	27.58	25.58	24.02	22.76	21.71	20.82	19.39
6.0	35.09	30.96	28.13	26.04	24.41	23.10	22.01	21.09	19.61
6.5	37.35	32.63	29.46	27.14	25.36	23.92	22.74	21.75	20.15
7.0	39.55	34.24	30.73	28.19	26.25	24.70	23.43	22.36	20.66
7.5	41.70	35.80	31.95	29.19	27.09	25.43	24.08	22.94	21.14
8.0	43.80	37.30	33.12	30.15	27.90	26.13	24.69	23.49	21.60
8.5	45.85	38.75	34.24	31.06	28.67	26.80	25.28	24.02	22.03
9.0	47.86	40.16	35.32	31.94	29.41	27.44	25.84	24.52	22.44
9.5	49.82	41.52	36.37	32.79	30.12	28.05	26.38	25.00	22.83
10.0	51.75	42.85	37.38	33.60	30.81	28.64	26.90	25.46	23.21
11.0	55.49	45.39	39.30	35.15	32.10	29.75	27.87	26.32	23.92
12.0	59.09	47.81	41.12	36.61	33.31	30.79	28.78	27.13	24.58
13.0	62.58	50.12	42.85	37.98	34.45	31.76	29.62	27.88	25.19
14.0	65.96	52.33	44.48	39.28	35.53	32.68	30.42	28.59	25.77
15.0	69.25	54.46	46.05	40.51	36.55	33.55	31.18	29.26	26.31
16.0	****	56.50	47.54	41.69	37.52	34.37	31.89	29.89	26.82
18.0	****	60.37	50.35	43.89	39.33	35.90	33.22	31.06	27.77
20.0	****	64.00	52.96	45.93	40.99	37.31	34.44	32.13	28.64

† For an explanation of the headings and the significance of asterisks and high returns which appear on some pages, see Introductory Section, pages xx-xxiii.

INTERNAL RATE OF RETURN (IRR) †
(Reflecting projected growth in Full Rental Value to next Review/Reversion)

RENTAL FACTOR †	YEARS TO REVIEW/REVERSION								
	1	2	3	4	5	6	7	8	10
1.1	9.30	9.72	10.08	10.38	10.62	10.81	10.96	11.07	11.22
1.2	10.06	10.45	10.77	11.02	11.21	11.36	11.47	11.54	11.62
1.3	10.82	11.16	11.43	11.63	11.78	11.88	11.95	11.99	12.01
1.4	11.57	11.85	12.06	12.21	12.31	12.37	12.40	12.41	12.37
1.5	12.31	12.53	12.68	12.78	12.83	12.85	12.84	12.81	12.71
1.6	13.04	13.19	13.28	13.32	13.32	13.30	13.25	13.19	13.04
1.7	13.76	13.84	13.87	13.85	13.80	13.73	13.65	13.56	13.35
1.8	14.47	14.48	14.43	14.36	14.26	14.15	14.03	13.91	13.65
1.9	15.18	15.10	14.99	14.85	14.71	14.55	14.40	14.24	13.94
2.0	15.87	15.71	15.52	15.33	15.14	14.94	14.75	14.57	14.21
2.2	17.24	16.89	16.56	16.25	15.96	15.68	15.42	15.18	14.73
2.4	18.58	18.04	17.55	17.12	16.73	16.38	16.05	15.75	15.22
2.5	19.24	18.59	18.03	17.54	17.10	16.71	16.35	16.03	15.45
2.6	19.89	19.14	18.50	17.95	17.47	17.03	16.65	16.29	15.67
2.8	21.18	20.21	19.42	18.74	18.16	17.66	17.21	16.80	16.10
3.0	22.44	21.25	20.29	19.50	18.83	18.25	17.74	17.28	16.51
3.2	23.68	22.26	21.14	20.23	19.47	18.81	18.24	17.74	16.89
3.4	24.90	23.24	21.96	20.93	20.08	19.35	18.73	18.18	17.26
3.5	25.50	23.72	22.36	21.27	20.37	19.62	18.96	18.39	17.44
3.6	26.09	24.19	22.75	21.60	20.66	19.87	19.19	18.60	17.61
3.8	27.27	25.12	23.52	22.26	21.23	20.37	19.64	19.00	17.94
4.0	28.43	26.03	24.26	22.89	21.77	20.85	20.06	19.38	18.27
4.2	29.57	26.92	24.98	23.50	22.30	21.31	20.48	19.76	18.57
4.4	30.70	27.78	25.69	24.09	22.81	21.76	20.87	20.11	18.87
4.5	31.25	28.21	26.03	24.38	23.06	21.98	21.07	20.29	19.02
4.6	31.80	28.63	26.37	24.66	23.30	22.19	21.26	20.46	19.16
4.8	32.90	29.46	27.04	25.22	23.78	22.61	21.63	20.79	19.43
5.0	33.97	30.27	27.69	25.76	24.25	23.01	21.99	21.12	19.70
5.2	35.04	31.07	28.33	26.29	24.70	23.41	22.34	21.43	19.96
5.4	36.09	31.85	28.95	26.81	25.14	23.79	22.68	21.73	20.21
5.5	36.61	32.23	29.26	27.06	25.35	23.98	22.84	21.88	20.34
5.6	37.12	32.62	29.56	27.31	25.57	24.16	23.01	22.03	20.46
5.8	38.15	33.37	30.15	27.80	25.98	24.53	23.33	22.32	20.69
6.0	39.16	34.11	30.74	28.28	26.39	24.88	23.64	22.60	20.93
6.5	41.64	35.90	32.14	29.43	27.37	25.73	24.39	23.26	21.48
7.0	44.04	37.63	33.48	30.53	28.30	26.53	25.09	23.89	21.99
7.5	46.39	39.29	34.76	31.57	29.17	27.29	25.76	24.49	22.48
8.0	48.68	40.89	35.99	32.57	30.01	28.01	26.39	25.05	22.94
8.5	50.91	42.43	37.17	33.52	30.81	28.70	27.00	25.59	23.38
9.0	53.10	43.93	38.31	34.44	31.58	29.36	27.57	26.10	23.80
9.5	55.23	45.38	39.41	35.32	32.31	29.99	28.12	26.59	24.20
10.0	57.32	46.79	40.47	36.17	33.02	30.59	28.65	27.06	24.58
11.0	61.39	49.50	42.49	37.78	34.36	31.74	29.65	27.94	25.30
12.0	65.30	52.07	44.40	39.30	35.61	32.80	30.58	28.77	25.97
13.0	69.08	54.52	46.21	40.72	36.79	33.80	31.45	29.53	26.59
14.0	****	56.86	47.92	42.07	37.90	34.75	32.27	30.26	27.18
15.0	****	59.11	49.56	43.35	38.95	35.64	33.04	30.94	27.73
16.0	****	61.27	51.12	44.57	39.95	36.48	33.77	31.58	28.25
18.0	****	65.36	54.06	46.86	41.82	38.06	35.13	32.78	29.21
20.0	****	69.20	56.78	48.97	43.53	39.50	36.37	33.87	30.09

† For an explanation of the headings and the significance of asterisks and high returns which appear on some pages, see Introductory Section, pages xx-xxiii.

INTERNAL RATE OF RETURN (IRR) †
(Reflecting projected growth in Full Rental Value to next Review/Reversion)

RENTAL FACTOR †	YEARS TO REVIEW/REVERSION								
	1	2	3	4	5	6	7	8	10
1.1	10.44	10.90	11.27	11.57	11.80	11.97	12.10	12.19	12.29
1.2	11.30	11.70	12.01	12.25	12.42	12.55	12.63	12.68	12.70
1.3	12.14	12.47	12.72	12.90	13.02	13.09	13.13	13.14	13.09
1.4	12.97	13.23	13.41	13.53	13.59	13.61	13.60	13.57	13.46
1.5	13.79	13.97	14.08	14.13	14.13	14.10	14.05	13.98	13.81
1.6	14.59	14.69	14.72	14.70	14.65	14.57	14.48	14.38	14.15
1.7	15.39	15.40	15.35	15.26	15.15	15.03	14.89	14.75	14.46
1.8	16.17	16.08	15.96	15.80	15.64	15.46	15.29	15.11	14.77
1.9	16.95	16.76	16.55	16.33	16.10	15.88	15.66	15.46	15.06
2.0	17.72	17.42	17.12	16.83	16.55	16.29	16.03	15.79	15.34
2.2	19.22	18.70	18.23	17.80	17.41	17.05	16.72	16.42	15.87
2.4	20.69	19.94	19.29	18.72	18.22	17.77	17.37	17.00	16.36
2.5	21.42	20.54	19.80	19.16	18.61	18.12	17.68	17.28	16.59
2.6	22.13	21.13	20.30	19.59	18.99	18.45	17.98	17.56	16.82
2.8	23.54	22.28	21.26	20.43	19.71	19.10	18.56	18.08	17.26
3.0	24.92	23.39	22.20	21.22	20.41	19.71	19.10	18.57	17.67
3.2	26.28	24.47	23.09	21.98	21.07	20.29	19.63	19.04	18.06
3.4	27.61	25.52	23.96	22.72	21.70	20.85	20.12	19.49	18.43
3.5	28.26	26.04	24.38	23.07	22.01	21.12	20.36	19.70	18.61
3.6	28.91	26.54	24.79	23.42	22.31	21.39	20.60	19.92	18.79
3.8	30.19	27.54	25.60	24.10	22.90	21.90	21.05	20.33	19.13
4.0	31.46	28.51	26.39	24.76	23.46	22.39	21.49	20.72	19.45
4.2	32.70	29.45	27.15	25.40	24.01	22.87	21.91	21.10	19.76
4.4	33.92	30.38	27.89	26.02	24.54	23.33	22.32	21.46	20.07
4.5	34.52	30.83	28.25	26.32	24.79	23.55	22.52	21.64	20.21
4.6	35.12	31.28	28.61	26.61	25.05	23.78	22.72	21.81	20.36
4.8	36.31	32.16	29.31	27.20	25.54	24.21	23.10	22.16	20.64
5.0	37.48	33.03	30.00	27.76	26.02	24.62	23.47	22.49	20.91
5.2	38.63	33.87	30.66	28.31	26.49	25.03	23.82	22.81	21.17
5.4	39.77	34.70	31.32	28.85	26.95	25.42	24.17	23.12	21.43
5.5	40.33	35.11	31.64	29.11	27.17	25.62	24.34	23.27	21.55
5.6	40.89	35.52	31.96	29.37	27.39	25.81	24.51	23.42	21.67
5.8	42.00	36.32	32.58	29.88	27.82	26.18	24.84	23.71	21.92
6.0	43.09	37.10	33.19	30.38	28.24	26.54	25.16	23.99	22.15
6.5	45.77	39.00	34.66	31.58	29.25	27.41	25.92	24.68	22.71
7.0	48.37	40.83	36.06	32.72	30.21	28.24	26.64	25.32	23.23
7.5	50.90	42.59	37.41	33.80	31.11	29.02	27.32	25.92	23.73
8.0	53.36	44.28	38.69	34.83	31.97	29.75	27.97	26.50	24.19
8.5	55.77	45.91	39.92	35.82	32.80	30.46	28.59	27.04	24.64
9.0	58.12	47.49	41.11	36.77	33.59	31.13	29.17	27.56	25.06
9.5	60.42	49.02	42.26	37.68	34.35	31.78	29.74	28.06	25.47
10.0	62.67	50.51	43.36	38.56	35.07	32.40	30.28	28.54	25.85
11.0	67.03	53.36	45.47	40.23	36.45	33.57	31.30	29.44	26.58
12.0	****	56.07	47.46	41.79	37.74	34.67	32.24	30.28	27.26
13.0	****	58.65	49.34	43.27	38.95	35.69	33.13	31.06	27.89
14.0	****	61.11	51.12	44.66	40.09	36.65	33.97	31.79	28.48
15.0	****	63.47	52.82	45.98	41.17	37.57	34.75	32.49	29.04
16.0	****	65.74	54.44	47.24	42.19	38.43	35.50	33.14	29.57
18.0	****	****	57.50	49.60	44.11	40.04	36.89	34.36	30.54
20.0	****	****	60.32	51.77	45.87	41.51	38.15	35.47	31.43

† For an explanation of the headings and the significance of asterisks and high returns which appear on some pages, see Introductory Section, pages xx-xxiii.

INTERNAL RATE OF RETURN (IRR) †

(Reflecting projected growth in Full Rental Value to next Review/Reversion)

RENTAL FACTOR †	YEARS TO REVIEW/REVERSION								
	1	2	3	4	5	6	7	8	10
1.1	11.59	12.07	12.44	12.74	12.95	13.11	13.21	13.28	13.33
1.2	12.52	12.93	13.24	13.46	13.61	13.71	13.76	13.78	13.75
1.3	13.45	13.77	14.00	14.15	14.23	14.27	14.27	14.25	14.15
1.4	14.35	14.59	14.73	14.81	14.83	14.81	14.76	14.69	14.52
1.5	15.25	15.39	15.44	15.44	15.39	15.32	15.22	15.12	14.87
1.6	16.13	16.16	16.13	16.05	15.94	15.81	15.67	15.52	15.21
1.7	17.00	16.92	16.79	16.64	16.46	16.28	16.09	15.90	15.53
1.8	17.86	17.66	17.44	17.20	16.96	16.73	16.49	16.27	15.84
1.9	18.70	18.39	18.07	17.75	17.45	17.16	16.88	16.62	16.13
2.0	19.54	19.09	18.68	18.28	17.92	17.58	17.26	16.96	16.42
2.2	21.17	20.47	19.85	19.30	18.81	18.37	17.97	17.60	16.95
2.4	22.77	21.78	20.96	20.26	19.65	19.11	18.63	18.20	17.45
2.5	23.56	22.43	21.50	20.72	20.05	19.47	18.95	18.48	17.69
2.6	24.33	23.05	22.03	21.17	20.44	19.81	19.26	18.76	17.92
2.8	25.86	24.28	23.04	22.04	21.19	20.47	19.85	19.29	18.36
3.0	27.35	25.47	24.02	22.87	21.91	21.10	20.41	19.80	18.77
3.2	28.82	26.62	24.96	23.66	22.60	21.70	20.94	20.27	19.17
3.4	30.25	27.73	25.87	24.42	23.25	22.27	21.45	20.73	19.54
3.5	30.96	28.28	26.31	24.79	23.57	22.55	21.69	20.95	19.73
3.6	31.66	28.81	26.75	25.16	23.88	22.82	21.93	21.16	19.90
3.8	33.04	29.87	27.60	25.86	24.48	23.35	22.40	21.58	20.25
4.0	34.40	30.89	28.42	26.55	25.07	23.86	22.85	21.98	20.58
4.2	35.74	31.89	29.21	27.21	25.63	24.35	23.28	22.37	20.89
4.4	37.05	32.87	29.99	27.85	26.17	24.82	23.69	22.74	21.20
4.5	37.70	33.35	30.37	28.16	26.44	25.05	23.89	22.92	21.34
4.6	38.35	33.82	30.74	28.47	26.70	25.27	24.09	23.10	21.49
4.8	39.62	34.76	31.47	29.07	27.21	25.72	24.48	23.44	21.77
5.0	40.88	35.67	32.19	29.65	27.70	26.14	24.86	23.78	22.05
5.2	42.11	36.56	32.89	30.22	28.18	26.56	25.22	24.10	22.31
5.4	43.33	37.44	33.57	30.78	28.65	26.96	25.58	24.42	22.57
5.5	43.94	37.87	33.90	31.05	28.88	27.16	25.75	24.57	22.70
5.6	44.54	38.29	34.23	31.32	29.11	27.35	25.92	24.72	22.82
5.8	45.72	39.14	34.88	31.85	29.55	27.74	26.26	25.02	23.06
6.0	46.90	39.96	35.52	32.36	29.98	28.11	26.58	25.31	23.30
6.5	49.76	41.96	37.05	33.60	31.02	29.00	27.36	26.00	23.86
7.0	52.54	43.88	38.51	34.78	32.00	29.84	28.10	26.65	24.39
7.5	55.25	45.73	39.90	35.89	32.93	30.63	28.79	27.27	24.89
8.0	57.88	47.50	41.24	36.96	33.82	31.39	29.45	27.85	25.37
8.5	60.45	49.22	42.52	37.98	34.66	32.11	30.08	28.41	25.82
9.0	62.96	50.87	43.75	38.96	35.47	32.80	30.67	28.94	26.24
9.5	65.41	52.48	44.94	39.90	36.25	33.46	31.25	29.44	26.65
10.0	67.81	54.04	46.09	40.80	36.99	34.09	31.80	29.93	27.05
11.0	****	57.02	48.27	42.52	38.41	35.29	32.83	30.84	27.78
12.0	****	59.85	50.33	44.13	39.73	36.41	33.80	31.69	28.47
13.0	****	62.55	52.27	45.65	40.96	37.45	34.70	32.48	29.10
14.0	****	65.12	54.12	47.08	42.13	38.43	35.55	33.23	29.70
15.0	****	67.59	55.88	48.44	43.24	39.36	36.35	33.93	30.26
16.0	****	69.96	57.56	49.73	44.29	40.24	37.11	34.60	30.80
18.0	****	****	60.71	52.16	46.24	41.88	38.52	35.83	31.78
20.0	****	****	63.64	54.39	48.04	43.39	39.81	36.96	32.68

† For an explanation of the headings and the significance of asterisks and high returns which appear on some pages, see Introductory Section, pages xx-xxiii.

CONVERSION TABLES

DECIMALS OF ONE POUND

s. d.	Decimal of Pound	s. d.	Decimal of Pound	s. d.	Decimal of Pound	s. d.	Decimal of Pound	s. d.	Decimal of Pound
0– 0½	·0020833	2– 0	·1000000	4– 0	·2000000	6– 0	·3000000	8– 0	·4000000
0– 1	·0041667	2– 1	·1041667	4– 1	·2041667	6– 1	·3041667	8– 1	·4041667
0– 2	·0083333	2– 2	·1083333	4– 2	·2083333	6– 2	·3083333	8– 2	·4083333
0– 3	·0125000	2– 3	·1125000	4– 3	·2125000	6– 3	·3125000	8– 3	·4125000
0– 4	·0166667	2– 4	·1166667	4– 4	·2166667	6– 4	·3166667	8– 4	·4166667
0– 5	·0208333	2– 5	·1208333	4– 5	·2208333	6– 5	·3208333	8– 5	·4208333
0– 6	·0250000	2– 6	·1250000	4– 6	·2250000	6– 6	·3250000	8– 6	·4250000
0– 7	·0291667	2– 7	·1291667	4– 7	·2291667	6– 7	·3291667	8– 7	·4291667
0– 8	·0333333	2– 8	·1333333	4– 8	·2333333	6– 8	·3333333	8– 8	·4333333
0– 9	·0375000	2– 9	·1375000	4– 9	·2375000	6– 9	·3375000	8– 9	·4375000
0–10	·0416667	2–10	·1416667	4–10	·2416667	6–10	·3416667	8–10	·4416667
0–11	·0458333	2–11	·1458333	4–11	·2458333	6–11	·3458333	8–11	·4458333
1– 0	·0500000	3– 0	·1500000	5– 0	·2500000	7– 0	·3500000	9– 0	·4500000
1– 1	·0541667	3– 1	·1541667	5– 1	·2541667	7– 1	·3541667	9– 1	·4541667
1– 2	·0583333	3– 2	·1583333	5– 2	·2583333	7– 2	·3583333	9– 2	·4583333
1– 3	·0625000	3– 3	·1625000	5– 3	·2625000	7– 3	·3625000	9– 3	·4625000
1– 4	·0666667	3– 4	·1666667	5– 4	·2666667	7– 4	·3666667	9– 4	·4666667
1– 5	·0708333	3– 5	·1708333	5– 5	·2708333	7– 5	·3708333	9– 5	·4708333
1– 6	·0750000	3– 6	·1750000	5– 6	·2750000	7– 6	·3750000	9– 6	·4750000
1– 7	·0791667	3– 7	·1791667	5– 7	·2791667	7– 7	·3791667	9– 7	·4791667
1– 8	·0833333	3– 8	·1833333	5– 8	·2833333	7– 8	·3833333	9– 8	·4833333
1– 9	·0875000	3– 9	·1875000	5– 9	·2875000	7– 9	·3875000	9– 9	·4875000
1–10	·0916667	3–10	·1916667	5–10	·2916667	7–10	·3916667	9–10	·4916667
1–11	·0958333	3–11	·1958333	5–11	·2958333	7–11	·3958333	9–11	·4958333

s. d.	Decimal of Pound	s. d.	Decimal of Pound	s. d.	Decimal of Pound	s. d.	Decimal of Pound	s. d.	Decimal of Pound
10– 0	·5000000	12– 0	·6000000	14– 0	·7000000	16– 0	·8000000	18– 0	·9000000
10– 1	·5041667	12– 1	·6041667	14– 1	·7041667	16– 1	·8041667	18– 1	·9041667
10– 2	·5083333	12– 2	·6083333	14– 2	·7083333	16– 2	·8083333	18– 2	·9083333
10– 3	·5125000	12– 3	·6125000	14– 3	·7125000	16– 3	·8125000	18– 3	·9125000
10– 4	·5166667	12– 4	·6166667	14– 4	·7166667	16– 4	·8166667	18– 4	·9166667
10– 5	·5208333	12– 5	·6208333	14– 5	·7208333	16– 5	·8208333	18– 5	·9208333
10– 6	·5250000	12– 6	·6250000	14– 6	·7250000	16– 6	·8250000	18– 6	·9250000
10– 7	·5291667	12– 7	·6291667	14– 7	·7291667	16– 7	·8291667	18– 7	·9291667
10– 8	·5333333	12– 8	·6333333	14– 8	·7333333	16– 8	·8333333	18– 8	·9333333
10– 9	·5375000	12– 9	·6375000	14– 9	·7375000	16– 9	·8375000	18– 9	·9375000
10–10	·5416667	12–10	·6416667	14–10	·7416667	16–10	·8416667	18–10	·9416667
10–11	·5458333	12–11	·6458333	14–11	·7458333	16–11	·8458333	18–11	·9458333
11– 0	·5500000	13– 0	·6500000	15– 0	·7500000	17– 0	·8500000	19– 0	·9500000
11– 1	·5541667	13– 1	·6541667	15– 1	·7541667	17– 1	·8541667	19– 1	·9541667
11– 2	·5583333	13– 2	·6583333	15– 2	·7583333	17– 2	·8583333	19– 2	·9583333
11– 3	·5625000	13– 3	·6625000	15– 3	·7625000	17– 3	·8625000	19– 3	·9625000
11– 4	·5666667	13– 4	·6666667	15– 4	·7666667	17– 4	·8666667	19– 4	·9666667
11– 5	·5708333	13– 5	·6708333	15– 5	·7708333	17– 5	·8708333	19– 5	·9708333
11– 6	·5750000	13– 6	·6750000	15– 6	·7750000	17– 6	·8750000	19– 6	·9750000
11– 7	·5791667	13– 7	·6791667	15– 7	·7791667	17– 7	·8791667	19– 7	·9791667
11– 8	·5833333	13– 8	·6833333	15– 8	·7833333	17– 8	·8833333	19– 8	·9833333
11– 9	·5875000	13– 9	·6875000	15– 9	·7875000	17– 9	·8875000	19– 9	·9875000
11–10	·5916667	13–10	·6916667	15–10	·7916667	17–10	·8916667	19–10	·9916667
11–11	·5958333	13–11	·6958333	15–11	·7958333	17–11	·8958333	19–11	·9958333

FEET to METRES

Feet	0	1	2	3	4	5	6	7	8	9
0		0.3048	0.6096	0.9144	1.2192	1.5240	1.8288	2.1336	2.4384	2.7432
10	3.0480	3.3528	3.6576	3.9624	4.2672	4.5720	4.8768	5.1816	5.4864	5.7912
20	6.0960	6.4008	6.7056	7.0104	7.3152	7.6200	7.9248	8.2296	8.5344	8.8392
30	9.1440	9.4488	9.7536	10.0584	10.3632	10.6680	10.9728	11.2776	11.5824	11.8872
40	12.1920	12.4968	12.8016	13.1064	13.4112	13.7160	14.0208	14.3256	14.6304	14.9352
50	15.2400	15.5448	15.8496	16.1544	16.4592	16.7640	17.0688	17.3736	17.6784	17.9832
60	18.2880	18.5928	18.8976	19.2024	19.5072	19.8120	20.1168	20.4216	20.7264	21.0312
70	21.3360	21.6408	21.9456	22.2504	22.5552	22.8600	23.1648	23.4696	23.7744	24.0792
80	24.3840	24.6888	24.9936	25.2984	25.6032	25.9080	26.2128	26.5176	26.8224	27.1272
90	27.4320	27.7368	28.0416	28.3464	28.6512	28.9560	29.2608	29.5656	29.8704	30.1752
100	30.4800	30.7848	31.0896	31.3944	31.6992	32.0040	32.3088	32.6136	32.9184	33.2232
110	33.5280	33.8328	34.1376	34.4424	34.7472	35.0520	35.3568	35.6616	35.9664	36.2712
120	36.5760	36.8808	37.1856	37.4904	37.7952	38.1000	38.4048	38.7096	39.0144	39.3192
130	39.6240	39.9288	40.2336	40.5384	40.8432	41.1480	41.4528	41.7576	42.0624	42.3672
140	42.6720	42.9768	43.2816	43.5864	43.8912	44.1960	44.5008	44.8056	45.1104	45.4152
150	45.7200	46.0248	46.3296	46.6344	46.9392	47.2440	47.5488	47.8536	48.1584	48.4632
160	48.7680	49.0728	49.3776	49.6824	49.9872	50.2920	50.5968	50.9016	51.2064	51.5112
170	51.8160	52.1208	52.4256	52.7304	53.0352	53.3400	53.6448	53.9496	54.2544	54.5592
180	54.8640	55.1688	55.4736	55.7784	56.0832	56.3880	56.6928	56.9976	57.3024	57.6072
190	57.9120	58.2168	58.5216	58.8264	59.1312	59.4360	59.7408	60.0456	60.3504	60.6552
200	60.9600	61.2648	61.5696	61.8744	62.1792	62.4840	62.7888	63.0936	63.3984	63.7032
210	64.0080	64.3128	64.6176	64.9224	65.2272	65.5320	65.8368	66.1416	66.4464	66.7512
220	67.0560	67.3608	67.6656	67.9704	68.2752	68.5800	68.8848	69.1896	69.4944	69.7992
230	70.1040	70.4088	70.7136	71.0184	71.3232	71.6280	71.9328	72.2376	72.5424	72.8472
240	73.1520	73.4568	73.7616	74.0664	74.3712	74.6760	74.9808	75.2856	75.5904	75.8952
250	76.2000	76.5048	76.8096	77.1144	77.4192	77.7240	78.0288	78.3336	78.6384	78.9432
260	79.2480	79.5528	79.8576	80.1624	80.4672	80.7720	81.0768	81.3816	81.6864	81.9912
270	82.2960	82.6008	82.9056	83.2104	83.5152	83.8200	84.1248	84.4296	84.7344	85.0392
280	85.3440	85.6488	85.9536	86.2584	86.5632	86.8680	87.1728	87.4776	87.7824	88.0872

FEET to METRES

Feet	0	1	2	3	4	5	6	7	8	9
300	91.4400	91.7448	92.0496	92.3544	92.6592	92.9640	93.2688	93.5736	93.8784	94.1832
310	94.4880	94.7928	95.0976	95.4024	95.7072	96.0120	96.3168	96.6216	96.9264	97.2312
320	97.5360	97.8408	98.1456	98.4504	98.7552	99.0600	99.3648	99.6696	99.9744	100.2792
330	100.5840	100.8888	101.1936	101.4984	101.8032	102.1080	102.4128	102.7176	103.0224	103.3272
340	103.6320	103.9368	104.2416	104.5464	104.8512	105.1560	105.4608	105.7656	106.0704	106.3752
350	106.6800	106.9848	107.2896	107.5944	107.8992	108.2040	108.5088	108.8136	109.1184	109.4232
360	109.7280	110.0328	110.3376	110.6424	110.9472	111.2520	111.5568	111.8616	112.1664	112.4712
370	112.7760	113.0808	113.3856	113.6904	113.9952	114.3000	114.6048	114.9096	115.2144	115.5192
380	115.8240	116.1288	116.4336	116.7384	117.0432	117.3480	117.6528	117.9576	118.2624	118.5672
390	118.8720	119.1768	119.4816	119.7864	120.0912	120.3960	120.7008	121.0056	121.3104	121.6152
400	121.9200	122.2248	122.5296	122.8344	123.1392	123.4440	123.7488	124.0536	124.3584	124.6632
410	124.9680	125.2728	125.5776	125.8824	126.1872	126.4920	126.7968	127.1016	127.4064	127.7112
420	128.0160	128.3208	128.6256	128.9304	129.2352	129.5400	129.8448	130.1496	130.4544	130.7592
430	131.0640	131.3688	131.6736	131.9784	132.2832	132.5880	132.8928	133.1976	133.5024	133.8072
440	134.1120	134.4168	134.7216	135.0264	135.3312	135.6360	135.9408	136.2456	136.5504	136.8552
450	137.1600	137.4648	137.7696	138.0744	138.3792	138.6840	138.9888	139.2936	139.5984	139.9032
460	140.2080	140.5128	140.8176	141.1224	141.4272	141.7320	142.0368	142.3416	142.6464	142.9512
470	143.2560	143.5608	143.8656	144.1704	144.4752	144.7800	145.0848	145.3896	145.6944	145.9992
480	146.3040	146.6088	146.9136	147.2184	147.5232	147.8280	148.1328	148.4376	148.7424	149.0472
490	149.3520	149.6568	149.9616	150.2664	150.5712	150.8760	151.1808	151.4856	151.7904	152.0952

SUPPLEMENTARY TABLE

Feet	500	600	700	800	900	1000
Metres	152.4000	182.8800	213.3600	243.8400	274.3200	304.8000

Basis : 1 ft. = 0.3048 m. (exactly)

243

METRES to FEET and INCHES

Metres	0 (Ft. Ins.)	1 (Ft. Ins.)	2 (Ft. Ins.)	3 (Ft. Ins.)	4 (Ft. Ins.)	5 (Ft. Ins.)	6 (Ft. Ins.)	7 (Ft. Ins.)	8 (Ft. Ins.)	9 (Ft. Ins.)
0	0	3 3$\frac{3}{8}$	6 6$\frac{3}{4}$	9 10$\frac{1}{8}$	13 1$\frac{1}{2}$	16 4$\frac{7}{8}$	19 8$\frac{1}{4}$	22 11$\frac{5}{8}$	26 3	29 6$\frac{3}{8}$
10	32 9$\frac{3}{4}$	36 1$\frac{1}{8}$	39 4$\frac{1}{2}$	42 7$\frac{3}{4}$	45 11$\frac{1}{8}$	49 2$\frac{1}{2}$	52 5$\frac{7}{8}$	55 9$\frac{1}{4}$	59 0$\frac{5}{8}$	62 4
20	65 7$\frac{3}{8}$	68 10$\frac{3}{4}$	72 2$\frac{1}{8}$	75 5$\frac{1}{2}$	78 8$\frac{7}{8}$	82 0$\frac{1}{4}$	85 3$\frac{5}{8}$	88 7	91 10$\frac{3}{8}$	95 1$\frac{3}{4}$
30	98 5$\frac{1}{8}$	101 8$\frac{1}{2}$	104 11$\frac{7}{8}$	108 3$\frac{1}{4}$	111 6$\frac{5}{8}$	114 10	118 1$\frac{3}{8}$	121 4$\frac{3}{4}$	124 8$\frac{1}{8}$	127 11$\frac{3}{8}$
40	131 2$\frac{3}{4}$	134 6$\frac{1}{8}$	137 9$\frac{1}{2}$	141 0$\frac{7}{8}$	144 4$\frac{1}{4}$	147 7$\frac{5}{8}$	150 11	154 2$\frac{3}{8}$	157 5$\frac{3}{4}$	160 9$\frac{1}{8}$
50	164 0$\frac{1}{2}$	167 3$\frac{7}{8}$	170 7$\frac{1}{4}$	173 10$\frac{5}{8}$	177 2	180 5$\frac{3}{8}$	183 8$\frac{3}{4}$	187 0$\frac{1}{8}$	190 3$\frac{1}{2}$	193 6$\frac{7}{8}$
60	196 10$\frac{1}{4}$	200 1$\frac{5}{8}$	203 5	206 8$\frac{3}{8}$	209 11$\frac{5}{8}$	213 3	216 6$\frac{3}{8}$	219 9$\frac{3}{4}$	223 1$\frac{1}{8}$	226 4$\frac{1}{2}$
70	229 7$\frac{7}{8}$	232 11$\frac{1}{4}$	236 2$\frac{5}{8}$	239 6	242 9$\frac{3}{8}$	246 0$\frac{3}{4}$	249 4$\frac{1}{8}$	252 7$\frac{1}{2}$	255 10$\frac{7}{8}$	259 2$\frac{1}{4}$
80	262 5$\frac{5}{8}$	265 9	269 0$\frac{3}{8}$	272 3$\frac{3}{4}$	275 7$\frac{1}{8}$	278 10$\frac{1}{2}$	282 1$\frac{7}{8}$	285 5$\frac{1}{4}$	288 8$\frac{5}{8}$	291 11$\frac{7}{8}$
90	295 3$\frac{1}{4}$	298 6$\frac{5}{8}$	301 10	305 1$\frac{3}{8}$	308 4$\frac{3}{4}$	311 8$\frac{1}{8}$	314 11$\frac{1}{2}$	318 2$\frac{7}{8}$	321 6$\frac{1}{4}$	324 9$\frac{5}{8}$
100	328 1	331 4$\frac{3}{8}$	334 7$\frac{3}{4}$	337 11$\frac{1}{8}$	341 2$\frac{1}{2}$	344 5$\frac{7}{8}$	347 9$\frac{1}{4}$	351 0$\frac{5}{8}$	354 4	357 7$\frac{3}{8}$
110	360 10$\frac{3}{4}$	364 2$\frac{1}{8}$	367 5$\frac{1}{2}$	370 8$\frac{7}{8}$	374 0$\frac{1}{4}$	377 3$\frac{1}{2}$	380 6$\frac{7}{8}$	383 10$\frac{1}{4}$	387 1$\frac{5}{8}$	390 5
120	393 8$\frac{3}{8}$	396 11$\frac{3}{4}$	400 3$\frac{1}{8}$	403 6$\frac{1}{2}$	406 9$\frac{7}{8}$	410 1$\frac{1}{4}$	413 4$\frac{5}{8}$	416 8	419 11$\frac{3}{8}$	423 2$\frac{3}{4}$
130	426 6$\frac{1}{8}$	429 9$\frac{1}{2}$	433 0$\frac{7}{8}$	436 4$\frac{1}{4}$	439 7$\frac{5}{8}$	442 11	446 2$\frac{3}{8}$	449 5$\frac{3}{4}$	452 9$\frac{1}{8}$	456 0$\frac{1}{2}$
140	459 3$\frac{3}{4}$	462 7$\frac{1}{8}$	465 10$\frac{1}{2}$	469 1$\frac{7}{8}$	472 5$\frac{1}{4}$	475 8$\frac{5}{8}$	479 0	482 3$\frac{3}{8}$	485 6$\frac{3}{4}$	488 10$\frac{1}{8}$
150	492 1$\frac{1}{2}$	495 4$\frac{7}{8}$	498 8$\frac{1}{4}$	501 11$\frac{5}{8}$	505 3	508 6$\frac{3}{8}$	511 9$\frac{3}{4}$	515 1$\frac{1}{8}$	518 4$\frac{1}{2}$	521 7$\frac{7}{8}$
160	524 11$\frac{1}{4}$	528 2$\frac{5}{8}$	531 6	534 9$\frac{3}{8}$	538 0$\frac{3}{4}$	541 4$\frac{1}{8}$	544 7$\frac{3}{8}$	547 10$\frac{3}{4}$	551 2$\frac{1}{8}$	554 5$\frac{1}{2}$
170	557 8$\frac{7}{8}$	561 0$\frac{1}{4}$	564 3$\frac{5}{8}$	567 7	570 10$\frac{3}{8}$	574 1$\frac{3}{4}$	577 5$\frac{1}{8}$	580 8$\frac{1}{2}$	583 11$\frac{7}{8}$	587 3$\frac{1}{4}$
180	590 6$\frac{5}{8}$	593 10	597 1$\frac{3}{8}$	600 4$\frac{3}{4}$	603 8$\frac{1}{8}$	606 11$\frac{1}{2}$	610 2$\frac{7}{8}$	613 6$\frac{1}{4}$	616 9$\frac{5}{8}$	620 1
190	623 4$\frac{3}{8}$	626 7$\frac{5}{8}$	629 11	633 2$\frac{3}{8}$	636 5$\frac{3}{4}$	639 9$\frac{1}{8}$	643 0$\frac{1}{2}$	646 3$\frac{7}{8}$	649 7$\frac{1}{4}$	652 10$\frac{5}{8}$
200	656 2	659 5$\frac{3}{8}$	662 8$\frac{3}{4}$	666 0$\frac{1}{8}$	669 3$\frac{1}{2}$	672 6$\frac{7}{8}$	675 10$\frac{1}{4}$	679 1$\frac{5}{8}$	682 5	685 8$\frac{3}{8}$
210	688 11$\frac{3}{4}$	692 3$\frac{1}{8}$	695 6$\frac{1}{2}$	698 9$\frac{7}{8}$	702 1$\frac{1}{4}$	705 4$\frac{5}{8}$	708 7$\frac{7}{8}$	711 11$\frac{1}{4}$	715 2$\frac{5}{8}$	718 6
220	721 9$\frac{3}{8}$	725 0$\frac{3}{4}$	728 4$\frac{1}{8}$	731 7$\frac{1}{2}$	734 10$\frac{7}{8}$	738 2$\frac{1}{4}$	741 5$\frac{5}{8}$	744 9	748 0$\frac{3}{8}$	751 3$\frac{3}{4}$
230	754 7$\frac{1}{8}$	757 10$\frac{1}{2}$	761 1$\frac{7}{8}$	764 5$\frac{1}{4}$	767 8$\frac{5}{8}$	771 0	774 3$\frac{3}{8}$	777 6$\frac{3}{4}$	780 10$\frac{1}{8}$	784 1$\frac{1}{2}$
240	787 4$\frac{7}{8}$	790 8$\frac{1}{4}$	793 11$\frac{1}{2}$	797 2$\frac{7}{8}$	800 6$\frac{1}{4}$	803 9$\frac{5}{8}$	807 1	810 4$\frac{3}{8}$	813 7$\frac{3}{4}$	816 11$\frac{1}{8}$

Metres	0	1	2	3	4	5	6	7	8	9
	Ft. Ins.	Ft. Ins.	Ft. Ins.	Ft. Ins.	Ft. Ins.	Ft. Ins.	Ft. Ins.	Ft. Ins.	Ft. Ins.	Ft. Ins.
250	820 2½	823 5⅞	826 9¼	830 0⅝	833 4	836 7⅜	839 10¾	843 2⅛	846 5½	849 8⅞
260	853 0¼	856 3⅝	859 7	862 10⅜	866 1¾	869 5⅛	872 8½	875 11¾	879 3⅛	882 6½
270	885 9⅞	889 1¼	892 4⅝	895 8	898 11⅜	902 2¾	905 6⅛	908 9½	912 0⅞	915 4¼
280	918 7⅝	921 11	925 2⅜	928 5¾	931 9⅛	935 0½	938 3⅞	941 7¼	944 10⅝	948 2
290	951 5⅜	954 8¾	958 0⅛	961 3⅜	964 6¾	967 10⅛	971 1½	974 4⅞	977 8¼	980 11⅝
300	984 3	987 6⅜	990 9¾	994 1⅛	997 4½	1000 7⅞	1003 11¼	1007 2⅝	1010 6	1013 9⅜
310	1017 0¾	1020 4⅛	1023 7½	1026 10⅞	1030 2¼	1033 5⅝	1036 9	1040 0⅜	1043 3⅝	1046 7
320	1049 10⅜	1053 1¾	1056 5⅛	1059 8½	1062 11⅞	1066 3¼	1069 6⅝	1072 10	1076 1⅜	1079 4¾
330	1082 8⅛	1085 11½	1089 2⅞	1092 6¼	1095 9⅝	1099 1	1102 4⅜	1105 7¾	1108 11⅛	1112 2½
340	1115 5⅞	1118 9¼	1122 0⅝	1125 3⅞	1128 7¼	1131 10⅝	1135 2	1138 5⅜	1141 8¾	1145 0⅛
350	1148 3½	1151 6⅞	1154 10¼	1158 1⅝	1161 5	1164 8⅜	1167 11¾	1171 3⅛	1174 6½	1177 9⅞
360	1181 1¼	1184 4⅝	1187 8	1190 11⅜	1194 2¾	1197 6⅛	1200 9½	1204 0⅞	1207 4¼	1210 7½
370	1213 10⅞	1217 2¼	1220 5⅝	1223 9	1227 0⅜	1230 3¾	1233 7⅛	1236 10½	1240 1⅞	1243 5¼
380	1246 8⅝	1250 0	1253 3⅜	1256 6¾	1259 10⅛	1263 1½	1266 4⅞	1269 8¼	1272 11⅝	1276 3
390	1279 6⅜	1282 9¾	1286 1⅛	1289 4½	1292 7¾	1295 11⅛	1299 2½	1302 5⅞	1305 9¼	1309 0⅝
400	1312 4	1315 7⅜	1318 10¾	1322 2⅛	1325 5½	1328 8⅞	1332 0¼	1335 3⅝	1338 7	1341 10⅜
410	1345 1¾	1348 5⅛	1351 8½	1354 11⅞	1358 3¼	1361 6⅝	1364 10	1368 1⅜	1371 4¾	1374 8⅛
420	1377 11⅜	1381 2¾	1384 6⅛	1387 9½	1391 0⅞	1394 4¼	1397 7⅝	1400 11	1404 2⅜	1407 5¾
430	1410 9⅛	1414 0½	1417 3⅞	1420 7¼	1423 10⅝	1427 2	1430 5⅜	1433 8¾	1437 0⅛	1440 3½
440	1443 6⅞	1446 10¼	1450 1⅝	1453 5	1456 8⅜	1459 11⅝	1463 3	1466 6⅜	1469 9¾	1473 1⅛
450	1476 4½	1479 7⅞	1482 11¼	1486 2⅝	1489 6	1492 9⅜	1496 0¾	1499 4⅛	1502 7½	1505 10⅞
460	1509 2¼	1512 5⅝	1515 9	1519 0⅜	1522 3¾	1525 7⅛	1528 10½	1532 1⅞	1535 5¼	1538 8⅝
470	1541 11⅞	1545 3¼	1548 6⅝	1551 10	1555 1⅜	1558 4¾	1561 8⅛	1564 11½	1568 2⅞	1571 6¼
480	1574 9⅝	1578 1	1581 4⅜	1584 7¾	1587 11⅛	1591 2½	1594 5⅞	1597 9¼	1601 0⅝	1604 4
490	1607 7⅜	1610 10¾	1614 2⅛	1617 5½	1620 8⅞	1624 0¼	1627 3½	1630 6⅞	1633 10¼	1637 1⅝

Basis 1 m. = 3.280839895 ft.

METRES to FEET and INCHES

YARDS to METRES

Yards	0	1	2	3	4	5	6	7	8	9
0	0	0.9144	1.8288	2.7432	3.6576	4.5720	5.4864	6.4008	7.3152	8.2296
10	9.1440	10.0584	10.9728	11.8872	12.8016	13.7160	14.6304	15.5448	16.4592	17.3736
20	18.2880	19.2024	20.1168	21.0312	21.9456	22.8600	23.7744	24.6888	25.6032	26.5176
30	27.4320	28.3464	29.2608	30.1752	31.0896	32.0040	32.9184	33.8328	34.7472	35.6616
40	36.5760	37.4904	38.4048	39.3192	40.2336	41.1480	42.0624	42.9768	43.8912	44.8056
50	45.7200	46.6344	47.5488	48.4632	49.3776	50.2920	51.2064	52.1208	53.0352	53.9496
60	54.8640	55.7784	56.6928	57.6072	58.5216	59.4360	60.3504	61.2648	62.1792	63.0936
70	64.0080	64.9224	65.8368	66.7512	67.6656	68.5800	69.4944	70.4088	71.3232	72.2376
80	73.1520	74.0664	74.9808	75.8952	76.8096	77.7240	78.6384	79.5528	80.4672	81.3816
90	82.2960	83.2104	84.1248	85.0392	85.9536	86.8680	87.7824	88.6968	89.6112	90.5256
100	91.4400	92.3544	93.2688	94.1832	95.0976	96.0120	96.9264	97.8408	98.7552	99.6696
110	100.5840	101.4984	102.4128	103.3272	104.2416	105.1560	106.0704	106.9848	107.8992	108.8136
120	109.7280	110.6424	111.5568	112.4712	113.3856	114.3000	115.2144	116.1288	117.0432	117.9576
130	118.8720	119.7864	120.7008	121.6152	122.5296	123.4440	124.3584	125.2728	126.1872	127.1016
140	128.0160	128.9304	129.8448	130.7592	131.6736	132.5880	133.5024	134.4168	135.3312	136.2456
150	137.1600	138.0744	138.9888	139.9032	140.8176	141.7320	142.6464	143.5608	144.4752	145.3896
160	146.3040	147.2184	148.1328	149.0472	149.9616	150.8760	151.7904	152.7048	153.6192	154.5336
170	155.4480	156.3624	157.2768	158.1912	159.1056	160.0200	160.9344	161.8488	162.7632	163.6776
180	164.5920	165.5064	166.4208	167.3352	168.2496	169.1640	170.0784	170.9928	171.9072	172.8216
190	173.7360	174.6504	175.5648	176.4792	177.3936	178.3080	179.2224	180.1368	181.0512	181.9656

SUPPLEMENTARY TABLE

Yards	200	300	400	500	600	700	800	900	1000
Metres	182.8800	274.3200	365.7600	457.2000	548.6400	640.0800	731.5200	822.9600	914.4000

Basis : 1 yd = 0.9144 m(exactly)

METRES to YARDS

Metres	0	1	2	3	4	5	6	7	8	9
0	0	1.0936	2.1872	3.2808	4.3745	5.4681	6.5617	7.6553	8.7489	9.8425
10	10.9361	12.0297	13.1234	14.2170	15.3106	16.4042	17.4978	18.5914	19.6850	20.7787
20	21.8723	22.9659	24.0595	25.1531	26.2467	27.3403	28.4339	29.5276	30.6212	31.7148
30	32.8084	33.9020	34.9956	36.0892	37.1829	38.2765	39.3701	40.4637	41.5573	42.6509
40	43.7445	44.8381	45.9318	47.0254	48.1190	49.2126	50.3062	51.3998	52.4934	53.5871
50	54.6807	55.7743	56.8679	57.9615	59.0551	60.1487	61.2423	62.3360	63.4296	64.5232
60	65.6168	66.7104	67.8040	68.8976	69.9913	71.0849	72.1785	73.2721	74.3657	75.4593
70	76.5529	77.6465	78.7402	79.8338	80.9274	82.0210	83.1146	84.2082	85.3018	86.3955
80	87.4891	88.5827	89.6763	90.7699	91.8635	92.9571	94.0507	95.1444	96.2380	97.3316
90	98.4252	99.5188	100.6124	101.7060	102.7997	103.8933	104.9869	106.0805	107.1741	108.2677
100	109.3613	110.4549	111.5486	112.6422	113.7358	114.8294	115.9230	117.0166	118.1102	119.2038
110	120.2975	121.3911	122.4847	123.5783	124.6719	125.7655	126.8591	127.9528	129.0464	130.1400
120	131.2336	132.3272	133.4208	134.5144	135.6080	136.7017	137.7953	138.8889	139.9825	141.0761
130	142.1697	143.2633	144.3570	145.4506	146.5442	147.6378	148.7314	149.8250	150.9186	152.0122
140	153.1059	154.1995	155.2931	156.3867	157.4803	158.5739	159.6675	160.7612	161.8548	162.9484
150	164.0420	165.1356	166.2292	167.3228	168.4164	169.5101	170.6037	171.6973	172.7909	173.8845
160	174.9781	176.0717	177.1654	178.2590	179.3526	180.4462	181.5398	182.6334	183.7270	184.8206
170	185.9143	187.0079	188.1015	189.1951	190.2887	191.3823	192.4759	193.5696	194.6632	195.7568
180	196.8504	197.9440	199.0376	200.1312	201.2248	202.3185	203.4121	204.5057	205.5993	206.6929
190	207.7865	208.8801	209.9738	211.0674	212.1610	213.2546	214.3482	215.4418	216.5354	217.6290

SUPPLEMENTARY TABLE

Metres	200	300	400	500	600	700	800	900	1000
Yards	218.7227	328.0840	437.4453	546.8066	656.1680	765.5293	874.8906	984.2520	1093.6133

Basis : 1 m. = 1.093613298 yds.

247

SQUARE FEET to SQUARE METRES

Sq. ft.	0	1	2	3	4	5	6	7	8	9
0		0.09290	0.18581	0.27871	0.37161	0.46452	0.55742	0.65032	0.74322	0.83613
10	0.92903	1.02193	1.11484	1.20774	1.30064	1.39355	1.48645	1.57935	1.67225	1.76516
20	1.85806	1.95096	2.04387	2.13677	2.22967	2.32258	2.41548	2.50838	2.60129	2.69419
30	2.78709	2.87999	2.97290	3.06580	3.15870	3.25161	3.34451	3.43741	3.53032	3.62322
40	3.71612	3.80902	3.90193	3.99483	4.08773	4.18064	4.27354	4.36644	4.45935	4.55225
50	4.64515	4.73806	4.83096	4.92386	5.01676	5.10967	5.20257	5.29547	5.38838	5.48128
60	5.57418	5.66709	5.75999	5.85289	5.94579	6.03870	6.13160	6.22450	6.31741	6.41031
70	6.50321	6.59612	6.68902	6.78192	6.87482	6.96773	7.06063	7.15353	7.24644	7.33934
80	7.43224	7.52515	7.61805	7.71095	7.80386	7.89676	7.98966	8.08256	8.17547	8.26837
90	8.36127	8.45418	8.54708	8.63998	8.73289	8.82579	8.91869	9.01159	9.10450	9.19740
100	9.29030	9.38321	9.47611	9.56901	9.66192	9.75482	9.84772	9.94063	10.03353	10.12643
110	10.21933	10.31224	10.40514	10.49804	10.59095	10.68385	10.77675	10.86966	10.96256	11.05546
120	11.14836	11.24127	11.33417	11.42707	11.51998	11.61288	11.70578	11.79869	11.89159	11.98449
130	12.07740	12.17030	12.26320	12.35610	12.44901	12.54191	12.63481	12.72772	12.82062	12.91352
140	13.00643	13.09933	13.19223	13.28513	13.37804	13.47094	13.56384	13.65675	13.74965	13.84255
150	13.93546	14.02836	14.12126	14.21417	14.30707	14.39997	14.49287	14.58578	14.67868	14.77158
160	14.86449	14.95739	15.05029	15.14320	15.23610	15.32900	15.42190	15.51481	15.60771	15.70061
170	15.79352	15.88642	15.97932	16.07223	16.16513	16.25803	16.35094	16.44384	16.53674	16.62964
180	16.72255	16.81545	16.90835	17.00126	17.09416	17.18706	17.27997	17.37287	17.46577	17.55867
190	17.65158	17.74448	17.83738	17.93029	18.02319	18.11609	18.20900	18.30190	18.39480	18.48770
200	18.58061	18.67351	18.76641	18.85932	18.95222	19.04512	19.13803	19.23093	19.32383	19.41674
210	19.50964	19.60254	19.69544	19.78835	19.88125	19.97415	20.06706	20.15996	20.25286	20.34577
220	20.43867	20.53157	20.62447	20.71738	20.81028	20.90318	20.99609	21.08899	21.18189	21.27480
230	21.36770	21.46060	21.55351	21.64641	21.73931	21.83221	21.92512	22.01802	22.11092	22.20383
240	22.29673	22.38963	22.48254	22.57544	22.66834	22.76124	22.85415	22.94705	23.03995	23.13286
250	23.22576	23.31866	23.41157	23.50447	23.59737	23.69028	23.78318	23.87608	23.96898	24.06189
260	24.15479	24.24769	24.34060	24.43350	24.52640	24.61931	24.71221	24.80511	24.89801	24.99092
270	25.08382	25.17672	25.26963	25.36253	25.45543	25.54834	25.64124	25.73414	25.82705	25.91995
280	26.01285	26.10575	26.19866	26.29156	26.38446	26.47737	26.57027	26.66317	26.75608	26.84898

SQUARE FEET to SQUARE METRES

Sq. ft.	0	1	2	3	4	5	6	7	8	9
300	27.87091	27.96382	28.05672	28.14962	28.24252	28.33543	28.42833	28.52123	28.61414	28.70704
310	28.79994	28.89285	28.98575	29.07865	29.17155	29.26446	29.35736	29.45026	29.54317	29.63607
320	29.72897	29.82188	29.91478	30.00768	30.10058	30.19349	30.28639	30.37929	30.47220	30.56510
330	30.65800	30.75091	30.84381	30.93671	31.02962	31.12252	31.21542	31.30832	31.40123	31.49413
340	31.58703	31.67994	31.77284	31.86574	31.95865	32.05155	32.14445	32.23735	32.33026	32.42316
350	32.51606	32.60897	32.70187	32.79477	32.88768	32.98058	33.07348	33.16639	33.25929	33.35219
360	33.44509	33.53800	33.63090	33.72380	33.81671	33.90961	34.00251	34.09542	34.18832	34.28122
370	34.37413	34.46703	34.55993	34.65283	34.74574	34.83864	34.93154	35.02445	35.11735	35.21025
380	35.30316	35.39606	35.48896	35.58186	35.67477	35.76767	35.86057	35.95348	36.04638	36.13928
390	36.23219	36.32509	36.41799	36.51089	36.60380	36.69670	36.78960	36.88251	36.97541	37.06831
400	37.16122	37.25412	37.34702	37.43993	37.53283	37.62573	37.71863	37.81154	37.90444	37.99734
410	38.09025	38.18315	38.27605	38.36896	38.46186	38.55476	38.64766	38.74057	38.83347	38.92637
420	39.01928	39.11218	39.20508	39.29799	39.39089	39.48379	39.57670	39.66960	39.76250	39.85540
430	39.94831	40.04121	40.13411	40.22702	40.31992	40.41282	40.50573	40.59863	40.69153	40.78443
440	40.87734	40.97024	41.06314	41.15605	41.24895	41.34185	41.43476	41.52766	41.62056	41.71347
450	41.80637	41.89927	41.99217	42.08508	42.17798	42.27088	42.36379	42.45669	42.54959	42.64250
460	42.73540	42.82830	42.92120	43.01411	43.10701	43.19991	43.29282	43.38572	43.47862	43.57153
470	43.66443	43.75733	43.85024	43.94314	44.03604	44.12894	44.22185	44.31475	44.40765	44.50056
480	44.59346	44.68636	44.77927	44.87217	44.96507	45.05797	45.15088	45.24378	45.33668	45.42959
490	45.52249	45.61539	45.70830	45.80120	45.89410	45.98700	46.07991	46.17281	46.26571	46.35862

SUPPLEMENTARY TABLE

Sq. ft.	500	600	700	800	900	1000
Sq. m.	46.45152	55.74182	65.03213	74.32243	83.61274	92.90304

Basis : 1 sq. ft. = 0.09290304 sq. m. (exactly)

SQUARE METRES to SQUARE FEET

Sq. m.	0	1	2	3	4	5	6	7	8	9
0	0	10.764	21.528	32.292	43.056	53.820	64.583	75.347	86.111	96.875
10	107.639	118.403	129.167	139.931	150.695	161.459	172.223	182.986	193.750	204.514
20	215.278	226.042	236.806	247.570	258.334	269.098	279.862	290.626	301.389	312.153
30	322.917	333.681	344.445	355.209	365.973	376.737	387.501	398.265	409.029	419.793
40	430.556	441.320	452.084	462.848	473.612	484.376	495.140	505.904	516.668	527.432
50	538.196	548.959	559.723	570.487	581.251	592.015	602.779	613.543	624.307	635.071
60	645.835	656.599	667.362	678.126	688.890	699.654	710.418	721.182	731.946	742.710
70	753.474	764.238	775.002	785.765	796.529	807.293	818.057	828.821	839.585	850.349
80	861.113	871.877	882.641	893.405	904.168	914.932	925.696	936.460	947.224	957.988
90	968.752	979.516	990.280	1001.044	1011.808	1022.571	1033.335	1044.099	1054.863	1065.627
100	1076.391	1087.155	1097.919	1108.683	1119.447	1130.211	1140.975	1151.738	1162.502	1173.266
110	1184.030	1194.794	1205.558	1216.322	1227.086	1237.850	1248.614	1259.378	1270.141	1280.905
120	1291.669	1302.433	1313.197	1323.961	1334.725	1345.489	1356.253	1367.017	1377.781	1388.544
130	1399.308	1410.072	1420.836	1431.600	1442.364	1453.128	1463.892	1474.656	1485.420	1496.184
140	1506.947	1517.711	1528.475	1539.239	1550.003	1560.767	1571.531	1582.295	1593.059	1603.823
150	1614.587	1625.350	1636.114	1646.878	1657.642	1668.406	1679.170	1689.934	1700.698	1711.462
160	1722.226	1732.990	1743.753	1754.517	1765.281	1776.045	1786.809	1797.573	1808.337	1819.101
170	1829.865	1840.629	1851.393	1862.157	1872.920	1883.684	1894.448	1905.212	1915.976	1926.740
180	1937.504	1948.268	1959.032	1969.796	1980.560	1991.323	2002.087	2012.851	2023.615	2034.379
190	2045.143	2055.907	2066.671	2077.435	2088.199	2098.963	2109.726	2120.490	2131.254	2142.018
200	2152.782	2163.546	2174.310	2185.074	2195.838	2206.602	2217.366	2228.129	2238.893	2249.657
210	2260.421	2271.185	2281.949	2292.713	2303.477	2314.241	2325.005	2335.769	2346.532	2357.296
220	2368.060	2378.824	2389.588	2400.352	2411.116	2421.880	2432.644	2443.408	2454.172	2464.936
230	2475.699	2486.463	2497.227	2507.991	2518.755	2529.519	2540.283	2551.047	2561.811	2572.575
240	2583.339	2594.102	2604.866	2615.630	2626.394	2637.158	2647.922	2658.686	2669.450	2680.214
250	2690.978	2701.742	2712.505	2723.269	2734.033	2744.797	2755.561	2766.325	2777.089	2787.853
260	2798.617	2809.381	2820.145	2830.908	2841.672	2852.436	2863.200	2873.964	2884.728	2895.492
270	2906.256	2917.020	2927.784	2938.548	2949.311	2960.075	2970.839	2981.603	2992.367	3003.131
280		3024.659	3035.423	3046.187	3056.951	3067.714	3078.478	3089.242	3100.006	3110.770

Sq. m.	0	1	2	3	4	5	6	7	8	9
300	3229.173	3239.937	3250.701	3261.465	3272.229	3282.993	3293.757	3304.520	3315.284	3326.048
310	3336.812	3347.576	3358.340	3369.104	3379.868	3390.632	3401.396	3412.160	3422.924	3433.687
320	3444.451	3455.215	3465.979	3476.743	3487.507	3498.271	3509.035	3519.799	3530.563	3541.327
330	3552.090	3562.854	3573.618	3584.382	3595.146	3605.910	3616.674	3627.438	3638.202	3648.966
340	3659.730	3670.493	3681.257	3692.021	3702.785	3713.549	3724.313	3735.077	3745.841	3756.605
350	3767.369	3778.133	3788.896	3799.660	3810.424	3821.188	3831.952	3842.716	3853.480	3864.244
360	3875.008	3885.772	3896.536	3907.299	3918.063	3928.827	3939.591	3950.355	3961.119	3971.883
370	3982.647	3993.411	4004.175	4014.939	4025.702	4036.466	4047.230	4057.994	4068.758	4079.522
380	4090.286	4101.050	4111.814	4122.578	4133.342	4144.106	4154.869	4165.633	4176.397	4187.161
390	4197.925	4208.689	4219.453	4230.217	4240.981	4251.745	4262.509	4273.272	4284.036	4294.800
400	4305.564	4316.328	4327.092	4337.856	4348.620	4359.384	4370.148	4380.912	4391.675	4402.439
410	4413.203	4423.967	4434.731	4445.495	4456.259	4467.023	4477.787	4488.551	4499.315	4510.078
420	4520.842	4531.606	4542.370	4553.134	4563.898	4574.662	4585.426	4596.190	4606.954	4617.718
430	4628.481	4639.245	4650.009	4660.773	4671.537	4682.301	4693.065	4703.829	4714.593	4725.357
440	4736.121	4746.884	4757.648	4768.412	4779.176	4789.940	4800.704	4811.468	4822.232	4832.996
450	4843.760	4854.524	4865.288	4876.051	4886.815	4897.579	4908.343	4919.107	4929.871	4940.635
460	4951.399	4962.163	4972.927	4983.691	4994.454	5005.218	5015.982	5026.746	5037.510	5048.274
470	5059.038	5069.802	5080.566	5091.330	5102.094	5112.857	5123.621	5134.385	5145.149	5155.913
480	5166.677	5177.441	5188.205	5198.969	5209.733	5220.497	5231.260	5242.024	5252.788	5263.552
490	5274.316	5285.080	5295.844	5306.608	5317.372	5328.136	5338.900	5349.663	5360.427	5371.191

SUPPLEMENTARY TABLE

Sq. m.	500	600	700	800	900	1000
Sq. ft.	5381.955	6458.346	7534.737	8611.128	9687.519	10763.910

Basis : 1 sq. m. = 10.76391042 sq. ft.

251

SQUARE YARDS to SQUARE METRES

Sq. yds.	0	1	2	3	4	5	6	7	8	9
0	0	0.8361	1.6723	2.5084	3.3445	4.1806	5.0168	5.8529	6.6890	7.5251
10	8.3613	9.1974	10.0335	10.8697	11.7058	12.5419	13.3780	14.2142	15.0503	15.8864
20	16.7225	17.5587	18.3948	19.2309	20.0671	20.9032	21.7393	22.5754	23.4116	24.2477
30	25.0838	25.9199	26.7561	27.5922	28.4283	29.2645	30.1006	30.9367	31.7728	32.6090
40	33.4451	34.2812	35.1173	35.9535	36.7896	37.6257	38.4619	39.2980	40.1341	40.9702
50	41.8064	42.6425	43.4786	44.3148	45.1509	45.9870	46.8231	47.6593	48.4954	49.3315
60	50.1676	51.0038	51.8399	52.6760	53.5122	54.3483	55.1844	56.0205	56.8567	57.6928
70	58.5289	59.3650	60.2012	61.0373	61.8734	62.7096	63.5457	64.3818	65.2179	66.0541
80	66.8902	67.7263	68.5624	69.3986	70.2347	71.0708	71.9070	72.7431	73.5792	74.4153
90	75.2515	76.0876	76.9237	77.7598	78.5960	79.4321	80.2682	81.1044	81.9405	82.7766
100	83.6127	84.4489	85.2850	86.1211	86.9572	87.7934	88.6295	89.4656	90.3018	91.1379
110	91.9740	92.8101	93.6463	94.4824	95.3185	96.1546	96.9908	97.8269	98.6630	99.4992
120	100.3353	101.1714	102.0075	102.8437	103.6798	104.5159	105.3520	106.1882	107.0243	107.8604
130	108.6966	109.5327	110.3688	111.2049	112.0411	112.8772	113.7133	114.5494	115.3856	116.2217
140	117.0578	117.8940	118.7301	119.5662	120.4023	121.2385	122.0746	122.9107	123.7468	124.5830
150	125.4191	126.2552	127.0914	127.9275	128.7636	129.5997	130.4359	131.2720	132.1081	132.9442
160	133.7804	134.6165	135.4526	136.2888	137.1249	137.9610	138.7971	139.6333	140.4694	141.3055
170	142.1416	142.9778	143.8139	144.6500	145.4862	146.3223	147.1584	147.9945	148.8307	149.6668
180	150.5029	151.3391	152.1752	153.0113	153.8474	154.6836	155.5197	156.3558	157.1919	158.0281
190	158.8642	159.7003	160.5365	161.3726	162.2087	163.0448	163.8810	164.7171	165.5532	166.3893

SUPPLEMENTARY TABLE

Sq. yds.	200	300	400	500	600	700	800	900	1000
Sq. m.	167.2255	250.8382	334.4509	418.0637	501.6764	585.2892	668.9019	752.5146	836.1274

252

Sq. m.	0	1	2	3	4	5	6	7	8	9
0	0	1.1960	2.3920	3.5880	4.7840	5.9800	7.1759	8.3719	9.5679	10.7639
10	11.9599	13.1559	14.3519	15.5479	16.7439	17.9399	19.1358	20.3318	21.5278	22.7238
20	23.9198	25.1158	26.3118	27.5078	28.7038	29.8998	31.0957	32.2917	33.4877	34.6837
30	35.8797	37.0757	38.2717	39.4677	40.6637	41.8597	43.0556	44.2516	45.4476	46.6436
40	47.8396	49.0356	50.2316	51.4276	52.6236	53.8196	55.0155	56.2115	57.4075	58.6035
50	59.7995	60.9955	62.1915	63.3875	64.5835	65.7795	66.9754	68.1714	69.3674	70.5634
60	71.7594	72.9554	74.1514	75.3474	76.5434	77.7394	78.9353	80.1313	81.3273	82.5233
70	83.7193	84.9153	86.1113	87.3073	88.5033	89.6993	90.8952	92.0912	93.2872	94.4832
80	95.6792	96.8752	98.0712	99.2672	100.4632	101.6592	102.8551	104.0511	105.2471	106.4431
90	107.6391	108.8351	110.0311	111.2271	112.4231	113.6191	114.8150	116.0110	117.2070	118.4030
100	119.5990	120.7950	121.9910	123.1870	124.3830	125.5790	126.7749	127.9709	129.1669	130.3629
110	131.5589	132.7549	133.9509	135.1469	136.3429	137.5389	138.7348	139.9308	141.1268	142.3228
120	143.5188	144.7148	145.9108	147.1068	148.3028	149.4988	150.6947	151.8907	153.0867	154.2827
130	155.4787	156.6747	157.8707	159.0667	160.2627	161.4587	162.6546	163.8506	165.0466	166.2426
140	167.4386	168.6346	169.8306	171.0266	172.2226	173.4186	174.6145	175.8105	177.0065	178.2025
150	179.3985	180.5945	181.7905	182.9865	184.1825	185.3785	186.5744	187.7704	188.9664	190.1624
160	191.3584	192.5544	193.7504	194.9464	196.1424	197.3384	198.5343	199.7303	200.9263	202.1223
170	203.3183	204.5143	205.7103	206.9063	208.1023	209.2983	210.4942	211.6902	212.8862	214.0822
180	215.2782	216.4742	217.6702	218.8662	220.0622	221.2582	222.4541	223.6501	224.8461	226.0421
190	227.2381	228.4341	229.6301	230.8261	232.0221	233.2181	234.4140	235.6100	236.8060	238.0020

SUPPLEMENTARY TABLE

Sq. m.	200	300	400	500	600	700	800	900	1000
Sq. yds.	239.1980	358.7970	478.3960	597.9950	717.5940	837.1930	956.7920	1076.3910	1195.9900

Basis : 1 sq. m. = 1.195990046 sq. yds.

ACRES to HECTARES

Acres	0	1	2	3	4	5	6	7	8	9
0	0	0.4047	0.8094	1.2141	1.6187	2.0234	2.4281	2.8328	3.2375	3.6422
10	4.0469	4.4515	4.8562	5.2609	5.6656	6.0703	6.4750	6.8797	7.2843	7.6890
20	8.0937	8.4984	8.9031	9.3078	9.7125	10.1171	10.5218	10.9265	11.3312	11.7359
30	12.1406	12.5453	12.9499	13.3546	13.7593	14.1640	14.5687	14.9734	15.3781	15.7827
40	16.1874	16.5921	16.9968	17.4015	17.8062	18.2109	18.6155	19.0202	19.4249	19.8296
50	20.2343	20.6390	21.0437	21.4483	21.8530	22.2577	22.6624	23.0671	23.4718	23.8765
60	24.2811	24.6858	25.0905	25.4952	25.8999	26.3046	26.7093	27.1139	27.5186	27.9233
70	28.3280	28.7327	29.1374	29.5421	29.9467	30.3514	30.7561	31.1608	31.5655	31.9702
80	32.3749	32.7795	33.1842	33.5889	33.9936	34.3983	34.8030	35.2077	35.6123	36.0170
90	36.4217	36.8264	37.2311	37.6358	38.0404	38.4451	38.8498	39.2545	39.6592	40.0639
100	40.4686	40.8732	41.2779	41.6826	42.0873	42.4920	42.8967	43.3014	43.7060	44.1107
110	44.5154	44.9201	45.3248	45.7295	46.1342	46.5388	46.9435	47.3482	47.7529	48.1576
120	48.5623	48.9670	49.3716	49.7763	50.1810	50.5857	50.9904	51.3951	51.7998	52.2044
130	52.6091	53.0138	53.4185	53.8232	54.2279	54.6326	55.0372	55.4419	55.8466	56.2513
140	56.6560	57.0607	57.4654	57.8700	58.2747	58.6794	59.0841	59.4888	59.8935	60.2982
150	60.7028	61.1075	61.5122	61.9169	62.3216	62.7263	63.1310	63.5356	63.9403	64.3450
160	64.7497	65.1544	65.5591	65.9638	66.3684	66.7731	67.1778	67.5825	67.9872	68.3919
170	68.7966	69.2012	69.6059	70.0106	70.4153	70.8200	71.2247	71.6294	72.0340	72.4387
180	72.8434	73.2481	73.6528	74.0575	74.4622	74.8668	75.2715	75.6762	76.0809	76.4856
190	76.8903	77.2950	77.6996	78.1043	78.5090	78.9137	79.3184	79.7231	80.1278	80.5324
200	80.9371	81.3418	81.7465	82.1512	82.5559	82.9606	83.3652	83.7699	84.1746	84.5793
210	84.9840	85.3887	85.7934	86.1980	86.6027	87.0074	87.4121	87.8168	88.2215	88.6262
220	89.0308	89.4355	89.8402	90.2449	90.6496	91.0543	91.4590	91.8636	92.2683	92.6730
230	93.0777	93.4824	93.8871	94.2918	94.6964	95.1011	95.5058	95.9105	96.3152	96.7199
240	97.1246	97.5292	97.9339	98.3386	98.7433	99.1480	99.5527	99.9574	100.3620	100.7667
250	101.1714	101.5761	101.9808	102.3855	102.7902	103.1948	103.5995	104.0042	104.4089	104.8136
260	105.2183	105.6230	106.0276	106.4323	106.8370	107.2417	107.6464	108.0511	108.4558	108.8604
270	109.2651	109.6698	110.0745	110.4792	110.8839	111.2885	111.6932	112.0979	112.5026	112.9073
280	113.3120	113.7167	114.1213	114.5260	114.9307	115.3354	115.7401	116.1448	116.5495	116.9541

ACRES to HECTARES

Acres	0	1	2	3	4	5	6	7	8	9
300	121.4057	121.8104	122.2151	122.6197	123.0244	123.4291	123.8338	124.2385	124.6432	125.0479
310	125.4525	125.8572	126.2619	126.6666	127.0713	127.4760	127.8807	128.2853	128.6900	129.0947
320	129.4994	129.9041	130.3088	130.7135	131.1181	131.5228	131.9275	132.3322	132.7369	133.1416
330	133.5463	133.9509	134.3556	134.7603	135.1650	135.5697	135.9744	136.3791	136.7837	137.1884
340	137.5931	137.9978	138.4025	138.8072	139.2119	139.6165	140.0212	140.4259	140.8306	141.2353
350	141.6400	142.0447	142.4493	142.8540	143.2587	143.6634	144.0681	144.4728	144.8775	145.2821
360	145.6868	146.0915	146.4962	146.9009	147.3056	147.7103	148.1149	148.5196	148.9243	149.3290
370	149.7337	150.1384	150.5431	150.9477	151.3524	151.7571	152.1618	152.5665	152.9712	153.3759
380	153.7805	154.1852	154.5899	154.9946	155.3993	155.8040	156.2087	156.6133	157.0180	157.4227
390	157.8274	158.2321	158.6368	159.0415	159.4461	159.8508	160.2555	160.6602	161.0649	161.4696
400	161.8743	162.2789	162.6836	163.0883	163.4930	163.8977	164.3024	164.7071	165.1117	165.5164
410	165.9211	166.3258	166.7305	167.1352	167.5399	167.9445	168.3492	168.7539	169.1586	169.5633
420	169.9680	170.3727	170.7773	171.1820	171.5867	171.9914	172.3961	172.8008	173.2054	173.6101
430	174.0148	174.4195	174.8242	175.2289	175.6336	176.0382	176.4429	176.8476	177.2523	177.6570
440	178.0617	178.4664	178.8710	179.2757	179.6804	180.0851	180.4898	180.8945	181.2992	181.7038
450	182.1085	182.5132	182.9179	183.3226	183.7273	184.1320	184.5366	184.9413	185.3460	185.7507
460	186.1554	186.5601	186.9648	187.3694	187.7741	188.1788	188.5835	188.9882	189.3929	189.7976
470	190.2022	190.6069	191.0116	191.4163	191.8210	192.2257	192.6304	193.0350	193.4397	193.8444
480	194.2491	194.6538	195.0585	195.4632	195.8678	196.2725	196.6772	197.0819	197.4866	197.8913
490	198.2960	198.7006	199.1053	199.5100	199.9147	200.3194	200.7241	201.1288	201.5334	201.9381

SUPPLEMENTARY TABLE

Acres	500	600	700	800	900	1000
Hectares	202.3428	242.8114	283.2799	323.7485	364.2171	404.6856

Basis : 1 acre = 0.404685642 hectares

HECTARES to ACRES

Hectares	0	1	2	3	4	5	6	7	8	9
0	0	2.471	4.942	7.413	9.884	12.355	14.826	17.297	19.768	22.239
10	24.711	27.182	29.653	32.124	34.595	37.066	39.537	42.008	44.479	46.950
20	49.421	51.892	54.363	56.834	59.305	61.776	64.247	66.718	69.190	71.661
30	74.132	76.603	79.074	81.545	84.016	86.487	88.958	91.429	93.900	96.371
40	98.842	101.313	103.784	106.255	108.726	111.197	113.668	116.140	118.611	121.082
50	123.553	126.024	128.495	130.966	133.437	135.908	138.379	140.850	143.321	145.792
60	148.263	150.734	153.205	155.676	158.147	160.618	163.090	165.561	168.032	170.503
70	172.974	175.445	177.916	180.387	182.858	185.329	187.800	190.271	192.742	195.213
80	197.684	200.155	202.626	205.097	207.569	210.040	212.511	214.982	217.453	219.924
90	222.395	224.866	227.337	229.808	232.279	234.750	237.221	239.692	242.163	244.634
100	247.105	249.576	252.047	254.519	256.990	259.461	261.932	264.403	266.874	269.345
110	271.816	274.287	276.758	279.229	281.700	284.171	286.642	289.113	291.584	294.055
120	296.526	298.998	301.469	303.940	306.411	308.882	311.353	313.824	316.295	318.766
130	321.237	323.708	326.179	328.650	331.121	333.592	336.063	338.534	341.005	343.476
140	345.948	348.419	350.890	353.361	355.832	358.303	360.774	363.245	365.716	368.187
150	370.658	373.129	375.600	378.071	380.542	383.013	385.484	387.955	390.427	392.898
160	395.369	397.840	400.311	402.782	405.253	407.724	410.195	412.666	415.137	417.608
170	420.079	422.550	425.021	427.492	429.963	432.434	434.905	437.377	439.848	442.319
180	444.790	447.261	449.732	452.203	454.674	457.145	459.616	462.087	464.558	467.029
190	469.500	471.971	474.442	476.913	479.384	481.855	484.327	486.798	489.269	491.740
200	494.211	496.682	499.153	501.624	504.095	506.566	509.037	511.508	513.979	516.450
210	518.921	521.392	523.863	526.334	528.806	531.277	533.748	536.219	538.690	541.161
220	543.632	546.103	548.574	551.045	553.516	555.987	558.458	560.929	563.400	565.871
230	568.342	570.813	573.284	575.756	578.227	580.698	583.169	585.640	588.111	590.582
240	593.053	595.524	597.995	600.466	602.937	605.408	607.879	610.350	612.821	615.292
250	617.763	620.235	622.706	625.177	627.648	630.119	632.590	635.061	637.532	640.003
260	642.474	644.945	647.416	649.887	652.358	654.829	657.300	659.771	662.242	664.713
270	667.185	669.656	672.127	674.598	677.069	679.540	682.011	684.482	686.953	689.424

HECTARES to ACRES

Hectares	0	1	2	3	4	5	6	7	8	9
300	741.316	743.787	746.258	748.729	751.200	753.671	756.142	758.614	761.085	763.556
310	766.027	768.498	770.969	773.440	775.911	778.382	780.853	783.324	785.795	788.266
320	790.737	793.208	795.679	798.150	800.621	803.092	805.564	808.035	810.506	812.977
330	815.448	817.919	820.390	822.861	825.332	827.803	830.274	832.745	835.216	837.687
340	840.158	842.629	845.100	847.571	850.043	852.514	854.985	857.456	859.927	862.398
350	864.869	867.340	869.811	872.282	874.753	877.224	879.695	882.166	884.637	887.108
360	889.579	892.050	894.521	896.993	899.464	901.935	904.406	906.877	909.348	911.819
370	914.290	916.761	919.232	921.703	924.174	926.645	929.116	931.587	934.058	936.529
380	939.000	941.472	943.943	946.414	948.885	951.356	953.827	956.298	958.769	961.240
390	963.711	966.182	968.653	971.124	973.595	976.066	978.537	981.008	983.479	985.950
400	988.422	990.893	993.364	995.835	998.306	1000.777	1003.248	1005.719	1008.190	1010.661
410	1013.132	1015.603	1018.074	1020.545	1023.016	1025.487	1027.958	1030.429	1032.901	1035.372
420	1037.843	1040.314	1042.785	1045.256	1047.727	1050.198	1052.669	1055.140	1057.611	1060.082
430	1062.553	1065.024	1067.495	1069.966	1072.437	1074.908	1077.379	1079.851	1082.322	1084.793
440	1087.264	1089.735	1092.206	1094.677	1097.148	1099.619	1102.090	1104.561	1107.032	1109.503
450	1111.974	1114.445	1116.916	1119.387	1121.858	1124.330	1126.801	1129.272	1131.743	1134.214
460	1136.685	1139.156	1141.627	1144.098	1146.569	1149.040	1151.511	1153.982	1156.453	1158.924
470	1161.395	1163.866	1166.337	1168.808	1171.280	1173.751	1176.222	1178.693	1181.164	1183.635
480	1186.106	1188.577	1191.048	1193.519	1195.990	1198.461	1200.932	1203.403	1205.874	1208.345
490	1210.816	1213.287	1215.758	1218.230	1220.701	1223.172	1225.643	1228.114	1230.585	1233.056

SUPPLEMENTARY TABLE

Hectares	500	600	700	800	900	1000
Acres	1235.527	1482.632	1729.738	1976.843	2223.948	2471.054

Basis : 1 hectare = 2.47105381 acres

257

CUBIC FEET to CUBIC METRES

Cu. ft.	0	1	2	3	4	5	6	7	8	9
0	0	0.02832	0.05663	0.08495	0.11327	0.14158	0.16990	0.19822	0.22653	0.25485
10	0.28317	0.31149	0.33980	0.36812	0.39644	0.42475	0.45307	0.48139	0.50970	0.53802
20	0.56634	0.59465	0.62297	0.65129	0.67960	0.70792	0.73624	0.76455	0.79287	0.82119
30	0.84951	0.87782	0.90614	0.93446	0.96277	0.99109	1.01941	1.04772	1.07604	1.10436
40	1.13267	1.16099	1.18931	1.21762	1.24594	1.27426	1.30257	1.33089	1.35921	1.38753
50	1.41584	1.44416	1.47248	1.50079	1.52911	1.55743	1.58574	1.61406	1.64238	1.67069
60	1.69901	1.72733	1.75564	1.78396	1.81228	1.84060	1.86891	1.89723	1.92555	1.95386
70	1.98218	2.01050	2.03881	2.06713	2.09545	2.12376	2.15208	2.18040	2.20871	2.23703
80	2.26535	2.29366	2.32198	2.35030	2.37862	2.40693	2.43525	2.46357	2.49188	2.52020
90	2.54852	2.57683	2.60515	2.63347	2.66178	2.69010	2.71842	2.74673	2.77505	2.80337
100	2.83168	2.86000	2.88832	2.91664	2.94495	2.97327	3.00159	3.02990	3.05822	3.08654
110	3.11485	3.14317	3.17149	3.19980	3.22812	3.25644	3.28475	3.31307	3.34139	3.36970
120	3.39802	3.42634	3.45466	3.48297	3.51129	3.53961	3.56792	3.59624	3.62456	3.65287
130	3.68119	3.70951	3.73782	3.76614	3.79446	3.82277	3.85109	3.87941	3.90772	3.93604
140	3.96436	3.99268	4.02099	4.04931	4.07763	4.10594	4.13426	4.16258	4.19089	4.21921
150	4.24753	4.27584	4.30416	4.33248	4.36079	4.38911	4.41743	4.44574	4.47406	4.50238
160	4.53070	4.55901	4.58733	4.61565	4.64396	4.67228	4.70060	4.72891	4.75723	4.78555
170	4.81386	4.84218	4.87050	4.89881	4.92713	4.95545	4.98376	5.01208	5.04040	5.06872
180	5.09703	5.12535	5.15367	5.18198	5.21030	5.23862	5.26693	5.29525	5.32357	5.35188
190	5.38020	5.40852	5.43683	5.46515	5.49347	5.52179	5.55010	5.57842	5.60674	5.63505

SUPPLEMENTARY TABLE

Cu. ft.	200	300	400	500	600	700	800	900	1000
Cu. m.	5.66337	8.49505	11.32674	14.15842	16.99011	19.82179	22.65348	25.48516	28.31685

CUBIC METRES to CUBIC FEET

Cu. m.	0	1	2	3	4	5	6	7	8	9
0	0	35.315	70.629	105.944	141.259	176.573	211.888	247.203	282.517	317.832
10	353.147	388.461	423.776	459.091	494.405	529.720	565.035	600.349	635.664	670.979
20	706.293	741.608	776.923	812.237	847.552	882.867	918.181	953.496	988.811	1024.125
30	1059.440	1094.755	1130.069	1165.384	1200.699	1236.013	1271.328	1306.643	1341.957	1377.272
40	1412.587	1447.901	1483.216	1518.531	1553.845	1589.160	1624.475	1659.789	1695.104	1730.419
50	1765.733	1801.048	1836.363	1871.677	1906.992	1942.307	1977.621	2012.936	2048.251	2083.565
60	2118.880	2154.195	2189.509	2224.824	2260.139	2295.453	2330.768	2366.083	2401.397	2436.712
70	2472.027	2507.341	2542.656	2577.971	2613.285	2648.600	2683.915	2719.229	2754.544	2789.859
80	2825.173	2860.488	2895.803	2931.117	2966.432	3001.747	3037.061	3072.376	3107.691	3143.005
90	3178.320	3213.635	3248.949	3284.264	3319.579	3354.893	3390.208	3425.523	3460.837	3496.152
100	3531.467	3566.781	3602.096	3637.411	3672.725	3708.040	3743.355	3778.669	3813.984	3849.299
110	3884.613	3919.928	3955.243	3990.557	4025.872	4061.187	4096.501	4131.816	4167.131	4202.445
120	4237.760	4273.075	4308.389	4343.704	4379.019	4414.333	4449.648	4484.963	4520.277	4555.592
130	4590.907	4626.221	4661.536	4696.851	4732.165	4767.480	4802.795	4838.109	4873.424	4908.739
140	4944.053	4979.368	5014.683	5049.997	5085.312	5120.627	5155.941	5191.256	5226.571	5261.885
150	5297.200	5332.515	5367.829	5403.144	5438.459	5473.773	5509.088	5544.403	5579.717	5615.032
160	5650.347	5685.661	5720.976	5756.291	5791.605	5826.920	5862.235	5897.549	5932.864	5968.179
170	6003.493	6038.808	6074.123	6109.437	6144.752	6180.067	6215.381	6250.696	6286.011	6321.325
180	6356.640	6391.955	6427.269	6462.584	6497.899	6533.213	6568.528	6603.843	6639.157	6674.472
190	6709.787	6745.101	6780.416	6815.731	6851.045	6886.360	6921.675	6956.989	6992.304	7027.619

SUPPLEMENTARY TABLE

Cu. m.	200	300	400	500	600	700	800	900	1000
Cu. ft.	7062.933	10594.400	14125.866	17657.333	21188.800	24720.266	28251.733	31783.200	35314.667

Basis : 1 cu. m. = 35.314666721 cu. ft.

259

PRICE PER SQUARE FOOT to PRICE PER SQUARE METRE

Price per sq. ft.	0	1p	2p	3p	4p	5p	6p	7p	8p	9p
	£'s per square metre									
	0.0	0.108	0.215	0.323	0.431	0.538	0.646	0.753	0.861	0.969
10p	1.076	1.184	1.292	1.399	1.507	1.615	1.722	1.830	1.938	2.045
20p	2.153	2.260	2.368	2.476	2.583	2.691	2.799	2.906	3.014	3.122
30p	3.229	3.337	3.444	3.552	3.660	3.767	3.875	3.983	4.090	4.198
40p	4.306	4.413	4.521	4.628	4.736	4.844	4.951	5.059	5.167	5.274
50p	5.382	5.490	5.597	5.705	5.813	5.920	6.028	6.135	6.243	6.351
60p	6.458	6.566	6.674	6.781	6.889	6.997	7.104	7.212	7.319	7.427
70p	7.535	7.642	7.750	7.858	7.965	8.073	8.181	8.288	8.396	8.503
80p	8.611	8.719	8.826	8.934	9.042	9.149	9.257	9.365	9.472	9.580
90p	9.688	9.795	9.903	10.010	10.118	10.226	10.333	10.441	10.549	10.656
£1.00	10.764	10.872	10.979	11.087	11.194	11.302	11.410	11.517	11.625	11.733
£1.10	11.840	11.948	12.056	12.163	12.271	12.378	12.486	12.594	12.701	12.809
£1.20	12.917	13.024	13.132	13.240	13.347	13.455	13.563	13.670	13.778	13.885
£1.30	13.993	14.101	14.208	14.316	14.424	14.531	14.639	14.747	14.854	14.962
£1.40	15.069	15.177	15.285	15.392	15.500	15.608	15.715	15.823	15.931	16.038
£1.50	16.146	16.254	16.361	16.469	16.576	16.684	16.792	16.899	17.007	17.115
£1.60	17.222	17.330	17.438	17.545	17.653	17.760	17.868	17.976	18.083	18.191
£1.70	18.299	18.406	18.514	18.622	18.729	18.837	18.944	19.052	19.160	19.267
£1.80	19.375	19.483	19.590	19.698	19.806	19.913	20.021	20.129	20.236	20.344
£1.90	20.451	20.559	20.667	20.774	20.882	20.990	21.097	21.205	21.313	21.420
£2.00	21.528	21.635	21.743	21.851	21.958	22.066	22.174	22.281	22.389	22.497
£2.10	22.604	22.712	22.819	22.927	23.035	23.142	23.250	23.358	23.465	23.573
£2.20	23.681	23.788	23.896	24.004	24.111	24.219	24.326	24.434	24.542	24.649
£2.30	24.757	24.865	24.972	25.080	25.188	25.295	25.403	25.510	25.618	25.726
£2.40	25.833	25.941	26.049	26.156	26.264	26.372	26.479	26.587	26.694	26.802

PRICE PER SQUARE FOOT to PRICE PER SQUARE METRE

Price per sq. ft.	0	1p	2p	3p	4p	5p	6p	7p	8p	9p
					£'s per square metre					
£2.50	26.910	27.017	27.125	27.233	27.340	27.448	27.556	27.663	27.771	27.879
£2.60	27.986	28.094	28.201	28.309	28.417	28.524	28.632	28.740	28.847	28.955
£2.70	29.063	29.170	29.278	29.385	29.493	29.601	29.708	29.816	29.924	30.031
£2.80	30.139	30.247	30.354	30.462	30.570	30.677	30.785	30.892	31.000	31.108
£2.90	31.215	31.323	31.431	31.538	31.646	31.754	31.861	31.969	32.076	32.184
£3.00	32.292	32.399	32.507	32.615	32.722	32.830	32.938	33.045	33.153	33.260
£3.10	33.368	33.476	33.583	33.691	33.799	33.906	34.014	34.122	34.229	34.337
£3.20	34.445	34.552	34.660	34.767	34.875	34.983	35.090	35.198	35.306	35.413
£3.30	35.521	35.629	35.736	35.844	35.951	36.059	36.167	36.274	36.382	36.490
£3.40	36.597	36.705	36.813	36.920	37.028	37.135	37.243	37.351	37.458	37.566
£3.50	37.674	37.781	37.889	37.997	38.104	38.212	38.320	38.427	38.535	38.642
£3.60	38.750	38.858	38.965	39.073	39.181	39.288	39.396	39.504	39.611	39.719
£3.70	39.826	39.934	40.042	40.149	40.257	40.365	40.472	40.580	40.688	40.795
£3.80	40.903	41.010	41.118	41.226	41.333	41.441	41.549	41.656	41.764	41.872
£3.90	41.979	42.087	42.195	42.302	42.410	42.517	42.625	42.733	42.840	42.948
£4.00	43.056	43.163	43.271	43.379	43.486	43.594	43.701	43.809	43.917	44.024

SUPPLEMENTARY TABLE

Price/sq. ft.	£5	£6	£7	£8	£9	£10	£11	£12	£13	£14	£15	£16
Price/sq.m.	53.820	64.583	75.347	86.111	96.875	107.639	118.403	129.167	139.931	150.695	161.459	172.223
Price/sq. ft.	£17	£18	£19	£20	£21	£22	£23	£24	£25	£26	£27	½p
Price/sq.m.	182.986	193.750	204.514	215.278	226.042	236.806	247.570	258.334	269.098	279.862	290.626	0.054

Basis: £1 per sq. ft. = £10.76391042 per sq. m.

261

PRICE PER CUBIC FOOT to PRICE PER CUBIC METRE

Price per cu. ft.	0	1p	2p	3p	4p	5p	6p	7p	8p	9p
					£'s per Cubic Metre					
10p	0.0	0.353	0.706	1.059	1.413	1.766	2.119	2.472	2.825	3.178
20p	3.531	3.885	4.238	4.591	4.944	5.297	5.650	6.003	6.357	6.710
30p	7.063	7.416	7.769	8.122	8.476	8.829	9.182	9.535	9.888	10.241
40p	10.594	10.948	11.301	11.654	12.007	12.360	12.713	13.066	13.420	13.773
	14.126	14.479	14.832	15.185	15.538	15.892	16.245	16.598	16.951	17.304
50p	17.657	18.010	18.364	18.717	19.070	19.423	19.776	20.129	20.483	20.836
60p	21.189	21.542	21.895	22.248	22.601	22.955	23.308	23.661	24.014	24.367
70p	24.720	25.073	25.427	25.780	26.133	26.486	26.839	27.192	27.545	27.899
80p	28.252	28.605	28.958	29.311	29.664	30.017	30.371	30.724	31.077	31.430
90p	31.783	32.136	32.489	32.843	33.196	33.549	33.902	34.255	34.608	34.962
£1.00	35.315	35.668	36.021	36.374	36.727	37.080	37.434	37.787	38.140	38.493
£1.10	38.846	39.199	39.552	39.906	40.259	40.612	40.965	41.318	41.671	42.024
£1.20	42.378	42.731	43.084	43.437	43.790	44.143	44.496	44.850	45.203	45.556
£1.30	45.909	46.262	46.615	46.969	47.322	47.675	48.028	48.381	48.734	49.087
£1.40	49.441	49.794	50.147	50.500	50.853	51.206	51.559	51.913	52.266	52.619
£1.50	52.972	53.325	53.678	54.031	54.385	54.738	55.091	55.444	55.797	56.150
£1.60	56.503	56.857	57.210	57.563	57.916	58.269	58.622	58.975	59.329	59.682
£1.70	60.035	60.388	60.741	61.094	61.448	61.801	62.154	62.507	62.860	63.213
£1.80	63.566	63.920	64.273	64.626	64.979	65.332	65.685	66.038	66.392	66.745
£1.90	67.098	67.451	67.804	68.157	68.510	68.864	69.217	69.570	69.923	70.276
£2.00	70.629	70.982	71.336	71.689	72.042	72.395	72.748	73.101	73.455	73.808
£2.10	74.161	74.514	74.867	75.220	75.573	75.927	76.280	76.633	76.986	77.339
£2.20	77.692	78.045	78.399	78.752	79.105	79.458	79.811	80.164	80.517	80.871
£2.30	81.224	81.577	81.930	82.283	82.636	82.989	83.343	83.696	84.049	84.402
£2.40	84.755	85.108	85.461	85.815	86.168	86.521	86.874	87.227	87.580	87.934

PRICE PER CUBIC FOOT to PRICE PER CUBIC METRE

Price per cu. ft.	0	1p	2p	3p	4p	5p	6p	7p	8p	9p
					£'s per Cubic Metre					
£2.50	88.287	88.640	88.993	89.346	89.699	90.052	90.406	90.759	91.112	91.465
£2.60	91.818	92.171	92.524	92.878	93.231	93.584	93.937	94.290	94.643	94.996
£2.70	95.350	95.703	96.056	96.409	96.762	97.115	97.468	97.822	98.175	98.528
£2.80	98.881	99.234	99.587	99.941	100.294	100.647	101.000	101.353	101.706	102.059
£2.90	102.413	102.766	103.119	103.472	103.825	104.178	104.531	104.885	105.238	105.591
£3.00	105.944	106.297	106.650	107.003	107.357	107.710	108.063	108.416	108.769	109.122
£3.10	109.475	109.829	110.182	110.535	110.888	111.241	111.594	111.947	112.301	112.654
£3.20	113.007	113.360	113.713	114.066	114.420	114.773	115.126	115.479	115.832	116.185
£3.30	116.538	116.892	117.245	117.598	117.951	118.304	118.657	119.010	119.364	119.717
£3.40	120.070	120.423	120.776	121.129	121.482	121.836	122.189	122.542	122.895	123.248
£3.50	123.601	123.954	124.308	124.661	125.014	125.367	125.720	126.073	126.427	126.780
£3.60	127.133	127.486	127.839	128.192	128.545	128.899	129.252	129.605	129.958	130.311
£3.70	130.664	131.017	131.371	131.724	132.077	132.430	132.783	133.136	133.489	133.843
£3.80	134.196	134.549	134.902	135.255	135.608	135.961	136.315	136.668	137.021	137.374
£3.90	137.727	138.080	138.433	138.787	139.140	139.493	139.846	140.199	140.552	140.906
£4.00	141.259	141.612	141.965	142.318	142.671	143.024	143.378	143.731	144.084	144.437

SUPPLEMENTARY TABLE

Price/cu. ft.	£5	£6	£7	£8	£9	£10	£11	£12	£13	£14	£15	£16
Price/cu.m.	176.573	211.888	247.203	282.517	317.832	353.147	388.461	423.776	459.091	494.405	529.720	565.035

Price/cu. ft.	£17	£18	£19	£20	£21	£22	£23	£24	£25	£26	£27	½p
Price/cu.m.	600.349	635.664	670.979	706.293	741.608	776.923	812.237	847.552	882.867	918.181	953.496	0.177

Basis: £1 per cu. ft. = £35.314666721 per cu. m.

PRICE PER ACRE to PRICE PER HECTARE

Price per acre	0	£1	£2	£3	£4	£5	£6	£7	£8	£9
					£'s per Hectare					
0	0	2.471	4.942	7.413	9.884	12.355	14.826	17.297	19.768	22.239
£10	24.711	27.182	29.653	32.124	34.595	37.066	39.537	42.008	44.479	46.950
£20	49.421	51.892	54.363	56.834	59.305	61.776	64.247	66.718	69.190	71.661
£30	74.132	76.603	79.074	81.545	84.016	86.487	88.958	91.429	93.900	96.371
£40	98.842	101.313	103.784	106.255	108.726	111.197	113.668	116.140	118.611	121.082
£50	123.553	126.024	128.495	130.966	133.437	135.908	138.379	140.850	143.321	145.792
£60	148.263	150.734	153.205	155.676	158.147	160.618	163.090	165.561	168.032	170.503
£70	172.974	175.445	177.916	180.387	182.858	185.329	187.800	190.271	192.742	195.213
£80	197.684	200.155	202.626	205.097	207.569	210.040	212.511	214.982	217.453	219.924
£90	222.395	224.866	227.337	229.808	232.279	234.750	237.221	239.692	242.163	244.634
£100	247.105	249.576	252.047	254.519	256.990	259.461	261.932	264.403	266.874	269.345
£110	271.816	274.287	276.758	279.229	281.700	284.171	286.642	289.113	291.584	294.055
£120	296.526	298.998	301.469	303.940	306.411	308.882	311.353	313.824	316.295	318.766
£130	321.237	323.708	326.179	328.650	331.121	333.592	336.063	338.534	341.005	343.476
£140	345.948	348.419	350.890	353.361	355.832	358.303	360.774	363.245	365.716	368.187
£150	370.658	373.129	375.600	378.071	380.542	383.013	385.484	387.955	390.426	392.898
£160	395.369	397.840	400.311	402.782	405.253	407.724	410.195	412.666	415.137	417.608
£170	420.079	422.550	425.021	427.492	429.963	432.434	434.905	437.377	439.848	442.319
£180	444.790	447.261	449.732	452.203	454.674	457.145	459.616	462.087	464.558	467.029
£190	469.500	471.971	474.442	476.913	479.384	481.855	484.327	486.798	489.269	491.740
£200	494.211	496.682	499.153	501.624	504.095	506.566	509.037	511.508	513.979	516.450

Price / acre	£300	£400	£500	£600	£700	£800	£900	£1000
Price / hectare	741.316	988.422	1235.527	1482.632	1729.738	1976.843	2223.948	2471.054

264

PRICE PER ACRE to PRICE PER HECTARE

Price per acre	0	1p	2p	3p	4p	5p	6p	7p	8p	9p
				£'s per Hectare						
	0.0	0.025	0.049	0.074	0.099	0.124	0.148	0.173	0.198	0.222
10p	0.247	0.272	0.297	0.321	0.346	0.371	0.395	0.420	0.445	0.470
20p	0.494	0.519	0.544	0.568	0.593	0.618	0.642	0.667	0.692	0.717
30p	0.741	0.766	0.791	0.815	0.840	0.865	0.890	0.914	0.939	0.964
40p	0.988	1.013	1.038	1.063	1.087	1.112	1.137	1.161	1.186	1.211
50p	1.236	1.260	1.285	1.310	1.334	1.359	1.384	1.409	1.433	1.458
60p	1.483	1.507	1.532	1.557	1.581	1.606	1.631	1.656	1.680	1.705
70p	1.730	1.754	1.779	1.804	1.829	1.853	1.878	1.903	1.927	1.952
80p	1.977	2.002	2.026	2.051	2.076	2.100	2.125	2.150	2.175	2.199
90p	2.224	2.249	2.273	2.298	2.323	2.348	2.372	2.397	2.422	2.446
£1.00	2.471	2.496	2.520	2.545	2.570	2.595	2.619	2.644	2.669	2.693
£1.10	2.718	2.743	2.768	2.792	2.817	2.842	2.866	2.891	2.916	2.941
£1.20	2.965	2.990	3.015	3.039	3.064	3.089	3.114	3.138	3.163	3.188
£1.30	3.212	3.237	3.262	3.287	3.311	3.336	3.361	3.385	3.410	3.435
£1.40	3.459	3.484	3.509	3.534	3.558	3.583	3.608	3.632	3.657	3.682
£1.50	3.707	3.731	3.756	3.781	3.805	3.830	3.855	3.880	3.904	3.929
£1.60	3.954	3.978	4.003	4.028	4.053	4.077	4.102	4.127	4.151	4.176
£1.70	4.201	4.226	4.250	4.275	4.300	4.324	4.349	4.374	4.398	4.423
£1.80	4.448	4.473	4.497	4.522	4.547	4.571	4.596	4.621	4.646	4.670
£1.90	4.695	4.720	4.744	4.769	4.794	4.819	4.843	4.868	4.893	4.917
£2.00	4.942	4.967	4.992	5.016	5.041	5.066	5.090	5.115	5.140	5.165
£2.10	5.189	5.214	5.239	5.263	5.288	5.313	5.337	5.362	5.387	5.412
£2.20	5.436	5.461	5.486	5.510	5.535	5.560	5.585	5.609	5.634	5.659
£2.30	5.683	5.708	5.733	5.758	5.782	5.807	5.832	5.856	5.881	5.906
£2.40	5.931	5.955	5.980	6.005	6.029	6.054	6.079	6.104	6.128	6.153
£2.50	6.178	6.202	6.227	6.252	6.276	6.301	6.326	6.351	6.375	6.400

Basis: £1 per acre = £2.47105381 per hectare

YEARS' PURCHASE

OR

PRESENT VALUE OF
ONE POUND PER ANNUM

for a

SINGLE LIFE

(Males and Females)

according to the

ENGLISH LIFE MORTALITY TABLE No. 15
ON THE SINGLE RATE % PRINCIPLE
AT RATES OF INTEREST FROM
5% TO 16%

YEARS' PURCHASE FOR A SINGLE LIFE

(ENGLISH LIFE MORTALITY TABLES - No 15)

				Rate Per Cent					
Age	5	5.5	6	6.5	7	7.5	8	8.5	Age
1	19.188	17.612	16.257	15.084	14.060	13.160	12.364	11.656	1
2	19.160	17.592	16.243	15.074	14.053	13.156	12.362	11.655	2
3	19.125	17.566	16.224	15.060	14.043	13.148	12.356	11.651	3
4	19.088	17.538	16.203	15.044	14.030	13.138	12.348	11.645	4
5	19.047	17.507	16.179	15.026	14.016	13.127	12.339	11.638	5
6	19.004	17.474	16.154	15.006	14.001	13.115	12.330	11.630	6
7	18.958	17.439	16.127	14.984	13.984	13.101	12.319	11.621	7
8	18.909	17.402	16.097	14.961	13.965	13.086	12.307	11.611	8
9	18.859	17.362	16.066	14.937	13.946	13.071	12.294	11.600	9
10	18.805	17.320	16.034	14.911	13.925	13.053	12.280	11.588	10
11	18.749	17.276	15.999	14.883	13.902	13.035	12.264	11.576	11
12	18.690	17.230	15.962	14.853	13.878	13.015	12.248	11.562	12
13	18.628	17.181	15.923	14.821	13.852	12.994	12.230	11.547	13
14	18.564	17.130	15.882	14.789	13.825	12.972	12.212	11.531	14
15	18.498	17.078	15.840	14.754	13.797	12.949	12.193	11.515	15
16	18.431	17.024	15.797	14.720	13.769	12.925	12.173	11.499	16
17	18.363	16.970	15.753	14.685	13.741	12.902	12.154	11.483	17
18	18.295	16.917	15.711	14.651	13.713	12.880	12.136	11.468	18
19	18.227	16.863	15.668	14.617	13.686	12.858	12.118	11.454	19
20	18.154	16.805	15.622	14.580	13.656	12.834	12.099	11.438	20
21	18.077	16.744	15.573	14.541	13.625	12.808	12.077	11.420	21
22	17.998	16.680	15.522	14.499	13.591	12.781	12.055	11.402	22
23	17.914	16.613	15.468	14.455	13.555	12.752	12.031	11.382	23
24	17.827	16.542	15.411	14.408	13.517	12.720	12.005	11.360	24
25	17.735	16.468	15.350	14.359	13.476	12.686	11.977	11.337	25
26	17.637	16.388	15.285	14.305	13.431	12.649	11.946	11.311	26
27	17.535	16.304	15.215	14.248	13.384	12.610	11.913	11.283	27
28	17.427	16.216	15.142	14.187	13.333	12.567	11.877	11.252	28
29	17.315	16.122	15.065	14.122	13.279	12.521	11.838	11.220	29
30	17.197	16.024	14.983	14.053	13.221	12.472	11.796	11.184	30
31	17.073	15.921	14.896	13.980	13.159	12.420	11.752	11.146	31
32	16.943	15.812	14.805	13.903	13.093	12.364	11.704	11.104	32
33	16.808	15.698	14.708	13.821	13.024	12.304	11.652	11.060	33
34	16.666	15.578	14.606	13.734	12.949	12.240	11.597	11.012	34
35	16.518	15.452	14.499	13.643	12.870	12.172	11.538	10.961	35
36	16.364	15.321	14.387	13.546	12.787	12.100	11.475	10.906	36
37	16.204	15.185	14.270	13.445	12.700	12.024	11.409	10.849	37
38	16.037	15.042	14.147	13.339	12.608	11.944	11.339	10.787	38
39	15.864	14.893	14.018	13.227	12.510	11.859	11.265	10.721	39
40	15.684	14.737	13.883	13.109	12.408	11.769	11.185	10.651	40
41	15.497	14.575	13.741	12.986	12.299	11.673	11.101	10.577	41
42	15.302	14.405	13.593	12.855	12.184	11.572	11.011	10.497	42
43	15.100	14.228	13.437	12.719	12.064	11.465	10.916	10.412	43
44	14.889	14.043	13.275	12.575	11.936	11.352	10.815	10.322	44
45	14.672	13.851	13.105	12.425	11.803	11.233	10.709	10.226	45
46	14.446	13.652	12.929	12.268	11.663	11.107	10.596	10.125	46
47	14.214	13.446	12.745	12.104	11.516	10.976	10.478	10.019	47
48	13.974	13.233	12.555	11.934	11.363	10.838	10.354	9.906	48
49	13.728	13.013	12.358	11.757	11.204	10.695	10.224	9.789	49
50	13.474	12.786	12.154	11.573	11.038	10.545	10.088	9.665	50

YEARS' PURCHASE FOR A SINGLE LIFE
(ENGLISH LIFE MORTALITY TABLES - No 15)

				Rate Per Cent					
Age	5	5.5	6	6.5	7	7.5	8	8.5	Age
51	13.214	12.552	11.943	11.383	10.866	10.388	9.946	9.535	51
52	12.947	12.311	11.726	11.186	10.687	10.226	9.798	9.400	52
53	12.673	12.064	11.502	10.982	10.502	10.057	9.643	9.258	53
54	12.393	11.809	11.270	10.772	10.310	9.881	9.482	9.110	54
55	12.106	11.548	11.033	10.555	10.111	9.698	9.314	8.955	55
56	11.814	11.282	10.789	10.331	9.905	9.509	9.140	8.794	56
57	11.516	11.009	10.539	10.101	9.694	9.314	8.959	8.628	57
58	11.213	10.731	10.283	9.866	9.477	9.113	8.773	8.455	58
59	10.906	10.449	10.023	9.625	9.254	8.907	8.582	8.277	59
60	10.595	10.162	9.758	9.380	9.027	8.696	8.385	8.093	60
61	10.282	9.872	9.489	9.131	8.795	8.480	8.184	7.905	61
62	9.967	9.580	9.218	8.878	8.560	8.260	7.978	7.713	62
63	9.652	9.287	8.945	8.624	8.322	8.038	7.770	7.518	63
64	9.338	8.994	8.672	8.368	8.083	7.814	7.560	7.320	64
65	9.025	8.702	8.399	8.113	7.843	7.589	7.348	7.121	65
66	8.714	8.411	8.126	7.857	7.603	7.362	7.135	6.920	66
67	8.405	8.121	7.853	7.601	7.361	7.135	6.921	6.717	67
68	8.097	7.832	7.582	7.345	7.120	6.907	6.705	6.513	68
69	7.792	7.544	7.310	7.088	6.878	6.678	6.488	6.307	69
70	7.487	7.257	7.038	6.831	6.634	6.447	6.269	6.099	70
71	7.183	6.969	6.766	6.573	6.389	6.214	6.047	5.888	71
72	6.882	6.684	6.495	6.315	6.144	5.981	5.825	5.677	72
73	6.586	6.402	6.228	6.061	5.902	5.750	5.605	5.466	73
74	6.296	6.126	5.965	5.810	5.663	5.521	5.386	5.257	74
75	6.010	5.854	5.704	5.562	5.425	5.294	5.169	5.049	75
76	5.728	5.584	5.446	5.314	5.188	5.067	4.951	4.840	76
77	5.451	5.319	5.193	5.071	4.955	4.843	4.736	4.633	77
78	5.182	5.061	4.945	4.834	4.726	4.624	4.525	4.429	78
79	4.921	4.810	4.704	4.602	4.503	4.409	4.317	4.230	79
80	4.668	4.567	4.469	4.376	4.286	4.199	4.115	4.034	80
81	4.423	4.331	4.242	4.156	4.073	3.994	3.917	3.843	81
82	4.184	4.100	4.019	3.941	3.865	3.792	3.722	3.654	82
83	3.951	3.875	3.801	3.730	3.661	3.595	3.531	3.468	83
84	3.727	3.658	3.591	3.526	3.464	3.403	3.344	3.288	84
85	3.513	3.450	3.389	3.330	3.273	3.218	3.165	3.113	85
86	3.307	3.250	3.195	3.142	3.090	3.040	2.991	2.944	86
87	3.114	3.062	3.013	2.964	2.917	2.872	2.827	2.784	87
88	2.932	2.885	2.840	2.797	2.754	2.713	2.672	2.633	88
89	2.757	2.715	2.674	2.635	2.596	2.559	2.522	2.487	89
90	2.585	2.547	2.510	2.475	2.440	2.406	2.373	2.341	90
91	2.412	2.379	2.346	2.314	2.283	2.252	2.223	2.194	91
92	2.243	2.213	2.184	2.155	2.128	2.100	2.074	2.048	92
93	2.085	2.058	2.032	2.007	1.982	1.957	1.933	1.910	93
94	1.942	1.918	1.894	1.871	1.849	1.827	1.806	1.785	94
95	1.812	1.790	1.769	1.748	1.728	1.708	1.689	1.670	95
96	1.692	1.672	1.654	1.635	1.617	1.599	1.582	1.565	96
97	1.579	1.561	1.544	1.527	1.511	1.495	1.479	1.464	97
98	1.470	1.454	1.439	1.424	1.409	1.395	1.381	1.367	98
99	1.369	1.355	1.341	1.328	1.315	1.302	1.289	1.276	99
100	1.270	1.258	1.245	1.233	1.222	1.210	1.199	1.187	100

YEARS' PURCHASE FOR A SINGLE LIFE
(ENGLISH LIFE MORTALITY TABLES - No 15)

	Rate Per Cent								
Age	9	10	11	12	13	14	15	16	Age
1	11.023	9.939	9.046	8.298	7.664	7.120	6.647	6.233	1
2	11.023	9.939	9.047	8.300	7.666	7.121	6.649	6.235	2
3	11.019	9.938	9.046	8.300	7.666	7.122	6.649	6.235	3
4	11.015	9.935	9.044	8.298	7.665	7.121	6.649	6.235	4
5	11.009	9.931	9.041	8.296	7.664	7.120	6.648	6.234	5
6	11.002	9.926	9.038	8.294	7.662	7.119	6.647	6.233	6
7	10.995	9.921	9.034	8.291	7.660	7.117	6.645	6.232	7
8	10.987	9.915	9.030	8.288	7.657	7.115	6.644	6.231	8
9	10.978	9.909	9.025	8.284	7.654	7.112	6.642	6.229	9
10	10.968	9.902	9.020	8.280	7.651	7.109	6.639	6.227	10
11	10.957	9.894	9.014	8.275	7.647	7.106	6.636	6.225	11
12	10.945	9.885	9.007	8.270	7.642	7.103	6.633	6.222	12
13	10.933	9.876	9.000	8.264	7.638	7.098	6.630	6.219	13
14	10.920	9.866	8.992	8.258	7.632	7.094	6.626	6.216	14
15	10.906	9.855	8.984	8.251	7.627	7.090	6.622	6.212	15
16	10.892	9.845	8.976	8.245	7.622	7.085	6.619	6.209	16
17	10.879	9.835	8.969	8.239	7.618	7.082	6.615	6.206	17
18	10.866	9.827	8.963	8.235	7.614	7.079	6.613	6.204	18
19	10.855	9.819	8.958	8.231	7.612	7.077	6.612	6.203	19
20	10.841	9.810	8.951	8.227	7.608	7.075	6.610	6.202	20
21	10.827	9.800	8.944	8.222	7.605	7.072	6.608	6.200	21
22	10.812	9.790	8.937	8.216	7.601	7.069	6.606	6.199	22
23	10.795	9.778	8.929	8.210	7.596	7.066	6.603	6.197	23
24	10.777	9.765	8.920	8.204	7.591	7.062	6.600	6.195	24
25	10.758	9.751	8.909	8.196	7.586	7.058	6.597	6.192	25
26	10.736	9.736	8.898	8.188	7.579	7.053	6.593	6.189	26
27	10.712	9.718	8.885	8.178	7.572	7.047	6.589	6.185	27
28	10.686	9.699	8.871	8.167	7.564	7.041	6.584	6.181	28
29	10.658	9.679	8.855	8.155	7.554	7.033	6.578	6.176	29
30	10.628	9.656	8.838	8.142	7.544	7.025	6.571	6.171	30
31	10.595	9.631	8.819	8.128	7.532	7.016	6.564	6.165	31
32	10.559	9.604	8.798	8.111	7.520	7.005	6.555	6.158	32
33	10.520	9.575	8.776	8.094	7.505	6.994	6.546	6.150	33
34	10.479	9.543	8.751	8.074	7.490	6.981	6.535	6.141	34
35	10.434	9.508	8.724	8.052	7.472	6.967	6.523	6.131	35
36	10.386	9.471	8.695	8.029	7.453	6.951	6.511	6.121	36
37	10.335	9.432	8.663	8.004	7.433	6.935	6.497	6.109	37
38	10.281	9.389	8.630	7.977	7.411	6.917	6.482	6.096	38
39	10.223	9.344	8.593	7.947	7.387	6.897	6.465	6.082	39
40	10.161	9.294	8.554	7.915	7.361	6.875	6.447	6.067	40
41	10.095	9.242	8.511	7.881	7.332	6.851	6.426	6.049	41
42	10.024	9.185	8.465	7.843	7.300	6.824	6.404	6.030	42
43	9.948	9.123	8.415	7.802	7.266	6.796	6.380	6.009	43
44	9.867	9.058	8.361	7.757	7.229	6.764	6.353	5.986	44
45	9.781	8.988	8.303	7.709	7.188	6.730	6.323	5.961	45
46	9.690	8.913	8.241	7.657	7.144	6.692	6.291	5.933	46
47	9.593	8.833	8.175	7.601	7.097	6.652	6.256	5.903	47
48	9.492	8.749	8.105	7.542	7.046	6.608	6.219	5.870	48
49	9.385	8.660	8.030	7.478	6.992	6.562	6.178	5.835	49
50	9.272	8.565	7.950	7.410	6.934	6.511	6.134	5.796	50

YEARS' PURCHASE FOR A SINGLE LIFE

(ENGLISH LIFE MORTALITY TABLES - No 15)

Age				Rate Per Cent					Age
	9	10	11	12	13	14	15	16	
51	9.153	8.466	7.866	7.338	6.872	6.458	6.087	5.755	51
52	9.029	8.361	7.777	7.262	6.806	6.400	6.037	5.711	52
53	8.899	8.251	7.682	7.180	6.735	6.339	5.983	5.663	53
54	8.763	8.134	7.582	7.094	6.660	6.273	5.925	5.611	54
55	8.620	8.012	7.477	7.002	6.580	6.202	5.863	5.556	55
56	8.471	7.884	7.366	6.906	6.495	6.127	5.796	5.497	56
57	8.317	7.751	7.250	6.804	6.406	6.048	5.725	5.433	57
58	8.156	7.611	7.128	6.697	6.311	5.964	5.650	5.366	58
59	7.990	7.466	7.001	6.585	6.212	5.875	5.571	5.294	59
60	7.819	7.316	6.869	6.468	6.108	5.782	5.487	5.219	60
61	7.643	7.162	6.732	6.346	5.999	5.685	5.399	5.139	61
62	7.463	7.003	6.591	6.221	5.886	5.583	5.308	5.056	62
63	7.279	6.840	6.446	6.091	5.770	5.478	5.212	4.969	63
64	7.093	6.675	6.299	5.959	5.651	5.370	5.114	4.880	64
65	6.905	6.508	6.149	5.824	5.529	5.260	5.014	4.788	65
66	6.716	6.338	5.996	5.686	5.404	5.146	4.910	4.693	66
67	6.524	6.166	5.841	5.546	5.277	5.030	4.804	4.596	67
68	6.331	5.992	5.684	5.404	5.147	4.912	4.696	4.496	68
69	6.136	5.816	5.524	5.258	5.014	4.790	4.584	4.393	69
70	5.938	5.636	5.361	5.109	4.878	4.665	4.468	4.286	70
71	5.737	5.453	5.194	4.956	4.737	4.535	4.349	4.176	71
72	5.535	5.269	5.025	4.801	4.594	4.403	4.226	4.062	72
73	5.333	5.085	4.856	4.645	4.450	4.270	4.102	3.947	73
74	5.133	4.901	4.687	4.489	4.306	4.136	3.977	3.830	74
75	4.934	4.717	4.516	4.331	4.159	3.999	3.850	3.711	75
76	4.733	4.531	4.345	4.171	4.010	3.860	3.721	3.590	76
77	4.534	4.347	4.173	4.012	3.861	3.721	3.590	3.467	77
78	4.338	4.164	4.003	3.853	3.713	3.582	3.459	3.344	78
79	4.145	3.985	3.835	3.696	3.565	3.443	3.328	3.221	79
80	3.956	3.808	3.670	3.540	3.419	3.305	3.199	3.098	80
81	3.771	3.634	3.507	3.387	3.275	3.169	3.070	2.976	81
82	3.588	3.462	3.345	3.234	3.131	3.033	2.940	2.853	82
83	3.408	3.293	3.185	3.083	2.987	2.897	2.811	2.731	83
84	3.232	3.127	3.028	2.934	2.846	2.763	2.684	2.609	84
85	3.062	2.966	2.875	2.790	2.708	2.632	2.559	2.489	85
86	2.898	2.810	2.727	2.649	2.574	2.503	2.436	2.372	86
87	2.743	2.662	2.586	2.514	2.446	2.381	2.319	2.260	87
88	2.595	2.522	2.452	2.387	2.324	2.264	2.207	2.153	88
89	2.452	2.385	2.322	2.262	2.205	2.150	2.098	2.048	89
90	2.310	2.249	2.192	2.137	2.085	2.035	1.988	1.942	90
91	2.165	2.111	2.059	2.010	1.962	1.917	1.874	1.832	91
92	2.023	1.974	1.927	1.882	1.839	1.799	1.759	1.722	92
93	1.887	1.843	1.801	1.761	1.722	1.685	1.650	1.616	93
94	1.764	1.725	1.687	1.650	1.615	1.582	1.549	1.518	94
95	1.652	1.616	1.581	1.548	1.517	1.486	1.457	1.429	95
96	1.548	1.515	1.484	1.454	1.426	1.398	1.371	1.345	96
97	1.449	1.420	1.391	1.364	1.338	1.313	1.289	1.265	97
98	1.353	1.327	1.301	1.277	1.253	1.231	1.209	1.187	98
99	1.264	1.240	1.218	1.195	1.174	1.153	1.133	1.114	99
100	1.176	1.155	1.134	1.114	1.095	1.077	1.058	1.041	100

YEARS' PURCHASE FOR A SINGLE LIFE
(ENGLISH LIFE MORTALITY TABLES - No 15)

				Rate Per Cent					
Age	5	5.5	6	6.5	7	7.5	8	8.5	Age
1	19.387	17.763	16.374	15.175	14.131	13.217	12.410	11.694	1
2	19.368	17.751	16.366	15.170	14.129	13.216	12.411	11.695	2
3	19.342	17.733	16.353	15.161	14.123	13.212	12.408	11.693	3
4	19.314	17.712	16.338	15.150	14.114	13.206	12.403	11.690	4
5	19.283	17.690	16.321	15.137	14.105	13.199	12.398	11.686	5
6	19.251	17.665	16.303	15.124	14.095	13.191	12.392	11.681	6
7	19.216	17.640	16.284	15.109	14.084	13.182	12.385	11.676	7
8	19.180	17.613	16.264	15.094	14.072	13.173	12.378	11.670	8
9	19.142	17.584	16.242	15.077	14.059	13.163	12.370	11.664	9
10	19.101	17.553	16.219	15.059	14.045	13.152	12.361	11.657	10
11	19.059	17.521	16.194	15.040	14.030	13.140	12.352	11.649	11
12	19.015	17.488	16.168	15.020	14.014	13.128	12.342	11.641	12
13	18.969	17.452	16.141	14.998	13.997	13.115	12.331	11.632	13
14	18.920	17.415	16.112	14.976	13.980	13.100	12.320	11.623	14
15	18.870	17.376	16.081	14.952	13.961	13.085	12.308	11.613	15
16	18.817	17.335	16.050	14.927	13.941	13.070	12.295	11.603	16
17	18.763	17.294	16.017	14.902	13.921	13.054	12.282	11.593	17
18	18.708	17.250	15.984	14.875	13.900	13.037	12.269	11.582	18
19	18.649	17.205	15.948	14.847	13.878	13.019	12.255	11.570	19
20	18.588	17.157	15.910	14.818	13.854	13.000	12.239	11.558	20
21	18.523	17.106	15.870	14.786	13.829	12.980	12.222	11.544	21
22	18.456	17.053	15.828	14.752	13.801	12.958	12.205	11.529	22
23	18.385	16.997	15.783	14.716	13.772	12.934	12.185	11.514	23
24	18.311	16.938	15.735	14.677	13.741	12.909	12.164	11.496	24
25	18.232	16.875	15.685	14.637	13.708	12.881	12.142	11.478	25
26	18.151	16.809	15.632	14.593	13.672	12.852	12.118	11.457	26
27	18.065	16.740	15.575	14.547	13.635	12.821	12.092	11.436	27
28	17.975	16.667	15.516	14.498	13.594	12.787	12.064	11.412	28
29	17.881	16.590	15.453	14.447	13.551	12.751	12.034	11.387	29
30	17.782	16.509	15.387	14.392	13.506	12.713	12.001	11.360	30
31	17.680	16.425	15.317	14.334	13.457	12.673	11.967	11.331	31
32	17.572	16.337	15.244	14.273	13.406	12.629	11.931	11.299	32
33	17.460	16.244	15.167	14.208	13.352	12.584	11.892	11.266	33
34	17.344	16.147	15.086	14.141	13.295	12.535	11.851	11.231	34
35	17.222	16.046	15.001	14.069	13.234	12.484	11.807	11.193	35
36	17.096	15.940	14.912	13.994	13.171	12.429	11.760	11.153	36
37	16.964	15.830	14.819	13.915	13.103	12.372	11.710	11.110	37
38	16.827	15.714	14.721	13.831	13.032	12.311	11.657	11.064	38
39	16.685	15.593	14.618	13.744	12.957	12.246	11.601	11.016	39
40	16.536	15.467	14.510	13.651	12.877	12.177	11.542	10.964	40
41	16.381	15.335	14.398	13.554	12.793	12.104	11.478	10.908	41
42	16.221	15.197	14.279	13.452	12.705	12.027	11.411	10.849	42
43	16.054	15.054	14.156	13.345	12.612	11.946	11.340	10.787	43
44	15.880	14.905	14.026	13.233	12.514	11.860	11.265	10.720	44
45	15.700	14.749	13.891	13.115	12.411	11.770	11.185	10.650	45
46	15.515	14.588	13.751	12.992	12.303	11.675	11.101	10.576	46
47	15.323	14.421	13.605	12.864	12.191	11.576	11.013	10.497	47
48	15.124	14.247	13.453	12.731	12.072	11.471	10.920	10.414	48
49	14.919	14.067	13.294	12.591	11.949	11.361	10.822	10.327	49
50	14.707	13.881	13.130	12.445	11.819	11.246	10.719	10.235	50

YEARS' PURCHASE FOR A SINGLE LIFE

(ENGLISH LIFE MORTALITY TABLES - No 15)

Age	5	5.5	6	6.5	7	7.5	8	8.5	Age
				Rate Per Cent					
51	14.488	13.687	12.959	12.293	11.684	11.125	10.611	10.137	51
52	14.262	13.487	12.781	12.135	11.543	10.999	10.497	10.035	52
53	14.028	13.280	12.596	11.970	11.395	10.866	10.378	9.927	53
54	13.788	13.065	12.404	11.798	11.240	10.727	10.252	9.813	54
55	13.539	12.843	12.205	11.619	11.079	10.581	10.120	9.693	55
56	13.284	12.614	11.999	11.433	10.911	10.429	9.982	9.567	56
57	13.023	12.379	11.787	11.241	10.737	10.270	9.838	9.436	57
58	12.755	12.138	11.569	11.043	10.557	10.106	9.688	9.299	58
59	12.482	11.891	11.344	10.839	10.371	9.937	9.533	9.156	59
60	12.204	11.638	11.115	10.630	10.180	9.761	9.372	9.008	60
61	11.921	11.381	10.880	10.415	9.984	9.581	9.206	8.856	61
62	11.634	11.118	10.640	10.196	9.782	9.396	9.035	8.698	62
63	11.340	10.850	10.394	9.970	9.574	9.204	8.858	8.534	63
64	11.043	10.578	10.144	9.739	9.361	9.007	8.676	8.365	64
65	10.744	10.303	9.890	9.505	9.145	8.807	8.490	8.193	65
66	10.442	10.023	9.632	9.267	8.924	8.602	8.300	8.015	66
67	10.133	9.738	9.368	9.022	8.696	8.390	8.102	7.831	67
68	9.821	9.449	9.100	8.772	8.463	8.173	7.899	7.641	68
69	9.506	9.156	8.827	8.517	8.226	7.951	7.692	7.447	69
70	9.187	8.858	8.549	8.258	7.983	7.723	7.478	7.246	70
71	8.862	8.555	8.265	7.991	7.733	7.488	7.257	7.038	71
72	8.534	8.247	7.976	7.720	7.478	7.248	7.030	6.824	72
73	8.209	7.941	7.689	7.449	7.222	7.007	6.803	6.609	73
74	7.887	7.639	7.403	7.180	6.968	6.767	6.575	6.393	74
75	7.562	7.332	7.114	6.906	6.709	6.521	6.343	6.172	75
76	7.234	7.022	6.820	6.627	6.444	6.270	6.103	5.945	76
77	6.906	6.710	6.524	6.346	6.177	6.015	5.861	5.713	77
78	6.582	6.402	6.230	6.067	5.910	5.761	5.618	5.481	78
79	6.262	6.097	5.939	5.789	5.645	5.507	5.375	5.249	79
80	5.948	5.797	5.653	5.515	5.383	5.256	5.135	5.018	80
81	5.641	5.504	5.372	5.246	5.125	5.009	4.897	4.790	81
82	5.340	5.215	5.095	4.980	4.869	4.763	4.661	4.562	82
83	5.043	4.930	4.821	4.716	4.615	4.519	4.425	4.335	83
84	4.751	4.649	4.550	4.455	4.364	4.276	4.191	4.109	84
85	4.467	4.375	4.286	4.200	4.118	4.038	3.961	3.886	85
86	4.196	4.113	4.033	3.956	3.881	3.809	3.739	3.671	86
87	3.942	3.867	3.795	3.725	3.657	3.592	3.528	3.467	87
88	3.695	3.628	3.563	3.500	3.440	3.380	3.323	3.268	88
89	3.453	3.393	3.335	3.278	3.224	3.170	3.119	3.069	89
90	3.223	3.169	3.117	3.067	3.017	2.970	2.923	2.878	90
91	3.007	2.959	2.913	2.867	2.823	2.780	2.739	2.698	91
92	2.805	2.762	2.720	2.679	2.640	2.601	2.564	2.528	92
93	2.617	2.578	2.541	2.504	2.469	2.434	2.401	2.368	93
94	2.440	2.405	2.372	2.339	2.307	2.276	2.246	2.216	94
95	2.275	2.244	2.214	2.185	2.156	2.129	2.101	2.075	95
96	2.128	2.100	2.073	2.047	2.021	1.996	1.971	1.947	96
97	1.993	1.968	1.944	1.920	1.897	1.874	1.852	1.830	97
98	1.859	1.837	1.815	1.794	1.773	1.753	1.733	1.713	98
99	1.725	1.705	1.686	1.667	1.648	1.630	1.612	1.595	99
100	1.600	1.582	1.565	1.548	1.531	1.515	1.499	1.483	100

YEARS' PURCHASE FOR A SINGLE LIFE
(ENGLISH LIFE MORTALITY TABLES - No 15)

Rate Per Cent

Age	9	10	11	12	13	14	15	16	Age
1	11.054	9.961	9.062	8.311	7.674	7.127	6.653	6.238	1
2	11.056	9.963	9.064	8.313	7.676	7.130	6.655	6.240	2
3	11.054	9.962	9.064	8.313	7.677	7.130	6.656	6.241	3
4	11.052	9.961	9.064	8.313	7.677	7.130	6.656	6.241	4
5	11.049	9.959	9.062	8.312	7.676	7.130	6.656	6.241	5
6	11.045	9.957	9.061	8.311	7.675	7.129	6.656	6.241	6
7	11.041	9.954	9.059	8.310	7.674	7.129	6.655	6.240	7
8	11.036	9.951	9.057	8.308	7.673	7.128	6.654	6.240	8
9	11.031	9.948	9.055	8.307	7.672	7.127	6.654	6.239	9
10	11.026	9.944	9.052	8.305	7.670	7.126	6.653	6.238	10
11	11.019	9.940	9.049	8.303	7.669	7.124	6.652	6.237	11
12	11.013	9.935	9.046	8.300	7.667	7.123	6.650	6.236	12
13	11.006	9.930	9.042	8.297	7.665	7.121	6.649	6.235	13
14	10.998	9.925	9.038	8.295	7.663	7.119	6.647	6.234	14
15	10.990	9.919	9.034	8.292	7.660	7.117	6.646	6.233	15
16	10.982	9.914	9.030	8.289	7.658	7.116	6.644	6.231	16
17	10.973	9.908	9.026	8.286	7.656	7.114	6.643	6.230	17
18	10.965	9.902	9.022	8.283	7.654	7.112	6.642	6.230	18
19	10.955	9.896	9.018	8.280	7.651	7.111	6.641	6.229	19
20	10.945	9.889	9.013	8.276	7.649	7.109	6.639	6.227	20
21	10.934	9.881	9.007	8.272	7.646	7.107	6.638	6.226	21
22	10.922	9.873	9.001	8.268	7.643	7.104	6.636	6.225	22
23	10.908	9.863	8.995	8.263	7.639	7.101	6.634	6.223	23
24	10.894	9.853	8.988	8.258	7.635	7.098	6.631	6.221	24
25	10.878	9.842	8.979	8.252	7.631	7.095	6.628	6.219	25
26	10.861	9.830	8.971	8.245	7.625	7.091	6.625	6.216	26
27	10.843	9.817	8.961	8.238	7.620	7.086	6.622	6.213	27
28	10.823	9.802	8.950	8.229	7.613	7.081	6.617	6.210	28
29	10.802	9.787	8.938	8.221	7.606	7.076	6.613	6.206	29
30	10.779	9.770	8.925	8.211	7.599	7.070	6.608	6.202	30
31	10.754	9.751	8.912	8.200	7.590	7.063	6.602	6.197	31
32	10.727	9.731	8.896	8.188	7.581	7.055	6.596	6.192	32
33	10.699	9.710	8.880	8.176	7.571	7.047	6.590	6.187	33
34	10.668	9.687	8.863	8.162	7.560	7.039	6.583	6.181	34
35	10.636	9.662	8.844	8.147	7.548	7.029	6.575	6.174	35
36	10.601	9.636	8.823	8.131	7.535	7.018	6.566	6.167	36
37	10.564	9.608	8.801	8.114	7.521	7.007	6.557	6.159	37
38	10.524	9.577	8.777	8.095	7.506	6.995	6.546	6.151	38
39	10.481	9.544	8.752	8.074	7.490	6.981	6.535	6.141	39
40	10.436	9.509	8.724	8.052	7.472	6.966	6.523	6.131	40
41	10.387	9.471	8.694	8.028	7.452	6.950	6.509	6.119	41
42	10.335	9.430	8.661	8.002	7.431	6.932	6.494	6.107	42
43	10.280	9.387	8.627	7.973	7.407	6.913	6.478	6.093	43
44	10.221	9.340	8.589	7.943	7.382	6.892	6.460	6.078	44
45	10.159	9.290	8.549	7.910	7.355	6.869	6.441	6.061	45
46	10.093	9.238	8.506	7.875	7.326	6.845	6.421	6.044	46
47	10.023	9.182	8.461	7.838	7.295	6.819	6.398	6.025	47
48	9.949	9.122	8.412	7.797	7.261	6.790	6.374	6.004	48
49	9.870	9.058	8.360	7.754	7.225	6.760	6.348	5.981	49
50	9.788	8.991	8.304	7.708	7.186	6.727	6.320	5.957	50

YEARS' PURCHASE FOR A SINGLE LIFE

(ENGLISH LIFE MORTALITY TABLES - No 15)

Rate Per Cent

Age	9	10	11	12	13	14	15	16	Age
51	9.700	8.919	8.245	7.658	7.144	6.691	6.289	5.931	51
52	9.607	8.843	8.182	7.605	7.100	6.653	6.256	5.902	52
53	9.510	8.762	8.114	7.549	7.051	6.611	6.220	5.871	53
54	9.406	8.676	8.042	7.487	6.999	6.567	6.182	5.837	54
55	9.297	8.585	7.965	7.422	6.943	6.518	6.139	5.800	55
56	9.182	8.489	7.884	7.352	6.883	6.466	6.094	5.760	56
57	9.062	8.387	7.797	7.279	6.819	6.411	6.045	5.717	57
58	8.936	8.281	7.707	7.201	6.752	6.352	5.994	5.671	58
59	8.805	8.169	7.611	7.118	6.680	6.289	5.938	5.622	59
60	8.669	8.053	7.511	7.032	6.605	6.223	5.880	5.571	60
61	8.528	7.933	7.407	6.941	6.526	6.154	5.819	5.516	61
62	8.382	7.807	7.299	6.847	6.443	6.080	5.754	5.458	62
63	8.230	7.676	7.185	6.747	6.355	6.003	5.685	5.396	63
64	8.073	7.540	7.066	6.643	6.263	5.921	5.612	5.331	64
65	7.913	7.400	6.944	6.535	6.168	5.837	5.536	5.263	65
66	7.747	7.256	6.817	6.424	6.069	5.748	5.457	5.192	66
67	7.575	7.105	6.684	6.306	5.964	5.654	5.373	5.116	67
68	7.398	6.949	6.546	6.183	5.854	5.556	5.284	5.036	68
69	7.215	6.787	6.402	6.055	5.739	5.453	5.191	4.951	69
70	7.026	6.620	6.253	5.921	5.619	5.344	5.093	4.862	70
71	6.830	6.445	6.096	5.780	5.492	5.229	4.988	4.766	71
72	6.627	6.263	5.933	5.633	5.358	5.107	4.877	4.664	72
73	6.424	6.080	5.768	5.483	5.222	4.983	4.763	4.560	73
74	6.220	5.896	5.601	5.332	5.085	4.857	4.648	4.454	74
75	6.010	5.706	5.429	5.174	4.941	4.726	4.527	4.343	75
76	5.793	5.509	5.249	5.010	4.790	4.587	4.398	4.224	76
77	5.572	5.307	5.064	4.840	4.633	4.442	4.264	4.100	77
78	5.350	5.104	4.877	4.668	4.474	4.294	4.127	3.972	78
79	5.127	4.899	4.688	4.493	4.312	4.144	3.988	3.842	79
80	4.906	4.695	4.500	4.318	4.150	3.993	3.846	3.709	80
81	4.687	4.492	4.311	4.143	3.986	3.840	3.703	3.575	81
82	4.468	4.288	4.122	3.966	3.821	3.686	3.558	3.439	82
83	4.249	4.084	3.931	3.788	3.654	3.528	3.410	3.300	83
84	4.030	3.880	3.739	3.608	3.485	3.369	3.260	3.157	84
85	3.814	3.677	3.549	3.429	3.316	3.209	3.109	3.014	85
86	3.606	3.481	3.364	3.254	3.150	3.053	2.961	2.873	86
87	3.408	3.294	3.188	3.087	2.993	2.903	2.818	2.738	87
88	3.214	3.111	3.014	2.923	2.836	2.755	2.677	2.603	88
89	3.020	2.928	2.840	2.757	2.678	2.604	2.533	2.466	89
90	2.835	2.751	2.672	2.596	2.525	2.457	2.393	2.332	90
91	2.659	2.583	2.511	2.443	2.379	2.317	2.258	2.203	91
92	2.492	2.424	2.359	2.297	2.239	2.183	2.129	2.079	92
93	2.336	2.274	2.216	2.160	2.107	2.056	2.007	1.961	93
94	2.187	2.132	2.079	2.029	1.980	1.934	1.890	1.848	94
95	2.049	1.999	1.951	1.905	1.861	1.819	1.779	1.741	95
96	1.924	1.878	1.835	1.793	1.754	1.716	1.679	1.644	96
97	1.809	1.768	1.728	1.691	1.654	1.620	1.586	1.554	97
98	1.694	1.657	1.621	1.587	1.555	1.523	1.493	1.464	98
99	1.578	1.544	1.513	1.482	1.452	1.424	1.397	1.370	99
100	1.468	1.438	1.409	1.382	1.355	1.330	1.305	1.281	100

YEARS' PURCHASE

OR

PRESENT VALUE OF
ONE POUND PER ANNUM

for the

JOINT CONTINUATION OF TWO LIVES

(Males and Females)

according to the

ENGLISH LIFE MORTALITY TABLE No. 15
AT RATES OF INTEREST FROM
5% to 14%

YEARS' PURCHASE FOR THE JOINT CONTINUATION OF TWO LIVES
(ENGLISH LIFE MORTALITY TABLES - No 15)

Ages		5	6	7	8	9	10	12	14	Ages	
						Rate Per Cent						
5	5	18.639	15.924	13.850	12.227	10.929	9.873	8.262	7.098	5	5
5	10	18.461	15.811	13.775	12.176	10.893	9.846	8.247	7.088	5	10
5	15	18.212	15.648	13.665	12.098	10.837	9.803	8.219	7.068	5	15
5	20	17.917	15.459	13.540	12.013	10.778	9.761	8.196	7.054	5	20
5	25	17.541	15.211	13.374	11.900	10.699	9.706	8.167	7.038	5	25
5	30	17.038	14.865	13.133	11.728	10.575	9.614	8.114	7.006	5	30
5	35	16.389	14.401	12.795	11.479	10.387	9.470	8.026	6.948	5	35
5	40	15.581	13.802	12.344	11.134	10.120	9.260	7.892	6.858	5	40
5	45	14.590	13.040	11.750	10.666	9.746	8.958	7.687	6.714	5	45
5	50	13.412	12.103	10.997	10.053	9.242	8.541	7.392	6.498	5	50
5	55	12.061	10.995	10.079	9.286	8.597	7.992	6.987	6.190	5	55
5	60	10.564	9.731	9.003	8.365	7.801	7.301	6.456	5.773	5	60
5	65	9.004	8.380	7.827	7.334	6.893	6.496	5.815	5.252	5	65
5	70	7.474	7.026	6.624	6.259	5.929	5.628	5.103	4.659	5	70
5	75	6.002	5.697	5.418	5.163	4.928	4.711	4.327	3.995	5	75
5	80	4.663	4.465	4.281	4.111	3.952	3.805	3.537	3.303	5	80
5	85	3.510	3.386	3.271	3.162	3.060	2.964	2.788	2.630	5	85
5	90	2.583	2.509	2.439	2.372	2.308	2.248	2.136	2.034	5	90
5	95	1.811	1.768	1.727	1.688	1.651	1.615	1.548	1.486	5	95
5	100	1.270	1.245	1.221	1.198	1.176	1.155	1.114	1.076	5	100
10	10	18.305	15.710	13.708	12.129	10.859	9.821	8.232	7.078	10	10
10	15	18.079	15.560	13.605	12.055	10.805	9.779	8.205	7.059	10	15
10	20	17.808	15.383	13.487	11.975	10.749	9.739	8.182	7.044	10	20
10	25	17.452	15.148	13.328	11.866	10.673	9.685	8.154	7.028	10	25
10	30	16.968	14.814	13.094	11.698	10.551	9.595	8.102	6.997	10	30
10	35	16.333	14.359	12.762	11.452	10.366	9.453	8.014	6.940	10	35
10	40	15.537	13.768	12.316	11.112	10.101	9.245	7.881	6.850	10	40
10	45	14.556	13.012	11.727	10.647	9.730	8.944	7.678	6.706	10	45
10	50	13.385	12.081	10.978	10.037	9.229	8.529	7.383	6.491	10	50
10	55	12.040	10.977	10.064	9.273	8.585	7.982	6.980	6.185	10	55
10	60	10.548	9.717	8.992	8.355	7.792	7.293	6.450	5.768	10	60
10	65	8.993	8.370	7.818	7.326	6.886	6.490	5.810	5.249	10	65
10	70	7.466	7.020	6.618	6.254	5.924	5.624	5.099	4.657	10	70
10	75	5.998	5.693	5.415	5.160	4.925	4.709	4.325	3.994	10	75
10	80	4.661	4.463	4.280	4.109	3.951	3.803	3.536	3.302	10	80
10	85	3.509	3.386	3.270	3.162	3.060	2.963	2.787	2.629	10	85
10	90	2.583	2.509	2.438	2.372	2.308	2.248	2.136	2.034	10	90
10	95	1.811	1.768	1.727	1.688	1.651	1.615	1.548	1.486	10	95
10	100	1.270	1.245	1.221	1.198	1.176	1.155	1.114	1.076	10	100
15	15	17.882	15.426	13.511	11.987	10.754	9.740	8.179	7.040	15	15
15	20	17.640	15.266	13.403	11.912	10.701	9.701	8.157	7.026	15	20
15	25	17.313	15.048	13.255	11.810	10.629	9.650	8.129	7.010	15	25
15	30	16.856	14.731	13.030	11.649	10.511	9.562	8.078	6.979	15	30
15	35	16.245	14.291	12.708	11.409	10.330	9.423	7.993	6.923	15	35
15	40	15.467	13.712	12.271	11.074	10.070	9.218	7.860	6.834	15	40
15	45	14.501	12.967	11.689	10.614	9.702	8.920	7.659	6.691	15	45
15	50	13.341	12.043	10.946	10.009	9.204	8.507	7.366	6.477	15	50
15	55	12.005	10.947	10.037	9.250	8.564	7.963	6.964	6.172	15	55
15	60	10.521	9.693	8.970	8.335	7.775	7.277	6.436	5.756	15	60
15	65	8.972	8.351	7.801	7.311	6.872	6.477	5.799	5.239	15	65

YEARS' PURCHASE FOR THE JOINT CONTINUATION OF TWO LIVES
(ENGLISH LIFE MORTALITY TABLES - No 15)

Ages		5	6	7	8	9	10	12	14	Ages	
15	70	7.450	7.005	6.604	6.242	5.913	5.614	5.090	4.649	15	70
15	75	5.986	5.682	5.405	5.150	4.916	4.701	4.317	3.987	15	75
15	80	4.653	4.456	4.273	4.103	3.945	3.797	3.531	3.297	15	80
15	85	3.504	3.381	3.266	3.157	3.056	2.960	2.784	2.626	15	85
15	90	2.580	2.506	2.436	2.369	2.306	2.246	2.134	2.032	15	90
15	95	1.809	1.766	1.726	1.687	1.649	1.614	1.546	1.484	15	95
15	100	1.269	1.244	1.220	1.197	1.175	1.154	1.113	1.075	15	100
20	20	17.432	15.127	13.307	11.845	10.652	9.665	8.136	7.012	20	20
20	25	17.143	14.931	13.173	11.751	10.586	9.617	8.110	6.997	20	25
20	30	16.722	14.636	12.962	11.599	10.474	9.534	8.061	6.967	20	30
20	35	16.144	14.218	12.654	11.368	10.299	9.399	7.977	6.912	20	35
20	40	15.395	13.657	12.229	11.042	10.044	9.197	7.846	6.823	20	40
20	45	14.450	12.927	11.658	10.589	9.681	8.903	7.646	6.682	20	45
20	50	13.307	12.015	10.923	9.990	9.188	8.494	7.356	6.469	20	50
20	55	11.981	10.927	10.020	9.235	8.552	7.953	6.956	6.165	20	55
20	60	10.505	9.679	8.958	8.324	7.765	7.269	6.429	5.750	20	60
20	65	8.961	8.341	7.792	7.302	6.864	6.470	5.793	5.234	20	65
20	70	7.443	6.998	6.598	6.236	5.907	5.608	5.085	4.644	20	70
20	75	5.981	5.677	5.400	5.146	4.912	4.697	4.314	3.984	20	75
20	80	4.649	4.452	4.269	4.099	3.941	3.794	3.528	3.295	20	80
20	85	3.501	3.378	3.263	3.155	3.053	2.957	2.781	2.624	20	85
20	90	2.578	2.504	2.434	2.367	2.304	2.244	2.132	2.030	20	90
20	95	1.808	1.765	1.724	1.686	1.648	1.613	1.545	1.483	20	95
20	100	1.268	1.243	1.219	1.196	1.174	1.153	1.113	1.075	20	100
25	25	16.897	14.763	13.055	11.668	10.526	9.574	8.086	6.983	25	25
25	30	16.524	14.498	12.864	11.528	10.422	9.495	8.039	6.954	25	30
25	35	15.992	14.109	12.575	11.311	10.256	9.366	7.958	6.900	25	35
25	40	15.285	13.577	12.169	10.997	10.010	9.171	7.830	6.813	25	40
25	45	14.375	12.871	11.615	10.556	9.656	8.883	7.634	6.674	25	45
25	50	13.258	11.978	10.893	9.967	9.170	8.479	7.346	6.462	25	50
25	55	11.952	10.903	10.001	9.220	8.540	7.943	6.949	6.160	25	55
25	60	10.488	9.665	8.946	8.315	7.757	7.262	6.425	5.747	25	60
25	65	8.952	8.334	7.786	7.297	6.860	6.467	5.790	5.232	25	65
25	70	7.438	6.994	6.595	6.233	5.905	5.606	5.084	4.643	25	70
25	75	5.979	5.676	5.398	5.144	4.911	4.696	4.313	3.983	25	75
25	80	4.648	4.451	4.269	4.099	3.941	3.794	3.528	3.294	25	80
25	85	3.501	3.378	3.263	3.155	3.053	2.957	2.781	2.624	25	85
25	90	2.578	2.504	2.434	2.367	2.304	2.244	2.132	2.030	25	90
25	95	1.808	1.765	1.724	1.686	1.648	1.613	1.545	1.483	25	95
25	100	1.268	1.243	1.219	1.196	1.174	1.153	1.113	1.075	25	100
30	30	16.206	14.269	12.696	11.404	10.329	9.424	7.995	6.926	30	30
30	35	15.736	13.921	12.435	11.205	10.175	9.303	7.918	6.875	30	35
30	40	15.088	13.429	12.057	10.910	9.942	9.118	7.796	6.790	30	40
30	45	14.232	12.760	11.529	10.488	9.602	8.840	7.605	6.654	30	45
30	50	13.159	11.900	10.831	9.917	9.129	8.445	7.323	6.446	30	50
30	55	11.887	10.851	9.958	9.185	8.510	7.918	6.931	6.147	30	55
30	60	10.447	9.631	8.918	8.291	7.737	7.245	6.412	5.737	30	60
30	65	8.927	8.313	7.768	7.282	6.846	6.455	5.781	5.225	30	65
30	70	7.424	6.982	6.584	6.223	5.896	5.599	5.078	4.638	30	70
30	75	5.971	5.669	5.392	5.139	4.906	4.691	4.309	3.980	30	75

YEARS' PURCHASE FOR THE JOINT CONTINUATION OF TWO LIVES
(ENGLISH LIFE MORTALITY TABLES - No 15)

Ages		Rate Per Cent								Ages	
		5	6	7	8	9	10	12	14		
30	80	4.644	4.448	4.265	4.096	3.938	3.791	3.526	3.292	30	80
30	85	3.499	3.376	3.261	3.153	3.052	2.956	2.780	2.623	30	85
30	90	2.577	2.503	2.433	2.366	2.303	2.243	2.131	2.030	30	90
30	95	1.807	1.765	1.724	1.685	1.648	1.612	1.545	1.483	30	95
30	100	1.268	1.243	1.219	1.196	1.174	1.153	1.112	1.075	30	100
35	35	15.338	13.621	12.206	11.028	10.037	9.194	7.847	6.826	35	35
35	40	14.767	13.183	11.866	10.760	9.823	9.022	7.732	6.746	35	40
35	45	13.986	12.568	11.377	10.367	9.503	8.759	7.550	6.615	35	45
35	50	12.982	11.757	10.716	9.823	9.051	8.381	7.277	6.412	35	50
35	55	11.765	10.750	9.875	9.115	8.451	7.868	6.894	6.119	35	55
35	60	10.367	9.564	8.861	8.242	7.695	7.208	6.384	5.715	35	60
35	65	8.877	8.270	7.730	7.249	6.817	6.429	5.761	5.208	35	65
35	70	7.394	6.955	6.560	6.202	5.877	5.581	5.063	4.626	35	70
35	75	5.953	5.653	5.378	5.125	4.894	4.680	4.300	3.972	35	75
35	80	4.634	4.438	4.256	4.088	3.931	3.784	3.520	3.287	35	80
35	85	3.493	3.371	3.256	3.148	3.047	2.952	2.776	2.620	35	85
35	90	2.574	2.500	2.430	2.364	2.301	2.241	2.129	2.028	35	90
35	95	1.806	1.763	1.722	1.684	1.646	1.611	1.544	1.482	35	95
35	100	1.267	1.242	1.218	1.195	1.173	1.152	1.112	1.074	35	100
40	40	14.288	12.809	11.571	10.525	9.634	8.868	7.627	6.671	40	40
40	45	13.604	12.265	11.133	10.170	9.342	8.627	7.457	6.547	40	45
40	50	12.693	11.524	10.525	9.665	8.921	8.271	7.198	6.354	40	50
40	55	11.560	10.580	9.733	8.996	8.350	7.782	6.831	6.070	40	55
40	60	10.229	9.447	8.761	8.156	7.621	7.144	6.335	5.677	40	60
40	65	8.789	8.193	7.663	7.190	6.766	6.383	5.725	5.179	40	65
40	70	7.340	6.907	6.517	6.164	5.843	5.550	5.038	4.605	40	70
40	75	5.921	5.624	5.351	5.102	4.872	4.660	4.283	3.958	40	75
40	80	4.616	4.422	4.241	4.073	3.917	3.772	3.509	3.278	40	80
40	85	3.483	3.361	3.247	3.140	3.039	2.944	2.770	2.614	40	85
40	90	2.568	2.495	2.425	2.359	2.296	2.236	2.125	2.024	40	90
40	95	1.803	1.760	1.720	1.681	1.644	1.608	1.542	1.480	40	95
40	100	1.265	1.241	1.217	1.194	1.172	1.151	1.110	1.073	40	100
45	45	13.034	11.804	10.758	9.861	9.086	8.412	7.303	6.433	45	45
45	50	12.243	11.154	10.219	9.410	8.706	8.089	7.064	6.253	45	50
45	55	11.225	10.300	9.497	8.796	8.179	7.635	6.719	5.985	45	55
45	60	9.994	9.246	8.589	8.008	7.491	7.031	6.247	5.607	45	60
45	65	8.633	8.057	7.544	7.085	6.673	6.301	5.659	5.125	45	65
45	70	7.241	6.820	6.439	6.094	5.780	5.493	4.991	4.566	45	70
45	75	5.862	5.570	5.303	5.057	4.831	4.622	4.251	3.930	45	75
45	80	4.582	4.390	4.212	4.046	3.892	3.749	3.488	3.260	45	80
45	85	3.464	3.344	3.231	3.124	3.024	2.930	2.757	2.602	45	85
45	90	2.558	2.485	2.416	2.350	2.288	2.228	2.118	2.017	45	90
45	95	1.797	1.755	1.715	1.676	1.639	1.604	1.537	1.476	45	95
45	100	1.262	1.238	1.214	1.191	1.169	1.148	1.108	1.070	45	100
50	50	11.591	10.611	9.763	9.024	8.377	7.807	6.852	6.089	50	50
50	55	10.717	9.871	9.131	8.482	7.908	7.399	6.539	5.843	50	55
50	60	9.623	8.927	8.312	7.766	7.280	6.845	6.100	5.490	50	60
50	65	8.377	7.833	7.347	6.910	6.517	6.162	5.546	5.033	50	65
50	70	7.074	6.670	6.305	5.973	5.671	5.394	4.909	4.497	50	70
50	75	5.759	5.476	5.217	4.978	4.759	4.556	4.194	3.881	50	75

YEARS' PURCHASE FOR THE JOINT CONTINUATION OF TWO LIVES
(ENGLISH LIFE MORTALITY TABLES - No 15)

Rate Per Cent

Ages		5	6	7	8	9	10	12	14	Ages	
50	80	4.521	4.334	4.159	3.998	3.847	3.706	3.451	3.227	50	80
50	85	3.429	3.311	3.200	3.095	2.997	2.904	2.734	2.581	50	85
50	90	2.538	2.466	2.398	2.333	2.271	2.213	2.104	2.004	50	90
50	95	1.787	1.745	1.705	1.667	1.630	1.595	1.529	1.468	50	95
50	100	1.256	1.232	1.209	1.186	1.164	1.143	1.103	1.066	50	100
55	55	10.007	9.262	8.606	8.026	7.510	7.049	6.265	5.625	55	55
55	60	9.080	8.454	7.899	7.403	6.959	6.560	5.873	5.305	55	60
55	65	7.986	7.488	7.040	6.637	6.273	5.942	5.367	4.884	55	65
55	70	6.809	6.433	6.091	5.779	5.495	5.234	4.775	4.383	55	70
55	75	5.590	5.322	5.076	4.849	4.640	4.446	4.100	3.800	55	75
55	80	4.419	4.239	4.072	3.916	3.771	3.635	3.389	3.171	55	80
55	85	3.371	3.256	3.148	3.046	2.951	2.860	2.694	2.546	55	85
55	90	2.505	2.435	2.368	2.305	2.244	2.187	2.080	1.982	55	90
55	95	1.769	1.728	1.688	1.651	1.615	1.580	1.515	1.455	55	95
55	100	1.247	1.222	1.199	1.177	1.155	1.135	1.095	1.058	55	100
60	60	8.338	7.801	7.321	6.889	6.500	6.148	5.538	5.028	60	60
60	65	7.428	6.989	6.594	6.236	5.910	5.614	5.094	4.655	60	65
60	70	6.413	6.074	5.765	5.483	5.224	4.986	4.564	4.203	60	70
60	75	5.327	5.080	4.853	4.644	4.450	4.270	3.947	3.666	60	75
60	80	4.254	4.085	3.929	3.782	3.645	3.517	3.284	3.079	60	80
60	85	3.272	3.163	3.060	2.963	2.872	2.786	2.627	2.485	60	85
60	90	2.449	2.381	2.316	2.255	2.197	2.141	2.038	1.944	60	90
60	95	1.738	1.698	1.659	1.623	1.588	1.554	1.491	1.432	60	95
60	100	1.229	1.206	1.183	1.161	1.140	1.119	1.081	1.045	60	100
65	65	6.712	6.345	6.011	5.707	5.428	5.173	4.723	4.339	65	65
65	70	5.883	5.591	5.323	5.078	4.851	4.642	4.270	3.948	65	70
65	75	4.959	4.741	4.539	4.353	4.179	4.018	3.727	3.472	65	75
65	80	4.014	3.861	3.719	3.585	3.460	3.343	3.129	2.939	65	80
65	85	3.123	3.022	2.927	2.837	2.752	2.672	2.524	2.390	65	85
65	90	2.360	2.296	2.235	2.178	2.122	2.070	1.972	1.882	65	90
65	95	1.687	1.649	1.613	1.578	1.544	1.512	1.451	1.395	65	95
65	100	1.200	1.177	1.155	1.134	1.114	1.094	1.057	1.022	65	100
70	70	5.245	5.005	4.784	4.580	4.391	4.216	3.900	3.625	70	70
70	75	4.500	4.315	4.144	3.984	3.835	3.696	3.444	3.221	70	75
70	80	3.705	3.572	3.447	3.329	3.219	3.115	2.925	2.756	70	80
70	85	2.927	2.837	2.751	2.670	2.593	2.520	2.386	2.264	70	85
70	90	2.241	2.182	2.126	2.073	2.022	1.974	1.883	1.800	70	90
70	95	1.619	1.583	1.549	1.516	1.484	1.454	1.397	1.344	70	95
70	100	1.161	1.139	1.118	1.098	1.078	1.060	1.024	0.991	70	100
75	75	3.938	3.791	3.653	3.524	3.403	3.289	3.082	2.897	75	75
75	80	3.309	3.198	3.094	2.996	2.904	2.817	2.656	2.511	75	80
75	85	2.665	2.587	2.514	2.444	2.377	2.314	2.197	2.090	75	85
75	90	2.076	2.024	1.974	1.927	1.882	1.838	1.757	1.683	75	90
75	95	1.521	1.489	1.458	1.428	1.399	1.371	1.319	1.271	75	95
75	100	1.103	1.083	1.063	1.045	1.027	1.009	0.976	0.945	75	100
80	80	2.842	2.756	2.675	2.598	2.524	2.455	2.326	2.210	80	80
80	85	2.341	2.278	2.218	2.161	2.106	2.054	1.957	1.869	80	85
80	90	1.862	1.819	1.777	1.736	1.698	1.661	1.592	1.528	80	90
80	95	1.390	1.362	1.334	1.308	1.283	1.259	1.213	1.170	80	95
80	100	1.022	1.005	0.987	0.970	0.954	0.939	0.909	0.880	80	100

YEARS' PURCHASE FOR THE JOINT CONTINUATION OF TWO LIVES
(ENGLISH LIFE MORTALITY TABLES - No 15)

Rate Per Cent

Ages		5	6	7	8	9	10	12	14	Ages	
5	5	18.802	16.029	13.920	12.275	10.964	9.898	8.277	7.108	5	5
5	10	18.604	15.906	13.840	12.221	10.926	9.871	8.261	7.097	5	10
5	15	18.335	15.732	13.724	12.140	10.868	9.827	8.234	7.078	5	15
5	20	18.022	15.532	13.592	12.052	10.807	9.783	8.210	7.063	5	20
5	25	17.628	15.274	13.421	11.936	10.726	9.727	8.180	7.047	5	25
5	30	17.111	14.920	13.174	11.760	10.599	9.633	8.127	7.014	5	30
5	35	16.449	14.447	12.830	11.506	10.409	9.488	8.038	6.957	5	35
5	40	15.630	13.840	12.374	11.158	10.139	9.276	7.902	6.865	5	40
5	45	14.629	13.071	11.775	10.686	9.762	8.971	7.697	6.721	5	45
5	50	13.442	12.127	11.016	10.069	9.256	8.552	7.400	6.503	5	50
5	55	12.082	11.012	10.093	9.299	8.607	8.001	6.994	6.195	5	55
5	60	10.578	9.743	9.013	8.373	7.809	7.307	6.461	5.776	5	60
5	65	9.013	8.388	7.834	7.340	6.898	6.501	5.818	5.255	5	65
5	70	7.479	7.031	6.628	6.263	5.932	5.631	5.105	4.661	5	70
5	75	6.005	5.700	5.420	5.165	4.930	4.713	4.328	3.997	5	75
5	80	4.665	4.466	4.283	4.112	3.953	3.806	3.538	3.304	5	80
5	85	3.511	3.387	3.272	3.163	3.061	2.965	2.788	2.630	5	85
5	90	2.584	2.509	2.439	2.372	2.309	2.248	2.136	2.034	5	90
5	95	1.811	1.768	1.727	1.688	1.651	1.615	1.548	1.486	5	95
5	100	1.270	1.245	1.221	1.198	1.176	1.155	1.114	1.076	5	100
10	10	18.508	15.846	13.802	12.196	10.909	9.859	8.255	7.093	10	10
10	15	18.257	15.682	13.691	12.118	10.852	9.816	8.228	7.074	10	15
10	20	17.960	15.491	13.565	12.033	10.793	9.773	8.204	7.059	10	20
10	25	17.581	15.241	13.398	11.919	10.714	9.718	8.175	7.043	10	25
10	30	17.075	14.894	13.155	11.746	10.589	9.625	8.122	7.011	10	30
10	35	16.422	14.427	12.815	11.495	10.400	9.481	8.034	6.953	10	35
10	40	15.610	13.825	12.362	11.149	10.132	9.270	7.898	6.862	10	40
10	45	14.615	13.060	11.766	10.678	9.756	8.967	7.693	6.718	10	45
10	50	13.431	12.119	11.009	10.063	9.251	8.548	7.397	6.501	10	50
10	55	12.074	11.006	10.088	9.294	8.603	7.998	6.991	6.193	10	55
10	60	10.573	9.738	9.010	8.370	7.806	7.305	6.459	5.775	10	60
10	65	9.009	8.385	7.831	7.337	6.896	6.499	5.817	5.254	10	65
10	70	7.477	7.029	6.626	6.261	5.931	5.630	5.104	4.660	10	70
10	75	6.004	5.698	5.419	5.164	4.929	4.712	4.328	3.996	10	75
10	80	4.664	4.466	4.282	4.112	3.953	3.805	3.538	3.303	10	80
10	85	3.510	3.387	3.271	3.163	3.061	2.965	2.788	2.630	10	85
10	90	2.583	2.509	2.439	2.372	2.309	2.248	2.136	2.034	10	90
10	95	1.811	1.768	1.727	1.688	1.651	1.615	1.548	1.486	10	95
10	100	1.270	1.245	1.221	1.198	1.176	1.155	1.114	1.076	10	100
15	15	18.135	15.603	13.638	12.081	10.826	9.796	8.216	7.066	15	15
15	20	17.860	15.424	13.519	12.000	10.769	9.756	8.193	7.052	15	20
15	25	17.502	15.187	13.359	11.891	10.693	9.702	8.165	7.036	15	25
15	30	17.014	14.851	13.123	11.722	10.570	9.611	8.113	7.004	15	30
15	35	16.376	14.393	12.789	11.475	10.384	9.468	8.025	6.947	15	35
15	40	15.576	13.799	12.341	11.132	10.118	9.259	7.891	6.857	15	40
15	45	14.589	13.040	11.750	10.665	9.745	8.957	7.686	6.713	15	45
15	50	13.413	12.103	10.996	10.053	9.242	8.540	7.391	6.496	15	50
15	55	12.061	10.995	10.078	9.286	8.596	7.991	6.986	6.189	15	55
15	60	10.563	9.730	9.002	8.364	7.800	7.300	6.455	5.771	15	60
15	65	9.002	8.379	7.825	7.332	6.891	6.495	5.813	5.251	15	65

YEARS' PURCHASE FOR THE JOINT CONTINUATION OF TWO LIVES
(ENGLISH LIFE MORTALITY TABLES - No 15)

Ages		5	6	7	8	9	10	12	14	Ages	
						Rate Per Cent					
15	70	7.472	7.025	6.622	6.258	5.927	5.627	5.101	4.658	15	70
15	75	6.000	5.695	5.417	5.161	4.926	4.710	4.325	3.994	15	75
15	80	4.662	4.464	4.280	4.110	3.951	3.803	3.536	3.302	15	80
15	85	3.509	3.386	3.270	3.162	3.060	2.963	2.787	2.629	15	85
15	90	2.583	2.508	2.438	2.371	2.308	2.248	2.136	2.034	15	90
15	95	1.810	1.768	1.727	1.688	1.651	1.615	1.548	1.485	15	95
15	100	1.269	1.245	1.221	1.198	1.176	1.154	1.114	1.076	15	100
20	20	17.718	15.330	13.455	11.956	10.738	9.733	8.181	7.045	20	20
20	25	17.388	15.109	13.305	11.853	10.666	9.681	8.153	7.029	20	25
20	30	16.926	14.789	13.079	11.691	10.547	9.593	8.102	6.998	20	30
20	35	16.311	14.346	12.755	11.449	10.365	9.453	8.016	6.941	20	35
20	40	15.529	13.764	12.315	11.112	10.103	9.247	7.883	6.851	20	40
20	45	14.557	13.015	11.730	10.650	9.733	8.948	7.680	6.708	20	45
20	50	13.391	12.086	10.983	10.042	9.233	8.532	7.386	6.493	20	50
20	55	12.047	10.983	10.069	9.278	8.589	7.986	6.982	6.186	20	55
20	60	10.554	9.722	8.996	8.358	7.795	7.296	6.452	5.769	20	60
20	65	8.997	8.374	7.821	7.329	6.888	6.492	5.811	5.249	20	65
20	70	7.469	7.022	6.619	6.255	5.925	5.625	5.100	4.657	20	70
20	75	5.998	5.694	5.415	5.160	4.925	4.709	4.324	3.993	20	75
20	80	4.661	4.463	4.279	4.109	3.950	3.803	3.536	3.301	20	80
20	85	3.508	3.385	3.269	3.161	3.059	2.963	2.786	2.629	20	85
20	90	2.582	2.508	2.438	2.371	2.308	2.247	2.135	2.033	20	90
20	95	1.810	1.767	1.727	1.688	1.650	1.615	1.547	1.485	20	95
20	100	1.269	1.245	1.221	1.198	1.176	1.154	1.114	1.076	20	100
25	25	17.209	14.986	13.219	11.791	10.621	9.648	8.134	7.017	25	25
25	30	16.784	14.689	13.007	11.638	10.508	9.564	8.084	6.986	25	30
25	35	16.202	14.267	12.697	11.406	10.332	9.428	8.000	6.931	25	35
25	40	15.449	13.704	12.270	11.078	10.076	9.226	7.869	6.842	25	40
25	45	14.500	12.971	11.696	10.623	9.711	8.930	7.669	6.700	25	45
25	50	13.352	12.055	10.958	10.022	9.217	8.519	7.377	6.486	25	50
25	55	12.021	10.962	10.052	9.264	8.577	7.976	6.975	6.181	25	55
25	60	10.538	9.709	8.985	8.349	7.787	7.289	6.446	5.765	25	60
25	65	8.987	8.366	7.814	7.323	6.883	6.487	5.808	5.246	25	65
25	70	7.463	7.017	6.615	6.252	5.922	5.622	5.097	4.655	25	70
25	75	5.995	5.691	5.413	5.157	4.923	4.707	4.323	3.992	25	75
25	80	4.659	4.461	4.278	4.108	3.949	3.802	3.535	3.301	25	80
25	85	3.507	3.384	3.269	3.160	3.058	2.962	2.786	2.628	25	85
25	90	2.582	2.508	2.437	2.371	2.307	2.247	2.135	2.033	25	90
25	95	1.810	1.767	1.727	1.688	1.650	1.614	1.547	1.485	25	95
25	100	1.269	1.244	1.221	1.198	1.176	1.154	1.114	1.076	25	100
30	30	16.558	14.526	12.888	11.549	10.441	9.512	8.052	6.965	30	30
30	35	16.023	14.135	12.598	11.331	10.274	9.382	7.971	6.911	30	35
30	40	15.313	13.601	12.191	11.016	10.027	9.187	7.843	6.824	30	40
30	45	14.400	12.893	11.635	10.574	9.672	8.898	7.646	6.685	30	45
30	50	13.281	11.999	10.912	9.984	9.186	8.493	7.358	6.473	30	50
30	55	11.973	10.922	10.019	9.236	8.554	7.956	6.960	6.170	30	55
30	60	10.506	9.682	8.962	8.329	7.770	7.274	6.435	5.756	30	60
30	65	8.967	8.348	7.799	7.309	6.871	6.477	5.799	5.240	30	65
30	70	7.451	7.006	6.605	6.243	5.914	5.615	5.091	4.650	30	70
30	75	5.988	5.684	5.407	5.152	4.918	4.702	4.319	3.989	30	75

YEARS' PURCHASE FOR THE JOINT CONTINUATION OF TWO LIVES
(ENGLISH LIFE MORTALITY TABLES - No 15)

						Rate Per Cent						
Ages		5	6	7	8	9	10	12	14	Ages	
30	80	4.655	4.457	4.274	4.104	3.946	3.799	3.532	3.298	30	80
30	85	3.505	3.382	3.267	3.158	3.057	2.961	2.785	2.627	30	85
30	90	2.580	2.506	2.436	2.370	2.306	2.246	2.134	2.032	30	90
30	95	1.809	1.767	1.726	1.687	1.650	1.614	1.547	1.485	30	95
30	100	1.269	1.244	1.220	1.197	1.175	1.154	1.113	1.076	30	100
35	35	15.748	13.930	12.442	11.211	10.180	9.308	7.922	6.878	35	35
35	40	15.097	13.436	12.063	10.915	9.947	9.122	7.800	6.794	35	40
35	45	14.237	12.765	11.534	10.493	9.606	8.844	7.609	6.657	35	45
35	50	13.164	11.904	10.835	9.921	9.133	8.449	7.327	6.449	35	50
35	55	11.892	10.855	9.963	9.189	8.514	7.922	6.935	6.150	35	55
35	60	10.452	9.636	8.922	8.295	7.741	7.249	6.415	5.740	35	60
35	65	8.932	8.318	7.772	7.286	6.850	6.459	5.785	5.228	35	65
35	70	7.429	6.986	6.588	6.227	5.900	5.602	5.081	4.641	35	70
35	75	5.975	5.672	5.396	5.142	4.909	4.694	4.312	3.983	35	75
35	80	4.647	4.450	4.268	4.098	3.941	3.793	3.528	3.294	35	80
35	85	3.501	3.378	3.263	3.155	3.053	2.957	2.782	2.624	35	85
35	90	2.578	2.504	2.434	2.367	2.304	2.244	2.132	2.031	35	90
35	95	1.808	1.765	1.725	1.686	1.649	1.613	1.546	1.484	35	95
35	100	1.268	1.243	1.220	1.197	1.175	1.153	1.113	1.075	35	100
40	40	14.769	13.183	11.865	10.759	9.822	9.021	7.731	6.745	40	40
40	45	13.984	12.566	11.374	10.364	9.501	8.757	7.548	6.613	40	45
40	50	12.977	11.753	10.712	9.819	9.048	8.378	7.275	6.411	40	50
40	55	11.760	10.746	9.871	9.112	8.449	7.866	6.893	6.118	40	55
40	60	10.363	9.560	8.858	8.240	7.693	7.206	6.383	5.715	40	60
40	65	8.875	8.268	7.728	7.247	6.816	6.428	5.760	5.208	40	65
40	70	7.393	6.954	6.559	6.201	5.877	5.581	5.064	4.627	40	70
40	75	5.953	5.653	5.378	5.126	4.894	4.680	4.300	3.973	40	75
40	80	4.635	4.439	4.257	4.088	3.931	3.785	3.520	3.288	40	80
40	85	3.494	3.371	3.257	3.149	3.048	2.952	2.777	2.620	40	85
40	90	2.574	2.500	2.430	2.364	2.301	2.241	2.130	2.028	40	90
40	95	1.806	1.763	1.723	1.684	1.647	1.611	1.544	1.482	40	95
40	100	1.267	1.242	1.219	1.196	1.174	1.152	1.112	1.074	40	100
45	45	13.600	12.259	11.126	10.162	9.334	8.619	7.450	6.541	45	45
45	50	12.684	11.514	10.515	9.656	8.912	8.263	7.191	6.347	45	50
45	55	11.547	10.569	9.722	8.986	8.341	7.773	6.823	6.064	45	55
45	60	10.216	9.435	8.750	8.147	7.612	7.136	6.328	5.671	45	60
45	65	8.778	8.183	7.654	7.182	6.758	6.377	5.719	5.174	45	65
45	70	7.331	6.900	6.510	6.157	5.837	5.545	5.034	4.601	45	70
45	75	5.916	5.619	5.347	5.097	4.868	4.656	4.280	3.955	45	75
45	80	4.613	4.418	4.238	4.071	3.915	3.770	3.507	3.276	45	80
45	85	3.481	3.360	3.246	3.138	3.038	2.943	2.768	2.612	45	85
45	90	2.567	2.494	2.424	2.358	2.295	2.235	2.124	2.023	45	90
45	95	1.802	1.760	1.719	1.681	1.643	1.608	1.541	1.479	45	95
45	100	1.265	1.240	1.217	1.194	1.172	1.150	1.110	1.072	45	100
50	50	12.250	11.158	10.220	9.409	8.703	8.085	7.060	6.248	50	50
50	55	11.224	10.298	9.493	8.791	8.174	7.629	6.714	5.979	50	55
50	60	9.988	9.240	8.582	8.001	7.485	7.024	6.241	5.602	50	60
50	65	8.624	8.049	7.537	7.078	6.666	6.294	5.653	5.120	50	65
50	70	7.233	6.812	6.432	6.087	5.773	5.487	4.986	4.561	50	70
50	75	5.856	5.564	5.297	5.052	4.826	4.618	4.247	3.927	50	75

YEARS' PURCHASE FOR THE JOINT CONTINUATION OF TWO LIVES
(ENGLISH LIFE MORTALITY TABLES - No 15)

						Rate Per Cent					
Ages		5	6	7	8	9	10	12	14	Ages	
50	80	4.577	4.386	4.208	4.043	3.889	3.745	3.485	3.257	50	80
50	85	3.461	3.341	3.228	3.122	3.022	2.928	2.755	2.600	50	85
50	90	2.556	2.483	2.414	2.348	2.286	2.227	2.116	2.016	50	90
50	95	1.796	1.754	1.714	1.675	1.638	1.603	1.536	1.475	50	95
50	100	1.262	1.237	1.213	1.191	1.169	1.148	1.107	1.070	50	100
55	55	10.743	9.891	9.147	8.493	7.917	7.406	6.542	5.844	55	55
55	60	9.637	8.937	8.320	7.772	7.284	6.847	6.101	5.489	55	60
55	65	8.382	7.836	7.348	6.911	6.517	6.161	5.545	5.031	55	65
55	70	7.073	6.669	6.303	5.971	5.669	5.392	4.906	4.494	55	70
55	75	5.756	5.473	5.214	4.976	4.756	4.553	4.192	3.879	55	75
55	80	4.518	4.331	4.157	3.995	3.845	3.704	3.449	3.225	55	80
55	85	3.427	3.309	3.198	3.094	2.995	2.903	2.732	2.580	55	85
55	90	2.537	2.465	2.397	2.332	2.270	2.212	2.103	2.003	55	90
55	95	1.786	1.744	1.704	1.666	1.630	1.595	1.529	1.468	55	95
55	100	1.256	1.232	1.208	1.186	1.164	1.143	1.103	1.066	55	100
60	60	9.125	8.493	7.931	7.430	6.982	6.579	5.886	5.314	60	60
60	65	8.015	7.512	7.061	6.654	6.287	5.954	5.375	4.890	60	65
60	70	6.825	6.445	6.101	5.788	5.502	5.240	4.779	4.386	60	70
60	75	5.596	5.327	5.080	4.852	4.642	4.448	4.101	3.800	60	75
60	80	4.420	4.240	4.072	3.916	3.770	3.634	3.388	3.170	60	80
60	85	3.369	3.255	3.147	3.045	2.949	2.859	2.693	2.544	60	85
60	90	2.504	2.434	2.367	2.303	2.243	2.185	2.078	1.981	60	90
60	95	1.768	1.727	1.687	1.650	1.614	1.579	1.514	1.454	60	95
60	100	1.246	1.222	1.199	1.176	1.155	1.134	1.094	1.058	60	100
65	65	7.509	7.062	6.659	6.294	5.963	5.661	5.133	4.687	65	65
65	70	6.472	6.127	5.813	5.526	5.263	5.021	4.594	4.228	65	70
65	75	5.365	5.115	4.885	4.672	4.476	4.294	3.968	3.684	65	75
65	80	4.276	4.106	3.947	3.799	3.661	3.532	3.297	3.089	65	80
65	85	3.284	3.173	3.070	2.973	2.881	2.794	2.634	2.491	65	85
65	90	2.454	2.386	2.321	2.260	2.201	2.145	2.042	1.947	65	90
65	95	1.740	1.700	1.662	1.625	1.590	1.556	1.493	1.434	65	95
65	100	1.230	1.206	1.184	1.162	1.141	1.120	1.081	1.045	65	100
70	70	5.982	5.683	5.408	5.157	4.925	4.711	4.330	4.001	70	70
70	75	5.034	4.811	4.605	4.414	4.236	4.071	3.774	3.515	70	75
70	80	4.066	3.910	3.765	3.629	3.501	3.382	3.164	2.971	70	80
70	85	3.157	3.055	2.958	2.866	2.780	2.698	2.548	2.412	70	85
70	90	2.381	2.316	2.255	2.196	2.140	2.087	1.988	1.897	70	90
70	95	1.699	1.661	1.624	1.589	1.555	1.522	1.461	1.404	70	95
70	100	1.207	1.184	1.162	1.141	1.120	1.100	1.063	1.027	70	100
75	75	4.564	4.376	4.201	4.039	3.887	3.746	3.489	3.263	75	75
75	80	3.755	3.619	3.492	3.372	3.260	3.155	2.961	2.789	75	80
75	85	2.963	2.871	2.783	2.701	2.623	2.549	2.412	2.289	75	85
75	90	2.265	2.205	2.148	2.094	2.043	1.993	1.901	1.817	75	90
75	95	1.633	1.597	1.563	1.529	1.497	1.467	1.409	1.355	75	95
75	100	1.169	1.148	1.127	1.106	1.087	1.068	1.032	0.998	75	100
80	80	3.319	3.209	3.105	3.007	2.915	2.827	2.666	2.522	80	80
80	85	2.676	2.598	2.524	2.454	2.388	2.325	2.207	2.100	80	85
80	90	2.086	2.033	1.984	1.936	1.891	1.847	1.766	1.691	80	90
80	95	1.528	1.496	1.464	1.434	1.405	1.378	1.325	1.276	80	95
80	100	1.107	1.087	1.068	1.049	1.031	1.013	0.980	0.949	80	100

YEARS' PURCHASE FOR THE JOINT CONTINUATION OF TWO LIVES
(ENGLISH LIFE MORTALITY TABLES - No 15)

		Rate Per Cent									
Ages		5	6	7	8	9	10	12	14	Ages	
5	5	18.802	16.029	13.920	12.275	10.964	9.898	8.277	7.108	5	5
5	10	18.688	15.960	13.877	12.247	10.945	9.885	8.271	7.104	5	10
5	15	18.521	15.856	13.810	12.203	10.915	9.864	8.258	7.096	5	15
5	20	18.298	15.717	13.721	12.144	10.875	9.836	8.244	7.088	5	20
5	25	17.994	15.520	13.590	12.056	10.814	9.793	8.221	7.074	5	25
5	30	17.587	15.247	13.403	11.925	10.720	9.724	8.181	7.049	5	30
5	35	17.063	14.883	13.146	11.738	10.583	9.620	8.119	7.010	5	35
5	40	16.407	14.412	12.801	11.482	10.388	9.471	8.026	6.948	5	40
5	45	15.597	13.810	12.347	11.134	10.117	9.256	7.886	6.852	5	45
5	50	14.625	13.064	11.766	10.676	9.752	8.961	7.687	6.711	5	50
5	55	13.476	12.154	11.036	10.084	9.267	8.560	7.404	6.504	5	55
5	60	12.157	11.076	10.147	9.344	8.645	8.033	7.016	6.211	5	60
5	65	10.712	9.863	9.121	8.470	7.895	7.385	6.523	5.827	5	65
5	70	9.166	8.530	7.967	7.464	7.013	6.608	5.912	5.337	5	70
5	75	7.549	7.102	6.699	6.333	6.001	5.698	5.168	4.720	5	75
5	80	5.940	5.646	5.377	5.129	4.901	4.690	4.314	3.989	5	80
5	85	4.463	4.282	4.114	3.957	3.811	3.674	3.426	3.207	5	85
5	90	3.221	3.115	3.015	2.921	2.833	2.749	2.595	2.456	5	90
5	95	2.274	2.213	2.155	2.100	2.048	1.998	1.904	1.818	5	95
5	100	1.599	1.564	1.530	1.498	1.467	1.437	1.381	1.329	5	100
10	10	18.508	15.846	13.802	12.196	10.909	9.859	8.255	7.093	10	10
10	15	18.363	15.754	13.742	12.155	10.881	9.838	8.243	7.086	10	15
10	20	18.164	15.627	13.659	12.101	10.843	9.812	8.229	7.078	10	20
10	25	17.883	15.444	13.537	12.017	10.785	9.770	8.207	7.064	10	25
10	30	17.498	15.184	13.357	11.890	10.693	9.703	8.168	7.040	10	30
10	35	16.992	14.831	13.107	11.708	10.559	9.601	8.106	7.000	10	35
10	40	16.351	14.370	12.768	11.456	10.367	9.454	8.014	6.939	10	40
10	45	15.553	13.776	12.319	11.111	10.098	9.241	7.875	6.844	10	45
10	50	14.590	13.036	11.743	10.657	9.736	8.947	7.677	6.704	10	50
10	55	13.449	12.131	11.017	10.068	9.253	8.548	7.395	6.497	10	55
10	60	12.136	11.058	10.132	9.331	8.634	8.023	7.009	6.205	10	60
10	65	10.696	9.849	9.109	8.459	7.886	7.377	6.517	5.822	10	65
10	70	9.154	8.520	7.958	7.456	7.006	6.602	5.907	5.333	10	70
10	75	7.542	7.096	6.693	6.328	5.996	5.694	5.165	4.717	10	75
10	80	5.936	5.643	5.374	5.126	4.898	4.688	4.312	3.987	10	80
10	85	4.461	4.281	4.112	3.956	3.810	3.673	3.425	3.206	10	85
10	90	3.220	3.114	3.015	2.921	2.832	2.749	2.594	2.455	10	90
10	95	2.274	2.213	2.155	2.100	2.047	1.997	1.904	1.818	10	95
10	100	1.599	1.564	1.530	1.498	1.467	1.437	1.381	1.329	10	100
15	15	18.135	15.603	13.638	12.081	10.826	9.796	8.216	7.066	15	15
15	20	17.963	15.491	13.564	12.032	10.792	9.772	8.203	7.059	15	20
15	25	17.713	15.326	13.452	11.954	10.737	9.732	8.181	7.046	15	25
15	30	17.357	15.083	13.283	11.834	10.649	9.668	8.143	7.022	15	30
15	35	16.880	14.748	13.043	11.659	10.519	9.569	8.083	6.983	15	35
15	40	16.262	14.301	12.714	11.413	10.332	9.424	7.992	6.922	15	40
15	45	15.483	13.720	12.273	11.073	10.067	9.214	7.855	6.828	15	45
15	50	14.535	12.991	11.705	10.624	9.708	8.923	7.658	6.688	15	50
15	55	13.405	12.094	10.985	10.040	9.229	8.526	7.377	6.483	15	55
15	60	12.101	11.028	10.105	9.307	8.612	8.004	6.993	6.192	15	60
15	65	10.668	9.824	9.087	8.439	7.868	7.360	6.503	5.811	15	65

YEARS' PURCHASE FOR THE JOINT CONTINUATION OF TWO LIVES
(ENGLISH LIFE MORTALITY TABLES - No 15)

Ages		5	6	7	8	9	10	12	14	Ages	
15	70	9.133	8.501	7.940	7.440	6.992	6.589	5.895	5.323	15	70
15	75	7.526	7.081	6.679	6.315	5.985	5.683	5.155	4.709	15	75
15	80	5.925	5.632	5.364	5.117	4.890	4.680	4.305	3.981	15	80
15	85	4.454	4.274	4.106	3.950	3.804	3.668	3.420	3.201	15	85
15	90	3.216	3.110	3.011	2.917	2.829	2.745	2.591	2.453	15	90
15	95	2.271	2.211	2.153	2.098	2.045	1.995	1.902	1.817	15	95
15	100	1.598	1.563	1.529	1.497	1.466	1.436	1.380	1.328	15	100
20	20	17.718	15.330	13.455	11.956	10.738	9.733	8.181	7.045	20	20
20	25	17.502	15.184	13.355	11.886	10.688	9.696	8.160	7.032	20	25
20	30	17.185	14.965	13.200	11.774	10.606	9.635	8.123	7.009	20	30
20	35	16.744	14.653	12.974	11.608	10.481	9.540	8.065	6.971	20	35
20	40	16.161	14.227	12.660	11.371	10.300	9.399	7.976	6.911	20	40
20	45	15.410	13.665	12.231	11.041	10.041	9.193	7.841	6.818	20	45
20	50	14.484	12.951	11.674	10.599	9.687	8.906	7.645	6.679	20	50
20	55	13.370	12.066	10.962	10.021	9.213	8.512	7.367	6.475	20	55
20	60	12.078	11.008	10.088	9.293	8.600	7.993	6.984	6.185	20	60
20	65	10.652	9.810	9.075	8.428	7.858	7.352	6.496	5.804	20	65
20	70	9.122	8.491	7.931	7.432	6.984	6.582	5.889	5.318	20	70
20	75	7.518	7.074	6.673	6.309	5.979	5.678	5.151	4.705	20	75
20	80	5.920	5.627	5.359	5.113	4.885	4.676	4.301	3.978	20	80
20	85	4.450	4.270	4.103	3.947	3.801	3.665	3.417	3.199	20	85
20	90	3.213	3.108	3.008	2.915	2.827	2.743	2.589	2.451	20	90
20	95	2.270	2.209	2.151	2.096	2.044	1.994	1.901	1.815	20	95
20	100	1.597	1.562	1.528	1.496	1.465	1.435	1.379	1.327	20	100
25	25	17.209	14.986	13.219	11.791	10.621	9.648	8.134	7.017	25	25
25	30	16.936	14.794	13.081	11.690	10.545	9.591	8.099	6.995	25	30
25	35	16.544	14.512	12.875	11.537	10.429	9.501	8.043	6.958	25	35
25	40	16.007	14.118	12.580	11.313	10.257	9.366	7.957	6.899	25	40
25	45	15.299	13.583	12.171	10.995	10.007	9.167	7.825	6.808	25	45
25	50	14.408	12.894	11.630	10.566	9.661	8.886	7.633	6.671	25	50
25	55	13.322	12.028	10.933	9.998	9.194	8.498	7.357	6.469	25	55
25	60	12.048	10.984	10.069	9.277	8.587	7.983	6.978	6.180	25	60
25	65	10.636	9.796	9.063	8.419	7.850	7.345	6.492	5.801	25	65
25	70	9.113	8.484	7.925	7.426	6.980	6.578	5.887	5.316	25	70
25	75	7.514	7.070	6.670	6.307	5.977	5.676	5.149	4.704	25	75
25	80	5.918	5.626	5.358	5.111	4.884	4.675	4.301	3.977	25	80
25	85	4.450	4.270	4.102	3.946	3.801	3.664	3.417	3.199	25	85
25	90	3.213	3.108	3.008	2.915	2.827	2.743	2.589	2.451	25	90
25	95	2.270	2.209	2.151	2.096	2.044	1.994	1.901	1.815	25	95
25	100	1.597	1.562	1.528	1.496	1.465	1.435	1.379	1.327	25	100
30	30	16.558	14.526	12.888	11.549	10.441	9.512	8.052	6.965	30	30
30	35	16.223	14.281	12.706	11.411	10.335	9.429	8.000	6.930	30	35
30	40	15.747	13.927	12.438	11.206	10.175	9.303	7.918	6.874	30	40
30	45	15.100	13.434	12.057	10.908	9.938	9.113	7.790	6.785	30	45
30	50	14.263	12.782	11.543	10.497	9.607	8.842	7.604	6.651	30	50
30	55	13.222	11.949	10.869	9.947	9.153	8.464	7.334	6.452	30	55
30	60	11.983	10.931	10.026	9.241	8.557	7.958	6.959	6.167	30	60
30	65	10.595	9.762	9.035	8.395	7.829	7.327	6.478	5.791	30	65
30	70	9.089	8.463	7.907	7.411	6.966	6.566	5.877	5.308	30	70
30	75	7.501	7.058	6.659	6.297	5.968	5.669	5.143	4.699	30	75

YEARS' PURCHASE FOR THE JOINT CONTINUATION OF TWO LIVES
(ENGLISH LIFE MORTALITY TABLES - No 15)

						Rate Per Cent						
Ages		5	6	7	8	9	10	12	14	Ages	
30	80	5.911	5.619	5.352	5.106	4.880	4.670	4.297	3.974	30	80
30	85	4.446	4.267	4.099	3.944	3.798	3.662	3.415	3.197	30	85
30	90	3.211	3.106	3.007	2.914	2.825	2.742	2.588	2.450	30	90
30	95	2.269	2.208	2.151	2.096	2.043	1.993	1.900	1.815	30	95
30	100	1.596	1.561	1.528	1.496	1.465	1.435	1.379	1.327	30	100
35	35	15.748	13.930	12.442	11.211	10.180	9.308	7.922	6.878	35	35
35	40	15.345	13.624	12.208	11.028	10.036	9.193	7.846	6.825	35	40
35	45	14.775	13.185	11.864	10.757	9.818	9.016	7.726	6.740	35	45
35	50	14.015	12.587	11.389	10.374	9.507	8.761	7.548	6.611	35	50
35	55	13.042	11.804	10.752	9.851	9.074	8.398	7.287	6.418	35	55
35	60	11.859	10.829	9.941	9.171	8.498	7.907	6.922	6.139	35	60
35	65	10.514	9.694	8.977	8.345	7.786	7.290	6.450	5.769	35	65
35	70	9.039	8.419	7.869	7.377	6.937	6.540	5.857	5.291	35	70
35	75	7.471	7.032	6.636	6.276	5.949	5.651	5.129	4.687	35	75
35	80	5.894	5.604	5.338	5.094	4.868	4.660	4.288	3.966	35	80
35	85	4.437	4.258	4.092	3.936	3.791	3.656	3.410	3.192	35	85
35	90	3.207	3.102	3.003	2.910	2.822	2.738	2.585	2.447	35	90
35	95	2.266	2.206	2.148	2.094	2.041	1.991	1.898	1.813	35	95
35	100	1.595	1.560	1.526	1.494	1.463	1.434	1.378	1.326	35	100
40	40	14.769	13.183	11.865	10.759	9.822	9.021	7.731	6.745	40	40
40	45	14.290	12.807	11.567	10.520	9.628	8.861	7.620	6.665	40	45
40	50	13.627	12.280	11.143	10.175	9.344	8.627	7.454	6.543	40	50
40	55	12.749	11.567	10.559	9.692	8.941	8.287	7.207	6.359	40	55
40	60	11.651	10.657	9.797	9.050	8.395	7.820	6.857	6.089	40	60
40	65	10.374	9.576	8.875	8.258	7.711	7.225	6.399	5.729	40	65
40	70	8.950	8.343	7.802	7.318	6.885	6.494	5.820	5.262	40	70
40	75	7.418	6.985	6.594	6.238	5.915	5.621	5.104	4.666	40	75
40	80	5.865	5.577	5.314	5.071	4.847	4.641	4.272	3.952	40	80
40	85	4.422	4.244	4.078	3.924	3.779	3.645	3.400	3.183	40	85
40	90	3.198	3.094	2.995	2.903	2.815	2.732	2.579	2.442	40	90
40	95	2.262	2.202	2.144	2.090	2.038	1.988	1.895	1.810	40	95
40	100	1.593	1.558	1.524	1.492	1.461	1.432	1.376	1.324	40	100
45	45	13.600	12.259	11.126	10.162	9.334	8.619	7.450	6.541	45	45
45	50	13.049	11.814	10.764	9.863	9.086	8.411	7.299	6.429	45	50
45	55	12.292	11.192	10.248	9.433	8.723	8.103	7.072	6.257	45	55
45	60	11.310	10.372	9.557	8.846	8.221	7.670	6.744	6.002	45	60
45	65	10.136	9.372	8.700	8.106	7.579	7.109	6.310	5.658	45	65
45	70	8.793	8.206	7.681	7.212	6.791	6.410	5.753	5.207	45	70
45	75	7.322	6.899	6.516	6.169	5.853	5.564	5.056	4.626	45	75
45	80	5.809	5.527	5.267	5.029	4.809	4.605	4.241	3.926	45	80
45	85	4.391	4.216	4.052	3.899	3.757	3.623	3.381	3.167	45	85
45	90	3.183	3.079	2.981	2.889	2.802	2.720	2.569	2.432	45	90
45	95	2.254	2.194	2.137	2.082	2.031	1.981	1.889	1.805	45	95
45	100	1.588	1.553	1.520	1.488	1.458	1.428	1.372	1.321	45	100
50	50	12.250	11.158	10.220	9.409	8.703	8.085	7.060	6.248	50	50
50	55	11.630	10.641	9.786	9.042	8.390	7.817	6.858	6.092	50	55
50	60	10.793	9.934	9.184	8.526	7.945	7.430	6.561	5.858	50	60
50	65	9.757	9.045	8.417	7.860	7.363	6.919	6.160	5.538	50	65
50	70	8.534	7.978	7.480	7.034	6.632	6.268	5.638	5.112	50	70
50	75	7.157	6.751	6.384	6.049	5.744	5.465	4.974	4.557	50	75

YEARS' PURCHASE FOR THE JOINT CONTINUATION OF TWO LIVES
(ENGLISH LIFE MORTALITY TABLES - No 15)

Ages		Rate Per Cent								Ages	
		5	6	7	8	9	10	12	14		
50	80	5.712	5.438	5.186	4.954	4.739	4.541	4.186	3.878	50	80
50	85	4.337	4.165	4.005	3.855	3.716	3.585	3.347	3.137	50	85
50	90	3.154	3.052	2.955	2.865	2.779	2.698	2.549	2.414	50	90
50	95	2.238	2.179	2.123	2.069	2.018	1.969	1.877	1.794	50	95
50	100	1.580	1.545	1.512	1.481	1.450	1.421	1.366	1.315	50	100
55	55	10.743	9.891	9.147	8.493	7.917	7.406	6.542	5.844	55	55
55	60	10.069	9.314	8.649	8.062	7.540	7.075	6.283	5.637	55	60
55	65	9.201	8.562	7.994	7.488	7.035	6.628	5.928	5.350	55	65
55	70	8.137	7.626	7.168	6.755	6.382	6.044	5.454	4.960	55	70
55	75	6.895	6.515	6.170	5.856	5.568	5.305	4.839	4.442	55	75
55	80	5.552	5.291	5.051	4.830	4.625	4.435	4.095	3.799	55	80
55	85	4.247	4.081	3.926	3.782	3.647	3.520	3.290	3.085	55	85
55	90	3.105	3.005	2.912	2.823	2.740	2.661	2.515	2.383	55	90
55	95	2.212	2.154	2.098	2.046	1.995	1.947	1.858	1.776	55	95
55	100	1.565	1.531	1.499	1.468	1.438	1.409	1.354	1.304	55	100
60	60	9.125	8.493	7.931	7.430	6.982	6.579	5.886	5.314	60	60
60	65	8.441	7.893	7.402	6.962	6.565	6.207	5.585	5.067	60	65
60	70	7.566	7.117	6.711	6.344	6.011	5.707	5.175	4.726	60	70
60	75	6.498	6.156	5.844	5.558	5.296	5.054	4.627	4.260	60	75
60	80	5.299	5.059	4.837	4.631	4.441	4.264	3.946	3.668	60	80
60	85	4.097	3.941	3.796	3.659	3.532	3.412	3.193	2.999	60	85
60	90	3.021	2.926	2.837	2.752	2.672	2.596	2.456	2.330	60	90
60	95	2.166	2.110	2.056	2.005	1.957	1.910	1.823	1.744	60	95
60	100	1.540	1.507	1.475	1.445	1.416	1.387	1.334	1.285	60	100
65	65	7.509	7.062	6.659	6.294	5.963	5.661	5.133	4.687	65	65
65	70	6.833	6.456	6.114	5.802	5.517	5.256	4.795	4.402	65	70
65	75	5.966	5.670	5.398	5.149	4.919	4.707	4.328	4.001	65	75
65	80	4.944	4.730	4.531	4.347	4.176	4.017	3.729	3.477	65	80
65	85	3.878	3.736	3.603	3.478	3.360	3.250	3.048	2.869	65	85
65	90	2.894	2.806	2.722	2.643	2.568	2.497	2.365	2.246	65	90
65	95	2.093	2.040	1.990	1.941	1.895	1.851	1.768	1.692	65	95
65	100	1.499	1.467	1.437	1.407	1.379	1.352	1.301	1.253	65	100
70	70	5.982	5.683	5.408	5.157	4.925	4.711	4.330	4.001	70	70
70	75	5.320	5.077	4.852	4.645	4.453	4.274	3.953	3.674	70	75
70	80	4.496	4.314	4.144	3.986	3.839	3.701	3.450	3.229	70	80
70	85	3.594	3.468	3.351	3.240	3.136	3.038	2.857	2.696	70	85
70	90	2.725	2.645	2.569	2.497	2.429	2.364	2.244	2.134	70	90
70	95	1.996	1.946	1.899	1.855	1.812	1.770	1.693	1.622	70	95
70	100	1.443	1.413	1.384	1.356	1.330	1.304	1.256	1.211	70	100
75	75	4.564	4.376	4.201	4.039	3.887	3.746	3.489	3.263	75	75
75	80	3.943	3.797	3.660	3.532	3.411	3.298	3.091	2.907	75	80
75	85	3.224	3.120	3.021	2.928	2.839	2.756	2.602	2.464	75	85
75	90	2.496	2.427	2.361	2.298	2.238	2.182	2.076	1.979	75	90
75	95	1.859	1.815	1.773	1.732	1.694	1.657	1.587	1.523	75	95
75	100	1.361	1.334	1.308	1.282	1.258	1.235	1.190	1.148	75	100
80	80	3.319	3.209	3.105	3.007	2.915	2.827	2.666	2.522	80	80
80	85	2.783	2.701	2.623	2.549	2.478	2.411	2.287	2.175	80	85
80	90	2.208	2.151	2.097	2.045	1.996	1.948	1.860	1.779	80	90
80	95	1.679	1.641	1.605	1.571	1.538	1.506	1.446	1.390	80	95
80	100	1.251	1.227	1.204	1.181	1.160	1.139	1.099	1.062	80	100

YEARS' PURCHASE FOR THE JOINT CONTINUATION OF TWO LIVES
(ENGLISH LIFE MORTALITY TABLES - No 15)

Ages		5	6	7	8	9	10	12	14	Ages	
5	5	18.987	16.145	13.996	12.327	11.001	9.925	8.293	7.117	5	5
5	10	18.854	16.067	13.948	12.296	10.980	9.911	8.286	7.113	5	10
5	15	18.666	15.952	13.875	12.249	10.948	9.889	8.273	7.105	5	15
5	20	18.423	15.801	13.780	12.187	10.906	9.860	8.258	7.097	5	20
5	25	18.099	15.594	13.643	12.095	10.843	9.815	8.235	7.083	5	25
5	30	17.675	15.311	13.450	11.960	10.747	9.745	8.194	7.058	5	30
5	35	17.137	14.938	13.187	11.770	10.607	9.640	8.132	7.018	5	35
5	40	16.468	14.458	12.837	11.510	10.410	9.488	8.038	6.956	5	40
5	45	15.646	13.849	12.377	11.158	10.136	9.272	7.897	6.860	5	45
5	50	14.664	13.095	11.791	10.696	9.768	8.975	7.696	6.718	5	50
5	55	13.506	12.178	11.056	10.101	9.280	8.571	7.412	6.510	5	55
5	60	12.179	11.094	10.162	9.357	8.656	8.042	7.023	6.216	5	60
5	65	10.727	9.875	9.131	8.478	7.902	7.391	6.528	5.831	5	65
5	70	9.174	8.538	7.973	7.469	7.018	6.613	5.915	5.339	5	70
5	75	7.554	7.107	6.702	6.336	6.004	5.701	5.170	4.722	5	75
5	80	5.943	5.649	5.379	5.131	4.903	4.692	4.315	3.990	5	80
5	85	4.464	4.283	4.115	3.958	3.812	3.675	3.427	3.207	5	85
5	90	3.221	3.115	3.016	2.922	2.833	2.750	2.595	2.456	5	90
5	95	2.274	2.213	2.156	2.100	2.048	1.998	1.904	1.819	5	95
5	100	1.599	1.564	1.531	1.498	1.467	1.437	1.381	1.329	5	100
10	10	18.738	15.997	13.904	12.268	10.962	9.898	8.279	7.109	10	10
10	15	18.569	15.892	13.837	12.224	10.931	9.876	8.267	7.101	10	15
10	20	18.344	15.751	13.746	12.164	10.891	9.849	8.252	7.093	10	20
10	25	18.037	15.552	13.615	12.075	10.829	9.805	8.229	7.079	10	25
10	30	17.627	15.278	13.427	11.943	10.735	9.736	8.189	7.055	10	30
10	35	17.100	14.912	13.168	11.756	10.597	9.632	8.127	7.015	10	35
10	40	16.440	14.439	12.822	11.499	10.402	9.481	8.033	6.953	10	40
10	45	15.626	13.834	12.365	11.148	10.129	9.266	7.893	6.857	10	45
10	50	14.650	13.084	11.782	10.689	9.762	8.970	7.693	6.715	10	50
10	55	13.496	12.170	11.049	10.095	9.276	8.567	7.409	6.508	10	55
10	60	12.172	11.088	10.157	9.352	8.652	8.039	7.020	6.214	10	60
10	65	10.721	9.870	9.128	8.475	7.899	7.389	6.526	5.829	10	65
10	70	9.171	8.535	7.971	7.467	7.016	6.611	5.914	5.338	10	70
10	75	7.552	7.105	6.701	6.335	6.003	5.700	5.169	4.721	10	75
10	80	5.942	5.648	5.378	5.130	4.902	4.691	4.315	3.990	10	80
10	85	4.464	4.283	4.115	3.958	3.812	3.675	3.426	3.207	10	85
10	90	3.221	3.115	3.016	2.922	2.833	2.749	2.595	2.456	10	90
10	95	2.274	2.213	2.156	2.100	2.048	1.998	1.904	1.819	10	95
10	100	1.599	1.564	1.531	1.498	1.467	1.438	1.381	1.329	10	100
15	15	18.422	15.798	13.776	12.182	10.902	9.856	8.255	7.094	15	15
15	20	18.220	15.670	13.693	12.127	10.864	9.829	8.241	7.086	15	20
15	25	17.936	15.485	13.569	12.043	10.806	9.787	8.218	7.072	15	25
15	30	17.547	15.223	13.388	11.915	10.714	9.719	8.179	7.048	15	30
15	35	17.039	14.868	13.136	11.732	10.578	9.617	8.117	7.008	15	35
15	40	16.394	14.404	12.796	11.479	10.386	9.469	8.025	6.947	15	40
15	45	15.592	13.807	12.344	11.132	10.116	9.255	7.885	6.851	15	45
15	50	14.624	13.064	11.766	10.676	9.751	8.960	7.686	6.710	15	50
15	55	13.477	12.154	11.037	10.084	9.266	8.559	7.403	6.503	15	55
15	60	12.158	11.076	10.147	9.344	8.644	8.032	7.015	6.210	15	60
15	65	10.712	9.862	9.120	8.469	7.894	7.383	6.522	5.826	15	65

YEARS' PURCHASE FOR THE JOINT CONTINUATION OF TWO LIVES
(ENGLISH LIFE MORTALITY TABLES - No 15)

Ages		5	6	7	8	9	10	12	14	Ages	
					Rate Per Cent						
15	70	9.164	8.529	7.965	7.462	7.012	6.607	5.910	5.335	15	70
15	75	7.547	7.100	6.697	6.331	5.999	5.697	5.166	4.719	15	75
15	80	5.938	5.645	5.375	5.127	4.899	4.689	4.313	3.988	15	80
15	85	4.462	4.281	4.113	3.956	3.810	3.673	3.425	3.206	15	85
15	90	3.220	3.114	3.015	2.921	2.832	2.748	2.594	2.455	15	90
15	95	2.274	2.213	2.155	2.100	2.047	1.997	1.904	1.818	15	95
15	100	1.599	1.564	1.530	1.498	1.467	1.437	1.381	1.329	15	100
20	20	18.045	15.557	13.618	12.076	10.829	9.804	8.227	7.078	20	20
20	25	17.791	15.389	13.504	11.998	10.774	9.764	8.205	7.065	20	25
20	30	17.432	15.144	13.334	11.877	10.686	9.699	8.167	7.041	20	30
20	35	16.950	14.806	13.092	11.700	10.555	9.600	8.107	7.002	20	35
20	40	16.328	14.357	12.761	11.453	10.366	9.454	8.015	6.941	20	40
20	45	15.544	13.772	12.318	11.112	10.100	9.243	7.877	6.846	20	45
20	50	14.591	13.039	11.746	10.660	9.739	8.951	7.679	6.706	20	50
20	55	13.455	12.137	11.023	10.073	9.257	8.551	7.397	6.499	20	55
20	60	12.144	11.065	10.137	9.336	8.638	8.026	7.011	6.207	20	60
20	65	10.703	9.854	9.114	8.463	7.889	7.379	6.519	5.823	20	65
20	70	9.159	8.524	7.961	7.458	7.008	6.604	5.908	5.333	20	70
20	75	7.544	7.097	6.694	6.329	5.997	5.695	5.165	4.717	20	75
20	80	5.937	5.643	5.374	5.126	4.898	4.687	4.312	3.987	20	80
20	85	4.461	4.280	4.112	3.955	3.809	3.672	3.424	3.205	20	85
20	90	3.219	3.114	3.014	2.920	2.832	2.748	2.594	2.455	20	90
20	95	2.273	2.212	2.155	2.099	2.047	1.997	1.903	1.818	20	95
20	100	1.599	1.564	1.530	1.498	1.467	1.437	1.381	1.329	20	100
25	25	17.572	15.242	13.403	11.927	10.723	9.727	8.184	7.052	25	25
25	30	17.251	15.020	13.246	11.814	10.641	9.665	8.148	7.029	25	30
25	35	16.807	14.705	13.019	11.647	10.515	9.570	8.089	6.990	25	35
25	40	16.218	14.277	12.703	11.409	10.333	9.428	8.000	6.930	25	40
25	45	15.464	13.711	12.272	11.077	10.073	9.221	7.864	6.837	25	45
25	50	14.534	12.994	11.712	10.633	9.717	8.933	7.668	6.698	25	50
25	55	13.416	12.106	10.998	10.053	9.241	8.538	7.388	6.493	25	55
25	60	12.118	11.044	10.120	9.321	8.626	8.016	7.004	6.201	25	60
25	65	10.686	9.841	9.102	8.453	7.881	7.372	6.513	5.819	25	65
25	70	9.149	8.516	7.954	7.452	7.003	6.599	5.904	5.330	25	70
25	75	7.539	7.093	6.690	6.325	5.994	5.692	5.163	4.715	25	75
25	80	5.934	5.640	5.371	5.124	4.896	4.686	4.310	3.986	25	80
25	85	4.459	4.279	4.111	3.954	3.808	3.672	3.424	3.205	25	85
25	90	3.219	3.113	3.013	2.920	2.831	2.748	2.593	2.454	25	90
25	95	2.273	2.212	2.154	2.099	2.047	1.997	1.903	1.818	25	95
25	100	1.598	1.563	1.530	1.498	1.467	1.437	1.381	1.329	25	100
30	30	16.975	14.825	13.107	11.713	10.565	9.608	8.113	7.006	30	30
30	35	16.579	14.541	12.899	11.558	10.448	9.518	8.057	6.969	30	35
30	40	16.038	14.144	12.603	11.333	10.275	9.382	7.970	6.911	30	40
30	45	15.326	13.607	12.192	11.014	10.024	9.182	7.838	6.819	30	45
30	50	14.433	12.916	11.650	10.584	9.678	8.901	7.645	6.682	30	50
30	55	13.345	12.049	10.951	10.015	9.210	8.512	7.369	6.479	30	55
30	60	12.069	11.003	10.087	9.293	8.602	7.996	6.989	6.190	30	60
30	65	10.655	9.814	9.079	8.434	7.863	7.357	6.502	5.810	30	65
30	70	9.129	8.498	7.938	7.439	6.991	6.589	5.896	5.324	30	70
30	75	7.527	7.082	6.680	6.317	5.986	5.685	5.157	4.711	30	75

YEARS' PURCHASE FOR THE JOINT CONTINUATION OF TWO LIVES

(ENGLISH LIFE MORTALITY TABLES - No 15)

					Rate Per Cent						
Ages		5	6	7	8	9	10	12	14	Ages	
30	80	5.927	5.634	5.366	5.119	4.891	4.681	4.306	3.982	30	80
30	85	4.456	4.275	4.108	3.951	3.805	3.669	3.421	3.203	30	85
30	90	3.217	3.111	3.012	2.918	2.830	2.746	2.592	2.453	30	90
30	95	2.272	2.211	2.153	2.098	2.046	1.996	1.903	1.817	30	95
30	100	1.598	1.563	1.529	1.497	1.466	1.436	1.380	1.328	30	100
35	35	16.240	14.293	12.715	11.419	10.341	9.435	8.004	6.934	35	35
35	40	15.760	13.937	12.446	11.212	10.180	9.308	7.922	6.877	35	40
35	45	15.108	13.441	12.063	10.913	9.943	9.117	7.794	6.788	35	45
35	50	14.269	12.787	11.548	10.502	9.611	8.846	7.607	6.654	35	50
35	55	13.226	11.953	10.874	9.951	9.157	8.468	7.337	6.455	35	55
35	60	11.987	10.935	10.030	9.245	8.562	7.962	6.963	6.170	35	60
35	65	10.600	9.767	9.039	8.399	7.834	7.331	6.482	5.794	35	65
35	70	9.094	8.468	7.912	7.415	6.970	6.570	5.881	5.311	35	70
35	75	7.505	7.063	6.663	6.301	5.972	5.672	5.146	4.702	35	75
35	80	5.915	5.623	5.355	5.109	4.883	4.673	4.300	3.976	35	80
35	85	4.449	4.269	4.102	3.946	3.800	3.664	3.417	3.199	35	85
35	90	3.213	3.108	3.009	2.915	2.827	2.743	2.589	2.451	35	90
35	95	2.270	2.209	2.152	2.097	2.044	1.994	1.901	1.816	35	95
35	100	1.597	1.562	1.528	1.496	1.465	1.435	1.379	1.327	35	100
40	40	15.353	13.628	12.209	11.028	10.035	9.192	7.845	6.824	40	40
40	45	14.777	13.185	11.864	10.755	9.817	9.015	7.725	6.739	40	45
40	50	14.013	12.585	11.386	10.371	9.505	8.759	7.546	6.610	40	50
40	55	13.037	11.800	10.749	9.848	9.071	8.396	7.285	6.417	40	55
40	60	11.854	10.824	9.937	9.167	8.495	7.905	6.921	6.138	40	60
40	65	10.510	9.690	8.973	8.343	7.784	7.289	6.449	5.768	40	65
40	70	9.036	8.417	7.867	7.376	6.936	6.539	5.856	5.291	40	70
40	75	7.470	7.031	6.635	6.276	5.949	5.651	5.129	4.687	40	75
40	80	5.894	5.604	5.338	5.094	4.868	4.660	4.288	3.967	40	80
40	85	4.438	4.259	4.092	3.937	3.792	3.656	3.410	3.193	40	85
40	90	3.207	3.102	3.003	2.910	2.822	2.739	2.586	2.447	40	90
40	95	2.267	2.206	2.149	2.094	2.042	1.992	1.899	1.814	40	95
40	100	1.595	1.560	1.527	1.495	1.464	1.434	1.378	1.326	40	100
45	45	14.293	12.806	11.563	10.514	9.622	8.855	7.614	6.659	45	45
45	50	13.624	12.274	11.136	10.168	9.337	8.619	7.447	6.537	45	50
45	55	12.740	11.558	10.549	9.683	8.932	8.279	7.200	6.352	45	55
45	60	11.638	10.645	9.786	9.039	8.386	7.811	6.850	6.083	45	60
45	65	10.361	9.563	8.864	8.248	7.702	7.216	6.393	5.723	45	65
45	70	8.939	8.332	7.792	7.310	6.877	6.487	5.814	5.257	45	70
45	75	7.410	6.977	6.587	6.232	5.910	5.615	5.099	4.662	45	75
45	80	5.859	5.572	5.309	5.067	4.843	4.637	4.268	3.950	45	80
45	85	4.418	4.241	4.075	3.921	3.777	3.642	3.398	3.182	45	85
45	90	3.197	3.092	2.994	2.901	2.814	2.731	2.578	2.441	45	90
45	95	2.261	2.201	2.144	2.089	2.037	1.987	1.895	1.810	45	95
45	100	1.592	1.557	1.524	1.492	1.461	1.431	1.376	1.324	45	100
50	50	13.066	11.825	10.769	9.866	9.086	8.409	7.296	6.425	50	50
50	55	12.300	11.196	10.249	9.432	8.721	8.099	7.068	6.252	50	55
50	60	11.310	10.369	9.553	8.841	8.216	7.664	6.739	5.997	50	60
50	65	10.129	9.364	8.693	8.099	7.572	7.103	6.304	5.653	50	65
50	70	8.784	8.197	7.673	7.204	6.783	6.403	5.747	5.201	50	70
50	75	7.313	6.891	6.509	6.162	5.846	5.558	5.051	4.622	50	75

YEARS' PURCHASE FOR THE JOINT CONTINUATION OF TWO LIVES
(ENGLISH LIFE MORTALITY TABLES - No 15)

Ages		5	6	7	8	9	10	12	14	Ages	
					Rate Per Cent						
50	80	5.803	5.521	5.262	5.023	4.804	4.600	4.237	3.922	50	80
50	85	4.387	4.212	4.048	3.896	3.753	3.620	3.378	3.164	50	85
50	90	3.180	3.077	2.979	2.887	2.800	2.718	2.567	2.430	50	90
50	95	2.252	2.192	2.135	2.081	2.029	1.980	1.888	1.803	50	95
50	100	1.587	1.553	1.519	1.487	1.457	1.427	1.372	1.320	50	100
55	55	11.669	10.672	9.810	9.060	8.404	7.828	6.864	6.095	55	55
55	60	10.820	9.955	9.200	8.538	7.955	7.437	6.564	5.860	55	60
55	65	9.771	9.056	8.425	7.866	7.367	6.922	6.161	5.538	55	65
55	70	8.538	7.981	7.482	7.035	6.632	6.267	5.636	5.111	55	70
55	75	7.156	6.750	6.382	6.047	5.742	5.463	4.972	4.554	55	75
55	80	5.709	5.435	5.183	4.951	4.737	4.538	4.184	3.876	55	80
55	85	4.334	4.163	4.002	3.853	3.713	3.583	3.345	3.135	55	85
55	90	3.152	3.050	2.954	2.863	2.778	2.697	2.547	2.413	55	90
55	95	2.237	2.178	2.122	2.068	2.017	1.968	1.877	1.793	55	95
55	100	1.579	1.545	1.512	1.480	1.450	1.421	1.365	1.314	55	100
60	60	10.133	9.367	8.694	8.099	7.572	7.101	6.302	5.651	60	60
60	65	9.248	8.601	8.027	7.516	7.058	6.647	5.942	5.360	60	65
60	70	8.166	7.651	7.189	6.773	6.397	6.056	5.463	4.966	60	70
60	75	6.909	6.527	6.180	5.864	5.575	5.310	4.843	4.445	60	75
60	80	5.556	5.295	5.054	4.832	4.627	4.436	4.095	3.799	60	80
60	85	4.246	4.080	3.926	3.781	3.646	3.519	3.289	3.084	60	85
60	90	3.103	3.004	2.910	2.822	2.738	2.659	2.513	2.381	60	90
60	95	2.210	2.152	2.097	2.044	1.994	1.946	1.857	1.774	60	95
60	100	1.564	1.530	1.498	1.467	1.437	1.408	1.354	1.303	60	100
65	65	8.547	7.987	7.486	7.037	6.632	6.267	5.634	5.107	65	65
65	70	7.650	7.192	6.778	6.404	6.065	5.756	5.215	4.759	65	70
65	75	6.557	6.209	5.891	5.601	5.335	5.090	4.657	4.285	65	75
65	80	5.335	5.091	4.867	4.659	4.466	4.287	3.965	3.685	65	80
65	85	4.116	3.959	3.812	3.675	3.546	3.425	3.205	3.009	65	85
65	90	3.030	2.935	2.845	2.760	2.679	2.603	2.462	2.335	65	90
65	95	2.170	2.114	2.060	2.009	1.960	1.913	1.826	1.746	65	95
65	100	1.542	1.509	1.477	1.446	1.417	1.389	1.335	1.286	65	100
70	70	6.957	6.570	6.220	5.900	5.608	5.340	4.868	4.466	70	70
70	75	6.066	5.763	5.485	5.229	4.994	4.777	4.390	4.056	70	75
70	80	5.017	4.798	4.596	4.408	4.233	4.070	3.777	3.519	70	80
70	85	3.927	3.782	3.646	3.518	3.399	3.287	3.082	2.899	70	85
70	90	2.923	2.834	2.749	2.669	2.592	2.520	2.387	2.266	70	90
70	95	2.111	2.057	2.006	1.957	1.910	1.865	1.781	1.705	70	95
70	100	1.509	1.477	1.446	1.417	1.388	1.361	1.309	1.261	70	100
75	75	5.398	5.150	4.922	4.711	4.515	4.334	4.007	3.723	75	75
75	80	4.561	4.375	4.202	4.041	3.891	3.751	3.496	3.270	75	80
75	85	3.641	3.514	3.394	3.281	3.175	3.076	2.892	2.728	75	85
75	90	2.757	2.676	2.598	2.525	2.456	2.390	2.268	2.157	75	90
75	95	2.016	1.966	1.918	1.873	1.829	1.787	1.709	1.637	75	95
75	100	1.455	1.425	1.396	1.368	1.341	1.315	1.266	1.220	75	100
80	80	3.950	3.804	3.668	3.540	3.420	3.308	3.101	2.917	80	80
80	85	3.236	3.131	3.032	2.939	2.851	2.767	2.613	2.474	80	85
80	90	2.507	2.438	2.372	2.309	2.249	2.192	2.086	1.989	80	90
80	95	1.867	1.823	1.781	1.740	1.702	1.665	1.595	1.530	80	95
80	100	1.368	1.340	1.314	1.288	1.264	1.240	1.195	1.153	80	100

YEARS' PURCHASE

OR

PRESENT VALUE OF
ONE POUND PER ANNUM

for the

LONGER OF TWO LIVES

A Rule is given on the following page to enable these figures of years' purchase to be derived from the preceding Life Tables.

YEARS' PURCHASE

FOR

THE LONGER OF TWO LIVES

RULE

For finding the Years' Purchase for the longer of two lives:
Add the years' purchase for the first single life to the years' purchase for the second single life and subtract the years' purchase for the joint continuation of the two lives.

Example 1—
What is the Years' Purchase for the longer of two lives aged 40 years (female) and 60 years (male) at 8%?

Answer—

	Y.P. for life aged 40 years (female) at 8%	...	=	11·542
Add	Y.P. for life aged 60 years (male) at 8% ...	...	=	8·385
		Sum	=	19·927

Deduct   Y.P. for joint continuation of two lives aged 40 (female) and 60 (male) at 8% ... ... = 8·240

Y.P. for longer of two lives aged 40 (female) and 60 (male) at 8% ... ... = 11·687

Example 2—
To find the Y.P. for the longer of two lives aged 22 years (male) and 44 years (male) at 7% the ages must be adjusted to the nearest age group available in the Y.P. for the joint continuation of two lives table. Each Y.P. figure is then taken at the appropriate adjusted age or age group as follows:

Add the Y.P. for life aged 20 years (male) at 7% to the Y.P. for life aged 45 years (male) at 7% *then subtract* the Y.P. for the joint continuation of two lives aged 20 years (male) and 45 years (male) at 7%.

Note—
The method shown in Example 2 will only give an approximate answer for the longer of the two lives considered (i.e., 22 years and 44 years).

TABLE OF
ANNUAL PREMIUMS

TO SECURE

ONE POUND

at the end of the year in which a person may die
according to the

ENGLISH LIFE MORTALITY TABLE No. 15
(Males and Females)

AT RATES OF INTEREST

3%, 4% and 5%

ANNUAL PREMIUMS TO SECURE £1 AT DEATH

(ENGLISH LIFE MORTALITY TABLES - No 15)

	Rate Per Cent				
Age	3	4	5	Age	
1	0.0042	0.0028	0.0019	1	
2	0.0043	0.0029	0.0020	2	
3	0.0045	0.0030	0.0021	3	
4	0.0046	0.0031	0.0022	4	
5	0.0048	0.0033	0.0023	5	
6	0.0049	0.0034	0.0024	6	
7	0.0051	0.0035	0.0025	7	
8	0.0053	0.0037	0.0026	8	
9	0.0055	0.0038	0.0027	9	
10	0.0056	0.0040	0.0029	10	
11	0.0058	0.0042	0.0030	11	
12	0.0060	0.0044	0.0032	12	
13	0.0063	0.0045	0.0033	13	
14	0.0065	0.0047	0.0035	14	
15	0.0067	0.0049	0.0037	15	
16	0.0069	0.0051	0.0038	16	
17	0.0072	0.0053	0.0040	17	
18	0.0074	0.0056	0.0042	18	
19	0.0077	0.0058	0.0044	19	
20	0.0079	0.0060	0.0046	20	
21	0.0082	0.0063	0.0048	21	
22	0.0085	0.0065	0.0050	22	
23	0.0088	0.0068	0.0053	23	
24	0.0091	0.0071	0.0055	24	
25	0.0094	0.0073	0.0058	25	
26	0.0098	0.0077	0.0060	26	
27	0.0101	0.0080	0.0063	27	
28	0.0105	0.0083	0.0066	28	
29	0.0109	0.0087	0.0070	29	
30	0.0113	0.0091	0.0073	30	
31	0.0118	0.0095	0.0077	31	
32	0.0122	0.0099	0.0081	32	
33	0.0127	0.0104	0.0085	33	
34	0.0133	0.0109	0.0090	34	
35	0.0138	0.0114	0.0095	35	
36	0.0144	0.0119	0.0100	36	
37	0.0150	0.0125	0.0105	37	
38	0.0156	0.0131	0.0111	38	
39	0.0162	0.0137	0.0117	39	
40	0.0169	0.0144	0.0123	40	
41	0.0177	0.0151	0.0130	41	
42	0.0185	0.0159	0.0137	42	
43	0.0193	0.0167	0.0145	43	
44	0.0202	0.0175	0.0153	44	
45	0.0211	0.0184	0.0162	45	
46	0.0221	0.0194	0.0171	46	
47	0.0231	0.0204	0.0181	47	
48	0.0242	0.0215	0.0192	48	
49	0.0253	0.0226	0.0203	49	
50	0.0266	0.0238	0.0215	50	

ANNUAL PREMIUMS TO SECURE £1 AT DEATH

(ENGLISH LIFE MORTALITY TABLES - No 15)

		Rate Per Cent				
	Age	3	4	5	Age	
	51	0.0279	0.0251	0.0227	51	
	52	0.0293	0.0265	0.0241	52	
	53	0.0307	0.0280	0.0255	53	
	54	0.0323	0.0295	0.0270	54	
	55	0.0339	0.0311	0.0287	55	
	56	0.0357	0.0329	0.0304	56	
	57	0.0376	0.0348	0.0323	57	
	58	0.0396	0.0368	0.0343	58	
	59	0.0417	0.0389	0.0364	59	
	60	0.0440	0.0412	0.0386	60	
	61	0.0464	0.0436	0.0410	61	
	62	0.0489	0.0461	0.0436	62	
	63	0.0516	0.0488	0.0463	63	
	64	0.0545	0.0517	0.0491	64	
	65	0.0575	0.0547	0.0521	65	
	66	0.0607	0.0579	0.0553	66	
	67	0.0641	0.0613	0.0587	67	
	68	0.0677	0.0649	0.0623	68	
	69	0.0715	0.0687	0.0661	69	
	70	0.0756	0.0728	0.0702	70	
	71	0.0800	0.0772	0.0746	71	
	72	0.0847	0.0819	0.0793	72	
	73	0.0896	0.0868	0.0842	73	
	74	0.0949	0.0921	0.0894	74	
	75	0.1005	0.0977	0.0950	75	
	76	0.1065	0.1037	0.1010	76	
	77	0.1129	0.1101	0.1074	77	
	78	0.1196	0.1168	0.1141	78	
	79	0.1268	0.1240	0.1213	79	
	80	0.1344	0.1315	0.1288	80	
	81	0.1424	0.1395	0.1368	81	
	82	0.1510	0.1481	0.1453	82	
	83	0.1601	0.1571	0.1543	83	
	84	0.1697	0.1667	0.1639	84	
	85	0.1798	0.1769	0.1740	85	
	86	0.1905	0.1875	0.1846	86	
	87	0.2015	0.1984	0.1955	87	
	88	0.2129	0.2097	0.2067	88	
	89	0.2249	0.2217	0.2186	89	
	90	0.2378	0.2345	0.2313	90	
	91	0.2521	0.2487	0.2454	91	
	92	0.2675	0.2640	0.2607	92	
	93	0.2834	0.2799	0.2765	93	
	94	0.2994	0.2958	0.2923	94	
	95	0.3153	0.3117	0.3081	95	
	96	0.3314	0.3276	0.3239	96	
	97	0.3479	0.3440	0.3402	97	
	98	0.3653	0.3612	0.3573	98	
	99	0.3828	0.3786	0.3745	99	
	100	0.4014	0.3971	0.3929	100	

ANNUAL PREMIUMS TO SECURE £1 AT DEATH

(ENGLISH LIFE MORTALITY TABLES - No 15)

		Rate Per Cent				
	Age	3	4	5	Age	
	1	0.0035	0.0022	0.0014	1	
	2	0.0036	0.0023	0.0015	2	
	3	0.0037	0.0024	0.0015	3	
	4	0.0038	0.0025	0.0016	4	
	5	0.0039	0.0026	0.0017	5	
	6	0.0041	0.0027	0.0018	6	
	7	0.0042	0.0028	0.0018	7	
	8	0.0043	0.0029	0.0019	8	
	9	0.0045	0.0030	0.0020	9	
	10	0.0046	0.0031	0.0021	10	
	11	0.0048	0.0033	0.0022	11	
	12	0.0050	0.0034	0.0023	12	
	13	0.0051	0.0035	0.0025	13	
	14	0.0053	0.0037	0.0026	14	
	15	0.0055	0.0038	0.0027	15	
	16	0.0057	0.0040	0.0028	16	
	17	0.0059	0.0042	0.0030	17	
	18	0.0061	0.0043	0.0031	18	
	19	0.0063	0.0045	0.0033	19	
	20	0.0065	0.0047	0.0034	20	
	21	0.0067	0.0049	0.0036	21	
	22	0.0070	0.0051	0.0038	22	
	23	0.0072	0.0053	0.0040	23	
	24	0.0075	0.0056	0.0042	24	
	25	0.0078	0.0058	0.0044	25	
	26	0.0080	0.0061	0.0046	26	
	27	0.0083	0.0063	0.0048	27	
	28	0.0086	0.0066	0.0051	28	
	29	0.0090	0.0069	0.0053	29	
	30	0.0093	0.0072	0.0056	30	
	31	0.0097	0.0075	0.0059	31	
	32	0.0100	0.0079	0.0062	32	
	33	0.0104	0.0082	0.0066	33	
	34	0.0108	0.0086	0.0069	34	
	35	0.0113	0.0090	0.0073	35	
	36	0.0117	0.0094	0.0076	36	
	37	0.0122	0.0099	0.0080	37	
	38	0.0127	0.0103	0.0085	38	
	39	0.0132	0.0108	0.0089	39	
	40	0.0137	0.0113	0.0094	40	
	41	0.0143	0.0119	0.0099	41	
	42	0.0149	0.0124	0.0105	42	
	43	0.0155	0.0130	0.0110	43	
	44	0.0162	0.0137	0.0116	44	
	45	0.0169	0.0143	0.0123	45	
	46	0.0176	0.0150	0.0129	46	
	47	0.0184	0.0158	0.0136	47	
	48	0.0192	0.0166	0.0144	48	
	49	0.0200	0.0174	0.0152	49	
	50	0.0209	0.0183	0.0160	50	

ANNUAL PREMIUMS TO SECURE £1 AT DEATH

(ENGLISH LIFE MORTALITY TABLES - No 15)

	Rate Per Cent					
Age	3	4	5	Age	
51	0.0219	0.0192	0.0169	51	
52	0.0229	0.0202	0.0179	52	
53	0.0239	0.0212	0.0189	53	
54	0.0251	0.0223	0.0200	54	
55	0.0263	0.0235	0.0212	55	
56	0.0275	0.0248	0.0224	56	
57	0.0289	0.0261	0.0237	57	
58	0.0303	0.0275	0.0251	58	
59	0.0318	0.0290	0.0266	59	
60	0.0334	0.0306	0.0281	60	
61	0.0351	0.0323	0.0298	61	
62	0.0369	0.0341	0.0315	62	
63	0.0388	0.0360	0.0334	63	
64	0.0408	0.0380	0.0354	64	
65	0.0430	0.0401	0.0375	65	
66	0.0453	0.0424	0.0398	66	
67	0.0477	0.0448	0.0422	67	
68	0.0503	0.0474	0.0448	68	
69	0.0531	0.0502	0.0476	69	
70	0.0561	0.0532	0.0505	70	
71	0.0594	0.0565	0.0538	71	
72	0.0629	0.0600	0.0573	72	
73	0.0666	0.0637	0.0610	73	
74	0.0706	0.0676	0.0649	74	
75	0.0748	0.0719	0.0692	75	
76	0.0795	0.0766	0.0738	76	
77	0.0846	0.0816	0.0789	77	
78	0.0900	0.0870	0.0843	78	
79	0.0958	0.0929	0.0901	79	
80	0.1020	0.0991	0.0963	80	
81	0.1087	0.1057	0.1030	81	
82	0.1159	0.1129	0.1101	82	
83	0.1237	0.1207	0.1179	83	
84	0.1321	0.1291	0.1263	84	
85	0.1411	0.1381	0.1353	85	
86	0.1507	0.1477	0.1448	86	
87	0.1607	0.1576	0.1547	87	
88	0.1714	0.1683	0.1654	88	
89	0.1830	0.1799	0.1769	89	
90	0.1953	0.1922	0.1892	90	
91	0.2081	0.2050	0.2019	91	
92	0.2215	0.2183	0.2152	92	
93	0.2353	0.2320	0.2289	93	
94	0.2497	0.2463	0.2431	94	
95	0.2644	0.2610	0.2577	95	
96	0.2789	0.2755	0.2721	96	
97	0.2936	0.2900	0.2865	97	
98	0.3094	0.3057	0.3021	98	
99	0.3268	0.3230	0.3193	99	
100	0.3448	0.3408	0.3370	100	

ENGLISH LIFE TABLES—No. 15

The l_x columns show the number of people who survive to any given age out of 100,000 born.

q_x is the probability that a person aged (x) will die within one year.

$\overset{\circ}{e}_x$ is the mean expectation of life showing the average duration of life at any age.

p_x—the probability that a person aged (x) will survive at least one year—can be obtained from the equation:

$$p_x = 1 - q_x$$

English Life Tables No. 15

Males

Age x	l_x	q_x	$\overset{o}{e}_x$	Age x	l_x	q_x	$\overset{o}{e}_x$
0	100000	.00814	73.413	55	91217	.00797	21.856
1	99186	.00062	73.019	56	90490	.00890	21.027
2	99124	.00038	72.064	57	89684	.00995	20.211
3	99086	.00030	71.091	58	88792	.01112	19.409
4	99056	.00024	70.113	59	87805	.01243	18.622
5	99032	.00022	69.130	60	86714	.01392	17.850
6	99010	.00020	68.145	61	85507	.01560	17.095
7	98990	.00019	67.158	62	84173	.01749	16.357
8	98972	.00018	66.171	63	82701	.01965	15.640
9	98953	.00018	65.183	64	81076	.02199	14.943
10	98935	.00018	64.195	65	79293	.02447	14.267
11	98917	.00018	63.206	66	77353	.02711	13.612
12	98899	.00019	62.218	67	75256	.02997	12.978
13	98880	.00023	61.230	68	73001	.03292	12.363
14	98857	.00029	60.244	69	70598	.03602	11.767
15	98828	.00040	59.261	70	68055	.03930	11.187
16	98789	.00052	58.285	71	65381	.04311	10.624
17	98737	.00075	57.315	72	62562	.04745	10.080
18	98663	.00087	56.358	73	59593	.05217	9.557
19	98577	.00083	55.406	74	56484	.05697	9.056
20	98496	.00084	54.452	75	53266	.06197	8.572
21	98413	.00086	53.497	76	49965	.06777	8.106
22	98328	.00089	52.543	77	46579	.07418	7.658
23	98241	.00089	51.589	78	43124	.08101	7.232
24	98154	.00088	50.635	79	39630	.08838	6.825
25	98067	.00086	49.679	80	36128	.09616	6.438
26	97983	.00085	48.721	81	32654	.10411	6.070
27	97900	.00085	47.762	82	29254	.11279	5.718
28	97817	.00087	46.802	83	25954	.12235	5.382
29	97732	.00090	45.842	84	22779	.13270	5.063
30	97645	.00091	44.883	85	19756	.14372	4.762
31	97556	.00094	43.923	86	16917	.15585	4.478
32	97465	.00097	42.964	87	14280	.16848	4.213
33	97370	.00099	42.005	88	11874	.18061	3.968
34	97273	.00106	41.046	89	9730	.19246	3.734
35	97170	.00116	40.090	90	7857	.20465	3.508
36	97057	.00127	39.136	91	6249	.21911	3.285
37	96933	.00138	38.185	92	4880	.23665	3.071
38	96800	.00149	37.237	93	3726	.25575	2.872
39	96655	.00160	36.292	94	2773	.27483	2.693
40	96500	.00172	35.349	95	2011	.29311	2.531
41	96334	.00186	34.409	96	1421	.31104	2.383
42	96155	.00201	33.473	97	979	.32919	2.244
43	95961	.00219	32.539	98	657	.34783	2.114
44	95751	.00240	31.609	99	428	.36712	1.991
45	95521	.00266	30.684	100	271	.38705	1.874
46	95266	.00297	29.765	101	166	.40760	1.764
47	94983	.00332	28.852	102	98	.42870	1.660
48	94668	.00371	27.947	103	56	.45030	1.562
49	94316	.00415	27.049	104	31	.47428	1.468
50	93925	.00464	26.159	105	16	.49634	1.384
51	93489	.00519	25.279	106	8	.51841	1.306
52	93004	.00577	24.408	107	4	.54041	1.234
53	92467	.00642	23.547	108	2	.56225	1.166
54	91873	.00714	22.696	109	1	.58385	1.104

For an explanation of headings see page 305

English Life Tables No. 15

Females

Age x	l_x	q_x	$\overset{\circ}{e}_x$	Age x	l_x	q_x	$\overset{\circ}{e}_x$
0	100000	.00632	78.956	55	94532	.00475	26.357
1	99368	.00055	78.462	56	94082	.00531	25.481
2	99313	.00030	77.505	57	93583	.00592	24.614
3	99283	.00022	76.528	58	93029	.00660	23.757
4	99261	.00018	75.545	59	92415	.00739	22.912
5	99243	.00016	74.559	60	91732	.00830	22.079
6	99228	.00015	73.570	61	90971	.00922	21.259
7	99213	.00014	72.581	62	90132	.01015	20.452
8	99199	.00014	71.591	63	89217	.01129	19.657
9	99185	.00013	70.601	64	88210	.01266	18.875
10	99172	.00013	69.610	65	87093	.01339	18.111
11	99159	.00014	68.620	66	85875	.01523	17.361
12	99145	.00014	67.629	67	84567	.01676	16.621
13	99131	.00015	66.638	68	83150	.01844	15.896
14	99116	.00018	65.649	69	81617	.02017	15.185
15	99098	.00022	64.660	70	79970	.02190	14.487
16	99077	.00026	63.674	71	78219	.02399	13.800
17	99051	.00031	62.691	72	76343	.02693	13.127
18	99020	.00031	61.710	73	74287	.03014	12.476
19	98989	.00032	60.729	74	72048	.03284	11.848
20	98957	.00031	59.748	75	69682	.03569	11.234
21	98926	.00032	58.767	76	67195	.03919	10.631
22	98894	.00033	57.786	77	64561	.04356	10.044
23	98862	.00033	56.805	78	61749	.04833	9.478
24	98829	.00033	55.823	79	58765	.05373	8.934
25	98797	.00034	54.842	80	55607	.05961	8.413
26	98763	.00035	53.860	81	52293	.06568	7.914
27	98729	.00036	52.878	82	48858	.07216	7.435
28	98694	.00038	51.897	83	45332	.07933	6.974
29	98656	.00040	50.917	84	41736	.08757	6.532
30	98617	.00043	49.937	85	38081	.09731	6.111
31	98574	.00047	48.958	86	34375	.10833	5.715
32	98528	.00052	47.981	87	30651	.11859	5.349
33	98477	.00057	47.006	88	27017	.12860	5.002
34	98420	.00063	46.032	89	23542	.14146	4.667
35	98359	.00069	45.061	90	20212	.15550	4.354
36	98291	.00075	44.092	91	17069	.17006	4.065
37	98217	.00082	43.124	92	14166	.18573	3.797
38	98136	.00090	42.160	93	11535	.20126	3.551
39	98048	.00098	41.197	94	9214	.21790	3.322
40	97952	.00107	40.237	95	7206	.23619	3.112
41	97847	.00117	39.279	96	5504	.25344	2.925
42	97733	.00129	38.325	97	4109	.26820	2.754
43	97607	.00142	37.374	98	3007	.28352	2.588
44	97469	.00158	36.426	99	2154	.30331	2.422
45	97315	.00177	35.483	100	1501	.32489	2.269
46	97142	.00198	34.545	101	1013	.34562	2.133
47	96950	.00219	33.612	102	663	.36186	2.011
48	96738	.00241	32.685	103	423	.37992	1.887
49	96504	.00266	31.763	104	262	.40045	1.758
50	96247	.00294	30.846	105	157	.43618	1.621
51	95964	.00326	29.936	106	89	.45994	1.518
52	95652	.00357	29.032	107	48	.48389	1.425
53	95310	.00390	28.134	108	25	.50791	1.338
54	94938	.00428	27.242	109	12	.53190	1.257
				110	6	.55574	1.183
				111	3	.57932	1.114
				112	1	.60255	1.050

For an explanation of headings see page 305

TAX ADJUSTMENT FACTORS

TAX ADJUSTMENT FACTORS

T_G = GROSS ADJUSTMENT FACTOR. A gross equivalent figure before tax may be found with this factor.

$$T_G = \frac{100}{100 - \text{Rate of Tax}}$$

T_N = NET ADJUSTMENT FACTOR. A net equivalent figure after tax may be found with this factor.

$$T_N = \frac{100 - \text{Rate of Tax}}{100}$$

Note—Examples are given in the introduction which illustrate the use of these factors.

TAX ADJUSTMENT FACTORS

Tax	T_G	T_N	Tax	T_G	T_N
1%	1.01010	.99000	51%	2.04082	.49000
2%	1.02041	.98000	52%	2.08333	.48000
3%	1.03093	.97000	53%	2.12766	.47000
4%	1.04167	.96000	54%	2.17391	.46000
5%	1.05263	.95000	55%	2.22222	.45000
6%	1.06383	.94000	56%	2.27273	.44000
7%	1.07527	.93000	57%	2.32558	.43000
8%	1.08696	.92000	58%	2.38095	.42000
9%	1.09890	.91000	59%	2.43902	.41000
10%	1.11111	.90000	60%	2.50000	.40000
11%	1.12360	.89000	61%	2.56410	.39000
12%	1.13636	.88000	62%	2.63158	.38000
13%	1.14943	.87000	63%	2.70270	.37000
14%	1.16279	.86000	64%	2.77778	.36000
15%	1.17647	.85000	65%	2.85714	.35000
16%	1.19048	.84000	66%	2.94118	.34000
17%	1.20482	.83000	67%	3.03030	.33000
18%	1.21951	.82000	68%	3.12500	.32000
19%	1.23457	.81000	69%	3.22581	.31000
20%	1.25000	.80000	70%	3.33333	.30000
21%	1.26582	.79000	71%	3.44828	.29000
22%	1.28205	.78000	72%	3.57143	.28000
23%	1.29870	.77000	73%	3.70370	.27000
24%	1.31579	.76000	74%	3.84615	.26000
25%	1.33333	.75000	75%	4.00000	.25000
26%	1.35135	.74000	76%	4.16667	.24000
27%	1.36986	.73000	77%	4.34783	.23000
28%	1.38889	.72000	78%	4.54545	.22000
29%	1.40845	.71000	79%	4.76190	.21000
30%	1.42857	.70000	80%	5.00000	.20000
31%	1.44928	.69000	81%	5.26316	.19000
32%	1.47059	.68000	82%	5.55556	.18000
33%	1.49254	.67000	83%	5.88235	.17000
34%	1.51515	.66000	84%	6.25000	.16000
35%	1.53846	.65000	85%	6.66667	.15000
36%	1.56250	.64000	86%	7.14286	.14000
37%	1.58730	.63000	87%	7.69231	.13000
38%	1.61290	.62000	88%	8.33333	.12000
39%	1.63934	.61000	89%	9.09091	.11000
40%	1.66667	.60000	90%	10.00000	.10000
41%	1.69492	.59000	91%	11.11111	.09000
42%	1.72414	.58000	92%	12.50000	.08000
43%	1.75439	.57000	93%	14.28571	.07000
44%	1.78571	.56000	94%	16.66667	.06000
45%	1.81818	.55000	95%	20.00000	.05000
46%	1.85185	.54000	96%	25.00000	.04000
47%	1.88679	.53000	97%	33.33333	.03000
48%	1.92308	.52000	98%	50.00000	.02000
49%	1.96078	.51000	99%	100.00000	.01000
50%	2.00000	.50000	100%	∞	.00000

YEARS' PURCHASE

(DUAL RATE % PRINCIPLE)

OR

PRESENT VALUE OF
ONE POUND PER ANNUM

*receivable quarterly in advance, allow-
ing for a sinking fund at a given rate
to replace the invested capital and for
the effect of income tax at* **10%** *on
that part of the income used to provide
the annual sinking fund instalment.*

AT RATES OF INTEREST*

FROM

4% to 20%

AND

ALLOWING FOR THE POSSIBLE INVESTMENT

OF SINKING FUNDS AT

3% and 4%

INCOME TAX at 10%

* *Note:*—In computing the quarterly in advance figures in these tables, all the rates of
interest quoted are effective rates, see the Introductory Section, pages xvi to
xvii.

Rate Per Cent*

Yrs.	4	4.5	5	5.5	6	6.25	6.5	6.75	Yrs.
1	0.8851	0.8814	0.8778	0.8742	0.8706	0.8689	0.8671	0.8654	1
2	1.7351	1.7209	1.7071	1.6935	1.6802	1.6737	1.6673	1.6609	2
3	2.5516	2.5210	2.4914	2.4626	2.4347	2.4210	2.4075	2.3942	3
4	3.3362	3.2842	3.2341	3.1857	3.1391	3.1164	3.0941	3.0722	4
5	4.0904	4.0125	3.9380	3.8665	3.7981	3.7649	3.7324	3.7005	5
6	4.8157	4.7081	4.6058	4.5084	4.4156	4.3708	4.3270	4.2842	6
7	5.5134	5.3728	5.2399	5.1143	4.9951	4.9379	4.8821	4.8277	7
8	6.1847	6.0083	5.8427	5.6869	5.5400	5.4696	5.4013	5.3348	8
9	6.8309	6.6163	6.4161	6.2286	6.0528	5.9690	5.8877	5.8087	9
10	7.4530	7.1983	6.9619	6.7418	6.5363	6.4386	6.3441	6.2526	10
11	8.0521	7.7557	7.4819	7.2283	6.9926	6.8809	6.7731	6.6689	11
12	8.6293	8.2897	7.9777	7.6900	7.4238	7.2980	7.1768	7.0599	12
13	9.1854	8.8017	8.4507	8.1285	7.8317	7.6919	7.5574	7.4279	13
14	9.7214	9.2926	8.9023	8.5455	8.2181	8.0642	7.9165	7.7745	14
15	10.2381	9.7637	9.3337	8.9422	8.5843	8.4166	8.2558	8.1015	15
16	10.7364	10.2158	9.7460	9.3200	8.9319	8.7505	8.5768	8.4103	16
17	11.2169	10.6499	10.1404	9.6800	9.2620	9.0670	8.8807	8.7024	17
18	11.6805	11.0669	10.5177	10.0233	9.5758	9.3676	9.1688	8.9788	18
19	12.1278	11.4676	10.8790	10.3508	9.8743	9.6531	9.4422	9.2408	19
20	12.5594	11.8528	11.2251	10.6636	10.1586	9.9246	9.7018	9.4893	20
21	12.9761	12.2232	11.5567	10.9625	10.4294	10.1829	9.9485	9.7253	21
22	13.3783	12.5795	11.8747	11.2482	10.6877	10.4290	10.1833	9.9495	22
23	13.7667	12.9223	12.1797	11.5216	10.9342	10.6636	10.4068	10.1627	23
24	14.1419	13.2523	12.4724	11.7831	11.1695	10.8872	10.6197	10.3657	24
25	14.5042	13.5700	12.7534	12.0336	11.3943	11.1007	10.8227	10.5590	25
26	14.8542	13.8759	13.0233	12.2736	11.6092	11.3046	11.0164	10.7433	26
27	15.1925	14.1706	13.2825	12.5036	11.8148	11.4995	11.2014	10.9192	27
28	15.5193	14.4546	13.5317	12.7241	12.0115	11.6858	11.3781	11.0870	28
29	15.8352	14.7282	13.7712	12.9357	12.1999	11.8640	11.5470	11.2473	29
30	16.1406	14.9920	14.0016	13.1388	12.3804	12.0346	11.7085	11.4005	30
31	16.4359	15.2464	14.2233	13.3338	12.5533	12.1979	11.8631	11.5470	31
32	16.7213	15.4918	14.4365	13.5210	12.7192	12.3545	12.0111	11.6872	32
33	16.9974	15.7284	14.6419	13.7010	12.8783	12.5045	12.1529	11.8214	33
34	17.2644	15.9568	14.8395	13.8739	13.0310	12.6484	12.2887	11.9499	34
35	17.5226	16.1772	15.0300	14.0402	13.1776	12.7865	12.4190	12.0731	35
36	17.7725	16.3899	15.2134	14.2002	13.3184	12.9190	12.5440	12.1911	36
37	18.0142	16.5952	15.3902	14.3540	13.4536	13.0463	12.6639	12.3044	37
38	18.2481	16.7935	15.5605	14.5021	13.5837	13.1685	12.7791	12.4131	38
39	18.4744	16.9850	15.7248	14.6447	13.7087	13.2859	12.8896	12.5174	39
40	18.6934	17.1699	15.8832	14.7820	13.8289	13.3989	12.9959	12.6175	40
41	18.9054	17.3486	16.0360	14.9143	13.9446	13.5074	13.0980	12.7138	41
42	19.1106	17.5213	16.1834	15.0417	14.0559	13.6118	13.1962	12.8062	42
43	19.3093	17.6881	16.3256	15.1645	14.1631	13.7123	13.2906	12.8951	43
44	19.5016	17.8494	16.4629	15.2828	14.2663	13.8090	13.3814	12.9806	44
45	19.6878	18.0052	16.5954	15.3970	14.3657	13.9022	13.4688	13.0629	45
46	19.8681	18.1559	16.7234	15.5070	14.4615	13.9918	13.5530	13.1420	46
47	20.0428	18.3017	16.8469	15.6132	14.5537	14.0782	13.6340	13.2182	47
48	20.2119	18.4426	16.9662	15.7156	14.6427	14.1614	13.7121	13.2915	48
49	20.3757	18.5788	17.0815	15.8145	14.7285	14.2416	13.7873	13.3622	49
50	20.5343	18.7107	17.1928	15.9099	14.8112	14.3190	13.8597	13.4302	50

* Note:-In computing the quarterly in advance figures in these tables, all the rates % quoted above are effective rates, see the Introductory Section, pages xvi-xvii

				Rate Per Cent*					
Yrs.	4	4.5	5	5.5	6	6.25	6.5	6.75	Yrs.
51	20.6880	18.8382	17.3005	16.0020	14.8910	14.3935	13.9296	13.4958	51
52	20.8369	18.9615	17.4044	16.0909	14.9680	14.4654	13.9969	13.5590	52
53	20.9811	19.0809	17.5050	16.1768	15.0423	14.5348	14.0618	13.6199	53
54	21.1209	19.1964	17.6021	16.2598	15.1139	14.6017	14.1245	13.6787	54
55	21.2563	19.3082	17.6961	16.3399	15.1832	14.6663	14.1849	13.7353	55
56	21.3875	19.4164	17.7869	16.4173	15.2500	14.7287	14.2432	13.7900	56
57	21.5146	19.5211	17.8748	16.4921	15.3145	14.7889	14.2995	13.8427	57
58	21.6378	19.6225	17.9597	16.5644	15.3768	14.8470	14.3538	13.8937	58
59	21.7572	19.7207	18.0419	16.6343	15.4371	14.9031	14.4063	13.9428	59
60	21.8730	19.8157	18.1214	16.7019	15.4952	14.9573	14.4569	13.9902	60
61	21.9852	19.9077	18.1984	16.7672	15.5514	15.0097	14.5058	14.0360	61
62	22.0939	19.9968	18.2728	16.8304	15.6058	15.0603	14.5531	14.0803	62
63	22.1993	20.0832	18.3449	16.8915	15.6583	15.1092	14.5987	14.1230	63
64	22.3015	20.1667	18.4146	16.9506	15.7091	15.1564	14.6429	14.1643	64
65	22.4005	20.2477	18.4821	17.0077	15.7581	15.2021	14.6855	14.2042	65
66	22.4966	20.3261	18.5474	17.0630	15.8056	15.2463	14.7267	14.2427	66
67	22.5897	20.4021	18.6106	17.1166	15.8515	15.2890	14.7666	14.2800	67
68	22.6800	20.4757	18.6719	17.1683	15.8959	15.3303	14.8051	14.3160	68
69	22.7675	20.5470	18.7311	17.2184	15.9388	15.3703	14.8423	14.3509	69
70	22.8524	20.6161	18.7885	17.2669	15.9804	15.4089	14.8783	14.3845	70
71	22.9347	20.6831	18.8441	17.3139	16.0206	15.4463	14.9132	14.4171	71
72	23.0145	20.7480	18.8980	17.3593	16.0595	15.4824	14.9469	14.4486	72
73	23.0918	20.8109	18.9501	17.4033	16.0971	15.5174	14.9795	14.4791	73
74	23.1669	20.8718	19.0006	17.4459	16.1336	15.5513	15.0110	14.5085	74
75	23.2397	20.9309	19.0496	17.4872	16.1688	15.5840	15.0416	14.5370	75
76	23.3103	20.9881	19.0970	17.5271	16.2030	15.6157	15.0711	14.5646	76
77	23.3787	21.0436	19.1429	17.5658	16.2360	15.6464	15.0997	14.5913	77
78	23.4452	21.0974	19.1874	17.6032	16.2680	15.6761	15.1274	14.6172	78
79	23.5096	21.1495	19.2305	17.6395	16.2990	15.7049	15.1542	14.6422	79
80	23.5720	21.2001	19.2723	17.6747	16.3290	15.7328	15.1801	14.6664	80
81	23.6326	21.2491	19.3128	17.7087	16.3581	15.7597	15.2052	14.6898	81
82	23.6914	21.2966	19.3520	17.7417	16.3862	15.7859	15.2295	14.7125	82
83	23.7484	21.3426	19.3901	17.7737	16.4135	15.8112	15.2531	14.7345	83
84	23.8038	21.3873	19.4269	17.8046	16.4399	15.8357	15.2759	14.7558	84
85	23.8574	21.4306	19.4627	17.8346	16.4655	15.8594	15.2979	14.7764	85
86	23.9095	21.4726	19.4973	17.8637	16.4902	15.8824	15.3193	14.7963	86
87	23.9600	21.5133	19.5309	17.8919	16.5142	15.9046	15.3400	14.8156	87
88	24.0090	21.5528	19.5634	17.9192	16.5375	15.9262	15.3601	14.8344	88
89	24.0565	21.5911	19.5949	17.9457	16.5600	15.9471	15.3795	14.8525	89
90	24.1026	21.6283	19.6255	17.9713	16.5819	15.9674	15.3984	14.8701	90
91	24.1474	21.6643	19.6552	17.9962	16.6031	15.9870	15.4166	14.8871	91
92	24.1908	21.6992	19.6839	18.0203	16.6236	16.0060	15.4343	14.9036	92
93	24.2329	21.7331	19.7118	18.0436	16.6434	16.0244	15.4515	14.9195	93
94	24.2738	21.7660	19.7389	18.0663	16.6627	16.0423	15.4681	14.9350	94
95	24.3134	21.7978	19.7651	18.0882	16.6814	16.0596	15.4841	14.9500	95
96	24.3519	21.8288	19.7905	18.1095	16.6995	16.0764	15.4997	14.9646	96
97	24.3892	21.8588	19.8151	18.1302	16.7170	16.0926	15.5149	14.9787	97
98	24.4254	21.8879	19.8390	18.1502	16.7340	16.1084	15.5295	14.9923	98
99	24.4606	21.9161	19.8622	18.1696	16.7505	16.1237	15.5437	15.0055	99
100	24.4947	21.9435	19.8847	18.1884	16.7665	16.1385	15.5575	15.0184	100

Note:-In computing the quarterly in advance figures in these tables, all the rates % quoted above are effective rates, see the
Introductory Section, pages xvi-xvii

				Rate Per Cent*					
Yrs.	7	7.25	7.5	8	8.5	9	9.5	10	Yrs.
1	0.8637	0.8620	0.8603	0.8569	0.8536	0.8503	0.8471	0.8439	1
2	1.6545	1.6483	1.6421	1.6299	1.6180	1.6063	1.5948	1.5836	2
3	2.3811	2.3682	2.3554	2.3304	2.3061	2.2824	2.2593	2.2368	3
4	3.0506	3.0294	3.0085	2.9679	2.9285	2.8904	2.8535	2.8177	4
5	3.6693	3.6386	3.6086	3.5502	3.4941	3.4400	3.3878	3.3375	5
6	4.2424	4.2015	4.1615	4.0841	4.0100	3.9389	3.8706	3.8051	6
7	4.7747	4.7230	4.6725	4.5751	4.4823	4.3936	4.3089	4.2278	7
8	5.2701	5.2071	5.1459	5.0280	4.9161	4.8096	4.7083	4.6116	8
9	5.7321	5.6577	5.5855	5.4469	5.3158	5.1915	5.0736	4.9616	9
10	6.1639	6.0780	5.9946	5.8353	5.6851	5.5432	5.4090	5.2818	10
11	6.5681	6.4706	6.3762	6.1963	6.0272	5.8679	5.7177	5.5758	11
12	6.9471	6.8381	6.7328	6.5325	6.3448	6.1686	6.0028	5.8466	12
13	7.3031	7.1827	7.0666	6.8463	6.6404	6.4476	6.2668	6.0967	13
14	7.6379	7.5064	7.3797	7.1397	6.9161	6.7072	6.5117	6.3283	14
15	7.9533	7.8108	7.6737	7.4145	7.1737	6.9492	6.7396	6.5433	15
16	8.2507	8.0974	7.9502	7.6724	7.4148	7.1752	6.9519	6.7433	16
17	8.5316	8.3678	8.2107	7.9147	7.6408	7.3867	7.1503	6.9297	17
18	8.7971	8.6231	8.4563	8.1427	7.8531	7.5850	7.3359	7.1039	18
19	9.0485	8.8645	8.6883	8.3576	8.0528	7.7711	7.5098	7.2669	19
20	9.2866	9.0929	8.9076	8.5603	8.2409	7.9461	7.6731	7.4197	20
21	9.5124	9.3093	9.1152	8.7519	8.4182	8.1108	7.8267	7.5632	21
22	9.7268	9.5145	9.3119	8.9330	8.5857	8.2662	7.9712	7.6981	22
23	9.9305	9.7094	9.4984	9.1045	8.7440	8.4128	8.1075	7.8251	23
24	10.1243	9.8945	9.6755	9.2671	8.8939	8.5515	8.2362	7.9449	24
25	10.3086	10.0705	9.8437	9.4213	9.0359	8.6826	8.3578	8.0580	25
26	10.4842	10.2380	10.0037	9.5678	9.1705	8.8069	8.4728	8.1649	26
27	10.6516	10.3975	10.1560	9.7070	9.2983	8.9247	8.5818	8.2661	27
28	10.8112	10.5496	10.3010	9.8394	9.4197	9.0365	8.6851	8.3619	28
29	10.9636	10.6946	10.4393	9.9655	9.5352	9.1427	8.7832	8.4527	29
30	11.1091	10.8331	10.5711	10.0855	9.6450	9.2436	8.8763	8.5390	30
31	11.2482	10.9653	10.6970	10.2000	9.7497	9.3397	8.9649	8.6209	31
32	11.3812	11.0916	10.8172	10.3093	9.8494	9.4312	9.0492	8.6988	32
33	11.5084	11.2124	10.9320	10.4135	9.9446	9.5184	9.1294	8.7729	33
34	11.6302	11.3279	11.0419	10.5131	10.0354	9.6016	9.2059	8.8435	34
35	11.7468	11.4386	11.1469	10.6083	10.1221	9.6809	9.2788	8.9108	35
36	11.8585	11.5445	11.2475	10.6994	10.2050	9.7567	9.3484	8.9749	36
37	11.9657	11.6460	11.3438	10.7865	10.2842	9.8291	9.4148	9.0362	37
38	12.0684	11.7433	11.4361	10.8700	10.3600	9.8983	9.4783	9.0946	38
39	12.1670	11.8366	11.5246	10.9499	10.4326	9.9645	9.5390	9.1505	39
40	12.2616	11.9262	11.6095	11.0264	10.5021	10.0279	9.5971	9.2039	40
41	12.3525	12.0121	11.6909	11.0999	10.5686	10.0886	9.6527	9.2550	41
42	12.4397	12.0946	11.7690	11.1703	10.6325	10.1467	9.7059	9.3039	42
43	12.5236	12.1739	11.8441	11.2379	10.6937	10.2025	9.7568	9.3508	43
44	12.6042	12.2500	11.9162	11.3027	10.7524	10.2559	9.8057	9.3956	44
45	12.6818	12.3233	11.9854	11.3650	10.8088	10.3072	9.8526	9.4386	45
46	12.7563	12.3937	12.0520	11.4249	10.8629	10.3564	9.8975	9.4799	46
47	12.8281	12.4614	12.1161	11.4824	10.9149	10.4036	9.9407	9.5195	47
48	12.8972	12.5266	12.1777	11.5377	10.9648	10.4490	9.9821	9.5574	48
49	12.9637	12.5893	12.2369	11.5909	11.0129	10.4926	10.0219	9.5939	49
50	13.0277	12.6497	12.2940	11.6421	11.0591	10.5345	10.0601	9.6289	50

* Note:-In computing the quarterly in advance figures in these tables, all the rates % quoted above are effective rates, see the
Introductory Section, pages xvi-xvii

				Rate Per Cent*					
Yrs.	7	7.25	7.5	8	8.5	9	9.5	10	Yrs.
51	13.0894	12.7078	12.3489	11.6913	11.1035	10.5748	10.0969	9.6626	51
52	13.1488	12.7638	12.4018	11.7387	11.1462	10.6136	10.1322	9.6950	52
53	13.2061	12.8178	12.4527	11.7844	11.1874	10.6509	10.1662	9.7261	53
54	13.2613	12.8698	12.5018	11.8283	11.2270	10.6868	10.1989	9.7560	54
55	13.3146	12.9200	12.5491	11.8707	11.2651	10.7213	10.2303	9.7848	55
56	13.3660	12.9683	12.5948	11.9115	11.3019	10.7546	10.2606	9.8125	56
57	13.4155	13.0150	12.6387	11.9508	11.3373	10.7867	10.2898	9.8392	57
58	13.4633	13.0600	12.6812	11.9887	11.3714	10.8176	10.3179	9.8649	58
59	13.5094	13.1034	12.7221	12.0253	11.4043	10.8473	10.3450	9.8896	59
60	13.5540	13.1453	12.7616	12.0606	11.4360	10.8760	10.3711	9.9134	60
61	13.5970	13.1857	12.7997	12.0946	11.4666	10.9037	10.3962	9.9364	61
62	13.6385	13.2247	12.8365	12.1274	11.4961	10.9304	10.4205	9.9586	62
63	13.6786	13.2624	12.8720	12.1591	11.5246	10.9561	10.4439	9.9799	63
64	13.7173	13.2988	12.9063	12.1897	11.5521	10.9809	10.4664	10.0005	64
65	13.7547	13.3340	12.9394	12.2192	11.5786	11.0049	10.4882	10.0204	65
66	13.7909	13.3680	12.9714	12.2478	11.6042	11.0280	10.5092	10.0396	66
67	13.8258	13.4008	13.0023	12.2753	11.6289	11.0503	10.5295	10.0581	67
68	13.8596	13.4325	13.0321	12.3019	11.6528	11.0719	10.5490	10.0759	68
69	13.8922	13.4632	13.0610	12.3276	11.6758	11.0927	10.5679	10.0932	69
70	13.9237	13.4928	13.0888	12.3525	11.6981	11.1128	10.5862	10.1098	70
71	13.9543	13.5214	13.1158	12.3765	11.7196	11.1323	10.6038	10.1259	71
72	13.9838	13.5491	13.1419	12.3997	11.7405	11.1510	10.6208	10.1414	72
73	14.0123	13.5759	13.1671	12.4221	11.7606	11.1692	10.6373	10.1564	73
74	14.0399	13.6018	13.1914	12.4438	11.7800	11.1867	10.6532	10.1709	74
75	14.0666	13.6269	13.2150	12.4648	11.7988	11.2036	10.6686	10.1849	75
76	14.0924	13.6511	13.2378	12.4850	11.8170	11.2200	10.6834	10.1985	76
77	14.1174	13.6746	13.2599	12.5047	11.8345	11.2358	10.6978	10.2115	77
78	14.1416	13.6973	13.2812	12.5236	11.8515	11.2512	10.7117	10.2242	78
79	14.1650	13.7192	13.3018	12.5420	11.8680	11.2660	10.7251	10.2364	79
80	14.1877	13.7405	13.3218	12.5597	11.8839	11.2803	10.7381	10.2482	80
81	14.2096	13.7611	13.3411	12.5769	11.8992	11.2942	10.7506	10.2597	81
82	14.2308	13.7810	13.3599	12.5936	11.9141	11.3076	10.7628	10.2707	82
83	14.2514	13.8002	13.3780	12.6096	11.9285	11.3205	10.7745	10.2814	83
84	14.2713	13.8189	13.3955	12.6252	11.9425	11.3331	10.7859	10.2918	84
85	14.2906	13.8370	13.4125	12.6403	11.9560	11.3452	10.7969	10.3018	85
86	14.3092	13.8545	13.4289	12.6549	11.9690	11.3570	10.8075	10.3115	86
87	14.3273	13.8714	13.4448	12.6690	11.9817	11.3684	10.8179	10.3209	87
88	14.3448	13.8878	13.4602	12.6827	11.9939	11.3794	10.8278	10.3300	88
89	14.3617	13.9037	13.4752	12.6960	12.0057	11.3901	10.8375	10.3388	89
90	14.3782	13.9191	13.4896	12.7088	12.0172	11.4004	10.8468	10.3473	90
91	14.3941	13.9340	13.5036	12.7212	12.0283	11.4104	10.8559	10.3555	91
92	14.4095	13.9484	13.5172	12.7333	12.0391	11.4201	10.8647	10.3635	92
93	14.4244	13.9624	13.5303	12.7449	12.0495	11.4295	10.8731	10.3712	93
94	14.4389	13.9760	13.5431	12.7562	12.0596	11.4385	10.8814	10.3787	94
95	14.4529	13.9891	13.5554	12.7672	12.0694	11.4473	10.8893	10.3859	95
96	14.4665	14.0019	13.5674	12.7778	12.0789	11.4559	10.8970	10.3929	96
97	14.4797	14.0142	13.5789	12.7880	12.0880	11.4641	10.9045	10.3997	97
98	14.4924	14.0261	13.5902	12.7980	12.0969	11.4721	10.9117	10.4063	98
99	14.5048	14.0377	13.6010	12.8076	12.1056	11.4799	10.9188	10.4127	99
100	14.5168	14.0490	13.6116	12.8170	12.1139	11.4874	10.9255	10.4189	100

Note:-In computing the quarterly in advance figures in these tables, all the rates % quoted above are effective rates, see the introductory Section, pages xvi-xvii

YEARS' PURCHASE*

Rate Per Cent*

Yrs.	11	12	13	14	15	16	18	20	Yrs.
1	0.8377	0.8316	0.8257	0.8199	0.8143	0.8088	0.7982	0.7880	1
2	1.5617	1.5407	1.5205	1.5010	1.4822	1.4641	1.4297	1.3976	2
3	2.1935	2.1523	2.1130	2.0756	2.0398	2.0057	1.9417	1.8828	3
4	2.7493	2.6849	2.6241	2.5666	2.5122	2.4605	2.3649	2.2782	4
5	3.2420	3.1528	3.0692	2.9909	2.9172	2.8478	2.7205	2.6064	5
6	3.6814	3.5668	3.4603	3.3610	3.2683	3.1814	3.0233	2.8831	6
7	4.0757	3.9357	3.8064	3.6866	3.5753	3.4716	3.2842	3.1194	7
8	4.4312	4.2662	4.1147	3.9751	3.8460	3.7263	3.5113	3.3235	8
9	4.7534	4.5641	4.3911	4.2324	4.0864	3.9515	3.7105	3.5015	9
10	5.0465	4.8336	4.6401	4.4633	4.3012	4.1520	3.8868	3.6580	10
11	5.3143	5.0787	4.8654	4.6714	4.4942	4.3316	4.0437	3.7966	11
12	5.5597	5.3024	5.0704	4.8600	4.6684	4.4933	4.1842	3.9203	12
13	5.7854	5.5073	5.2574	5.0316	4.8265	4.6395	4.3108	4.0312	13
14	5.9935	5.6956	5.4287	5.1883	4.9705	4.7724	4.4253	4.1311	14
15	6.1860	5.8691	5.5861	5.3319	5.1022	4.8937	4.5294	4.2217	15
16	6.3645	6.0295	5.7313	5.4640	5.2230	5.0047	4.6243	4.3040	16
17	6.5303	6.1782	5.8654	5.5857	5.3342	5.1067	4.7112	4.3792	17
18	6.6847	6.3162	5.9897	5.6984	5.4368	5.2006	4.7911	4.4481	18
19	6.8289	6.4448	6.1052	5.8028	5.5317	5.2875	4.8647	4.5115	19
20	6.9636	6.5647	6.2127	5.8998	5.6198	5.3679	4.9327	4.5699	20
21	7.0899	6.6767	6.3129	5.9901	5.7018	5.4426	4.9957	4.6240	21
22	7.2083	6.7816	6.4066	6.0744	5.7781	5.5121	5.0542	4.6740	22
23	7.3196	6.8800	6.4944	6.1533	5.8494	5.5769	5.1086	4.7206	23
24	7.4243	6.9725	6.5767	6.2271	5.9161	5.6375	5.1594	4.7639	24
25	7.5229	7.0594	6.6540	6.2964	5.9785	5.6942	5.2069	4.8043	25
26	7.6160	7.1413	6.7267	6.3614	6.0372	5.7474	5.2513	4.8421	26
27	7.7040	7.2186	6.7952	6.4227	6.0923	5.7973	5.2930	4.8775	27
28	7.7871	7.2915	6.8598	6.4804	6.1442	5.8443	5.3321	4.9107	28
29	7.8659	7.3605	6.9209	6.5348	6.1931	5.8885	5.3689	4.9419	29
30	7.9405	7.4258	6.9786	6.5862	6.2393	5.9303	5.4036	4.9713	30
31	8.0113	7.4877	7.0332	6.6349	6.2829	5.9697	5.4362	4.9989	31
32	8.0785	7.5464	7.0850	6.6809	6.3242	6.0069	5.4671	5.0250	32
33	8.1424	7.6021	7.1340	6.7245	6.3633	6.0422	5.4963	5.0497	33
34	8.2032	7.6551	7.1807	6.7659	6.4003	6.0756	5.5239	5.0730	34
35	8.2610	7.7054	7.2249	6.8052	6.4355	6.1072	5.5501	5.0951	35
36	8.3161	7.7534	7.2671	6.8426	6.4689	6.1373	5.5749	5.1160	36
37	8.3687	7.7990	7.3072	6.8781	6.5006	6.1659	5.5985	5.1358	37
38	8.4188	7.8425	7.3453	6.9120	6.5308	6.1930	5.6209	5.1546	38
39	8.4667	7.8841	7.3817	6.9442	6.5596	6.2189	5.6422	5.1725	39
40	8.5124	7.9237	7.4165	6.9749	6.5870	6.2435	5.6624	5.1896	40
41	8.5561	7.9615	7.4496	7.0042	6.6131	6.2670	5.6817	5.2058	41
42	8.5978	7.9977	7.4813	7.0322	6.6381	6.2894	5.7001	5.2212	42
43	8.6378	8.0323	7.5115	7.0589	6.6619	6.3108	5.7177	5.2359	43
44	8.6761	8.0654	7.5404	7.0844	6.6846	6.3312	5.7344	5.2500	44
45	8.7128	8.0970	7.5681	7.1089	6.7064	6.3507	5.7504	5.2634	45
46	8.7479	8.1274	7.5946	7.1322	6.7272	6.3693	5.7657	5.2762	46
47	8.7816	8.1564	7.6200	7.1546	6.7471	6.3871	5.7803	5.2884	47
48	8.8139	8.1843	7.6443	7.1761	6.7661	6.4042	5.7943	5.3001	48
49	8.8449	8.2110	7.6676	7.1966	6.7844	6.4206	5.8077	5.3113	49
50	8.8747	8.2367	7.6900	7.2163	6.8019	6.4362	5.8205	5.3220	50

* Note:-In computing the quarterly in advance figures in these tables, all the rates % quoted above are effective rates, see the Introductory Section, pages xvi-xvii

Rate Per Cent*

Yrs.	11	12	13	14	15	16	18	20	Yrs.
51	8.9032	8.2613	7.7114	7.2352	6.8186	6.4513	5.8328	5.3323	51
52	8.9307	8.2849	7.7320	7.2533	6.8347	6.4657	5.8445	5.3421	52
53	8.9571	8.3076	7.7518	7.2707	6.8502	6.4795	5.8558	5.3516	53
54	8.9825	8.3295	7.7708	7.2874	6.8650	6.4928	5.8667	5.3606	54
55	9.0069	8.3504	7.7891	7.3035	6.8793	6.5055	5.8771	5.3693	55
56	9.0303	8.3706	7.8066	7.3189	6.8930	6.5177	5.8871	5.3776	56
57	9.0529	8.3900	7.8235	7.3337	6.9061	6.5295	5.8967	5.3856	57
58	9.0747	8.4087	7.8397	7.3480	6.9188	6.5408	5.9059	5.3933	58
59	9.0956	8.4267	7.8553	7.3617	6.9309	6.5517	5.9147	5.4007	59
60	9.1158	8.4440	7.8704	7.3749	6.9426	6.5621	5.9232	5.4078	60
61	9.1352	8.4606	7.8848	7.3876	6.9539	6.5722	5.9314	5.4146	61
62	9.1539	8.4767	7.8988	7.3999	6.9647	6.5819	5.9393	5.4212	62
63	9.1720	8.4922	7.9122	7.4116	6.9752	6.5912	5.9469	5.4275	63
64	9.1894	8.5071	7.9252	7.4230	6.9852	6.6002	5.9542	5.4336	64
65	9.2061	8.5214	7.9376	7.4339	6.9949	6.6088	5.9613	5.4395	65
66	9.2223	8.5353	7.9497	7.4445	7.0042	6.6172	5.9680	5.4451	66
67	9.2379	8.5487	7.9613	7.4547	7.0132	6.6252	5.9746	5.4505	67
68	9.2530	8.5616	7.9724	7.4645	7.0219	6.6329	5.9809	5.4558	68
69	9.2675	8.5740	7.9832	7.4739	7.0303	6.6404	5.9869	5.4608	69
70	9.2816	8.5860	7.9936	7.4830	7.0384	6.6476	5.9928	5.4657	70
71	9.2951	8.5976	8.0037	7.4918	7.0461	6.6545	5.9984	5.4704	71
72	9.3082	8.6088	8.0134	7.5003	7.0537	6.6612	6.0039	5.4749	72
73	9.3208	8.6196	8.0227	7.5085	7.0609	6.6677	6.0091	5.4793	73
74	9.3330	8.6300	8.0318	7.5165	7.0679	6.6740	6.0142	5.4835	74
75	9.3448	8.6401	8.0405	7.5241	7.0747	6.6800	6.0191	5.4876	75
76	9.3562	8.6499	8.0489	7.5315	7.0812	6.6858	6.0238	5.4915	76
77	9.3672	8.6593	8.0571	7.5386	7.0875	6.6914	6.0284	5.4953	77
78	9.3779	8.6684	8.0650	7.5455	7.0936	6.6969	6.0328	5.4990	78
79	9.3882	8.6772	8.0726	7.5522	7.0995	6.7021	6.0371	5.5025	79
80	9.3981	8.6857	8.0799	7.5586	7.1052	6.7072	6.0412	5.5059	80
81	9.4077	8.6939	8.0870	7.5648	7.1107	6.7121	6.0451	5.5092	81
82	9.4170	8.7018	8.0939	7.5708	7.1160	6.7168	6.0490	5.5124	82
83	9.4260	8.7095	8.1006	7.5767	7.1211	6.7214	6.0527	5.5155	83
84	9.4347	8.7169	8.1070	7.5823	7.1261	6.7258	6.0563	5.5185	84
85	9.4431	8.7241	8.1132	7.5877	7.1309	6.7301	6.0597	5.5213	85
86	9.4513	8.7311	8.1192	7.5930	7.1355	6.7342	6.0631	5.5241	86
87	9.4592	8.7378	8.1250	7.5981	7.1400	6.7382	6.0663	5.5268	87
88	9.4668	8.7443	8.1306	7.6030	7.1444	6.7421	6.0695	5.5294	88
89	9.4742	8.7506	8.1361	7.6077	7.1486	6.7458	6.0725	5.5319	89
90	9.4813	8.7567	8.1414	7.6123	7.1526	6.7494	6.0754	5.5344	90
91	9.4882	8.7626	8.1465	7.6168	7.1566	6.7529	6.0783	5.5367	91
92	9.4949	8.7683	8.1514	7.6211	7.1604	6.7563	6.0810	5.5390	92
93	9.5014	8.7738	8.1562	7.6253	7.1641	6.7596	6.0837	5.5412	93
94	9.5077	8.7792	8.1608	7.6293	7.1676	6.7628	6.0863	5.5433	94
95	9.5138	8.7843	8.1653	7.6332	7.1711	6.7659	6.0888	5.5454	95
96	9.5196	8.7894	8.1696	7.6370	7.1744	6.7688	6.0912	5.5474	96
97	9.5253	8.7942	8.1738	7.6407	7.1777	6.7717	6.0935	5.5493	97
98	9.5309	8.7989	8.1779	7.6442	7.1808	6.7745	6.0958	5.5512	98
99	9.5362	8.8035	8.1818	7.6477	7.1838	6.7772	6.0979	5.5530	99
100	9.5414	8.8079	8.1856	7.6510	7.1868	6.7798	6.1001	5.5548	100

Note:-In computing the quarterly in advance figures in these tables, all the rates % quoted above are effective rates, see the introductory Section, pages xvi-xvii

YEARS' PURCHASE*

Rate Per Cent*

Yrs.	4	4.5	5	5.5	6	6.25	6.5	6.75	Yrs.
1	0.8903	0.8866	0.8829	0.8793	0.8757	0.8739	0.8721	0.8704	1
2	1.7529	1.7385	1.7243	1.7105	1.6970	1.6903	1.6837	1.6772	2
3	2.5884	2.5570	2.5266	2.4970	2.4682	2.4542	2.4403	2.4266	3
4	3.3975	3.3436	3.2917	3.2416	3.1933	3.1699	3.1468	3.1241	4
5	4.1808	4.0994	4.0216	3.9472	3.8758	3.8413	3.8074	3.7743	5
6	4.9389	4.8257	4.7183	4.6161	4.5189	4.4720	4.4262	4.3814	6
7	5.6724	5.5237	5.3834	5.2508	5.1253	5.0651	5.0064	4.9492	7
8	6.3821	6.1944	6.0185	5.8533	5.6978	5.6234	5.5512	5.4810	8
9	7.0685	6.8390	6.6253	6.4256	6.2387	6.1496	6.0633	5.9797	9
10	7.7323	7.4585	7.2050	6.9695	6.7501	6.6460	6.5453	6.4479	10
11	8.3740	8.0539	7.7590	7.4866	7.2341	7.1146	6.9994	6.8881	11
12	8.9943	8.6261	8.2887	7.9786	7.6924	7.5575	7.4276	7.3024	12
13	9.5939	9.1760	8.7952	8.4468	8.1267	7.9763	7.8317	7.6927	13
14	10.1732	9.7046	9.2797	8.8926	8.5386	8.3727	8.2135	8.0608	14
15	10.7328	10.2125	9.7431	9.3173	8.9294	8.7481	8.5745	8.4082	15
16	11.2734	10.7008	10.1865	9.7220	9.3004	9.1039	8.9161	8.7363	16
17	11.7955	11.1701	10.6109	10.1078	9.6529	9.4414	9.2395	9.0467	17
18	12.2996	11.6212	11.0171	10.4758	9.9879	9.7616	9.5460	9.3403	18
19	12.7864	12.0547	11.4060	10.8268	10.3065	10.0657	9.8366	9.6183	19
20	13.2562	12.4715	11.7784	11.1618	10.6096	10.3547	10.1123	9.8818	20
21	13.7097	12.8721	12.1351	11.4816	10.8982	10.6293	10.3741	10.1316	21
22	14.1473	13.2571	12.4767	11.7869	11.1729	10.8905	10.6228	10.3686	22
23	14.5696	13.6272	12.8040	12.0786	11.4347	11.1390	10.8591	10.5937	23
24	14.9771	13.9830	13.1176	12.3573	11.6841	11.3756	11.0838	10.8075	24
25	15.3701	14.3250	13.4181	12.6237	11.9220	11.6009	11.2976	11.0106	25
26	15.7492	14.6538	13.7061	12.8783	12.1488	11.8156	11.5011	11.2038	26
27	16.1148	14.9698	13.9822	13.1217	12.3652	12.0202	11.6949	11.3876	27
28	16.4674	15.2736	14.2469	13.3545	12.5717	12.2153	11.8795	11.5626	28
29	16.8074	15.5656	14.5007	13.5773	12.7689	12.4014	12.0554	11.7292	29
30	17.1352	15.8464	14.7440	13.7904	12.9572	12.5790	12.2231	11.8879	30
31	17.4512	16.1163	14.9774	13.9943	13.1371	12.7484	12.3831	12.0391	31
32	17.7558	16.3757	15.2012	14.1895	13.3090	12.9102	12.5357	12.1833	32
33	18.0494	16.6251	15.4159	14.3764	13.4733	13.0647	12.6813	12.3208	33
34	18.3324	16.8649	15.6218	14.5554	13.6303	13.2124	12.8204	12.4520	34
35	18.6051	17.0954	15.8194	14.7267	13.7805	13.3534	12.9531	12.5772	35
36	18.8678	17.3170	16.0090	14.8909	13.9241	13.4882	13.0800	12.6968	36
37	19.1210	17.5300	16.1909	15.0481	14.0615	13.6171	13.2011	12.8109	37
38	19.3649	17.7348	16.3654	15.1988	14.1930	13.7404	13.3169	12.9200	38
39	19.5999	17.9317	16.5329	15.3432	14.3188	13.8583	13.4277	13.0241	39
40	19.8263	18.1210	16.6937	15.4816	14.4393	13.9711	13.5335	13.1237	40
41	20.0444	18.3030	16.8480	15.6142	14.5546	14.0790	13.6348	13.2189	41
42	20.2544	18.4780	16.9962	15.7414	14.6650	14.1823	13.7316	13.3099	42
43	20.4567	18.6462	17.1384	15.8633	14.7708	14.2812	13.8243	13.3970	43
44	20.6515	18.8079	17.2749	15.9802	14.8721	14.3759	13.9130	13.4802	44
45	20.8391	18.9634	17.4060	16.0923	14.9691	14.4665	13.9979	13.5599	45
46	21.0198	19.1128	17.5318	16.1998	15.0621	14.5533	14.0792	13.6362	46
47	21.1937	19.2566	17.6527	16.3029	15.1512	14.6365	14.1570	13.7092	47
48	21.3612	19.3947	17.7687	16.4018	15.2366	14.7162	14.2315	13.7791	48
49	21.5224	19.5276	17.8802	16.4967	15.3185	14.7925	14.3029	13.8460	49
50	21.6777	19.6553	17.9872	16.5878	15.3970	14.8657	14.3713	13.9101	50

* Note:-In computing the quarterly in advance figures in these tables, all the rates % quoted above are effective rates, see the Introductory Section, pages xvi-xvii

Rate Per Cent*

Yrs.	4	4.5	5	5.5	6	6.25	6.5	6.75	Yrs.
51	21.8271	19.7781	18.0900	16.6751	15.4722	14.9359	14.4369	13.9715	51
52	21.9710	19.8961	18.1887	16.7590	15.5444	15.0031	14.4997	14.0303	52
53	22.1095	20.0096	18.2835	16.8394	15.6136	15.0675	14.5599	14.0866	53
54	22.2428	20.1188	18.3746	16.9167	15.6799	15.1293	14.6175	14.1406	54
55	22.3711	20.2237	18.4620	16.9908	15.7436	15.1886	14.6729	14.1924	55
56	22.4946	20.3246	18.5461	17.0619	15.8046	15.2454	14.7259	14.2420	56
57	22.6135	20.4215	18.6268	17.1302	15.8632	15.2999	14.7767	14.2895	57
58	22.7279	20.5148	18.7043	17.1958	15.9194	15.3522	14.8255	14.3351	58
59	22.8380	20.6045	18.7788	17.2587	15.9734	15.4024	14.8723	14.3788	59
60	22.9440	20.6907	18.8504	17.3192	16.0251	15.4505	14.9171	14.4208	60
61	23.0459	20.7735	18.9192	17.3772	16.0748	15.4967	14.9602	14.4610	61
62	23.1440	20.8532	18.9853	17.4330	16.1225	15.5410	15.0014	14.4996	62
63	23.2385	20.9299	19.0488	17.4865	16.1683	15.5835	15.0411	14.5366	63
64	23.3293	21.0035	19.1098	17.5379	16.2122	15.6243	15.0791	14.5721	64
65	23.4167	21.0744	19.1684	17.5872	16.2543	15.6634	15.1155	14.6061	65
66	23.5008	21.1425	19.2247	17.6346	16.2948	15.7010	15.1505	14.6388	66
67	23.5818	21.2079	19.2788	17.6802	16.3337	15.7371	15.1841	14.6702	67
68	23.6596	21.2709	19.3308	17.7239	16.3710	15.7717	15.2164	14.7003	68
69	23.7346	21.3314	19.3808	17.7659	16.4069	15.8050	15.2473	14.7291	69
70	23.8066	21.3896	19.4289	17.8062	16.4413	15.8369	15.2770	14.7569	70
71	23.8760	21.4456	19.4750	17.8450	16.4743	15.8676	15.3056	14.7835	71
72	23.9427	21.4994	19.5194	17.8823	16.5060	15.8970	15.3330	14.8090	72
73	24.0069	21.5511	19.5620	17.9180	16.5365	15.9253	15.3593	14.8336	73
74	24.0686	21.6009	19.6030	17.9524	16.5658	15.9524	15.3845	14.8571	74
75	24.1280	21.6487	19.6424	17.9854	16.5939	15.9785	15.4087	14.8797	75
76	24.1851	21.6947	19.6802	18.0172	16.6209	16.0035	15.4320	14.9014	76
77	24.2401	21.7389	19.7166	18.0476	16.6468	16.0276	15.4544	14.9223	77
78	24.2930	21.7814	19.7516	18.0769	16.6718	16.0507	15.4759	14.9423	78
79	24.3438	21.8223	19.7852	18.1051	16.6957	16.0729	15.4965	14.9615	79
80	24.3928	21.8616	19.8175	18.1321	16.7187	16.0942	15.5163	14.9800	80
81	24.4398	21.8994	19.8485	18.1581	16.7408	16.1147	15.5353	14.9977	81
82	24.4851	21.9357	19.8784	18.1831	16.7620	16.1343	15.5536	15.0148	82
83	24.5286	21.9707	19.9071	18.2071	16.7824	16.1532	15.5712	15.0311	83
84	24.5705	22.0043	19.9346	18.2301	16.8020	16.1714	15.5880	15.0468	84
85	24.6108	22.0366	19.9611	18.2523	16.8208	16.1888	15.6042	15.0619	85
86	24.6495	22.0676	19.9866	18.2736	16.8389	16.2056	15.6198	15.0764	86
87	24.6868	22.0975	20.0111	18.2941	16.8563	16.2217	15.6347	15.0904	87
88	24.7226	22.1262	20.0346	18.3138	16.8730	16.2371	15.6491	15.1037	88
89	24.7571	22.1538	20.0573	18.3327	16.8891	16.2520	15.6629	15.1166	89
90	24.7903	22.1803	20.0790	18.3508	16.9045	16.2663	15.6762	15.1290	90
91	24.8221	22.2059	20.1000	18.3683	16.9193	16.2800	15.6889	15.1408	91
92	24.8528	22.2304	20.1201	18.3851	16.9335	16.2932	15.7012	15.1522	92
93	24.8823	22.2540	20.1394	18.4012	16.9472	16.3059	15.7129	15.1632	93
94	24.9107	22.2767	20.1580	18.4167	16.9604	16.3180	15.7243	15.1737	94
95	24.9380	22.2985	20.1758	18.4317	16.9730	16.3297	15.7351	15.1838	95
96	24.9642	22.3195	20.1930	18.4460	16.9852	16.3410	15.7456	15.1936	96
97	24.9894	22.3397	20.2095	18.4598	16.9969	16.3518	15.7556	15.2029	97
98	25.0137	22.3591	20.2254	18.4730	17.0081	16.3622	15.7652	15.2119	98
99	25.0370	22.3777	20.2406	18.4857	17.0189	16.3722	15.7745	15.2205	99
100	25.0595	22.3956	20.2553	18.4980	17.0292	16.3818	15.7834	15.2288	100

Note:-In computing the quarterly in advance figures in these tables, all the rates % quoted above are effective rates, see the
introductory Section, pages xvi-xvii

YEARS' PURCHASE*

Rate Per Cent*

Yrs.	7	7.25	7.5	8	8.5	9	9.5	10	Yrs.
1	0.8686	0.8669	0.8652	0.8618	0.8584	0.8551	0.8519	0.8487	1
2	1.6708	1.6644	1.6581	1.6456	1.6335	1.6215	1.6099	1.5984	2
3	2.4132	2.3999	2.3868	2.3611	2.3361	2.3118	2.2881	2.2651	3
4	3.1018	3.0799	3.0583	3.0163	2.9757	2.9363	2.8982	2.8613	4
5	3.7418	3.7100	3.6787	3.6181	3.5598	3.5036	3.4495	3.3974	5
6	4.3377	4.2950	4.2532	4.1724	4.0950	4.0209	3.9498	3.8815	6
7	4.8935	4.8392	4.7862	4.6841	4.5868	4.4940	4.4054	4.3207	7
8	5.4127	5.3464	5.2818	5.1577	5.0400	4.9282	4.8218	4.7205	8
9	5.8985	5.8198	5.7433	5.5969	5.4586	5.3277	5.2036	5.0858	9
10	6.3537	6.2624	6.1740	6.0051	5.8461	5.6962	5.5546	5.4206	10
11	6.7807	6.6768	6.5764	6.3852	6.2057	6.0370	5.8782	5.7283	11
12	7.1818	7.0654	6.9530	6.7396	6.5400	6.3529	6.1772	6.0119	12
13	7.5589	7.4301	7.3059	7.0707	6.8513	6.6463	6.4542	6.2740	13
14	7.9140	7.7729	7.6371	7.3804	7.1417	6.9192	6.7113	6.5167	14
15	8.2486	8.0954	7.9482	7.6706	7.4131	7.1736	6.9504	6.7419	15
16	8.5642	8.3992	8.2409	7.9428	7.6670	7.4112	7.1732	6.9513	16
17	8.8622	8.6856	8.5164	8.1984	7.9050	7.6333	7.3811	7.1463	17
18	9.1438	8.9559	8.7762	8.4388	8.1282	7.8413	7.5754	7.3283	18
19	9.4101	9.2112	9.0212	8.6652	8.3380	8.0363	7.7572	7.4983	19
20	9.6621	9.4526	9.2526	8.8784	8.5353	8.2194	7.9277	7.6575	20
21	9.9008	9.6810	9.4712	9.0796	8.7210	8.3915	8.0877	7.8067	21
22	10.1271	9.8972	9.6781	9.2695	8.8961	8.5535	8.2380	7.9466	22
23	10.3416	10.1020	9.8738	9.4489	9.0612	8.7060	8.3795	8.0782	23
24	10.5453	10.2962	10.0593	9.6186	9.2171	8.8499	8.5126	8.2019	24
25	10.7386	10.4804	10.2351	9.7792	9.3645	8.9857	8.6382	8.3184	25
26	10.9223	10.6553	10.4018	9.9313	9.5039	9.1139	8.7567	8.4282	26
27	11.0969	10.8214	10.5601	10.0755	9.6358	9.2352	8.8685	8.5318	27
28	11.2630	10.9793	10.7103	10.2122	9.7608	9.3499	8.9743	8.6296	28
29	11.4210	11.1294	10.8531	10.3419	9.8793	9.4585	9.0743	8.7220	29
30	11.5714	11.2722	10.9889	10.4651	9.9916	9.5615	9.1690	8.8095	30
31	11.7146	11.4081	11.1180	10.5821	10.0982	9.6591	9.2587	8.8923	31
32	11.8511	11.5375	11.2408	10.6934	10.1995	9.7517	9.3438	8.9707	32
33	11.9812	11.6607	11.3578	10.7992	10.2957	9.8396	9.4245	9.0450	33
34	12.1052	11.7782	11.4692	10.8998	10.3871	9.9231	9.5010	9.1155	34
35	12.2235	11.8901	11.5753	10.9956	10.4741	10.0024	9.5737	9.1825	35
36	12.3364	11.9969	11.6765	11.0869	10.5569	10.0779	9.6429	9.2460	36
37	12.4441	12.0988	11.7730	11.1738	10.6357	10.1497	9.7086	9.3064	37
38	12.5470	12.1960	11.8650	11.2567	10.7107	10.2180	9.7710	9.3638	38
39	12.6452	12.2888	11.9528	11.3357	10.7822	10.2830	9.8305	9.4184	39
40	12.7391	12.3774	12.0366	11.4111	10.8504	10.3450	9.8871	9.4704	40
41	12.8288	12.4620	12.1167	11.4830	10.9154	10.4041	9.9411	9.5198	41
42	12.9145	12.5429	12.1931	11.5516	10.9774	10.4604	9.9925	9.5669	42
43	12.9964	12.6202	12.2661	11.6171	11.0365	10.5141	10.0414	9.6118	43
44	13.0748	12.6940	12.3359	11.6797	11.0930	10.5653	10.0882	9.6546	44
45	13.1497	12.7647	12.4026	11.7394	11.1469	10.6142	10.1327	9.6954	45
46	13.2214	12.8322	12.4663	11.7965	11.1983	10.6608	10.1752	9.7344	46
47	13.2900	12.8968	12.5273	11.8511	11.2475	10.7054	10.2158	9.7715	47
48	13.3557	12.9587	12.5856	11.9033	11.2945	10.7480	10.2546	9.8070	48
49	13.4185	13.0178	12.6414	11.9532	11.3394	10.7886	10.2916	9.8408	49
50	13.4787	13.0745	12.6948	12.0010	11.3824	10.8275	10.3270	9.8731	50

* Note:-In computing the quarterly in advance figures in these tables, all the rates % quoted above are effective rates, see the Introductory Section, pages xvi-xvii

				Rate Per Cent*					
Yrs.	7	7.25	7.5	8	8.5	9	9.5	10	Yrs.
51	13.5364	13.1287	12.7460	12.0466	11.4235	10.8647	10.3608	9.9040	51
52	13.5916	13.1806	12.7949	12.0903	11.4627	10.9002	10.3931	9.9335	52
53	13.6444	13.2303	12.8417	12.1321	11.5003	10.9342	10.4240	9.9617	53
54	13.6951	13.2779	12.8866	12.1722	11.5363	10.9667	10.4535	9.9887	54
55	13.7436	13.3236	12.9296	12.2105	11.5707	10.9978	10.4817	10.0145	55
56	13.7901	13.3673	12.9707	12.2472	11.6037	11.0275	10.5088	10.0392	56
57	13.8347	13.4092	13.0101	12.2823	11.6352	11.0560	10.5346	10.0628	57
58	13.8774	13.4493	13.0479	12.3160	11.6654	11.0833	10.5594	10.0854	58
59	13.9184	13.4878	13.0841	12.3483	11.6944	11.1094	10.5831	10.1070	59
60	13.9577	13.5247	13.1189	12.3792	11.7221	11.1344	10.6058	10.1277	60
61	13.9954	13.5600	13.1521	12.4088	11.7486	11.1584	10.6275	10.1475	61
62	14.0315	13.5939	13.1840	12.4372	11.7741	11.1814	10.6484	10.1665	62
63	14.0661	13.6265	13.2146	12.4644	11.7985	11.2033	10.6683	10.1847	63
64	14.0994	13.6577	13.2439	12.4905	11.8218	11.2244	10.6874	10.2021	64
65	14.1313	13.6876	13.2721	12.5155	11.8443	11.2446	10.7057	10.2188	65
66	14.1618	13.7163	13.2990	12.5395	11.8657	11.2640	10.7233	10.2348	66
67	14.1912	13.7438	13.3249	12.5625	11.8863	11.2825	10.7401	10.2501	67
68	14.2194	13.7702	13.3497	12.5846	11.9061	11.3003	10.7562	10.2648	68
69	14.2464	13.7955	13.3736	12.6057	11.9250	11.3174	10.7717	10.2788	69
70	14.2723	13.8199	13.3964	12.6260	11.9432	11.3338	10.7865	10.2923	70
71	14.2972	13.8432	13.4183	12.6455	11.9606	11.3494	10.8007	10.3053	71
72	14.3211	13.8656	13.4394	12.6642	11.9773	11.3645	10.8143	10.3177	72
73	14.3440	13.8871	13.4596	12.6821	11.9934	11.3789	10.8274	10.3296	73
74	14.3661	13.9077	13.4790	12.6993	12.0088	11.3928	10.8399	10.3410	74
75	14.3872	13.9276	13.4976	12.7159	12.0235	11.4061	10.8520	10.3519	75
76	14.4075	13.9466	13.5154	12.7317	12.0377	11.4188	10.8635	10.3624	76
77	14.4270	13.9648	13.5326	12.7469	12.0513	11.4311	10.8746	10.3725	77
78	14.4457	13.9824	13.5491	12.7615	12.0644	11.4428	10.8852	10.3822	78
79	14.4637	13.9992	13.5649	12.7756	12.0769	11.4541	10.8954	10.3915	79
80	14.4809	14.0154	13.5800	12.7890	12.0889	11.4649	10.9052	10.4004	80
81	14.4975	14.0309	13.5946	12.8019	12.1005	11.4753	10.9146	10.4089	81
82	14.5134	14.0458	13.6086	12.8143	12.1115	11.4853	10.9236	10.4171	82
83	14.5287	14.0601	13.6220	12.8263	12.1222	11.4948	10.9323	10.4250	83
84	14.5434	14.0739	13.6349	12.8377	12.1324	11.5040	10.9406	10.4326	84
85	14.5575	14.0871	13.6473	12.8487	12.1422	11.5128	10.9486	10.4398	85
86	14.5710	14.0998	13.6592	12.8592	12.1516	11.5213	10.9562	10.4468	86
87	14.5840	14.1119	13.6707	12.8694	12.1607	11.5294	10.9636	10.4535	87
88	14.5965	14.1236	13.6817	12.8791	12.1694	11.5373	10.9707	10.4599	88
89	14.6086	14.1349	13.6922	12.8885	12.1777	11.5448	10.9774	10.4660	89
90	14.6201	14.1457	13.7024	12.8974	12.1857	11.5520	10.9839	10.4720	90
91	14.6312	14.1561	13.7121	12.9061	12.1934	11.5589	10.9902	10.4777	91
92	14.6418	14.1660	13.7214	12.9144	12.2008	11.5655	10.9962	10.4831	92
93	14.6521	14.1756	13.7304	12.9223	12.2079	11.5719	11.0020	10.4884	93
94	14.6619	14.1848	13.7391	12.9300	12.2148	11.5780	11.0075	10.4934	94
95	14.6713	14.1937	13.7474	12.9373	12.2213	11.5839	11.0129	10.4982	95
96	14.6804	14.2022	13.7553	12.9444	12.2276	11.5896	11.0180	10.5029	96
97	14.6891	14.2103	13.7630	12.9511	12.2337	11.5950	11.0229	10.5073	97
98	14.6975	14.2182	13.7703	12.9577	12.2395	11.6003	11.0276	10.5116	98
99	14.7056	14.2257	13.7774	12.9639	12.2451	11.6053	11.0321	10.5158	99
100	14.7133	14.2329	13.7842	12.9699	12.2504	11.6101	11.0365	10.5197	100

Note:-In computing the quarterly in advance figures in these tables, all the rates % quoted above are effective rates, see the introductory Section, pages xvi-xvii

YEARS' PURCHASE*

Rate Per Cent*

Yrs.	11	12	13	14	15	16	18	20	Yrs.
1	0.8424	0.8362	0.8302	0.8244	0.8187	0.8131	0.8024	0.7922	1
2	1.5762	1.5548	1.5342	1.5144	1.4952	1.4768	1.4418	1.4091	2
3	2.2207	2.1785	2.1382	2.0999	2.0633	2.0284	1.9629	1.9028	3
4	2.7908	2.7245	2.6619	2.6027	2.5468	2.4937	2.3955	2.3066	4
5	3.2985	3.2062	3.1198	3.0389	2.9629	2.8913	2.7602	2.6428	5
6	3.7529	3.6339	3.5234	3.4205	3.3245	3.2347	3.0714	2.9268	6
7	4.1619	4.0160	3.8815	3.7570	3.6415	3.5340	3.3400	3.1697	7
8	4.5317	4.3592	4.2012	4.0557	3.9214	3.7971	3.5740	3.3797	8
9	4.8673	4.6689	4.4881	4.3225	4.1703	4.0299	3.7796	3.5629	9
10	5.1731	4.9496	4.7468	4.5619	4.3928	4.2373	3.9614	3.7240	10
11	5.4526	5.2049	4.9811	4.7780	4.5927	4.4231	4.1233	3.8667	11
12	5.7090	5.4380	5.1942	4.9737	4.7733	4.5903	4.2682	3.9939	12
13	5.9448	5.6516	5.3887	5.1517	4.9370	4.7415	4.3987	4.1079	13
14	6.1622	5.8477	5.5667	5.3142	5.0860	4.8788	4.5166	4.2106	14
15	6.3632	6.0284	5.7302	5.4630	5.2222	5.0039	4.6236	4.3034	15
16	6.5494	6.1953	5.8808	5.5997	5.3469	5.1183	4.7212	4.3878	16
17	6.7223	6.3497	6.0198	5.7256	5.4616	5.2233	4.8103	4.4647	17
18	6.8830	6.4930	6.1484	5.8418	5.5672	5.3199	4.8921	4.5351	18
19	7.0329	6.6261	6.2677	5.9494	5.6648	5.4089	4.9673	4.5996	19
20	7.1727	6.7501	6.3785	6.0491	5.7552	5.4913	5.0367	4.6590	20
21	7.3034	6.8658	6.4817	6.1418	5.8391	5.5676	5.1008	4.7138	21
22	7.4258	6.9738	6.5779	6.2282	5.9170	5.6384	5.1602	4.7645	22
23	7.5405	7.0749	6.6677	6.3087	5.9896	5.7043	5.2153	4.8115	23
24	7.6482	7.1696	6.7518	6.3839	6.0574	5.7657	5.2666	4.8551	24
25	7.7494	7.2584	6.8305	6.4542	6.1207	5.8230	5.3144	4.8957	25
26	7.8446	7.3419	6.9044	6.5201	6.1799	5.8766	5.3590	4.9335	26
27	7.9342	7.4204	6.9737	6.5819	6.2354	5.9268	5.4007	4.9688	27
28	8.0188	7.4943	7.0390	6.6400	6.2875	5.9738	5.4397	5.0019	28
29	8.0985	7.5639	7.1004	6.6946	6.3365	6.0180	5.4763	5.0328	29
30	8.1739	7.6296	7.1582	6.7460	6.3825	6.0595	5.5106	5.0618	30
31	8.2451	7.6916	7.2128	6.7944	6.4258	6.0985	5.5429	5.0890	31
32	8.3125	7.7502	7.2643	6.8401	6.4667	6.1353	5.5733	5.1146	32
33	8.3763	7.8056	7.3129	6.8833	6.5052	6.1700	5.6019	5.1387	33
34	8.4367	7.8581	7.3590	6.9240	6.5416	6.2027	5.6289	5.1613	34
35	8.4940	7.9078	7.4025	6.9626	6.5760	6.2336	5.6543	5.1827	35
36	8.5484	7.9548	7.4438	6.9990	6.6085	6.2629	5.6783	5.2029	36
37	8.5999	7.9995	7.4829	7.0336	6.6393	6.2905	5.7011	5.2220	37
38	8.6489	8.0419	7.5199	7.0663	6.6685	6.3167	5.7225	5.2400	38
39	8.6955	8.0821	7.5551	7.0974	6.6961	6.3415	5.7429	5.2571	39
40	8.7398	8.1204	7.5885	7.1269	6.7224	6.3650	5.7622	5.2732	40
41	8.7819	8.1567	7.6202	7.1548	6.7472	6.3873	5.7804	5.2885	41
42	8.8220	8.1913	7.6504	7.1814	6.7709	6.4085	5.7978	5.3030	42
43	8.8601	8.2242	7.6791	7.2067	6.7933	6.4286	5.8142	5.3168	43
44	8.8965	8.2555	7.7064	7.2307	6.8147	6.4477	5.8299	5.3299	44
45	8.9311	8.2853	7.7323	7.2536	6.8350	6.4659	5.8447	5.3423	45
46	8.9641	8.3137	7.7571	7.2753	6.8543	6.4832	5.8588	5.3541	46
47	8.9956	8.3408	7.7806	7.2961	6.8727	6.4996	5.8723	5.3653	47
48	9.0257	8.3666	7.8031	7.3158	6.8902	6.5153	5.8851	5.3759	48
49	9.0543	8.3912	7.8245	7.3346	6.9069	6.5302	5.8972	5.3861	49
50	9.0817	8.4147	7.8449	7.3526	6.9228	6.5444	5.9088	5.3958	50

* Note:-In computing the quarterly in advance figures in these tables, all the rates % quoted above are effective rates, see the Introductory Section, pages xvi-xvii

Rate Per Cent*

Yrs.	11	12	13	14	15	16	18	20	Yrs.
51	9.1078	8.4371	7.8644	7.3697	6.9380	6.5580	5.9199	5.4050	51
52	9.1328	8.4585	7.8830	7.3860	6.9525	6.5709	5.9304	5.4138	52
53	9.1566	8.4790	7.9008	7.4016	6.9663	6.5833	5.9405	5.4221	53
54	9.1794	8.4985	7.9177	7.4165	6.9795	6.5950	5.9500	5.4301	54
55	9.2012	8.5172	7.9339	7.4307	6.9920	6.6063	5.9592	5.4377	55
56	9.2220	8.5350	7.9494	7.4443	7.0041	6.6170	5.9679	5.4450	56
57	9.2419	8.5521	7.9642	7.4572	7.0155	6.6272	5.9762	5.4519	57
58	9.2610	8.5684	7.9784	7.4696	7.0265	6.6370	5.9842	5.4586	58
59	9.2792	8.5840	7.9919	7.4815	7.0370	6.6464	5.9918	5.4649	59
60	9.2966	8.5989	8.0048	7.4928	7.0470	6.6553	5.9991	5.4709	60
61	9.3133	8.6132	8.0172	7.5037	7.0566	6.6639	6.0060	5.4767	61
62	9.3293	8.6269	8.0290	7.5140	7.0658	6.6721	6.0127	5.4822	62
63	9.3446	8.6400	8.0404	7.5240	7.0746	6.6799	6.0190	5.4875	63
64	9.3593	8.6525	8.0512	7.5335	7.0830	6.6874	6.0251	5.4926	64
65	9.3733	8.6645	8.0616	7.5426	7.0910	6.6945	6.0309	5.4974	65
66	9.3868	8.6760	8.0715	7.5513	7.0987	6.7014	6.0365	5.5020	66
67	9.3996	8.6870	8.0811	7.5596	7.1061	6.7080	6.0418	5.5064	67
68	9.4120	8.6975	8.0902	7.5676	7.1131	6.7142	6.0469	5.5107	68
69	9.4238	8.7076	8.0989	7.5752	7.1199	6.7203	6.0518	5.5147	69
70	9.4352	8.7173	8.1073	7.5826	7.1263	6.7260	6.0565	5.5186	70
71	9.4460	8.7266	8.1153	7.5896	7.1325	6.7315	6.0609	5.5223	71
72	9.4565	8.7355	8.1230	7.5963	7.1385	6.7368	6.0652	5.5259	72
73	9.4665	8.7440	8.1304	7.6028	7.1442	6.7419	6.0693	5.5293	73
74	9.4760	8.7522	8.1375	7.6089	7.1496	6.7468	6.0733	5.5326	74
75	9.4852	8.7600	8.1443	7.6149	7.1549	6.7514	6.0771	5.5357	75
76	9.4941	8.7675	8.1508	7.6206	7.1599	6.7559	6.0807	5.5387	76
77	9.5025	8.7748	8.1570	7.6260	7.1647	6.7602	6.0841	5.5416	77
78	9.5106	8.7817	8.1630	7.6312	7.1693	6.7643	6.0875	5.5443	78
79	9.5184	8.7883	8.1687	7.6362	7.1737	6.7682	6.0907	5.5470	79
80	9.5259	8.7947	8.1742	7.6410	7.1780	6.7720	6.0937	5.5495	80
81	9.5331	8.8008	8.1795	7.6457	7.1820	6.7756	6.0966	5.5520	81
82	9.5399	8.8067	8.1845	7.6501	7.1859	6.7791	6.0995	5.5543	82
83	9.5465	8.8123	8.1894	7.6543	7.1897	6.7824	6.1022	5.5565	83
84	9.5529	8.8177	8.1941	7.6584	7.1933	6.7856	6.1047	5.5587	84
85	9.5590	8.8229	8.1985	7.6623	7.1967	6.7887	6.1072	5.5607	85
86	9.5648	8.8278	8.2028	7.6661	7.2000	6.7916	6.1096	5.5627	86
87	9.5704	8.8326	8.2070	7.6697	7.2032	6.7945	6.1119	5.5646	87
88	9.5758	8.8372	8.2109	7.6731	7.2063	6.7972	6.1141	5.5664	88
89	9.5809	8.8416	8.2147	7.6764	7.2092	6.7998	6.1162	5.5682	89
90	9.5859	8.8458	8.2184	7.6796	7.2120	6.8023	6.1182	5.5698	90
91	9.5907	8.8499	8.2219	7.6827	7.2147	6.8047	6.1202	5.5715	91
92	9.5952	8.8538	8.2252	7.6856	7.2173	6.8070	6.1220	5.5730	92
93	9.5996	8.8575	8.2285	7.6884	7.2198	6.8092	6.1238	5.5745	93
94	9.6039	8.8611	8.2316	7.6911	7.2222	6.8113	6.1255	5.5759	94
95	9.6079	8.8646	8.2345	7.6937	7.2244	6.8134	6.1272	5.5773	95
96	9.6118	8.8679	8.2374	7.6962	7.2266	6.8153	6.1288	5.5786	96
97	9.6155	8.8711	8.2401	7.6986	7.2288	6.8172	6.1303	5.5798	97
98	9.6191	8.8741	8.2428	7.7009	7.2308	6.8190	6.1317	5.5810	98
99	9.6226	8.8770	8.2453	7.7031	7.2327	6.8207	6.1331	5.5822	99
100	9.6259	8.8799	8.2477	7.7053	7.2346	6.8224	6.1345	5.5833	100

* Note:-In computing the quarterly in advance figures in these tables, all the rates % quoted above are effective rates, see the introductory Section, pages xvi-xvii

YEARS' PURCHASE

(DUAL RATE % PRINCIPLE)

OR

PRESENT VALUE OF
ONE POUND PER ANNUM

*receivable quarterly in advance, allow-
ing for a sinking fund at a given rate
to replace the invested capital and for
the effect of income tax at* **20%** *on
that part of the income used to provide
the annual sinking fund instalment.*

AT RATES OF INTEREST*

FROM

4% to 20%

AND

ALLOWING FOR THE POSSIBLE INVESTMENT

OF SINKING FUNDS AT

3% and 4%

INCOME TAX at 20%

* *Note:*—In computing the quarterly in advance figures in these tables, all the rates of
interest quoted are effective rates, see the Introductory Section, pages xvi to
xvii.

Rate Per Cent*

Yrs.	4	4.5	5	5.5	6	6.25	6.5	6.75	Yrs.
1	0.7898	0.7869	0.7840	0.7811	0.7783	0.7769	0.7755	0.7741	1
2	1.5540	1.5426	1.5315	1.5206	1.5098	1.5046	1.4994	1.4942	2
3	2.2934	2.2687	2.2447	2.2213	2.1986	2.1874	2.1764	2.1655	3
4	3.0090	2.9667	2.9257	2.8861	2.8478	2.8291	2.8107	2.7926	4
5	3.7016	3.6377	3.5763	3.5173	3.4605	3.4330	3.4059	3.3794	5
6	4.3719	4.2830	4.1982	4.1171	4.0396	4.0021	3.9653	3.9294	6
7	5.0208	4.9040	4.7931	4.6877	4.5874	4.5391	4.4919	4.4458	7
8	5.6490	5.5015	5.3623	5.2308	5.1062	5.0464	4.9882	4.9314	8
9	6.2572	6.0768	5.9074	5.7481	5.5981	5.5263	5.4565	5.3887	9
10	6.8461	6.6307	6.4295	6.2413	6.0648	5.9806	5.8990	5.8198	10
11	7.4164	7.1642	6.9300	6.7118	6.5081	6.4113	6.3176	6.2268	11
12	7.9687	7.6782	7.4098	7.1609	6.9296	6.8199	6.7139	6.6115	12
13	8.5035	8.1736	7.8701	7.5899	7.3305	7.2079	7.0896	6.9755	13
14	9.0216	8.6511	8.3119	8.0000	7.7123	7.5767	7.4461	7.3204	14
15	9.5234	9.1115	8.7359	8.3921	8.0761	7.9275	7.7847	7.6473	15
16	10.0095	9.5555	9.1433	8.7673	8.4230	8.2615	8.1065	7.9576	16
17	10.4804	9.9837	9.5346	9.1265	8.7540	8.5797	8.4126	8.2524	17
18	10.9366	10.3969	9.9107	9.4705	9.0700	8.8830	8.7041	8.5327	18
19	11.3787	10.7956	10.2724	9.8002	9.3720	9.1724	8.9818	8.7994	19
20	11.8070	11.1804	10.6202	10.1163	9.6606	9.4488	9.2466	9.0534	20
21	12.2220	11.5519	10.9548	10.4194	9.9367	9.7127	9.4992	9.2955	21
22	12.6242	11.9106	11.2768	10.7104	10.2009	9.9650	9.7404	9.5263	22
23	13.0140	12.2569	11.5868	10.9896	10.4540	10.2063	9.9708	9.7466	23
24	13.3918	12.5914	11.8853	11.2578	10.6963	10.4372	10.1911	9.9569	24
25	13.7580	12.9146	12.1729	11.5154	10.9286	10.6583	10.4017	10.1579	25
26	14.1129	13.2268	12.4499	11.7630	11.1514	10.8701	10.6033	10.3501	26
27	14.4569	13.5285	12.7168	12.0010	11.3651	11.0730	10.7964	10.5339	27
28	14.7903	13.8201	12.9741	12.2299	11.5702	11.2676	10.9813	10.7099	28
29	15.1136	14.1020	13.2222	12.4501	11.7671	11.4542	11.1585	10.8784	29
30	15.4270	14.3745	13.4615	12.6620	11.9562	11.6333	11.3284	11.0398	30
31	15.7309	14.6379	13.6922	12.8660	12.1379	11.8053	11.4914	11.1945	31
32	16.0255	14.8926	13.9149	13.0624	12.3125	11.9704	11.6478	11.3429	32
33	16.3111	15.1390	14.1297	13.2515	12.4804	12.1291	11.7979	11.4853	33
34	16.5880	15.3773	14.3371	13.4337	12.6419	12.2816	11.9421	11.6219	34
35	16.8566	15.6078	14.5373	13.6093	12.7973	12.4281	12.0807	11.7531	35
36	17.1170	15.8308	14.7305	13.7786	12.9468	12.5691	12.2139	11.8791	36
37	17.3695	16.0465	14.9171	13.9417	13.0908	12.7048	12.3419	12.0002	37
38	17.6144	16.2553	15.0974	14.0991	13.2294	12.8353	12.4650	12.1165	38
39	17.8519	16.4574	15.2715	14.2508	13.3629	12.9609	12.5835	12.2285	39
40	18.0822	16.6529	15.4398	14.3972	13.4915	13.0819	12.6975	12.3361	40
41	18.3056	16.8422	15.6023	14.5384	13.6155	13.1984	12.8073	12.4396	41
42	18.5222	17.0254	15.7595	14.6748	13.7350	13.3107	12.9129	12.5393	42
43	18.7324	17.2028	15.9113	14.8064	13.8502	13.4189	13.0147	12.6353	43
44	18.9362	17.3745	16.0581	14.9334	13.9613	13.5231	13.1128	12.7277	44
45	19.1339	17.5408	16.2001	15.0561	14.0685	13.6237	13.2073	12.8167	45
46	19.3257	17.7019	16.3373	15.1746	14.1719	13.7206	13.2984	12.9024	46
47	19.5117	17.8578	16.4701	15.2890	14.2717	13.8141	13.3862	12.9851	47
48	19.6921	18.0088	16.5985	15.3996	14.3680	13.9043	13.4709	13.0648	48
49	19.8672	18.1551	16.7227	15.5065	14.4609	13.9913	13.5525	13.1416	49
50	20.0370	18.2968	16.8428	15.6097	14.5507	14.0754	13.6313	13.2157	50

* Note:-In computing the quarterly in advance figures in these tables, all the rates % quoted above are effective rates, see the Introductory Section, pages xvi-xvii

Yrs.	4	4.5	5	5.5	6	6.25	6.5	6.75	Yrs.
51	20.2017	18.4341	16.9591	15.7095	14.6374	14.1565	13.7074	13.2871	51
52	20.3615	18.5671	17.0716	15.8060	14.7211	14.2347	13.7808	13.3561	52
53	20.5166	18.6959	17.1804	15.8993	14.8020	14.3104	13.8516	13.4226	53
54	20.6670	18.8208	17.2858	15.9895	14.8801	14.3834	13.9200	13.4869	54
55	20.8130	18.9417	17.3878	16.0767	14.9556	14.4539	13.9861	13.5489	55
56	20.9546	19.0590	17.4865	16.1610	15.0286	14.5221	14.0499	13.6087	56
57	21.0920	19.1725	17.5821	16.2426	15.0991	14.5879	14.1115	13.6665	57
58	21.2253	19.2826	17.6746	16.3216	15.1673	14.6516	14.1711	13.7224	58
59	21.3546	19.3893	17.7642	16.3979	15.2333	14.7131	14.2286	13.7763	59
60	21.4801	19.4927	17.8510	16.4718	15.2970	14.7725	14.2842	13.8285	60
61	21.6019	19.5930	17.9350	16.5434	15.3587	14.8300	14.3380	13.8788	61
62	21.7201	19.6901	18.0163	16.6126	15.4183	14.8856	14.3899	13.9275	62
63	21.8347	19.7843	18.0952	16.6796	15.4760	14.9394	14.4402	13.9746	63
64	21.9460	19.8756	18.1715	16.7444	15.5318	14.9914	14.4888	14.0201	64
65	22.0539	19.9641	18.2455	16.8072	15.5858	15.0417	14.5357	14.0640	65
66	22.1587	20.0499	18.3171	16.8680	15.6381	15.0904	14.5812	14.1066	66
67	22.2604	20.1331	18.3865	16.9268	15.6886	15.1374	14.6251	14.1477	67
68	22.3590	20.2138	18.4538	16.9838	15.7376	15.1830	14.6676	14.1875	68
69	22.4548	20.2920	18.5190	17.0390	15.7850	15.2271	14.7088	14.2260	69
70	22.5477	20.3679	18.5821	17.0924	15.8308	15.2698	14.7486	14.2632	70
71	22.6379	20.4414	18.6433	17.1442	15.8752	15.3111	14.7871	14.2992	71
72	22.7254	20.5127	18.7026	17.1943	15.9182	15.3510	14.8244	14.3341	72
73	22.8103	20.5819	18.7601	17.2429	15.9598	15.3897	14.8605	14.3679	73
74	22.8927	20.6490	18.8158	17.2900	16.0001	15.4272	14.8954	14.4005	74
75	22.9727	20.7140	18.8698	17.3356	16.0392	15.4635	14.9293	14.4321	75
76	23.0503	20.7771	18.9222	17.3797	16.0770	15.4986	14.9620	14.4627	76
77	23.1257	20.8383	18.9729	17.4225	16.1136	15.5327	14.9937	14.4923	77
78	23.1988	20.8977	19.0221	17.4640	16.1490	15.5656	15.0244	14.5210	78
79	23.2698	20.9552	19.0698	17.5042	16.1834	15.5975	15.0542	14.5488	79
80	23.3386	21.0111	19.1160	17.5431	16.2167	15.6285	15.0830	14.5757	80
81	23.4055	21.0652	19.1608	17.5809	16.2489	15.6584	15.1108	14.6017	81
82	23.4704	21.1178	19.2043	17.6175	16.2802	15.6874	15.1379	14.6270	82
83	23.5333	21.1688	19.2464	17.6529	16.3104	15.7155	15.1640	14.6514	83
84	23.5945	21.2182	19.2873	17.6873	16.3398	15.7428	15.1894	14.6751	84
85	23.6538	21.2662	19.3269	17.7206	16.3682	15.7691	15.2140	14.6980	85
86	23.7114	21.3127	19.3654	17.7529	16.3958	15.7947	15.2378	14.7202	86
87	23.7673	21.3578	19.4026	17.7842	16.4225	15.8195	15.2608	14.7417	87
88	23.8215	21.4016	19.4388	17.8146	16.4484	15.8435	15.2832	14.7626	88
89	23.8742	21.4441	19.4738	17.8440	16.4734	15.8668	15.3048	14.7828	89
90	23.9253	21.4854	19.5078	17.8725	16.4978	15.8893	15.3258	14.8024	90
91	23.9749	21.5254	19.5408	17.9002	16.5213	15.9112	15.3462	14.8213	91
92	24.0230	21.5642	19.5727	17.9270	16.5442	15.9324	15.3659	14.8397	92
93	24.0698	21.6018	19.6038	17.9530	16.5663	15.9529	15.3850	14.8575	93
94	24.1151	21.6383	19.6338	17.9783	16.5878	15.9729	15.4035	14.8748	94
95	24.1592	21.6738	19.6630	18.0027	16.6086	15.9922	15.4215	14.8916	95
96	24.2019	21.7082	19.6913	18.0265	16.6288	16.0109	15.4389	14.9078	96
97	24.2434	21.7416	19.7188	18.0495	16.6484	16.0290	15.4557	14.9235	97
98	24.2837	21.7740	19.7454	18.0718	16.6674	16.0466	15.4721	14.9388	98
99	24.3228	21.8054	19.7713	18.0934	16.6858	16.0637	15.4880	14.9536	99
100	24.3607	21.8359	19.7963	18.1144	16.7037	16.0802	15.5033	14.9679	100

* Note:-In computing the quarterly in advance figures in these tables, all the rates % quoted above are effective rates, see the introductory Section, pages xvi-xvii

				Rate Per Cent*					
Yrs.	7	7.25	7.5	8	8.5	9	9.5	10	Yrs.
1	0.7727	0.7713	0.7700	0.7673	0.7646	0.7620	0.7594	0.7569	1
2	1.4891	1.4840	1.4790	1.4691	1.4594	1.4498	1.4405	1.4313	2
3	2.1548	2.1442	2.1337	2.1132	2.0931	2.0736	2.0545	2.0359	3
4	2.7747	2.7572	2.7399	2.7061	2.6734	2.6416	2.6107	2.5807	4
5	3.3533	3.3277	3.3025	3.2536	3.2064	3.1607	3.1166	3.0740	5
6	3.8942	3.8597	3.8259	3.7604	3.6974	3.6369	3.5786	3.5225	6
7	4.4008	4.3568	4.3138	4.2307	4.1512	4.0750	4.0020	3.9320	7
8	4.8761	4.8222	4.7695	4.6681	4.5715	4.4793	4.3913	4.3071	8
9	5.3227	5.2585	5.1960	5.0758	4.9618	4.8534	4.7502	4.6518	9
10	5.7429	5.6682	5.5957	5.4566	5.3250	5.2004	5.0820	4.9696	10
11	6.1389	6.0536	5.9709	5.8129	5.6638	5.5229	5.3897	5.2634	11
12	6.5124	6.4166	6.3238	6.1467	5.9803	5.8235	5.6755	5.5357	12
13	6.8654	6.7589	6.6560	6.4602	6.2766	6.1041	5.9417	5.7886	13
14	7.1991	7.0822	6.9693	6.7549	6.5544	6.3665	6.1901	6.0241	14
15	7.5151	7.3877	7.2650	7.0323	6.8153	6.6124	6.4223	6.2438	15
16	7.8146	7.6770	7.5445	7.2939	7.0606	6.8431	6.6397	6.4491	16
17	8.0987	7.9510	7.8090	7.5408	7.2918	7.0600	6.8437	6.6414	17
18	8.3684	8.2108	8.0595	7.7741	7.5097	7.2641	7.0353	6.8217	18
19	8.6248	8.4575	8.2970	7.9949	7.7155	7.4565	7.2157	6.9911	19
20	8.8687	8.6919	8.5224	8.2040	7.9101	7.6381	7.3856	7.1505	20
21	9.1008	8.9147	8.7366	8.4022	8.0943	7.8097	7.5459	7.3007	21
22	9.3220	9.1268	8.9402	8.5904	8.2687	7.9720	7.6973	7.4423	22
23	9.5328	9.3288	9.1339	8.7691	8.4342	8.1257	7.8405	7.5761	23
24	9.7340	9.5214	9.3184	8.9390	8.5913	8.2713	7.9760	7.7026	24
25	9.9260	9.7050	9.4942	9.1007	8.7405	8.4096	8.1045	7.8223	25
26	10.1094	9.8803	9.6619	9.2547	8.8824	8.5409	8.2263	7.9358	26
27	10.2847	10.0477	9.8219	9.4014	9.0175	8.6656	8.3420	8.0434	27
28	10.4524	10.2076	9.9747	9.5413	9.1461	8.7844	8.4520	8.1456	28
29	10.6128	10.3605	10.1207	9.6748	9.2687	8.8974	8.5566	8.2427	29
30	10.7664	10.5069	10.2603	9.8022	9.3856	9.0051	8.6561	8.3350	30
31	10.9135	10.6469	10.3938	9.9240	9.4972	9.1078	8.7510	8.4229	31
32	11.0544	10.7811	10.5216	10.0405	9.6038	9.2058	8.8414	8.5066	32
33	11.1896	10.9096	10.6440	10.1518	9.7057	9.2993	8.9277	8.5864	33
34	11.3193	11.0328	10.7612	10.2584	9.8030	9.3887	9.0100	8.6626	34
35	11.4437	11.1509	10.8736	10.3605	9.8962	9.4741	9.0886	8.7353	35
36	11.5631	11.2643	10.9814	10.4583	9.9854	9.5558	9.1638	8.8047	36
37	11.6778	11.3731	11.0848	10.5520	10.0708	9.6340	9.2357	8.8710	37
38	11.7880	11.4776	11.1840	10.6419	10.1527	9.7089	9.3045	8.9344	38
39	11.8938	11.5780	11.2793	10.7281	10.2311	9.7806	9.3703	8.9951	39
40	11.9956	11.6744	11.3708	10.8109	10.3063	9.8493	9.4334	9.0532	40
41	12.0935	11.7671	11.4587	10.8903	10.3785	9.9152	9.4938	9.1089	41
42	12.1877	11.8563	11.5432	10.9667	10.4478	9.9784	9.5518	9.1622	42
43	12.2784	11.9420	11.6245	11.0400	10.5143	10.0391	9.6073	9.2134	43
44	12.3656	12.0245	11.7027	11.1105	10.5783	10.0973	9.6607	9.2624	44
45	12.4496	12.1039	11.7779	11.1782	10.6397	10.1533	9.7119	9.3094	45
46	12.5305	12.1804	11.8502	11.2434	10.6987	10.2070	9.7610	9.3546	46
47	12.6084	12.2540	11.9199	11.3061	10.7555	10.2587	9.8083	9.3980	47
48	12.6835	12.3250	11.9870	11.3665	10.8101	10.3084	9.8537	9.4396	48
49	12.7559	12.3933	12.0517	11.4246	10.8626	10.3561	9.8973	9.4797	49
50	12.8257	12.4592	12.1140	11.4805	10.9132	10.4021	9.9393	9.5182	50

* Note:-In computing the quarterly in advance figures in these tables, all the rates % quoted above are effective rates, see the Introductory Section, pages xvi-xvii

Rate Per Cent*

Yrs.	7	7.25	7.5	8	8.5	9	9.5	10	Yrs.
51	12.8930	12.5227	12.1740	11.5344	10.9619	10.4463	9.9796	9.5552	51
52	12.9579	12.5839	12.2318	11.5863	11.0088	10.4889	10.0185	9.5908	52
53	13.0206	12.6429	12.2876	11.6364	11.0539	10.5299	10.0559	9.6250	53
54	13.0810	12.6999	12.3414	11.6846	11.0974	10.5694	10.0919	9.6580	54
55	13.1393	12.7549	12.3933	11.7311	11.1394	10.6074	10.1265	9.6898	55
56	13.1956	12.8079	12.4434	11.7760	11.1798	10.6441	10.1599	9.7204	56
57	13.2499	12.8591	12.4917	11.8193	11.2188	10.6794	10.1921	9.7498	57
58	13.3024	12.9085	12.5383	11.8610	11.2564	10.7135	10.2232	9.7782	58
59	13.3531	12.9563	12.5834	11.9013	11.2927	10.7463	10.2531	9.8056	59
60	13.4021	13.0023	12.6268	11.9402	11.3277	10.7780	10.2819	9.8319	60
61	13.4494	13.0469	12.6688	11.9777	11.3615	10.8086	10.3097	9.8574	61
62	13.4951	13.0899	12.7094	12.0139	11.3941	10.8381	10.3366	9.8819	62
63	13.5393	13.1314	12.7485	12.0489	11.4255	10.8665	10.3625	9.9056	63
64	13.5820	13.1716	12.7864	12.0827	11.4559	10.8940	10.3875	9.9284	64
65	13.6232	13.2104	12.8230	12.1154	11.4853	10.9206	10.4116	9.9505	65
66	13.6631	13.2479	12.8583	12.1469	11.5136	10.9462	10.4349	9.9717	66
67	13.7017	13.2842	12.8925	12.1774	11.5410	10.9709	10.4574	9.9923	67
68	13.7391	13.3193	12.9255	12.2069	11.5675	10.9949	10.4791	10.0121	68
69	13.7751	13.3532	12.9574	12.2354	11.5930	11.0180	10.5001	10.0312	69
70	13.8101	13.3860	12.9883	12.2629	11.6178	11.0403	10.5203	10.0497	70
71	13.8438	13.4177	13.0182	12.2895	11.6417	11.0619	10.5399	10.0676	71
72	13.8765	13.4484	13.0471	12.3153	11.6648	11.0827	10.5589	10.0849	72
73	13.9081	13.4781	13.0750	12.3402	11.6871	11.1029	10.5772	10.1016	73
74	13.9387	13.5068	13.1021	12.3642	11.7087	11.1224	10.5948	10.1177	74
75	13.9683	13.5347	13.1282	12.3875	11.7296	11.1412	10.6119	10.1333	75
76	13.9970	13.5616	13.1536	12.4101	11.7498	11.1594	10.6285	10.1484	76
77	14.0247	13.5876	13.1781	12.4319	11.7693	11.1771	10.6445	10.1630	77
78	14.0516	13.6128	13.2018	12.4530	11.7882	11.1941	10.6599	10.1771	78
79	14.0776	13.6372	13.2247	12.4734	11.8065	11.2106	10.6749	10.1907	79
80	14.1028	13.6608	13.2469	12.4932	11.8242	11.2266	10.6894	10.2039	80
81	14.1272	13.6837	13.2684	12.5123	11.8414	11.2420	10.7034	10.2166	81
82	14.1508	13.7059	13.2893	12.5308	11.8580	11.2570	10.7169	10.2290	82
83	14.1736	13.7273	13.3094	12.5487	11.8740	11.2714	10.7300	10.2409	83
84	14.1958	13.7481	13.3290	12.5661	11.8896	11.2854	10.7427	10.2525	84
85	14.2172	13.7682	13.3479	12.5829	11.9046	11.2990	10.7550	10.2637	85
86	14.2380	13.7877	13.3662	12.5992	11.9192	11.3121	10.7669	10.2745	86
87	14.2582	13.8066	13.3839	12.6150	11.9333	11.3248	10.7784	10.2850	87
88	14.2777	13.8249	13.4011	12.6302	11.9469	11.3371	10.7895	10.2951	88
89	14.2966	13.8426	13.4178	12.6450	11.9602	11.3490	10.8003	10.3049	89
90	14.3149	13.8598	13.4339	12.6593	11.9730	11.3606	10.8108	10.3144	90
91	14.3326	13.8764	13.4495	12.6732	11.9854	11.3717	10.8209	10.3237	91
92	14.3498	13.8925	13.4647	12.6866	11.9974	11.3826	10.8307	10.3326	92
93	14.3665	13.9081	13.4793	12.6997	12.0091	11.3930	10.8402	10.3412	93
94	14.3826	13.9233	13.4936	12.7123	12.0203	11.4032	10.8494	10.3496	94
95	14.3983	13.9379	13.5073	12.7245	12.0313	11.4130	10.8583	10.3577	95
96	14.4135	13.9522	13.5207	12.7364	12.0419	11.4226	10.8669	10.3655	96
97	14.4282	13.9659	13.5336	12.7478	12.0521	11.4318	10.8753	10.3731	97
98	14.4424	13.9793	13.5462	12.7590	12.0621	11.4408	10.8834	10.3805	98
99	14.4562	13.9922	13.5583	12.7698	12.0717	11.4494	10.8912	10.3876	99
100	14.4696	14.0048	13.5701	12.7802	12.0810	11.4578	10.8988	10.3946	100

* Note:-In computing the quarterly in advance figures in these tables, all the rates % quoted above are effective rates, see the introductory Section, pages xvi-xvii

YEARS' PURCHASE*

Yrs.	11	12	13	14	15	16	18	20	Yrs.
				Rate Per Cent*					
1	0.7518	0.7469	0.7421	0.7375	0.7329	0.7285	0.7198	0.7116	1
2	1.4135	1.3962	1.3796	1.3636	1.3480	1.3330	1.3044	1.2776	2
3	2.0000	1.9657	1.9329	1.9015	1.8714	1.8426	1.7885	1.7384	3
4	2.5232	2.4689	2.4174	2.3685	2.3221	2.2779	2.1957	2.1207	4
5	2.9928	2.9166	2.8450	2.7775	2.7139	2.6537	2.5428	2.4429	5
6	3.4163	3.3174	3.2251	3.1386	3.0576	2.9815	2.8422	2.7179	6
7	3.8001	3.6781	3.5649	3.4596	3.3615	3.2697	3.1029	2.9553	7
8	4.1493	4.0043	3.8705	3.7467	3.6319	3.5249	3.3319	3.1623	8
9	4.4683	4.3006	4.1467	4.0049	3.8740	3.7525	3.5345	3.3443	9
10	4.7608	4.5709	4.3974	4.2383	4.0919	3.9567	3.7150	3.5055	10
11	5.0297	4.8182	4.6259	4.4501	4.2890	4.1407	3.8768	3.6492	11
12	5.2778	5.0454	4.8348	4.6432	4.4681	4.3073	4.0225	3.7780	12
13	5.5072	5.2547	5.0267	4.8199	4.6314	4.4589	4.1544	3.8941	13
14	5.7199	5.4480	5.2033	4.9820	4.7809	4.5974	4.2744	3.9993	14
15	5.9176	5.6270	5.3664	5.1313	4.9183	4.7242	4.3838	4.0949	15
16	6.1018	5.7933	5.5174	5.2692	5.0448	4.8408	4.4840	4.1823	16
17	6.2736	5.9479	5.6575	5.3969	5.1617	4.9484	4.5762	4.2623	17
18	6.4343	6.0922	5.7878	5.5153	5.2699	5.0478	4.6611	4.3358	18
19	6.5848	6.2269	5.9093	5.6256	5.3705	5.1399	4.7395	4.4037	19
20	6.7260	6.3530	6.0228	5.7283	5.4640	5.2256	4.8122	4.4664	20
21	6.8587	6.4713	6.1290	5.8243	5.5513	5.3053	4.8798	4.5245	21
22	6.9835	6.5823	6.2285	5.9140	5.6328	5.3797	4.9427	4.5785	22
23	7.1012	6.6868	6.3219	5.9982	5.7091	5.4493	5.0013	4.6288	23
24	7.2122	6.7851	6.4097	6.0772	5.7806	5.5144	5.0561	4.6757	24
25	7.3171	6.8778	6.4924	6.1515	5.8478	5.5755	5.1074	4.7195	25
26	7.4163	6.9654	6.5704	6.2215	5.9110	5.6329	5.1556	4.7606	26
27	7.5102	7.0482	6.6440	6.2874	5.9705	5.6869	5.2008	4.7991	27
28	7.5992	7.1265	6.7136	6.3497	6.0266	5.7378	5.2433	4.8353	28
29	7.6836	7.2007	6.7794	6.4085	6.0796	5.7858	5.2834	4.8694	29
30	7.7638	7.2711	6.8417	6.4642	6.1297	5.8312	5.3212	4.9015	30
31	7.8400	7.3379	6.9009	6.5170	6.1771	5.8740	5.3568	4.9317	31
32	7.9125	7.4014	6.9570	6.5670	6.2220	5.9146	5.3906	4.9603	32
33	7.9815	7.4617	7.0102	6.6144	6.2646	5.9531	5.4225	4.9873	33
34	8.0473	7.5192	7.0609	6.6595	6.3050	5.9896	5.4528	5.0129	34
35	8.1099	7.5738	7.1091	6.7024	6.3434	6.0243	5.4815	5.0372	35
36	8.1697	7.6260	7.1550	6.7432	6.3800	6.0572	5.5088	5.0602	36
37	8.2268	7.6757	7.1988	6.7820	6.4147	6.0885	5.5346	5.0820	37
38	8.2814	7.7231	7.2405	6.8190	6.4478	6.1183	5.5593	5.1028	38
39	8.3335	7.7685	7.2803	6.8543	6.4794	6.1467	5.5827	5.1225	39
40	8.3833	7.8118	7.3183	6.8880	6.5095	6.1738	5.6050	5.1413	40
41	8.4310	7.8532	7.3546	6.9202	6.5382	6.1997	5.6263	5.1592	41
42	8.4767	7.8928	7.3894	6.9509	6.5656	6.2243	5.6466	5.1763	42
43	8.5204	7.9307	7.4226	6.9803	6.5918	6.2479	5.6660	5.1926	43
44	8.5624	7.9670	7.4544	7.0084	6.6169	6.2704	5.6845	5.2081	44
45	8.6025	8.0018	7.4848	7.0353	6.6409	6.2919	5.7022	5.2229	45
46	8.6411	8.0351	7.5140	7.0611	6.6638	6.3125	5.7191	5.2371	46
47	8.6781	8.0671	7.5420	7.0858	6.6858	6.3322	5.7353	5.2507	47
48	8.7136	8.0978	7.5688	7.1094	6.7069	6.3511	5.7508	5.2637	48
49	8.7477	8.1272	7.5945	7.1321	6.7270	6.3692	5.7656	5.2761	49
50	8.7805	8.1555	7.6192	7.1539	6.7464	6.3866	5.7798	5.2880	50

* Note:-In computing the quarterly in advance figures in these tables, all the rates % quoted above are effective rates, see the Introductory Section, pages xvi-xvii

				Rate Per Cent*					
Yrs.	11	12	13	14	15	16	18	20	Yrs.
51	8.8120	8.1826	7.6429	7.1748	6.7650	6.4032	5.7935	5.2994	51
52	8.8422	8.2087	7.6656	7.1948	6.7828	6.4192	5.8065	5.3103	52
53	8.8714	8.2338	7.6875	7.2141	6.7999	6.4345	5.8191	5.3208	53
54	8.8994	8.2579	7.7085	7.2326	6.8164	6.4492	5.8311	5.3309	54
55	8.9263	8.2811	7.7287	7.2504	6.8322	6.4634	5.8427	5.3405	55
56	8.9523	8.3035	7.7482	7.2675	6.8474	6.4770	5.8538	5.3498	56
57	8.9772	8.3250	7.7669	7.2840	6.8620	6.4900	5.8644	5.3587	57
58	9.0013	8.3457	7.7849	7.2998	6.8760	6.5026	5.8747	5.3673	58
59	9.0245	8.3656	7.8022	7.3150	6.8895	6.5147	5.8846	5.3755	59
60	9.0468	8.3848	7.8189	7.3297	6.9025	6.5263	5.8941	5.3835	60
61	9.0684	8.4033	7.8350	7.3438	6.9151	6.5375	5.9032	5.3911	61
62	9.0891	8.4211	7.8505	7.3574	6.9271	6.5483	5.9120	5.3984	62
63	9.1091	8.4383	7.8654	7.3706	6.9388	6.5587	5.9204	5.4055	63
64	9.1284	8.4548	7.8798	7.3832	6.9500	6.5687	5.9286	5.4122	64
65	9.1471	8.4708	7.8937	7.3954	6.9607	6.5783	5.9364	5.4188	65
66	9.1650	8.4862	7.9071	7.4071	6.9711	6.5876	5.9440	5.4251	66
67	9.1824	8.5011	7.9200	7.4184	6.9812	6.5966	5.9513	5.4312	67
68	9.1991	8.5154	7.9324	7.4294	6.9909	6.6052	5.9583	5.4370	68
69	9.2153	8.5293	7.9444	7.4399	7.0002	6.6135	5.9651	5.4427	69
70	9.2309	8.5427	7.9560	7.4501	7.0092	6.6216	5.9716	5.4481	70
71	9.2460	8.5556	7.9672	7.4599	7.0179	6.6293	5.9779	5.4533	71
72	9.2605	8.5680	7.9780	7.4694	7.0263	6.6368	5.9840	5.4584	72
73	9.2746	8.5801	7.9885	7.4785	7.0344	6.6440	5.9899	5.4633	73
74	9.2882	8.5917	7.9986	7.4874	7.0422	6.6510	5.9956	5.4680	74
75	9.3013	8.6030	8.0083	7.4959	7.0497	6.6577	6.0010	5.4726	75
76	9.3141	8.6138	8.0177	7.5041	7.0570	6.6642	6.0063	5.4770	76
77	9.3263	8.6243	8.0268	7.5121	7.0641	6.6705	6.0114	5.4812	77
78	9.3382	8.6345	8.0356	7.5198	7.0709	6.6766	6.0164	5.4853	78
79	9.3497	8.6443	8.0441	7.5273	7.0775	6.6825	6.0211	5.4893	79
80	9.3608	8.6538	8.0523	7.5344	7.0838	6.6881	6.0257	5.4931	80
81	9.3715	8.6629	8.0603	7.5414	7.0900	6.6936	6.0302	5.4968	81
82	9.3819	8.6718	8.0680	7.5481	7.0959	6.6989	6.0345	5.5003	82
83	9.3919	8.6804	8.0754	7.5546	7.1017	6.7040	6.0386	5.5038	83
84	9.4017	8.6887	8.0826	7.5609	7.1072	6.7090	6.0426	5.5071	84
85	9.4111	8.6967	8.0895	7.5670	7.1126	6.7138	6.0465	5.5104	85
86	9.4202	8.7045	8.0962	7.5729	7.1178	6.7184	6.0503	5.5135	86
87	9.4290	8.7120	8.1027	7.5786	7.1228	6.7229	6.0539	5.5165	87
88	9.4375	8.7193	8.1090	7.5841	7.1277	6.7272	6.0574	5.5194	88
89	9.4458	8.7263	8.1151	7.5894	7.1324	6.7314	6.0608	5.5222	89
90	9.4537	8.7332	8.1210	7.5946	7.1369	6.7355	6.0641	5.5250	90
91	9.4615	8.7398	8.1267	7.5996	7.1413	6.7394	6.0673	5.5276	91
92	9.4690	8.7462	8.1323	7.6044	7.1456	6.7432	6.0704	5.5302	92
93	9.4762	8.7523	8.1376	7.6091	7.1497	6.7469	6.0734	5.5326	93
94	9.4833	8.7583	8.1428	7.6136	7.1537	6.7504	6.0762	5.5350	94
95	9.4901	8.7641	8.1478	7.6180	7.1576	6.7539	6.0790	5.5373	95
96	9.4966	8.7698	8.1527	7.6222	7.1614	6.7572	6.0817	5.5396	96
97	9.5030	8.7752	8.1574	7.6263	7.1650	6.7604	6.0844	5.5418	97
98	9.5092	8.7805	8.1619	7.6303	7.1685	6.7636	6.0869	5.5439	98
99	9.5152	8.7856	8.1663	7.6342	7.1719	6.7666	6.0893	5.5459	99
100	9.5210	8.7905	8.1706	7.6379	7.1752	6.7695	6.0917	5.5479	100

Note:-In computing the quarterly in advance figures in these tables, all the rates % quoted above are effective rates, see the troductory Section, pages xvi-xvii

Yrs.	4	4.5	5	5.5	6	6.25	6.5	6.75	Yrs.
					Rate Per Cent*				
1	0.7945	0.7915	0.7885	0.7856	0.7828	0.7814	0.7799	0.7785	1
2	1.5701	1.5585	1.5471	1.5360	1.5250	1.5197	1.5143	1.5091	2
3	2.3270	2.3015	2.2768	2.2528	2.2293	2.2179	2.2065	2.1954	3
4	3.0652	3.0212	2.9788	2.9377	2.8980	2.8787	2.8596	2.8409	4
5	3.7849	3.7181	3.6540	3.5924	3.5332	3.5045	3.4763	3.4486	5
6	4.4862	4.3926	4.3034	4.2183	4.1369	4.0976	4.0591	4.0214	6
7	5.1693	5.0455	4.9282	4.8169	4.7110	4.6601	4.6104	4.5618	7
8	5.8344	5.6772	5.5291	5.3894	5.2573	5.1939	5.1322	5.0721	8
9	6.4818	6.2883	6.1071	5.9371	5.7772	5.7007	5.6265	5.5544	9
10	7.1116	6.8794	6.6631	6.4612	6.2722	6.1822	6.0950	6.0105	10
11	7.7241	7.4509	7.1979	6.9628	6.7439	6.6399	6.5394	6.4422	11
12	8.3195	8.0034	7.7122	7.4430	7.1933	7.0752	6.9612	6.8512	12
13	8.8981	8.5375	8.2069	7.9027	7.6219	7.4894	7.3618	7.2388	13
14	9.4602	9.0536	8.6827	8.3430	8.0306	7.8836	7.7424	7.6065	14
15	10.0060	9.5523	9.1404	8.7646	8.4205	8.2591	8.1042	7.9554	15
16	10.5359	10.0341	9.5805	9.1685	8.7927	8.6168	8.4483	8.2868	16
17	11.0501	10.4994	10.0038	9.5555	9.1479	8.9577	8.7758	8.6016	17
18	11.5490	10.9488	10.4110	9.9263	9.4872	9.2828	9.0876	8.9010	18
19	12.0329	11.3827	10.8026	10.2816	9.8113	9.5929	9.3845	9.1856	19
20	12.5020	11.8017	11.1792	10.6222	10.1210	9.8887	9.6674	9.4565	20
21	12.9567	12.2060	11.5414	10.9487	10.4169	10.1710	9.9371	9.7144	21
22	13.3974	12.5963	11.8897	11.2617	10.6999	10.4406	10.1943	9.9600	22
23	13.8242	12.9730	12.2247	11.5618	10.9704	10.6980	10.4396	10.1940	23
24	14.2377	13.3364	12.5469	11.8496	11.2292	10.9439	10.6736	10.4171	24
25	14.6380	13.6870	12.8567	12.1256	11.4767	11.1789	10.8970	10.6298	25
26	15.0255	14.0252	13.1547	12.3903	11.7136	11.4035	11.1104	10.8327	26
27	15.4005	14.3515	13.4413	12.6442	11.9403	11.6183	11.3141	11.0262	27
28	15.7634	14.6661	13.7169	12.8878	12.1572	11.8236	11.5087	11.2110	28
29	16.1144	14.9695	13.9819	13.1214	12.3650	12.0200	11.6947	11.3874	29
30	16.4539	15.2620	14.2368	13.3457	12.5639	12.2079	11.8725	11.5559	30
31	16.7822	15.5440	14.4819	13.5608	12.7544	12.3877	12.0425	11.7169	31
32	17.0996	15.8159	14.7176	13.7673	12.9369	12.5598	12.2050	11.8707	32
33	17.4064	16.0780	14.9443	13.9655	13.1117	12.7245	12.3605	12.0178	33
34	17.7028	16.3306	15.1623	14.1557	13.2792	12.8822	12.5093	12.1583	34
35	17.9893	16.5741	15.3720	14.3382	13.4397	13.0332	12.6516	12.2928	35
36	18.2660	16.8087	15.5736	14.5134	13.5936	13.1778	12.7878	12.4213	36
37	18.5332	17.0347	15.7674	14.6817	13.7411	13.3164	12.9183	12.5444	37
38	18.7913	17.2525	15.9539	14.8432	13.8824	13.4491	13.0431	12.6621	38
39	19.0406	17.4624	16.1331	14.9982	14.0180	13.5763	13.1627	12.7747	39
40	19.2812	17.6645	16.3055	15.1471	14.1480	13.6981	13.2773	12.8826	40
41	19.5134	17.8593	16.4713	15.2901	14.2726	13.8150	13.3870	12.9859	41
42	19.7376	18.0468	16.6307	15.4274	14.3921	13.9269	13.4921	13.0848	42
43	19.9539	18.2275	16.7841	15.5592	14.5068	14.0343	13.5928	13.1795	43
44	20.1626	18.4015	16.9315	15.6858	14.6168	14.1372	13.6894	13.2702	44
45	20.3639	18.5691	17.0732	15.8074	14.7224	14.2359	13.7819	13.3571	45
46	20.5582	18.7305	17.2096	15.9242	14.8236	14.3306	13.8706	13.4404	46
47	20.7455	18.8858	17.3407	16.0364	14.9208	14.4213	13.9556	13.5202	47
48	20.9262	19.0355	17.4667	16.1441	15.0140	14.5084	14.0371	13.5968	48
49	21.1004	19.1795	17.5879	16.2476	15.1035	14.5920	14.1153	13.6701	49
50	21.2684	19.3182	17.7045	16.3471	15.1894	14.6721	14.1903	13.7404	50

* Note:-In computing the quarterly in advance figures in these tables, all the rates % quoted above are effective rates, see the
Introductory Section, pages xvi-xvii

Rate Per Cent*

Yrs.	4	4.5	5	5.5	6	6.25	6.5	6.75	Yrs.
51	21.4304	19.4518	17.8166	16.4426	15.2718	14.7490	14.2622	13.8078	51
52	21.5865	19.5803	17.9244	16.5343	15.3509	14.8228	14.3312	13.8725	52
53	21.7370	19.7041	18.0280	16.6225	15.4269	14.8936	14.3974	13.9345	53
54	21.8821	19.8232	18.1277	16.7072	15.4998	14.9616	14.4609	13.9939	54
55	22.0219	19.9378	18.2235	16.7886	15.5698	15.0268	14.5218	14.0510	55
56	22.1566	20.0482	18.3157	16.8667	15.6370	15.0894	14.5803	14.1057	56
57	22.2864	20.1544	18.4043	16.9419	15.7016	15.1495	14.6364	14.1582	57
58	22.4115	20.2566	18.4895	17.0140	15.7636	15.2072	14.6902	14.2086	58
59	22.5320	20.3550	18.5714	17.0834	15.8231	15.2626	14.7419	14.2569	59
60	22.6481	20.4497	18.6502	17.1501	15.8802	15.3157	14.7915	14.3033	60
61	22.7599	20.5409	18.7260	17.2141	15.9351	15.3668	14.8391	14.3478	61
62	22.8676	20.6285	18.7989	17.2756	15.9879	15.4158	14.8848	14.3906	62
63	22.9714	20.7129	18.8689	17.3348	16.0385	15.4629	14.9287	14.4316	63
64	23.0713	20.7941	18.9363	17.3916	16.0871	15.5081	14.9708	14.4710	64
65	23.1675	20.8723	19.0010	17.4462	16.1339	15.5515	15.0113	14.5088	65
66	23.2602	20.9474	19.0633	17.4987	16.1787	15.5932	15.0501	14.5450	66
67	23.3494	21.0198	19.1232	17.5492	16.2219	15.6333	15.0874	14.5799	67
68	23.4353	21.0894	19.1808	17.5977	16.2633	15.6717	15.1233	14.6133	68
69	23.5180	21.1564	19.2362	17.6443	16.3031	15.7087	15.1577	14.6455	69
70	23.5977	21.2208	19.2894	17.6891	16.3413	15.7442	15.1907	14.6763	70
71	23.6743	21.2828	19.3406	17.7321	16.3780	15.7783	15.2224	14.7059	71
72	23.7481	21.3424	19.3899	17.7735	16.4133	15.8110	15.2529	14.7344	72
73	23.8192	21.3998	19.4372	17.8133	16.4472	15.8425	15.2822	14.7617	73
74	23.8876	21.4550	19.4827	17.8515	16.4798	15.8727	15.3103	14.7879	74
75	23.9534	21.5081	19.5265	17.8882	16.5111	15.9018	15.3374	14.8131	75
76	24.0168	21.5591	19.5686	17.9236	16.5412	15.9297	15.3633	14.8373	76
77	24.0778	21.6083	19.6091	17.9575	16.5701	15.9565	15.3882	14.8606	77
78	24.1365	21.6555	19.6480	17.9901	16.5979	15.9822	15.4122	14.8829	78
79	24.1930	21.7010	19.6854	18.0215	16.6246	16.0070	15.4352	14.9044	79
80	24.2474	21.7447	19.7214	18.0517	16.6503	16.0308	15.4573	14.9250	80
81	24.2997	21.7868	19.7560	18.0806	16.6749	16.0536	15.4786	14.9448	81
82	24.3500	21.8273	19.7893	18.1085	16.6986	16.0756	15.4990	14.9639	82
83	24.3985	21.8662	19.8212	18.1353	16.7214	16.0967	15.5186	14.9821	83
84	24.4451	21.9036	19.8520	18.1610	16.7433	16.1170	15.5375	14.9997	84
85	24.4900	21.9397	19.8816	18.1858	16.7643	16.1364	15.5556	15.0166	85
86	24.5331	21.9743	19.9100	18.2096	16.7845	16.1552	15.5730	15.0328	86
87	24.5747	22.0076	19.9374	18.2324	16.8040	16.1732	15.5897	15.0484	87
88	24.6146	22.0397	19.9637	18.2544	16.8226	16.1905	15.6058	15.0634	88
89	24.6531	22.0705	19.9890	18.2756	16.8406	16.2071	15.6212	15.0778	89
90	24.6901	22.1001	20.0133	18.2959	16.8578	16.2231	15.6361	15.0916	90
91	24.7257	22.1286	20.0366	18.3154	16.8744	16.2384	15.6503	15.1049	91
92	24.7599	22.1560	20.0591	18.3342	16.8904	16.2532	15.6640	15.1176	92
93	24.7928	22.1824	20.0807	18.3523	16.9057	16.2674	15.6772	15.1299	93
94	24.8245	22.2078	20.1015	18.3696	16.9204	16.2810	15.6899	15.1417	94
95	24.8550	22.2322	20.1215	18.3863	16.9346	16.2941	15.7021	15.1531	95
96	24.8843	22.2556	20.1407	18.4024	16.9482	16.3067	15.7138	15.1640	96
97	24.9126	22.2782	20.1592	18.4178	16.9613	16.3188	15.7250	15.1744	97
98	24.9397	22.2999	20.1770	18.4326	16.9738	16.3305	15.7358	15.1845	98
99	24.9658	22.3208	20.1940	18.4469	16.9859	16.3417	15.7462	15.1942	99
100	24.9909	22.3408	20.2105	18.4606	16.9975	16.3524	15.7562	15.2035	100

Note:-In computing the quarterly in advance figures in these tables, all the rates % quoted above are effective rates, see the introductory Section, pages xvi-xvii

YEARS' PURCHASE*

Rate Per Cent*

Yrs.	7	7.25	7.5	8	8.5	9	9.5	10	Yrs.
1	0.7772	0.7758	0.7744	0.7717	0.7690	0.7663	0.7637	0.7611	1
2	1.5038	1.4987	1.4936	1.4835	1.4736	1.4639	1.4543	1.4450	2
3	2.1843	2.1734	2.1627	2.1416	2.1210	2.1010	2.0814	2.0623	3
4	2.8224	2.8043	2.7864	2.7515	2.7176	2.6848	2.6529	2.6219	4
5	3.4215	3.3948	3.3687	3.3178	3.2687	3.2213	3.1755	3.1312	5
6	3.9846	3.9485	3.9131	3.8446	3.7788	3.7156	3.6548	3.5963	6
7	4.5145	4.4682	4.4230	4.3357	4.2522	4.1723	4.0958	4.0225	7
8	5.0136	4.9566	4.9011	4.7940	4.6922	4.5951	4.5025	4.4140	8
9	5.4843	5.4162	5.3499	5.2226	5.1020	4.9874	4.8785	4.7748	9
10	5.9285	5.8490	5.7718	5.6239	5.4843	5.3521	5.2269	5.1080	10
11	6.3482	6.2570	6.1688	6.0002	5.8415	5.6918	5.5503	5.4165	11
12	6.7449	6.6421	6.5427	6.3534	6.1757	6.0086	5.8513	5.7027	12
13	7.1203	7.0058	6.8953	6.6854	6.4889	6.3047	6.1317	5.9688	13
14	7.4757	7.3496	7.2281	6.9978	6.7828	6.5818	6.3935	6.2165	14
15	7.8125	7.6749	7.5425	7.2920	7.0589	6.8415	6.6382	6.4477	15
16	8.1318	7.9829	7.8397	7.5694	7.3186	7.0851	6.8673	6.6636	16
17	8.4347	8.2746	8.1209	7.8313	7.5631	7.3140	7.0821	6.8657	17
18	8.7223	8.5512	8.3872	8.0786	7.7935	7.5293	7.2838	7.0551	18
19	8.9955	8.8137	8.6395	8.3124	8.0109	7.7320	7.4733	7.2327	19
20	9.2552	9.0627	8.8787	8.5336	8.2161	7.9230	7.6516	7.3996	20
21	9.5020	9.2993	9.1057	8.7431	8.4101	8.1033	7.8196	7.5566	21
22	9.7369	9.5242	9.3211	8.9415	8.5936	8.2734	7.9780	7.7044	22
23	9.9604	9.7379	9.5258	9.1297	8.7672	8.4343	8.1274	7.8437	23
24	10.1733	9.9413	9.7203	9.3082	8.9317	8.5864	8.2686	7.9751	24
25	10.3760	10.1348	9.9052	9.4776	9.0876	8.7304	8.4020	8.0991	25
26	10.5692	10.3191	10.0811	9.6386	9.2355	8.8668	8.5283	8.2164	26
27	10.7535	10.4946	10.2486	9.7915	9.3758	8.9961	8.6478	8.3273	27
28	10.9291	10.6618	10.4080	9.9370	9.5091	9.1187	8.7611	8.4322	28
29	11.0967	10.8213	10.5599	10.0753	9.6357	9.2351	8.8684	8.5316	29
30	11.2567	10.9733	10.7046	10.2070	9.7561	9.3456	8.9703	8.6259	30
31	11.4093	11.1184	10.8426	10.3324	9.8706	9.4506	9.0670	8.7152	31
32	11.5552	11.2568	10.9742	10.4518	9.9795	9.5504	9.1588	8.8001	32
33	11.6944	11.3889	11.0998	10.5656	10.0832	9.6453	9.2461	8.8806	33
34	11.8275	11.5151	11.2196	10.6741	10.1820	9.7357	9.3291	8.9571	34
35	11.9547	11.6356	11.3340	10.7776	10.2761	9.8217	9.4080	9.0299	35
36	12.0762	11.7507	11.4432	10.8763	10.3658	9.9036	9.4832	9.0991	36
37	12.1925	11.8608	11.5475	10.9705	10.4513	9.9816	9.5547	9.1649	37
38	12.3037	11.9659	11.6472	11.0604	10.5329	10.0560	9.6228	9.2276	38
39	12.4100	12.0665	11.7424	11.1463	10.6107	10.1269	9.6878	9.2873	39
40	12.5118	12.1627	11.8335	11.2283	10.6850	10.1946	9.7497	9.3442	40
41	12.6092	12.2547	11.9206	11.3067	10.7560	10.2592	9.8087	9.3984	41
42	12.7024	12.3427	12.0039	11.3816	10.8237	10.3208	9.8650	9.4501	42
43	12.7916	12.4270	12.0835	11.4532	10.8885	10.3796	9.9188	9.4994	43
44	12.8771	12.5076	12.1597	11.5216	10.9503	10.4358	9.9701	9.5464	44
45	12.9589	12.5848	12.2327	11.5871	11.0094	10.4895	10.0190	9.5913	45
46	13.0373	12.6587	12.3025	11.6497	11.0660	10.5408	10.0658	9.6342	46
47	13.1124	12.7295	12.3694	11.7097	11.1200	10.5898	10.1105	9.6751	47
48	13.1843	12.7973	12.4334	11.7670	11.1717	10.6367	10.1533	9.7142	48
49	13.2533	12.8622	12.4947	11.8219	11.2212	10.6816	10.1941	9.7516	49
50	13.3194	12.9245	12.5534	11.8745	11.2685	10.7244	10.2332	9.7874	50

* Note:-In computing the quarterly in advance figures in these tables, all the rates % quoted above are effective rates, see the Introductory Section, pages xvi-xvii

YEARS' PURCHASE*

				Rate Per Cent*					
Yrs.	7	7.25	7.5	8	8.5	9	9.5	10	Yrs.
51	13.3827	12.9841	12.6096	11.9248	11.3138	10.7655	10.2705	9.8215	51
52	13.4434	13.0413	12.6635	11.9730	11.3572	10.8047	10.3062	9.8542	52
53	13.5016	13.0960	12.7152	12.0191	11.3987	10.8423	10.3404	9.8854	53
54	13.5575	13.1486	12.7647	12.0633	11.4385	10.8783	10.3731	9.9153	54
55	13.6110	13.1989	12.8121	12.1057	11.4766	10.9127	10.4044	9.9439	55
56	13.6623	13.2472	12.8576	12.1463	11.5131	10.9457	10.4344	9.9713	56
57	13.7116	13.2935	12.9012	12.1852	11.5480	10.9773	10.4631	9.9975	57
58	13.7588	13.3379	12.9430	12.2225	11.5815	11.0075	10.4906	10.0226	58
59	13.8042	13.3805	12.9831	12.2583	11.6136	11.0365	10.5169	10.0466	59
60	13.8476	13.4213	13.0216	12.2925	11.6444	11.0643	10.5421	10.0696	60
61	13.8894	13.4605	13.0585	12.3254	11.6738	11.0909	10.5663	10.0917	61
62	13.9294	13.4981	13.0939	12.3569	11.7021	11.1164	10.5895	10.1128	62
63	13.9678	13.5342	13.1278	12.3871	11.7292	11.1409	10.6117	10.1330	63
64	14.0047	13.5688	13.1604	12.4161	11.7552	11.1643	10.6329	10.1524	64
65	14.0401	13.6020	13.1916	12.4440	11.7801	11.1868	10.6533	10.1710	65
66	14.0741	13.6339	13.2216	12.4706	11.8041	11.2084	10.6729	10.1888	66
67	14.1067	13.6645	13.2504	12.4962	11.8270	11.2291	10.6916	10.2059	67
68	14.1380	13.6939	13.2780	12.5208	11.8490	11.2489	10.7096	10.2223	68
69	14.1681	13.7221	13.3045	12.5444	11.8701	11.2679	10.7268	10.2380	69
70	14.1969	13.7492	13.3300	12.5670	11.8904	11.2862	10.7434	10.2531	70
71	14.2247	13.7752	13.3544	12.5887	11.9098	11.3037	10.7592	10.2675	71
72	14.2513	13.8001	13.3779	12.6096	11.9284	11.3205	10.7745	10.2814	72
73	14.2768	13.8241	13.4004	12.6296	11.9463	11.3366	10.7891	10.2947	73
74	14.3014	13.8471	13.4220	12.6488	11.9635	11.3521	10.8031	10.3074	74
75	14.3250	13.8692	13.4428	12.6672	11.9800	11.3669	10.8165	10.3197	75
76	14.3476	13.8904	13.4627	12.6849	11.9958	11.3812	10.8294	10.3314	76
77	14.3693	13.9108	13.4819	12.7019	12.0110	11.3948	10.8418	10.3427	77
78	14.3902	13.9304	13.5002	12.7182	12.0256	11.4080	10.8537	10.3535	78
79	14.4103	13.9492	13.5179	12.7339	12.0396	11.4206	10.8651	10.3639	79
80	14.4296	13.9672	13.5348	12.7489	12.0531	11.4327	10.8761	10.3739	80
81	14.4481	13.9846	13.5511	12.7634	12.0660	11.4443	10.8866	10.3834	81
82	14.4659	14.0012	13.5668	12.7773	12.0784	11.4555	10.8967	10.3926	82
83	14.4829	14.0173	13.5818	12.7906	12.0903	11.4662	10.9064	10.4014	83
84	14.4994	14.0326	13.5962	12.8034	12.1018	11.4765	10.9157	10.4099	84
85	14.5151	14.0474	13.6101	12.8157	12.1127	11.4863	10.9246	10.4180	85
86	14.5303	14.0616	13.6234	12.8275	12.1233	11.4958	10.9332	10.4258	86
87	14.5448	14.0752	13.6362	12.8388	12.1334	11.5049	10.9414	10.4333	87
88	14.5588	14.0883	13.6485	12.8497	12.1432	11.5137	10.9493	10.4405	88
89	14.5723	14.1009	13.6603	12.8602	12.1525	11.5221	10.9569	10.4474	89
90	14.5852	14.1130	13.6717	12.8703	12.1615	11.5302	10.9642	10.4541	90
91	14.5976	14.1246	13.6826	12.8799	12.1701	11.5379	10.9713	10.4604	91
92	14.6095	14.1358	13.6931	12.8892	12.1784	11.5454	10.9780	10.4666	92
93	14.6210	14.1465	13.7031	12.8981	12.1864	11.5525	10.9845	10.4724	93
94	14.6320	14.1568	13.7128	12.9067	12.1940	11.5594	10.9907	10.4781	94
95	14.6426	14.1667	13.7221	12.9149	12.2014	11.5660	10.9966	10.4835	95
96	14.6528	14.1763	13.7310	12.9229	12.2084	11.5724	11.0024	10.4887	96
97	14.6625	14.1854	13.7396	12.9305	12.2152	11.5785	11.0079	10.4937	97
98	14.6719	14.1942	13.7479	12.9378	12.2217	11.5843	11.0132	10.4985	98
99	14.6810	14.2027	13.7558	12.9448	12.2280	11.5899	11.0183	10.5032	99
100	14.6897	14.2108	13.7634	12.9515	12.2340	11.5953	11.0232	10.5076	100

Note:-In computing the quarterly in advance figures in these tables, all the rates % quoted above are effective rates, see the introductory Section, pages xvi-xvii

YEARS' PURCHASE*

Rate Per Cent*

Yrs.	11	12	13	14	15	16	18	20	Yrs.
1	0.7560	0.7511	0.7463	0.7415	0.7369	0.7324	0.7237	0.7154	1
2	1.4268	1.4092	1.3923	1.3759	1.3601	1.3449	1.3158	1.2885	2
3	2.0254	1.9902	1.9566	1.9245	1.8937	1.8642	1.8088	1.7576	3
4	2.5626	2.5065	2.4535	2.4031	2.3553	2.3099	2.2254	2.1485	4
5	3.0470	2.9681	2.8939	2.8242	2.7584	2.6963	2.5819	2.4789	5
6	3.4857	3.3828	3.2868	3.1971	3.1131	3.0342	2.8901	2.7616	6
7	3.8845	3.7572	3.6391	3.5295	3.4274	3.3320	3.1590	3.0062	7
8	4.2485	4.0966	3.9567	3.8274	3.7076	3.5963	3.3955	3.2196	8
9	4.5817	4.4055	4.2441	4.0958	3.9589	3.8322	3.6051	3.4074	9
10	4.8877	4.6877	4.5054	4.3385	4.1852	4.0439	3.7919	3.5738	10
11	5.1694	4.9462	4.7437	4.5591	4.3901	4.2348	3.9592	3.7221	11
12	5.4294	5.1838	4.9618	4.7602	4.5763	4.4078	4.1100	3.8551	12
13	5.6701	5.4027	5.1620	4.9441	4.7460	4.5651	4.2464	3.9748	13
14	5.8932	5.6049	5.3463	5.1129	4.9013	4.7086	4.3704	4.0832	14
15	6.1005	5.7921	5.5163	5.2682	5.0439	4.8400	4.4833	4.1817	15
16	6.2935	5.9658	5.6736	5.4115	5.1751	4.9607	4.5867	4.2714	16
17	6.4734	6.1272	5.8195	5.5441	5.2962	5.0718	4.6815	4.3536	17
18	6.6415	6.2776	5.9549	5.6669	5.4081	5.1744	4.7688	4.4289	18
19	6.7987	6.4179	6.0810	5.7809	5.5119	5.2693	4.8493	4.4983	19
20	6.9460	6.5489	6.1986	5.8871	5.6083	5.3574	4.9238	4.5623	20
21	7.0841	6.6716	6.3083	5.9860	5.6980	5.4392	4.9928	4.6215	21
22	7.2138	6.7865	6.4110	6.0784	5.7816	5.5153	5.0569	4.6764	22
23	7.3358	6.8944	6.5072	6.1647	5.8597	5.5863	5.1165	4.7273	23
24	7.4506	6.9957	6.5973	6.2456	5.9328	5.6527	5.1721	4.7747	24
25	7.5588	7.0910	6.6820	6.3214	6.0011	5.7147	5.2240	4.8189	25
26	7.6608	7.1807	6.7616	6.3926	6.0653	5.7728	5.2726	4.8602	26
27	7.7571	7.2652	6.8365	6.4596	6.1255	5.8274	5.3180	4.8988	27
28	7.8481	7.3450	6.9071	6.5225	6.1821	5.8786	5.3606	4.9349	28
29	7.9341	7.4203	6.9737	6.5819	6.2354	5.9267	5.4006	4.9688	29
30	8.0156	7.4915	7.0365	6.6378	6.2855	5.9720	5.4382	5.0006	30
31	8.0927	7.5588	7.0959	6.6906	6.3329	6.0147	5.4736	5.0305	31
32	8.1658	7.6225	7.1520	6.7405	6.3775	6.0550	5.5070	5.0587	32
33	8.2351	7.6829	7.2051	6.7876	6.4197	6.0931	5.5384	5.0852	33
34	8.3009	7.7401	7.2554	6.8323	6.4596	6.1290	5.5681	5.1102	34
35	8.3633	7.7944	7.3031	6.8745	6.4974	6.1630	5.5961	5.1338	35
36	8.4226	7.8459	7.3482	6.9145	6.5331	6.1951	5.6226	5.1561	36
37	8.4790	7.8948	7.3911	6.9525	6.5670	6.2256	5.6476	5.1771	37
38	8.5326	7.9412	7.4318	6.9885	6.5991	6.2544	5.6714	5.1971	38
39	8.5836	7.9854	7.4705	7.0227	6.6296	6.2818	5.6939	5.2160	39
40	8.6322	8.0274	7.5073	7.0551	6.6585	6.3077	5.7152	5.2339	40
41	8.6784	8.0674	7.5422	7.0860	6.6860	6.3324	5.7354	5.2508	41
42	8.7225	8.1054	7.5755	7.1153	6.7121	6.3558	5.7546	5.2669	42
43	8.7645	8.1417	7.6071	7.1433	6.7370	6.3781	5.7729	5.2822	43
44	8.8045	8.1762	7.6373	7.1698	6.7606	6.3993	5.7902	5.2967	44
45	8.8427	8.2091	7.6660	7.1951	6.7831	6.4194	5.8067	5.3105	45
46	8.8791	8.2405	7.6933	7.2192	6.8045	6.4386	5.8224	5.3236	46
47	8.9139	8.2704	7.7194	7.2422	6.8249	6.4568	5.8373	5.3361	47
48	8.9471	8.2990	7.7443	7.2641	6.8443	6.4742	5.8516	5.3480	48
49	8.9788	8.3263	7.7680	7.2850	6.8629	6.4908	5.8651	5.3593	49
50	9.0091	8.3523	7.7907	7.3049	6.8805	6.5066	5.8780	5.3701	50

* Note:-In computing the quarterly in advance figures in these tables, all the rates % quoted above are effective rates, see the Introductory Section, pages xvi-xvii

				Rate Per Cent*					
Yrs.	11	12	13	14	15	16	18	20	Yrs.
51	9.0380	8.3772	7.8123	7.3239	6.8974	6.5217	5.8903	5.3803	51
52	9.0656	8.4009	7.8330	7.3421	6.9135	6.5361	5.9020	5.3901	52
53	9.0921	8.4236	7.8527	7.3594	6.9289	6.5498	5.9132	5.3994	53
54	9.1174	8.4453	7.8715	7.3759	6.9435	6.5629	5.9239	5.4084	54
55	9.1415	8.4661	7.8896	7.3918	6.9575	6.5755	5.9341	5.4168	55
56	9.1647	8.4859	7.9068	7.4069	6.9709	6.5874	5.9439	5.4250	56
57	9.1868	8.5049	7.9233	7.4213	6.9837	6.5989	5.9532	5.4327	57
58	9.2080	8.5230	7.9390	7.4351	6.9960	6.6098	5.9620	5.4401	58
59	9.2283	8.5404	7.9541	7.4484	7.0077	6.6202	5.9705	5.4472	59
60	9.2477	8.5570	7.9685	7.4610	7.0189	6.6302	5.9787	5.4539	60
61	9.2663	8.5729	7.9823	7.4731	7.0296	6.6398	5.9864	5.4604	61
62	9.2841	8.5882	7.9955	7.4847	7.0398	6.6489	5.9938	5.4666	62
63	9.3011	8.6028	8.0081	7.4958	7.0496	6.6576	6.0010	5.4725	63
64	9.3175	8.6167	8.0203	7.5064	7.0590	6.6660	6.0077	5.4781	64
65	9.3331	8.6301	8.0319	7.5165	7.0680	6.6740	6.0143	5.4835	65
66	9.3481	8.6429	8.0430	7.5262	7.0766	6.6817	6.0205	5.4887	66
67	9.3625	8.6552	8.0536	7.5356	7.0848	6.6890	6.0264	5.4937	67
68	9.3763	8.6670	8.0638	7.5445	7.0927	6.6960	6.0321	5.4984	68
69	9.3895	8.6783	8.0736	7.5530	7.1003	6.7028	6.0376	5.5030	69
70	9.4022	8.6891	8.0829	7.5612	7.1075	6.7092	6.0428	5.5073	70
71	9.4143	8.6995	8.0919	7.5691	7.1144	6.7154	6.0479	5.5115	71
72	9.4260	8.7094	8.1005	7.5766	7.1211	6.7213	6.0527	5.5155	72
73	9.4371	8.7190	8.1088	7.5838	7.1275	6.7270	6.0573	5.5193	73
74	9.4479	8.7281	8.1167	7.5908	7.1336	6.7325	6.0617	5.5230	74
75	9.4581	8.7369	8.1243	7.5974	7.1394	6.7377	6.0659	5.5265	75
76	9.4680	8.7453	8.1315	7.6038	7.1451	6.7427	6.0700	5.5298	76
77	9.4775	8.7534	8.1385	7.6099	7.1504	6.7475	6.0739	5.5331	77
78	9.4866	8.7611	8.1452	7.6157	7.1556	6.7521	6.0776	5.5362	78
79	9.4953	8.7686	8.1516	7.6213	7.1606	6.7565	6.0812	5.5391	79
80	9.5036	8.7757	8.1578	7.6267	7.1653	6.7607	6.0846	5.5420	80
81	9.5117	8.7826	8.1637	7.6319	7.1699	6.7648	6.0879	5.5447	81
82	9.5194	8.7891	8.1694	7.6368	7.1743	6.7687	6.0910	5.5473	82
83	9.5268	8.7954	8.1748	7.6416	7.1785	6.7724	6.0941	5.5498	83
84	9.5339	8.8015	8.1801	7.6462	7.1825	6.7760	6.0970	5.5522	84
85	9.5407	8.8073	8.1851	7.6506	7.1864	6.7795	6.0998	5.5545	85
86	9.5472	8.8129	8.1899	7.6548	7.1901	6.7828	6.1024	5.5568	86
87	9.5535	8.8182	8.1945	7.6588	7.1936	6.7859	6.1050	5.5589	87
88	9.5595	8.8234	8.1990	7.6627	7.1971	6.7890	6.1075	5.5609	88
89	9.5653	8.8283	8.2032	7.6664	7.2003	6.7919	6.1098	5.5629	89
90	9.5709	8.8330	8.2073	7.6700	7.2035	6.7947	6.1121	5.5648	90
91	9.5762	8.8376	8.2113	7.6734	7.2065	6.7974	6.1143	5.5666	91
92	9.5814	8.8420	8.2150	7.6767	7.2094	6.8000	6.1164	5.5683	92
93	9.5863	8.8462	8.2186	7.6799	7.2122	6.8025	6.1184	5.5700	93
94	9.5910	8.8502	8.2221	7.6829	7.2149	6.8049	6.1203	5.5716	94
95	9.5956	8.8541	8.2255	7.6858	7.2175	6.8071	6.1222	5.5731	95
96	9.5999	8.8578	8.2287	7.6886	7.2199	6.8093	6.1239	5.5746	96
97	9.6041	8.8613	8.2318	7.6913	7.2223	6.8115	6.1256	5.5760	97
98	9.6082	8.8648	8.2347	7.6939	7.2246	6.8135	6.1273	5.5774	98
99	9.6120	8.8681	8.2376	7.6964	7.2268	6.8154	6.1289	5.5787	99
100	9.6158	8.8712	8.2403	7.6988	7.2289	6.8173	6.1304	5.5799	100

ote:-In computing the quarterly in advance figures in these tables, all the rates % quoted above are effective rates, see the
roductory Section, pages xvi-xvii

YEARS' PURCHASE

(DUAL RATE % PRINCIPLE)

OR

PRESENT VALUE OF
ONE POUND PER ANNUM

*receivable quarterly in advance, allow-
ing for a sinking fund at a given rate
to replace the invested capital and for
the effect of income tax at 25% on that
part of the income used to provide the
annual sinking fund instalment.*

AT RATES OF INTEREST*

FROM

4% to 20%

AND

ALLOWING FOR THE POSSIBLE INVESTMENT

OF SINKING FUNDS AT

3% and 4%

INCOME TAX at 25%

Note:—In computing the quarterly in advance figures in these tables, all the rates
of interest quoted are effective rates, see the Introductory Section, pages
xvi-xvii.

				Rate per Cent*					
Yrs.	4	4.5	5	5.5	6	6.25	6.5	6.75	Yrs.
1	0.7419	0.7393	0.7367	0.7342	0.7317	0.7304	0.7292	0.7280	1
2	1.4624	1.4523	1.4425	1.4328	1.4233	1.4186	1.4139	1.4093	2
3	2.1622	2.1402	2.1188	2.0980	2.0777	2.0677	2.0578	2.0481	3
4	2.8418	2.8040	2.7674	2.7319	2.6976	2.6808	2.6643	2.6480	4
5	3.5019	3.4446	3.3895	3.3365	3.2854	3.2605	3.2361	3.2121	5
6	4.1429	4.0630	3.9865	3.9134	3.8432	3.8093	3.7760	3.7434	6
7	4.7654	4.6600	4.5597	4.4643	4.3732	4.3293	4.2863	4.2444	7
8	5.3700	5.2365	5.1102	4.9906	4.8771	4.8225	4.7693	4.7174	8
9	5.9571	5.7933	5.6391	5.4938	5.3566	5.2908	5.2268	5.1646	9
10	6.5273	6.3311	6.1475	5.9752	5.8132	5.7359	5.6607	5.5877	10
11	7.0810	6.8507	6.6362	6.4359	6.2484	6.1591	6.0725	5.9886	11
12	7.6187	7.3528	7.1063	6.8771	6.6634	6.5619	6.4638	6.3688	12
13	8.1409	7.8381	7.5585	7.2997	7.0595	6.9457	6.8358	6.7296	13
14	8.6480	8.3070	7.9937	7.7049	7.4377	7.3115	7.1898	7.0725	14
15	9.1405	8.7604	8.4127	8.0934	7.7991	7.6604	7.5270	7.3985	15
16	9.6187	9.1988	8.8161	8.4661	8.1446	7.9935	7.8483	7.7087	16
17	10.0831	9.6226	9.2047	8.8238	8.4751	8.3116	8.1547	8.0041	17
18	10.5341	10.0325	9.5791	9.1672	8.7914	8.6156	8.4472	8.2857	18
19	10.9720	10.4289	9.9398	9.4971	9.0944	8.9064	8.7265	8.5543	19
20	11.3973	10.8124	10.2876	9.8140	9.3846	9.1845	8.9934	8.8106	20
21	11.8103	11.1834	10.6228	10.1187	9.6628	9.4509	9.2486	9.0553	21
22	12.2113	11.5423	10.9462	10.4116	9.9296	9.7059	9.4927	9.2892	22
23	12.6007	11.8896	11.2580	10.6934	10.1855	9.9503	9.7264	9.5129	23
24	12.9788	12.2257	11.5589	10.9645	10.4312	10.1846	9.9501	9.7268	24
25	13.3460	12.5509	11.8492	11.2254	10.6671	10.4094	10.1645	9.9316	25
26	13.7025	12.8657	12.1294	11.4765	10.8936	10.6250	10.3700	10.1277	26
27	14.0487	13.1705	12.4000	11.7184	11.1113	10.8320	10.5671	10.3156	27
28	14.3849	13.4655	12.6611	11.9514	11.3206	11.0308	10.7562	10.4957	28
29	14.7114	13.7512	12.9133	12.1759	11.5218	11.2217	10.9377	10.6684	29
30	15.0284	14.0277	13.1569	12.3922	11.7153	11.4052	11.1119	10.8341	30
31	15.3362	14.2956	13.3922	12.6008	11.9015	11.5816	11.2793	10.9932	31
32	15.6351	14.5549	13.6196	12.8018	12.0807	11.7513	11.4401	11.1459	32
33	15.9253	14.8061	13.8393	12.9957	12.2533	11.9144	11.5948	11.2926	33
34	16.2071	15.0494	14.0516	13.1828	12.4194	12.0715	11.7434	11.4336	34
35	16.4807	15.2850	14.2568	13.3633	12.5795	12.2226	11.8864	11.5691	35
36	16.7464	15.5133	14.4552	13.5374	12.7337	12.3682	12.0240	11.6994	36
37	17.0044	15.7344	14.6470	13.7055	12.8823	12.5083	12.1564	11.8248	37
38	17.2549	15.9487	14.8325	13.8678	13.0256	12.6433	12.2839	11.9453	38
39	17.4981	16.1562	15.0119	14.0245	13.1637	12.7734	12.4067	12.0614	39
40	17.7343	16.3574	15.1854	14.1758	13.2969	12.8988	12.5250	12.1732	40
41	17.9636	16.5523	15.3532	14.3219	13.4254	13.0197	12.6389	12.2808	41
42	18.1863	16.7412	15.5156	14.4631	13.5494	13.1363	12.7487	12.3844	42
43	18.4025	16.9242	15.6727	14.5995	13.6690	13.2487	12.8546	12.4843	43
44	18.6124	17.1016	15.8247	14.7313	13.7845	13.3572	12.9567	12.5806	44
45	18.8162	17.2735	15.9718	14.8587	13.8960	13.4618	13.0551	12.6734	45
46	19.0142	17.4402	16.1142	14.9819	14.0037	13.5628	13.1501	12.7629	46
47	19.2063	17.6017	16.2520	15.1009	14.1076	13.6603	13.2417	12.8491	47
48	19.3929	17.7583	16.3854	15.2160	14.2080	13.7545	13.3302	12.9324	48
49	19.5741	17.9101	16.5145	15.3273	14.3050	13.8453	13.4155	13.0127	49
50	19.7500	18.0572	16.6396	15.4350	14.3988	13.9331	13.4979	13.0902	50

Note:—In computing the quarterly in advance figures in these tables, all the rates % quoted above are effective rates, see the Introductory Section, pages xvi-xvii.

				Rate per Cent*					
Yrs.	4	4.5	5	5.5	6	6.25	6.5	6.75	Yrs.
51	19.9208	18.1999	16.7606	15.5391	14.4893	14.0179	13.5775	13.1650	51
52	20.0866	18.3382	16.8779	15.6398	14.5769	14.0998	13.6543	13.2373	52
53	20.2476	18.4723	16.9914	15.7373	14.6615	14.1790	13.7285	13.3070	53
54	20.4040	18.6024	17.1014	15.8315	14.7433	14.2555	13.8002	13.3743	54
55	20.5558	18.7285	17.2079	15.9228	14.8224	14.3294	13.8695	13.4394	55
56	20.7032	18.8507	17.3111	16.0111	14.8988	14.4009	13.9364	13.5022	56
57	20.8463	18.9693	17.4110	16.0965	14.9728	14.4700	14.0011	13.5630	57
58	20.9852	19.0843	17.5078	16.1792	15.0444	14.5368	14.0637	13.6216	58
59	21.1201	19.1958	17.6016	16.2593	15.1136	14.6014	14.1241	13.6784	59
60	21.2511	19.3039	17.6925	16.3368	15.1805	14.6639	14.1826	13.7332	60
61	21.3783	19.4088	17.7806	16.4119	15.2453	14.7243	14.2391	13.7862	61
62	21.5018	19.5105	17.8659	16.4846	15.3080	14.7828	14.2938	13.8374	62
63	21.6217	19.6092	17.9486	16.5549	15.3687	14.8394	14.3467	13.8870	63
64	21.7381	19.7049	18.0287	16.6231	15.4274	14.8941	14.3978	13.9349	64
65	21.8511	19.7977	18.1064	16.6891	15.4842	14.9471	14.4473	13.9813	65
66	21.9608	19.8878	18.1817	16.7531	15.5393	14.9983	14.4952	14.0261	66
67	22.0674	19.9751	18.2547	16.8150	15.5925	15.0479	14.5416	14.0695	67
68	22.1708	20.0598	18.3254	16.8750	15.6441	15.0960	14.5864	14.1115	68
69	22.2712	20.1420	18.3940	16.9331	15.6940	15.1425	14.6298	14.1521	69
70	22.3688	20.2217	18.4604	16.9894	15.7424	15.1875	14.6718	14.1914	70
71	22.4634	20.2991	18.5249	17.0440	15.7892	15.2311	14.7125	14.2295	71
72	22.5554	20.3741	18.5873	17.0969	15.8346	15.2733	14.7519	14.2663	72
73	22.6446	20.4469	18.6479	17.1481	15.8785	15.3142	14.7900	14.3019	73
74	22.7313	20.5176	18.7066	17.1977	15.9211	15.3537	14.8269	14.3365	74
75	22.8154	20.5861	18.7636	17.2458	15.9623	15.3921	14.8627	14.3699	75
76	22.8971	20.6526	18.8188	17.2925	16.0023	15.4292	14.8973	14.4023	76
77	22.9764	20.7171	18.8723	17.3377	16.0410	15.4652	14.9308	14.4336	77
78	23.0534	20.7797	18.9243	17.3815	16.0785	15.5000	14.9633	14.4639	78
79	23.1282	20.8404	18.9746	17.4240	16.1148	15.5338	14.9948	14.4933	79
80	23.2008	20.8993	19.0234	17.4651	16.1500	15.5665	15.0253	14.5218	80
81	23.2713	20.9565	19.0708	17.5050	16.1841	15.5982	15.0548	14.5494	81
82	23.3397	21.0119	19.1167	17.5437	16.2172	15.6289	15.0834	14.5761	82
83	23.4061	21.0658	19.1613	17.5812	16.2492	15.6587	15.1111	14.6020	83
84	23.4706	21.1180	19.2045	17.6176	16.2803	15.6875	15.1380	14.6271	84
85	23.5333	21.1687	19.2464	17.6529	16.3104	15.7155	15.1640	14.6514	85
86	23.5941	21.2179	19.2870	17.6871	16.3396	15.7426	15.1892	14.6749	86
87	23.6531	21.2656	19.3265	17.7202	16.3679	15.7688	15.2137	14.6977	87
88	23.7104	21.3119	19.3647	17.7524	16.3953	15.7943	15.2374	14.7198	88
89	23.7661	21.3569	19.4018	17.7836	16.4219	15.8190	15.2603	14.7413	89
90	23.8201	21.4005	19.4378	17.8138	16.4477	15.8429	15.2826	14.7620	90
91	23.8726	21.4428	19.4727	17.8431	16.4727	15.8661	15.3042	14.7822	91
92	23.9235	21.4839	19.5066	17.8715	16.4969	15.8886	15.3251	14.8017	92
93	23.9730	21.5238	19.5395	17.8991	16.5204	15.9104	15.3454	14.8206	93
94	24.0210	21.5625	19.5714	17.9259	16.5432	15.9315	15.3650	14.8389	94
95	24.0676	21.6000	19.6023	17.9518	16.5653	15.9520	15.3841	14.8567	95
96	24.1128	21.6365	19.6323	17.9770	16.5867	15.9719	15.4026	14.8739	96
97	24.1568	21.6719	19.6614	18.0014	16.6075	15.9911	15.4205	14.8907	97
98	24.1994	21.7062	19.6897	18.0251	16.6277	16.0098	15.4378	14.9069	98
99	24.2409	21.7395	19.7171	18.0481	16.6472	16.0279	15.4547	14.9226	99
100	24.2811	21.7718	19.7437	18.0703	16.6662	16.0455	15.4710	14.9378	100

*Note:—In computing the quarterly in advance figures in these tables, all the rates % quoted above are effective rates, see the Introductory Section, pages xvi-xvii.

Rate per Cent*

Yrs.	7	7.25	7.5	8	8.5	9	9.5	10	Yrs.
1	0.7268	0.7255	0.7243	0.7220	0.7196	0.7173	0.7150	0.7127	1
2	1.4048	1.4003	1.3958	1.3870	1.3783	1.3698	1.3615	1.3533	2
3	2.0385	2.0290	2.0197	2.0012	1.9833	1.9657	1.9486	1.9318	3
4	2.6319	2.6161	2.6006	2.5701	2.5406	2.5118	2.4839	2.4567	4
5	3.1885	3.1654	3.1426	3.0983	3.0554	3.0140	2.9738	2.9350	5
6	3.7114	3.6801	3.6494	3.5897	3.5323	3.4770	3.4237	3.3723	6
7	4.2033	4.1632	4.1239	4.0479	3.9750	3.9051	3.8380	3.7736	7
8	4.6668	4.6173	4.5691	4.4759	4.3870	4.3020	4.2207	4.1429	8
9	5.1039	5.0449	4.9873	4.8765	4.7712	4.6708	4.5752	4.4839	9
10	5.5168	5.4479	5.3808	5.2521	5.1301	5.0143	4.9042	4.7994	10
11	5.9072	5.8283	5.7516	5.6048	5.4660	5.3348	5.2103	5.0922	11
12	6.2768	6.1877	6.1014	5.9364	5.7810	5.6344	5.4957	5.3645	12
13	6.6270	6.5278	6.4318	6.2487	6.0768	5.9149	5.7624	5.6183	13
14	6.9592	6.8499	6.7442	6.5432	6.3549	6.1782	6.0119	5.8552	14
15	7.2746	7.1552	7.0400	6.8213	6.6169	6.4255	6.2458	6.0769	15
16	7.5744	7.4450	7.3203	7.0842	6.8640	6.6582	6.4655	6.2846	16
17	7.8594	7.7202	7.5863	7.3329	7.0972	6.8775	6.6720	6.4796	17
18	8.1307	7.9819	7.8387	7.5685	7.3177	7.0843	6.8666	6.6629	18
19	8.3892	8.2308	8.0787	7.7920	7.5264	7.2797	7.0500	6.8355	19
20	8.6355	8.4678	8.3069	8.0041	7.7241	7.4645	7.2232	6.9982	20
21	8.8705	8.6936	8.5241	8.2056	7.9116	7.6395	7.3869	7.1517	21
22	9.0949	8.9090	8.7311	8.3972	8.0896	7.8053	7.5418	7.2968	22
23	9.3091	9.1145	8.9284	8.5795	8.2586	7.9626	7.6885	7.4341	23
24	9.5139	9.3107	9.1166	8.7531	8.4194	8.1119	7.8277	7.5641	24
25	9.7097	9.4982	9.2962	8.9186	8.5724	8.2538	7.9597	7.6874	25
26	9.8971	9.6774	9.4678	9.0764	8.7181	8.3888	8.0852	7.8043	26
27	10.0764	9.8488	9.6318	9.2270	8.8570	8.5173	8.2045	7.9154	27
28	10.2482	10.0128	9.7887	9.3709	8.9894	8.6397	8.3180	8.0211	28
29	10.4129	10.1699	9.9387	9.5083	9.1158	8.7565	8.4262	8.1216	29
30	10.5707	10.3204	10.0824	9.6397	9.2366	8.8678	8.5292	8.2172	30
31	10.7220	10.4647	10.2200	9.7655	9.3519	8.9741	8.6275	8.3084	31
32	10.8673	10.6030	10.3519	9.8858	9.4622	9.0756	8.7213	8.3954	32
33	11.0067	10.7356	10.4783	10.0010	9.5677	9.1726	8.8108	8.4783	33
34	11.1406	10.8630	10.5996	10.1114	9.6687	9.2654	8.8964	8.5575	34
35	11.2692	10.9852	10.7160	10.2173	9.7655	9.3542	8.9782	8.6332	35
36	11.3928	11.1026	10.8277	10.3188	9.8581	9.4392	9.0565	8.7056	36
37	11.5116	11.2154	10.9349	10.4162	9.9470	9.5206	9.1314	8.7748	37
38	11.6258	11.3239	11.0380	10.5096	10.0322	9.5986	9.2032	8.8410	38
39	11.7358	11.4281	11.1370	10.5994	10.1139	9.6734	9.2719	8.9044	39
40	11.8415	11.5284	11.2322	10.6856	10.1924	9.7452	9.3378	8.9652	40
41	11.9433	11.6249	11.3238	10.7684	10.2677	9.8140	9.4010	9.0234	41
42	12.0414	11.7177	11.4118	10.8480	10.3401	9.8801	9.4616	9.0793	42
43	12.1358	11.8071	11.4966	10.9246	10.4096	9.9436	9.5198	9.1328	43
44	12.2267	11.8931	11.5782	10.9982	10.4764	10.0045	9.5757	9.1842	44
45	12.3143	11.9760	11.6567	11.0691	10.5407	10.0631	9.6294	9.2336	45
46	12.3988	12.0559	11.7324	11.1373	10.6025	10.1195	9.6809	9.2810	46
47	12.4802	12.1329	11.8053	11.2029	10.6620	10.1737	9.7305	9.3266	47
48	12.5587	12.2071	11.8755	11.2661	10.7193	10.2258	9.7782	9.3703	48
49	12.6345	12.2786	11.9432	11.3270	10.7744	10.2759	9.8240	9.4124	49
50	12.7075	12.3476	12.0085	11.3857	10.8275	10.3242	9.8681	9.4529	50

Note:—In computing the quarterly in advance figures in these tables, all the rates % quoted above are effective rates, see the Introductory Section, pages xvi-xvii.

			Rate per Cent*						
Yrs.	7	7.25	7.5	8	8.5	9	9.5	10	Yrs.
51	12.7780	12.4141	12.0714	11.4423	10.8786	10.3707	9.9106	9.4919	51
52	12.8460	12.4783	12.1321	11.4968	10.9279	10.4154	9.9514	9.5293	52
53	12.9117	12.5403	12.1906	11.5494	10.9754	10.4586	9.9908	9.5654	53
54	12.9751	12.6001	12.2471	11.6001	11.0211	10.5001	10.0287	9.6002	54
55	13.0363	12.6578	12.3017	11.6490	11.0653	10.5402	10.0653	9.6337	55
56	13.0955	12.7135	12.3543	11.6962	11.1078	10.5788	10.1005	9.6659	56
57	13.1526	12.7674	12.4051	11.7417	11.1489	10.6160	10.1344	9.6970	57
58	13.2077	12.8193	12.4542	11.7857	11.1885	10.6519	10.1671	9.7269	58
59	13.2610	12.8696	12.5016	11.8281	11.2268	10.6866	10.1987	9.7558	59
60	13.3126	12.9181	12.5473	11.8691	11.2637	10.7200	10.2291	9.7837	60
61	13.3624	12.9650	12.5916	11.9086	11.2993	10.7523	10.2585	9.8106	61
62	13.4105	13.0103	12.6343	11.9468	11.3337	10.7834	10.2869	9.8365	62
63	13.4570	13.0541	12.6756	11.9838	11.3669	10.8135	10.3142	9.8615	63
64	13.5020	13.0964	12.7155	12.0194	11.3990	10.8426	10.3406	9.8856	64
65	13.5456	13.1374	12.7541	12.0539	11.4300	10.8706	10.3662	9.9089	65
66	13.5877	13.1769	12.7914	12.0872	11.4600	10.8977	10.3908	9.9315	66
67	13.6284	13.2152	12.8275	12.1194	11.4889	10.9239	10.4146	9.9532	67
68	13.6677	13.2523	12.8624	12.1506	11.5169	10.9491	10.4376	9.9742	68
69	13.7059	13.2881	12.8961	12.1807	11.5439	10.9736	10.4598	9.9945	69
70	13.7427	13.3227	12.9288	12.2098	11.5701	10.9972	10.4812	10.0140	70
71	13.7784	13.3563	12.9603	12.2379	11.5954	11.0200	10.5020	10.0330	71
72	13.8129	13.3887	12.9909	12.2652	11.6198	11.0421	10.5220	10.0513	72
73	13.8464	13.4201	13.0204	12.2915	11.6434	11.0635	10.5414	10.0690	73
74	13.8787	13.4505	13.0490	12.3170	11.6663	11.0841	10.5601	10.0861	74
75	13.9100	13.4799	13.0767	12.3417	11.6884	11.1041	10.5783	10.1026	75
76	13.9404	13.5084	13.1035	12.3655	11.7098	11.1234	10.5958	10.1186	76
77	13.9697	13.5359	13.1295	12.3886	11.7305	11.1421	10.6127	10.1340	77
78	13.9981	13.5626	13.1546	12.4110	11.7506	11.1602	10.6291	10.1490	78
79	14.0257	13.5885	13.1789	12.4326	11.7700	11.1777	10.6450	10.1634	79
80	14.0523	13.6135	13.2024	12.4536	11.7888	11.1946	10.6604	10.1774	80
81	14.0782	13.6377	13.2252	12.4738	11.8069	11.2110	10.6752	10.1910	81
82	14.1032	13.6612	13.2473	12.4935	11.8245	11.2268	10.6896	10.2041	82
83	14.1274	13.6839	13.2687	12.5125	11.8415	11.2422	10.7035	10.2168	83
84	14.1509	13.7060	13.2894	12.5309	11.8580	11.2570	10.7170	10.2290	84
85	14.1736	13.7273	13.3094	12.5487	11.8740	11.2714	10.7300	10.2409	85
86	14.1957	13.7480	13.3288	12.5660	11.8895	11.2854	10.7426	10.2524	86
87	14.2170	13.7680	13.3477	12.5827	11.9044	11.2988	10.7549	10.2635	87
88	14.2377	13.7874	13.3659	12.5989	11.9189	11.3119	10.7667	10.2743	88
89	14.2577	13.8062	13.3836	12.6146	11.9330	11.3246	10.7782	10.2847	89
90	14.2772	13.8244	13.4007	12.6298	11.9466	11.3368	10.7893	10.2949	90
91	14.2960	13.8421	13.4173	12.6446	11.9598	11.3487	10.8000	10.3046	91
92	14.3142	13.8592	13.4333	12.6588	11.9725	11.3602	10.8104	10.3141	92
93	14.3319	13.8758	13.4489	12.6727	11.9849	11.3713	10.8205	10.3233	93
94	14.3491	13.8918	13.4640	12.6861	11.9969	11.3821	10.8303	10.3322	94
95	14.3657	13.9074	13.4786	12.6991	12.0085	11.3926	10.8397	10.3408	95
96	14.3818	13.9225	13.4928	12.7116	12.0198	11.4027	10.8489	10.3492	96
97	14.3974	13.9371	13.5066	12.7238	12.0307	11.4125	10.8578	10.3572	97
98	14.4126	13.9513	13.5199	12.7357	12.0412	11.4220	10.8664	10.3651	98
99	14.4273	13.9651	13.5328	12.7471	12.0515	11.4312	10.8747	10.3727	99
100	14.4415	13.9784	13.5453	12.7582	12.0614	11.4402	10.8828	10.3800	100

Note:—In computing the quarterly in advance figures in these tables, all the rates % quoted above are effective rates,
see the Introductory Section, pages xvi-xvii.

Rate per Cent*

Yrs.	11	12	13	14	15	16	18	20	Yrs.
1	0.7083	0.7039	0.6997	0.6955	0.6915	0.6875	0.6798	0.6724	1
2	1.3373	1.3219	1.3070	1.2925	1.2786	1.2651	1.2393	1.2151	2
3	1.8994	1.8685	1.8388	1.8104	1.7831	1.7570	1.7077	1.6620	3
4	2.4046	2.3552	2.3083	2.2637	2.2212	2.1807	2.1053	2.0363	4
5	2.8609	2.7912	2.7255	2.6635	2.6050	2.5495	2.4470	2.3543	5
6	3.2748	3.1838	3.0987	3.0188	2.9438	2.8731	2.7436	2.6276	6
7	3.6519	3.5391	3.4342	3.3364	3.2450	3.1594	3.0034	2.8649	7
8	3.9968	3.8620	3.7375	3.6219	3.5144	3.4142	3.2328	3.0729	8
9	4.3132	4.1567	4.0127	3.8798	3.7568	3.6425	3.4367	3.2566	9
10	4.6044	4.4265	4.2636	4.1139	3.9758	3.8480	3.6191	3.4199	10
11	4.8732	4.6744	4.4931	4.3271	4.1746	4.0340	3.7831	3.5660	11
12	5.1220	4.9028	4.7038	4.5222	4.3559	4.2030	3.9314	3.6975	12
13	5.3528	5.1139	4.8977	4.7012	4.5217	4.3572	4.0660	3.8163	13
14	5.5675	5.3095	5.0768	4.8660	4.6739	4.4983	4.1886	3.9241	14
15	5.7675	5.4911	5.2426	5.0181	4.8141	4.6280	4.3009	4.0225	15
16	5.9543	5.6602	5.3965	5.1589	4.9436	4.7476	4.4039	4.1125	16
17	6.1291	5.8179	5.5397	5.2896	5.0634	4.8580	4.4988	4.1951	17
18	6.2928	5.9652	5.6731	5.4111	5.1747	4.9603	4.5864	4.2711	18
19	6.4465	6.1031	5.7977	5.5243	5.2782	5.0553	4.6675	4.3414	19
20	6.5910	6.2325	5.9143	5.6301	5.3746	5.1437	4.7428	4.4065	20
21	6.7271	6.3540	6.0236	5.7291	5.4647	5.2262	4.8128	4.4668	21
22	6.8553	6.4683	6.1263	5.8218	5.5491	5.3033	4.8781	4.5230	22
23	6.9763	6.5759	6.2227	5.9089	5.6281	5.3754	4.9390	4.5754	23
24	7.0907	6.6774	6.3136	5.9907	5.7023	5.4431	4.9961	4.6243	24
25	7.1989	6.7733	6.3992	6.0678	5.7721	5.5066	5.0496	4.6701	25
26	7.3014	6.8640	6.4801	6.1404	5.8378	5.5664	5.0998	4.7130	26
27	7.3985	6.9498	6.5565	6.2090	5.8997	5.6227	5.1470	4.7533	27
28	7.4907	7.0310	6.6288	6.2738	5.9582	5.6757	5.1914	4.7912	28
29	7.5783	7.1081	6.6973	6.3351	6.0134	5.7259	5.2333	4.8269	29
30	7.6615	7.1813	6.7622	6.3932	6.0657	5.7733	5.2729	4.8605	30
31	7.7407	7.2509	6.8238	6.4482	6.1153	5.8181	5.3103	4.8922	31
32	7.8162	7.3170	6.8824	6.5005	6.1623	5.8606	5.3457	4.9223	32
33	7.8880	7.3799	6.9380	6.5501	6.2068	5.9009	5.3792	4.9507	33
34	7.9565	7.4399	6.9910	6.5973	6.2492	5.9392	5.4110	4.9776	34
35	8.0219	7.4970	7.0414	6.6422	6.2894	5.9756	5.4411	5.0031	35
36	8.0844	7.5515	7.0894	6.6849	6.3278	6.0101	5.4698	5.0273	36
37	8.1440	7.6035	7.1353	6.7256	6.3642	6.0430	5.4970	5.0503	37
38	8.2010	7.6532	7.1790	6.7645	6.3990	6.0744	5.5230	5.0722	38
39	8.2556	7.7007	7.2208	6.8015	6.4322	6.1043	5.5476	5.0930	39
40	8.3078	7.7461	7.2607	6.8369	6.4638	6.1327	5.5712	5.1128	40
41	8.3577	7.7895	7.2988	6.8708	6.4940	6.1599	5.5936	5.1317	41
42	8.4056	7.8311	7.3353	6.9031	6.5229	6.1859	5.6150	5.1497	42
43	8.4515	7.8709	7.3702	6.9340	6.5505	6.2107	5.6354	5.1669	43
44	8.4955	7.9091	7.4037	6.9636	6.5769	6.2345	5.6550	5.1833	44
45	8.5377	7.9457	7.4357	6.9919	6.6022	6.2572	5.6737	5.1990	45
46	8.5783	7.9807	7.4664	7.0191	6.6264	6.2789	5.6915	5.2140	46
47	8.6172	8.0144	7.4959	7.0451	6.6496	6.2997	5.7086	5.2283	47
48	8.6545	8.0467	7.5241	7.0701	6.6718	6.3197	5.7250	5.2421	48
49	8.6904	8.0777	7.5513	7.0940	6.6931	6.3388	5.7407	5.2552	49
50	8.7249	8.1075	7.5773	7.1170	6.7136	6.3571	5.7557	5.2678	50

*Note:—In computing the quarterly in advance figures in these tables, all the rates % quoted above are effective rates, see the Introductory Section, pages xvi-xvii.

YEARS' PURCHASE*

Rate per Cent*

Yrs.	11	12	13	14	15	16	18	20	Yrs.
51	8.7581	8.1362	7.6023	7.1390	6.7332	6.3747	5.7701	5.2799	51
52	8.7900	8.1637	7.6263	7.1602	6.7520	6.3916	5.7840	5.2914	52
53	8.8207	8.1902	7.6494	7.1806	6.7701	6.4078	5.7972	5.3026	53
54	8.8502	8.2156	7.6716	7.2001	6.7875	6.4234	5.8100	5.3132	54
55	8.8787	8.2401	7.6930	7.2189	6.8042	6.4384	5.8222	5.3235	55
56	8.9061	8.2637	7.7135	7.2370	6.8203	6.4527	5.8340	5.3333	56
57	8.9324	8.2864	7.7333	7.2544	6.8358	6.4666	5.8453	5.3427	57
58	8.9578	8.3083	7.7524	7.2712	6.8506	6.4799	5.8562	5.3518	58
59	8.9823	8.3293	7.7707	7.2873	6.8649	6.4927	5.8666	5.3606	59
60	9.0059	8.3496	7.7884	7.3029	6.8787	6.5050	5.8767	5.3690	60
61	9.0287	8.3692	7.8054	7.3178	6.8920	6.5169	5.8864	5.3770	61
62	9.0507	8.3881	7.8218	7.3322	6.9048	6.5283	5.8957	5.3848	62
63	9.0718	8.4062	7.8376	7.3461	6.9171	6.5393	5.9047	5.3923	63
64	9.0923	8.4238	7.8528	7.3595	6.9290	6.5499	5.9133	5.3995	64
65	9.1120	8.4407	7.8675	7.3724	6.9404	6.5602	5.9216	5.4065	65
66	9.1310	8.4570	7.8817	7.3849	6.9514	6.5700	5.9297	5.4131	66
67	9.1494	8.4728	7.8954	7.3969	6.9621	6.5795	5.9374	5.4196	67
68	9.1671	8.4880	7.9086	7.4085	6.9723	6.5887	5.9449	5.4258	68
69	9.1842	8.5027	7.9213	7.4196	6.9822	6.5975	5.9521	5.4318	69
70	9.2008	8.5168	7.9336	7.4304	6.9918	6.6061	5.9590	5.4376	70
71	9.2168	8.5305	7.9455	7.4409	7.0010	6.6143	5.9657	5.4432	71
72	9.2322	8.5438	7.9570	7.4509	7.0099	6.6222	5.9722	5.4485	72
73	9.2471	8.5565	7.9681	7.4606	7.0185	6.6299	5.9784	5.4537	73
74	9.2615	8.5689	7.9788	7.4700	7.0268	6.6373	5.9844	5.4588	74
75	9.2755	8.5808	7.9891	7.4791	7.0349	6.6445	5.9903	5.4636	75
76	9.2889	8.5923	7.9991	7.4878	7.0426	6.6514	5.9959	5.4683	76
77	9.3020	8.6035	8.0088	7.4963	7.0501	6.6581	6.0013	5.4728	77
78	9.3146	8.6142	8.0181	7.5045	7.0573	6.6645	6.0065	5.4771	78
79	9.3267	8.6247	8.0271	7.5124	7.0643	6.6707	6.0116	5.4813	79
80	9.3385	8.6347	8.0359	7.5200	7.0711	6.6768	6.0165	5.4854	80
81	9.3499	8.6445	8.0443	7.5274	7.0776	6.6826	6.0212	5.4893	81
82	9.3609	8.6539	8.0525	7.5346	7.0839	6.6882	6.0258	5.4931	82
83	9.3716	8.6630	8.0603	7.5415	7.0900	6.6937	6.0302	5.4968	83
84	9.3819	8.6718	8.0680	7.5481	7.0959	6.6989	6.0345	5.5004	84
85	9.3919	8.6804	8.0754	7.5546	7.1016	6.7040	6.0386	5.5038	85
86	9.4016	8.6886	8.0825	7.5609	7.1072	6.7089	6.0426	5.5071	86
87	9.4110	8.6966	8.0894	7.5669	7.1125	6.7137	6.0465	5.5103	87
88	9.4200	8.7044	8.0961	7.5728	7.1177	6.7183	6.0502	5.5134	88
89	9.4288	8.7119	8.1026	7.5784	7.1227	6.7228	6.0538	5.5164	89
90	9.4373	8.7191	8.1089	7.5839	7.1275	6.7271	6.0573	5.5193	90
91	9.4455	8.7261	8.1149	7.5892	7.1322	6.7313	6.0607	5.5222	91
92	9.4535	8.7329	8.1208	7.5944	7.1368	6.7353	6.0640	5.5249	92
93	9.4612	8.7395	8.1265	7.5994	7.1412	6.7392	6.0672	5.5275	93
94	9.4687	8.7459	8.1320	7.6042	7.1454	6.7430	6.0702	5.5301	94
95	9.4759	8.7521	8.1374	7.6088	7.1495	6.7467	6.0732	5.5325	95
96	9.4829	8.7580	8.1425	7.6134	7.1535	6.7502	6.0761	5.5349	96
97	9.4897	8.7638	8.1475	7.6177	7.1574	6.7537	6.0789	5.5372	97
98	9.4963	8.7694	8.1524	7.6220	7.1611	6.7570	6.0816	5.5395	98
99	9.5026	8.7749	8.1571	7.6261	7.1648	6.7602	6.0842	5.5416	99
100	9.5088	8.7801	8.1616	7.6300	7.1683	6.7634	6.0867	5.5437	100

Note:—In computing the quarterly in advance figures in these tables, all the rates % quoted above are effective rates,
see the Introductory Section, pages xvi-xvii.

Rate per Cent*

Yrs.	4	4.5	5	5.5	6	6.25	6.5	6.75	Yrs.
1	0.7463	0.7436	0.7410	0.7385	0.7359	0.7347	0.7334	0.7322	1
2	1.4776	1.4673	1.4573	1.4474	1.4377	1.4329	1.4281	1.4234	2
3	2.1940	2.1714	2.1493	2.1279	2.1070	2.0967	2.0866	2.0766	3
4	2.8952	2.8560	2.8180	2.7813	2.7457	2.7283	2.7112	2.6943	4
5	3.5814	3.5215	3.4639	3.4086	3.3552	3.3293	3.3039	3.2789	5
6	4.2523	4.1682	4.0878	4.0109	3.9373	3.9016	3.8667	3.8325	6
7	4.9081	4.7964	4.6902	4.5893	4.4931	4.4468	4.4015	4.3572	7
8	5.5488	5.4064	5.2719	5.1447	5.0242	4.9663	4.9098	4.8548	8
9	6.1743	5.9985	5.8334	5.6781	5.5316	5.4615	5.3933	5.3270	9
10	6.7848	6.5731	6.3754	6.1903	6.0166	5.9338	5.8534	5.7754	10
11	7.3804	7.1306	6.8985	6.6823	6.4804	6.3843	6.2914	6.2014	11
12	7.9611	7.6712	7.4032	7.1548	6.9238	6.8143	6.7085	6.6063	12
13	8.5270	8.1953	7.8902	7.6087	7.3480	7.2248	7.1060	6.9913	13
14	9.0784	8.7034	8.3601	8.0446	7.7538	7.6167	7.4848	7.3577	14
15	9.6153	9.1956	8.8132	8.4634	8.1421	7.9911	7.8460	7.7065	15
16	10.1379	9.6725	9.2503	8.8657	8.5138	8.3488	8.1905	8.0386	16
17	10.6465	10.1343	9.6719	9.2522	8.8695	8.6906	8.5193	8.3551	17
18	11.1410	10.5815	10.0783	9.6234	9.2102	9.0174	8.8331	8.6557	18
19	11.6219	11.0143	10.4702	9.9801	9.5364	9.3299	9.1327	8.9442	19
20	12.0893	11.4332	10.8480	10.3228	9.8488	9.6287	9.4188	9.2185	20
21	12.5434	11.8385	11.2122	10.6521	10.1481	9.9145	9.6922	9.4802	21
22	12.9843	12.2306	11.5633	10.9684	10.4348	10.1880	9.9534	9.7299	22
23	13.4125	12.6097	11.9016	11.2724	10.7095	10.4498	10.2031	9.9684	23
24	13.8281	12.9764	12.2277	11.5645	10.9728	10.7003	10.4418	10.1961	24
25	14.2313	13.3308	12.5419	11.8451	11.2252	10.9402	10.6700	10.4136	25
26	14.6223	13.6733	12.8447	12.1148	11.4671	11.1698	10.8884	10.6215	26
27	15.0016	14.0044	13.1364	12.3740	11.6990	11.3898	11.0973	10.8202	27
28	15.3692	14.3242	13.4174	12.6230	11.9214	11.6004	11.2971	11.0102	28
29	15.7254	14.6332	13.6881	12.8624	12.1346	11.8022	11.4885	11.1918	29
30	16.0706	14.9316	13.9489	13.0924	12.3391	11.9956	11.6716	11.3655	30
31	16.4049	15.2198	14.2001	13.3134	12.5353	12.1809	11.8469	11.5317	31
32	16.7287	15.4981	14.4420	13.5258	12.7234	12.3585	12.0149	11.6908	32
33	17.0421	15.7667	14.6750	13.7300	12.9039	12.5287	12.1757	11.8430	33
34	17.3454	16.0260	14.8994	13.9262	13.0771	12.6919	12.3297	11.9887	34
35	17.6389	16.2762	15.1154	14.1148	13.2432	12.8483	12.4773	12.1282	35
36	17.9229	16.5177	15.3235	14.2960	13.4027	12.9983	12.6188	12.2617	36
37	18.1976	16.7507	15.5238	14.4702	13.5557	13.1422	12.7543	12.3897	37
38	18.4632	16.9755	15.7167	14.6377	13.7025	13.2802	12.8842	12.5122	38
39	18.7200	17.1924	15.9024	14.7986	13.8434	13.4125	13.0087	12.6296	39
40	18.9682	17.4015	16.0812	14.9533	13.9787	13.5394	13.1281	12.7421	40
41	19.2081	17.6032	16.2533	15.1020	14.1086	13.6612	13.2426	12.8499	41
42	19.4399	17.7977	16.4189	15.2449	14.2332	13.7781	13.3524	12.9533	42
43	19.6639	17.9852	16.5784	15.3823	14.3529	13.8902	13.4576	13.0523	43
44	19.8802	18.1660	16.7319	15.5144	14.4678	13.9978	13.5586	13.1473	44
45	20.0891	18.3403	16.8796	15.6413	14.5782	14.1010	13.6554	13.2383	45
46	20.2908	18.5083	17.0218	15.7633	14.6841	14.2001	13.7484	13.3256	46
47	20.4856	18.6702	17.1587	15.8806	14.7858	14.2953	13.8375	13.4094	47
48	20.6736	18.8262	17.2904	15.9934	14.8835	14.3866	13.9230	13.4897	48
49	20.8551	18.9766	17.4171	16.1018	14.9774	14.4742	14.0051	13.5667	49
50	21.0302	19.1215	17.5391	16.2060	15.0675	14.5583	14.0839	13.6406	50

*Note:—In computing the quarterly in advance figures in these tables, all the rates % quoted above are effective rates, see the Introductory Section, pages xvi-xvii.

				Rate per Cent*					
Yrs.	4	4.5	5	5.5	6	6.25	6.5	6.75	Yrs.
51	21.1992	19.2611	17.6565	16.3061	15.1540	14.6391	14.1594	13.7115	51
52	21.3622	19.3956	17.7694	16.4024	15.2371	14.7167	14.2320	13.7795	52
53	21.5195	19.5251	17.8781	16.4950	15.3170	14.7912	14.3016	13.8448	53
54	21.6712	19.6499	17.9827	16.5840	15.3937	14.8627	14.3685	13.9074	54
55	21.8175	19.7702	18.0833	16.6695	15.4674	14.9313	14.4327	13.9675	55
56	21.9586	19.8860	18.1802	16.7518	15.5381	14.9973	14.4943	14.0252	56
57	22.0947	19.9975	18.2733	16.8308	15.6061	15.0606	14.5534	14.0806	57
58	22.2258	20.1049	18.3630	16.9068	15.6715	15.1215	14.6102	14.1337	58
59	22.3523	20.2083	18.4492	16.9799	15.7342	15.1799	14.6647	14.1848	59
60	22.4742	20.3078	18.5322	17.0502	15.7945	15.2360	14.7171	14.2338	60
61	22.5917	20.4037	18.6120	17.1177	15.8525	15.2899	14.7674	14.2808	61
62	22.7049	20.4960	18.6887	17.1826	15.9081	15.3417	14.8157	14.3260	62
63	22.8140	20.5849	18.7626	17.2450	15.9616	15.3914	14.8621	14.3693	63
64	22.9192	20.6705	18.8337	17.3051	16.0130	15.4392	14.9066	14.4110	64
65	23.0205	20.7529	18.9020	17.3627	16.0624	15.4851	14.9494	14.4510	65
66	23.1181	20.8322	18.9678	17.4182	16.1099	15.5292	14.9905	14.4894	66
67	23.2121	20.9085	19.0311	17.4715	16.1555	15.5716	15.0300	14.5262	67
68	23.3027	20.9820	19.0919	17.5228	16.1993	15.6123	15.0679	14.5617	68
69	23.3900	21.0527	19.1504	17.5721	16.2414	15.6515	15.1044	14.5957	69
70	23.4740	21.1207	19.2067	17.6195	16.2819	15.6890	15.1394	14.6284	70
71	23.5550	21.1862	19.2609	17.6651	16.3208	15.7252	15.1730	14.6598	71
72	23.6329	21.2493	19.3130	17.7089	16.3582	15.7599	15.2053	14.6899	72
73	23.7080	21.3100	19.3631	17.7510	16.3941	15.7932	15.2364	14.7189	73
74	23.7803	21.3684	19.4113	17.7915	16.4287	15.8253	15.2662	14.7467	74
75	23.8499	21.4245	19.4576	17.8304	16.4619	15.8561	15.2948	14.7735	75
76	23.9169	21.4786	19.5022	17.8679	16.4938	15.8857	15.3224	14.7992	76
77	23.9814	21.5306	19.5451	17.9039	16.5244	15.9141	15.3488	14.8238	77
78	24.0436	21.5807	19.5864	17.9385	16.5539	15.9414	15.3743	14.8476	78
79	24.1034	21.6289	19.6260	17.9717	16.5822	15.9677	15.3987	14.8703	79
80	24.1609	21.6752	19.6642	18.0037	16.6095	15.9929	15.4222	14.8922	80
81	24.2164	21.7198	19.7009	18.0345	16.6356	16.0172	15.4447	14.9133	81
82	24.2697	21.7627	19.7362	18.0640	16.6608	16.0405	15.4664	14.9335	82
83	24.3211	21.8040	19.7701	18.0925	16.6850	16.0629	15.4873	14.9529	83
84	24.3705	21.8437	19.8028	18.1198	16.7082	16.0845	15.5073	14.9716	84
85	24.4181	21.8819	19.8342	18.1461	16.7306	16.1052	15.5265	14.9895	85
86	24.4638	21.9187	19.8644	18.1714	16.7521	16.1251	15.5450	15.0068	86
87	24.5079	21.9540	19.8934	18.1957	16.7727	16.1442	15.5628	15.0233	87
88	24.5503	21.9881	19.9213	18.2190	16.7926	16.1626	15.5799	15.0392	88
89	24.5911	22.0208	19.9482	18.2415	16.8116	16.1803	15.5963	15.0545	89
90	24.6304	22.0523	19.9740	18.2631	16.8300	16.1973	15.6121	15.0693	90
91	24.6681	22.0825	19.9989	18.2838	16.8476	16.2136	15.6273	15.0834	91
92	24.7045	22.1117	20.0227	18.3038	16.8646	16.2293	15.6418	15.0970	92
93	24.7395	22.1397	20.0457	18.3230	16.8809	16.2444	15.6559	15.1100	93
94	24.7731	22.1666	20.0678	18.3415	16.8965	16.2589	15.6693	15.1226	94
95	24.8055	22.1926	20.0891	18.3592	16.9116	16.2728	15.6823	15.1346	95
96	24.8367	22.2175	20.1095	18.3763	16.9261	16.2863	15.6947	15.1462	96
97	24.8667	22.2415	20.1291	18.3927	16.9400	16.2991	15.7067	15.1574	97
98	24.8955	22.2646	20.1480	18.4085	16.9534	16.3115	15.7182	15.1681	98
99	24.9233	22.2868	20.1662	18.4236	16.9662	16.3234	15.7293	15.1784	99
100	24.9500	22.3081	20.1837	18.4382	16.9786	16.3349	15.7399	15.1883	100

*Note:—In computing the quarterly in advance figures in these tables, all the rates % quoted above are effective rates,
see the Introductory Section, pages xvi-xvii.

Rate per Cent*

Yrs.	7	7.25	7.5	8	8.5	9	9.5	10	Yrs.
1	0.7310	0.7297	0.7285	0.7261	0.7237	0.7214	0.7191	0.7168	1
2	1.4188	1.4142	1.4096	1.4006	1.3918	1.3832	1.3746	1.3663	2
3	2.0667	2.0570	2.0474	2.0284	2.0100	1.9920	1.9743	1.9572	3
4	2.6777	2.6614	2.6453	2.6138	2.5832	2.5535	2.5246	2.4966	4
5	3.2543	3.2302	3.2065	3.1604	3.1158	3.0727	3.0310	2.9906	5
6	3.7990	3.7662	3.7340	3.6716	3.6115	3.5538	3.4981	3.4445	6
7	4.3140	4.2717	4.2304	4.1504	4.0738	4.0005	3.9301	3.8625	7
8	4.8012	4.7489	4.6979	4.5995	4.5056	4.4160	4.3304	4.2485	8
9	5.2625	5.1998	5.1387	5.0211	4.9095	4.8033	4.7022	4.6058	9
10	5.6997	5.6261	5.5546	5.4176	5.2879	5.1649	5.0482	4.9372	10
11	6.1141	6.0296	5.9476	5.7907	5.6427	5.5029	5.3706	5.2452	11
12	6.5074	6.4117	6.3190	6.1422	5.9760	5.8194	5.6717	5.5320	12
13	6.8807	6.7738	6.6704	6.4737	6.2894	6.1162	5.9532	5.7995	13
14	7.2352	7.1171	7.0031	6.7867	6.5843	6.3947	6.2168	6.0494	14
15	7.5722	7.4429	7.3184	7.0823	6.8622	6.6566	6.4639	6.2832	15
16	7.8927	7.7523	7.6172	7.3618	7.1243	6.9029	6.6960	6.5022	16
17	8.1975	8.0462	7.9008	7.6263	7.3718	7.1350	6.9141	6.7077	17
18	8.4876	8.3255	8.1699	7.8768	7.6056	7.3537	7.1194	6.9007	18
19	8.7639	8.5912	8.4256	8.1142	7.8266	7.5602	7.3127	7.0822	19
20	9.0270	8.8439	8.6685	8.3393	8.0358	7.7553	7.4951	7.2531	20
21	9.2778	9.0845	8.8995	8.5529	8.2340	7.9396	7.6671	7.4141	21
22	9.5169	9.3136	9.1193	8.7556	8.4217	8.1141	7.8297	7.5660	22
23	9.7449	9.5318	9.3284	8.9483	8.5998	8.2792	7.9833	7.7094	23
24	9.9624	9.7398	9.5276	9.1313	8.7687	8.4357	8.1287	7.8449	24
25	10.1700	9.9381	9.7173	9.3054	8.9292	8.5841	8.2664	7.9731	25
26	10.3682	10.1273	9.8980	9.4711	9.0816	8.7248	8.3969	8.0943	26
27	10.5574	10.3078	10.0703	9.6287	9.2264	8.8584	8.5206	8.2092	27
28	10.7382	10.4800	10.2347	9.7788	9.3642	8.9854	8.6379	8.3181	28
29	10.9109	10.6444	10.3914	9.9219	9.4952	9.1060	8.7493	8.4213	29
30	11.0759	10.8015	10.5411	10.0582	9.6200	9.2206	8.8551	8.5193	30
31	11.2337	10.9515	10.6839	10.1881	9.7388	9.3297	8.9557	8.6124	31
32	11.3846	11.0948	10.8202	10.3120	9.8520	9.4336	9.0513	8.7008	32
33	11.5289	11.2318	10.9505	10.4303	9.9599	9.5324	9.1423	8.7848	33
34	11.6669	11.3628	11.0749	10.5431	10.0627	9.6266	9.2289	8.8647	34
35	11.7989	11.4880	11.1939	10.6509	10.1608	9.7163	9.3113	8.9408	35
36	11.9253	11.6078	11.3076	10.7538	10.2544	9.8019	9.3898	9.0131	36
37	12.0463	11.7224	11.4163	10.8520	10.3437	9.8834	9.4647	9.0821	37
38	12.1621	11.8320	11.5203	10.9459	10.4290	9.9613	9.5360	9.1478	38
39	12.2730	11.9370	11.6197	11.0357	10.5104	10.0355	9.6041	9.2104	39
40	12.3792	12.0374	11.7149	11.1215	10.5882	10.1064	9.6690	9.2700	40
41	12.4810	12.1336	11.8059	11.2035	10.6626	10.1742	9.7310	9.3270	41
42	12.5784	12.2257	11.8931	11.2820	10.7336	10.2388	9.7901	9.3813	42
43	12.6718	12.3139	11.9766	11.3570	10.8015	10.3006	9.8466	9.4331	43
44	12.7613	12.3984	12.0565	11.4289	10.8665	10.3597	9.9005	9.4826	44
45	12.8471	12.4793	12.1330	11.4976	10.9286	10.4161	9.9521	9.5299	45
46	12.9293	12.5568	12.2063	11.5634	10.9880	10.4701	10.0013	9.5751	46
47	13.0081	12.6312	12.2765	11.6264	11.0449	10.5217	10.0484	9.6182	47
48	13.0836	12.7024	12.3438	11.6867	11.0993	10.5711	10.0934	9.6595	48
49	13.1561	12.7707	12.4082	11.7445	11.1514	10.6183	10.1365	9.6989	49
50	13.2255	12.8361	12.4700	11.7998	11.2013	10.6635	10.1777	9.7366	50

*Note:—In computing the quarterly in advance figures in these tables, all the rates % quoted above are effective rates, see the Introductory Section, pages xvi-xvii.

				Rate per Cent*					
Yrs.	7	7.25	7.5	8	8.5	9	9.5	10	Yrs.
51	13.2922	12.8989	12.5292	11.8528	11.2491	10.7068	10.2171	9.7727	51
52	13.3561	12.9591	12.5860	11.9036	11.2948	10.7482	10.2548	9.8072	52
53	13.4174	13.0168	12.6404	11.9523	11.3386	10.7879	10.2909	9.8402	53
54	13.4762	13.0721	12.6926	11.9990	11.3806	10.8259	10.3255	9.8718	54
55	13.5327	13.1252	12.7427	12.0437	11.4208	10.8623	10.3586	9.9020	55
56	13.5868	13.1761	12.7907	12.0866	11.4594	10.8971	10.3903	9.9310	56
57	13.6388	13.2250	12.8367	12.1277	11.4963	10.9305	10.4206	9.9587	57
58	13.6886	13.2719	12.8809	12.1671	11.5317	10.9625	10.4497	9.9853	58
59	13.7365	13.3169	12.9233	12.2049	11.5657	10.9932	10.4776	10.0107	59
60	13.7824	13.3600	12.9639	12.2411	11.5982	11.0226	10.5043	10.0351	60
61	13.8265	13.4015	13.0029	12.2759	11.6294	11.0508	10.5299	10.0585	61
62	13.8689	13.4412	13.0403	12.3093	11.6594	11.0778	10.5544	10.0809	62
63	13.9095	13.4794	13.0763	12.3413	11.6881	11.1038	10.5780	10.1023	63
64	13.9485	13.5160	13.1107	12.3720	11.7156	11.1286	10.6005	10.1229	64
65	13.9860	13.5512	13.1438	12.4014	11.7420	11.1524	10.6221	10.1426	65
66	14.0219	13.5850	13.1756	12.4297	11.7674	11.1753	10.6429	10.1615	66
67	14.0565	13.6174	13.2061	12.4568	11.7917	11.1972	10.6627	10.1796	67
68	14.0897	13.6485	13.2354	12.4829	11.8150	11.2183	10.6818	10.1970	68
69	14.1215	13.6784	13.2635	12.5079	11.8374	11.2384	10.7001	10.2137	69
70	14.1521	13.7071	13.2904	12.5319	11.8589	11.2578	10.7177	10.2297	70
71	14.1815	13.7347	13.3164	12.5549	11.8795	11.2764	10.7345	10.2450	71
72	14.2097	13.7611	13.3412	12.5770	11.8993	11.2942	10.7507	10.2597	72
73	14.2368	13.7866	13.3651	12.5982	11.9183	11.3113	10.7662	10.2739	73
74	14.2628	13.8110	13.3881	12.6186	11.9366	11.3278	10.7811	10.2874	74
75	14.2879	13.8344	13.4101	12.6382	11.9541	11.3435	10.7954	10.3004	75
76	14.3119	13.8570	13.4313	12.6570	11.9709	11.3587	10.8091	10.3129	76
77	14.3350	13.8786	13.4516	12.6750	11.9870	11.3732	10.8222	10.3249	77
78	14.3571	13.8994	13.4711	12.6924	12.0025	11.3872	10.8349	10.3364	78
79	14.3784	13.9193	13.4899	12.7090	12.0174	11.4006	10.8470	10.3474	79
80	14.3989	13.9385	13.5079	12.7250	12.0317	11.4134	10.8586	10.3580	80
81	14.4186	13.9569	13.5252	12.7404	12.0454	11.4258	10.8698	10.3682	81
82	14.4375	13.9747	13.5418	12.7551	12.0586	11.4376	10.8805	10.3779	82
83	14.4556	13.9917	13.5578	12.7693	12.0713	11.4490	10.8909	10.3873	83
84	14.4731	14.0080	13.5731	12.7829	12.0834	11.4600	10.9008	10.3963	84
85	14.4898	14.0237	13.5879	12.7960	12.0951	11.4705	10.9103	10.4050	85
86	14.5059	14.0388	13.6020	12.8085	12.1063	11.4806	10.9194	10.4133	86
87	14.5214	14.0533	13.6156	12.8206	12.1171	11.4903	10.9282	10.4213	87
88	14.5363	14.0672	13.6287	12.8322	12.1275	11.4996	10.9366	10.4289	88
89	14.5506	14.0806	13.6413	12.8433	12.1374	11.5085	10.9447	10.4363	89
90	14.5643	14.0935	13.6534	12.8540	12.1470	11.5171	10.9524	10.4433	90
91	14.5775	14.1058	13.6650	12.8643	12.1562	11.5254	10.9599	10.4501	91
92	14.5902	14.1177	13.6761	12.8742	12.1650	11.5333	10.9671	10.4566	92
93	14.6024	14.1291	13.6868	12.8837	12.1735	11.5409	10.9740	10.4629	93
94	14.6141	14.1401	13.6971	12.8928	12.1816	11.5482	10.9806	10.4689	94
95	14.6254	14.1507	13.7070	12.9016	12.1894	11.5553	10.9869	10.4747	95
96	14.6362	14.1608	13.7165	12.9100	12.1970	11.5620	10.9931	10.4802	96
97	14.6466	14.1705	13.7257	12.9181	12.2042	11.5685	10.9989	10.4856	97
98	14.6566	14.1799	13.7344	12.9259	12.2111	11.5748	11.0046	10.4907	98
99	14.6662	14.1889	13.7429	12.9333	12.2178	11.5808	11.0100	10.4956	99
100	14.6755	14.1975	13.7510	12.9405	12.2242	11.5865	11.0152	10.5004	100

*Note:—In computing the quarterly in advance figures in these tables, all the rates % quoted above are effective rates,
see the Introductory Section, pages xvi-xvii.

Rate per Cent*

Yrs.	11	12	13	14	15	16	18	20	Yrs.
1	0.7123	0.7079	0.7036	0.6994	0.6953	0.6912	0.6835	0.6760	1
2	1.3500	1.3343	1.3191	1.3044	1.2902	1.2764	1.2502	1.2256	2
3	1.9239	1.8921	1.8617	1.8326	1.8047	1.7779	1.7274	1.6807	3
4	2.4427	2.3918	2.3434	2.2974	2.2537	2.2121	2.1345	2.0636	4
5	2.9137	2.8415	2.7734	2.7093	2.6487	2.5914	2.4855	2.3899	5
6	3.3428	3.2481	3.1595	3.0765	2.9986	2.9254	2.7912	2.6712	6
7	3.7352	3.6172	3.5077	3.4058	3.3106	3.2215	3.0595	2.9159	7
8	4.0950	3.9537	3.8232	3.7024	3.5902	3.4856	3.2968	3.1307	8
9	4.4259	4.2613	4.1101	3.9708	3.8420	3.7226	3.5079	3.3205	9
10	4.7311	4.5434	4.3720	4.2147	4.0699	3.9361	3.6969	3.4893	10
11	5.0131	4.8030	4.6118	4.4371	4.2769	4.1294	3.8669	3.6404	11
12	5.2745	5.0423	4.8320	4.6406	4.4657	4.3051	4.0206	3.7763	12
13	5.5171	5.2636	5.0349	4.8274	4.6384	4.4654	4.1600	3.8990	13
14	5.7427	5.4686	5.2222	4.9993	4.7968	4.6121	4.2871	4.0104	14
15	5.9530	5.6590	5.3955	5.1579	4.9427	4.7467	4.4032	4.1118	15
16	6.1493	5.8361	5.5562	5.3046	5.0772	4.8707	4.5096	4.2045	16
17	6.3327	6.0011	5.7055	5.4406	5.2016	4.9851	4.6075	4.2895	17
18	6.5045	6.1551	5.8446	5.5669	5.3170	5.0909	4.6978	4.3676	18
19	6.6655	6.2991	5.9743	5.6844	5.4241	5.1890	4.7812	4.4396	19
20	6.8167	6.4339	6.0954	5.7939	5.5237	5.2801	4.8585	4.5062	20
21	6.9587	6.5603	6.2087	5.8962	5.6166	5.3650	4.9302	4.5678	21
22	7.0923	6.6789	6.3149	5.9919	5.7034	5.4440	4.9969	4.6250	22
23	7.2182	6.7904	6.4145	6.0815	5.7845	5.5179	5.0591	4.6782	23
24	7.3369	6.8953	6.5080	6.1655	5.8604	5.5870	5.1171	4.7278	24
25	7.4488	6.9941	6.5960	6.2444	5.9316	5.6517	5.1713	4.7740	25
26	7.5546	7.0873	6.6787	6.3185	5.9985	5.7123	5.2220	4.8172	26
27	7.6546	7.1752	6.7568	6.3883	6.0614	5.7693	5.2696	4.8577	27
28	7.7491	7.2582	6.8303	6.4540	6.1205	5.8229	5.3143	4.8956	28
29	7.8387	7.3367	6.8998	6.5160	6.1762	5.8733	5.3562	4.9312	29
30	7.9235	7.4110	6.9654	6.5745	6.2288	5.9208	5.3957	4.9646	30
31	8.0039	7.4813	7.0275	6.6298	6.2784	5.9656	5.4329	4.9961	31
32	8.0802	7.5479	7.0863	6.6821	6.3252	6.0079	5.4679	5.0257	32
33	8.1526	7.6111	7.1419	6.7315	6.3695	6.0478	5.5010	5.0536	33
34	8.2214	7.6710	7.1946	6.7784	6.4114	6.0856	5.5322	5.0800	34
35	8.2868	7.7279	7.2446	6.8227	6.4511	6.1213	5.5617	5.1048	35
36	8.3489	7.7819	7.2921	6.8648	6.4887	6.1551	5.5896	5.1284	36
37	8.4080	7.8332	7.3372	6.9047	6.5244	6.1872	5.6161	5.1506	37
38	8.4643	7.8820	7.3800	6.9426	6.5582	6.2176	5.6411	5.1717	38
39	8.5179	7.9284	7.4206	6.9786	6.5903	6.2465	5.6649	5.1916	39
40	8.5689	7.9726	7.4593	7.0128	6.6208	6.2739	5.6874	5.2105	40
41	8.6175	8.0147	7.4962	7.0453	6.6498	6.2999	5.7088	5.2285	41
42	8.6639	8.0548	7.5312	7.0763	6.6774	6.3246	5.7291	5.2455	42
43	8.7081	8.0930	7.5646	7.1057	6.7036	6.3482	5.7484	5.2616	43
44	8.7502	8.1294	7.5964	7.1338	6.7285	6.3705	5.7667	5.2770	44
45	8.7905	8.1641	7.6267	7.1605	6.7523	6.3918	5.7842	5.2916	45
46	8.8289	8.1972	7.6556	7.1860	6.7749	6.4121	5.8008	5.3055	46
47	8.8656	8.2288	7.6831	7.2103	6.7965	6.4315	5.8166	5.3187	47
48	8.9006	8.2590	7.7094	7.2334	6.8171	6.4499	5.8316	5.3313	48
49	8.9341	8.2878	7.7345	7.2555	6.8367	6.4674	5.8460	5.3433	49
50	8.9660	8.3153	7.7585	7.2766	6.8554	6.4842	5.8597	5.3547	50

Note:—In computing the quarterly in advance figures in these tables, all the rates % quoted above are effective rates,
see the Introductory Section, pages xvi-xvii.

352

Yrs.	11	12	13	14	15	16	18	20	Yrs.
				Rate per Cent*					
51	8.9966	8.3416	7.7814	7.2967	6.8733	6.5001	5.8727	5.3656	51
52	9.0258	8.3667	7.8032	7.3159	6.8903	6.5154	5.8851	5.3760	52
53	9.0538	8.3908	7.8241	7.3343	6.9066	6.5299	5.8970	5.3859	53
54	9.0805	8.4137	7.8441	7.3518	6.9222	6.5438	5.9084	5.3954	54
55	9.1061	8.4357	7.8632	7.3686	6.9370	6.5571	5.9192	5.4044	55
56	9.1306	8.4567	7.8814	7.3846	6.9512	6.5698	5.9295	5.4130	56
57	9.1541	8.4768	7.8989	7.3999	6.9648	6.5819	5.9394	5.4212	57
58	9.1765	8.4960	7.9156	7.4146	6.9778	6.5935	5.9488	5.4291	58
59	9.1980	8.5144	7.9316	7.4286	6.9902	6.6046	5.9578	5.4366	59
60	9.2186	8.5321	7.9469	7.4420	7.0021	6.6152	5.9665	5.4438	60
61	9.2383	8.5490	7.9615	7.4549	7.0134	6.6254	5.9747	5.4507	61
62	9.2571	8.5651	7.9755	7.4672	7.0243	6.6351	5.9826	5.4572	62
63	9.2752	8.5806	7.9889	7.4789	7.0347	6.6443	5.9902	5.4635	63
64	9.2926	8.5954	8.0018	7.4902	7.0447	6.6532	5.9974	5.4695	64
65	9.3092	8.6096	8.0141	7.5010	7.0542	6.6617	6.0043	5.4753	65
66	9.3251	8.6233	8.0259	7.5113	7.0634	6.6699	6.0109	5.4808	66
67	9.3404	8.6363	8.0372	7.5212	7.0721	6.6777	6.0173	5.4860	67
68	9.3550	8.6488	8.0480	7.5307	7.0805	6.6852	6.0233	5.4911	68
69	9.3690	8.6608	8.0584	7.5398	7.0885	6.6923	6.0291	5.4959	69
70	9.3825	8.6723	8.0684	7.5485	7.0962	6.6992	6.0347	5.5005	70
71	9.3954	8.6833	8.0779	7.5568	7.1036	6.7058	6.0400	5.5050	71
72	9.4078	8.6939	8.0871	7.5649	7.1107	6.7121	6.0452	5.5092	72
73	9.4196	8.7040	8.0958	7.5725	7.1175	6.7181	6.0501	5.5133	73
74	9.4310	8.7138	8.1043	7.5799	7.1240	6.7239	6.0548	5.5172	74
75	9.4420	8.7231	8.1123	7.5869	7.1302	6.7295	6.0593	5.5209	75
76	9.4524	8.7320	8.1201	7.5937	7.1362	6.7348	6.0636	5.5245	76
77	9.4625	8.7406	8.1275	7.6002	7.1419	6.7399	6.0677	5.5280	77
78	9.4722	8.7489	8.1346	7.6064	7.1474	6.7448	6.0717	5.5313	78
79	9.4814	8.7568	8.1414	7.6124	7.1527	6.7495	6.0755	5.5344	79
80	9.4903	8.7644	8.1480	7.6181	7.1578	6.7540	6.0791	5.5374	80
81	9.4989	8.7716	8.1543	7.6236	7.1626	6.7583	6.0826	5.5403	81
82	9.5071	8.7786	8.1603	7.6289	7.1673	6.7625	6.0860	5.5431	82
83	9.5149	8.7853	8.1661	7.6340	7.1717	6.7665	6.0892	5.5458	83
84	9.5225	8.7918	8.1717	7.6389	7.1760	6.7703	6.0923	5.5484	84
85	9.5297	8.7980	8.1770	7.6435	7.1802	6.7739	6.0953	5.5508	85
86	9.5367	8.8039	8.1822	7.6480	7.1841	6.7775	6.0981	5.5532	86
87	9.5434	8.8096	8.1871	7.6523	7.1879	6.7808	6.1009	5.5555	87
88	9.5498	8.8151	8.1918	7.6564	7.1915	6.7841	6.1035	5.5576	88
89	9.5560	8.8203	8.1964	7.6604	7.1950	6.7872	6.1060	5.5597	89
90	9.5619	8.8254	8.2007	7.6642	7.1984	6.7902	6.1084	5.5617	90
91	9.5676	8.8302	8.2049	7.6679	7.2016	6.7931	6.1108	5.5637	91
92	9.5731	8.8349	8.2089	7.6714	7.2047	6.7958	6.1130	5.5655	92
93	9.5783	8.8394	8.2128	7.6747	7.2077	6.7984	6.1151	5.5673	93
94	9.5833	8.8436	8.2165	7.6780	7.2105	6.8010	6.1172	5.5690	94
95	9.5882	8.8478	8.2200	7.6811	7.2133	6.8034	6.1192	5.5706	95
96	9.5928	8.8517	8.2235	7.6841	7.2159	6.8058	6.1210	5.5722	96
97	9.5973	8.8555	8.2267	7.6869	7.2185	6.8080	6.1229	5.5737	97
98	9.6016	8.8592	8.2299	7.6897	7.2209	6.8102	6.1246	5.5751	98
99	9.6057	8.8627	8.2329	7.6923	7.2232	6.8123	6.1263	5.5765	99
100	9.6097	8.8661	8.2358	7.6949	7.2255	6.8142	6.1279	5.5779	100

Note:—In computing the quarterly in advance figures in these tables, all the rates % quoted above are effective rates,
see the Introductory Section, pages xvi-xvii.

YEARS' PURCHASE

(DUAL RATE % PRINCIPLE)

OR

PRESENT VALUE OF
ONE POUND PER ANNUM

*receivable quarterly in advance, allow-
ing for a sinking fund at a given rate
to replace the invested capital and for
the effect of income tax at 30% on that
part of the income used to provide the
annual sinking fund instalment.*

AT RATES OF INTEREST*

FROM

4% to 20%

AND

ALLOWING FOR THE POSSIBLE INVESTMENT

OF SINKING FUNDS AT

3% and 4%

INCOME TAX at 30%

*Note:—In computing the quarterly in advance figures in these tables, all the rates
of interest quoted are effective rates, see the Introductory Section, pages
xvi-xvii.

Rate per Cent*

Yrs.	4	4.5	5	5.5	6	6.25	6.5	6.75	Yrs.
1	0.6938	0.6915	0.6892	0.6870	0.6848	0.6837	0.6827	0.6816	1
2	1.3701	1.3613	1.3526	1.3441	1.3357	1.3316	1.3275	1.3234	2
3	2.0295	2.0101	1.9912	1.9728	1.9548	1.9460	1.9373	1.9286	3
4	2.6721	2.6387	2.6062	2.5747	2.5442	2.5293	2.5145	2.5000	4
5	3.2985	3.2476	3.1986	3.1513	3.1057	3.0835	3.0616	3.0401	5
6	3.9088	3.8376	3.7694	3.7039	3.6410	3.6105	3.5806	3.5512	6
7	4.5035	4.4093	4.3194	4.2337	4.1517	4.1121	4.0733	4.0354	7
8	5.0830	4.9632	4.8497	4.7418	4.6392	4.5898	4.5416	4.4945	8
9	5.6475	5.5000	5.3609	5.2294	5.1050	5.0452	4.9870	4.9302	9
10	6.1974	6.0203	5.8540	5.6976	5.5501	5.4795	5.4109	5.3442	10
11	6.7330	6.5245	6.3296	6.1471	5.9758	5.8941	5.8148	5.7378	11
12	7.2546	7.0131	6.7885	6.5790	6.3832	6.2900	6.1998	6.1123	12
13	7.7626	7.4868	7.2313	6.9941	6.7732	6.6684	6.5671	6.4690	13
14	8.2573	7.9459	7.6587	7.3932	7.1468	7.0302	6.9177	6.8090	14
15	8.7390	8.3909	8.0714	7.7770	7.5048	7.3764	7.2526	7.1332	15
16	9.2079	8.8223	8.4698	8.1462	7.8481	7.7077	7.5726	7.4426	16
17	9.6645	9.2406	8.8545	8.5015	8.1773	8.0250	7.8787	7.7380	17
18	10.1089	9.6461	9.2262	8.8435	8.4933	8.3291	8.1716	8.0204	18
19	10.5415	10.0392	9.5852	9.1728	8.7966	8.6206	8.4520	8.2903	19
20	10.9626	10.4204	9.9320	9.4900	9.0879	8.9001	8.7205	8.5485	20
21	11.3724	10.7899	10.2672	9.7955	9.3677	9.1684	8.9779	8.7957	21
22	11.7712	11.1483	10.5912	10.0900	9.6366	9.4258	9.2246	9.0324	22
23	12.1593	11.4958	10.9044	10.3738	9.8952	9.6730	9.4612	9.2591	23
24	12.5369	11.8328	11.2071	10.6474	10.1439	9.9105	9.6883	9.4765	24
25	12.9044	12.1596	11.4998	10.9113	10.3831	10.1387	9.9063	9.6849	25
26	13.2619	12.4765	11.7829	11.1658	10.6133	10.3581	10.1156	9.8849	26
27	13.6097	12.7838	12.0566	11.4113	10.8349	10.5691	10.3167	10.0769	27
28	13.9480	13.0819	12.3214	11.6482	11.0482	10.7720	10.5100	10.2612	28
29	14.2771	13.3710	12.5775	11.8769	11.2537	10.9673	10.6958	10.4382	29
30	14.5973	13.6514	12.8253	12.0976	11.4517	11.1552	10.8745	10.6083	30
31	14.9087	13.9234	13.0651	12.3107	11.6425	11.3361	11.0463	10.7718	31
32	15.2115	14.1872	13.2971	12.5165	11.8263	11.5104	11.2117	10.9290	32
33	15.5061	14.4431	13.5216	12.7153	12.0036	11.6783	11.3710	11.0802	33
34	15.7926	14.6913	13.7390	12.9072	12.1746	11.8400	11.5243	11.2258	34
35	16.0711	14.9321	13.9493	13.0927	12.3395	11.9959	11.6719	11.3658	35
36	16.3420	15.1657	14.1530	13.2720	12.4985	12.1462	11.8141	11.5006	36
37	16.6054	15.3923	14.3501	13.4452	12.6520	12.2911	11.9512	11.6305	37
38	16.8616	15.6121	14.5410	13.6126	12.8002	12.4309	12.0833	11.7555	38
39	17.1106	15.8253	14.7258	13.7744	12.9432	12.5657	12.2106	11.8760	39
40	17.3527	16.0322	14.9048	13.9309	13.0812	12.6958	12.3334	11.9922	40
41	17.5881	16.2329	15.0781	14.0822	13.2146	12.8213	12.4519	12.1041	41
42	17.8169	16.4277	15.2460	14.2285	13.3433	12.9425	12.5661	12.2121	42
43	18.0394	16.6166	15.4086	14.3701	13.4677	13.0595	12.6764	12.3162	43
44	18.2557	16.7999	15.5661	14.5069	13.5879	13.1725	12.7828	12.4166	44
45	18.4659	16.9778	15.7187	14.6394	13.7040	13.2816	12.8855	12.5135	45
46	18.6702	17.1504	15.8665	14.7675	13.8162	13.3869	12.9847	12.6070	46
47	18.8688	17.3178	16.0097	14.8915	13.9247	13.4887	13.0804	12.6972	47
48	19.0619	17.4803	16.1485	15.0115	14.0295	13.5871	13.1729	12.7843	48
49	19.2495	17.6380	16.2829	15.1276	14.1309	13.6822	13.2623	12.8685	49
50	19.4319	17.7910	16.4132	15.2400	14.2289	13.7741	13.3486	12.9497	50

*Note:—In computing the quarterly in advance figures in these tables, all the rates % quoted above are effective rates,
see the Introductory Section, pages xvi-xvii.

				Rate per Cent*					
Yrs.	4	4.5	5	5.5	6	6.25	6.5	6.75	Yrs.
51	19.6091	17.9394	16.5395	15.3488	14.3237	13.8629	13.4320	13.0282	51
52	19.7814	18.0835	16.6618	15.4541	14.4154	13.9487	13.5126	13.1040	52
53	19.9488	18.2232	16.7804	15.5561	14.5041	14.0318	13.5905	13.1772	53
54	20.1114	18.3589	16.8954	15.6549	14.5899	14.1120	13.6658	13.2480	54
55	20.2695	18.4905	17.0068	15.7505	14.6729	14.1897	13.7386	13.3164	55
56	20.4231	18.6183	17.1148	15.8431	14.7533	14.2648	13.8090	13.3826	56
57	20.5724	18.7422	17.2195	15.9327	14.8310	14.3375	13.8770	13.4465	57
58	20.7174	18.8625	17.3210	16.0196	14.9062	14.4078	13.9429	13.5083	58
59	20.8583	18.9793	17.4194	16.1037	14.9790	14.4758	14.0066	13.5681	59
60	20.9953	19.0926	17.5148	16.1852	15.0495	14.5416	14.0682	13.6259	60
61	21.1283	19.2026	17.6073	16.2642	15.1178	14.6053	14.1278	13.6818	61
62	21.2576	19.3093	17.6970	16.3407	15.1838	14.6670	14.1855	13.7359	62
63	21.3832	19.4129	17.7840	16.4148	15.2478	14.7266	14.2413	13.7882	63
64	21.5052	19.5134	17.8683	16.4866	15.3098	14.7844	14.2953	13.8389	64
65	21.6238	19.6109	17.9500	16.5562	15.3697	14.8403	14.3476	13.8879	65
66	21.7389	19.7056	18.0293	16.6236	15.4278	14.8945	14.3982	13.9353	66
67	21.8508	19.7975	18.1062	16.6890	15.4841	14.9469	14.4472	13.9812	67
68	21.9595	19.8867	18.1808	16.7523	15.5386	14.9977	14.4947	14.0256	68
69	22.0651	19.9733	18.2531	16.8137	15.5914	15.0469	14.5406	14.0686	69
70	22.1677	20.0573	18.3233	16.8732	15.6426	15.0945	14.5851	14.1102	70
71	22.2674	20.1389	18.3913	16.9309	15.6921	15.1407	14.6281	14.1505	71
72	22.3642	20.2180	18.4573	16.9868	15.7401	15.1854	14.6699	14.1896	72
73	22.4582	20.2948	18.5213	17.0410	15.7867	15.2287	14.7103	14.2274	73
74	22.5496	20.3694	18.5834	17.0935	15.8317	15.2706	14.7494	14.2640	74
75	22.6383	20.4418	18.6436	17.1445	15.8754	15.3113	14.7873	14.2994	75
76	22.7245	20.5120	18.7020	17.1938	15.9178	15.3506	14.8240	14.3338	76
77	22.8082	20.5802	18.7587	17.2417	15.9588	15.3888	14.8596	14.3670	77
78	22.8895	20.6464	18.8137	17.2882	15.9986	15.4258	14.8941	14.3993	78
79	22.9685	20.7106	18.8670	17.3332	16.0371	15.4616	14.9275	14.4305	79
80	23.0453	20.7730	18.9187	17.3768	16.0745	15.4963	14.9599	14.4607	80
81	23.1198	20.8335	18.9689	17.4192	16.1107	15.5300	14.9912	14.4900	81
82	23.1921	20.8923	19.0176	17.4602	16.1458	15.5626	15.0216	14.5184	82
83	23.2624	20.9493	19.0649	17.5000	16.1799	15.5943	15.0511	14.5459	83
84	23.3307	21.0047	19.1107	17.5387	16.2129	15.6249	15.0796	14.5726	84
85	23.3970	21.0584	19.1552	17.5761	16.2449	15.6546	15.1073	14.5985	85
86	23.4614	21.1106	19.1983	17.6124	16.2759	15.6834	15.1342	14.6235	86
87	23.5240	21.1612	19.2402	17.6477	16.3060	15.7114	15.1602	14.6478	87
88	23.5848	21.2103	19.2808	17.6818	16.3351	15.7384	15.1854	14.6713	88
89	23.6438	21.2581	19.3202	17.7150	16.3634	15.7647	15.2098	14.6941	89
90	23.7011	21.3044	19.3585	17.7471	16.3908	15.7901	15.2335	14.7162	90
91	23.7567	21.3493	19.3956	17.7783	16.4174	15.8148	15.2565	14.7377	91
92	23.8108	21.3930	19.4316	17.8086	16.4432	15.8388	15.2787	14.7585	92
93	23.8633	21.4353	19.4665	17.8379	16.4682	15.8620	15.3003	14.7786	93
94	23.9142	21.4765	19.5005	17.8664	16.4925	15.8845	15.3213	14.7981	94
95	23.9637	21.5164	19.5334	17.8940	16.5160	15.9063	15.3416	14.8171	95
96	24.0118	21.5551	19.5653	17.9208	16.5389	15.9275	15.3613	14.8354	96
97	24.0585	21.5927	19.5963	17.9468	16.5610	15.9480	15.3804	14.8533	97
98	24.1039	21.6293	19.6264	17.9720	16.5825	15.9679	15.3989	14.8705	98
99	24.1479	21.6647	19.6555	17.9965	16.6033	15.9872	15.4168	14.8873	99
100	24.1906	21.6991	19.6839	18.0202	16.6235	16.0060	15.4343	14.9035	100

Note:—In computing the quarterly in advance figures in these tables, all the rates % quoted above are effective rates, see the Introductory Section, pages xvi-xvii.

				Rate per Cent*					
Yrs.	7	7.25	7.5	8	8.5	9	9.5	10	Yrs.
1	0.6805	0.6795	0.6784	0.6763	0.6743	0.6722	0.6702	0.6682	1
2	1.3194	1.3154	1.3115	1.3037	1.2960	1.2885	1.2811	1.2739	2
3	1.9201	1.9117	1.9034	1.8870	1.8710	1.8554	1.8401	1.8252	3
4	2.4857	2.4716	2.4578	2.4305	2.4041	2.3783	2.3533	2.3289	4
5	3.0190	2.9983	2.9778	2.9380	2.8994	2.8621	2.8258	2.7907	5
6	3.5225	3.4942	3.4665	3.4126	3.3607	3.3106	3.2623	3.2156	6
7	3.9983	3.9619	3.9263	3.8574	3.7911	3.7275	3.6663	3.6075	7
8	4.4485	4.4036	4.3596	4.2748	4.1936	4.1159	4.0414	3.9700	8
9	4.8749	4.8210	4.7684	4.6671	4.5705	4.4783	4.3903	4.3062	9
10	5.2793	5.2161	5.1546	5.0364	4.9241	4.8173	4.7156	4.6187	10
11	5.6630	5.5904	5.5198	5.3845	5.2563	5.1348	5.0194	4.9097	11
12	6.0276	5.9454	5.8656	5.7130	5.5689	5.4327	5.3037	5.1814	12
13	6.3742	6.2823	6.1933	6.0234	5.8635	5.7127	5.5702	5.4355	13
14	6.7040	6.6024	6.5042	6.3171	6.1414	5.9761	5.8204	5.6735	14
15	7.0180	6.9068	6.7994	6.5952	6.4039	6.2244	6.0557	5.8968	15
16	7.3173	7.1965	7.0800	6.8588	6.6522	6.4587	6.2772	6.1066	16
17	7.6027	7.4724	7.3468	7.1089	6.8872	6.6801	6.4861	6.3041	17
18	7.8751	7.7353	7.6008	7.3465	7.1100	6.8894	6.6833	6.4903	18
19	8.1351	7.9861	7.8428	7.5724	7.3213	7.0877	6.8697	6.6659	19
20	8.3836	8.2254	8.0736	7.7872	7.5220	7.2756	7.0461	6.8318	20
21	8.6212	8.4540	8.2937	7.9918	7.7127	7.4538	7.2131	6.9888	21
22	8.8485	8.6725	8.5038	8.1867	7.8940	7.6231	7.3716	7.1374	22
23	9.0660	8.8813	8.7045	8.3725	8.0667	7.7840	7.5219	7.2782	23
24	9.2743	9.0811	8.8963	8.5499	8.2312	7.9371	7.6647	7.4119	24
25	9.4738	9.2723	9.0798	8.7192	8.3880	8.0828	7.8005	7.5388	25
26	9.6651	9.4555	9.2553	8.8809	8.5376	8.2216	7.9297	7.6594	26
27	9.8485	9.6310	9.4234	9.0356	8.6804	8.3539	8.0528	7.7741	27
28	10.0245	9.7992	9.5844	9.1835	8.8168	8.4802	8.1700	7.8834	28
29	10.1934	9.9605	9.7386	9.3250	8.9472	8.6007	8.2819	7.9874	29
30	10.3556	10.1153	9.8865	9.4605	9.0719	8.7159	8.3886	8.0867	30
31	10.5113	10.2638	10.0284	9.5904	9.1912	8.8260	8.4905	8.1813	31
32	10.6610	10.4065	10.1645	9.7148	9.3054	8.9312	8.5879	8.2717	32
33	10.8048	10.5435	10.2952	9.8341	9.4148	9.0320	8.6810	8.3580	33
34	10.9431	10.6752	10.4207	9.9485	9.5197	9.1284	8.7701	8.4406	34
35	11.0762	10.8017	10.5413	10.0584	9.6202	9.2208	8.8553	8.5195	35
36	11.2042	10.9234	10.6572	10.1638	9.7166	9.3094	8.9369	8.5950	36
37	11.3274	11.0405	10.7686	10.2651	9.8091	9.3942	9.0151	8.6673	37
38	11.4460	11.1531	10.8757	10.3624	9.8979	9.4757	9.0901	8.7366	38
39	11.5602	11.2615	10.9787	10.4559	9.9832	9.5538	9.1620	8.8030	39
40	11.6702	11.3659	11.0779	10.5458	10.0652	9.6288	9.2309	8.8666	40
41	11.7762	11.4664	11.1734	10.6323	10.1439	9.7009	9.2971	8.9277	41
42	11.8783	11.5633	11.2653	10.7155	10.2196	9.7701	9.3607	8.9863	42
43	11.9768	11.6565	11.3538	10.7956	10.2924	9.8366	9.4217	9.0425	43
44	12.0717	11.7465	11.4391	10.8727	10.3625	9.9006	9.4804	9.0965	44
45	12.1633	11.8331	11.5213	10.9469	10.4299	9.9621	9.5368	9.1484	45
46	12.2516	11.9167	11.6005	11.0184	10.4947	10.0212	9.5910	9.1983	46
47	12.3368	11.9973	11.6769	11.0872	10.5572	10.0782	9.6431	9.2462	47
48	12.4191	12.0751	11.7505	11.1536	10.6174	10.1330	9.6933	9.2924	48
49	12.4985	12.1501	11.8216	11.2176	10.6753	10.1858	9.7416	9.3367	49
50	12.5751	12.2225	11.8901	11.2793	10.7312	10.2366	9.7881	9.3794	50

*Note:—In computing the quarterly in advance figures in these tables, all the rates % quoted above are effective rates, see the Introductory Section, pages xvi-xvii.

Rate per Cent*

Yrs.	7	7.25	7.5	8	8.5	9	9.5	10	Yrs.
51	12.6491	12.2924	11.9562	11.3388	10.7850	10.2856	9.8328	9.4205	51
52	12.7205	12.3599	12.0200	11.3961	10.8369	10.3328	9.8759	9.4601	52
53	12.7895	12.4250	12.0816	11.4515	10.8870	10.3782	9.9175	9.4982	53
54	12.8562	12.4879	12.1411	11.5049	10.9352	10.4221	9.9575	9.5349	54
55	12.9206	12.5487	12.1986	11.5565	10.9818	10.4644	9.9961	9.5703	55
56	12.9828	12.6074	12.2540	11.6063	11.0267	10.5052	10.0333	9.6044	56
57	13.0430	12.6641	12.3076	11.6543	11.0701	10.5445	10.0692	9.6373	57
58	13.1012	12.7189	12.3594	11.7007	11.1119	10.5825	10.1039	9.6690	58
59	13.1574	12.7719	12.4094	11.7455	11.1524	10.6192	10.1373	9.6996	59
60	13.2117	12.8231	12.4577	11.7888	11.1914	10.6545	10.1695	9.7291	60
61	13.2643	12.8726	12.5044	11.8307	11.2291	10.6887	10.2006	9.7576	61
62	13.3151	12.9205	12.5496	11.8711	11.2655	10.7217	10.2306	9.7851	62
63	13.3643	12.9668	12.5933	11.9101	11.3007	10.7535	10.2597	9.8116	63
64	13.4118	13.0115	12.6355	11.9479	11.3347	10.7843	10.2877	9.8372	64
65	13.4579	13.0548	12.6763	11.9844	11.3675	10.8140	10.3147	9.8619	65
66	13.5024	13.0967	12.7158	12.0197	11.3993	10.8428	10.3408	9.8858	66
67	13.5455	13.1373	12.7540	12.0538	11.4299	10.8705	10.3661	9.9089	67
68	13.5872	13.1765	12.7910	12.0868	11.4596	10.8974	10.3905	9.9312	68
69	13.6275	13.2144	12.8267	12.1188	11.4883	10.9233	10.4141	9.9527	69
70	13.6666	13.2512	12.8613	12.1496	11.5161	10.9484	10.4369	9.9735	70
71	13.7044	13.2867	12.8948	12.1795	11.5429	10.9727	10.4589	9.9937	71
72	13.7410	13.3211	12.9272	12.2084	11.5689	10.9961	10.4802	10.0131	72
73	13.7764	13.3544	12.9586	12.2364	11.5940	11.0188	10.5008	10.0319	73
74	13.8108	13.3867	12.9890	12.2635	11.6183	11.0407	10.5208	10.0501	74
75	13.8440	13.4179	13.0183	12.2897	11.6418	11.0620	10.5400	10.0677	75
76	13.8762	13.4481	13.0468	12.3150	11.6645	11.0825	10.5587	10.0847	76
77	13.9073	13.4774	13.0744	12.3396	11.6865	11.1024	10.5767	10.1012	77
78	13.9375	13.5057	13.1010	12.3633	11.7079	11.1216	10.5942	10.1171	78
79	13.9668	13.5332	13.1269	12.3863	11.7285	11.1402	10.6111	10.1325	79
80	13.9951	13.5598	13.1519	12.4086	11.7485	11.1582	10.6274	10.1474	80
81	14.0226	13.5856	13.1761	12.4302	11.7678	11.1757	10.6432	10.1618	81
82	14.0492	13.6105	13.1996	12.4511	11.7865	11.1926	10.6585	10.1758	82
83	14.0749	13.6347	13.2224	12.4713	11.8046	11.2089	10.6734	10.1893	83
84	14.0999	13.6581	13.2444	12.4909	11.8222	11.2247	10.6877	10.2024	84
85	14.1241	13.6808	13.2657	12.5099	11.8392	11.2401	10.7016	10.2150	85
86	14.1475	13.7028	13.2864	12.5283	11.8557	11.2549	10.7151	10.2273	86
87	14.1703	13.7241	13.3065	12.5461	11.8716	11.2693	10.7281	10.2392	87
88	14.1923	13.7448	13.3259	12.5634	11.8871	11.2832	10.7407	10.2506	88
89	14.2136	13.7648	13.3447	12.5801	11.9021	11.2967	10.7529	10.2618	89
90	14.2343	13.7842	13.3629	12.5963	11.9166	11.3098	10.7648	10.2726	90
91	14.2544	13.8030	13.3806	12.6120	11.9306	11.3224	10.7762	10.2830	91
92	14.2738	13.8213	13.3977	12.6272	11.9442	11.3347	10.7873	10.2931	92
93	14.2927	13.8389	13.4143	12.6419	11.9574	11.3466	10.7981	10.3029	93
94	14.3109	13.8561	13.4304	12.6562	11.9702	11.3581	10.8085	10.3124	94
95	14.3286	13.8727	13.4460	12.6701	11.9826	11.3692	10.8186	10.3216	95
96	14.3458	13.8888	13.4611	12.6835	11.9946	11.3800	10.8284	10.3305	96
97	14.3625	13.9044	13.4758	12.6965	12.0062	11.3905	10.8379	10.3391	97
98	14.3786	13.9195	13.4900	12.7091	12.0175	11.4007	10.8471	10.3475	98
99	14.3943	13.9342	13.5038	12.7214	12.0285	11.4105	10.8560	10.3556	99
100	14.4095	13.9484	13.5172	12.7332	12.0391	11.4201	10.8646	10.3635	100

Note:—In computing the quarterly in advance figures in these tables, all the rates % quoted above are effective rates, see the Introductory Section, pages xvi-xvii.

				Rate per Cent*					
Yrs.	11	12	13	14	15	16	18	20	Yrs.
1	0.6643	0.6605	0.6567	0.6530	0.6495	0.6460	0.6392	0.6327	1
2	1.2597	1.2460	1.2328	1.2199	1.2075	1.1954	1.1724	1.1507	2
3	1.7962	1.7685	1.7419	1.7164	1.6919	1.6683	1.6238	1.5824	3
4	2.2820	2.2374	2.1950	2.1547	2.1162	2.0794	2.0107	1.9477	4
5	2.7237	2.6604	2.6007	2.5442	2.4907	2.4400	2.3459	2.2605	5
6	3.1268	3.0438	2.9658	2.8926	2.8236	2.7586	2.6389	2.5314	6
7	3.4961	3.3926	3.2961	3.2059	3.1214	3.0421	2.8972	2.7682	7
8	3.8356	3.7113	3.5961	3.4890	3.3892	3.2959	3.1265	2.9768	8
9	4.1485	4.0035	3.8698	3.7461	3.6312	3.5243	3.3314	3.1619	9
10	4.4377	4.2723	4.1203	3.9803	3.8509	3.7309	3.5154	3.3271	10
11	4.7058	4.5201	4.3504	4.1946	4.0512	3.9186	3.6815	3.4756	11
12	4.9548	4.7494	4.5624	4.3914	4.2344	4.0897	3.8321	3.6095	12
13	5.1866	4.9620	4.7582	4.5725	4.4025	4.2464	3.9694	3.7310	13
14	5.4029	5.1596	4.9396	4.7398	4.5574	4.3903	4.0948	3.8417	14
15	5.6050	5.3436	5.1080	4.8946	4.7004	4.5228	4.2099	3.9428	15
16	5.7943	5.5154	5.2648	5.0383	4.8328	4.6453	4.3157	4.0355	16
17	5.9718	5.6760	5.4109	5.1720	4.9556	4.7587	4.4135	4.1208	17
18	6.1386	5.8264	5.5475	5.2966	5.0699	4.8640	4.5039	4.1995	18
19	6.2955	5.9676	5.6753	5.4130	5.1765	4.9619	4.5878	4.2724	19
20	6.4433	6.1002	5.7951	5.5219	5.2760	5.0533	4.6658	4.3399	20
21	6.5827	6.2250	5.9076	5.6240	5.3691	5.1387	4.7384	4.4027	21
22	6.7144	6.3427	6.0135	5.7199	5.4564	5.2185	4.8063	4.4612	22
23	6.8389	6.4537	6.1131	5.8100	5.5383	5.2934	4.8697	4.5159	23
24	6.9567	6.5585	6.2071	5.8948	5.6153	5.3638	4.9292	4.5670	24
25	7.0684	6.6577	6.2959	5.9748	5.6879	5.4299	4.9850	4.6148	25
26	7.1743	6.7516	6.3798	6.0503	5.7563	5.4922	5.0375	4.6597	26
27	7.2749	6.8406	6.4592	6.1217	5.8208	5.5510	5.0869	4.7020	27
28	7.3705	6.9250	6.5344	6.1892	5.8819	5.6064	5.1334	4.7417	28
29	7.4614	7.0052	6.6058	6.2532	5.9396	5.6589	5.1773	4.7792	29
30	7.5479	7.0814	6.6735	6.3138	5.9943	5.7085	5.2188	4.8145	30
31	7.6303	7.1539	6.7378	6.3714	6.0462	5.7555	5.2581	4.8479	31
32	7.7089	7.2229	6.7990	6.4261	6.0954	5.8001	5.2953	4.8795	32
33	7.7838	7.2886	6.8573	6.4781	6.1421	5.8424	5.3305	4.9094	33
34	7.8553	7.3513	6.9127	6.5275	6.1866	5.8826	5.3640	4.9378	34
35	7.9236	7.4111	6.9656	6.5746	6.2289	5.9209	5.3957	4.9647	35
36	7.9889	7.4682	7.0160	6.6195	6.2691	5.9572	5.4259	4.9902	36
37	8.0514	7.5227	7.0641	6.6623	6.3075	5.9919	5.4547	5.0145	37
38	8.1111	7.5749	7.1100	6.7032	6.3441	6.0249	5.4820	5.0376	38
39	8.1683	7.6247	7.1539	6.7422	6.3791	6.0564	5.5081	5.0596	39
40	8.2231	7.6724	7.1959	6.7795	6.4124	6.0865	5.5329	5.0806	40
41	8.2755	7.7181	7.2361	6.8151	6.4443	6.1152	5.5567	5.1006	41
42	8.3259	7.7618	7.2745	6.8492	6.4748	6.1426	5.5793	5.1196	42
43	8.3741	7.8038	7.3113	6.8818	6.5039	6.1688	5.6009	5.1378	43
44	8.4204	7.8440	7.3466	6.9131	6.5318	6.1939	5.6216	5.1552	44
45	8.4649	7.8825	7.3804	6.9430	6.5585	6.2179	5.6414	5.1719	45
46	8.5076	7.9195	7.4128	6.9717	6.5841	6.2409	5.6603	5.1878	46
47	8.5486	7.9550	7.4439	6.9992	6.6087	6.2630	5.6784	5.2030	47
48	8.5880	7.9891	7.4738	7.0256	6.6322	6.2841	5.6958	5.2176	48
49	8.6258	8.0219	7.5025	7.0509	6.6547	6.3044	5.7124	5.2315	49
50	8.6623	8.0534	7.5300	7.0752	6.6764	6.3238	5.7284	5.2449	50

*Note:—In computing the quarterly in advance figures in these tables, all the rates % quoted above are effective rates, see the Introductory Section, pages xvi-xvii.

Yrs.	11	12	13	14	15	16	18	20	Yrs.
				Rate per Cent*					
51	8.6973	8.0837	7.5565	7.0986	6.6972	6.3424	5.7437	5.2577	51
52	8.7310	8.1128	7.5819	7.1210	6.7172	6.3604	5.7584	5.2700	52
53	8.7635	8.1408	7.6064	7.1426	6.7364	6.3776	5.7725	5.2818	53
54	8.7947	8.1678	7.6299	7.1634	6.7548	6.3941	5.7860	5.2932	54
55	8.8248	8.1937	7.6525	7.1833	6.7726	6.4100	5.7990	5.3041	55
56	8.8538	8.2187	7.6743	7.2025	6.7896	6.4253	5.8115	5.3145	56
57	8.8818	8.2428	7.6953	7.2210	6.8060	6.4400	5.8235	5.3246	57
58	8.9087	8.2660	7.7155	7.2388	6.8218	6.4541	5.8351	5.3342	58
59	8.9347	8.2883	7.7350	7.2559	6.8371	6.4677	5.8462	5.3435	59
60	8.9597	8.3099	7.7537	7.2724	6.8517	6.4808	5.8569	5.3525	60
61	8.9838	8.3306	7.7718	7.2883	6.8658	6.4935	5.8673	5.3611	61
62	9.0071	8.3506	7.7892	7.3036	6.8794	6.5056	5.8772	5.3694	62
63	9.0296	8.3700	7.8060	7.3184	6.8925	6.5173	5.8867	5.3773	63
64	9.0513	8.3886	7.8222	7.3326	6.9051	6.5286	5.8959	5.3850	64
65	9.0722	8.4066	7.8379	7.3464	6.9173	6.5395	5.9048	5.3924	65
66	9.0924	8.4239	7.8529	7.3596	6.9291	6.5500	5.9134	5.3996	66
67	9.1119	8.4407	7.8675	7.3724	6.9404	6.5601	5.9216	5.4064	67
68	9.1308	8.4568	7.8815	7.3847	6.9513	6.5699	5.9296	5.4131	68
69	9.1490	8.4724	7.8951	7.3966	6.9619	6.5793	5.9373	5.4195	69
70	9.1666	8.4875	7.9082	7.4081	6.9720	6.5884	5.9447	5.4256	70
71	9.1836	8.5021	7.9209	7.4192	6.9819	6.5972	5.9518	5.4316	71
72	9.2000	8.5162	7.9331	7.4299	6.9914	6.6057	5.9587	5.4373	72
73	9.2159	8.5298	7.9449	7.4403	7.0005	6.6138	5.9653	5.4429	73
74	9.2312	8.5429	7.9563	7.4503	7.0094	6.6217	5.9718	5.4482	74
75	9.2461	8.5556	7.9673	7.4599	7.0179	6.6294	5.9780	5.4534	75
76	9.2604	8.5679	7.9779	7.4693	7.0262	6.6367	5.9840	5.4584	76
77	9.2743	8.5798	7.9882	7.4783	7.0342	6.6439	5.9898	5.4632	77
78	9.2877	8.5913	7.9982	7.4870	7.0419	6.6507	5.9954	5.4678	78
79	9.3007	8.6024	8.0078	7.4955	7.0493	6.6574	6.0008	5.4723	79
80	9.3132	8.6131	8.0171	7.5036	7.0566	6.6638	6.0060	5.4767	80
81	9.3254	8.6235	8.0261	7.5115	7.0635	6.6700	6.0110	5.4809	81
82	9.3371	8.6335	8.0348	7.5191	7.0703	6.6761	6.0159	5.4849	82
83	9.3485	8.6433	8.0432	7.5265	7.0768	6.6819	6.0206	5.4889	83
84	9.3595	8.6527	8.0514	7.5336	7.0831	6.6875	6.0252	5.4926	84
85	9.3702	8.6618 .	8.0593	7.5405	7.0892	6.6929	6.0296	5.4963	85
86	9.3805	8.6706	8.0669	7.5472	7.0951	6.6982	6.0339	5.4999	86
87	9.3905	8.6791	8.0743	7.5537	7.1008	6.7033	6.0380	5.5033	87
88	9.4001	8.6874	8.0814	7.5599	7.1063	6.7082	6.0420	5.5066	88
89	9.4095	8.6954	8.0883	7.5660	7.1117	6.7130	6.0459	5.5098	89
90	9.4185	8.7031	8.0950	7.5718	7.1169	6.7176	6.0496	5.5129	90
91	9.4273	8.7106	8.1015	7.5775	7.1219	6.7220	6.0532	5.5159	91
92	9.4358	8.7179	8.1078	7.5830	7.1267	6.7264	6.0567	5.5188	92
93	9.4441	8.7249	8.1139	7.5883	7.1314	6.7305	6.0601	5.5217	93
94	9.4520	8.7317	8.1198	7.5934	7.1360	6.7346	6.0634	5.5244	94
95	9.4597	8.7383	8.1255	7.5984	7.1404	6.7385	6.0666	5.5270	95
96	9.4672	8.7447	8.1310	7.6033	7.1446	6.7423	6.0697	5.5296	96
97	9.4745	8.7509	8.1363	7.6079	7.1487	6.7460	6.0726	5.5320	97
98	9.4815	8.7568	8.1415	7.6125	7.1527	6.7495	6.0755	5.5344	98
99	9.4883	8.7626	8.1465	7.6168	7.1566	6.7530	6.0783	5.5368	99
100	9.4949	8.7683	8.1514	7.6211	7.1604	6.7563	6.0810	5.5390	100

*Note:—In computing the quarterly in advance figures in these tables, all the rates % quoted above are effective rates, see the Introductory Section, pages xvi-xvii.

Rate per Cent*

Yrs.	4	4.5	5	5.5	6	6.25	6.5	6.75	Yrs.
1	0.6979	0.6956	0.6933	0.6910	0.6888	0.6877	0.6866	0.6855	1
2	1.3844	1.3754	1.3665	1.3578	1.3493	1.3451	1.3409	1.3368	2
3	2.0595	2.0395	2.0201	2.0011	1.9826	1.9736	1.9646	1.9557	3
4	2.7227	2.6880	2.6543	2.6217	2.5900	2.5746	2.5593	2.5443	4
5	3.3740	3.3209	3.2696	3.2203	3.1726	3.1494	3.1266	3.1042	5
6	4.0132	3.9382	3.8664	3.7975	3.7314	3.6994	3.6680	3.6372	6
7	4.6402	4.5402	4.4450	4.3542	4.2675	4.2257	4.1848	4.1447	7
8	5.2547	5.1268	5.0058	4.8909	4.7819	4.7294	4.6782	4.6282	8
9	5.8568	5.6984	5.5492	5.4084	5.2754	5.2116	5.1495	5.0890	9
10	6.4463	6.2549	6.0756	5.9073	5.7489	5.6732	5.5997	5.5283	10
11	7.0232	6.7966	6.5855	6.3881	6.2034	6.1153	6.0300	5.9472	11
12	7.5875	7.3237	7.0791	6.8516	6.6395	6.5388	6.4413	6.3470	12
13	8.1391	7.8364	7.5570	7.2983	7.0581	6.9444	6.8345	6.7284	13
14	8.6782	8.3348	8.0195	7.7288	7.4599	7.3330	7.2106	7.0926	14
15	9.2046	8.8193	8.4669	8.1435	7.8456	7.7053	7.5703	7.4404	15
16	9.7184	9.2899	8.8998	8.5432	8.2159	8.0622	7.9145	7.7726	16
17	10.2198	9.7470	9.3184	8.9282	8.5714	8.4042	8.2439	8.0900	17
18	10.7087	10.1908	9.7232	9.2992	8.9127	8.7321	8.5591	8.3934	18
19	11.1854	10.6215	10.1146	9.6565	9.2404	9.0464	8.8609	8.6834	19
20	11.6498	11.0394	10.4928	10.0007	9.5551	9.3478	9.1499	8.9607	20
21	12.1021	11.4447	10.8584	10.3322	9.8573	9.6368	9.4266	9.2259	21
22	12.5425	11.8377	11.2115	10.6514	10.1475	9.9140	9.6916	9.4797	22
23	12.9710	12.2187	11.5527	10.9589	10.4262	10.1798	9.9455	9.7224	23
24	13.3879	12.5880	11.8822	11.2550	10.6938	10.4348	10.1888	9.9548	24
25	13.7933	12.9457	12.2005	11.5401	10.9509	10.6795	10.4219	10.1772	25
26	14.1873	13.2922	12.5078	11.8147	11.1978	10.9142	10.6453	10.3901	26
27	14.5702	13.6277	12.8044	12.0790	11.4350	11.1394	10.8594	10.5940	27
28	14.9421	13.9525	13.0908	12.3335	11.6628	11.3554	11.0647	10.7892	28
29	15.3032	14.2669	13.3671	12.5785	11.8817	11.5628	11.2615	10.9763	29
30	15.6538	14.5711	13.6338	12.8144	12.0919	11.7618	11.4502	11.1554	30
31	15.9939	14.8654	13.8911	13.0414	12.2939	11.9528	11.6311	11.3271	31
32	16.3239	15.1501	14.1394	13.2600	12.4879	12.1362	11.8047	11.4917	32
33	16.6440	15.4253	14.3788	13.4704	12.6744	12.3122	11.9711	11.6493	33
34	16.9542	15.6915	14.6098	13.6729	12.8535	12.4812	12.1308	11.8005	34
35	17.2549	15.9487	14.8326	13.8678	13.0256	12.6434	12.2840	11.9454	35
36	17.5463	16.1973	15.0474	14.0554	13.1910	12.7991	12.4309	12.0843	36
37	17.8286	16.4376	15.2545	14.2359	13.3498	12.9486	12.5719	12.2175	37
38	18.1019	16.6697	15.4542	14.4097	13.5025	13.0922	12.7072	12.3453	38
39	18.3666	16.8938	15.6467	14.5769	13.6492	13.2301	12.8371	12.4678	39
40	18.6228	17.1103	15.8322	14.7378	13.7902	13.3625	12.9617	12.5853	40
41	18.8707	17.3194	16.0110	14.8927	13.9257	13.4897	13.0813	12.6981	41
42	19.1106	17.5212	16.1834	15.0416	14.0559	13.6118	13.1961	12.8062	42
43	19.3426	17.7161	16.3494	15.1850	14.1810	13.7291	13.3064	12.9100	43
44	19.5670	17.9041	16.5095	15.3230	14.3012	13.8418	13.4122	13.0096	44
45	19.7839	18.0856	16.6637	15.4557	14.4168	13.9500	13.5138	13.1051	45
46	19.9937	18.2607	16.8122	15.5834	14.5279	14.0540	13.6113	13.1968	46
47	20.1964	18.4297	16.9553	15.7063	14.6346	14.1538	13.7049	13.2848	47
48	20.3923	18.5927	17.0932	15.8245	14.7372	14.2498	13.7949	13.3693	48
49	20.5816	18.7499	17.2260	15.9383	14.8358	14.3419	13.8812	13.4504	49
50	20.7644	18.9015	17.3538	16.0477	14.9305	14.4305	13.9641	13.5283	50

*Note:—In computing the quarterly in advance figures in these tables, all the rates % quoted above are effective rates, see the Introductory Section, pages xvi-xvii.

Yrs.	4	4.5	5	5.5	6	6.25	6.5	6.75	Yrs.
51	20.9410	19.0477	17.4770	16.1529	15.0216	14.5155	14.0438	13.6030	51
52	21.1115	19.1887	17.5956	16.2542	15.1092	14.5973	14.1203	13.6747	52
53	21.2762	19.3246	17.7099	16.3516	15.1933	14.6758	14.1937	13.7436	53
54	21.4351	19.4557	17.8199	16.4454	15.2742	14.7512	14.2643	13.8098	54
55	21.5885	19.5820	17.9258	16.5355	15.3519	14.8237	14.3321	13.8733	55
56	21.7366	19.7037	18.0277	16.6223	15.4267	14.8934	14.3972	13.9343	56
57	21.8795	19.8211	18.1259	16.7057	15.4985	14.9604	14.4598	13.9929	57
58	22.0174	19.9341	18.2204	16.7859	15.5676	15.0247	14.5198	14.0492	58
59	22.1504	20.0431	18.3114	16.8631	15.6339	15.0865	14.5776	14.1032	59
60	22.2787	20.1481	18.3990	16.9374	15.6977	15.1459	14.6330	14.1551	60
61	22.4024	20.2492	18.4833	17.0088	15.7591	15.2030	14.6863	14.2049	61
62	22.5217	20.3467	18.5645	17.0775	15.8180	15.2579	14.7375	14.2528	62
63	22.6368	20.4406	18.6426	17.1436	15.8747	15.3106	14.7867	14.2988	63
64	22.7478	20.5310	18.7178	17.2072	15.9292	15.3613	14.8339	14.3430	64
65	22.8547	20.6181	18.7901	17.2683	15.9816	15.4100	14.8794	14.3855	65
66	22.9578	20.7019	18.8598	17.3271	16.0319	15.4568	14.9230	14.4262	66
67	23.0572	20.7827	18.9268	17.3836	16.0803	15.5018	14.9649	14.4654	67
68	23.1530	20.8605	18.9913	17.4380	16.1268	15.5450	15.0052	14.5031	68
69	23.2453	20.9354	19.0534	17.4904	16.1716	15.5866	15.0439	14.5392	69
70	23.3343	21.0076	19.1131	17.5407	16.2146	15.6265	15.0811	14.5740	70
71	23.4200	21.0770	19.1706	17.5891	16.2559	15.6649	15.1169	14.6074	71
72	23.5026	21.1439	19.2259	17.6356	16.2957	15.7018	15.1513	14.6395	72
73	23.5822	21.2083	19.2791	17.6804	16.3339	15.7373	15.1843	14.6703	73
74	23.6588	21.2702	19.3303	17.7234	16.3706	15.7714	15.2160	14.6999	74
75	23.7326	21.3299	19.3795	17.7648	16.4059	15.8042	15.2465	14.7284	75
76	23.8038	21.3873	19.4269	17.8046	16.4399	15.8357	15.2759	14.7558	76
77	23.8723	21.4426	19.4725	17.8429	16.4725	15.8659	15.3040	14.7821	77
78	23.9382	21.4958	19.5164	17.8798	16.5039	15.8951	15.3311	14.8073	78
79	24.0018	21.5470	19.5586	17.9152	16.5341	15.9230	15.3572	14.8316	79
80	24.0629	21.5963	19.5992	17.9492	16.5631	15.9499	15.3822	14.8549	80
81	24.1218	21.6437	19.6383	17.9820	16.5910	15.9758	15.4062	14.8774	81
82	24.1786	21.6894	19.6759	18.0135	16.6178	16.0007	15.4293	14.8989	82
83	24.2332	21.7333	19.7120	18.0438	16.6436	16.0246	15.4516	14.9196	83
84	24.2858	21.7756	19.7468	18.0729	16.6684	16.0475	15.4729	14.9396	84
85	24.3364	21.8163	19.7802	18.1009	16.6922	16.0696	15.4935	14.9587	85
86	24.3851	21.8555	19.8124	18.1279	16.7151	16.0909	15.5132	14.9771	86
87	24.4320	21.8931	19.8434	18.1538	16.7371	16.1113	15.5322	14.9948	87
88	24.4772	21.9294	19.8731	18.1787	16.7583	16.1309	15.5504	15.0118	88
89	24.5206	21.9643	19.9018	18.2027	16.7787	16.1498	15.5679	15.0281	89
90	24.5625	21.9978	19.9293	18.2257	16.7982	16.1679	15.5848	15.0438	90
91	24.6027	22.0301	19.9558	18.2479	16.8171	16.1853	15.6010	15.0589	91
92	24.6415	22.0612	19.9813	18.2692	16.8352	16.2021	15.6166	15.0734	92
93	24.6788	22.0911	20.0058	18.2897	16.8526	16.2182	15.6315	15.0874	93
94	24.7147	22.1198	20.0294	18.3094	16.8693	16.2337	15.6459	15.1008	94
95	24.7492	22.1475	20.0521	18.3283	16.8854	16.2486	15.6598	15.1137	95
96	24.7824	22.1741	20.0739	18.3466	16.9008	16.2629	15.6731	15.1260	96
97	24.8144	22.1997	20.0949	18.3641	16.9157	16.2767	15.6858	15.1380	97
98	24.8452	22.2243	20.1151	18.3809	16.9300	16.2899	15.6981	15.1494	98
99	24.8748	22.2480	20.1345	18.3971	16.9437	16.3026	15.7099	15.1604	99
100	24.9033	22.2708	20.1531	18.4127	16.9570	16.3149	15.7213	15.1710	100

*Note:—In computing the quarterly in advance figures in these tables, all the rates % quoted above are effective rates, see the Introductory Section, pages xvi-xvii.

Rate per Cent*

Yrs.	7	7.25	7.5	8	8.5	9	9.5	10	Yrs.
1	0.6845	0.6834	0.6823	0.6802	0.6781	0.6761	0.6740	0.6720	1
2	1.3327	1.3286	1.3246	1.3166	1.3088	1.3012	1.2936	1.2862	2
3	1.9470	1.9383	1.9297	1.9129	1.8965	1.8804	1.8648	1.8494	3
4	2.5295	2.5149	2.5005	2.4723	2.4450	2.4184	2.3925	2.3672	4
5	3.0822	3.0606	3.0393	2.9978	2.9577	2.9188	2.8812	2.8447	5
6	3.6070	3.5774	3.5484	3.4920	3.4376	3.3852	3.3347	3.2859	6
7	4.1056	4.0673	4.0298	3.9572	3.8875	3.8206	3.7564	3.6946	7
8	4.5795	4.5319	4.4854	4.3956	4.3098	4.2278	4.1492	4.0740	8
9	5.0301	4.9727	4.9168	4.8091	4.7066	4.6089	4.5158	4.4268	9
10	5.4589	5.3913	5.3257	5.1996	5.0799	4.9664	4.8583	4.7555	10
11	5.8670	5.7891	5.7134	5.5685	5.4315	5.3019	5.1790	5.0623	11
12	6.2556	6.1671	6.0813	5.9174	5.7630	5.6173	5.4795	5.3490	12
13	6.6259	6.5267	6.4307	6.2477	6.0758	5.9140	5.7615	5.6174	13
14	6.9787	6.8688	6.7625	6.5605	6.3712	6.1935	6.0264	5.8690	14
15	7.3152	7.1944	7.0780	6.8569	6.6504	6.4571	6.2757	6.1051	15
16	7.6360	7.5046	7.3779	7.1381	6.9146	6.7058	6.5104	6.3270	16
17	7.9422	7.8001	7.6634	7.4049	7.1647	6.9408	6.7316	6.5358	17
18	8.2344	8.0817	7.9350	7.6583	7.4016	7.1629	6.9403	6.7324	18
19	8.5133	8.3502	8.1938	7.8990	7.6262	7.3730	7.1375	6.9177	19
20	8.7797	8.6064	8.4402	8.1278	7.8393	7.5720	7.3238	7.0925	20
21	9.0342	8.8508	8.6751	8.3454	8.0415	7.7605	7.5000	7.2577	21
22	9.2773	9.0840	8.8991	8.5525	8.2336	7.9393	7.6668	7.4138	22
23	9.5097	9.3067	9.1127	8.7496	8.4161	8.1089	7.8248	7.5615	23
24	9.7319	9.5194	9.3165	8.9373	8.5897	8.2698	7.9746	7.7013	24
25	9.9443	9.7225	9.5110	9.1161	8.7547	8.4227	8.1167	7.8337	25
26	10.1475	9.9167	9.6967	9.2866	8.9118	8.5681	8.2516	7.9592	26
27	10.3419	10.1022	9.8741	9.4491	9.0614	8.7062	8.3796	8.0783	27
28	10.5279	10.2797	10.0435	9.6042	9.2039	8.8377	8.5013	8.1914	28
29	10.7059	10.4493	10.2054	9.7521	9.3397	8.9628	8.6170	8.2987	29
30	10.8763	10.6116	10.3601	9.8933	9.4691	9.0819	8.7271	8.4008	30
31	11.0394	10.7668	10.5080	10.0281	9.5925	9.1954	8.8318	8.4977	31
32	11.1957	10.9153	10.6495	10.1568	9.7102	9.3035	8.9315	8.5900	32
33	11.3453	11.0575	10.7847	10.2798	9.8226	9.4066	9.0265	8.6778	33
34	11.4886	11.1936	10.9142	10.3973	9.9298	9.5049	9.1169	8.7614	34
35	11.6259	11.3239	11.0380	10.5096	10.0322	9.5986	9.2032	8.8410	35
36	11.7574	11.4487	11.1565	10.6170	10.1300	9.6881	9.2854	8.9169	36
37	11.8835	11.5682	11.2700	10.7197	10.2234	9.7736	9.3639	8.9892	37
38	12.0043	11.6826	11.3786	10.8179	10.3127	9.8552	9.4387	9.0582	38
39	12.1201	11.7923	11.4826	10.9119	10.3981	9.9331	9.5102	9.1240	39
40	12.2312	11.8974	11.5822	11.0018	10.4797	10.0075	9.5784	9.1868	40
41	12.3376	11.9981	11.6776	11.0879	10.5578	10.0787	9.6436	9.2467	41
42	12.4397	12.0946	11.7690	11.1703	10.6324	10.1467	9.7059	9.3039	42
43	12.5376	12.1871	11.8566	11.2491	10.7039	10.2118	9.7653	9.3586	43
44	12.6315	12.2758	11.9405	11.3247	10.7722	10.2740	9.8222	9.4108	44
45	12.7216	12.3609	12.0210	11.3970	10.8377	10.3335	9.8766	9.4607	45
46	12.8080	12.4424	12.0981	11.4663	10.9003	10.3904	9.9286	9.5084	46
47	12.8909	12.5206	12.1720	11.5327	10.9603	10.4449	9.9783	9.5540	47
48	12.9704	12.5956	12.2429	11.5963	11.0177	10.4970	10.0259	9.5976	48
49	13.0467	12.6676	12.3109	11.6573	11.0727	10.5470	10.0714	9.6393	49
50	13.1199	12.7366	12.3761	11.7157	11.1255	10.5948	10.1150	9.6792	50

**Note:*—In computing the quarterly in advance figures in these tables, all the rates % quoted above are effective rates, see the Introductory Section, pages xvi-xvii.

				Rate per Cent*					
Yrs.	7	7.25	7.5	8	8.5	9	9.5	10	Yrs.
51	13.1902	12.8028	12.4386	11.7717	11.1759	10.6405	10.1567	9.7174	51
52	13.2577	12.8664	12.4986	11.8254	11.2243	10.6844	10.1967	9.7540	52
53	13.3224	12.9273	12.5561	11.8769	11.2707	10.7264	10.2349	9.7890	53
54	13.3846	12.9858	12.6113	11.9262	11.3152	10.7667	10.2716	9.8225	54
55	13.4442	13.0420	12.6642	11.9736	11.3578	10.8052	10.3067	9.8546	55
56	13.5015	13.0959	12.7150	12.0190	11.3986	10.8422	10.3403	9.8853	56
57	13.5565	13.1476	12.7638	12.0626	11.4378	10.8776	10.3725	9.9148	57
58	13.6093	13.1973	12.8106	12.1043	11.4753	10.9116	10.4034	9.9430	58
59	13.6600	13.2450	12.8555	12.1444	11.5114	10.9442	10.4330	9.9700	59
60	13.7087	13.2907	12.8986	12.1829	11.5459	10.9754	10.4614	9.9959	60
61	13.7554	13.3347	12.9400	12.2198	11.5791	11.0053	10.4886	10.0208	61
62	13.8003	13.3768	12.9797	12.2552	11.6109	11.0341	10.5147	10.0446	62
63	13.8434	13.4174	13.0179	12.2892	11.6414	11.0616	10.5397	10.0674	63
64	13.8848	13.4563	13.0545	12.3218	11.6707	11.0880	10.5637	10.0893	64
65	13.9246	13.4936	13.0896	12.3532	11.6987	11.1134	10.5867	10.1103	65
66	13.9628	13.5295	13.1234	12.3832	11.7257	11.1377	10.6088	10.1304	66
67	13.9995	13.5639	13.1558	12.4121	11.7516	11.1611	10.6299	10.1497	67
68	14.0348	13.5970	13.1869	12.4398	11.7764	11.1834	10.6503	10.1682	68
69	14.0687	13.6288	13.2168	12.4664	11.8002	11.2049	10.6698	10.1860	69
70	14.1012	13.6594	13.2455	12.4919	11.8231	11.2256	10.6885	10.2030	70
71	14.1325	13.6887	13.2731	12.5164	11.8451	11.2454	10.7064	10.2194	71
72	14.1625	13.7169	13.2996	12.5400	11.8662	11.2644	10.7236	10.2351	72
73	14.1913	13.7439	13.3250	12.5626	11.8864	11.2826	10.7402	10.2502	73
74	14.2191	13.7699	13.3495	12.5843	11.9059	11.3001	10.7560	10.2646	74
75	14.2457	13.7949	13.3730	12.6052	11.9245	11.3170	10.7713	10.2785	75
76	14.2713	13.8189	13.3955	12.6252	11.9425	11.3331	10.7859	10.2918	76
77	14.2959	13.8420	13.4172	12.6445	11.9597	11.3486	10.7999	10.3046	77
78	14.3195	13.8641	13.4380	12.6630	11.9762	11.3635	10.8134	10.3169	78
79	14.3422	13.8854	13.4580	12.6807	11.9921	11.3778	10.8264	10.3286	79
80	14.3640	13.9058	13.4772	12.6978	12.0073	11.3915	10.8388	10.3399	80
81	14.3850	13.9255	13.4957	12.7141	12.0220	11.4047	10.8507	10.3508	81
82	14.4052	13.9444	13.5134	12.7299	12.0361	11.4174	10.8622	10.3612	82
83	14.4245	13.9625	13.5304	12.7450	12.0496	11.4295	10.8732	10.3713	83
84	14.4431	13.9800	13.5468	12.7595	12.0626	11.4412	10.8838	10.3809	84
85	14.4610	13.9967	13.5625	12.7735	12.0750	11.4524	10.8939	10.3901	85
86	14.4782	14.0128	13.5777	12.7869	12.0870	11.4632	10.9037	10.3990	86
87	14.4947	14.0283	13.5922	12.7998	12.0985	11.4736	10.9130	10.4075	87
88	14.5106	14.0432	13.6062	12.8122	12.1096	11.4835	10.9220	10.4157	88
89	14.5259	14.0575	13.6196	12.8241	12.1202	11.4931	10.9307	10.4236	89
90	14.5406	14.0712	13.6325	12.8355	12.1304	11.5023	10.9390	10.4311	90
91	14.5547	14.0844	13.6449	12.8465	12.1403	11.5111	10.9470	10.4384	91
92	14.5682	14.0971	13.6568	12.8571	12.1497	11.5196	10.9546	10.4453	92
93	14.5812	14.1093	13.6682	12.8672	12.1587	11.5277	10.9620	10.4520	93
94	14.5938	14.1210	13.6792	12.8769	12.1675	11.5355	10.9691	10.4585	94
95	14.6058	14.1323	13.6898	12.8863	12.1758	11.5430	10.9759	10.4646	95
96	14.6174	14.1431	13.7000	12.8953	12.1839	11.5503	10.9824	10.4706	96
97	14.6285	14.1535	13.7097	12.9040	12.1916	11.5572	10.9887	10.4763	97
98	14.6392	14.1636	13.7191	12.9123	12.1990	11.5639	10.9947	10.4818	98
99	14.6495	14.1732	13.7281	12.9203	12.2061	11.5703	11.0005	10.4870	99
100	14.6593	14.1824	13.7368	12.9280	12.2130	11.5764	11.0061	10.4921	100

Note:—In computing the quarterly in advance figures in these tables, all the rates % quoted above are effective rates, see the Introductory Section, pages xvi-xvii.

Rate per Cent*

Yrs.	11	12	13	14	15	16	18	20	Yrs.
1	0.6680	0.6642	0.6604	0.6567	0.6531	0.6495	0.6427	0.6361	1
2	1.2718	1.2578	1.2443	1.2312	1.2186	1.2063	1.1828	1.1608	2
3	1.8197	1.7912	1.7640	1.7378	1.7127	1.6885	1.6429	1.6006	3
4	2.3188	2.2728	2.2291	2.1874	2.1478	2.1099	2.0392	1.9744	4
5	2.7750	2.7094	2.6475	2.5890	2.5336	2.4811	2.3839	2.2958	5
6	3.1933	3.1067	3.0256	2.9494	2.8777	2.8102	2.6861	2.5748	6
7	3.5779	3.4696	3.3687	3.2745	3.1864	3.1038	2.9532	2.8192	7
8	3.9326	3.8021	3.6813	3.5691	3.4647	3.3673	3.1907	3.0349	8
9	4.2603	4.1076	3.9670	3.8370	3.7166	3.6047	3.4031	3.2264	9
10	4.5639	4.3891	4.2289	4.0816	3.9456	3.8197	3.5941	3.3976	10
11	4.8458	4.6491	4.4698	4.3055	4.1545	4.0152	3.7666	3.5513	11
12	5.1078	4.8899	4.6918	4.5112	4.3456	4.1935	3.9231	3.6901	12
13	5.3520	5.1132	4.8971	4.7006	4.5212	4.3566	4.0655	3.8159	13
14	5.5799	5.3208	5.0872	4.8755	4.6827	4.5065	4.1957	3.9303	14
15	5.7930	5.5142	5.2637	5.0373	4.8318	4.6444	4.3150	4.0348	15
16	5.9924	5.6945	5.4278	5.1874	4.9698	4.7717	4.4247	4.1306	16
17	6.1793	5.8631	5.5807	5.3269	5.0977	4.8895	4.5258	4.2185	17
18	6.3547	6.0208	5.7234	5.4568	5.2165	4.9987	4.6192	4.2996	18
19	6.5196	6.1686	5.8568	5.5779	5.3270	5.1001	4.7057	4.3744	19
20	6.6747	6.3072	5.9816	5.6910	5.4301	5.1945	4.7859	4.4437	20
21	6.8207	6.4375	6.0987	5.7969	5.5264	5.2826	4.8606	4.5080	21
22	6.9584	6.5600	6.2085	5.8960	5.6164	5.3648	4.9301	4.5677	22
23	7.0884	6.6754	6.3117	5.9890	5.7008	5.4417	4.9949	4.6233	23
24	7.2111	6.7841	6.4088	6.0764	5.7799	5.5137	5.0556	4.6752	24
25	7.3271	6.8867	6.5003	6.1586	5.8542	5.5813	5.1123	4.7237	25
26	7.4368	6.9835	6.5865	6.2359	5.9240	5.6447	5.1655	4.7690	26
27	7.5406	7.0750	6.6678	6.3088	5.9897	5.7044	5.2154	4.8116	27
28	7.6390	7.1616	6.7447	6.3775	6.0516	5.7605	5.2622	4.8514	28
29	7.7323	7.2435	6.8173	6.4424	6.1100	5.8134	5.3064	4.8889	29
30	7.8208	7.3211	6.8860	6.5037	6.1652	5.8633	5.3479	4.9241	30
31	7.9048	7.3946	6.9510	6.5617	6.2172	5.9103	5.3870	4.9573	31
32	7.9846	7.4644	7.0126	6.6166	6.2665	5.9548	5.4240	4.9885	32
33	8.0604	7.5306	7.0710	6.6685	6.3131	5.9969	5.4588	5.0180	33
34	8.1325	7.5935	7.1264	6.7178	6.3572	6.0367	5.4918	5.0459	34
35	8.2010	7.6532	7.1790	6.7645	6.3990	6.0744	5.5230	5.0722	35
36	8.2663	7.7100	7.2290	6.8088	6.4387	6.1101	5.5525	5.0971	36
37	8.3284	7.7640	7.2764	6.8509	6.4763	6.1440	5.5804	5.1206	37
38	8.3876	7.8154	7.3216	6.8909	6.5120	6.1761	5.6069	5.1429	38
39	8.4439	7.8644	7.3645	6.9289	6.5460	6.2066	5.6321	5.1640	39
40	8.4977	7.9110	7.4053	6.9650	6.5782	6.2356	5.6559	5.1841	40
41	8.5489	7.9554	7.4442	6.9994	6.6089	6.2632	5.6786	5.2031	41
42	8.5978	7.9977	7.4813	7.0322	6.6381	6.2894	5.7001	5.2212	42
43	8.6445	8.0380	7.5166	7.0634	6.6658	6.3143	5.7206	5.2384	43
44	8.6890	8.0765	7.5502	7.0931	6.6923	6.3380	5.7401	5.2547	44
45	8.7315	8.1132	7.5823	7.1214	6.7175	6.3606	5.7586	5.2702	45
46	8.7721	8.1483	7.6129	7.1484	6.7415	6.3822	5.7762	5.2850	46
47	8.8110	8.1818	7.6421	7.1741	6.7644	6.4027	5.7930	5.2990	47
48	8.8480	8.2137	7.6700	7.1987	6.7862	6.4222	5.8090	5.3124	48
49	8.8835	8.2443	7.6966	7.2221	6.8070	6.4409	5.8243	5.3252	49
50	8.9174	8.2734	7.7220	7.2445	6.8269	6.4587	5.8388	5.3373	50

*Note:—In computing the quarterly in advance figures in these tables, all the rates % quoted above are effective rates, see the Introductory Section, pages xvi-xvii.

				Rate per Cent*					
Yrs.	11	12	13	14	15	16	18	20	Yrs.
51	8.9498	8.3013	7.7463	7.2659	6.8459	6.4757	5.8527	5.3489	51
52	8.9808	8.3280	7.7695	7.2863	6.8640	6.4919	5.8660	5.3600	52
53	9.0104	8.3535	7.7917	7.3058	6.8813	6.5074	5.8786	5.3705	53
54	9.0388	8.3779	7.8129	7.3245	6.8979	6.5222	5.8907	5.3806	54
55	9.0660	8.4012	7.8332	7.3423	6.9137	6.5363	5.9022	5.3902	55
56	9.0920	8.4236	7.8526	7.3593	6.9288	6.5498	5.9132	5.3994	56
57	9.1169	8.4449	7.8712	7.3757	6.9433	6.5627	5.9237	5.4082	57
58	9.1408	8.4654	7.8890	7.3913	6.9571	6.5751	5.9338	5.4166	58
59	9.1636	8.4850	7.9060	7.4062	6.9703	6.5869	5.9434	5.4246	59
60	9.1855	8.5037	7.9223	7.4205	6.9830	6.5982	5.9526	5.4323	60
61	9.2065	8.5217	7.9379	7.4341	6.9951	6.6090	5.9614	5.4396	61
62	9.2265	8.5389	7.9528	7.4472	7.0067	6.6193	5.9698	5.4466	62
63	9.2458	8.5554	7.9671	7.4598	7.0178	6.6292	5.9779	5.4533	63
64	9.2643	8.5712	7.9808	7.4718	7.0284	6.6387	5.9856	5.4597	64
65	9.2820	8.5864	7.9939	7.4833	7.0386	6.6478	5.9930	5.4658	65
66	9.2989	8.6009	8.0065	7.4943	7.0483	6.6565	6.0000	5.4717	66
67	9.3152	8.6148	8.0186	7.5049	7.0577	6.6648	6.0068	5.4773	67
68	9.3308	8.6281	8.0301	7.5150	7.0666	6.6728	6.0133	5.4827	68
69	9.3457	8.6409	8.0412	7.5247	7.0752	6.6804	6.0195	5.4879	69
70	9.3601	8.6532	8.0518	7.5340	7.0834	6.6878	6.0254	5.4928	70
71	9.3738	8.6649	8.0620	7.5429	7.0913	6.6948	6.0311	5.4976	71
72	9.3870	8.6762	8.0718	7.5515	7.0988	6.7015	6.0366	5.5021	72
73	9.3997	8.6870	8.0811	7.5596	7.1061	6.7080	6.0418	5.5065	73
74	9.4119	8.6974	8.0901	7.5675	7.1130	6.7142	6.0469	5.5106	74
75	9.4235	8.7074	8.0987	7.5750	7.1197	6.7201	6.0517	5.5146	75
76	9.4347	8.7169	8.1070	7.5823	7.1261	6.7258	6.0563	5.5185	76
77	9.4455	8.7261	8.1149	7.5892	7.1322	6.7313	6.0607	5.5221	77
78	9.4558	8.7349	8.1225	7.5959	7.1381	6.7365	6.0649	5.5257	78
79	9.4657	8.7433	8.1298	7.6022	7.1437	6.7415	6.0690	5.5290	79
80	9.4752	8.7514	8.1368	7.6084	7.1491	6.7463	6.0729	5.5323	80
81	9.4843	8.7592	8.1436	7.6143	7.1543	6.7509	6.0767	5.5354	81
82	9.4930	8.7667	8.1500	7.6199	7.1593	6.7554	6.0803	5.5384	82
83	9.5015	8.7739	8.1562	7.6253	7.1641	6.7596	6.0837	5.5412	83
84	9.5095	8.7807	8.1622	7.6305	7.1687	6.7637	6.0870	5.5440	84
85	9.5173	8.7873	8.1679	7.6355	7.1731	6.7676	6.0902	5.5466	85
86	9.5247	8.7937	8.1733	7.6403	7.1773	6.7714	6.0932	5.5491	86
87	9.5319	8.7998	8.1786	7.6449	7.1814	6.7750	6.0962	5.5516	87
88	9.5387	8.8056	8.1837	7.6493	7.1853	6.7785	6.0990	5.5539	88
89	9.5453	8.8113	8.1885	7.6535	7.1890	6.7818	6.1017	5.5561	89
90	9.5517	8.8166	8.1932	7.6576	7.1926	6.7850	6.1043	5.5583	90
91	9.5577	8.8218	8.1976	7.6615	7.1960	6.7881	6.1067	5.5603	91
92	9.5636	8.8268	8.2019	7.6653	7.1994	6.7910	6.1091	5.5623	92
93	9.5692	8.8316	8.2061	7.6689	7.2025	6.7939	6.1114	5.5642	93
94	9.5746	8.8362	8.2100	7.6723	7.2056	6.7966	6.1136	5.5660	94
95	9.5798	8.8406	8.2138	7.6757	7.2085	6.7992	6.1157	5.5678	95
96	9.5847	8.8448	8.2175	7.6789	7.2113	6.8017	6.1177	5.5695	96
97	9.5895	8.8489	8.2210	7.6819	7.2140	6.8041	6.1197	5.5711	97
98	9.5941	8.8528	8.2244	7.6849	7.2166	6.8064	6.1216	5.5726	98
99	9.5985	8.8566	8.2276	7.6877	7.2191	6.8086	6.1234	5.5741	99
100	9.6028	8.8602	8.2307	7.6904	7.2215	6.8108	6.1251	5.5755	100

*Note:—In computing the quarterly in advance figures in these tables, all the rates % quoted above are effective rates, see the Introductory Section, pages xvi-xvii.

YEARS' PURCHASE
(DUAL RATE % PRINCIPLE)

OR

PRESENT VALUE OF
ONE POUND PER ANNUM

*receivable quarterly in advance, allow-
ing for a sinking fund at a given rate
to replace the invested capital and for
the effect of income tax at 35% on that
part of the income used to provide the
annual sinking fund instalment.*

AT RATES OF INTEREST*

FROM

4% to 20%

AND

ALLOWING·FOR THE POSSIBLE INVESTMENT
OF SINKING FUNDS AT

3% and 4%

INCOME TAX at 35%

Note:—In computing the quarterly in advance figures in these tables, all the rates
of interest quoted are effective rates, see the Introductory Section, pages
xvi-xvii.

Rate per Cent*

Yrs.	4	4.5	5	5.5	6	6.25	6.5	6.75	Yrs.
1	0.6455	0.6435	0.6415	0.6396	0.6377	0.6368	0.6358	0.6349	1
2	1.2771	1.2695	1.2619	1.2545	1.2472	1.2436	1.2400	1.2365	2
3	1.8952	1.8783	1.8618	1.8457	1.8300	1.8222	1.8146	1.8070	3
4	2.4999	2.4706	2.4421	2.4144	2.3876	2.3744	2.3614	2.3486	4
5	3.0913	3.0466	3.0034	2.9617	2.9213	2.9017	2.8823	2.8633	5
6	3.6696	3.6068	3.5464	3.4884	3.4326	3.4054	3.3788	3.3527	6
7	4.2350	4.1516	4.0718	3.9955	3.9224	3.8870	3.8524	3.8185	7
8	4.7878	4.6814	4.5802	4.4839	4.3921	4.3477	4.3044	4.2621	8
9	5.3280	5.1966	5.0722	4.9543	4.8425	4.7886	4.7362	4.6850	9
10	5.8559	5.6975	5.5484	5.4076	5.2746	5.2108	5.1488	5.0883	10
11	6.3716	6.1846	6.0092	5.8445	5.6895	5.6153	5.5433	5.4733	11
12	6.8755	6.6582	6.4554	6.2657	6.0878	6.0030	5.9208	5.8410	12
13	7.3676	7.1186	6.8873	6.6718	6.4705	6.3748	6.2821	6.1923	13
14	7.8482	7.5663	7.3055	7.0635	6.8383	6.7314	6.6282	6.5283	14
15	8.3174	8.0015	7.7104	7.4413	7.1918	7.0737	6.9598	6.8498	15
16	8.7755	8.4246	8.1025	7.8059	7.5317	7.4023	7.2777	7.1575	16
17	9.2226	8.8358	8.4822	8.1577	7.8588	7.7180	7.5826	7.4522	17
18	9.6591	9.2356	8.8500	8.4973	8.1735	8.0213	7.8751	7.7346	18
19	10.0849	9.6242	9.2062	8.8251	8.4763	8.3128	8.1559	8.0053	19
20	10.5004	10.0019	9.5512	9.1417	8.7680	8.5931	8.4255	8.2649	20
21	10.9058	10.3691	9.8854	9.4474	9.0488	8.8627	8.6846	8.5140	21
22	11.3013	10.7259	10.2092	9.7427	9.3194	9.1221	8.9335	8.7531	22
23	11.6869	11.0727	10.5229	10.0280	9.5801	9.3717	9.1728	8.9827	23
24	12.0630	11.4097	10.8269	10.3037	9.8314	9.6120	9.4029	9.2032	24
25	12.4298	11.7373	11.1214	10.5700	10.0736	9.8434	9.6242	9.4151	25
26	12.7873	12.0556	11.4068	10.8275	10.3072	10.0664	9.8372	9.6189	26
27	13.1359	12.3650	11.6834	11.0764	10.5325	10.2811	10.0422	9.8148	27
28	13.4757	12.6656	11.9514	11.3170	10.7498	10.4881	10.2396	10.0032	28
29	13.8069	12.9577	12.2111	11.5497	10.9595	10.6876	10.4297	10.1846	29
30	14.1296	13.2415	12.4629	11.7746	11.1619	10.8800	10.6128	10.3591	30
31	14.4441	13.5174	12.7069	11.9922	11.3572	11.0655	10.7892	10.5272	31
32	14.7505	13.7854	12.9435	12.2027	11.5458	11.2445	10.9593	10.6890	32
33	15.0491	14.0458	13.1728	12.4063	11.7279	11.4171	11.1232	10.8449	33
34	15.3399	14.2988	13.3951	12.6033	11.9038	11.5837	11.2813	10.9951	34
35	15.6232	14.5446	13.6106	12.7939	12.0736	11.7445	11.4338	11.1399	35
36	15.8991	14.7834	13.8195	12.9783	12.2378	11.8998	11.5809	11.2795	36
37	16.1678	15.0155	14.0221	13.1568	12.3964	12.0497	11.7228	11.4140	37
38	16.4295	15.2409	14.2185	13.3296	12.5496	12.1944	11.8597	11.5438	38
39	16.6843	15.4600	14.4089	13.4968	12.6977	12.3342	11.9919	11.6691	39
40	16.9323	15.6727	14.5936	13.6587	12.8409	12.4693	12.1196	11.7899	40
41	17.1739	15.8794	14.7726	13.8154	12.9794	12.5998	12.2428	11.9065	41
42	17.4090	16.0803	14.9463	13.9672	13.1132	12.7259	12.3618	12.0190	42
43	17.6379	16.2753	15.1147	14.1141	13.2426	12.8478	12.4768	12.1277	43
44	17.8607	16.4648	15.2780	14.2564	13.3678	12.9656	12.5879	12.2326	44
45	18.0775	16.6489	15.4364	14.3942	13.4889	13.0794	12.6952	12.3339	45
46	18.2885	16.8278	15.5900	14.5277	13.6061	13.1896	12.7989	12.4318	46
47	18.4939	17.0015	15.7389	14.6570	13.7194	13.2960	12.8991	12.5263	47
48	18.6937	17.1702	15.8834	14.7822	13.8291	13.3990	12.9960	12.6177	48
49	18.8882	17.3341	16.0236	14.9035	13.9352	13.4986	13.0897	12.7060	49
50	19.0774	17.4933	16.1595	15.0211	14.0379	13.5950	13.1803	12.7913	50

*Note:—In computing the quarterly in advance figures in these tables, all the rates % quoted above are effective rates, see the Introductory Section, pages xvi-xvii.

Rate per Cent*

Yrs.	4	4.5	5	5.5	6	6.25	6.5	6.75	Yrs.
51	19.2614	17.6480	16.2914	15.1350	14.1373	13.6882	13.2679	12.8738	51
52	19.4405	17.7982	16.4193	15.2453	14.2335	13.7784	13.3526	12.9535	52
53	19.6147	17.9441	16.5434	15.3522	14.3267	13.8656	13.4346	13.0306	53
54	19.7841	18.0858	16.6638	15.4558	14.4169	13.9501	13.5138	13.1052	54
55	19.9489	18.2234	16.7806	15.5562	14.5042	14.0318	13.5905	13.1773	55
56	20.1092	18.3571	16.8939	15.6535	14.5888	14.1110	13.6648	13.2471	56
57	20.2652	18.4869	17.0038	15.7478	14.6707	14.1876	13.7366	13.3146	57
58	20.4168	18.6130	17.1104	15.8393	14.7500	14.2617	13.8061	13.3798	58
59	20.5642	18.7355	17.2138	15.9279	14.8268	14.3335	13.8733	13.4430	59
60	20.7076	18.8544	17.3142	16.0138	14.9012	14.4030	13.9385	13.5041	60
61	20.8471	18.9700	17.4116	16.0970	14.9732	14.4704	14.0015	13.5633	61
62	20.9827	19.0822	17.5060	16.1777	15.0430	14.5355	14.0625	13.6206	62
63	21.1145	19.1911	17.5977	16.2560	15.1107	14.5987	14.1216	13.6760	63
64	21.2427	19.2970	17.6867	16.3319	15.1762	14.6599	14.1788	13.7297	64
65	21.3673	19.3998	17.7730	16.4054	15.2397	14.7191	14.2342	13.7816	65
66	21.4884	19.4996	17.8567	16.4767	15.3012	14.7765	14.2879	13.8319	66
67	21.6062	19.5965	17.9379	16.5459	15.3609	14.8321	14.3399	13.8806	67
68	21.7207	19.6906	18.0168	16.6130	15.4187	14.8859	14.3902	13.9278	68
69	21.8320	19.7821	18.0933	16.6780	15.4747	14.9381	14.4390	13.9735	69
70	21.9402	19.8709	18.1676	16.7410	15.5289	14.9887	14.4862	14.0177	70
71	22.0454	19.9571	18.2396	16.8022	15.5815	15.0377	14.5320	14.0605	71
72	22.1476	20.0408	18.3095	16.8615	15.6325	15.0852	14.5764	14.1021	72
73	22.2469	20.1221	18.3774	16.9190	15.6820	15.1312	14.6193	14.1423	73
74	22.3435	20.2011	18.4432	16.9748	15.7299	15.1758	14.6610	14.1812	74
75	22.4373	20.2778	18.5071	17.0289	15.7763	15.2191	14.7013	14.2190	75
76	22.5285	20.3522	18.5691	17.0814	15.8214	15.2610	14.7404	14.2555	76
77	22.6172	20.4245	18.6293	17.1323	15.8650	15.3016	14.7783	14.2910	77
78	22.7033	20.4947	18.6877	17.1817	15.9074	15.3410	14.8150	14.3253	78
79	22.7870	20.5629	18.7443	17.2296	15.9484	15.3791	14.8506	14.3586	79
80	22.8683	20.6291	18.7994	17.2761	15.9882	15.4162	14.8851	14.3909	80
81	22.9474	20.6934	18.8527	17.3211	16.0268	15.4520	14.9186	14.4221	81
82	23.0242	20.7559	18.9045	17.3649	16.0642	15.4868	14.9510	14.4524	82
83	23.0988	20.8165	18.9548	17.4073	16.1005	15.5205	14.9824	14.4818	83
84	23.1713	20.8754	19.0036	17.4484	16.1357	15.5533	15.0129	14.5103	84
85	23.2418	20.9326	19.0510	17.4884	16.1699	15.5850	15.0424	14.5379	85
86	23.3103	20.9881	19.0970	17.5271	16.2030	15.6157	15.0711	14.5646	86
87	23.3768	21.0420	19.1416	17.5647	16.2351	15.6455	15.0989	14.5906	87
88	23.4414	21.0943	19.1849	17.6011	16.2662	15.6745	15.1258	14.6157	88
89	23.5042	21.1452	19.2269	17.6365	16.2964	15.7025	15.1519	14.6401	89
90	23.5652	21.1945	19.2677	17.6708	16.3257	15.7297	15.1772	14.6637	90
91	23.6244	21.2424	19.3073	17.7041	16.3541	15.7561	15.2018	14.6867	91
92	23.6820	21.2890	19.3458	17.7364	16.3817	15.7817	15.2256	14.7089	92
93	23.7379	21.3341	19.3831	17.7678	16.4085	15.8065	15.2487	14.7304	93
94	23.7923	21.3780	19.4193	17.7982	16.4344	15.8306	15.2711	14.7513	94
95	23.8450	21.4206	19.4544	17.8277	16.4596	15.8539	15.2928	14.7716	95
96	23.8963	21.4620	19.4885	17.8564	16.4840	15.8766	15.3139	14.7913	96
97	23.9461	21.5022	19.5216	17.8842	16.5077	15.8985	15.3344	14.8103	97
98	23.9945	21.5412	19.5538	17.9111	16.5306	15.9198	15.3542	14.8288	98
99	24.0415	21.5790	19.5850	17.9373	16.5529	15.9405	15.3734	14.8468	99
100	24.0871	21.6158	19.6153	17.9627	16.5746	15.9606	15.3921	14.8642	100

*Note:—In computing the quarterly in advance figures in these tables, all the rates % quoted above are effective rates, see the Introductory Section, pages xvi-xvii.

				Rate per Cent*					
Yrs.	7	7.25	7.5	8	8.5	9	9.5	10	Yrs.
1	0.6340	0.6331	0.6321	0.6303	0.6285	0.6268	0.6250	0.6233	1
2	1.2330	1.2295	1.2260	1.2192	1.2125	1.2060	1.1995	1.1931	2
3	1.7995	1.7921	1.7848	1.7704	1.7563	1.7426	1.7291	1.7159	3
4	2.3360	2.3236	2.3113	2.2872	2.2638	2.2409	2.2187	2.1970	4
5	2.8445	2.8261	2.8079	2.7725	2.7381	2.7048	2.6724	2.6410	5
6	3.3270	3.3018	3.2771	3.2289	3.1823	3.1374	3.0939	3.0519	6
7	3.7852	3.7526	3.7207	3.6587	3.5991	3.5417	3.4864	3.4331	7
8	4.2207	4.1802	4.1407	4.0640	3.9906	3.9201	3.8525	3.7876	8
9	4.6350	4.5862	4.5386	4.4467	4.3589	4.2750	4.1948	4.1179	9
10	5.0294	4.9721	4.9162	4.8085	4.7060	4.6084	4.5152	4.4263	10
11	5.4052	5.3390	5.2746	5.1509	5.0335	4.9219	4.8158	4.7148	11
12	5.7635	5.6883	5.6153	5.4752	5.3428	5.2173	5.0982	4.9851	12
13	6.1054	6.0210	5.9393	5.7828	5.6353	5.4958	5.3639	5.2388	13
14	6.4317	6.3382	6.2477	6.0748	5.9122	5.7589	5.6141	5.4773	14
15	6.7435	6.6408	6.5414	6.3522	6.1746	6.0076	5.8502	5.7018	15
16	7.0415	6.9296	6.8215	6.6159	6.4235	6.2429	6.0732	5.9134	16
17	7.3266	7.2055	7.0887	6.8669	6.6599	6.4660	6.2841	6.1131	17
18	7.5993	7.4691	7.3437	7.1060	6.8845	6.6775	6.4837	6.3018	18
19	7.8605	7.7213	7.5873	7.3338	7.0981	6.8783	6.6728	6.4804	19
20	8.1107	7.9625	7.8201	7.5511	7.3015	7.0691	6.8522	6.6494	20
21	8.3504	8.1935	8.0427	7.7585	7.4952	7.2505	7.0226	6.8097	21
22	8.5803	8.4147	8.2558	7.9566	7.6799	7.4232	7.1845	6.9618	22
23	8.8008	8.6266	8.4597	8.1458	7.8561	7.5877	7.3384	7.1063	23
24	9.0124	8.8298	8.6550	8.3268	8.0242	7.7444	7.4850	7.2436	24
25	9.2155	9.0247	8.8422	8.4999	8.1849	7.8940	7.6245	7.3743	25
26	9.4106	9.2118	9.0217	8.6656	8.3384	8.0367	7.7576	7.4987	26
27	9.5981	9.3913	9.1938	8.8243	8.4852	8.1730	7.8845	7.6172	27
28	9.7782	9.5637	9.3590	8.9763	8.6257	8.3033	8.0057	7.7302	28
29	9.9514	9.7293	9.5175	9.1221	8.7602	8.4278	8.1214	7.8381	29
30	10.1180	9.8885	9.6698	9.2619	8.8890	8.5470	8.2320	7.9410	30
31	10.2782	10.0415	9.8160	9.3960	9.0125	8.6611	8.3378	8.0394	31
32	10.4324	10.1886	9.9566	9.5247	9.1308	8.7703	8.4390	8.1335	32
33	10.5809	10.3302	10.0917	9.6483	9.2444	8.8750	8.5359	8.2234	33
34	10.7238	10.4664	10.2217	9.7670	9.3533	8.9753	8.6286	8.3095	34
35	10.8615	10.5975	10.3467	9.8810	9.4579	9.0716	8.7176	8.3919	35
36	10.9942	10.7237	10.4670	9.9907	9.5583	9.1639	8.8028	8.4709	36
37	11.1220	10.8453	10.5828	10.0961	9.6547	9.2525	8.8846	8.5466	37
38	11.2452	10.9624	10.6943	10.1976	9.7475	9.3377	8.9630	8.6191	38
39	11.3640	11.0753	10.8016	10.2952	9.8366	9.4194	9.0383	8.6888	39
40	11.4785	11.1840	10.9051	10.3891	9.9223	9.4980	9.1106	8.7556	40
41	11.5890	11.2889	11.0048	10.4795	10.0047	9.5735	9.1801	8.8197	41
42	11.6956	11.3900	11.1008	10.5666	10.0841	9.6461	9.2468	8.8813	42
43	11.7985	11.4876	11.1935	10.6505	10.1605	9.7160	9.3110	8.9405	43
44	11.8977	11.5817	11.2828	10.7313	10.2340	9.7832	9.3727	8.9974	44
45	11.9936	11.6724	11.3689	10.8092	10.3048	9.8479	9.4321	9.0521	45
46	12.0861	11.7601	11.4520	10.8843	10.3730	9.9102	9.4892	9.1047	46
47	12.1754	11.8446	11.5322	10.9567	10.4388	9.9702	9.5442	9.1553	47
48	12.2617	11.9263	11.6096	11.0265	10.5022	10.0280	9.5972	9.2040	48
49	12.3451	12.0051	11.6843	11.0939	10.5632	10.0837	9.6482	9.2509	49
50	12.4256	12.0813	11.7564	11.1589	10.6222	10.1374	9.6973	9.2960	50

*Note:—In computing the quarterly in advance figures in these tables, all the rates % quoted above are effective rates,
  see the Introductory Section, pages xvi-xvii.

				Rate per Cent*					
Yrs.	7	7.25	7.5	8	8.5	9	9.5	10	Yrs.
51	12.5035	12.1548	11.8261	11.2216	10.6790	10.1891	9.7446	9.3395	51
52	12.5787	12.2259	11.8933	11.2822	10.7338	10.2390	9.7902	9.3814	52
53	12.6514	12.2946	11.9583	11.3406	10.7867	10.2871	9.8342	9.4218	53
54	12.7216	12.3609	12.0211	11.3971	10.8377	10.3335	9.8766	9.4607	54
55	12.7896	12.4251	12.0817	11.4516	10.8870	10.3783	9.9175	9.4982	55
56	12.8553	12.4871	12.1403	11.5042	10.9346	10.4215	9.9570	9.5344	56
57	12.9188	12.5470	12.1970	11.5551	10.9805	10.4632	9.9951	9.5693	57
58	12.9803	12.6050	12.2517	11.6042	11.0249	10.5035	10.0318	9.6030	58
59	13.0397	12.6610	12.3047	11.6517	11.0677	10.5424	10.0673	9.6355	59
60	13.0972	12.7152	12.3559	11.6976	11.1091	10.5800	10.1015	9.6669	60
61	13.1529	12.7677	12.4054	11.7420	11.1491	10.6162	10.1346	9.6972	61
62	13.2067	12.8184	12.4533	11.7848	11.1878	10.6513	10.1665	9.7264	62
63	13.2588	12.8675	12.4996	11.8263	11.2252	10.6852	10.1974	9.7546	63
64	13.3093	12.9150	12.5444	11.8664	11.2613	10.7179	10.2272	9.7819	64
65	13.3581	12.9609	12.5878	11.9052	11.2962	10.7495	10.2560	9.8082	65
66	13.4053	13.0054	12.6297	11.9427	11.3300	10.7801	10.2838	9.8337	66
67	13.4511	13.0484	12.6703	11.9790	11.3627	10.8097	10.3107	9.8583	67
68	13.4953	13.0901	12.7096	12.0141	11.3942	10.8382	10.3367	9.8820	68
69	13.5382	13.1305	12.7476	12.0481	11.4248	10.8659	10.3619	9.9050	69
70	13.5798	13.1695	12.7844	12.0810	11.4543	10.8926	10.3862	9.9272	70
71	13.6200	13.2073	12.8201	12.1128	11.4829	10.9185	10.4097	9.9487	71
72	13.6589	13.2439	12.8546	12.1436	11.5106	10.9435	10.4324	9.9695	72
73	13.6966	13.2794	12.8880	12.1734	11.5374	10.9677	10.4544	9.9895	73
74	13.7332	13.3138	12.9203	12.2022	11.5633	10.9911	10.4757	10.0090	74
75	13.7686	13.3470	12.9516	12.2302	11.5884	11.0138	10.4963	10.0278	75
76	13.8029	13.3792	12.9820	12.2572	11.6127	11.0357	10.5162	10.0459	76
77	13.8361	13.4104	13.0114	12.2834	11.6362	11.0569	10.5354	10.0635	77
78	13.8683	13.4407	13.0398	12.3088	11.6589	11.0775	10.5541	10.0805	78
79	13.8995	13.4700	13.0674	12.3333	11.6810	11.0973	10.5721	10.0970	79
80	13.9297	13.4984	13.0941	12.3571	11.7023	11.1166	10.5896	10.1129	80
81	13.9590	13.5259	13.1200	12.3802	11.7230	11.1353	10.6065	10.1284	81
82	13.9874	13.5525	13.1450	12.4025	11.7430	11.1533	10.6229	10.1433	82
83	14.0149	13.5783	13.1693	12.4241	11.7624	11.1708	10.6388	10.1578	83
84	14.0415	13.6034	13.1929	12.4451	11.7811	11.1877	10.6541	10.1718	84
85	14.0674	13.6276	13.2157	12.4654	11.7993	11.2041	10.6690	10.1853	85
86	14.0924	13.6511	13.2378	12.4850	11.8169	11.2200	10.6834	10.1984	86
87	14.1167	13.6739	13.2592	12.5041	11.8340	11.2354	10.6974	10.2112	87
88	14.1402	13.6960	13.2800	12.5226	11.8506	11.2503	10.7109	10.2235	88
89	14.1631	13.7174	13.3001	12.5404	11.8666	11.2647	10.7240	10.2354	89
90	14.1852	13.7381	13.3196	12.5578	11.8821	11.2787	10.7366	10.2469	90
91	14.2066	13.7583	13.3385	12.5746	11.8972	11.2923	10.7489	10.2581	91
92	14.2274	13.7778	13.3569	12.5909	11.9117	11.3054	10.7608	10.2690	92
93	14.2476	13.7967	13.3746	12.6067	11.9259	11.3182	10.7724	10.2795	93
94	14.2672	13.8150	13.3919	12.6220	11.9396	11.3305	10.7835	10.2896	94
95	14.2861	13.8328	13.4086	12.6368	11.9528	11.3424	10.7944	10.2995	95
96	14.3045	13.8500	13.4248	12.6512	11.9657	11.3540	10.8049	10.3091	96
97	14.3223	13.8668	13.4405	12.6652	11.9782	11.3653	10.8150	10.3183	97
98	14.3396	13.8830	13.4557	12.6787	11.9903	11.3762	10.8249	10.3273	98
99	14.3564	13.8987	13.4705	12.6918	12.0020	11.3867	10.8344	10.3360	99
100	14.3727	13.9139	13.4848	12.7045	12.0134	11.3969	10.8437	10.3444	100

*Note:—In computing the quarterly in advance figures in these tables, all the rates % quoted above are effective rates,
see the Introductory Section, pages xvi-xvii.

				Rate per Cent*					
Yrs.	11	12	13	14	15	16	18	20	Yrs.
1	0.6199	0.6165	0.6133	0.6101	0.6069	0.6039	0.5979	0.5922	1
2	1.1807	1.1686	1.1570	1.1456	1.1347	1.1240	1.1036	1.0844	2
3	1.6903	1.6657	1.6421	1.6194	1.5975	1.5765	1.5367	1.4996	3
4	2.1552	2.1154	2.0774	2.0413	2.0067	1.9736	1.9116	1.8545	4
5	2.5808	2.5240	2.4702	2.4192	2.3707	2.3247	2.2391	2.1613	5
6	2.9718	2.8967	2.8261	2.7595	2.6967	2.6373	2.5277	2.4289	6
7	3.3321	3.2380	3.1499	3.0675	2.9900	2.9172	2.7837	2.6644	7
8	3.6651	3.5514	3.4458	3.3474	3.2554	3.1692	3.0123	2.8730	8
9	3.9735	3.8403	3.7171	3.6028	3.4964	3.3972	3.2175	3.0591	9
10	4.2598	4.1071	3.9665	3.8366	3.7163	3.6044	3.4028	3.2261	10
11	4.5264	4.3544	4.1966	4.0515	3.9175	3.7934	3.5708	3.3767	11
12	4.7750	4.5839	4.4095	4.2495	4.1023	3.9664	3.7237	3.5132	12
13	5.0072	4.7976	4.6068	4.4325	4.2726	4.1254	3.8634	3.6373	13
14	5.2247	4.9968	4.7902	4.6020	4.4299	4.2719	3.9916	3.7507	14
15	5.4285	5.1830	4.9610	4.7595	4.5756	4.4072	4.1095	3.8546	15
16	5.6200	5.3572	5.1205	4.9060	4.7109	4.5326	4.2183	3.9502	16
17	5.8001	5.5207	5.2696	5.0427	4.8368	4.6490	4.3190	4.0383	17
18	5.9698	5.6741	5.4092	5.1705	4.9542	4.7574	4.4123	4.1198	18
19	6.1297	5.8184	5.5402	5.2900	5.0639	4.8584	4.4991	4.1954	19
20	6.2808	5.9544	5.6633	5.4022	5.1665	4.9528	4.5800	4.2656	20
21	6.4236	6.0826	5.7792	5.5075	5.2628	5.0412	4.6555	4.3310	21
22	6.5588	6.2037	5.8884	5.6066	5.3532	5.1241	4.7260	4.3920	22
23	6.6869	6.3181	5.9914	5.6999	5.4382	5.2019	4.7922	4.4491	23
24	6.8083	6.4264	6.0887	5.7879	5.5182	5.2751	4.8542	4.5025	24
25	6.9236	6.5291	6.1808	5.8710	5.5937	5.3441	4.9126	4.5527	25
26	7.0332	6.6264	6.2679	5.9496	5.6650	5.4091	4.9675	4.5998	26
27	7.1373	6.7188	6.3505	6.0240	5.7324	5.4705	5.0192	4.6441	27
28	7.2365	6.8066	6.4289	6.0944	5.7962	5.5286	5.0680	4.6859	28
29	7.3309	6.8900	6.5033	6.1613	5.8566	5.5835	5.1142	4.7253	29
30	7.4209	6.9695	6.5740	6.2247	5.9139	5.6356	5.1578	4.7625	30
31	7.5067	7.0451	6.6413	6.2850	5.9683	5.6849	5.1991	4.7977	31
32	7.5887	7.1173	6.7054	6.3423	6.0200	5.7318	5.2383	4.8311	32
33	7.6669	7.1860	6.7664	6.3969	6.0691	5.7763	5.2755	4.8627	33
34	7.7417	7.2517	6.8245	6.4489	6.1159	5.8187	5.3108	4.8926	34
35	7.8132	7.3144	6.8800	6.4984	6.1604	5.8590	5.3443	4.9211	35
36	7.8816	7.3743	6.9330	6.5457	6.2029	5.8973	5.3762	4.9481	36
37	7.9471	7.4316	6.9836	6.5908	6.2433	5.9339	5.4066	4.9739	37
38	8.0098	7.4864	7.0320	6.6338	6.2820	5.9688	5.4356	4.9983	38
39	8.0698	7.5389	7.0783	6.6750	6.3189	6.0021	5.4632	5.0217	39
40	8.1274	7.5891	7.1226	6.7143	6.3541	6.0339	5.4895	5.0439	40
41	8.1827	7.6372	7.1650	6.7520	6.3878	6.0643	5.5146	5.0651	41
42	8.2357	7.6834	7.2056	6.7880	6.4201	6.0934	5.5387	5.0854	42
43	8.2866	7.7277	7.2445	6.8226	6.4510	6.1212	5.5616	5.1048	43
44	8.3354	7.7701	7.2818	6.8556	6.4805	6.1478	5.5836	5.1232	44
45	8.3823	7.8109	7.3176	6.8873	6.5089	6.1733	5.6046	5.1409	45
46	8.4274	7.8500	7.3519	6.9178	6.5360	6.1977	5.6247	5.1579	46
47	8.4708	7.8876	7.3849	6.9469	6.5621	6.2211	5.6440	5.1741	47
48	8.5124	7.9237	7.4165	6.9749	6.5870	6.2436	5.6625	5.1896	48
49	8.5525	7.9585	7.4469	7.0018	6.6110	6.2651	5.6802	5.2045	49
50	8.5911	7.9919	7.4762	7.0277	6.6340	6.2858	5.6972	5.2187	50

Note:—In computing the quarterly in advance figures in these tables, all the rates % quoted above are effective rates, see the Introductory Section, pages xvi-xvii.

Rate per Cent*

Yrs.	11	12	13	14	15	16	18	20	Yrs.
51	8.6282	8.0240	7.5043	7.0525	6.6562	6.3056	5.7135	5.2324	51
52	8.6640	8.0549	7.5313	7.0764	6.6774	6.3247	5.7291	5.2455	52
53	8.6984	8.0846	7.5573	7.0993	6.6979	6.3430	5.7442	5.2581	53
54	8.7316	8.1133	7.5823	7.1214	6.7175	6.3606	5.7586	5.2702	54
55	8.7635	8.1409	7.6064	7.1426	6.7364	6.3776	5.7725	5.2818	55
56	8.7943	8.1674	7.6296	7.1631	6.7546	6.3939	5.7858	5.2930	56
57	8.8240	8.1930	7.6519	7.1828	6.7721	6.4096	5.7987	5.3038	57
58	8.8526	8.2177	7.6734	7.2017	6.7889	6.4247	5.8110	5.3141	58
59	8.8803	8.2415	7.6942	7.2200	6.8052	6.4392	5.8229	5.3240	59
60	8.9069	8.2644	7.7142	7.2376	6.8208	6.4532	5.8343	5.3336	60
61	8.9326	8.2865	7.7334	7.2545	6.8358	6.4667	5.8454	5.3428	61
62	8.9574	8.3079	7.7520	7.2709	6.8504	6.4796	5.8560	5.3517	62
63	8.9813	8.3285	7.7699	7.2867	6.8643	6.4922	5.8662	5.3602	63
64	9.0044	8.3483	7.7872	7.3019	6.8778	6.5042	5.8760	5.3684	64
65	9.0267	8.3675	7.8039	7.3165	6.8909	6.5159	5.8855	5.3763	65
66	9.0483	8.3860	7.8200	7.3307	6.9034	6.5271	5.8947	5.3840	66
67	9.0691	8.4039	7.8356	7.3443	6.9155	6.5379	5.9035	5.3913	67
68	9.0892	8.4212	7.8506	7.3575	6.9272	6.5484	5.9120	5.3984	68
69	9.1087	8.4379	7.8651	7.3702	6.9385	6.5584	5.9202	5.4053	69
70	9.1274	8.4540	7.8791	7.3825	6.9494	6.5682	5.9282	5.4119	70
71	9.1456	8.4695	7.8926	7.3944	6.9599	6.5776	5.9358	5.4183	71
72	9.1631	8.4846	7.9056	7.4059	6.9700	6.5866	5.9432	5.4244	72
73	9.1801	8.4991	7.9183	7.4169	6.9799	6.5954	5.9503	5.4304	73
74	9.1965	8.5132	7.9305	7.4276	6.9893	6.6038	5.9572	5.4361	74
75	9.2123	8.5268	7.9423	7.4380	6.9985	6.6120	5.9639	5.4416	75
76	9.2277	8.5399	7.9536	7.4480	7.0073	6.6199	5.9703	5.4470	76
77	9.2425	8.5526	7.9647	7.4576	7.0159	6.6275	5.9765	5.4521	77
78	9.2569	8.5649	7.9753	7.4670	7.0242	6.6349	5.9825	5.4571	78
79	9.2708	8.5768	7.9856	7.4760	7.0321	6.6421	5.9883	5.4620	79
80	9.2842	8.5883	7.9956	7.4848	7.0399	6.6489	5.9939	5.4666	80
81	9.2972	8.5994	8.0052	7.4932	7.0473	6.6556	5.9993	5.4711	81
82	9.3098	8.6102	8.0146	7.5014	7.0546	6.6621	6.0046	5.4755	82
83	9.3220	8.6206	8.0236	7.5093	7.0616	6.6683	6.0096	5.4797	83
84	9.3337	8.6307	8.0323	7.5169	7.0683	6.6743	6.0145	5.4838	84
85	9.3452	8.6404	8.0408	7.5243	7.0749	6.6802	6.0192	5.4877	85
86	9.3562	8.6499	8.0489	7.5315	7.0812	6.6858	6.0238	5.4915	86
87	9.3669	8.6590	8.0569	7.5384	7.0873	6.6913	6.0283	5.4952	87
88	9.3773	8.6679	8.0645	7.5451	7.0933	6.6965	6.0326	5.4988	88
89	9.3873	8.6764	8.0719	7.5516	7.0990	6.7017	6.0367	5.5022	89
90	9.3970	8.6847	8.0791	7.5579	7.1045	6.7066	6.0407	5.5055	90
91	9.4064	8.6928	8.0861	7.5640	7.1099	6.7114	6.0446	5.5088	91
92	9.4155	8.7005	8.0928	7.5699	7.1151	6.7160	6.0484	5.5119	92
93	9.4244	8.7081	8.0993	7.5756	7.1202	6.7205	6.0520	5.5149	93
94	9.4329	8.7154	8.1056	7.5811	7.1251	6.7249	6.0555	5.5178	94
95	9.4412	8.7224	8.1118	7.5865	7.1298	6.7291	6.0589	5.5207	95
96	9.4492	8.7293	8.1177	7.5916	7.1344	6.7332	6.0623	5.5234	96
97	9.4570	8.7359	8.1234	7.5967	7.1388	6.7371	6.0655	5.5261	97
98	9.4645	8.7424	8.1290	7.6015	7.1431	6.7409	6.0686	5.5286	98
99	9.4718	8.7486	8.1344	7.6062	7.1472	6.7446	6.0716	5.5311	99
100	9.4789	8.7546	8.1396	7.6108	7.1513	6.7482	6.0745	5.5336	100

*Note:—In computing the quarterly in advance figures in these tables, all the rates % quoted above are effective rates, see the Introductory Section, pages xvi-xvii.

Rate per Cent*

Yrs.	4	4.5	5	5.5	6	6.25	6.5	6.75	Yrs.
1	0.6493	0.6473	0.6453	0.6434	0.6415	0.6405	0.6396	0.6386	1
2	1.2905	1.2827	1.2750	1.2674	1.2599	1.2563	1.2526	1.2490	2
3	1.9234	1.9060	1.8890	1.8724	1.8562	1.8483	1.8404	1.8326	3
4	2.5476	2.5172	2.4876	2.4589	2.4311	2.4174	2.4040	2.3907	4
5	3.1628	3.1160	3.0709	3.0273	2.9851	2.9646	2.9444	2.9245	5
6	3.7687	3.7025	3.6389	3.5779	3.5192	3.4907	3.4627	3.4352	6
7	4.3652	4.2766	4.1920	4.1112	4.0338	3.9964	3.9598	3.9240	7
8	4.9519	4.8382	4.7302	4.6276	4.5298	4.4827	4.4367	4.3917	8
9	5.5287	5.3873	5.2538	5.1274	5.0077	4.9502	4.8941	4.8395	9
10	6.0954	5.9240	5.7629	5.6112	5.4682	5.3997	5.3330	5.2682	10
11	6.6518	6.4482	6.2578	6.0794	5.9118	5.8318	5.7541	5.6787	11
12	7.1978	6.9600	6.7387	6.5322	6.3392	6.2473	6.1582	6.0720	12
13	7.7332	7.4594	7.2058	6.9703	6.7509	6.6467	6.5460	6.4486	13
14	8.2581	7.9466	7.6594	7.3938	7.1474	7.0308	6.9182	6.8095	14
15	8.7722	8.4215	8.0997	7.8033	7.5293	7.4000	7.2754	7.1553	15
16	9.2756	8.8844	8.5270	8.1991	7.8972	7.7550	7.6183	7.4867	16
17	9.7681	9.3353	8.9414	8.5816	8.2514	8.0963	7.9474	7.8043	17
18	10.2498	9.7743	9.3434	8.9511	8.5925	8.4245	8.2634	8.1088	18
19	10.7207	10.2016	9.7331	9.3082	8.9210	8.7400	8.5668	8.4007	19
20	11.1808	10.6173	10.1108	9.6531	9.2373	9.0434	8.8580	8.6806	20
21	11.6300	11.0216	10.4768	9.9861	9.5418	9.3351	9.1377	8.9490	21
22	12.0686	11.4147	10.8313	10.3077	9.8350	9.6155	9.4062	9.2064	22
23	12.4964	11.7967	11.1747	10.6182	10.1173	9.8852	9.6641	9.4533	23
24	12.9136	12.1678	11.5071	10.9179	10.3890	10.1444	9.9117	9.6901	24
25	13.3202	12.5281	11.8289	11.2071	10.6506	10.3937	10.1496	9.9173	25
26	13.7164	12.8780	12.1403	11.4863	10.9024	10.6333	10.3780	10.1353	26
27	14.1023	13.2175	12.4416	11.7556	11.1448	10.8638	10.5974	10.3444	27
28	14.4779	13.5469	12.7331	12.0155	11.3781	11.0853	10.8081	10.5451	28
29	14.8434	13.8664	13.0149	12.2662	11.6026	11.2983	11.0105	10.7377	29
30	15.1989	14.1762	13.2875	12.5080	11.8187	11.5032	11.2049	10.9225	30
31	15.5446	14.4765	13.5509	12.7411	12.0267	11.7001	11.3917	11.0999	31
32	15.8806	14.7675	13.8056	12.9660	12.2268	11.8894	11.5711	11.2702	32
33	16.2071	15.0494	14.0516	13.1828	12.4195	12.0715	11.7434	11.4336	33
34	16.5242	15.3224	14.2894	13.3919	12.6048	12.2465	11.9090	11.5905	34
35	16.8321	15.5868	14.5191	13.5934	12.7832	12.4149	12.0681	11.7412	35
36	17.1310	15.8428	14.7409	13.7876	12.9548	12.5767	12.2210	11.8858	36
37	17.4210	16.0905	14.9551	13.9749	13.1200	12.7323	12.3679	12.0247	37
38	17.7023	16.3301	15.1619	14.1553	13.2789	12.8819	12.5090	12.1581	38
39	17.9751	16.5620	15.3616	14.3292	13.4318	13.0257	12.6446	12.2861	39
40	18.2395	16.7862	15.5543	14.4967	13.5789	13.1640	12.7749	12.4091	40
41	18.4958	17.0031	15.7403	14.6582	13.7205	13.2970	12.9001	12.5272	41
42	18.7441	17.2127	15.9198	14.8137	13.8566	13.4249	13.0204	12.6406	42
43	18.9847	17.4154	16.0930	14.9636	13.9877	13.5478	13.1360	12.7496	43
44	19.2176	17.6112	16.2601	15.1079	14.1137	13.6661	13.2471	12.8542	44
45	19.4432	17.8004	16.4212	15.2469	14.2350	13.7797	13.3539	12.9547	45
46	19.6615	17.9832	16.5767	15.3808	14.3516	13.8890	13.4565	13.0513	46
47	19.8727	18.1598	16.7266	15.5098	14.4639	13.9941	13.5551	13.1440	47
48	20.0771	18.3303	16.8711	15.6340	14.5718	14.0951	13.6499	13.2331	48
49	20.2748	18.4949	17.0105	15.7536	14.6757	14.1923	13.7410	13.3187	49
50	20.4659	18.6539	17.1449	15.8688	14.7756	14.2857	13.8285	13.4009	50

Note:—In computing the quarterly in advance figures in these tables, all the rates % quoted above are effective rates,
see the Introductory Section, pages xvi-xvii.

			Rate per Cent*						
Yrs.	4	4.5	5	5.5	6	6.25	6.5	6.75	Yrs.
51	20.6508	18.8073	17.2744	15.9797	14.8717	14.3755	13.9127	13.4799	51
52	20.8295	18.9554	17.3993	16.0865	14.9641	14.4619	13.9935	13.5558	52
53	21.0022	19.0983	17.5196	16.1893	15.0531	14.5449	14.0713	13.6288	53
54	21.1690	19.2362	17.6356	16.2883	15.1386	14.6247	14.1460	13.6989	54
55	21.3303	19.3692	17.7473	16.3836	15.2209	14.7015	14.2178	13.7662	55
56	21.4860	19.4976	17.8550	16.4753	15.3000	14.7753	14.2868	13.8309	56
57	21.6364	19.6213	17.9588	16.5636	15.3761	14.8463	14.3532	13.8931	57
58	21.7817	19.7407	18.0587	16.6486	15.4493	14.9145	14.4170	13.9528	58
59	21.9219	19.8558	18.1550	16.7304	15.5198	14.9802	14.4783	14.0102	59
60	22.0573	19.9668	18.2478	16.8091	15.5875	15.0433	14.5372	14.0654	60
61	22.1879	20.0739	18.3371	16.8849	15.6526	15.1039	14.5938	14.1184	61
62	22.3141	20.1770	18.4231	16.9578	15.7153	15.1623	14.6483	14.1694	62
63	22.4357	20.2765	18.5060	17.0280	15.7755	15.2183	14.7006	14.2183	63
64	22.5532	20.3723	18.5858	17.0956	15.8335	15.2723	14.7509	14.2654	64
65	22.6664	20.4647	18.6627	17.1606	15.8892	15.3241	14.7993	14.3106	65
66	22.7757	20.5537	18.7367	17.2231	15.9429	15.3740	14.8458	14.3541	66
67	22.8810	20.6395	18.8079	17.2833	15.9944	15.4219	14.8905	14.3959	67
68	22.9827	20.7221	18.8765	17.3412	16.0440	15.4680	14.9335	14.4360	68
69	23.0806	20.8018	18.9426	17.3970	16.0917	15.5123	14.9748	14.4746	69
70	23.1751	20.8785	19.0062	17.4506	16.1376	15.5550	15.0145	14.5117	70
71	23.2662	20.9524	19.0674	17.5022	16.1817	15.5959	15.0527	14.5474	71
72	23.3540	21.0235	19.1263	17.5518	16.2241	15.6353	15.0894	14.5817	72
73	23.4386	21.0921	19.1830	17.5996	16.2649	15.6732	15.1247	14.6146	73
74	23.5202	21.1581	19.2376	17.6455	16.3041	15.7097	15.1586	14.6463	74
75	23.5988	21.2217	19.2902	17.6897	16.3419	15.7447	15.1912	14.6767	75
76	23.6745	21.2829	19.3408	17.7322	16.3781	15.7784	15.2225	14.7060	76
77	23.7475	21.3419	19.3895	17.7732	16.4130	15.8107	15.2527	14.7341	77
78	23.8178	21.3987	19.4363	17.8125	16.4466	15.8419	15.2817	14.7612	78
79	23.8856	21.4533	19.4814	17.8504	16.4789	15.8718	15.3095	14.7872	79
80	23.9508	21.5060	19.5248	17.8868	16.5099	15.9006	15.3363	14.8121	80
81	24.0137	21.5566	19.5665	17.9218	16.5398	15.9283	15.3620	14.8362	81
82	24.0742	21.6054	19.6067	17.9555	16.5685	15.9549	15.3868	14.8593	82
83	24.1326	21.6524	19.6454	17.9880	16.5961	15.9805	15.4106	14.8814	83
84	24.1887	21.6976	19.6826	18.0191	16.6226	16.0051	15.4335	14.9028	84
85	24.2428	21.7411	19.7184	18.0491	16.6481	16.0288	15.4555	14.9233	85
86	24.2949	21.7830	19.7528	18.0780	16.6727	16.0515	15.4766	14.9430	86
87	24.3450	21.8233	19.7860	18.1057	16.6963	16.0734	15.4970	14.9620	87
88	24.3933	21.8621	19.8178	18.1324	16.7190	16.0944	15.5165	14.9802	88
89	24.4398	21.8994	19.8485	18.1581	16.7408	16.1147	15.5353	14.9977	89
90	24.4846	21.9353	19.8780	18.1828	16.7618	16.1341	15.5534	15.0146	90
91	24.5277	21.9699	19.9064	18.2066	16.7820	16.1528	15.5708	15.0308	91
92	24.5691	22.0032	19.9337	18.2294	16.8014	16.1708	15.5875	15.0463	92
93	24.6091	22.0352	19.9600	18.2514	16.8200	16.1881	15.6035	15.0613	93
94	24.6475	22.0660	19.9853	18.2725	16.8380	16.2047	15.6190	15.0757	94
95	24.6845	22.0957	20.0096	18.2928	16.8552	16.2207	15.6338	15.0895	95
96	24.7201	22.1242	20.0330	18.3124	16.8718	16.2361	15.6481	15.1028	96
97	24.7544	22.1516	20.0555	18.3312	16.8878	16.2508	15.6618	15.1156	97
98	24.7874	22.1780	20.0772	18.3493	16.9031	16.2650	15.6750	15.1279	98
99	24.8191	22.2035	20.0980	18.3667	16.9179	16.2787	15.6877	15.1397	99
100	24.8497	22.2279	20.1180	18.3834	16.9321	16.2918	15.6999	15.1511	100

*Note:—In computing the quarterly in advance figures in these tables, all the rates % quoted above are effective rates, see the Introductory Section, pages xvi-xvii.

YEARS' PURCHASE*

				Rate per Cent*					
Yrs.	7	7.25	7.5	8	8.5	9	9.5	10	Yrs.
1	0.6377	0.6367	0.6358	0.6340	0.6322	0.6304	0.6286	0.6268	1
2	1.2454	1.2419	1.2384	1.2314	1.2246	1.2179	1.2113	1.2048	2
3	1.8249	1.8173	1.8098	1.7950	1.7805	1.7664	1.7525	1.7389	3
4	2.3776	2.3647	2.3520	2.3271	2.3028	2.2792	2.2562	2.2337	4
5	2.9050	2.8857	2.8668	2.8299	2.7941	2.7594	2.7257	2.6930	5
6	3.4083	3.3819	3.3559	3.3054	3.2566	3.2096	3.1641	3.1202	6
7	3.8888	3.8545	3.8208	3.7554	3.6926	3.6322	3.5741	3.5182	7
8	4.3478	4.3048	4.2629	4.1817	4.1040	4.0295	3.9581	3.8896	8
9	4.7862	4.7342	4.6835	4.5857	4.4924	4.4033	4.3182	4.2368	9
10	5.2051	5.1437	5.0839	4.9688	4.8595	4.7554	4.6563	4.5618	10
11	5.6055	5.5343	5.4652	5.3324	5.2067	5.0874	4.9742	4.8664	11
12	5.9883	5.9072	5.8284	5.6777	5.5354	5.4008	5.2733	5.1523	12
13	6.3544	6.2631	6.1746	6.0057	5.8467	5.6967	5.5551	5.4210	13
14	6.7045	6.6029	6.5047	6.3175	6.1418	5.9766	5.8208	5.6738	14
15	7.0394	6.9276	6.8195	6.6141	6.4217	6.2413	6.0716	5.9119	15
16	7.3599	7.2377	7.1199	6.8962	6.6874	6.4919	6.3086	6.1363	16
17	7.6667	7.5342	7.4065	7.1648	6.9397	6.7294	6.5326	6.3481	17
18	7.9603	7.8176	7.6802	7.4207	7.1794	6.9546	6.7446	6.5481	18
19	8.2414	8.0885	7.9416	7.6644	7.4073	7.1682	6.9454	6.7371	19
20	8.5107	8.3477	8.1913	7.8967	7.6241	7.3710	7.1356	6.9159	20
21	8.7685	8.5956	8.4299	8.1182	7.8303	7.5637	7.3159	7.0852	21
22	9.0155	8.8328	8.6579	8.3294	8.0267	7.7467	7.4871	7.2456	22
23	9.2521	9.0598	8.8759	8.5310	8.2137	7.9208	7.6495	7.3977	23
24	9.4788	9.2771	9.0843	8.7234	8.3919	8.0864	7.8039	7.5419	24
25	9.6961	9.4851	9.2837	8.9071	8.5618	8.2440	7.9506	7.6788	25
26	9.9043	9.6843	9.4744	9.0825	8.7237	8.3940	8.0900	7.8088	26
27	10.1039	9.8751	9.6570	9.2501	8.8782	8.5370	8.2227	7.9324	27
28	10.2953	10.0578	9.8316	9.4102	9.0256	8.6732	8.3490	8.0499	28
29	10.4788	10.2328	9.9988	9.5633	9.1663	8.8030	8.4693	8.1616	29
30	10.6548	10.4006	10.1589	9.7096	9.3007	8.9269	8.5839	8.2680	30
31	10.8235	10.5613	10.3122	9.8496	9.4290	9.0450	8.6930	8.3692	31
32	10.9853	10.7153	10.4590	9.9834	9.5516	9.1578	8.7971	8.4657	32
33	11.1406	10.8630	10.5996	10.1115	9.6688	9.2654	8.8964	8.5576	33
34	11.2895	11.0045	10.7343	10.2340	9.7807	9.3682	8.9911	8.6451	34
35	11.4324	11.1402	10.8634	10.3513	9.8878	9.4664	9.0815	8.7287	35
36	11.5695	11.2704	10.9871	10.4635	9.9902	9.5602	9.1678	8.8084	36
37	11.7010	11.3952	11.1057	10.5710	10.0881	9.6498	9.2502	8.8844	37
38	11.8272	11.5148	11.2194	10.6739	10.1818	9.7355	9.3289	8.9570	38
39	11.9484	11.6296	11.3283	10.7725	10.2714	9.8174	9.4041	9.0263	39
40	12.0647	11.7398	11.4328	10.8669	10.3573	9.8958	9.4760	9.0925	40
41	12.1763	11.8454	11.5330	10.9574	10.4394	9.9708	9.5447	9.1558	41
42	12.2834	11.9468	11.6290	11.0441	10.5180	10.0425	9.6104	9.2162	42
43	12.3863	12.0441	11.7212	11.1271	10.5934	10.1111	9.6733	9.2740	43
44	12.4850	12.1374	11.8095	11.2068	10.6655	10.1768	9.7334	9.3292	44
45	12.5798	12.2270	11.8943	11.2831	10.7346	10.2397	9.7909	9.3820	45
46	12.6708	12.3129	11.9757	11.3562	10.8008	10.2999	9.8460	9.4326	46
47	12.7582	12.3954	12.0537	11.4264	10.8643	10.3576	9.8987	9.4809	47
48	12.8421	12.4747	12.1286	11.4937	10.9251	10.4129	9.9491	9.5272	48
49	12.9227	12.5507	12.2005	11.5582	10.9833	10.4658	9.9974	9.5715	49
50	13.0001	12.6237	12.2694	11.6201	11.0392	10.5165	10.0437	9.6139	50

*Note:—In computing the quarterly in advance figures in these tables, all the rates % quoted above are effective rates, see the Introductory Section, pages xvi-xvii.

YEARS' PURCHASE*

Rate per Cent*

Yrs.	7	7.25	7.5	8	8.5	9	9.5	10	Yrs.
51	13.0745	12.6938	12.3356	11.6794	11.0927	10.5651	10.0880	9.6545	51
52	13.1459	12.7611	12.3992	11.7364	11.1441	10.6117	10.1304	9.6933	52
53	13.2144	12.8257	12.4601	11.7910	11.1933	10.6563	10.1711	9.7306	53
54	13.2803	12.8877	12.5187	11.8434	11.2406	10.6991	10.2101	9.7663	54
55	13.3436	12.9473	12.5749	11.8937	11.2859	10.7 01	10.2474	9.8004	55
56	13.4044	13.0045	12.6289	11.9420	11.3293	10.7795	10.2833	9.8332	56
57	13.4628	13.0595	12.6807	11.9883	11.3710	10.8172	10.3176	9.8646	57
58	13.5189	13.1122	12.7304	12.0327	11.4110	10.8534	10.3505	9.8946	58
59	13.5727	13.1629	12.7782	12.0754	11.4494	10.8881	10.3821	9.9235	59
60	13.6245	13.2116	12.8241	12.1164	11.4862	10.9214	10.4123	9.9511	60
61	13.6743	13.2584	12.8681	12.1557	11.5215	10.9533	10.4414	9.9776	61
62	13.7221	13.3033	12.9105	12.1935	11.5554	10.9840	10.4692	10.0031	62
63	13.7680	13.3465	12.9511	12.2297	11.5880	11.0134	10.4959	10.0274	63
64	13.8121	13.3879	12.9901	12.2645	11.6192	11.0416	10.5215	10.0508	64
65	13.8545	13.4277	13.0276	12.2979	11.6492	11.0687	10.5461	10.0733	65
66	13.8952	13.4660	13.0637	12.3300	11.6780	11.0947	10.5697	10.0948	66
67	13.9344	13.5028	13.0983	12.3608	11.7056	11.1196	10.5923	10.1154	67
68	13.9720	13.5381	13.1315	12.3904	11.7322	11.1436	10.6141	10.1352	68
69	14.0082	13.5720	13.1634	12.4189	11.7577	11.1665	10.6349	10.1543	69
70	14.0429	13.6047	13.1941	12.4462	11.7821	11.1886	10.6549	10.1725	70
71	14.0763	13.6360	13.2236	12.4724	11.8056	11.2098	10.6741	10.1900	71
72	14.1084	13.6661	13.2519	12.4976	11.8282	11.2301	10.6926	10.2068	72
73	14.1392	13.6950	13.2791	12.5218	11.8498	11.2497	10.7103	10.2229	73
74	14.1689	13.7228	13.3052	12.5450	11.8707	11.2684	10.7273	10.2384	74
75	14.1974	13.7496	13.3303	12.5673	11.8906	11.2864	10.7436	10.2533	75
76	14.2247	13.7752	13.3545	12.5888	11.9098	11.3037	10.7593	10.2676	76
77	14.2511	13.7999	13.3777	12.6094	11.9283	11.3203	10.7743	10.2813	77
78	14.2763	13.8236	13.4000	12.6292	11.9460	11.3363	10.7888	10.2944	78
79	14.3007	13.8464	13.4214	12.6482	11.9630	11.3516	10.8027	10.3071	79
80	14.3240	13.8683	13.4420	12.6665	11.9794	11.3663	10.8160	10.3192	80
81	14.3465	13.8894	13.4617	12.6840	11.9951	11.3805	10.8288	10.3308	81
82	14.3681	13.9096	13.4807	12.7009	12.0102	11.3940	10.8411	10.3420	82
83	14.3888	13.9291	13.4990	12.7171	12.0247	11.4071	10.8529	10.3528	83
84	14.4088	13.9478	13.5166	12.7327	12.0386	11.4196	10.8642	10.3631	84
85	14.4279	13.9657	13.5334	12.7477	12.0520	11.4317	10.8751	10.3730	85
86	14.4464	13.9830	13.5496	12.7621	12.0648	11.4432	10.8856	10.3825	86
87	14.4641	13.9996	13.5652	12.7759	12.0772	11.4544	10.8957	10.3917	87
88	14.4811	14.0155	13.5802	12.7892	12.0891	11.4650	10.9053	10.4005	88
89	14.4975	14.0309	13.5946	12.8019	12.1005	11.4753	10.9146	10.4089	89
90	14.5132	14.0456	13.6084	12.8142	12.1114	11.4851	10.9235	10.4170	90
91	14.5284	14.0598	13.6217	12.8260	12.1220	11.4946	10.9321	10.4248	91
92	14.5429	14.0734	13.6345	12.8373	12.1321	11.5037	10.9403	10.4323	92
93	14.5569	14.0865	13.6468	12.8482	12.1418	11.5125	10.9482	10.4395	93
94	14.5703	14.0991	13.6586	12.8587	12.1512	11.5209	10.9558	10.4464	94
95	14.5833	14.1112	13.6700	12.8688	12.1601	11.5289	10.9631	10.4531	95
96	14.5957	14.1228	13.6809	12.8784	12.1688	11.5367	10.9702	10.4594	96
97	14.6076	14.1340	13.6914	12.8877	12.1771	11.5442	10.9769	10.4656	97
98	14.6191	14.1448	13.7015	12.8967	12.1851	11.5513	10.9834	10.4715	98
99	14.6301	14.1551	13.7112	12.9053	12.1927	11.5582	10.9896	10.4771	99
100	14.6407	14.1650	13.7205	12.9135	12.2001	11.5649	10.9956	10.4826	100

*Note:—In computing the quarterly in advance figures in these tables, all the rates % quoted above are effective rates,
see the Introductory Section, pages xvi-xvii.

Rate per Cent*

Yrs.	11	12	13	14	15	16	18	20	Yrs.
1	0.6234	0.6200	0.6167	0.6135	0.6103	0.6072	0.6012	0.5955	1
2	1.1921	1.1798	1.1679	1.1564	1.1452	1.1344	1.1136	1.0940	2
3	1.7126	1.6874	1.6632	1.6399	1.6175	1.5960	1.5552	1.5172	3
4	2.1905	2.1494	2.1103	2.0730	2.0373	2.0032	1.9394	1.8807	4
5	2.6305	2.5715	2.5156	2.4627	2.4126	2.3649	2.2764	2.1960	5
6	3.0365	2.9581	2.8845	2.8152	2.7498	2.6881	2.5743	2.4719	6
7	3.4122	3.3135	3.2214	3.1352	3.0543	2.9784	2.8394	2.7153	7
8	3.7605	3.6410	3.5300	3.4268	3.3304	3.2403	3.0765	2.9313	8
9	4.0840	3.9435	3.8137	3.6934	3.5817	3.4777	3.2897	3.1243	9
10	4.3852	4.2235	4.0750	3.9380	3.8113	3.6937	3.4823	3.2975	10
11	4.6660	4.4834	4.3164	4.1630	4.0216	3.8910	3.6571	3.4538	11
12	4.9282	4.7250	4.5398	4.3705	4.2149	4.0716	3.8162	3.5954	12
13	5.1735	4.9500	4.7472	4.5623	4.3931	4.2376	3.9617	3.7242	13
14	5.4032	5.1599	4.9399	4.7400	4.5576	4.3905	4.0950	3.8418	14
15	5.6187	5.3560	5.1194	4.9050	4.7100	4.5317	4.2176	3.9495	15
16	5.8210	5.5396	5.2868	5.0585	4.8513	4.6624	4.3305	4.0484	16
17	6.0112	5.7116	5.4432	5.2015	4.9827	4.7837	4.4349	4.1395	17
18	6.1903	5.8730	5.5896	5.3351	5.1051	4.8964	4.5316	4.2236	18
19	6.3589	6.0246	5.7268	5.4599	5.2193	5.0013	4.6214	4.3015	19
20	6.5180	6.1672	5.8555	5.5768	5.3260	5.0992	4.7048	4.3737	20
21	6.6682	6.3015	5.9764	5.6863	5.4258	5.1906	4.7826	4.4408	21
22	6.8101	6.4280	6.0901	5.7892	5.5194	5.2762	4.8551	4.5033	22
23	6.9442	6.5474	6.1972	5.8858	5.6072	5.3564	4.9229	4.5616	23
24	7.0712	6.6601	6.2981	5.9768	5.6897	5.4316	4.9864	4.6160	24
25	7.1914	6.7667	6.3933	6.0624	5.7672	5.5022	5.0459	4.6669	25
26	7.3053	6.8674	6.4832	6.1432	5.8403	5.5687	5.1017	4.7146	26
27	7.4133	6.9628	6.5681	6.2194	5.9091	5.6312	5.1542	4.7594	27
28	7.5158	7.0532	6.6484	6.2914	5.9741	5.6902	5.2035	4.8014	28
29	7.6132	7.1388	6.7245	6.3594	6.0354	5.7458	5.2500	4.8410	29
30	7.7056	7.2200	6.7965	6.4238	6.0933	5.7983	5.2938	4.8782	30
31	7.7935	7.2971	6.8648	6.4848	6.1482	5.8479	5.3351	4.9133	31
32	7.8770	7.3703	6.9295	6.5425	6.2000	5.8948	5.3741	4.9463	32
33	7.9565	7.4399	6.9910	6.5973	6.2492	5.9392	5.4110	4.9776	33
34	8.0322	7.5060	7.0493	6.6492	6.2958	5.9813	5.4459	5.0071	34
35	8.1043	7.5689	7.1048	6.6985	6.3400	6.0211	5.4789	5.0350	35
36	8.1729	7.6288	7.1575	6.7454	6.3819	6.0590	5.5102	5.0614	36
37	8.2384	7.6857	7.2076	6.7899	6.4217	6.0948	5.5399	5.0864	37
38	8.3007	7.7400	7.2553	6.8322	6.4596	6.1289	5.5680	5.1101	38
39	8.3602	7.7917	7.3007	6.8724	6.4955	6.1613	5.5947	5.1326	39
40	8.4170	7.8410	7.3440	6.9107	6.5297	6.1921	5.6201	5.1540	40
41	8.4712	7.8880	7.3852	6.9472	6.5623	6.2213	5.6442	5.1742	41
42	8.5229	7.9328	7.4244	6.9820	6.5933	6.2492	5.6671	5.1935	42
43	8.5723	7.9755	7.4619	7.0151	6.6228	6.2757	5.6889	5.2118	43
44	8.6194	8.0164	7.4976	7.0466	6.6509	6.3009	5.7096	5.2292	44
45	8.6645	8.0553	7.5317	7.0767	6.6777	6.3250	5.7294	5.2457	45
46	8.7076	8.0926	7.5642	7.1054	6.7033	6.3479	5.7482	5.2615	46
47	8.7488	8.1281	7.5953	7.1328	6.7277	6.3698	5.7661	5.2765	47
48	8.7882	8.1621	7.6250	7.1590	6.7509	6.3906	5.7832	5.2908	48
49	8.8258	8.1946	7.6533	7.1840	6.7731	6.4105	5.7994	5.3044	49
50	8.8619	8.2257	7.6804	7.2078	6.7944	6.4295	5.8150	5.3174	50

*Note:—In computing the quarterly in advance figures in these tables, all the rates % quoted above are effective rates, see the Introductory Section, pages xvi-xvii.

				Rate per Cent*					
Yrs.	11	12	13	14	15	16	18	20	Yrs.
51	8.8963	8.2553	7.7063	7.2306	6.8146	6.4476	5.8298	5.3298	51
52	8.9293	8.2838	7.7310	7.2524	6.8339	6.4650	5.8440	5.3416	52
53	8.9609	8.3109	7.7547	7.2732	6.8524	6.4815	5.8575	5.3529	53
54	8.9912	8.3369	7.7773	7.2931	6.8701	6.4973	5.8704	5.3637	54
55	9.0201	8.3618	7.7990	7.3122	6.8870	6.5124	5.8827	5.3740	55
56	9.0479	8.3857	7.8197	7.3304	6.9032	6.5269	5.8945	5.3838	56
57	9.0744	8.4085	7.8395	7.3478	6.9186	6.5407	5.9058	5.3932	57
58	9.0999	8.4303	7.8585	7.3645	6.9334	6.5539	5.9165	5.4022	58
59	9.1243	8.4512	7.8767	7.3805	6.9475	6.5665	5.9268	5.4108	59
60	9.1476	8.4713	7.8941	7.3957	6.9611	6.5786	5.9367	5.4190	60
61	9.1700	8.4905	7.9108	7.4104	6.9740	6.5902	5.9461	5.4268	61
62	9.1915	8.5089	7.9267	7.4244	6.9864	6.6013	5.9551	5.4344	62
63	9.2121	8.5265	7.9421	7.4378	6.9983	6.6119	5.9638	5.4415	63
64	9.2318	8.5434	7.9567	7.4507	7.0097	6.6220	5.9720	5.4484	64
65	9.2507	8.5596	7.9708	7.4630	7.0206	6.6318	5.9799	5.4550	65
66	9.2689	8.5752	7.9842	7.4748	7.0311	6.6411	5.9875	5.4613	66
67	9.2863	8.5901	7.9971	7.4861	7.0411	6.6500	5.9948	5.4673	67
68	9.3030	8.6043	8.0095	7.4970	7.0507	6.6586	6.0017	5.4731	68
69	9.3190	8.6180	8.0214	7.5074	7.0599	6.6668	6.0084	5.4787	69
70	9.3344	8.6312	8.0328	7.5173	7.0687	6.6746	6.0148	5.4840	70
71	9.3491	8.6438	8.0437	7.5269	7.0771	6.6822	6.0209	5.4891	71
72	9.3632	8.6559	8.0542	7.5360	7.0852	6.6894	6.0267	5.4939	72
73	9.3768	8.6675	8.0642	7.5448	7.0930	6.6963	6.0324	5.4986	73
74	9.3898	8.6786	8.0738	7.5533	7.1005	6.7030	6.0378	5.5031	74
75	9.4023	8.6893	8.0831	7.5614	7.1076	6.7093	6.0429	5.5074	75
76	9.4144	8.6995	8.0919	7.5691	7.1145	6.7154	6.0479	5.5115	76
77	9.4259	8.7094	8.1004	7.5766	7.1210	6.7213	6.0526	5.5154	77
78	9.4369	8.7188	8.1086	7.5837	7.1273	6.7269	6.0572	5.5192	78
79	9.4475	8.7279	8.1165	7.5906	7.1334	6.7323	6.0616	5.5228	79
80	9.4577	8.7366	8.1240	7.5971	7.1392	6.7375	6.0658	5.5263	80
81	9.4675	8.7449	8.1312	7.6034	7.1448	6.7425	6.0698	5.5297	81
82	9.4769	8.7529	8.1381	7.6095	7.1501	6.7472	6.0736	5.5329	82
83	9.4859	8.7606	8.1448	7.6153	7.1553	6.7518	6.0773	5.5359	83
84	9.4946	8.7680	8.1512	7.6209	7.1602	6.7562	6.0809	5.5389	84
85	9.5029	8.7751	8.1573	7.6263	7.1649	6.7604	6.0843	5.5417	85
86	9.5109	8.7819	8.1632	7.6314	7.1695	6.7644	6.0876	5.5444	86
87	9.5186	8.7885	8.1688	7.6364	7.1738	6.7683	6.0907	5.5471	87
88	9.5260	8.7948	8.1743	7.6411	7.1780	6.7720	6.0937	5.5496	88
89	9.5331	8.8008	8.1795	7.6457	7.1820	6.7756	6.0966	5.5520	89
90	9.5399	8.8066	8.1845	7.6500	7.1859	6.7791	6.0994	5.5543	90
91	9.5464	8.8122	8.1893	7.6542	7.1896	6.7824	6.1021	5.5565	91
92	9.5527	8.8175	8.1939	7.6583	7.1932	6.7855	6.1047	5.5586	92
93	9.5587	8.8226	8.1984	7.6621	7.1966	6.7886	6.1071	5.5606	93
94	9.5645	8.8276	8.2026	7.6659	7.1999	6.7915	6.1095	5.5626	94
95	9.5701	8.8323	8.2067	7.6694	7.2030	6.7943	6.1118	5.5645	95
96	9.5754	8.8369	8.2106	7.6729	7.2061	6.7970	6.1139	5.5663	96
97	9.5805	8.8413	8.2144	7.6762	7.2090	6.7996	6.1160	5.5680	97
98	9.5855	8.8455	8.2180	7.6793	7.2118	6.8021	6.1180	5.5697	98
99	9.5902	8.8495	8.2215	7.6824	7.2144	6.8045	6.1200	5.5713	99
100	9.5948	8.8534	8.2249	7.6853	7.2170	6.8067	6.1218	5.5728	100

Note:—In computing the quarterly in advance figures in these tables, all the rates % quoted above are effective rates, see the Introductory Section, pages xvi-xvii.

YEARS' PURCHASE
(DUAL RATE % PRINCIPLE)
OR
PRESENT VALUE OF
ONE POUND PER ANNUM

receivable quarterly in advance, allow-
ing for a sinking fund at a given rate
to replace the invested capital and for
the effect of income tax at **40%** *on that*
part of the income used to provide the
annual sinking fund instalment.

AT RATES OF INTEREST*

FROM

4% to 20%

AND

ALLOWING FOR THE POSSIBLE INVESTMENT

OF SINKING FUNDS AT

3% and 4%

INCOME TAX at 40%

Note:—In computing the quarterly in advance figures in these tables, all the rates
of interest quoted are effective rates, see the Introductory Section, pages
xvi-xvii.

Rate per Cent*

Yrs.	4	4.5	5	5.5	6	6.25	6.5	6.75	Yrs.
1	0.5970	0.5953	0.5936	0.5920	0.5903	0.5895	0.5887	0.5879	1
2	1.1834	1.1768	1.1703	1.1639	1.1577	1.1546	1.1515	1.1484	2
3	1.7595	1.7449	1.7306	1.7167	1.7031	1.6964	1.6897	1.6832	3
4	2.3250	2.2997	2.2750	2.2509	2.2276	2.2161	2.2048	2.1936	4
5	2.8802	2.8414	2.8038	2.7674	2.7321	2.7149	2.6980	2.6813	5
6	3.4251	3.3703	3.3175	3.2667	3.2177	3.1938	3.1704	3.1474	6
7	3.9596	3.8866	3.8166	3.7494	3.6850	3.6538	3.6231	3.5931	7
8	4.4839	4.3905	4.3014	4.2163	4.1350	4.0957	4.0573	4.0196	8
9	4.9981	4.8823	4.7723	4.6678	4.5684	4.5205	4.4737	4.4280	9
10	5.5021	5.3621	5.2298	5.1046	4.9859	4.9289	4.8733	4.8191	10
11	5.9962	5.8303	5.6742	5.5271	5.3882	5.3217	5.2569	5.1939	11
12	6.4804	6.2870	6.1059	5.9359	5.7760	5.6996	5.6254	5.5533	12
13	6.9547	6.7324	6.5252	6.3314	6.1499	6.0633	5.9794	5.8980	13
14	7.4193	7.1669	6.9325	6.7142	6.5103	6.4134	6.3196	6.2288	14
15	7.8742	7.5905	7.3281	7.0846	6.8580	6.7506	6.6468	6.5464	15
16	8.3196	8.0036	7.7123	7.4431	7.1935	7.0753	6.9614	6.8513	16
17	8.7556	8.4063	8.0856	7.7902	7.5171	7.3882	7.2640	7.1443	17
18	9.1823	8.7988	8.4481	8.1261	7.8295	7.6897	7.5553	7.4258	18
19	9.5998	9.1815	8.8002	8.4514	8.1310	7.9804	7.8357	7.6965	19
20	10.0082	9.5544	9.1422	8.7663	8.4221	8.2606	8.1057	7.9569	20
21	10.4077	9.9177	9.4744	9.0713	8.7032	8.5309	8.3657	8.2073	21
22	10.7983	10.2718	9.7970	9.3666	8.9747	8.7915	8.6162	8.4483	22
23	11.1802	10.6168	10.1104	9.6526	9.2369	9.0430	8.8577	8.6803	23
24	11.5535	10.9529	10.4147	9.9296	9.4903	9.2857	9.0904	8.9036	24
25	11.9184	11.2803	10.7102	10.1980	9.7351	9.5200	9.3148	9.1188	25
26	12.2749	11.5991	10.9973	10.4579	9.9717	9.7461	9.5311	9.3260	26
27	12.6233	11.9097	11.2761	10.7097	10.2003	9.9644	9.7398	9.5258	27
28	12.9636	12.2121	11.5468	10.9536	10.4214	10.1753	9.9412	9.7183	28
29	13.2959	12.5066	11.8098	11.1899	10.6351	10.3789	10.1355	9.9038	29
30	13.6205	12.7934	12.0651	11.4190	10.8417	10.5756	10.3230	10.0828	30
31	13.9374	13.0726	12.3131	11.6409	11.0416	10.7657	10.5040	10.2554	31
32	14.2468	13.3444	12.5540	11.8559	11.2348	10.9493	10.6788	10.4220	32
33	14.5488	13.6090	12.7879	12.0643	11.4218	11.1268	10.8475	10.5826	33
34	14.8435	13.8665	13.0150	12.2662	11.6027	11.2984	11.0105	10.7377	34
35	15.1311	14.1172	13.2356	12.4620	11.7777	11.4643	11.1680	10.8874	35
36	15.4117	14.3612	13.4498	12.6517	11.9470	11.6246	11.3201	11.0320	36
37	15.6855	14.5986	13.6578	12.8356	12.1108	11.7797	11.4671	11.1715	37
38	15.9525	14.8296	13.8599	13.0139	12.2694	11.9297	11.6092	11.3063	38
39	16.2130	15.0544	14.0560	13.1867	12.4229	12.0747	11.7465	11.4365	39
40	16.4670	15.2732	14.2466	13.3542	12.5715	12.2151	11.8793	11.5623	40
41	16.7146	15.4860	14.4316	13.5166	12.7153	12.3508	12.0076	11.6839	41
42	16.9561	15.6931	14.6112	13.6741	12.8546	12.4822	12.1317	11.8014	42
43	17.1915	15.8945	14.7857	13.8268	12.9894	12.6093	12.2518	11.9149	43
44	17.4209	16.0904	14.9551	13.9748	13.1200	12.7323	12.3678	12.0247	44
45	17.6446	16.2810	15.1196	14.1184	13.2464	12.8513	12.4801	12.1308	45
46	17.8625	16.4664	15.2793	14.2576	13.3689	12.9665	12.5888	12.2334	46
47	18.0749	16.6467	15.4344	14.3925	13.4875	13.0781	12.6939	12.3327	47
48	18.2818	16.8220	15.5851	14.5234	13.6023	13.1860	12.7956	12.4287	48
49	18.4834	16.9926	15.7313	14.6504	13.7136	13.2906	12.8940	12.5215	49
50	18.6797	17.1584	15.8734	14.7735	13.8214	13.3918	12.9893	12.6113	50

Note:—In computing the quarterly in advance figures in these tables, all the rates % quoted above are effective rates, see the Introductory Section, pages xvi-xvii.

				Rate per Cent*					
Yrs.	4	4.5	5	5.5	6	6.25	6.5	6.75	Yrs.
51	18.8710	17.3197	16.0113	14.8929	13.9259	13.4899	13.0815	12.6982	51
52	19.0573	17.4765	16.1452	15.0087	14.0271	13.5848	13.1707	12.7823	52
53	19.2388	17.6290	16.2752	15.1210	14.1251	13.6767	13.2572	12.8637	53
54	19.4155	17.7772	16.4015	15.2299	14.2201	13.7658	13.3408	12.9424	54
55	19.5875	17.9213	16.5241	15.3356	14.3122	13.8521	13.4218	13.0187	55
56	19.7550	18.0615	16.6432	15.4381	14.4014	13.9356	13.5003	13.0924	56
57	19.9181	18.1977	16.7588	15.5375	14.4879	14.0166	13.5762	13.1639	57
58	20.0769	18.3301	16.8710	15.6339	14.5717	14.0950	13.6498	13.2330	58
59	20.2314	18.4589	16.9800	15.7275	14.6530	14.1710	13.7211	13.3000	59
60	20.3819	18.5840	17.0859	15.8182	14.7317	14.2447	13.7901	13.3648	60
61	20.5283	18.7056	17.1886	15.9063	14.8081	14.3160	13.8570	13.4276	61
62	20.6708	18.8239	17.2884	15.9917	14.8821	14.3852	13.9217	13.4885	62
63	20.8094	18.9388	17.3853	16.0746	14.9538	14.4522	13.9845	13.5474	63
64	20.9444	19.0505	17.4794	16.1549	15.0233	14.5172	14.0453	13.6044	64
65	21.0756	19.1590	17.5707	16.2329	15.0908	14.5801	14.1042	13.6597	65
66	21.2034	19.2646	17.6594	16.3086	15.1562	14.6411	14.1613	13.7132	66
67	21.3277	19.3671	17.7455	16.3820	15.2195	14.7003	14.2166	13.7651	67
68	21.4486	19.4667	17.8291	16.4533	15.2810	14.7576	14.2703	13.8154	68
69	21.5662	19.5636	17.9103	16.5224	15.3406	14.8132	14.3222	13.8641	69
70	21.6806	19.6577	17.9892	16.5895	15.3984	14.8671	14.3726	13.9113	70
71	21.7919	19.7491	18.0657	16.6545	15.4545	14.9193	14.4214	13.9570	71
72	21.9001	19.8380	18.1400	16.7177	15.5088	14.9700	14.4687	14.0013	72
73	22.0054	19.9243	18.2122	16.7790	15.5615	15.0191	14.5146	14.0443	73
74	22.1077	20.0082	18.2823	16.8384	15.6127	15.0667	14.5591	14.0859	74
75	22.2073	20.0897	18.3503	16.8961	15.6623	15.1129	14.6022	14.1263	75
76	22.3041	20.1689	18.4164	16.9521	15.7104	15.1577	14.6440	14.1654	76
77	22.3983	20.2459	18.4805	17.0064	15.7570	15.2011	14.6845	14.2033	77
78	22.4898	20.3206	18.5428	17.0592	15.8023	15.2432	14.7238	14.2400	78
79	22.5788	20.3933	18.6032	17.1103	15.8462	15.2840	14.7619	14.2757	79
80	22.6654	20.4638	18.6620	17.1600	15.8887	15.3236	14.7989	14.3102	80
81	22.7495	20.5324	18.7190	17.2082	15.9300	15.3620	14.8347	14.3437	81
82	22.8313	20.5990	18.7743	17.2549	15.9701	15.3993	14.8694	14.3762	82
83	22.9108	20.6637	18.8280	17.3003	16.0090	15.4354	14.9031	14.4077	83
84	22.9881	20.7266	18.8802	17.3443	16.0467	15.4705	14.9358	14.4382	84
85	23.0633	20.7876	18.9309	17.3871	16.0832	15.5045	14.9675	14.4678	85
86	23.1363	20.8469	18.9801	17.4286	16.1187	15.5375	14.9982	14.4965	86
87	23.2073	20.9046	19.0278	17.4688	16.1532	15.5694	15.0280	14.5244	87
88	23.2763	20.9606	19.0742	17.5079	16.1866	15.6005	15.0569	14.5514	88
89	23.3434	21.0149	19.1192	17.5458	16.2190	15.6306	15.0849	14.5775	89
90	23.4086	21.0677	19.1629	17.5826	16.2504	15.6598	15.1121	14.6029	90
91	23.4719	21.1191	19.2053	17.6183	16.2809	15.6881	15.1385	14.6276	91
92	23.5335	21.1689	19.2466	17.6530	16.3105	15.7156	15.1641	14.6515	92
93	23.5934	21.2173	19.2866	17.6867	16.3393	15.7423	15.1889	14.6746	93
94	23.6515	21.2643	19.3254	17.7193	16.3671	15.7681	15.2130	14.6971	94
95	23.7080	21.3100	19.3631	17.7510	16.3942	15.7932	15.2364	14.7189	95
96	23.7629	21.3543	19.3997	17.7818	16.4204	15.8176	15.2590	14.7401	96
97	23.8163	21.3974	19.4353	17.8117	16.4459	15.8412	15.2810	14.7606	97
98	23.8682	21.4393	19.4698	17.8406	16.4706	15.8641	15.3024	14.7805	98
99	23.9185	21.4799	19.5033	17.8688	16.4946	15.8864	15.3230	14.7998	99
100	23.9675	21.5194	19.5359	17.8961	16.5178	15.9080	15.3431	14.8185	100

*Note:—In computing the quarterly in advance figures in these tables, all the rates % quoted above are effective rates, see the Introductory Section, pages xvi-xvii.

| Income Tax 40% | | | | | | | | | Quarterly |
| Sinking Fund 3% | | | **YEARS' PURCHASE*** | | | | | | in Advance |

Rate per Cent*

Yrs.	7	7.25	7.5	8	8.5	9	9.5	10	Yrs.
1	0.5871	0.5863	0.5856	0.5840	0.5825	0.5809	0.5794	0.5779	1
2	1.1454	1.1424	1.1394	1.1335	1.1278	1.1221	1.1165	1.1109	2
3	1.6767	1.6703	1.6639	1.6514	1.6391	1.6271	1.6153	1.6038	3
4	2.1826	2.1718	2.1610	2.1400	2.1194	2.0994	2.0798	2.0608	4
5	2.6648	2.6486	2.6327	2.6015	2.5712	2.5418	2.5132	2.4854	5
6	3.1247	3.1025	3.0806	3.0380	2.9968	2.9569	2.9183	2.8808	6
7	3.5636	3.5348	3.5064	3.4513	3.3982	3.3470	3.2976	3.2499	7
8	3.9828	3.9467	3.9114	3.8430	3.7772	3.7141	3.6533	3.5949	8
9	4.3833	4.3397	4.2970	4.2145	4.1356	4.0600	3.9876	3.9180	9
10	4.7663	4.7147	4.6644	4.5674	4.4748	4.3865	4.3020	4.2212	10
11	5.1326	5.0729	5.0147	4.9027	4.7962	4.6949	4.5982	4.5060	11
12	5.4833	5.4152	5.3489	5.2217	5.1011	4.9866	4.8777	4.7740	12
13	5.8191	5.7424	5.6680	5.5253	5.3905	5.2627	5.1416	5.0266	13
14	6.1408	6.0555	5.9728	5.8146	5.6654	5.5245	5.3912	5.2649	14
15	6.4492	6.3552	6.2642	6.0904	5.9269	5.7729	5.6275	5.4899	15
16	6.7450	6.6422	6.5428	6.3535	6.1758	6.0087	5.8513	5.7028	16
17	7.0288	6.9172	6.8095	6.6047	6.4129	6.2329	6.0637	5.9043	17
18	7.3011	7.1808	7.0648	6.8446	6.6388	6.4461	6.2653	6.0954	18
19	7.5626	7.4337	7.3094	7.0739	6.8543	6.6491	6.4569	6.2765	19
20	7.8138	7.6762	7.5438	7.2932	7.0600	6.8425	6.6392	6.4486	20
21	8.0552	7.9091	7.7685	7.5030	7.2565	7.0269	6.8126	6.6121	21
22	8.2872	8.1326	7.9841	7.7040	7.4443	7.2028	6.9778	6.7677	22
23	8.5103	8.3474	8.1910	7.8964	7.6238	7.3708	7.1354	6.9157	23
24	8.7249	8.5537	8.3896	8.0808	7.7956	7.5312	7.2856	7.0568	24
25	8.9314	8.7521	8.5803	8.2576	7.9600	7.6846	7.4290	7.1912	25
26	9.1301	8.9428	8.7636	8.4272	8.1174	7.8312	7.5660	7.3195	26
27	9.3215	9.1263	8.9397	8.5900	8.2683	7.9716	7.6969	7.4420	27
28	9.5057	9.3029	9.1091	8.7462	8.4130	8.1060	7.8221	7.5589	28
29	9.6832	9.4728	9.2719	8.8962	8.5517	8.2347	7.9419	7.6707	29
30	9.8542	9.6364	9.4286	9.0404	8.6848	8.3580	8.0566	7.7777	30
31	10.0190	9.7939	9.5794	9.1789	8.8126	8.4763	8.1664	7.8800	31
32	10.1779	9.9457	9.7245	9.3121	8.9353	8.5897	8.2717	7.9779	32
33	10.3311	10.0920	9.8643	9.4401	9.0531	8.6986	8.3726	8.0717	33
34	10.4789	10.2329	9.9989	9.5633	9.1664	8.8031	8.4693	8.1617	34
35	10.6214	10.3688	10.1286	9.6819	9.2753	8.9035	8.5622	8.2479	35
36	10.7589	10.4998	10.2535	9.7960	9.3800	8.9999	8.6513	8.3305	36
37	10.8916	10.6261	10.3740	9.9059	9.4807	9.0925	8.7369	8.4099	37
38	11.0197	10.7480	10.4901	10.0118	9.5776	9.1816	8.8192	8.4860	38
39	11.1434	10.8656	10.6021	10.1137	9.6708	9.2673	8.8982	8.5592	39
40	11.2627	10.9791	10.7101	10.2120	9.7606	9.3498	8.9741	8.6294	40
41	11.3781	11.0886	10.8144	10.3067	9.8471	9.4291	9.0472	8.6970	41
42	11.4894	11.1944	10.9149	10.3980	9.9304	9.5054	9.1175	8.7619	42
43	11.5970	11.2965	11.0120	10.4861	10.0107	9.5790	9.1851	8.8243	43
44	11.7010	11.3951	11.1057	10.5710	10.0881	9.6498	9.2502	8.8844	44
45	11.8015	11.4904	11.1961	10.6529	10.1627	9.7180	9.3129	8.9422	45
46	11.8986	11.5824	11.2835	10.7320	10.2346	9.7838	9.3732	8.9978	46
47	11.9924	11.6713	11.3679	10.8083	10.3040	9.8471	9.4314	9.0514	47
48	12.0832	11.7573	11.4494	10.8819	10.3709	9.9082	9.4874	9.1030	48
49	12.1709	11.8403	11.5281	10.9530	10.4354	9.9671	9.5414	9.1527	49
50	12.2557	11.9206	11.6042	11.0217	10.4977	10.0240	9.5935	9.2006	50

*Note:—In computing the quarterly in advance figures in these tables, all the rates % quoted above are effective rates, see the Introductory Section, pages xvi-xvii.

				Rate per Cent*					
Yrs.	7	7.25	7.5	8	8.5	9	9.5	10	Yrs.
51	12.3378	11.9982	11.6777	11.0880	10.5579	10.0788	9.6437	9.2468	51
52	12.4171	12.0733	11.7488	11.1521	10.6160	10.1317	9.6921	9.2913	52
53	12.4939	12.1458	11.8175	11.2139	10.6720	10.1828	9.7388	9.3342	53
54	12.5682	12.2160	11.8840	11.2737	10.7262	10.2320	9.7839	9.3756	54
55	12.6401	12.2839	11.9482	11.3315	10.7785	10.2796	9.8274	9.4155	55
56	12.7096	12.3496	12.0103	11.3874	10.8290	10.3256	9.8694	9.4541	56
57	12.7769	12.4131	12.0704	11.4414	10.8778	10.3700	9.9099	9.4913	57
58	12.8421	12.4746	12.1285	11.4936	10.9250	10.4128	9.9491	9.5272	58
59	12.9051	12.5341	12.1848	11.5441	10.9706	10.4542	9.9869	9.5618	59
60	12.9662	12.5916	12.2392	11.5929	11.0147	10.4943	10.0234	9.5953	60
61	13.0253	12.6474	12.2918	11.6401	11.0573	10.5329	10.0587	9.6276	61
62	13.0825	12.7013	12.3427	11.6858	11.0985	10.5703	10.0928	9.6588	62
63	13.1379	12.7535	12.3921	11.7300	11.1384	10.6065	10.1257	9.6890	63
64	13.1915	12.8041	12.4398	11.7728	11.1769	10.6414	10.1575	9.7182	64
65	13.2435	12.8530	12.4860	11.8141	11.2142	10.6752	10.1883	9.7463	65
66	13.2938	12.9004	12.5307	11.8542	11.2502	10.7079	10.2181	9.7736	66
67	13.3426	12.9463	12.5740	11.8929	11.2851	10.7395	10.2468	9.7999	67
68	13.3898	12.9908	12.6159	11.9304	11.3189	10.7700	10.2747	9.8253	68
69	13.4355	13.0338	12.6565	11.9667	11.3516	10.7996	10.3016	9.8499	69
70	13.4798	13.0755	12.6958	12.0018	11.3832	10.8282	10.3276	9.8737	70
71	13.5228	13.1159	12.7339	12.0359	11.4138	10.8559	10.3528	9.8968	71
72	13.5644	13.1551	12.7708	12.0688	11.4434	10.8827	10.3772	9.9190	72
73	13.6047	13.1930	12.8065	12.1007	11.4721	10.9086	10.4007	9.9406	73
74	13.6438	13.2297	12.8411	12.1316	11.4998	10.9337	10.4236	9.9614	74
75	13.6816	13.2653	12.8747	12.1615	11.5267	10.9580	10.4456	9.9816	75
76	13.7183	13.2998	12.9071	12.1905	11.5528	10.9816	10.4670	10.0011	76
77	13.7539	13.3332	12.9386	12.2186	11.5780	11.0043	10.4877	10.0200	77
78	13.7883	13.3656	12.9691	12.2458	11.6024	11.0264	10.5077	10.0382	78
79	13.8217	13.3970	12.9987	12.2721	11.6260	11.0477	10.5271	10.0559	79
80	13.8541	13.4274	13.0273	12.2976	11.6489	11.0684	10.5459	10.0731	80
81	13.8855	13.4569	13.0550	12.3223	11.6711	11.0884	10.5641	10.0896	81
82	13.9159	13.4854	13.0819	12.3463	11.6926	11.1078	10.5817	10.1057	82
83	13.9454	13.5131	13.1080	12.3695	11.7134	11.1266	10.5987	10.1212	83
84	13.9740	13.5400	13.1333	12.3920	11.7336	11.1448	10.6152	10.1363	84
85	14.0018	13.5660	13.1578	12.4138	11.7531	11.1625	10.6312	10.1509	85
86	14.0286	13.5913	13.1815	12.4350	11.7721	11.1795	10.6467	10.1650	86
87	14.0547	13.6157	13.2045	12.4554	11.7904	11.1961	10.6617	10.1787	87
88	14.0800	13.6395	13.2268	12.4753	11.8082	11.2121	10.6763	10.1919	88
89	14.1045	13.6625	13.2485	12.4945	11.8255	11.2277	10.6904	10.2048	89
90	14.1283	13.6848	13.2694	12.5132	11.8422	11.2427	10.7040	10.2172	90
91	14.1513	13.7064	13.2898	12.5313	11.8584	11.2573	10.7172	10.2293	91
92	14.1737	13.7274	13.3095	12.5488	11.8741	11.2715	10.7301	10.2410	92
93	14.1954	13.7477	13.3286	12.5658	11.8893	11.2852	10.7425	10.2523	93
94	14.2164	13.7674	13.3472	12.5823	11.9040	11.2985	10.7545	10.2632	94
95	14.2368	13.7866	13.3651	12.5982	11.9183	11.3114	10.7662	10.2739	95
96	14.2566	13.8051	13.3826	12.6137	11.9322	11.3238	10.7775	10.2842	96
97	14.2758	13.8231	13.3995	12.6288	11.9456	11.3359	10.7885	10.2941	97
98	14.2944	13.8406	13.4159	12.6433	11.9587	11.3477	10.7991	10.3038	98
99	14.3125	13.8575	13.4318	12.6574	11.9713	11.3591	10.8094	10.3132	99
100	14.3300	13.8739	13.4472	12.6711	11.9835	11.3701	10.8194	10.3223	100

*Note:—In computing the quarterly in advance figures in these tables, all the rates % quoted above are effective rates, see the Introductory Section, pages xvi-xvii.

Rate per Cent*

Yrs.	11	12	13	14	15	16	18	20	Yrs.
1	0.5750	0.5721	0.5693	0.5666	0.5639	0.5612	0.5561	0.5512	1
2	1.1001	1.0897	1.0795	1.0697	1.0601	1.0508	1.0330	1.0161	2
3	1.5814	1.5599	1.5392	1.5192	1.5000	1.4814	1.4462	1.4133	3
4	2.0240	1.9888	1.9553	1.9232	1.8924	1.8630	1.8076	1.7565	4
5	2.4320	2.3815	2.3335	2.2879	2.2446	2.2033	2.1263	2.0559	5
6	2.8094	2.7422	2.6788	2.6189	2.5622	2.5085	2.4092	2.3193	6
7	3.1592	3.0745	2.9950	2.9203	2.8501	2.7838	2.6620	2.5526	7
8	3.4843	3.3815	3.2856	3.1960	3.1120	3.0331	2.8891	2.7608	8
9	3.7870	3.6659	3.5534	3.4488	3.3512	3.2600	3.0942	2.9474	9
10	4.0695	3.9299	3.8010	3.6816	3.5706	3.4672	3.2802	3.1158	10
11	4.3336	4.1757	4.0304	3.8964	3.7723	3.6571	3.4497	3.2683	11
12	4.5810	4.4049	4.2435	4.0952	3.9583	3.8317	3.6046	3.4070	12
13	4.8130	4.6190	4.4419	4.2796	4.1304	3.9927	3.7468	3.5337	13
14	5.0311	4.8194	4.6270	4.4512	4.2899	4.1416	3.8776	3.6498	14
15	5.2362	5.0074	4.7999	4.6110	4.4382	4.2796	3.9983	3.7566	15
16	5.4295	5.1839	4.9618	4.7602	4.5763	4.4078	4.1101	3.8551	16
17	5.6119	5.3498	5.1137	4.8998	4.7052	4.5273	4.2137	3.9461	17
18	5.7842	5.5062	5.2564	5.0307	4.8257	4.6387	4.3101	4.0306	18
19	5.9471	5.6536	5.3906	5.1534	4.9386	4.7429	4.3999	4.1090	19
20	6.1013	5.7928	5.5170	5.2689	5.0445	4.8405	4.4838	4.1820	20
21	6.2475	5.9244	5.6362	5.3775	5.1440	4.9321	4.5622	4.2502	21
22	6.3862	6.0490	5.7489	5.4799	5.2376	5.0181	4.6357	4.3139	22
23	6.5178	6.1670	5.8553	5.5766	5.3259	5.0991	4.7047	4.3736	23
24	6.6430	6.2789	5.9561	5.6680	5.4091	5.1753	4.7696	4.4296	24
25	6.7620	6.3852	6.0517	5.7544	5.4878	5.2473	4.8306	4.4822	25
26	6.8753	6.4861	6.1422	5.8362	5.5622	5.3153	4.8882	4.5317	26
27	6.9832	6.5821	6.2282	5.9138	5.6326	5.3795	4.9425	4.5784	27
28	7.0861	6.6734	6.3100	5.9875	5.6994	5.4404	4.9938	4.6224	28
29	7.1843	6.7604	6.3877	6.0574	5.7627	5.4981	5.0424	4.6639	29
30	7.2780	6.8433	6.4617	6.1239	5.8228	5.5528	5.0884	4.7033	30
31	7.3675	6.9224	6.5321	6.1871	5.8800	5.6047	5.1320	4.7405	31
32	7.4531	6.9979	6.5993	6.2474	5.9343	5.6541	5.1733	4.7758	32
33	7.5349	7.0700	6.6634	6.3047	5.9861	5.7011	5.2126	4.8092	33
34	7.6132	7.1388	6.7245	6.3595	6.0354	5.7458	5.2500	4.8410	34
35	7.6881	7.2047	6.7829	6.4117	6.0824	5.7884	5.2855	4.8712	35
36	7.7599	7.2677	6.8387	6.4615	6.1273	5.8290	5.3193	4.8999	36
37	7.8287	7.3280	6.8921	6.5092	6.1701	5.8677	5.3516	4.9272	37
38	7.8947	7.3858	6.9432	6.5547	6.2110	5.9047	5.3823	4.9533	38
39	7.9580	7.4411	6.9921	6.5982	6.2501	5.9400	5.4116	4.9781	39
40	8.0187	7.4942	7.0389	6.6399	6.2874	5.9738	5.4396	5.0018	40
41	8.0769	7.5451	7.0837	6.6798	6.3232	6.0060	5.4664	5.0244	41
42	8.1329	7.5939	7.1268	6.7181	6.3575	6.0369	5.4920	5.0460	42
43	8.1867	7.6407	7.1680	6.7547	6.3903	6.0665	5.5164	5.0667	43
44	8.2383	7.6857	7.2076	6.7899	6.4217	6.0948	5.5399	5.0864	44
45	8.2880	7.7289	7.2456	6.8236	6.4519	6.1220	5.5623	5.1053	45
46	8.3358	7.7705	7.2821	6.8559	6.4808	6.1480	5.5838	5.1234	46
47	8.3818	7.8104	7.3171	6.8870	6.5085	6.1730	5.6043	5.1407	47
48	8.4260	7.8488	7.3508	6.9168	6.5351	6.1969	5.6241	5.1573	48
49	8.4685	7.8857	7.3832	6.9455	6.5607	6.2199	5.6430	5.1732	49
50	8.5095	7.9212	7.4143	6.9730	6.5853	6.2420	5.6612	5.1885	50

Note:—In computing the quarterly in advance figures in these tables, all the rates % quoted above are effective rates,
see the Introductory Section, pages xvi-xvii.

Rate per Cent*

Yrs.	11	12	13	14	15	16	18	20	Yrs.
51	8.5490	7.9554	7.4443	6.9995	6.6089	6.2632	5.6786	5.2032	51
52	8.5870	7.9883	7.4731	7.0250	6.6316	6.2836	5.6954	5.2172	52
53	8.6237	8.0200	7.5008	7.0495	6.6535	6.3032	5.7115	5.2307	53
54	8.6590	8.0506	7.5275	7.0731	6.6745	6.3221	5.7269	5.2437	54
55	8.6931	8.0800	7.5533	7.0958	6.6947	6.3402	5.7418	5.2562	55
56	8.7259	8.1084	7.5780	7.1176	6.7141	6.3576	5.7561	5.2682	56
57	8.7576	8.1357	7.6019	7.1387	6.7329	6.3744	5.7699	5.2797	57
58	8.7881	8.1621	7.6249	7.1590	6.7509	6.3906	5.7831	5.2908	58
59	8.8176	8.1875	7.6471	7.1785	6.7683	6.4062	5.7959	5.3014	59
60	8.8461	8.2120	7.6685	7.1974	6.7851	6.4212	5.8082	5.3117	60
61	8.8735	8.2357	7.6891	7.2155	6.8012	6.4357	5.8200	5.3216	61
62	8.9001	8.2585	7.7090	7.2331	6.8168	6.4496	5.8314	5.3311	62
63	8.9257	8.2806	7.7282	7.2500	6.8318	6.4630	5.8424	5.3403	63
64	8.9504	8.3019	7.7468	7.2663	6.8463	6.4760	5.8530	5.3492	64
65	8.9743	8.3224	7.7647	7.2820	6.8602	6.4885	5.8632	5.3577	65
66	8.9974	8.3423	7.7819	7.2972	6.8737	6.5005	5.8730	5.3659	66
67	9.0197	8.3614	7.7986	7.3119	6.8867	6.5122	5.8825	5.3738	67
68	9.0412	8.3800	7.8147	7.3260	6.8993	6.5234	5.8917	5.3815	68
69	9.0621	8.3978	7.8303	7.3397	6.9114	6.5342	5.9005	5.3888	69
70	9.0822	8.4151	7.8453	7.3529	6.9231	6.5447	5.9090	5.3960	70
71	9.1017	8.4318	7.8598	7.3657	6.9344	6.5548	5.9173	5.4028	71
72	9.1205	8.4480	7.8739	7.3780	6.9453	6.5646	5.9252	5.4095	72
73	9.1387	8.4636	7.8874	7.3899	6.9559	6.5740	5.9329	5.4158	73
74	9.1563	8.4787	7.9006	7.4014	6.9661	6.5831	5.9403	5.4220	74
75	9.1733	8.4933	7.9132	7.4125	6.9760	6.5919	5.9475	5.4280	75
76	9.1898	8.5075	7.9255	7.4233	6.9855	6.6004	5.9544	5.4338	76
77	9.2058	8.5211	7.9374	7.4337	6.9947	6.6086	5.9611	5.4393	77
78	9.2212	8.5343	7.9488	7.4437	7.0036	6.6166	5.9676	5.4447	78
79	9.2361	8.5471	7.9599	7.4535	7.0122	6.6243	5.9738	5.4499	79
80	9.2506	8.5595	7.9706	7.4629	7.0205	6.6317	5.9799	5.4549	80
81	9.2645	8.5715	7.9810	7.4720	7.0286	6.6389	5.9857	5.4598	81
82	9.2781	8.5830	7.9911	7.4808	7.0364	6.6458	5.9914	5.4645	82
83	9.2912	8.5943	8.0008	7.4893	7.0439	6.6525	5.9968	5.4690	83
84	9.3039	8.6051	8.0102	7.4975	7.0512	6.6590	6.0021	5.4734	84
85	9.3162	8.6156	8.0193	7.5055	7.0582	6.6653	6.0072	5.4777	85
86	9.3281	8.6258	8.0281	7.5132	7.0651	6.6714	6.0121	5.4818	86
87	9.3396	8.6356	8.0366	7.5207	7.0717	6.6773	6.0169	5.4858	87
88	9.3507	8.6452	8.0449	7.5279	7.0781	6.6830	6.0216	5.4896	88
89	9.3615	8.6544	8.0529	7.5349	7.0843	6.6885	6.0260	5.4933	89
90	9.3720	8.6634	8.0606	7.5417	7.0902	6.6939	6.0304	5.4969	90
91	9.3821	8.6720	8.0681	7.5483	7.0960	6.6990	6.0346	5.5004	91
92	9.3920	8.6804	8.0754	7.5546	7.1017	6.7040	6.0386	5.5038	92
93	9.4015	8.6885	8.0824	7.5608	7.1071	6.7089	6.0426	5.5071	93
94	9.4107	8.6964	8.0892	7.5668	7.1124	6.7136	6.0464	5.5102	94
95	9.4196	8.7040	8.0958	7.5725	7.1175	6.7181	6.0501	5.5133	95
96	9.4283	8.7114	8.1022	7.5781	7.1224	6.7225	6.0536	5.5163	96
97	9.4367	8.7186	8.1084	7.5835	7.1272	6.7268	6.0571	5.5191	97
98	9.4448	8.7255	8.1144	7.5888	7.1318	6.7309	6.0604	5.5219	98
99	9.4527	8.7323	8.1203	7.5939	7.1363	6.7349	6.0637	5.5246	99
100	9.4603	8.7388	8.1259	7.5988	7.1407	6.7388	6.0668	5.5272	100

*Note:—In computing the quarterly in advance figures in these tables, all the rates % quoted above are effective rates, see the Introductory Section, pages xvi-xvii.

Rate per Cent*

Yrs.	4	4.5	5	5.5	6	6.25	6.5	6.75	Yrs.
1	0.6005	0.5988	0.5971	0.5955	0.5938	0.5930	0.5922	0.5914	1
2	1.1959	1.1891	1.1825	1.1760	1.1696	1.1664	1.1633	1.1602	2
3	1.7858	1.7708	1.7561	1.7417	1.7277	1.7208	1.7140	1.7072	3
4	2.3697	2.3434	2.3178	2.2928	2.2686	2.2567	2.2450	2.2334	4
5	2.9475	2.9068	2.8675	2.8294	2.7926	2.7746	2.7569	2.7395	5
6	3.5187	3.4609	3.4052	3.3517	3.3001	3.2751	3.2504	3.2262	6
7	4.0829	4.0053	3.9310	3.8598	3.7916	3.7585	3.7261	3.6944	7
8	4.6400	4.5400	4.4448	4.3540	4.2674	4.2255	4.1846	4.1446	8
9	5.1896	5.0648	4.9466	4.8344	4.7279	4.6765	4.6265	4.5776	9
10	5.7314	5.5796	5.4365	5.3013	5.1734	5.1121	5.0523	4.9941	10
11	6.2652	6.0843	5.9145	5.7549	5.6045	5.5325	5.4626	5.3946	11
12	6.7909	6.5788	6.3807	6.1953	6.0214	5.9384	5.8579	5.7798	12
13	7.3080	7.0630	6.8353	6.6229	6.4245	6.3302	6.2388	6.1502	13
14	7.8166	7.5370	7.2782	7.0379	6.8143	6.7082	6.6057	6.5065	14
15	8.3164	8.0006	7.7096	7.4405	7.1911	7.0730	6.9591	6.8491	15
16	8.8073	8.4539	8.1296	7.8310	7.5552	7.4250	7.2996	7.1786	16
17	9.2891	8.8969	8.5384	8.2097	7.9070	7.7645	7.6275	7.4955	17
18	9.7618	9.3295	8.9361	8.5767	8.2469	8.0920	7.9433	7.8003	18
19	10.2252	9.7519	9.3229	8.9323	8.5752	8.4078	8.2474	8.0934	19
20	10.6792	10.1640	9.6989	9.2769	8.8922	8.7124	8.5402	8.3752	20
21	11.1238	10.5660	10.0642	9.6106	9.1984	9.0061	8.8223	8.6463	21
22	11.5590	10.9578	10.4191	9.9337	9.4940	9.2893	9.0938	8.9069	22
23	11.9848	11.3397	10.7638	10.2465	9.7793	9.5623	9.3552	9.1576	23
24	12.4010	11.7116	11.0984	10.5492	10.0547	9.8254	9.6070	9.3986	24
25	12.8078	12.0738	11.4230	10.8422	10.3204	10.0790	9.8493	9.6304	25
26	13.2051	12.4262	11.7380	11.1255	10.5769	10.3235	10.0826	9.8533	26
27	13.5930	12.7691	12.0435	11.3996	10.8243	10.5590	10.3072	10.0677	27
28	13.9715	13.1026	12.3397	11.6646	11.0629	10.7860	10.5233	10.2739	28
29	14.3407	13.4267	12.6268	11.9208	11.2931	11.0047	10.7314	10.4721	29
30	14.7006	13.7417	12.9050	12.1685	11.5152	11.2154	10.9317	10.6627	30
31	15.0513	14.0477	13.1745	12.4078	11.7293	11.4184	11.1245	10.8461	31
32	15.3929	14.3449	13.4355	12.6391	11.9357	11.6140	11.3100	11.0223	32
33	15.7256	14.6333	13.6882	12.8625	12.1347	11.8023	11.4885	11.1919	33
34	16.0493	14.9133	13.9329	13.0782	12.3266	11.9837	11.6604	11.3549	34
35	16.3643	15.1848	14.1696	13.2866	12.5116	12.1585	11.8257	11.5116	35
36	16.6706	15.4482	14.3987	13.4878	12.6898	12.3268	11.9849	11.6624	36
37	16.9684	15.7036	14.6203	13.6821	12.8616	12.4888	12.1380	11.8073	37
38	17.2577	15.9511	14.8346	13.8696	13.0272	12.6449	12.2854	11.9467	38
39	17.5388	16.1910	15.0419	14.0506	13.1867	12.7951	12.4272	12.0808	39
40	17.8118	16.4233	15.2422	14.2253	13.3405	12.9398	12.5636	12.2096	40
41	18.0768	16.6484	15.4359	14.3938	13.4886	13.0791	12.6949	12.3336	41
42	18.3340	16.8663	15.6230	14.5564	13.6312	13.2132	12.8212	12.4528	42
43	18.5835	17.0772	15.8038	14.7132	13.7687	13.3423	12.9427	12.5674	43
44	18.8255	17.2813	15.9785	14.8645	13.9011	13.4666	13.0596	12.6776	44
45	19.0601	17.4788	16.1472	15.0104	14.0286	13.5862	13.1721	12.7836	45
46	19.2875	17.6699	16.3101	15.1511	14.1514	13.7014	13.2803	12.8855	46
47	19.5079	17.8547	16.4674	15.2867	14.2697	13.8122	13.3844	12.9834	47
48	19.7214	18.0333	16.6193	15.4175	14.3836	13.9189	13.4846	13.0777	48
49	19.9282	18.2061	16.7659	15.5436	14.4932	14.0216	13.5809	13.1683	49
50	20.1284	18.3730	16.9074	15.6651	14.5988	14.1204	13.6736	13.2554	50

*Note:—In computing the quarterly in advance figures in these tables, all the rates % quoted above are effective rates, see the Introductory Section, pages xvi-xvii.

Rate per Cent*

Yrs.	4	4.5	5	5.5	6	6.25	6.5	6.75	Yrs.
51	20.3222	18.5344	17.0439	15.7823	14.7005	14.2155	13.7628	13.3392	51
52	20.5098	18.6903	17.1756	15.8952	14.7984	14.3070	13.8485	13.4197	52
53	20.6913	18.8409	17.3027	16.0040	14.8927	14.3951	13.9310	13.4972	53
54	20.8668	18.9863	17.4253	16.1088	14.9834	14.4799	14.0104	13.5717	54
55	21.0366	19.1268	17.5436	16.2098	15.0708	14.5614	14.0867	13.6433	55
56	21.2008	19.2624	17.6576	16.3071	15.1548	14.6399	14.1602	13.7122	56
57	21.3595	19.3934	17.7676	16.4008	15.2358	14.7154	14.2308	13.7784	57
58	21.5130	19.5198	17.8736	16.4911	15.3137	14.7881	14.2987	13.8421	58
59	21.6612	19.6417	17.9758	16.5781	15.3887	14.8580	14.3641	13.9033	59
60	21.8045	19.7595	18.0744	16.6619	15.4608	14.9252	14.4269	13.9622	60
61	21.9429	19.8730	18.1694	16.7426	15.5303	14.9900	14.4874	14.0188	61
62	22.0765	19.9826	18.2609	16.8203	15.5971	15.0522	14.5455	14.0732	62
63	22.2056	20.0883	18.3492	16.8951	15.6614	15.1121	14.6015	14.1256	63
64	22.3303	20.1903	18.4342	16.9672	15.7233	15.1697	14.6553	14.1759	64
65	22.4506	20.2886	18.5161	17.0366	15.7829	15.2252	14.7070	14.2243	65
66	22.5668	20.3834	18.5951	17.1034	15.8402	15.2785	14.7568	14.2708	66
67	22.6789	20.4748	18.6711	17.1677	15.8954	15.3298	14.8046	14.3156	67
68	22.7871	20.5630	18.7444	17.2296	15.9484	15.3792	14.8506	14.3586	68
69	22.8914	20.6479	18.8150	17.2892	15.9995	15.4266	14.8949	14.4000	69
70	22.9921	20.7298	18.8829	17.3466	16.0486	15.4723	14.9375	14.4398	70
71	23.0893	20.8088	18.9484	17.4019	16.0959	15.5162	14.9784	14.4780	71
72	23.1830	20.8849	19.0115	17.4550	16.1414	15.5585	15.0178	14.5148	72
73	23.2734	20.9582	19.0722	17.5062	16.1851	15.5992	15.0557	14.5502	73
74	23.3605	21.0288	19.1307	17.5555	16.2272	15.6383	15.0921	14.5842	74
75	23.4445	21.0969	19.1870	17.6029	16.2677	15.6759	15.1271	14.6169	75
76	23.5255	21.1624	19.2412	17.6485	16.3067	15.7120	15.1608	14.6484	76
77	23.6036	21.2256	19.2934	17.6924	16.3442	15.7468	15.1932	14.6786	77
78	23.6789	21.2865	19.3437	17.7347	16.3802	15.7803	15.2243	14.7077	78
79	23.7514	21.3451	19.3921	17.7754	16.4149	15.8125	15.2543	14.7356	79
80	23.8214	21.4015	19.4387	17.8145	16.4483	15.8434	15.2831	14.7625	80
81	23.8887	21.4559	19.4835	17.8521	16.4804	15.8732	15.3108	14.7884	81
82	23.9537	21.5082	19.5267	17.8884	16.5113	15.9019	15.3375	14.8132	82
83	24.0162	21.5587	19.5682	17.9232	16.5410	15.9294	15.3631	14.8371	83
84	24.0765	21.6072	19.6082	17.9568	16.5695	15.9559	15.3877	14.8601	84
85	24.1346	21.6540	19.6467	17.9891	16.5970	15.9814	15.4114	14.8822	85
86	24.1905	21.6990	19.6837	18.0201	16.6234	16.0059	15.4342	14.9035	86
87	24.2444	21.7423	19.7194	18.0500	16.6488	16.0294	15.4561	14.9239	87
88	24.2962	21.7840	19.7537	18.0787	16.6733	16.0521	15.4772	14.9435	88
89	24.3462	21.8242	19.7867	18.1064	16.6968	16.0739	15.4974	14.9624	89
90	24.3943	21.8629	19.8185	18.1330	16.7194	16.0949	15.5169	14.9806	90
91	24.4407	21.9001	19.8491	18.1586	16.7412	16.1150	15.5357	14.9980	91
92	24.4853	21.9359	19.8785	18.1832	16.7621	16.1344	15.5537	15.0148	92
93	24.5283	21.9704	19.9068	18.2069	16.7823	16.1531	15.5710	15.0310	93
94	24.5697	22.0036	19.9341	18.2297	16.8016	16.1710	15.5877	15.0465	94
95	24.6095	22.0355	19.9603	18.2516	16.8202	16.1883	15.6037	15.0614	95
96	24.6478	22.0663	19.9855	18.2727	16.8381	16.2048	15.6191	15.0758	96
97	24.6848	22.0958	20.0098	18.2930	16.8554	16.2208	15.6339	15.0896	97
98	24.7203	22.1243	20.0331	18.3125	16.8719	16.2361	15.6482	15.1029	98
99	24.7545	22.1517	20.0556	18.3312	16.8878	16.2509	15.6619	15.1156	99
100	24.7874	22.1781	20.0772	18.3493	16.9032	16.2651	15.6751	15.1279	100

*Note:—In computing the quarterly in advance figures in these tables, all the rates % quoted above are effective rates, see the Introductory Section, pages xvi-xvii.

Rate per Cent*

Yrs.	7	7.25	7.5	8	8.5	9	9.5	10	Yrs.
1	0.5906	0.5898	0.5890	0.5874	0.5858	0.5843	0.5828	0.5813	1
2	1.1571	1.1540	1.1510	1.1450	1.1391	1.1332	1.1275	1.1219	2
3	1.7006	1.6939	1.6874	1.6745	1.6619	1.6496	1.6375	1.6257	3
4	2.2220	2.2107	2.1996	2.1778	2.1565	2.1358	2.1156	2.0958	4
5	2.7223	2.7054	2.6888	2.6563	2.6247	2.5940	2.5643	2.5353	5
6	3.2025	3.1791	3.1562	3.1114	3.0682	3.0264	2.9859	2.9468	6
7	3.6632	3.6327	3.6028	3.5446	3.4886	3.4347	3.3826	3.3325	7
8	4.1054	4.0671	4.0297	3.9570	3.8874	3.8205	3.7563	3.6945	8
9	4.5299	4.4833	4.4378	4.3499	4.2659	4.1855	4.1085	4.0347	9
10	4.9373	4.8820	4.8281	4.7242	4.6253	4.5309	4.4409	4.3548	10
11	5.3285	5.2641	5.2015	5.0811	4.9668	4.8582	4.7548	4.6562	11
12	5.7039	5.6303	5.5587	5.4214	5.2915	5.1684	5.0515	4.9404	12
13	6.0644	5.9812	5.9005	5.7461	5.6004	5.4626	5.3322	5.2086	13
14	6.4105	6.3176	6.2277	6.0559	5.8942	5.7419	5.5980	5.4619	14
15	6.7429	6.6402	6.5408	6.3516	6.1740	6.0071	5.8498	5.7013	15
16	7.0620	6.9494	6.8407	6.6340	6.4405	6.2590	6.0884	5.9278	16
17	7.3685	7.2460	7.1279	6.9037	6.6945	6.4986	6.3149	6.1422	17
18	7.6628	7.5304	7.4029	7.1614	6.9365	6.7264	6.5298	6.3454	18
19	7.9454	7.8032	7.6664	7.4077	7.1673	6.9432	6.7339	6.5380	19
20	8.2169	8.0649	7.9188	7.6431	7.3875	7.1496	6.9279	6.7207	20
21	8.4776	8.3159	8.1607	7.8682	7.5975	7.3462	7.1123	6.8941	21
22	8.7281	8.5567	8.3925	8.0835	7.7981	7.5336	7.2878	7.0588	22
23	8.9686	8.7878	8.6147	8.2894	7.9895	7.7121	7.4547	7.2153	23
24	9.1997	9.0096	8.8277	8.4864	8.1724	7.8824	7.6137	7.3641	24
25	9.4217	9.2224	9.0318	8.6750	8.3471	8.0448	7.7651	7.5057	25
26	9.6349	9.4266	9.2276	8.8555	8.5140	8.1997	7.9094	7.6404	26
27	9.8398	9.6226	9.4154	9.0282	8.6736	8.3476	8.0469	7.7687	27
28	10.0366	9.8108	9.5954	9.1937	8.8262	8.4889	8.1781	7.8909	28
29	10.2257	9.9914	9.7681	9.3521	8.9721	8.6238	8.3032	8.0073	29
30	10.4074	10.1648	9.9338	9.5038	9.1117	8.7526	8.4226	8.1183	30
31	10.5820	10.3312	10.0927	9.6492	9.2452	8.8758	8.5366	8.2241	31
32	10.7498	10.4911	10.2452	9.7885	9.3730	8.9935	8.6454	8.3250	32
33	10.9109	10.6445	10.3915	9.9219	9.4953	9.1060	8.7494	8.4214	33
34	11.0658	10.7919	10.5319	10.0498	9.6124	9.2136	8.8487	8.5134	34
35	11.2146	10.9334	10.6666	10.1724	9.7245	9.3166	8.9436	8.6012	35
36	11.3576	11.0693	10.7959	10.2900	9.8318	9.4151	9.0343	8.6850	36
37	11.4951	11.1998	10.9200	10.4026	9.9346	9.5093	9.1210	8.7652	37
38	11.6271	11.3251	11.0391	10.5107	10.0331	9.5995	9.2040	8.8418	38
39	11.7541	11.4455	11.1535	10.6143	10.1275	9.6859	9.2833	8.9150	39
40	11.8761	11.5611	11.2633	10.7137	10.2179	9.7685	9.3593	8.9850	40
41	11.9933	11.6722	11.3687	10.8090	10.3046	9.8477	9.4319	9.0519	41
42	12.1059	11.7788	11.4698	10.9004	10.3877	9.9236	9.5015	9.1159	42
43	12.2142	11.8813	11.5670	10.9881	10.4673	9.9962	9.5680	9.1772	43
44	12.3183	11.9798	11.6603	11.0723	10.5436	10.0658	9.6318	9.2358	44
45	12.4183	12.0744	11.7499	11.1530	10.6168	10.1325	9.6928	9.2919	45
46	12.5145	12.1652	11.8359	11.2305	10.6870	10.1964	9.7513	9.3457	46
47	12.6069	12.2525	11.9185	11.3049	10.7543	10.2577	9.8073	9.3971	47
48	12.6957	12.3364	11.9979	11.3762	10.8189	10.3164	9.8610	9.4464	48
49	12.7811	12.4170	12.0741	11.4447	10.8808	10.3727	9.9124	9.4935	49
50	12.8631	12.4944	12.1473	11.5105	10.9402	10.4267	9.9617	9.5387	50

*Note:—In computing the quarterly in advance figures in these tables, all the rates % quoted above are effective rates, see the Introductory Section, pages xvi-xvii.

Rate per Cent*

Yrs.	7	7.25	7.5	8	8.5	9	9.5	10	Yrs.
51	12.9420	12.5689	12.2176	11.5736	10.9972	10.4784	10.0089	9.5820	51
52	13.0178	12.6403	12.2852	11.6342	11.0519	10.5281	10.0542	9.6235	52
53	13.0907	12.7090	12.3500	11.6924	11.1044	10.5757	10.0976	9.6633	53
54	13.1607	12.7751	12.4124	11.7482	11.1548	10.6214	10.1393	9.7014	54
55	13.2281	12.8385	12.4723	11.8018	11.2031	10.6652	10.1792	9.7380	55
56	13.2928	12.8995	12.5298	11.8533	11.2495	10.7072	10.2175	9.7730	56
57	13.3550	12.9581	12.5851	11.9028	11.2941	10.7476	10.2542	9.8066	57
58	13.4149	13.0144	12.6382	11.9503	11.3368	10.7863	10.2894	9.8388	58
59	13.4724	13.0685	12.6892	11.9959	11.3778	10.8234	10.3232	9.8697	59
60	13.5276	13.1205	12.7382	12.0397	11.4172	10.8590	10.3556	9.8994	60
61	13.5808	13.1705	12.7853	12.0818	11.4551	10.8933	10.3868	9.9278	61
62	13.6319	13.2185	12.8306	12.1222	11.4914	10.9261	10.4166	9.9550	62
63	13.6810	13.2647	12.8741	12.1610	11.5263	10.9576	10.4453	9.9812	63
64	13.7282	13.3091	12.9159	12.1983	11.5598	10.9879	10.4728	10.0063	64
65	13.7736	13.3517	12.9561	12.2341	11.5919	11.0170	10.4992	10.0304	65
66	13.8172	13.3927	12.9947	12.2685	11.6228	11.0449	10.5245	10.0535	66
67	13.8591	13.4321	13.0318	12.3016	11.6525	11.0716	10.5488	10.0757	67
68	13.8995	13.4700	13.0674	12.3334	11.6810	11.0974	10.5722	10.0970	68
69	13.9382	13.5064	13.1017	12.3639	11.7084	11.1221	10.5946	10.1175	69
70	13.9755	13.5414	13.1346	12.3932	11.7346	11.1458	10.6161	10.1371	70
71	14.0114	13.5750	13.1662	12.4214	11.7599	11.1686	10.6368	10.1559	71
72	14.0458	13.6074	13.1967	12.4484	11.7842	11.1904	10.6566	10.1740	72
73	14.0789	13.6385	13.2259	12.4744	11.8075	11.2115	10.6757	10.1914	73
74	14.1108	13.6683	13.2540	12.4994	11.8298	11.2316	10.6940	10.2081	74
75	14.1414	13.6971	13.2810	12.5234	11.8514	11.2510	10.7115	10.2241	75
76	14.1708	13.7247	13.3069	12.5465	11.8720	11.2696	10.7284	10.2394	76
77	14.1991	13.7512	13.3319	12.5687	11.8919	11.2875	10.7446	10.2542	77
78	14.2263	13.7767	13.3559	12.5900	11.9110	11.3047	10.7602	10.2684	78
79	14.2525	13.8012	13.3789	12.6105	11.9293	11.3212	10.7751	10.2820	79
80	14.2776	13.8248	13.4011	12.6302	11.9469	11.3371	10.7895	10.2951	80
81	14.3018	13.8475	13.4224	12.6491	11.9638	11.3523	10.8033	10.3077	81
82	14.3250	13.8693	13.4428	12.6673	11.9801	11.3670	10.8166	10.3197	82
83	14.3474	13.8902	13.4625	12.6847	11.9957	11.3810	10.8293	10.3313	83
84	14.3689	13.9104	13.4814	12.7015	12.0107	11.3946	10.8415	10.3425	84
85	14.3895	13.9297	13.4996	12.7177	12.0252	11.4075	10.8533	10.3532	85
86	14.4094	13.9483	13.5171	12.7332	12.0390	11.4200	10.8646	10.3634	86
87	14.4285	13.9662	13.5339	12.7481	12.0524	11.4320	10.8754	10.3733	87
88	14.4469	13.9834	13.5501	12.7624	12.0652	11.4435	10.8859	10.3828	88
89	14.4645	14.0000	13.5656	12.7762	12.0775	11.4546	10.8959	10.3919	89
90	14.4815	14.0159	13.5805	12.7894	12.0893	11.4653	10.9055	10.4007	90
91	14.4978	14.0312	13.5949	12.8022	12.1007	11.4755	10.9148	10.4091	91
92	14.5135	14.0459	13.6087	12.8144	12.1116	11.4853	10.9237	10.4172	92
93	14.5286	14.0600	13.6219	12.8262	12.1221	11.4948	10.9322	10.4249	93
94	14.5431	14.0736	13.6347	12.8375	12.1322	11.5038	10.9404	10.4324	94
95	14.5570	14.0866	13.6469	12.8483	12.1419	11.5126	10.9483	10.4396	95
96	14.5704	14.0992	13.6587	12.8588	12.1512	11.5209	10.9559	10.4465	96
97	14.5833	14.1113	13.6701	12.8688	12.1602	11.5290	10.9632	10.4531	97
98	14.5957	14.1229	13.6809	12.8785	12.1688	11.5367	10.9702	10.4595	98
99	14.6076	14.1340	13.6914	12.8878	12.1771	11.5442	10.9769	10.4656	99
100	14.6191	14.1448	13.7015	12.8967	12.1851	11.5513	10.9834	10.4715	100

**Note:*—In computing the quarterly in advance figures in these tables, all the rates % quoted above are effective rates,
see the Introductory Section, pages xvi-xvii.

Rate per Cent*

Yrs.	11	12	13	14	15	16	18	20	Yrs.
1	0.5783	0.5754	0.5725	0.5698	0.5670	0.5644	0.5592	0.5542	1
2	1.1109	1.1002	1.0899	1.0798	1.0701	1.0606	1.0424	1.0252	2
3	1.6027	1.5805	1.5593	1.5388	1.5191	1.5000	1.4639	1.4302	3
4	2.0578	2.0214	1.9868	1.9537	1.9220	1.8916	1.8345	1.7819	4
5	2.4798	2.4273	2.3775	2.3302	2.2852	2.2424	2.1627	2.0900	5
6	2.8721	2.8018	2.7357	2.6733	2.6142	2.5584	2.4551	2.3618	6
7	3.2372	3.1483	3.0650	2.9869	2.9134	2.8442	2.7172	2.6033	7
8	3.5778	3.4695	3.3686	3.2745	3.1864	3.1038	2.9531	2.8191	8
9	3.8960	3.7678	3.6492	3.5389	3.4363	3.3404	3.1665	3.0130	9
10	4.1936	4.0455	3.9090	3.7828	3.6657	3.5568	3.3604	3.1880	10
11	4.4724	4.3044	4.1502	4.0082	3.8770	3.7554	3.5371 ·	3.3466	11
12	4.7340	4.5462	4.3745	4.2170	4.0721	3.9381	3.6987	3.4909	12
13	4.9797	4.7723	4.5835	4.4109	4.2525	4.1067	3.8470	3.6227	13
14	5.2107	4.9840	4.7785	4.5912	4.4199	4.2625	3.9834	3.7435	14
15	5.4281	5.1826	4.9607	4.7592	4.5753	4.4069	4.1093	3.8544	15
16	5.6331	5.3691	5.1313	4.9160	4.7201	4.5411	4.2257	3.9566	16
17	5.8263	5.5444	5.2912	5.0625	4.8550	4.6658	4.3335	4.0510	17
18	6.0088	5.7094	5.4413	5.1998	4.9811	4.7821	4.4336	4.1384	18
19	6.1813	5.8649	5.5823	5.3284	5.0990	4.8907	4.5268	4.2194	19
20	6.3443	6.0114	5.7149	5.4491	5.2094	4.9922	4.6137	4.2948	20
21	6.4986	6.1498	5.8398	5.5626	5.3130	5.0873	4.6947	4.3650	21
22	6.6448	6.2805	5.9576	5.6693	5.4103	5.1764	4.7705	4.4304	22
23	6.7833	6.4042	6.0687	5.7698	5.5018	5.2601	4.8415	4.4916	23
24	6.9147	6.5211	6.1736	5.8646	5.5879	5.3387	4.9081	4.5488	24
25	7.0393	6.6319	6.2728	5.9540	5.6690	5.4128	4.9705	4.6024	25
26	7.1577	6.7368	6.3666	6.0385	5.7455	5.4825	5.0293	4.6527	26
27	7.2702	6.8364	6.4555	6.1183	5.8178	5.5482	5.0845	4.7000	27
28	7.3770	6.9308	6.5396	6.1938	5.8860	5.6102	5.1366	4.7444	28
29	7.4787	7.0204	6.6194	6.2653	5.9506	5.6688	5.1857	4.7863	29
30	7.5754	7.1056	6.6950	6.3331	6.0116	5.7242	5.2320	4.8257	30
31	7.6675	7.1866	6.7668	6.3973	6.0695	5.7767	5.2757	4.8629	31
32	7.7552	7.2635	6.8350	6.4582	6.1243	5.8263	5.3171	4.8980	32
33	7.8387	7.3368	6.8998	6.5161	6.1763	5.8733	5.3562	4.9312	33
34	7.9183	7.4065	6.9615	6.5710	6.2256	5.9179	5.3933	4.9626	34
35	7.9942	7.4728	7.0201	6.6232	6.2724	5.9602	5.4284	4.9923	35
36	8.0667	7.5361	7.0758	6.6728	6.3169	6.0003	5.4617	5.0204	36
37	8.1357	7.5963	7.1289	6.7200	6.3592	6.0385	5.4933	5.0471	37
38	8.2017	7.6538	7.1795	6.7649	6.3994	6.0747	5.5233	5.0724	38
39	8.2646	7.7086	7.2277	6.8077	6.4377	6.1092	5.5517	5.0964	39
40	8.3247	7.7609	7.2736	6.8484	6.4741	6.1420	5.5788	5.1192	40
41	8.3822	7.8107	7.3174	6.8873	6.5088	6.1732	5.6045	5.1409	41
42	8.4371	7.8584	7.3592	6.9243	6.5418	6.2029	5.6290	5.1615	42
43	8.4895	7.9039	7.3991	6.9595	6.5733	6.2312	5.6523	5.1811	43
44	8.5397	7.9473	7.4372	6.9932	6.6033	6.2582	5.6745	5.1997	44
45	8.5876	7.9888	7.4735	7.0253	6.6320	6.2839	5.6956	5.2174	45
46	8.6335	8.0285	7.5082	7.0560	6.6593	6.3084	5.7158	5.2343	46
47	8.6773	8.0664	7.5414	7.0853	6.6854	6.3318	5.7350	5.2504	47
48	8.7193	8.1027	7.5731	7.1133	6.7103	6.3542	5.7533	5.2658	48
49	8.7595	8.1374	7.6034	7.1400	6.7340	6.3755	5.7707	5.2804	49
50	8.7980	8.1706	7.6323	7.1655	6.7567	6.3958	5.7874	5.2943	50

*Note:—In computing the quarterly in advance figures in these tables, all the rates % quoted above are effective rates,
see the Introductory Section, pages xvi-xvii.

Rate per Cent*

Yrs.	11	12	13	14	15	16	18	20	Yrs.
51	8.8348	8.2023	7.6600	7.1899	6.7784	6.4153	5.8033	5.3077	51
52	8.8701	8.2327	7.6865	7.2133	6.7992	6.4338	5.8185	5.3204	52
53	8.9039	8.2618	7.7119	7.2356	6.8190	6.4516	5.8330	5.3325	53
54	8.9362	8.2897	7.7361	7.2569	6.8380	6.4686	5.8469	5.3441	54
55	8.9672	8.3163	7.7594	7.2774	6.8561	6.4848	5.8602	5.3552	55
56	8.9969	8.3419	7.7816	7.2969	6.8734	6.5003	5.8728	5.3657	56
57	9.0254	8.3663	7.8029	7.3156	6.8900	6.5151	5.8849	5.3758	57
58	9.0526	8.3898	7.8233	7.3335	6.9059	6.5293	5.8965	5.3855	58
59	9.0788	8.4122	7.8428	7.3507	6.9211	6.5429	5.9076	5.3948	59
60	9.1039	8.4337	7.8615	7.3671	6.9357	6.5559	5.9182	5.4036	60
61	9.1279	8.4544	7.8794	7.3828	6.9496	6.5684	5.9284	5.4121	61
62	9.1509	8.4741	7.8966	7.3979	6.9630	6.5803	5.9381	5.4202	62
63	9.1731	8.4931	7.9130	7.4123	6.9758	6.5918	5.9474	5.4279	63
64	9.1943	8.5113	7.9288	7.4262	6.9880	6.6027	5.9563	5.4353	64
65	9.2146	8.5287	7.9439	7.4394	6.9998	6.6132	5.9648	5.4424	65
66	9.2341	8.5454	7.9584	7.4522	7.0110	6.6232	5.9730	5.4492	66
67	9.2528	8.5614	7.9723	7.4643	7.0218	6.6328	5.9808	5.4557	67
68	9.2708	8.5768	7.9856	7.4760	7.0322	6.6421	5.9883	5.4620	68
69	9.2880	8.5915	7.9984	7.4872	7.0421	6.6509	5.9955	5.4679	69
70	9.3045	8.6057	8.0107	7.4980	7.0516	6.6594	6.0024	5.4737	70
71	9.3204	8.6193	8.0224	7.5083	7.0607	6.6675	6.0090	5.4792	71
72	9.3356	8.6323	8.0337	7.5181	7.0694	6.6753	6.0153	5.4844	72
73	9.3503	8.6448	8.0445	7.5276	7.0778	6.6828	6.0214	5.4895	73
74	9.3643	8.6568	8.0549	7.5367	7.0858	6.6899	6.0272	5.4943	74
75	9.3778	8.6683	8.0649	7.5454	7.0935	6.6968	6.0328	5.4989	75
76	9.3907	8.6793	8.0745	7.5538	7.1009	6.7034	6.0381	5.5034	76
77	9.4031	8.6899	8.0836	7.5618	7.1080	6.7097	6.0432	5.5076	77
78	9.4150	8.7001	8.0924	7.5696	7.1148	6.7158	6.0482	5.5117	78
79	9.4265	8.7099	8.1009	7.5770	7.1214	6.7216	6.0529	5.5156	79
80	9.4375	8.7193	8.1090	7.5841	7.1277	6.7272	6.0574	5.5194	80
81	9.4480	8.7283	8.1168	7.5909	7.1337	6.7326	6.0618	5.5230	81
82	9.4582	8.7369	8.1243	7.5974	7.1395	6.7377	6.0659	5.5265	82
83	9.4679	8.7453	8.1315	7.6037	7.1450	6.7426	6.0699	5.5298	83
84	9.4773	8.7532	8.1384	7.6097	7.1503	6.7474	6.0738	5.5330	84
85	9.4863	8.7609 ·	8.1450	7.6155	7.1554	6.7519	6.0775	5.5361	85
86	9.4949	8.7683	8.1514	7.6211	7.1603	6.7563	6.0810	5.5390	86
87	9.5032	8.7753	8.1575	7.6264	7.1651	6.7605	6.0844	5.5418	87
88	9.5111	8.7821	8.1633	7.6315	7.1696	6.7645	6.0877	5.5445	88
89	9.5188	8.7886	8.1690	7.6365	7.1739	6.7684	6.0908	5.5471	89
90	9.5261	8.7949	8.1744	7.6412	7.1781	6.7721	6.0938	5.5496	90
91	9.5332	8.8009	8.1796	7.6457	7.1821	6.7757	6.0967	5.5520	91
92	9.5400	8.8067	8.1846	7.6501	7.1860	6.7791	6.0995	5.5543	92
93	9.5465	8.8122	8.1894	7.6543	7.1897	6.7824	6.1021	5.5565	93
94	9.5527	8.8176	8.1940	7.6583	7.1932	6.7856	6.1047	5.5586	94
95	9.5588	8.8227	8.1984	7.6622	7.1966	6.7886	6.1072	5.5607	95
96	9.5645	8.8276	8.2026	7.6659	7.1999	6.7915	6.1095	5.5626	96
97	9.5701	8.8324	8.2067	7.6695	7.2030	6.7943	6.1118	5.5645	97
98	9.5754	8.8369	8.2107	7.6729	7.2061	6.7970	6.1140	5.5663	98
99	9.5806	8.8413	8.2144	7.6762	7.2090	6.7996	6.1160	5.5680	99
100	9.5855	8.8455	8.2181	7.6793	7.2118	6.8021	6.1180	5.5697	100

*Note:—In computing the quarterly in advance figures in these tables, all the rates % quoted above are effective rates, see the Introductory Section, pages xvi-xvii.

YEARS' PURCHASE

(DUAL RATE % PRINCIPLE)

OR

PRESENT VALUE OF
ONE POUND PER ANNUM

*receivable quarterly in advance, allow-
ing for a sinking fund at a given rate
to replace the invested capital and for
the effect of income tax at* **50%** *on that
part of the income used to provide the
annual sinking fund instalment.*

AT RATES OF INTEREST*

FROM

4% to 20%

AND

ALLOWING FOR THE POSSIBLE INVESTMENT

OF SINKING FUNDS AT

3% and 4%

INCOME TAX at 50%

**Note:*—In computing the quarterly in advance figures in these tables, all the rates
of interest quoted are effective rates, see the Introductory Section, pages
xvi-xvii.

	Rate per Cent*								
Yrs.	4	4.5	5	5.5	6	6.25	6.5	6.75	Yrs.
1	0.4994	0.4982	0.4971	0.4959	0.4948	0.4942	0.4936	0.4931	1
2	0.9939	0.9892	0.9846	0.9801	0.9756	0.9734	0.9712	0.9690	2
3	1.4832	1.4728	1.4627	1.4527	1.4429	1.4381	1.4333	1.4286	3
4	1.9673	1.9491	1.9313	1.9140	1.8971	1.8887	1.8805	1.8724	4
5	2.4460	2.4179	2.3907	2.3641	2.3384	2.3258	2.3133	2.3010	5
6	2.9192	2.8794	2.8408	2.8034	2.7672	2.7496	2.7322	2.7151	6
7	3.3869	3.3333	3.2817	3.2320	3.1840	3.1606	3.1377	3.1151	7
8	3.8489	3.7798	3.7136	3.6500	3.5889	3.5593	3.5302	3.5017	8
9	4.3050	4.2188	4.1365	4.0578	3.9824	3.9459	3.9102	3.8753	9
10	4.7553	4.6504	4.5505	4.4554	4.3647	4.3210	4.2782	4.2364	10
11	5.1997	5.0744	4.9558	4.8432	4.7362	4.6847	4.6345	4.5855	11
12	5.6380	5.4910	5.3524	5.2213	5.0972	5.0376	4.9795	4.9230	12
13	6.0702	5.9002	5.7404	5.5899	5.4479	5.3799	5.3137	5.2494	13
14	6.4962	6.3019	6.1200	5.9492	5.7886	5.7119	5.6374	5.5650	14
15	6.9161	6.6963	6.4912	6.2994	6.1197	6.0340	5.9509	5.8703	15
16	7.3297	7.0833	6.8542	6.6407	6.4413	6.3464	6.2545	6.1656	16
17	7.7370	7.4629	7.2091	6.9733	6.7537	6.6495	6.5487	6.4513	17
18	8.1380	7.8354	7.5560	7.2974	7.0573	6.9436	6.8337	6.7277	18
19	8.5327	8.2005	7.8951	7.6132	7.3522	7.2288	7.1099	6.9951	19
20	8.9210	8.5586	8.2264	7.9208	7.6387	7.5056	7.3775	7.2540	20
21	9.3029	8.9095	8.5500	8.2204	7.9170	7.7741	7.6367	7.5045	21
22	9.6784	9.2533	8.8662	8.5122	8.1873	8.0346	7.8880	7.7470	22
23	10.0476	9.5902	9.1750	8.7965	8.4499	8.2874	8.1314	7.9817	23
24	10.4103	9.9201	9.4766	9.0733	8.7050	8.5326	8.3674	8.2089	24
25	10.7667	10.2432	9.7710	9.3428	8.9528	8.7706	8.5961	8.4289	25
26	11.1168	10.5596	10.0584	9.6053	9.1936	9.0015	8.8178	8.6420	26
27	11.4605	10.8692	10.3390	9.8608	9.4274	9.2255	9.0327	8.8483	27
28	11.7978	11.1722	10.6128	10.1096	9.6545	9.4429	9.2410	9.0480	28
29	12.1290	11.4687	10.8800	10.3517	9.8751	9.6538	9.4429	9.2415	29
30	12.4538	11.7587	11.1406	10.5874	10.0894	9.8585	9.6386	9.4289	30
31	12.7725	12.0424	11.3949	10.8168	10.2975	10.0571	9.8284	9.6105	31
32	13.0849	12.3198	11.6430	11.0401	10.4997	10.2499	10.0124	9.7863	32
33	13.3913	12.5910	11.8849	11.2574	10.6960	10.4369	10.1908	9.9567	33
34	13.6916	12.8561	12.1209	11.4689	10.8867	10.6184	10.3637	10.1217	34
35	13.9858	13.1152	12.3509	11.6746	11.0719	10.7945	10.5315	10.2816	35
36	14.2741	13.3683	12.5752	11.8748	11.2518	10.9655	10.6941	10.4366	36
37	14.5565	13.6157	12.7938	12.0696	11.4265	11.1313	10.8518	10.5867	37
38	14.8330	13.8573	13.0069	12.2590	11.5962	11.2923	11.0047	10.7322	38
39	15.1037	14.0933	13.2146	12.4434	11.7610	11.4485	11.1530	10.8732	39
40	15.3687	14.3238	13.4170	12.6227	11.9211	11.6001	11.2969	11.0099	40
41	15.6280	14.5488	13.6143	12.7971	12.0765	11.7473	11.4364	11.1424	41
42	15.8818	14.7685	13.8064	12.9668	12.2275	11.8901	11.5717	11.2708	42
43	16.1300	14.9829	13.9936	13.1318	12.3741	12.0287	11.7029	11.3952	43
44	16.3728	15.1922	14.1760	13.2922	12.5165	12.1632	11.8302	11.5158	44
45	16.6102	15.3964	14.3537	13.4483	12.6548	12.2937	11.9537	11.6328	45
46	16.8424	15.5956	14.5267	13.6001	12.7891	12.4204	12.0734	11.7462	46
47	17.0693	15.7900	14.6952	13.7476	12.9195	12.5434	12.1896	11.8561	47
48	17.2911	15.9796	14.8593	13.8911	13.0462	12.6628	12.3023	11.9627	48
49	17.5078	16.1645	15.0190	14.0307	13.1692	12.7786	12.4116	12.0660	49
50	17.7195	16.3448	15.1746	14.1663	13.2886	12.8910	12.5176	12.1662	50

Note:—In computing the quarterly in advance figures in these tables, all the rates % quoted above are effective rates, see the Introductory Section, pages xvi-xvii.

Rate per Cent*

Yrs.	4	4.5	5	5.5	6	6.25	6.5	6.75	Yrs.
51	17.9264	16.5207	15.3260	14.2982	13.4046	13.0002	12.6205	12.2634	51
52	18.1284	16.6921	15.4735	14.4265	13.5172	13.1061	12.7203	12.3576	52
53	18.3257	16.8592	15.6170	14.5511	13.6266	13.2089	12.8171	12.4489	53
54	18.5183	17.0221	15.7566	14.6723	13.7329	13.3087	12.9110	12.5375	54
55	18.7064	17.1809	15.8926	14.7901	13.8360	13.4055	13.0022	12.6235	55
56	18.8900	17.3356	16.0249	14.9047	13.9362	13.4995	13.0906	12.7068	56
57	19.0691	17.4864	16.1536	15.0160	14.0335	13.5908	13.1764	12.7876	57
58	19.2440	17.6333	16.2789	15.1242	14.1279	13.6794	13.2596	12.8660	58
59	19.4145	17.7764	16.4008	15.2293	14.2196	13.7653	13.3404	12.9420	59
60	19.5810	17.9158	16.5194	15.3315	14.3087	13.8488	13.4187	13.0157	60
61	19.7433	18.0516	16.6348	15.4309	14.3952	13.9298	13.4948	13.0873	61
62	19.9016	18.1839	16.7471	15.5274	14.4792	14.0084	13.5686	13.1567	62
63	20.0560	18.3127	16.8563	15.6213	14.5607	14.0847	13.6402	13.2240	63
64	20.2066	18.4381	16.9625	15.7124	14.6399	14.1588	13.7096	13.2892	64
65	20.3534	18.5603	17.0658	15.8011	14.7168	14.2307	13.7770	13.3526	65
66	20.4965	18.6792	17.1663	15.8872	14.7915	14.3005	13.8425	13.4140	66
67	20.6359	18.7950	17.2640	15.9708	14.8640	14.3683	13.9059	13.4736	67
68	20.7719	18.9077	17.3591	16.0521	14.9344	14.4341	13.9675	13.5314	68
69	20.9044	19.0174	17.4515	16.1312	15.0028	14.4979	14.0273	13.5875	69
70	21.0335	19.1242	17.5414	16.2079	15.0691	14.5599	14.0853	13.6420	70
71	21.1593	19.2281	17.6288	16.2825	15.1336	14.6201	14.1416	13.6948	71
72	21.2818	19.3293	17.7138	16.3550	15.1962	14.6785	14.1963	13.7460	72
73	21.4012	19.4277	17.7964	16.4254	15.2570	14.7352	14.2493	13.7957	73
74	21.5175	19.5235	17.8768	16.4938	15.3160	14.7902	14.3007	13.8439	74
75	21.6308	19.6167	17.9549	16.5603	15.3733	14.8436	14.3507	13.8907	75
76	21.7411	19.7074	18.0308	16.6249	15.4289	14.8955	14.3992	13.9361	76
77	21.8485	19.7956	18.1046	16.6876	15.4829	14.9458	14.4462	13.9802	77
78	21.9531	19.8814	18.1764	16.7486	15.5354	14.9947	14.4919	14.0230	78
79	22.0549	19.9649	18.2461	16.8078	15.5863	15.0422	14.5362	14.0644	79
80	22.1541	20.0461	18.3140	16.8653	15.6358	15.0882	14.5792	14.1047	80
81	22.2506	20.1251	18.3799	16.9212	15.6838	15.1329	14.6209	14.1438	81
82	22.3446	20.2020	18.4439	16.9754	15.7304	15.1763	14.6614	14.1817	82
83	22.4360	20.2767	18.5062	17.0282	15.7757	15.2185	14.7007	14.2184	83
84	22.5250	20.3494	18.5667	17.0794	15.8196	15.2594	14.7389	14.2541	84
85	22.6116	20.4200	18.6255	17.1292	15.8623	15.2991	14.7759	14.2888	85
86	22.6959	20.4888	18.6827	17.1775	15.9038	15.3376	14.8119	14.3224	86
87	22.7780	20.5556	18.7382	17.2244	15.9440	15.3750	14.8468	14.3550	87
88	22.8578	20.6206	18.7922	17.2700	15.9831	15.4114	14.8807	14.3867	88
89	22.9355	20.6837	18.8447	17.3143	16.0210	15.4466	14.9135	14.4174	89
90	23.0110	20.7452	18.8957	17.3574	16.0578	15.4809	14.9454	14.4472	90
91	23.0845	20.8049	18.9452	17.3992	16.0936	15.5141	14.9764	14.4762	91
92	23.1560	20.8630	18.9933	17.4397	16.1283	15.5464	15.0065	14.5043	92
93	23.2256	20.9194	19.0401	17.4792	16.1620	15.5777	15.0357	14.5315	93
94	23.2932	20.9743	19.0855	17.5175	16.1947	15.6081	15.0640	14.5580	94
95	23.3590	21.0276	19.1297	17.5547	16.2265	15.6376	15.0915	14.5836	95
96	23.4230	21.0795	19.1726	17.5908	16.2574	15.6663	15.1182	14.6086	96
97	23.4853	21.1299	19.2143	17.6259	16.2873	15.6941	15.1441	14.6328	97
98	23.5458	21.1789	19.2548	17.6599	16.3164	15.7211	15.1692	14.6562	98
99	23.6047	21.2265	19.2941	17.6930	16.3447	15.7473	15.1936	14.6790	99
100	23.6619	21.2727	19.3324	17.7252	16.3721	15.7728	15.2173	14.7011	100

Note:—In computing the quarterly in advance figures in these tables, all the rates % quoted above are effective rates, see the Introductory Section, pages xvi-xvii.

Rate per Cent*

Yrs.	7	7.25	7.5	8	8.5	9	9.5	10	Yrs.
1	0.4925	0.4920	0.4914	0.4903	0.4892	0.4881	0.4871	0.4860	1
2	0.9669	0.9647	0.9626	0.9584	0.9543	0.9502	0.9462	0.9422	2
3	1.4239	1.4193	1.4147	1.4056	1.3967	1.3880	1.3794	1.3710	3
4	1.8644	1.8564	1.8486	1.8331	1.8180	1.8033	1.7888	1.7747	4
5	2.2889	2.2769	2.2652	2.2420	2.2195	2.1975	2.1761	2.1552	5
6	2.6982	2.6816	2.6653	2.6333	2.6023	2.5721	2.5429	2.5144	6
7	3.0930	3.0712	3.0497	3.0080	2.9675	2.9284	2.8905	2.8538	7
8	3.4737	3.4462	3.4193	3.3668	3.3163	3.2675	3.2204	3.1749	8
9	3.8410	3.8075	3.7746	3.7108	3.6495	3.5905	3.5337	3.4790	9
10	4.1955	4.1555	4.1164	4.0406	3.9680	3.8984	3.8315	3.7673	10
11	4.5376	4.4909	4.4452	4.3570	4.2727	4.1920	4.1148	4.0408	11
12	4.8678	4.8141	4.7617	4.6606	4.5643	4.4724	4.3846	4.3006	12
13	5.1867	5.1257	5.0663	4.9521	4.8434	4.7401	4.6416	4.5476	13
14	5.4946	5.4262	5.3597	5.2320	5.1109	4.9959	4.8867	4.7826	14
15	5.7920	5.7161	5.6423	5.5010	5.3673	5.2406	5.1205	5.0064	15
16	6.0793	5.9957	5.9146	5.7595	5.6131	5.4747	5.3437	5.2196	16
17	6.3569	6.2655	6.1770	6.0080	5.8489	5.6988	5.5570	5.4229	17
18	6.6251	6.5259	6.4300	6.2470	6.0752	5.9134	5.7609	5.6169	18
19	6.8843	6.7773	6.6739	6.4770	6.2924	6.1191	5.9559	5.8021	19
20	7.1349	7.0200	6.9091	6.6983	6.5011	6.3162	6.1425	5.9791	20
21	7.3771	7.2544	7.1360	6.9113	6.7016	6.5053	6.3212	6.1482	21
22	7.6113	7.4807	7.3549	7.1165	6.8943	6.6867	6.4924	6.3100	22
23	7.8378	7.6993	7.5661	7.3141	7.0796	6.8609	6.6565	6.4649	23
24	8.0568	7.9106	7.7700	7.5044	7.2578	7.0281	6.8138	6.6132	24
25	8.2686	8.1147	7.9668	7.6879	7.4292	7.1888	6.9646	6.7552	25
26	8.4735	8.3119	8.1569	7.8647	7.5942	7.3432	7.1095	6.8914	26
27	8.6717	8.5026	8.3404	8.0352	7.7531	7.4916	7.2485	7.0219	27
28	8.8635	8.6869	8.5177	8.1996	7.9060	7.6343	7.3820	7.1472	28
29	9.0491	8.8651	8.6889	8.3582	8.0533	7.7716	7.5103	7.2673	29
30	9.2287	9.0374	8.8544	8.5111	8.1953	7.9037	7.6336	7.3827	30
31	9.4026	9.2040	9.0143	8.6588	8.3321	8.0308	7.7521	7.4936	31
32	9.5708	9.3652	9.1688	8.8013	8.4639	8.1532	7.8661	7.6000	32
33	9.7337	9.5211	9.3182	8.9388	8.5911	8.2711	7.9758	7.7024	33
34	9.8914	9.6719	9.4626	9.0716	8.7137	8.3847	8.0814	7.8008	34
35	10.0440	9.8178	9.6022	9.1999	8.8319	8.4942	8.1830	7.8954	35
36	10.1919	9.9590	9.7372	9.3237	8.9460	8.5996	8.2809	7.9865	36
37	10.3350	10.0956	9.8678	9.4434	9.0561	8.7013	8.3751	8.0741	37
38	10.4736	10.2279	9.9941	9.5590	9.1624	8.7994	8.4659	8.1585	38
39	10.6079	10.3559	10.1163	9.6707	9.2649	8.8940	8.5534	8.2397	39
40	10.7379	10.4798	10.2345	9.7786	9.3640	8.9852	8.6377	8.3179	40
41	10.8639	10.5997	10.3488	9.8830	9.4596	9.0732	8.7191	8.3933	41
42	10.9859	10.7158	10.4595	9.9839	9.5520	9.1582	8.7975	8.4660	42
43	11.1041	10.8283	10.5666	10.0814	9.6413	9.2402	8.8731	8.5360	43
44	11.2186	10.9372	10.6702	10.1757	9.7275	9.3193	8.9461	8.6035	44
45	11.3296	11.0426	10.7706	10.2669	9.8108	9.3958	9.0165	8.6686	45
46	11.4371	11.1447	10.8677	10.3551	9.8913	9.4696	9.0845	8.7314	46
47	11.5413	11.2436	10.9617	10.4405	9.9692	9.5409	9.1501	8.7920	47
48	11.6423	11.3394	11.0528	10.5230	10.0444	9.6098	9.2135	8.8505	48
49	11.7401	11.4322	11.1409	10.6029	10.1171	9.6764	9.2746	8.9069	49
50	11.8350	11.5221	11.2263	10.6802	10.1875	9.7407	9.3337	8.9614	50

*Note:—In computing the quarterly in advance figures in these tables, all the rates % quoted above are effective rates, see the Introductory Section, pages xvi-xvii.

				Rate per Cent*					
Yrs.	7	7.25	7.5	8	8.5	9	9.5	10	Yrs.
51	11.9269	11.6092	11.3090	10.7550	10.2555	9.8029	9.3908	9.0140	51
52	12.0160	11.6936	11.3890	10.8274	10.3213	9.8630	9.4459	9.0648	52
53	12.1023	11.7754	11.4666	10.8975	10.3850	9.9211	9.4992	9.1139	53
54	12.1860	11.8547	11.5417	10.9653	10.4466	9.9773	9.5507	9.1613	54
55	12.2672	11.9314	11.6145	11.0310	10.5062	10.0316	9.6005	9.2071	55
56	12.3459	12.0059	11.6850	11.0945	10.5638	10.0842	9.6486	9.2513	56
57	12.4221	12.0780	11.7533	11.1561	10.6196	10.1350	9.6952	9.2941	57
58	12.4961	12.1479	11.8195	11.2157	10.6736	10.1842	9.7401	9.3354	58
59	12.5678	12.2156	11.8836	11.2734	10.7259	10.2318	9.7837	9.3754	59
60	12.6373	12.2813	11.9457	11.3293	10.7765	10.2778	9.8257	9.4140	60
61	12.7047	12.3450	12.0060	11.3835	10.8255	10.3224	9.8664	9.4514	61
62	12.7701	12.4067	12.0643	11.4360	10.8729	10.3655	9.9058	9.4875	62
63	12.8335	12.4665	12.1209	11.4868	10.9188	10.4072	9.9439	9.5225	63
64	12.8950	12.5245	12.1757	11.5360	10.9633	10.4476	9.9808	9.5563	64
65	12.9546	12.5808	12.2289	11.5837	11.0064	10.4867	10.0165	9.5890	65
66	13.0124	12.6353	12.2804	11.6299	11.0481	10.5246	10.0510	9.6206	66
67	13.0685	12.6882	12.3303	11.6747	11.0885	10.5612	10.0844	9.6512	67
68	13.1229	12.7394	12.3787	11.7181	11.1276	10.5967	10.1168	9.6809	68
69	13.1757	12.7891	12.4257	11.7601	11.1655	10.6311	10.1481	9.7095	69
70	13.2268	12.8373	12.4712	11.8009	11.2022	10.6644	10.1785	9.7373	70
71	13.2765	12.8841	12.5153	11.8404	11.2378	10.6966	10.2078	9.7642	71
72	13.3246	12.9294	12.5580	11.8786	11.2723	10.7278	10.2363	9.7902	72
73	13.3713	12.9734	12.5995	11.9157	11.3057	10.7581	10.2638	9.8154	73
74	13.4166	13.0160	12.6397	11.9517	11.3381	10.7874	10.2905	9.8398	74
75	13.4606	13.0574	12.6787	11.9866	11.3694	10.8158	10.3163	9.8634	75
76	13.5032	13.0975	12.7166	12.0204	11.3998	10.8433	10.3413	9.8863	76
77	13.5446	13.1364	12.7532	12.0531	11.4293	10.8700	10.3656	9.9084	77
78	13.5847	13.1742	12.7888	12.0849	11.4579	10.8958	10.3891	9.9299	78
79	13.6236	13.2108	12.8233	12.1157	11.4855	10.9208	10.4118	9.9507	79
80	13.6614	13.2463	12.8567	12.1455	11.5124	10.9451	10.4338	9.9708	80
81	13.6980	13.2807	12.8892	12.1745	11.5384	10.9686	10.4552	9.9903	81
82	13.7336	13.3141	12.9207	12.2026	11.5636	10.9914	10.4759	10.0092	82
83	13.7681	13.3466	12.9512	12.2298	11.5880	11.0134	10.4960	10.0275	83
84	13.8015	13.3780	12.9808	12.2562	11.6117	11.0348	10.5154	10.0452	84
85	13.8340	13.4085	13.0095	12.2818	11.6347	11.0556	10.5342	10.0624	85
86	13.8655	13.4381	13.0374	12.3066	11.6570	11.0757	10.5525	10.0791	86
87	13.8961	13.4668	13.0644	12.3307	11.6786	11.0952	10.5702	10.0952	87
88	13.9258	13.4947	13.0906	12.3541	11.6995	11.1141	10.5874	10.1109	88
89	13.9546	13.5217	13.1161	12.3767	11.7199	11.1324	10.6040	10.1260	89
90	13.9825	13.5479	13.1407	12.3987	11.7396	11.1502	10.6201	10.1408	90
91	14.0096	13.5734	13.1647	12.4200	11.7587	11.1674	10.6357	10.1550	91
92	14.0359	13.5981	13.1879	12.4407	11.7772	11.1841	10.6509	10.1688	92
93	14.0614	13.6220	13.2104	12.4607	11.7951	11.2004	10.6656	10.1822	93
94	14.0862	13.6453	13.2323	12.4801	11.8126	11.2161	10.6798	10.1952	94
95	14.1102	13.6678	13.2535	12.4990	11.8295	11.2313	10.6936	10.2078	95
96	14.1336	13.6897	13.2741	12.5173	11.8459	11.2461	10.7070	10.2200	96
97	14.1562	13.7110	13.2941	12.5351	11.8618	11.2604	10.7200	10.2318	97
98	14.1782	13.7316	13.3134	12.5523	11.8772	11.2743	10.7326	10.2433	98
99	14.1995	13.7516	13.3322	12.5690	11.8921	11.2878	10.7448	10.2544	99
100	14.2202	13.7710	13.3505	12.5852	11.9067	11.3008	10.7567	10.2652	100

Note:—In computing the quarterly in advance figures in these tables, all the rates % quoted above are effective rates, see the Introductory Section, pages xvi-xvii.

YEARS' PURCHASE*

Rate per Cent*

Yrs.	11	12	13	14	15	16	18	20	Yrs.
1	0.4839	0.4819	0.4799	0.4780	0.4760	0.4742	0.4705	0.4670	1
2	0.9344	0.9269	0.9195	0.9124	0.9054	0.8986	0.8855	0.8731	2
3	1.3546	1.3388	1.3235	1.3087	1.2944	1.2806	1.2542	1.2294	3
4	1.7473	1.7211	1.6959	1.6717	1.6484	1.6260	1.5837	1.5444	4
5	2.1150	2.0767	2.0401	2.0052	1.9718	1.9399	1.8799	1.8247	5
6	2.4598	2.4081	2.3591	2.3125	2.2682	2.2261	2.1475	2.0757	6
7	2.7837	2.7177	2.6554	2.5965	2.5408	2.4880	2.3903	2.3017	7
8	3.0883	3.0073	2.9312	2.8596	2.7922	2.7286	2.6115	2.5062	8
9	3.3753	3.2787	3.1885	3.1040	3.0247	2.9502	2.8138	2.6919	9
10	3.6460	3.5336	3.4290	3.3315	3.2403	3.1550	2.9994	2.8613	10
11	3.9017	3.7732	3.6542	3.5436	3.4407	3.3446	3.1703	3.0164	11
12	4.1434	3.9987	3.8653	3.7419	3.6273	3.5206	3.3280	3.1589	12
13	4.3721	4.2114	4.0637	3.9275	3.8014	3.6845	3.4741	3.2901	13
14	4.5889	4.4122	4.2503	4.1015	3.9642	3.8372	3.6095	3.4114	14
15	4.7945	4.6019	4.4261	4.2650	4.1168	3.9799	3.7356	3.5237	15
16	4.9897	4.7815	4.5920	4.4188	4.2598	4.1135	3.8530	3.6280	16
17	5.1752	4.9515	4.7486	4.5636	4.3943	4.2387	3.9626	3.7251	17
18	5.3516	5.1128	4.8967	4.7002	4.5208	4.3563	4.0652	3.8156	18
19	5.5194	5.2658	5.0368	4.8292	4.6400	4.4669	4.1614	3.9002	19
20	5.6793	5.4111	5.1697	4.9512	4.7525	4.5711	4.2516	3.9794	20
21	5.8318	5.5493	5.2957	5.0666	4.8588	4.6693	4.3365	4.0536	21
22	5.9771	5.6808	5.4153	5.1760	4.9593	4.7620	4.4164	4.1233	22
23	6.1159	5.8060	5.5289	5.2797	5.0544	4.8497	4.4917	4.1889	23
24	6.2484	5.9253	5.6370	5.3782	5.1446	4.9327	4.5627	4.2506	24
25	6.3751	6.0391	5.7399	5.4718	5.2302	5.0113	4.6299	4.3089	25
26	6.4962	6.1477	5.8379	5.5608	5.3114	5.0858	4.6935	4.3639	26
27	6.6121	6.2513	5.9313	5.6455	5.3886	5.1566	4.7537	4.4159	27
28	6.7230	6.3504	6.0204	5.7261	5.4621	5.2238	4.8107	4.4651	28
29	6.8293	6.4451	6.1055	5.8030	5.5320	5.2877	4.8649	4.5117	29
30	6.9311	6.5357	6.1867	5.8764	5.5986	5.3485	4.9163	4.5559	30
31	7.0287	6.6224	6.2643	5.9464	5.6621	5.4064	4.9652	4.5978	31
32	7.1222	6.7054	6.3386	6.0132	5.7227	5.4616	5.0117	4.6377	32
33	7.2121	6.7850	6.4096	6.0771	5.7805	5.5143	5.0560	4.6756	33
34	7.2983	6.8612	6.4776	6.1382	5.8358	5.5646	5.0983	4.7117	34
35	7.3810	6.9343	6.5427	6.1967	5.8886	5.6125	5.1385	4.7461	35
36	7.4606	7.0045	6.6051	6.2526	5.9391	5.6584	5.1769	4.7788	36
37	7.5370	7.0718	6.6650	6.3062	5.9874	5.7023	5.2136	4.8101	37
38	7.6104	7.1364	6.7223	6.3575	6.0337	5.7442	5.2487	4.8399	38
39	7.6811	7.1985	6.7774	6.4067	6.0780	5.7844	5.2822	4.8683	39
40	7.7490	7.2581	6.8302	6.4540	6.1204	5.8228	5.3142	4.8955	40
41	7.8144	7.3155	6.8810	6.4992	6.1612	5.8596	5.3449	4.9216	41
42	7.8773	7.3706	6.9297	6.5427	6.2002	5.8950	5.3742	4.9465	42
43	7.9379	7.4236	6.9766	6.5845	6.2377	5.9288	5.4024	4.9703	43
44	7.9963	7.4746	7.0216	6.6246	6.2737	5.9613	5.4293	4.9931	44
45	8.0525	7.5237	7.0649	6.6631	6.3082	5.9925	5.4552	5.0150	45
46	8.1067	7.5710	7.1066	6.7001	6.3414	6.0225	5.4800	5.0359	46
47	8.1589	7.6165	7.1467	6.7358	6.3733	6.0512	5.5038	5.0560	47
48	8.2092	7.6603	7.1853	6.7700	6.4040	6.0789	5.5267	5.0753	48
49	8.2577	7.7026	7.2224	6.8030	6.4335	6.1054	5.5486	5.0938	49
50	8.3045	7.7433	7.2582	6.8347	6.4619	6.1310	5.5697	5.1116	50

*Note:—In computing the quarterly in advance figures in these tables, all the rates % quoted above are effective rates, see the Introductory Section, pages xvi-xvii.

				Rate per Cent*					
Yrs.	11	12	13	14	15	16	18	20	Yrs.
51	8.3497	7.7825	7.2927	6.8653	6.4892	6.1556	5.5900	5.1286	51
52	8.3932	7.8204	7.3259	6.8947	6.5154	6.1792	5.6095	5.1450	52
53	8.4353	7.8569	7.3579	6.9231	6.5408	6.2020	5.6282	5.1608	53
54	8.4759	7.8920	7.3888	6.9504	6.5651	6.2239	5.6463	5.1760	54
55	8.5151	7.9260	7.4185	6.9767	6.5886	6.2450	5.6636	5.1906	55
56	8.5529	7.9588	7.4472	7.0021	6.6112	6.2653	5.6803	5.2046	56
57	8.5894	7.9904	7.4749	7.0266	6.6331	6.2849	5.6964	5.2181	57
58	8.6247	8.0209	7.5016	7.0502	6.6541	6.3038	5.7119	5.2311	58
59	8.6588	8.0504	7.5274	7.0729	6.6744	6.3220	5.7269	5.2436	59
60	8.6918	8.0789	7.5523	7.0949	6.6939	6.3395	5.7413	5.2557	60
61	8.7236	8.1064	7.5763	7.1161	6.7128	6.3564	5.7551	5.2673	61
62	8.7544	8.1330	7.5995	7.1366	6.7310	6.3727	5.7685	5.2785	62
63	8.7841	8.1586	7.6219	7.1563	6.7486	6.3885	5.7814	5.2893	63
64	8.8129	8.1834	7.6436	7.1754	6.7655	6.4037	5.7939	5.2997	64
65	8.8407	8.2074	7.6645	7.1938	6.7819	6.4184	5.8059	5.3098	65
66	8.8676	8.2306	7.6847	7.2116	6.7977	6.4325	5.8174	5.3195	66
67	8.8936	8.2530	7.7042	7.2288	6.8130	6.4462	5.8286	5.3288	67
68	8.9187	8.2746	7.7231	7.2454	6.8277	6.4594	5.8394	5.3378	68
69	8.9431	8.2956	7.7413	7.2615	6.8420	6.4722	5.8498	5.3465	69
70	8.9666	8.3158	7.7589	7.2770	6.8558	6.4845	5.8599	5.3550	70
71	8.9894	8.3354	7.7760	7.2920	6.8691	6.4964	5.8696	5.3631	71
72	9.0115	8.3544	7.7925	7.3065	6.8819	6.5079	5.8790	5.3709	72
73	9.0328	8.3727	7.8084	7.3205	6.8944	6.5190	5.8881	5.3785	73
74	9.0534	8.3905	7.8239	7.3341	6.9064	6.5298	5.8969	5.3858	74
75	9.0734	8.4076	7.8388	7.3472	6.9180	6.5402	5.9053	5.3929	75
76	9.0928	8.4242	7.8532	7.3599	6.9293	6.5502	5.9135	5.3997	76
77	9.1115	8.4403	7.8672	7.3721	6.9401	6.5599	5.9215	5.4063	77
78	9.1297	8.4559	7.8807	7.3840	6.9507	6.5693	5.9291	5.4127	78
79	9.1472	8.4709	7.8938	7.3955	6.9608	6.5784	5.9365	5.4188	79
80	9.1642	8.4855	7.9065	7.4066	6.9707	6.5872	5.9437	5.4248	80
81	9.1807	8.4997	7.9187	7.4174	6.9802	6.5957	5.9506	5.4306	81
82	9.1967	8.5133	7.9306	7.4278	6.9894	6.6039	5.9573	5.4362	82
83	9.2121	8.5266	7.9421	7.4378	6.9984	6.6119	5.9638	5.4416	83
84	9.2271	8.5394	7.9532	7.4476	7.0070	6.6196	5.9700	5.4468	84
85	9.2416	8.5518	7.9640	7.4570	7.0154	6.6271	5.9761	5.4518	85
86	9.2557	8.5638	7.9744	7.4662	7.0235	6.6343	5.9820	5.4567	86
87	9.2693	8.5755	7.9845	7.4750	7.0313	6.6413	5.9877	5.4614	87
88	9.2825	8.5868	7.9943	7.4836	7.0389	6.6481	5.9932	5.4660	88
89	9.2952	8.5977	8.0038	7.4919	7.0462	6.6546	5.9985	5.4704	89
90	9.3076	8.6083	8.0130	7.5000	7.0533	6.6610	6.0037	5.4747	90
91	9.3196	8.6186	8.0219	7.5078	7.0602	6.6671	6.0086	5.4789	91
92	9.3313	8.6285	8.0305	7.5153	7.0669	6.6731	6.0135	5.4829	92
93	9.3425	8.6382	8.0388	7.5226	7.0734	6.6788	6.0182	5.4868	93
94	9.3535	8.6475	8.0469	7.5297	7.0796	6.6844	6.0227	5.4906	94
95	9.3641	8.6566	8.0548	7.5366	7.0857	6.6898	6.0271	5.4942	95
96	9.3743	8.6653	8.0624	7.5432	7.0916	6.6950	6.0313	5.4977	96
97	9.3843	8.6738	8.0697	7.5497	7.0973	6.7001	6.0355	5.5012	97
98	9.3939	8.6821	8.0768	7.5559	7.1028	6.7050	6.0394	5.5045	98
99	9.4033	8.6901	8.0838	7.5620	7.1081	6.7098	6.0433	5.5077	99
100	9.4124	8.6978	8.0905	7.5678	7.1133	6.7144	6.0471	5.5108	100

Note:—In computing the quarterly in advance figures in these tables, all the rates % quoted above are effective rates, see the Introductory Section, pages xvi-xvii.

YEARS' PURCHASE*

Rate per Cent*

Yrs.	4	4.5	5	5.5	6	6.25	6.5	6.75	Yrs.
1	0.5024	0.5012	0.5000	0.4988	0.4977	0.4971	0.4965	0.4960	1
2	1.0044	0.9996	0.9949	0.9903	0.9858	0.9835	0.9813	0.9791	2
3	1.5056	1.4949	1.4845	1.4742	1.4641	1.4592	1.4543	1.4494	3
4	2.0057	1.9868	1.9683	1.9503	1.9328	1.9241	1.9156	1.9072	4
5	2.5043	2.4748	2.4463	2.4185	2.3916	2.3784	2.3653	2.3525	5
6	3.0009	2.9588	2.9180	2.8786	2.8405	2.8219	2.8036	2.7856	6
7	3.4953	3.4382	3.3833	3.3305	3.2796	3.2548	3.2305	3.2065	7
8	3.9870	3.9129	3.8420	3.7740	3.7087	3.6771	3.6461	3.6156	8
9	4.4757	4.3826	4.2938	4.2091	4.1280	4.0889	4.0505	4.0130	9
10	4.9611	4.8470	4.7386	4.6356	4.5375	4.4902	4.4440	4.3989	10
11	5.4428	5.3058	5.1762	5.0535	4.9372	4.8812	4.8267	4.7736	11
12	5.9206	5.7587	5.6064	5.4628	5.3271	5.2620	5.1987	5.1371	12
13	6.3940	6.2057	6.0291	5.8633	5.7073	5.6327	5.5602	5.4898	13
14	6.8628	6.6463	6.4442	6.2552	6.0779	5.9934	5.9114	5.8318	14
15	7.3267	7.0805	6.8516	6.6383	6.4390	6.3442	6.2524	6.1635	15
16	7.7855	7.5080	7.2511	7.0126	6.7906	6.6852	6.5834	6.4849	16
17	8.2388	7.9287	7.6428	7.3783	7.1329	7.0168	6.9046	6.7964	17
18	8.6864	8.3424	8.0265	7.7353	7.4660	7.3388	7.2163	7.0981	18
19	9.1281	8.7490	8.4022	8.0836	7.7900	7.6517	7.5185	7.3903	19
20	9.5637	9.1484	8.7699	8.4234	8.1051	7.9554	7.8116	7.6733	20
21	9.9929	9.5404	9.1295	8.7546	8.4113	8.2502	8.0956	7.9472	21
22	10.4157	9.9250	9.4810	9.0774	8.7088	8.5362	8.3709	8.2123	22
23	10.8317	10.3021	9.8245	9.3918	8.9978	8.8137	8.6375	8.4687	23
24	11.2409	10.6716	10.1600	9.6979	9.2783	9.0827	8.8958	8.7168	24
25	11.6432	11.0334	10.4875	9.9958	9.5507	9.3435	9.1458	8.9568	25
26	12.0383	11.3876	10.8070	10.2856	9.8149	9.5963	9.3878	9.1888	26
27	12.4262	11.7341	11.1185	10.5675	10.0712	9.8412	9.6221	9.4131	27
28	12.8068	12.0729	11.4223	10.8415	10.3198	10.0784	9.8487	9.6299	28
29	13.1800	12.4040	11.7182	11.1077	10.5608	10.3081	10.0680	9.8394	29
30	13.5458	12.7275	12.0065	11.3664	10.7943	10.5305	10.2800	10.0418	30
31	13.9041	13.0432	12.2871	11.6176	11.0206	10.7458	10.4850	10.2374	31
32	14.2548	13.3514	12.5602	11.8614	11.2398	10.9540	10.6832	10.4262	32
33	14.5979	13.6520	12.8258	12.0981	11.4521	11.1556	10.8748	10.6086	33
34	14.9335	13.9450	13.0841	12.3276	11.6576	11.3505	11.0600	10.7847	34
35	15.2615	14.2306	13.3352	12.5503	11.8565	11.5389	11.2388	10.9548	35
36	15.5819	14.5088	13.5792	12.7661	12.0490	11.7212	11.4116	11.1189	36
37	15.8947	14.7797	13.8162	12.9754	12.2352	11.8973	11.5786	11.2773	37
38	16.2001	15.0433	14.0463	13.1782	12.4153	12.0676	11.7397	11.4301	38
39	16.4979	15.2998	14.2697	13.3746	12.5895	12.2321	11.8954	11.5776	39
40	16.7883	15.5493	14.4865	13.5648	12.7579	12.3910	12.0456	11.7199	40
41	17.0714	15.7918	14.6968	13.7490	12.9207	12.5445	12.1906	11.8571	41
42	17.3472	16.0275	14.9007	13.9273	13.0781	12.6928	12.3306	11.9895	42
43	17.6157	16.2564	15.0984	14.0999	13.2301	12.8360	12.4657	12.1172	43
44	17.8771	16.4788	15.2900	14.2669	13.3770	12.9742	12.5960	12.2403	44
45	18.1314	16.6947	15.4757	14.4284	13.5189	13.1077	12.7218	12.3590	45
46	18.3788	16.9042	15.6555	14.5846	13.6560	13.2364	12.8431	12.4734	46
47	18.6193	17.1074	15.8297	14.7356	13.7883	13.3607	12.9600	12.5837	47
48	18.8531	17.3045	15.9983	14.8817	13.9161	13.4807	13.0729	12.6901	48
49	19.0802	17.4957	16.1616	15.0228	14.0394	13.5964	13.1817	12.7926	49
50	19.3007	17.6810	16.3195	15.1592	14.1585	13.7080	13.2866	12.8913	50

Note:—In computing the quarterly in advance figures in these tables, all the rates % quoted above are effective rates, see the Introductory Section, pages xvi-xvii.

Rate per Cent*

Yrs.	4	4.5	5	5.5	6	6.25	6.5	6.75	Yrs.
51	19.5149	17.8605	16.4724	15.2910	14.2734	13.8157	13.3877	12.9865	51
52	19.7227	18.0345	16.6202	15.4183	14.3843	13.9196	13.4852	13.0782	52
53	19.9244	18.2029	16.7632	15.5413	14.4912	14.0197	13.5792	13.1666	53
54	20.1200	18.3661	16.9015	15.6601	14.5944	14.1163	13.6697	13.2518	54
55	20.3097	18.5240	17.0351	15.7747	14.6940	14.2094	13.7570	13.3338	55
56	20.4936	18.6768	17.1643	15.8854	14.7900	14.2991	13.8411	13.4128	56
57	20.6718	18.8247	17.2891	15.9923	14.8826	14.3857	13.9222	13.4889	57
58	20.8444	18.9678	17.4097	16.0954	14.9718	14.4691	14.0003	13.5622	58
59	21.0116	19.1061	17.5262	16.1949	15.0579	14.5494	14.0755	13.6328	59
60	21.1735	19.2399	17.6387	16.2910	15.1409	14.6269	14.1480	13.7007	60
61	21.3303	19.3693	17.7474	16.3836	15.2209	14.7015	14.2178	13.7662	61
62	21.4820	19.4943	17.8523	16.4730	15.2980	14.7735	14.2851	13.8292	62
63	21.6288	19.6151	17.9535	16.5592	15.3723	14.8427	14.3498	13.8899	63
64	21.7709	19.7319	18.0513	16.6423	15.4439	14.9095	14.4122	13.9484	64
65	21.9083	19.8447	18.1457	16.7225	15.5129	14.9738	14.4723	14.0047	65
66	22.0411	19.9536	18.2367	16.7997	15.5794	15.0357	14.5302	14.0588	66
67	22.1696	20.0588	18.3245	16.8743	15.6435	15.0954	14.5859	14.1110	67
68	22.2937	20.1604	18.4093	16.9461	15.7052	15.1529	14.6395	14.1612	68
69	22.4137	20.2585	18.4910	17.0153	15.7647	15.2082	14.6912	14.2095	69
70	22.5297	20.3531	18.5699	17.0821	15.8219	15.2615	14.7409	14.2560	70
71	22.6417	20.4445	18.6459	17.1464	15.8771	15.3128	14.7888	14.3008	71
72	22.7499	20.5327	18.7192	17.2084	15.9302	15.3622	14.8348	14.3439	72
73	22.8544	20.6178	18.7899	17.2681	15.9814	15.4098	14.8792	14.3853	73
74	22.9553	20.6999	18.8581	17.3256	16.0307	15.4556	14.9219	14.4252	74
75	23.0527	20.7791	18.9238	17.3811	16.0781	15.4997	14.9630	14.4637	75
76	23.1468	20.8554	18.9871	17.4345	16.1238	15.5422	15.0026	14.5006	76
77	23.2375	20.9291	19.0481	17.4859	16.1678	15.5830	15.0407	14.5362	77
78	23.3251	21.0001	19.1069	17.5355	16.2102	15.6224	15.0773	14.5704	78
79	23.4096	21.0686	19.1636	17.5832	16.2509	15.6603	15.1126	14.6034	79
80	23.4912	21.1346	19.2182	17.6292	16.2902	15.6967	15.1465	14.6351	80
81	23.5699	21.1983	19.2709	17.6735	16.3280	15.7318	15.1792	14.6655	81
82	23.6457	21.2597	19.3216	17.7161	16.3644	15.7656	15.2106	14.6949	82
83	23.7189	21.3188	19.3704	17.7571	16.3994	15.7981	15.2409	14.7231	83
84	23.7895	21.3758	19.4174	17.7967	16.4331	15.8293	15.2700	14.7503	84
85	23.8576	21.4307	19.4628	17.8347	16.4655	15.8594	15.2980	14.7764	85
86	23.9232	21.4837	19.5064	17.8714	16.4968	15.8884	15.3249	14.8016	86
87	23.9864	21.5347	19.5484	17.9066	16.5268	15.9163	15.3509	14.8257	87
88	24.0474	21.5838	19.5889	17.9406	16.5557	15.9431	15.3758	14.8490	88
89	24.1062	21.6311	19.6279	17.9733	16.5836	15.9689	15.3998	14.8714	89
90	24.1628	21.6767	19.6654	18.0047	16.6103	15.9938	15.4229	14.8929	90
91	24.2174	21.7206	19.7016	18.0350	16.6361	16.0177	15.4451	14.9137	91
92	24.2700	21.7629	19.7364	18.0642	16.6609	16.0406	15.4665	14.9336	92
93	24.3207	21.8037	19.7699	18.0922	16.6848	16.0628	15.4871	14.9528	93
94	24.3695	21.8429	19.8021	18.1193	16.7078	16.0841	15.5069	14.9712	94
95	24.4165	21.8807	19.8332	18.1452	16.7299	16.1045	15.5259	14.9889	95
96	24.4618	21.9171	19.8630	18.1703	16.7511	16.1242	15.5442	15.0060	96
97	24.5055	21.9521	19.8918	18.1943	16.7716	16.1432	15.5618	15.0224	97
98	24.5475	21.9858	19.9195	18.2175	16.7913	16.1614	15.5788	15.0382	98
99	24.5880	22.0183	19.9462	18.2398	16.8102	16.1790	15.5951	15.0534	99
100	24.6270	22.0496	19.9718	18.2612	16.8284	16.1958	15.6107	15.0680	100

*Note:—In computing the quarterly in advance figures in these tables, all the rates % quoted above are effective rates, see the Introductory Section, pages xvi-xvii.

				Rate per Cent*					
Yrs.	7	7.25	7.5	8	8.5	9	9.5	10	Yrs.
1	0.4954	0.4948	0.4943	0.4932	0.4921	0.4910	0.4899	0.4888	1
2	0.9769	0.9747	0.9725	0.9682	0.9640	0.9598	0.9557	0.9517	2
3	1.4446	1.4398	1.4351	1.4258	1.4166	1.4077	1.3988	1.3902	3
4	1.8988	1.8906	1.8825	1.8665	1.8508	1.8355	1.8206	1.8059	4
5	2.3398	2.3273	2.3150	2.2909	2.2673	2.2444	2.2221	2.2003	5
6	2.7678	2.7504	2.7332	2.6996	2.6670	2.6353	2.6046	2.5747	6
7	3.1831	3.1600	3.1373	3.0931	3.0504	3.0091	2.9691	2.9303	7
8	3.5858	3.5566	3.5279	3.4721	3.4183	3.3665	3.3165	3.2683	8
9	3.9763	3.9404	3.9052	3.8369	3.7714	3.7084	3.6479	3.5896	9
10	4.3549	4.3118	4.2697	4.1882	4.1103	4.0356	3.9640	3.8953	10
11	4.7217	4.6711	4.6217	4.5264	4.4355	4.3487	4.2656	4.1862	11
12	5.0771	5.0186	4.9617	4.8520	4.7477	4.6484	4.5536	4.4631	12
13	5.4213	5.3547	5.2899	5.1655	5.0474	4.9353	4.8286	4.7270	13
14	5.7546	5.6796	5.6068	5.4672	5.3351	5.2100	5.0912	4.9784	14
15	6.0773	5.9937	5.9127	5.7576	5.6113	5.4730	5.3422	5.2181	15
16	6.3896	6.2973	6.2078	6.0372	5.8765	5.7250	5.5820	5.4466	16
17	6.6917	6.5906	6.4927	6.3062	6.1311	5.9664	5.8112	5.6647	17
18	6.9841	6.8739	6.7675	6.5652	6.3756	6.1977	6.0304	5.8728	18
19	7.2668	7.1476	7.0327	6.8144	6.6104	6.4194	6.2400	6.0714	19
20	7.5402	7.4120	7.2884	7.0542	6.8359	6.6318	6.4406	6.2611	20
21	7.8045	7.6672	7.5351	7.2851	7.0524	6.8354	6.6324	6.4422	21
22	8.0600	7.9137	7.7730	7.5072	7.2604	7.0306	6.8160	6.6153	22
23	8.3069	8.1516	8.0024	7.7210	7.4601	7.2177	6.9918	6.7808	23
24	8.5455	8.3812	8.2235	7.9266	7.6520	7.3971	7.1600	6.9389	24
25	8.7759	8.6028	8.4368	8.1246	7.8363	7.5692	7.3211	7.0901	25
26	8.9986	8.8166	8.6423	8.3150	8.0133	7.7342	7.4754	7.2347	26
27	9.2136	9.0229	8.8404	8.4982	8.1833	7.8925	7.6232	7.3730	27
28	9.4212	9.2219	9.0314	8.6745	8.3467	8.0444	7.7648	7.5054	28
29	9.6216	9.4138	9.2154	8.8442	8.5036	8.1901	7.9004	7.6320	29
30	9.8151	9.5989	9.3927	9.0074	8.6544	8.3298	8.0304	7.7532	30
31	10.0018	9.7775	9.5636	9.1644	8.7992	8.4639	8.1549	7.8693	31
32	10.1820	9.9496	9.7282	9.3155	8.9384	8.5926	8.2743	7.9804	32
33	10.3559	10.1156	9.8868	9.4608	9.0721	8.7161	8.3888	8.0869	33
34	10.5236	10.2756	10.0396	9.6006	9.2006	8.8347	8.4985	8.1888	34
35	10.6855	10.4298	10.1868	9.7351	9.3241	8.9484	8.6038	8.2864	35
36	10.8415	10.5785	10.3286	9.8645	9.4427	9.0576	8.7047	8.3800	36
37	10.9921	10.7217	10.4651	9.9890	9.5567	9.1625	8.8015	8.4697	37
38	11.1372	10.8598	10.5966	10.1087	9.6662	9.2631	8.8943	8.5556	38
39	11.2772	10.9928	10.7232	10.2239	9.7715	9.3597	8.9833	8.6379	39
40	11.4122	11.1210	10.8452	10.3347	9.8727	9.4525	9.0688	8.7169	40
41	11.5423	11.2445	10.9626	10.4413	9.9699	9.5416	9.1507	8.7926	41
42	11.6677	11.3635	11.0757	10.5438	10.0633	9.6271	9.2294	8.8652	42
43	11.7885	11.4782	11.1845	10.6424	10.1531	9.7093	9.3048	8.9348	43
44	11.9050	11.5886	11.2893	10.7372	10.2394	9.7881	9.3773	9.0015	44
45	12.0173	11.6949	11.3902	10.8285	10.3223	9.8639	9.4468	9.0656	45
46	12.1255	11.7973	11.4874	10.9162	10.4020	9.9367	9.5135	9.1270	46
47	12.2297	11.8960	11.5809	11.0006	10.4786	10.0065	9.5775	9.1859	47
48	12.3301	11.9909	11.6709	11.0818	10.5523	10.0737	9.6390	9.2425	48
49	12.4268	12.0824	11.7575	11.1599	10.6230	10.1381	9.6980	9.2967	49
50	12.5200	12.1705	11.8409	11.2350	10.6911	10.2001	9.7547	9.3488	50

**Note:*—In computing the quarterly in advance figures in these tables, all the rates % quoted above are effective rates,
see the Introductory Section, pages xvi-xvii.

				Rate per Cent*					
Yrs.	7	7.25	7.5	8	8.5	9	9.5	10	Yrs.
51	12.6098	12.2553	11.9211	11.3072	10.7564	10.2596	9.8091	9.3987	51
52	12.6962	12.3369	11.9984	11.3767	10.8193	10.3167	9.8613	9.4467	52
53	12.7795	12.4155	12.0727	11.4435	10.8797	10.3717	9.9115	9.4927	53
54	12.8597	12.4912	12.1443	11.5077	10.9378	10.4244	9.9596	9.5369	54
55	12.9369	12.5641	12.2131	11.5695	10.9936	10.4751	10.0059	9.5793	55
56	13.0113	12.6342	12.2793	11.6290	11.0472	10.5238	10.0503	9.6200	56
57	13.0829	12.7017	12.3431	11.6861	11.0988	10.5706	10.0930	9.6591	57
58	13.1518	12.7667	12.4044	11.7411	11.1484	10.6155	10.1340	9.6966	58
59	13.2182	12.8292	12.4635	11.7940	11.1960	10.6587	10.1733	9.7326	59
60	13.2821	12.8894	12.5203	11.8448	11.2418	10.7003	10.2111	9.7672	60
61	13.3436	12.9473	12.5749	11.8937	11.2859	10.7401	10.2475	9.8004	61
62	13.4028	13.0030	12.6275	11.9407	11.3282	10.7785	10.2823	9.8323	62
63	13.4598	13.0567	12.6781	11.9860	11.3689	10.8153	10.3159	9.8630	63
64	13.5147	13.1083	12.7268	12.0295	11.4080	10.8507	10.3481	9.8924	64
65	13.5675	13.1580	12.7736	12.0713	11.4456	10.8847	10.3790	9.9207	65
66	13.6184	13.2058	12.8186	12.1115	11.4818	10.9174	10.4087	9.9478	66
67	13.6673	13.2518	12.8620	12.1502	11.5166	10.9488	10.4373	9.9739	67
68	13.7144	13.2961	12.9037	12.1874	11.5500	10.9790	10.4647	9.9990	68
69	13.7597	13.3387	12.9438	12.2232	11.5821	11.0081	10.4911	10.0230	69
70	13.8033	13.3796	12.9823	12.2576	11.6130	11.0360	10.5164	10.0462	70
71	13.8453	13.4191	13.0195	12.2907	11.6427	11.0628	10.5408	10.0684	71
72	13.8856	13.4570	13.0552	12.3225	11.6712	11.0885	10.5642	10.0897	72
73	13.9245	13.4935	13.0895	12.3531	11.6987	11.1133	10.5866	10.1102	73
74	13.9619	13.5286	13.1226	12.3825	11.7250	11.1371	10.6082	10.1299	74
75	13.9979	13.5624	13.1543	12.4108	11.7504	11.1600	10.6290	10.1488	75
76	14.0325	13.5949	13.1849	12.4380	11.7748	11.1820	10.6489	10.1670	76
77	14.0658	13.6261	13.2143	12.4641	11.7982	11.2031	10.6681	10.1845	77
78	14.0978	13.6562	13.2426	12.4893	11.8208	11.2234	10.6865	10.2013	78
79	14.1287	13.6851	13.2698	12.5135	11.8424	11.2430	10.7042	10.2174	79
80	14.1583	13.7130	13.2959	12.5368	11.8633	11.2618	10.7213	10.2329	80
81	14.1869	13.7397	13.3211	12.5591	11.8833	11.2798	10.7376	10.2478	81
82	14.2143	13.7655	13.3453	12.5806	11.9026	11.2972	10.7533	10.2621	82
83	14.2408	13.7903	13.3686	12.6013	11.9211	11.3138	10.7684	10.2759	83
84	14.2662	13.8141	13.3910	12.6212	11.9389	11.3299	10.7830	10.2891	84
85	14.2906	13.8370	13.4125	12.6403	11.9560	11.3453	10.7969	10.3018	85
86	14.3141	13.8591	13.4332	12.6587	11.9724	11.3601	10.8103	10.3141	86
87	14.3367	13.8803	13.4532	12.6764	11.9883	11.3743	10.8232	10.3258	87
88	14.3585	13.9007	13.4723	12.6934	12.0035	11.3880	10.8356	10.3371	88
89	14.3794	13.9203	13.4907	12.7098	12.0181	11.4012	10.8476	10.3479	89
90	14.3996	13.9391	13.5085	12.7255	12.0322	11.4138	10.8590	10.3583	90
91	14.4189	13.9573	13.5255	12.7406	12.0457	11.4260	10.8700	10.3684	91
92	14.4376	13.9747	13.5419	12.7552	12.0587	11.4377	10.8806	10.3780	92
93	14.4555	13.9915	13.5577	12.7692	12.0712	11.4489	10.8908	10.3872	93
94	14.4727	14.0077	13.5728	12.7826	12.0832	11.4598	10.9006	10.3961	94
95	14.4893	14.0232	13.5874	12.7955	12.0947	11.4701	10.9100	10.4047	95
96	14.5052	14.0381	13.6014	12.8080	12.1059	11.4801	10.9190	10.4129	96
97	14.5206	14.0525	13.6149	12.8199	12.1165	11.4897	10.9277	10.4208	97
98	14.5353	14.0663	13.6279	12.8314	12.1268	11.4990	10.9360	10.4284	98
99	14.5495	14.0796	13.6403	12.8425	12.1367	11.5079	10.9441	10.4357	99
100	14.5632	14.0924	13.6523	12.8531	12.1462	11.5164	10.9518	10.4427	100

Note:—In computing the quarterly in advance figures in these tables, all the rates % quoted above are effective rates, see the Introductory Section, pages xvi-xvii.

				Rate per Cent*					
Yrs.	11	12	13	14	15	16	18	20	Yrs.
1	0.4867	0.4847	0.4827	0.4807	0.4787	0.4768	0.4731	0.4696	1
2	0.9437	0.9360	0.9285	0.9212	0.9141	0.9072	0.8939	0.8812	2
3	1.3733	1.3571	1.3414	1.3262	1.3115	1.2973	1.2702	1.2448	3
4	1.7776	1.7504	1.7244	1.6994	1.6753	1.6522	1.6085	1.5679	4
5	2.1584	2.1185	2.0805	2.0442	2.0095	1.9763	1.9141	1.8569	5
6	2.5175	2.4634	2.4121	2.3635	2.3172	2.2732	2.1913	2.1167	6
7	2.8565	2.7870	2.7215	2.6597	2.6013	2.5460	2.4437	2.3513	7
8	3.1766	3.0909	3.0106	2.9352	2.8642	2.7973	2.6743	2.5640	8
9	3.4794	3.3768	3.2812	3.1918	3.1080	3.0294	2.8857	2.7577	9
10	3.7658	3.6460	3.5347	3.4312	3.3346	3.2442	3.0800	2.9346	10
11	4.0370	3.8996	3.7726	3.6549	3.5455	3.4435	3.2591	3.0967	11
12	4.2940	4.1389	3.9961	3.8643	3.7422	3.6288	3.4245	3.2457	12
13	4.5377	4.3648	4.2063	4.0605	3.9259	3.8013	3.5778	3.3830	13
14	4.7688	4.5783	4.4043	4.2447	4.0978	3.9622	3.7200	3.5098	14
15	4.9883	4.7802	4.5908	4.4177	4.2588	4.1126	3.8522	3.6273	15
16	5.1968	4.9713	4.7668	4.5804	4.4099	4.2532	3.9753	3.7363	16
17	5.3949	5.1523	4.9330	4.7336	4.5517	4.3850	4.0902	3.8376	17
18	5.5833	5.3239	5.0900	4.8781	4.6851	4.5087	4.1976	3.9320	18
19	5.7626	5.4866	5.2386	5.0143	4.8107	4.6249	4.2981	4.0201	19
20	5.9332	5.6411	5.3792	5.1430	4.9290	4.7341	4.3923	4.1024	20
21	6.0956	5.7877	5.5123	5.2646	5.0406	4.8370	4.4807	4.1794	21
22	6.2504	5.9270	5.6386	5.3797	5.1459	4.9339	4.5638	4.2515	22
23	6.3978	6.0595	5.7583	5.4885	5.2455	5.0253	4.6419	4.3193	23
24	6.5384	6.1854	5.8720	5.5917	5.3396	5.1116	4.7155	4.3829	24
25	6.6725	6.3053	5.9799	5.6895	5.4287	5.1932	4.7848	4.4427	25
26	6.8004	6.4194	6.0824	5.7822	5.5131	5.2704	4.8502	4.4991	26
27	6.9225	6.5281	6.1799	5.8702	5.5930	5.3434	4.9120	4.5522	27
28	7.0390	6.6316	6.2726	5.9538	5.6688	5.4126	4.9704	4.6023	28
29	7.1503	6.7303	6.3608	6.0332	5.7408	5.4781	5.0256	4.6496	29
30	7.2566	6.8244	6.4448	6.1087	5.8091	5.5403	5.0779	4.6943	30
31	7.3582	6.9142	6.5248	6.1806	5.8740	5.5993	5.1274	4.7366	31
32	7.4553	6.9998	6.6010	6.2489	5.9357	5.6554	5.1744	4.7766	32
33	7.5481	7.0815	6.6736	6.3140	5.9944	5.7086	5.2189	4.8146	33
34	7.6368	7.1596	6.7429	6.3759	6.0502	5.7592	5.2612	4.8505	34
35	7.7217	7.2341	6.8090	6:4350	6.1034	5.8073	5.3013	4.8846	35
36	7.8028	7.3053	6.8720	6.4912	6.1540	5.8531	5.3395	4.9170	36
37	7.8805	7.3734	6.9322	6.5449	6.2022	5.8967	5.3757	4.9477	37
38	7.9548	7.4384	6.9897	6.5961	6.2481	5.9383	5.4102	4.9769	38
39	8.0260	7.5006	7.0445	6.6450	6.2920	5.9778	5.4430	5.0047	39
40	8.0941	7.5600	7.0970	6.6916	6.3337	6.0155	5.4743	5.0311	40
41	8.1593	7.6169	7.1471	6.7361	6.3736	6.0515	5.5040	5.0562	41
42	8.2218	7.6713	7.1949	6.7786	6.4117	6.0858	5.5324	5.0801	42
43	8.2816	7.7234	7.2407	6.8192	6.4480	6.1185	5.5594	5.1029	43
44	8.3390	7.7732	7.2845	6.8581	6.4827	6.1497	5.5852	5.1246	44
45	8.3939	7.8209	7.3264	6.8952	6.5158	6.1796	5.6098	5.1453	45
46	8.4465	7.8666	7.3664	6.9306	6.5475	6.2080	5.6332	5.1650	46
47	8.4970	7.9103	7.4048	6.9646	6.5778	6.2352	5.6556	5.1838	47
48	8.5453	7.9522	7.4415	6.9970	6.6067	6.2612	5.6770	5.2018	48
49	8.5917	7.9923	7.4766	7.0281	6.6344	6.2861	5.6974	5.2189	49
50	8.6361	8.0308	7.5102	7.0578	6.6609	6.3098	5.7169	5.2353	50

Note:—In computing the quarterly in advance figures in these tables, all the rates % quoted above are effective rates, see the Introductory Section, pages xvi-xvii.

				Rate per Cent*					
Yrs.	11	12	13	14	15	16	18	20	Yrs.
51	8.6787	8.0676	7.5424	7.0862	6.6862	6.3326	5.7356	5.2509	51
52	8.7196	8.1029	7.5733	7.1134	6.7104	6.3543	5.7534	5.2659	52
53	8.7588	8.1368	7.6028	7.1395	6.7336	6.3751	5.7704	5.2801	53
54	8.7964	8.1692	7.6311	7.1644	6.7558	6.3950	5.7867	5.2938	54
55	8.8324	8.2003	7.6583	7.1884	6.7770	6.4140	5.8023	5.3068	55
56	8.8670	8.2301	7.6843	7.2112	6.7974	6.4322	5.8172	5.3193	56
57	8.9002	8.2587	7.7092	7.2332	6.8169	6.4497	5.8315	5.3312	57
58	8.9321	8.2861	7.7331	7.2542	6.8356	6.4664	5.8451	5.3426	58
59	8.9627	8.3124	7.7560	7.2744	6.8534	6.4824	5.8582	5.3535	59
60	8.9920	8.3376	7.7779	7.2937	6.8706	6.4977	5.8707	5.3640	60
61	9.0201	8.3618	7.7990	7.3122	6.8870	6.5124	5.8827	5.3740	61
62	9.0472	8.3851	7.8192	7.3299	6.9027	6.5265	5.8942	5.3836	62
63	9.0731	8.4073	7.8385	7.3469	6.9178	6.5400	5.9052	5.3927	63
64	9.0980	8.4287	7.8571	7.3633	6.9323	6.5529	5.9157	5.4015	64
65	9.1219	8.4492	7.8749	7.3789	6.9462	6.5653	5.9258	5.4099	65
66	9.1449	8.4689	7.8920	7.3939	6.9595	6.5772	5.9355	5.4180	66
67	9.1669	8.4878	7.9084	7.4083	6.9722	6.5886	5.9448	5.4257	67
68	9.1881	8.5059	7.9242	7.4221	6.9845	6.5995	5.9537	5.4331	68
69	9.2084	8.5233	7.9393	7.4354	6.9962	6.6100	5.9622	5.4402	69
70	9.2279	8.5401	7.9538	7.4481	7.0074	6.6200	5.9704	5.4470	70
71	9.2466	8.5561	7.9677	7.4603	7.0182	6.6297	5.9782	5.4536	71
72	9.2646	8.5715	7.9811	7.4720	7.0286	6.6389	5.9857	5.4598	72
73	9.2819	8.5863	7.9939	7.4833	7.0386	6.6478	5.9929	5.4658	73
74	9.2985	8.6005	8.0062	7.4940	7.0481	6.6563	5.9999	5.4716	74
75	9.3144	8.6141	8.0180	7.5044	7.0573	6.6644	6.0065	5.4771	75
76	9.3298	8.6272	8.0294	7.5143	7.0660	6.6723	6.0129	5.4824	76
77	9.3445	8.6398	8.0403	7.5239	7.0745	6.6798	6.0190	5.4875	77
78	9.3586	8.6519	8.0507	7.5330	7.0826	6.6870	6.0248	5.4923	78
79	9.3722	8.6635	8.0608	7.5418	7.0903	6.6940	6.0304	5.4970	79
80	9.3852	8.6747	8.0704	7.5503	7.0978	6.7006	6.0358	5.5015	80
81	9.3978	8.6854	8.0797	7.5584	7.1050	6.7070	6.0410	5.5058	81
82	9.4098	8.6956	8.0886	7.5662	7.1119	6.7131	6.0460	5.5099	82
83	9.4214	8.7055	8.0971	7.5736	7.1185	6.7190	6.0508	5.5139	83
84	9.4325	8.7150	8.1053	7.5808	7.1248	6.7247	6.0554	5.5177	84
85	9.4432	8.7241	8.1132	7.5877	7.1309	6.7301	6.0598	5.5213	85
86	9.4534	8.7329	8.1208	7.5943	7.1367	6.7353	6.0640	5.5249	86
87	9.4633	8.7413	8.1281	7.6007	7.1424	6.7403	6.0680	5.5282	87
88	9.4728	8.7494	8.1351	7.6068	7.1478	6.7451	6.0719	5.5315	88
89	9.4819	8.7571	8.1418	7.6127	7.1529	6.7497	6.0757	5.5346	89
90	9.4906	8.7646	8.1482	7.6183	7.1579	6.7542	6.0793	5.5375	90
91	9.4990	8.7718	8.1544	7.6237	7.1627	6.7584	6.0827	5.5404	91
92	9.5071	8.7787	8.1604	7.6290	7.1673	6.7625	6.0860	5.5431	92
93	9.5149	8.7853	8.1661	7.6340	7.1717	6.7664	6.0892	5.5458	93
94	9.5223	8.7917	8.1716	7.6388	7.1760	6.7702	6.0923	5.5483	94
95	9.5295	8.7978	8.1769	7.6434	7.1800	6.7738	6.0952	5.5508	95
96	9.5364	8.8036	8.1819	7.6478	7.1839	6.7773	6.0980	5.5531	96
97	9.5430	8.8093	8.1868	7.6521	7.1877	6.7807	6.1007	5.5553	97
98	9.5494	8.8147	8.1915	7.6562	7.1913	6.7839	6.1033	5.5575	98
99	9.5555	8.8199	8.1960	7.6601	7.1948	6.7870	6.1058	5.5596	99
100	9.5614	8.8249	8.2003	7.6639	7.1981	6.7899	6.1082	5.5616	100

Note:—In computing the quarterly in advance figures in these tables, all the rates % quoted above are effective rates, see the Introductory Section, pages xvi-xvii.

YEARS' PURCHASE

(DUAL RATE % PRINCIPLE)

OR

PRESENT VALUE OF
ONE POUND PER ANNUM

*receivable at the end of each year after
allowing for a sinking fund at a given
rate to replace the invested capital and
for the effect of income tax at* **10%** *on
that part of the income used to provide
the annual sinking fund instalment.*

AT RATES OF INTEREST FROM

4% to **20%**

AND

ALLOWING FOR THE POSSIBLE INVESTMENT

OF SINKING FUNDS AT

2·5%, 3% and 4%

INCOME TAX at 10%

Note:—Tables of Years' Purchase in which no allowance has been made for the effect
of income tax on that part of the income used to provide the annual sinking
fund instalment will be found on pages 53 to 71.

YEARS' PURCHASE

				Rate Per Cent					
Yrs.	4	4.5	5	5.5	6	6.25	6.5	6.75	Yrs
1	0.8687	0.8650	0.8612	0.8576	0.8539	0.8521	0.8503	0.8485	1
2	1.6987	1.6844	1.6703	1.6565	1.6429	1.6361	1.6295	1.6229	2
3	2.4921	2.4615	2.4315	2.4023	2.3738	2.3598	2.3460	2.3323	3
4	3.2512	3.1992	3.1489	3.1001	3.0527	3.0296	3.0068	2.9844	4
5	3.9780	3.9004	3.8258	3.7540	3.6848	3.6512	3.6181	3.5857	5
6	4.6741	4.5674	4.4654	4.3679	4.2745	4.2293	4.1851	4.1417	6
7	5.3414	5.2024	5.0706	4.9452	4.8259	4.7683	4.7122	4.6573	7
8	5.9814	5.8077	5.6438	5.4889	5.3423	5.2719	5.2033	5.1365	8
9	6.5955	6.3849	6.1874	6.0017	5.8269	5.7432	5.6619	5.5829	9
10	7.1851	6.9359	6.7035	6.4861	6.2823	6.1852	6.0910	5.9996	10
11	7.7515	7.4623	7.1939	6.9441	6.7111	6.6004	6.4932	6.3895	11
12	8.2959	7.9655	7.6604	7.3778	7.1153	6.9910	6.8709	6.7549	12
13	8.8194	8.4469	8.1046	7.7890	7.4970	7.3591	7.2261	7.0979	13
14	9.3229	8.9077	8.5279	8.1791	7.8578	7.7064	7.5607	7.4204	14
15	9.8075	9.3491	8.9315	8.5497	8.1992	8.0345	7.8763	7.7242	15
16	10.2741	9.7721	9.3169	8.9022	8.5228	8.3450	8.1745	8.0108	16
17	10.7235	10.1778	9.6849	9.2376	8.8298	8.6391	8.4564	8.2814	17
18	11.1566	10.5671	10.0368	9.5572	9.1213	8.9179	8.7235	8.5373	18
19	11.5740	10.9408	10.3734	9.8619	9.3984	9.1827	8.9766	8.7796	19
20	11.9765	11.2998	10.6956	10.1526	9.6621	9.4342	9.2169	9.0093	20
21	12.3648	11.6449	11.0041	10.4303	9.9133	9.6735	9.4451	9.2272	21
22	12.7395	11.9766	11.2999	10.6956	10.1527	9.9014	9.6622	9.4343	22
23	13.1011	12.2957	11.5836	10.9494	10.3811	10.1185	9.8688	9.6312	23
24	13.4503	12.6028	11.8557	11.1922	10.5991	10.3255	10.0657	9.8186	24
25	13.7876	12.8984	12.1170	11.4248	10.8074	10.5231	10.2534	9.9971	25
26	14.1135	13.1832	12.3679	11.6477	11.0067	10.7119	10.4325	10.1673	26
27	14.4284	13.4576	12.6091	11.8613	11.1973	10.8923	10.6036	10.3298	27
28	14.7329	13.7220	12.8410	12.0663	11.3797	11.0649	10.7671	10.4849	28
29	15.0272	13.9771	13.0641	12.2630	11.5546	11.2302	10.9235	10.6331	29
30	15.3119	14.2230	13.2787	12.4520	11.7222	11.3884	11.0732	10.7749	30
31	15.5874	14.4604	13.4854	12.6335	11.8829	11.5401	11.2165	10.9105	31
32	15.8539	14.6895	13.6844	12.8080	12.0372	11.6855	11.3538	11.0405	32
33	16.1119	14.9107	13.8762	12.9759	12.1853	11.8251	11.4855	11.1650	33
34	16.3616	15.1243	14.0610	13.1374	12.3276	11.9591	11.6119	11.2843	34
35	16.6035	15.3307	14.2392	13.2928	12.4644	12.0877	11.7332	11.3988	35
36	16.8377	15.5302	14.4112	13.4426	12.5959	12.2114	11.8497	11.5087	36
37	17.0646	15.7230	14.5770	13.5868	12.7225	12.3303	11.9616	11.6143	37
38	17.2844	15.9095	14.7372	13.7258	12.8443	12.4447	12.0692	11.7157	38
39	17.4974	16.0898	14.8917	13.8598	12.9615	12.5547	12.1727	11.8132	39
40	17.7039	16.2642	15.0411	13.9890	13.0745	12.6607	12.2722	11.9069	40
41	17.9041	16.4330	15.1853	14.1137	13.1834	12.7627	12.3681	11.9971	41
42	18.0982	16.5964	15.3247	14.2340	13.2883	12.8610	12.4604	12.0840	42
43	18.2864	16.7545	15.4594	14.3502	13.3895	12.9558	12.5493	12.1676	43
44	18.4690	16.9076	15.5897	14.4624	13.4871	13.0472	12.6351	12.2482	44
45	18.6461	17.0559	15.7157	14.5708	13.5813	13.1353	12.7177	12.3258	45
46	18.8179	17.1996	15.8376	14.6755	13.6723	13.2204	12.7974	12.4007	46
47	18.9847	17.3388	15.9556	14.7767	13.7601	13.3025	12.8743	12.4729	47
48	19.1465	17.4737	16.0697	14.8746	13.8449	13.3817	12.9485	12.5425	48
49	19.3036	17.6044	16.1802	14.9692	13.9268	13.4582	13.0202	12.6097	49
50	19.4561	17.7312	16.2872	15.0607	14.0060	13.5322	13.0894	12.6746	50

YEARS' PURCHASE

Rate Per Cent

Yrs.	4	4.5	5	5.5	6	6.25	6.5	6.75	Yrs.
51	19.6041	17.8541	16.3908	15.1493	14.0826	13.6037	13.1562	12.7373	51
52	19.7479	17.9732	16.4912	15.2350	14.1566	13.6727	13.2208	12.7978	52
53	19.8875	18.0888	16.5885	15.3180	14.2282	13.7395	13.2833	12.8563	53
54	20.0231	18.2009	16.6827	15.3983	14.2975	13.8041	13.3436	12.9129	54
55	20.1549	18.3097	16.7741	15.4761	14.3646	13.8666	13.4020	12.9675	55
56	20.2829	18.4153	16.8626	15.5514	14.4295	13.9271	13.4585	13.0204	56
57	20.4072	18.5177	16.9485	15.6244	14.4923	13.9856	13.5131	13.0715	57
58	20.5281	18.6172	17.0318	15.6952	14.5531	14.0422	13.5660	13.1210	58
59	20.6455	18.7137	17.1125	15.7637	14.6120	14.0971	13.6172	13.1689	59
60	20.7596	18.8074	17.1909	15.8302	14.6691	14.1502	13.6667	13.2152	60
61	20.8706	18.8985	17.2669	15.8946	14.7244	14.2016	13.7147	13.2601	61
62	20.9784	18.9868	17.3406	15.9571	14.7780	14.2515	13.7612	13.3035	62
63	21.0833	19.0727	17.4122	16.0177	14.8300	14.2998	13.8062	13.3456	63
64	21.1852	19.1561	17.4817	16.0765	14.8803	14.3466	13.8499	13.3864	64
65	21.2843	19.2371	17.5491	16.1335	14.9292	14.3920	13.8922	13.4259	65
66	21.3807	19.3158	17.6146	16.1888	14.9765	14.4360	13.9332	13.4642	66
67	21.4745	19.3923	17.6782	16.2425	15.0225	14.4787	13.9729	13.5013	67
68	21.5657	19.4666	17.7399	16.2946	15.0670	14.5201	14.0115	13.5373	68
69	21.6544	19.5389	17.7999	16.3452	15.1103	14.5603	14.0489	13.5722	69
70	21.7406	19.6091	17.8582	16.3943	15.1523	14.5992	14.0851	13.6060	70
71	21.8246	19.6773	17.9148	16.4420	15.1930	14.6370	14.1203	13.6389	71
72	21.9063	19.7437	17.9698	16.4883	15.2325	14.6737	14.1545	13.6707	72
73	21.9857	19.8082	18.0232	16.5333	15.2709	14.7093	14.1876	13.7016	73
74	22.0630	19.8710	18.0751	16.5770	15.3081	14.7439	14.2198	13.7316	74
75	22.1383	19.9320	18.1256	16.6194	15.3443	14.7775	14.2510	13.7607	75
76	22.2115	19.9913	18.1746	16.6606	15.3795	14.8100	14.2813	13.7890	76
77	22.2828	20.0490	18.2223	16.7007	15.4136	14.8417	14.3107	13.8164	77
78	22.3521	20.1052	18.2687	16.7396	15.4468	14.8724	14.3393	13.8430	78
79	22.4196	20.1598	18.3138	16.7775	15.4790	14.9023	14.3670	13.8689	79
80	22.4854	20.2129	18.3576	16.8142	15.5103	14.9313	14.3940	13.8940	80
81	22.5493	20.2646	18.4002	16.8500	15.5407	14.9595	14.4202	13.9184	81
82	22.6116	20.3149	18.4417	16.8847	15.5702	14.9869	14.4456	13.9421	82
83	22.6723	20.3638	18.4820	16.9185	15.5990	15.0135	14.4704	13.9652	83
84	22.7313	20.4114	18.5212	16.9514	15.6269	15.0394	14.4944	13.9875	84
85	22.7888	20.4578	18.5593	16.9833	15.6541	15.0645	14.5177	14.0093	85
86	22.8448	20.5029	18.5965	17.0144	15.6804	15.0889	14.5404	14.0304	86
87	22.8993	20.5467	18.6326	17.0446	15.7061	15.1127	14.5625	14.0510	87
88	22.9524	20.5895	18.6677	17.0740	15.7311	15.1358	14.5840	14.0709	88
89	23.0041	20.6311	18.7019	17.1026	15.7553	15.1583	14.6048	14.0903	89
90	23.0544	20.6716	18.7351	17.1304	15.7789	15.1801	14.6251	14.1092	90
91	23.1034	20.7110	18.7675	17.1575	15.8019	15.2014	14.6448	14.1276	91
92	23.1512	20.7493	18.7990	17.1838	15.8242	15.2220	14.6640	14.1454	92
93	23.1977	20.7867	18.8297	17.2094	15.8459	15.2421	14.6826	14.1628	93
94	23.2430	20.8231	18.8595	17.2344	15.8671	15.2617	14.7008	14.1796	94
95	23.2872	20.8585	18.8886	17.2586	15.8876	15.2807	14.7184	14.1961	95
96	23.3302	20.8930	18.9169	17.2822	15.9076	15.2992	14.7356	14.2120	96
97	23.3721	20.9266	18.9444	17.3052	15.9271	15.3172	14.7523	14.2276	97
98	23.4129	20.9593	18.9712	17.3276	15.9460	15.3347	14.7685	14.2427	98
99	23.4527	20.9912	18.9973	17.3493	15.9645	15.3518	14.7844	14.2574	99
100	23.4914	21.0222	19.0227	17.3705	15.9824	15.3684	14.7997	14.2717	100

YEARS' PURCHASE

Rate Per Cent

Yrs.	7	7.25	7.5	8	8.5	9	9.5	10	Yrs.
1	0.8467	0.8449	0.8431	0.8396	0.8360	0.8326	0.8291	0.8257	1
2	1.6163	1.6098	1.6033	1.5906	1.5780	1.5657	1.5535	1.5416	2
3	2.3188	2.3054	2.2922	2.2662	2.2408	2.2160	2.1917	2.1680	3
4	2.9623	2.9405	2.9191	2.8771	2.8363	2.7966	2.7580	2.7205	4
5	3.5538	3.5225	3.4918	3.4319	3.3740	3.3180	3.2639	3.2115	5
6	4.0993	4.0577	4.0170	3.9379	3.8618	3.7887	3.7182	3.6504	6
7	4.6037	4.5513	4.5001	4.4011	4.3063	4.2155	4.1285	4.0450	7
8	5.0714	5.0079	4.9459	4.8266	4.7128	4.6043	4.5007	4.4017	8
9	5.5060	5.4313	5.3585	5.2187	5.0860	4.9599	4.8398	4.7255	9
10	5.9110	5.8249	5.7413	5.5811	5.4296	5.2861	5.1500	5.0207	10
11	6.2890	6.1917	6.0973	5.9169	5.7469	5.5864	5.4346	5.2908	11
12	6.6427	6.5342	6.4292	6.2289	6.0408	5.8637	5.6967	5.5389	12
13	6.9741	6.8546	6.7391	6.5195	6.3137	6.1204	5.9387	5.7674	13
14	7.2853	7.1550	7.0292	6.7906	6.5676	6.3588	6.1628	5.9786	14
15	7.5779	7.4370	7.3013	7.0441	6.8044	6.5806	6.3709	6.1743	15
16	7.8535	7.7022	7.5567	7.2816	7.0258	6.7874	6.5646	6.3560	16
17	8.1134	7.9521	7.7971	7.5045	7.2331	6.9807	6.7452	6.5252	17
18	8.3589	8.1878	8.0235	7.7141	7.4276	7.1616	6.9140	6.6830	18
19	8.5910	8.4104	8.2372	7.9113	7.6103	7.3313	7.0721	6.8306	19
20	8.8108	8.6209	8.4390	8.0974	7.7823	7.4908	7.2204	6.9688	20
21	9.0192	8.8203	8.6300	8.2730	7.9444	7.6409	7.3597	7.0985	21
22	9.2169	9.0093	8.8109	8.4391	8.0974	7.7823	7.4908	7.2204	22
23	9.4047	9.1887	8.9824	8.5963	8.2420	7.9158	7.6144	7.3352	23
24	9.5834	9.3591	9.1451	8.7453	8.3789	8.0420	7.7311	7.4434	24
25	9.7534	9.5212	9.2998	8.8866	8.5086	8.1613	7.8414	7.5455	25
26	9.9153	9.6755	9.4470	9.0209	8.6315	8.2744	7.9457	7.6421	26
27	10.0697	9.8224	9.5870	9.1485	8.7483	8.3817	8.0446	7.7335	27
28	10.2171	9.9626	9.7205	9.2699	8.8593	8.4835	8.1383	7.8201	28
29	10.3578	10.0963	9.8478	9.3856	8.9649	8.5803	8.2273	7.9023	29
30	10.4922	10.2241	9.9692	9.4959	9.0655	8.6724	8.3120	7.9803	30
31	10.6208	10.3461	10.0853	9.6011	9.1613	8.7601	8.3925	8.0545	31
32	10.7439	10.4629	10.1962	9.7016	9.2528	8.8436	8.4691	8.1251	32
33	10.8618	10.5746	10.3023	9.7976	9.3400	8.9233	8.5422	8.1923	33
34	10.9747	10.6816	10.4038	9.8894	9.4234	8.9994	8.6119	8.2564	34
35	11.0830	10.7842	10.5011	9.9772	9.5031	9.0721	8.6784	8.3175	35
36	11.1869	10.8825	10.5943	10.0613	9.5794	9.1415	8.7420	8.3759	36
37	11.2866	10.9768	10.6836	10.1419	9.6524	9.2080	8.8027	8.4316	37
38	11.3823	11.0674	10.7694	10.2191	9.7224	9.2716	8.8609	8.4850	38
39	11.4743	11.1543	10.8517	10.2932	9.7894	9.3326	8.9165	8.5360	39
40	11.5627	11.2379	10.9308	10.3643	9.8537	9.3910	8.9698	8.5848	40
41	11.6478	11.3182	11.0068	10.4326	9.9154	9.4470	9.0209	8.6316	41
42	11.7296	11.3955	11.0798	10.4982	9.9746	9.5008	9.0700	8.6765	42
43	11.8084	11.4698	11.1501	10.5613	10.0316	9.5524	9.1170	8.7195	43
44	11.8843	11.5414	11.2177	10.6219	10.0863	9.6020	9.1621	8.7608	44
45	11.9574	11.6103	11.2828	10.6803	10.1388	9.6497	9.2055	8.8004	45
46	12.0278	11.6767	11.3455	10.7364	10.1894	9.6955	9.2472	8.8385	46
47	12.0957	11.7407	11.4059	10.7905	10.2381	9.7396	9.2873	8.8752	47
48	12.1612	11.8024	11.4641	10.8426	10.2850	9.7820	9.3258	8.9104	48
49	12.2244	11.8619	11.5202	10.8928	10.3302	9.8228	9.3630	8.9442	49
50	12.2853	11.9193	11.5744	10.9412	10.3737	9.8621	9.3987	8.9768	50

YEARS' PURCHASE

Rate Per Cent

Yrs.	7	7.25	7.5	8	8.5	9	9.5	10	Yrs.
51	12.3442	11.9747	11.6266	10.9878	10.4156	9.9000	9.4331	9.0082	51
52	12.4011	12.0282	11.6770	11.0329	10.4561	9.9366	9.4663	9.0385	52
53	12.4560	12.0798	11.7257	11.0763	10.4951	9.9718	9.4982	9.0676	53
54	12.5090	12.1297	11.7727	11.1183	10.5327	10.0058	9.5291	9.0957	54
55	12.5603	12.1779	11.8181	11.1588	10.5691	10.0386	9.5588	9.1228	55
56	12.6099	12.2245	11.8620	11.1979	10.6042	10.0702	9.5875	9.1489	56
57	12.6579	12.2696	11.9044	11.2357	10.6380	10.1008	9.6152	9.1741	57
58	12.7042	12.3132	11.9455	11.2722	10.6708	10.1303	9.6419	9.1985	58
59	12.7491	12.3553	11.9851	11.3075	10.7024	10.1588	9.6677	9.2220	59
60	12.7926	12.3961	12.0235	11.3417	10.7330	10.1864	9.6927	9.2447	60
61	12.8346	12.4356	12.0606	11.3747	10.7626	10.2130	9.7168	9.2666	61
62	12.8753	12.4738	12.0966	11.4067	10.7912	10.2388	9.7401	9.2878	62
63	12.9147	12.5108	12.1314	11.4376	10.8189	10.2637	9.7627	9.3083	63
64	12.9529	12.5466	12.1650	11.4675	10.8457	10.2878	9.7845	9.3281	64
65	12.9899	12.5813	12.1977	11.4965	10.8716	10.3111	9.8056	9.3473	65
66	13.0257	12.6149	12.2293	11.5246	10.8967	10.3337	9.8260	9.3658	66
67	13.0605	12.6475	12.2599	11.5518	10.9210	10.3555	9.8457	9.3838	67
68	13.0941	12.6791	12.2895	11.5781	10.9445	10.3767	9.8649	9.4011	68
69	13.1268	12.7097	12.3183	11.6036	10.9673	10.3972	9.8834	9.4180	69
70	13.1584	12.7394	12.3462	11.6283	10.9894	10.4170	9.9013	9.4342	70
71	13.1892	12.7681	12.3732	11.6523	11.0108	10.4362	9.9187	9.4500	71
72	13.2189	12.7961	12.3994	11.6755	11.0316	10.4549	9.9355	9.4653	72
73	13.2478	12.8231	12.4248	11.6981	11.0517	10.4729	9.9518	9.4801	73
74	13.2759	12.8494	12.4495	11.7199	11.0712	10.4905	9.9676	9.4944	74
75	13.3031	12.8749	12.4734	11.7411	11.0901	10.5074	9.9830	9.5084	75
76	13.3295	12.8996	12.4966	11.7617	11.1084	10.5239	9.9978	9.5218	76
77	13.3551	12.9236	12.5191	11.7817	11.1262	10.5399	10.0122	9.5349	77
78	13.3800	12.9469	12.5410	11.8010	11.1435	10.5554	10.0262	9.5476	78
79	13.4042	12.9695	12.5622	11.8198	11.1602	10.5704	10.0398	9.5599	79
80	13.4276	12.9915	12.5828	11.8380	11.1765	10.5850	10.0529	9.5718	80
81	13.4504	13.0128	12.6028	11.8558	11.1923	10.5991	10.0657	9.5834	81
82	13.4725	13.0336	12.6223	11.8730	11.2076	10.6129	10.0781	9.5946	82
83	13.4940	13.0537	12.6411	11.8897	11.2225	10.6262	10.0901	9.6055	83
84	13.5149	13.0732	12.6595	11.9059	11.2369	10.6392	10.1018	9.6161	84
85	13.5352	13.0922	12.6773	11.9216	11.2510	10.6518	10.1131	9.6264	85
86	13.5550	13.1107	12.6946	11.9369	11.2646	10.6640	10.1242	9.6364	86
87	13.5741	13.1286	12.7114	11.9518	11.2778	10.6758	10.1348	9.6460	87
88	13.5928	13.1460	12.7277	11.9662	11.2907	10.6874	10.1452	9.6554	88
89	13.6109	13.1630	12.7436	11.9803	11.3032	10.6986	10.1553	9.6646	89
90	13.6285	13.1795	12.7591	11.9939	11.3153	10.7094	10.1651	9.6735	90
91	13.6456	13.1955	12.7741	12.0072	11.3271	10.7200	10.1746	9.6821	91
92	13.6623	13.2110	12.7887	12.0201	11.3386	10.7303	10.1839	9.6905	92
93	13.6785	13.2262	12.8028	12.0326	11.3498	10.7403	10.1929	9.6986	93
94	13.6942	13.2409	12.8166	12.0448	11.3606	10.7500	10.2016	9.7065	94
95	13.7095	13.2552	12.8300	12.0566	11.3711	10.7594	10.2101	9.7142	95
96	13.7244	13.2691	12.8431	12.0681	11.3814	10.7686	10.2184	9.7217	96
97	13.7389	13.2827	12.8558	12.0793	11.3913	10.7775	10.2264	9.7289	97
98	13.7530	13.2958	12.8681	12.0902	11.4010	10.7861	10.2342	9.7360	98
99	13.7667	13.3087	12.8801	12.1008	11.4104	10.7946	10.2418	9.7429	99
100	13.7800	13.3211	12.8918	12.1111	11.4196	10.8028	10.2492	9.7496	100

YEARS' PURCHASE

Rate Per Cent

Yrs.	11	12	13	14	15	16	18	20	Yrs.
1	0.8189	0.8123	0.8057	0.7993	0.7930	0.7867	0.7745	0.7627	1
2	1.5181	1.4954	1.4734	1.4520	1.4312	1.4110	1.3723	1.3357	2
3	2.1220	2.0779	2.0356	1.9950	1.9559	1.9184	1.8475	1.7817	3
4	2.6485	2.5801	2.5152	2.4535	2.3948	2.3388	2.2343	2.1387	4
5	3.1115	3.0176	2.9292	2.8459	2.7671	2.6926	2.5550	2.4308	5
6	3.5218	3.4020	3.2901	3.1853	3.0869	2.9945	2.8253	2.6742	6
7	3.8878	3.7423	3.6073	3.4817	3.3645	3.2550	3.0561	2.8800	7
8	4.2161	4.0455	3.8882	3.7427	3.6077	3.4821	3.2554	3.0564	8
9	4.5123	4.3174	4.1388	3.9743	3.8224	3.6816	3.4291	3.2090	9
10	4.7807	4.5625	4.3635	4.1810	4.0132	3.8584	3.5820	3.3425	10
11	5.0250	4.7845	4.5661	4.3667	4.1840	4.0160	3.7174	3.4601	11
12	5.2482	4.9865	4.7497	4.5343	4.3376	4.1573	3.8382	3.5645	12
13	5.4530	5.1710	4.9167	4.6863	4.4765	4.2847	3.9465	3.6578	13
14	5.6413	5.3401	5.0694	4.8248	4.6027	4.4002	4.0443	3.7416	14
15	5.8152	5.4956	5.2093	4.9514	4.7178	4.5053	4.1329	3.8173	15
16	5.9761	5.6391	5.3381	5.0676	4.8232	4.6013	4.2135	3.8860	16
17	6.1255	5.7719	5.4569	5.1746	4.9200	4.6893	4.2872	3.9486	17
18	6.2644	5.8951	5.5669	5.2733	5.0092	4.7702	4.3548	4.0059	18
19	6.3938	6.0096	5.6689	5.3648	5.0916	4.8449	4.4169	4.0584	19
20	6.5148	6.1163	5.7638	5.4497	5.1680	4.9141	4.4743	4.1068	20
21	6.6280	6.2160	5.8522	5.5287	5.2390	4.9782	4.5275	4.1515	21
22	6.7342	6.3093	5.9349	5.6024	5.3051	5.0379	4.5767	4.1929	22
23	6.8339	6.3968	6.0122	5.6712	5.3668	5.0935	4.6226	4.2314	23
24	6.9277	6.4789	6.0847	5.7357	5.4245	5.1454	4.6653	4.2672	24
25	7.0161	6.5561	6.1528	5.7961	5.4786	5.1940	4.7052	4.3005	25
26	7.0995	6.6289	6.2168	5.8529	5.5293	5.2396	4.7426	4.3317	26
27	7.1784	6.6976	6.2772	5.9064	5.5770	5.2824	4.7777	4.3610	27
28	7.2529	6.7624	6.3341	5.9568	5.6219	5.3227	4.8106	4.3884	28
29	7.3235	6.8238	6.3879	6.0044	5.6643	5.3606	4.8415	4.4141	29
30	7.3905	6.8819	6.4388	6.0493	5.7042	5.3964	4.8707	4.4384	30
31	7.4541	6.9370	6.4870	6.0918	5.7420	5.4302	4.8983	4.4612	31
32	7.5145	6.9893	6.5327	6.1321	5.7778	5.4622	4.9243	4.4828	32
33	7.5720	7.0390	6.5761	6.1703	5.8117	5.4925	4.9489	4.5032	33
34	7.6267	7.0862	6.6173	6.2066	5.8439	5.5212	4.9722	4.5225	34
35	7.6788	7.1312	6.6565	6.2411	5.8745	5.5485	4.9943	4.5407	35
36	7.7285	7.1741	6.6939	6.2739	5.9035	5.5744	5.0153	4.5581	36
37	7.7760	7.2150	6.7294	6.3051	5.9312	5.5991	5.0352	4.5745	37
38	7.8213	7.2540	6.7634	6.3349	5.9575	5.6225	5.0542	4.5902	38
39	7.8646	7.2912	6.7957	6.3633	5.9826	5.6449	5.0722	4.6051	39
40	7.9061	7.3268	6.8267	6.3904	6.0066	5.6662	5.0895	4.6193	40
41	7.9458	7.3609	6.8562	6.4163	6.0294	5.6866	5.1059	4.6328	41
42	7.9838	7.3935	6.8845	6.4411	6.0513	5.7060	5.1215	4.6457	42
43	8.0202	7.4247	6.9115	6.4647	6.0722	5.7246	5.1365	4.6580	43
44	8.0551	7.4546	6.9375	6.4874	6.0922	5.7423	5.1508	4.6697	44
45	8.0886	7.4833	6.9623	6.5091	6.1113	5.7594	5.1645	4.6810	45
46	8.1208	7.5108	6.9861	6.5299	6.1297	5.7756	5.1776	4.6917	46
47	8.1517	7.5373	7.0090	6.5499	6.1473	5.7913	5.1901	4.7020	47
48	8.1814	7.5626	7.0309	6.5691	6.1641	5.8062	5.2021	4.7119	48
49	8.2099	7.5870	7.0520	6.5874	6.1803	5.8206	5.2137	4.7213	49
50	8.2374	7.6105	7.0722	6.6051	6.1959	5.8344	5.2247	4.7304	50

YEARS' PURCHASE

				Rate Per Cent					
Yrs.	**11**	**12**	**13**	**14**	**15**	**16**	**18**	**20**	Yrs.
51	8.2638	7.6330	7.0917	6.6221	6.2108	5.8476	5.2353	4.7391	**51**
52	8.2892	7.6547	7.1104	6.6384	6.2252	5.8603	5.2455	4.7475	**52**
53	8.3137	7.6756	7.1285	6.6541	6.2390	5.8726	5.2553	4.7555	**53**
54	8.3373	7.6957	7.1458	6.6692	6.2523	5.8844	5.2648	4.7632	**54**
55	8.3601	7.7151	7.1625	6.6838	6.2650	5.8957	5.2738	4.7706	**55**
56	8.3820	7.7338	7.1786	6.6978	6.2774	5.9066	5.2825	4.7778	**56**
57	8.4032	7.7518	7.1941	6.7113	6.2892	5.9171	5.2909	4.7846	**57**
58	8.4236	7.7692	7.2091	6.7243	6.3006	5.9272	5.2990	4.7912	**58**
59	8.4433	7.7859	7.2235	6.7369	6.3117	5.9369	5.3068	4.7976	**59**
60	8.4624	7.8021	7.2374	6.7490	6.3223	5.9463	5.3143	4.8038	**60**
61	8.4807	7.8177	7.2509	6.7607	6.3325	5.9554	5.3216	4.8097	**61**
62	8.4985	7.8328	7.2638	6.7719	6.3424	5.9642	5.3286	4.8154	**62**
63	8.5156	7.8474	7.2764	6.7828	6.3520	5.9726	5.3353	4.8209	**63**
64	8.5322	7.8615	7.2885	6.7934	6.3612	5.9808	5.3418	4.8262	**64**
65	8.5483	7.8751	7.3002	6.8035	6.3701	5.9886	5.3481	4.8313	**65**
66	8.5638	7.8882	7.3115	6.8133	6.3787	5.9962	5.3541	4.8363	**66**
67	8.5788	7.9010	7.3224	6.8228	6.3870	6.0036	5.3600	4.8410	**67**
68	8.5933	7.9133	7.3330	6.8320	6.3951	6.0107	5.3657	4.8457	**68**
69	8.6073	7.9252	7.3432	6.8409	6.4029	6.0176	5.3711	4.8501	**69**
70	8.6209	7.9367	7.3531	6.8495	6.4104	6.0242	5.3764	4.8544	**70**
71	8.6341	7.9479	7.3627	6.8578	6.4177	6.0306	5.3816	4.8586	**71**
72	8.6468	7.9587	7.3720	6.8658	6.4247	6.0369	5.3865	4.8627	**72**
73	8.6592	7.9691	7.3809	6.8736	6.4315	6.0429	5.3913	4.8666	**73**
74	8.6712	7.9793	7.3896	6.8811	6.4381	6.0487	5.3959	4.8703	**74**
75	8.6828	7.9891	7.3981	6.8884	6.4445	6.0543	5.4004	4.8740	**75**
76	8.6940	7.9986	7.4062	6.8955	6.4507	6.0598	5.4048	4.8775	**76**
77	8.7049	8.0078	7.4141	6.9024	6.4567	6.0651	5.4090	4.8810	**77**
78	8.7155	8.0168	7.4218	6.9090	6.4625	6.0702	5.4131	4.8843	**78**
79	8.7257	8.0254	7.4292	6.9154	6.4681	6.0752	5.4170	4.8875	**79**
80	8.7357	8.0338	7.4364	6.9217	6.4736	6.0800	5.4208	4.8906	**80**
81	8.7453	8.0420	7.4434	6.9277	6.4789	6.0847	5.4245	4.8936	**81**
82	8.7546	8.0499	7.4502	6.9336	6.4840	6.0892	5.4281	4.8966	**82**
83	8.7637	8.0576	7.4567	6.9393	6.4890	6.0936	5.4316	4.8994	**83**
84	8.7725	8.0650	7.4631	6.9448	6.4938	6.0979	5.4350	4.9021	**84**
85	8.7811	8.0722	7.4693	6.9502	6.4985	6.1020	5.4383	4.9048	**85**
86	8.7894	8.0793	7.4753	6.9554	6.5031	6.1060	5.4415	4.9074	**86**
87	8.7974	8.0861	7.4811	6.9604	6.5075	6.1099	5.4446	4.9099	**87**
88	8.8053	8.0927	7.4868	6.9653	6.5118	6.1136	5.4476	4.9124	**88**
89	8.8129	8.0991	7.4923	6.9701	6.5159	6.1173	5.4505	4.9147	**89**
90	8.8202	8.1053	7.4976	6.9747	6.5199	6.1209	5.4533	4.9170	**90**
91	8.8274	8.1114	7.5028	6.9792	6.5239	6.1243	5.4560	4.9192	**91**
92	8.8344	8.1173	7.5078	6.9835	6.5277	6.1277	5.4587	4.9214	**92**
93	8.8411	8.1230	7.5127	6.9877	6.5314	6.1309	5.4613	4.9235	**93**
94	8.8477	8.1285	7.5175	6.9919	6.5349	6.1341	5.4638	4.9255	**94**
95	8.8541	8.1339	7.5221	6.9958	6.5384	6.1372	5.4662	4.9275	**95**
96	8.8603	8.1392	7.5266	6.9997	6.5418	6.1401	5.4686	4.9294	**96**
97	8.8663	8.1442	7.5309	7.0035	6.5451	6.1430	5.4709	4.9313	**97**
98	8.8722	8.1492	7.5351	7.0071	6.5483	6.1458	5.4731	4.9331	**98**
99	8.8779	8.1540	7.5393	7.0107	6.5514	6.1486	5.4753	4.9349	**99**
100	8.8835	8.1587	7.5433	7.0142	6.5544	6.1512	5.4774	4.9366	**100**

YEARS' PURCHASE

Rate Per Cent

Yrs.	4	4.5	5	5.5	6	6.25	6.5	6.75	Yrs.
1	0.8687	0.8650	0.8612	0.8576	0.8539	0.8521	0.8503	0.8485	1
2	1.7026	1.6882	1.6741	1.6602	1.6465	1.6398	1.6331	1.6264	2
3	2.5033	2.4723	2.4421	2.4127	2.3839	2.3698	2.3558	2.3420	3
4	3.2724	3.2197	3.1687	3.1193	3.0714	3.0480	3.0249	3.0022	4
5	4.0115	3.9326	3.8568	3.7838	3.7136	3.6794	3.6459	3.6129	5
6	4.7220	4.6131	4.5091	4.4097	4.3145	4.2685	4.2234	4.1793	6
7	5.4052	5.2630	5.1280	4.9998	4.8779	4.8191	4.7617	4.7057	7
8	6.0624	5.8840	5.7159	5.5571	5.4068	5.3347	5.2645	5.1961	8
9	6.6947	6.4779	6.2747	6.0838	5.9042	5.8183	5.7349	5.6538	9
10	7.3034	7.0461	6.8063	6.5823	6.3726	6.2726	6.1758	6.0819	10
11	7.8894	7.5900	7.3125	7.0545	6.8142	6.7000	6.5897	6.4829	11
12	8.4537	8.1109	7.7948	7.5024	7.2311	7.1027	6.9788	6.8591	12
13	8.9973	8.6100	8.2546	7.9275	7.6252	7.4826	7.3452	7.2127	13
14	9.5212	9.0885	8.6934	8.3313	7.9981	7.8413	7.6906	7.5455	14
15	10.0260	9.5474	9.1124	8.7153	8.3514	8.1806	8.0166	7.8591	15
16	10.5127	9.9877	9.5127	9.0807	8.6863	8.5017	8.3248	8.1551	16
17	10.9820	10.4103	9.8953	9.4288	9.0043	8.8060	8.6164	8.4347	17
18	11.4346	10.8162	10.2612	9.7605	9.3063	9.0947	8.8925	8.6991	18
19	11.8712	11.2060	10.6115	10.0768	9.5935	9.3688	9.1544	8.9495	19
20	12.2924	11.5807	10.9468	10.3787	9.8667	9.6292	9.4028	9.1869	20
21	12.6990	11.9408	11.2681	10.6671	10.1269	9.8769	9.6389	9.4121	21
22	13.0914	12.2871	11.5760	10.9426	10.3750	10.1127	9.8633	9.6259	22
23	13.4703	12.6203	11.8712	11.2060	10.6115	10.3372	10.0768	9.8292	23
24	13.8361	12.9408	12.1544	11.4580	10.8372	10.5513	10.2801	10.0226	24
25	14.1893	13.2493	12.4262	11.6993	11.0527	10.7555	10.4739	10.2066	25
26	14.5306	13.5464	12.6871	11.9303	11.2587	10.9505	10.6587	10.3820	26
27	14.8603	13.8325	12.9377	12.1516	11.4556	11.1367	10.8350	10.5492	27
28	15.1788	14.1081	13.1785	12.3638	11.6440	11.3146	11.0034	10.7088	28
29	15.4866	14.3736	13.4099	12.5673	11.8243	11.4848	11.1642	10.8611	29
30	15.7841	14.6296	13.6324	12.7625	11.9969	11.6476	11.3180	11.0066	30
31	16.0717	14.8763	13.8464	12.9498	12.1623	11.8034	11.4651	11.1457	31
32	16.3498	15.1142	14.0523	13.1298	12.3209	11.9527	11.6059	11.2787	32
33	16.6186	15.3437	14.2504	13.3026	12.4730	12.0958	11.7407	11.4060	33
34	16.8786	15.5650	14.4411	13.4686	12.6188	12.2329	11.8699	11.5278	34
35	17.1300	15.7786	14.6248	13.6283	12.7588	12.3645	11.9937	11.6446	35
36	17.3732	15.9847	14.8017	13.7817	12.8933	12.4907	12.1124	11.7564	36
37	17.6085	16.1837	14.9721	13.9294	13.0224	12.6118	12.2263	11.8637	37
38	17.8361	16.3757	15.1364	14.0714	13.1465	12.7282	12.3356	11.9666	38
39	18.0564	16.5612	15.2947	14.2082	13.2658	12.8399	12.4406	12.0653	39
40	18.2695	16.7403	15.4474	14.3398	13.3804	12.9473	12.5414	12.1601	40
41	18.4758	16.9133	15.5946	14.4666	13.4907	13.0506	12.6382	12.2512	41
42	18.6754	17.0805	15.7366	14.5887	13.5969	13.1499	12.7313	12.3386	42
43	18.8687	17.2420	15.8735	14.7063	13.6990	13.2454	12.8209	12.4227	43
44	19.0557	17.3981	16.0057	14.8197	13.7974	13.3373	12.9070	12.5035	44
45	19.2369	17.5489	16.1333	14.9290	13.8921	13.4258	12.9898	12.5812	45
46	19.4122	17.6948	16.2565	15.0344	13.9833	13.5110	13.0695	12.6560	46
47	19.5820	17.8357	16.3754	15.1361	14.0712	13.5930	13.1463	12.7280	47
48	19.7465	17.9720	16.4902	15.2342	14.1559	13.6720	13.2202	12.7972	48
49	19.9057	18.1039	16.6011	15.3288	14.2375	13.7482	13.2914	12.8639	49
50	20.0600	18.2314	16.7083	15.4201	14.3163	13.8216	13.3600	12.9282	50

YEARS' PURCHASE

				Rate Per Cent					
Yrs.	4	4.5	5	5.5	6	6.25	6.5	6.75	Yrs.
51	20.2094	18.3547	16.8118	15.5082	14.3922	13.8924	13.4261	12.9900	51
52	20.3541	18.4740	16.9118	15.5933	14.4655	13.9606	13.4898	13.0497	52
53	20.4943	18.5894	17.0085	15.6754	14.5361	14.0264	13.5512	13.1072	53
54	20.6301	18.7011	17.1019	15.7548	14.6043	14.0899	13.6105	13.1626	54
55	20.7617	18.8091	17.1923	15.8314	14.6701	14.1512	13.6676	13.2160	55
56	20.8892	18.9137	17.2796	15.9054	14.7337	14.2103	13.7228	13.2676	56
57	21.0127	19.0150	17.3641	15.9770	14.7951	14.2673	13.7760	13.3173	57
58	21.1325	19.1130	17.4458	16.0461	14.8543	14.3224	13.8273	13.3653	58
59	21.2485	19.2078	17.5248	16.1129	14.9115	14.3756	13.8769	13.4116	59
60	21.3609	19.2997	17.6012	16.1775	14.9668	14.4270	13.9248	13.4563	60
61	21.4699	19.3886	17.6751	16.2399	15.0203	14.4766	13.9710	13.4995	61
62	21.5756	19.4747	17.7466	16.3003	15.0719	14.5246	14.0157	13.5412	62
63	21.6780	19.5581	17.8158	16.3586	15.1218	14.5709	14.0588	13.5815	63
64	21.7772	19.6388	17.8828	16.4151	15.1700	14.6157	14.1005	13.6203	64
65	21.8734	19.7170	17.9477	16.4697	15.2166	14.6590	14.1408	13.6579	65
66	21.9667	19.7928	18.0104	16.5225	15.2617	14.7008	14.1797	13.6942	66
67	22.0571	19.8662	18.0712	16.5736	15.3053	14.7413	14.2173	13.7293	67
68	22.1448	19.9373	18.1300	16.6231	15.3475	14.7804	14.2537	13.7632	68
69	22.2298	20.0061	18.1869	16.6709	15.3883	14.8182	14.2888	13.7960	69
70	22.3122	20.0729	18.2420	16.7172	15.4277	14.8548	14.3229	13.8277	70
71	22.3921	20.1375	18.2954	16.7621	15.4659	14.8901	14.3557	13.8584	71
72	22.4696	20.2002	18.3471	16.8055	15.5028	14.9244	14.3876	13.8880	72
73	22.5448	20.2609	18.3972	16.8475	15.5385	14.9575	14.4183	13.9167	73
74	22.6177	20.3197	18.4457	16.8881	15.5731	14.9895	14.4481	13.9444	74
75	22.6883	20.3768	18.4927	16.9275	15.6066	15.0205	14.4769	13.9713	75
76	22.7569	20.4320	18.5382	16.9656	15.6390	15.0505	14.5048	13.9972	76
77	22.8233	20.4856	18.5823	17.0025	15.6703	15.0796	14.5318	14.0223	77
78	22.8878	20.5375	18.6250	17.0383	15.7007	15.1077	14.5579	14.0466	78
79	22.9503	20.5879	18.6664	17.0729	15.7301	15.1349	14.5831	14.0702	79
80	23.0110	20.6366	18.7065	17.1064	15.7586	15.1613	14.6076	14.0929	80
81	23.0698	20.6839	18.7453	17.1389	15.7861	15.1868	14.6313	14.1150	81
82	23.1269	20.7298	18.7830	17.1704	15.8128	15.2115	14.6542	14.1363	82
83	23.1822	20.7743	18.8195	17.2009	15.8387	15.2354	14.6764	14.1570	83
84	23.2359	20.8174	18.8548	17.2304	15.8637	15.2586	14.6979	14.1770	84
85	23.2880	20.8592	18.8891	17.2591	15.8880	15.2810	14.7187	14.1964	85
86	23.3385	20.8997	18.9223	17.2868	15.9115	15.3028	14.7389	14.2151	86
87	23.3875	20.9390	18.9545	17.3137	15.9343	15.3238	14.7584	14.2333	87
88	23.4351	20.9771	18.9858	17.3397	15.9563	15.3442	14.7774	14.2509	88
89	23.4812	21.0141	19.0160	17.3650	15.9777	15.3640	14.7957	14.2679	89
90	23.5260	21.0499	19.0454	17.3894	15.9984	15.3832	14.8135	14.2845	90
91	23.5694	21.0847	19.0738	17.4131	16.0185	15.4017	14.8307	14.3005	91
92	23.6115	21.1184	19.1014	17.4361	16.0379	15.4197	14.8473	14.3159	92
93	23.6524	21.1511	19.1282	17.4584	16.0568	15.4371	14.8635	14.3310	93
94	23.6921	21.1828	19.1541	17.4800	16.0751	15.4540	14.8791	14.3455	94
95	23.7306	21.2135	19.1792	17.5010	16.0928	15.4704	14.8943	14.3596	95
96	23.7679	21.2434	19.2036	17.5213	16.1099	15.4862	14.9090	14.3733	96
97	23.8041	21.2723	19.2273	17.5409	16.1266	15.5016	14.9233	14.3865	97
98	23.8393	21.3004	19.2502	17.5600	16.1427	15.5165	14.9371	14.3994	98
99	23.8734	21.3276	19.2724	17.5785	16.1583	15.5309	14.9504	14.4118	99
100	23.9065	21.3540	19.2940	17.5965	16.1735	15.5449	14.9634	14.4238	100

YEARS' PURCHASE

Rate Per Cent

Yrs.	7	7.25	7.5	8	8.5	9	9.5	10	Yrs.
1	0.8467	0.8449	0.8431	0.8396	0.8360	0.8326	0.8291	0.8257	1
2	1.6198	1.6133	1.6068	1.5940	1.5814	1.5690	1.5568	1.5448	2
3	2.3284	2.3149	2.3016	2.2754	2.2498	2.2248	2.2003	2.1764	3
4	2.9799	2.9578	2.9361	2.8936	2.8524	2.8123	2.7733	2.7353	4
5	3.5806	3.5488	3.5176	3.4568	3.3981	3.3413	3.2864	3.2333	5
6	4.1361	4.0937	4.0523	3.9718	3.8945	3.8201	3.7485	3.6795	6
7	4.6510	4.5976	4.5453	4.4443	4.3477	4.2552	4.1665	4.0815	7
8	5.1295	5.0645	5.0012	4.8792	4.7630	4.6522	4.5465	4.4454	8
9	5.5750	5.4984	5.4238	5.2806	5.1448	5.0158	4.8931	4.7762	9
10	5.9908	5.9024	5.8166	5.6522	5.4968	5.3498	5.2104	5.0781	10
11	6.3795	6.2793	6.1823	5.9969	5.8223	5.6576	5.5020	5.3547	11
12	6.7435	6.6317	6.5235	6.3175	6.1240	5.9421	5.7706	5.6088	12
13	7.0850	6.9617	6.8426	6.6162	6.4043	6.2056	6.0189	5.8430	13
14	7.4058	7.2712	7.1414	6.8952	6.6654	6.4504	6.2489	6.0595	14
15	7.7077	7.5620	7.4217	7.1561	6.9089	6.6782	6.4624	6.2601	15
16	7.9921	7.8356	7.6850	7.4007	7.1366	6.8907	6.6612	6.4465	16
17	8.2605	8.0933	7.9328	7.6302	7.3498	7.0893	6.8466	6.6200	17
18	8.5140	8.3365	8.1663	7.8460	7.5498	7.2752	7.0198	6.7818	18
19	8.7537	8.5662	8.3866	8.0491	7.7377	7.4495	7.1820	6.9330	19
20	8.9806	8.7834	8.5947	8.2406	7.9145	7.6132	7.3340	7.0746	20
21	9.1957	8.9891	8.7915	8.4213	8.0810	7.7672	7.4768	7.2074	21
22	9.3997	9.1839	8.9778	8.5921	8.2382	7.9123	7.6112	7.3321	22
23	9.5935	9.3688	9.1544	8.7537	8.3866	8.0491	7.7377	7.4495	23
24	9.7776	9.5443	9.3218	8.9067	8.5270	8.1783	7.8570	7.5600	24
25	9.9527	9.7111	9.4809	9.0518	8.6599	8.3004	7.9697	7.6643	25
26	10.1194	9.8697	9.6320	9.1895	8.7858	8.4161	8.0762	7.7628	26
27	10.2782	10.0207	9.7758	9.3202	8.9052	8.5256	8.1770	7.8559	27
28	10.4296	10.1645	9.9126	9.4445	9.0186	8.6295	8.2726	7.9440	28
29	10.5740	10.3016	10.0430	9.5628	9.1264	8.7281	8.3632	8.0275	29
30	10.7118	10.4325	10.1673	9.6754	9.2289	8.8219	8.4492	8.1067	30
31	10.8435	10.5573	10.2858	9.7827	9.3265	8.9110	8.5309	8.1819	31
32	10.9694	10.6766	10.3990	9.8850	9.4195	8.9958	8.6086	8.2534	32
33	11.0897	10.7906	10.5071	9.9827	9.5081	9.0766	8.6826	8.3213	33
34	11.2049	10.8996	10.6105	10.0759	9.5926	9.1536	8.7530	8.3860	34
35	11.3152	11.0039	10.7093	10.1650	9.6733	9.2271	8.8201	8.4476	35
36	11.4208	11.1037	10.8038	10.2501	9.7504	9.2972	8.8842	8.5063	36
37	11.5220	11.1994	10.8944	10.3316	9.8241	9.3641	8.9453	8.5623	37
38	11.6190	11.2910	10.9811	10.4095	9.8945	9.4281	9.0037	8.6158	38
39	11.7121	11.3789	11.0641	10.4842	9.9619	9.4893	9.0594	8.6669	39
40	11.8014	11.4632	11.1438	10.5557	10.0265	9.5478	9.1128	8.7157	40
41	11.8871	11.5440	11.2202	10.6242	10.0883	9.6039	9.1638	8.7623	41
42	11.9694	11.6216	11.2935	10.6899	10.1475	9.6575	9.2127	8.8070	42
43	12.0485	11.6962	11.3639	10.7529	10.2043	9.7089	9.2594	8.8497	43
44	12.1245	11.7678	11.4315	10.8134	10.2588	9.7582	9.3043	8.8907	44
45	12.1976	11.8366	11.4964	10.8715	10.3110	9.8055	9.3472	8.9299	45
46	12.2678	11.9028	11.5588	10.9273	10.3612	9.8509	9.3884	8.9675	46
47	12.3354	11.9664	11.6188	10.9809	10.4094	9.8944	9.4280	9.0036	47
48	12.4005	12.0276	11.6765	11.0324	10.4557	9.9362	9.4659	9.0382	48
49	12.4631	12.0865	11.7320	11.0820	10.5001	9.9764	9.5024	9.0714	49
50	12.5234	12.1432	11.7854	11.1296	10.5429	10.0150	9.5374	9.1033	50

YEARS' PURCHASE

Yrs.	7	7.25	7.5	8	8.5	9	9.5	10	Yrs.
				Rate Per Cent					
51	12.5815	12.1978	11.8368	11.1754	10.5840	10.0521	9.5710	9.1339	51
52	12.6374	12.2504	11.8863	11.2195	10.6236	10.0877	9.6034	9.1634	52
53	12.6913	12.3010	11.9340	11.2620	10.6616	10.1221	9.6345	9.1917	53
54	12.7433	12.3498	11.9799	11.3029	10.6983	10.1551	9.6644	9.2189	54
55	12.7933	12.3969	12.0242	11.3423	10.7336	10.1869	9.6932	9.2451	55
56	12.8416	12.4422	12.0669	11.3802	10.7676	10.2175	9.7209	9.2703	56
57	12.8882	12.4859	12.1080	11.4168	10.8003	10.2469	9.7475	9.2945	57
58	12.9332	12.5281	12.1476	11.4521	10.8318	10.2753	9.7732	9.3179	58
59	12.9765	12.5688	12.1859	11.4860	10.8622	10.3027	9.7980	9.3404	59
60	13.0184	12.6081	12.2228	11.5188	10.8915	10.3290	9.8218	9.3620	60
61	13.0588	12.6459	12.2584	11.5504	10.9198	10.3545	9.8448	9.3829	61
62	13.0978	12.6825	12.2928	11.5809	10.9471	10.3790	9.8669	9.4030	62
63	13.1355	12.7178	12.3259	11.6104	10.9734	10.4026	9.8883	9.4224	63
64	13.1718	12.7519	12.3580	11.6388	10.9987	10.4254	9.9089	9.4411	64
65	13.2070	12.7849	12.3889	11.6662	11.0232	10.4474	9.9288	9.4592	65
66	13.2409	12.8167	12.4187	11.6927	11.0469	10.4686	9.9479	9.4766	66
67	13.2737	12.8474	12.4476	11.7183	11.0697	10.4891	9.9664	9.4934	67
68	13.3054	12.8771	12.4755	11.7430	11.0917	10.5089	9.9843	9.5096	68
69	13.3361	12.9058	12.5024	11.7668	11.1130	10.5280	10.0015	9.5252	69
70	13.3657	12.9335	12.5284	11.7899	11.1336	10.5465	10.0182	9.5403	70
71	13.3943	12.9603	12.5536	11.8122	11.1534	10.5643	10.0343	9.5549	71
72	13.4220	12.9863	12.5779	11.8337	11.1726	10.5815	10.0498	9.5690	72
73	13.4488	13.0113	12.6014	11.8545	11.1912	10.5981	10.0648	9.5826	73
74	13.4747	13.0356	12.6242	11.8746	11.2091	10.6142	10.0793	9.5957	74
75	13.4997	13.0590	12.6461	11.8941	11.2264	10.6298	10.0933	9.6084	75
76	13.5240	13.0817	12.6674	11.9129	11.2432	10.6448	10.1069	9.6207	76
77	13.5474	13.1036	12.6880	11.9311	11.2594	10.6593	10.1199	9.6325	77
78	13.5701	13.1248	12.7079	11.9487	11.2751	10.6733	10.1326	9.6440	78
79	13.5921	13.1454	12.7271	11.9657	11.2902	10.6869	10.1448	9.6551	79
80	13.6133	13.1653	12.7458	11.9821	11.3049	10.7001	10.1567	9.6658	80
81	13.6339	13.1845	12.7638	11.9981	11.3190	10.7128	10.1681	9.6762	81
82	13.6538	13.2031	12.7812	12.0135	11.3328	10.7250	10.1792	9.6862	82
83	13.6731	13.2211	12.7981	12.0284	11.3460	10.7369	10.1899	9.6959	83
84	13.6917	13.2386	12.8145	12.0429	11.3589	10.7484	10.2002	9.7053	84
85	13.7098	13.2555	12.8303	12.0568	11.3713	10.7596	10.2103	9.7143	85
86	13.7273	13.2718	12.8456	12.0704	11.3834	10.7703	10.2200	9.7231	86
87	13.7442	13.2877	12.8604	12.0835	11.3950	10.7808	10.2294	9.7316	87
88	13.7606	13.3030	12.8748	12.0961	11.4063	10.7909	10.2385	9.7398	88
89	13.7765	13.3178	12.8887	12.1084	11.4172	10.8006	10.2472	9.7478	89
90	13.7919	13.3322	12.9022	12.1203	11.4278	10.8101	10.2558	9.7555	90
91	13.8068	13.3462	12.9152	12.1318	11.4380	10.8192	10.2640	9.7630	91
92	13.8213	13.3597	12.9279	12.1430	11.4479	10.8281	10.2720	9.7702	92
93	13.8353	13.3727	12.9401	12.1538	11.4575	10.8367	10.2797	9.7772	93
94	13.8488	13.3854	12.9520	12.1642	11.4668	10.8450	10.2872	9.7840	94
95	13.8620	13.3977	12.9635	12.1744	11.4758	10.8531	10.2944	9.7905	95
96	13.8747	13.4096	12.9746	12.1842	11.4845	10.8609	10.3015	9.7969	96
97	13.8871	13.4211	12.9854	12.1937	11.4930	10.8684	10.3083	9.8030	97
98	13.8990	13.4323	12.9959	12.2029	11.5012	10.8758	10.3149	9.8090	98
99	13.9106	13.4431	13.0060	12.2119	11.5091	10.8829	10.3212	9.8147	99
100	13.9218	13.4536	13.0158	12.2205	11.5168	10.8897	10.3274	9.8203	100

YEARS' PURCHASE

				Rate Per Cent					
Yrs.	11	12	13	14	15	16	18	20	Yrs.
1	0.8189	0.8123	0.8057	0.7993	0.7930	0.7867	0.7745	0.7627	1
2	1.5213	1.4985	1.4764	1.4549	1.4340	1.4137	1.3749	1.3381	2
3	2.1300	2.0856	2.0430	2.0021	1.9628	1.9250	1.8536	1.7874	3
4	2.6625	2.5935	2.5279	2.4656	2.4062	2.3497	2.2442	2.1478	4
5	3.1320	3.0369	2.9474	2.8630	2.7833	2.7080	2.5688	2.4433	5
6	3.5489	3.4273	3.3137	3.2074	3.1078	3.0141	2.8427	2.6898	6
7	3.9215	3.7735	3.6363	3.5087	3.3897	3.2786	3.0769	2.8985	7
8	4.2562	4.0824	3.9223	3.7743	3.6370	3.5094	3.2792	3.0774	8
9	4.5585	4.3597	4.1776	4.0101	3.8555	3.7124	3.4558	3.2324	9
10	4.8327	4.6099	4.4068	4.2208	4.0498	3.8922	3.6111	3.3679	10
11	5.0825	4.8367	4.6136	4.4101	4.2238	4.0526	3.7488	3.4873	11
12	5.3109	5.0431	4.8010	4.5810	4.3804	4.1965	3.8716	3.5934	12
13	5.5205	5.2317	4.9716	4.7361	4.5219	4.3263	3.9818	3.6881	13
14	5.7133	5.4046	5.1274	4.8774	4.6505	4.4439	4.0811	3.7732	14
15	5.8913	5.5636	5.2704	5.0065	4.7678	4.5508	4.1712	3.8500	15
16	6.0561	5.7103	5.4018	5.1250	4.8751	4.6485	4.2531	3.9197	16
17	6.2089	5.8460	5.5231	5.2340	4.9737	4.7380	4.3279	3.9831	17
18	6.3511	5.9718	5.6353	5.3346	5.0645	4.8203	4.3965	4.0412	18
19	6.4835	6.0887	5.7393	5.4278	5.1483	4.8963	4.4596	4.0944	19
20	6.6072	6.1977	5.8360	5.5142	5.2260	4.9665	4.5177	4.1433	20
21	6.7229	6.2994	5.9261	5.5945	5.2981	5.0315	4.5715	4.1885	21
22	6.8313	6.3944	6.0101	5.6694	5.3652	5.0920	4.6214	4.2304	22
23	6.9330	6.4835	6.0887	5.7393	5.4278	5.1483	4.6677	4.2692	23
24	7.0286	6.5671	6.1624	5.8047	5.4862	5.2009	4.7109	4.3052	24
25	7.1187	6.6456	6.2315	5.8659	5.5409	5.2500	4.7511	4.3389	25
26	7.2036	6.7195	6.2964	5.9235	5.5922	5.2960	4.7888	4.3702	26
27	7.2837	6.7892	6.3575	5.9775	5.6404	5.3392	4.8241	4.3996	27
28	7.3594	6.8549	6.4151	6.0284	5.6857	5.3798	4.8572	4.4271	28
29	7.4310	6.9170	6.4695	6.0764	5.7283	5.4179	4.8883	4.4529	29
30	7.4988	6.9757	6.5208	6.1216	5.7685	5.4539	4.9175	4.4772	30
31	7.5631	7.0313	6.5694	6.1644	5.8065	5.4878	4.9451	4.5000	31
32	7.6241	7.0840	6.6154	6.2049	5.8424	5.5199	4.9711	4.5216	32
33	7.6821	7.1340	6.6590	6.2432	5.8764	5.5502	4.9957	4.5419	33
34	7.7371	7.1815	6.7003	6.2796	5.9085	5.5789	5.0189	4.5611	34
35	7.7896	7.2266	6.7396	6.3140	5.9391	5.6061	5.0409	4.5792	35
36	7.8395	7.2696	6.7769	6.3468	5.9680	5.6319	5.0618	4.5964	36
37	7.8870	7.3104	6.8124	6.3779	5.9955	5.6564	5.0815	4.6127	37
38	7.9324	7.3494	6.8462	6.4075	6.0217	5.6797	5.1003	4.6282	38
39	7.9756	7.3865	6.8784	6.4357	6.0466	5.7018	5.1182	4.6429	39
40	8.0169	7.4219	6.9091	6.4626	6.0703	5.7229	5.1352	4.6569	40
41	8.0564	7.4557	6.9384	6.4882	6.0929	5.7430	5.1513	4.6702	41
42	8.0941	7.4880	6.9664	6.5127	6.1145	5.7622	5.1667	4.6828	42
43	8.1302	7.5189	6.9931	6.5360	6.1350	5.7804	5.1814	4.6949	43
44	8.1648	7.5484	7.0186	6.5583	6.1547	5.7979	5.1954	4.7064	44
45	8.1978	7.5767	7.0431	6.5797	6.1735	5.8145	5.2088	4.7173	45
46	8.2295	7.6038	7.0664	6.6001	6.1914	5.8304	5.2216	4.7278	46
47	8.2599	7.6297	7.0888	6.6196	6.2086	5.8457	5.2338	4.7378	47
48	8.2890	7.6545	7.1103	6.6383	6.2250	5.8602	5.2454	4.7474	48
49	8.3169	7.6783	7.1308	6.6562	6.2408	5.8742	5.2566	4.7565	49
50	8.3437	7.7012	7.1505	6.6733	6.2558	5.8875	5.2673	4.7653	50

YEARS' PURCHASE

				Rate Per Cent					
Yrs.	11	12	13	14	15	16	18	20	Yrs.
51	8.3695	7.7231	7.1694	6.6898	6.2703	5.9003	5.2775	4.7737	51
52	8.3942	7.7441	7.1875	6.7055	6.2842	5.9126	5.2874	4.7817	52
53	8.4179	7.7643	7.2049	6.7207	6.2975	5.9244	5.2968	4.7894	53
54	8.4408	7.7837	7.2216	6.7352	6.3102	5.9357	5.3058	4.7968	54
55	8.4627	7.8024	7.2377	6.7492	6.3225	5.9465	5.3145	4.8039	55
56	8.4838	7.8203	7.2531	6.7626	6.3343	5.9569	5.3228	4.8107	56
57	8.5041	7.8376	7.2680	6.7755	6.3456	5.9669	5.3308	4.8172	57
58	8.5237	7.8542	7.2822	6.7879	6.3564	5.9766	5.3384	4.8234	58
59	8.5425	7.8702	7.2960	6.7998	6.3669	5.9858	5.3458	4.8295	59
60	8.5606	7.8855	7.3092	6.8113	6.3770	5.9947	5.3529	4.8353	60
61	8.5780	7.9003	7.3219	6.8224	6.3866	6.0032	5.3597	4.8408	61
62	8.5949	7.9146	7.3341	6.8330	6.3960	6.0115	5.3663	4.8462	62
63	8.6111	7.9283	7.3459	6.8432	6.4049	6.0194	5.3726	4.8513	63
64	8.6267	7.9416	7.3573	6.8531	6.4136	6.0270	5.3787	4.8563	64
65	8.6417	7.9543	7.3682	6.8626	6.4219	6.0344	5.3845	4.8610	65
66	8.6563	7.9666	7.3788	6.8717	6.4299	6.0414	5.3902	4.8656	66
67	8.6703	7.9785	7.3890	6.8806	6.4376	6.0483	5.3956	4.8700	67
68	8.6838	7.9899	7.3988	6.8891	6.4451	6.0548	5.4008	4.8743	68
69	8.6968	8.0010	7.4083	6.8973	6.4523	6.0612	5.4059	4.8784	69
70	8.7094	8.0116	7.4174	6.9052	6.4592	6.0673	5.4107	4.8824	70
71	8.7216	8.0219	7.4262	6.9128	6.4659	6.0732	5.4154	4.8862	71
72	8.7333	8.0318	7.4347	6.9202	6.4723	6.0789	5.4199	4.8899	72
73	8.7446	8.0414	7.4429	6.9273	6.4785	6.0843	5.4243	4.8934	73
74	8.7556	8.0507	7.4508	6.9342	6.4845	6.0896	5.4285	4.8968	74
75	8.7661	8.0596	7.4585	6.9408	6.4903	6.0948	5.4326	4.9001	75
76	8.7763	8.0682	7.4659	6.9472	6.4959	6.0997	5.4365	4.9033	76
77	8.7862	8.0766	7.4730	6.9534	6.5013	6.1045	5.4403	4.9064	77
78	8.7957	8.0846	7.4799	6.9594	6.5065	6.1091	5.4439	4.9094	78
79	8.8050	8.0924	7.4866	6.9651	6.5116	6.1135	5.4474	4.9123	79
80	8.8139	8.1000	7.4930	6.9707	6.5165	6.1178	5.4509	4.9150	80
81	8.8225	8.1072	7.4992	6.9761	6.5212	6.1219	5.4541	4.9177	81
82	8.8308	8.1143	7.5053	6.9813	6.5257	6.1260	5.4573	4.9203	82
83	8.8389	8.1211	7.5111	6.9863	6.5301	6.1298	5.4604	4.9228	83
84	8.8467	8.1276	7.5167	6.9912	6.5344	6.1336	5.4634	4.9252	84
85	8.8542	8.1340	7.5222	6.9959	6.5385	6.1372	5.4663	4.9276	85
86	8.8615	8.1402	7.5274	7.0005	6.5425	6.1407	5.4690	4.9298	86
87	8.8686	8.1461	7.5325	7.0049	6.5463	6.1441	5.4717	4.9320	87
88	8.8754	8.1519	7.5374	7.0091	6.5500	6.1474	5.4743	4.9341	88
89	8.8820	8.1575	7.5422	7.0133	6.5536	6.1505	5.4768	4.9361	89
90	8.8884	8.1629	7.5468	7.0172	6.5571	6.1536	5.4793	4.9381	90
91	8.8946	8.1681	7.5513	7.0211	6.5605	6.1566	5.4816	4.9400	91
92	8.9006	8.1731	7.5556	7.0248	6.5637	6.1594	5.4839	4.9419	92
93	8.9064	8.1780	7.5598	7.0284	6.5669	6.1622	5.4861	4.9437	93
94	8.9120	8.1828	7.5638	7.0319	6.5699	6.1649	5.4882	4.9454	94
95	8.9174	8.1873	7.5677	7.0353	6.5729	6.1675	5.4903	4.9471	95
96	8.9227	8.1918	7.5715	7.0386	6.5758	6.1700	5.4923	4.9487	96
97	8.9278	8.1961	7.5752	7.0418	6.5785	6.1725	5.4942	4.9503	97
98	8.9328	8.2002	7.5788	7.0449	6.5812	6.1748	5.4961	4.9518	98
99	8.9375	8.2043	7.5822	7.0478	6.5838	6.1771	5.4979	4.9533	99
100	8.9422	8.2082	7.5855	7.0507	6.5863	6.1793	5.4997	4.9547	100

YEARS' PURCHASE

Rate Per Cent

Yrs.	4	4.5	5	5.5	6	6.25	6.5	6.75	Yrs.
1	0.8687	0.8650	0.8612	0.8576	0.8539	0.8521	0.8503	0.8485	1
2	1.7104	1.6959	1.6816	1.6676	1.6538	1.6470	1.6403	1.6336	2
3	2.5256	2.4941	2.4634	2.4334	2.4042	2.3898	2.3756	2.3616	3
4	3.3150	3.2610	3.2087	3.1580	3.1089	3.0849	3.0613	3.0381	4
5	4.0793	3.9977	3.9194	3.8441	3.7716	3.7363	3.7018	3.6678	5
6	4.8190	4.7056	4.5974	4.4941	4.3953	4.3476	4.3008	4.2551	6
7	5.5347	5.3857	5.2445	5.1105	4.9831	4.9218	4.8620	4.8036	7
8	6.2272	6.0391	5.8621	5.6952	5.5375	5.4619	5.3883	5.3167	8
9	6.8969	6.6670	6.4519	6.2503	6.0609	5.9704	5.8826	5.7974	9
10	7.5446	7.2703	7.0153	6.7776	6.5554	6.4497	6.3474	6.2482	10
11	8.1707	7.8500	7.5536	7.2787	7.0231	6.9019	6.7848	6.6717	11
12	8.7760	8.4071	8.0680	7.7551	7.4656	7.3289	7.1970	7.0698	12
13	9.3610	8.9424	8.5597	8.2084	7.8848	7.7324	7.5857	7.4446	13
14	9.9262	9.4569	9.0299	8.6398	8.2820	8.1140	7.9527	7.7977	14
15	10.4723	9.9512	9.4796	9.0506	8.6587	8.4753	8.2994	8.1307	15
16	10.9998	10.4263	9.9097	9.4419	9.0162	8.8175	8.6273	8.4452	16
17	11.5092	10.8829	10.3213	9.8148	9.3557	9.1418	8.9376	8.7422	17
18	12.0011	11.3217	10.7151	10.1703	9.6781	9.4495	9.2314	9.0232	18
19	12.4760	11.7434	11.0921	10.5093	9.9846	9.7415	9.5099	9.2890	19
20	12.9344	12.1487	11.4530	10.8327	10.2761	10.0187	9.7739	9.5408	20
21	13.3769	12.5383	11.7986	11.1414	10.5535	10.2822	10.0245	9.7794	21
22	13.8039	12.9127	12.1296	11.4360	10.8175	10.5326	10.2624	10.0057	22
23	14.2160	13.2726	12.4466	11.7174	11.0689	10.7708	10.4884	10.2204	23
24	14.6135	13.6184	12.7503	11.9861	11.3084	10.9975	10.7032	10.4243	24
25	14.9970	13.9509	13.0412	12.2429	11.5367	11.2133	10.9075	10.6180	25
26	15.3669	14.2705	13.3200	12.4883	11.7544	11.4188	11.1019	10.8021	26
27	15.7237	14.5776	13.5873	12.7229	11.9620	11.6146	11.2869	10.9771	27
28	16.0677	14.8728	13.8434	12.9472	12.1600	11.8013	11.4631	11.1437	28
29	16.3994	15.1566	14.0889	13.1618	12.3491	11.9792	11.6309	11.3023	29
30	16.7193	15.4294	14.3243	13.3670	12.5296	12.1490	11.7909	11.4533	30
31	17.0276	15.6916	14.5501	13.5633	12.7019	12.3110	11.9434	11.5971	31
32	17.3248	15.9437	14.7665	13.7513	12.8666	12.4656	12.0889	11.7342	32
33	17.6113	16.1860	14.9742	13.9311	13.0239	12.6132	12.2277	11.8650	33
34	17.8874	16.4189	15.1733	14.1033	13.1743	12.7542	12.3601	11.9896	34
35	18.1535	16.6428	15.3643	14.2682	13.3181	12.8889	12.4866	12.1086	35
36	18.4098	16.8581	15.5476	14.4261	13.4556	13.0177	12.6074	12.2221	36
37	18.6569	17.0650	15.7234	14.5774	13.5870	13.1407	12.7227	12.3305	37
38	18.8949	17.2639	15.8921	14.7223	13.7128	13.2583	12.8330	12.4340	38
39	19.1242	17.4551	16.0540	14.8611	13.8332	13.3708	12.9383	12.5329	39
40	19.3451	17.6389	16.2094	14.9941	13.9484	13.4784	13.0390	12.6274	40
41	19.5578	17.8157	16.3585	15.1216	14.0587	13.5813	13.1354	12.7177	41
42	19.7628	17.9855	16.5016	15.2439	14.1643	13.6798	13.2275	12.8041	42
43	19.9601	18.1489	16.6390	15.3610	14.2654	13.7741	13.3156	12.8866	43
44	20.1502	18.3059	16.7708	15.4733	14.3622	13.8644	13.3999	12.9656	44
45	20.3333	18.4568	16.8975	15.5811	14.4549	13.9508	13.4806	13.0411	45
46	20.5095	18.6019	17.0190	15.6843	14.5438	14.0335	13.5579	13.1134	46
47	20.6793	18.7415	17.1357	15.7834	14.6289	14.1128	13.6318	13.1826	47
48	20.8427	18.8756	17.2478	15.8784	14.7105	14.1887	13.7027	13.2488	48
49	21.0000	19.0045	17.3554	15.9696	14.7887	14.2615	13.7705	13.3122	49
50	21.1515	19.1285	17.4587	16.0570	14.8637	14.3312	13.8355	13.3729	50

YEARS' PURCHASE

Rate Per Cent

Yrs.	4	4.5	5	5.5	6	6.25	6.5	6.75	Yrs.
51	21.2973	19.2477	17.5579	16.1409	14.9356	14.3980	13.8977	13.4311	51
52	21.4377	19.3623	17.6532	16.2214	15.0045	14.4620	13.9574	13.4868	52
53	21.5728	19.4725	17.7448	16.2987	15.0705	14.5234	14.0145	13.5401	53
54	21.7029	19.5784	17.8327	16.3728	15.1339	14.5822	14.0693	13.5912	54
55	21.8281	19.6802	17.9171	16.4440	15.1947	14.6386	14.1218	13.6402	55
56	21.9486	19.7781	17.9982	16.5123	15.2530	14.6927	14.1721	13.6872	56
57	22.0646	19.8722	18.0762	16.5778	15.3089	14.7446	14.2204	13.7322	57
58	22.1762	19.9627	18.1510	16.6408	15.3626	14.7944	14.2667	13.7754	58
59	22.2836	20.0497	18.2229	16.7012	15.4140	14.8421	14.3111	13.8167	59
60	22.3870	20.1334	18.2920	16.7592	15.4634	14.8879	14.3536	13.8564	60
61	22.4865	20.2138	18.3584	16.8149	15.5108	14.9318	14.3945	13.8945	61
62	22.5823	20.2912	18.4221	16.8684	15.5563	14.9740	14.4336	13.9310	62
63	22.6744	20.3655	18.4834	16.9197	15.6000	15.0144	14.4712	13.9660	63
64	22.7630	20.4370	18.5423	16.9690	15.6419	15.0532	14.5073	13.9995	64
65	22.8483	20.5057	18.5988	17.0164	15.6821	15.0905	14.5419	14.0318	65
66	22.9304	20.5718	18.6532	17.0619	15.7207	15.1263	14.5751	14.0627	66
67	23.0094	20.6353	18.7054	17.1055	15.7578	15.1606	14.6069	14.0923	67
68	23.0853	20.6964	18.7556	17.1475	15.7934	15.1935	14.6375	14.1208	68
69	23.1584	20.7552	18.8038	17.1878	15.8276	15.2251	14.6669	14.1481	69
70	23.2288	20.8116	18.8501	17.2265	15.8604	15.2555	14.6951	14.1743	70
71	23.2964	20.8659	18.8946	17.2637	15.8919	15.2847	14.7221	14.1995	71
72	23.3615	20.9181	18.9374	17.2994	15.9222	15.3127	14.7481	14.2237	72
73	23.4241	20.9683	18.9786	17.3337	15.9513	15.3395	14.7730	14.2468	73
74	23.4844	21.0166	19.0181	17.3667	15.9792	15.3654	14.7970	14.2691	74
75	23.5423	21.0630	19.0561	17.3984	16.0060	15.3901	14.8199	14.2905	75
76	23.5981	21.1076	19.0926	17.4288	16.0317	15.4139	14.8420	14.3110	76
77	23.6517	21.1505	19.1277	17.4580	16.0565	15.4368	14.8632	14.3307	77
78	23.7033	21.1917	19.1614	17.4861	16.0802	15.4588	14.8836	14.3496	78
79	23.7529	21.2314	19.1938	17.5131	16.1030	15.4799	14.9031	14.3678	79
80	23.8007	21.2695	19.2250	17.5390	16.1250	15.5001	14.9219	14.3853	80
81	23.8466	21.3062	19.2549	17.5640	16.1460	15.5196	14.9399	14.4020	81
82	23.8907	21.3414	19.2837	17.5879	16.1663	15.5383	14.9572	14.4181	82
83	23.9332	21.3753	19.3114	17.6109	16.1857	15.5562	14.9739	14.4336	83
84	23.9741	21.4079	19.3380	17.6330	16.2044	15.5735	14.9899	14.4484	84
85	24.0134	21.4392	19.3635	17.6543	16.2223	15.5901	15.0052	14.4627	85
86	24.0512	21.4694	19.3881	17.6747	16.2396	15.6060	15.0200	14.4764	86
87	24.0876	21.4983	19.4117	17.6943	16.2561	15.6213	15.0342	14.4896	87
88	24.1225	21.5262	19.4344	17.7132	16.2721	15.6360	15.0478	14.5022	88
89	24.1562	21.5530	19.4563	17.7313	16.2874	15.6501	15.0609	14.5144	89
90	24.1885	21.5787	19.4773	17.7488	16.3021	15.6637	15.0734	14.5260	90
91	24.2196	21.6035	19.4974	17.7655	16.3162	15.6767	15.0855	14.5372	91
92	24.2495	21.6273	19.5168	17.7816	16.3298	15.6893	15.0971	14.5480	92
93	24.2783	21.6502	19.5354	17.7971	16.3428	15.7013	15.1083	14.5584	93
94	24.3060	21.6722	19.5534	17.8119	16.3553	15.7129	15.1190	14.5683	94
95	24.3326	21.6933	19.5706	17.8262	16.3674	15.7240	15.1293	14.5779	95
96	24.3582	21.7137	19.5871	17.8400	16.3790	15.7347	15.1392	14.5871	96
97	24.3829	21.7333	19.6031	17.8532	16.3901	15.7449	15.1487	14.5959	97
98	24.4065	21.7521	19.6184	17.8659	16.4008	15.7548	15.1578	14.6044	98
99	24.4293	21.7702	19.6331	17.8781	16.4111	15.7643	15.1666	14.6125	99
100	24.4512	21.7875	19.6472	17.8898	16.4210	15.7734	15.1750	14.6204	100

YEARS' PURCHASE

Rate Per Cent

Yrs.	7	7.25	7.5	8	8.5	9	9.5	10	Yrs.
1	0.8467	0.8449	0.8431	0.8396	0.8360	0.8326	0.8291	0.8257	1
2	1.6269	1.6203	1.6138	1.6009	1.5882	1.5756	1.5633	1.5512	2
3	2.3477	2.3340	2.3205	2.2939	2.2679	2.2424	2.2176	2.1933	3
4	3.0152	2.9926	2.9704	2.9269	2.8847	2.8437	2.8038	2.7651	4
5	3.6345	3.6018	3.5696	3.5070	3.4466	3.3882	3.3318	3.2772	5
6	4.2103	4.1664	4.1235	4.0402	3.9602	3.8833	3.8093	3.7381	6
7	4.7466	4.6909	4.6366	4.5315	4.4311	4.3351	4.2431	4.1549	7
8	5.2470	5.1790	5.1128	4.9854	4.8641	4.7486	4.6385	4.5334	8
9	5.7145	5.6340	5.5558	5.4056	5.2634	5.1284	5.0002	4.8782	9
10	6.1521	6.0589	5.9685	5.7956	5.6324	5.4781	5.3320	5.1936	10
11	6.5622	6.4563	6.3537	6.1581	5.9742	5.8009	5.6374	5.4828	11
12	6.9470	6.8284	6.7138	6.4957	6.2914	6.0995	5.9190	5.7489	12
13	7.3085	7.1774	7.0509	6.8108	6.5865	6.3765	6.1795	5.9943	13
14	7.6486	7.5051	7.3668	7.1051	6.8614	6.6338	6.4208	6.2211	14
15	7.9688	7.8131	7.6634	7.3806	7.1179	6.8733	6.6450	6.4313	15
16	8.2705	8.1030	7.9421	7.6388	7.3578	7.0967	6.8535	6.6264	16
17	8.5553	8.3761	8.2043	7.8810	7.5822	7.3053	7.0478	6.8079	17
18	8.8241	8.6336	8.4512	8.1086	7.7927	7.5004	7.2293	6.9771	18
19	9.0782	8.8767	8.6840	8.3227	7.9902	7.6832	7.3990	7.1350	19
20	9.3185	9.1064	8.9037	8.5242	8.1757	7.8547	7.5578	7.2826	20
21	9.5460	9.3235	9.1111	8.7142	8.3503	8.0157	7.7068	7.4208	21
22	9.7615	9.5290	9.3072	8.8934	8.5148	8.1671	7.8466	7.5504	22
23	9.9658	9.7235	9.4928	9.0626	8.6698	8.3096	7.9781	7.6720	23
24	10.1595	9.9079	9.6684	9.2226	8.8160	8.4438	8.1018	7.7864	24
25	10.3434	10.0827	9.8348	9.3738	8.9542	8.5705	8.2183	7.8939	25
26	10.5180	10.2485	9.9925	9.5170	9.0847	8.6900	8.3281	7.9952	26
27	10.6839	10.4060	10.1422	9.6527	9.2082	8.8029	8.4318	8.0907	27
28	10.8417	10.5556	10.2842	9.7812	9.3252	8.9097	8.5298	8.1809	28
29	10.9917	10.6977	10.4191	9.9032	9.4359	9.0108	8.6223	8.2660	29
30	11.1345	10.8329	10.5473	10.0189	9.5410	9.1065	8.7100	8.3465	30
31	11.2704	10.9615	10.6692	10.1288	9.6406	9.1972	8.7929	8.4226	31
32	11.3998	11.0839	10.7851	10.2333	9.7351	9.2833	8.8715	8.4947	32
33	11.5232	11.2005	10.8954	10.3325	9.8249	9.3649	8.9460	8.5630	33
34	11.6407	11.3115	11.0005	10.4270	9.9103	9.4424	9.0167	8.6277	34
35	11.7528	11.4174	11.1005	10.5168	9.9914	9.5160	9.0838	8.6892	35
36	11.8598	11.5182	11.1959	10.6023	10.0686	9.5860	9.1476	8.7475	36
37	11.9618	11.6145	11.2867	10.6838	10.1420	9.6526	9.2081	8.8029	37
38	12.0592	11.7063	11.3734	10.7614	10.2120	9.7159	9.2657	8.8555	38
39	12.1522	11.7939	11.4561	10.8354	10.2786	9.7761	9.3205	8.9055	39
40	12.2410	11.8775	11.5350	10.9060	10.3420	9.8335	9.3727	8.9531	40
41	12.3258	11.9574	11.6103	10.9733	10.4025	9.8882	9.4224	8.9984	41
42	12.4069	12.0337	11.6822	11.0375	10.4602	9.9403	9.4697	9.0416	42
43	12.4844	12.1066	11.7509	11.0988	10.5153	9.9900	9.5148	9.0827	43
44	12.5585	12.1762	11.8165	11.1573	10.5678	10.0374	9.5577	9.1218	44
45	12.6294	12.2428	11.8792	11.2132	10.6179	10.0826	9.5987	9.1591	45
46	12.6971	12.3065	11.9392	11.2666	10.6658	10.1258	9.6378	9.1947	46
47	12.7620	12.3674	11.9965	11.3176	10.7115	10.1670	9.6751	9.2287	47
48	12.8240	12.4257	12.0513	11.3664	10.7552	10.2063	9.7108	9.2611	48
49	12.8834	12.4814	12.1037	11.4130	10.7969	10.2439	9.7448	9.2920	49
50	12.9403	12.5348	12.1539	11.4576	10.8368	10.2798	9.7773	9.3216	50

YEARS' PURCHASE

Rate Per Cent

Yrs.	7	7.25	7.5	8	8.5	9	9.5	10	Yrs.
51	12.9947	12.5859	12.2019	11.5003	10.8750	10.3141	9.8083	9.3498	51
52	13.0469	12.6347	12.2479	11.5411	10.9115	10.3470	9.8380	9.3767	52
53	13.0968	12.6816	12.2919	11.5802	10.9464	10.3783	9.8663	9.4025	53
54	13.1446	12.7264	12.3340	11.6175	10.9797	10.4083	9.8935	9.4271	54
55	13.1904	12.7694	12.3743	11.6533	11.0117	10.4371	9.9194	9.4507	55
56	13.2343	12.8105	12.4130	11.6876	11.0423	10.4645	9.9442	9.4732	56
57	13.2764	12.8499	12.4500	11.7204	11.0716	10.4908	9.9680	9.4947	57
58	13.3168	12.8877	12.4854	11.7518	11.0996	10.5160	9.9907	9.5154	58
59	13.3554	12.9239	12.5194	11.7819	11.1264	10.5401	10.0124	9.5351	59
60	13.3925	12.9586	12.5520	11.8107	11.1522	10.5632	10.0332	9.5540	60
61	13.4280	12.9919	12.5832	11.8384	11.1768	10.5852	10.0532	9.5720	61
62	13.4621	13.0238	12.6131	11.8649	11.2004	10.6064	10.0723	9.5893	62
63	13.4948	13.0544	12.6418	11.8902	11.2230	10.6267	10.0905	9.6059	63
64	13.5261	13.0837	12.6693	11.9146	11.2447	10.6461	10.1081	9.6218	64
65	13.5562	13.1119	12.6957	11.9379	11.2655	10.6647	10.1249	9.6370	65
66	13.5851	13.1388	12.7210	11.9603	11.2854	10.6826	10.1409	9.6516	66
67	13.6127	13.1647	12.7453	11.9817	11.3045	10.6997	10.1564	9.6655	67
68	13.6393	13.1896	12.7685	12.0023	11.3228	10.7161	10.1711	9.6789	68
69	13.6648	13.2134	12.7909	12.0220	11.3403	10.7318	10.1853	9.6917	69
70	13.6892	13.2363	12.8123	12.0409	11.3572	10.7469	10.1989	9.7040	70
71	13.7127	13.2582	12.8328	12.0591	11.3733	10.7614	10.2119	9.7158	71
72	13.7352	13.2793	12.8526	12.0765	11.3888	10.7752	10.2244	9.7271	72
73	13.7569	13.2995	12.8715	12.0932	11.4037	10.7885	10.2364	9.7380	73
74	13.7776	13.3189	12.8897	12.1093	11.4179	10.8013	10.2478	9.7483	74
75	13.7975	13.3375	12.9071	12.1246	11.4316	10.8135	10.2589	9.7583	75
76	13.8167	13.3554	12.9239	12.1394	11.4447	10.8253	10.2694	9.7679	76
77	13.8350	13.3725	12.9399	12.1536	11.4573	10.8366	10.2796	9.7771	77
78	13.8527	13.3890	12.9553	12.1672	11.4694	10.8474	10.2893	9.7859	78
79	13.8696	13.4048	12.9702	12.1803	11.4810	10.8578	10.2987	9.7943	79
80	13.8859	13.4200	12.9844	12.1928	11.4922	10.8677	10.3076	9.8024	80
81	13.9015	13.4346	12.9980	12.2048	11.5029	10.8773	10.3162	9.8102	81
82	13.9165	13.4486	13.0111	12.2164	11.5131	10.8865	10.3245	9.8177	82
83	13.9309	13.4620	13.0237	12.2275	11.5230	10.8953	10.3324	9.8248	83
84	13.9447	13.4750	13.0358	12.2381	11.5325	10.9037	10.3400	9.8317	84
85	13.9580	13.4874	13.0474	12.2484	11.5416	10.9119	10.3473	9.8383	85
86	13.9708	13.4993	13.0586	12.2582	11.5503	10.9197	10.3543	9.8447	86
87	13.9830	13.5107	13.0693	12.2676	11.5587	10.9271	10.3611	9.8507	87
88	13.9948	13.5217	13.0796	12.2767	11.5667	10.9343	10.3675	9.8566	88
89	14.0061	13.5323	13.0895	12.2854	11.5744	10.9412	10.3737	9.8622	89
90	14.0170	13.5424	13.0990	12.2938	11.5819	10.9479	10.3797	9.8676	90
91	14.0274	13.5522	13.1081	12.3018	11.5890	10.9542	10.3854	9.8728	91
92	14.0375	13.5615	13.1168	12.3095	11.5958	10.9604	10.3909	9.8777	92
93	14.0471	13.5705	13.1253	12.3169	11.6024	10.9662	10.3962	9.8825	93
94	14.0564	13.5792	13.1333	12.3241	11.6087	10.9719	10.4013	9.8871	94
95	14.0653	13.5875	13.1411	12.3309	11.6148	10.9773	10.4061	9.8915	95
96	14.0738	13.5955	13.1486	12.3375	11.6206	10.9825	10.4108	9.8957	96
97	14.0820	13.6031	13.1557	12.3438	11.6262	10.9875	10.4153	9.8998	97
98	14.0899	13.6105	13.1626	12.3499	11.6316	10.9923	10.4196	9.9037	98
99	14.0975	13.6176	13.1693	12.3557	11.6368	10.9969	10.4238	9.9074	99
100	14.1048	13.6244	13.1756	12.3613	11.6417	11.0014	10.4278	9.9110	100

YEARS' PURCHASE

Rate Per Cent

Yrs.	11	12	13	14	15	16	18	20	Yrs.
1	0.8189	0.8123	0.8057	0.7993	0.7930	0.7867	0.7745	0.7627	1
2	1.5275	1.5045	1.4822	1.4606	1.4395	1.4191	1.3800	1.3429	2
3	2.1462	2.1011	2.0579	2.0164	1.9765	1.9382	1.8659	1.7987	3
4	2.6907	2.6202	2.5533	2.4897	2.4292	2.3716	2.2642	2.1661	4
5	3.1732	3.0756	2.9838	2.8974	2.8158	2.7387	2.5965	2.4683	5
6	3.6034	3.4781	3.3612	3.2519	3.1495	3.0533	2.8776	2.7210	6
7	3.9892	3.8362	3.6944	3.5628	3.4402	3.3258	3.1184	2.9353	7
8	4.3368	4.1565	3.9906	3.8375	3.6957	3.5640	3.3268	3.1193	8
9	4.6513	4.4446	4.2555	4.0818	3.9217	3.7737	3.5089	3.2788	9
10	4.9372	4.7049	4.4935	4.3002	4.1229	3.9597	3.6691	3.4183	10
11	5.1978	4.9410	4.7084	4.4966	4.3032	4.1256	3.8112	3.5412	11
12	5.4364	5.1561	4.9032	4.6741	4.4653	4.2745	3.9378	3.6503	12
13	5.6553	5.3526	5.0806	4.8350	4.6120	4.4087	4.0514	3.7478	13
14	5.8567	5.5327	5.2426	4.9815	4.7451	4.5301	4.1538	3.8352	14
15	6.0427	5.6983	5.3911	5.1153	4.8664	4.6406	4.2465	3.9140	15
16	6.2146	5.8510	5.5276	5.2380	4.9773	4.7413	4.3307	3.9855	16
17	6.3740	5.9921	5.6533	5.3508	5.0790	4.8336	4.4075	4.0504	17
18	6.5221	6.1227	5.7695	5.4548	5.1726	4.9182	4.4778	4.1097	18
19	6.6598	6.2440	5.8770	5.5508	5.2589	4.9962	4.5423	4.1640	19
20	6.7883	6.3568	5.9768	5.6397	5.3387	5.0681	4.6017	4.2138	20
21	6.9082	6.4618	6.0696	5.7223	5.4126	5.1346	4.6565	4.2597	21
22	7.0203	6.5598	6.1560	5.7990	5.4812	5.1963	4.7071	4.3021	22
23	7.1254	6.6514	6.2366	5.8705	5.5450	5.2537	4.7541	4.3413	23
24	7.2239	6.7372	6.3119	5.9372	5.6044	5.3070	4.7978	4.3777	24
25	7.3164	6.8176	6.3824	5.9995	5.6600	5.3568	4.8384	4.4115	25
26	7.4033	6.8930	6.4485	6.0579	5.7118	5.4032	4.8763	4.4430	26
27	7.4851	6.9639	6.5105	6.1125	5.7604	5.4467	4.9116	4.4723	27
28	7.5622	7.0305	6.5687	6.1638	5.8060	5.4874	4.9447	4.4997	28
29	7.6349	7.0933	6.6235	6.2120	5.8487	5.5255	4.9757	4.5253	29
30	7.7035	7.1525	6.6751	6.2574	5.8889	5.5614	5.0047	4.5494	30
31	7.7683	7.2083	6.7237	6.3001	5.9267	5.5951	5.0320	4.5719	31
32	7.8296	7.2611	6.7695	6.3403	5.9623	5.6268	5.0576	4.5930	32
33	7.8876	7.3109	6.8128	6.3783	5.9959	5.6567	5.0818	4.6129	33
34	7.9425	7.3581	6.8538	6.4142	6.0275	5.6849	5.1045	4.6317	34
35	7.9945	7.4027	6.8925	6.4480	6.0575	5.7115	5.1259	4.6493	35
36	8.0438	7.4450	6.9291	6.4801	6.0857	5.7366	5.1462	4.6659	36
37	8.0906	7.4851	6.9638	6.5104	6.1125	5.7604	5.1653	4.6817	37
38	8.1351	7.5231	6.9967	6.5392	6.1378	5.7829	5.1834	4.6965	38
39	8.1773	7.5592	7.0279	6.5664	6.1618	5.8042	5.2005	4.7105	39
40	8.2174	7.5934	7.0575	6.5923	6.1846	5.8244	5.2167	4.7238	40
41	8.2556	7.6260	7.0856	6.6168	6.2062	5.8435	5.2320	4.7364	41
42	8.2919	7.6570	7.1124	6.6401	6.2266	5.8617	5.2466	4.7483	42
43	8.3264	7.6864	7.1378	6.6622	6.2461	5.8789	5.2604	4.7596	43
44	8.3593	7.7144	7.1619	6.6833	6.2646	5.8953	5.2735	4.7704	44
45	8.3906	7.7411	7.1849	6.7033	6.2822	5.9108	5.2860	4.7806	45
46	8.4205	7.7665	7.2068	6.7223	6.2989	5.9257	5.2978	4.7902	46
47	8.4490	7.7907	7.2276	6.7405	6.3148	5.9397	5.3090	4.7994	47
48	8.4761	7.8138	7.2475	6.7577	6.3300	5.9531	5.3198	4.8082	48
49	8.5020	7.8358	7.2664	6.7742	6.3444	5.9659	5.3299	4.8165	49
50	8.5267	7.8568	7.2845	6.7899	6.3582	5.9781	5.3397	4.8244	50

YEARS' PURCHASE

Rate Per Cent

Yrs.	11	12	13	14	15	16	18	20	Yrs.
51	8.5503	7.8769	7.3017	6.8048	6.3713	5.9897	5.3489	4.8320	51
52	8.5729	7.8960	7.3181	6.8191	6.3838	6.0007	5.3577	4.8392	52
53	8.5944	7.9142	7.3338	6.8327	6.3957	6.0113	5.3661	4.8460	53
54	8.6150	7.9317	7.3488	6.8457	6.4071	6.0213	5.3741	4.8526	54
55	8.6346	7.9483	7.3631	6.8581	6.4180	6.0309	5.3818	4.8588	55
56	8.6534	7.9643	7.3768	6.8700	6.4283	6.0401	5.3891	4.8647	56
57	8.6714	7.9795	7.3898	6.8813	6.4383	6.0488	5.3960	4.8704	57
58	8.6886	7.9940	7.4023	6.8921	6.4477	6.0572	5.4027	4.8758	58
59	8.7050	8.0079	7.4142	6.9025	6.4568	6.0652	5.4090	4.8810	59
60	8.7208	8.0213	7.4256	6.9123	6.4654	6.0728	5.4151	4.8859	60
61	8.7358	8.0340	7.4365	6.9218	6.4737	6.0801	5.4209	4.8907	61
62	8.7502	8.0462	7.4470	6.9308	6.4816	6.0871	5.4264	4.8952	62
63	8.7640	8.0578	7.4570	6.9395	6.4892	6.0937	5.4318	4.8995	63
64	8.7773	8.0690	7.4665	6.9478	6.4964	6.1001	5.4368	4.9036	64
65	8.7899	8.0797	7.4757	6.9557	6.5034	6.1062	5.4417	4.9076	65
66	8.8020	8.0899	7.4845	6.9633	6.5100	6.1121	5.4463	4.9113	66
67	8.8136	8.0998	7.4928	6.9706	6.5163	6.1177	5.4508	4.9150	67
68	8.8248	8.1091	7.5009	6.9775	6.5224	6.1230	5.4550	4.9184	68
69	8.8354	8.1181	7.5086	6.9842	6.5282	6.1282	5.4591	4.9217	69
70	8.8456	8.1268	7.5160	6.9906	6.5338	6.1331	5.4630	4.9249	70
71	8.8554	8.1350	7.5230	6.9967	6.5392	6.1378	5.4667	4.9279	71
72	8.8648	8.1430	7.5298	7.0025	6.5443	6.1423	5.4703	4.9308	72
73	8.8738	8.1506	7.5363	7.0082	6.5492	6.1466	5.4737	4.9336	73
74	8.8825	8.1578	7.5425	7.0135	6.5539	6.1508	5.4770	4.9363	74
75	8.8907	8.1648	7.5485	7.0187	6.5584	6.1547	5.4802	4.9388	75
76	8.8987	8.1715	7.5542	7.0236	6.5627	6.1585	5.4832	4.9413	76
77	8.9063	8.1779	7.5597	7.0284	6.5668	6.1622	5.4861	4.9436	77
78	8.9136	8.1841	7.5650	7.0329	6.5708	6.1657	5.4888	4.9459	78
79	8.9206	8.1900	7.5700	7.0373	6.5746	6.1690	5.4915	4.9480	79
80	8.9273	8.1957	7.5749	7.0415	6.5783	6.1722	5.4940	4.9501	80
81	8.9338	8.2011	7.5795	7.0455	6.5818	6.1753	5.4965	4.9521	81
82	8.9400	8.2063	7.5840	7.0493	6.5851	6.1783	5.4988	4.9540	82
83	8.9459	8.2113	7.5882	7.0530	6.5884	6.1811	5.5011	4.9558	83
84	8.9516	8.2161	7.5923	7.0566	6.5914	6.1838	5.5032	4.9576	84
85	8.9571	8.2207	7.5963	7.0600	6.5944	6.1865	5.5053	4.9592	85
86	8.9623	8.2252	7.6001	7.0632	6.5973	6.1890	5.5073	4.9609	86
87	8.9674	8.2294	7.6037	7.0664	6.6000	6.1914	5.5092	4.9624	87
88	8.9722	8.2335	7.6072	7.0694	6.6026	6.1937	5.5110	4.9639	88
89	8.9769	8.2374	7.6105	7.0723	6.6051	6.1959	5.5128	4.9653	89
90	8.9813	8.2412	7.6137	7.0750	6.6076	6.1980	5.5144	4.9667	90
91	8.9856	8.2448	7.6168	7.0777	6.6099	6.2001	5.5161	4.9680	91
92	8.9897	8.2482	7.6198	7.0803	6.6121	6.2020	5.5176	4.9692	92
93	8.9937	8.2516	7.6226	7.0827	6.6142	6.2039	5.5191	4.9705	93
94	8.9975	8.2548	7.6253	7.0851	6.6163	6.2057	5.5205	4.9716	94
95	9.0011	8.2578	7.6279	7.0873	6.6183	6.2074	5.5219	4.9727	95
96	9.0046	8.2608	7.6304	7.0895	6.6202	6.2091	5.5232	4.9738	96
97	9.0080	8.2636	7.6329	7.0916	6.6220	6.2107	5.5245	4.9748	97
98	9.0112	8.2663	7.6352	7.0936	6.6237	6.2122	5.5257	4.9758	98
99	9.0143	8.2689	7.6374	7.0955	6.6254	6.2137	5.5269	4.9767	99
100	9.0173	8.2715	7.6395	7.0973	6.6270	6.2151	5.5280	4.9777	100

YEARS' PURCHASE

(DUAL RATE % PRINCIPLE)

OR

PRESENT VALUE OF
ONE POUND PER ANNUM

*receivable at the end of each year after
allowing for a sinking fund at a given
rate to replace the invested capital and
for the effect of income tax at* **20%** *on
that part of the income used to provide
the annual sinking fund instalment.*

AT RATES OF INTEREST FROM

4% to 20%

AND

ALLOWING FOR THE POSSIBLE INVESTMENT

OF SINKING FUNDS AT

2·5%, 3% and 4%

INCOME TAX at 20%

Note:–Tables of Years' Purchase in which no allowance has been made for the effect of
income tax on that part of the income used to provide the annual sinking fund
instalment will be found on pages 53 to 71.

YEARS' PURCHASE

Rate Per Cent

Yrs.	4	4.5	5	5.5	6	6.25	6.5	6.75	Yrs.
1	0.7752	0.7722	0.7692	0.7663	0.7634	0.7619	0.7605	0.7590	1
2	1.5214	1.5099	1.4986	1.4875	1.4765	1.4711	1.4657	1.4603	2
3	2.2400	2.2152	2.1910	2.1672	2.1440	2.1326	2.1212	2.1101	3
4	2.9324	2.8900	2.8488	2.8088	2.7699	2.7509	2.7321	2.7135	4
5	3.5996	3.5360	3.4745	3.4152	3.3579	3.3299	3.3024	3.2754	5
6	4.2429	4.1548	4.0702	3.9890	3.9110	3.8732	3.8360	3.7996	6
7	4.8634	4.7479	4.6378	4.5327	4.4322	4.3837	4.3362	4.2896	7
8	5.4620	5.3168	5.1791	5.0484	4.9241	4.8642	4.8057	4.7487	8
9	6.0397	5.8627	5.6957	5.5380	5.3888	5.3171	5.2474	5.1794	9
10	6.5975	6.3868	6.1891	6.0034	5.8284	5.7447	5.6634	5.5843	10
11	7.1361	6.8903	6.6608	6.4461	6.2448	6.1488	6.0557	5.9654	11
12	7.6564	7.3741	7.1119	6.8677	6.6397	6.5313	6.4264	6.3248	12
13	8.1592	7.8394	7.5437	7.2695	7.0146	6.8937	6.7769	6.6640	13
14	8.6452	8.2870	7.9573	7.6528	7.3708	7.2374	7.1088	6.9847	14
15	9.1151	8.7178	8.3537	8.0187	7.7096	7.5638	7.4235	7.2882	15
16	9.5695	9.1325	8.7337	8.3683	8.0322	7.8741	7.7221	7.5758	16
17	10.0090	9.5320	9.0984	8.7025	8.3396	8.1693	8.0058	7.8487	17
18	10.4343	9.9169	9.4484	9.0222	8.6328	8.4504	8.2756	8.1078	18
19	10.8459	10.2880	9.7847	9.3283	8.9126	8.7183	8.5324	8.3542	19
20	11.2443	10.6458	10.1078	9.6215	9.1799	8.9739	8.7770	8.5886	20
21	11.6300	10.9909	10.4184	9.9025	9.4354	9.2179	9.0103	8.8118	21
22	12.0036	11.3240	10.7172	10.1721	9.6798	9.4511	9.2329	9.0246	22
23	12.3655	11.6454	11.0047	10.4307	9.9137	9.6739	9.4455	9.2276	23
24	12.7160	11.9559	11.2815	10.6791	10.1378	9.8872	9.6487	9.4214	24
25	13.0557	12.2557	11.5480	10.9176	10.3525	10.0913	9.8430	9.6066	25
26	13.3849	12.5453	11.8049	11.1469	10.5584	10.2869	10.0290	9.7837	26
27	13.7041	12.8253	12.0524	11.3674	10.7560	10.4744	10.2071	9.9531	27
28	14.0135	13.0959	12.2911	11.5794	10.9457	10.6542	10.3778	10.1153	28
29	14.3135	13.3575	12.5213	11.7835	11.1279	10.8267	10.5414	10.2707	29
30	14.6045	13.6106	12.7434	11.9801	11.3030	10.9924	10.6984	10.4197	30
31	14.8868	13.8555	12.9578	12.1693	11.4713	11.1515	10.8491	10.5626	31
32	15.1606	14.0924	13.1648	12.3517	11.6333	11.3045	10.9938	10.6997	32
33	15.4263	14.3217	13.3647	12.5275	11.7891	11.4516	11.1329	10.8314	33
34	15.6842	14.5437	13.5578	12.6971	11.9391	11.5931	11.2665	10.9579	34
35	15.9345	14.7586	13.7444	12.8606	12.0836	11.7292	11.3951	11.0795	35
36	16.1774	14.9668	13.9248	13.0184	12.2228	11.8604	11.5188	11.1964	36
37	16.4133	15.1685	14.0992	13.1707	12.3569	11.9867	11.6379	11.3089	37
38	16.6424	15.3639	14.2678	13.3178	12.4863	12.1083	11.7526	11.4171	38
39	16.8648	15.5533	14.4310	13.4598	12.6111	12.2257	11.8631	11.5214	39
40	17.0808	15.7368	14.5889	13.5971	12.7315	12.3388	11.9696	11.6218	40
41	17.2906	15.9147	14.7417	13.7297	12.8477	12.4479	12.0722	11.7185	41
42	17.4945	16.0873	14.8896	13.8579	12.9599	12.5532	12.1712	11.8118	42
43	17.6925	16.2546	15.0328	13.9819	13.0683	12.6548	12.2668	11.9018	43
44	17.8849	16.4169	15.1715	14.1018	13.1730	12.7530	12.3590	11.9885	44
45	18.0719	16.5743	15.3059	14.2178	13.2742	12.8478	12.4480	12.0723	45
46	18.2537	16.7270	15.4360	14.3300	13.3719	12.9394	12.5339	12.1531	46
47	18.4303	16.8753	15.5622	14.4387	13.4665	13.0279	12.6170	12.2312	47
48	18.6021	17.0191	15.6844	14.5439	13.5579	13.1135	12.6972	12.3066	48
49	18.7690	17.1587	15.8029	14.6457	13.6464	13.1962	12.7748	12.3794	49
50	18.9313	17.2943	15.9178	14.7444	13.7320	13.2762	12.8497	12.4498	50

YEARS' PURCHASE

Rate Per Cent

Yrs.	4	4.5	5	5.5	6	6.25	6.5	6.75	Yrs.
51	19.0891	17.4259	16.0293	14.8399	13.8149	13.3537	12.9223	12.5179	51
52	19.2426	17.5537	16.1373	14.9325	13.8950	13.4286	12.9924	12.5837	52
53	19.3918	17.6778	16.2422	15.0222	13.9727	13.5011	13.0603	12.6473	53
54	19.5370	17.7984	16.3439	15.1092	14.0479	13.5713	13.1259	12.7089	54
55	19.6782	17.9155	16.4426	15.1935	14.1208	13.6393	13.1895	12.7685	55
56	19.8155	18.0292	16.5383	15.2752	14.1913	13.7051	13.2511	12.8262	56
57	19.9491	18.1397	16.6313	15.3545	14.2597	13.7689	13.3107	12.8820	57
58	20.0791	18.2472	16.7216	15.4314	14.3260	13.8307	13.3684	12.9361	58
59	20.2056	18.3515	16.8092	15.5060	14.3903	13.8906	13.4244	12.9885	59
60	20.3286	18.4530	16.8942	15.5783	14.4526	13.9486	13.4786	13.0392	60
61	20.4484	18.5516	16.9769	15.6486	14.5130	14.0049	13.5311	13.0884	61
62	20.5649	18.6475	17.0571	15.7167	14.5716	14.0594	13.5821	13.1360	62
63	20.6783	18.7407	17.1351	15.7829	14.6285	14.1124	13.6314	13.1822	63
64	20.7887	18.8313	17.2108	15.8471	14.6836	14.1637	13.6793	13.2270	64
65	20.8961	18.9194	17.2844	15.9094	14.7371	14.2135	13.7257	13.2704	65
66	21.0007	19.0051	17.3558	15.9700	14.7891	14.2618	13.7708	13.3125	66
67	21.1025	19.0884	17.4253	16.0288	14.8395	14.3087	13.8145	13.3533	67
68	21.2016	19.1695	17.4928	16.0859	14.8884	14.3542	13.8569	13.3929	68
69	21.2981	19.2483	17.5585	16.1414	14.9359	14.3983	13.8980	13.4314	69
70	21.3920	19.3250	17.6223	16.1953	14.9821	14.4412	13.9380	13.4687	70
71	21.4835	19.3996	17.6843	16.2476	15.0269	14.4828	13.9768	13.5049	71
72	21.5725	19.4722	17.7446	16.2985	15.0704	14.5232	14.0144	13.5400	72
73	21.6593	19.5429	17.8032	16.3480	15.1127	14.5625	14.0509	13.5741	73
74	21.7437	19.6116	17.8602	16.3961	15.1537	14.6006	14.0864	13.6072	74
75	21.8260	19.6785	17.9157	16.4428	15.1936	14.6377	14.1209	13.6394	75
76	21.9061	19.7436	17.9696	16.4882	15.2324	14.6736	14.1544	13.6706	76
77	21.9841	19.8069	18.0221	16.5324	15.2701	14.7086	14.1869	13.7010	77
78	22.0601	19.8686	18.0731	16.5753	15.3067	14.7426	14.2185	13.7305	78
79	22.1341	19.9286	18.1228	16.6170	15.3423	14.7756	14.2492	13.7591	79
80	22.2062	19.9870	18.1711	16.6576	15.3769	14.8077	14.2791	13.7869	80
81	22.2764	20.0439	18.2181	16.6971	15.4106	14.8389	14.3081	13.8139	81
82	22.3448	20.0992	18.2638	16.7355	15.4433	14.8692	14.3363	13.8402	82
83	22.4114	20.1531	18.3083	16.7729	15.4751	14.8987	14.3637	13.8658	83
84	22.4763	20.2056	18.3516	16.8092	15.5060	14.9273	14.3903	13.8906	84
85	22.5396	20.2567	18.3937	16.8446	15.5361	14.9552	14.4162	13.9147	85
86	22.6012	20.3065	18.4347	16.8789	15.5653	14.9823	14.4414	13.9382	86
87	22.6613	20.3549	18.4747	16.9124	15.5938	15.0087	14.4659	13.9610	87
88	22.7198	20.4021	18.5135	16.9450	15.6214	15.0343	14.4897	13.9832	88
89	22.7768	20.4481	18.5514	16.9767	15.6484	15.0592	14.5129	14.0047	89
90	22.8323	20.4928	18.5882	17.0075	15.6746	15.0835	14.5354	14.0257	90
91	22.8864	20.5364	18.6240	17.0375	15.7001	15.1071	14.5573	14.0461	91
92	22.9392	20.5789	18.6590	17.0667	15.7249	15.1301	14.5786	14.0660	92
93	22.9906	20.6202	18.6929	17.0951	15.7490	15.1524	14.5994	14.0853	93
94	23.0406	20.6605	18.7260	17.1228	15.7725	15.1741	14.6195	14.1041	94
95	23.0895	20.6997	18.7583	17.1498	15.7953	15.1953	14.6392	14.1223	95
96	23.1370	20.7380	18.7897	17.1760	15.8176	15.2159	14.6583	14.1401	96
97	23.1834	20.7752	18.8202	17.2015	15.8392	15.2359	14.6769	14.1574	97
98	23.2286	20.8115	18.8500	17.2264	15.8603	15.2554	14.6950	14.1743	98
99	23.2726	20.8468	18.8790	17.2506	15.8808	15.2744	14.7126	14.1906	99
100	23.3155	20.8813	18.9072	17.2742	15.9008	15.2929	14.7297	14.2066	100

433

YEARS' PURCHASE

Rate Per Cent

Yrs.	7	7.25	7.5	8	8.5	9	9.5	10	Yrs.
1	0.7576	0.7561	0.7547	0.7519	0.7491	0.7463	0.7435	0.7407	1
2	1.4550	1.4497	1.4445	1.4341	1.4239	1.4139	1.4039	1.3941	2
3	2.0990	2.0880	2.0772	2.0558	2.0349	2.0144	1.9943	1.9746	3
4	2.6953	2.6772	2.6594	2.6245	2.5905	2.5574	2.5251	2.4936	4
5	3.2488	3.2226	3.1968	3.1465	3.0978	3.0506	3.0047	2.9603	5
6	3.7638	3.7287	3.6943	3.6273	3.5627	3.5003	3.4401	3.3819	6
7	4.2441	4.1996	4.1559	4.0713	3.9901	3.9121	3.8370	3.7648	7
8	4.6930	4.6386	4.5854	4.4826	4.3843	4.2903	4.2002	4.1138	8
9	5.1132	5.0487	4.9858	4.8645	4.7490	4.6388	4.5337	4.4332	9
10	5.5074	5.4326	5.3598	5.2199	5.0872	4.9610	4.8409	4.7265	10
11	5.8778	5.7927	5.7100	5.5515	5.4015	5.2595	5.1247	4.9967	11
12	6.2263	6.1309	6.0383	5.8614	5.6945	5.5368	5.3877	5.2463	12
13	6.5548	6.4491	6.3468	6.1516	5.9680	5.7951	5.6319	5.4776	13
14	6.8648	6.7490	6.6370	6.4238	6.2239	6.0361	5.8592	5.6925	14
15	7.1578	7.0319	6.9105	6.6797	6.4638	6.2614	6.0713	5.8925	15
16	7.4350	7.2993	7.1685	6.9205	6.6890	6.4725	6.2696	6.0791	16
17	7.6977	7.5523	7.4124	7.1475	6.9008	6.6707	6.4554	6.2535	17
18	7.9468	7.7919	7.6431	7.3617	7.1004	6.8569	6.6296	6.4169	18
19	8.1833	8.0192	7.8616	7.5643	7.2886	7.0323	6.7934	6.5703	19
20	8.4080	8.2349	8.0688	7.7559	7.4664	7.1977	6.9476	6.7144	20
21	8.6219	8.4399	8.2655	7.9375	7.6345	7.3538	7.0930	6.8501	21
22	8.8255	8.6350	8.4525	8.1098	7.7937	7.5014	7.2302	6.9780	22
23	9.0195	8.8206	8.6303	8.2733	7.9447	7.6411	7.3600	7.0987	23
24	9.2046	8.9976	8.7996	8.4288	8.0879	7.7736	7.4827	7.2129	24
25	9.3813	9.1663	8.9610	8.5767	8.2240	7.8992	7.5991	7.3209	25
26	9.5501	9.3274	9.1149	8.7176	8.3535	8.0185	7.7095	7.4233	26
27	9.7115	9.4813	9.2617	8.8518	8.4767	8.1320	7.8143	7.5204	27
28	9.8658	9.6284	9.4020	8.9799	8.5940	8.2400	7.9139	7.6127	28
29	10.0136	9.7691	9.5362	9.1022	8.7059	8.3428	8.0087	7.7004	29
30	10.1552	9.9037	9.6644	9.2190	8.8127	8.4408	8.0990	7.7838	30
31	10.2908	10.0327	9.7872	9.3306	8.9147	8.5343	8.1851	7.8633	31
32	10.4210	10.1564	9.9049	9.4375	9.0122	8.6236	8.2672	7.9390	32
33	10.5458	10.2749	10.0176	9.5398	9.1055	8.7090	8.3456	8.0113	33
34	10.6657	10.3887	10.1257	9.6378	9.1947	8.7906	8.4205	8.0803	34
35	10.7809	10.4979	10.2294	9.7317	9.2801	8.8686	8.4921	8.1462	35
36	10.8915	10.6028	10.3290	9.8218	9.3620	8.9434	8.5606	8.2092	36
37	10.9979	10.7036	10.4247	9.9082	9.4405	9.0150	8.6262	8.2695	37
38	11.1003	10.8006	10.5166	9.9912	9.5159	9.0837	8.6890	8.3273	38
39	11.1988	10.8938	10.6050	10.0710	9.5882	9.1495	8.7493	8.3826	39
40	11.2937	10.9835	10.6900	10.1476	9.6576	9.2127	8.8071	8.4356	40
41	11.3850	11.0699	10.7718	10.2213	9.7243	9.2734	8.8625	8.4865	41
42	11.4730	11.1531	10.8506	10.2922	9.7885	9.3318	8.9158	8.5353	42
43	11.5579	11.2333	10.9264	10.3604	9.8502	9.3878	8.9669	8.5821	43
44	11.6397	11.3106	10.9995	10.4261	9.9095	9.4417	9.0161	8.6272	44
45	11.7186	11.3851	11.0700	10.4894	9.9667	9.4936	9.0634	8.6704	45
46	11.7948	11.4569	11.1379	10.5504	10.0217	9.5435	9.1088	8.7121	46
47	11.8683	11.5263	11.2034	10.6091	10.0747	9.5916	9.1526	8.7521	47
48	11.9392	11.5932	11.2667	10.6658	10.1258	9.6379	9.1948	8.7906	48
49	12.0078	11.6578	11.3277	10.7205	10.1751	9.6825	9.2354	8.8277	49
50	12.0740	11.7202	11.3866	10.7732	10.2226	9.7255	9.2745	8.8635	50

YEARS' PURCHASE

				Rate Per Cent					
Yrs.	7	7.25	7.5	8	8.5	9	9.5	10	Yrs.
51	12.1380	11.7805	11.4435	10.8242	10.2684	9.7670	9.3122	8.8979	51
52	12.1999	11.8388	11.4985	10.8733	10.3127	9.8070	9.3486	8.9311	52
53	12.2597	11.8951	11.5516	10.9208	10.3554	9.8456	9.3837	8.9631	53
54	12.3175	11.9496	11.6029	10.9667	10.3966	9.8829	9.4175	8.9940	54
55	12.3735	12.0022	11.6526	11.0111	10.4365	9.9189	9.4502	9.0238	55
56	12.4277	12.0532	11.7006	11.0539	10.4750	9.9537	9.4818	9.0526	56
57	12.4801	12.1025	11.7471	11.0954	10.5122	9.9873	9.5123	9.0804	57
58	12.5308	12.1502	11.7920	11.1355	10.5482	10.0197	9.5417	9.1072	58
59	12.5800	12.1964	11.8355	11.1743	10.5830	10.0511	9.5702	9.1331	59
60	12.6276	12.2411	11.8776	11.2118	10.6166	10.0815	9.5977	9.1582	60
61	12.6737	12.2845	11.9184	11.2481	10.6492	10.1108	9.6243	9.1824	61
62	12.7183	12.3264	11.9579	11.2833	10.6807	10.1393	9.6500	9.2059	62
63	12.7616	12.3671	11.9962	11.3174	10.7112	10.1667	9.6749	9.2285	63
64	12.8036	12.4065	12.0332	11.3503	10.7408	10.1934	9.6990	9.2504	64
65	12.8443	12.4447	12.0692	11.3823	10.7694	10.2191	9.7224	9.2716	65
66	12.8837	12.4817	12.1040	11.4132	10.7971	10.2441	9.7449	9.2922	66
67	12.9219	12.5176	12.1377	11.4432	10.8239	10.2682	9.7668	9.3120	67
68	12.9590	12.5524	12.1704	11.4723	10.8500	10.2916	9.7880	9.3313	68
69	12.9950	12.5861	12.2022	11.5005	10.8752	10.3143	9.8085	9.3499	69
70	13.0299	12.6189	12.2330	11.5279	10.8996	10.3363	9.8284	9.3680	70
71	13.0638	12.6506	12.2628	11.5544	10.9233	10.3576	9.8476	9.3855	71
72	13.0967	12.6815	12.2918	11.5801	10.9463	10.3783	9.8663	9.4025	72
73	13.1286	12.7114	12.3199	11.6050	10.9686	10.3983	9.8844	9.4189	73
74	13.1596	12.7404	12.3472	11.6292	10.9902	10.4177	9.9019	9.4348	74
75	13.1897	12.7686	12.3736	11.6527	11.0112	10.4366	9.9190	9.4503	75
76	13.2189	12.7960	12.3993	11.6755	11.0315	10.4548	9.9355	9.4653	76
77	13.2472	12.8226	12.4243	11.6976	11.0513	10.4726	9.9515	9.4798	77
78	13.2748	12.8484	12.4485	11.7191	11.0704	10.4898	9.9670	9.4939	78
79	13.3015	12.8735	12.4721	11.7400	11.0890	10.5065	9.9821	9.5076	79
80	13.3275	12.8978	12.4949	11.7602	11.1071	10.5227	9.9967	9.5209	80
81	13.3528	12.9215	12.5171	11.7799	11.1246	10.5385	10.0110	9.5337	81
82	13.3774	12.9445	12.5387	11.7990	11.1417	10.5537	10.0247	9.5462	82
83	13.4012	12.9668	12.5596	11.8175	11.1582	10.5686	10.0381	9.5584	83
84	13.4244	12.9885	12.5800	11.8355	11.1743	10.5830	10.0511	9.5702	84
85	13.4469	13.0096	12.5998	11.8531	11.1899	10.5970	10.0638	9.5816	85
86	13.4688	13.0301	12.6190	11.8701	11.2051	10.6106	10.0760	9.5927	86
87	13.4901	13.0500	12.6377	11.8866	11.2198	10.6238	10.0879	9.6035	87
88	13.5109	13.0694	12.6559	11.9027	11.2341	10.6366	10.0995	9.6140	88
89	13.5310	13.0882	12.6736	11.9183	11.2480	10.6491	10.1108	9.6242	89
90	13.5506	13.1066	12.6907	11.9335	11.2616	10.6613	10.1217	9.6341	90
91	13.5696	13.1244	12.7074	11.9483	11.2747	10.6730	10.1323	9.6438	91
92	13.5881	13.1417	12.7237	11.9626	11.2875	10.6845	10.1426	9.6531	92
93	13.6062	13.1586	12.7395	11.9766	11.2999	10.6956	10.1527	9.6622	93
94	13.6237	13.1750	12.7548	11.9902	11.3120	10.7065	10.1624	9.6710	94
95	13.6407	13.1909	12.7698	12.0034	11.3238	10.7170	10.1719	9.6796	95
96	13.6573	13.2064	12.7843	12.0162	11.3352	10.7272	10.1811	9.6880	96
97	13.6735	13.2215	12.7985	12.0287	11.3463	10.7372	10.1901	9.6961	97
98	13.6892	13.2362	12.8122	12.0409	11.3571	10.7469	10.1988	9.7040	98
99	13.7045	13.2505	12.8256	12.0527	11.3676	10.7563	10.2073	9.7117	99
100	13.7193	13.2644	12.8386	12.0642	11.3779	10.7654	10.2156	9.7191	100

YEARS' PURCHASE

Rate Per Cent

Yrs.	11	12	13	14	15	16	18	20	Yrs
1	0.7353	0.7299	0.7246	0.7194	0.7143	0.7092	0.6993	0.6897	1
2	1.3750	1.3563	1.3382	1.3205	1.3033	1.2865	1.2543	1.2236	2
3	1.9364	1.8996	1.8642	1.8301	1.7972	1.7655	1.7053	1.6490	3
4	2.4330	2.3752	2.3201	2.2675	2.2172	2.1691	2.0789	1.9959	4
5	2.8751	2.7948	2.7188	2.6468	2.5786	2.5138	2.3934	2.2841	5
6	3.2713	3.1677	3.0704	2.9790	2.8928	2.8115	2.6618	2.5272	6
7	3.6282	3.5012	3.3827	3.2720	3.1684	3.0711	2.8934	2.7351	7
8	3.9513	3.8011	3.6619	3.5325	3.4120	3.2994	3.0952	2.9147	8
9	4.2450	4.0721	3.9128	3.7655	3.6288	3.5018	3.2726	3.0715	9
10	4.5132	4.3183	4.1395	3.9750	3.8230	3.6822	3.4297	3.2095	10
11	4.7589	4.5427	4.3453	4.1644	3.9979	3.8442	3.5697	3.3319	11
12	4.9848	4.7481	4.5329	4.3363	4.1561	3.9903	3.6954	3.4411	12
13	5.1932	4.9368	4.7045	4.4932	4.3000	4.1227	3.8086	3.5391	13
14	5.3859	5.1106	4.8621	4.6367	4.4312	4.2432	3.9113	3.6275	14
15	5.5646	5.2713	5.0073	4.7685	4.5515	4.3534	4.0047	3.7077	15
16	5.7307	5.4201	5.1414	4.8900	4.6620	4.4544	4.0900	3.7807	16
17	5.8855	5.5583	5.2657	5.0023	4.7640	4.5473	4.1682	3.8475	17
18	6.0300	5.6871	5.3810	5.1063	4.8582	4.6331	4.2402	3.9087	18
19	6.1652	5.8072	5.4885	5.2029	4.9456	4.7125	4.3066	3.9651	19
20	6.2919	5.9195	5.5887	5.2929	5.0268	4.7862	4.3681	4.0171	20
21	6.4109	6.0247	5.6823	5.3768	5.1025	4.8547	4.4251	4.0653	21
22	6.5228	6.1234	5.7701	5.4553	5.1731	4.9186	4.4781	4.1100	22
23	6.6282	6.2162	5.8524	5.5288	5.2392	4.9783	4.5275	4.1516	23
24	6.7276	6.3035	5.9298	5.5978	5.3011	5.0342	4.5737	4.1904	24
25	6.8215	6.3859	6.0026	5.6627	5.3592	5.0866	4.6169	4.2266	25
26	6.9103	6.4637	6.0712	5.7237	5.4139	5.1358	4.6574	4.2606	26
27	6.9944	6.5372	6.1361	5.7813	5.4653	5.1821	4.6955	4.2924	27
28	7.0741	6.6068	6.1973	5.8357	5.5139	5.2258	4.7313	4.3223	28
29	7.1498	6.6727	6.2553	5.8871	5.5598	5.2669	4.7650	4.3504	29
30	7.2217	6.7353	6.3103	5.9357	5.6031	5.3058	4.7968	4.3769	30
31	7.2900	6.7947	6.3624	5.9818	5.6442	5.3426	4.8269	4.4019	31
32	7.3551	6.8512	6.4119	6.0255	5.6831	5.3775	4.8553	4.4256	32
33	7.4171	6.9049	6.4589	6.0671	5.7200	5.4105	4.8822	4.4479	33
34	7.4762	6.9561	6.5037	6.1066	5.7551	5.4419	4.9078	4.4691	34
35	7.5326	7.0049	6.5463	6.1441	5.7885	5.4717	4.9320	4.4892	35
36	7.5864	7.0515	6.5870	6.1799	5.8202	5.5001	4.9550	4.5083	36
37	7.6379	7.0959	6.6258	6.2140	5.8505	5.5271	4.9770	4.5264	37
38	7.6871	7.1384	6.6628	6.2466	5.8793	5.5529	4.9978	4.5436	38
39	7.7342	7.1790	6.6981	6.2777	5.9068	5.5774	5.0177	4.5601	39
40	7.7794	7.2179	6.7320	6.3073	5.9331	5.6008	5.0366	4.5757	40
41	7.8226	7.2551	6.7643	6.3357	5.9582	5.6232	5.0547	4.5906	41
42	7.8640	7.2907	6.7953	6.3629	5.9823	5.6446	5.0720	4.6049	42
43	7.9038	7.3249	6.8250	6.3889	6.0052	5.6650	5.0885	4.6185	43
44	7.9420	7.3577	6.8534	6.4138	6.0273	5.6846	5.1043	4.6315	44
45	7.9787	7.3891	6.8807	6.4377	6.0483	5.7034	5.1194	4.6439	45
46	8.0139	7.4193	6.9069	6.4606	6.0686	5.7214	5.1339	4.6559	46
47	8.0477	7.4483	6.9320	6.4826	6.0880	5.7386	5.1478	4.6673	47
48	8.0803	7.4762	6.9562	6.5037	6.1066	5.7551	5.1611	4.6782	48
49	8.1117	7.5030	6.9794	6.5240	6.1245	5.7710	5.1739	4.6887	49
50	8.1418	7.5288	7.0017	6.5435	6.1417	5.7863	5.1861	4.6987	50

YEARS' PURCHASE

				Rate Per Cent					
Yrs.	11	12	13	14	15	16	18	20	Yrs.
51	8.1709	7.5537	7.0232	6.5623	6.1582	5.8009	5.1979	4.7084	51
52	8.1989	7.5776	7.0438	6.5803	6.1741	5.8150	5.2092	4.7177	52
53	8.2258	7.6006	7.0637	6.5977	6.1893	5.8286	5.2201	4.7266	53
54	8.2518	7.6228	7.0829	6.6144	6.2041	5.8416	5.2305	4.7352	54
55	8.2769	7.6442	7.1014	6.6305	6.2182	5.8542	5.2406	4.7434	55
56	8.3011	7.6649	7.1192	6.6460	6.2319	5.8663	5.2503	4.7514	56
57	8.3245	7.6848	7.1364	6.6610	6.2450	5.8779	5.2596	4.7590	57
58	8.3470	7.7040	7.1529	6.6754	6.2577	5.8892	5.2686	4.7664	58
59	8.3688	7.7225	7.1689	6.6894	6.2699	5.9000	5.2773	4.7735	59
60	8.3898	7.7404	7.1843	6.7028	6.2817	5.9105	5.2856	4.7803	60
61	8.4102	7.7577	7.1992	6.7158	6.2931	5.9205	5.2937	4.7869	61
62	8.4298	7.7744	7.2136	6.7283	6.3041	5.9303	5.3015	4.7933	62
63	8.4488	7.7906	7.2275	6.7404	6.3147	5.9397	5.3090	4.7994	63
64	8.4672	7.8062	7.2410	6.7521	6.3250	5.9487	5.3162	4.8053	64
65	8.4849	7.8213	7.2540	6.7633	6.3349	5.9575	5.3232	4.8110	65
66	8.5021	7.8359	7.2665	6.7743	6.3445	5.9660	5.3300	4.8166	66
67	8.5188	7.8500	7.2787	6.7848	6.3537	5.9742	5.3365	4.8219	67
68	8.5349	7.8637	7.2904	6.7950	6.3627	5.9821	5.3428	4.8270	68
69	8.5505	7.8770	7.3018	6.8049	6.3714	5.9897	5.3489	4.8320	69
70	8.5656	7.8898	7.3128	6.8145	6.3797	5.9971	5.3549	4.8368	70
71	8.5802	7.9022	7.3235	6.8237	6.3878	6.0043	5.3606	4.8415	71
72	8.5944	7.9142	7.3338	6.8327	6.3957	6.0112	5.3661	4.8460	72
73	8.6081	7.9258	7.3438	6.8414	6.4033	6.0179	5.3714	4.8504	73
74	8.6214	7.9371	7.3535	6.8498	6.4107	6.0244	5.3766	4.8546	74
75	8.6343	7.9481	7.3628	6.8579	6.4178	6.0307	5.3816	4.8587	75
76	8.6468	7.9587	7.3719	6.8658	6.4247	6.0368	5.3865	4.8626	76
77	8.6589	7.9689	7.3808	6.8734	6.4314	6.0428	5.3912	4.8665	77
78	8.6707	7.9789	7.3893	6.8809	6.4379	6.0485	5.3958	4.8702	78
79	8.6821	7.9885	7.3976	6.8880	6.4442	6.0540	5.4002	4.8738	79
80	8.6932	7.9979	7.4056	6.8950	6.4503	6.0594	5.4045	4.8773	80
81	8.7039	8.0070	7.4134	6.9018	6.4562	6.0646	5.4086	4.8807	81
82	8.7144	8.0158	7.4210	6.9083	6.4619	6.0697	5.4126	4.8839	82
83	8.7245	8.0244	7.4283	6.9147	6.4675	6.0746	5.4165	4.8871	83
84	8.7343	8.0327	7.4354	6.9208	6.4729	6.0794	5.4203	4.8902	84
85	8.7438	8.0408	7.4423	6.9268	6.4781	6.0840	5.4240	4.8932	85
86	8.7531	8.0486	7.4490	6.9326	6.4832	6.0885	5.4275	4.8961	86
87	8.7621	8.0562	7.4556	6.9383	6.4881	6.0928	5.4310	4.8989	87
88	8.7708	8.0636	7.4619	6.9437	6.4929	6.0970	5.4344	4.9016	88
89	8.7793	8.0707	7.4680	6.9491	6.4975	6.1011	5.4376	4.9043	89
90	8.7875	8.0777	7.4740	6.9542	6.5021	6.1051	5.4408	4.9068	90
91	8.7955	8.0845	7.4798	6.9592	6.5064	6.1090	5.4438	4.9093	91
92	8.8033	8.0910	7.4854	6.9641	6.5107	6.1127	5.4468	4.9117	92
93	8.8109	8.0974	7.4909	6.9688	6.5148	6.1164	5.4497	4.9141	93
94	8.8182	8.1036	7.4962	6.9734	6.5188	6.1199	5.4525	4.9164	94
95	8.8254	8.1097	7.5013	6.9779	6.5227	6.1233	5.4552	4.9186	95
96	8.8323	8.1155	7.5063	6.9822	6.5265	6.1267	5.4579	4.9208	96
97	8.8390	8.1212	7.5112	6.9864	6.5302	6.1299	5.4605	4.9229	97
98	8.8456	8.1267	7.5159	6.9905	6.5338	6.1331	5.4630	4.9249	98
99	8.8520	8.1321	7.5206	6.9945	6.5373	6.1361	5.4654	4.9269	99
100	8.8582	8.1374	7.5250	6.9984	6.5407	6.1391	5.4678	4.9288	100

Sinking Fund 3%

YEARS' PURCHASE

Rate Per Cent

Yrs.	4	4.5	5	5.5	6	6.25	6.5	6.75	Yrs.
1	0.7752	0.7722	0.7692	0.7663	0.7634	0.7619	0.7605	0.7590	1
2	1.5249	1.5134	1.5020	1.4908	1.4798	1.4744	1.4689	1.4636	2
3	2.2502	2.2251	2.2006	2.1767	2.1533	2.1417	2.1303	2.1190	3
4	2.9517	2.9088	2.8671	2.8266	2.7872	2.7679	2.7489	2.7301	4
5	3.6305	3.5658	3.5033	3.4430	3.3847	3.3563	3.3284	3.3009	5
6	4.2873	4.1973	4.1110	4.0282	3.9487	3.9101	3.8723	3.8351	6
7	4.9229	4.8046	4.6919	4.5844	4.4816	4.4320	4.3834	4.3359	7
8	5.5380	5.3888	5.2474	5.1132	4.9858	4.9244	4.8645	4.8061	8
9	6.1334	5.9509	5.7789	5.6166	5.4632	5.3896	5.3180	5.2482	9
10	6.7097	6.4919	6.2878	6.0961	5.9158	5.8296	5.7459	5.6645	10
11	7.2676	7.0128	6.7752	6.5532	6.3453	6.2462	6.1502	6.0571	11
12	7.8078	7.5144	7.2423	6.9892	6.7532	6.6411	6.5326	6.4277	12
13	8.3308	7.9976	7.6901	7.4054	7.1410	7.0157	6.8948	6.7780	13
14	8.8372	8.4632	8.1197	7.8029	7.5099	7.3715	7.2381	7.1094	14
15	9.3276	8.9120	8.5318	8.1828	7.8611	7.7096	7.5638	7.4235	15
16	9.8026	9.3446	8.9275	8.5460	8.1958	8.0313	7.8732	7.7212	16
17	10.2627	9.7617	9.3075	8.8936	8.5149	8.3375	8.1672	8.0038	17
18	10.7083	10.1641	9.6725	9.2263	8.8194	8.6292	8.4470	8.2723	18
19	11.1399	10.5522	10.0233	9.5450	9.1102	8.9073	8.7133	8.5275	19
20	11.5581	10.9266	10.3606	9.8503	9.3879	9.1727	8.9670	8.7704	20
21	11.9632	11.2880	10.6849	10.1430	9.6535	9.4260	9.2090	9.0017	21
22	12.3557	11.6368	10.9970	10.4238	9.9074	9.6680	9.4398	9.2222	22
23	12.7360	11.9736	11.2972	10.6932	10.1505	9.8993	9.6602	9.4324	23
24	13.1046	12.2987	11.5862	10.9518	10.3832	10.1205	9.8708	9.6331	24
25	13.4617	12.6128	11.8645	11.2001	10.6062	10.3322	10.0720	9.8246	25
26	13.8078	12.9161	12.1326	11.4387	10.8198	10.5349	10.2645	10.0077	26
27	14.1432	13.2091	12.3908	11.6679	11.0247	10.7290	10.4487	10.1828	27
28	14.4683	13.4923	12.6396	11.8883	11.2213	10.9151	10.6251	10.3502	28
29	14.7834	13.7659	12.8794	12.1002	11.4099	11.0934	10.7941	10.5105	29
30	15.0889	14.0303	13.1106	12.3040	11.5910	11.2646	10.9560	10.6639	30
31	15.3849	14.2860	13.3336	12.5002	11.7649	11.4287	11.1113	10.8110	31
32	15.6719	14.5331	13.5486	12.6890	11.9320	11.5864	11.2602	10.9519	32
33	15.9502	14.7721	13.7561	12.8708	12.0926	11.7378	11.4031	11.0871	33
34	16.2200	15.0032	13.9563	13.0459	12.2470	11.8832	11.5404	11.2167	34
35	16.4815	15.2267	14.1495	13.2146	12.3955	12.0230	11.6721	11.3412	35
36	16.7351	15.4429	14.3359	13.3771	12.5384	12.1574	11.7987	11.4607	36
37	16.9809	15.6520	14.5160	13.5337	12.6760	12.2866	11.9204	11.5755	37
38	17.2194	15.8543	14.6899	13.6847	12.8083	12.4109	12.0374	11.6858	38
39	17.4505	16.0501	14.8578	13.8303	12.9358	12.5306	12.1500	11.7918	39
40	17.6747	16.2396	15.0200	13.9708	13.0586	12.6457	12.2582	11.8937	40
41	17.8921	16.4229	15.1767	14.1063	13.1769	12.7566	12.3624	11.9918	41
42	18.1030	16.6004	15.3281	14.2370	13.2909	12.8635	12.4627	12.0861	42
43	18.3074	16.7722	15.4745	14.3631	13.4008	12.9664	12.5592	12.1769	43
44	18.5057	16.9384	15.6159	14.4849	13.5067	13.0655	12.6523	12.2643	44
45	18.6981	17.0994	15.7526	14.6025	13.6089	13.1611	12.7419	12.3485	45
46	18.8846	17.2553	15.8848	14.7160	13.7074	13.2533	12.8282	12.4296	46
47	19.0655	17.4062	16.0126	14.8257	13.8025	13.3421	12.9115	12.5077	47
48	19.2410	17.5524	16.1363	14.9316	13.8943	13.4278	12.9917	12.5830	48
49	19.4113	17.6940	16.2558	15.0339	13.9828	13.5105	13.0691	12.6556	49
50	19.5764	17.8311	16.3715	15.1327	14.0683	13.5903	13.1437	12.7256	50

YEARS' PURCHASE

Rate Per Cent

Yrs.	4	4.5	5	5.5	6	6.25	6.5	6.75	Yrs.
51	19.7366	17.9639	16.4834	15.2283	14.1508	13.6673	13.2158	12.7931	51
52	19.8920	18.0925	16.5916	15.3206	14.2305	13.7416	13.2852	12.8582	52
53	20.0427	18.2171	16.6963	15.4099	14.3075	13.8134	13.3523	12.9210	53
54	20.1890	18.3379	16.7977	15.4962	14.3819	13.8827	13.4171	12.9816	54
55	20.3308	18.4548	16.8958	15.5796	14.4537	13.9497	13.4796	13.0401	55
56	20.4685	18.5682	16.9907	15.6603	14.5231	14.0143	13.5399	13.0966	56
57	20.6020	18.6780	17.0826	15.7384	14.5902	14.0768	13.5982	13.1512	57
58	20.7316	18.7844	17.1716	15.8139	14.6551	14.1372	13.6546	13.2038	58
59	20.8573	18.8876	17.2578	15.8869	14.7178	14.1955	13.7090	13.2547	59
60	20.9792	18.9875	17.3412	15.9576	14.7784	14.2519	13.7616	13.3039	60
61	21.0976	19.0844	17.4220	16.0259	14.8370	14.3064	13.8124	13.3513	61
62	21.2124	19.1783	17.5002	16.0921	14.8937	14.3591	13.8615	13.3972	62
63	21.3238	19.2693	17.5759	16.1561	14.9486	14.4100	13.9090	13.4416	63
64	21.4319	19.3575	17.6493	16.2181	15.0016	14.4593	13.9549	13.4844	64
65	21.5367	19.4430	17.7204	16.2781	15.0529	14.5070	13.9993	13.5259	65
66	21.6385	19.5260	17.7892	16.3362	15.1026	14.5531	14.0422	13.5660	66
67	21.7373	19.6063	17.8559	16.3924	15.1506	14.5977	14.0837	13.6047	67
68	21.8331	19.6843	17.9205	16.4468	15.1971	14.6409	14.1239	13.6422	68
69	21.9261	19.7598	17.9831	16.4995	15.2421	14.6826	14.1628	13.6784	69
70	22.0164	19.8331	18.0438	16.5506	15.2857	14.7230	14.2004	13.7135	70
71	22.1039	19.9041	18.1026	16.6000	15.3278	14.7621	14.2367	13.7474	71
72	22.1889	19.9730	18.1595	16.6479	15.3686	14.8000	14.2719	13.7803	72
73	22.2714	20.0398	18.2147	16.6943	15.4082	14.8366	14.3060	13.8120	73
74	22.3514	20.1046	18.2682	16.7392	15.4464	14.8721	14.3390	13.8428	74
75	22.4291	20.1674	18.3201	16.7828	15.4835	14.9065	14.3709	13.8725	75
76	22.5045	20.2283	18.3703	16.8249	15.5194	14.9397	14.4018	13.9013	76
77	22.5776	20.2874	18.4190	16.8658	15.5541	14.9719	14.4318	13.9292	77
78	22.6486	20.3447	18.4663	16.9054	15.5878	15.0031	14.4607	13.9562	78
79	22.7175	20.4003	18.5120	16.9437	15.6204	15.0333	14.4888	13.9823	79
80	22.7844	20.4542	18.5564	16.9809	15.6520	15.0626	14.5160	14.0076	80
81	22.8493	20.5065	18.5995	17.0169	15.6826	15.0909	14.5423	14.0321	81
82	22.9123	20.5572	18.6412	17.0518	15.7122	15.1184	14.5678	14.0559	82
83	22.9734	20.6064	18.6816	17.0857	15.7410	15.1450	14.5925	14.0788	83
84	23.0328	20.6542	18.7208	17.1185	15.7688	15.1707	14.6164	14.1011	84
85	23.0904	20.7004	18.7589	17.1503	15.7958	15.1957	14.6395	14.1227	85
86	23.1462	20.7454	18.7957	17.1811	15.8219	15.2199	14.6620	14.1436	86
87	23.2005	20.7889	18.8315	17.2110	15.8472	15.2433	14.6837	14.1638	87
88	23.2532	20.8312	18.8662	17.2399	15.8718	15.2660	14.7048	14.1834	88
89	23.3043	20.8722	18.8998	17.2680	15.8956	15.2880	14.7252	14.2024	89
90	23.3539	20.9120	18.9324	17.2952	15.9186	15.3094	14.7450	14.2208	90
91	23.4020	20.9506	18.9641	17.3216	15.9410	15.3301	14.7642	14.2387	91
92	23.4488	20.9880	18.9947	17.3472	15.9627	15.3501	14.7828	14.2559	92
93	23.4941	21.0244	19.0245	17.3720	15.9837	15.3695	14.8008	14.2727	93
94	23.5382	21.0596	19.0533	17.3961	16.0040	15.3884	14.8183	14.2889	94
95	23.5809	21.0938	19.0813	17.4194	16.0238	15.4066	14.8352	14.3047	95
96	23.6224	21.1270	19.1085	17.4420	16.0429	15.4243	14.8516	14.3199	96
97	23.6626	21.1592	19.1348	17.4640	16.0615	15.4415	14.8675	14.3347	97
98	23.7017	21.1905	19.1604	17.4853	16.0795	15.4581	14.8829	14.3491	98
99	23.7397	21.2208	19.1852	17.5059	16.0969	15.4742	14.8979	14.3630	99
100	23.7765	21.2502	19.2092	17.5259	16.1139	15.4899	14.9124	14.3764	100

YEARS' PURCHASE

				Rate Per Cent					
Yrs.	7	7.25	7.5	8	8.5	9	9.5	10	Yr
1	0.7576	0.7561	0.7547	0.7519	0.7491	0.7463	0.7435	0.7407	1
2	1.4582	1.4529	1.4477	1.4373	1.4270	1.4169	1.4069	1.3971	2
3	2.1079	2.0968	2.0859	2.0644	2.0433	2.0226	2.0024	1.9825	3
4	2.7116	2.6934	2.6753	2.6400	2.6056	2.5721	2.5395	2.5076	4
5	3.2739	3.2474	3.2212	3.1701	3.1207	3.0727	3.0262	2.9811	5
6	3.7987	3.7630	3.7279	3.6597	3.5939	3.5305	3.4692	3.4101	6
7	4.2894	4.2439	4.1993	4.1130	4.0301	3.9505	3.8740	3.8004	7
8	4.7490	4.6933	4.6389	4.5337	4.4332	4.3371	4.2450	4.1568	8
9	5.1802	5.1140	5.0494	4.9251	4.8067	4.6939	4.5863	4.4835	9
10	5.5854	5.5085	5.4337	5.2899	5.1536	5.0242	4.9010	4.7838	10
11	5.9667	5.8790	5.7938	5.6307	5.4765	5.3306	5.1922	5.0608	11
12	6.3260	6.2275	6.1320	5.9496	5.7778	5.6155	5.4622	5.3170	12
13	6.6650	6.5558	6.4501	6.2486	6.0592	5.8811	5.7131	5.5544	13
14	6.9853	6.8654	6.7496	6.5292	6.3228	6.1290	5.9468	5.7751	14
15	7.2882	7.1578	7.0319	6.7931	6.5699	6.3610	6.1649	5.9806	15
16	7.5750	7.4342	7.2985	7.0416	6.8021	6.5784	6.3689	6.1723	16
17	7.8468	7.6958	7.5506	7.2759	7.0205	6.7824	6.5599	6.3516	17
18	8.1047	7.9437	7.7890	7.4970	7.2262	6.9742	6.7392	6.5195	18
19	8.3495	8.1788	8.0149	7.7061	7.4202	7.1547	6.9076	6.6770	19
20	8.5822	8.4020	8.2291	7.9039	7.6034	7.3250	7.0662	6.8250	20
21	8.8036	8.6140	8.4324	8.0913	7.7767	7.4856	7.2155	6.9643	21
22	9.0143	8.8157	8.6256	8.2690	7.9407	7.6374	7.3565	7.0955	22
23	9.2151	9.0076	8.8092	8.4376	8.0960	7.7810	7.4897	7.2193	23
24	9.4065	9.1904	8.9840	8.5978	8.2434	7.9171	7.6156	7.3363	24
25	9.5891	9.3646	9.1504	8.7501	8.3833	8.0460	7.7349	7.4469	25
26	9.7634	9.5308	9.3090	8.8950	8.5162	8.1684	7.8479	7.5516	26
27	9.9300	9.6894	9.4603	9.0330	8.6426	8.2846	7.9551	7.6508	27
28	10.0891	9.8409	9.6046	9.1645	8.7630	8.3951	8.0569	7.7449	28
29	10.2414	9.9857	9.7425	9.2899	8.8776	8.5003	8.1537	7.8343	29
30	10.3870	10.1241	9.8742	9.4096	8.9868	8.6004	8.2458	7.9193	30
31	10.5265	10.2566	10.0001	9.5239	9.0910	8.6958	8.3334	8.0001	31
32	10.6600	10.3833	10.1206	9.6331	9.1905	8.7867	8.4169	8.0770	32
33	10.7881	10.5047	10.2359	9.7376	9.2855	8.8735	8.4965	8.1503	33
34	10.9108	10.6211	10.3463	9.8374	9.3762	8.9564	8.5725	8.2201	34
35	11.0285	10.7326	10.4521	9.9330	9.4631	9.0355	8.6450	8.2868	35
36	11.1415	10.8396	10.5536	10.0246	9.5461	9.1112	8.7142	8.3504	36
37	11.2499	10.9422	10.6508	10.1123	9.6256	9.1836	8.7804	8.4112	37
38	11.3541	11.0407	10.7441	10.1964	9.7018	9.2529	8.8438	8.4693	38
39	11.4541	11.1353	10.8337	10.2770	9.7747	9.3192	8.9043	8.5248	39
40	11.5503	11.2261	10.9197	10.3543	9.8447	9.3828	8.9623	8.5779	40
41	11.6427	11.3134	11.0022	10.4286	9.9117	9.4437	9.0179	8.6288	41
42	11.7316	11.3974	11.0816	10.4998	9.9761	9.5021	9.0712	8.6776	42
43	11.8172	11.4781	11.1579	10.5683	10.0379	9.5582	9.1222	8.7243	43
44	11.8995	11.5557	11.2312	10.6341	10.0972	9.6119	9.1712	8.7691	44
45	11.9787	11.6304	11.3018	10.6973	10.1542	9.6636	9.2182	8.8120	45
46	12.0550	11.7023	11.3697	10.7581	10.2090	9.7132	9.2633	8.8532	46
47	12.1285	11.7715	11.4350	10.8166	10.2616	9.7608	9.3066	8.8928	47
48	12.1993	11.8382	11.4979	10.8728	10.3122	9.8066	9.3482	8.9308	48
49	12.2675	11.9024	11.5585	10.9270	10.3609	9.8506	9.3882	8.9673	49
50	12.3332	11.9643	11.6168	10.9791	10.4078	9.8930	9.4267	9.0024	50

YEARS' PURCHASE

				Rate Per Cent					
Yrs.	7	7.25	7.5	8	8.5	9	9.5	10	Yrs.
51	12.3966	12.0240	11.6731	11.0293	10.4529	9.9337	9.4637	9.0361	51
52	12.4577	12.0815	11.7273	11.0777	10.4963	9.9729	9.4992	9.0685	52
53	12.5167	12.1369	11.7795	11.1243	10.5381	10.0107	9.5335	9.0997	53
54	12.5736	12.1904	11.8298	11.1692	10.5784	10.0470	9.5665	9.1298	54
55	12.6284	12.2419	11.8784	11.2125	10.6173	10.0820	9.5982	9.1587	55
56	12.6814	12.2917	11.9253	11.2542	10.6547	10.1158	9.6288	9.1865	56
57	12.7325	12.3397	11.9705	11.2945	10.6907	10.1483	9.6582	9.2133	57
58	12.7819	12.3861	12.0141	11.3333	10.7255	10.1796	9.6866	9.2391	58
59	12.8296	12.4309	12.0562	11.3708	10.7591	10.2098	9.7139	9.2640	59
60	12.8756	12.4741	12.0968	11.4069	10.7914	10.2390	9.7403	9.2880	60
61	12.9201	12.5158	12.1361	11.4418	10.8226	10.2671	9.7657	9.3111	61
62	12.9631	12.5561	12.1740	11.4755	10.8528	10.2942	9.7903	9.3334	62
63	13.0046	12.5951	12.2106	11.5080	10.8819	10.3203	9.8139	9.3549	63
64	13.0447	12.6327	12.2460	11.5394	10.9099	10.3456	9.8368	9.3756	64
65	13.0835	12.6691	12.2801	11.5698	10.9371	10.3700	9.8588	9.3956	65
66	13.1210	12.7042	12.3132	11.5991	10.9632	10.3935	9.8801	9.4150	66
67	13.1572	12.7382	12.3451	11.6274	10.9885	10.4162	9.9006	9.4336	67
68	13.1923	12.7711	12.3759	11.6547	11.0130	10.4382	9.9204	9.4516	68
69	13.2262	12.8028	12.4058	11.6812	11.0366	10.4594	9.9396	9.4690	69
70	13.2589	12.8335	12.4346	11.7068	11.0594	10.4799	9.9581	9.4858	70
71	13.2907	12.8633	12.4625	11.7315	11.0815	10.4997	9.9760	9.5020	71
72	13.3213	12.8920	12.4895	11.7554	11.1028	10.5188	9.9932	9.5177	72
73	13.3510	12.9198	12.5155	11.7785	11.1234	10.5373	10.0099	9.5328	73
74	13.3797	12.9467	12.5408	11.8008	11.1433	10.5552	10.0261	9.5475	74
75	13.4075	12.9727	12.5652	11.8224	11.1626	10.5725	10.0417	9.5616	75
76	13.4344	12.9979	12.5888	11.8433	11.1812	10.5892	10.0568	9.5753	76
77	13.4605	13.0223	12.6117	11.8636	11.1993	10.6054	10.0713	9.5885	77
78	13.4857	13.0458	12.6338	11.8831	11.2167	10.6210	10.0854	9.6013	78
79	13.5101	13.0687	12.6552	11.9021	11.2336	10.6362	10.0991	9.6136	79
80	13.5337	13.0908	12.6759	11.9204	11.2499	10.6508	10.1123	9.6256	80
81	13.5566	13.1122	12.6960	11.9382	11.2657	10.6650	10.1250	9.6372	81
82	13.5787	13.1329	12.7154	11.9553	11.2810	10.6787	10.1374	9.6483	82
83	13.6002	13.1530	12.7342	11.9720	11.2958	10.6919	10.1493	9.6592	83
84	13.6209	13.1724	12.7524	11.9880	11.3101	10.7048	10.1609	9.6696	84
85	13.6411	13.1912	12.7701	12.0036	11.3240	10.7172	10.1721	9.6798	85
86	13.6605	13.2094	12.7871	12.0187	11.3374	10.7292	10.1829	9.6896	86
87	13.6794	13.2271	12.8037	12.0333	11.3504	10.7408	10.1934	9.6991	87
88	13.6977	13.2442	12.8197	12.0475	11.3630	10.7521	10.2036	9.7083	88
89	13.7154	13.2607	12.8352	12.0612	11.3752	10.7630	10.2134	9.7172	89
90	13.7326	13.2768	12.8503	12.0745	11.3870	10.7736	10.2229	9.7258	90
91	13.7492	13.2923	12.8648	12.0873	11.3984	10.7838	10.2321	9.7341	91
92	13.7653	13.3074	12.8789	12.0998	11.4095	10.7938	10.2411	9.7422	92
93	13.7810	13.3220	12.8926	12.1118	11.4202	10.8034	10.2497	9.7500	93
94	13.7961	13.3361	12.9059	12.1235	11.4306	10.8127	10.2581	9.7576	94
95	13.8108	13.3498	12.9187	12.1349	11.4407	10.8217	10.2662	9.7649	95
96	13.8250	13.3631	12.9311	12.1458	11.4505	10.8304	10.2740	9.7720	96
97	13.8388	13.3760	12.9432	12.1565	11.4599	10.8389	10.2816	9.7789	97
98	13.8521	13.3885	12.9549	12.1668	11.4691	10.8470	10.2890	9.7856	98
99	13.8651	13.4006	12.9662	12.1768	11.4779	10.8550	10.2962	9.7921	99
100	13.8776	13.4123	12.9772	12.1865	11.4866	10.8627	10.3031	9.7983	100

YEARS' PURCHASE

				Rate Per Cent					
Yrs.	11	12	13	14	15	16	18	20	Yr
1	0.7353	0.7299	0.7246	0.7194	0.7143	0.7092	0.6993	0.6897	1
2	1.3779	1.3591	1.3409	1.3232	1.3059	1.2891	1.2567	1.2258	2
3	1.9440	1.9069	1.8712	1.8368	1.8037	1.7718	1.7111	1.6545	3
4	2.4463	2.3879	2.3322	2.2790	2.2282	2.1797	2.0886	2.0049	4
5	2.8948	2.8134	2.7364	2.6635	2.5944	2.5288	2.4071	2.2965	5
6	3.2976	3.1924	3.0936	3.0008	2.9134	2.8309	2.6792	2.5429	6
7	3.6612	3.5319	3.4114	3.2989	3.1935	3.0947	2.9143	2.7538	7
8	3.9909	3.8377	3.6959	3.5642	3.4415	3.3270	3.1194	2.9362	8
9	4.2911	4.1145	3.9519	3.8017	3.6624	3.5330	3.2999	3.0956	9
10	4.5654	4.3661	4.1834	4.0154	3.8604	3.7169	3.4598	3.2358	10
11	4.8170	4.5957	4.3937	4.2088	4.0388	3.8820	3.6023	3.3603	11
12	5.0485	4.8059	4.5855	4.3845	4.2003	4.0310	3.7303	3.4713	12
13	5.2621	4.9991	4.7611	4.5447	4.3471	4.1660	3.8456	3.5710	13
14	5.4598	5.1771	4.9223	4.6914	4.4811	4.2889	3.9501	3.6609	14
15	5.6431	5.3416	5.0708	4.8261	4.6039	4.4012	4.0452	3.7424	15
16	5.8135	5.4941	5.2080	4.9502	4.7167	4.5042	4.1320	3.8166	16
17	5.9723	5.6357	5.3350	5.0648	4.8207	4.5990	4.2116	3.8844	17
18	6.1205	5.7675	5.4530	5.1710	4.9168	4.6863	4.2847	3.9465	18
19	6.2591	5.8904	5.5627	5.2696	5.0058	4.7672	4.3522	4.0037	19
20	6.3890	6.0053	5.6651	5.3614	5.0886	4.8422	4.4146	4.0565	20
21	6.5109	6.1129	5.7607	5.4469	5.1656	4.9118	4.4725	4.1053	21
22	6.6254	6.2137	5.8502	5.5269	5.2374	4.9768	4.5262	4.1505	22
23	6.7332	6.3085	5.9341	5.6017	5.3046	5.0373	4.5763	4.1926	23
24	6.8348	6.3976	6.0129	5.6719	5.3674	5.0940	4.6230	4.2317	24
25	6.9307	6.4815	6.0870	5.7377	5.4264	5.1471	4.6667	4.2683	25
26	7.0213	6.5607	6.1568	5.7997	5.4818	5.1969	4.7076	4.3025	26
27	7.1071	6.6355	6.2226	5.8581	5.5339	5.2437	4.7460	4.3345	27
28	7.1882	6.7062	6.2847	5.9131	5.5830	5.2877	4.7820	4.3646	28
29	7.2652	6.7731	6.3434	5.9650	5.6293	5.3293	4.8160	4.3928	29
30	7.3382	6.8365	6.3990	6.0142	5.6730	5.3684	4.8479	4.4194	30
31	7.4075	6.8966	6.4517	6.0607	5.7143	5.4054	4.8781	4.4445	31
32	7.4734	6.9537	6.5016	6.1047	5.7535	5.4405	4.9066	4.4681	32
33	7.5361	7.0079	6.5490	6.1465	5.7906	5.4736	4.9335	4.4904	33
34	7.5958	7.0595	6.5940	6.1861	5.8257	5.5050	4.9590	4.5116	34
35	7.6526	7.1086	6.6368	6.2238	5.8591	5.5348	4.9832	4.5316	35
36	7.7068	7.1554	6.6776	6.2596	5.8909	5.5631	5.0061	4.5505	36
37	7.7586	7.2000	6.7164	6.2937	5.9210	5.5900	5.0279	4.5685	37
38	7.8080	7.2425	6.7534	6.3261	5.9498	5.6156	5.0486	4.5856	38
39	7.8552	7.2831	6.7886	6.3571	5.9771	5.6400	5.0683	4.6018	39
40	7.9003	7.3218	6.8223	6.3866	6.0032	5.6632	5.0870	4.6173	40
41	7.9434	7.3589	6.8545	6.4148	6.0281	5.6854	5.1049	4.6320	41
42	7.9847	7.3943	6.8852	6.4417	6.0518	5.7065	5.1219	4.6460	42
43	8.0242	7.4282	6.9145	6.4674	6.0745	5.7266	5.1381	4.6593	43
44	8.0621	7.4606	6.9426	6.4919	6.0962	5.7459	5.1536	4.6721	44
45	8.0984	7.4917	6.9695	6.5154	6.1169	5.7643	5.1685	4.6842	45
46	8.1332	7.5214	6.9953	6.5379	6.1367	5.7819	5.1826	4.6959	46
47	8.1666	7.5500	7.0200	6.5595	6.1557	5.7988	5.1961	4.7070	47
48	8.1986	7.5774	7.0436	6.5802	6.1739	5.8149	5.2091	4.7176	48
49	8.2293	7.6036	7.0663	6.5999	6.1913	5.8303	5.2215	4.7278	49
50	8.2589	7.6288	7.0881	6.6189	6.2080	5.8452	5.2334	4.7375	50

YEARS' PURCHASE

Rate Per Cent

Yrs.	11	12	13	14	15	16	18	20	Yrs.
51	8.2873	7.6530	7.1090	6.6371	6.2240	5.8594	5.2447	4.7468	51
52	8.3145	7.6763	7.1290	6.6546	6.2394	5.8730	5.2556	4.7558	52
53	8.3407	7.6986	7.1483	6.6714	6.2542	5.8860	5.2661	4.7643	53
54	8.3660	7.7201	7.1668	6.6875	6.2683	5.8986	5.2762	4.7725	54
55	8.3902	7.7408	7.1846	6.7030	6.2819	5.9106	5.2858	4.7804	55
56	8.4136	7.7606	7.2017	6.7179	6.2950	5.9222	5.2951	4.7880	56
57	8.4360	7.7797	7.2182	6.7322	6.3076	5.9333	5.3039	4.7953	57
58	8.4577	7.7981	7.2340	6.7460	6.3197	5.9440	5.3125	4.8023	58
59	8.4785	7.8159	7.2493	6.7593	6.3313	5.9543	5.3207	4.8090	59
60	8.4986	7.8329	7.2639	6.7720	6.3425	5.9642	5.3286	4.8154	60
61	8.5180	7.8494	7.2781	6.7843	6.3533	5.9738	5.3362	4.8216	61
62	8.5366	7.8652	7.2917	6.7961	6.3637	5.9829	5.3435	4.8276	62
63	8.5546	7.8805	7.3048	6.8075	6.3736	5.9918	5.3506	4.8333	63
64	8.5720	7.8952	7.3175	6.8185	6.3833	6.0003	5.3573	4.8389	64
65	8.5887	7.9094	7.3296	6.8291	6.3925	6.0085	5.3639	4.8442	65
66	8.6048	7.9231	7.3414	6.8393	6.4015	6.0163	5.3702	4.8493	66
67	8.6204	7.9363	7.3527	6.8491	6.4101	6.0240	5.3762	4.8543	67
68	8.6354	7.9490	7.3637	6.8586	6.4184	6.0313	5.3821	4.8590	68
69	8.6499	7.9613	7.3742	6.8678	6.4264	6.0384	5.3877	4.8636	69
70	8.6640	7.9732	7.3844	6.8766	6.4341	6.0452	5.3931	4.8681	70
71	8.6775	7.9846	7.3942	6.8851	6.4416	6.0518	5.3984	4.8723	71
72	8.6905	7.9957	7.4037	6.8933	6.4488	6.0581	5.4034	4.8764	72
73	8.7032	8.0064	7.4129	6.9013	6.4558	6.0643	5.4083	4.8804	73
74	8.7154	8.0167	7.4217	6.9089	6.4625	6.0702	5.4130	4.8842	74
75	8.7271	8.0267	7.4303	6.9163	6.4689	6.0759	5.4176	4.8879	75
76	8.7385	8.0363	7.4385	6.9235	6.4752	6.0814	5.4219	4.8915	76
77	8.7495	8.0456	7.4465	6.9304	6.4812	6.0867	5.4262	4.8950	77
78	8.7602	8.0546	7.4542	6.9371	6.4871	6.0919	5.4303	4.8983	78
79	8.7705	8.0633	7.4616	6.9435	6.4927	6.0969	5.4342	4.9015	79
80	8.7804	8.0717	7.4688	6.9498	6.4982	6.1017	5.4380	4.9046	80
81	8.7900	8.0798	7.4758	6.9558	6.5034	6.1063	5.4417	4.9076	81
82	8.7994	8.0877	7.4825	6.9616	6.5085	6.1108	5.4453	4.9105	82
83	8.8084	8.0953	7.4890	6.9673	6.5134	6.1151	5.4487	4.9133	83
84	8.8171	8.1026	7.4953	6.9727	6.5182	6.1193	5.4521	4.9160	84
85	8.8255	8.1098	7.5014	6.9780	6.5228	6.1234	5.4553	4.9186	85
86	8.8336	8.1166	7.5073	6.9831	6.5273	6.1273	5.4584	4.9212	86
87	8.8415	8.1233	7.5130	6.9880	6.5316	6.1311	5.4614	4.9236	87
88	8.8492	8.1298	7.5185	6.9928	6.5357	6.1348	5.4643	4.9260	88
89	8.8566	8.1360	7.5239	6.9974	6.5398	6.1383	5.4672	4.9283	89
90	8.8637	8.1420	7.5290	7.0018	6.5437	6.1418	5.4699	4.9305	90
91	8.8706	8.1479	7.5340	7.0062	6.5474	6.1451	5.4725	4.9326	91
92	8.8774	8.1535	7.5389	7.0104	6.5511	6.1483	5.4751	4.9347	92
93	8.8838	8.1590	7.5435	7.0144	6.5546	6.1514	5.4775	4.9367	93
94	8.8901	8.1643	7.5481	7.0183	6.5581	6.1544	5.4799	4.9387	94
95	8.8962	8.1695	7.5525	7.0221	6.5614	6.1574	5.4822	4.9405	95
96	8.9021	8.1744	7.5567	7.0258	6.5646	6.1602	5.4845	4.9424	96
97	8.9078	8.1792	7.5608	7.0293	6.5677	6.1629	5.4866	4.9441	97
98	8.9134	8.1839	7.5648	7.0328	6.5707	6.1656	5.4887	4.9458	98
99	8.9187	8.1884	7.5687	7.0361	6.5736	6.1681	5.4908	4.9475	99
100	8.9239	8.1928	7.5724	7.0394	6.5764	6.1706	5.4927	4.9491	100

YEARS' PURCHASE

				Rate Per Cent					
Yrs.	4	4.5	5	5.5	6	6.25	6.5	6.75	Yrs.
1	0.7752	0.7722	0.7692	0.7663	0.7634	0.7619	0.7605	0.7590	1
2	1.5320	1.5203	1.5089	1.4976	1.4864	1.4809	1.4755	1.4701	2
3	2.2705	2.2450	2.2201	2.1957	2.1719	2.1601	2.1485	2.1370	3
4	2.9908	2.9467	2.9039	2.8624	2.8220	2.8022	2.7827	2.7635	4
5	3.6930	3.6260	3.5615	3.4991	3.4390	3.4097	3.3808	3.3525	5
6	4.3773	4.2835	4.1937	4.1076	4.0249	3.9848	3.9455	3.9070	6
7	5.0438	4.9198	4.8016	4.6891	4.5816	4.5298	4.4790	4.4294	7
8	5.6928	5.5353	5.3862	5.2449	5.1109	5.0464	4.9836	4.9222	8
9	6.3245	6.1306	5.9483	5.7765	5.6143	5.5366	5.4610	5.3875	9
10	6.9390	6.7063	6.4887	6.2848	6.0933	6.0019	5.9132	5.8270	10
11	7.5366	7.2629	7.0084	6.7711	6.5494	6.4439	6.3417	6.2427	11
12	8.1175	7.8009	7.5081	7.2364	6.9837	6.8639	6.7481	6.6361	12
13	8.6821	8.3209	7.9885	7.6817	7.3976	7.2632	7.1337	7.0087	13
14	9.2305	8.8233	8.4505	8.1079	7.7920	7.6431	7.4998	7.3618	14
15	9.7631	9.3087	8.8947	8.5160	8.1682	8.0047	7.8477	7.6967	15
16	10.2801	9.7776	9.3218	8.9067	8.5270	8.3490	8.1783	8.0144	16
17	10.7819	10.2304	9.7325	9.2809	8.8693	8.6769	8.4927	8.3161	17
18	11.2687	10.6676	10.1274	9.6393	9.1961	8.9894	8.7919	8.6028	18
19	11.7408	11.0898	10.5072	9.9827	9.5081	9.2874	9.0766	8.8752	19
20	12.1985	11.4973	10.8723	10.3117	9.8061	9.5715	9.3478	9.1343	20
21	12.6422	11.8906	11.2233	10.6270	10.0908	9.8425	9.6061	9.3809	21
22	13.0721	12.2702	11.5609	10.9291	10.3629	10.1012	9.8524	9.6155	22
23	13.4887	12.6364	11.8855	11.2188	10.6229	10.3481	10.0871	9.8390	23
24	13.8921	12.9898	12.1976	11.4964	10.8715	10.5839	10.3110	10.0519	24
25	14.2827	13.3307	12.4977	11.7626	11.1093	10.8091	10.5247	10.2548	25
26	14.6608	13.6595	12.7862	12.0179	11.3367	11.0242	10.7286	10.4483	26
27	15.0267	13.9766	13.0637	12.2627	11.5543	11.2299	10.9232	10.6328	27
28	15.3808	14.2824	13.3305	12.4975	11.7625	11.4265	11.1091	10.8089	28
29	15.7233	14.5773	13.5870	12.7227	11.9617	11.6144	11.2867	10.9770	29
30	16.0546	14.8616	13.8336	12.9387	12.1525	11.7942	11.4564	11.1374	30
31	16.3749	15.1356	14.0708	13.1459	12.3351	11.9661	11.6186	11.2906	31
32	16.6845	15.3998	14.2988	13.3448	12.5100	12.1307	11.7736	11.4370	32
33	16.9839	15.6545	14.5181	13.5356	12.6776	12.2881	11.9219	11.5768	33
34	17.2731	15.8999	14.7290	13.7187	12.8380	12.4388	12.0637	11.7105	34
35	17.5526	16.1364	14.9317	13.8944	12.9918	12.5831	12.1993	11.8383	35
36	17.8226	16.3643	15.1266	14.0630	13.1391	12.7213	12.3291	11.9605	36
37	18.0834	16.5839	15.3141	14.2249	13.2803	12.8536	12.4534	12.0774	37
38	18.3352	16.7955	15.4943	14.3802	13.4156	12.9803	12.5723	12.1892	38
39	18.5784	16.9993	15.6676	14.5294	13.5454	13.1017	12.6862	12.2962	39
40	18.8131	17.1956	15.8342	14.6726	13.6697	13.2180	12.7952	12.3986	40
41	19.0397	17.3847	15.9944	14.8100	13.7890	13.3295	12.8996	12.4966	41
42	19.2585	17.5669	16.1485	14.9420	13.9033	13.4363	12.9996	12.5905	42
43	19.4695	17.7423	16.2966	15.0688	14.0130	13.5387	13.0955	12.6803	43
44	19.6732	17.9113	16.4391	15.1905	14.1182	13.6369	13.1873	12.7664	44
45	19.8696	18.0740	16.5760	15.3074	14.2191	13.7310	13.2753	12.8488	45
46	20.0592	18.2307	16.7077	15.4196	14.3159	13.8212	13.3596	12.9278	46
47	20.2419	18.3816	16.8343	15.5274	14.4087	13.9077	13.4404	13.0035	47
48	20.4182	18.5268	16.9561	15.6309	14.4978	13.9907	13.5179	13.0760	48
49	20.5882	18.6667	17.0732	15.7303	14.5833	14.0704	13.5922	13.1455	49
50	20.7522	18.8013	17.1857	15.8259	14.6654	14.1467	13.6635	13.2122	50

YEARS' PURCHASE

				Rate Per Cent					
Yrs.	4	4.5	5	5.5	6	6.25	6.5	6.75	Yrs.
51	20.9102	18.9309	17.2940	15.9176	14.7441	14.2200	13.7318	13.2761	51
52	21.0625	19.0557	17.3981	16.0057	14.8197	14.2903	13.7974	13.3373	52
53	21.2094	19.1758	17.4981	16.0904	14.8923	14.3577	13.8602	13.3960	53
54	21.3509	19.2915	17.5944	16.1717	14.9619	14.4224	13.9205	13.4524	54
55	21.4873	19.4028	17.6869	16.2498	15.0288	14.4846	13.9784	13.5064	55
56	21.6188	19.5099	17.7759	16.3249	15.0930	14.5442	14.0339	13.5582	56
57	21.7454	19.6130	17.8614	16.3970	15.1546	14.6014	14.0872	13.6079	57
58	21.8675	19.7122	17.9437	16.4663	15.2138	14.6563	14.1383	13.6556	58
59	21.9851	19.8077	18.0227	16.5329	15.2706	14.7090	14.1873	13.7014	59
60	22.0983	19.8996	18.0988	16.5969	15.3251	14.7596	14.2344	13.7453	60
61	22.2074	19.9880	18.1719	16.6583	15.3775	14.8082	14.2796	13.7874	61
62	22.3125	20.0731	18.2422	16.7174	15.4278	14.8549	14.3230	13.8278	62
63	22.4138	20.1550	18.3098	16.7742	15.4762	14.8997	14.3646	13.8667	63
64	22.5112	20.2338	18.3748	16.8287	15.5226	14.9427	14.4046	13.9039	64
65	22.6051	20.3096	18.4373	16.8811	15.5672	14.9840	14.4430	13.9397	65
66	22.6955	20.3826	18.4974	16.9315	15.6100	15.0237	14.4798	13.9740	66
67	22.7826	20.4528	18.5552	16.9799	15.6511	15.0618	14.5152	14.0069	67
68	22.8664	20.5203	18.6108	17.0264	15.6906	15.0984	14.5492	14.0386	68
69	22.9471	20.5853	18.6642	17.0711	15.7286	15.1335	14.5819	14.0690	69
70	23.0249	20.6478	18.7156	17.1141	15.7651	15.1673	14.6132	14.0981	70
71	23.0997	20.7079	18.7650	17.1554	15.8001	15.1997	14.6433	14.1262	71
72	23.1717	20.7658	18.8125	17.1951	15.8338	15.2309	14.6722	14.1531	72
73	23.2410	20.8215	18.8582	17.2332	15.8661	15.2608	14.7000	14.1789	73
74	23.3078	20.8750	18.9021	17.2699	15.8972	15.2895	14.7266	14.2037	74
75	23.3720	20.9265	18.9443	17.3052	15.9271	15.3172	14.7523	14.2275	75
76	23.4338	20.9761	18.9849	17.3390	15.9557	15.3437	14.7769	14.2504	76
77	23.4933	21.0237	19.0240	17.3716	15.9833	15.3692	14.8005	14.2724	77
78	23.5506	21.0696	19.0615	17.4029	16.0098	15.3937	14.8232	14.2935	78
79	23.6057	21.1137	19.0976	17.4330	16.0353	15.4172	14.8450	14.3138	79
80	23.6588	21.1561	19.1323	17.4619	16.0597	15.4398	14.8660	14.3333	80
81	23.7098	21.1970	19.1657	17.4897	16.0832	15.4615	14.8861	14.3520	81
82	23.7590	21.2362	19.1978	17.5164	16.1058	15.4824	14.9055	14.3700	82
83	23.8062	21.2740	19.2286	17.5421	16.1275	15.5025	14.9241	14.3873	83
84	23.8517	21.3103	19.2583	17.5668	16.1484	15.5218	14.9420	14.4039	84
85	23.8955	21.3452	19.2868	17.5905	16.1684	15.5403	14.9591	14.4198	85
86	23.9376	21.3788	19.3143	17.6133	16.1877	15.5581	14.9756	14.4352	86
87	23.9782	21.4112	19.3406	17.6352	16.2062	15.5752	14.9915	14.4499	87
88	24.0171	21.4422	19.3660	17.6563	16.2240	15.5916	15.0067	14.4641	88
89	24.0547	21.4721	19.3904	17.6766	16.2412	15.6074	15.0213	14.4777	89
90	24.0908	21.5009	19.4138	17.6961	16.2576	15.6226	15.0354	14.4907	90
91	24.1255	21.5286	19.4364	17.7148	16.2734	15.6372	15.0489	14.5033	91
92	24.1589	21.5552	19.4580	17.7328	16.2886	15.6513	15.0619	14.5153	92
93	24.1910	21.5807	19.4789	17.7501	16.3032	15.6647	15.0744	14.5269	93
94	24.2220	21.6053	19.4989	17.7668	16.3172	15.6777	15.0864	14.5381	94
95	24.2517	21.6290	19.5182	17.7828	16.3307	15.6902	15.0979	14.5488	95
96	24.2803	21.6518	19.5367	17.7981	16.3437	15.7021	15.1090	14.5591	96
97	24.3078	21.6736	19.5546	17.8129	16.3562	15.7136	15.1197	14.5690	97
98	24.3343	21.6947	19.5717	17.8271	16.3682	15.7247	15.1299	14.5785	98
99	24.3598	21.7149	19.5882	17.8408	16.3797	15.7353	15.1398	14.5876	99
100	24.3843	21.7344	19.6040	17.8540	16.3908	15.7456	15.1492	14.5964	100

YEARS' PURCHASE

Rate Per Cent

Yrs.	7	7.25	7.5	8	8.5	9	9.5	10	Yrs
1	0.7576	0.7561	0.7547	0.7519	0.7491	0.7463	0.7435	0.7407	1
2	1.4647	1.4593	1.4540	1.4435	1.4332	1.4230	1.4129	1.4030	2
3	2.1257	2.1145	2.1033	2.0814	2.0600	2.0390	2.0184	1.9983	3
4	2.7445	2.7258	2.7074	2.6712	2.6360	2.6017	2.5683	2.5357	4
5	3.3246	3.2972	3.2703	3.2177	3.1667	3.1174	3.0695	3.0231	5
6	3.8692	3.8321	3.7958	3.7251	3.6569	3.5913	3.5279	3.4668	6
7	4.3809	4.3335	4.2870	4.1971	4.1108	4.0280	3.9485	3.8720	7
8	4.8624	4.8040	4.7470	4.6369	4.5319	4.4314	4.3354	4.2434	8
9	5.3159	5.2461	5.1782	5.0475	4.9233	4.8050	4.6923	4.5847	9
10	5.7434	5.6621	5.5830	5.4314	5.2878	5.1516	5.0223	4.8992	10
11	6.1468	6.0538	5.9635	5.7908	5.6279	5.4739	5.3280	5.1898	11
12	6.5278	6.4230	6.3215	6.1278	5.9456	5.7740	5.6120	5.4588	12
13	6.8880	6.7714	6.6587	6.4441	6.2430	6.0540	5.8761	5.7084	13
14	7.2288	7.1004	6.9766	6.7414	6.5216	6.3157	6.1223	5.9405	14
15	7.5514	7.4114	7.2766	7.0212	6.7830	6.5605	6.3522	6.1566	15
16	7.8570	7.7056	7.5600	7.2846	7.0286	6.7900	6.5671	6.3583	16
17	8.1468	7.9841	7.8279	7.5331	7.2596	7.0053	6.7683	6.5467	17
18	8.4216	8.2480	8.0814	7.7675	7.4771	7.2076	6.9569	6.7231	18
19	8.6826	8.4981	8.3213	7.9889	7.6821	7.3979	7.1340	6.8883	19
20	8.9304	8.7354	8.5487	8.1983	7.8754	7.5771	7.3005	7.0434	20
21	9.1659	8.9606	8.7642	8.3963	8.0580	7.7459	7.4571	7.1891	21
22	9.3898	9.1744	8.9687	8.5838	8.2306	7.9052	7.6046	7.3261	22
23	9.6028	9.3777	9.1629	8.7615	8.3937	8.0557	7.7438	7.4551	23
24	9.8055	9.5709	9.3472	8.9299	8.5482	8.1978	7.8750	7.5767	24
25	9.9985	9.7547	9.5225	9.0897	8.6945	8.3323	7.9990	7.6914	25
26	10.1823	9.9296	9.6891	9.2414	8.8332	8.4596	8.1163	7.7998	26
27	10.3575	10.0961	9.8475	9.3854	8.9647	8.5801	8.2272	7.9021	27
28	10.5245	10.2547	9.9984	9.5223	9.0896	8.6944	8.3322	7.9990	28
29	10.6838	10.4058	10.1420	9.6525	9.2081	8.8028	8.4317	8.0906	29
30	10.8357	10.5499	10.2788	9.7764	9.3207	8.9057	8.5260	8.1774	30
31	10.9807	10.6873	10.4092	9.8942	9.4278	9.0034	8.6156	8.2597	31
32	11.1191	10.8183	10.5334	10.0064	9.5296	9.0962	8.7005	8.3378	32
33	11.2512	10.9434	10.6520	10.1133	9.6265	9.1845	8.7812	8.4119	33
34	11.3774	11.0627	10.7650	10.2152	9.7188	9.2684	8.8579	8.4822	34
35	11.4980	11.1767	10.8729	10.3123	9.8066	9.3483	8.9308	8.5491	35
36	11.6132	11.2856	10.9759	10.4049	9.8904	9.4243	9.0002	8.6126	36
37	11.7234	11.3896	11.0743	10.4932	9.9701	9.4967	9.0662	8.6731	37
38	11.8287	11.4890	11.1682	10.5775	10.0462	9.5657	9.1291	8.7306	38
39	11.9295	11.5840	11.2580	10.6580	10.1188	9.6315	9.1890	8.7853	39
40	12.0258	11.6748	11.3437	10.7349	10.1880	9.6942	9.2460	8.8375	40
41	12.1180	11.7617	11.4257	10.8083	10.2541	9.7540	9.3004	8.8872	41
42	12.2062	11.8448	11.5041	10.8784	10.3172	9.8111	9.3523	8.9345	42
43	12.2907	11.9243	11.5791	10.9454	10.3775	9.8656	9.4018	8.9797	43
44	12.3715	12.0004	11.6508	11.0095	10.4351	9.9176	9.4491	9.0228	44
45	12.4489	12.0732	11.7195	11.0708	10.4901	9.9673	9.4941	9.0639	45
46	12.5231	12.1429	11.7851	11.1293	10.5427	10.0148	9.5372	9.1031	46
47	12.5941	12.2097	11.8480	11.1854	10.5930	10.0601	9.5783	9.1406	47
48	12.6621	12.2736	11.9082	11.2390	10.6410	10.1035	9.6176	9.1763	48
49	12.7273	12.3348	11.9658	11.2903	10.6870	10.1449	9.6552	9.2105	49
50	12.7897	12.3935	12.0210	11.3394	10.7310	10.1846	9.6911	9.2432	50

YEARS' PURCHASE

Rate Per Cent

Yrs.	7	7.25	7.5	8	8.5	9	9.5	10	Yrs.
51	12.8496	12.4496	12.0739	11.3865	10.7731	10.2225	9.7254	9.2744	51
52	12.9069	12.5035	12.1245	11.4315	10.8134	10.2588	9.7582	9.3043	52
53	12.9619	12.5551	12.1730	11.4746	10.8520	10.2935	9.7896	9.3328	53
54	13.0147	12.6046	12.2195	11.5159	10.8889	10.3267	9.8197	9.3601	54
55	13.0652	12.6520	12.2641	11.5555	10.9243	10.3585	9.8484	9.3862	55
56	13.1137	12.6974	12.3068	11.5934	10.9582	10.3890	9.8759	9.4112	56
57	13.1602	12.7410	12.3477	11.6297	10.9906	10.4181	9.9023	9.4352	57
58	13.2048	12.7828	12.3870	11.6645	11.0217	10.4460	9.9275	9.4581	58
59	13.2476	12.8229	12.4246	11.6979	11.0515	10.4728	9.9517	9.4800	59
60	13.2886	12.8614	12.4607	11.7299	11.0800	10.4984	9.9748	9.5010	60
61	13.3280	12.8982	12.4953	11.7606	11.1074	10.5230	9.9970	9.5211	61
62	13.3658	12.9336	12.5285	11.7900	11.1336	10.5465	10.0182	9.5404	62
63	13.4020	12.9676	12.5604	11.8182	11.1588	10.5691	10.0386	9.5588	63
64	13.4368	13.0001	12.5909	11.8452	11.1829	10.5907	10.0581	9.5765	64
65	13.4702	13.0314	12.6202	11.8712	11.2060	10.6115	10.0768	9.5935	65
66	13.5023	13.0614	12.6484	11.8961	11.2282	10.6313	10.0947	9.6097	66
67	13.5331	13.0902	12.6754	11.9199	11.2495	10.6504	10.1119	9.6253	67
68	13.5626	13.1178	12.7013	11.9428	11.2699	10.6687	10.1284	9.6402	68
69	13.5909	13.1443	12.7261	11.9648	11.2894	10.6862	10.1442	9.6545	69
70	13.6182	13.1698	12.7500	11.9859	11.3082	10.7030	10.1594	9.6683	70
71	13.6443	13.1942	12.7729	12.0061	11.3262	10.7192	10.1739	9.6814	71
72	13.6694	13.2177	12.7949	12.0256	11.3435	10.7347	10.1879	9.6940	72
73	13.6935	13.2402	12.8160	12.0442	11.3601	10.7495	10.2012	9.7062	73
74	13.7166	13.2619	12.8363	12.0621	11.3760	10.7638	10.2141	9.7178	74
75	13.7389	13.2826	12.8557	12.0793	11.3913	10.7775	10.2264	9.7289	75
76	13.7602	13.3026	12.8744	12.0958	11.4060	10.7906	10.2382	9.7396	76
77	13.7807	13.3217	12.8924	12.1116	11.4201	10.8032	10.2495	9.7499	77
78	13.8004	13.3401	12.9096	12.1268	11.4336	10.8153	10.2604	9.7597	78
79	13.8193	13.3578	12.9261	12.1414	11.4465	10.8269	10.2709	9.7692	79
80	13.8375	13.3748	12.9420	12.1555	11.4590	10.8380	10.2809	9.7783	80
81	13.8549	13.3911	12.9573	12.1689	11.4710	10.8487	10.2905	9.7870	81
82	13.8717	13.4067	12.9720	12.1818	11.4825	10.8590	10.2998	9.7953	82
83	13.8878	13.4218	12.9860	12.1943	11.4935	10.8689	10.3087	9.8034	83
84	13.9032	13.4362	12.9996	12.2062	11.5041	10.8784	10.3172	9.8111	84
85	13.9181	13.4501	13.0126	12.2176	11.5143	10.8875	10.3254	9.8185	85
86	13.9324	13.4634	13.0250	12.2286	11.5240	10.8962	10.3332	9.8256	86
87	13.9461	13.4762	13.0370	12.2392	11.5334	10.9046	10.3408	9.8324	87
88	13.9593	13.4886	13.0485	12.2494	11.5424	10.9126	10.3480	9.8389	88
89	13.9719	13.5004	13.0596	12.2591	11.5511	10.9204	10.3550	9.8452	89
90	13.9841	13.5117	13.0702	12.2685	11.5594	10.9278	10.3617	9.8513	90
91	13.9958	13.5227	13.0805	12.2775	11.5674	10.9349	10.3681	9.8571	91
92	14.0070	13.5332	13.0903	12.2861	11.5751	10.9418	10.3742	9.8627	92
93	14.0178	13.5432	13.0997	12.2944	11.5824	10.9484	10.3802	9.8680	93
94	14.0282	13.5529	13.1088	12.3024	11.5895	10.9547	10.3858	9.8731	94
95	14.0382	13.5622	13.1175	12.3101	11.5963	10.9608	10.3913	9.8781	95
96	14.0478	13.5712	13.1258	12.3175	11.6029	10.9666	10.3966	9.8828	96
97	14.0570	13.5798	13.1339	12.3245	11.6091	10.9723	10.4016	9.8874	97
98	14.0658	13.5880	13.1416	12.3313	11.6152	10.9776	10.4065	9.8918	98
99	14.0743	13.5960	13.1490	12.3379	11.6210	10.9828	10.4111	9.8960	99
100	14.0825	13.6036	13.1562	12.3442	11.6266	10.9878	10.4156	9.9000	100

YEARS' PURCHASE

Rate Per Cent

Yrs.	11	12	13	14	15	16	18	20	Yrs
1	0.7353	0.7299	0.7246	0.7194	0.7143	0.7092	0.6993	0.6897	1
2	1.3836	1.3647	1.3464	1.3285	1.3111	1.2941	1.2614	1.2304	2
3	1.9591	1.9215	1.8852	1.8504	1.8167	1.7843	1.7228	1.6655	3
4	2.4730	2.4133	2.3565	2.3022	2.2504	2.2009	2.1081	2.0228	4
5	2.9344	2.8508	2.7717	2.6970	2.6262	2.5590	2.4344	2.3213	5
6	3.3506	3.2420	3.1402	3.0446	2.9546	2.8698	2.7141	2.5743	6
7	3.7277	3.5937	3.4691	3.3528	3.2440	3.1421	2.9563	2.7913	7
8	4.0707	3.9114	3.7642	3.6277	3.5007	3.3823	3.1680	2.9792	8
9	4.3837	4.1996	4.0304	3.8742	3.7297	3.5956	3.3544	3.1435	9
10	4.6704	4.4620	4.2714	4.0965	3.9352	3.7862	3.5197	3.2882	10
11	4.9337	4.7018	4.4906	4.2976	4.1205	3.9575	3.6672	3.4166	11
12	5.1762	4.9215	4.6906	4.4805	4.2883	4.1120	3.7995	3.5312	12
13	5.4002	5.1235	4.8738	4.6473	4.4409	4.2521	3.9188	3.6340	13
14	5.6074	5.3097	5.0419	4.7999	4.5801	4.3795	4.0268	3.7267	14
15	5.7996	5.4817	5.1968	4.9401	4.7075	4.4959	4.1250	3.8106	15
16	5.9782	5.6410	5.3397	5.0691	4.8245	4.6025	4.2145	3.8869	16
17	6.1445	5.7888	5.4720	5.1881	4.9322	4.7004	4.2965	3.9565	17
18	6.2995	5.9262	5.5947	5.2982	5.0317	4.7906	4.3717	4.0202	18
19	6.4444	6.0542	5.7086	5.4003	5.1237	4.8739	4.4410	4.0787	19
20	6.5799	6.1737	5.8147	5.4952	5.2090	4.9511	4.5050	4.1326	20
21	6.7069	6.2853	5.9137	5.5835	5.2882	5.0226	4.5641	4.1823	21
22	6.8260	6.3898	6.0061	5.6658	5.3620	5.0891	4.6190	4.2284	22
23	6.9379	6.4878	6.0925	5.7426	5.4308	5.1510	4.6699	4.2710	23
24	7.0431	6.5797	6.1735	5.8145	5.4950	5.2088	4.7173	4.3107	24
25	7.1421	6.6660	6.2494	5.8818	5.5551	5.2627	4.7616	4.3475	25
26	7.2354	6.7472	6.3207	5.9450	5.6114	5.3132	4.8029	4.3819	26
27	7.3234	6.8237	6.3878	6.0043	5.6642	5.3605	4.8415	4.4141	27
28	7.4065	6.8958	6.4509	6.0600	5.7138	5.4049	4.8777	4.4441	28
29	7.4850	6.9638	6.5104	6.1125	5.7604	5.4466	4.9116	4.4723	29
30	7.5593	7.0280	6.5665	6.1619	5.8042	5.4858	4.9435	4.4987	30
31	7.6296	7.0887	6.6195	6.2085	5.8456	5.5228	4.9734	4.5235	31
32	7.6961	7.1461	6.6695	6.2525	5.8846	5.5575	5.0016	4.5468	32
33	7.7592	7.2005	6.7168	6.2941	5.9214	5.5904	5.0282	4.5687	33
34	7.8190	7.2520	6.7616	6.3334	5.9562	5.6213	5.0532	4.5894	34
35	7.8758	7.3008	6.8040	6.3706	5.9890	5.6506	5.0769	4.6089	35
36	7.9297	7.3471	6.8442	6.4058	6.0202	5.6783	5.0992	4.6273	36
37	7.9809	7.3910	6.8823	6.4392	6.0496	5.7045	5.1203	4.6447	37
38	8.0296	7.4327	6.9185	6.4708	6.0776	5.7293	5.1403	4.6611	38
39	8.0758	7.4724	6.9528	6.5009	6.1040	5.7529	5.1593	4.6767	39
40	8.1199	7.5101	6.9855	6.5294	6.1292	5.7752	5.1772	4.6914	40
41	8.1618	7.5459	7.0165	6.5564	6.1530	5.7964	5.1942	4.7054	41
42	8.2017	7.5801	7.0460	6.5822	6.1757	5.8165	5.2104	4.7186	42
43	8.2398	7.6125	7.0740	6.6067	6.1972	5.8356	5.2257	4.7312	43
44	8.2760	7.6435	7.1007	6.6300	6.2177	5.8537	5.2402	4.7431	44
45	8.3106	7.6729	7.1262	6.6521	6.2372	5.8710	5.2541	4.7545	45
46	8.3436	7.7010	7.1504	6.6732	6.2558	5.8875	5.2672	4.7653	46
47	8.3750	7.7278	7.1735	6.6933	6.2734	5.9031	5.2798	4.7755	47
48	8.4051	7.7534	7.1955	6.7125	6.2903	5.9180	5.2917	4.7852	48
49	8.4337	7.7778	7.2165	6.7308	6.3063	5.9322	5.3030	4.7945	49
50	8.4611	7.8011	7.2365	6.7482	6.3216	5.9457	5.3138	4.8034	50

YEARS' PURCHASE

				Rate Per Cent					
Yrs.	11	12	13	14	15	16	18	20	Yrs.
51	8.4873	7.8233	7.2557	6.7648	6.3362	5.9586	5.3241	4.8118	51
52	8.5123	7.8445	7.2739	6.7807	6.3501	5.9709	5.3340	4.8198	52
53	8.5361	7.8648	7.2913	6.7958	6.3634	5.9827	5.3433	4.8274	53
54	8.5590	7.8842	7.3080	6.8103	6.3761	5.9939	5.3523	4.8347	54
55	8.5808	7.9027	7.3239	6.8241	6.3882	6.0046	5.3608	4.8417	55
56	8.6017	7.9204	7.3391	6.8373	6.3998	6.0148	5.3690	4.8483	56
57	8.6217	7.9373	7.3537	6.8499	6.4108	6.0246	5.3767	4.8547	57
58	8.6408	7.9536	7.3676	6.8620	6.4214	6.0339	5.3842	4.8607	58
59	8.6591	7.9691	7.3809	6.8735	6.4315	6.0428	5.3913	4.8665	59
60	8.6766	7.9839	7.3936	6.8846	6.4411	6.0514	5.3980	4.8721	60
61	8.6934	7.9981	7.4058	6.8951	6.4504	6.0595	5.4045	4.8773	61
62	8.7094	8.0117	7.4174	6.9052	6.4592	6.0673	5.4107	4.8824	62
63	8.7248	8.0247	7.4286	6.9149	6.4677	6.0748	5.4167	4.8872	63
64	8.7396	8.0371	7.4392	6.9241	6.4758	6.0819	5.4223	4.8918	64
65	8.7537	8.0491	7.4495	6.9330	6.4835	6.0887	5.4278	4.8963	65
66	8.7672	8.0605	7.4593	6.9415	6.4909	6.0953	5.4330	4.9005	66
67	8.7802	8.0715	7.4686	6.9496	6.4980	6.1015	5.4379	4.9045	67
68	8.7926	8.0820	7.4776	6.9574	6.5048	6.1075	5.4427	4.9084	68
69	8.8045	8.0920	7.4862	6.9648	6.5113	6.1133	5.4473	4.9121	69
70	8.8159	8.1017	7.4945	6.9720	6.5176	6.1188	5.4516	4.9157	70
71	8.8269	8.1109	7.5024	6.9788	6.5236	6.1240	5.4558	4.9191	71
72	8.8373	8.1198	7.5100	6.9854	6.5293	6.1291	5.4598	4.9223	72
73	8.8474	8.1283	7.5172	6.9917	6.5348	6.1339	5.4637	4.9254	73
74	8.8571	8.1364	7.5242	6.9977	6.5400	6.1386	5.4673	4.9284	74
75	8.8663	8.1442	7.5309	7.0035	6.5451	6.1430	5.4709	4.9313	75
76	8.8752	8.1517	7.5373	7.0090	6.5499	6.1473	5.4743	4.9340	76
77	8.8837	8.1589	7.5435	7.0143	6.5546	6.1514	5.4775	4.9367	77
78	8.8919	8.1658	7.5493	7.0194	6.5590	6.1553	5.4806	4.9392	78
79	8.8998	8.1724	7.5550	7.0243	6.5633	6.1591	5.4836	4.9416	79
80	8.9073	8.1788	7.5604	7.0290	6.5674	6.1627	5.4864	4.9439	80
81	8.9145	8.1849	7.5656	7.0335	6.5713	6.1661	5.4892	4.9462	81
82	8.9215	8.1907	7.5706	7.0378	6.5751	6.1694	5.4918	4.9483	82
83	8.9281	8.1963	7.5754	7.0420	6.5787	6.1726	5.4943	4.9504	83
84	8.9345	8.2017	7.5800	7.0459	6.5822	6.1757	5.4968	4.9523	84
85	8.9406	8.2069	7.5844	7.0498	6.5855	6.1786	5.4991	4.9542	85
86	8.9465	8.2119	7.5887	7.0534	6.5887	6.1814	5.5013	4.9560	86
87	8.9522	8.2166	7.5927	7.0569	6.5918	6.1841	5.5034	4.9577	87
88	8.9576	8.2212	7.5967	7.0603	6.5947	6.1867	5.5055	4.9594	88
89	8.9628	8.2256	7.6004	7.0635	6.5975	6.1892	5.5075	4.9610	89
90	8.9678	8.2298	7.6040	7.0667	6.6002	6.1916	5.5093	4.9625	90
91	8.9726	8.2338	7.6075	7.0696	6.6028	6.1939	5.5112	4.9640	91
92	8.9773	8.2377	7.6108	7.0725	6.6053	6.1961	5.5129	4.9654	92
93	8.9817	8.2415	7.6140	7.0753	6.6077	6.1982	5.5146	4.9668	93
94	8.9859	8.2451	7.6170	7.0779	6.6100	6.2002	5.5162	4.9681	94
95	8.9900	8.2485	7.6200	7.0804	6.6123	6.2022	5.5177	4.9693	95
96	8.9940	8.2518	7.6228	7.0829	6.6144	6.2040	5.5192	4.9705	96
97	8.9977	8.2550	7.6255	7.0852	6.6164	6.2058	5.5206	4.9717	97
98	9.0014	8.2580	7.6281	7.0875	6.6184	6.2075	5.5220	4.9728	98
99	9.0049	8.2610	7.6306	7.0896	6.6203	6.2092	5.5233	4.9739	99
100	9.0082	8.2638	7.6330	7.0917	6.6221	6.2108	5.5246	4.9749	100

YEARS' PURCHASE

(DUAL RATE % PRINCIPLE)

OR

PRESENT VALUE OF
ONE POUND PER ANNUM

*receivable at the end of each year after
allowing for a sinking fund at a given
rate to replace the invested capital and
for the effect of income tax at* **25%** *on
that part of the income used to provide
the annual sinking fund instalment.*

AT RATES OF INTEREST FROM

4% to 20%

AND

ALLOWING FOR THE POSSIBLE INVESTMENT

OF SINKING FUNDS AT

2·5%, 3% and 4%

INCOME TAX at 25%

Note:—Tables of Years' Purchase in which no allowance has been made for the effect
of income tax on that part of the income used to provide the annual sinking
fund instalment will be found on pages 53 to 71.

YEARS' PURCHASE

Rate per Cent

Yrs.	4	4.5	5	5.5	6	6.25	6.5	6.75	Yrs.
1	0.7282	0.7255	0.7229	0.7203	0.7177	0.7164	0.7151	0.7139	1
2	1.4318	1.4216	1.4116	1.4017	1.3919	1.3871	1.3823	1.3775	2
3	2.1119	2.0898	2.0682	2.0470	2.0263	2.0161	2.0060	1.9959	3
4	2.7694	2.7316	2.6948	2.6589	2.6240	2.6069	2.5901	2.5734	4
5	3.4053	3.3483	3.2931	3.2398	3.1881	3.1629	3.1381	3.1137	5
6	4.0204	3.9411	3.8650	3.7917	3.7212	3.6869	3.6532	3.6201	6
7	4.6155	4.5114	4.4119	4.3167	4.2255	4.1813	4.1380	4.0957	7
8	5.1915	5.0601	4.9353	4.8164	4.7032	4.6485	4.5951	4.5429	8
9	5.7490	5.5884	5.4365	5.2926	5.1562	5.0905	5.0266	4.9642	9
10	6.2888	6.0971	5.9168	5.7467	5.5862	5.5093	5.4344	5.3616	10
11	6.8116	6.5873	6.3772	6.1802	5.9949	5.9064	5.8205	5.7370	11
12	7.3180	7.0597	6.8190	6.5942	6.3837	6.2834	6.1862	6.0920	12
13	7.8086	7.5152	7.2430	6.9899	6.7538	6.6417	6.5332	6.4282	13
14	8.2839	7.9545	7.6502	7.3684	7.1065	6.9825	6.8627	6.7469	14
15	8.7447	8.3783	8.0415	7.7306	7.4430	7.3070	7.1759	7.0494	15
16	9.1913	8.7875	8.4176	8.0776	7.7641	7.6162	7.4739	7.3368	16
17	9.6243	9.1824	8.7793	8.4102	8.0708	7.9112	7.7577	7.6101	17
18	10.0442	9.5639	9.1274	8.7290	8.3640	8.1927	8.0283	7.8703	18
19	10.4514	9.9324	9.4625	9.0350	8.6445	8.4616	8.2863	8.1181	19
20	10.8464	10.2885	9.7851	9.3287	8.9130	8.7187	8.5327	8.3545	20
21	11.2297	10.6327	10.0959	9.6108	9.1701	8.9646	8.7681	8.5800	21
22	11.6015	10.9655	10.3955	9.8819	9.4166	9.2000	8.9932	8.7954	22
23	11.9624	11.2873	10.6843	10.1425	9.6530	9.4255	9.2085	9.0013	23
24	12.3127	11.5986	10.9629	10.3932	9.8798	9.6416	9.4147	9.1982	24
25	12.6527	11.8999	11.2316	10.6344	10.0975	9.8489	9.6122	9.3866	25
26	12.9828	12.1914	11.4910	10.8666	10.3066	10.0477	9.8015	9.5671	26
27	13.3033	12.4736	11.7413	11.0903	10.5076	10.2386	9.9831	9.7400	27
28	13.6146	12.7469	11.9831	11.3057	10.7008	10.4220	10.1574	9.9058	28
29	13.9169	13.0115	12.2167	11.5134	10.8867	10.5983	10.3247	10.0649	29
30	14.2106	13.2678	12.4424	11.7137	11.0656	10.7677	10.4855	10.2176	30
31	14.4958	13.5162	12.6606	11.9068	11.2378	10.9307	10.6400	10.3643	31
32	14.7730	13.7568	12.8715	12.0932	11.4037	11.0876	10.7885	10.5052	32
33	15.0423	13.9901	13.0755	12.2731	11.5635	11.2386	10.9314	10.6407	33
34	15.3040	14.2162	13.2728	12.4467	11.7175	11.3840	11.0690	10.7709	34
35	15.5584	14.4354	13.4636	12.6145	11.8660	11.5242	11.2015	10.8963	35
36	15.8056	14.6480	13.6484	12.7765	12.0093	11.6593	11.3290	11.0170	36
37	16.0459	14.8542	13.8272	12.9331	12.1475	11.7895	11.4520	11.1332	37
38	16.2795	15.0542	14.0003	13.0844	12.2810	11.9151	11.5705	11.2452	38
39	16.5067	15.2482	14.1680	13.2307	12.4098	12.0364	11.6848	11.3531	39
40	16.7275	15.4365	14.3304	13.3723	12.5342	12.1534	11.7950	11.4572	40
41	16.9423	15.6192	14.4878	13.5092	12.6544	12.2664	11.9014	11.5575	41
42	17.1512	15.7965	14.6402	13.6416	12.7706	12.3755	12.0041	11.6543	42
43	17.3543	15.9687	14.7880	13.7698	12.8829	12.4809	12.1032	11.7478	43
44	17.5519	16.1358	14.9312	13.8939	12.9914	12.5828	12.1990	11.8380	44
45	17.7441	16.2981	15.0701	14.0141	13.0964	12.6812	12.2916	11.9251	45
46	17.9311	16.4558	15.2047	14.1305	13.1980	12.7764	12.3810	12.0093	46
47	18.1130	16.6088	15.3353	14.2432	13.2963	12.8685	12.4674	12.0906	47
48	18.2900	16.7575	15.4620	14.3524	13.3914	12.9576	12.5510	12.1692	48
49	18.4622	16.9020	15.5849	14.4583	13.4835	13.0438	12.6319	12.2452	49
50	18.6298	17.0423	15.7042	14.5608	13.5727	13.1273	12.7101	12.3187	50

YEARS' PURCHASE

				Rate per Cent					
Yrs.	4	4.5	5	5.5	6	6.25	6.5	6.75	Yrs.
51	18.7929	17.1787	15.8199	14.6603	13.6590	13.2080	12.7858	12.3898	51
52	18.9516	17.3112	15.9322	14.7567	13.7427	13.2862	12.8591	12.4586	52
53	19.1061	17.4400	16.0412	14.8502	13.8237	13.3620	12.9300	12.5252	53
54	19.2565	17.5652	16.1471	14.9408	13.9023	13.4353	12.9987	12.5896	54
55	19.4028	17.6869	16.2499	15.0288	13.9784	13.5064	13.0652	12.6520	55
56	19.5453	17.8052	16.3497	15.1141	14.0522	13.5753	13.1297	12.7124	56
57	19.6840	17.9203	16.4466	15.1969	14.1238	13.6421	13.1921	12.7709	57
58	19.8190	18.0321	16.5408	15.2773	14.1931	13.7068	13.2527	12.8277	58
59	19.9505	18.1409	16.6323	15.3553	14.2604	13.7695	13.3113	12.8826	59
60	20.0785	18.2467	16.7211	15.4310	14.3257	13.8304	13.3682	12.9359	60
61	20.2031	18.3495	16.8075	15.5045	14.3891	13.8894	13.4233	12.9875	61
62	20.3245	18.4496	16.8914	15.5759	14.4505	13.9467	13.4768	13.0375	62
63	20.4427	18.5470	16.9730	15.6452	14.5102	14.0022	13.5287	13.0861	63
64	20.5578	18.6417	17.0522	15.7126	14.5681	14.0561	13.5790	13.1331	64
65	20.6699	18.7338	17.1293	15.7780	14.6243	14.1085	13.6278	13.1788	65
66	20.7791	18.8234	17.2042	15.8415	14.6788	14.1592	13.6752	13.2231	66
67	20.8854	18.9106	17.2770	15.9032	14.7318	14.2085	13.7211	13.2661	67
68	20.9890	18.9955	17.3479	15.9632	14.7833	14.2564	13.7658	13.3078	68
69	21.0899	19.0781	17.4167	16.0215	14.8333	14.3029	13.8091	13.3483	69
70	21.1882	19.1585	17.4837	16.0782	14.8818	14.3480	13.8512	13.3876	70
71	21.2839	19.2367	17.5488	16.1332	14.9290	14.3918	13.8920	13.4257	71
72	21.3772	19.3129	17.6122	16.1868	14.9748	14.4344	13.9317	13.4628	72
73	21.4680	19.3870	17.6738	16.2388	15.0193	14.4758	13.9702	13.4988	73
74	21.5565	19.4592	17.7338	16.2894	15.0626	14.5160	14.0076	13.5337	74
75	21.6428	19.5294	17.7921	16.3386	15.1047	14.5550	14.0440	13.5676	75
76	21.7268	19.5978	17.8488	16.3864	15.1455	14.5930	14.0793	13.6006	76
77	21.8087	19.6644	17.9040	16.4330	15.1853	14.6299	14.1137	13.6327	77
78	21.8885	19.7292	17.9578	16.4782	15.2239	14.6657	14.1470	13.6638	78
79	21.9662	19.7924	18.0101	16.5222	15.2615	14.7006	14.1795	13.6940	79
80	22.0419	19.8539	18.0610	16.5651	15.2980	14.7345	14.2110	13.7234	80
81	22.1158	19.9137	18.1105	16.6067	15.3335	14.7674	14.2416	13.7520	81
82	22.1877	19.9720	18.1587	16.6472	15.3681	14.7995	14.2714	13.7798	82
83	22.2578	20.0288	18.2056	16.6867	15.4016	14.8306	14.3004	13.8068	83
84	22.3261	20.0841	18.2513	16.7250	15.4343	14.8609	14.3286	13.8330	84
85	22.3927	20.1380	18.2958	16.7624	15.4661	14.8904	14.3560	13.8586	85
86	22.4576	20.1904	18.3391	16.7987	15.4970	14.9190	14.3826	13.8834	86
87	22.5208	20.2415	18.3812	16.8341	15.5271	14.9469	14.4085	13.9075	87
88	22.5824	20.2913	18.4222	16.8685	15.5564	14.9740	14.4337	13.9310	88
89	22.6425	20.3398	18.4622	16.9020	15.5849	15.0004	14.4582	13.9539	89
90	22.7011	20.3870	18.5011	16.9346	15.6126	15.0261	14.4821	13.9761	90
91	22.7581	20.4331	18.5390	16.9663	15.6396	15.0511	14.5053	13.9977	91
92	22.8138	20.4779	18.5759	16.9972	15.6658	15.0754	14.5279	14.0187	92
93	22.8680	20.5216	18.6119	17.0273	15.6914	15.0991	14.5499	14.0392	93
94	22.9209	20.5641	18.6469	17.0566	15.7163	15.1221	14.5712	14.0591	94
95	22.9724	20.6056	18.6809	17.0851	15.7405	15.1445	14.5920	14.0785	95
96	23.0226	20.6460	18.7142	17.1129	15.7640	15.1663	14.6123	14.0973	96
97	23.0716	20.6854	18.7465	17.1399	15.7870	15.1876	14.6320	14.1157	97
98	23.1194	20.7238	18.7780	17.1663	15.8093	15.2082	14.6512	14.1335	98
99	23.1659	20.7611	18.8087	17.1919	15.8311	15.2284	14.6699	14.1509	99
100	23.2113	20.7976	18.8386	17.2169	15.8523	15.2480	14.6881	14.1678	100

YEARS' PURCHASE

Rate per Cent

Yrs.	7	7.25	7.5	8	8.5	9	9.5	10	Yrs.
1	0.7126	0.7113	0.7101	0.7075	0.7051	0.7026	0.7001	0.6977	1
2	1.3728	1.3681	1.3634	1.3542	1.3451	1.3361	1.3273	1.3185	2
3	1.9860	1.9762	1.9665	1.9474	1.9286	1.9102	1.8921	1.8744	3
4	2.5570	2.5407	2.5247	2.4932	2.4625	2.4326	2.4033	2.3748	4
5	3.0896	3.0660	3.0426	2.9970	2.9528	2.9098	2.8681	2.8276	5
6	3.5877	3.5558	3.5244	3.4634	3.4044	3.3475	3.2924	3.2390	6
7	4.0542	4.0135	3.9736	3.8962	3.8217	3.7501	3.6811	3.6145	7
8	4.4919	4.4420	4.3932	4.2988	4.2083	4.1216	4.0384	3.9585	8
9	4.9033	4.8440	4.7860	4.6742	4.5674	4.4654	4.3679	4.2746	9
10	5.2907	5.2216	5.1543	5.0248	4.9017	4.7844	4.6726	4.5660	10
11	5.6559	5.5770	5.5003	5.3531	5.2135	5.0811	4.9552	4.8354	11
12	6.0006	5.9119	5.8258	5.6609	5.5051	5.3576	5.2179	5.0852	12
13	6.3265	6.2280	6.1325	5.9501	5.7782	5.6159	5.4626	5.3173	13
14	6.6350	6.5268	6.4220	6.2222	6.0344	5.8577	5.6910	5.5336	14
15	6.9274	6.8094	6.6954	6.4786	6.2753	6.0844	5.9047	5.7354	15
16	7.2047	7.0772	6.9542	6.7205	6.5020	6.2973	6.1051	5.9242	16
17	7.4681	7.3312	7.1992	6.9491	6.7158	6.4976	6.2931	6.1011	17
18	7.7184	7.5723	7.4316	7.1654	6.9175	6.6863	6.4700	6.2672	18
19	7.9567	7.8015	7.6522	7.3702	7.1083	6.8643	6.6365	6.4234	19
20	8.1836	8.0195	7.8619	7.5645	7.2888	7.0325	6.7937	6.5705	20
21	8.3998	8.2271	8.0613	7.7489	7.4599	7.1917	6.9420	6.7092	21
22	8.6062	8.4249	8.2511	7.9242	7.6222	7.3424	7.0824	6.8402	22
23	8.8032	8.6136	8.4320	8.0909	7.7763	7.4853	7.2153	6.9640	23
24	8.9914	8.7938	8.6046	8.2497	7.9229	7.6210	7.3412	7.0813	24
25	9.1714	8.9658	8.7693	8.4009	8.0623	7.7499	7.4608	7.1925	25
26	9.3436	9.1303	8.9266	8.5452	8.1950	7.8725	7.5743	7.2979	26
27	9.5085	9.2877	9.0770	8.6829	8.3216	7.9892	7.6823	7.3981	27
28	9.6665	9.4384	9.2208	8.8144	8.4423	8.1004	7.7851	7.4934	28
29	9.8179	9.5827	9.3585	8.9401	8.5576	8.2065	7.8830	7.5841	29
30	9.9631	9.7210	9.4904	9.0604	8.6678	8.3077	7.9764	7.6705	30
31	10.1025	9.8536	9.6167	9.1755	8.7731	8.4044	8.0655	7.7528	31
32	10.2363	9.9809	9.7379	9.2858	8.8738	8.4968	8.1506	7.8314	32
33	10.3649	10.1031	9.8542	9.3915	8.9703	8.5852	8.2319	7.9064	33
34	10.4885	10.2205	9.9659	9.4929	9.0627	8.6698	8.3096	7.9782	34
35	10.6074	10.3333	10.0731	9.5901	9.1513	8.7509	8.3840	8.0467	35
36	10.7217	10.4418	10.1762	9.6835	9.2363	8.8286	8.4553	8.1124	36
37	10.8317	10.5462	10.2753	9.7731	9.3178	8.9030	8.5236	8.1752	37
38	10.9377	10.6466	10.3706	9.8593	9.3961	8.9745	8.5891	8.2354	38
39	11.0398	10.7433	10.4623	9.9422	9.4714	9.0431	8.6519	8.2931	39
40	11.1381	10.8364	10.5506	10.0219	9.5437	9.1090	8.7122	8.3485	40
41	11.2330	10.9261	10.6356	10.0986	9.6132	9.1723	8.7701	8.4017	41
42	11.3244	11.0126	10.7175	10.1724	9.6801	9.2332	8.8257	8.4527	42
43	11.4126	11.0960	10.7965	10.2435	9.7445	9.2917	8.8792	8.5018	43
44	11.4977	11.1765	10.8727	10.3121	9.8064	9.3481	8.9307	8.5489	44
45	11.5799	11.2541	10.9461	10.3781	9.8661	9.4023	8.9802	8.5943	45
46	11.6592	11.3290	11.0170	10.4418	9.9237	9.4546	9.0278	8.6379	46
47	11.7359	11.4013	11.0854	10.5032	9.9792	9.5049	9.0737	8.6799	47
48	11.8099	11.4712	11.1514	10.5625	10.0326	9.5534	9.1179	8.7203	48
49	11.8815	11.5387	11.2152	10.6197	10.0842	9.6002	9.1605	8.7593	49
50	11.9507	11.6040	11.2768	10.6749	10.1340	9.6453	9.2015	8.7968	50

YEARS' PURCHASE

				Rate per Cent					
Yrs.	7	7.25	7.5	8	8.5	9	9.5	10	Yrs.
51	12.0176	11.6670	11.3364	10.7283	10.1821	9.6888	9.2412	8.8330	51
52	12.0823	11.7280	11.3939	10.7798	10.2285	9.7308	9.2794	8.8679	52
53	12.1449	11.7870	11.4496	10.8296	10.2733	9.7714	9.3163	8.9016	53
54	12.2054	11.8440	11.5034	10.8778	10.3167	9.8106	9.3519	8.9341	54
55	12.2641	11.8992	11.5555	10.9243	10.3585	9.8484	9.3862	8.9655	55
56	12.3208	11.9527	11.6059	10.9693	10.3990	9.8850	9.4195	8.9958	56
57	12.3758	12.0044	11.6546	11.0129	10.4381	9.9204	9.4516	9.0250	57
58	12.4291	12.0545	11.7018	11.0550	10.4760	9.9545	9.4826	9.0533	58
59	12.4806	12.1030	11.7476	11.0958	10.5126	9.9876	9.5126	9.0807	59
60	12.5306	12.1500	11.7918	11.1353	10.5480	10.0196	9.5416	9.1071	60
61	12.5791	12.1955	11.8347	11.1735	10.5823	10.0505	9.5696	9.1326	61
62	12.6260	12.2397	11.8762	11.2106	10.6155	10.0805	9.5968	9.1574	62
63	12.6715	12.2824	11.9165	11.2464	10.6477	10.1095	9.6230	9.1813	63
64	12.7156	12.3239	11.9555	11.2812	10.6788	10.1375	9.6485	9.2044	64
65	12.7584	12.3641	11.9934	11.3148	10.7090	10.1647	9.6731	9.2268	65
66	12.7999	12.4031	12.0300	11.3475	10.7382	10.1911	9.6969	9.2485	66
67	12.8402	12.4409	12.0656	11.3791	10.7665	10.2166	9.7200	9.2695	67
68	12.8793	12.4775	12.1001	11.4098	10.7940	10.2413	9.7424	9.2899	68
69	12.9172	12.5131	12.1336	11.4395	10.8206	10.2652	9.7641	9.3096	69
70	12.9540	12.5477	12.1660	11.4684	10.8464	10.2885	9.7851	9.3287	70
71	12.9897	12.5812	12.1975	11.4964	10.8715	10.3110	9.8055	9.3472	71
72	13.0244	12.6137	12.2281	11.5235	10.8958	10.3328	9.8252	9.3651	72
73	13.0581	12.6453	12.2578	11.5499	10.9193	10.3540	9.8444	9.3825	73
74	13.0908	12.6759	12.2866	11.5755	10.9422	10.3746	9.8629	9.3994	74
75	13.1225	12.7057	12.3145	11.6003	10.9643	10.3945	9.8810	9.4158	75
76	13.1534	12.7346	12.3417	11.6244	10.9859	10.4138	9.8984	9.4316	76
77	13.1833	12.7627	12.3681	11.6478	11.0068	10.4326	9.9154	9.4470	77
78	13.2125	12.7900	12.3937	11.6705	11.0270	10.4508	9.9319	9.4620	78
79	13.2407	12.8165	12.4186	11.6926	11.0467	10.4685	9.9478	9.4765	79
80	13.2682	12.8422	12.4428	11.7140	11.0659	10.4857	9.9633	9.4905	80
81	13.2949	12.8673	12.4662	11.7348	11.0844	10.5024	9.9784	9.5042	81
82	13.3209	12.8916	12.4891	11.7550	11.1025	10.5186	9.9930	9.5175	82
83	13.3461	12.9152	12.5112	11.7747	11.1200	10.5343	10.0072	9.5303	83
84	13.3707	12.9382	12.5328	11.7938	11.1370	10.5496	10.0210	9.5428	84
85	13.3945	12.9605	12.5538	11.8123	11.1536	10.5644	10.0344	9.5550	85
86	13.4177	12.9822	12.5741	11.8303	11.1696	10.5788	10.0474	9.5668	86
87	13.4402	13.0033	12.5939	11.8479	11.1853	10.5928	10.0600	9.5782	87
88	13.4622	13.0239	12.6132	11.8649	11.2004	10.6065	10.0723	9.5894	88
89	13.4835	13.0438	12.6319	11.8815	11.2152	10.6197	10.0842	9.6002	89
90	13.5042	13.0632	12.6501	11.8976	11.2295	10.6326	10.0958	9.6107	90
91	13.5244	13.0821	12.6678	11.9132	11.2435	10.6451	10.1071	9.6209	91
92	13.5441	13.1005	12.6850	11.9285	11.2571	10.6572	10.1181	9.6308	92
93	13.5631	13.1183	12.7018	11.9433	11.2702	10.6690	10.1287	9.6405	93
94	13.5817	13.1357	12.7181	11.9577	11.2831	10.6805	10.1391	9.6499	94
95	13.5998	13.1526	12.7339	11.9717	11.2955	10.6917	10.1491	9.6590	95
96	13.6174	13.1691	12.7493	11.9853	11.3077	10.7026	10.1589	9.6679	96
97	13.6345	13.1851	12.7643	11.9986	11.3195	10.7131	10.1685	9.6765	97
98	13.6512	13.2007	12.7789	12.0115	11.3310	10.7234	10.1777	9.6849	98
99	13.6674	13.2158	12.7931	12.0240	11.3421	10.7334	10.1867	9.6930	99
100	13.6832	13.2306	12.8070	12.0362	11.3530	10.7432	10.1955	9.7010	100

YEARS' PURCHASE

Rate per Cent

Yrs.	11	12	13	14	15	16	18	20	Yrs.
1	0.6928	0.6881	0.6834	0.6787	0.6742	0.6696	0.6608	0.6522	1
2	1.3013	1.2846	1.2683	1.2524	1.2370	1.2218	1.1927	1.1649	2
3	1.8399	1.8066	1.7746	1.7436	1.7137	1.6849	1.6299	1.5785	3
4	2.3197	2.2671	2.2169	2.1688	2.1227	2.0786	1.9956	1.9191	4
5	2.7498	2.6762	2.6065	2.5402	2.4773	2.4174	2.3059	2.2043	5
6	3.1374	3.0420	2.9522	2.8675	2.7876	2.7120	2.5725	2.4466	6
7	3.4884	3.3709	3.2609	3.1580	3.0613	2.9704	2.8038	2.6549	7
8	3.8077	3.6681	3.5383	3.4174	3.3044	3.1987	3.0064	2.8359	8
9	4.0993	3.9379	3.7887	3.6504	3.5218	3.4020	3.1853	2.9945	9
10	4.3666	4.1839	4.0159	3.8608	3.7173	3.5841	3.3444	3.1347	10
11	4.6124	4.4090	4.2228	4.0517	3.8940	3.7480	3.4867	3.2594	11
12	4.8391	4.6158	4.4121	4.2257	4.0543	3.8964	3.6147	3.3710	12
13	5.0489	4.8062	4.5858	4.3847	4.2005	4.0312	3.7304	3.4714	13
14	5.2434	4.9822	4.7457	4.5307	4.3343	4.1543	3.8356	3.5623	14
15	5.4243	5.1452	4.8934	4.6652	4.4572	4.2670	3.9315	3.6449	15
16	5.5929	5.2966	5.0302	4.7893	4.5704	4.3707	4.0193	3.7203	16
17	5.7503	5.4376	5.1572	4.9043	4.6750	4.4662	4.1000	3.7893	17
18	5.8976	5.5692	5.2754	5.0110	4.7719	4.5546	4.1743	3.8527	18
19	6.0357	5.6921	5.3856	5.1104	4.8619	4.6365	4.2430	3.9111	19
20	6.1654	5.8073	5.4886	5.2030	4.9457	4.7126	4.3067	3.9652	20
21	6.2873	5.9154	5.5850	5.2896	5.0239	4.7835	4.3659	4.0153	21
22	6.4022	6.0170	5.6755	5.3707	5.0970	4.8498	4.4210	4.0618	22
23	6.5106	6.1126	5.7605	5.4468	5.1654	4.9117	4.4724	4.1052	23
24	6.6130	6.2028	5.8405	5.5182	5.2297	4.9698	4.5204	4.1456	24
25	6.7099	6.2879	5.9159	5.5855	5.2900	5.0243	4.5655	4.1835	25
26	6.8016	6.3684	5.9871	5.6489	5.3469	5.0755	4.6078	4.2190	26
27	6.8885	6.4446	6.0544	5.7088	5.4005	5.1238	4.6475	4.2523	27
28	6.9710	6.5168	6.1181	5.7653	5.4511	5.1693	4.6849	4.2836	28
29	7.0495	6.5852	6.1784	5.8189	5.4989	5.2123	4.7202	4.3130	29
30	7.1240	6.6503	6.2356	5.8696	5.5442	5.2529	4.7535	4.3408	30
31	7.1950	6.7121	6.2899	5.9177	5.5870	5.2914	4.7850	4.3671	31
32	7.2626	6.7709	6.3415	5.9633	5.6277	5.3279	4.8148	4.3919	32
33	7.3271	6.8269	6.3906	6.0068	5.6664	5.3625	4.8431	4.4154	33
34	7.3887	6.8803	6.4374	6.0481	5.7031	5.3954	4.8699	4.4377	34
35	7.4475	6.9313	6.4820	6.0874	5.7381	5.4267	4.8954	4.4588	35
36	7.5036	6.9799	6.5245	6.1249	5.7714	5.4565	4.9196	4.4789	36
37	7.5574	7.0264	6.5651	6.1606	5.8031	5.4848	4.9426	4.4980	37
38	7.6088	7.0708	6.6038	6.1948	5.8334	5.5119	4.9646	4.5162	38
39	7.6581	7.1133	6.6409	6.2274	5.8623	5.5377	4.9855	4.5335	39
40	7.7052	7.1540	6.6764	6.2585	5.8899	5.5623	5.0055	4.5500	40
41	7.7505	7.1930	6.7103	6.2884	5.9163	5.5858	5.0245	4.5657	41
42	7.7939	7.2304	6.7429	6.3169	5.9416	5.6084	5.0427	4.5807	42
43	7.8356	7.2663	6.7740	6.3443	5.9658	5.6299	5.0602	4.5951	43
44	7.8756	7.3007	6.8039	6.3705	5.9890	5.6506	5.0768	4.6089	44
45	7.9141	7.3337	6.8326	6.3956	6.0112	5.6703	5.0928	4.6220	45
46	7.9511	7.3655	6.8602	6.4198	6.0325	5.6893	5.1081	4.6346	46
47	7.9867	7.3960	6.8866	6.4429	6.0529	5.7075	5.1227	4.6466	47
48	8.0209	7.4253	6.9121	6.4652	6.0726	5.7249	5.1368	4.6582	48
49	8.0538	7.4535	6.9365	6.4866	6.0914	5.7417	5.1503	4.6693	49
50	8.0856	7.4807	6.9600	6.5071	6.1096	5.7578	5.1632	4.6800	50

YEARS' PURCHASE

Rate per Cent

Yrs.	11	12	13	14	15	16	18	20	Yrs.
51	8.1161	7.5069	6.9827	6.5269	6.1270	5.7733	5.1757	4.6902	51
52	8.1456	7.5321	7.0045	6.5460	6.1438	5.7882	5.1876	4.7000	52
53	8.1740	7.5563	7.0255	6.5643	6.1599	5.8025	5.1991	4.7094	53
54	8.2014	7.5797	7.0457	6.5820	6.1755	5.8163	5.2102	4.7185	54
55	8.2278	7.6023	7.0652	6.5990	6.1905	5.8296	5.2209	4.7273	55
56	8.2533	7.6241	7.0840	6.6154	6.2049	5.8424	5.2311	4.7357	56
57	8.2780	7.6451	7.1021	6.6312	6.2188	5.8547	5.2410	4.7438	57
58	8.3017	7.6654	7.1196	6.6464	6.2322	5.8666	5.2505	4.7516	58
59	8.3247	7.6850	7.1365	6.6612	6.2452	5.8781	5.2597	4.7591	59
60	8.3469	7.7039	7.1528	6.6754	6.2576	5.8891	5.2686	4.7663	60
61	8.3684	7.7222	7.1686	6.6891	6.2697	5.8998	5.2771	4.7733	61
62	8.3891	7.7398	7.1838	6.7023	6.2813	5.9101	5.2854	4.7801	62
63	8.4092	7.7569	7.1985	6.7151	6.2926	5.9201	5.2933	4.7866	63
64	8.4286	7.7734	7.2128	6.7275	6.3034	5.9297	5.3010	4.7929	64
65	8.4474	7.7894	7.2265	6.7395	6.3139	5.9390	5.3084	4.7989	65
66	8.4656	7.8049	7.2398	6.7510	6.3241	5.9479	5.3156	4.8048	66
67	8.4832	7.8198	7.2527	6.7622	6.3339	5.9566	5.3225	4.8105	67
68	8.5002	7.8343	7.2651	6.7730	6.3434	5.9650	5.3292	4.8159	68
69	8.5167	7.8483	7.2772	6.7835	6.3526	5.9731	5.3357	4.8212	69
70	8.5327	7.8619	7.2888	6.7937	6.3615	5.9810	5.3420	4.8263	70
71	8.5482	7.8750	7.3001	6.8035	6.3701	5.9886	5.3481	4.8313	71
72	8.5632	7.8877	7.3111	6.8130	6.3784	5.9960	5.3539	4.8361	72
73	8.5777	7.9001	7.3217	6.8222	6.3865	6.0031	5.3596	4.8407	73
74	8.5918	7.9120	7.3319	6.8311	6.3943	6.0100	5.3651	4.8452	74
75	8.6055	7.9236	7.3419	6.8397	6.4019	6.0167	5.3704	4.8495	75
76	8.6188	7.9349	7.3515	6.8481	6.4092	6.0232	5.3756	4.8538	76
77	8.6316	7.9458	7.3609	6.8562	6.4163	6.0294	5.3806	4.8578	77
78	8.6441	7.9563	7.3699	6.8641	6.4232	6.0355	5.3854	4.8618	78
79	8.6562	7.9666	7.3787	6.8717	6.4299	6.0414	5.3901	4.8656	79
80	8.6679	7.9765	7.3873	6.8791	6.4363	6.0471	5.3947	4.8693	80
81	8.6793	7.9862	7.3955	6.8863	6.4426	6.0527	5.3991	4.8729	81
82	8.6904	7.9955	7.4036	6.8932	6.4487	6.0580	5.4034	4.8764	82
83	8.7011	8.0046	7.4114	6.9000	6.4546	6.0632	5.4075	4.8798	83
84	8.7115	8.0134	7.4189	6.9065	6.4603	6.0683	5.4115	4.8830	84
85	8.7216	8.0220	7.4262	6.9129	6.4659	6.0732	5.4154	4.8862	85
86	8.7315	8.0303	7.4334	6.9191	6.4713	6.0780	5.4192	4.8893	86
87	8.7410	8.0384	7.4403	6.9250	6.4765	6.0826	5.4229	4.8923	87
88	8.7503	8.0462	7.4470	6.9309	6.4816	6.0871	5.4265	4.8952	88
89	8.7593	8.0538	7.4535	6.9365	6.4866	6.0914	5.4299	4.8980	89
90	8.7680	8.0612	7.4599	6.9420	6.4914	6.0957	5.4333	4.9007	90
91	8.7765	8.0684	7.4660	6.9473	6.4960	6.0998	5.4365	4.9034	91
92	8.7848	8.0754	7.4720	6.9525	6.5005	6.1038	5.4397	4.9060	92
93	8.7928	8.0822	7.4778	6.9575	6.5049	6.1076	5.4428	4.9085	93
94	8.8006	8.0888	7.4834	6.9624	6.5092	6.1114	5.4458	4.9109	94
95	8.8082	8.0952	7.4889	6.9672	6.5134	6.1151	5.4487	4.9133	95
96	8.8156	8.1014	7.4943	6.9718	6.5174	6.1186	5.4515	4.9156	96
97	8.8228	8.1075	7.4994	6.9763	6.5213	6.1221	5.4542	4.9178	97
98	8.8297	8.1133	7.5045	6.9806	6.5251	6.1254	5.4569	4.9200	98
99	8.8365	8.1191	7.5094	6.9849	6.5288	6.1287	5.4595	4.9221	99
100	8.8431	8.1246	7.5141	6.9890	6.5324	6.1319	5.4620	4.9241	100

YEARS' PURCHASE

Rate per Cent

Yrs.	4	4.5	5	5.5	6	6.25	6.5	6.75	Yrs.
1	0.7282	0.7255	0.7229	0.7203	0.7177	0.7164	0.7151	0.7139	1
2	1.4351	1.4249	1.4148	1.4049	1.3951	1.3902	1.3854	1.3806	2
3	2.1215	2.0992	2.0774	2.0560	2.0351	2.0248	2.0146	2.0045	3
4	2.7878	2.7495	2.7122	2.6759	2.6406	2.6233	2.6062	2.5893	4
5	3.4348	3.3768	3.3207	3.2665	3.2140	3.1884	3.1632	3.1383	5
6	4.0629	3.9820	3.9043	3.8295	3.7576	3.7226	3.6883	3.6546	6
7	4.6727	4.5660	4.4641	4.3667	4.2733	4.2282	4.1840	4.1406	7
8	5.2648	5.1297	5.0015	4.8794	4.7632	4.7072	4.6524	4.5989	8
9	5.8396	5.6739	5.5174	5.3693	5.2289	5.1614	5.0957	5.0316	9
10	6.3977	6.1993	6.0130	5.8375	5.6719	5.5926	5.5155	5.4405	10
11	6.9395	6.7068	6.4892	6.2852	6.0937	6.0023	5.9135	5.8274	11
12	7.4655	7.1969	6.9469	6.7137	6.4956	6.3918	6.2913	6.1939	12
13	7.9762	7.6703	7.3870	7.1239	6.8789	6.7626	6.6501	6.5414	13
14	8.4721	8.1278	7.8104	7.5168	7.2445	7.1157	6.9913	6.8712	14
15	8.9535	8.5698	8.2177	7.8934	7.5937	7.4522	7.3159	7.1845	15
16	9.4208	8.9970	8.6097	8.2544	7.9272	7.7732	7.6250	7.4824	16
17	9.8746	9.4100	8.9871	8.6007	8.2461	8.0795	7.9195	7.7658	17
18	10.3151	9.8092	9.3506	8.9330	8.5510	8.3721	8.2004	8.0357	18
19	10.7428	10.1952	9.7007	9.2520	8.8429	8.6516	8.4685	8.2929	19
20	11.1581	10.5685	10.0380	9.5583	9.1223	8.9189	8.7244	8.5382	20
21	11.5613	10.9295	10.3632	9.8526	9.3901	9.1747	8.9690	8.7723	21
22	11.9527	11.2786	10.6766	10.1355	9.6466	9.4195	9.2027	8.9958	22
23	12.3327	11.6164	10.9787	10.4074	9.8926	9.6539	9.4264	9.2094	23
24	12.7017	11.9432	11.2702	10.6690	10.1286	9.8785	9.6404	9.4135	24
25	13.0599	12.2593	11.5513	10.9205	10.3551	10.0938	9.8454	9.6089	25
26	13.4076	12.5653	11.8225	11.1627	10.5726	10.3003	10.0417	9.7958	26
27	13.7453	12.8614	12.0843	11.3957	10.7814	10.4984	10.2299	9.9748	27
28	14.0731	13.1479	12.3369	11.6201	10.9821	10.6886	10.4104	10.1463	28
29	14.3913	13.4253	12.5808	11.8362	11.1749	10.8712	10.5836	10.3107	29
30	14.7003	13.6938	12.8163	12.0445	11.3603	11.0466	10.7497	10.4684	30
31	15.0003	13.9538	13.0437	12.2451	11.5387	11.2151	10.9093	10.6196	31
32	15.2916	14.2055	13.2634	12.4385	11.7102	11.3771	11.0625	10.7648	32
33	15.5743	14.4492	13.4756	12.6250	11.8753	11.5329	11.2097	10.9042	33
34	15.8489	14.6852	13.6806	12.8048	12.0343	11.6828	11.3513	11.0380	34
35	16.1154	14.9137	13.8788	12.9782	12.1873	11.8270	11.4873	11.1666	35
36	16.3742	15.1351	14.0703	13.1455	12.3348	11.9658	11.6182	11.2903	36
37	16.6254	15.3495	14.2554	13.3069	12.4768	12.0994	11.7441	11.4092	37
38	16.8693	15.5572	14.4344	13.4627	12.6137	12.2281	11.8653	11.5235	38
39	17.1062	15.7583	14.6074	13.6131	12.7456	12.3520	11.9820	11.6335	39
40	17.3361	15.9532	14.7747	13.7583	12.8728	12.4714	12.0944	11.7394	40
41	17.5593	16.1421	14.9365	13.8986	12.9955	12.5866	12.2026	11.8413	41
42	17.7760	16.3250	15.0931	14.0340	13.1138	12.6975	12.3068	11.9395	42
43	17.9864	16.5023	15.2445	14.1648	13.2280	12.8045	12.4073	12.0341	43
44	18.1907	16.6741	15.3910	14.2912	13.3381	12.9077	12.5042	12.1252	44
45	18.3890	16.8406	15.5327	14.4133	13.4444	13.0072	12.5976	12.2130	45
46	18.5816	17.0020	15.6699	14.5314	13.5471	13.1033	12.6877	12.2976	46
47	18.7685	17.1583	15.8026	14.6454	13.6462	13.1960	12.7745	12.3792	47
48	18.9500	17.3099	15.9311	14.7557	13.7419	13.2854	12.8584	12.4579	48
49	19.1262	17.4568	16.0554	14.8623	13.8343	13.3718	12.9393	12.5338	49
50	19.2973	17.5992	16.1758	14.9654	13.9236	13.4552	13.0173	12.6071	50

YEARS' PURCHASE

				Rate per Cent					
Yrs.	4	4.5	5	5.5	6	6.25	6.5	6.75	Yrs.
51	19.4634	17.7373	16.2924	15.0651	14.0098	13.5358	13.0927	12.6777	51
52	19.6247	17.8711	16.4052	15.1616	14.0932	13.6135	13.1655	12.7460	52
53	19.7812	18.0008	16.5145	15.2548	14.1738	13.6887	13.2358	12.8118	53
54	19.9332	18.1266	16.6203	15.3451	14.2516	13.7613	13.3036	12.8754	54
55	20.0808	18.2486	16.7228	15.4324	14.3269	13.8315	13.3692	12.9368	55
56	20.2241	18.3668	16.8220	15.5169	14.3997	13.8993	13.4326	12.9961	56
57	20.3632	18.4815	16.9181	15.5986	14.4701	13.9649	13.4938	13.0534	57
58	20.4983	18.5927	17.0112	15.6778	14.5381	14.0283	13.5530	13.1088	58
59	20.6294	18.7005	17.1015	15.7543	14.6040	14.0896	13.6102	13.1623	59
60	20.7567	18.8050	17.1888	15.8285	14.6676	14.1488	13.6654	13.2140	60
61	20.8803	18.9064	17.2735	15.9002	14.7293	14.2061	13.7189	13.2640	61
62	21.0003	19.0047	17.3555	15.9697	14.7889	14.2616	13.7706	13.3123	62
63	21.1168	19.1001	17.4350	16.0370	14.8465	14.3152	13.8206	13.3590	63
64	21.2299	19.1926	17.5121	16.1022	14.9024	14.3671	13.8690	13.4042	64
65	21.3397	19.2823	17.5867	16.1653	14.9564	14.4173	13.9157	13.4479	65
66	21.4463	19.3693	17.6591	16.2264	15.0087	14.4659	13.9610	13.4902	66
67	21.5498	19.4537	17.7292	16.2855	15.0593	14.5129	14.0048	13.5310	67
68	21.6503	19.5355	17.7971	16.3429	15.1083	14.5584	14.0472	13.5706	68
69	21.7478	19.6149	17.8630	16.3984	15.1557	14.6025	14.0882	13.6089	69
70	21.8426	19.6920	17.9269	16.4522	15.2017	14.6451	14.1279	13.6459	70
71	21.9345	19.7667	17.9888	16.5043	15.2462	14.6864	14.1663	13.6817	71
72	22.0238	19.8391	18.0488	16.5548	15.2893	14.7264	14.2035	13.7164	72
73	22.1105	19.9095	18.1070	16.6037	15.3310	14.7651	14.2395	13.7500	73
74	22.1947	19.9777	18.1634	16.6512	15.3714	14.8026	14.2743	13.7825	74
75	22.2764	20.0438	18.2180	16.6971	15.4105	14.8389	14.3081	13.8139	75
76	22.3557	20.1081	18.2711	16.7416	15.4485	14.8740	14.3408	13.8444	76
77	22.4327	20.1703	18.3225	16.7848	15.4852	14.9081	14.3724	13.8739	77
78	22.5075	20.2308	18.3723	16.8266	15.5208	14.9411	14.4031	13.9025	78
79	22.5801	20.2894	18.4207	16.8672	15.5553	14.9730	14.4328	13.9301	79
80	22.6506	20.3463	18.4676	16.9065	15.5887	15.0040	14.4615	13.9569	80
81	22.7190	20.4015	18.5130	16.9446	15.6211	15.0340	14.4894	13.9829	81
82	22.7854	20.4551	18.5571	16.9815	15.6525	15.0630	14.5164	14.0080	82
83	22.8499	20.5070	18.5999	17.0173	15.6829	15.0912	14.5425	14.0324	83
84	22.9126	20.5574	18.6414	17.0520	15.7124	15.1185	14.5679	14.0560	84
85	22.9734	20.6064	18.6816	17.0856	15.7409	15.1449	14.5924	14.0788	85
86	23.0324	20.6539	18.7206	17.1183	15.7686	15.1706	14.6162	14.1010	86
87	23.0897	20.6999	18.7584	17.1499	15.7955	15.1954	14.6393	14.1224	87
88	23.1453	20.7446	18.7951	17.1806	15.8215	15.2195	14.6616	14.1432	88
89	23.1994	20.7880	18.8307	17.2103	15.8467	15.2428	14.6833	14.1634	89
90	23.2518	20.8301	18.8653	17.2392	15.8711	15.2654	14.7043	14.1829	90
91	23.3027	20.8710	18.8988	17.2671	15.8949	15.2874	14.7246	14.2018	91
92	23.3522	20.9106	18.9313	17.2943	15.9178	15.3086	14.7443	14.2202	92
93	23.4002	20.9491	18.9628	17.3206	15.9401	15.3293	14.7635	14.2380	93
94	23.4468	20.9864	18.9934	17.3461	15.9617	15.3492	14.7820	14.2552	94
95	23.4920	21.0227	19.0231	17.3709	15.9827	15.3686	14.8000	14.2719	95
96	23.5359	21.0578	19.0519	17.3949	16.0030	15.3874	14.8174	14.2881	96
97	23.5786	21.0920	19.0798	17.4181	16.0227	15.4056	14.8343	14.3038	97
98	23.6200	21.1251	19.1069	17.4407	16.0418	15.4233	14.8507	14.3190	98
99	23.6602	21.1572	19.1332	17.4626	16.0603	15.4404	14.8665	14.3338	99
100	23.6992	21.1884	19.1587	17.4839	16.0783	15.4570	14.8819	14.3481	100

YEARS' PURCHASE

Rate per Cent

Yrs.	7	7.25	7.5	8	8.5	9	9.5	10	Yrs.
1	0.7126	0.7113	0.7101	0.7075	0.7051	0.7026	0.7001	0.6977	1
2	1.3759	1.3712	1.3665	1.3572	1.3480	1.3390	1.3301	1.3213	2
3	1.9945	1.9846	1.9748	1.9555	1.9366	1.9180	1.8998	1.8819	3
4	2.5727	2.5562	2.5400	2.5081	2.4771	2.4468	2.4172	2.3883	4
5	3.1139	3.0899	3.0662	3.0199	2.9750	2.9314	2.8890	2.8479	5
6	3.6215	3.5890	3.5571	3.4949	3.4349	3.3769	3.3208	3.2666	6
7	4.0982	4.0567	4.0159	3.9369	3.8609	3.7878	3.7174	3.6495	7
8	4.5467	4.4956	4.4456	4.3489	4.2564	4.1677	4.0826	4.0009	8
9	4.9691	4.9081	4.8486	4.7338	4.6244	4.5199	4.4200	4.3244	9
10	5.3675	5.2964	5.2272	5.0941	4.9675	4.8471	4.7324	4.6231	10
11	5.7437	5.6624	5.5834	5.4317	5.2881	5.1519	5.0225	4.8995	11
12	6.0994	6.0078	5.9189	5.7488	5.5882	5.4363	5.2924	5.1560	12
13	6.4361	6.3342	6.2355	6.0469	5.8695	5.7021	5.5441	5.3945	13
14	6.7552	6.6430	6.5344	6.3277	6.1336	5.9511	5.7792	5.6169	14
15	7.0577	6.9354	6.8172	6.5924	6.3821	6.1847	5.9992	5.8245	15
16	7.3450	7.2125	7.0848	6.8424	6.6160	6.4042	6.2055	6.0187	16
17	7.6179	7.4755	7.3384	7.0786	6.8367	6.6107	6.3992	6.2008	17
18	7.8774	7.7253	7.5789	7.3022	7.0450	6.8053	6.5813	6.3717	18
19	8.1245	7.9627	7.8073	7.5140	7.2419	6.9888	6.7529	6.5323	19
20	8.3597	8.1886	8.0243	7.7148	7.4283	7.1622	6.9146	6.6836	20
21	8.5840	8.4037	8.2307	7.9054	7.6048	7.3262	7.0674	6.8261	21
22	8.7979	8.6086	8.4272	8.0865	7.7722	7.4815	7.2117	6.9607	22
23	9.0021	8.8040	8.6144	8.2586	7.9311	7.6286	7.3483	7.0879	23
24	9.1971	8.9904	8.7928	8.4225	8.0821	7.7682	7.4778	7.2083	24
25	9.3835	9.1684	8.9629	8.5785	8.2257	7.9007	7.6005	7.3222	25
26	9.5617	9.3384	9.1254	8.7272	8.3623	8.0267	7.7170	7.4303	26
27	9.7321	9.5010	9.2805	8.8690	8.4924	8.1465	7.8276	7.5328	27
28	9.8953	9.6565	9.4288	9.0043	8.6164	8.2605	7.9329	7.6302	28
29	10.0516	9.8052	9.5706	9.1336	8.7347	8.3692	8.0330	7.7228	29
30	10.2014	9.9477	9.7063	9.2571	8.8475	8.4727	8.1284	7.8109	30
31	10.3450	10.0842	9.8362	9.3751	8.9553	8.5715	8.2193	7.8948	31
32	10.4827	10.2150	9.9606	9.4881	9.0583	8.6658	8.3060	7.9748	32
33	10.6148	10.3404	10.0798	9.5962	9.1568	8.7559	8.3887	8.0510	33
34	10.7416	10.4607	10.1941	9.6997	9.2510	8.8420	8.4677	8.1237	34
35	10.8634	10.5761	10.3037	9.7989	9.3412	8.9244	8.5432	8.1932	35
36	10.9804	10.6870	10.4089	9.8940	9.4276	9.0032	8.6154	8.2596	36
37	11.0928	10.7934	10.5099	9.9851	9.5103	9.0786	8.6844	8.3230	37
38	11.2008	10.8957	10.6068	10.0726	9.5896	9.1509	8.7505	8.3837	38
39	11.3047	10.9940	10.6999	10.1566	9.6657	9.2201	8.8138	8.4418	39
40	11.4047	11.0885	10.7894	10.2372	9.7387	9.2865	8.8744	8.4974	40
41	11.5009	11.1794	10.8755	10.3146	9.8087	9.3502	8.9326	8.5507	41
42	11.5935	11.2669	10.9582	10.3890	9.8760	9.4113	8.9883	8.6017	42
43	11.6826	11.3511	11.0378	10.4605	9.9406	9.4699	9.0418	8.6507	43
44	11.7684	11.4321	11.1144	10.5293	10.0027	9.5262	9.0931	8.6977	44
45	11.8511	11.5101	11.1882	10.5954	10.0624	9.5804	9.1424	8.7428	45
46	11.9308	11.5852	11.2591	10.6591	10.1197	9.6324	9.1898	8.7861	46
47	12.0076	11.6576	11.3275	10.7203	10.1749	9.6824	9.2353	8.8276	47
48	12.0816	11.7274	11.3934	10.7793	10.2280	9.7304	9.2790	8.8676	48
49	12.1530	11.7947	11.4568	10.8361	10.2792	9.7767	9.3210	8.9060	49
50	12.2219	11.8595	11.5180	10.8908	10.3284	9.8212	9.3615	8.9429	50

YEARS' PURCHASE

				Rate per Cent					
Yrs.	7	7.25	7.5	8	8.5	9	9.5	10	Yrs.
51	12.2883	11.9220	11.5770	10.9435	10.3758	9.8640	9.4004	8.9784	51
52	12.3524	11.9823	11.6338	10.9943	10.4214	9.9053	9.4379	9.0126	52
53	12.4142	12.0405	11.6887	11.0433	10.4654	9.9450	9.4739	9.0454	53
54	12.4739	12.0967	11.7416	11.0905	10.5078	9.9833	9.5086	9.0771	54
55	12.5315	12.1509	11.7926	11.1360	10.5487	10.0202	9.5421	9.1076	55
56	12.5872	12.2032	11.8419	11.1799	10.5881	10.0557	9.5743	9.1369	56
57	12.6409	12.2537	11.8894	11.2223	10.6261	10.0900	9.6054	9.1652	57
58	12.6928	12.3024	11.9354	11.2632	10.6627	10.1230	9.6353	9.1925	58
59	12.7430	12.3496	11.9797	11.3027	10.6981	10.1549	9.6642	9.2188	59
60	12.7914	12.3951	12.0225	11.3408	10.7322	10.1857	9.6921	9.2441	60
61	12.8383	12.4390	12.0639	11.3776	10.7652	10.2153	9.7189	9.2685	61
62	12.8835	12.4815	12.1038	11.4131	10.7970	10.2440	9.7448	9.2921	62
63	12.9273	12.5226	12.1424	11.4474	10.8277	10.2716	9.7698	9.3148	63
64	12.9696	12.5623	12.1798	11.4806	10.8574	10.2983	9.7940	9.3368	64
65	13.0105	12.6006	12.2158	11.5126	10.8860	10.3241	9.8173	9.3579	65
66	13.0500	12.6377	12.2507	11.5436	10.9137	10.3490	9.8398	9.3784	66
67	13.0883	12.6736	12.2844	11.5735	10.9404	10.3730	9.8615	9.3981	67
68	13.1253	12.7083	12.3170	11.6024	10.9663	10.3962	9.8825	9.4172	68
69	13.1611	12.7418	12.3485	11.6304	10.9912	10.4187	9.9028	9.4356	69
70	13.1957	12.7743	12.3790	11.6574	11.0154	10.4404	9.9224	9.4534	70
71	13.2292	12.8057	12.4085	11.6836	11.0387	10.4613	9.9413	9.4706	71
72	13.2617	12.8361	12.4370	11.7089	11.0613	10.4816	9.9596	9.4872	72
73	13.2930	12.8655	12.4646	11.7333	11.0831	10.5012	9.9773	9.5032	73
74	13.3234	12.8939	12.4913	11.7570	11.1042	10.5201	9.9944	9.5187	74
75	13.3528	12.9215	12.5171	11.7799	11.1246	10.5384	10.0110	9.5337	75
76	13.3813	12.9481	12.5421	11.8020	11.1444	10.5562	10.0269	9.5482	76
77	13.4088	12.9739	12.5663	11.8234	11.1635	10.5733	10.0424	9.5623	77
78	13.4355	12.9989	12.5898	11.8442	11.1820	10.5899	10.0574	9.5758	78
79	13.4613	13.0231	12.6124	11.8643	11.1999	10.6059	10.0718	9.5889	79
80	13.4864	13.0465	12.6344	11.8837	11.2172	10.6215	10.0858	9.6016	80
81	13.5106	13.0692	12.6557	11.9025	11.2339	10.6365	10.0994	9.6139	81
82	13.5341	13.0911	12.6763	11.9207	11.2502	10.6510	10.1125	9.6258	82
83	13.5568	13.1124	12.6962	11.9383	11.2659	10.6651	10.1252	9.6373	83
84	13.5788	13.1330	12.7155	11.9554	11.2811	10.6787	10.1374	9.6484	84
85	13.6001	13.1529	12.7342	11.9719	11.2958	10.6919	10.1493	9.6592	85
86	13.6208	13.1723	12.7523	11.9879	11.3100	10.7047	10.1608	9.6696	86
87	13.6408	13.1910	12.7699	12.0035	11.3238	10.7170	10.1720	9.6797	87
88	13.6602	13.2091	12.7869	12.0185	11.3372	10.7290	10.1828	9.6894	88
89	13.6790	13.2267	12.8033	12.0330	11.3501	10.7406	10.1932	9.6989	89
90	13.6972	13.2437	12.8193	12.0471	11.3627	10.7518	10.2033	9.7080	90
91	13.7149	13.2602	12.8348	12.0608	11.3748	10.7627	10.2131	9.7169	91
92	13.7320	13.2762	12.8497	12.0740	11.3866	10.7732	10.2226	9.7255	92
93	13.7486	13.2917	12.8643	12.0868	11.3980	10.7834	10.2318	9.7338	93
94	13.7647	13.3067	12.8783	12.0992	11.4090	10.7933	10.2407	9.7419	94
95	13.7802	13.3213	12.8920	12.1113	11.4197	10.8029	10.2493	9.7497	95
96	13.7953	13.3354	12.9052	12.1229	11.4301	10.8122	10.2576	9.7572	96
97	13.8100	13.3491	12.9180	12.1342	11.4402	10.8212	10.2657	9.7645	97
98	13.8242	13.3624	12.9304	12.1452	11.4499	10.8299	10.2736	9.7716	98
99	13.8379	13.3752	12.9424	12.1558	11.4593	10.8383	10.2812	9.7785	99
100	13.8513	13.3877	12.9541	12.1661	11.4685	10.8465	10.2885	9.7852	100

YEARS' PURCHASE

Rate per Cent

Yrs.	11	12	13	14	15	16	18	20	Yrs.
1	0.6928	0.6881	0.6834	0.6787	0.6742	0.6696	0.6608	0.6522	1
2	1.3041	1.2873	1.2709	1.2550	1.2394	1.2243	1.1950	1.1671	2
3	1.8472	1.8137	1.7813	1.7502	1.7201	1.6910	1.6357	1.5838	3
4	2.3326	2.2794	2.2286	2.1801	2.1335	2.0890	2.0052	1.9279	4
5	2.7690	2.6944	2.6237	2.5566	2.4929	2.4323	2.3194	2.2166	5
6	3.1633	3.0663	2.9750	2.8891	2.8080	2.7313	2.5898	2.4623	6
7	3.5210	3.4013	3.2894	3.1846	3.0863	2.9939	2.8248	2.6737	7
8	3.8470	3.7045	3.5722	3.4490	3.3340	3.2264	3.0308	2.8576	8
9	4.1452	3.9802	3.8278	3.6867	3.5556	3.4335	3.2129	3.0189	9
10	4.4188	4.2318	4.0600	3.9016	3.7551	3.6192	3.3749	3.1615	10
11	4.6706	4.4622	4.2716	4.0966	3.9354	3.7864	3.5198	3.2884	11
12	4.9032	4.6740	4.4653	4.2744	4.0992	3.9378	3.6503	3.4019	12
13	5.1184	4.8692	4.6431	4.4371	4.2486	4.0754	3.7683	3.5042	13
14	5.3182	5.0496	4.8069	4.5864	4.3853	4.2011	3.8754	3.5967	14
15	5.5039	5.2168	4.9581	4.7239	4.5108	4.3161	3.9732	3.6807	15
16	5.6771	5.3721	5.0982	4.8509	4.6265	4.4219	4.0626	3.7573	16
17	5.8387	5.5166	5.2282	4.9684	4.7333	4.5194	4.1447	3.8275	17
18	5.9900	5.6515	5.3492	5.0776	4.8322	4.6095	4.2204	3.8919	18
19	6.1318	5.7775	5.4619	5.1791	4.9240	4.6930	4.2903	3.9512	19
20	6.2648	5.8955	5.5673	5.2737	5.0095	4.7705	4.3550	4.0061	20
21	6.3900	6.0062	5.6659	5.3621	5.0892	4.8427	4.4151	4.0569	21
22	6.5077	6.1101	5.7583	5.4447	5.1636	4.9101	4.4710	4.1040	22
23	6.6188	6.2079	5.8450	5.5223	5.2333	4.9730	4.5231	4.1479	23
24	6.7236	6.3000	5.9266	5.5950	5.2986	5.0320	4.5718	4.1888	24
25	6.8227	6.3869	6.0035	5.6635	5.3599	5.0872	4.6174	4.2271	25
26	6.9164	6.4690	6.0759	5.7279	5.4176	5.1392	4.6602	4.2629	26
27	7.0051	6.5465	6.1443	5.7886	5.4719	5.1880	4.7003	4.2964	27
28	7.0893	6.6200	6.2090	5.8460	5.5231	5.2340	4.7380	4.3279	28
29	7.1692	6.6896	6.2701	5.9002	5.5715	5.2774	4.7736	4.3576	29
30	7.2450	6.7556	6.3281	5.9515	5.6172	5.3184	4.8071	4.3855	30
31	7.3171	6.8182	6.3830	6.0001	5.6604	5.3572	4.8387	4.4118	31
32	7.3858	6.8778	6.4352	6.0461	5.7014	5.3939	4.8687	4.4366	32
33	7.4511	6.9344	6.4847	6.0898	5.7403	5.4286	4.8970	4.4601	33
34	7.5134	6.9883	6.5319	6.1314	5.7771	5.4616	4.9238	4.4824	34
35	7.5728	7.0397	6.5767	6.1708	5.8122	5.4929	4.9492	4.5034	35
36	7.6294	7.0886	6.6194	6.2084	5.8455	5.5227	4.9733	4.5234	36
37	7.6835	7.1353	6.6601	6.2442	5.8772	5.5510	4.9963	4.5424	37
38	7.7352	7.1798	6.6989	6.2783	5.9074	5.5779	5.0181	4.5604	38
39	7.7846	7.2224	6.7359	6.3108	5.9362	5.6035	5.0388	4.5775	39
40	7.8319	7.2631	6.7713	6.3418	5.9636	5.6280	5.0586	4.5938	40
41	7.8771	7.3019	6.8050	6.3715	5.9898	5.6513	5.0774	4.6094	41
42	7.9204	7.3391	6.8373	6.3998	6.0148	5.6736	5.0954	4.6242	42
43	7.9619	7.3748	6.8683	6.4268	6.0387	5.6948	5.1125	4.6383	43
44	8.0017	7.4089	6.8978	6.4527	6.0616	5.7152	5.1289	4.6517	44
45	8.0399	7.4416	6.9262	6.4775	6.0835	5.7346	5.1446	4.6646	45
46	8.0765	7.4729	6.9533	6.5012	6.1044	5.7532	5.1595	4.6769	46
47	8.1116	7.5030	6.9793	6.5240	6.1244	5.7710	5.1738	4.6887	47
48	8.1453	7.5318	7.0043	6.5458	6.1436	5.7880	5.1875	4.6999	48
49	8.1777	7.5595	7.0282	6.5667	6.1620	5.8044	5.2006	4.7107	49
50	8.2088	7.5861	7.0512	6.5867	6.1797	5.8200	5.2132	4.7210	50

YEARS' PURCHASE

				Rate per Cent					
Yrs.	11	12	13	14	15	16	18	20	Yrs.
51	8.2387	7.6116	7.0732	6.6060	6.1966	5.8350	5.2252	4.7309	51
52	8.2675	7.6361	7.0944	6.6244	6.2129	5.8494	5.2368	4.7403	52
53	8.2951	7.6597	7.1148	6.6422	6.2285	5.8633	5.2479	4.7494	53
54	8.3217	7.6824	7.1343	6.6592	6.2435	5.8766	5.2585	4.7581	54
55	8.3473	7.7042	7.1531	6.6756	6.2579	5.8893	5.2687	4.7665	55
56	8.3720	7.7252	7.1712	6.6914	6.2717	5.9016	5.2786	4.7745	56
57	8.3957	7.7454	7.1886	6.7065	6.2850	5.9134	5.2880	4.7822	57
58	8.4186	7.7649	7.2054	6.7211	6.2978	5.9247	5.2970	4.7896	58
59	8.4406	7.7836	7.2215	6.7352	6.3102	5.9356	5.3058	4.7967	59
60	8.4619	7.8017	7.2371	6.7487	6.3220	5.9461	5.3141	4.8036	60
61	8.4823	7.8191	7.2520	6.7617	6.3334	5.9562	5.3222	4.8102	61
62	8.5021	7.8359	7.2665	6.7742	6.3444	5.9659	5.3300	4.8165	62
63	8.5211	7.8520	7.2804	6.7863	6.3550	5.9753	5.3374	4.8226	63
64	8.5395	7.8676	7.2938	6.7979	6.3652	5.9843	5.3446	4.8285	64
65	8.5572	7.8826	7.3067	6.8092	6.3751	5.9930	5.3516	4.8342	65
66	8.5743	7.8971	7.3191	6.8200	6.3845	6.0014	5.3582	4.8396	66
67	8.5908	7.9111	7.3312	6.8304	6.3937	6.0095	5.3647	4.8449	67
68	8.6067	7.9246	7.3427	6.8405	6.4025	6.0173	5.3709	4.8499	68
69	8.6221	7.9377	7.3539	6.8502	6.4110	6.0248	5.3769	4.8548	69
70	8.6369	7.9503	7.3647	6.8595	6.4192	6.0320	5.3826	4.8595	70
71	8.6513	7.9624	7.3752	6.8686	6.4271	6.0390	5.3882	4.8640	71
72	8.6651	7.9741	7.3852	6.8773	6.4348	6.0458	5.3936	4.8684	72
73	8.6785	7.9855	7.3950	6.8858	6.4422	6.0523	5.3988	4.8726	73
74	8.6914	7.9964	7.4043	6.8939	6.4493	6.0586	5.4038	4.8767	74
75	8.7039	8.0070	7.4134	6.9018	6.4562	6.0646	5.4086	4.8807	75
76	8.7160	8.0172	7.4222	6.9094	6.4628	6.0705	5.4133	4.8845	76
77	8.7277	8.0271	7.4306	6.9167	6.4692	6.0762	5.4178	4.8881	77
78	8.7390	8.0367	7.4388	6.9238	6.4754	6.0816	5.4221	4.8917	78
79	8.7499	8.0459	7.4467	6.9306	6.4814	6.0869	5.4263	4.8951	79
80	8.7605	8.0548	7.4544	6.9373	6.4872	6.0920	5.4304	4.8984	80
81	8.7707	8.0635	7.4618	6.9437	6.4928	6.0970	5.4343	4.9016	81
82	8.7806	8.0718	7.4689	6.9499	6.4982	6.1017	5.4381	4.9047	82
83	8.7901	8.0799	7.4759	6.9559	6.5035	6.1064	5.4418	4.9076	83
84	8.7994	8.0877	7.4826	6.9616	6.5085	6.1108	5.4453	4.9105	84
85	8.8083	8.0953	7.4890	6.9672	6.5134	6.1151	5.4487	4.9133	85
86	8.8170	8.1026	7.4953	6.9727	6.5182	6.1193	5.4521	4.9160	86
87	8.8254	8.1097	7.5013	6.9779	6.5228	6.1233	5.4553	4.9186	87
88	8.8335	8.1165	7.5072	6.9830	6.5272	6.1273	5.4584	4.9211	88
89	8.8414	8.1232	7.5129	6.9879	6.5315	6.1310	5.4614	4.9236	89
90	8.8490	8.1296	7.5184	6.9926	6.5356	6.1347	5.4643	4.9259	90
91	8.8563	8.1358	7.5237	6.9972	6.5396	6.1382	5.4671	4.9282	91
92	8.8635	8.1418	7.5288	7.0017	6.5435	6.1417	5.4698	4.9304	92
93	8.8704	8.1477	7.5338	7.0060	6.5473	6.1450	5.4724	4.9326	93
94	8.8771	8.1533	7.5386	7.0102	6.5509	6.1482	5.4750	4.9346	94
95	8.8835	8.1588	7.5433	7.0142	6.5545	6.1513	5.4774	4.9366	95
96	8.8898	8.1640	7.5478	7.0181	6.5579	6.1543	5.4798	4.9386	96
97	8.8959	8.1692	7.5522	7.0219	6.5612	6.1572	5.4821	4.9404	97
98	8.9018	8.1741	7.5565	7.0256	6.5644	6.1600	5.4843	4.9422	98
99	8.9075	8.1789	7.5606	7.0291	6.5675	6.1628	5.4865	4.9440	99
100	8.9130	8.1836	7.5646	7.0326	6.5705	6.1654	5.4886	4.9457	100

YEARS' PURCHASE

Rate per Cent

Yrs.	4	4.5	5	5.5	6	6.25	6.5	6.75	Yrs.
1	0.7282	0.7255	0.7229	0.7203	0.7177	0.7164	0.7151	0.7139	1
2	1.4418	1.4314	1.4213	1.4112	1.4014	1.3965	1.3916	1.3868	2
3	2.1407	2.1181	2.0959	2.0741	2.0528	2.0424	2.0320	2.0217	3
4	2.8250	2.7856	2.7474	2.7101	2.6739	2.6561	2.6386	2.6213	4
5	3.4944	3.4344	3.3764	3.3204	3.2662	3.2397	3.2137	3.1881	5
6	4.1491	4.0648	3.9838	3.9060	3.8312	3.7948	3.7592	3.7242	6
7	4.7890	4.6770	4.5701	4.4680	4.3704	4.3231	4.2769	4.2317	7
8	5.4141	5.2714	5.1360	5.0074	4.8851	4.8262	4.7686	4.7125	8
9	6.0244	5.8483	5.6821	5.5251	5.3766	5.3053	5.2359	5.1682	9
10	6.6201	6.4080	6.2091	6.0221	5.8461	5.7619	5.6801	5.6005	10
11	7.2012	6.9509	6.7175	6.4992	6.2946	6.1971	6.1026	6.0109	11
12	7.7678	7.4774	7.2079	6.9572	6.7233	6.6122	6.5046	6.4006	12
13	8.3200	7.9878	7.6810	7.3969	7.1331	7.0081	6.8874	6.7709	13
14	8.8580	8.4823	8.1372	7.8191	7.5249	7.3860	7.2521	7.1229	14
15	9.3819	8.9615	8.5772	8.2245	7.8996	7.7466	7.5995	7.4578	15
16	9.8918	9.4257	9.0014	8.6138	8.2581	8.0911	7.9306	7.7765	16
17	10.3880	9.8751	9.4105	8.9876	8.6011	8.4200	8.2464	8.0798	17
18	10.8706	10.3102	9.8048	9.3466	8.9293	8.7343	8.5477	8.3688	18
19	11.3398	10.7314	10.1849	9.6913	9.2434	9.0347	8.8351	8.6442	19
20	11.7958	11.1389	10.5512	10.0225	9.5442	9.3218	9.1095	8.9066	20
21	12.2389	11.5331	10.9043	10.3405	9.8322	9.5963	9.3715	9.1569	21
22	12.6692	11.9144	11.2446	10.6460	10.1080	9.8588	9.6217	9.3957	22
23	13.0869	12.2832	11.5725	10.9395	10.3721	10.1100	9.8608	9.6235	23
24	13.4924	12.6397	11.8884	11.2214	10.6252	10.3503	10.0892	9.8410	24
25	13.8858	12.9843	12.1928	11.4921	10.8677	10.5802	10.3076	10.0486	25
26	14.2674	13.3174	12.4860	11.7523	11.1000	10.8003	10.5164	10.2470	26
27	14.6374	13.6392	12.7684	12.0022	11.3227	11.0110	10.7160	10.4364	27
28	14.9961	13.9501	13.0405	12.2423	11.5362	11.2128	10.9070	10.6175	28
29	15.3437	14.2504	13.3026	12.4730	11.7408	11.4060	11.0898	10.7906	29
30	15.6805	14.5405	13.5550	12.6946	11.9370	11.5910	11.2646	10.9561	30
31	16.0067	14.8206	13.7981	12.9076	12.1251	11.7683	11.4320	11.1143	31
32	16.3226	15.0910	14.0322	13.1122	12.3055	11.9382	11.5922	11.2657	32
33	16.6284	15.3520	14.2576	13.3088	12.4785	12.1010	11.7456	11.4106	33
34	16.9244	15.6039	14.4746	13.4978	12.6444	12.2570	11.8925	11.5492	34
35	17.2108	15.8471	14.6836	13.6793	12.8036	12.4065	12.0332	11.6818	35
36	17.4879	16.0817	14.8848	13.8538	12.9563	12.5498	12.1680	11.8088	36
37	17.7559	16.3080	15.0785	14.0214	13.1028	12.6872	12.2972	11.9304	37
38	18.0150	16.5264	15.2650	14.1825	13.2434	12.8190	12.4209	12.0469	38
39	18.2656	16.7370	15.4445	14.3374	13.3783	12.9454	12.5395	12.1584	39
40	18.5078	16.9402	15.6174	14.4862	13.5078	13.0665	12.6532	12.2652	40
41	18.7419	17.1361	15.7837	14.6292	13.6321	13.1828	12.7622	12.3676	41
42	18.9680	17.3249	15.9438	14.7666	13.7513	13.2943	12.8667	12.4657	42
43	19.1865	17.5070	16.0979	14.8987	13.8658	13.4013	12.9668	12.5597	43
44	19.3976	17.6826	16.2462	15.0257	13.9757	13.5039	13.0629	12.6498	44
45	19.6015	17.8518	16.3890	15.1477	14.0812	13.6024	13.1550	12.7362	45
46	19.7983	18.0150	16.5264	15.2650	14.1825	13.6969	13.2434	12.8190	46
47	19.9883	18.1722	16.6586	15.3777	14.2798	13.7876	13.3281	12.8984	47
48	20.1718	18.3237	16.7858	15.4861	14.3731	13.8746	13.4095	12.9745	48
49	20.3488	18.4697	16.9082	15.5902	14.4628	13.9581	13.4875	13.0475	49
50	20.5197	18.6103	17.0260	15.6903	14.5489	14.0383	13.5623	13.1176	50

YEARS' PURCHASE

				Rate per Cent					
Yrs.	4	4.5	5	5.5	6	6.25	6.5	6.75	Yrs.
51	20.6846	18.7459	17.1394	15.7865	14.6316	14.1153	13.6342	13.1848	51
52	20.8437	18.8764	17.2485	15.8790	14.7110	14.1892	13.7031	13.2492	52
53	20.9971	19.0022	17.3534	15.9679	14.7873	14.2601	13.7693	13.3111	53
54	21.1452	19.1233	17.4544	16.0534	14.8606	14.3283	13.8328	13.3704	54
55	21.2879	19.2400	17.5516	16.1355	14.9309	14.3937	13.8937	13.4273	55
56	21.4256	19.3524	17.6450	16.2145	14.9985	14.4565	13.9522	13.4820	56
57	21.5583	19.4606	17.7350	16.2904	15.0635	14.5168	14.0084	13.5344	57
58	21.6863	19.5649	17.8215	16.3634	15.1258	14.5747	14.0623	13.5847	58
59	21.8097	19.6652	17.9047	16.4335	15.1858	14.6303	14.1141	13.6331	59
60	21.9286	19.7619	17.9848	16.5010	15.2433	14.6838	14.1638	13.6794	60
61	22.0433	19.8549	18.0619	16.5658	15.2986	14.7351	14.2115	13.7240	61
62	22.1538	19.9445	18.1360	16.6281	15.3518	14.7844	14.2574	13.7667	62
63	22.2602	20.0308	18.2073	16.6880	15.4028	14.8317	14.3014	13.8077	63
64	22.3628	20.1138	18.2758	16.7456	15.4519	14.8772	14.3437	13.8471	64
65	22.4617	20.1938	18.3418	16.8010	15.4990	14.9209	14.3843	13.8850	65
66	22.5569	20.2707	18.4053	16.8542	15.5443	14.9628	14.4233	13.9213	66
67	22.6487	20.3448	18.4663	16.9054	15.5878	15.0031	14.4608	13.9562	67
68	22.7371	20.4161	18.5250	16.9546	15.6296	15.0419	14.4967	13.9897	68
69	22.8222	20.4847	18.5815	17.0019	15.6698	15.0791	14.5313	14.0219	69
70	22.9042	20.5507	18.6358	17.0474	15.7084	15.1149	14.5645	14.0528	70
71	22.9832	20.6143	18.6881	17.0911	15.7455	15.1492	14.5964	14.0825	71
72	23.0593	20.6755	18.7383	17.1331	15.7812	15.1822	14.6270	14.1110	72
73	23.1325	20.7343	18.7867	17.1735	15.8155	15.2139	14.6565	14.1384	73
74	23.2030	20.7910	18.8332	17.2124	15.8484	15.2444	14.6848	14.1647	74
75	23.2710	20.8455	18.8779	17.2497	15.8801	15.2737	14.7119	14.1900	75
76	23.3364	20.8980	18.9209	17.2856	15.9105	15.3018	14.7380	14.2143	76
77	23.3993	20.9484	18.9623	17.3201	15.9397	15.3289	14.7631	14.2377	77
78	23.4599	20.9970	19.0021	17.3533	15.9678	15.3549	14.7872	14.2601	78
79	23.5183	21.0437	19.0403	17.3852	15.9949	15.3799	14.8104	14.2816	79
80	23.5745	21.0887	19.0771	17.4159	16.0208	15.4039	14.8327	14.3023	80
81	23.6285	21.1320	19.1125	17.4454	16.0458	15.4269	14.8541	14.3222	81
82	23.6806	21.1736	19.1466	17.4738	16.0698	15.4491	14.8746	14.3413	82
83	23.7307	21.2136	19.1793	17.5010	16.0928	15.4704	14.8944	14.3597	83
84	23.7789	21.2522	19.2108	17.5272	16.1150	15.4909	14.9133	14.3773	84
85	23.8253	21.2892	19.2411	17.5524	16.1363	15.5106	14.9316	14.3943	85
86	23.8700	21.3249	19.2702	17.5767	16.1568	15.5295	14.9491	14.4106	86
87	23.9130	21.3592	19.2982	17.6000	16.1764	15.5477	14.9660	14.4262	87
88	23.9544	21.3922	19.3251	17.6224	16.1954	15.5652	14.9822	14.4413	88
89	23.9942	21.4239	19.3511	17.6439	16.2136	15.5820	14.9977	14.4557	89
90	24.0325	21.4545	19.3760	17.6646	16.2310	15.5981	15.0127	14.4696	90
91	24.0694	21.4838	19.3999	17.6845	16.2478	15.6136	15.0271	14.4830	91
92	24.1048	21.5121	19.4230	17.7037	16.2640	15.6285	15.0409	14.4958	92
93	24.1390	21.5393	19.4451	17.7221	16.2795	15.6429	15.0542	14.5081	93
94	24.1718	21.5654	19.4664	17.7398	16.2945	15.6567	15.0669	14.5200	94
95	24.2034	21.5906	19.4869	17.7568	16.3088	15.6699	15.0792	14.5314	95
96	24.2338	21.6148	19.5066	17.7731	16.3226	15.6827	15.0910	14.5424	96
97	24.2631	21.6380	19.5256	17.7889	16.3359	15.6949	15.1023	14.5529	97
98	24.2912	21.6604	19.5438	17.8040	16.3486	15.7067	15.1132	14.5630	98
99	24.3183	21.6819	19.5613	17.8185	16.3609	15.7180	15.1237	14.5727	99
100	24.3443	21.7026	19.5782	17.8325	16.3727	15.7289	15.1338	14.5821	100

YEARS' PURCHASE

Rate per Cent

Yrs.	7	7.25	7.5	8	8.5	9	9.5	10	Yrs.
1	0.7126	0.7113	0.7101	0.7075	0.7051	0.7026	0.7001	0.6977	1
2	1.3820	1.3772	1.3725	1.3632	1.3539	1.3448	1.3358	1.3270	2
3	2.0115	2.0015	1.9915	1.9719	1.9526	1.9337	1.9152	1.8971	3
4	2.6043	2.5874	2.5708	2.5382	2.5063	2.4753	2.4451	2.4155	4
5	3.1629	3.1380	3.1136	3.0659	3.0196	2.9747	2.9311	2.8888	5
6	3.6898	3.6561	3.6230	3.5585	3.4963	3.4362	3.3782	3.3221	6
7	4.1874	4.1440	4.1015	4.0191	3.9399	3.8638	3.7906	3.7201	7
8	4.6576	4.6040	4.5516	4.4503	4.3534	4.2607	4.1718	4.0866	8
9	5.1023	5.0380	4.9754	4.8546	4.7395	4.6298	4.5251	4.4250	9
10	5.5232	5.4480	5.3748	5.2341	5.1006	4.9738	4.8531	4.7381	10
11	5.9219	5.8355	5.7516	5.5908	5.4388	5.2948	5.1582	5.0285	11
12	6.2998	6.2021	6.1074	5.9264	5.7559	5.5948	5.4426	5.2984	12
13	6.6582	6.5491	6.4436	6.2425	6.0536	5.8757	5.7080	5.5496	13
14	6.9983	6.8780	6.7617	6.5406	6.3334	6.1390	5.9562	5.7840	14
15	7.3213	7.1897	7.0627	6.8218	6.5968	6.3862	6.1886	6.0028	15
16	7.6282	7.4854	7.3479	7.0875	6.8449	6.6184	6.4064	6.2076	16
17	7.9199	7.7661	7.6182	7.3387	7.0789	6.8369	6.6109	6.3994	17
18	8.1973	8.0327	7.8746	7.5763	7.2997	7.0427	6.8031	6.5793	18
19	8.4613	8.2860	8.1179	7.8012	7.5084	7.2367	6.9840	6.7483	19
20	8.7126	8.5269	8.3489	8.0144	7.7056	7.4197	7.1543	6.9072	20
21	8.9520	8.7560	8.5685	8.2165	7.8922	7.5926	7.3149	7.0568	21
22	9.1801	8.9741	8.7772	8.4082	8.0690	7.7560	7.4665	7.1978	22
23	9.3974	9.1817	8.9757	8.5902	8.2364	7.9106	7.6097	7.3307	23
24	9.6047	9.3795	9.1646	8.7630	8.3952	8.0570	7.7450	7.4562	24
25	9.8024	9.5679	9.3444	8.9273	8.5458	8.1956	7.8730	7.5748	25
26	9.9910	9.7476	9.5157	9.0835	8.6889	8.3271	7.9943	7.6870	26
27	10.1711	9.9189	9.6789	9.2321	8.8247	8.4518	8.1091	7.7931	27
28	10.3430	10.0823	9.8344	9.3735	8.9538	8.5702	8.2180	7.8937	28
29	10.5072	10.2382	9.9827	9.5081	9.0766	8.6826	8.3213	7.9889	29
30	10.6640	10.3871	10.1242	9.6364	9.1934	8.7894	8.4194	8.0793	30
31	10.8139	10.5292	10.2592	9.7586	9.3046	8.8910	8.5125	8.1650	31
32	10.9571	10.6650	10.3880	9.8751	9.4105	8.9876	8.6011	8.2464	32
33	11.0941	10.7947	10.5110	9.9862	9.5113	9.0795	8.6852	8.3238	33
34	11.2251	10.9187	10.6285	10.0922	9.6074	9.1670	8.7653	8.3973	34
35	11.3503	11.0372	10.7408	10.1934	9.6990	9.2504	8.8415	8.4672	35
36	11.4702	11.1504	10.8480	10.2899	9.7864	9.3299	8.9140	8.5337	36
37	11.5849	11.2588	10.9506	10.3821	9.8698	9.4056	8.9832	8.5970	37
38	11.6946	11.3624	11.0486	10.4702	9.9493	9.4778	9.0490	8.6573	38
39	11.7997	11.4616	11.1423	10.5543	10.0253	9.5467	9.1118	8.7148	39
40	11.9003	11.5565	11.2320	10.6348	10.0978	9.6125	9.1717	8.7695	40
41	11.9967	11.6473	11.3178	10.7116	10.1671	9.6753	9.2288	8.8217	41
42	12.0889	11.7343	11.3999	10.7851	10.2333	9.7352	9.2833	8.8715	42
43	12.1773	11.8176	11.4784	10.8554	10.2966	9.7924	9.3353	8.9190	43
44	12.2620	11.8973	11.5537	10.9227	10.3570	9.8471	9.3850	8.9644	44
45	12.3431	11.9737	11.6257	10.9870	10.4149	9.8994	9.4325	9.0077	45
46	12.4209	12.0468	11.6946	11.0486	10.4702	9.9493	9.4778	9.0490	46
47	12.4954	12.1169	11.7607	11.1075	10.5231	9.9971	9.5212	9.0885	47
48	12.5669	12.1841	11.8239	11.1639	10.5737	10.0428	9.5626	9.1262	48
49	12.6354	12.2485	11.8846	11.2180	10.6222	10.0865	9.6022	9.1623	49
50	12.7011	12.3102	11.9426	11.2697	10.6685	10.1283	9.6401	9.1968	50

YEARS' PURCHASE

Rate per Cent

Yrs.	7	7.25	7.5	8	8.5	9	9.5	10	Yrs.
51	12.7640	12.3693	11.9983	11.3192	10.7129	10.1683	9.6763	9.2298	51
52	12.8244	12.4260	12.0517	11.3667	10.7554	10.2066	9.7110	9.2613	52
53	12.8824	12.4804	12.1028	11.4122	10.7962	10.2432	9.7442	9.2915	53
54	12.9379	12.5326	12.1518	11.4558	10.8352	10.2783	9.7759	9.3203	54
55	12.9912	12.5826	12.1988	11.4976	10.8725	10.3119	9.8063	9.3480	55
56	13.0424	12.6305	12.2439	11.5376	10.9083	10.3441	9.8354	9.3744	56
57	13.0914	12.6766	12.2872	11.5760	10.9426	10.3750	9.8633	9.3998	57
58	13.1385	12.7207	12.3286	11.6128	10.9755	10.4045	9.8900	9.4240	58
59	13.1837	12.7631	12.3684	11.6481	11.0070	10.4328	9.9156	9.4472	59
60	13.2271	12.8037	12.4066	11.6819	11.0372	10.4600	9.9401	9.4695	60
61	13.2687	12.8427	12.4432	11.7144	11.0662	10.4860	9.9636	9.4908	61
62	13.3087	12.8801	12.4783	11.7455	11.0940	10.5109	9.9861	9.5112	62
63	13.3470	12.9160	12.5120	11.7754	11.1206	10.5348	10.0077	9.5308	63
64	13.3838	12.9505	12.5444	11.8040	11.1462	10.5578	10.0284	9.5495	64
65	13.4192	12.9836	12.5754	11.8315	11.1707	10.5797	10.0482	9.5675	65
66	13.4531	13.0154	12.6052	11.8579	11.1942	10.6008	10.0672	9.5848	66
67	13.4857	13.0459	12.6338	11.8832	11.2167	10.6210	10.0855	9.6013	67
68	13.5170	13.0751	12.6613	11.9074	11.2383	10.6404	10.1029	9.6171	68
69	13.5470	13.1032	12.6876	11.9308	11.2591	10.6591	10.1197	9.6323	69
70	13.5759	13.1302	12.7129	11.9531	11.2790	10.6769	10.1358	9.6469	70
71	13.6036	13.1562	12.7372	11.9746	11.2982	10.6940	10.1512	9.6609	71
72	13.6302	13.1810	12.7606	11.9952	11.3165	10.7105	10.1661	9.6743	72
73	13.6558	13.2049	12.7830	12.0150	11.3341	10.7263	10.1803	9.6872	73
74	13.6803	13.2279	12.8045	12.0340	11.3510	10.7414	10.1939	9.6995	74
75	13.7039	13.2499	12.8251	12.0523	11.3673	10.7559	10.2070	9.7114	75
76	13.7265	13.2711	12.8450	12.0698	11.3828	10.7699	10.2196	9.7227	76
77	13.7483	13.2915	12.8640	12.0866	11.3978	10.7833	10.2316	9.7337	77
78	13.7692	13.3110	12.8823	12.1027	11.4122	10.7961	10.2432	9.7441	78
79	13.7893	13.3298	12.8999	12.1183	11.4259	10.8085	10.2543	9.7542	79
80	13.8086	13.3478	12.9168	12.1332	11.4392	10.8203	10.2650	9.7638	80
81	13.8271	13.3651	12.9330	12.1475	11.4519	10.8317	10.2752	9.7731	81
82	13.8449	13.3817	12.9486	12.1612	11.4641	10.8426	10.2850	9.7820	82
83	13.8620	13.3977	12.9635	12.1744	11.4759	10.8531	10.2945	9.7905	83
84	13.8785	13.4131	12.9779	12.1871	11.4871	10.8632	10.3035	9.7987	84
85	13.8943	13.4278	12.9917	12.1993	11.4979	10.8729	10.3122	9.8066	85
86	13.9094	13.4420	13.0050	12.2110	11.5083	10.8822	10.3206	9.8142	86
87	13.9240	13.4556	13.0177	12.2222	11.5183	10.8911	10.3286	9.8214	87
88	13.9381	13.4687	13.0300	12.2330	11.5279	10.8997	10.3363	9.8284	88
89	13.9515	13.4813	13.0418	12.2434	11.5371	10.9079	10.3437	9.8351	89
90	13.9645	13.4934	13.0531	12.2533	11.5460	10.9158	10.3509	9.8415	90
91	13.9769	13.5050	13.0639	12.2629	11.5545	10.9234	10.3577	9.8477	91
92	13.9889	13.5162	13.0744	12.2721	11.5626	10.9307	10.3643	9.8536	92
93	14.0003	13.5269	13.0844	12.2810	11.5705	10.9377	10.3706	9.8593	93
94	14.0114	13.5372	13.0941	12.2895	11.5780	10.9444	10.3766	9.8648	94
95	14.0220	13.5471	13.1033	12.2976	11.5853	10.9509	10.3824	9.8701	95
96	14.0322	13.5566	13.1122	12.3055	11.5922	10.9571	10.3880	9.8751	96
97	14.0420	13.5658	13.1208	12.3130	11.5989	10.9631	10.3934	9.8800	97
98	14.0514	13.5746	13.1290	12.3203	11.6053	10.9689	10.3986	9.8846	98
99	14.0605	13.5830	13.1369	12.3272	11.6115	10.9744	10.4035	9.8891	99
100	14.0692	13.5911	13.1445	12.3339	11.6175	10.9797	10.4083	9.8934	100

YEARS' PURCHASE

Rate per Cent

Yrs.	11	12	13	14	15	16	18	20	Yrs.
1	0.6928	0.6881	0.6834	0.6787	0.6742	0.6696	0.6608	0.6522	1
2	1.3096	1.2927	1.2762	1.2601	1.2444	1.2291	1.1996	1.1715	2
3	1.8617	1.8277	1.7949	1.7633	1.7327	1.7032	1.6471	1.5946	3
4	2.3586	2.3042	2.2523	2.2027	2.1552	2.1098	2.0243	1.9456	4
5	2.8077	2.7310	2.6584	2.5895	2.5242	2.4620	2.3465	2.2413	5
6	3.2153	3.1151	3.0210	2.9324	2.8489	2.7700	2.6246	2.4937	6
7	3.5866	3.4625	3.3466	3.2382	3.1366	3.0412	2.8669	2.7114	7
8	3.9261	3.7778	3.6403	3.5124	3.3932	3.2819	3.0797	2.9010	8
9	4.2375	4.0652	3.9064	3.7595	3.6233	3.4966	3.2681	3.0676	9
10	4.5238	4.3280	4.1484	3.9832	3.8306	3.6893	3.4358	3.2149	10
11	4.7878	4.5690	4.3694	4.1865	4.0182	3.8630	3.5860	3.3460	11
12	5.0318	4.7907	4.5717	4.3718	4.1887	4.0203	3.7211	3.4634	12
13	5.2579	4.9952	4.7576	4.5415	4.3442	4.1633	3.8433	3.5690	13
14	5.4677	5.1842	4.9287	4.6972	4.4865	4.2938	3.9543	3.6645	14
15	5.6629	5.3594	5.0868	4.8405	4.6171	4.4133	4.0553	3.7511	15
16	5.8448	5.5220	5.2330	4.9728	4.7372	4.5230	4.1478	3.8300	16
17	6.0145	5.6733	5.3687	5.0952	4.8481	4.6240	4.2325	3.9022	17
18	6.1732	5.8143	5.4948	5.2086	4.9507	4.7172	4.3105	3.9684	18
19	6.3217	5.9458	5.6122	5.3139	5.0458	4.8034	4.3824	4.0293	19
20	6.4610	6.0689	5.7216	5.4120	5.1341	4.8834	4.4489	4.0854	20
21	6.5917	6.1840	5.8239	5.5034	5.2163	4.9577	4.5105	4.1372	21
22	6.7145	6.2920	5.9196	5.5887	5.2929	5.0268	4.5676	4.1853	22
23	6.8300	6.3934	6.0092	5.6685	5.3645	5.0913	4.6208	4.2299	23
24	6.9389	6.4886	6.0933	5.7433	5.4314	5.1516	4.6704	4.2714	24
25	7.0415	6.5783	6.1722	5.8134	5.4940	5.2079	4.7166	4.3100	25
26	7.1383	6.6627	6.2465	5.8792	5.5528	5.2607	4.7599	4.3461	26
27	7.2297	6.7423	6.3164	5.9411	5.6080	5.3102	4.8004	4.3799	27
28	7.3161	6.8174	6.3823	5.9994	5.6598	5.3566	4.8383	4.4114	28
29	7.3979	6.8883	6.4444	6.0542	5.7086	5.4003	4.8739	4.4410	29
30	7.4753	6.9554	6.5031	6.1060	5.7546	5.4415	4.9074	4.4688	30
31	7.5487	7.0188	6.5585	6.1548	5.7980	5.4802	4.9389	4.4949	31
32	7.6182	7.0789	6.6109	6.2010	5.8389	5.5168	4.9686	4.5195	32
33	7.6841	7.1358	6.6605	6.2446	5.8776	5.5513	4.9966	4.5426	33
34	7.7468	7.1898	6.7075	6.2859	5.9141	5.5839	5.0229	4.5644	34
35	7.8062	7.2410	6.7521	6.3250	5.9487	5.6147	5.0479	4.5850	35
36	7.8627	7.2896	6.7943	6.3620	5.9815	5.6439	5.0714	4.6044	36
37	7.9164	7.3357	6.8344	6.3972	6.0125	5.6715	5.0937	4.6228	37
38	7.9675	7.3796	6.8724	6.4305	6.0420	5.6977	5.1148	4.6402	38
39	8.0162	7.4213	6.9086	6.4621	6.0699	5.7225	5.1348	4.6566	39
40	8.0625	7.4609	6.9429	6.4922	6.0964	5.7461	5.1538	4.6722	40
41	8.1066	7.4987	6.9756	6.5208	6.1216	5.7685	5.1718	4.6870	41
42	8.1486	7.5346	7.0067	6.5479	6.1455	5.7897	5.1889	4.7010	42
43	8.1887	7.5689	7.0363	6.5738	6.1683	5.8099	5.2051	4.7143	43
44	8.2269	7.6015	7.0645	6.5984	6.1899	5.8291	5.2205	4.7270	44
45	8.2633	7.6326	7.0914	6.6218	6.2105	5.8474	5.2351	4.7390	45
46	8.2981	7.6623	7.1170	6.6441	6.2302	5.8648	5.2491	4.7504	46
47	8.3313	7.6906	7.1414	6.6654	6.2489	5.8813	5.2623	4.7612	47
48	8.3630	7.7176	7.1646	6.6856	6.2667	5.8971	5.2750	4.7716	48
49	8.3933	7.7434	7.1869	6.7050	6.2837	5.9122	5.2870	4.7814	49
50	8.4222	7.7680	7.2081	6.7234	6.2999	5.9265	5.2985	4.7908	50

YEARS' PURCHASE

Rate per Cent

Yrs.	11	12	13	14	15	16	18	20	Yrs.
51	8.4499	7.7915	7.2283	6.7410	6.3153	5.9402	5.3094	4.7997	51
52	8.4763	7.8140	7.2476	6.7578	6.3301	5.9532	5.3198	4.8082	52
53	8.5016	7.8354	7.2661	6.7739	6.3442	5.9657	5.3298	4.8164	53
54	8.5257	7.8559	7.2837	6.7892	6.3576	5.9776	5.3393	4.8241	54
55	8.5488	7.8756	7.3006	6.8039	6.3704	5.9889	5.3483	4.8315	55
56	8.5710	7.8943	7.3167	6.8179	6.3827	5.9998	5.3570	4.8386	56
57	8.5921	7.9123	7.3321	6.8313	6.3944	6.0101	5.3652	4.8453	57
58	8.6124	7.9295	7.3469	6.8441	6.4057	6.0200	5.3731	4.8517	58
59	8.6318	7.9459	7.3610	6.8563	6.4164	6.0295	5.3807	4.8579	59
60	8.6503	7.9616	7.3745	6.8680	6.4266	6.0386	5.3879	4.8638	60
61	8.6681	7.9767	7.3874	6.8792	6.4364	6.0472	5.3948	4.8694	61
62	8.6851	7.9911	7.3998	6.8899	6.4458	6.0555	5.4013	4.8747	62
63	8.7015	8.0049	7.4116	6.9002	6.4548	6.0634	5.4077	4.8799	63
64	8.7171	8.0181	7.4230	6.9100	6.4634	6.0710	5.4137	4.8848	64
65	8.7321	8.0308	7.4338	6.9194	6.4716	6.0783	5.4195	4.8895	65
66	8.7464	8.0430	7.4442	6.9285	6.4795	6.0852	5.4250	4.8940	66
67	8.7602	8.0546	7.4542	6.9371	6.4871	6.0919	5.4303	4.8983	67
68	8.7734	8.0657	7.4637	6.9454	6.4943	6.0983	5.4353	4.9024	68
69	8.7860	8.0764	7.4729	6.9533	6.5012	6.1044	5.4402	4.9064	69
70	8.7982	8.0867	7.4817	6.9609	6.5079	6.1102	5.4448	4.9101	70
71	8.8098	8.0965	7.4901	6.9682	6.5142	6.1158	5.4493	4.9138	71
72	8.8209	8.1059	7.4981	6.9751	6.5203	6.1212	5.4536	4.9172	72
73	8.8316	8.1150	7.5059	6.9818	6.5262	6.1264	5.4576	4.9206	73
74	8.8419	8.1236	7.5133	6.9882	6.5318	6.1313	5.4616	4.9237	74
75	8.8518	8.1319	7.5204	6.9944	6.5371	6.1360	5.4653	4.9268	75
76	8.8612	8.1399	7.5272	7.0003	6.5423	6.1406	5.4689	4.9297	76
77	8.8703	8.1476	7.5337	7.0059	6.5472	6.1449	5.4724	4.9325	77
78	8.8790	8.1549	7.5400	7.0114	6.5520	6.1491	5.4757	4.9352	78
79	8.8873	8.1619	7.5460	7.0166	6.5565	6.1531	5.4788	4.9378	79
80	8.8953	8.1687	7.5518	7.0215	6.5609	6.1569	5.4819	4.9403	80
81	8.9030	8.1752	7.5573	7.0263	6.5651	6.1606	5.4848	4.9426	81
82	8.9104	8.1814	7.5627	7.0309	6.5691	6.1641	5.4876	4.9449	82
83	8.9175	8.1874	7.5678	7.0353	6.5729	6.1675	5.4903	4.9471	83
84	8.9243	8.1931	7.5727	7.0396	6.5766	6.1708	5.4929	4.9492	84
85	8.9308	8.1986	7.5774	7.0436	6.5802	6.1739	5.4953	4.9512	85
86	8.9371	8.2039	7.5819	7.0475	6.5836	6.1769	5.4977	4.9531	86
87	8.9431	8.2089	7.5862	7.0513	6.5868	6.1798	5.5000	4.9550	87
88	8.9489	8.2138	7.5904	7.0549	6.5900	6.1825	5.5022	4.9567	88
89	8.9544	8.2185	7.5944	7.0583	6.5930	6.1852	5.5043	4.9584	89
90	8.9597	8.2230	7.5982	7.0616	6.5959	6.1877	5.5063	4.9601	90
91	8.9649	8.2273	7.6019	7.0648	6.5986	6.1902	5.5082	4.9616	91
92	8.9698	8.2314	7.6054	7.0679	6.6013	6.1925	5.5101	4.9631	92
93	8.9745	8.2354	7.6088	7.0708	6.6039	6.1948	5.5119	4.9646	93
94	8.9790	8.2392	7.6121	7.0736	6.6063	6.1969	5.5136	4.9660	94
95	8.9834	8.2429	7.6152	7.0763	6.6087	6.1990	5.5152	4.9673	95
96	8.9876	8.2464	7.6182	7.0789	6.6109	6.2010	5.5168	4.9686	96
97	8.9916	8.2498	7.6211	7.0814	6.6131	6.2029	5.5183	4.9698	97
98	8.9955	8.2531	7.6239	7.0838	6.6152	6.2047	5.5198	4.9710	98
99	8.9992	8.2562	7.6265	7.0861	6.6172	6.2065	5.5212	4.9721	99
100	9.0027	8.2592	7.6291	7.0883	6.6191	6.2082	5.5225	4.9732	100

YEARS' PURCHASE

(DUAL RATE % PRINCIPLE)

OR

PRESENT VALUE OF
ONE POUND PER ANNUM

*receivable at the end of each year after
allowing for a sinking fund at a given
rate to replace the invested capital and
for the effect of income tax at* **30%** *on
that part of the income used to provide
the annual sinking fund instalment.*

AT RATES OF INTEREST FROM

4% to 20%

AND

ALLOWING FOR THE POSSIBLE INVESTMENT

OF SINKING FUNDS AT

2·5%, 3% and 4%

INCOME TAX at 30%

Note:—Tables of Years' Purchase in which no allowance has been made for the effect
of income tax on that part of the income used to provide the annual sinking
fund instalment will be found on pages 53 to 71.

YEARS' PURCHASE

				Rate per Cent					
Yrs.	4	4.5	5	5.5	6	6.25	6.5	6.75	Yrs.
1	0.6809	0.6786	0.6763	0.6740	0.6718	0.6707	0.6695	0.6684	1
2	1.3414	1.3325	1.3237	1.3150	1.3064	1.3021	1.2979	1.2937	2
3	1.9822	1.9628	1.9437	1.9250	1.9066	1.8976	1.8886	1.8798	3
4	2.6040	2.5705	2.5379	2.5061	2.4751	2.4599	2.4448	2.4300	4
5	3.2074	3.1568	3.1077	3.0602	3.0140	2.9915	2.9693	2.9474	5
6	3.7930	3.7224	3.6544	3.5888	3.5256	3.4948	3.4645	3.4347	6
7	4.3615	4.2684	4.1792	4.0937	4.0116	3.9717	3.9327	3.8944	7
8	4.9134	4.7956	4.6833	4.5761	4.4738	4.4243	4.3759	4.3285	8
9	5.4493	5.3048	5.1677	5.0375	4.9138	4.8541	4.7959	4.7391	9
10	5.9697	5.7967	5.6334	5.4791	5.3330	5.2628	5.1945	5.1279	10
11	6.4751	6.2721	6.0814	5.9019	5.7327	5.6517	5.5730	5.4964	11
12	6.9661	6.7316	6.5124	6.3070	6.1142	6.0222	5.9329	5.8461	12
13	7.4430	7.1759	6.9274	6.6955	6.4786	6.3753	6.2753	6.1784	13
14	7.9063	7.6057	7.3270	7.0681	6.8268	6.7123	6.6015	6.4943	14
15	8.3566	8.0214	7.7121	7.4258	7.1599	7.0340	6.9125	6.7950	15
16	8.7941	8.4237	8.0832	7.7692	7.4787	7.3415	7.2091	7.0815	16
17	9.2193	8.8130	8.4411	8.0992	7.7840	7.6354	7.4924	7.3547	17
18	9.6326	9.1900	8.7862	8.4165	8.0766	7.9168	7.7631	7.6153	18
19	10.0343	9.5549	9.1193	8.7216	8.3571	8.1861	8.0219	7.8642	19
20	10.4249	9.9084	9.4407	9.0151	8.6263	8.4442	8.2696	8.1021	20
21	10.8046	10.2508	9.7510	9.2977	8.8847	8.6916	8.5068	8.3296	21
22	11.1738	10.5826	10.0507	9.5698	9.1328	8.9290	8.7340	8.5474	22
23	11.5328	10.9040	10.3403	9.8320	9.3713	9.1567	8.9518	8.7559	23
24	11.8820	11.2156	10.6201	10.0846	9.6005	9.3755	9.1608	8.9557	24
25	12.2215	11.5177	10.8905	10.3282	9.8210	9.5856	9.3613	9.1472	25
26	12.5518	11.8106	11.1520	10.5630	10.0331	9.7876	9.5539	9.3310	26
27	12.8731	12.0946	11.4049	10.7897	10.2374	9.9819	9.7389	9.5074	27
28	13.1857	12.3701	11.6496	11.0084	10.4341	10.1688	9.9167	9.6768	28
29	13.4897	12.6374	11.8863	11.2195	10.6236	10.3487	10.0877	9.8396	29
30	13.7856	12.8967	12.1154	11.4234	10.8062	10.5219	10.2523	9.9961	30
31	14.0735	13.1483	12.3372	11.6204	10.9823	10.6888	10.4106	10.1465	31
32	14.3536	13.3924	12.5519	11.8107	11.1521	10.8496	10.5631	10.2913	32
33	14.6262	13.6294	12.7599	11.9946	11.3160	11.0047	10.7100	10.4307	33
34	14.8915	13.8595	12.9613	12.1725	11.4741	11.1542	10.8516	10.5650	34
35	15.1497	14.0829	13.1565	12.3445	11.6268	11.2984	10.9880	10.6943	35
36	15.4010	14.2998	13.3456	12.5108	11.7743	11.4376	11.1197	10.8189	36
37	15.6456	14.5105	13.5289	12.6718	11.9167	11.5720	11.2466	10.9391	37
38	15.8838	14.7151	13.7066	12.8275	12.0544	11.7017	11.3692	11.0549	38
39	16.1156	14.9139	13.8789	12.9783	12.1874	11.8271	11.4874	11.1667	39
40	16.3413	15.1070	14.0460	13.1243	12.3161	11.9482	11.6017	11.2746	40
41	16.5610	15.2946	14.2080	13.2657	12.4405	12.0653	11.7120	11.3788	41
42	16.7750	15.4769	14.3652	13.4026	12.5608	12.1784	11.8186	11.4794	42
43	16.9833	15.6540	14.5177	13.5352	12.6773	12.2878	11.9216	11.5766	43
44	17.1862	15.8262	14.6657	13.6638	12.7900	12.3937	12.0212	11.6705	44
45	17.3837	15.9936	14.8093	13.7883	12.8991	12.4961	12.1175	11.7612	45
46	17.5761	16.1563	14.9487	13.9091	13.0047	12.5952	12.2107	11.8490	46
47	17.7635	16.3145	15.0840	14.0262	13.1070	12.6911	12.3008	11.9338	47
48	17.9460	16.4683	15.2154	14.1397	13.2061	12.7840	12.3881	12.0159	48
49	18.1237	16.6178	15.3430	14.2498	13.3020	12.8739	12.4725	12.0953	49
50	18.2968	16.7632	15.4669	14.3566	13.3951	12.9610	12.5542	12.1722	50

YEARS' PURCHASE

				Rate per Cent					
Yrs.	4	4.5	5	5.5	6	6.25	6.5	6.75	Yrs.
51	18.4654	16.9047	15.5872	14.4602	13.4852	13.0454	12.6334	12.2466	51
52	18.6297	17.0422	15.7041	14.5607	13.5726	13.1272	12.7101	12.3186	52
53	18.7897	17.1760	15.8176	14.6583	13.6573	13.2064	12.7843	12.3884	53
54	18.9456	17.3062	15.9279	14.7530	13.7395	13.2832	12.8563	12.4560	54
55	19.0974	17.4328	16.0351	14.8449	13.8192	13.3577	12.9261	12.5214	55
56	19.2453	17.5560	16.1393	14.9342	13.8965	13.4299	12.9937	12.5849	56
57	19.3895	17.6759	16.2405	15.0208	13.9715	13.5000	13.0592	12.6463	57
58	19.5299	17.7925	16.3389	15.1049	14.0443	13.5679	13.1228	12.7059	58
59	19.6668	17.9060	16.4346	15.1867	14.1149	13.6338	13.1844	12.7637	59
60	19.8001	18.0164	16.5276	15.2660	14.1834	13.6977	13.2442	12.8197	60
61	19.9300	18.1239	16.6180	15.3432	14.2500	13.7598	13.3022	12.8740	61
62	20.0566	18.2286	16.7059	15.4181	14.3146	13.8200	13.3585	12.9268	62
63	20.1799	18.3304	16.7914	15.4909	14.3773	13.8785	13.4131	12.9779	63
64	20.3002	18.4295	16.8746	15.5616	14.4382	13.9352	13.4661	13.0275	64
65	20.4173	18.5261	16.9555	15.6304	14.4974	13.9903	13.5175	13.0756	65
66	20.5315	18.6200	17.0341	15.6972	14.5548	14.0438	13.5675	13.1224	66
67	20.6427	18.7115	17.1106	15.7621	14.6107	14.0958	13.6160	13.1677	67
68	20.7512	18.8005	17.1851	15.8253	14.6649	14.1463	13.6631	13.2118	68
69	20.8569	18.8872	17.2575	15.8867	14.7176	14.1953	13.7088	13.2545	69
70	20.9599	18.9717	17.3280	15.9464	14.7688	14.2430	13.7532	13.2961	70
71	21.0603	19.0539	17.3965	16.0044	14.8186	14.2892	13.7964	13.3364	71
72	21.1582	19.1340	17.4633	16.0609	14.8670	14.3342	13.8383	13.3756	72
73	21.2535	19.2119	17.5282	16.1158	14.9140	14.3779	13.8791	13.4136	73
74	21.3465	19.2879	17.5914	16.1692	14.9598	14.4204	13.9187	13.4506	74
75	21.4372	19.3618	17.6529	16.2211	15.0042	14.4617	13.9571	13.4865	75
76	21.5255	19.4339	17.7127	16.2717	15.0474	14.5019	13.9945	13.5215	76
77	21.6116	19.5041	17.7710	16.3208	15.0895	14.5409	14.0309	13.5554	77
78	21.6956	19.5724	17.8277	16.3687	15.1303	14.5789	14.0662	13.5884	78
79	21.7774	19.6390	17.8830	16.4152	15.1701	14.6158	14.1006	13.6204	79
80	21.8572	19.7039	17.9367	16.4605	15.2088	14.6517	14.1340	13.6516	80
81	21.9350	19.7670	17.9891	16.5046	15.2464	14.6866	14.1665	13.6819	81
82	22.0108	19.8286	18.0401	16.5475	15.2830	14.7206	14.1981	13.7114	82
83	22.0847	19.8886	18.0897	16.5892	15.3186	14.7536	14.2288	13.7400	83
84	22.1568	19.9470	18.1380	16.6299	15.3532	14.7857	14.2587	13.7679	84
85	22.2271	20.0039	18.1851	16.6694	15.3869	14.8170	14.2877	13.7950	85
86	22.2956	20.0594	18.2309	16.7079	15.4197	14.8474	14.3160	13.8213	86
87	22.3624	20.1135	18.2755	16.7454	15.4517	14.8770	14.3435	13.8470	87
88	22.4275	20.1661	18.3190	16.7819	15.4827	14.9058	14.3703	13.8719	88
89	22.4910	20.2175	18.3614	16.8174	15.5130	14.9338	14.3963	13.8962	89
90	22.5529	20.2675	18.4026	16.8520	15.5424	14.9611	14.4217	13.9198	90
91	22.6133	20.3162	18.4428	16.8857	15.5710	14.9876	14.4463	13.9428	91
92	22.6722	20.3637	18.4819	16.9185	15.5989	15.0134	14.4703	13.9651	92
93	22.7296	20.4100	18.5200	16.9504	15.6261	15.0386	14.4937	13.9869	93
94	22.7855	20.4551	18.5572	16.9815	15.6525	15.0631	14.5164	14.0081	94
95	22.8401	20.4991	18.5934	17.0118	15.6782	15.0869	14.5385	14.0287	95
96	22.8933	20.5419	18.6286	17.0413	15.7033	15.1101	14.5601	14.0487	96
97	22.9452	20.5837	18.6629	17.0701	15.7277	15.1327	14.5811	14.0682	97
98	22.9958	20.6244	18.6964	17.0980	15.7515	15.1547	14.6015	14.0872	98
99	23.0451	20.6641	18.7290	17.1253	15.7746	15.1761	14.6214	14.1057	99
100	23.0933	20.7028	18.7608	17.1519	15.7971	15.1969	14.6407	14.1238	100

YEARS' PURCHASE

				Rate per Cent					
Yrs.	7	7.25	7.5	8	8.5	9	9.5	10	Yrs.
1	0.6673	0.6662	0.6651	0.6629	0.6607	0.6585	0.6564	0.6542	1
2	1.2895	1.2854	1.2813	1.2731	1.2651	1.2571	1.2493	1.2415	2
3	1.8710	1.8623	1.8536	1.8366	1.8199	1.8035	1.7874	1.7715	3
4	2.4153	2.4008	2.3865	2.3583	2.3309	2.3040	2.2778	2.2521	4
5	2.9258	2.9046	2.8837	2.8427	2.8028	2.7641	2.7264	2.6898	5
6	3.4055	3.3767	3.3485	3.2933	3.2400	3.1883	3.1383	3.0898	6
7	3.8568	3.8200	3.7839	3.7136	3.6459	3.5806	3.5177	3.4569	7
8	4.2822	4.2368	4.1924	4.1064	4.0237	3.9444	3.8681	3.7947	8
9	4.6836	4.6294	4.5765	4.4741	4.3762	4.2825	4.1927	4.1066	9
10	5.0630	4.9997	4.9380	4.8190	4.7056	4.5974	4.4941	4.3954	10
11	5.4219	5.3494	5.2788	5.1431	5.0141	4.8915	4.7747	4.6634	11
12	5.7619	5.6801	5.6006	5.4480	5.3035	5.1665	5.0364	4.9127	12
13	6.0844	5.9932	5.9048	5.7354	5.5755	5.4243	5.2811	5.1452	13
14	6.3906	6.2901	6.1927	6.0067	5.8316	5.6663	5.5102	5.3625	14
15	6.6815	6.5718	6.4655	6.2631	6.0729	5.8939	5.7252	5.5659	15
16	6.9583	6.8393	6.7244	6.5056	6.3007	6.1083	5.9272	5.7566	16
17	7.2219	7.0938	6.9702	6.7354	6.5160	6.3104	6.1174	5.9358	17
18	7.4730	7.3360	7.2039	6.9534	6.7198	6.5013	6.2967	6.1045	18
19	7.7126	7.5667	7.4262	7.1603	6.9129	6.6819	6.4659	6.2634	19
20	7.9413	7.7867	7.6380	7.3570	7.0960	6.8529	6.6258	6.4134	20
21	8.1597	7.9966	7.8399	7.5441	7.2699	7.0149	6.7772	6.5551	21
22	8.3685	8.1970	8.0324	7.7223	7.4352	7.1687	6.9206	6.6892	22
23	8.5683	8.3886	8.2163	7.8921	7.5925	7.3148	7.0567	6.8162	23
24	8.7595	8.5718	8.3920	8.0540	7.7423	7.4537	7.1859	6.9367	24
25	8.9427	8.7472	8.5600	8.2086	7.8850	7.5859	7.3087	7.0511	25
26	9.1183	8.9151	8.7207	8.3563	8.0212	7.7119	7.4256	7.1598	26
27	9.2867	9.0760	8.8746	8.4975	8.1512	7.8320	7.5369	7.2632	27
28	9.4482	9.2302	9.0220	8.6326	8.2754	7.9466	7.6429	7.3616	28
29	9.6033	9.3782	9.1634	8.7619	8.3942	8.0560	7.7441	7.4554	29
30	9.7523	9.5202	9.2989	8.8858	8.5078	8.1606	7.8407	7.5449	30
31	9.8955	9.6566	9.4290	9.0045	8.6166	8.2607	7.9330	7.6303	31
32	10.0332	9.7877	9.5539	9.1183	8.7208	8.3564	8.0212	7.7119	32
33	10.1656	9.9137	9.6739	9.2276	8.8206	8.4480	8.1057	7.7899	33
34	10.2931	10.0349	9.7893	9.3325	8.9164	8.5359	8.1865	7.8646	34
35	10.4158	10.1515	9.9002	9.4333	9.0084	8.6201	8.2639	7.9360	35
36	10.5340	10.2637	10.0069	9.5301	9.0966	8.7009	8.3381	8.0044	36
37	10.6479	10.3718	10.1096	9.6232	9.1814	8.7784	8.4093	8.0700	37
38	10.7576	10.4759	10.2085	9.7128	9.2629	8.8529	8.4776	8.1329	38
39	10.8635	10.5762	10.3038	9.7990	9.3413	8.9245	8.5432	8.1933	39
40	10.9656	10.6730	10.3956	9.8819	9.4167	8.9932	8.6063	8.2512	40
41	11.0641	10.7663	10.4841	9.9619	9.4892	9.0594	8.6668	8.3068	41
42	11.1592	10.8563	10.5694	10.0389	9.5591	9.1230	8.7251	8.3603	42
43	11.2510	10.9432	10.6518	10.1131	9.6264	9.1843	8.7811	8.4118	43
44	11.3396	11.0270	10.7312	10.1847	9.6912	9.2433	8.8350	8.4612	44
45	11.4253	11.1080	10.8079	10.2538	9.7537	9.3002	8.8869	8.5088	45
46	11.5081	11.1863	10.8819	10.3204	9.8140	9.3549	8.9369	8.5547	46
47	11.5881	11.2619	10.9535	10.3847	9.8721	9.4078	8.9851	8.5988	47
48	11.6655	11.3349	11.0226	10.4468	9.9282	9.4587	9.0316	8.6413	48
49	11.7403	11.4056	11.0894	10.5068	9.9824	9.5078	9.0764	8.6823	49
50	11.8127	11.4739	11.1539	10.5648	10.0347	9.5553	9.1196	8.7219	50

YEARS' PURCHASE

				Rate per Cent					
Yrs.	7	7.25	7.5	8	8.5	9	9.5	10	Yrs.
51	11.8828	11.5400	11.2164	10.6208	10.0852	9.6010	9.1613	8.7600	51
52	11.9506	11.6039	11.2768	10.6749	10.1340	9.6453	9.2015	8.7968	52
53	12.0162	11.6658	11.3352	10.7272	10.1812	9.6880	9.2404	8.8323	53
54	12.0798	11.7257	11.3917	10.7779	10.2267	9.7293	9.2779	8.8666	54
55	12.1414	11.7837	11.4465	10.8268	10.2708	9.7691	9.3142	8.8997	55
56	12.2010	11.8398	11.4995	10.8742	10.3135	9.8077	9.3492	8.9317	56
57	12.2588	11.8942	11.5508	10.9201	10.3547	9.8450	9.3831	8.9626	57
58	12.3147	11.9469	11.6005	10.9645	10.3946	9.8811	9.4159	8.9925	58
59	12.3690	11.9980	11.6486	11.0075	10.4333	9.9160	9.4476	9.0214	59
60	12.4216	12.0475	11.6952	11.0491	10.4707	9.9498	9.4782	9.0494	60
61	12.4726	12.0955	11.7404	11.0895	10.5069	9.9825	9.5079	9.0764	61
62	12.5221	12.1420	11.7843	11.1286	10.5420	10.0141	9.5366	9.1026	62
63	12.5701	12.1871	11.8267	11.1664	10.5760	10.0448	9.5644	9.1279	63
64	12.6166	12.2308	11.8679	11.2031	10.6089	10.0745	9.5913	9.1524	64
65	12.6617	12.2732	11.9079	11.2387	10.6408	10.1033	9.6174	9.1762	65
66	12.7056	12.3144	11.9466	11.2732	10.6717	10.1311	9.6427	9.1992	66
67	12.7481	12.3543	11.9842	11.3067	10.7017	10.1581	9.6671	9.2214	67
68	12.7894	12.3931	12.0207	11.3392	10.7308	10.1843	9.6909	9.2430	68
69	12.8294	12.4307	12.0561	11.3706	10.7590	10.2097	9.7138	9.2639	69
70	12.8683	12.4672	12.0904	11.4012	10.7863	10.2343	9.7361	9.2842	70
71	12.9061	12.5027	12.1238	11.4308	10.8128	10.2582	9.7577	9.3038	71
72	12.9428	12.5371	12.1561	11.4596	10.8386	10.2814	9.7787	9.3229	72
73	12.9784	12.5706	12.1875	11.4875	10.8635	10.3039	9.7990	9.3413	73
74	13.0130	12.6030	12.2181	11.5146	10.8878	10.3257	9.8187	9.3593	74
75	13.0467	12.6346	12.2477	11.5410	10.9113	10.3468	9.8379	9.3766	75
76	13.0793	12.6652	12.2765	11.5665	10.9342	10.3674	9.8564	9.3935	76
77	13.1111	12.6950	12.3045	11.5913	10.9563	10.3873	9.8745	9.4099	77
78	13.1419	12.7239	12.3316	11.6154	10.9779	10.4067	9.8919	9.4258	78
79	13.1719	12.7520	12.3580	11.6389	10.9988	10.4255	9.9089	9.4412	79
80	13.2011	12.7793	12.3837	11.6616	11.0191	10.4437	9.9254	9.4561	80
81	13.2294	12.8059	12.4086	11.6837	11.0388	10.4614	9.9414	9.4707	81
82	13.2569	12.8317	12.4328	11.7052	11.0580	10.4786	9.9570	9.4848	82
83	13.2837	12.8568	12.4564	11.7261	11.0766	10.4954	9.9721	9.4985	83
84	13.3098	12.8811	12.4793	11.7464	11.0947	10.5116	9.9867	9.5118	84
85	13.3351	12.9049	12.5015	11.7661	11.1123	10.5274	10.0010	9.5247	85
86	13.3597	12.9279	12.5232	11.7852	11.1294	10.5428	10.0148	9.5373	86
87	13.3837	12.9504	12.5442	11.8039	11.1460	10.5577	10.0283	9.5495	87
88	13.4070	12.9722	12.5647	11.8220	11.1622	10.5722	10.0414	9.5613	88
89	13.4296	12.9934	12.5846	11.8396	11.1779	10.5862	10.0541	9.5728	89
90	13.4517	13.0140	12.6040	11.8568	11.1932	10.5999	10.0664	9.5840	90
91	13.4731	13.0341	12.6228	11.8734	11.2080	10.6133	10.0784	9.5949	91
92	13.4940	13.0536	12.6411	11.8896	11.2225	10.6262	10.0901	9.6055	92
93	13.5143	13.0727	12.6589	11.9054	11.2365	10.6388	10.1015	9.6158	93
94	13.5341	13.0911	12.6763	11.9207	11.2502	10.6510	10.1125	9.6258	94
95	13.5533	13.1091	12.6932	11.9356	11.2635	10.6630	10.1232	9.6355	95
96	13.5720	13.1267	12.7096	11.9502	11.2764	10.6745	10.1337	9.6450	96
97	13.5903	13.1437	12.7255	11.9643	11.2890	10.6858	10.1438	9.6542	97
98	13.6080	13.1603	12.7411	11.9780	11.3012	10.6968	10.1537	9.6631	98
99	13.6253	13.1764	12.7562	11.9914	11.3131	10.7074	10.1633	9.6718	99
100	13.6421	13.1921	12.7710	12.0044	11.3247	10.7178	10.1727	9.6803	100

YEARS' PURCHASE

					Rate per Cent				
Yrs.	11	12	13	14	15	16	18	20	Yrs.
1	0.6500	0.6458	0.6416	0.6375	0.6335	0.6295	0.6217	0.6140	1
2	1.2263	1.2114	1.1969	1.1828	1.1690	1.1554	1.1293	1.1044	2
3	1.7407	1.7109	1.6821	1.6543	1.6274	1.6013	1.5516	1.5049	3
4	2.2025	2.1551	2.1096	2.0660	2.0242	1.9840	1.9083	1.8381	4
5	2.6193	2.5524	2.4889	2.4285	2.3709	2.3160	2.2135	2.1196	5
6	2.9972	2.9100	2.8277	2.7500	2.6764	2.6066	2.4774	2.3605	6
7	3.3414	3.2333	3.1321	3.0369	2.9474	2.8630	2.7080	2.5689	7
8	3.6560	3.5270	3.4069	3.2946	3.1895	3.0910	2.9110	2.7508	8
9	3.9446	3.7949	3.6562	3.5272	3.4070	3.2948	3.0911	2.9111	9
10	4.2103	4.0402	3.8833	3.7381	3.6034	3.4781	3.2519	3.0533	10
11	4.4556	4.2655	4.0910	3.9302	3.7816	3.6438	3.3963	3.1803	11
12	4.6827	4.4732	4.2817	4.1059	3.9439	3.7943	3.5267	3.2943	12
13	4.8935	4.6652	4.4572	4.2670	4.0924	3.9315	3.6449	3.3973	13
14	5.0896	4.8431	4.6193	4.4154	4.2287	4.0571	3.7526	3.4906	14
15	5.2724	5.0084	4.7695	4.5524	4.3541	4.1725	3.8511	3.5757	15
16	5.4433	5.1623	4.9089	4.6792	4.4700	4.2788	3.9415	3.6535	16
17	5.6032	5.3059	5.0386	4.7969	4.5773	4.3770	4.0247	3.7248	17
18	5.7533	5.4403	5.1596	4.9064	4.6770	4.4680	4.1015	3.7905	18
19	5.8942	5.5661	5.2726	5.0086	4.7697	4.5525	4.1726	3.8512	19
20	6.0268	5.6843	5.3785	5.1040	4.8561	4.6312	4.2386	3.9074	20
21	6.1518	5.7953	5.4779	5.1934	4.9370	4.7047	4.3001	3.9596	21
22	6.2698	5.8999	5.5712	5.2772	5.0127	4.7734	4.3574	4.0081	22
23	6.3812	5.9985	5.6590	5.3559	5.0836	4.8377	4.4109	4.0534	23
24	6.4867	6.0916	5.7418	5.4300	5.1504	4.8981	4.4611	4.0957	24
25	6.5866	6.1796	5.8200	5.4999	5.2131	4.9548	4.5081	4.1353	25
26	6.6814	6.2629	5.8938	5.5658	5.2723	5.0083	4.5523	4.1724	26
27	6.7713	6.3419	5.9637	5.6281	5.3282	5.0586	4.5939	4.2073	27
28	6.8568	6.4168	6.0299	5.6870	5.3810	5.1062	4.6331	4.2402	28
29	6.9382	6.4880	6.0927	5.7428	5.4309	5.1512	4.6701	4.2711	29
30	7.0156	6.5557	6.1524	5.7958	5.4783	5.1937	4.7050	4.3003	30
31	7.0894	6.6201	6.2090	5.8460	5.5232	5.2341	4.7381	4.3280	31
32	7.1598	6.6814	6.2630	5.8938	5.5658	5.2723	4.7694	4.3541	32
33	7.2270	6.7399	6.3143	5.9393	5.6063	5.3087	4.7991	4.3788	33
34	7.2912	6.7957	6.3632	5.9826	5.6449	5.3432	4.8274	4.4023	34
35	7.3525	6.8489	6.4099	6.0238	5.6816	5.3761	4.8542	4.4246	35
36	7.4112	6.8998	6.4545	6.0631	5.7165	5.4074	4.8797	4.4458	36
37	7.4674	6.9485	6.4971	6.1007	5.7499	5.4373	4.9040	4.4660	37
38	7.5212	6.9951	6.5378	6.1366	5.7818	5.4658	4.9271	4.4852	38
39	7.5728	7.0397	6.5767	6.1709	5.8122	5.4929	4.9492	4.5035	39
40	7.6223	7.0824	6.6140	6.2037	5.8413	5.5189	4.9703	4.5209	40
41	7.6697	7.1234	6.6497	6.2351	5.8691	5.5438	4.9905	4.5376	41
42	7.7153	7.1627	6.6839	6.2652	5.8958	5.5675	5.0097	4.5535	42
43	7.7591	7.2004	6.7168	6.2940	5.9213	5.5903	5.0281	4.5687	43
44	7.8011	7.2366	6.7483	6.3217	5.9458	5.6121	5.0458	4.5832	44
45	7.8416	7.2714	6.7785	6.3482	5.9693	5.6330	5.0626	4.5972	45
46	7.8805	7.3048	6.8076	6.3737	5.9918	5.6531	5.0788	4.6105	46
47	7.9180	7.3370	6.8355	6.3981	6.0134	5.6723	5.0944	4.6233	47
48	7.9540	7.3680	6.8623	6.4217	6.0342	5.6908	5.1093	4.6356	48
49	7.9887	7.3977	6.8882	6.4443	6.0541	5.7085	5.1236	4.6473	49
50	8.0222	7.4264	6.9130	6.4660	6.0733	5.7256	5.1373	4.6587	50

YEARS' PURCHASE

				Rate per Cent					
Yrs.	11	12	13	14	15	16	18	20	Yrs.
51	8.0544	7.4540	6.9370	6.4870	6.0918	5.7420	5.1505	4.6695	51
52	8.0855	7.4807	6.9600	6.5071	6.1096	5.7578	5.1632	4.6799	52
53	8.1155	7.5063	6.9822	6.5265	6.1267	5.7730	5.1754	4.6900	53
54	8.1445	7.5311	7.0036	6.5452	6.1432	5.7876	5.1872	4.6996	54
55	8.1724	7.5550	7.0243	6.5633	6.1590	5.8017	5.1985	4.7089	55
56	8.1994	7.5780	7.0442	6.5807	6.1743	5.8153	5.2094	4.7179	56
57	8.2254	7.6003	7.0634	6.5974	6.1891	5.8284	5.2199	4.7265	57
58	8.2506	7.6217	7.0820	6.6136	6.2033	5.8410	5.2300	4.7348	58
59	8.2749	7.6425	7.0999	6.6292	6.2171	5.8532	5.2398	4.7428	59
60	8.2984	7.6626	7.1172	6.6443	6.2303	5.8649	5.2492	4.7505	60
61	8.3212	7.6819	7.1339	6.6589	6.2431	5.8763	5.2583	4.7579	61
62	8.3431	7.7007	7.1501	6.6729	6.2555	5.8872	5.2671	4.7651	62
63	8.3644	7.7188	7.1657	6.6865	6.2675	5.8978	5.2755	4.7720	63
64	8.3850	7.7363	7.1808	6.6997	6.2790	5.9080	5.2837	4.7787	64
65	8.4049	7.7533	7.1954	6.7124	6.2902	5.9179	5.2916	4.7852	65
66	8.4242	7.7697	7.2095	6.7247	6.3010	5.9275	5.2993	4.7914	66
67	8.4429	7.7855	7.2232	6.7366	6.3114	5.9367	5.3066	4.7975	67
68	8.4610	7.8009	7.2364	6.7481	6.3215	5.9457	5.3138	4.8033	68
69	8.4785	7.8158	7.2492	6.7592	6.3313	5.9543	5.3207	4.8089	69
70	8.4954	7.8302	7.2616	6.7700	6.3407	5.9627	5.3274	4.8144	70
71	8.5119	7.8442	7.2736	6.7805	6.3499	5.9708	5.3338	4.8197	71
72	8.5278	7.8577	7.2853	6.7906	6.3588	5.9786	5.3401	4.8248	72
73	8.5433	7.8709	7.2966	6.8004	6.3674	5.9862	5.3461	4.8297	73
74	8.5583	7.8836	7.3075	6.8099	6.3757	5.9935	5.3520	4.8345	74
75	8.5728	7.8959	7.3181	6.8191	6.3837	6.0007	5.3577	4.8391	75
76	8.5869	7.9079	7.3283	6.8280	6.3916	6.0076	5.3632	4.8436	76
77	8.6006	7.9195	7.3383	6.8366	6.3991	6.0143	5.3685	4.8480	77
78	8.6138	7.9307	7.3480	6.8450	6.4065	6.0207	5.3737	4.8522	78
79	8.6267	7.9416	7.3573	6.8531	6.4136	6.0270	5.3787	4.8563	79
80	8.6392	7.9522	7.3664	6.8610	6.4205	6.0331	5.3835	4.8602	80
81	8.6513	7.9625	7.3752	6.8686	6.4272	6.0390	5.3882	4.8641	81
82	8.6631	7.9724	7.3838	6.8761	6.4337	6.0448	5.3928	4.8678	82
83	8.6745	7.9821	7.3921	6.8833	6.4400	6.0503	5.3972	4.8714	83
84	8.6856	7.9915	7.4001	6.8902	6.4461	6.0557	5.4015	4.8749	84
85	8.6964	8.0006	7.4079	6.8970	6.4520	6.0610	5.4057	4.8783	85
86	8.7069	8.0095	7.4155	6.9036	6.4578	6.0661	5.4097	4.8816	86
87	8.7170	8.0181	7.4229	6.9100	6.4634	6.0710	5.4137	4.8848	87
88	8.7269	8.0265	7.4301	6.9162	6.4688	6.0758	5.4175	4.8879	88
89	8.7365	8.0346	7.4370	6.9222	6.4741	6.0804	5.4212	4.8909	89
90	8.7458	8.0425	7.4438	6.9281	6.4792	6.0849	5.4248	4.8938	90
91	8.7549	8.0501	7.4504	6.9338	6.4842	6.0893	5.4282	4.8966	91
92	8.7637	8.0576	7.4567	6.9393	6.4890	6.0936	5.4316	4.8994	92
93	8.7723	8.0648	7.4629	6.9447	6.4937	6.0977	5.4349	4.9021	93
94	8.7806	8.0718	7.4690	6.9499	6.4983	6.1017	5.4381	4.9047	94
95	8.7887	8.0787	7.4748	6.9549	6.5027	6.1057	5.4412	4.9072	95
96	8.7966	8.0853	7.4805	6.9599	6.5070	6.1094	5.4442	4.9096	96
97	8.8042	8.0918	7.4860	6.9647	6.5112	6.1131	5.4472	4.9120	97
98	8.8116	8.0981	7.4914	6.9693	6.5152	6.1167	5.4500	4.9143	98
99	8.8189	8.1042	7.4966	6.9738	6.5192	6.1202	5.4528	4.9166	99
100	8.8259	8.1101	7.5017	6.9782	6.5230	6.1236	5.4555	4.9188	100

YEARS' PURCHASE

				Rate per Cent					
Yrs.	4	4.5	5	5.5	6	6.25	6.5	6.75	Yrs.
1	0.6809	0.6786	0.6763	0.6740	0.6718	0.6707	0.6695	0.6684	1
2	1.3446	1.3356	1.3267	1.3180	1.3094	1.3051	1.3008	1.2966	2
3	1.9913	1.9717	1.9524	1.9335	1.9150	1.9059	1.8969	1.8879	3
4	2.6215	2.5875	2.5545	2.5223	2.4909	2.4754	2.4602	2.4452	4
5	3.2354	3.1839	3.1340	3.0857	3.0388	3.0159	2.9933	2.9711	5
6	3.8336	3.7615	3.6920	3.6251	3.5606	3.5292	3.4983	3.4680	6
7	4.4162	4.3208	4.2294	4.1419	4.0578	4.0171	3.9771	3.9380	7
8	4.9838	4.8626	4.7472	4.6371	4.5320	4.4813	4.4316	4.3830	8
9	5.5365	5.3874	5.2460	5.1120	4.9846	4.9232	4.8633	4.8049	9
10	6.0748	5.8957	5.7269	5.5675	5.4167	5.3443	5.2738	5.2052	10
11	6.5990	6.3882	6.1904	6.0046	5.8296	5.7458	5.6645	5.5854	11
12	7.1093	6.8653	6.6375	6.4243	6.2243	6.1289	6.0365	5.9467	12
13	7.6063	7.3276	7.0686	6.8273	6.6019	6.4947	6.3910	6.2905	13
14	8.0900	7.7755	7.4845	7.2145	6.9633	6.8442	6.7291	6.6177	14
15	8.5610	8.2095	7.8859	7.5867	7.3094	7.1783	7.0517	6.9296	15
16	9.0194	8.6302	8.2732	7.9445	7.6410	7.4978	7.3598	7.2269	16
17	9.4655	9.0378	8.6470	8.2887	7.9588	7.8036	7.6542	7.5105	17
18	9.8998	9.4329	9.0080	8.6198	8.2636	8.0964	7.9357	7.7813	18
19	10.3224	9.8158	9.3566	8.9384	8.5560	8.3768	8.2050	8.0401	19
20	10.7336	10.1869	9.6932	9.2451	8.8366	8.6456	8.4627	8.2874	20
21	11.1338	10.5467	10.0184	9.5405	9.1061	8.9034	8.7095	8.5239	21
22	11.5231	10.8954	10.3325	9.8249	9.3649	9.1506	8.9460	8.7503	22
23	11.9019	11.2334	10.6361	10.0990	9.6135	9.3879	9.1726	8.9670	23
24	12.2705	11.5612	10.9294	10.3631	9.8526	9.6157	9.3900	9.1746	24
25	12.6290	11.8789	11.2129	10.6177	10.0824	9.8345	9.5985	9.3736	25
26	12.9778	12.1870	11.4870	10.8631	10.3035	10.0447	9.7987	9.5644	26
27	13.3170	12.4857	11.7520	11.0998	10.5162	10.2468	9.9908	9.7474	27
28	13.6470	12.7753	12.0083	11.3281	10.7209	10.4410	10.1754	9.9230	28
29	13.9680	13.0561	12.2560	11.5484	10.9179	10.6279	10.3528	10.0916	29
30	14.2801	13.3284	12.4957	11.7609	11.1077	10.8076	10.5233	10.2535	30
31	14.5837	13.5925	12.7275	11.9660	11.2905	10.9806	10.6872	10.4091	31
32	14.8789	13.8486	12.9518	12.1641	11.4666	11.1471	10.8449	10.5586	32
33	15.1659	14.0970	13.1688	12.3552	11.6364	11.3074	10.9966	10.7024	33
34	15.4451	14.3378	13.3787	12.5399	11.8000	11.4619	11.1426	10.8406	34
35	15.7165	14.5714	13.5819	12.7182	11.9578	11.6107	11.2832	10.9736	35
36	15.9804	14.7980	13.7785	12.8904	12.1099	11.7541	11.4185	11.1016	36
37	16.2369	15.0177	13.9688	13.0569	12.2567	11.8923	11.5489	11.2249	37
38	16.4864	15.2309	14.1530	13.2177	12.3983	12.0256	11.6746	11.3435	38
39	16.7289	15.4376	14.3314	13.3731	12.5350	12.1541	11.7957	11.4578	39
40	16.9646	15.6381	14.5041	13.5233	12.6668	12.2780	11.9124	11.5679	40
41	17.1938	15.8327	14.6712	13.6686	12.7942	12.3976	12.0249	11.6740	41
42	17.4165	16.0214	14.8331	13.8090	12.9171	12.5130	12.1335	11.7762	42
43	17.6331	16.2044	14.9899	13.9447	13.0358	12.6244	12.2382	11.8749	43
44	17.8436	16.3820	15.1417	14.0761	13.1505	12.7319	12.3392	11.9699	44
45	18.0481	16.5543	15.2888	14.2031	13.2613	12.8358	12.4367	12.0617	45
46	18.2470	16.7214	15.4312	14.3259	13.3683	12.9360	12.5308	12.1501	46
47	18.4402	16.8835	15.5692	14.4448	13.4718	13.0328	12.6216	12.2355	47
48	18.6280	17.0408	15.7029	14.5597	13.5717	13.1264	12.7093	12.3179	48
49	18.8106	17.1935	15.8324	14.6710	13.6684	13.2167	12.7940	12.3975	49
50	18.9879	17.3415	15.9579	14.7787	13.7618	13.3041	12.8758	12.4743	50

YEARS' PURCHASE

Rate per Cent

Yrs.	4	4.5	5	5.5	6	6.25	6.5	6.75	Yrs.
51	19.1603	17.4852	16.0795	14.8829	13.8521	13.3885	12.9548	12.5484	51
52	19.3278	17.6246	16.1973	14.9838	13.9395	13.4700	13.0312	12.6201	52
53	19.4906	17.7599	16.3114	15.0814	14.0239	13.5489	13.1050	12.6893	53
54	19.6488	17.8911	16.4221	15.1760	14.1056	13.6251	13.1763	12.7561	54
55	19.8025	18.0184	16.5293	15.2675	14.1847	13.6989	13.2453	12.8207	55
56	19.9518	18.1420	16.6332	15.3561	14.2611	13.7702	13.3119	12.8832	56
57	20.0970	18.2619	16.7339	15.4419	14.3351	13.8391	13.3764	12.9435	57
58	20.2380	18.3783	16.8316	15.5250	14.4067	13.9059	13.4387	13.0019	58
59	20.3749	18.4912	16.9262	15.6055	14.4760	13.9704	13.4989	13.0583	59
60	20.5080	18.6007	17.0180	15.6835	14.5430	14.0328	13.5572	13.1128	60
61	20.6373	18.7070	17.1069	15.7590	14.6080	14.0933	13.6136	13.1655	61
62	20.7630	18.8102	17.1932	15.8321	14.6708	14.1517	13.6682	13.2166	62
63	20.8850	18.9103	17.2768	15.9030	14.7316	14.2083	13.7210	13.2659	63
64	21.0036	19.0075	17.3578	15.9717	14.7905	14.2631	13.7720	13.3137	64
65	21.1188	19.1018	17.4364	16.0382	14.8476	14.3162	13.8215	13.3598	65
66	21.2307	19.1933	17.5127	16.1027	14.9028	14.3675	13.8693	13.4045	66
67	21.3394	19.2821	17.5866	16.1651	14.9563	14.4172	13.9156	13.4478	67
68	21.4450	19.3683	17.6582	16.2256	15.0081	14.4653	13.9605	13.4897	68
69	21.5476	19.4519	17.7277	16.2843	15.0582	14.5119	14.0039	13.5302	69
70	21.6473	19.5331	17.7951	16.3412	15.1068	14.5571	14.0459	13.5694	70
71	21.7441	19.6119	17.8605	16.3963	15.1539	14.6008	14.0866	13.6074	71
72	21.8381	19.6883	17.9239	16.4497	15.1995	14.6431	14.1260	13.6441	72
73	21.9295	19.7625	17.9854	16.5014	15.2437	14.6841	14.1642	13.6797	73
74	22.0182	19.8346	18.0450	16.5516	15.2865	14.7239	14.2011	13.7142	74
75	22.1044	19.9045	18.1028	16.6003	15.3280	14.7623	14.2369	13.7476	75
76	22.1881	19.9723	18.1589	16.6474	15.3682	14.7996	14.2716	13.7799	76
77	22.2694	20.0382	18.2134	16.6932	15.4072	14.8358	14.3052	13.8113	77
78	22.3483	20.1021	18.2662	16.7375	15.4450	14.8708	14.3377	13.8416	78
79	22.4250	20.1641	18.3174	16.7805	15.4816	14.9047	14.3693	13.8710	79
80	22.4995	20.2243	18.3670	16.8222	15.5170	14.9376	14.3998	13.8994	80
81	22.5719	20.2828	18.4152	16.8626	15.5514	14.9694	14.4294	13.9270	81
82	22.6422	20.3395	18.4620	16.9018	15.5847	15.0003	14.4581	13.9537	82
83	22.7104	20.3946	18.5073	16.9398	15.6170	15.0302	14.4859	13.9796	83
84	22.7767	20.4480	18.5513	16.9766	15.6484	15.0592	14.5128	14.0047	84
85	22.8411	20.4999	18.5940	17.0124	15.6787	15.0873	14.5390	14.0290	85
86	22.9036	20.5503	18.6354	17.0470	15.7082	15.1146	14.5643	14.0526	86
87	22.9644	20.5991	18.6756	17.0807	15.7367	15.1410	14.5888	14.0754	87
88	23.0233	20.6466	18.7146	17.1133	15.7644	15.1666	14.6126	14.0976	88
89	23.0806	20.6926	18.7524	17.1449	15.7912	15.1915	14.6356	14.1190	89
90	23.1362	20.7373	18.7891	17.1756	15.8172	15.2156	14.6580	14.1398	90
91	23.1903	20.7807	18.8248	17.2053	15.8425	15.2389	14.6796	14.1600	91
92	23.2427	20.8228	18.8593	17.2342	15.8669	15.2615	14.7006	14.1795	92
93	23.2937	20.8637	18.8928	17.2622	15.8906	15.2835	14.7210	14.1985	93
94	23.3432	20.9034	18.9254	17.2893	15.9137	15.3048	14.7408	14.2168	94
95	23.3912	20.9419	18.9569	17.3157	15.9360	15.3254	14.7599	14.2347	95
96	23.4379	20.9793	18.9876	17.3412	15.9576	15.3454	14.7785	14.2519	96
97	23.4832	21.0156	19.0173	17.3660	15.9786	15.3648	14.7965	14.2687	97
98	23.5272	21.0509	19.0462	17.3901	15.9990	15.3837	14.8139	14.2849	98
99	23.5699	21.0851	19.0742	17.4134	16.0187	15.4019	14.8309	14.3006	99
100	23.6114	21.1183	19.1013	17.4361	16.0379	15.4196	14.8473	14.3159	100

YEARS' PURCHASE

				Rate per Cent					
Yrs.	7	7.25	7.5	8	8.5	9	9.5	10	Yrs.
1	0.6673	0.6662	0.6651	0.6629	0.6607	0.6585	0.6564	0.6542	1
2	1.2924	1.2883	1.2841	1.2760	1.2679	1.2599	1.2520	1.2442	2
3	1.8790	1.8703	1.8616	1.8444	1.8275	1.8110	1.7947	1.7788	3
4	2.4303	2.4157	2.4011	2.3727	2.3448	2.3177	2.2911	2.2652	4
5	2.9492	2.9276	2.9063	2.8647	2.8242	2.7849	2.7467	2.7095	5
6	3.4382	3.4089	3.3801	3.3239	3.2695	3.2169	3.1660	3.1167	6
7	3.8996	3.8619	3.8250	3.7532	3.6841	3.6175	3.5532	3.4912	7
8	4.3355	4.2890	4.2435	4.1554	4.0708	3.9896	3.9116	3.8365	8
9	4.7479	4.6922	4.6378	4.5327	4.4322	4.3361	4.2441	4.1559	9
10	5.1383	5.0732	5.0096	4.8872	4.7707	4.6595	4.5534	4.4521	10
11	5.5085	5.4336	5.3608	5.2209	5.0880	4.9618	4.8417	4.7273	11
12	5.8596	5.7750	5.6928	5.5353	5.3862	5.2449	5.1109	4.9836	12
13	6.1931	6.0986	6.0071	5.8319	5.6667	5.5105	5.3628	5.2227	13
14	6.5100	6.4058	6.3048	6.1121	5.9309	5.7601	5.5988	5.4464	14
15	6.8116	6.6975	6.5872	6.3772	6.1801	5.9949	5.8204	5.6558	15
16	7.0986	6.9748	6.8553	6.6281	6.4155	6.2161	6.0287	5.8523	16
17	7.3721	7.2387	7.1100	6.8659	6.6381	6.4248	6.2248	6.0370	17
18	7.6329	7.4899	7.3523	7.0916	6.8487	6.6220	6.4097	6.2107	18
19	7.8817	7.7294	7.5828	7.3058	7.0484	6.8084	6.5843	6.3744	19
20	8.1192	7.9576	7.8024	7.5095	7.2377	6.9849	6.7492	6.5289	20
21	8.3461	8.1755	8.0117	7.7032	7.4175	7.1522	6.9053	6.6748	21
22	8.5630	8.3835	8.2114	7.8876	7.5883	7.3109	7.0531	6.8128	22
23	8.7704	8.5822	8.4020	8.0632	7.7507	7.4616	7.1932	6.9435	23
24	8.9689	8.7722	8.5840	8.2307	7.9054	7.6048	7.3262	7.0673	24
25	9.1590	8.9539	8.7579	8.3905	8.0527	7.7410	7.4525	7.1848	25
26	9.3410	9.1279	8.9242	8.5430	8.1930	7.8706	7.5726	7.2964	26
27	9.5155	9.2944	9.0833	8.6887	8.3270	7.9941	7.6869	7.4024	27
28	9.6828	9.4539	9.2357	8.8280	8.4548	8.1119	7.7957	7.5032	28
29	9.8433	9.6069	9.3815	8.9612	8.5769	8.2242	7.8994	7.5992	29
30	9.9972	9.7535	9.5213	9.0886	8.6936	8.3314	7.9982	7.6907	30
31	10.1451	9.8941	9.6553	9.2107	8.8052	8.4338	8.0926	7.7779	31
32	10.2871	10.0291	9.7838	9.3275	8.9119	8.5317	8.1827	7.8611	32
33	10.4235	10.1587	9.9071	9.4395	9.0141	8.6254	8.2687	7.9405	33
34	10.5546	10.2832	10.0255	9.5469	9.1120	8.7149	8.3510	8.0163	34
35	10.6806	10.4028	10.1392	9.6499	9.2058	8.8007	8.4297	8.0888	35
36	10.8018	10.5178	10.2483	9.7488	9.2957	8.8828	8.5051	8.1581	36
37	10.9185	10.6283	10.3532	9.8437	9.3819	8.9615	8.5772	8.2245	37
38	11.0307	10.7347	10.4541	9.9348	9.4647	9.0370	8.6463	8.2880	38
39	11.1387	10.8369	10.5511	10.0224	9.5441	9.1094	8.7126	8.3489	39
40	11.2427	10.9354	10.6444	10.1065	9.6204	9.1788	8.7761	8.4072	40
41	11.3429	11.0302	10.7342	10.1874	9.6936	9.2455	8.8370	8.4631	41
42	11.4395	11.1214	10.8206	10.2652	9.7640	9.3095	8.8955	8.5167	42
43	11.5325	11.2093	10.9037	10.3400	9.8317	9.3711	8.9516	8.5681	43
44	11.6221	11.2940	10.9839	10.4120	9.8968	9.4302	9.0055	8.6175	44
45	11.7086	11.3756	11.0610	10.4814	9.9594	9.4870	9.0574	8.6650	45
46	11.7919	11.4543	11.1354	10.5481	10.0197	9.5417	9.1072	8.7105	46
47	11.8724	11.5301	11.2071	10.6124	10.0777	9.5942	9.1551	8.7543	47
48	11.9499	11.6033	11.2762	10.6743	10.1335	9.6448	9.2011	8.7964	48
49	12.0248	11.6738	11.3428	10.7340	10.1873	9.6935	9.2454	8.8369	49
50	12.0970	11.7419	11.4071	10.7916	10.2391	9.7404	9.2881	8.8759	50

YEARS' PURCHASE

				Rate per Cent					
Yrs.	7	7.25	7.5	8	8.5	9	9.5	10	Yrs.
51	12.1668	11.8076	11.4691	10.8470	10.2890	9.7856	9.3291	8.9134	51
52	12.2341	11.8710	11.5289	10.9005	10.3371	9.8291	9.3687	8.9494	52
53	12.2991	11.9322	11.5866	10.9521	10.3835	9.8710	9.4067	8.9842	53
54	12.3619	11.9913	11.6423	11.0019	10.4282	9.9114	9.4434	9.0176	54
55	12.4226	12.0484	11.6961	11.0499	10.4713	9.9504	9.4788	9.0499	55
56	12.4812	12.1035	11.7480	11.0962	10.5130	9.9879	9.5129	9.0809	56
57	12.5378	12.1568	11.7982	11.1410	10.5531	10.0242	9.5457	9.1109	57
58	12.5925	12.2082	11.8466	11.1842	10.5919	10.0591	9.5774	9.1398	58
59	12.6454	12.2579	11.8934	11.2259	10.6293	10.0929	9.6080	9.1676	59
60	12.6966	12.3060	11.9387	11.2662	10.6654	10.1254	9.6375	9.1944	60
61	12.7460	12.3524	11.9824	11.3051	10.7002	10.1568	9.6660	9.2203	61
62	12.7938	12.3973	12.0246	11.3427	10.7339	10.1872	9.6934	9.2453	62
63	12.8401	12.4407	12.0655	11.3790	10.7664	10.2165	9.7200	9.2695	63
64	12.8848	12.4827	12.1049	11.4141	10.7979	10.2448	9.7456	9.2927	64
65	12.9281	12.5233	12.1431	11.4480	10.8282	10.2721	9.7703	9.3152	65
66	12.9699	12.5626	12.1800	11.4808	10.8576	10.2985	9.7942	9.3369	66
67	13.0104	12.6006	12.2157	11.5126	10.8859	10.3240	9.8172	9.3579	67
68	13.0496	12.6373	12.2503	11.5432	10.9134	10.3487	9.8395	9.3782	68
69	13.0875	12.6729	12.2837	11.5729	10.9399	10.3725	9.8611	9.3977	69
70	13.1242	12.7073	12.3160	11.6016	10.9655	10.3955	9.8819	9.4166	70
71	13.1597	12.7406	12.3473	11.6293	10.9903	10.4178	9.9020	9.4349	71
72	13.1941	12.7728	12.3775	11.6562	11.0142	10.4393	9.9215	9.4526	72
73	13.2274	12.8040	12.4068	11.6821	11.0374	10.4602	9.9403	9.4696	73
74	13.2596	12.8342	12.4352	11.7073	11.0599	10.4803	9.9585	9.4861	74
75	13.2908	12.8634	12.4626	11.7316	11.0816	10.4998	9.9761	9.5021	75
76	13.3210	12.8917	12.4892	11.7551	11.1026	10.5186	9.9931	9.5175	76
77	13.3503	12.9191	12.5149	11.7779	11.1229	10.5369	10.0095	9.5325	77
78	13.3786	12.9456	12.5398	11.8000	11.1426	10.5545	10.0255	9.5469	78
79	13.4061	12.9713	12.5639	11.8213	11.1616	10.5716	10.0409	9.5609	79
80	13.4327	12.9962	12.5873	11.8420	11.1800	10.5881	10.0558	9.5744	80
81	13.4584	13.0203	12.6099	11.8620	11.1978	10.6041	10.0702	9.5875	81
82	13.4834	13.0437	12.6318	11.8814	11.2151	10.6196	10.0842	9.6001	82
83	13.5076	13.0663	12.6530	11.9001	11.2318	10.6346	10.0977	9.6124	83
84	13.5310	13.0882	12.6736	11.9183	11.2480	10.6491	10.1108	9.6242	84
85	13.5537	13.1095	12.6935	11.9359	11.2637	10.6632	10.1234	9.6357	85
86	13.5757	13.1300	12.7127	11.9530	11.2789	10.6768	10.1357	9.6468	86
87	13.5970	13.1500	12.7314	11.9695	11.2936	10.6900	10.1476	9.6576	87
88	13.6176	13.1693	12.7495	11.9855	11.3078	10.7027	10.1591	9.6680	88
89	13.6377	13.1880	12.7671	12.0010	11.3216	10.7151	10.1702	9.6781	89
90	13.6571	13.2062	12.7841	12.0160	11.3350	10.7271	10.1810	9.6878	90
91	13.6759	13.2237	12.8006	12.0306	11.3480	10.7387	10.1914	9.6973	91
92	13.6941	13.2408	12.8165	12.0447	11.3605	10.7499	10.2016	9.7065	92
93	13.7118	13.2573	12.8320	12.0583	11.3727	10.7608	10.2114	9.7153	93
94	13.7289	13.2733	12.8470	12.0716	11.3845	10.7713	10.2209	9.7239	94
95	13.7455	13.2888	12.8616	12.0844	11.3959	10.7815	10.2301	9.7323	95
96	13.7616	13.3039	12.8756	12.0969	11.4069	10.7914	10.2390	9.7403	96
97	13.7772	13.3185	12.8893	12.1089	11.4177	10.8010	10.2476	9.7481	97
98	13.7923	13.3326	12.9026	12.1206	11.4280	10.8103	10.2560	9.7557	98
99	13.8070	13.3463	12.9154	12.1320	11.4381	10.8194	10.2641	9.7631	99
100	13.8212	13.3596	12.9279	12.1429	11.4479	10.8281	10.2720	9.7702	100

YEARS' PURCHASE

				Rate per Cent					
Yrs.	11	12	13	14	15	16	18	20	Yrs.
1	0.6500	0.6458	0.6416	0.6375	0.6335	0.6295	0.6217	0.6140	1
2	1.2289	1.2140	1.1994	1.1852	1.1713	1.1578	1.1316	1.1065	2
3	1.7477	1.7177	1.6887	1.6606	1.6335	1.6072	1.5572	1.5101	3
4	2.2150	2.1670	2.1210	2.0770	2.0347	1.9941	1.9177	1.8468	4
5	2.6380	2.5702	2.5058	2.4445	2.3862	2.3306	2.2268	2.1318	5
6	3.0225	2.9338	2.8502	2.7712	2.6965	2.6257	2.4947	2.3761	6
7	3.3734	3.2633	3.1602	3.0634	2.9723	2.8865	2.7290	2.5877	7
8	3.6948	3.5631	3.4405	3.3261	3.2190	3.1186	2.9355	2.7728	8
9	3.9901	3.8370	3.6952	3.5635	3.4409	3.3265	3.1190	2.9358	9
10	4.2623	4.0881	3.9275	3.7791	3.6415	3.5135	3.2828	3.0806	10
11	4.5139	4.3189	4.1401	3.9755	3.8235	3.6827	3.4301	3.2099	11
12	4.7470	4.5319	4.3354	4.1552	3.9895	3.8364	3.5630	3.3260	12
13	4.9635	4.7288	4.5153	4.3202	4.1413	3.9766	3.6836	3.4309	13
14	5.1650	4.9114	4.6814	4.4721	4.2807	4.1049	3.7935	3.5260	14
15	5.3530	5.0811	4.8354	4.6123	4.4090	4.2228	3.8939	3.6126	15
16	5.5288	5.2391	4.9783	4.7422	4.5275	4.3314	3.9861	3.6918	16
17	5.6933	5.3866	5.1113	4.8627	4.6372	4.4317	4.0709	3.7644	17
18	5.8475	5.5245	5.2353	4.9748	4.7391	4.5246	4.1492	3.8312	18
19	5.9924	5.6536	5.3511	5.0793	4.8338	4.6109	4.2216	3.8929	19
20	6.1288	5.7748	5.4595	5.1769	4.9221	4.6912	4.2888	3.9500	20
21	6.2572	5.8887	5.5612	5.2682	5.0046	4.7661	4.3513	4.0029	21
22	6.3783	5.9959	5.6567	5.3538	5.0818	4.8360	4.4095	4.0522	22
23	6.4927	6.0968	5.7465	5.4342	5.1541	4.9015	4.4639	4.0980	23
24	6.6008	6.1921	5.8310	5.5098	5.2220	4.9629	4.5147	4.1409	24
25	6.7032	6.2821	5.9108	5.5809	5.2859	5.0205	4.5624	4.1809	25
26	6.8002	6.3672	5.9861	5.6480	5.3460	5.0747	4.6071	4.2184	26
27	6.8922	6.4478	6.0572	5.7113	5.4027	5.1258	4.6492	4.2537	27
28	6.9795	6.5242	6.1246	5.7711	5.4563	5.1739	4.6888	4.2868	28
29	7.0625	6.5966	6.1884	5.8278	5.5068	5.2194	4.7261	4.3179	29
30	7.1415	6.6654	6.2489	5.8814	5.5547	5.2624	4.7613	4.3473	30
31	7.2166	6.7308	6.3064	5.9323	5.6000	5.3031	4.7946	4.3750	31
32	7.2881	6.7930	6.3609	5.9805	5.6430	5.3416	4.8260	4.4012	32
33	7.3563	6.8523	6.4128	6.0264	5.6838	5.3782	4.8558	4.4260	33
34	7.4214	6.9087	6.4622	6.0700	5.7226	5.4128	4.8841	4.4495	34
35	7.4835	6.9625	6.5092	6.1114	5.7595	5.4458	4.9109	4.4717	35
36	7.5428	7.0138	6.5541	6.1509	5.7945	5.4771	4.9364	4.4928	36
37	7.5995	7.0627	6.5968	6.1886	5.8279	5.5070	4.9606	4.5129	37
38	7.6537	7.1095	6.6376	6.2245	5.8597	5.5354	4.9836	4.5319	38
39	7.7055	7.1543	6.6766	6.2587	5.8901	5.5624	5.0056	4.5501	39
40	7.7552	7.1970	6.7138	6.2914	5.9190	5.5883	5.0265	4.5673	40
41	7.8027	7.2380	6.7494	6.3227	5.9467	5.6129	5.0464	4.5838	41
42	7.8483	7.2771	6.7835	6.3526	5.9731	5.6364	5.0654	4.5995	42
43	7.8919	7.3147	6.8161	6.3811	5.9984	5.6589	5.0836	4.6144	43
44	7.9338	7.3506	6.8473	6.4085	6.0225	5.6804	5.1009	4.6287	44
45	7.9740	7.3851	6.8772	6.4347	6.0457	5.7010	5.1175	4.6424	45
46	8.0126	7.4182	6.9059	6.4598	6.0678	5.7207	5.1334	4.6554	46
47	8.0496	7.4499	6.9334	6.4839	6.0890	5.7396	5.1486	4.6679	47
48	8.0852	7.4804	6.9598	6.5069	6.1094	5.7576	5.1631	4.6798	48
49	8.1194	7.5097	6.9851	6.5291	6.1289	5.7750	5.1770	4.6913	49
50	8.1523	7.5378	7.0094	6.5503	6.1476	5.7916	5.1904	4.7022	50

YEARS' PURCHASE

				Rate per Cent					
Yrs.	11	12	13	14	15	16	18	20	Yrs.
51	8.1839	7.5648	7.0328	6.5707	6.1656	5.8075	5.2032	4.7127	51
52	8.2143	7.5908	7.0552	6.5903	6.1828	5.8228	5.2154	4.7228	52
53	8.2436	7.6158	7.0768	6.6091	6.1994	5.8375	5.2272	4.7325	53
54	8.2717	7.6398	7.0975	6.6272	6.2153	5.8516	5.2385	4.7417	54
55	8.2988	7.6629	7.1175	6.6446	6.2306	5.8651	5.2494	4.7506	55
56	8.3250	7.6852	7.1367	6.6613	6.2453	5.8782	5.2598	4.7592	56
57	8.3501	7.7066	7.1552	6.6774	6.2594	5.8907	5.2698	4.7674	57
58	8.3744	7.7273	7.1730	6.6929	6.2730	5.9028	5.2795	4.7753	58
59	8.3977	7.7471	7.1901	6.7078	6.2861	5.9144	5.2888	4.7829	59
60	8.4202	7.7663	7.2066	6.7222	6.2988	5.9255	5.2977	4.7902	60
61	8.4420	7.7848	7.2225	6.7360	6.3109	5.9363	5.3063	4.7972	61
62	8.4629	7.8026	7.2378	6.7493	6.3226	5.9466	5.3146	4.8039	62
63	8.4831	7.8198	7.2526	6.7622	6.3339	5.9566	5.3225	4.8104	63
64	8.5026	7.8363	7.2669	6.7746	6.3447	5.9662	5.3302	4.8167	64
65	8.5214	7.8523	7.2806	6.7865	6.3552	5.9755	5.3376	4.8227	65
66	8.5396	7.8677	7.2939	6.7980	6.3653	5.9844	5.3447	4.8285	66
67	8.5571	7.8826	7.3067	6.8091	6.3750	5.9930	5.3516	4.8341	67
68	8.5741	7.8970	7.3190	6.8198	6.3844	6.0013	5.3582	4.8395	68
69	8.5904	7.9108	7.3309	6.8302	6.3935	6.0093	5.3646	4.8448	69
70	8.6062	7.9242	7.3424	6.8402	6.4022	6.0170	5.3707	4.8498	70
71	8.6215	7.9372	7.3535	6.8498	6.4107	6.0245	5.3766	4.8546	71
72	8.6362	7.9497	7.3642	6.8591	6.4188	6.0317	5.3824	4.8593	72
73	8.6505	7.9617	7.3746	6.8681	6.4267	6.0386	5.3879	4.8638	73
74	8.6642	7.9734	7.3846	6.8768	6.4343	6.0453	5.3932	4.8681	74
75	8.6775	7.9847	7.3943	6.8852	6.4416	6.0518	5.3984	4.8723	75
76	8.6904	7.9956	7.4036	6.8933	6.4487	6.0581	5.4034	4.8764	76
77	8.7029	8.0061	7.4126	6.9011	6.4556	6.0641	5.4082	4.8803	77
78	8.7149	8.0163	7.4214	6.9087	6.4622	6.0699	5.4128	4.8841	78
79	8.7265	8.0261	7.4298	6.9160	6.4686	6.0756	5.4173	4.8878	79
80	8.7378	8.0357	7.4380	6.9230	6.4748	6.0810	5.4217	4.8913	80
81	8.7487	8.0449	7.4459	6.9299	6.4808	6.0863	5.4259	4.8947	81
82	8.7592	8.0538	7.4535	6.9365	6.4865	6.0914	5.4299	4.8980	82
83	8.7694	8.0624	7.4609	6.9429	6.4921	6.0963	5.4338	4.9012	83
84	8.7793	8.0707	7.4680	6.9491	6.4975	6.1011	5.4376	4.9043	84
85	8.7888	8.0788	7.4749	6.9550	6.5028	6.1057	5.4413	4.9072	85
86	8.7981	8.0866	7.4816	6.9608	6.5078	6.1102	5.4448	4.9101	86
87	8.8070	8.0942	7.4881	6.9664	6.5127	6.1145	5.4482	4.9129	87
88	8.8157	8.1015	7.4943	6.9718	6.5175	6.1187	5.4515	4.9156	88
89	8.8241	8.1086	7.5004	6.9771	6.5220	6.1227	5.4548	4.9182	89
90	8.8322	8.1154	7.5063	6.9822	6.5265	6.1266	5.4579	4.9207	90
91	8.8400	8.1221	7.5119	6.9871	6.5308	6.1304	5.4609	4.9232	91
92	8.8477	8.1285	7.5174	6.9918	6.5349	6.1341	5.4638	4.9255	92
93	8.8550	8.1347	7.5228	6.9964	6.5389	6.1376	5.4666	4.9278	93
94	8.8622	8.1407	7.5279	7.0009	6.5428	6.1410	5.4693	4.9300	94
95	8.8691	8.1466	7.5329	7.0052	6.5466	6.1444	5.4719	4.9322	95
96	8.8758	8.1522	7.5377	7.0094	6.5502	6.1476	5.4745	4.9342	96
97	8.8823	8.1577	7.5424	7.0134	6.5538	6.1507	5.4769	4.9362	97
98	8.8886	8.1630	7.5469	7.0173	6.5572	6.1537	5.4793	4.9382	98
99	8.8947	8.1681	7.5513	7.0211	6.5605	6.1566	5.4816	4.9401	99
100	8.9006	8.1731	7.5556	7.0248	6.5637	6.1594	5.4839	4.9419	100

YEARS' PURCHASE

				Rate Per Cent					
Yrs.	4	4.5	5	5.5	6	6.25	6.5	6.75	Yrs.
1	0.6809	0.6786	0.6763	0.6740	0.6718	0.6707	0.6695	0.6684	1
2	1.3508	1.3418	1.3328	1.3240	1.3153	1.3110	1.3067	1.3025	2
3	2.0095	1.9895	1.9699	1.9507	1.9318	1.9226	1.9134	1.9043	3
4	2.6566	2.6218	2.5879	2.5548	2.5226	2.5068	2.4912	2.4758	4
5	3.2921	3.2388	3.1872	3.1372	3.0888	3.0651	3.0418	3.0188	5
6	3.9158	3.8406	3.7683	3.6986	3.6314	3.5988	3.5667	3.5351	6
7	4.5275	4.4273	4.3314	4.2396	4.1516	4.1090	4.0672	4.0262	7
8	5.1272	4.9990	4.8771	4.7610	4.6503	4.5969	4.5446	4.4936	8
9	5.7146	5.5559	5.4057	5.2634	5.1285	5.0635	5.0003	4.9385	9
10	6.2898	6.0980	5.9176	5.7476	5.5870	5.5100	5.4352	5.3623	10
11	6.8527	6.6257	6.4132	6.2140	6.0267	5.9373	5.8504	5.7661	11
12	7.4033	7.1391	6.8930	6.6634	6.4485	6.3462	6.2471	6.1510	12
13	7.9416	7.6383	7.3573	7.0962	6.8531	6.7377	6.6260	6.5181	13
14	8.4675	8.1236	7.8065	7.5132	7.2412	7.1124	6.9882	6.8682	14
15	8.9811	8.5952	8.2410	7.9149	7.6136	7.4714	7.3344	7.2023	15
16	9.4825	9.0533	8.6612	8.3017	7.9708	7.8151	7.6654	7.5212	16
17	9.9717	9.4982	9.0675	8.6743	8.3137	8.1444	7.9819	7.8257	17
18	10.4488	9.9300	9.4603	9.0330	8.6427	8.4599	8.2847	8.1166	18
19	10.9139	10.3491	9.8399	9.3785	8.9584	8.7622	8.5744	8.3944	19
20	11.3670	10.7557	10.2068	9.7112	9.2615	9.0519	8.8516	8.6600	20
21	11.8083	11.1500	10.5612	10.0315	9.5524	9.3296	9.1169	8.9138	21
22	12.2380	11.5324	10.9036	10.3399	9.8316	9.5958	9.3710	9.1565	22
23	12.6562	11.9029	11.2343	10.6368	10.0997	9.8510	9.6142	9.3885	23
24	13.0629	12.2620	11.5537	10.9227	10.3570	10.0956	9.8471	9.6105	24
25	13.4584	12.6099	11.8620	11.1979	10.6041	10.3303	10.0702	9.8229	25
26	13.8429	12.9468	12.1597	11.4627	10.8414	10.5553	10.2839	10.0262	26
27	14.2165	13.2730	12.4470	11.7177	11.0692	10.7711	10.4887	10.2207	27
28	14.5794	13.5888	12.7243	11.9632	11.2880	10.9782	10.6849	10.4069	28
29	14.9318	13.8944	12.9918	12.1994	11.4980	11.1768	10.8730	10.5852	29
30	15.2738	14.1901	13.2500	12.4267	11.6998	11.3673	11.0532	10.7560	30
31	15.6057	14.4762	13.4991	12.6456	11.8936	11.5501	11.2260	10.9195	31
32	15.9277	14.7528	13.7393	12.8562	12.0797	11.7256	11.3916	11.0762	32
33	16.2400	15.0203	13.9711	13.0588	12.2584	11.8939	11.5505	11.2263	33
34	16.5427	15.2789	14.1945	13.2539	12.4301	12.0555	11.7028	11.3701	34
35	16.8361	15.5289	14.4100	13.4416	12.5951	12.2106	11.8489	11.5080	35
36	17.1204	15.7704	14.6178	13.6222	12.7535	12.3594	11.9890	11.6401	36
37	17.3958	16.0038	14.8181	13.7959	12.9057	12.5023	12.1234	11.7668	37
38	17.6625	16.2293	15.0112	13.9632	13.0519	12.6395	12.2523	11.8882	38
39	17.9208	16.4470	15.1973	14.1241	13.1924	12.7712	12.3761	12.0046	39
40	18.1707	16.6574	15.3767	14.2789	13.3274	12.8976	12.4948	12.1163	40
41	18.4126	16.8604	15.5496	14.4278	13.4570	13.0190	12.6087	12.2234	41
42	18.6467	17.0564	15.7161	14.5711	13.5816	13.1356	12.7180	12.3261	42
43	18.8731	17.2457	15.8767	14.7090	13.7013	13.2476	12.8229	12.4246	43
44	19.0920	17.4283	16.0313	14.8417	13.8164	13.3551	12.9236	12.5191	44
45	19.3037	17.6045	16.1803	14.9693	13.9269	13.4583	13.0202	12.6098	45
46	19.5084	17.7746	16.3238	15.0920	14.0331	13.5575	13.1130	12.6968	46
47	19.7062	17.9387	16.4621	15.2102	14.1352	13.6527	13.2021	12.7803	47
48	19.8973	18.0969	16.5953	15.3238	14.2332	13.7442	13.2876	12.8604	48
49	20.0820	18.2495	16.7236	15.4331	14.3275	13.8320	13.3697	12.9373	49
50	20.2604	18.3967	16.8471	15.5382	14.4181	13.9164	13.4486	13.0111	50

YEARS' PURCHASE

Yrs.	4	4.5	5	5.5	6	6.25	6.5	6.75	Yrs.
				Rate Per Cent					
51	20.4327	18.5387	16.9661	15.6394	14.5051	13.9975	13.5243	13.0819	51
52	20.5991	18.6756	17.0806	15.7367	14.5888	14.0754	13.5970	13.1500	52
53	20.7597	18.8075	17.1909	15.8302	14.6692	14.1502	13.6668	13.2152	53
54	20.9148	18.9347	17.2971	15.9203	14.7464	14.2221	13.7338	13.2779	54
55	21.0645	19.0573	17.3994	16.0069	14.8207	14.2912	13.7982	13.3381	55
56	21.2090	19.1755	17.4979	16.0902	14.8921	14.3575	13.8601	13.3959	56
57	21.3484	19.2894	17.5927	16.1703	14.9607	14.4213	13.9195	13.4514	57
58	21.4829	19.3992	17.6839	16.2473	15.0266	14.4826	13.9765	13.5046	58
59	21.6127	19.5049	17.7717	16.3214	15.0900	14.5414	14.0313	13.5558	59
60	21.7379	19.6068	17.8563	16.3927	15.1509	14.5980	14.0840	13.6050	60
61	21.8586	19.7050	17.9377	16.4613	15.2095	14.6523	14.1346	13.6521	61
62	21.9751	19.7996	18.0160	16.5272	15.2657	14.7046	14.1832	13.6975	62
63	22.0873	19.8907	18.0914	16.5907	15.3198	14.7547	14.2298	13.7410	63
64	22.1956	19.9784	18.1640	16.6517	15.3718	14.8030	14.2747	13.7828	64
65	22.3000	20.0629	18.2338	16.7104	15.4218	14.8493	14.3178	13.8230	65
66	22.4006	20.1443	18.3010	16.7668	15.4699	14.8939	14.3592	13.8616	66
67	22.4975	20.2227	18.3657	16.8211	15.5161	14.9367	14.3990	13.8987	67
68	22.5910	20.2982	18.4279	16.8733	15.5605	14.9778	14.4372	13.9343	68
69	22.6811	20.3709	18.4878	16.9235	15.6032	15.0174	14.4740	13.9685	69
70	22.7679	20.4409	18.5455	16.9717	15.6442	15.0554	14.5093	14.0014	70
71	22.8515	20.5083	18.6009	17.0182	15.6836	15.0919	14.5432	14.0330	71
72	22.9321	20.5732	18.6543	17.0628	15.7215	15.1270	14.5758	14.0633	72
73	23.0097	20.6356	18.7056	17.1058	15.7580	15.1607	14.6071	14.0925	73
74	23.0845	20.6958	18.7550	17.1471	15.7930	15.1932	14.6372	14.1205	74
75	23.1566	20.7537	18.8025	17.1868	15.8267	15.2243	14.6661	14.1474	75
76	23.2260	20.8094	18.8483	17.2250	15.8591	15.2543	14.6939	14.1733	76
77	23.2928	20.8630	18.8923	17.2617	15.8902	15.2831	14.7207	14.1981	77
78	23.3572	20.9146	18.9346	17.2970	15.9202	15.3108	14.7463	14.2220	78
79	23.4192	20.9643	18.9753	17.3310	15.9489	15.3374	14.7710	14.2450	79
80	23.4788	21.0121	19.0145	17.3637	15.9766	15.3630	14.7948	14.2671	80
81	23.5363	21.0582	19.0521	17.3951	16.0032	15.3876	14.8176	14.2883	81
82	23.5917	21.1025	19.0884	17.4253	16.0288	15.4112	14.8395	14.3086	82
83	23.6450	21.1451	19.1233	17.4543	16.0533	15.4339	14.8605	14.3282	83
84	23.6963	21.1861	19.1568	17.4823	16.0770	15.4558	14.8808	14.3470	84
85	23.7456	21.2256	19.1891	17.5092	16.0997	15.4768	14.9002	14.3651	85
86	23.7932	21.2636	19.2201	17.5350	16.1215	15.4969	14.9190	14.3825	86
87	23.8390	21.3001	19.2500	17.5598	16.1425	15.5164	14.9369	14.3992	87
88	23.8830	21.3353	19.2787	17.5837	16.1627	15.5350	14.9542	14.4153	88
89	23.9254	21.3691	19.3063	17.6067	16.1821	15.5529	14.9708	14.4307	89
90	23.9662	21.4017	19.3329	17.6288	16.2008	15.5702	14.9868	14.4456	90
91	24.0055	21.4330	19.3584	17.6500	16.2187	15.5867	15.0022	14.4598	91
92	24.0433	21.4631	19.3830	17.6705	16.2360	15.6027	15.0169	14.4735	92
93	24.0797	21.4921	19.4067	17.6901	16.2526	15.6180	15.0311	14.4867	93
94	24.1147	21.5200	19.4294	17.7090	16.2685	15.6327	15.0447	14.4994	94
95	24.1484	21.5468	19.4513	17.7272	16.2839	15.6469	15.0579	14.5116	95
96	24.1809	21.5726	19.4723	17.7447	16.2986	15.6605	15.0705	14.5233	96
97	24.2121	21.5975	19.4925	17.7615	16.3128	15.6736	15.0826	14.5345	97
98	24.2421	21.6214	19.5120	17.7776	16.3264	15.6861	15.0942	14.5453	98
99	24.2710	21.6444	19.5307	17.7931	16.3395	15.6982	15.1054	14.5557	99
100	24.2988	21.6665	19.5487	17.8081	16.3521	15.7099	15.1162	14.5657	100

YEARS' PURCHASE

Rate Per Cent

Yrs.	7	7.25	7.5	8	8.5	9	9.5	10	Yrs.
1	0.6673	0.6662	0.6651	0.6629	0.6607	0.6585	0.6564	0.6542	1
2	1.2982	1.2940	1.2899	1.2816	1.2734	1.2654	1.2574	1.2496	2
3	1.8952	1.8863	1.8774	1.8600	1.8428	1.8260	1.8095	1.7933	3
4	2.4605	2.4455	2.4306	2.4015	2.3730	2.3451	2.3180	2.2914	4
5	2.9962	2.9740	2.9520	2.9091	2.8674	2.8268	2.7874	2.7491	5
6	3.5042	3.4737	3.4438	3.3855	3.3292	3.2747	3.2219	3.1708	6
7	3.9861	3.9468	3.9082	3.8333	3.7612	3.6918	3.6249	3.5604	7
8	4.4437	4.3948	4.3471	4.2546	4.1660	4.0810	3.9994	3.9210	8
9	4.8783	4.8195	4.7621	4.6514	4.5457	4.4446	4.3480	4.2555	9
10	5.2914	5.2223	5.1550	5.0255	4.9023	4.7850	4.6732	4.5665	10
11	5.6842	5.6045	5.5271	5.3784	5.2376	5.1039	4.9769	4.8561	11
12	6.0579	5.9675	5.8798	5.7119	5.5533	5.4032	5.2611	5.1262	12
13	6.4136	6.3123	6.2143	6.0270	5.8507	5.6844	5.5273	5.3787	13
14	6.7523	6.6402	6.5317	6.3252	6.1313	5.9489	5.7771	5.6149	14
15	7.0749	6.9520	6.8332	6.6074	6.3961	6.1979	6.0116	5.8362	15
16	7.3824	7.2486	7.1196	6.8749	6.6464	6.4326	6.2322	6.0439	16
17	7.6756	7.5311	7.3919	7.1284	6.8831	6.6541	6.4398	6.2389	17
18	7.9551	7.8000	7.6508	7.3689	7.1071	6.8632	6.6355	6.4224	18
19	8.2219	8.0563	7.8972	7.5972	7.3192	7.0608	6.8200	6.5951	19
20	8.4764	8.3006	8.1318	7.8141	7.5203	7.2477	6.9943	6.7579	20
21	8.7195	8.5335	8.3552	8.0202	7.7109	7.4247	7.1589	6.9115	21
22	8.9515	8.7556	8.5681	8.2161	7.8919	7.5923	7.3146	7.0565	22
23	9.1732	8.9676	8.7709	8.4024	8.0637	7.7512	7.4620	7.1936	23
24	9.3850	9.1699	8.9644	8.5798	8.2269	7.9019	7.6015	7.3232	24
25	9.5875	9.3631	9.1489	8.7487	8.3820	8.0449	7.7338	7.4459	25
26	9.7810	9.5475	9.3250	8.9095	8.5296	8.1807	7.8592	7.5621	26
27	9.9660	9.7238	9.4930	9.0628	8.6700	8.3097	7.9782	7.6722	27
28	10.1430	9.8922	9.6534	9.2090	8.8036	8.4324	8.0913	7.7767	28
29	10.3123	10.0531	9.8067	9.3483	8.9309	8.5491	8.1986	7.8758	29
30	10.4743	10.2070	9.9531	9.4812	9.0521	8.6601	8.3007	7.9699	30
31	10.6294	10.3542	10.0929	9.6081	9.1677	8.7658	8.3978	8.0594	31
32	10.7778	10.4950	10.2267	9.7292	9.2778	8.8665	8.4901	8.1444	32
33	10.9198	10.6296	10.3545	9.8448	9.3829	8.9625	8.5781	8.2253	33
34	11.0559	10.7585	10.4767	9.9552	9.4832	9.0539	8.6618	8.3022	34
35	11.1862	10.8818	10.5936	10.0607	9.5789	9.1411	8.7416	8.3755	35
36	11.3110	10.9999	10.7055	10.1616	9.6703	9.2243	8.8176	8.4452	36
37	11.4305	11.1130	10.8126	10.2580	9.7575	9.3036	8.8901	8.5117	37
38	11.5451	11.2212	10.9150	10.3501	9.8409	9.3794	8.9592	8.5751	38
39	11.6549	11.3249	11.0131	10.4383	9.9205	9.4517	9.0252	8.6355	39
40	11.7601	11.4242	11.1070	10.5226	9.9966	9.5208	9.0881	8.6931	40
41	11.8609	11.5193	11.1969	10.6033	10.0694	9.5868	9.1482	8.7481	41
42	11.9576	11.6105	11.2830	10.6805	10.1390	9.6498	9.2057	8.8006	42
43	12.0503	11.6979	11.3655	10.7544	10.2056	9.7101	9.2605	8.8507	43
44	12.1392	11.7816	11.4445	10.8251	10.2693	9.7677	9.3129	8.8985	44
45	12.2244	11.8619	11.5203	10.8928	10.3302	9.8228	9.3630	8.9443	45
46	12.3062	11.9389	11.5928	10.9577	10.3885	9.8756	9.4109	8.9880	46
47	12.3846	12.0127	11.6624	11.0198	10.4443	9.9260	9.4567	9.0297	47
48	12.4598	12.0834	11.7291	11.0793	10.4978	9.9743	9.5005	9.0696	48
49	12.5320	12.1513	11.7930	11.1364	10.5490	10.0205	9.5424	9.1078	49
50	12.6012	12.2164	11.8543	11.1910	10.5980	10.0647	9.5825	9.1443	50

YEARS' PURCHASE

				Rate Per Cent					
Yrs.	7	7.25	7.5	8	8.5	9	9.5	10	Yrs.
51	12.6677	12.2788	11.9131	11.2434	10.6450	10.1070	9.6208	9.1793	51
52	12.7314	12.3387	11.9695	11.2936	10.6899	10.1476	9.6576	9.2127	52
53	12.7926	12.3962	12.0235	11.3417	10.7330	10.1864	9.6927	9.2447	53
54	12.8513	12.4513	12.0754	11.3878	10.7744	10.2236	9.7264	9.2753	54
55	12.9077	12.5042	12.1252	11.4321	10.8139	10.2592	9.7586	9.3046	55
56	12.9618	12.5550	12.1729	11.4745	10.8519	10.2934	9.7895	9.3327	56
57	13.0137	12.6037	12.2187	11.5152	10.8883	10.3261	9.8191	9.3596	57
58	13.0636	12.6504	12.2626	11.5542	10.9232	10.3575	9.8475	9.3854	58
59	13.1115	12.6953	12.3048	11.5916	10.9566	10.3876	9.8747	9.4101	59
60	13.1574	12.7384	12.3453	11.6276	10.9887	10.4164	9.9007	9.4337	60
61	13.2016	12.7798	12.3841	11.6620	11.0195	10.4440	9.9257	9.4564	61
62	13.2440	12.8195	12.4214	11.6951	11.0490	10.4705	9.9496	9.4781	62
63	13.2847	12.8576	12.4572	11.7268	11.0773	10.4960	9.9726	9.4989	63
64	13.3237	12.8942	12.4916	11.7572	11.1044	10.5203	9.9946	9.5189	64
65	13.3613	12.9294	12.5246	11.7865	11.1305	10.5437	10.0157	9.5381	65
66	13.3973	12.9632	12.5562	11.8145	11.1555	10.5662	10.0360	9.5564	66
67	13.4320	12.9956	12.5866	11.8414	11.1795	10.5877	10.0554	9.5740	67
68	13.4652	13.0267	12.6158	11.8673	11.2025	10.6083	10.0740	9.5909	68
69	13.4972	13.0566	12.6439	11.8921	11.2247	10.6282	10.0919	9.6071	69
70	13.5279	13.0853	12.6708	11.9159	11.2459	10.6472	10.1090	9.6226	70
71	13.5573	13.1129	12.6967	11.9388	11.2662	10.6654	10.1255	9.6376	71
72	13.5857	13.1394	12.7215	11.9607	11.2858	10.6830	10.1413	9.6519	72
73	13.6129	13.1648	12.7454	11.9818	11.3046	10.6998	10.1564	9.6656	73
74	13.6390	13.1893	12.7683	12.0021	11.3226	10.7159	10.1710	9.6788	74
75	13.6641	13.2128	12.7903	12.0215	11.3399	10.7314	10.1849	9.6914	75
76	13.6883	13.2353	12.8114	12.0402	11.3565	10.7463	10.1983	9.7035	76
77	13.7115	13.2570	12.8317	12.0581	11.3725	10.7606	10.2112	9.7152	77
78	13.7337	13.2778	12.8513	12.0753	11.3878	10.7743	10.2235	9.7264	78
79	13.7551	13.2979	12.8700	12.0919	11.4025	10.7875	10.2354	9.7371	79
80	13.7757	13.3171	12.8880	12.1078	11.4166	10.8001	10.2468	9.7474	80
81	13.7955	13.3356	12.9053	12.1230	11.4302	10.8123	10.2577	9.7573	81
82	13.8145	13.3533	12.9219	12.1377	11.4432	10.8239	10.2682	9.7668	82
83	13.8327	13.3704	12.9379	12.1518	11.4558	10.8351	10.2783	9.7759	83
84	13.8503	13.3867	12.9532	12.1653	11.4678	10.8459	10.2880	9.7847	84
85	13.8671	13.4025	12.9680	12.1783	11.4793	10.8562	10.2973	9.7931	85
86	13.8833	13.4176	12.9822	12.1908	11.4904	10.8662	10.3062	9.8012	86
87	13.8989	13.4322	12.9958	12.2028	11.5011	10.8757	10.3148	9.8089	87
88	13.9139	13.4461	13.0088	12.2144	11.5114	10.8849	10.3230	9.8164	88
89	13.9282	13.4596	13.0214	12.2255	11.5212	10.8937	10.3309	9.8235	89
90	13.9421	13.4725	13.0335	12.2361	11.5307	10.9021	10.3386	9.8304	90
91	13.9554	13.4849	13.0451	12.2463	11.5397	10.9102	10.3459	9.8370	91
92	13.9681	13.4968	13.0563	12.2562	11.5485	10.9180	10.3529	9.8433	92
93	13.9804	13.5083	13.0670	12.2656	11.5569	10.9255	10.3596	9.8494	93
94	13.9922	13.5193	13.0773	12.2747	11.5649	10.9327	10.3661	9.8553	94
95	14.0035	13.5299	13.0872	12.2834	11.5727	10.9397	10.3723	9.8609	95
96	14.0144	13.5400	13.0967	12.2918	11.5801	10.9463	10.3783	9.8663	96
97	14.0249	13.5498	13.1059	12.2999	11.5873	10.9527	10.3840	9.8715	97
98	14.0350	13.5592	13.1147	12.3076	11.5941	10.9588	10.3896	9.8765	98
99	14.0447	13.5683	13.1231	12.3151	11.6007	10.9647	10.3949	9.8813	99
100	14.0540	13.5769	13.1312	12.3222	11.6071	10.9704	10.4000	9.8859	100

YEARS' PURCHASE

				Rate Per Cent					
Yrs.	11	12	13	14	15	16	18	20	Yrs.
1	0.6500	0.6458	0.6416	0.6375	0.6335	0.6295	0.6217	0.6140	1
2	1.2341	1.2191	1.2044	1.1901	1.1761	1.1624	1.1360	1.1108	2
3	1.7617	1.7312	1.7017	1.6732	1.6457	1.6191	1.5683	1.5206	3
4	2.2401	2.1910	2.1440	2.0990	2.0559	2.0144	1.9364	1.8642	4
5	2.6756	2.6058	2.5397	2.4768	2.4169	2.3599	2.2535	2.1563	5
6	3.0734	2.9817	2.8954	2.8139	2.7369	2.6640	2.5292	2.4075	6
7	3.4380	3.3237	3.2168	3.1165	3.0223	2.9337	2.7711	2.6256	7
8	3.7730	3.6358	3.5083	3.3894	3.2783	3.1742	2.9847	2.8166	8
9	4.0818	3.9217	3.7737	3.6365	3.5089	3.3899	3.1747	2.9852	9
10	4.3671	4.1843	4.0163	3.8612	3.7177	3.5844	3.3446	3.1349	10
11	4.6312	4.4262	4.2386	4.0662	3.9074	3.7604	3.4974	3.2688	11
12	4.8763	4.6496	4.4430	4.2540	4.0804	3.9204	3.6354	3.3890	12
13	5.1041	4.8563	4.6314	4.4264	4.2387	4.0664	3.7605	3.4975	13
14	5.3164	5.0480	4.8054	4.5851	4.3841	4.1999	3.8745	3.5958	14
15	5.5144	5.2262	4.9666	4.7316	4.5178	4.3226	3.9786	3.6854	15
16	5.6994	5.3921	5.1162	4.8672	4.6413	4.4354	4.0740	3.7671	16
17	5.8726	5.5468	5.2553	4.9929	4.7555	4.5396	4.1617	3.8420	17
18	6.0348	5.6914	5.3849	5.1097	4.8613	4.6360	4.2426	3.9108	18
19	6.1871	5.8266	5.5058	5.2185	4.9597	4.7253	4.3173	3.9741	19
20	6.3302	5.9533	5.6188	5.3199	5.0512	4.8083	4.3865	4.0327	20
21	6.4647	6.0722	5.7246	5.4146	5.1365	4.8855	4.4507	4.0869	21
22	6.5914	6.1838	5.8237	5.5032	5.2161	4.9575	4.5103	4.1371	22
23	6.7108	6.2888	5.9167	5.5862	5.2906	5.0248	4.5659	4.1839	23
24	6.8235	6.3876	6.0041	5.6640	5.3604	5.0877	4.6178	4.2274	24
25	6.9299	6.4808	6.0863	5.7371	5.4259	5.1466	4.6663	4.2680	25
26	7.0304	6.5686	6.1637	5.8059	5.4873	5.2019	4.7117	4.3059	26
27	7.1255	6.6516	6.2367	5.8706	5.5451	5.2537	4.7542	4.3414	27
28	7.2155	6.7299	6.3056	5.9316	5.5994	5.3025	4.7941	4.3746	28
29	7.3008	6.8040	6.3706	5.9891	5.6506	5.3484	4.8316	4.4058	29
30	7.3816	6.8742	6.4320	6.0433	5.6989	5.3917	4.8669	4.4351	30
31	7.4583	6.9406	6.4902	6.0946	5.7445	5.4324	4.9001	4.4627	31
32	7.5310	7.0036	6.5452	6.1431	5.7876	5.4709	4.9314	4.4887	32
33	7.6001	7.0633	6.5973	6.1890	5.8283	5.5073	4.9609	4.5131	33
34	7.6658	7.1200	6.6467	6.2325	5.8668	5.5417	4.9888	4.5362	34
35	7.7282	7.1738	6.6936	6.2737	5.9033	5.5743	5.0151	4.5580	35
36	7.7876	7.2249	6.7381	6.3127	5.9379	5.6051	5.0401	4.5785	36
37	7.8441	7.2735	6.7803	6.3498	5.9707	5.6343	5.0637	4.5980	37
38	7.8978	7.3197	6.8205	6.3850	6.0018	5.6620	5.0860	4.6164	38
39	7.9491	7.3637	6.8587	6.4184	6.0313	5.6882	5.1072	4.6339	39
40	7.9979	7.4056	6.8950	6.4502	6.0594	5.7132	5.1273	4.6504	40
41	8.0444	7.4454	6.9295	6.4804	6.0860	5.7369	5.1464	4.6661	41
42	8.0887	7.4834	6.9624	6.5092	6.1114	5.7594	5.1645	4.6810	42
43	8.1310	7.5196	6.9937	6.5366	6.1355	5.7808	5.1817	4.6952	43
44	8.1714	7.5541	7.0236	6.5626	6.1585	5.8012	5.1981	4.7086	44
45	8.2099	7.5871	7.0520	6.5875	6.1803	5.8206	5.2137	4.7214	45
46	8.2467	7.6185	7.0791	6.6111	6.2012	5.8391	5.2285	4.7335	46
47	8.2819	7.6484	7.1050	6.6337	6.2210	5.8567	5.2426	4.7451	47
48	8.3155	7.6771	7.1297	6.6552	6.2399	5.8734	5.2560	4.7561	48
49	8.3475	7.7044	7.1533	6.6758	6.2580	5.8894	5.2688	4.7665	49
50	8.3782	7.7305	7.1758	6.6953	6.2752	5.9047	5.2810	4.7765	50

YEARS' PURCHASE

				Rate Per Cent					
Yrs.	11	12	13	14	15	16	18	20	Yrs.
51	8.4075	7.7555	7.1973	6.7141	6.2916	5.9192	5.2927	4.7860	51
52	8.4356	7.7793	7.2178	6.7319	6.3073	5.9331	5.3037	4.7951	52
53	8.4624	7.8021	7.2375	6.7490	6.3223	5.9464	5.3143	4.8038	53
54	8.4880	7.8239	7.2562	6.7653	6.3366	5.9590	5.3244	4.8120	54
55	8.5126	7.8448	7.2741	6.7809	6.3503	5.9711	5.3341	4.8199	55
56	8.5361	7.8647	7.2913	6.7958	6.3634	5.9827	5.3433	4.8274	56
57	8.5586	7.8838	7.3077	6.8100	6.3758	5.9937	5.3521	4.8346	57
58	8.5801	7.9021	7.3234	6.8237	6.3878	6.0043	5.3605	4.8415	58
59	8.6007	7.9196	7.3384	6.8367	6.3992	6.0143	5.3686	4.8480	59
60	8.6205	7.9363	7.3528	6.8492	6.4101	6.0240	5.3763	4.8543	60
61	8.6394	7.9524	7.3666	6.8611	6.4206	6.0332	5.3836	4.8603	61
62	8.6575	7.9677	7.3797	6.8726	6.4306	6.0421	5.3907	4.8660	62
63	8.6749	7.9825	7.3924	6.8835	6.4402	6.0505	5.3974	4.8715	63
64	8.6916	7.9965	7.4044	6.8940	6.4494	6.0586	5.4038	4.8768	64
65	8.7075	8.0101	7.4160	6.9040	6.4581	6.0664	5.4100	4.8818	65
66	8.7228	8.0230	7.4271	6.9136	6.4666	6.0738	5.4159	4.8866	66
67	8.7375	8.0354	7.4377	6.9228	6.4746	6.0809	5.4215	4.8912	67
68	8.7516	8.0473	7.4479	6.9317	6.4823	6.0877	5.4270	4.8956	68
69	8.7650	8.0587	7.4577	6.9401	6.4897	6.0942	5.4321	4.8998	69
70	8.7780	8.0696	7.4671	6.9482	6.4968	6.1005	5.4371	4.9038	70
71	8.7904	8.0801	7.4760	6.9560	6.5036	6.1065	5.4419	4.9077	71
72	8.8023	8.0902	7.4846	6.9634	6.5101	6.1122	5.4464	4.9114	72
73	8.8137	8.0998	7.4929	6.9706	6.5164	6.1177	5.4508	4.9150	73
74	8.8246	8.1090	7.5008	6.9774	6.5223	6.1230	5.4550	4.9184	74
75	8.8351	8.1179	7.5084	6.9840	6.5281	6.1280	5.4590	4.9216	75
76	8.8452	8.1264	7.5157	6.9903	6.5336	6.1329	5.4628	4.9248	76
77	8.8549	8.1346	7.5227	6.9963	6.5389	6.1375	5.4665	4.9278	77
78	8.8642	8.1424	7.5294	7.0021	6.5439	6.1420	5.4701	4.9306	78
79	8.8731	8.1500	7.5358	7.0077	6.5488	6.1463	5.4735	4.9334	79
80	8.8817	8.1572	7.5420	7.0130	6.5534	6.1504	5.4767	4.9360	80
81	8.8899	8.1641	7.5479	7.0182	6.5579	6.1543	5.4798	4.9386	81
82	8.8978	8.1707	7.5536	7.0231	6.5622	6.1581	5.4828	4.9410	82
83	8.9053	8.1771	7.5590	7.0278	6.5663	6.1617	5.4857	4.9433	83
84	8.9126	8.1833	7.5643	7.0323	6.5703	6.1652	5.4885	4.9456	84
85	8.9196	8.1891	7.5693	7.0367	6.5741	6.1685	5.4911	4.9477	85
86	8.9263	8.1948	7.5741	7.0408	6.5777	6.1717	5.4936	4.9498	86
87	8.9327	8.2002	7.5787	7.0448	6.5812	6.1748	5.4961	4.9518	87
88	8.9389	8.2054	7.5832	7.0487	6.5845	6.1778	5.4984	4.9537	88
89	8.9448	8.2104	7.5875	7.0524	6.5878	6.1806	5.5007	4.9555	89
90	8.9505	8.2152	7.5916	7.0559	6.5909	6.1833	5.5028	4.9572	90
91	8.9560	8.2198	7.5955	7.0593	6.5938	6.1859	5.5049	4.9589	91
92	8.9613	8.2243	7.5993	7.0626	6.5967	6.1884	5.5069	4.9605	92
93	8.9663	8.2285	7.6029	7.0657	6.5994	6.1908	5.5088	4.9621	93
94	8.9712	8.2326	7.6064	7.0687	6.6020	6.1932	5.5106	4.9636	94
95	8.9758	8.2365	7.6097	7.0716	6.6046	6.1954	5.5124	4.9650	95
96	8.9803	8.2403	7.6130	7.0744	6.6070	6.1975	5.5140	4.9664	96
97	8.9846	8.2439	7.6160	7.0771	6.6093	6.1996	5.5157	4.9677	97
98	8.9887	8.2474	7.6190	7.0796	6.6115	6.2015	5.5172	4.9689	98
99	8.9927	8.2507	7.6219	7.0821	6.6137	6.2034	5.5187	4.9701	99
100	8.9965	8.2539	7.6246	7.0844	6.6158	6.2052	5.5202	4.9713	100

YEARS' PURCHASE

(DUAL RATE % PRINCIPLE)

OR

PRESENT VALUE OF
ONE POUND PER ANNUM

*receivable at the end of each year after
allowing for a sinking fund at a given
rate to replace the invested capital and
for the effect of income tax at* **35%** *on
that part of the income used to provide
the annual sinking fund instalment.*

AT RATES OF INTEREST FROM

4% to 20%

AND

ALLOWING FOR THE POSSIBLE INVESTMENT

OF SINKING FUNDS AT

2·5%, 3% and 4%

INCOME TAX at 35%

Note:–Tables of Years' Purchase in which no allowance has been made for the effect of
income tax on that part of the income used to provide the annual sinking fund
instalment will be found on pages 53 to 71.

YEARS' PURCHASE

Rate Per Cent

Yrs.	4	4.5	5	5.5	6	6.25	6.5	6.75	Yrs.
1	0.6335	0.6315	0.6295	0.6276	0.6256	0.6246	0.6237	0.6227	1
2	1.2504	1.2426	1.2350	1.2274	1.2199	1.2162	1.2125	1.2088	2
3	1.8511	1.8342	1.8175	1.8011	1.7850	1.7771	1.7693	1.7615	3
4	2.4361	2.4068	2.3782	2.3502	2.3229	2.3095	2.2963	2.2832	4
5	3.0058	2.9613	2.9181	2.8761	2.8354	2.8154	2.7957	2.7763	5
6	3.5607	3.4984	3.4382	3.3801	3.3240	3.2966	3.2696	3.2431	6
7	4.1011	4.0187	3.9395	3.8634	3.7902	3.7546	3.7197	3.6854	7
8	4.6274	4.5228	4.4228	4.3271	4.2354	4.1911	4.1476	4.1050	8
9	5.1401	5.0113	4.8888	4.7722	4.6609	4.6073	4.5548	4.5035	9
10	5.6395	5.4848	5.3384	5.1996	5.0679	5.0045	4.9426	4.8823	10
11	6.1260	5.9439	5.7723	5.6104	5.4573	5.3839	5.3124	5.2427	11
12	6.5998	6.3890	6.1912	6.0053	5.8303	5.7465	5.6651	5.5860	12
13	7.0615	6.8207	6.5958	6.3852	6.1876	6.0934	6.0019	5.9132	13
14	7.5113	7.2394	6.9865	6.7507	6.5303	6.4254	6.3238	6.2254	14
15	7.9495	7.6456	7.3641	7.1025	6.8590	6.7433	6.6315	6.5234	15
16	8.3764	8.0397	7.7290	7.4414	7.1745	7.0481	6.9260	6.8081	16
17	8.7924	8.4221	8.0818	7.7679	7.4775	7.3403	7.2080	7.0804	17
18	9.1977	8.7933	8.4230	8.0826	7.7686	7.6206	7.4781	7.3409	18
19	9.5926	9.1536	8.7530	8.3859	8.0485	7.8897	7.7371	7.5903	19
20	9.9774	9.5033	9.0722	8.6786	8.3176	8.1482	7.9855	7.8292	20
21	10.3524	9.8429	9.3812	8.9609	8.5766	8.3966	8.2240	8.0583	21
22	10.7178	10.1727	9.6803	9.2334	8.8259	8.6354	8.4529	8.2780	22
23	11.0739	10.4929	9.9699	9.4965	9.0660	8.8651	8.6729	8.4888	23
24	11.4210	10.8040	10.2503	9.7506	9.2973	9.0861	8.8843	8.6913	24
25	11.7592	11.1062	10.5219	9.9960	9.5202	9.2989	9.0876	8.8857	25
26	12.0888	11.3998	10.7850	10.2332	9.7351	9.5038	9.2832	9.0727	26
27	12.4101	11.6850	11.0400	10.4625	9.9424	9.7012	9.4715	9.2524	27
28	12.7232	11.9622	11.2871	10.6841	10.1423	9.8915	9.6528	9.4254	28
29	13.0283	12.2315	11.5266	10.8985	10.3353	10.0750	9.8274	9.5918	29
30	13.3258	12.4934	11.7588	11.1059	10.5216	10.2519	9.9957	9.7521	30
31	13.6157	12.7478	11.9840	11.3065	10.7015	10.4227	10.1580	9.9064	31
32	13.8983	12.9952	12.2024	11.5007	10.8753	10.5875	10.3145	10.0552	32
33	14.1738	13.2358	12.4142	11.6887	11.0433	10.7466	10.4654	10.1986	33
34	14.4423	13.4696	12.6197	11.8707	11.2056	10.9002	10.6111	10.3369	34
35	14.7040	13.6970	12.8191	12.0469	11.3625	11.0487	10.7517	10.4703	35
36	14.9592	13.9182	13.0126	12.2177	11.5143	11.1921	10.8875	10.5990	36
37	15.2079	14.1332	13.2004	12.3831	11.6611	11.3308	11.0186	10.7233	37
38	15.4504	14.3424	13.3827	12.5434	11.8031	11.4648	11.1454	10.8433	38
39	15.6868	14.5459	13.5597	12.6987	11.9406	11.5945	11.2679	10.9592	39
40	15.9172	14.7438	13.7316	12.8493	12.0737	11.7199	11.3863	11.0711	40
41	16.1419	14.9364	13.8984	12.9954	12.2025	11.8412	11.5008	11.1794	41
42	16.3609	15.1238	14.0605	13.1370	12.3272	11.9587	11.6115	11.2840	42
43	16.5745	15.3060	14.2179	13.2743	12.4481	12.0724	11.7187	11.3852	43
44	16.7827	15.4834	14.3709	13.4075	12.5652	12.1825	11.8224	11.4830	44
45	16.9857	15.6560	14.5195	13.5367	12.6786	12.2891	11.9228	11.5777	45
46	17.1836	15.8240	14.6638	13.6621	12.7885	12.3923	12.0200	11.6693	46
47	17.3766	15.9875	14.8041	13.7838	12.8951	12.4924	12.1141	11.7580	47
48	17.5647	16.1467	14.9405	13.9020	12.9984	12.5893	12.2052	11.8438	48
49	17.7482	16.3016	15.0730	14.0166	13.0986	12.6833	12.2935	11.9269	49
50	17.9270	16.4523	15.2018	14.1280	13.1958	12.7744	12.3791	12.0075	50

YEARS' PURCHASE

			Rate	Per	Cent				
Yrs.	4	4.5	5	5.5	6	6.25	6.5	6.75	Yrs.
51	18.1015	16.5991	15.3270	14.2361	13.2901	12.8627	12.4620	12.0854	51
52	18.2715	16.7420	15.4488	14.3410	13.3815	12.9483	12.5423	12.1610	52
53	18.4374	16.8811	15.5672	14.4430	13.4702	13.0314	12.6203	12.2343	53
54	18.5991	17.0166	15.6823	14.5420	13.5564	13.1120	12.6958	12.3052	54
55	18.7567	17.1485	15.7943	14.6383	13.6399	13.1902	12.7691	12.3741	55
56	18.9105	17.2769	15.9031	14.7317	13.7211	13.2660	12.8402	12.4408	56
57	19.0604	17.4020	16.0090	14.8226	13.7998	13.3396	12.9091	12.5055	57
58	19.2067	17.5238	16.1121	14.9108	13.8763	13.4111	12.9760	12.5683	58
59	19.3492	17.6424	16.2123	14.9966	13.9506	13.4804	13.0409	12.6292	59
60	19.4883	17.7579	16.3098	15.0800	14.0227	13.5478	13.1039	12.6883	60
61	19.6239	17.8704	16.4046	15.1611	14.0928	13.6132	13.1651	12.7456	61
62	19.7561	17.9800	16.4969	15.2399	14.1608	13.6767	13.2245	12.8013	62
63	19.8850	18.0868	16.5868	15.3165	14.2270	13.7383	13.2821	12.8553	63
64	20.0108	18.1907	16.6742	15.3910	14.2912	13.7982	13.3381	12.9077	64
65	20.1334	18.2920	16.7592	15.4634	14.3537	13.8564	13.3925	12.9586	65
66	20.2530	18.3907	16.8420	15.5339	14.4143	13.9130	13.4453	13.0081	66
67	20.3697	18.4868	16.9226	15.6024	14.4733	13.9679	13.4966	13.0561	67
68	20.4834	18.5805	17.0010	15.6691	14.5307	14.0213	13.5465	13.1027	68
69	20.5943	18.6717	17.0774	15.7339	14.5864	14.0732	13.5949	13.1480	69
70	20.7025	18.7606	17.1517	15.7970	14.6406	14.1236	13.6420	13.1920	70
71	20.8081	18.8472	17.2241	15.8583	14.6933	14.1727	13.6877	13.2348	71
72	20.9110	18.9316	17.2945	15.9180	14.7445	14.2203	13.7322	13.2764	72
73	21.0113	19.0138	17.3631	15.9761	14.7944	14.2667	13.7754	13.3168	73
74	21.1092	19.0939	17.4299	16.0327	14.8428	14.3118	13.8174	13.3560	74
75	21.2047	19.1720	17.4949	16.0877	14.8900	14.3556	13.8582	13.3942	75
76	21.2978	19.2481	17.5583	16.1412	14.9358	14.3982	13.8979	13.4313	76
77	21.3886	19.3223	17.6200	16.1933	14.9804	14.4396	13.9365	13.4673	77
78	21.4772	19.3945	17.6800	16.2441	15.0238	14.4800	13.9741	13.5024	78
79	21.5636	19.4649	17.7385	16.2934	15.0660	14.5192	14.0106	13.5365	79
80	21.6479	19.5336	17.7955	16.3415	15.1071	14.5573	14.0461	13.5696	80
81	21.7301	19.6005	17.8510	16.3883	15.1471	14.5944	14.0807	13.6019	81
82	21.8102	19.6657	17.9051	16.4338	15.1860	14.6306	14.1143	13.6333	82
83	21.8884	19.7292	17.9577	16.4782	15.2239	14.6657	14.1470	13.6638	83
84	21.9647	19.7911	18.0090	16.5214	15.2607	14.6999	14.1788	13.6934	84
85	22.0391	19.8515	18.0590	16.5634	15.2966	14.7332	14.2098	13.7223	85
86	22.1116	19.9104	18.1077	16.6044	15.3315	14.7656	14.2399	13.7504	86
87	22.1824	19.9677	18.1551	16.6442	15.3655	14.7971	14.2692	13.7777	87
88	22.2514	20.0236	18.2013	16.6831	15.3986	14.8278	14.2978	13.8043	88
89	22.3187	20.0781	18.2464	16.7209	15.4308	14.8576	14.3255	13.8302	89
90	22.3844	20.1313	18.2902	16.7577	15.4622	14.8867	14.3526	13.8554	90
91	22.4484	20.1831	18.3330	16.7936	15.4927	14.9150	14.3789	13.8799	91
92	22.5109	20.2335	18.3746	16.8285	15.5224	14.9426	14.4045	13.9038	92
93	22.5719	20.2828	18.4152	16.8626	15.5514	14.9694	14.4294	13.9270	93
94	22.6313	20.3308	18.4548	16.8957	15.5796	14.9955	14.4537	13.9496	94
95	22.6893	20.3775	18.4933	16.9280	15.6070	15.0210	14.4773	13.9716	95
96	22.7459	20.4232	18.5309	16.9595	15.6338	15.0457	14.5003	13.9931	96
97	22.8010	20.4676	18.5675	16.9901	15.6598	15.0699	14.5227	14.0139	97
98	22.8549	20.5110	18.6031	17.0200	15.6852	15.0933	14.5445	14.0342	98
99	22.9074	20.5533	18.6379	17.0491	15.7099	15.1162	14.5658	14.0540	99
100	22.9586	20.5945	18.6718	17.0775	15.7340	15.1385	14.5865	14.0733	100

YEARS' PURCHASE

Rate Per Cent

Yrs.	7	7.25	7.5	8	8.5	9	9.5	10	Yrs.
1	0.6217	0.6207	0.6198	0.6179	0.6160	0.6141	0.6122	0.6103	1
2	1.2052	1.2016	1.1980	1.1909	1.1838	1.1768	1.1700	1.1632	2
3	1.7537	1.7461	1.7385	1.7235	1.7088	1.6943	1.6801	1.6661	3
4	2.2702	2.2574	2.2447	2.2198	2.1954	2.1716	2.1483	2.1254	4
5	2.7572	2.7383	2.7197	2.6832	2.6477	2.6131	2.5794	2.5466	5
6	3.2170	3.1914	3.1661	3.1168	3.0689	3.0226	2.9776	2.9339	6
7	3.6518	3.6187	3.5863	3.5231	3.4621	3.4032	3.3463	3.2912	7
8	4.0633	4.0225	3.9824	3.9047	3.8299	3.7579	3.6886	3.6218	8
9	4.4534	4.4043	4.3564	4.2635	4.1745	4.0892	4.0072	3.9285	9
10	4.8234	4.7660	4.7098	4.6015	4.4980	4.3991	4.3044	4.2137	10
11	5.1749	5.1088	5.0444	4.9203	4.8022	4.6896	4.5821	4.4795	11
12	5.5091	5.4342	5.3614	5.2214	5.0886	4.9623	4.8422	4.7277	12
13	5.8271	5.7434	5.6621	5.5062	5.3587	5.2189	5.0861	4.9600	13
14	6.1300	6.0374	5.9477	5.7759	5.6138	5.4605	5.3154	5.1778	14
15	6.4187	6.3173	6.2191	6.0316	5.8550	5.6885	5.5311	5.3823	15
16	6.6942	6.5840	6.4774	6.2742	6.0834	5.9038	5.7345	5.5747	16
17	6.9572	6.8383	6.7234	6.5047	6.2998	6.1074	5.9265	5.7559	17
18	7.2086	7.0810	6.9578	6.7239	6.5052	6.3003	6.1079	5.9269	18
19	7.4490	7.3128	7.1815	6.9325	6.7003	6.4831	6.2795	6.0884	19
20	7.6789	7.5343	7.3950	7.1313	6.8858	6.6566	6.4422	6.2412	20
21	7.8992	7.7462	7.5990	7.3209	7.0624	6.8215	6.5965	6.3859	21
22	8.1101	7.9490	7.7941	7.5017	7.2305	6.9782	6.7430	6.5231	22
23	8.3124	8.1432	7.9807	7.6745	7.3909	7.1275	6.8822	6.6533	23
24	8.5064	8.3293	8.1594	7.8396	7.5439	7.2697	7.0147	6.7770	24
25	8.6926	8.5078	8.3306	7.9974	7.6899	7.4052	7.1408	6.8947	25
26	8.8714	8.6790	8.4947	8.1486	7.8296	7.5346	7.2610	7.0067	26
27	9.0432	8.8433	8.6520	8.2933	7.9631	7.6582	7.3757	7.1134	27
28	9.2084	9.0012	8.8031	8.4319	8.0908	7.7762	7.4852	7.2152	28
29	9.3672	9.1528	8.9481	8.5649	8.2132	7.8892	7.5898	7.3123	29
30	9.5200	9.2986	9.0874	8.6924	8.3304	7.9973	7.6898	7.4051	30
31	9.6670	9.4389	9.2213	8.8149	8.4428	8.1008	7.7855	7.4937	31
32	9.8086	9.5738	9.3501	8.9325	8.5506	8.2000	7.8770	7.5786	32
33	9.9450	9.7038	9.4739	9.0454	8.6540	8.2951	7.9648	7.6597	33
34	10.0765	9.8289	9.5931	9.1541	8.7534	8.3864	8.0489	7.7375	34
35	10.2032	9.9494	9.7079	9.2585	8.8489	8.4740	8.1295	7.8120	35
36	10.3254	10.0656	9.8185	9.3590	8.9407	8.5581	8.2069	7.8834	36
37	10.4433	10.1776	9.9250	9.4558	9.0289	8.6389	8.2812	7.9520	37
38	10.5571	10.2856	10.0278	9.5490	9.1138	8.7166	8.3526	8.0178	38
39	10.6669	10.3898	10.1268	9.6387	9.1956	8.7914	8.4212	8.0809	39
40	10.7730	10.4904	10.2223	9.7253	9.2743	8.8633	8.4872	8.1417	40
41	10.8754	10.5876	10.3145	9.8087	9.3501	8.9325	8.5506	8.2000	41
42	10.9744	10.6813	10.4035	9.8891	9.4232	8.9992	8.6117	8.2562	42
43	11.0701	10.7720	10.4895	9.9667	9.4936	9.0634	8.6705	8.3102	43
44	11.1626	10.8595	10.5725	10.0417	9.5616	9.1253	8.7271	8.3622	44
45	11.2520	10.9441	10.6527	10.1140	9.6271	9.1850	8.7817	8.4123	45
46	11.3385	11.0260	10.7302	10.1838	9.6904	9.2426	8.8343	8.4606	46
47	11.4222	11.1051	10.8051	10.2513	9.7515	9.2981	8.8850	8.5071	47
48	11.5032	11.1816	10.8776	10.3165	9.8104	9.3517	8.9340	8.5520	48
49	11.5816	11.2557	10.9476	10.3795	9.8674	9.4035	8.9812	8.5952	49
50	11.6575	11.3274	11.0154	10.4404	9.9224	9.4534	9.0268	8.6369	50

YEARS' PURCHASE

Rate Per Cent

Yrs.	7	7.25	7.5	8	8.5	9	9.5	10	Yrs.
51	11.7310	11.3968	11.0811	10.4993	9.9756	9.5017	9.0708	8.6772	51
52	11.8022	11.4640	11.1446	10.5563	10.0271	9.5484	9.1133	8.7161	52
53	11.8712	11.5290	11.2060	10.6115	10.0768	9.5935	9.1544	8.7537	53
54	11.9380	11.5920	11.2656	10.6648	10.1249	9.6371	9.1940	8.7900	54
55	12.0028	11.6531	11.3232	10.7165	10.1715	9.6792	9.2324	8.8250	55
56	12.0655	11.7123	11.3791	10.7665	10.2165	9.7200	9.2695	8.8589	56
57	12.1264	11.7696	11.4332	10.8149	10.2601	9.7595	9.3054	8.8917	57
58	12.1854	11.8252	11.4856	10.8619	10.3023	9.7976	9.3401	8.9234	58
59	12.2427	11.8791	11.5365	10.9073	10.3432	9.8346	9.3737	8.9540	59
60	12.2982	11.9313	11.5858	10.9514	10.3828	9.8704	9.4062	8.9837	60
61	12.3520	11.9820	11.6335	10.9940	10.4212	9.9051	9.4377	9.0124	61
62	12.4043	12.0312	11.6799	11.0354	10.4584	9.9386	9.4681	9.0402	62
63	12.4550	12.0789	11.7248	11.0755	10.4944	9.9712	9.4977	9.0671	63
64	12.5042	12.1252	11.7684	11.1144	10.5293	10.0027	9.5263	9.0931	64
65	12.5520	12.1701	11.8107	11.1522	10.5632	10.0332	9.5540	9.1184	65
66	12.5984	12.2137	11.8518	11.1888	10.5960	10.0629	9.5808	9.1428	66
67	12.6434	12.2560	11.8916	11.2243	10.6278	10.0916	9.6068	9.1665	67
68	12.6871	12.2971	11.9303	11.2587	10.6587	10.1194	9.6321	9.1895	68
69	12.7296	12.3370	11.9679	11.2922	10.6887	10.1464	9.6565	9.2117	69
70	12.7709	12.3757	12.0043	11.3246	10.7177	10.1726	9.6802	9.2333	70
71	12.8109	12.4134	12.0397	11.3561	10.7460	10.1980	9.7032	9.2543	71
72	12.8499	12.4499	12.0741	11.3867	10.7733	10.2227	9.7256	9.2746	72
73	12.8877	12.4854	12.1075	11.4164	10.7999	10.2466	9.7472	9.2943	73
74	12.9245	12.5199	12.1400	11.4452	10.8257	10.2698	9.7682	9.3134	74
75	12.9602	12.5535	12.1715	11.4732	10.8508	10.2924	9.7886	9.3319	75
76	12.9949	12.5860	12.2021	11.5004	10.8751	10.3143	9.8084	9.3499	76
77	13.0287	12.6177	12.2318	11.5269	10.8987	10.3355	9.8276	9.3673	77
78	13.0615	12.6485	12.2608	11.5526	10.9217	10.3562	9.8463	9.3843	78
79	13.0934	12.6784	12.2889	11.5775	10.9440	10.3762	9.8644	9.4008	79
80	13.1244	12.7075	12.3162	11.6017	10.9656	10.3957	9.8820	9.4167	80
81	13.1546	12.7357	12.3428	11.6253	10.9867	10.4146	9.8991	9.4323	81
82	13.1839	12.7632	12.3686	11.6482	11.0071	10.4330	9.9157	9.4473	82
83	13.2124	12.7900	12.3937	11.6705	11.0270	10.4508	9.9318	9.4620	83
84	13.2402	12.8160	12.4181	11.6921	11.0463	10.4682	9.9475	9.4762	84
85	13.2672	12.8413	12.4418	11.7132	11.0651	10.4850	9.9627	9.4900	85
86	13.2934	12.8659	12.4649	11.7336	11.0834	10.5014	9.9775	9.5034	86
87	13.3190	12.8898	12.4874	11.7535	11.1011	10.5174	9.9919	9.5165	87
88	13.3438	12.9131	12.5092	11.7729	11.1184	10.5329	10.0059	9.5292	88
89	13.3680	12.9357	12.5305	11.7917	11.1352	10.5479	10.0195	9.5415	89
90	13.3915	12.9577	12.5511	11.8100	11.1515	10.5626	10.0327	9.5535	90
91	13.4144	12.9792	12.5713	11.8278	11.1674	10.5768	10.0456	9.5651	91
92	13.4367	13.0000	12.5908	11.8451	11.1828	10.5907	10.0580	9.5764	92
93	13.4584	13.0203	12.6099	11.8620	11.1978	10.6041	10.0702	9.5875	93
94	13.4795	13.0401	12.6284	11.8784	11.2125	10.6172	10.0820	9.5982	94
95	13.5001	13.0593	12.6464	11.8943	11.2267	10.6300	10.0935	9.6086	95
96	13.5201	13.0780	12.6640	11.9099	11.2405	10.6424	10.1047	9.6187	96
97	13.5396	13.0963	12.6811	11.9250	11.2540	10.6544	10.1156	9.6286	97
98	13.5585	13.1140	12.6977	11.9397	11.2671	10.6662	10.1261	9.6381	98
99	13.5770	13.1313	12.7139	11.9540	11.2798	10.6776	10.1364	9.6475	99
100	13.5949	13.1481	12.7297	11.9679	11.2922	10.6887	10.1464	9.6565	100

YEARS' PURCHASE

Rate Per Cent

Yrs.	11	12	13	14	15	16	18	20	Yrs.
1	0.6066	0.6030	0.5994	0.5958	0.5923	0.5888	0.5819	0.5752	1
2	1.1498	1.1367	1.1239	1.1114	1.0992	1.0873	1.0641	1.0420	2
3	1.6388	1.6124	1.5868	1.5620	1.5380	1.5147	1.4701	1.4281	3
4	2.0812	2.0388	1.9980	1.9589	1.9213	1.8851	1.8166	1.7529	4
5	2.4833	2.4231	2.3658	2.3111	2.2589	2.2090	2.1156	2.0297	5
6	2.8503	2.7713	2.6965	2.6257	2.5586	2.4947	2.3762	2.2684	6
7	3.1863	3.0880	2.9955	2.9083	2.8261	2.7485	2.6053	2.4762	7
8	3.4952	3.3772	3.2669	3.1635	3.0665	2.9753	2.8082	2.6588	8
9	3.7800	3.6423	3.5143	3.3950	3.2835	3.1792	2.9891	2.8205	9
10	4.0433	3.8862	3.7408	3.6059	3.4804	3.3634	3.1514	2.9645	10
11	4.2874	4.1112	3.9488	3.7988	3.6598	3.5306	3.2977	3.0937	11
12	4.5143	4.3193	4.1405	3.9758	3.8238	3.6830	3.4303	3.2101	12
13	4.7256	4.5124	4.3176	4.1389	3.9744	3.8224	3.5510	3.3155	13
14	4.9229	4.6919	4.4816	4.2894	4.1130	3.9505	3.6612	3.4114	14
15	5.1074	4.8592	4.6340	4.4288	4.2410	4.0684	3.7623	3.4990	15
16	5.2803	5.0155	4.7759	4.5582	4.3595	4.1774	3.8553	3.5793	16
17	5.4426	5.1617	4.9083	4.6787	4.4696	4.2783	3.9411	3.6532	17
18	5.5952	5.2988	5.0321	4.7910	4.5720	4.3721	4.0205	3.7213	18
19	5.7390	5.4275	5.1481	4.8960	4.6675	4.4594	4.0942	3.7843	19
20	5.8745	5.5486	5.2569	4.9943	4.7568	4.5408	4.1627	3.8428	20
21	6.0025	5.6626	5.3592	5.0866	4.8404	4.6169	4.2266	3.8972	21
22	6.1236	5.7703	5.4555	5.1732	4.9188	4.6882	4.2863	3.9478	22
23	6.2382	5.8719	5.5462	5.2548	4.9925	4.7551	4.3421	3.9952	23
24	6.3469	5.9681	5.6320	5.3317	5.0618	4.8179	4.3945	4.0395	24
25	6.4500	6.0591	5.7130	5.4042	5.1272	4.8771	4.4437	4.0810	25
26	6.5479	6.1455	5.7897	5.4728	5.1888	4.9329	4.4899	4.1200	26
27	6.6410	6.2274	5.8624	5.5377	5.2471	4.9855	4.5335	4.1566	27
28	6.7296	6.3053	5.9313	5.5992	5.3023	5.0353	4.5746	4.1912	28
29	6.8140	6.3793	5.9968	5.6575	5.3546	5.0824	4.6135	4.2238	29
30	6.8945	6.4498	6.0590	5.7129	5.4042	5.1271	4.6502	4.2545	30
31	6.9713	6.5170	6.1183	5.7655	5.4512	5.1694	4.6851	4.2837	31
32	7.0447	6.5811	6.1747	5.8156	5.4960	5.2097	4.7181	4.3113	32
33	7.1148	6.6422	6.2285	5.8633	5.5385	5.2479	4.7494	4.3374	33
34	7.1818	6.7006	6.2798	5.9087	5.5791	5.2843	4.7792	4.3622	34
35	7.2459	6.7564	6.3288	5.9521	5.6177	5.3189	4.8075	4.3858	35
36	7.3073	6.8097	6.3756	5.9935	5.6546	5.3519	4.8345	4.4082	36
37	7.3662	6.8608	6.4203	6.0330	5.6897	5.3834	4.8601	4.4296	37
38	7.4226	6.9097	6.4632	6.0708	5.7233	5.4135	4.8846	4.4499	38
39	7.4768	6.9566	6.5042	6.1069	5.7555	5.4422	4.9080	4.4693	39
40	7.5287	7.0016	6.5434	6.1416	5.7862	5.4697	4.9304	4.4878	40
41	7.5786	7.0447	6.5811	6.1747	5.8156	5.4960	4.9517	4.5055	41
42	7.6265	7.0861	6.6172	6.2065	5.8438	5.5212	4.9721	4.5224	42
43	7.6726	7.1259	6.6519	6.2370	5.8708	5.5453	4.9917	4.5386	43
44	7.7169	7.1641	6.6852	6.2662	5.8967	5.5684	5.0104	4.5540	44
45	7.7596	7.2008	6.7171	6.2943	5.9216	5.5906	5.0283	4.5689	45
46	7.8006	7.2362	6.7479	6.3213	5.9455	5.6118	5.0455	4.5831	46
47	7.8401	7.2701	6.7774	6.3472	5.9684	5.6323	5.0620	4.5967	47
48	7.8782	7.3029	6.8059	6.3722	5.9905	5.6519	5.0779	4.6097	48
49	7.9149	7.3344	6.8332	6.3962	6.0116	5.6707	5.0931	4.6223	49
50	7.9503	7.3648	6.8596	6.4192	6.0320	5.6889	5.1077	4.6343	50

YEARS' PURCHASE

				Rate Per Cent					
Yrs.	11	12	13	14	15	16	18	20	Yrs.
51	7.9844	7.3940	6.8850	6.4415	6.0517	5.7063	5.1218	4.6459	51
52	8.0173	7.4223	6.9094	6.4629	6.0705	5.7231	5.1353	4.6570	52
53	8.0491	7.4495	6.9330	6.4835	6.0887	5.7393	5.1483	4.6677	53
54	8.0798	7.4757	6.9557	6.5034	6.1063	5.7549	5.1609	4.6780	54
55	8.1094	7.5011	6.9777	6.5226	6.1232	5.7699	5.1729	4.6879	55
56	8.1380	7.5255	6.9988	6.5410	6.1395	5.7843	5.1846	4.6975	56
57	8.1656	7.5492	7.0193	6.5589	6.1552	5.7983	5.1958	4.7067	57
58	8.1923	7.5720	7.0390	6.5761	6.1704	5.8117	5.2066	4.7155	58
59	8.2182	7.5941	7.0581	6.5928	6.1850	5.8247	5.2170	4.7241	59
60	8.2431	7.6154	7.0765	6.6088	6.1991	5.8373	5.2270	4.7323	60
61	8.2673	7.6360	7.0943	6.6243	6.2128	5.8494	5.2367	4.7403	61
62	8.2907	7.6560	7.1115	6.6393	6.2260	5.8611	5.2461	4.7479	62
63	8.3133	7.6752	7.1281	6.6538	6.2387	5.8724	5.2552	4.7554	63
64	8.3352	7.6939	7.1442	6.6679	6.2511	5.8833	5.2639	4.7625	64
65	8.3564	7.7120	7.1598	6.6814	6.2630	5.8938	5.2724	4.7694	65
66	8.3769	7.7294	7.1749	6.6945	6.2745	5.9040	5.2805	4.7761	66
67	8.3968	7.7464	7.1895	6.7072	6.2856	5.9139	5.2884	4.7826	67
68	8.4161	7.7628	7.2036	6.7195	6.2964	5.9235	5.2960	4.7888	68
69	8.4348	7.7786	7.2172	6.7314	6.3069	5.9327	5.3034	4.7949	69
70	8.4529	7.7940	7.2305	6.7429	6.3170	5.9417	5.3106	4.8007	70
71	8.4704	7.8089	7.2433	6.7541	6.3268	5.9503	5.3175	4.8063	71
72	8.4874	7.8234	7.2557	6.7649	6.3363	5.9587	5.3242	4.8118	72
73	8.5039	7.8374	7.2678	6.7754	6.3454	5.9668	5.3307	4.8171	73
74	8.5199	7.8510	7.2795	6.7855	6.3543	5.9747	5.3370	4.8222	74
75	8.5354	7.8641	7.2908	6.7954	6.3630	5.9823	5.3430	4.8272	75
76	8.5504	7.8769	7.3018	6.8049	6.3713	5.9897	5.3489	4.8320	76
77	8.5650	7.8893	7.3124	6.8141	6.3794	5.9969	5.3546	4.8367	77
78	8.5792	7.9013	7.3227	6.8231	6.3873	6.0038	5.3602	4.8412	78
79	8.5930	7.9130	7.3328	6.8318	6.3949	6.0105	5.3655	4.8456	79
80	8.6063	7.9243	7.3425	6.8402	6.4023	6.0171	5.3707	4.8498	80
81	8.6193	7.9353	7.3519	6.8484	6.4095	6.0234	5.3758	4.8539	81
82	8.6318	7.9460	7.3611	6.8564	6.4164	6.0295	5.3807	4.8579	82
83	8.6441	7.9563	7.3699	6.8641	6.4232	6.0355	5.3854	4.8618	83
84	8.6559	7.9664	7.3786	6.8715	6.4297	6.0413	5.3900	4.8655	84
85	8.6675	7.9761	7.3869	6.8788	6.4361	6.0469	5.3945	4.8692	85
86	8.6787	7.9856	7.3951	6.8859	6.4423	6.0523	5.3988	4.8727	86
87	8.6895	7.9948	7.4030	6.8927	6.4482	6.0576	5.4030	4.8761	87
88	8.7001	8.0038	7.4106	6.8994	6.4541	6.0628	5.4071	4.8795	88
89	8.7104	8.0125	7.4181	6.9058	6.4597	6.0678	5.4111	4.8827	89
90	8.7204	8.0209	7.4253	6.9121	6.4652	6.0726	5.4149	4.8858	90
91	8.7301	8.0291	7.4324	6.9182	6.4705	6.0773	5.4187	4.8889	91
92	8.7395	8.0371	7.4392	6.9241	6.4757	6.0819	5.4223	4.8918	92
93	8.7487	8.0449	7.4459	6.9299	6.4808	6.0863	5.4258	4.8947	93
94	8.7576	8.0524	7.4523	6.9355	6.4857	6.0906	5.4293	4.8975	94
95	8.7663	8.0597	7.4586	6.9409	6.4904	6.0948	5.4326	4.9002	95
96	8.7747	8.0669	7.4647	6.9462	6.4950	6.0989	5.4358	4.9028	96
97	8.7829	8.0738	7.4706	6.9513	6.4995	6.1029	5.4390	4.9054	97
98	8.7909	8.0805	7.4764	6.9563	6.5039	6.1067	5.4420	4.9079	98
99	8.7986	8.0871	7.4820	6.9612	6.5081	6.1104	5.4450	4.9103	99
100	8.8062	8.0934	7.4875	6.9659	6.5123	6.1141	5.4479	4.9126	100

YEARS' PURCHASE

Rate Per Cent

Yrs.	4	4.5	5	5.5	6	6.25	6.5	6.75	Yrs.
1	0.6335	0.6315	0.6295	0.6276	0.6256	0.6246	0.6237	0.6227	1
2	1.2533	1.2455	1.2378	1.2302	1.2227	1.2190	1.2153	1.2116	2
3	1.8596	1.8425	1.8257	1.8092	1.7930	1.7850	1.7770	1.7692	3
4	2.4526	2.4229	2.3939	2.3656	2.3379	2.3243	2.3109	2.2976	4
5	3.0324	2.9871	2.9431	2.9004	2.8590	2.8387	2.8187	2.7990	5
6	3.5992	3.5355	3.4741	3.4148	3.3575	3.3295	3.3020	3.2750	6
7	4.1532	4.0687	3.9876	3.9096	3.8347	3.7983	3.7625	3.7275	7
8	4.6946	4.5869	4.4841	4.3858	4.2917	4.2461	4.2015	4.1578	8
9	5.2237	5.0907	4.9643	4.8441	4.7295	4.6743	4.6203	4.5675	9
10	5.7405	5.5803	5.4289	5.2854	5.1493	5.0839	5.0201	4.9578	10
11	6.2453	6.0562	5.8782	5.7104	5.5519	5.4759	5.4019	5.3299	11
12	6.7384	6.5188	6.3130	6.1198	5.9381	5.8513	5.7669	5.6849	12
13	7.2199	6.9683	6.7337	6.5144	6.3089	6.2109	6.1159	6.0238	13
14	7.6899	7.4052	7.1408	6.8946	6.6649	6.5556	6.4499	6.3476	14
15	8.1488	7.8298	7.5348	7.2612	7.0068	6.8862	6.7697	6.6570	15
16	8.5966	8.2424	7.9161	7.6147	7.3354	7.2033	7.0759	6.9529	16
17	9.0337	8.6433	8.2853	7.9557	7.6513	7.5077	7.3694	7.2361	17
18	9.4602	9.0330	8.6426	8.2846	7.9551	7.8000	7.6508	7.5072	18
19	9.8763	9.4116	8.9886	8.6020	8.2473	8.0807	7.9207	7.7669	19
20	10.2823	9.7795	9.3236	8.9083	8.5284	8.3504	8.1796	8.0157	20
21	10.6782	10.1370	9.6480	9.2040	8.7990	8.6096	8.4282	8.2543	21
22	11.0643	10.4843	9.9621	9.4894	9.0596	8.8589	8.6670	8.4832	22
23	11.4409	10.8218	10.2663	9.7651	9.3105	9.0987	8.8963	8.7028	23
24	11.8080	11.1497	10.5610	10.0313	9.5522	9.3294	9.1167	8.9136	24
25	12.1659	11.4683	10.8464	10.2884	9.7850	9.5514	9.3286	9.1160	25
26	12.5149	11.7779	11.1228	10.5368	10.0095	9.7651	9.5324	9.3105	26
27	12.8549	12.0786	11.3907	10.7769	10.2259	9.9710	9.7285	9.4975	27
28	13.1864	12.3708	11.6502	11.0089	10.4345	10.1692	9.9171	9.6772	28
29	13.5094	12.6546	11.9016	11.2331	10.6357	10.3603	10.0987	9.8500	29
30	13.8241	12.9304	12.1452	11.4499	10.8299	10.5444	10.2736	10.0163	30
31	14.1308	13.1983	12.3812	11.6594	11.0171	10.7218	10.4419	10.1763	31
32	14.4295	13.4585	12.6099	11.8620	11.1979	10.8930	10.6042	10.3303	32
33	14.7205	13.7113	12.8316	12.0580	11.3724	11.0580	10.7605	10.4786	33
34	15.0039	13.9569	13.0465	12.2475	11.5408	11.2172	10.9112	10.6214	34
35	15.2800	14.1955	13.2547	12.4308	11.7034	11.3707	11.0564	10.7590	35
36	15.5488	14.4272	13.4565	12.6082	11.8605	11.5189	11.1965	10.8916	36
37	15.8106	14.6523	13.6521	12.7798	12.0122	11.6620	11.3316	11.0194	37
38	16.0655	14.8710	13.8418	12.9458	12.1588	11.8001	11.4620	11.1427	38
39	16.3137	15.0834	14.0256	13.1065	12.3004	11.9334	11.5877	11.2615	39
40	16.5553	15.2897	14.2038	13.2620	12.4372	12.0622	11.7091	11.3761	40
41	16.7905	15.4901	14.3766	13.4125	12.5695	12.1866	11.8263	11.4867	41
42	17.0194	15.6847	14.5441	13.5581	12.6974	12.3067	11.9394	11.5933	42
43	17.2422	15.8737	14.7065	13.6992	12.8210	12.4228	12.0486	11.6963	43
44	17.4591	16.0574	14.8640	13.8357	12.9405	12.5350	12.1541	11.7957	44
45	17.6702	16.2357	15.0167	13.9679	13.0561	12.6434	12.2560	11.8917	45
46	17.8755	16.4089	15.1648	14.0959	13.1679	12.7482	12.3545	11.9843	46
47	18.0754	16.5772	15.3083	14.2199	13.2760	12.8495	12.4496	12.0738	47
48	18.2698	16.7406	15.4476	14.3400	13.3806	12.9475	12.5415	12.1603	48
49	18.4590	16.8993	15.5826	14.4563	13.4818	13.0422	12.6304	12.2438	49
50	18.6431	17.0534	15.7136	14.5689	13.5797	13.1338	12.7163	12.3245	50

YEARS' PURCHASE

				Rate Per Cent					
Yrs.	4	4.5	5	5.5	6	6.25	6.5	6.75	Yrs.
51	18.8221	17.2031	15.8406	14.6780	13.6745	13.2224	12.7994	12.4025	51
52	18.9963	17.3485	15.9638	14.7838	13.7662	13.3082	12.8797	12.4779	52
53	19.1657	17.4897	16.0833	14.8862	13.8549	13.3911	12.9573	12.5508	53
54	19.3305	17.6268	16.1991	14.9854	13.9408	13.4713	13.0324	12.6212	54
55	19.4908	17.7600	16.3115	15.0815	14.0240	13.5490	13.1051	12.6893	55
56	19.6467	17.8893	16.4206	15.1747	14.1045	13.6241	13.1754	12.7552	56
57	19.7983	18.0150	16.5263	15.2650	14.1825	13.6969	13.2434	12.8190	57
58	19.9457	18.1369	16.6289	15.3525	14.2580	13.7673	13.3092	12.8806	58
59	20.0891	18.2554	16.7285	15.4373	14.3311	13.8354	13.3729	12.9402	59
60	20.2285	18.3704	16.8250	15.5194	14.4019	13.9014	13.4345	12.9979	60
61	20.3640	18.4821	16.9187	15.5991	14.4705	13.9653	13.4941	13.0538	61
62	20.4958	18.5906	17.0095	15.6763	14.5369	14.0271	13.5519	13.1078	62
63	20.6239	18.6960	17.0977	15.7512	14.6012	14.0870	13.6078	13.1601	63
64	20.7485	18.7983	17.1832	15.8237	14.6635	14.1450	13.6619	13.2107	64
65	20.8696	18.8976	17.2662	15.8940	14.7239	14.2012	13.7143	13.2597	65
66	20.9873	18.9941	17.3467	15.9622	14.7824	14.2556	13.7650	13.3071	66
67	21.1018	19.0878	17.4248	16.0284	14.8391	14.3083	13.8142	13.3530	67
68	21.2130	19.1788	17.5006	16.0925	14.8941	14.3594	13.8618	13.3975	68
69	21.3211	19.2671	17.5741	16.1546	14.9473	14.4088	13.9079	13.4405	69
70	21.4262	19.3529	17.6455	16.2149	14.9989	14.4568	13.9525	13.4822	70
71	21.5284	19.4363	17.7147	16.2733	15.0489	14.5032	13.9958	13.5226	71
72	21.6277	19.5171	17.7819	16.3300	15.0973	14.5482	14.0377	13.5617	72
73	21.7242	19.5957	17.8471	16.3850	15.1443	14.5918	14.0782	13.5996	73
74	21.8180	19.6720	17.9103	16.4383	15.1898	14.6341	14.1176	13.6363	74
75	21.9092	19.7461	17.9717	16.4899	15.2339	14.6750	14.1557	13.6718	75
76	21.9977	19.8180	18.0313	16.5401	15.2767	14.7147	14.1926	13.7063	76
77	22.0838	19.8878	18.0891	16.5887	15.3182	14.7532	14.2284	13.7397	77
78	22.1675	19.9556	18.1452	16.6359	15.3584	14.7905	14.2631	13.7720	78
79	22.2488	20.0215	18.1996	16.6816	15.3973	14.8266	14.2967	13.8033	79
80	22.3278	20.0854	18.2524	16.7260	15.4351	14.8616	14.3293	13.8337	80
81	22.4045	20.1475	18.3037	16.7690	15.4718	14.8956	14.3608	13.8631	81
82	22.4791	20.2078	18.3534	16.8107	15.5073	14.9285	14.3914	13.8916	82
83	22.5516	20.2664	18.4017	16.8512	15.5417	14.9605	14.4211	13.9193	83
84	22.6220	20.3232	18.4485	16.8905	15.5752	14.9914	14.4499	13.9461	84
85	22.6904	20.3784	18.4940	16.9286	15.6076	15.0214	14.4777	13.9720	85
86	22.7568	20.4320	18.5381	16.9656	15.6390	15.0505	14.5048	13.9972	86
87	22.8214	20.4840	18.5810	17.0015	15.6694	15.0788	14.5310	14.0216	87
88	22.8842	20.5346	18.6225	17.0363	15.6990	15.1061	14.5564	14.0453	88
89	22.9451	20.5836	18.6629	17.0700	15.7277	15.1327	14.5810	14.0682	89
90	23.0043	20.6313	18.7020	17.1028	15.7555	15.1584	14.6049	14.0904	90
91	23.0618	20.6775	18.7400	17.1345	15.7824	15.1833	14.6281	14.1120	91
92	23.1177	20.7225	18.7769	17.1654	15.8086	15.2075	14.6505	14.1329	92
93	23.1720	20.7661	18.8127	17.1953	15.8339	15.2310	14.6723	14.1532	93
94	23.2248	20.8084	18.8475	17.2243	15.8585	15.2538	14.6935	14.1728	94
95	23.2760	20.8495	18.8812	17.2525	15.8824	15.2759	14.7139	14.1919	95
96	23.3258	20.8894	18.9139	17.2798	15.9056	15.2973	14.7338	14.2104	96
97	23.3741	20.9282	18.9457	17.3063	15.9280	15.3181	14.7531	14.2283	97
98	23.4211	20.9658	18.9765	17.3320	15.9498	15.3382	14.7718	14.2457	98
99	23.4667	21.0024	19.0065	17.3570	15.9710	15.3578	14.7899	14.2626	99
100	23.5110	21.0379	19.0355	17.3812	15.9915	15.3767	14.8075	14.2789	100

YEARS' PURCHASE

Rate Per Cent

Yrs.	7	7.25	7.5	8	8.5	9	9.5	10	Yrs.
1	0.6217	0.6207	0.6198	0.6179	0.6160	0.6141	0.6122	0.6103	1
2	1.2079	1.2043	1.2007	1.1935	1.1864	1.1794	1.1725	1.1657	2
3	1.7614	1.7537	1.7460	1.7309	1.7160	1.7014	1.6871	1.6730	3
4	2.2845	2.2715	2.2587	2.2335	2.2088	2.1847	2.1611	2.1380	4
5	2.7795	2.7603	2.7414	2.7043	2.6683	2.6331	2.5989	2.5656	5
6	3.2484	3.2222	3.1965	3.1462	3.0975	3.0502	3.0044	2.9600	6
7	3.6930	3.6593	3.6261	3.5615	3.4992	3.4390	3.3809	3.3247	7
8	4.1151	4.0732	4.0321	3.9524	3.8758	3.8021	3.7312	3.6629	8
9	4.5160	4.4655	4.4162	4.3208	4.2295	4.1419	4.0578	3.9771	9
10	4.8971	4.8379	4.7801	4.6685	4.5620	4.4603	4.3630	4.2698	10
11	5.2599	5.1916	5.1251	4.9970	4.8752	4.7592	4.6486	4.5430	11
12	5.6053	5.5278	5.4525	5.3078	5.1705	5.0402	4.9163	4.7984	12
13	5.9345	5.8477	5.7635	5.6020	5.4494	5.3048	5.1678	5.0376	13
14	6.2484	6.1523	6.0591	5.8810	5.7130	5.5543	5.4042	5.2620	14
15	6.5480	6.4426	6.3404	6.1456	5.9624	5.7898	5.6269	5.4729	15
16	6.8341	6.7193	6.6083	6.3970	6.1987	6.0123	5.8369	5.6714	16
17	7.1075	6.9834	6.8636	6.6359	6.4228	6.2229	6.0351	5.8584	17
18	7.3689	7.2356	7.1070	6.8632	6.6354	6.4224	6.2226	6.0348	18
19	7.6189	7.4765	7.3393	7.0795	6.8375	6.6115	6.3999	6.2015	19
20	7.8582	7.7068	7.5612	7.2857	7.0296	6.7909	6.5679	6.3591	20
21	8.0874	7.9271	7.7731	7.4823	7.2125	6.9614	6.7273	6.5083	21
22	8.3070	8.1380	7.9757	7.6699	7.3866	7.1235	6.8785	6.6498	22
23	8.5175	8.3399	8.1695	7.8489	7.5525	7.2777	7.0222	6.7840	23
24	8.7193	8.5333	8.3550	8.0200	7.7108	7.4245	7.1588	6.9114	24
25	8.9129	8.7186	8.5327	8.1835	7.8618	7.5645	7.2888	7.0325	25
26	9.0988	8.8964	8.7028	8.3399	8.0061	7.6979	7.4126	7.1477	26
27	9.2772	9.0669	8.8659	8.4896	8.1439	7.8253	7.5306	7.2574	27
28	9.4486	9.2306	9.0224	8.6329	8.2757	7.9469	7.6432	7.3618	28
29	9.6133	9.3877	9.1724	8.7702	8.4018	8.0630	7.7506	7.4614	29
30	9.7716	9.5386	9.3164	8.9018	8.5224	8.1741	7.8532	7.5564	30
31	9.9238	9.6836	9.4547	9.0279	8.6380	8.2804	7.9512	7.6472	31
32	10.0702	9.8229	9.5875	9.1489	8.7487	8.3821	8.0449	7.7338	32
33	10.2111	9.9569	9.7151	9.2651	8.8548	8.4794	8.1345	7.8166	33
34	10.3467	10.0858	9.8378	9.3765	8.9566	8.5727	8.2204	7.8958	34
35	10.4772	10.2098	9.9557	9.4836	9.0543	8.6621	8.3025	7.9716	35
36	10.6029	10.3291	10.0691	9.5865	9.1480	8.7479	8.3813	8.0442	36
37	10.7240	10.4440	10.1783	9.6854	9.2380	8.8301	8.4568	8.1137	37
38	10.8407	10.5546	10.2833	9.7804	9.3244	8.9091	8.5291	8.1803	38
39	10.9531	10.6612	10.3844	9.8718	9.4075	8.9849	8.5986	8.2441	39
40	11.0615	10.7638	10.4818	9.9598	9.4873	9.0577	8.6652	8.3054	40
41	11.1660	10.8628	10.5756	10.0444	9.5641	9.1276	8.7292	8.3642	41
42	11.2668	10.9581	10.6659	10.1259	9.6380	9.1949	8.7907	8.4206	42
43	11.3640	11.0501	10.7530	10.2044	9.7090	9.2595	8.8498	8.4748	43
44	11.4578	11.1387	10.8370	10.2800	9.7774	9.3217	8.9066	8.5268	44
45	11.5483	11.2243	10.9179	10.3528	9.8432	9.3815	8.9612	8.5769	45
46	11.6357	11.3068	10.9960	10.4229	9.9066	9.4391	9.0137	8.6250	46
47	11.7201	11.3864	11.0713	10.4906	9.9677	9.4945	9.0642	8.6712	47
48	11.8015	11.4633	11.1439	10.5558	10.0266	9.5479	9.1129	8.7157	48
49	11.8801	11.5375	11.2140	10.6186	10.0833	9.5993	9.1597	8.7586	49
50	11.9561	11.6091	11.2817	10.6793	10.1380	9.6489	9.2048	8.7998	50

YEARS' PURCHASE

Rate Per Cent

Yrs.	7	7.25	7.5	8	8.5	9	9.5	10	Yrs.
51	12.0295	11.6783	11.3470	10.7378	10.1907	9.6966	9.2482	8.8395	51
52	12.1004	11.7451	11.4101	10.7943	10.2415	9.7426	9.2901	8.8777	52
53	12.1689	11.8097	11.4710	10.8488	10.2906	9.7870	9.3304	8.9145	53
54	12.2352	11.8720	11.5298	10.9014	10.3379	9.8298	9.3693	8.9500	54
55	12.2992	11.9323	11.5866	10.9522	10.3835	9.8711	9.4068	8.9842	55
56	12.3611	11.9905	11.6416	11.0012	10.4276	9.9109	9.4429	9.0172	56
57	12.4209	12.0468	11.6946	11.0486	10.4702	9.9493	9.4778	9.0490	57
58	12.4788	12.1012	11.7459	11.0943	10.5113	9.9864	9.5115	9.0797	58
59	12.5347	12.1539	11.7955	11.1385	10.5509	10.0222	9.5440	9.1093	59
60	12.5889	12.2047	11.8434	11.1813	10.5893	10.0568	9.5753	9.1378	60
61	12.6412	12.2540	11.8897	11.2226	10.6263	10.0902	9.6056	9.1654	61
62	12.6919	12.3016	11.9345	11.2625	10.6621	10.1224	9.6348	9.1920	62
63	12.7409	12.3476	11.9779	11.3010	10.6966	10.1536	9.6630	9.2177	63
64	12.7883	12.3921	12.0198	11.3383	10.7300	10.1837	9.6903	9.2425	64
65	12.8342	12.4352	12.0603	11.3744	10.7623	10.2128	9.7166	9.2664	65
66	12.8787	12.4769	12.0995	11.4093	10.7936	10.2409	9.7420	9.2895	66
67	12.9217	12.5173	12.1375	11.4430	10.8237	10.2681	9.7666	9.3119	67
68	12.9633	12.5564	12.1742	11.4757	10.8529	10.2943	9.7904	9.3335	68
69	13.0036	12.5942	12.2097	11.5072	10.8812	10.3197	9.8134	9.3544	69
70	13.0426	12.6308	12.2441	11.5378	10.9085	10.3443	9.8356	9.3746	70
71	13.0804	12.6662	12.2774	11.5673	10.9349	10.3680	9.8570	9.3941	71
72	13.1170	12.7005	12.3097	11.5960	10.9605	10.3910	9.8778	9.4129	72
73	13.1524	12.7337	12.3409	11.6236	10.9852	10.4132	9.8979	9.4312	73
74	13.1867	12.7659	12.3711	11.6504	11.0091	10.4347	9.9173	9.4488	74
75	13.2200	12.7970	12.4003	11.6764	11.0323	10.4555	9.9361	9.4658	75
76	13.2522	12.8272	12.4287	11.7015	11.0547	10.4757	9.9543	9.4823	76
77	13.2834	12.8564	12.4561	11.7258	11.0764	10.4952	9.9719	9.4983	77
78	13.3136	12.8847	12.4827	11.7493	11.0974	10.5140	9.9889	9.5137	78
79	13.3429	12.9122	12.5084	11.7721	11.1177	10.5323	10.0054	9.5287	79
80	13.3713	12.9387	12.5333	11.7942	11.1374	10.5499	10.0213	9.5431	80
81	13.3987	12.9645	12.5575	11.8156	11.1565	10.5670	10.0367	9.5571	81
82	13.4254	12.9894	12.5809	11.8363	11.1750	10.5836	10.0517	9.5707	82
83	13.4512	13.0136	12.6035	11.8564	11.1928	10.5996	10.0662	9.5838	83
84	13.4762	13.0370	12.6255	11.8758	11.2102	10.6152	10.0802	9.5965	84
85	13.5005	13.0597	12.6468	11.8946	11.2269	10.6302	10.0937	9.6088	85
86	13.5240	13.0817	12.6674	11.9129	11.2432	10.6448	10.1068	9.6207	86
87	13.5467	13.1030	12.6874	11.9305	11.2589	10.6589	10.1196	9.6322	87
88	13.5688	13.1236	12.7067	11.9477	11.2742	10.6725	10.1319	9.6434	88
89	13.5902	13.1437	12.7255	11.9643	11.2889	10.6858	10.1438	9.6542	89
90	13.6110	13.1631	12.7437	11.9803	11.3033	10.6986	10.1554	9.6646	90
91	13.6311	13.1819	12.7613	11.9959	11.3171	10.7110	10.1666	9.6748	91
92	13.6506	13.2001	12.7784	12.0110	11.3306	10.7231	10.1774	9.6846	92
93	13.6695	13.2178	12.7950	12.0257	11.3436	10.7347	10.1879	9.6941	93
94	13.6878	13.2350	12.8111	12.0399	11.3562	10.7460	10.1981	9.7033	94
95	13.7056	13.2516	12.8266	12.0536	11.3685	10.7570	10.2080	9.7123	95
96	13.7229	13.2677	12.8417	12.0669	11.3803	10.7676	10.2175	9.7209	96
97	13.7396	13.2833	12.8564	12.0799	11.3918	10.7779	10.2268	9.7293	97
98	13.7558	13.2985	12.8706	12.0924	11.4029	10.7879	10.2358	9.7374	98
99	13.7715	13.3132	12.8843	12.1045	11.4137	10.7975	10.2445	9.7453	99
100	13.7868	13.3274	12.8977	12.1163	11.4242	10.8069	10.2529	9.7529	100

YEARS' PURCHASE

				Rate Per Cent					
Yrs.	11	12	13	14	15	16	18	20	Yrs.
1	0.6066	0.6030	0.5994	0.5958	0.5923	0.5888	0.5819	0.5752	1
2	1.1523	1.1391	1.1263	1.1138	1.1015	1.0895	1.0663	1.0440	2
3	1.6454	1.6188	1.5930	1.5680	1.5438	1.5204	1.4755	1.4332	3
4	2.0932	2.0503	2.0091	1.9695	1.9315	1.8949	1.8257	1.7614	4
5	2.5014	2.4404	2.3822	2.3268	2.2739	2.2233	2.1287	2.0417	5
6	2.8749	2.7945	2.7186	2.6466	2.5784	2.5136	2.3932	2.2839	6
7	3.2177	3.1174	3.0232	2.9345	2.8508	2.7718	2.6262	2.4951	7
8	3.5334	3.4129	3.3002	3.1948	3.0959	3.0029	2.8328	2.6809	8
9	3.8250	3.6841	3.5532	3.4313	3.3174	3.2109	3.0172	2.8455	9
10	4.0950	3.9339	3.7850	3.6470	3.5186	3.3990	3.1827	2.9922	10
11	4.3456	4.1646	3.9981	3.8444	3.7021	3.5699	3.3320	3.1238	11
12	4.5787	4.3782	4.1946	4.0257	3.8699	3.7257	3.4674	3.2425	12
13	4.7960	4.5765	4.3762	4.1927	4.0240	3.8684	3.5906	3.3500	13
14	4.9990	4.7610	4.5446	4.3471	4.1660	3.9994	3.7031	3.4478	14
15	5.1889	4.9330	4.7011	4.4900	4.2970	4.1200	3.8064	3.5371	15
16	5.3670	5.0936	4.8467	4.6227	4.4184	4.2315	3.9013	3.6189	16
17	5.5342	5.2439	4.9827	4.7462	4.5311	4.3347	3.9889	3.6942	17
18	5.6913	5.3849	5.1097	4.8613	4.6359	4.4305	4.0699	3.7636	18
19	5.8393	5.5172	5.2287	4.9689	4.7337	4.5197	4.1450	3.8277	19
20	5.9789	5.6416	5.3403	5.0696	4.8250	4.6029	4.2149	3.8872	20
21	6.1106	5.7587	5.4452	5.1640	4.9104	4.6806	4.2799	3.9425	21
22	6.2352	5.8692	5.5438	5.2526	4.9905	4.7533	4.3406	3.9939	22
23	6.3530	5.9735	5.6368	5.3360	5.0657	4.8215	4.3974	4.0419	23
24	6.4646	6.0721	5.7245	5.4145	5.1364	4.8855	4.4506	4.0868	24
25	6.5704	6.1654	5.8073	5.4886	5.2030	4.9457	4.5005	4.1289	25
26	6.6709	6.2537	5.8856	5.5585	5.2658	5.0024	4.5474	4.1683	26
27	6.7663	6.3375	5.9598	5.6246	5.3251	5.0558	4.5916	4.2054	27
28	6.8570	6.4170	6.0301	5.6871	5.3811	5.1063	4.6332	4.2402	28
29	6.9434	6.4926	6.0967	5.7464	5.4341	5.1540	4.6724	4.2731	29
30	7.0256	6.5644	6.1600	5.8026	5.4843	5.1992	4.7095	4.3041	30
31	7.1039	6.6327	6.2202	5.8559	5.5320	5.2420	4.7446	4.3334	31
32	7.1786	6.6978	6.2774	5.9066	5.5772	5.2826	4.7778	4.3611	32
33	7.2499	6.7598	6.3318	5.9548	5.6201	5.3211	4.8093	4.3873	33
34	7.3180	6.8190	6.3837	6.0006	5.6609	5.3576	4.8391	4.4121	34
35	7.3831	6.8754	6.4331	6.0443	5.6998	5.3924	4.8675	4.4357	35
36	7.4453	6.9294	6.4803	6.0859	5.7368	5.4255	4.8944	4.4580	36
37	7.5048	6.9809	6.5253	6.1256	5.7721	5.4571	4.9201	4.4793	37
38	7.5617	7.0301	6.5684	6.1635	5.8057	5.4871	4.9445	4.4995	38
39	7.6162	7.0772	6.6095	6.1997	5.8378	5.5158	4.9678	4.5188	39
40	7.6685	7.1223	6.6488	6.2343	5.8684	5.5431	4.9899	4.5371	40
41	7.7186	7.1655	6.6864	6.2673	5.8977	5.5692	5.0111	4.5546	41
42	7.7666	7.2069	6.7224	6.2990	5.9257	5.5942	5.0313	4.5713	42
43	7.8127	7.2465	6.7569	6.3292	5.9525	5.6181	5.0506	4.5872	43
44	7.8569	7.2846	6.7899	6.3582	5.9781	5.6409	5.0690	4.6024	44
45	7.8994	7.3210	6.8216	6.3860	6.0027	5.6628	5.0867	4.6170	45
46	7.9401	7.3561	6.8520	6.4126	6.0262	5.6837	5.1035	4.6309	46
47	7.9793	7.3897	6.8812	6.4382	6.0487	5.7037	5.1197	4.6442	47
48	8.0170	7.4220	6.9092	6.4627	6.0704	5.7230	5.1352	4.6569	48
49	8.0532	7.4530	6.9361	6.4862	6.0911	5.7414	5.1500	4.6691	49
50	8.0880	7.4828	6.9619	6.5088	6.1110	5.7591	5.1642	4.6808	50

YEARS' PURCHASE

				Rate Per Cent					
Yrs.	11	12	13	14	15	16	18	20	Yrs.
51	8.1216	7.5115	6.9867	6.5304	6.1301	5.7760	5.1779	4.6920	51
52	8.1538	7.5391	7.0106	6.5513	6.1485	5.7923	5.1910	4.7027	52
53	8.1849	7.5656	7.0335	6.5713	6.1661	5.8080	5.2036	4.7131	53
54	8.2148	7.5912	7.0556	6.5906	6.1831	5.8230	5.2156	4.7230	54
55	8.2436	7.6158	7.0768	6.6091	6.1994	5.8375	5.2272	4.7325	55
56	8.2714	7.6395	7.0973	6.6269	6.2151	5.8514	5.2384	4.7416	56
57	8.2981	7.6623	7.1170	6.6441	6.2302	5.8648	5.2491	4.7504	57
58	8.3239	7.6843	7.1359	6.6606	6.2447	5.8776	5.2594	4.7588	58
59	8.3488	7.7054	7.1542	6.6765	6.2587	5.8900	5.2693	4.7669	59
60	8.3727	7.7259	7.1718	6.6919	6.2721	5.9020	5.2788	4.7747	60
61	8.3959	7.7456	7.1887	6.7066	6.2851	5.9134	5.2880	4.7823	61
62	8.4182	7.7645	7.2051	6.7209	6.2976	5.9245	5.2969	4.7895	62
63	8.4397	7.7829	7.2209	6.7346	6.3096	5.9352	5.3054	4.7965	63
64	8.4605	7.8005	7.2361	6.7478	6.3213	5.9454	5.3136	4.8032	64
65	8.4806	7.8176	7.2508	6.7606	6.3325	5.9553	5.3215	4.8096	65
66	8.4999	7.8341	7.2649	6.7729	6.3433	5.9649	5.3291	4.8158	66
67	8.5187	7.8499	7.2786	6.7847	6.3537	5.9741	5.3365	4.8218	67
68	8.5367	7.8653	7.2918	6.7962	6.3637	5.9830	5.3436	4.8276	68
69	8.5542	7.8801	7.3045	6.8073	6.3734	5.9915	5.3504	4.8332	69
70	8.5711	7.8944	7.3168	6.8179	6.3828	5.9998	5.3570	4.8386	70
71	8.5874	7.9082	7.3287	6.8283	6.3918	6.0078	5.3634	4.8438	71
72	8.6031	7.9216	7.3402	6.8382	6.4005	6.0155	5.3695	4.8488	72
73	8.6183	7.9345	7.3512	6.8478	6.4090	6.0230	5.3754	4.8536	73
74	8.6331	7.9470	7.3619	6.8571	6.4171	6.0301	5.3812	4.8583	74
75	8.6473	7.9591	7.3723	6.8661	6.4250	6.0371	5.3867	4.8628	75
76	8.6611	7.9707	7.3823	6.8748	6.4326	6.0438	5.3920	4.8671	76
77	8.6744	7.9820	7.3920	6.8832	6.4399	6.0503	5.3972	4.8713	77
78	8.6873	7.9929	7.4013	6.8913	6.4470	6.0565	5.4022	4.8754	78
79	8.6997	8.0034	7.4104	6.8991	6.4538	6.0626	5.4070	4.8793	79
80	8.7118	8.0136	7.4191	6.9067	6.4605	6.0684	5.4116	4.8831	80
81	8.7234	8.0235	7.4276	6.9140	6.4669	6.0741	5.4161	4.8868	81
82	8.7347	8.0330	7.4357	6.9211	6.4731	6.0796	5.4205	4.8903	82
83	8.7456	8.0423	7.4436	6.9280	6.4791	6.0848	5.4247	4.8937	83
84	8.7562	8.0512	7.4513	6.9346	6.4849	6.0900	5.4287	4.8970	84
85	8.7664	8.0599	7.4587	6.9410	6.4905	6.0949	5.4327	4.9002	85
86	8.7763	8.0682	7.4659	6.9472	6.4959	6.0997	5.4365	4.9033	86
87	8.7859	8.0763	7.4728	6.9532	6.5012	6.1043	5.4401	4.9063	87
88	8.7952	8.0842	7.4795	6.9590	6.5063	6.1088	5.4437	4.9092	88
89	8.8042	8.0918	7.4860	6.9646	6.5112	6.1131	5.4471	4.9120	89
90	8.8129	8.0991	7.4923	6.9701	6.5159	6.1173	5.4505	4.9147	90
91	8.8213	8.1062	7.4984	6.9754	6.5205	6.1214	5.4537	4.9173	91
92	8.8295	8.1131	7.5043	6.9805	6.5250	6.1253	5.4568	4.9199	92
93	8.8374	8.1198	7.5100	6.9854	6.5293	6.1291	5.4598	4.9223	93
94	8.8451	8.1263	7.5155	6.9902	6.5335	6.1328	5.4628	4.9247	94
95	8.8525	8.1325	7.5209	6.9948	6.5375	6.1364	5.4656	4.9270	95
96	8.8597	8.1386	7.5261	6.9993	6.5415	6.1398	5.4683	4.9292	96
97	8.8666	8.1445	7.5311	7.0037	6.5453	6.1432	5.4710	4.9314	97
98	8.8734	8.1502	7.5360	7.0079	6.5489	6.1464	5.4736	4.9335	98
99	8.8799	8.1557	7.5407	7.0120	6.5525	6.1495	5.4760	4.9355	99
100	8.8863	8.1610	7.5453	7.0159	6.5559	6.1526	5.4785	4.9375	100

YEARS' PURCHASE

				Rate Per Cent					
Yrs.	4	4.5	5	5.5	6	6.25	6.5	6.75	Yrs.
1	0.6335	0.6315	0.6295	0.6276	0.6256	0.6246	0.6237	0.6227	1
2	1.2592	1.2513	1.2436	1.2359	1.2283	1.2245	1.2208	1.2171	2
3	1.8767	1.8593	1.8422	1.8253	1.8088	1.8007	1.7926	1.7846	3
4	2.4858	2.4552	2.4255	2.3964	2.3680	2.3541	2.3403	2.3267	4
5	3.0860	3.0391	2.9936	2.9495	2.9066	2.8857	2.8650	2.8446	5
6	3.6773	3.6109	3.5468	3.4850	3.4253	3.3963	3.3677	3.3396	6
7	4.2592	4.1704	4.0852	4.0035	3.9249	3.8868	3.8494	3.8127	7
8	4.8317	4.7177	4.6090	4.5052	4.4059	4.3579	4.3110	4.2650	8
9	5.3945	5.2528	5.1184	4.9907	4.8692	4.8106	4.7534	4.6976	9
10	5.9474	5.7757	5.6136	5.4603	5.3152	5.2455	5.1776	5.1114	10
11	6.4903	6.2863	6.0948	5.9145	5.7446	5.6633	5.5842	5.5074	11
12	7.0231	6.7848	6.5622	6.3537	6.1581	6.0647	5.9741	5.8862	12
13	7.5455	7.2712	7.0161	6.7783	6.5561	6.4504	6.3480	6.2489	13
14	8.0576	7.7456	7.4568	7.1888	6.9393	6.8210	6.7066	6.5960	14
15	8.5593	8.2080	7.8844	7.5854	7.3082	7.1771	7.0506	6.9284	15
16	9.0504	8.6586	8.2993	7.9686	7.6633	7.5192	7.3805	7.2468	16
17	9.5310	9.0975	8.7016	8.3388	8.0051	7.8480	7.6970	7.5517	17
18	10.0010	9.5247	9.0918	8.6964	8.3340	8.1639	8.0007	7.8438	18
19	10.4605	9.9406	9.4699	9.0418	8.6507	8.4676	8.2920	8.1236	19
20	10.9094	10.3451	9.8363	9.3752	8.9554	8.7593	8.5716	8.3918	20
21	11.3477	10.7385	10.1913	9.6971	9.2487	9.0397	8.8399	8.6488	21
22	11.7756	11.1208	10.5350	10.0079	9.5310	9.3091	9.0974	8.8951	22
23	12.1931	11.4924	10.8679	10.3078	9.8026	9.5681	9.3446	9.1313	23
24	12.6001	11.8533	11.1901	10.5972	10.0640	9.8170	9.5818	9.3577	24
25	12.9969	12.2038	11.5020	10.8765	10.3155	10.0562	9.8095	9.5747	25
26	13.3835	12.5441	11.8037	11.1459	10.5575	10.2861	10.0282	9.7829	26
27	13.7600	12.8742	12.0956	11.4058	10.7904	10.5070	10.2381	9.9826	27
28	14.1265	13.1945	12.3779	11.6565	11.0145	10.7194	10.4396	10.1741	28
29	14.4831	13.5051	12.6509	11.8982	11.2301	10.9235	10.6331	10.3578	29
30	14.8300	13.8063	12.9147	12.1314	11.4376	11.1196	10.8189	10.5340	30
31	15.1673	14.0981	13.1698	12.3562	11.6372	11.3082	10.9973	10.7031	31
32	15.4952	14.3810	13.4163	12.5729	11.8292	11.4895	11.1687	10.8653	32
33	15.8137	14.6550	13.6544	12.7818	12.0140	11.6637	11.3332	11.0210	33
34	16.1231	14.9203	13.8845	12.9832	12.1917	11.8311	11.4913	11.1703	34
35	16.4235	15.1772	14.1067	13.1773	12.3627	11.9921	11.6430	11.3137	35
36	16.7152	15.4259	14.3213	13.3643	12.5273	12.1468	11.7888	11.4513	36
37	16.9981	15.6666	14.5285	13.5446	12.6855	12.2956	11.9289	11.5834	37
38	17.2726	15.8994	14.7286	13.7183	12.8378	12.4385	12.0634	11.7103	38
39	17.5387	16.1247	14.9217	13.8857	12.9842	12.5760	12.1926	11.8320	39
40	17.7968	16.3425	15.1080	14.0469	13.1251	12.7081	12.3168	11.9489	40
41	18.0468	16.5532	15.2879	14.2023	13.2606	12.8351	12.4361	12.0611	41
42	18.2891	16.7568	15.4614	14.3519	13.3910	12.9572	12.5506	12.1688	42
43	18.5239	16.9536	15.6288	14.4960	13.5164	13.0746	12.6607	12.2723	43
44	18.7511	17.1438	15.7903	14.6348	13.6370	13.1874	12.7665	12.3716	44
45	18.9712	17.3276	15.9460	14.7685	13.7530	13.2958	12.8681	12.4670	45
46	19.1842	17.5051	16.0963	14.8973	13.8646	13.4001	12.9658	12.5587	46
47	19.3903	17.6765	16.2411	15.0213	13.9719	13.5004	13.0596	12.6467	47
48	19.5897	17.8421	16.3808	15.1407	14.0752	13.5967	13.1497	12.7312	48
49	19.7826	18.0020	16.5154	15.2557	14.1745	13.6894	13.2364	12.8124	49
50	19.9692	18.1563	16.6452	15.3664	14.2700	13.7784	13.3196	12.8904	50

YEARS' PURCHASE

				Rate Per Cent					
Yrs.	4	4.5	5	5.5	6	6.25	6.5	6.75	Yrs.
51	20.1495	18.3053	16.7704	15.4729	14.3618	13.8640	13.3996	12.9653	51
52	20.3239	18.4491	16.8910	15.5755	14.4502	13.9464	13.4765	13.0373	52
53	20.4924	18.5878	17.0072	15.6743	14.5352	14.0255	13.5504	13.1064	53
54	20.6552	18.7217	17.1192	15.7694	14.6169	14.1016	13.6214	13.1728	54
55	20.8125	18.8508	17.2271	15.8609	14.6955	14.1747	13.6896	13.2366	55
56	20.9645	18.9754	17.3311	15.9490	14.7711	14.2451	13.7552	13.2979	56
57	21.1112	19.0956	17.4313	16.0338	14.8438	14.3127	13.8182	13.3568	57
58	21.2529	19.2114	17.5278	16.1154	14.9137	14.3777	13.8788	13.4134	58
59	21.3898	19.3232	17.6207	16.1940	14.9810	14.4402	13.9370	13.4678	59
60	21.5219	19.4309	17.7103	16.2696	15.0457	14.5002	13.9930	13.5200	60
61	21.6494	19.5348	17.7965	16.3423	15.1079	14.5580	14.0468	13.5702	61
62	21.7724	19.6349	17.8796	16.4124	15.1677	14.6135	14.0985	13.6185	62
63	21.8911	19.7314	17.9596	16.4797	15.2252	14.6669	14.1482	13.6648	63
64	22.0057	19.8244	18.0366	16.5446	15.2805	14.7183	14.1959	13.7094	64
65	22.1162	19.9141	18.1108	16.6070	15.3337	14.7676	14.2418	13.7522	65
66	22.2228	20.0005	18.1822	16.6670	15.3849	14.8151	14.2860	13.7933	66
67	22.3256	20.0837	18.2510	16.7248	15.4341	14.8607	14.3284	13.8329	67
68	22.4248	20.1639	18.3172	16.7804	15.4814	14.9046	14.3692	13.8709	68
69	22.5204	20.2412	18.3809	16.8338	15.5269	14.9467	14.4084	13.9074	69
70	22.6126	20.3156	18.4423	16.8853	15.5707	14.9873	14.4460	13.9425	70
71	22.7014	20.3873	18.5014	16.9348	15.6128	15.0263	14.4822	13.9762	71
72	22.7871	20.4564	18.5582	16.9824	15.6533	15.0638	14.5171	14.0087	72
73	22.8697	20.5229	18.6130	17.0282	15.6922	15.0998	14.5505	14.0398	73
74	22.9493	20.5870	18.6656	17.0723	15.7296	15.1345	14.5827	14.0698	74
75	23.0260	20.6487	18.7163	17.1147	15.7656	15.1678	14.6136	14.0986	75
76	23.0999	20.7081	18.7651	17.1555	15.8002	15.1998	14.6434	14.1262	76
77	23.1711	20.7653	18.8121	17.1948	15.8335	15.2306	14.6720	14.1528	77
78	23.2397	20.8204	18.8573	17.2325	15.8655	15.2602	14.6994	14.1784	78
79	23.3058	20.8734	18.9008	17.2688	15.8963	15.2887	14.7258	14.2030	79
80	23.3695	20.9245	18.9427	17.3038	15.9259	15.3161	14.7512	14.2266	80
81	23.4308	20.9737	18.9829	17.3374	15.9543	15.3424	14.7757	14.2493	81
82	23.4899	21.0210	19.0217	17.3697	15.9817	15.3677	14.7991	14.2711	82
83	23.5468	21.0665	19.0590	17.4008	16.0080	15.3920	14.8217	14.2921	83
84	23.6016	21.1104	19.0949	17.4307	16.0333	15.4154	14.8434	14.3123	84
85	23.6543	21.1526	19.1294	17.4595	16.0577	15.4379	14.8642	14.3317	85
86	23.7052	21.1932	19.1626	17.4871	16.0811	15.4596	14.8843	14.3503	86
87	23.7541	21.2323	19.1946	17.5137	16.1036	15.4804	14.9036	14.3682	87
88	23.8012	21.2700	19.2253	17.5393	16.1252	15.5004	14.9221	14.3855	88
89	23.8466	21.3062	19.2549	17.5640	16.1460	15.5196	14.9399	14.4020	89
90	23.8902	21.3410	19.2834	17.5876	16.1660	15.5381	14.9571	14.4179	90
91	23.9323	21.3746	19.3108	17.6104	16.1853	15.5558	14.9735	14.4332	91
92	23.9728	21.4069	19.3371	17.6323	16.2038	15.5729	14.9894	14.4479	92
93	24.0117	21.4379	19.3625	17.6534	16.2216	15.5894	15.0046	14.4621	93
94	24.0492	21.4678	19.3868	17.6737	16.2387	15.6052	15.0192	14.4757	94
95	24.0853	21.4966	19.4103	17.6932	16.2551	15.6204	15.0333	14.4888	95
96	24.1201	21.5242	19.4329	17.7119	16.2709	15.6350	15.0468	14.5013	96
97	24.1535	21.5509	19.4546	17.7299	16.2862	15.6490	15.0598	14.5134	97
98	24.1857	21.5765	19.4754	17.7473	16.3008	15.6625	15.0723	14.5250	98
99	24.2167	21.6011	19.4955	17.7639	16.3149	15.6755	15.0844	14.5362	99
100	24.2465	21.6249	19.5148	17.7800	16.3284	15.6880	15.0959	14.5469	100

YEARS' PURCHASE

Rate Per Cent

Yrs.	7	7.25	7.5	8	8.5	9	9.5	10	Yrs.
1	0.6217	0.6207	0.6198	0.6179	0.6160	0.6141	0.6122	0.6103	1
2	1.2134	1.2097	1.2061	1.1988	1.1917	1.1846	1.1777	1.1708	2
3	1.7767	1.7688	1.7610	1.7457	1.7306	1.7157	1.7011	1.6868	3
4	2.3132	2.2999	2.2868	2.2609	2.2357	2.2110	2.1868	2.1631	4
5	2.8245	2.8047	2.7852	2.7469	2.7097	2.6735	2.6382	2.6039	5
6	3.3119	3.2847	3.2580	3.2057	3.1552	3.1062	3.0587	3.0126	6
7	3.7767	3.7413	3.7067	3.6392	3.5742	3.5114	3.4508	3.3923	7
8	4.2200	4.1760	4.1328	4.0491	3.9688	3.8916	3.8173	3.7458	8
9	4.6431	4.5898	4.5377	4.4371	4.3408	4.2486	4.1602	4.0754	9
10	5.0469	4.9841	4.9227	4.8045	4.6918	4.5842	4.4815	4.3833	10
11	5.4326	5.3598	5.2889	5.1526	5.0232	4.9001	4.7830	4.6713	11
12	5.8009	5.7179	5.6374	5.4828	5.3365	5.1978	5.0662	4.9410	12
13	6.1528	6.0595	5.9691	5.7961	5.6329	5.4786	5.3325	5.1940	13
14	6.4890	6.3854	6.2851	6.0936	5.9134	5.7436	5.5833	5.4317	14
15	6.8105	6.6965	6.5862	6.3762	6.1792	5.9940	5.8196	5.6551	15
16	7.1178	6.9934	6.8732	6.6449	6.4312	6.2308	6.0426	5.8654	16
17	7.4118	7.2769	7.1469	6.9003	6.6702	6.4549	6.2531	6.0635	17
18	7.6929	7.5478	7.4080	7.1434	6.8970	6.6671	6.4520	6.2504	18
19	7.9619	7.8065	7.6571	7.3747	7.1125	6.8682	6.6402	6.4268	19
20	8.2193	8.0538	7.8949	7.5951	7.3172	7.0589	6.8183	6.5935	20
21	8.4657	8.2903	8.1219	7.8050	7.5118	7.2399	6.9870	6.7511	21
22	8.7016	8.5163	8.3388	8.0050	7.6970	7.4117	7.1469	6.9003	22
23	8.9275	8.7326	8.5460	8.1958	7.8732	7.5750	7.2985	7.0416	23
24	9.1437	8.9394	8.7440	8.3777	8.0409	7.7301	7.4424	7.1754	24
25	9.3509	9.1373	8.9332	8.5513	8.2007	7.8776	7.5791	7.3024	25
26	9.5494	9.3267	9.1142	8.7169	8.3529	8.0180	7.7090	7.4229	26
27	9.7395	9.5080	9.2872	8.8751	8.4980	8.1516	7.8324	7.5372	27
28	9.9217	9.6816	9.4528	9.0262	8.6364	8.2789	7.9498	7.6459	28
29	10.0963	9.8477	9.6111	9.1704	8.7684	8.4001	8.0615	7.7492	29
30	10.2637	10.0069	9.7627	9.3083	8.8944	8.5156	8.1679	7.8474	30
31	10.4241	10.1594	9.9077	9.4401	9.0146	8.6258	8.2692	7.9408	31
32	10.5779	10.3054	10.0466	9.5661	9.1294	8.7309	8.3657	8.0298	32
33	10.7254	10.4454	10.1795	9.6865	9.2390	8.8311	8.4576	8.1145	33
34	10.8669	10.5795	10.3069	9.8017	9.3438	8.9268	8.5453	8.1952	34
35	11.0025	10.7080	10.4288	9.9120	9.4439	9.0181	8.6290	8.2721	35
36	11.1326	10.8312	10.5456	10.0174	9.5396	9.1053	8.7088	8.3454	36
37	11.2574	10.9493	10.6576	10.1184	9.6311	9.1886	8.7850	8.4154	37
38	11.3772	11.0625	10.7648	10.2150	9.7186	9.2682	8.8578	8.4821	38
39	11.4921	11.1711	10.8676	10.3075	9.8023	9.3443	8.9272	8.5458	39
40	11.6023	11.2752	10.9661	10.3961	9.8824	9.4171	8.9936	8.6066	40
41	11.7080	11.3751	11.0606	10.4809	9.9590	9.4866	9.0570	8.6647	41
42	11.8095	11.4709	11.1511	10.5622	10.0324	9.5532	9.1177	8.7201	42
43	11.9070	11.5628	11.2379	10.6401	10.1026	9.6168	9.1756	8.7731	43
44	12.0005	11.6509	11.3212	10.7147	10.1698	9.6777	9.2310	8.8238	44
45	12.0902	11.7355	11.4010	10.7862	10.2342	9.7360	9.2841	8.8722	45
46	12.1764	11.8167	11.4776	10.8547	10.2959	9.7918	9.3348	8.9185	46
47	12.2591	11.8945	11.5511	10.9204	10.3550	9.8452	9.3833	8.9628	47
48	12.3385	11.9693	11.6215	10.9833	10.4116	9.8964	9.4298	9.0052	48
49	12.4147	12.0410	11.6891	11.0437	10.4658	9.9454	9.4742	9.0457	49
50	12.4879	12.1099	11.7540	11.1016	10.5178	9.9923	9.5168	9.0845	50

YEARS' PURCHASE

				Rate Per Cent					
Yrs.	7	7.25	7.5	8	8.5	9	9.5	10	Yrs.
51	12.5582	12.1760	11.8163	11.1571	10.5676	10.0372	9.5576	9.1217	51
52	12.6257	12.2394	11.8760	11.2104	10.6153	10.0803	9.5966	9.1572	52
53	12.6906	12.3003	11.9334	11.2614	10.6611	10.1216	9.6340	9.1913	53
54	12.7528	12.3588	11.9884	11.3104	10.7050	10.1612	9.6699	9.2239	54
55	12.8126	12.4149	12.0412	11.3574	10.7471	10.1991	9.7042	9.2551	55
56	12.8700	12.4689	12.0919	11.4025	10.7875	10.2354	9.7371	9.2851	56
57	12.9252	12.5206	12.1406	11.4458	10.8262	10.2703	9.7687	9.3137	57
58	12.9782	12.5703	12.1873	11.4873	10.8634	10.3037	9.7989	9.3412	58
59	13.0291	12.6181	12.2322	11.5272	10.8990	10.3358	9.8279	9.3676	59
60	13.0780	12.6639	12.2753	11.5655	10.9332	10.3665	9.8557	9.3928	60
61	13.1250	12.7080	12.3167	11.6022	10.9660	10.3960	9.8823	9.4170	61
62	13.1701	12.7503	12.3564	11.6374	10.9975	10.4243	9.9079	9.4402	62
63	13.2134	12.7909	12.3946	11.6713	11.0277	10.4514	9.9324	9.4625	63
64	13.2551	12.8299	12.4312	11.7037	11.0567	10.4775	9.9559	9.4838	64
65	13.2951	12.8674	12.4664	11.7349	11.0845	10.5025	9.9785	9.5043	65
66	13.3335	12.9034	12.5002	11.7649	11.1113	10.5265	10.0001	9.5239	66
67	13.3705	12.9380	12.5327	11.7936	11.1369	10.5495	10.0209	9.5428	67
68	13.4060	12.9713	12.5638	11.8212	11.1615	10.5716	10.0408	9.5608	68
69	13.4401	13.0032	12.5938	11.8478	11.1852	10.5928	10.0599	9.5782	69
70	13.4729	13.0339	12.6226	11.8732	11.2079	10.6131	10.0783	9.5948	70
71	13.5044	13.0633	12.6502	11.8977	11.2296	10.6326	10.0959	9.6108	71
72	13.5346	13.0917	12.6768	11.9212	11.2506	10.6514	10.1128	9.6261	72
73	13.5637	13.1189	12.7023	11.9437	11.2707	10.6694	10.1290	9.6408	73
74	13.5917	13.1450	12.7268	11.9654	11.2899	10.6867	10.1446	9.6549	74
75	13.6186	13.1702	12.7503	11.9862	11.3085	10.7033	10.1596	9.6684	75
76	13.6444	13.1943	12.7730	12.0062	11.3263	10.7192	10.1739	9.6815	76
77	13.6692	13.2175	12.7947	12.0254	11.3434	10.7345	10.1877	9.6939	77
78	13.6930	13.2398	12.8156	12.0439	11.3598	10.7492	10.2010	9.7059	78
79	13.7160	13.2612	12.8357	12.0616	11.3756	10.7634	10.2137	9.7174	79
80	13.7380	13.2818	12.8550	12.0786	11.3907	10.7769	10.2259	9.7285	80
81	13.7592	13.3016	12.8735	12.0950	11.4053	10.7899	10.2376	9.7391	81
82	13.7795	13.3206	12.8913	12.1107	11.4192	10.8025	10.2489	9.7493	82
83	13.7991	13.3389	12.9084	12.1258	11.4327	10.8145	10.2597	9.7591	83
84	13.8179	13.3565	12.9249	12.1403	11.4456	10.8260	10.2701	9.7685	84
85	13.8359	13.3734	12.9407	12.1543	11.4580	10.8371	10.2801	9.7775	85
86	13.8533	13.3896	12.9559	12.1677	11.4699	10.8478	10.2897	9.7862	86
87	13.8700	13.4052	12.9705	12.1806	11.4813	10.8580	10.2989	9.7945	87
88	13.8861	13.4202	12.9845	12.1929	11.4923	10.8678	10.3077	9.8025	88
89	13.9015	13.4346	12.9980	12.2048	11.5029	10.8773	10.3162	9.8102	89
90	13.9163	13.4484	13.0110	12.2163	11.5130	10.8864	10.3244	9.8176	90
91	13.9306	13.4618	13.0235	12.2272	11.5228	10.8951	10.3322	9.8247	91
92	13.9443	13.4745	13.0354	12.2378	11.5322	10.9035	10.3398	9.8315	92
93	13.9575	13.4868	13.0469	12.2480	11.5412	10.9115	10.3470	9.8380	93
94	13.9701	13.4987	13.0580	12.2577	11.5498	10.9193	10.3540	9.8443	94
95	13.9823	13.5100	13.0686	12.2671	11.5581	10.9267	10.3606	9.8504	95
96	13.9940	13.5210	13.0789	12.2761	11.5661	10.9338	10.3671	9.8562	96
97	14.0052	13.5315	13.0887	12.2847	11.5738	10.9407	10.3732	9.8618	97
98	14.0161	13.5416	13.0981	12.2931	11.5812	10.9473	10.3792	9.8671	98
99	14.0265	13.5513	13.1072	12.3011	11.5883	10.9536	10.3849	9.8723	99
100	14.0365	13.5606	13.1159	12.3087	11.5951	10.9597	10.3904	9.8772	100

YEARS' PURCHASE

				Rate Per Cent					
Yrs.	11	12	13	14	15	16	18	20	Yrs.
1	0.6066	0.6030	0.5994	0.5958	0.5923	0.5888	0.5819	0.5752	1
2	1.1572	1.1440	1.1310	1.1184	1.1060	1.0939	1.0705	1.0481	2
3	1.6588	1.6317	1.6055	1.5802	1.5556	1.5318	1.4862	1.4433	3
4	2.1173	2.0734	2.0313	1.9909	1.9520	1.9146	1.8440	1.7784	4
5	2.5378	2.4750	2.4152	2.3583	2.3039	2.2520	2.1550	2.0659	5
6	2.9245	2.8414	2.7629	2.6886	2.6182	2.5514	2.4275	2.3151	6
7	3.2810	3.1768	3.0790	2.9870	2.9004	2.8186	2.6682	2.5330	7
8	3.6106	3.4847	3.3674	3.2577	3.1549	3.0584	2.8821	2.7250	8
9	3.9158	3.7683	3.6314	3.5042	3.3855	3.2747	3.0734	2.8954	9
10	4.1992	4.0300	3.8739	3.7294	3.5953	3.4705	3.2453	3.0475	10
11	4.4628	4.2721	4.0971	3.9358	3.7868	3.6486	3.4005	3.1839	11
12	4.7084	4.4966	4.3031	4.1256	3.9622	3.8112	3.5412	3.3070	12
13	4.9376	4.7052	4.4938	4.3005	4.1232	3.9599	3.6693	3.4185	13
14	5.1518	4.8994	4.6706	4.4622	4.2716	4.0966	3.7864	3.5198	14
15	5.3524	5.0805	4.8348	4.6119	4.4085	4.2224	3.8936	3.6123	15
16	5.5404	5.2496	4.9877	4.7508	4.5353	4.3385	3.9921	3.6970	16
17	5.7169	5.4077	5.1303	4.8799	4.6529	4.4460	4.0830	3.7747	17
18	5.8827	5.5559	5.2634	5.0003	4.7621	4.5457	4.1668	3.8463	18
19	6.0387	5.6948	5.3880	5.1125	4.8639	4.6383	4.2445	3.9124	19
20	6.1857	5.8253	5.5047	5.2175	4.9587	4.7245	4.3166	3.9735	20
21	6.3242	5.9480	5.6141	5.3157	5.0474	4.8048	4.3836	4.0303	21
22	6.4549	6.0635	5.7169	5.4077	5.1303	4.8799	4.4460	4.0829	22
23	6.5783	6.1723	5.8135	5.4941	5.2080	4.9502	4.5042	4.1320	23
24	6.6950	6.2749	5.9044	5.5752	5.2808	5.0159	4.5586	4.1777	24
25	6.8054	6.3718	5.9901	5.6516	5.3493	5.0777	4.6095	4.2205	25
26	6.9099	6.4633	6.0709	5.7235	5.4136	5.1356	4.6572	4.2604	26
27	7.0090	6.5499	6.1472	5.7912	5.4742	5.1901	4.7020	4.2978	27
28	7.1028	6.6318	6.2193	5.8552	5.5313	5.2414	4.7441	4.3330	28
29	7.1919	6.7093	6.2875	5.9155	5.5852	5.2897	4.7836	4.3659	29
30	7.2764	6.7828	6.3520	5.9726	5.6360	5.3353	4.8209	4.3969	30
31	7.3567	6.8525	6.4131	6.0266	5.6840	5.3783	4.8560	4.4261	31
32	7.4329	6.9187	6.4710	6.0777	5.7295	5.4190	4.8891	4.4536	32
33	7.5055	6.9815	6.5259	6.1261	5.7725	5.4574	4.9204	4.4796	33
34	7.5745	7.0411	6.5780	6.1720	5.8132	5.4938	4.9499	4.5040	34
35	7.6401	7.0978	6.6274	6.2155	5.8518	5.5283	4.9779	4.5272	35
36	7.7026	7.1517	6.6744	6.2568	5.8884	5.5609	5.0044	4.5491	36
37	7.7622	7.2030	6.7191	6.2960	5.9231	5.5919	5.0294	4.5698	37
38	7.8189	7.2519	6.7615	6.3333	5.9561	5.6213	5.0532	4.5894	38
39	7.8730	7.2984	6.8020	6.3688	5.9874	5.6492	5.0757	4.6079	39
40	7.9246	7.3427	6.8404	6.4025	6.0172	5.6757	5.0971	4.6256	40
41	7.9738	7.3849	6.8770	6.4345	6.0455	5.7009	5.1174	4.6423	41
42	8.0207	7.4252	6.9119	6.4651	6.0725	5.7248	5.1367	4.6582	42
43	8.0655	7.4636	6.9452	6.4942	6.0981	5.7476	5.1551	4.6732	43
44	8.1083	7.5002	6.9769	6.5219	6.1226	5.7693	5.1725	4.6876	44
45	8.1492	7.5351	7.0071	6.5483	6.1458	5.7900	5.1891	4.7012	45
46	8.1882	7.5685	7.0360	6.5735	6.1680	5.8097	5.2049	4.7142	46
47	8.2256	7.6004	7.0635	6.5975	6.1892	5.8285	5.2200	4.7265	47
48	8.2612	7.6308	7.0898	6.6204	6.2094	5.8463	5.2343	4.7383	48
49	8.2954	7.6599	7.1149	6.6423	6.2286	5.8634	5.2480	4.7495	49
50	8.3280	7.6877	7.1389	6.6632	6.2470	5.8797	5.2610	4.7602	50

YEARS' PURCHASE

Rate Per Cent

Yrs.	11	12	13	14	15	16	18	20	Yrs.
51	8.3592	7.7143	7.1618	6.6832	6.2645	5.8952	5.2735	4.7703	51
52	8.3890	7.7397	7.1837	6.7023	6.2813	5.9101	5.2853	4.7800	52
53	8.4176	7.7641	7.2047	6.7205	6.2973	5.9242	5.2966	4.7893	53
54	8.4449	7.7873	7.2247	6.7379	6.3126	5.9377	5.3075	4.7981	54
55	8.4711	7.8096	7.2439	6.7546	6.3272	5.9507	5.3178	4.8066	55
56	8.4962	7.8309	7.2622	6.7705	6.3412	5.9630	5.3277	4.8146	56
57	8.5202	7.8513	7.2797	6.7857	6.3545	5.9749	5.3371	4.8223	57
58	8.5432	7.8708	7.2965	6.8003	6.3673	5.9861	5.3461	4.8297	58
59	8.5652	7.8895	7.3125	6.8142	6.3795	5.9970	5.3547	4.8367	59
60	8.5863	7.9074	7.3279	6.8276	6.3912	6.0073	5.3630	4.8435	60
61	8.6065	7.9245	7.3426	6.8404	6.4024	6.0172	5.3708	4.8499	61
62	8.6259	7.9409	7.3567	6.8526	6.4131	6.0266	5.3784	4.8560	62
63	8.6445	7.9567	7.3703	6.8643	6.4234	6.0357	5.3856	4.8619	63
64	8.6623	7.9718	7.3832	6.8756	6.4332	6.0444	5.3925	4.8675	64
65	8.6794	7.9862	7.3956	6.8863	6.4426	6.0527	5.3991	4.8729	65
66	8.6957	8.0001	7.4075	6.8966	6.4517	6.0607	5.4054	4.8781	66
67	8.7114	8.0134	7.4189	6.9065	6.4603	6.0683	5.4115	4.8830	67
68	8.7265	8.0261	7.4298	6.9159	6.4686	6.0756	5.4173	4.8877	68
69	8.7409	8.0383	7.4402	6.9250	6.4765	6.0826	5.4229	4.8923	69
70	8.7548	8.0500	7.4503	6.9337	6.4841	6.0893	5.4282	4.8966	70
71	8.7681	8.0613	7.4599	6.9420	6.4914	6.0957	5.4333	4.9008	71
72	8.7808	8.0720	7.4691	6.9500	6.4984	6.1019	5.4382	4.9047	72
73	8.7931	8.0824	7.4780	6.9577	6.5051	6.1078	5.4429	4.9086	73
74	8.8048	8.0923	7.4865	6.9650	6.5115	6.1134	5.4474	4.9122	74
75	8.8161	8.1018	7.4946	6.9721	6.5177	6.1189	5.4517	4.9157	75
76	8.8269	8.1109	7.5024	6.9788	6.5236	6.1241	5.4558	4.9191	76
77	8.8373	8.1197	7.5099	6.9853	6.5292	6.1291	5.4598	4.9223	77
78	8.8472	8.1281	7.5171	6.9915	6.5347	6.1338	5.4636	4.9254	78
79	8.8568	8.1362	7.5240	6.9975	6.5399	6.1384	5.4672	4.9283	79
80	8.8660	8.1439	7.5306	7.0032	6.5449	6.1429	5.4707	4.9312	80
81	8.8748	8.1514	7.5370	7.0087	6.5497	6.1471	5.4741	4.9339	81
82	8.8832	8.1585	7.5431	7.0140	6.5543	6.1511	5.4773	4.9365	82
83	8.8914	8.1654	7.5490	7.0191	6.5587	6.1550	5.4804	4.9390	83
84	8.8992	8.1719	7.5546	7.0239	6.5630	6.1588	5.4834	4.9414	84
85	8.9067	8.1783	7.5600	7.0286	6.5670	6.1624	5.4862	4.9438	85
86	8.9139	8.1843	7.5652	7.0331	6.5710	6.1658	5.4889	4.9460	86
87	8.9208	8.1901	7.5701	7.0374	6.5747	6.1691	5.4915	4.9481	87
88	8.9274	8.1957	7.5749	7.0415	6.5783	6.1723	5.4941	4.9501	88
89	8.9338	8.2011	7.5795	7.0455	6.5818	6.1753	5.4965	4.9521	89
90	8.9399	8.2063	7.5839	7.0493	6.5851	6.1783	5.4988	4.9540	90
91	8.9458	8.2112	7.5881	7.0530	6.5883	6.1811	5.5010	4.9558	91
92	8.9514	8.2160	7.5922	7.0565	6.5913	6.1838	5.5032	4.9575	92
93	8.9569	8.2206	7.5961	7.0598	6.5943	6.1863	5.5052	4.9592	93
94	8.9621	8.2249	7.5999	7.0631	6.5971	6.1888	5.5072	4.9608	94
95	8.9671	8.2292	7.6035	7.0662	6.5998	6.1912	5.5091	4.9623	95
96	8.9719	8.2332	7.6069	7.0692	6.6024	6.1935	5.5109	4.9638	96
97	8.9765	8.2371	7.6102	7.0720	6.6049	6.1957	5.5126	4.9652	97
98	8.9810	8.2408	7.6134	7.0748	6.6073	6.1978	5.5143	4.9666	98
99	8.9852	8.2444	7.6165	7.0774	6.6097	6.1999	5.5159	4.9679	99
100	8.9893	8.2479	7.6194	7.0800	6.6119	6.2018	5.5175	4.9691	100

YEARS' PURCHASE

(DUAL RATE % PRINCIPLE)

OR

PRESENT VALUE OF
ONE POUND PER ANNUM

receivable at the end of each year after allowing for a sinking fund at a given rate to replace the invested capital and for the effect of income tax at **40%** *on that part of the income used to provide the annual sinking fund instalment.*

AT RATES OF INTEREST FROM

4% to **20%**

AND

ALLOWING FOR THE POSSIBLE INVESTMENT

OF SINKING FUNDS AT

2·5%, 3% and 4%

INCOME TAX at 40%

Note:—Tables of Years' Purchase in which no allowance has been made for the effect of income tax on that part of the income used to provide the annual sinking fund instalment will be found on pages 53 to 71.

YEARS' PURCHASE

				Rate Per Cent					
Yrs.	4	4.5	5	5.5	6	6.25	6.5	6.75	Yrs.
1	0.5859	0.5842	0.5825	0.5808	0.5792	0.5783	0.5775	0.5766	1
2	1.1587	1.1520	1.1454	1.1389	1.1324	1.1292	1.1261	1.1229	2
3	1.7185	1.7039	1.6895	1.6753	1.6614	1.6545	1.6477	1.6410	3
4	2.2657	2.2403	2.2155	2.1912	2.1675	2.1558	2.1443	2.1328	4
5	2.8005	2.7618	2.7242	2.6876	2.6520	2.6345	2.6173	2.6003	5
6	3.3232	3.2689	3.2163	3.1654	3.1161	3.0920	3.0683	3.0449	6
7	3.8340	3.7619	3.6924	3.6255	3.5609	3.5295	3.4986	3.4683	7
8	4.3332	4.2413	4.1532	4.0687	3.9876	3.9482	3.9096	3.8718	8
9	4.8209	4.7075	4.5992	4.4958	4.3970	4.3492	4.3024	4.2566	9
10	5.2976	5.1609	5.0311	4.9076	4.7901	4.7334	4.6780	4.6240	10
11	5.7634	5.6019	5.4493	5.3048	5.1677	5.1018	5.0375	4.9749	11
12	6.2184	6.0309	5.8544	5.6879	5.5306	5.4552	5.3818	5.3103	12
13	6.6631	6.4483	6.2469	6.0577	5.8796	5.7944	5.7117	5.6313	13
14	7.0975	6.8543	6.6272	6.4146	6.2153	6.1202	6.0279	5.9384	14
15	7.5220	7.2493	6.9957	6.7593	6.5383	6.4332	6.3314	6.2327	15
16	7.9366	7.6337	7.3530	7.0923	6.8494	6.7341	6.6226	6.5147	16
17	8.3417	8.0077	7.6994	7.4140	7.1490	7.0235	6.9023	6.7852	17
18	8.7374	8.3717	8.0353	7.7250	7.4377	7.3019	7.1710	7.0447	18
19	9.1240	8.7259	8.3611	8.0256	7.7160	7.5700	7.4294	7.2939	19
20	9.5016	9.0707	8.6771	8.3163	7.9843	7.8281	7.6778	7.5332	20
21	9.8705	9.4063	8.9837	8.5975	8.2432	8.0767	7.9169	7.7632	21
22	10.2308	9.7329	9.2812	8.8696	8.4930	8.3164	8.1470	7.9844	22
23	10.5827	10.0509	9.5699	9.1329	8.7341	8.5474	8.3686	8.1971	23
24	10.9264	10.3604	9.8501	9.3878	8.9669	8.7703	8.5821	8.4019	24
25	11.2621	10.6618	10.1222	9.6345	9.1918	8.9853	8.7879	8.5990	25
26	11.5900	10.9552	10.3862	9.8735	9.4090	9.1928	8.9862	8.7888	26
27	11.9102	11.2408	10.6427	10.1049	9.6189	9.3931	9.1776	8.9717	27
28	12.2230	11.5190	10.8917	10.3292	9.8219	9.5865	9.3621	9.1480	28
29	12.5284	11.7898	11.1335	10.5464	10.0182	9.7734	9.5403	9.3180	29
30	12.8266	12.0536	11.3685	10.7570	10.2080	9.9539	9.7123	9.4820	30
31	13.1179	12.3105	11.5967	10.9611	10.3916	10.1285	9.8783	9.6403	31
32	13.4023	12.5606	11.8184	11.1590	10.5693	10.2972	10.0388	9.7930	32
33	13.6801	12.8043	12.0338	11.3509	10.7413	10.4604	10.1938	9.9405	33
34	13.9513	13.0416	12.2432	11.5370	10.9078	10.6182	10.3436	10.0829	34
35	14.2161	13.2727	12.4467	11.7175	11.0690	10.7709	10.4885	10.2205	35
36	14.4747	13.4978	12.6445	11.8926	11.2251	10.9187	10.6286	10.3535	36
37	14.7272	13.7171	12.8367	12.0625	11.3764	11.0618	10.7641	10.4820	37
38	14.9738	13.9308	13.0236	12.2274	11.5229	11.2003	10.8952	10.6063	38
39	15.2145	14.1389	13.2053	12.3874	11.6649	11.3344	11.0221	10.7265	39
40	15.4495	14.3416	13.3820	12.5428	11.8026	11.4643	11.1449	10.8428	40
41	15.6790	14.5392	13.5539	12.6936	11.9361	11.5902	11.2638	10.9553	41
42	15.9030	14.7316	13.7209	12.8401	12.0655	11.7122	11.3790	11.0642	42
43	16.1217	14.9191	13.8835	12.9823	12.1909	11.8304	11.4905	11.1697	43
44	16.3353	15.1018	14.0415	13.1204	12.3127	11.9450	11.5986	11.2718	44
45	16.5437	15.2798	14.1953	13.2545	12.4307	12.0561	11.7033	11.3706	45
46	16.7473	15.4533	14.3449	13.3849	12.5453	12.1638	11.8048	11.4664	46
47	16.9460	15.6223	14.4904	13.5115	12.6564	12.2683	11.9032	11.5592	47
48	17.1399	15.7870	14.6320	13.6345	12.7643	12.3696	11.9986	11.6491	48
49	17.3293	15.9475	14.7698	13.7541	12.8691	12.4679	12.0910	11.7363	49
50	17.5141	16.1039	14.9038	13.8702	12.9707	12.5633	12.1807	11.8208	50

YEARS' PURCHASE

				Rate Per Cent					
Yrs.	4	4.5	5	5.5	6	6.25	6.5	6.75	Yrs.
51	17.6946	16.2563	15.0343	13.9832	13.0694	12.6559	12.2678	11.9027	51
52	17.8707	16.4049	15.1613	14.0930	13.1653	12.7458	12.3522	11.9822	52
53	18.0427	16.5497	15.2849	14.1997	13.2584	12.8330	12.4341	12.0592	53
54	18.2105	16.6908	15.4052	14.3034	13.3488	12.9177	12.5136	12.1340	54
55	18.3744	16.8283	15.5222	14.4043	13.4366	12.9999	12.5907	12.2065	55
56	18.5343	16.9624	15.6362	14.5024	13.5219	13.0798	12.6656	12.2769	56
57	18.6904	17.0930	15.7472	14.5978	13.6048	13.1573	12.7383	12.3452	57
58	18.8428	17.2204	15.8552	14.6906	13.6854	13.2326	12.8089	12.4115	58
59	18.9915	17.3445	15.9604	14.7808	13.7637	13.3058	12.8775	12.4758	59
60	19.1367	17.4655	16.0628	14.8686	13.8397	13.3769	12.9440	12.5383	60
61	19.2784	17.5835	16.1625	14.9540	13.9137	13.4460	13.0087	12.5990	61
62	19.4167	17.6985	16.2596	15.0371	13.9856	13.5131	13.0715	12.6579	62
63	19.5517	17.8105	16.3542	15.1180	14.0555	13.5784	13.1326	12.7151	63
64	19.6834	17.9198	16.4462	15.1966	14.1235	13.6418	13.1919	12.7707	64
65	19.8120	18.0263	16.5359	15.2731	14.1896	13.7034	13.2495	12.8247	65
66	19.9375	18.1302	16.6233	15.3476	14.2538	13.7634	13.3055	12.8772	66
67	20.0600	18.2314	16.7083	15.4201	14.3163	13.8216	13.3600	12.9282	67
68	20.1796	18.3301	16.7912	15.4907	14.3771	13.8783	13.4129	12.9777	68
69	20.2963	18.4264	16.8719	15.5593	14.4362	13.9334	13.4644	13.0259	69
70	20.4102	18.5202	16.9505	15.6262	14.4938	13.9870	13.5144	13.0727	70
71	20.5213	18.6116	17.0271	15.6912	14.5497	14.0391	13.5630	13.1182	71
72	20.6298	18.7008	17.1017	15.7546	14.6042	14.0897	13.6103	13.1625	72
73	20.7356	18.7878	17.1744	15.8162	14.6571	14.1390	13.6563	13.2055	73
74	20.8390	18.8725	17.2452	15.8763	14.7087	14.1870	13.7011	13.2473	74
75	20.9398	18.9552	17.3142	15.9347	14.7588	14.2337	13.7446	13.2880	75
76	21.0382	19.0358	17.3814	15.9917	14.8077	14.2791	13.7869	13.3275	76
77	21.1342	19.1144	17.4469	16.0471	14.8552	14.3232	13.8281	13.3660	77
78	21.2279	19.1910	17.5108	16.1011	14.9014	14.3662	13.8681	13.4034	78
79	21.3194	19.2657	17.5730	16.1536	14.9464	14.4081	13.9071	13.4398	79
80	21.4087	19.3386	17.6336	16.2048	14.9902	14.4488	13.9450	13.4753	80
81	21.4958	19.4096	17.6926	16.2547	15.0329	14.4884	13.9819	13.5097	81
82	21.5808	19.4789	17.7501	16.3032	15.0744	14.5270	14.0179	13.5432	82
83	21.6637	19.5465	17.8062	16.3505	15.1148	14.5645	14.0528	13.5759	83
84	21.7447	19.6123	17.8609	16.3966	15.1542	14.6010	14.0868	13.6076	84
85	21.8237	19.6766	17.9141	16.4415	15.1925	14.6366	14.1199	13.6385	85
86	21.9007	19.7392	17.9660	16.4852	15.2298	14.6712	14.1522	13.6686	86
87	21.9760	19.8003	18.0166	16.5278	15.2662	14.7050	14.1835	13.6978	87
88	22.0494	19.8599	18.0659	16.5693	15.3016	14.7378	14.2141	13.7263	88
89	22.1210	19.9180	18.1140	16.6097	15.3360	14.7698	14.2438	13.7540	89
90	22.1909	19.9746	18.1609	16.6491	15.3696	14.8009	14.2728	13.7810	90
91	22.2591	20.0299	18.2065	16.6874	15.4023	14.8312	14.3010	13.8073	91
92	22.3257	20.0838	18.2510	16.7248	15.4341	14.8607	14.3284	13.8329	92
93	22.3906	20.1363	18.2944	16.7612	15.4652	14.8895	14.3551	13.8578	93
94	22.4540	20.1876	18.3367	16.7967	15.4954	14.9175	14.3812	13.8821	94
95	22.5159	20.2376	18.3779	16.8313	15.5248	14.9448	14.4065	13.9057	95
96	22.5762	20.2863	18.4181	16.8650	15.5535	14.9713	14.4312	13.9287	96
97	22.6351	20.3338	18.4573	16.8979	15.5814	14.9972	14.4552	13.9511	97
98	22.6926	20.3802	18.4955	16.9299	15.6086	15.0224	14.4786	13.9729	98
99	22.7487	20.4254	18.5327	16.9611	15.6351	15.0470	14.5015	13.9941	99
100	22.8034	20.4695	18.5690	16.9914	15.6609	15.0709	14.5237	14.0148	100

YEARS' PURCHASE

				Rate Per Cent					
Yrs.	7	7.25	7.5	8	8.5	9	9.5	10	Yrs.
1	0.5758	0.5750	0.5742	0.5725	0.5709	0.5693	0.5676	0.5660	1
2	1.1198	1.1166	1.1135	1.1074	1.1013	1.0952	1.0893	1.0834	2
3	1.6343	1.6276	1.6210	1.6080	1.5952	1.5825	1.5701	1.5579	3
4	2.1215	2.1103	2.0992	2.0774	2.0561	2.0352	2.0147	1.9946	4
5	2.5835	2.5669	2.5505	2.5184	2.4871	2.4565	2.4267	2.3976	5
6	3.0219	2.9993	2.9769	2.9333	2.8909	2.8497	2.8096	2.7707	6
7	3.4385	3.4092	3.3804	3.3242	3.2698	3.2172	3.1663	3.1170	7
8	3.8347	3.7983	3.7625	3.6930	3.6261	3.5615	3.4992	3.4390	8
9	4.2118	4.1679	4.1249	4.0416	3.9615	3.8846	3.8106	3.7393	9
10	4.5711	4.5195	4.4690	4.3713	4.2778	4.1882	4.1023	4.0199	10
11	4.9138	4.8541	4.7959	4.6836	4.5765	4.4741	4.3762	4.2825	11
12	5.2408	5.1730	5.1069	4.9798	4.8588	4.7436	4.6337	4.5287	12
13	5.5531	5.4770	5.4031	5.2609	5.1261	4.9980	4.8761	4.7601	13
14	5.8516	5.7672	5.6852	5.5281	5.3794	5.2385	5.1048	4.9777	14
15	6.1371	6.0443	5.9544	5.7822	5.6197	5.4662	5.3207	5.1828	15
16	6.4103	6.3092	6.2112	6.0242	5.8480	5.6819	5.5249	5.3764	16
17	6.6720	6.5626	6.4566	6.2547	6.0650	5.8865	5.7182	5.5593	17
18	6.9228	6.8050	6.6912	6.4746	6.2716	6.0809	5.9014	5.7323	18
19	7.1633	7.0372	6.9156	6.6844	6.4683	6.2656	6.0753	5.8962	19
20	7.3940	7.2598	7.1304	6.8849	6.6558	6.4414	6.2404	6.0516	20
21	7.6154	7.4732	7.3361	7.0765	6.8347	6.6089	6.3975	6.1992	21
22	7.8281	7.6779	7.5333	7.2598	7.0055	6.7684	6.5469	6.3394	22
23	8.0325	7.8744	7.7224	7.4353	7.1688	6.9207	6.6892	6.4727	23
24	8.2290	8.0631	7.9038	7.6033	7.3249	7.0661	6.8249	6.5997	24
25	8.4180	8.2445	8.0780	7.7644	7.4742	7.2050	6.9544	6.7207	25
26	8.5998	8.4188	8.2453	7.9188	7.6172	7.3378	7.0781	6.8361	26
27	8.7749	8.5865	8.4061	8.0670	7.7543	7.4648	7.1962	6.9463	27
28	8.9435	8.7479	8.5607	8.2093	7.8856	7.5865	7.3092	7.0515	28
29	9.1059	8.9032	8.7094	8.3459	8.0116	7.7030	7.4174	7.1521	29
30	9.2625	9.0528	8.8525	8.4773	8.1325	7.8148	7.5209	7.2483	30
31	9.4134	9.1970	8.9902	8.6035	8.2487	7.9219	7.6201	7.3404	31
32	9.5590	9.3359	9.1229	8.7249	8.3602	8.0248	7.7152	7.4287	32
33	9.6994	9.4698	9.2508	8.8418	8.4675	8.1235	7.8065	7.5132	33
34	9.8350	9.5990	9.3740	8.9543	8.5706	8.2184	7.8940	7.5943	34
35	9.9659	9.7236	9.4928	9.0627	8.6698	8.3096	7.9781	7.6721	35
36	10.0922	9.8439	9.6074	9.1671	8.7653	8.3973	8.0589	7.7468	36
37	10.2143	9.9600	9.7180	9.2677	8.8573	8.4817	8.1366	7.8185	37
38	10.3323	10.0722	9.8248	9.3647	8.9459	8.5629	8.2113	7.8875	38
39	10.4464	10.1805	9.9278	9.4583	9.0312	8.6410	8.2832	7.9537	39
40	10.5566	10.2852	10.0274	9.5486	9.1135	8.7163	8.3523	8.0175	40
41	10.6633	10.3864	10.1235	9.6358	9.1929	8.7889	8.4189	8.0789	41
42	10.7664	10.4842	10.2165	9.7199	9.2695	8.8589	8.4831	8.1379	42
43	10.8662	10.5789	10.3063	9.8012	9.3433	8.9263	8.5450	8.1948	43
44	10.9628	10.6704	10.3931	9.8797	9.4147	8.9914	8.6046	8.2496	44
45	11.0563	10.7590	10.4771	9.9556	9.4835	9.0542	8.6621	8.3025	45
46	11.1469	10.8447	10.5584	10.0290	9.5501	9.1148	8.7175	8.3534	46
47	11.2345	10.9276	10.6370	10.0999	9.6144	9.1734	8.7711	8.4026	47
48	11.3195	11.0080	10.7131	10.1685	9.6765	9.2299	8.8228	8.4500	48
49	11.4018	11.0858	10.7868	10.2348	9.7365	9.2846	8.8727	8.4958	49
50	11.4815	11.1611	10.8581	10.2990	9.7946	9.3374	8.9209	8.5399	50

YEARS' PURCHASE

				Rate Per Cent					
Yrs.	7	7.25	7.5	8	8.5	9	9.5	10	Yrs.
51	11.5588	11.2341	10.9272	10.3611	9.8508	9.3884	8.9674	8.5826	51
52	11.6337	11.3049	10.9942	10.4213	9.9052	9.4378	9.0125	8.6239	52
53	11.7063	11.3734	11.0590	10.4795	9.9578	9.4855	9.0560	8.6637	53
54	11.7767	11.4399	11.1218	10.5359	10.0087	9.5317	9.0981	8.7022	54
55	11.8450	11.5043	11.1827	10.5906	10.0580	9.5764	9.1388	8.7395	55
56	11.9113	11.5668	11.2418	10.6435	10.1057	9.6196	9.1782	8.7755	56
57	11.9756	11.6274	11.2990	10.6948	10.1519	9.6615	9.2163	8.8103	57
58	12.0379	11.6862	11.3545	10.7445	10.1967	9.7021	9.2532	8.8440	58
59	12.0985	11.7433	11.4083	10.7927	10.2401	9.7414	9.2889	8.8767	59
60	12.1572	11.7986	11.4606	10.8394	10.2822	9.7794	9.3235	8.9082	60
61	12.2143	11.8523	11.5112	10.8848	10.3229	9.8163	9.3570	8.9388	61
62	12.2696	11.9045	11.5604	10.9287	10.3625	9.8520	9.3895	8.9684	62
63	12.3234	11.9551	11.6081	10.9713	10.4008	9.8866	9.4209	8.9971	63
64	12.3756	12.0042	11.6544	11.0127	10.4380	9.9202	9.4514	9.0249	64
65	12.4263	12.0519	11.6994	11.0529	10.4740	9.9528	9.4810	9.0519	65
66	12.4756	12.0982	11.7431	11.0918	10.5090	9.9844	9.5096	9.0780	66
67	12.5234	12.1432	11.7855	11.1296	10.5429	10.0150	9.5374	9.1033	67
68	12.5699	12.1869	11.8266	11.1663	10.5759	10.0447	9.5643	9.1278	68
69	12.6151	12.2294	11.8666	11.2020	10.6078	10.0735	9.5905	9.1516	69
70	12.6590	12.2707	11.9054	11.2366	10.6388	10.1015	9.6158	9.1747	70
71	12.7017	12.3107	11.9432	11.2702	10.6690	10.1287	9.6404	9.1971	71
72	12.7431	12.3497	11.9798	11.3028	10.6982	10.1550	9.6643	9.2188	72
73	12.7835	12.3876	12.0155	11.3345	10.7266	10.1806	9.6875	9.2399	73
74	12.8226	12.4244	12.0501	11.3653	10.7542	10.2054	9.7100	9.2604	74
75	12.8607	12.4601	12.0837	11.3952	10.7810	10.2296	9.7318	9.2802	75
76	12.8978	12.4949	12.1164	11.4243	10.8070	10.2530	9.7530	9.2995	76
77	12.9338	12.5287	12.1482	11.4526	10.8323	10.2757	9.7736	9.3182	77
78	12.9689	12.5616	12.1791	11.4800	10.8569	10.2978	9.7936	9.3364	78
79	13.0029	12.5936	12.2092	11.5067	10.8807	10.3193	9.8130	9.3540	79
80	13.0361	12.6247	12.2384	11.5327	10.9039	10.3402	9.8319	9.3712	80
81	13.0683	12.6549	12.2668	11.5579	10.9265	10.3605	9.8502	9.3878	81
82	13.0997	12.6843	12.2944	11.5824	10.9484	10.3802	9.8680	9.4040	82
83	13.1302	12.7129	12.3213	11.6063	10.9697	10.3993	9.8853	9.4197	83
84	13.1599	12.7408	12.3475	11.6295	10.9904	10.4179	9.9021	9.4350	84
85	13.1888	12.7678	12.3729	11.6520	11.0106	10.4360	9.9185	9.4498	85
86	13.2169	12.7942	12.3976	11.6740	11.0302	10.4536	9.9344	9.4643	86
87	13.2443	12.8198	12.4217	11.6953	11.0492	10.4707	9.9498	9.4783	87
88	13.2709	12.8448	12.4451	11.7161	11.0677	10.4874	9.9648	9.4919	88
89	13.2968	12.8690	12.4679	11.7363	11.0858	10.5036	9.9795	9.5052	89
90	13.3221	12.8927	12.4901	11.7559	11.1033	10.5193	9.9937	9.5181	90
91	13.3466	12.9157	12.5117	11.7750	11.1203	10.5346	10.0075	9.5306	91
92	13.3705	12.9380	12.5327	11.7936	11.1369	10.5495	10.0209	9.5428	92
93	13.3938	12.9598	12.5531	11.8117	11.1531	10.5640	10.0340	9.5546	93
94	13.4164	12.9810	12.5730	11.8294	11.1688	10.5780	10.0467	9.5661	94
95	13.4385	13.0017	12.5924	11.8465	11.1840	10.5918	10.0590	9.5773	95
96	13.4600	13.0218	12.6112	11.8632	11.1989	10.6051	10.0711	9.5882	96
97	13.4809	13.0414	12.6296	11.8794	11.2134	10.6181	10.0828	9.5989	97
98	13.5012	13.0604	12.6475	11.8952	11.2275	10.6307	10.0942	9.6092	98
99	13.5211	13.0790	12.6649	11.9106	11.2412	10.6430	10.1052	9.6192	99
100	13.5404	13.0970	12.6818	11.9256	11.2545	10.6549	10.1160	9.6290	100

YEARS' PURCHASE

				Rate Per Cent					
Yrs.	11	12	13	14	15	16	18	20	Yrs.
1	0.5629	0.5597	0.5566	0.5535	0.5505	0.5474	0.5415	0.5357	1
2	1.0718	1.0604	1.0493	1.0384	1.0277	1.0172	0.9970	0.9775	2
3	1.5340	1.5108	1.4883	1.4665	1.4453	1.4247	1.3852	1.3479	3
4	1.9556	1.9180	1.8820	1.8472	1.8137	1.7814	1.7201	1.6629	4
5	2.3415	2.2879	2.2367	2.1878	2.1410	2.0961	2.0118	1.9339	5
6	2.6960	2.6252	2.5581	2.4943	2.4336	2.3758	2.2680	2.1696	6
7	3.0227	2.9341	2.8504	2.7714	2.6967	2.6259	2.4948	2.3763	7
8	3.3247	3.2177	3.1174	3.0232	2.9345	2.8508	2.6970	2.5590	8
9	3.6045	3.4791	3.3622	3.2528	3.1503	3.0541	2.8783	2.7216	9
10	3.8645	3.7207	3.5873	3.4630	3.3471	3.2387	3.0417	2.8673	10
11	4.1066	3.9446	3.7949	3.6562	3.5272	3.4070	3.1897	2.9984	11
12	4.3325	4.1526	3.9871	3.8342	3.6926	3.5611	3.3243	3.1171	12
13	4.5438	4.3463	4.1653	3.9987	3.8450	3.7026	3.4473	3.2250	13
14	4.7417	4.5271	4.3310	4.1512	3.9857	3.8330	3.5601	3.3234	14
15	4.9275	4.6961	4.4854	4.2929	4.1162	3.9534	3.6638	3.4136	15
16	5.1021	4.8544	4.6297	4.4248	4.2373	4.0651	3.7594	3.4965	16
17	5.2665	5.0030	4.7646	4.5479	4.3501	4.1688	3.8479	3.5730	17
18	5.4215	5.1427	4.8912	4.6631	4.4553	4.2653	3.9300	3.6436	18
19	5.5679	5.2742	5.0100	4.7710	4.5537	4.3554	4.0064	3.7092	19
20	5.7063	5.3983	5.1218	4.8722	4.6459	4.4396	4.0776	3.7701	20
21	5.8373	5.5153	5.2271	4.9674	4.7323	4.5185	4.1440	3.8268	21
22	5.9615	5.6261	5.3264	5.0570	4.8136	4.5925	4.2062	3.8798	22
23	6.0793	5.7309	5.4202	5.1415	4.8901	4.6621	4.2645	3.9294	23
24	6.1911	5.8302	5.5090	5.2213	4.9623	4.7277	4.3193	3.9758	24
25	6.2975	5.9244	5.5931	5.2968	5.0304	4.7894	4.3708	4.0194	25
26	6.3987	6.0139	5.6728	5.3682	5.0947	4.8478	4.4193	4.0604	26
27	6.4951	6.0990	5.7484	5.4359	5.1557	4.9029	4.4651	4.0990	27
28	6.5870	6.1800	5.8203	5.5002	5.2134	4.9551	4.5083	4.1354	28
29	6.6747	6.2571	5.8886	5.5612	5.2682	5.0045	4.5492	4.1698	29
30	6.7585	6.3306	5.9537	5.6192	5.3202	5.0515	4.5879	4.2023	30
31	6.8385	6.4008	6.0157	5.6743	5.3697	5.0960	4.6247	4.2331	31
32	6.9150	6.4677	6.0748	5.7269	5.4167	5.1384	4.6595	4.2623	32
33	6.9882	6.5317	6.1312	5.7770	5.4615	5.1787	4.6927	4.2900	33
34	7.0583	6.5929	6.1851	5.8249	5.5042	5.2171	4.7242	4.3163	34
35	7.1254	6.6515	6.2366	5.8705	5.5450	5.2537	4.7541	4.3414	35
36	7.1898	6.7075	6.2859	5.9142	5.5839	5.2886	4.7827	4.3652	36
37	7.2516	6.7613	6.3331	5.9559	5.6211	5.3219	4.8100	4.3879	37
38	7.3108	6.8128	6.3782	5.9958	5.6566	5.3538	4.8360	4.4095	38
39	7.3677	6.8621	6.4215	6.0340	5.6906	5.3842	4.8608	4.4301	39
40	7.4224	6.9096	6.4630	6.0706	5.7232	5.4134	4.8846	4.4498	40
41	7.4750	6.9551	6.5028	6.1058	5.7544	5.4413	4.9073	4.4687	41
42	7.5255	6.9988	6.5410	6.1394	5.7843	5.4680	4.9290	4.4867	42
43	7.5741	7.0409	6.5777	6.1718	5.8130	5.4937	4.9498	4.5039	43
44	7.6209	7.0813	6.6130	6.2028	5.8405	5.5182	4.9698	4.5204	44
45	7.6660	7.1202	6.6469	6.2326	5.8670	5.5418	4.9889	4.5363	45
46	7.7094	7.1576	6.6795	6.2613	5.8924	5.5645	5.0072	4.5514	46
47	7.7513	7.1937	6.7109	6.2889	5.9168	5.5862	5.0248	4.5660	47
48	7.7916	7.2284	6.7411	6.3154	5.9402	5.6072	5.0418	4.5799	48
49	7.8305	7.2619	6.7702	6.3409	5.9628	5.6273	5.0580	4.5934	49
50	7.8680	7.2941	6.7982	6.3655	5.9846	5.6466	5.0737	4.6062	50

YEARS' PURCHASE

Yrs.	11	12	13	14	15	16	18	20	Yrs.
				Rate Per Cent					
51	7.9042	7.3252	6.8253	6.3892	6.0055	5.6653	5.0887	4.6186	51
52	7.9392	7.3552	6.8513	6.4120	6.0256	5.6832	5.1031	4.6305	52
53	7.9729	7.3842	6.8764	6.4340	6.0451	5.7005	5.1171	4.6420	53
54	8.0056	7.4122	6.9007	6.4552	6.0638	5.7171	5.1305	4.6530	54
55	8.0371	7.4392	6.9241	6.4757	6.0818	5.7332	5.1434	4.6637	55
56	8.0675	7.4652	6.9467	6.4954	6.0993	5.7486	5.1559	4.6739	56
57	8.0969	7.4904	6.9685	6.5145	6.1161	5.7636	5.1679	4.6838	57
58	8.1254	7.5148	6.9895	6.5329	6.1323	5.7780	5.1794	4.6933	58
59	8.1529	7.5383	7.0099	6.5507	6.1480	5.7919	5.1906	4.7024	59
60	8.1796	7.5611	7.0296	6.5679	6.1631	5.8053	5.2014	4.7113	60
61	8.2054	7.5831	7.0486	6.5845	6.1777	5.8183	5.2118	4.7198	61
62	8.2303	7.6044	7.0670	6.6006	6.1919	5.8308	5.2219	4.7281	62
63	8.2545	7.6251	7.0848	6.6161	6.2055	5.8429	5.2316	4.7360	63
64	8.2779	7.6450	7.1021	6.6311	6.2187	5.8547	5.2410	4.7437	64
65	8.3005	7.6643	7.1187	6.6456	6.2315	5.8660	5.2500	4.7512	65
66	8.3225	7.6831	7.1349	6.6597	6.2439	5.8769	5.2588	4.7584	66
67	8.3437	7.7012	7.1505	6.6733	6.2559	5.8875	5.2673	4.7653	67
68	8.3644	7.7187	7.1656	6.6865	6.2674	5.8978	5.2755	4.7720	68
69	8.3843	7.7357	7.1803	6.6993	6.2786	5.9077	5.2835	4.7785	69
70	8.4037	7.7522	7.1945	6.7116	6.2895	5.9173	5.2911	4.7848	70
71	8.4225	7.7682	7.2083	6.7236	6.3000	5.9266	5.2986	4.7909	71
72	8.4407	7.7837	7.2216	6.7352	6.3102	5.9356	5.3058	4.7968	72
73	8.4584	7.7987	7.2345	6.7464	6.3201	5.9444	5.3128	4.8025	73
74	8.4755	7.8133	7.2471	6.7573	6.3296	5.9528	5.3195	4.8080	74
75	8.4921	7.8274	7.2592	6.7679	6.3389	5.9610	5.3261	4.8133	75
76	8.5083	7.8411	7.2710	6.7782	6.3479	5.9690	5.3324	4.8185	76
77	8.5239	7.8544	7.2824	6.7881	6.3566	5.9767	5.3386	4.8235	77
78	8.5391	7.8673	7.2935	6.7977	6.3651	5.9842	5.3445	4.8284	78
79	8.5539	7.8799	7.3043	6.8071	6.3733	5.9914	5.3503	4.8331	79
80	8.5682	7.8920	7.3147	6.8162	6.3812	5.9984	5.3559	4.8377	80
81	8.5822	7.9038	7.3249	6.8250	6.3889	6.0053	5.3613	4.8421	81
82	8.5957	7.9153	7.3347	6.8335	6.3964	6.0119	5.3666	4.8464	82
83	8.6088	7.9264	7.3443	6.8418	6.4037	6.0183	5.3717	4.8506	83
84	8.6216	7.9372	7.3536	6.8499	6.4107	6.0245	5.3767	4.8546	84
85	8.6340	7.9477	7.3626	6.8577	6.4176	6.0306	5.3815	4.8586	85
86	8.6460	7.9579	7.3713	6.8653	6.4242	6.0364	5.3862	4.8624	86
87	8.6577	7.9679	7.3798	6.8727	6.4307	6.0421	5.3907	4.8661	87
88	8.6691	7.9775	7.3881	6.8798	6.4370	6.0477	5.3951	4.8697	88
89	8.6801	7.9868	7.3961	6.8868	6.4431	6.0531	5.3994	4.8732	89
90	8.6909	7.9959	7.4039	6.8935	6.4490	6.0583	5.4036	4.8765	90
91	8.7013	8.0048	7.4115	6.9001	6.4547	6.0633	5.4076	4.8798	91
92	8.7115	8.0134	7.4189	6.9065	6.4603	6.0683	5.4115	4.8830	92
93	8.7213	8.0217	7.4260	6.9127	6.4657	6.0731	5.4153	4.8861	93
94	8.7309	8.0298	7.4330	6.9187	6.4710	6.0777	5.4190	4.8891	94
95	8.7403	8.0377	7.4398	6.9246	6.4761	6.0822	5.4226	4.8921	95
96	8.7493	8.0454	7.4463	6.9303	6.4811	6.0866	5.4261	4.8949	96
97	8.7582	8.0529	7.4527	6.9358	6.4860	6.0909	5.4295	4.8977	97
98	8.7668	8.0601	7.4589	6.9412	6.4907	6.0951	5.4328	4.9003	98
99	8.7751	8.0672	7.4650	6.9464	6.4953	6.0991	5.4360	4.9030	99
100	8.7832	8.0741	7.4709	6.9515	6.4997	6.1030	5.4391	4.9055	100

YEARS' PURCHASE

				Rate Per Cent					
Yrs.	4	4.5	5	5.5	6	6.25	6.5	6.75	Yrs.
1	0.5859	0.5842	0.5825	0.5808	0.5792	0.5783	0.5775	0.5766	1
2	1.1614	1.1547	1.1481	1.1415	1.1351	1.1318	1.1286	1.1255	2
3	1.7265	1.7117	1.6972	1.6829	1.6688	1.6619	1.6550	1.6482	3
4	2.2811	2.2554	2.2303	2.2057	2.1816	2.1698	2.1581	2.1465	4
5	2.8255	2.7861	2.7478	2.7106	2.6743	2.6566	2.6390	2.6218	5
6	3.3595	3.3040	3.2503	3.1983	3.1480	3.1234	3.0992	3.0754	6
7	3.8833	3.8094	3.7382	3.6696	3.6035	3.5713	3.5397	3.5086	7
8	4.3970	4.3024	4.2118	4.1249	4.0416	4.0012	3.9615	3.9227	8
9	4.9006	4.7834	4.6717	4.5650	4.4632	4.4139	4.3657	4.3186	9
10	5.3942	5.2525	5.1181	4.9904	4.8689	4.8104	4.7532	4.6974	10
11	5.8779	5.7101	5.5516	5.4016	5.2596	5.1913	5.1248	5.0600	11
12	6.3518	6.1562	5.9724	5.7992	5.6358	5.5575	5.4813	5.4073	12
13	6.8159	6.5913	6.3810	6.1837	5.9982	5.9096	5.8236	5.7400	13
14	7.2704	7.0154	6.7776	6.5555	6.3474	6.2483	6.1522	6.0590	14
15	7.7154	7.4288	7.1628	6.9151	6.6840	6.5741	6.4678	6.3649	15
16	8.1510	7.8318	7.5367	7.2630	7.0085	6.8878	6.7712	6.6585	16
17	8.5772	8.2245	7.8997	7.5995	7.3213	7.1897	7.0628	6.9402	17
18	8.9943	8.6072	8.2521	7.9251	7.6230	7.4805	7.3432	7.2108	18
19	9.4023	8.9802	8.5943	8.2402	7.9141	7.7606	7.6129	7.4707	19
20	9.8014	9.3435	8.9265	8.5451	8.1950	8.0304	7.8724	7.7204	20
21	10.1916	9.6975	9.2490	8.8402	8.4660	8.2905	8.1222	7.9605	21
22	10.5732	10.0423	9.5621	9.1258	8.7276	8.5412	8.3627	8.1914	22
23	10.9461	10.3781	9.8662	9.4023	8.9802	8.7830	8.5943	8.4135	23
24	11.3106	10.7052	10.1613	9.6700	9.2240	9.0161	8.8174	8.6272	24
25	11.6668	11.0238	10.4479	9.9292	9.4596	9.2410	9.0324	8.8329	25
26	12.0148	11.3340	10.7261	10.1801	9.6871	9.4580	9.2395	9.0309	26
27	12.3548	11.6360	10.9962	10.4231	9.9068	9.6674	9.4393	9.2217	27
28	12.6868	11.9300	11.2585	10.6585	10.1192	9.8695	9.6319	9.4054	28
29	13.0110	12.2163	11.5131	10.8864	10.3244	10.0646	9.8176	9.5824	29
30	13.3276	12.4950	11.7603	11.1071	10.5228	10.2530	9.9968	9.7530	30
31	13.6367	12.7662	12.0003	11.3210	10.7145	10.4350	10.1697	9.9175	31
32	13.9384	13.0303	12.2333	11.5281	10.8999	10.6107	10.3365	10.0761	32
33	14.2328	13.2872	12.4595	11.7288	11.0791	10.7805	10.4976	10.2291	33
34	14.5201	13.5373	12.6791	11.9232	11.2524	10.9445	10.6530	10.3767	34
35	14.8005	13.7807	12.8923	12.1116	11.4200	11.1030	10.8032	10.5191	35
36	15.0739	14.0174	13.0994	12.2941	11.5822	11.2562	10.9481	10.6565	36
37	15.3407	14.2478	13.3003	12.4710	11.7390	11.4043	11.0882	10.7891	37
38	15.6009	14.4720	13.4955	12.6424	11.8908	11.5475	11.2235	10.9172	38
39	15.8546	14.6901	13.6849	12.8085	12.0376	11.6859	11.3542	11.0408	39
40	16.1020	14.9022	13.8689	12.9695	12.1797	11.8198	11.4805	11.1602	40
41	16.3432	15.1086	14.0474	13.1255	12.3172	11.9492	11.6026	11.2756	41
42	16.5784	15.3094	14.2208	13.2768	12.4503	12.0745	11.7207	11.3870	42
43	16.8076	15.5046	14.3891	13.4234	12.5791	12.1956	11.8348	11.4947	43
44	17.0310	15.6945	14.5526	13.5655	12.7038	12.3128	11.9451	11.5987	44
45	17.2487	15.8792	14.7112	13.7033	12.8246	12.4262	12.0518	11.6993	45
46	17.4609	16.0589	14.8653	13.8368	12.9415	12.5359	12.1550	11.7965	46
47	17.6676	16.2335	15.0148	13.9663	13.0547	12.6421	12.2548	11.8905	47
48	17.8690	16.4034	15.1600	14.0919	13.1643	12.7449	12.3513	11.9814	48
49	18.0651	16.5686	15.3010	14.2136	13.2705	12.8444	12.4447	12.0692	49
50	18.2562	16.7292	15.4379	14.3316	13.3733	12.9407	12.5351	12.1542	50

YEARS' PURCHASE

				Rate Per Cent					
Yrs.	4	4.5	5	5.5	6	6.25	6.5	6.75	Yrs.
51	18.4423	16.8853	15.5707	14.4461	13.4729	13.0339	12.6226	12.2365	51
52	18.6236	17.0371	15.6997	14.5570	13.5694	13.1242	12.7072	12.3160	52
53	18.8001	17.1847	15.8250	14.6647	13.6628	13.2116	12.7892	12.3929	53
54	18.9720	17.3282	15.9466	14.7690	13.7534	13.2962	12.8685	12.4674	54
55	19.1393	17.4677	16.0647	14.8702	13.8411	13.3782	12.9452	12.5394	55
56	19.3022	17.6033	16.1793	14.9684	13.9261	13.4576	13.0196	12.6092	56
57	19.4608	17.7351	16.2906	15.0636	14.0085	13.5345	13.0915	12.6766	57
58	19.6152	17.8633	16.3986	15.1559	14.0883	13.6090	13.1612	12.7420	58
59	19.7655	17.9878	16.5035	15.2455	14.1657	13.6812	13.2287	12.8052	59
60	19.9118	18.1089	16.6053	15.3323	14.2406	13.7511	13.2941	12.8664	60
61	20.0541	18.2265	16.7042	15.4166	14.3133	13.8188	13.3574	12.9257	61
62	20.1926	18.3409	16.8002	15.4983	14.3837	13.8844	13.4187	12.9831	62
63	20.3274	18.4520	16.8934	15.5776	14.4520	13.9480	13.4781	13.0387	63
64	20.4585	18.5600	16.9839	15.6545	14.5181	14.0097	13.5356	13.0925	64
65	20.5861	18.6650	17.0717	15.7291	14.5823	14.0694	13.5913	13.1447	65
66	20.7103	18.7669	17.1570	15.8015	14.6445	14.1273	13.6453	13.1952	66
67	20.8311	18.8661	17.2398	15.8717	14.7048	14.1833	13.6977	13.2441	67
68	20.9486	18.9624	17.3202	15.9398	14.7632	14.2377	13.7484	13.2915	68
69	21.0628	19.0560	17.3983	16.0059	14.8199	14.2904	13.7975	13.3374	69
70	21.1740	19.1469	17.4741	16.0700	14.8748	14.3415	13.8451	13.3819	70
71	21.2821	19.2353	17.5476	16.1322	14.9281	14.3910	13.8912	13.4250	71
72	21.3873	19.3212	17.6191	16.1926	14.9798	14.4390	13.9360	13.4668	72
73	21.4896	19.4046	17.6884	16.2511	15.0299	14.4856	13.9793	13.5073	73
74	21.5890	19.4856	17.7557	16.3079	15.0784	14.5307	14.0213	13.5465	74
75	21.6857	19.5644	17.8211	16.3631	15.1256	14.5744	14.0621	13.5845	75
76	21.7798	19.6409	17.8846	16.4165	15.1712	14.6169	14.1016	13.6214	76
77	21.8712	19.7152	17.9462	16.4684	15.2156	14.6580	14.1398	13.6571	77
78	21.9601	19.7875	18.0060	16.5188	15.2585	14.6979	14.1769	13.6917	78
79	22.0466	19.8576	18.0641	16.5677	15.3002	14.7365	14.2129	13.7252	79
80	22.1306	19.9258	18.1205	16.6151	15.3407	14.7741	14.2478	13.7578	80
81	22.2123	19.9920	18.1752	16.6611	15.3799	14.8104	14.2816	13.7893	81
82	22.2918	20.0563	18.2284	16.7058	15.4179	14.8457	14.3144	13.8199	82
83	22.3690	20.1188	18.2800	16.7491	15.4548	14.8799	14.3462	13.8495	83
84	22.4441	20.1795	18.3301	16.7911	15.4906	14.9131	14.3771	13.8782	84
85	22.5170	20.2385	18.3787	16.8319	15.5253	14.9453	14.4070	13.9061	85
86	22.5879	20.2958	18.4259	16.8715	15.5590	14.9765	14.4360	13.9331	86
87	22.6569	20.3514	18.4718	16.9100	15.5917	15.0067	14.4641	13.9593	87
88	22.7239	20.4054	18.5163	16.9473	15.6234	15.0361	14.4914	13.9847	88
89	22.7890	20.4579	18.5595	16.9835	15.6542	15.0646	14.5178	14.0094	89
90	22.8523	20.5089	18.6014	17.0186	15.6840	15.0922	14.5435	14.0333	90
91	22.9138	20.5585	18.6422	17.0527	15.7129	15.1190	14.5684	14.0564	91
92	22.9736	20.6066	18.6817	17.0858	15.7410	15.1450	14.5925	14.0789	92
93	23.0317	20.6533	18.7201	17.1179	15.7683	15.1703	14.6159	14.1007	93
94	23.0881	20.6987	18.7574	17.1490	15.7947	15.1947	14.6387	14.1218	94
95	23.1430	20.7427	18.7936	17.1793	15.8204	15.2185	14.6607	14.1423	95
96	23.1963	20.7856	18.8287	17.2086	15.8453	15.2415	14.6821	14.1622	96
97	23.2481	20.8271	18.8628	17.2371	15.8694	15.2639	14.7028	14.1815	97
98	23.2984	20.8675	18.8960	17.2648	15.8929	15.2855	14.7229	14.2002	98
99	23.3473	20.9068	18.9281	17.2916	15.9156	15.3066	14.7424	14.2184	99
100	23.3949	20.9448	18.9593	17.3177	15.9377	15.3270	14.7614	14.2360	100

YEARS' PURCHASE

				Rate Per Cent					
Yrs.	7	7.25	7.5	8	8.5	9	9.5	10	Yrs.
1	0.5758	0.5750	0.5742	0.5725	0.5709	0.5693	0.5676	0.5660	1
2	1.1223	1.1192	1.1160	1.1099	1.1037	1.0977	1.0917	1.0858	2
3	1.6415	1.6347	1.6281	1.6149	1.6020	1.5893	1.5767	1.5644	3
4	2.1350	2.1237	2.1125	2.0904	2.0688	2.0476	2.0268	2.0065	4
5	2.6047	2.5878	2.5712	2.5386	2.5067	2.4757	2.4454	2.4159	5
6	3.0519	3.0288	3.0060	2.9615	2.9183	2.8764	2.8356	2.7959	6
7	3.4781	3.4481	3.4187	3.3612	3.3057	3.2519	3.1999	3.1495	7
8	3.8846	3.8472	3.8106	3.7393	3.6707	3.6045	3.5407	3.4791	8
9	4.2725	4.2273	4.1831	4.0974	4.0152	3.9361	3.8602	3.7871	9
10	4.6429	4.5896	4.5375	4.4369	4.3406	4.2484	4.1600	4.0752	10
11	4.9968	4.9351	4.8750	4.7590	4.6484	4.5428	4.4419	4.3454	11
12	5.3351	5.2649	5.1965	5.0649	4.9398	4.8207	4.7073	4.5990	12
13	5.6588	5.5799	5.5031	5.3557	5.2161	5.0835	4.9575	4.8376	13
14	5.9686	5.8808	5.7956	5.6324	5.4781	5.3321	5.1936	5.0622	14
15	6.2652	6.1686	6.0749	5.8958	5.7270	5.5676	5.4168	5.2740	15
16	6.5494	6.4439	6.3418	6.1469	5.9636	5.7909	5.6279	5.4739	16
17	6.8219	6.7075	6.5968	6.3862	6.1886	6.0028	5.8279	5.6629	17
18	7.0831	6.9599	6.8408	6.6146	6.4028	6.2042	6.0175	5.8418	18
19	7.3337	7.2017	7.0743	6.8326	6.6069	6.3956	6.1975	6.0112	19
20	7.5743	7.4335	7.2979	7.0410	6.8015	6.5778	6.3684	6.1718	20
21	7.8052	7.6558	7.5120	7.2401	6.9872	6.7513	6.5308	6.3243	21
22	8.0270	7.8691	7.7173	7.4306	7.1644	6.9166	6.6854	6.4692	22
23	8.2402	8.0739	7.9141	7.6129	7.3337	7.0743	6.8326	6.6069	23
24	8.4451	8.2705	8.1029	7.7874	7.4956	7.2248	6.9729	6.7380	24
25	8.6421	8.4593	8.2841	7.9546	7.6503	7.3685	7.1067	6.8628	25
26	8.8315	8.6408	8.4581	8.1149	7.7985	7.5058	7.2343	6.9818	26
27	9.0139	8.8152	8.6251	8.2685	7.9403	7.6371	7.3562	7.0952	27
28	9.1893	8.9829	8.7856	8.4159	8.0761	7.7626	7.4726	7.2035	28
29	9.3582	9.1443	8.9399	8.5574	8.2063	7.8828	7.5839	7.3069	29
30	9.5209	9.2995	9.0883	8.6932	8.3311	7.9979	7.6904	7.4056	30
31	9.6776	9.4490	9.2309	8.8237	8.4508	8.1082	7.7923	7.5001	31
32	9.8286	9.5928	9.3682	8.9490	8.5657	8.2139	7.8899	7.5905	32
33	9.9740	9.7314	9.5003	9.0695	8.6760	8.3153	7.9834	7.6769	33
34	10.1143	9.8649	9.6274	9.1853	8.7819	8.4126	8.0730	7.7598	34
35	10.2495	9.9935	9.7499	9.2967	8.8837	8.5059	8.1589	7.8391	35
36	10.3799	10.1174	9.8678	9.4038	8.9815	8.5955	8.2413	7.9152	36
37	10.5057	10.2369	9.9814	9.5070	9.0756	8.6816	8.3204	7.9881	37
38	10.6271	10.3521	10.0909	9.6062	9.1660	8.7643	8.3964	8.0581	38
39	10.7442	10.4632	10.1965	9.7019	9.2530	8.8438	8.4693	8.1253	39
40	10.8573	10.5704	10.2982	9.7939	9.3367	8.9203	8.5394	8.1897	40
41	10.9664	10.6738	10.3964	9.8827	9.4173	8.9938	8.6068	8.2517	41
42	11.0718	10.7736	10.4910	9.9682	9.4949	9.0646	8.6716	8.3112	42
43	11.1736	10.8699	10.5824	10.0506	9.5697	9.1327	8.7339	8.3684	43
44	11.2719	10.9629	10.6705	10.1300	9.6417	9.1982	8.7938	8.4234	44
45	11.3668	11.0527	10.7555	10.2067	9.7111	9.2614	8.8515	8.4764	45
46	11.4586	11.1395	10.8377	10.2806	9.7780	9.3222	8.9070	8.5273	46
47	11.5472	11.2232	10.9169	10.3519	9.8424	9.3808	8.9605	8.5763	47
48	11.6329	11.3042	10.9935	10.4207	9.9046	9.4373	9.0120	8.6234	48
49	11.7157	11.3824	11.0674	10.4871	9.9646	9.4917	9.0616	8.6689	49
50	11.7958	11.4579	11.1389	10.5512	10.0225	9.5442	9.1095	8.7126	50

YEARS' PURCHASE

				Rate Per Cent					
Yrs.	7	7.25	7.5	8	8.5	9	9.5	10	Yrs.
51	11.8732	11.5310	11.2079	10.6131	10.0783	9.5948	9.1556	8.7548	51
52	11.9481	11.6016	11.2746	10.6729	10.1322	9.6436	9.2000	8.7954	52
53	12.0205	11.6698	11.3390	10.7306	10.1842	9.6908	9.2429	8.8346	53
54	12.0905	11.7358	11.4013	10.7864	10.2344	9.7362	9.2842	8.8724	54
55	12.1583	11.7996	11.4615	10.8403	10.2829	9.7801	9.3241	8.9088	55
56	12.2238	11.8613	11.5197	10.8924	10.3298	9.8225	9.3626	8.9439	56
57	12.2872	11.9210	11.5761	10.9427	10.3750	9.8634	9.3998	8.9779	57
58	12.3486	11.9788	11.6305	10.9913	10.4187	9.9029	9.4357	9.0106	58
59	12.4080	12.0347	11.6832	11.0384	10.4610	9.9410	9.4703	9.0421	59
60	12.4655	12.0887	11.7341	11.0838	10.5018	9.9779	9.5038	9.0726	60
61	12.5211	12.1411	11.7834	11.1278	10.5413	10.0135	9.5361	9.1021	61
62	12.5750	12.1917	11.8311	11.1703	10.5794	10.0479	9.5673	9.1305	62
63	12.6271	12.2407	11.8772	11.2114	10.6163	10.0812	9.5974	9.1580	63
64	12.6776	12.2881	11.9219	11.2512	10.6520	10.1133	9.6266	9.1845	64
65	12.7265	12.3341	11.9651	11.2897	10.6865	10.1444	9.6547	9.2101	65
66	12.7738	12.3785	12.0069	11.3269	10.7198	10.1745	9.6819	9.2349	66
67	12.8197	12.4216	12.0474	11.3630	10.7521	10.2035	9.7082	9.2588	67
68	12.8641	12.4632	12.0866	11.3978	10.7833	10.2316	9.7337	9.2820	68
69	12.9071	12.5036	12.1246	11.4316	10.8135	10.2588	9.7583	9.3043	69
70	12.9487	12.5427	12.1614	11.4642	10.8427	10.2851	9.7821	9.3259	70
71	12.9891	12.5806	12.1969	11.4959	10.8710	10.3106	9.8051	9.3469	71
72	13.0282	12.6172	12.2314	11.5265	10.8984	10.3352	9.8274	9.3671	72
73	13.0660	12.6527	12.2648	11.5561	10.9249	10.3590	9.8489	9.3867	73
74	13.1028	12.6872	12.2971	11.5848	10.9505	10.3821	9.8697	9.4056	74
75	13.1383	12.7205	12.3284	11.6126	10.9753	10.4044	9.8899	9.4239	75
76	13.1728	12.7528	12.3588	11.6395	10.9994	10.4260	9.9094	9.4416	76
77	13.2062	12.7841	12.3882	11.6656	11.0227	10.4469	9.9283	9.4588	77
78	13.2385	12.8144	12.4166	11.6908	11.0452	10.4671	9.9466	9.4753	78
79	13.2699	12.8438	12.4442	11.7153	11.0670	10.4867	9.9643	9.4914	79
80	13.3003	12.8723	12.4710	11.7390	11.0882	10.5057	9.9814	9.5069	80
81	13.3298	12.8999	12.4969	11.7619	11.1086	10.5241	9.9980	9.5220	81
82	13.3583	12.9266	12.5220	11.7842	11.1285	10.5419	10.0141	9.5366	82
83	13.3860	12.9526	12.5463	11.8057	11.1477	10.5591	10.0296	9.5507	83
84	13.4129	12.9777	12.5699	11.8266	11.1663	10.5758	10.0447	9.5643	84
85	13.4389	13.0021	12.5927	11.8468	11.1843	10.5920	10.0593	9.5775	85
86	13.4641	13.0257	12.6149	11.8664	11.2018	10.6077	10.0734	9.5904	86
87	13.4886	13.0486	12.6364	11.8854	11.2187	10.6229	10.0871	9.6028	87
88	13.5123	13.0708	12.6572	11.9038	11.2351	10.6376	10.1003	9.6148	88
89	13.5353	13.0923	12.6774	11.9217	11.2510	10.6518	10.1132	9.6264	89
90	13.5576	13.1132	12.6969	11.9390	11.2664	10.6656	10.1256	9.6377	90
91	13.5792	13.1334	12.7159	11.9557	11.2814	10.6790	10.1377	9.6486	91
92	13.6002	13.1530	12.7343	11.9720	11.2958	10.6920	10.1494	9.6592	92
93	13.6206	13.1720	12.7521	11.9878	11.3099	10.7045	10.1607	9.6695	93
94	13.6403	13.1905	12.7694	12.0030	11.3235	10.7167	10.1717	9.6794	94
95	13.6594	13.2084	12.7862	12.0178	11.3366	10.7285	10.1823	9.6890	95
96	13.6780	13.2257	12.8024	12.0322	11.3494	10.7399	10.1926	9.6984	96
97	13.6960	13.2425	12.8182	12.0461	11.3618	10.7510	10.2026	9.7074	97
98	13.7134	13.2588	12.8335	12.0596	11.3738	10.7618	10.2123	9.7162	98
99	13.7303	13.2747	12.8483	12.0727	11.3854	10.7722	10.2217	9.7247	99
100	13.7468	13.2900	12.8627	12.0854	11.3967	10.7823	10.2308	9.7329	100

YEARS' PURCHASE

Yrs.	11	12	13	14	15	16	18	20	Yrs.
1	0.5629	0.5597	0.5566	0.5535	0.5505	0.5474	0.5415	0.5357	1
2	1.0741	1.0627	1.0515	1.0406	1.0298	1.0193	0.9990	0.9794	2
3	1.5403	1.5170	1.4943	1.4723	1.4509	1.4302	1.3904	1.3528	3
4	1.9670	1.9291	1.8926	1.8574	1.8236	1.7909	1.7290	1.6712	4
5	2.3589	2.3045	2.2526	2.2030	2.1555	2.1100	2.0246	1.9458	5
6	2.7199	2.6479	2.5796	2.5147	2.4530	2.3943	2.2849	2.1850	6
7	3.0533	2.9629	2.8776	2.7971	2.7210	2.6489	2.5157	2.3951	7
8	3.3622	3.2528	3.1503	3.0541	2.9636	2.8783	2.7216	2.5811	8
9	3.6489	3.5204	3.4007	3.2889	3.1841	3.0859	2.9065	2.7468	9
10	3.9157	3.7681	3.6313	3.5040	3.3854	3.2746	3.0733	2.8953	10
11	4.1644	3.9979	3.8442	3.7019	3.5698	3.4467	3.2245	3.0291	11
12	4.3968	4.2116	4.0414	3.8844	3.7392	3.6044	3.3621	3.1502	12
13	4.6143	4.4108	4.2245	4.0532	3.8954	3.7493	3.4878	3.2603	13
14	4.8183	4.5968	4.3948	4.2097	4.0397	3.8828	3.6030	3.3608	14
15	5.0097	4.7707	4.5535	4.3552	4.1734	4.0062	3.7090	3.4529	15
16	5.1898	4.9338	4.7018	4.4907	4.2977	4.1206	3.8068	3.5375	16
17	5.3594	5.0868	4.8406	4.6171	4.4133	4.2268	3.8973	3.6155	17
18	5.5193	5.2306	4.9706	4.7353	4.5212	4.3256	3.9812	3.6876	18
19	5.6703	5.3661	5.0928	4.8460	4.6220	4.4178	4.0592	3.7544	19
20	5.8131	5.4937	5.2076	4.9499	4.7164	4.5040	4.1318	3.8164	20
21	5.9481	5.6142	5.3158	5.0475	4.8049	4.5846	4.1996	3.8742	21
22	6.0761	5.7281	5.4177	5.1393	4.8881	4.6603	4.2630	3.9281	22
23	6.1975	5.8358	5.5140	5.2258	4.9663	4.7313	4.3223	3.9784	23
24	6.3126	5.9378	5.6050	5.3075	5.0400	4.7982	4.3780	4.0256	24
25	6.4221	6.0345	5.6911	5.3846	5.1095	4.8611	4.4304	4.0698	25
26	6.5261	6.1263	5.7727	5.4576	5.1752	4.9205	4.4797	4.1113	26
27	6.6251	6.2135	5.8500	5.5267	5.2372	4.9766	4.5261	4.1504	27
28	6.7194	6.2964	5.9234	5.5922	5.2960	5.0296	4.5699	4.1872	28
29	6.8093	6.3752	5.9931	5.6543	5.3517	5.0798	4.6113	4.2219	29
30	6.8950	6.4503	6.0594	5.7132	5.4045	5.1274	4.6505	4.2547	30
31	6.9768	6.5218	6.1225	5.7693	5.4546	5.1725	4.6875	4.2857	31
32	7.0550	6.5900	6.1826	5.8226	5.5022	5.2153	4.7227	4.3151	32
33	7.1296	6.6551	6.2399	5.8734	5.5475	5.2560	4.7560	4.3429	33
34	7.2010	6.7173	6.2945	5.9217	5.5907	5.2946	4.7877	4.3693	34
35	7.2693	6.7767	6.3466	5.9678	5.6317	5.3315	4.8178	4.3943	35
36	7.3346	6.8334	6.3963	6.0118	5.6709	5.3665	4.8464	4.4181	36
37	7.3972	6.8877	6.4439	6.0538	5.7082	5.4000	4.8736	4.4408	37
38	7.4572	6.9397	6.4893	6.0939	5.7439	5.4319	4.8996	4.4623	38
39	7.5147	6.9894	6.5328	6.1322	5.7779	5.4623	4.9243	4.4828	39
40	7.5698	7.0371	6.5745	6.1689	5.8104	5.4914	4.9479	4.5024	40
41	7.6227	7.0828	6.6143	6.2040	5.8416	5.5191	4.9705	4.5211	41
42	7.6735	7.1266	6.6525	6.2375	5.8713	5.5457	4.9920	4.5389	42
43	7.7222	7.1686	6.6891	6.2697	5.8998	5.5711	5.0126	4.5559	43
44	7.7690	7.2090	6.7242	6.3005	5.9271	5.5955	5.0323	4.5721	44
45	7.8140	7.2477	6.7579	6.3301	5.9533	5.6188	5.0511	4.5877	45
46	7.8573	7.2849	6.7902	6.3585	5.9783	5.6411	5.0692	4.6025	46
47	7.8988	7.3206	6.8212	6.3857	6.0024	5.6625	5.0864	4.6168	47
48	7.9388	7.3549	6.8511	6.4118	6.0254	5.6830	5.1030	4.6304	48
49	7.9773	7.3880	6.8797	6.4369	6.0476	5.7027	5.1189	4.6435	49
50	8.0144	7.4197	6.9072	6.4610	6.0689	5.7216	5.1341	4.6560	50

YEARS' PURCHASE

			Rate Per Cent						
Yrs.	11	12	13	14	15	16	18	20	Yrs.
51	8.0500	7.4503	6.9337	6.4841	6.0893	5.7398	5.1487	4.6680	51
52	8.0844	7.4797	6.9592	6.5064	6.1089	5.7572	5.1627	4.6796	52
53	8.1175	7.5080	6.9837	6.5278	6.1278	5.7740	5.1762	4.6906	53
54	8.1493	7.5353	7.0073	6.5484	6.1459	5.7901	5.1892	4.7013	54
55	8.1801	7.5615	7.0300	6.5682	6.1634	5.8056	5.2016	4.7115	55
56	8.2097	7.5868	7.0518	6.5873	6.1802	5.8205	5.2136	4.7213	56
57	8.2382	7.6112	7.0729	6.6057	6.1964	5.8348	5.2251	4.7307	57
58	8.2658	7.6347	7.0932	6.6234	6.2119	5.8486	5.2361	4.7398	58
59	8.2923	7.6574	7.1127	6.6404	6.2269	5.8619	5.2468	4.7485	59
60	8.3180	7.6792	7.1316	6.6568	6.2414	5.8747	5.2570	4.7569	60
61	8.3427	7.7003	7.1497	6.6727	6.2553	5.8870	5.2669	4.7650	61
62	8.3666	7.7206	7.1673	6.6879	6.2687	5.8989	5.2764	4.7727	62
63	8.3896	7.7403	7.1842	6.7027	6.2816	5.9104	5.2856	4.7802	63
64	8.4119	7.7592	7.2005	6.7168	6.2941	5.9214	5.2944	4.7875	64
65	8.4334	7.7775	7.2162	6.7305	6.3061	5.9320	5.3029	4.7944	65
66	8.4541	7.7951	7.2314	6.7438	6.3177	5.9423	5.3111	4.8011	66
67	8.4742	7.8122	7.2461	6.7565	6.3289	5.9522	5.3190	4.8076	67
68	8.4936	7.8286	7.2603	6.7688	6.3397	5.9618	5.3266	4.8138	68
69	8.5123	7.8446	7.2739	6.7807	6.3501	5.9710	5.3340	4.8198	69
70	8.5304	7.8599	7.2872	6.7922	6.3602	5.9799	5.3411	4.8256	70
71	8.5479	7.8748	7.2999	6.8033	6.3699	5.9885	5.3479	4.8312	71
72	8.5648	7.8891	7.3123	6.8140	6.3793	5.9968	5.3546	4.8366	72
73	8.5812	7.9030	7.3242	6.8243	6.3884	6.0048	5.3609	4.8418	73
74	8.5970	7.9164	7.3357	6.8343	6.3971	6.0125	5.3671	4.8468	74
75	8.6123	7.9294	7.3468	6.8440	6.4056	6.0200	5.3731	4.8517	75
76	8.6271	7.9419	7.3576	6.8533	6.4138	6.0272	5.3788	4.8564	76
77	8.6414	7.9540	7.3680	6.8624	6.4217	6.0342	5.3844	4.8609	77
78	8.6552	7.9658	7.3781	6.8711	6.4293	6.0409	5.3898	4.8653	78
79	8.6686	7.9771	7.3878	6.8795	6.4367	6.0475	5.3950	4.8695	79
80	8.6816	7.9881	7.3972	6.8877	6.4439	6.0538	5.4000	4.8736	80
81	8.6941	7.9987	7.4063	6.8956	6.4508	6.0599	5.4048	4.8776	81
82	8.7063	8.0090	7.4151	6.9032	6.4575	6.0658	5.4095	4.8814	82
83	8.7180	8.0189	7.4236	6.9106	6.4639	6.0715	5.4140	4.8851	83
84	8.7294	8.0286	7.4319	6.9178	6.4702	6.0770	5.4184	4.8887	84
85	8.7404	8.0379	7.4399	6.9247	6.4762	6.0823	5.4227	4.8921	85
86	8.7511	8.0469	7.4476	6.9314	6.4821	6.0875	5.4268	4.8954	86
87	8.7614	8.0556	7.4551	6.9379	6.4877	6.0925	5.4307	4.8987	87
88	8.7714	8.0641	7.4623	6.9441	6.4932	6.0973	5.4346	4.9018	88
89	8.7811	8.0723	7.4693	6.9502	6.4985	6.1020	5.4383	4.9048	89
90	8.7905	8.0802	7.4761	6.9561	6.5037	6.1065	5.4419	4.9078	90
91	8.7996	8.0879	7.4827	6.9618	6.5086	6.1109	5.4454	4.9106	91
92	8.8084	8.0953	7.4891	6.9673	6.5135	6.1152	5.4488	4.9133	92
93	8.8169	8.1025	7.4952	6.9726	6.5181	6.1193	5.4520	4.9160	93
94	8.8252	8.1095	7.5012	6.9778	6.5226	6.1232	5.4552	4.9185	94
95	8.8332	8.1162	7.5070	6.9828	6.5270	6.1271	5.4582	4.9210	95
96	8.8409	8.1228	7.5126	6.9876	6.5312	6.1308	5.4612	4.9234	96
97	8.8484	8.1291	7.5180	6.9923	6.5353	6.1344	5.4641	4.9258	97
98	8.8557	8.1353	7.5232	6.9969	6.5393	6.1379	5.4668	4.9280	98
99	8.8628	8.1412	7.5283	7.0013	6.5432	6.1413	5.4695	4.9302	99
100	8.8696	8.1470	7.5333	7.0055	6.5469	6.1446	5.4721	4.9323	100

YEARS' PURCHASE

				Rate Per Cent					
Yrs.	4	4.5	5	5.5	6	6.25	6.5	6.75	Yrs.
1	0.5859	0.5842	0.5825	0.5808	0.5792	0.5783	0.5775	0.5766	1
2	1.1669	1.1601	1.1534	1.1468	1.1403	1.1370	1.1338	1.1306	2
3	1.7424	1.7274	1.7126	1.6980	1.6837	1.6767	1.6697	1.6627	3
4	2.3122	2.2858	2.2600	2.2347	2.2100	2.1979	2.1859	2.1740	4
5	2.8759	2.8352	2.7955	2.7570	2.7195	2.7012	2.6830	2.6652	5
6	3.4332	3.3753	3.3193	3.2651	3.2126	3.1870	3.1619	3.1371	6
7	3.9838	3.9060	3.8312	3.7592	3.6898	3.6561	3.6230	3.5905	7
8	4.5274	4.4271	4.3313	4.2394	4.1514	4.1088	4.0670	4.0261	8
9	5.0636	4.9386	4.8195	4.7061	4.5979	4.5457	4.4946	4.4447	9
10	5.5923	5.4402	5.2961	5.1595	5.0297	4.9673	4.9063	4.8469	10
11	6.1132	5.9318	5.7610	5.5997	5.4472	5.3740	5.3027	5.2334	11
12	6.6260	6.4135	6.2143	6.0270	5.8507	5.7663	5.6844	5.6047	12
13	7.1307	6.8852	6.6560	6.4417	6.2407	6.1448	6.0518	5.9616	13
14	7.6269	7.3467	7.0864	6.8439	6.6175	6.5098	6.4055	6.3046	14
15	8.1146	7.7982	7.5055	7.2340	6.9815	6.8618	6.7460	6.6342	15
16	8.5935	8.2395	7.9135	7.6123	7.3332	7.2012	7.0738	6.9509	16
17	9.0636	8.6707	8.3104	7.9789	7.6728	7.5284	7.3893	7.2553	17
18	9.5248	9.0918	8.6965	8.3341	8.0007	7.8438	7.6930	7.5478	18
19	9.9770	9.5029	9.0719	8.6782	8.3173	8.1479	7.9852	7.8290	19
20	10.4200	9.9040	9.4367	9.0115	8.6230	8.4410	8.2665	8.0992	20
21	10.8538	10.2951	9.7911	9.3341	8.9179	8.7235	8.5373	8.3589	21
22	11.2784	10.6764	10.1353	9.6465	9.2026	8.9957	8.7978	8.6085	22
23	11.6938	11.0479	10.4695	9.9488	9.4773	9.2580	9.0485	8.8484	23
24	12.1000	11.4097	10.7939	10.2412	9.7423	9.5107	9.2898	9.0790	24
25	12.4969	11.7619	11.1087	10.5241	9.9980	9.7542	9.5220	9.3006	25
26	12.8846	12.1047	11.4139	10.7977	10.2446	9.9888	9.7454	9.5136	26
27	13.2630	12.4382	11.7099	11.0622	10.4824	10.2148	9.9604	9.7184	27
28	13.6323	12.7624	11.9969	11.3180	10.7118	10.4324	10.1673	9.9152	28
29	13.9926	13.0776	12.2750	11.5652	10.9330	10.6421	10.3663	10.1044	29
30	14.3437	13.3839	12.5444	11.8040	11.1462	10.8440	10.5578	10.2863	30
31	14.6860	13.6813	12.8054	12.0348	11.3517	11.0385	10.7420	10.4611	31
32	15.0193	13.9702	13.0581	12.2578	11.5499	11.2257	10.9193	10.6291	32
33	15.3439	14.2506	13.3027	12.4731	11.7409	11.4061	11.0898	10.7907	33
34	15.6598	14.5227	13.5395	12.6810	11.9249	11.5797	11.2539	10.9460	34
35	15.9671	14.7866	13.7686	12.8818	12.1023	11.7469	11.4118	11.0952	35
36	16.2659	15.0425	13.9903	13.0756	12.2732	11.9079	11.5636	11.2387	36
37	16.5565	15.2907	14.2047	13.2627	12.4379	12.0628	11.7097	11.3767	37
38	16.8388	15.5312	14.4120	13.4433	12.5966	12.2120	11.8502	11.5093	38
39	17.1131	15.7642	14.6125	13.6175	12.7495	12.3556	11.9854	11.6367	39
40	17.3795	15.9900	14.8062	13.7857	12.8967	12.4939	12.1155	11.7593	40
41	17.6380	16.2086	14.9935	13.9479	13.0386	12.6270	12.2406	11.8771	41
42	17.8890	16.4203	15.1744	14.1043	13.1752	12.7550	12.3609	11.9904	42
43	18.1324	16.6252	15.3492	14.2552	13.3068	12.8783	12.4766	12.0992	43
44	18.3685	16.8234	15.5181	14.4007	13.4335	12.9970	12.5880	12.2039	44
45	18.5975	17.0153	15.6812	14.5411	13.5555	13.1112	12.6951	12.3045	45
46	18.8194	17.2008	15.8386	14.6764	13.6730	13.2211	12.7981	12.4013	46
47	19.0344	17.3803	15.9907	14.8068	13.7862	13.3269	12.8972	12.4943	47
48	19.2427	17.5538	16.1374	14.9326	13.8951	13.4286	12.9925	12.5837	48
49	19.4445	17.7215	16.2791	15.0538	14.0000	13.5266	13.0841	12.6697	49
50	19.6398	17.8836	16.4158	15.1706	14.1010	13.6208	13.1723	12.7523	50

YEARS' PURCHASE

					Rate Per Cent				
Yrs.	4	4.5	5	5.5	6	6.25	6.5	6.75	Yrs.
51	19.8289	18.0403	16.5477	15.2832	14.1982	13.7115	13.2571	12.8318	51
52	20.0119	18.1917	16.6749	15.3917	14.2918	13.7988	13.3386	12.9082	52
53	20.1890	18.3379	16.7977	15.4962	14.3819	13.8827	13.4171	12.9816	53
54	20.3603	18.4791	16.9161	15.5969	14.4686	13.9635	13.4925	13.0522	54
55	20.5260	18.6155	17.0303	15.6940	14.5521	14.0412	13.5651	13.1201	55
56	20.6862	18.7471	17.1405	15.7875	14.6324	14.1160	13.6349	13.1854	56
57	20.8411	18.8743	17.2467	15.8775	14.7097	14.1880	13.7020	13.2482	57
58	20.9908	18.9970	17.3491	15.9642	14.7841	14.2572	13.7665	13.3085	58
59	21.1354	19.1154	17.4478	16.0478	14.8558	14.3238	13.8286	13.3665	59
60	21.2752	19.2296	17.5429	16.1282	14.9247	14.3879	13.8883	13.4223	60
61	21.4102	19.3399	17.6346	16.2057	14.9910	14.4495	13.9457	13.4759	61
62	21.5407	19.4462	17.7230	16.2803	15.0548	14.5088	14.0009	13.5274	62
63	21.6666	19.5488	17.8082	16.3522	15.1163	14.5658	14.0540	13.5770	63
64	21.7882	19.6478	17.8903	16.4214	15.1754	14.6207	14.1051	13.6247	64
65	21.9056	19.7432	17.9693	16.4880	15.2322	14.6734	14.1542	13.6705	65
66	22.0190	19.8352	18.0455	16.5521	15.2869	14.7242	14.2014	13.7145	66
67	22.1284	19.9240	18.1189	16.6138	15.3396	14.7730	14.2469	13.7569	67
68	22.2339	20.0095	18.1897	16.6733	15.3902	14.8200	14.2906	13.7976	68
69	22.3358	20.0919	18.2578	16.7305	15.4390	14.8652	14.3326	13.8368	69
70	22.4340	20.1714	18.3234	16.7855	15.4858	14.9087	14.3730	13.8744	70
71	22.5288	20.2480	18.3866	16.8385	15.5310	14.9505	14.4118	13.9106	71
72	22.6203	20.3218	18.4474	16.8896	15.5743	14.9907	14.4492	13.9454	72
73	22.7084	20.3930	18.5060	16.9387	15.6161	15.0293	14.4851	13.9789	73
74	22.7935	20.4615	18.5624	16.9859	15.6563	15.0665	14.5196	14.0111	74
75	22.8754	20.5276	18.6168	17.0314	15.6949	15.1023	14.5529	14.0420	75
76	22.9545	20.5912	18.6691	17.0752	15.7321	15.1367	14.5848	14.0717	76
77	23.0307	20.6525	18.7195	17.1173	15.7678	15.1698	14.6155	14.1003	77
78	23.1041	20.7115	18.7680	17.1579	15.8022	15.2017	14.6451	14.1278	78
79	23.1749	20.7684	18.8146	17.1969	15.8353	15.2323	14.6735	14.1543	79
80	23.2431	20.8232	18.8596	17.2344	15.8671	15.2617	14.7008	14.1797	80
81	23.3089	20.8759	18.9028	17.2705	15.8977	15.2900	14.7271	14.2041	81
82	23.3722	20.9267	18.9445	17.3053	15.9272	15.3173	14.7523	14.2276	82
83	23.4333	20.9756	18.9846	17.3387	15.9555	15.3435	14.7766	14.2502	83
84	23.4921	21.0227	19.0231	17.3709	15.9827	15.3686	14.8000	14.2719	84
85	23.5487	21.0681	19.0603	17.4018	16.0089	15.3929	14.8225	14.2928	85
86	23.6033	21.1118	19.0960	17.4316	16.0341	15.4162	14.8441	14.3129	86
87	23.6559	21.1538	19.1304	17.4603	16.0584	15.4386	14.8648	14.3322	87
88	23.7065	21.1943	19.1635	17.4879	16.0817	15.4601	14.8848	14.3508	88
89	23.7552	21.2332	19.1953	17.5144	16.1041	15.4808	14.9040	14.3686	89
90	23.8022	21.2707	19.2260	17.5399	16.1257	15.5008	14.9225	14.3858	90
91	23.8474	21.3069	19.2555	17.5644	16.1464	15.5199	14.9403	14.4023	91
92	23.8910	21.3416	19.2839	17.5880	16.1664	15.5384	14.9573	14.4182	92
93	23.9329	21.3751	19.3112	17.6107	16.1856	15.5561	14.9738	14.4335	93
94	23.9733	21.4073	19.3374	17.6326	16.2040	15.5731	14.9895	14.4481	94
95	24.0121	21.4382	19.3627	17.6536	16.2218	15.5895	15.0047	14.4622	95
96	24.0495	21.4681	19.3870	17.6738	16.2388	15.6053	15.0193	14.4758	96
97	24.0856	21.4968	19.4104	17.6933	16.2552	15.6204	15.0334	14.4888	97
98	24.1202	21.5244	19.4330	17.7120	16.2710	15.6350	15.0469	14.5014	98
99	24.1536	21.5510	19.4546	17.7300	16.2862	15.6490	15.0599	14.5134	99
100	24.1857	21.5765	19.4755	17.7473	16.3008	15.6625	15.0723	14.5250	100

YEARS' PURCHASE

Rate Per Cent

Yrs.	7	7.25	7.5	8	8.5	9	9.5	10	Yrs.
1	0.5758	0.5750	0.5742	0.5725	0.5709	0.5693	0.5676	0.5660	1
2	1.1274	1.1242	1.1211	1.1148	1.1087	1.1025	1.0965	1.0905	2
3	1.6559	1.6490	1.6423	1.6289	1.6157	1.6028	1.5900	1.5775	3
4	2.1622	2.1506	2.1391	2.1165	2.0943	2.0726	2.0514	2.0305	4
5	2.6475	2.6301	2.6129	2.5792	2.5464	2.5144	2.4832	2.4527	5
6	3.1126	3.0886	3.0649	3.0187	2.9738	2.9302	2.8879	2.8468	6
7	3.5585	3.5271	3.4963	3.4362	3.3782	3.3221	3.2678	3.2153	7
8	3.9860	3.9466	3.9081	3.8332	3.7611	3.6917	3.6248	3.5602	8
9	4.3958	4.3480	4.3013	4.2107	4.1239	4.0406	3.9606	3.8837	9
10	4.7889	4.7322	4.6769	4.5700	4.4679	4.3703	4.2768	4.1873	10
11	5.1658	5.0999	5.0357	4.9120	4.7943	4.6820	4.5749	4.4726	11
12	5.5273	5.4520	5.3786	5.2378	5.1041	4.9771	4.8562	4.7411	12
13	5.8741	5.7891	5.7065	5.5482	5.3984	5.2565	5.1219	4.9940	13
14	6.2067	6.1119	6.0199	5.8440	5.6781	5.5214	5.3730	5.2325	14
15	6.5259	6.4212	6.3197	6.1261	5.9441	5.7725	5.6106	5.4575	15
16	6.8322	6.7174	6.6065	6.3952	6.1971	6.0108	5.8354	5.6700	16
17	7.1260	7.0013	6.8809	6.6520	6.4379	6.2371	6.0485	5.8709	17
18	7.4080	7.2733	7.1434	6.8971	6.6672	6.4521	6.2504	6.0610	18
19	7.6787	7.5340	7.3948	7.1311	6.8856	6.6564	6.4420	6.2410	19
20	7.9384	7.7839	7.6354	7.3546	7.0937	6.8507	6.6239	6.4115	20
21	8.1878	8.0235	7.8657	7.5681	7.2922	7.0356	6.7965	6.5732	21
22	8.4271	8.2532	8.0864	7.7721	7.4814	7.2116	6.9607	6.7265	22
23	8.6569	8.4735	8.2977	7.9672	7.6619	7.3793	7.1167	6.8721	23
24	8.8775	8.6847	8.5002	8.1536	7.8342	7.5389	7.2651	7.0104	24
25	9.0893	8.8873	8.6941	8.3319	7.9987	7.6911	7.4063	7.1418	25
26	9.2926	9.0816	8.8800	8.5025	8.1558	7.8362	7.5408	7.2668	26
27	9.4879	9.2680	9.0582	8.6657	8.3058	7.9746	7.6689	7.3857	27
28	9.6754	9.4469	9.2289	8.8218	8.4492	8.1067	7.7909	7.4988	28
29	9.8555	9.6185	9.3926	8.9713	8.5862	8.2327	7.9072	7.6065	29
30	10.0284	9.7831	9.5496	9.1144	8.7171	8.3530	8.0182	7.7091	30
31	10.1945	9.9411	9.7000	9.2514	8.8423	8.4680	8.1240	7.8069	31
32	10.3540	10.0928	9.8444	9.3825	8.9621	8.5777	8.2250	7.9001	32
33	10.5072	10.2383	9.9828	9.5082	9.0767	8.6826	8.3214	7.9890	33
34	10.6544	10.3780	10.1155	9.6285	9.1863	8.7829	8.4134	8.0738	34
35	10.7958	10.5121	10.2429	9.7438	9.2912	8.8787	8.5013	8.1547	35
36	10.9316	10.6408	10.3650	9.8543	9.3916	8.9704	8.5853	8.2319	36
37	11.0620	10.7643	10.4823	9.9602	9.4877	9.0580	8.6656	8.3057	37
38	11.1874	10.8830	10.5947	10.0617	9.5798	9.1419	8.7423	8.3762	38
39	11.3078	10.9969	10.7027	10.1590	9.6679	9.2221	8.8156	8.4435	39
40	11.4235	11.1063	10.8062	10.2523	9.7524	9.2989	8.8858	8.5078	40
41	11.5346	11.2113	10.9056	10.3417	9.8333	9.3725	8.9529	8.5693	41
42	11.6414	11.3122	11.0011	10.4275	9.9108	9.4428	9.0171	8.6281	42
43	11.7440	11.4090	11.0926	10.5097	9.9850	9.5102	9.0785	8.6843	43
44	11.8426	11.5021	11.1806	10.5886	10.0562	9.5748	9.1374	8.7381	44
45	11.9373	11.5914	11.2650	10.6643	10.1245	9.6366	9.1936	8.7896	45
46	12.0284	11.6772	11.3460	10.7369	10.1899	9.6959	9.2476	8.8389	46
47	12.1159	11.7597	11.4238	10.8065	10.2526	9.7526	9.2992	8.8860	47
48	12.1999	11.8388	11.4985	10.8734	10.3127	9.8070	9.3486	8.9311	48
49	12.2807	11.9149	11.5703	10.9375	10.3704	9.8592	9.3960	8.9744	49
50	12.3583	11.9880	11.6391	10.9990	10.4257	9.9091	9.4414	9.0158	50

YEARS' PURCHASE

					Rate Per Cent				
Yrs.	7	7.25	7.5	8	8.5	9	9.5	10	Yrs.
51	12.4330	12.0582	11.7053	11.0581	10.4787	9.9570	9.4848	9.0554	51
52	12.5047	12.1256	11.7688	11.1148	10.5296	10.0030	9.5265	9.0934	52
53	12.5736	12.1904	11.8299	11.1692	10.5784	10.0470	9.5665	9.1298	53
54	12.6398	12.2526	11.8885	11.2214	10.6253	10.0893	9.6047	9.1646	54
55	12.7035	12.3124	11.9448	11.2716	10.6702	10.1298	9.6415	9.1980	55
56	12.7646	12.3699	11.9988	11.3197	10.7134	10.1687	9.6767	9.2301	56
57	12.8234	12.4251	12.0508	11.3659	10.7547	10.2059	9.7104	9.2608	57
58	12.8800	12.4782	12.1007	11.4103	10.7945	10.2417	9.7428	9.2902	58
59	12.9343	12.5291	12.1486	11.4529	10.8326	10.2760	9.7738	9.3185	59
60	12.9865	12.5781	12.1947	11.4938	10.8692	10.3090	9.8036	9.3455	60
61	13.0367	12.6252	12.2389	11.5331	10.9043	10.3406	9.8322	9.3715	61
62	13.0849	12.6704	12.2814	11.5709	10.9381	10.3709	9.8596	9.3964	62
63	13.1313	12.7139	12.3223	11.6071	10.9705	10.4000	9.8859	9.4203	63
64	13.1759	12.7557	12.3615	11.6419	11.0015	10.4279	9.9112	9.4432	64
65	13.2187	12.7959	12.3992	11.6754	11.0314	10.4547	9.9354	9.4652	65
66	13.2599	12.8344	12.4354	11.7075	11.0601	10.4805	9.9586	9.4863	66
67	13.2995	12.8715	12.4703	11.7383	11.0876	10.5052	9.9810	9.5065	67
68	13.3375	12.9072	12.5037	11.7680	11.1140	10.5289	10.0024	9.5260	68
69	13.3741	12.9414	12.5359	11.7965	11.1394	10.5517	10.0229	9.5446	69
70	13.4093	12.9744	12.5667	11.8238	11.1638	10.5736	10.0427	9.5625	70
71	13.4431	13.0060	12.5964	11.8501	11.1872	10.5946	10.0616	9.5797	71
72	13.4756	13.0364	12.6250	11.8753	11.2097	10.6148	10.0798	9.5962	72
73	13.5069	13.0657	12.6524	11.8996	11.2313	10.6342	10.0973	9.6120	73
74	13.5369	13.0938	12.6787	11.9229	11.2521	10.6528	10.1141	9.6272	74
75	13.5658	13.1208	12.7041	11.9453	11.2721	10.6707	10.1302	9.6418	75
76	13.5935	13.1467	12.7284	11.9668	11.2912	10.6878	10.1456	9.6558	76
77	13.6202	13.1717	12.7518	11.9875	11.3096	10.7043	10.1605	9.6693	77
78	13.6459	13.1957	12.7743	12.0074	11.3273	10.7201	10.1748	9.6822	78
79	13.6705	13.2188	12.7959	12.0264	11.3443	10.7354	10.1885	9.6946	79
80	13.6942	13.2409	12.8167	12.0448	11.3606	10.7500	10.2016	9.7065	80
81	13.7170	13.2622	12.8366	12.0624	11.3763	10.7640	10.2143	9.7180	81
82	13.7389	13.2827	12.8558	12.0794	11.3914	10.7775	10.2264	9.7290	82
83	13.7600	13.3024	12.8743	12.0956	11.4058	10.7905	10.2381	9.7395	83
84	13.7803	13.3213	12.8920	12.1113	11.4198	10.8029	10.2493	9.7497	84
85	13.7997	13.3395	12.9090	12.1263	11.4331	10.8149	10.2601	9.7594	85
86	13.8185	13.3570	12.9254	12.1408	11.4460	10.8264	10.2704	9.7688	86
87	13.8365	13.3738	12.9412	12.1547	11.4583	10.8374	10.2804	9.7778	87
88	13.8538	13.3900	12.9563	12.1680	11.4702	10.8480	10.2899	9.7864	88
89	13.8704	13.4055	12.9708	12.1809	11.4816	10.8582	10.2991	9.7947	89
90	13.8864	13.4205	12.9848	12.1932	11.4925	10.8680	10.3079	9.8027	90
91	13.9018	13.4349	12.9983	12.2051	11.5031	10.8775	10.3164	9.8103	91
92	13.9166	13.4487	13.0112	12.2165	11.5132	10.8865	10.3245	9.8177	92
93	13.9308	13.4619	13.0236	12.2274	11.5229	10.8952	10.3323	9.8248	93
94	13.9444	13.4747	13.0356	12.2379	11.5323	10.9036	10.3399	9.8316	94
95	13.9576	13.4870	13.0471	12.2481	11.5413	10.9116	10.3471	9.8381	95
96	13.9702	13.4988	13.0581	12.2578	11.5499	10.9193	10.3540	9.8444	96
97	13.9824	13.5101	13.0687	12.2671	11.5582	10.9267	10.3607	9.8504	97
98	13.9940	13.5210	13.0789	12.2761	11.5662	10.9339	10.3671	9.8562	98
99	14.0053	13.5315	13.0887	12.2848	11.5739	10.9407	10.3733	9.8618	99
100	14.0161	13.5416	13.0981	12.2931	11.5812	10.9473	10.3792	9.8671	100

YEARS' PURCHASE

Rate Per Cent

Yrs.	11	12	13	14	15	16	18	20	Yrs.
1	0.5629	0.5597	0.5566	0.5535	0.5505	0.5474	0.5415	0.5357	1
2	1.0788	1.0672	1.0560	1.0449	1.0341	1.0235	1.0030	0.9833	2
3	1.5530	1.5293	1.5062	1.4839	1.4622	1.4411	1.4007	1.3626	3
4	1.9901	1.9513	1.9139	1.8780	1.8434	1.8100	1.7468	1.6878	4
5	2.3940	2.3380	2.2846	2.2336	2.1848	2.1381	2.0504	1.9696	5
6	2.7680	2.6935	2.6228	2.5558	2.4921	2.4315	2.3187	2.2160	6
7	3.1151	3.0210	2.9324	2.8489	2.7700	2.6953	2.5574	2.4330	7
8	3.4378	3.3236	3.2167	3.1164	3.0222	2.9336	2.7710	2.6255	8
9	3.7385	3.6038	3.4784	3.3615	3.2522	3.1497	2.9631	2.7973	9
10	4.0190	3.8637	3.7200	3.5866	3.4624	3.3465	3.1366	2.9514	10
11	4.2812	4.1054	3.9435	3.7939	3.6552	3.5263	3.2940	3.0904	11
12	4.5265	4.3305	4.1508	3.9853	3.8326	3.6911	3.4374	3.2163	12
13	4.7565	4.5405	4.3433	4.1625	3.9962	3.8426	3.5684	3.3307	13
14	4.9723	4.7368	4.5225	4.3269	4.1474	3.9822	3.6885	3.4351	14
15	5.1750	4.9204	4.6897	4.4796	4.2875	4.1112	3.7989	3.5306	15
16	5.3658	5.0925	4.8457	4.6218	4.4176	4.2307	3.9007	3.6184	16
17	5.5454	5.2540	4.9917	4.7544	4.5386	4.3416	3.9947	3.6992	17
18	5.7146	5.4057	5.1285	4.8783	4.6514	4.4447	4.0818	3.7737	18
19	5.8744	5.5484	5.2568	4.9942	4.7567	4.5407	4.1627	3.8427	19
20	6.0252	5.6828	5.3772	5.1028	4.8551	4.6303	4.2378	3.9067	20
21	6.1678	5.8094	5.4905	5.2047	4.9472	4.7140	4.3079	3.9662	21
22	6.3026	5.9289	5.5971	5.3004	5.0336	4.7924	4.3732	4.0215	22
23	6.4302	6.0417	5.6975	5.3904	5.1147	4.8658	4.4343	4.0731	23
24	6.5512	6.1484	5.7922	5.4751	5.1909	4.9347	4.4915	4.1213	24
25	6.6658	6.2492	5.8817	5.5549	5.2626	4.9995	4.5450	4.1663	25
26	6.7745	6.3447	5.9661	5.6302	5.3301	5.0604	4.5953	4.2085	26
27	6.8777	6.4351	6.0460	5.7013	5.3938	5.1178	4.6426	4.2481	27
28	6.9757	6.5208	6.1216	5.7685	5.4539	5.1718	4.6870	4.2853	28
29	7.0688	6.6021	6.1932	5.8320	5.5107	5.2228	4.7289	4.3203	29
30	7.1573	6.6793	6.2611	5.8922	5.5643	5.2710	4.7683	4.3532	30
31	7.2415	6.7525	6.3254	5.9491	5.6151	5.3165	4.8056	4.3842	31
32	7.3217	6.8222	6.3865	6.0031	5.6631	5.3596	4.8407	4.4134	32
33	7.3979	6.8884	6.4444	6.0543	5.7087	5.4004	4.8739	4.4410	33
34	7.4706	6.9513	6.4995	6.1028	5.7518	5.4390	4.9054	4.4671	34
35	7.5398	7.0112	6.5518	6.1490	5.7928	5.4756	4.9351	4.4918	35
36	7.6058	7.0682	6.6016	6.1928	5.8316	5.5103	4.9633	4.5151	36
37	7.6688	7.1225	6.6490	6.2344	5.8686	5.5433	4.9900	4.5372	37
38	7.7288	7.1743	6.6940	6.2741	5.9037	5.5746	5.0154	4.5582	38
39	7.7861	7.2236	6.7370	6.3117	5.9370	5.6043	5.0394	4.5780	39
40	7.8407	7.2707	6.7779	6.3476	5.9688	5.6326	5.0623	4.5969	40
41	7.8929	7.3155	6.8168	6.3818	5.9990	5.6594	5.0840	4.6148	41
42	7.9428	7.3583	6.8540	6.4144	6.0277	5.6850	5.1046	4.6318	42
43	7.9904	7.3992	6.8894	6.4454	6.0551	5.7094	5.1243	4.6479	43
44	8.0359	7.4382	6.9232	6.4750	6.0812	5.7326	5.1430	4.6633	44
45	8.0795	7.4755	6.9555	6.5032	6.1061	5.7547	5.1607	4.6779	45
46	8.1211	7.5111	6.9863	6.5301	6.1298	5.7758	5.1777	4.6918	46
47	8.1608	7.5451	7.0157	6.5558	6.1525	5.7959	5.1938	4.7051	47
48	8.1989	7.5776	7.0439	6.5803	6.1741	5.8150	5.2092	4.7177	48
49	8.2353	7.6087	7.0707	6.6038	6.1947	5.8333	5.2239	4.7297	49
50	8.2701	7.6384	7.0964	6.6262	6.2144	5.8508	5.2379	4.7412	50

YEARS' PURCHASE

				Rate Per Cent					
Yrs.	11	12	13	14	15	16	18	20	Yrs.
51	8.3035	7.6669	7.1209	6.6475	6.2332	5.8675	5.2512	4.7521	51
52	8.3354	7.6941	7.1444	6.6680	6.2512	5.8834	5.2640	4.7626	52
53	8.3660	7.7201	7.1668	6.6875	6.2683	5.8986	5.2762	4.7725	53
54	8.3952	7.7450	7.1883	6.7062	6.2848	5.9131	5.2878	4.7821	54
55	8.4233	7.7689	7.2088	6.7241	6.3005	5.9270	5.2989	4.7911	55
56	8.4501	7.7917	7.2285	6.7412	6.3155	5.9403	5.3095	4.7998	56
57	8.4759	7.8136	7.2473	6.7576	6.3298	5.9530	5.3197	4.8081	57
58	8.5005	7.8345	7.2653	6.7732	6.3436	5.9652	5.3294	4.8160	58
59	8.5241	7.8546	7.2826	6.7882	6.3567	5.9768	5.3386	4.8236	59
60	8.5468	7.8738	7.2991	6.8026	6.3693	5.9879	5.3475	4.8308	60
61	8.5685	7.8922	7.3149	6.8163	6.3814	5.9986	5.3560	4.8378	61
62	8.5893	7.9099	7.3301	6.8295	6.3929	6.0088	5.3641	4.8444	62
63	8.6093	7.9268	7.3446	6.8421	6.4039	6.0185	5.3719	4.8507	63
64	8.6284	7.9430	7.3586	6.8542	6.4145	6.0279	5.3793	4.8568	64
65	8.6468	7.9586	7.3719	6.8658	6.4247	6.0368	5.3865	4.8626	65
66	8.6644	7.9735	7.3847	6.8769	6.4344	6.0454	5.3933	4.8682	66
67	8.6812	7.9878	7.3970	6.8875	6.4437	6.0536	5.3998	4.8735	67
68	8.6974	8.0015	7.4087	6.8977	6.4526	6.0615	5.4061	4.8786	68
69	8.7130	8.0147	7.4200	6.9075	6.4612	6.0690	5.4121	4.8835	69
70	8.7279	8.0273	7.4308	6.9168	6.4693	6.0763	5.4178	4.8882	70
71	8.7422	8.0394	7.4412	6.9258	6.4772	6.0832	5.4234	4.8927	71
72	8.7559	8.0510	7.4511	6.9344	6.4847	6.0898	5.4286	4.8970	72
73	8.7691	8.0621	7.4607	6.9427	6.4920	6.0962	5.4337	4.9011	73
74	8.7818	8.0728	7.4698	6.9506	6.4989	6.1023	5.4386	4.9050	74
75	8.7939	8.0831	7.4786	6.9582	6.5055	6.1082	5.4432	4.9088	75
76	8.8056	8.0929	7.4870	6.9655	6.5119	6.1138	5.4477	4.9124	76
77	8.8168	8.1024	7.4951	6.9725	6.5180	6.1192	5.4520	4.9159	77
78	8.8275	8.1115	7.5029	6.9792	6.5239	6.1244	5.4561	4.9193	78
79	8.8378	8.1202	7.5103	6.9857	6.5295	6.1293	5.4600	4.9225	79
80	8.8477	8.1285	7.5175	6.9919	6.5349	6.1341	5.4638	4.9255	80
81	8.8572	8.1366	7.5243	6.9978	6.5401	6.1387	5.4674	4.9285	81
82	8.8664	8.1443	7.5309	7.0035	6.5451	6.1430	5.4709	4.9313	82
83	8.8751	8.1517	7.5373	7.0090	6.5499	6.1472	5.4742	4.9340	83
84	8.8836	8.1588	7.5433	7.0142	6.5545	6.1513	5.4774	4.9366	84
85	8.8916	8.1656	7.5492	7.0193	6.5589	6.1552	5.4805	4.9391	85
86	8.8994	8.1721	7.5548	7.0241	6.5631	6.1589	5.4834	4.9415	86
87	8.9069	8.1784	7.5601	7.0287	6.5672	6.1625	5.4863	4.9438	87
88	8.9140	8.1845	7.5653	7.0332	6.5711	6.1659	5.4890	4.9460	88
89	8.9209	8.1903	7.5703	7.0375	6.5748	6.1692	5.4916	4.9481	89
90	8.9275	8.1959	7.5750	7.0416	6.5784	6.1723	5.4941	4.9502	90
91	8.9339	8.2012	7.5796	7.0456	6.5818	6.1754	5.4965	4.9521	91
92	8.9400	8.2064	7.5840	7.0494	6.5851	6.1783	5.4988	4.9540	92
93	8.9459	8.2113	7.5882	7.0530	6.5883	6.1811	5.5010	4.9558	93
94	8.9515	8.2160	7.5923	7.0565	6.5914	6.1838	5.5032	4.9575	94
95	8.9569	8.2206	7.5962	7.0599	6.5943	6.1864	5.5052	4.9592	95
96	8.9621	8.2250	7.5999	7.0631	6.5971	6.1889	5.5072	4.9608	96
97	8.9671	8.2292	7.6035	7.0662	6.5998	6.1912	5.5091	4.9623	97
98	8.9719	8.2332	7.6069	7.0692	6.6024	6.1935	5.5109	4.9638	98
99	8.9765	8.2371	7.6103	7.0721	6.6049	6.1957	5.5126	4.9652	99
100	8.9810	8.2409	7.6134	7.0748	6.6073	6.1978	5.5143	4.9666	100

YEARS' PURCHASE

(DUAL RATE % PRINCIPLE)

OR

PRESENT VALUE OF
ONE POUND PER ANNUM

*receivable at the end of each year after
allowing for a sinking fund at a given
rate to replace the invested capital and
for the effect of income tax at* **50%** *on
that part of the income used to provide
the annual sinking fund instalment.*

AT RATES OF INTEREST FROM

4% to 20%

AND

ALLOWING FOR THE POSSIBLE INVESTMENT

• OF SINKING FUNDS AT

2·5%, 3% and 4%

INCOME TAX at 50%

Note:—Tables of Years' Purchase in which no allowance has been made for the effect
of income tax on that part of the income used to provide the annual sinking
fund instalment will be found on pages 53 to 71.

YEARS' PURCHASE

				Rate Per Cent					
Yrs.	4	4.5	5	5.5	6	6.25	6.5	6.75	Yrs.
1	0.4902	0.4890	0.4878	0.4866	0.4854	0.4848	0.4843	0.4837	1
2	0.9731	0.9684	0.9637	0.9591	0.9545	0.9522	0.9500	0.9477	2
3	1.4487	1.4383	1.4280	1.4179	1.4079	1.4030	1.3981	1.3932	3
4	1.9170	1.8988	1.8810	1.8635	1.8463	1.8378	1.8294	1.8210	4
5	2.3782	2.3502	2.3229	2.2962	2.2702	2.2574	2.2447	2.2322	5
6	2.8321	2.7925	2.7541	2.7167	2.6802	2.6624	2.6448	2.6274	6
7	3.2788	3.2259	3.1747	3.1251	3.0770	3.0535	3.0304	3.0076	7
8	3.7184	3.6505	3.5851	3.5219	3.4610	3.4313	3.4021	3.3734	8
9	4.1509	4.0665	3.9854	3.9076	3.8327	3.7963	3.7606	3.7256	9
10	4.5763	4.4739	4.3760	4.2823	4.1926	4.1491	4.1065	4.0648	10
11	4.9947	4.8730	4.7571	4.6466	4.5411	4.4901	4.4403	4.3915	11
12	5.4062	5.2639	5.1289	5.0006	4.8787	4.8199	4.7625	4.7065	12
13	5.8107	5.6466	5.4916	5.3448	5.2057	5.1388	5.0737	5.0101	13
14	6.2084	6.0214	5.8455	5.6795	5.5226	5.4474	5.3742	5.3030	14
15	6.5992	6.3884	6.1907	6.0048	5.8298	5.7460	5.6647	5.5856	15
16	6.9833	6.7477	6.5275	6.3212	6.1275	6.0351	5.9454	5.8583	16
17	7.3608	7.0995	6.8561	6.6289	6.4162	6.3149	6.2168	6.1216	17
18	7.7316	7.4438	7.1767	6.9281	6.6961	6.5859	6.4792	6.3759	18
19	8.0958	7.7808	7.4894	7.2191	6.9676	6.8483	6.7330	6.6216	19
20	8.4535	8.1107	7.7946	7.5022	7.2310	7.1026	6.9786	6.8590	20
21	8.8048	8.4335	8.0923	7.7776	7.4865	7.3489	7.2163	7.0884	21
22	9.1497	8.7494	8.3827	8.0455	7.7344	7.5876	7.4464	7.3103	22
23	9.4883	9.0586	8.6661	8.3061	7.9749	7.8191	7.6691	7.5249	23
24	9.8207	9.3610	8.9425	8.5598	8.2085	8.0434	7.8848	7.7324	24
25	10.1469	9.6570	9.2122	8.8066	8.4351	8.2609	8.0938	7.9332	25
26	10.4671	9.9465	9.4753	9.0467	8.6552	8.4719	8.2962	8.1276	26
27	10.7812	10.2298	9.7320	9.2804	8.8689	8.6765	8.4923	8.3157	27
28	11.0894	10.5069	9.9824	9.5079	9.0764	8.8750	8.6824	8.4979	28
29	11.3918	10.7779	10.2268	9.7293	9.2779	9.0676	8.8666	8.6743	29
30	11.6883	11.0430	10.4651	9.9448	9.4737	9.2545	9.0452	8.8452	30
31	11.9792	11.3022	10.6977	10.1546	9.6639	9.4359	9.2185	9.0108	31
32	12.2644	11.5558	10.9246	10.3588	9.8487	9.6120	9.3864	9.1712	32
33	12.5441	11.8038	11.1459	10.5576	10.0282	9.7829	9.5494	9.3267	33
34	12.8183	12.0462	11.3619	10.7511	10.2027	9.9489	9.7075	9.4775	34
35	13.0871	12.2833	11.5726	10.9396	10.3722	10.1101	9.8608	9.6236	35
36	13.3506	12.5152	11.7781	11.1231	10.5371	10.2666	10.0097	9.7653	36
37	13.6088	12.7418	11.9787	11.3018	10.6973	10.4186	10.1542	9.9028	37
38	13.8619	12.9634	12.1743	11.4758	10.8530	10.5663	10.2944	10.0361	38
39	14.1099	13.1800	12.3652	11.6452	11.0045	10.7098	10.4305	10.1655	39
40	14.3529	13.3918	12.5514	11.8102	11.1517	10.8492	10.5627	10.2910	40
41	14.5909	13.5988	12.7331	11.9709	11.2949	10.9847	10.6911	10.4128	41
42	14.8241	13.8012	12.9103	12.1275	11.4341	11.1164	10.8158	10.5310	42
43	15.0526	13.9990	13.0832	12.2799	11.5696	11.2443	10.9369	10.6458	43
44	15.2763	14.1923	13.2519	12.4284	11.7013	11.3687	11.0545	10.7572	44
45	15.4955	14.3812	13.4165	12.5731	11.8294	11.4896	11.1688	10.8654	45
46	15.7101	14.5659	13.5771	12.7140	11.9541	11.6072	11.2799	10.9705	46
47	15.9202	14.7464	13.7337	12.8513	12.0753	11.7215	11.3878	11.0726	47
48	16.1259	14.9227	13.8866	12.9850	12.1933	11.8326	11.4927	11.1717	48
49	16.3273	15.0950	14.0357	13.1153	12.3082	11.9407	11.5946	11.2680	49
50	16.5245	15.2634	14.1812	13.2422	12.4199	12.0459	11.6937	11.3616	50

YEARS' PURCHASE

					Rate Per Cent				
Yrs.	4	4.5	5	5.5	6	6.25	6.5	6.75	Yrs.
51	16.7175	15.4280	14.3231	13.3659	12.5286	12.1481	11.7900	11.4525	51
52	16.9065	15.5887	14.4615	13.4864	12.6344	12.2476	11.8837	11.5408	52
53	17.0914	15.7458	14.5966	13.6038	12.7374	12.3443	11.9748	11.6267	53
54	17.2724	15.8993	14.7284	13.7182	12.8376	12.4384	12.0633	11.7102	54
55	17.4495	16.0492	14.8570	13.8297	12.9352	12.5300	12.1494	11.7913	55
56	17.6227	16.1957	14.9824	13.9383	13.0302	12.6191	12.2332	11.8702	56
57	17.7923	16.3388	15.1048	14.0441	13.1227	12.7058	12.3147	11.9468	57
58	17.9582	16.4786	15.2242	14.1473	13.2127	12.7902	12.3939	12.0214	58
59	18.1205	16.6151	15.3407	14.2478	13.3003	12.8723	12.4710	12.0939	59
60	18.2793	16.7485	15.4543	14.3458	13.3857	12.9522	12.5460	12.1644	60
61	18.4346	16.8788	15.5652	14.4413	13.4688	13.0300	12.6190	12.2330	61
62	18.5865	17.0061	15.6734	14.5344	13.5497	13.1057	12.6900	12.2997	62
63	18.7351	17.1304	15.7789	14.6251	13.6285	13.1794	12.7590	12.3646	63
64	18.8804	17.2518	15.8818	14.7135	13.7052	13.2512	12.8263	12.4278	64
65	19.0225	17.3704	15.9823	14.7996	13.7799	13.3210	12.8917	12.4892	65
66	19.1615	17.4862	16.0803	14.8836	13.8527	13.3890	12.9554	12.5489	66
67	19.2974	17.5993	16.1759	14.9655	13.9236	13.4553	13.0174	12.6071	67
68	19.4303	17.7098	16.2692	15.0453	13.9927	13.5197	13.0777	12.6637	68
69	19.5602	17.8176	16.3602	15.1231	14.0599	13.5825	13.1364	12.7187	69
70	19.6873	17.9230	16.4489	15.1989	14.1254	13.6436	13.1936	12.7723	70
71	19.8115	18.0259	16.5355	15.2728	14.1893	13.7032	13.2493	12.8245	71
72	19.9329	18.1263	16.6200	15.3449	14.2514	13.7612	13.3035	12.8753	72
73	20.0516	18.2244	16.7025	15.4151	14.3120	13.8176	13.3562	12.9247	73
74	20.1676	18.3202	16.7829	15.4836	14.3710	13.8726	13.4076	12.9728	74
75	20.2810	18.4138	16.8614	15.5504	14.4285	13.9262	13.4577	13.0196	75
76	20.3919	18.5051	16.9379	15.6155	14.4845	13.9784	13.5064	13.0652	76
77	20.5002	18.5943	17.0126	15.6789	14.5391	14.0292	13.5538	13.1096	77
78	20.6061	18.6814	17.0855	15.7408	14.5923	14.0787	13.6000	13.1528	78
79	20.7096	18.7664	17.1566	15.8011	14.6441	14.1269	13.6450	13.1949	79
80	20.8107	18.8494	17.2259	15.8599	14.6946	14.1739	13.6889	13.2359	80
81	20.9096	18.9304	17.2936	15.9172	14.7438	14.2197	13.7316	13.2758	81
82	21.0062	19.0096	17.3596	15.9731	14.7918	14.2643	13.7731	13.3147	82
83	21.1005	19.0868	17.4240	16.0277	14.8385	14.3078	13.8136	13.3525	83
84	21.1927	19.1622	17.4868	16.0808	14.8841	14.3501	13.8531	13.3894	84
85	21.2828	19.2359	17.5481	16.1326	14.9284	14.3913	13.8916	13.4253	85
86	21.3709	19.3078	17.6079	16.1832	14.9717	14.4315	13.9290	13.4603	86
87	21.4569	19.3779	17.6663	16.2324	15.0139	14.4707	13.9655	13.4943	87
88	21.5409	19.4464	17.7232	16.2805	15.0550	14.5089	14.0010	13.5275	88
89	21.6230	19.5133	17.7787	16.3273	15.0950	14.5461	14.0357	13.5599	89
90	21.7032	19.5786	17.8329	16.3730	15.1340	14.5823	14.0694	13.5914	90
91	21.7815	19.6423	17.8857	16.4175	15.1721	14.6176	14.1023	13.6220	91
92	21.8581	19.7045	17.9373	16.4610	15.2092	14.6521	14.1343	13.6519	92
93	21.9328	19.7653	17.9876	16.5033	15.2453	14.6856	14.1656	13.6811	93
94	22.0058	19.8245	18.0367	16.5446	15.2806	14.7183	14.1960	13.7094	94
95	22.0771	19.8824	18.0846	16.5849	15.3149	14.7502	14.2256	13.7371	95
96	22.1468	19.9389	18.1313	16.6242	15.3484	14.7813	14.2545	13.7640	96
97	22.2148	19.9940	18.1769	16.6625	15.3811	14.8115	14.2827	13.7903	97
98	22.2813	20.0478	18.2213	16.6999	15.4129	14.8410	14.3101	13.8158	98
99	22.3462	20.1004	18.2647	16.7363	15.4439	14.8698	14.3368	13.8408	99
100	22.4096	20.1516	18.3071	16.7718	15.4742	14.8979	14.3629	13.8651	100

YEARS' PURCHASE

				Rate Per Cent					
Yrs.	7	7.25	7.5	8	8.5	9	9.5	10	Yrs.
1	0.4831	0.4825	0.4819	0.4808	0.4796	0.4785	0.4773	0.4762	1
2	0.9455	0.9433	0.9410	0.9366	0.9323	0.9279	0.9237	0.9194	2
3	1.3884	1.3836	1.3788	1.3693	1.3600	1.3509	1.3418	1.3328	3
4	1.8128	1.8046	1.7965	1.7805	1.7648	1.7494	1.7342	1.7193	4
5	2.2198	2.2075	2.1954	2.1716	2.1483	2.1254	2.1031	2.0812	5
6	2.6103	2.5934	2.5767	2.5439	2.5119	2.4808	2.4504	2.4207	6
7	2.9852	2.9630	2.9413	2.8986	2.8572	2.8170	2.7778	2.7398	7
8	3.3452	3.3175	3.2902	3.2369	3.1854	3.1354	3.0870	3.0401	8
9	3.6912	3.6575	3.6243	3.5598	3.4976	3.4374	3.3794	3.3232	9
10	4.0239	3.9838	3.9445	3.8682	3.7948	3.7242	3.6561	3.5904	10
11	4.3438	4.2972	4.2515	4.1630	4.0781	3.9966	3.9183	3.8430	11
12	4.6517	4.5982	4.5460	4.4450	4.3483	4.2558	4.1671	4.0821	12
13	4.9481	4.8877	4.8287	4.7148	4.6062	4.5025	4.4034	4.3086	13
14	5.2336	5.1660	5.1001	4.9733	4.8526	4.7377	4.6281	4.5234	14
15	5.5086	5.4338	5.3610	5.2210	5.0882	4.9620	4.8418	4.7274	15
16	5.7737	5.6916	5.6117	5.4586	5.3136	5.1760	5.0455	4.9213	16
17	6.0293	5.9398	5.8529	5.6865	5.5293	5.3805	5.2396	5.1058	17
18	6.2759	6.1789	6.0849	5.9053	5.7359	5.5760	5.4248	5.2815	18
19	6.5138	6.4094	6.3083	6.1154	5.9340	5.7630	5.6016	5.4490	19
20	6.7433	6.6316	6.5234	6.3173	6.1239	5.9420	5.7705	5.6087	20
21	6.9650	6.8458	6.7306	6.5115	6.3062	6.1134	5.9321	5.7612	21
22	7.1791	7.0525	6.9303	6.6982	6.4812	6.2777	6.0867	5.9069	22
23	7.3859	7.2520	7.1229	6.8779	6.6493	6.4353	6.2347	6.0462	23
24	7.5858	7.4446	7.3086	7.0509	6.8108	6.5865	6.3765	6.1795	24
25	7.7790	7.6306	7.4877	7.2175	6.9661	6.7317	6.5125	6.3071	25
26	7.9657	7.8102	7.6606	7.3780	7.1155	6.8711	6.6429	6.4293	26
27	8.1464	7.9838	7.8276	7.5327	7.2593	7.0051	6.7680	6.5465	27
28	8.3211	8.1516	7.9888	7.6819	7.3978	7.1339	6.8882	6.6589	28
29	8.4902	8.3138	8.1445	7.8258	7.5311	7.2578	7.0037	6.7667	29
30	8.6539	8.4706	8.2949	7.9646	7.6596	7.3771	7.1146	6.8702	30
31	8.8123	8.6223	8.4404	8.0986	7.7834	7.4919	7.2214	6.9697	31
32	8.9657	8.7691	8.5810	8.2280	7.9029	7.6024	7.3240	7.0653	32
33	9.1142	8.9112	8.7170	8.3529	8.0180	7.7090	7.4229	7.1572	33
34	9.2581	9.0487	8.8485	8.4736	8.1292	7.8117	7.5180	7.2457	34
35	9.3975	9.1818	8.9758	8.5902	8.2365	7.9107	7.6097	7.3308	35
36	9.5326	9.3107	9.0989	8.7030	8.3401	8.0062	7.6980	7.4127	36
37	9.6635	9.4356	9.2181	8.8120	8.4401	8.0984	7.7832	7.4917	37
38	9.7905	9.5566	9.3336	8.9174	8.5368	8.1873	7.8653	7.5677	38
39	9.9135	9.6738	9.4453	9.0194	8.6302	8.2732	7.9446	7.6410	39
40	10.0329	9.7874	9.5536	9.1181	8.7205	8.3561	8.0210	7.7117	40
41	10.1486	9.8975	9.6585	9.2136	8.8078	8.4363	8.0948	7.7799	41
42	10.2609	10.0043	9.7601	9.3060	8.8922	8.5137	8.1661	7.8458	42
43	10.3698	10.1078	9.8586	9.3955	8.9739	8.5886	8.2349	7.9093	43
44	10.4755	10.2082	9.9541	9.4822	9.0530	8.6610	8.3015	7.9706	44
45	10.5781	10.3056	10.0467	9.5662	9.1295	8.7310	8.3657	8.0299	45
46	10.6777	10.4000	10.1365	9.6475	9.2036	8.7987	8.4279	8.0871	46
47	10.7743	10.4917	10.2236	9.7264	9.2753	8.8642	8.4880	8.1424	47
48	10.8681	10.5807	10.3080	9.8028	9.3447	8.9276	8.5461	8.1959	48
49	10.9593	10.6670	10.3899	9.8768	9.4120	8.9890	8.6024	8.2476	49
50	11.0478	10.7508	10.4694	9.9487	9.4772	9.0485	8.6568	8.2976	50

YEARS' PURCHASE

				Rate Per Cent					
Yrs.	7	7.25	7.5	8	8.5	9	9.5	10	Yrs.
51	11.1337	10.8322	10.5466	10.0183	9.5404	9.1060	8.7095	8.3460	51
52	11.2172	10.9112	10.6215	10.0858	9.6016	9.1618	8.7605	8.3929	52
53	11.2983	10.9879	10.6942	10.1514	9.6610	9.2158	8.8099	8.4382	53
54	11.3771	11.0624	10.7647	10.2149	9.7186	9.2682	8.8577	8.4821	54
55	11.4537	11.1348	10.8333	10.2766	9.7744	9.3189	8.9041	8.5245	55
56	11.5281	11.2051	10.8998	10.3365	9.8285	9.3681	8.9490	8.5657	56
57	11.6004	11.2734	10.9644	10.3946	9.8810	9.4158	8.9925	8.6055	57
58	11.6707	11.3398	11.0272	10.4510	9.9320	9.4621	9.0347	8.6442	58
59	11.7390	11.4043	11.0882	10.5057	9.9814	9.5070	9.0756	8.6816	59
60	11.8054	11.4670	11.1474	10.5589	10.0294	9.5505	9.1152	8.7179	60
61	11.8700	11.5279	11.2050	10.6105	10.0760	9.5927	9.1537	8.7531	61
62	11.9328	11.5871	11.2609	10.6607	10.1212	9.6337	9.1910	8.7872	62
63	11.9939	11.6447	11.3153	10.7094	10.1651	9.6734	9.2272	8.8202	63
64	12.0533	11.7007	11.3682	10.7567	10.2077	9.7120	9.2623	8.8523	64
65	12.1110	11.7551	11.4195	10.8027	10.2491	9.7495	9.2963	8.8834	65
66	12.1672	11.8081	11.4695	10.8474	10.2893	9.7859	9.3294	8.9136	66
67	12.2219	11.8595	11.5180	10.8908	10.3284	9.8212	9.3615	8.9429	67
68	12.2751	11.9096	11.5652	10.9330	10.3663	9.8555	9.3927	8.9713	68
69	12.3268	11.9583	11.6111	10.9740	10.4032	9.8888	9.4229	8.9989	69
70	12.3771	12.0056	11.6558	11.0139	10.4390	9.9212	9.4523	9.0257	70
71	12.4261	12.0517	11.6992	11.0527	10.4739	9.9526	9.4808	9.0518	71
72	12.4738	12.0965	11.7415	11.0904	10.5077	9.9832	9.5086	9.0770	72
73	12.5201	12.1401	11.7825	11.1270	10.5406	10.0129	9.5355	9.1016	73
74	12.5653	12.1826	11.8225	11.1627	10.5726	10.0417	9.5617	9.1254	74
75	12.6092	12.2239	11.8614	11.1973	10.6036	10.0698	9.5871	9.1485	75
76	12.6520	12.2641	11.8992	11.2310	10.6339	10.0970	9.6118	9.1710	76
77	12.6936	12.3032	11.9360	11.2638	10.6633	10.1235	9.6358	9.1929	77
78	12.7341	12.3412	11.9719	11.2957	10.6918	10.1493	9.6591	9.2141	78
79	12.7736	12.3783	12.0067	11.3267	10.7196	10.1743	9.6818	9.2347	79
80	12.8120	12.4143	12.0406	11.3569	10.7467	10.1987	9.7038	9.2548	80
81	12.8493	12.4494	12.0737	11.3863	10.7730	10.2223	9.7253	9.2743	81
82	12.8858	12.4836	12.1058	11.4149	10.7985	10.2454	9.7461	9.2932	82
83	12.9212	12.5169	12.1371	11.4427	10.8234	10.2678	9.7664	9.3117	83
84	12.9557	12.5493	12.1675	11.4697	10.8476	10.2896	9.7861	9.3296	84
85	12.9893	12.5808	12.1972	11.4961	10.8712	10.3107	9.8052	9.3470	85
86	13.0221	12.6115	12.2260	11.5217	10.8941	10.3314	9.8239	9.3639	86
87	13.0540	12.6414	12.2541	11.5467	10.9164	10.3514	9.8420	9.3804	87
88	13.0850	12.6705	12.2815	11.5710	10.9381	10.3709	9.8597	9.3964	88
89	13.1153	12.6989	12.3081	11.5946	10.9593	10.3899	9.8768	9.4120	89
90	13.1447	12.7265	12.3341	11.6176	10.9798	10.4084	9.8935	9.4272	90
91	13.1734	12.7534	12.3593	11.6400	10.9998	10.4264	9.9098	9.4419	91
92	13.2014	12.7796	12.3839	11.6618	11.0193	10.4439	9.9256	9.4563	92
93	13.2286	12.8051	12.4079	11.6831	11.0383	10.4609	9.9410	9.4703	93
94	13.2551	12.8300	12.4312	11.7038	11.0567	10.4775	9.9559	9.4838	94
95	13.2810	12.8542	12.4540	11.7239	11.0747	10.4937	9.9705	9.4971	95
96	13.3061	12.8778	12.4761	11.7435	11.0922	10.5094	9.9847	9.5099	96
97	13.3307	12.9007	12.4977	11.7626	11.1093	10.5247	9.9985	9.5225	97
98	13.3546	12.9231	12.5187	11.7812	11.1259	10.5396	10.0119	9.5346	98
99	13.3779	12.9449	12.5391	11.7994	11.1420	10.5541	10.0250	9.5465	99
100	13.4006	12.9662	12.5591	11.8170	11.1578	10.5682	10.0378	9.5581	100

YEARS' PURCHASE

				Rate Per Cent					
Yrs.	11	12	13	14	15	16	18	20	Yrs.
1	0.4739	0.4717	0.4695	0.4673	0.4651	0.4630	0.4587	0.4545	1
2	0.9110	0.9028	0.8947	0.8868	0.8790	0.8713	0.8564	0.8420	2
3	1.3153	1.2982	1.2816	1.2654	1.2496	1.2341	1.2044	1.1761	3
4	1.6902	1.6621	1.6350	1.6087	1.5832	1.5585	1.5114	1.4671	4
5	2.0388	1.9980	1.9589	1.9213	1.8850	1.8502	1.7841	1.7227	5
6	2.3635	2.3089	2.2568	2.2070	2.1594	2.1137	2.0280	1.9489	6
7	2.6667	2.5975	2.5317	2.4692	2.4097	2.3530	2.2472	2.1506	7
8	2.9504	2.8659	2.7860	2.7105	2.6390	2.5711	2.4454	2.3314	8
9	3.2163	3.1161	3.0219	2.9333	2.8497	2.7707	2.6253	2.4943	9
10	3.4660	3.3499	3.2413	3.1395	3.0440	2.9541	2.7893	2.6419	10
11	3.7008	3.5687	3.4458	3.3310	3.2236	3.1229	2.9393	2.7761	11
12	3.9220	3.7740	3.6367	3.5091	3.3901	3.2790	3.0772	2.8988	12
13	4.1306	3.9667	3.8154	3.6752	3.5449	3.4235	3.2041	3.0112	13
14	4.3276	4.1481	3.9829	3.8303	3.6890	3.5578	3.3215	3.1146	14
15	4.5140	4.3190	4.1402	3.9756	3.8236	3.6828	3.4301	3.2099	15
16	4.6905	4.4803	4.2882	4.1119	3.9495	3.7994	3.5311	3.2982	16
17	4.8578	4.6327	4.4276	4.2399	4.0674	3.9085	3.6251	3.3800	17
18	5.0166	4.7769	4.5591	4.3603	4.1782	4.0106	3.7128	3.4561	18
19	5.1674	4.9135	4.6834	4.4738	4.2823	4.1064	3.7948	3.5271	19
20	5.3108	5.0430	4.8009	4.5810	4.3803	4.1965	3.8716	3.5933	20
21	5.4474	5.1660	4.9122	4.6822	4.4728	4.2813	3.9436	3.6553	21
22	5.5775	5.2828	5.0177	4.7780	4.5601	4.3612	4.0113	3.7134	22
23	5.7015	5.3940	5.1179	4.8687	4.6427	4.4367	4.0751	3.7680	23
24	5.8199	5.4998	5.2131	4.9548	4.7209	4.5080	4.1352	3.8193	24
25	5.9329	5.6006	5.3036	5.0365	4.7950	4.5756	4.1920	3.8677	25
26	6.0409	5.6968	5.3897	5.1141	4.8653	4.6396	4.2456	3.9133	26
27	6.1443	5.7886	5.4718	5.1880	4.9321	4.7003	4.2964	3.9564	27
28	6.2431	5.8763	5.5501	5.2583	4.9956	4.7579	4.3445	3.9972	28
29	6.3378	5.9601	5.6248	5.3253	5.0561	4.8127	4.3902	4.0358	29
30	6.4286	6.0403	5.6962	5.3892	5.1136	4.8649	4.4335	4.0724	30
31	6.5156	6.1170	5.7644	5.4502	5.1685	4.9145	4.4747	4.1071	31
32	6.5991	6.1905	5.8297	5.5085	5.2209	4.9619	4.5139	4.1402	32
33	6.6792	6.2610	5.8921	5.5642	5.2710	5.0070	4.5513	4.1716	33
34	6.7561	6.3286	5.9519	5.6176	5.3188	5.0502	4.5869	4.2014	34
35	6.8301	6.3934	6.0092	5.6686	5.3645	5.0914	4.6208	4.2299	35
36	6.9012	6.4556	6.0642	5.7174	5.4082	5.1308	4.6533	4.2571	36
37	6.9695	6.5154	6.1169	5.7643	5.4501	5.1684	4.6842	4.2830	37
38	7.0353	6.5729	6.1675	5.8092	5.4903	5.2045	4.7139	4.3077	38
39	7.0986	6.6281	6.2161	5.8523	5.5288	5.2391	4.7422	4.3314	39
40	7.1596	6.6813	6.2628	5.8937	5.5657	5.2722	4.7693	4.3540	40
41	7.2184	6.7324	6.3077	5.9335	5.6011	5.3040	4.7953	4.3757	41
42	7.2750	6.7816	6.3509	5.9717	5.6352	5.3345	4.8203	4.3964	42
43	7.3296	6.8290	6.3925	6.0084	5.6678	5.3638	4.8442	4.4163	43
44	7.3822	6.8747	6.4325	6.0437	5.6993	5.3920	4.8671	4.4354	44
45	7.4330	6.9187	6.4710	6.0777	5.7295	5.4190	4.8891	4.4536	45
46	7.4820	6.9612	6.5081	6.1105	5.7586	5.4450	4.9103	4.4712	46
47	7.5294	7.0021	6.5439	6.1420	5.7866	5.4701	4.9306	4.4881	47
48	7.5751	7.0417	6.5784	6.1724	5.8135	5.4941	4.9502	4.5043	48
49	7.6192	7.0798	6.6117	6.2017	5.8395	5.5173	4.9690	4.5198	49
50	7.6619	7.1166	6.6438	6.2299	5.8645	5.5397	4.9871	4.5348	50

YEARS' PURCHASE

				Rate Per Cent					
Yrs.	11	12	13	14	15	16	18	20	Yrs.
51	7.7031	7.1522	6.6748	6.2571	5.8887	5.5612	5.0046	4.5492	51
52	7.7430	7.1865	6.7047	6.2834	5.9120	5.5819	5.0214	4.5631	52
53	7.7816	7.2197	6.7336	6.3088	5.9344	5.6020	5.0376	4.5765	53
54	7.8189	7.2518	6.7615	6.3333	5.9561	5.6213	5.0532	4.5893	54
55	7.8549	7.2829	6.7885	6.3569	5.9770	5.6399	5.0682	4.6018	55
56	7.8899	7.3129	6.8145	6.3798	5.9972	5.6579	5.0827	4.6137	56
57	7.9237	7.3419	6.8398	6.4019	6.0167	5.6752	5.0967	4.6253	57
58	7.9564	7.3700	6.8641	6.4232	6.0356	5.6920	5.1103	4.6364	58
59	7.9881	7.3972	6.8877	6.4439	6.0538	5.7082	5.1233	4.6471	59
60	8.0188	7.4235	6.9105	6.4638	6.0714	5.7239	5.1359	4.6575	60
61	8.0486	7.4490	6.9326	6.4832	6.0884	5.7390	5.1481	4.6675	61
62	8.0774	7.4737	6.9540	6.5018	6.1049	5.7537	5.1599	4.6772	62
63	8.1053	7.4976	6.9747	6.5199	6.1209	5.7678	5.1713	4.6866	63
64	8.1324	7.5208	6.9947	6.5374	6.1363	5.7815	5.1823	4.6956	64
65	8.1587	7.5432	7.0141	6.5544	6.1512	5.7948	5.1929	4.7043	65
66	8.1841	7.5650	7.0329	6.5708	6.1657	5.8076	5.2032	4.7128	66
67	8.2088	7.5861	7.0512	6.5867	6.1797	5.8200	5.2132	4.7210	67
68	8.2328	7.6065	7.0688	6.6021	6.1933	5.8321	5.2229	4.7289	68
69	8.2560	7.6264	7.0860	6.6171	6.2064	5.8437	5.2322	4.7366	69
70	8.2785	7.6456	7.1026	6.6316	6.2191	5.8550	5.2412	4.7440	70
71	8.3004	7.6643	7.1187	6.6456	6.2315	5.8659	5.2500	4.7511	71
72	8.3217	7.6824	7.1343	6.6592	6.2434	5.8765	5.2585	4.7581	72
73	8.3423	7.6999	7.1494	6.6724	6.2550	5.8868	5.2667	4.7648	73
74	8.3623	7.7170	7.1641	6.6852	6.2663	5.8968	5.2747	4.7713	74
75	8.3817	7.7335	7.1784	6.6976	6.2772	5.9064	5.2824	4.7777	75
76	8.4006	7.7496	7.1922	6.7096	6.2878	5.9158	5.2899	4.7838	76
77	8.4189	7.7652	7.2057	6.7213	6.2980	5.9249	5.2972	4.7897	77
78	8.4367	7.7803	7.2187	6.7327	6.3080	5.9337	5.3042	4.7955	78
79	8.4540	7.7950	7.2314	6.7437	6.3176	5.9422	5.3110	4.8011	79
80	8.4708	7.8093	7.2436	6.7544	6.3270	5.9505	5.3177	4.8065	80
81	8.4872	7.8232	7.2556	6.7648	6.3361	5.9586	5.3241	4.8117	81
82	8.5030	7.8367	7.2672	6.7748	6.3450	5.9664	5.3303	4.8168	82
83	8.5185	7.8498	7.2784	6.7846	6.3536	5.9740	5.3364	4.8218	83
84	8.5334	7.8625	7.2894	6.7941	6.3619	5.9814	5.3423	4.8266	84
85	8.5480	7.8749	7.3000	6.8034	6.3700	5.9885	5.3480	4.8312	85
86	8.5622	7.8869	7.3103	6.8123	6.3778	5.9955	5.3535	4.8358	86
87	8.5760	7.8986	7.3204	6.8210	6.3855	6.0022	5.3589	4.8402	87
88	8.5893	7.9099	7.3301	6.8295	6.3929	6.0088	5.3641	4.8444	88
89	8.6024	7.9210	7.3396	6.8377	6.4001	6.0151	5.3692	4.8486	89
90	8.6150	7.9317	7.3488	6.8457	6.4071	6.0213	5.3741	4.8526	90
91	8.6273	7.9422	7.3578	6.8535	6.4139	6.0273	5.3789	4.8565	91
92	8.6393	7.9523	7.3665	6.8611	6.4206	6.0332	5.3836	4.8603	92
93	8.6510	7.9622	7.3750	6.8684	6.4270	6.0389	5.3881	4.8640	93
94	8.6623	7.9718	7.3832	6.8756	6.4332	6.0444	5.3925	4.8675	94
95	8.6733	7.9811	7.3912	6.8825	6.4393	6.0498	5.3968	4.8710	95
96	8.6841	7.9902	7.3990	6.8893	6.4452	6.0550	5.4009	4.8744	96
97	8.6945	7.9990	7.4066	6.8958	6.4510	6.0601	5.4050	4.8777	97
98	8.7047	8.0076	7.4140	6.9022	6.4566	6.0650	5.4089	4.8809	98
99	8.7146	8.0160	7.4211	6.9084	6.4620	6.0698	5.4127	4.8840	99
100	8.7242	8.0242	7.4281	6.9145	6.4673	6.0745	5.4164	4.8870	100

YEARS' PURCHASE

				Rate Per Cent					
Yrs.	4	4.5	5	5.5	6	6.25	6.5	6.75	Yrs.
1	0.4902	0.4890	0.4878	0.4866	0.4854	0.4848	0.4843	0.4837	1
2	0.9754	0.9707	0.9660	0.9613	0.9567	0.9545	0.9522	0.9499	2
3	1.4555	1.4450	1.4346	1.4244	1.4143	1.4093	1.4044	1.3995	3
4	1.9303	1.9118	1.8937	1.8760	1.8585	1.8500	1.8414	1.8330	4
5	2.3998	2.3713	2.3435	2.3164	2.2899	2.2768	2.2639	2.2512	5
6	2.8637	2.8233	2.7840	2.7458	2.7086	2.6904	2.6724	2.6547	6
7	3.3221	3.2678	3.2153	3.1644	3.1151	3.0911	3.0674	3.0440	7
8	3.7748	3.7049	3.6375	3.5725	3.5098	3.4793	3.4493	3.4198	8
9	4.2218	4.1345	4.0508	3.9703	3.8931	3.8555	3.8187	3.7826	9
10	4.6629	4.5566	4.4551	4.3580	4.2651	4.2201	4.1760	4.1329	10
11	5.0980	4.9713	4.8507	4.7359	4.6263	4.5734	4.5217	4.4712	11
12	5.5272	5.3785	5.2377	5.1040	4.9770	4.9158	4.8562	4.7979	12
13	5.9503	5.7784	5.6161	5.4627	5.3175	5.2477	5.1798	5.1135	13
14	6.3673	6.1708	5.9861	5.8122	5.6480	5.5694	5.4929	5.4185	14
15	6.7781	6.5560	6.3479	6.1526	5.9690	5.8812	5.7960	5.7132	15
16	7.1828	6.9338	6.7014	6.4842	6.2806	6.1835	6.0893	5.9980	16
17	7.5812	7.3043	7.0470	6.8071	6.5831	6.4765	6.3733	6.2733	17
18	7.9734	7.6677	7.3846	7.1216	6.8768	6.7605	6.6482	6.5395	18
19	8.3593	8.0239	7.7144	7.4279	7.1619	7.0359	6.9143	6.7968	19
20	8.7389	8.3730	8.0366	7.7261	7.4387	7.3029	7.1720	7.0457	20
21	9.1122	8.7151	8.3512	8.0164	7.7075	7.5618	7.4215	7.2863	21
22	9.4791	9.0502	8.6584	8.2991	7.9685	7.8128	7.6631	7.5191	22
23	9.8398	9.3784	8.9583	8.5743	8.2218	8.0562	7.8972	7.7443	23
24	10.1942	9.6998	9.2511	8.8421	8.4678	8.2922	8.1238	7.9621	24
25	10.5423	10.0144	9.5369	9.1028	8.7066	8.5211	8.3434	8.1729	25
26	10.8842	10.3224	9.8158	9.3566	8.9384	8.7430	8.5560	8.3769	26
27	11.2198	10.6238	10.0879	9.6035	9.1635	8.9583	8.7621	8.5742	27
28	11.5491	10.9186	10.3534	9.8438	9.3821	9.1670	8.9617	8.7653	28
29	11.8723	11.2071	10.6124	10.0777	9.5942	9.3695	9.1550	8.9502	29
30	12.1894	11.4892	10.8650	10.3052	9.8002	9.5659	9.3424	9.1292	30
31	12.5003	11.7650	11.1114	10.5266	10.0002	9.7563	9.5240	9.3025	31
32	12.8052	12.0347	11.3516	10.7419	10.1944	9.9410	9.7000	9.4703	32
33	13.1041	12.2983	11.5858	10.9514	10.3829	10.1202	9.8705	9.6328	33
34	13.3969	12.5559	11.8142	11.1552	10.5659	10.2940	10.0357	9.7901	34
35	13.6839	12.8076	12.0368	11.3535	10.7436	10.4626	10.1959	9.9425	35
36	13.9650	13.0535	12.2538	11.5463	10.9161	10.6261	10.3512	10.0900	36
37	14.2403	13.2938	12.4652	11.7339	11.0836	10.7848	10.5016	10.2330	37
38	14.5099	13.5284	12.6713	11.9163	11.2462	10.9387	10.6475	10.3714	38
39	14.7737	13.7575	12.8720	12.0937	11.4041	11.0880	10.7889	10.5056	39
40	15.0320	13.9812	13.0677	12.2662	11.5574	11.2328	10.9260	10.6355	40
41	15.2847	14.1995	13.2582	12.4340	11.7062	11.3734	11.0589	10.7614	41
42	15.5320	14.4127	13.4439	12.5971	11.8507	11.5097	11.1878	10.8834	42
43	15.7738	14.6207	13.6247	12.7557	11.9910	11.6420	11.3127	11.0016	43
44	16.0103	14.8237	13.8008	12.9099	12.1271	11.7703	11.4338	11.1161	44
45	16.2416	15.0217	13.9723	13.0599	12.2594	11.8948	11.5513	11.2271	45
46	16.4677	15.2149	14.1393	13.2057	12.3877	12.0156	11.6652	11.3346	46
47	16.6886	15.4033	14.3019	13.3474	12.5124	12.1328	11.7756	11.4389	47
48	16.9046	15.5871	14.4602	13.4852	12.6334	12.2466	11.8828	11.5399	48
49	17.1156	15.7663	14.6143	13.6191	12.7508	12.3569	11.9866	11.6379	49
50	17.3217	15.9411	14.7643	13.7493	12.8649	12.4640	12.0874	11.7328	50

YEARS' PURCHASE

				Rate Per Cent					
Yrs.	4	4.5	5	5.5	6	6.25	6.5	6.75	Yrs.
51	17.5231	16.1115	14.9103	13.8759	12.9756	12.5679	12.1851	11.8249	51
52	17.7197	16.2775	15.0525	13.9989	13.0831	12.6688	12.2798	11.9141	52
53	17.9117	16.4394	15.1908	14.1184	13.1875	12.7666	12.3717	12.0006	53
54	18.0992	16.5972	15.3254	14.2346	13.2888	12.8615	12.4609	12.0844	54
55	18.2822	16.7509	15.4564	14.3476	13.3872	12.9537	12.5473	12.1657	55
56	18.4608	16.9008	15.5839	14.4574	13.4827	13.0431	12.6312	12.2446	56
57	18.6351	17.0467	15.7079	14.5640	13.5755	13.1299	12.7126	12.3210	57
58	18.8051	17.1889	15.8286	14.6677	13.6655	13.2141	12.7915	12.3951	58
59	18.9711	17.3275	15.9459	14.7685	13.7529	13.2958	12.8680	12.4670	59
60	19.1329	17.4624	16.0602	14.8664	13.8378	13.3751	12.9423	12.5367	60
61	19.2908	17.5938	16.1712	14.9615	13.9202	13.4520	13.0144	12.6043	61
62	19.4448	17.7218	16.2793	15.0540	14.0002	13.5267	13.0843	12.6698	62
63	19.5949	17.8464	16.3844	15.1438	14.0778	13.5992	13.1521	12.7334	63
64	19.7413	17.9678	16.4866	15.2311	14.1532	13.6696	13.2179	12.7951	64
65	19.8840	18.0859	16.5861	15.3159	14.2264	13.7378	13.2817	12.8549	65
66	20.0231	18.2009	16.6827	15.3983	14.2975	13.8041	13.3436	12.9129	66
67	20.1587	18.3129	16.7768	15.4784	14.3665	13.8684	13.4037	12.9691	67
68	20.2909	18.4219	16.8682	15.5562	14.4335	13.9308	13.4620	13.0237	68
69	20.4197	18.5280	16.9571	15.6318	14.4986	13.9914	13.5186	13.0766	69
70	20.5452	18.6313	17.0435	15.7052	14.5617	14.0502	13.5734	13.1280	70
71	20.6674	18.7317	17.1276	15.7765	14.6230	14.1073	13.6267	13.1778	71
72	20.7865	18.8295	17.2093	15.8458	14.6825	14.1627	13.6784	13.2261	72
73	20.9025	18.9247	17.2888	15.9132	14.7403	14.2164	13.7285	13.2730	73
74	21.0155	19.0173	17.3660	15.9786	14.7964	14.2686	13.7772	13.3184	74
75	21.1256	19.1073	17.4411	16.0421	14.8509	14.3193	13.8244	13.3626	75
76	21.2328	19.1950	17.5141	16.1038	14.9038	14.3684	13.8702	13.4054	76
77	21.3371	19.2802	17.5850	16.1638	14.9551	14.4162	13.9147	13.4469	77
78	21.4388	19.3632	17.6540	16.2221	15.0050	14.4625	13.9578	13.4872	78
79	21.5377	19.4438	17.7210	16.2786	15.0534	14.5074	13.9997	13.5263	79
80	21.6340	19.5223	17.7862	16.3336	15.1004	14.5511	14.0403	13.5642	80
81	21.7278	19.5986	17.8495	16.3870	15.1460	14.5934	14.0797	13.6010	81
82	21.8190	19.6728	17.9110	16.4388	15.1903	14.6345	14.1180	13.6367	82
83	21.9079	19.7450	17.9708	16.4892	15.2333	14.6744	14.1551	13.6713	83
84	21.9943	19.8152	18.0290	16.5381	15.2750	14.7132	14.1912	13.7050	84
85	22.0785	19.8835	18.0855	16.5857	15.3156	14.7508	14.2262	13.7376	85
86	22.1603	19.9499	18.1404	16.6318	15.3549	14.7873	14.2601	13.7692	86
87	22.2400	20.0144	18.1937	16.6767	15.3931	14.8227	14.2931	13.8000	87
88	22.3175	20.0772	18.2456	16.7202	15.4302	14.8571	14.3250	13.8298	88
89	22.3929	20.1382	18.2959	16.7625	15.4662	14.8905	14.3561	13.8587	89
90	22.4663	20.1975	18.3449	16.8036	15.5012	14.9229	14.3862	13.8867	90
91	22.5377	20.2552	18.3924	16.8435	15.5351	14.9544	14.4154	13.9140	91
92	22.6071	20.3112	18.4387	16.8822	15.5681	14.9849	14.4438	13.9404	92
93	22.6746	20.3657	18.4836	16.9199	15.6001	15.0145	14.4713	13.9661	93
94	22.7403	20.4187	18.5272	16.9564	15.6312	15.0433	14.4981	13.9910	94
95	22.8042	20.4702	18.5696	16.9919	15.6613	15.0712	14.5240	14.0151	95
96	22.8663	20.5202	18.6107	17.0264	15.6906	15.0984	14.5492	14.0386	96
97	22.9268	20.5689	18.6508	17.0599	15.7190	15.1247	14.5736	14.0613	97
98	22.9855	20.6162	18.6896	17.0924	15.7466	15.1502	14.5973	14.0834	98
99	23.0427	20.6621	18.7274	17.1240	15.7734	15.1750	14.6204	14.1048	99
100	23.0982	20.7068	18.7641	17.1546	15.7995	15.1991	14.6427	14.1256	100

YEARS' PURCHASE

				Rate Per Cent					
Yrs.	7	7.25	7.5	8	8.5	9	9.5	10	Yrs.
1	0.4831	0.4825	0.4819	0.4808	0.4796	0.4785	0.4773	0.4762	1
2	0.9477	0.9454	0.9432	0.9388	0.9344	0.9300	0.9257	0.9215	2
3	1.3946	1.3897	1.3849	1.3754	1.3660	1.3567	1.3476	1.3386	3
4	1.8246	1.8164	1.8081	1.7919	1.7760	1.7604	1.7450	1.7299	4
5	2.2386	2.2261	2.2138	2.1896	2.1659	2.1427	2.1200	2.0977	5
6	2.6372	2.6199	2.6028	2.5694	2.5368	2.5050	2.4741	2.4438	6
7	3.0210	2.9984	2.9761	2.9324	2.8901	2.8489	2.8089	2.7700	7
8	3.3908	3.3623	3.3343	3.2796	3.2267	3.1755	3.1259	3.0777	8
9	3.7472	3.7124	3.6783	3.6118	3.5478	3.4859	3.4262	3.3685	9
10	4.0906	4.0492	4.0086	3.9299	3.8541	3.7813	3.7111	3.6435	10
11	4.4217	4.3734	4.3261	4.2345	4.1467	4.0625	3.9816	3.9039	11
12	4.7410	4.6855	4.6313	4.5264	4.4263	4.3304	4.2387	4.1507	12
13	5.0490	4.9861	4.9247	4.8063	4.6935	4.5859	4.4831	4.3848	13
14	5.3461	5.2756	5.2069	5.0748	4.9492	4.8297	4.7158	4.6072	14
15	5.6328	5.5545	5.4785	5.3324	5.1939	5.0624	4.9375	4.8185	15
16	5.9094	5.8234	5.7398	5.5797	5.4282	5.2848	5.1488	5.0195	16
17	6.1765	6.0825	5.9914	5.8172	5.6527	5.4974	5.3503	5.2109	17
18	6.4343	6.3324	6.2337	6.0453	5.8679	5.7007	5.5427	5.3932	18
19	6.6833	6.5734	6.4671	6.2646	6.0743	5.8953	5.7265	5.5671	19
20	6.9237	6.8059	6.6920	6.4754	6.2723	6.0816	5.9021	5.7329	20
21	7.1560	7.0302	6.9088	6.6781	6.4623	6.2600	6.0700	5.8912	21
22	7.3804	7.2467	7.1177	6.8731	6.6447	6.4311	6.2307	6.0425	22
23	7.5972	7.4556	7.3192	7.0608	6.8200	6.5951	6.3846	6.1871	23
24	7.8067	7.6573	7.5134	7.2414	6.9884	6.7524	6.5319	6.3253	24
25	8.0092	7.8520	7.7009	7.4153	7.1502	6.9034	6.6731	6.4576	25
26	8.2050	8.0401	7.8817	7.5828	7.3059	7.0484	6.8084	6.5843	26
27	8.3943	8.2218	8.0562	7.7442	7.4555	7.1876	6.9383	6.7056	27
28	8.5773	8.3973	8.2246	7.8997	7.5996	7.3214	7.0628	6.8219	28
29	8.7543	8.5668	8.3872	8.0496	7.7382	7.4499	7.1824	6.9334	29
30	8.9255	8.7307	8.5442	8.1941	7.8716	7.5735	7.2972	7.0403	30
31	9.0911	8.8891	8.6958	8.3335	8.0001	7.6924	7.4075	7.1430	31
32	9.2513	9.0421	8.8423	8.4679	8.1239	7.8068	7.5135	7.2415	32
33	9.4063	9.1901	8.9837	8.5975	8.2432	7.9169	7.6154	7.3361	33
34	9.5562	9.3332	9.1204	8.7227	8.3581	8.0229	7.7134	7.4270	34
35	9.7013	9.4716	9.2525	8.8434	8.4689	8.1249	7.8077	7.5144	35
36	9.8418	9.6054	9.3802	8.9600	8.5758	8.2232	7.8984	7.5983	36
37	9.9777	9.7349	9.5036	9.0725	8.6788	8.3179	7.9857	7.6791	37
38	10.1093	9.8601	9.6229	9.1812	8.7782	8.4091	8.0698	7.7568	38
39	10.2367	9.9813	9.7383	9.2861	8.8741	8.4971	8.1508	7.8316	39
40	10.3600	10.0985	9.8498	9.3875	8.9666	8.5819	8.2288	7.9036	40
41	10.4795	10.2119	9.9577	9.4854	9.0559	8.6636	8.3039	7.9729	41
42	10.5951	10.3217	10.0620	9.5801	9.1422	8.7425	8.3764	8.0397	42
43	10.7071	10.4279	10.1630	9.6715	9.2254	8.8186	8.4462	8.1040	43
44	10.8155	10.5308	10.2607	9.7599	9.3058	8.8921	8.5136	8.1660	44
45	10.9206	10.6303	10.3551	9.8454	9.3835	8.9630	8.5785	8.2257	45
46	11.0223	10.7267	10.4466	9.9280	9.4585	9.0314	8.6412	8.2833	46
47	11.1209	10.8201	10.5351	10.0079	9.5310	9.0974	8.7016	8.3388	47
48	11.2164	10.9104	10.6207	10.0852	9.6010	9.1612	8.7600	8.3924	48
49	11.3089	10.9979	10.7036	10.1599	9.6687	9.2229	8.8163	8.4441	49
50	11.3985	11.0827	10.7839	10.2322	9.7342	9.2824	8.8707	8.4939	50

YEARS' PURCHASE

				Rate Per Cent					
Yrs.	7	7.25	7.5	8	8.5	9	9.5	10	Yrs.
51	11.4853	11.1647	10.8616	10.3021	9.7974	9.3399	8.9232	8.5421	51
52	11.5695	11.2442	10.9368	10.3697	9.8586	9.3955	8.9739	8.5885	52
53	11.6510	11.3213	11.0096	10.4352	9.9177	9.4492	9.0229	8.6334	53
54	11.7300	11.3959	11.0802	10.4986	9.9749	9.5011	9.0702	8.6767	54
55	11.8066	11.4681	11.1485	10.5599	10.0303	9.5513	9.1159	8.7185	55
56	11.8809	11.5382	11.2147	10.6192	10.0838	9.5998	9.1601	8.7589	56
57	11.9528	11.6060	11.2788	10.6767	10.1356	9.6467	9.2028	8.7980	57
58	12.0226	11.6717	11.3408	10.7323	10.1857	9.6921	9.2441	8.8357	58
59	12.0902	11.7355	11.4010	10.7861	10.2342	9.7360	9.2840	8.8722	59
60	12.1557	11.7972	11.4592	10.8382	10.2811	9.7784	9.3226	8.9074	60
61	12.2192	11.8570	11.5157	10.8887	10.3265	9.8195	9.3600	8.9415	61
62	12.2808	11.9150	11.5704	10.9376	10.3705	9.8592	9.3961	8.9744	62
63	12.3406	11.9712	11.6234	10.9849	10.4130	9.8977	9.4310	9.0063	63
64	12.3985	12.0257	11.6747	11.0308	10.4542	9.9349	9.4647	9.0371	64
65	12.4546	12.0785	11.7245	11.0752	10.4941	9.9709	9.4974	9.0669	65
66	12.5090	12.1297	11.7727	11.1183	10.5327	10.0058	9.5291	9.0957	66
67	12.5618	12.1793	11.8195	11.1599	10.5701	10.0395	9.5597	9.1236	67
68	12.6130	12.2275	11.8648	11.2003	10.6063	10.0722	9.5893	9.1505	68
69	12.6627	12.2741	11.9087	11.2394	10.6414	10.1038	9.6179	9.1766	69
70	12.7108	12.3193	11.9513	11.2774	10.6754	10.1345	9.6457	9.2019	70
71	12.7575	12.3632	11.9925	11.3141	10.7083	10.1641	9.6726	9.2263	71
72	12.8028	12.4057	12.0325	11.3497	10.7402	10.1928	9.6986	9.2500	72
73	12.8467	12.4469	12.0713	11.3842	10.7711	10.2207	9.7237	9.2729	73
74	12.8893	12.4869	12.1089	11.4176	10.8010	10.2476	9.7481	9.2951	74
75	12.9306	12.5257	12.1454	11.4500	10.8300	10.2737	9.7717	9.3165	75
76	12.9707	12.5633	12.1807	11.4815	10.8581	10.2990	9.7946	9.3373	76
77	13.0095	12.5998	12.2150	11.5119	10.8853	10.3235	9.8168	9.3575	77
78	13.0473	12.6351	12.2482	11.5414	10.9117	10.3472	9.8382	9.3770	78
79	13.0838	12.6694	12.2805	11.5700	10.9373	10.3702	9.8590	9.3958	79
80	13.1193	12.7027	12.3117	11.5978	10.9621	10.3925	9.8791	9.4141	80
81	13.1537	12.7350	12.3420	11.6247	10.9861	10.4141	9.8986	9.4318	81
82	13.1871	12.7663	12.3714	11.6507	11.0094	10.4350	9.9175	9.4490	82
83	13.2195	12.7966	12.3999	11.6760	11.0320	10.4553	9.9358	9.4656	83
84	13.2510	12.8261	12.4276	11.7005	11.0538	10.4749	9.9536	9.4817	84
85	13.2814	12.8546	12.4544	11.7243	11.0751	10.4939	9.9708	9.4973	85
86	13.3110	12.8823	12.4804	11.7473	11.0956	10.5124	9.9874	9.5124	86
87	13.3397	12.9092	12.5056	11.7697	11.1156	10.5303	10.0036	9.5271	87
88	13.3676	12.9353	12.5301	11.7914	11.1349	10.5476	10.0193	9.5413	88
89	13.3946	12.9606	12.5538	11.8124	11.1536	10.5645	10.0344	9.5550	89
90	13.4208	12.9851	12.5769	11.8328	11.1718	10.5808	10.0491	9.5684	90
91	13.4463	13.0089	12.5992	11.8525	11.1894	10.5966	10.0634	9.5813	91
92	13.4709	13.0321	12.6209	11.8717	11.2065	10.6119	10.0772	9.5938	92
93	13.4949	13.0545	12.6419	11.8903	11.2231	10.6268	10.0906	9.6060	93
94	13.5181	13.0762	12.6623	11.9083	11.2391	10.6412	10.1036	9.6177	94
95	13.5407	13.0973	12.6821	11.9258	11.2547	10.6551	10.1162	9.6291	95
96	13.5626	13.1178	12.7013	11.9428	11.2698	10.6687	10.1284	9.6402	96
97	13.5838	13.1376	12.7199	11.9593	11.2845	10.6818	10.1402	9.6509	97
98	13.6044	13.1569	12.7379	11.9752	11.2987	10.6945	10.1517	9.6613	98
99	13.6244	13.1756	12.7555	11.9907	11.3125	10.7069	10.1628	9.6714	99
100	13.6438	13.1938	12.7725	12.0058	11.3259	10.7189	10.1736	9.6812	100

YEARS' PURCHASE

Rate Per Cent

Yrs.	11	12	13	14	15	16	18	20	Yrs.
1	0.4739	0.4717	0.4695	0.4673	0.4651	0.4630	0.4587	0.4545	1
2	0.9131	0.9048	0.8967	0.8887	0.8809	0.8732	0.8582	0.8437	2
3	1.3209	1.3037	1.2869	1.2705	1.2546	1.2391	1.2091	1.1806	3
4	1.7005	1.6721	1.6446	1.6180	1.5922	1.5673	1.5196	1.4748	4
5	2.0546	2.0133	1.9735	1.9353	1.8986	1.8632	1.7963	1.7340	5
6	2.3855	2.3299	2.2769	2.2262	2.1777	2.1313	2.0442	1.9639	6
7	2.6953	2.6246	2.5575	2.4937	2.4330	2.3752	2.2675	2.1691	7
8	2.9859	2.8993	2.8176	2.7404	2.6673	2.5980	2.4697	2.3534	8
9	3.2587	3.1559	3.0593	2.9685	2.8829	2.8022	2.6535	2.5197	9
10	3.5154	3.3960	3.2845	3.1800	3.0820	2.9899	2.8212	2.6705	10
11	3.7572	3.6212	3.4946	3.3766	3.2663	3.1630	2.9748	2.8078	11
12	3.9853	3.8325	3.6911	3.5597	3.4373	3.3231	3.1160	2.9332	12
13	4.2006	4.0313	3.8751	3.7305	3.5964	3.4715	3.2461	3.0482	13
14	4.4043	4.2185	4.0477	3.8903	3.7446	3.6094	3.3664	3.1541	14
15	4.5970	4.3950	4.2099	4.0399	3.8830	3.7379	3.4779	3.2517	15
16	4.7796	4.5616	4.3626	4.1802	4.0125	3.8577	3.5814	3.3420	16
17	4.9528	4.7191	4.5064	4.3121	4.1339	3.9697	3.6778	3.4258	17
18	5.1173	4.8681	4.6422	4.4362	4.2478	4.0747	3.7677	3.5036	18
19	5.2735	5.0093	4.7704	4.5532	4.3549	4.1731	3.8517	3.5762	19
20	5.4221	5.1432	4.8916	4.6635	4.4557	4.2656	3.9303	3.6439	20
21	5.5635	5.2703	5.0064	4.7677	4.5508	4.3527	4.0041	3.7072	21
22	5.6982	5.3910	5.1152	4.8663	4.6405	4.4347	4.0734	3.7666	22
23	5.8266	5.5058	5.2184	4.9596	4.7253	4.5121	4.1386	3.8222	23
24	5.9490	5.6150	5.3165	5.0481	4.8055	4.5852	4.2000	3.8745	24
25	6.0659	5.7190	5.4096	5.1320	4.8815	4.6543	4.2579	3.9238	25
26	6.1775	5.8181	5.4982	5.2117	4.9535	4.7197	4.3126	3.9702	26
27	6.2842	5.9127	5.5826	5.2874	5.0219	4.7818	4.3644	4.0140	27
28	6.3862	6.0029	5.6629	5.3594	5.0868	4.8406	4.4133	4.0554	28
29	6.4839	6.0890	5.7396	5.4280	5.1486	4.8965	4.4597	4.0945	29
30	6.5773	6.1714	5.8127	5.4933	5.2073	4.9496	4.5037	4.1316	30
31	6.6668	6.2501	5.8824	5.5556	5.2632	5.0001	4.5455	4.1667	31
32	6.7525	6.3254	5.9491	5.6150	5.3165	5.0481	4.5852	4.2000	32
33	6.8347	6.3975	6.0128	5.6718	5.3673	5.0939	4.6229	4.2317	33
34	6.9135	6.4665	6.0737	5.7259	5.4158	5.1376	4.6589	4.2618	34
35	6.9892	6.5326	6.1320	5.7777	5.4621	5.1792	4.6931	4.2904	35
36	7.0618	6.5960	6.1878	5.8272	5.5064	5.2190	4.7257	4.3176	36
37	7.1315	6.6568	6.2413	5.8746	5.5487	5.2570	4.7568	4.3436	37
38	7.1985	6.7151	6.2925	5.9200	5.5891	5.2933	4.7866	4.3684	38
39	7.2628	6.7710	6.3416	5.9635	5.6279	5.3280	4.8149	4.3920	39
40	7.3247	6.8248	6.3888	6.0051	5.6649	5.3612	4.8420	4.4145	40
41	7.3842	6.8764	6.4340	6.0450	5.7004	5.3930	4.8680	4.4361	41
42	7.4414	6.9260	6.4774	6.0833	5.7345	5.4235	4.8928	4.4567	42
43	7.4965	6.9737	6.5191	6.1201	5.7671	5.4527	4.9165	4.4764	43
44	7.5495	7.0195	6.5591	6.1554	5.7985	5.4807	4.9393	4.4952	44
45	7.6005	7.0636	6.5976	6.1893	5.8285	5.5075	4.9610	4.5132	45
46	7.6496	7.1061	6.6346	6.2218	5.8574	5.5333	4.9819	4.5305	46
47	7.6970	7.1469	6.6702	6.2531	5.8851	5.5580	5.0020	4.5471	47
48	7.7426	7.1862	6.7044	6.2832	5.9117	5.5817	5.0212	4.5630	48
49	7.7866	7.2241	6.7374	6.3121	5.9373	5.6046	5.0397	4.5782	49
50	7.8290	7.2605	6.7691	6.3399	5.9619	5.6265	5.0574	4.5928	50

YEARS' PURCHASE

				Rate Per Cent					
Yrs.	11	12	13	14	15	16	18	20	Yrs.
51	7.8698	7.2957	6.7996	6.3667	5.9856	5.6476	5.0744	4.6069	51
52	7.9092	7.3295	6.8290	6.3925	6.0084	5.6678	5.0908	4.6203	52
53	7.9473	7.3622	6.8573	6.4173	6.0303	5.6873	5.1065	4.6333	53
54	7.9840	7.3937	6.8846	6.4412	6.0514	5.7061	5.1216	4.6457	54
55	8.0194	7.4240	6.9109	6.4642	6.0717	5.7242	5.1362	4.6577	55
56	8.0535	7.4533	6.9363	6.4864	6.0913	5.7416	5.1502	4.6692	56
57	8.0865	7.4815	6.9608	6.5078	6.1101	5.7583	5.1636	4.6803	57
58	8.1184	7.5088	6.9844	6.5284	6.1283	5.7744	5.1766	4.6909	58
59	8.1492	7.5351	7.0071	6.5483	6.1458	5.7900	5.1891	4.7012	59
60	8.1789	7.5605	7.0291	6.5675	6.1627	5.8050	5.2011	4.7111	60
61	8.2076	7.5851	7.0503	6.5860	6.1790	5.8194	5.2127	4.7206	61
62	8.2354	7.6087	7.0708	6.6038	6.1947	5.8334	5.2239	4.7297	62
63	8.2622	7.6316	7.0905	6.6210	6.2099	5.8468	5.2347	4.7386	63
64	8.2881	7.6537	7.1096	6.6377	6.2245	5.8598	5.2451	4.7471	64
65	8.3131	7.6751	7.1280	6.6537	6.2386	5.8723	5.2551	4.7553	65
66	8.3373	7.6957	7.1458	6.6692	6.2523	5.8844	5.2648	4.7632	66
67	8.3608	7.7157	7.1630	6.6842	6.2654	5.8960	5.2741	4.7708	67
68	8.3834	7.7350	7.1796	6.6987	6.2781	5.9073	5.2831	4.7782	68
69	8.4053	7.7536	7.1957	6.7127	6.2904	5.9181	5.2918	4.7853	69
70	8.4265	7.7716	7.2112	6.7262	6.3023	5.9286	5.3002	4.7922	70
71	8.4470	7.7891	7.2262	6.7392	6.3137	5.9388	5.3083	4.7988	71
72	8.4668	7.8059	7.2407	6.7518	6.3248	5.9486	5.3161	4.8052	72
73	8.4860	7.8222	7.2547	6.7640	6.3355	5.9580	5.3236	4.8114	73
74	8.5046	7.8380	7.2683	6.7758	6.3458	5.9672	5.3310	4.8173	74
75	8.5225	7.8532	7.2814	6.7872	6.3558	5.9760	5.3380	4.8231	75
76	8.5399	7.8680	7.2941	6.7982	6.3655	5.9845	5.3448	4.8287	76
77	8.5568	7.8823	7.3064	6.8089	6.3748	5.9928	5.3514	4.8340	77
78	8.5731	7.8961	7.3183	6.8192	6.3839	6.0008	5.3578	4.8392	78
79	8.5888	7.9095	7.3298	6.8292	6.3926	6.0085	5.3639	4.8443	79
80	8.6041	7.9225	7.3409	6.8388	6.4011	6.0160	5.3699	4.8491	80
81	8.6189	7.9350	7.3516	6.8482	6.4093	6.0232	5.3756	4.8538	81
82	8.6332	7.9471	7.3621	6.8572	6.4172	6.0302	5.3812	4.8583	82
83	8.6471	7.9589	7.3721	6.8660	6.4248	6.0370	5.3866	4.8627	83
84	8.6605	7.9703	7.3819	6.8744	6.4323	6.0435	5.3918	4.8670	84
85	8.6736	7.9813	7.3914	6.8826	6.4394	6.0499	5.3969	4.8711	85
86	8.6862	7.9920	7.4005	6.8906	6.4464	6.0560	5.4017	4.8751	86
87	8.6984	8.0023	7.4094	6.8983	6.4531	6.0619	5.4065	4.8789	87
88	8.7102	8.0123	7.4180	6.9057	6.4596	6.0677	5.4110	4.8826	88
89	8.7217	8.0220	7.4263	6.9129	6.4659	6.0732	5.4154	4.8862	89
90	8.7328	8.0314	7.4343	6.9199	6.4720	6.0786	5.4197	4.8897	90
91	8.7435	8.0405	7.4421	6.9266	6.4779	6.0838	5.4239	4.8931	91
92	8.7540	8.0493	7.4497	6.9332	6.4837	6.0889	5.4279	4.8963	92
93	8.7641	8.0579	7.4570	6.9395	6.4892	6.0938	5.4318	4.8995	93
94	8.7739	8.0662	7.4641	6.9457	6.4946	6.0985	5.4355	4.9026	94
95	8.7834	8.0742	7.4710	6.9516	6.4998	6.1031	5.4392	4.9055	95
96	8.7926	8.0820	7.4776	6.9574	6.5048	6.1075	5.4427	4.9084	96
97	8.8015	8.0895	7.4841	6.9630	6.5097	6.1118	5.4461	4.9112	97
98	8.8101	8.0968	7.4903	6.9684	6.5144	6.1160	5.4494	4.9139	98
99	8.8185	8.1039	7.4964	6.9736	6.5190	6.1200	5.4526	4.9165	99
100	8.8266	8.1107	7.5023	6.9787	6.5234	6.1239	5.4557	4.9190	100

YEARS' PURCHASE

Yrs.	4	4.5	5	5.5	6	6.25	6.5	6.75	Yrs.
				Rate Per Cent					
1	0.4902	0.4890	0.4878	0.4866	0.4854	0.4848	0.4843	0.4837	1
2	0.9800	0.9752	0.9705	0.9658	0.9612	0.9589	0.9566	0.9543	2
3	1.4691	1.4584	1.4478	1.4374	1.4272	1.4221	1.4170	1.4120	3
4	1.9570	1.9381	1.9195	1.9012	1.8833	1.8745	1.8657	1.8571	4
5	2.4435	2.4140	2.3852	2.3571	2.3296	2.3161	2.3028	2.2896	5
6	2.9281	2.8858	2.8448	2.8049	2.7661	2.7471	2.7283	2.7099	6
7	3.4104	3.3532	3.2979	3.2444	3.1927	3.1674	3.1425	3.1180	7
8	3.8902	3.8160	3.7445	3.6757	3.6094	3.5771	3.5454	3.5143	8
9	4.3671	4.2738	4.1843	4.0986	4.0163	3.9764	3.9372	3.8988	9
10	4.8407	4.7263	4.6172	4.5130	4.4134	4.3653	4.3181	4.2720	10
11	5.3107	5.1734	5.0429	4.9189	4.8008	4.7439	4.6883	4.6340	11
12	5.7769	5.6147	5.4614	5.3162	5.1785	5.1124	5.0478	4.9849	12
13	6.2388	6.0501	5.8724	5.7049	5.5467	5.4708	5.3970	5.3252	13
14	6.6962	6.4793	6.2760	6.0850	5.9054	5.8194	5.7360	5.6549	14
15	7.1489	6.9022	6.6719	6.4565	6.2546	6.1583	6.0649	5.9744	15
16	7.5965	7.3185	7.0602	6.8194	6.5946	6.4876	6.3841	6.2838	16
17	8.0388	7.7282	7.4406	7.1738	6.9253	6.8075	6.6936	6.5834	17
18	8.4755	8.1310	7.8133	7.5196	7.2471	7.1181	6.9937	6.8735	18
19	8.9065	8.5268	8.1781	7.8569	7.5599	7.4197	7.2845	7.1542	19
20	9.3315	8.9156	8.5351	8.1858	7.8639	7.7123	7.5664	7.4259	20
21	9.7504	9.2971	8.8841	8.5063	8.1593	7.9961	7.8394	7.6887	21
22	10.1628	9.6714	9.2253	8.8185	8.4461	8.2715	8.1039	7.9430	22
23	10.5688	10.0383	9.5586	9.1226	8.7246	8.5384	8.3599	8.1888	23
24	10.9681	10.3979	9.8840	9.4185	8.9949	8.7971	8.6078	8.4265	24
25	11.3605	10.7499	10.2016	9.7065	9.2572	9.0478	8.8477	8.6562	25
26	11.7461	11.0945	10.5114	9.9865	9.5116	9.2907	9.0798	8.8783	26
27	12.1246	11.4316	10.8135	10.2588	9.7583	9.5259	9.3043	9.0928	27
28	12.4959	11.7611	11.1079	10.5234	9.9974	9.7536	9.5215	9.3001	28
29	12.8601	12.0832	11.3947	10.7805	10.2291	9.9741	9.7314	9.5003	29
30	13.2170	12.3977	11.6740	11.0302	10.4537	10.1874	9.9344	9.6937	30
31	13.5666	12.7048	11.9459	11.2726	10.6711	10.3939	10.1306	9.8804	31
32	13.9088	13.0044	12.2104	11.5079	10.8817	10.5935	10.3202	10.0606	32
33	14.2436	13.2966	12.4677	11.7361	11.0856	10.7867	10.5034	10.2347	33
34	14.5710	13.5815	12.7179	11.9575	11.2829	10.9734	10.6804	10.4026	34
35	14.8910	13.8591	12.9610	12.1722	11.4739	11.1539	10.8513	10.5647	35
36	15.2036	14.1295	13.1972	12.3803	11.6586	11.3284	11.0164	10.7211	36
37	15.5089	14.3928	13.4266	12.5819	11.8372	11.4970	11.1758	10.8720	37
38	15.8068	14.6490	13.6493	12.7773	12.0100	11.6599	11.3297	11.0176	38
39	16.0974	14.8983	13.8655	12.9665	12.1771	11.8173	11.4782	11.1580	39
40	16.3808	15.1407	14.0752	13.1498	12.3385	11.9693	11.6216	11.2934	40
41	16.6570	15.3764	14.2786	13.3272	12.4946	12.1161	11.7599	11.4240	41
42	16.9261	15.6054	14.4759	13.4988	12.6454	12.2578	11.8934	11.5500	42
43	17.1881	15.8279	14.6671	13.6650	12.7910	12.3947	12.0222	11.6714	43
44	17.4432	16.0439	14.8524	13.8257	12.9317	12.5268	12.1464	11.7884	44
45	17.6913	16.2536	15.0320	13.9811	13.0676	12.6542	12.2662	11.9012	45
46	17.9327	16.4571	15.2059	14.1315	13.1989	12.7773	12.3817	12.0100	46
47	18.1674	16.6545	15.3743	14.2768	13.3256	12.8959	12.4932	12.1148	47
48	18.3954	16.8460	15.5373	14.4173	13.4479	13.0104	12.6006	12.2158	48
49	18.6170	17.0316	15.6951	14.5530	13.5659	13.1209	12.7042	12.3131	49
50	18.8322	17.2116	15.8478	14.6842	13.6798	13.2274	12.8040	12.4069	50

YEARS' PURCHASE

Yrs.	4	4.5	5	5.5	6	6.25	6.5	6.75	Yrs.
					Rate Per Cent				
51	19.0412	17.3860	15.9955	14.8109	13.7897	13.3302	12.9003	12.4972	51
52	19.2440	17.5549	16.1383	14.9333	13.8958	13.4293	12.9930	12.5843	52
53	19.4408	17.7185	16.2765	15.0516	13.9981	13.5248	13.0824	12.6681	53
54	19.6316	17.8769	16.4101	15.1657	14.0968	13.6169	13.1686	12.7489	54
55	19.8167	18.0302	16.5392	15.2759	14.1919	13.7057	13.2516	12.8267	55
56	19.9961	18.1786	16.6640	15.3823	14.2837	13.7913	13.3316	12.9016	56
57	20.1700	18.3222	16.7845	15.4850	14.3722	13.8737	13.4087	12.9738	57
58	20.3384	18.4611	16.9010	15.5841	14.4575	13.9532	13.4829	13.0432	58
59	20.5016	18.5954	17.0135	15.6797	14.5398	14.0298	13.5544	13.1102	59
60	20.6596	18.7253	17.1222	15.7720	14.6191	14.1036	13.6233	13.1746	60
61	20.8125	18.8509	17.2271	15.8609	14.6955	14.1748	13.6896	13.2366	61
62	20.9606	18.9722	17.3284	15.9468	14.7692	14.2433	13.7535	13.2964	62
63	21.1038	19.0895	17.4262	16.0296	14.8402	14.3093	13.8151	13.3539	63
64	21.2424	19.2029	17.5206	16.1094	14.9086	14.3729	13.8743	13.4092	64
65	21.3765	19.3123	17.6117	16.1864	14.9745	14.4341	13.9314	13.4625	65
66	21.5061	19.4181	17.6996	16.2606	15.0380	14.4931	13.9863	13.5138	66
67	21.6314	19.5202	17.7844	16.3321	15.0991	14.5499	14.0392	13.5632	67
68	21.7526	19.6188	17.8662	16.4011	15.1580	14.6046	14.0902	13.6107	68
69	21.8697	19.7140	17.9451	16.4676	15.2148	14.6573	14.1392	13.6565	69
70	21.9828	19.8059	18.0212	16.5316	15.2695	14.7080	14.1864	13.7005	70
71	22.0921	19.8945	18.0946	16.5934	15.3221	14.7569	14.2318	13.7429	71
72	22.1977	19.9801	18.1654	16.6528	15.3728	14.8039	14.2756	13.7836	72
73	22.2996	20.0627	18.2336	16.7102	15.4217	14.8492	14.3177	13.8229	73
74	22.3981	20.1423	18.2994	16.7654	15.4687	14.8928	14.3582	13.8607	74
75	22.4931	20.2192	18.3628	16.8186	15.5140	14.9347	14.3972	13.8970	75
76	22.5849	20.2933	18.4239	16.8698	15.5576	14.9751	14.4347	13.9320	76
77	22.6735	20.3648	18.4828	16.9192	15.5995	15.0140	14.4709	13.9656	77
78	22.7589	20.4337	18.5395	16.9668	15.6400	15.0514	14.5056	13.9980	78
79	22.8414	20.5001	18.5942	17.0126	15.6789	15.0875	14.5391	14.0292	79
80	22.9210	20.5642	18.6469	17.0567	15.7163	15.1222	14.5713	14.0591	80
81	22.9977	20.6260	18.6977	17.0991	15.7524	15.1555	14.6023	14.0880	81
82	23.0718	20.6855	18.7466	17.1400	15.7871	15.1876	14.6321	14.1157	82
83	23.1432	20.7429	18.7937	17.1794	15.8205	15.2186	14.6608	14.1424	83
84	23.2121	20.7982	18.8391	17.2173	15.8526	15.2483	14.6884	14.1681	84
85	23.2785	20.8515	18.8828	17.2538	15.8836	15.2769	14.7149	14.1928	85
86	23.3425	20.9028	18.9249	17.2890	15.9133	15.3045	14.7405	14.2166	86
87	23.4042	20.9523	18.9655	17.3228	15.9420	15.3310	14.7651	14.2395	87
88	23.4637	21.0000	19.0045	17.3554	15.9696	15.3565	14.7887	14.2615	88
89	23.5210	21.0459	19.0421	17.3867	15.9961	15.3810	14.8115	14.2826	89
90	23.5763	21.0901	19.0783	17.4169	16.0217	15.4046	14.8334	14.3030	90
91	23.6295	21.1328	19.1132	17.4459	16.0462	15.4274	14.8544	14.3226	91
92	23.6809	21.1738	19.1467	17.4739	16.0699	15.4492	14.8747	14.3414	92
93	23.7303	21.2133	19.1791	17.5008	16.0926	15.4702	14.8942	14.3595	93
94	23.7779	21.2514	19.2102	17.5267	16.1145	15.4905	14.9130	14.3770	94
95	23.8238	21.2880	19.2401	17.5516	16.1356	15.5099	14.9310	14.3937	95
96	23.8681	21.3233	19.2689	17.5756	16.1559	15.5287	14.9484	14.4098	96
97	23.9106	21.3573	19.2967	17.5987	16.1754	15.5467	14.9650	14.4254	97
98	23.9517	21.3900	19.3234	17.6209	16.1941	15.5640	14.9811	14.4403	98
99	23.9912	21.4215	19.3491	17.6423	16.2122	15.5807	14.9966	14.4546	99
100	24.0292	21.4519	19.3738	17.6629	16.2296	15.5967	15.0114	14.4684	100

YEARS' PURCHASE

				Rate Per Cent					
Yrs.	7	7.25	7.5	8	8.5	9	9.5	10	Yrs.
1	0.4831	0.4825	0.4819	0.4808	0.4796	0.4785	0.4773	0.4762	1
2	0.9520	0.9498	0.9475	0.9430	0.9386	0.9342	0.9299	0.9256	2
3	1.4071	1.4021	1.3972	1.3875	1.3780	1.3686	1.3593	1.3501	3
4	1.8485	1.8400	1.8316	1.8149	1.7986	1.7826	1.7668	1.7514	4
5	2.2766	2.2637	2.2510	2.2259	2.2014	2.1774	2.1540	2.1310	5
6	2.6916	2.6736	2.6559	2.6211	2.5872	2.5541	2.5219	2.4905	6
7	3.0939	3.0701	3.0467	3.0010	2.9567	2.9136	2.8718	2.8311	7
8	3.4836	3.4536	3.4240	3.3664	3.3106	3.2567	3.2046	3.1540	8
9	3.8612	3.8243	3.7881	3.7177	3.6498	3.5844	3.5213	3.4604	9
10	4.2269	4.1827	4.1394	4.0554	3.9748	3.8974	3.8229	3.7512	10
11	4.5809	4.5290	4.4783	4.3802	4.2864	4.1964	4.1102	4.0274	11
12	4.9236	4.8637	4.8053	4.6925	4.5850	4.4822	4.3840	4.2899	12
13	5.2552	5.1871	5.1207	4.9928	4.8712	4.7554	4.6450	4.5395	13
14	5.5761	5.4994	5.4248	5.2816	5.1457	5.0166	4.8939	4.7770	14
15	5.8864	5.8011	5.7181	5.5592	5.4088	5.2664	5.1313	5.0029	15
16	6.1866	6.0924	6.0010	5.8261	5.6612	5.5054	5.3579	5.2181	16
17	6.4768	6.3736	6.2736	6.0828	5.9033	5.7340	5.5742	5.4231	17
18	6.7574	6.6451	6.5365	6.3296	6.1355	5.9529	5.7808	5.6184	18
19	7.0285	6.9072	6.7899	6.5670	6.3582	6.1623	5.9781	5.8046	19
20	7.2906	7.1601	7.0342	6.7952	6.5719	6.3628	6.1666	5.9822	20
21	7.5437	7.4041	7.2695	7.0146	6.7769	6.5548	6.3468	6.1516	21
22	7.7883	7.6396	7.4964	7.2256	6.9736	6.7387	6.5190	6.3132	22
23	8.0245	7.8667	7.7150	7.4284	7.1624	6.9148	6.6837	6.4675	23
24	8.2526	8.0858	7.9256	7.6235	7.3436	7.0835	6.8412	6.6149	24
25	8.4729	8.2971	8.1285	7.8110	7.5174	7.2451	6.9918	6.7557	25
26	8.6855	8.5009	8.3240	7.9914	7.6843	7.4000	7.1360	6.8901	26
27	8.8907	8.6974	8.5123	8.1648	7.8445	7.5485	7.2739	7.0187	27
28	9.0888	8.8868	8.6937	8.3315	7.9983	7.6908	7.4060	7.1415	28
29	9.2799	9.0695	8.8684	8.4919	8.1460	7.8272	7.5324	7.2590	29
30	9.4643	9.2455	9.0367	8.6460	8.2877	7.9580	7.6534	7.3714	30
31	9.6422	9.4152	9.1987	8.7942	8.4238	8.0834	7.7694	7.4788	31
32	9.8138	9.5788	9.3548	8.9368	8.5545	8.2036	7.8804	7.5817	32
33	9.9793	9.7364	9.5051	9.0738	8.6800	8.3190	7.9868	7.6801	33
34	10.1389	9.8883	9.6498	9.2056	8.8005	8.4296	8.0887	7.7743	34
35	10.2929	10.0347	9.7891	9.3323	8.9163	8.5357	8.1863	7.8644	35
36	10.4413	10.1757	9.9232	9.4541	9.0274	8.6375	8.2799	7.9508	36
37	10.5843	10.3115	10.0524	9.5713	9.1342	8.7352	8.3697	8.0335	37
38	10.7223	10.4424	10.1767	9.6839	9.2367	8.8289	8.4557	8.1127	38
39	10.8552	10.5684	10.2964	9.7922	9.3352	8.9189	8.5381	8.1886	39
40	10.9833	10.6898	10.4116	9.8964	9.4298	9.0052	8.6172	8.2613	40
41	11.1068	10.8068	10.5225	9.9965	9.5207	9.0880	8.6930	8.3309	41
42	11.2258	10.9194	10.6292	10.0928	9.6080	9.1675	8.7657	8.3977	42
43	11.3405	11.0278	10.7319	10.1854	9.6918	9.2439	8.8355	8.4617	43
44	11.4509	11.1323	10.8308	10.2744	9.7724	9.3171	8.9024	8.5230	44
45	11.5574	11.2328	10.9260	10.3600	9.8498	9.3875	8.9666	8.5819	45
46	11.6599	11.3296	11.0176	10.4423	9.9242	9.4550	9.0282	8.6383	46
47	11.7587	11.4229	11.1057	10.5215	9.9956	9.5198	9.0873	8.6923	47
48	11.8538	11.5126	11.1905	10.5976	10.0643	9.5821	9.1440	8.7442	48
49	11.9454	11.5990	11.2721	10.6707	10.1302	9.6419	9.1984	8.7940	49
50	12.0336	11.6822	11.3507	10.7411	10.1936	9.6993	9.2507	8.8417	50

YEARS' PURCHASE

				Rate Per Cent					
Yrs.	7	7.25	7.5	8	8.5	9	9.5	10	Yrs.
51	12.1186	11.7623	11.4263	10.8087	10.2545	9.7544	9.3008	8.8875	51
52	12.2004	11.8393	11.4990	10.8738	10.3131	9.8074	9.3489	8.9314	52
53	12.2792	11.9135	11.5689	10.9363	10.3693	9.8582	9.3951	8.9736	53
54	12.3551	11.9849	11.6363	10.9965	10.4234	9.9071	9.4395	9.0140	54
55	12.4282	12.0536	11.7010	11.0543	10.4753	9.9540	9.4820	9.0528	55
56	12.4985	12.1198	11.7634	11.1099	10.5252	9.9990	9.5229	9.0901	56
57	12.5662	12.1834	11.8233	11.1634	10.5732	10.0423	9.5622	9.1259	57
58	12.6314	12.2447	11.8810	11.2148	10.6193	10.0839	9.5999	9.1602	58
59	12.6941	12.3036	11.9365	11.2642	10.6636	10.1238	9.6361	9.1931	59
60	12.7545	12.3604	11.9899	11.3117	10.7062	10.1622	9.6708	9.2248	60
61	12.8126	12.4150	12.0412	11.3574	10.7471	10.1991	9.7042	9.2551	61
62	12.8686	12.4675	12.0906	11.4014	10.7865	10.2345	9.7363	9.2843	62
63	12.9224	12.5180	12.1382	11.4436	10.8243	10.2685	9.7671	9.3123	63
64	12.9743	12.5667	12.1839	11.4843	10.8606	10.3013	9.7967	9.3392	64
65	13.0242	12.6135	12.2279	11.5233	10.8956	10.3327	9.8251	9.3650	65
66	13.0722	12.6585	12.2702	11.5609	10.9292	10.3629	9.8524	9.3898	66
67	13.1184	12.7018	12.3109	11.5970	10.9614	10.3919	9.8786	9.4136	67
68	13.1628	12.7435	12.3500	11.6318	10.9924	10.4198	9.9038	9.4365	68
69	13.2056	12.7836	12.3877	11.6651	11.0223	10.4465	9.9280	9.4585	69
70	13.2468	12.8221	12.4239	11.6973	11.0509	10.4723	9.9512	9.4796	70
71	13.2864	12.8592	12.4587	11.7281	11.0785	10.4970	9.9736	9.4998	71
72	13.3245	12.8949	12.4922	11.7578	11.1050	10.5208	9.9950	9.5193	72
73	13.3612	12.9293	12.5245	11.7864	11.1304	10.5437	10.0156	9.5380	73
74	13.3964	12.9623	12.5555	11.8138	11.1549	10.5656	10.0355	9.5560	74
75	13.4304	12.9941	12.5853	11.8402	11.1784	10.5867	10.0545	9.5732	75
76	13.4631	13.0247	12.6139	11.8656	11.2010	10.6070	10.0728	9.5898	76
77	13.4945	13.0541	12.6415	11.8900	11.2228	10.6265	10.0904	9.6057	77
78	13.5247	13.0824	12.6680	11.9134	11.2437	10.6452	10.1073	9.6210	78
79	13.5538	13.1096	12.6936	11.9360	11.2638	10.6632	10.1235	9.6358	79
80	13.5818	13.1357	12.7181	11.9577	11.2831	10.6805	10.1391	9.6499	80
81	13.6087	13.1609	12.7417	11.9786	11.3017	10.6972	10.1541	9.6635	81
82	13.6346	13.1851	12.7644	11.9986	11.3195	10.7132	10.1685	9.6765	82
83	13.6595	13.2084	12.7862	12.0179	11.3367	10.7285	10.1823	9.6891	83
84	13.6834	13.2308	12.8072	12.0364	11.3532	10.7433	10.1956	9.7011	84
85	13.7065	13.2524	12.8274	12.0543	11.3690	10.7575	10.2084	9.7127	85
86	13.7287	13.2731	12.8468	12.0714	11.3843	10.7712	10.2207	9.7238	86
87	13.7500	13.2930	12.8655	12.0879	11.3989	10.7843	10.2325	9.7345	87
88	13.7705	13.3122	12.8834	12.1037	11.4130	10.7969	10.2439	9.7448	88
89	13.7902	13.3306	12.9007	12.1190	11.4266	10.8090	10.2548	9.7547	89
90	13.8092	13.3484	12.9173	12.1336	11.4396	10.8207	10.2653	9.7641	90
91	13.8275	13.3654	12.9333	12.1477	11.4521	10.8319	10.2754	9.7733	91
92	13.8450	13.3818	12.9486	12.1613	11.4642	10.8427	10.2851	9.7820	92
93	13.8619	13.3976	12.9634	12.1743	11.4758	10.8530	10.2944	9.7905	93
94	13.8781	13.4128	12.9776	12.1868	11.4869	10.8630	10.3034	9.7986	94
95	13.8938	13.4274	12.9913	12.1989	11.4976	10.8725	10.3120	9.8063	95
96	13.9088	13.4414	13.0044	12.2105	11.5079	10.8817	10.3202	9.8138	96
97	13.9232	13.4549	13.0170	12.2216	11.5178	10.8906	10.3282	9.8210	97
98	13.9371	13.4679	13.0292	12.2323	11.5273	10.8991	10.3358	9.8279	98
99	13.9505	13.4804	13.0409	12.2426	11.5364	10.9073	10.3432	9.8346	99
100	13.9634	13.4924	13.0521	12.2525	11.5452	10.9151	10.3503	9.8410	100

YEARS' PURCHASE

					Rate Per Cent				
Yrs.	11	12	13	14	15	16	18	20	Yrs.
1	0.4739	0.4717	0.4695	0.4673	0.4651	0.4630	0.4587	0.4545	1
2	0.9171	0.9088	0.9006	0.8925	0.8846	0.8769	0.8618	0.8472	2
3	1.3321	1.3146	1.2975	1.2809	1.2647	1.2489	1.2185	1.1895	3
4	1.7212	1.6921	1.6639	1.6367	1.6104	1.5848	1.5361	1.4904	4
5	2.0866	2.0439	2.0030	1.9637	1.9258	1.8895	1.8207	1.7567	5
6	2.4300	2.3723	2.3174	2.2649	2.2147	2.1667	2.0767	1.9939	6
7	2.7532	2.6794	2.6095	2.5431	2.4800	2.4200	2.3083	2.2064	7
8	3.0576	2.9669	2.8814	2.8007	2.7244	2.6521	2.5185	2.3978	8
9	3.3446	3.2364	3.1349	3.0396	2.9500	2.8654	2.7101	2.5708	9
10	3.6156	3.4894	3.3718	3.2618	3.1587	3.0620	2.8853	2.7279	10
11	3.8715	3.7272	3.5933	3.4686	3.3524	3.2436	3.0460	2.8711	11
12	4.1135	3.9509	3.8008	3.6616	3.5323	3.4118	3.1938	3.0021	12
13	4.3424	4.1617	3.9954	3.8419	3.6998	3.5678	3.3301	3.1222	13
14	4.5592	4.3604	4.1782	4.0106	3.8560	3.7128	3.4562	3.2327	14
15	4.7646	4.5479	4.3501	4.1687	4.0019	3.8479	3.5729	3.3346	15
16	4.9593	4.7250	4.5118	4.3170	4.1384	3.9739	3.6813	3.4289	16
17	5.1441	4.8924	4.6642	4.4564	4.2663	4.0917	3.7822	3.5162	17
18	5.3195	5.0508	4.8080	4.5874	4.3862	4.2019	3.8762	3.5973	18
19	5.4861	5.2008	4.9437	4.7108	4.4989	4.3052	3.9639	3.6727	19
20	5.6445	5.3429	5.0719	4.8271	4.6048	4.4021	4.0459	3.7430	20
21	5.7951	5.4776	5.1932	4.9368	4.7045	4.4932	4.1227	3.8087	21
22	5.9383	5.6055	5.3079	5.0404	4.7985	4.5788	4.1947	3.8700	22
23	6.0747	5.7268	5.4166	5.1383	4.8872	4.6594	4.2622	3.9275	23
24	6.2045	5.8420	5.5196	5.2308	4.9708	4.7354	4.3258	3.9813	24
25	6.3282	5.9515	5.6172	5.3185	5.0499	4.8071	4.3855	4.0319	25
26	6.4460	6.0557	5.7099	5.4015	5.1247	4.8748	4.4418	4.0794	26
27	6.5584	6.1547	5.7979	5.4801	5.1954	4.9388	4.4948	4.1241	27
28	6.6655	6.2490	5.8815	5.5548	5.2624	4.9994	4.5449	4.1662	28
29	6.7677	6.3387	5.9609	5.6256	5.3260	5.0566	4.5922	4.2059	29
30	6.8653	6.4242	6.0365	5.6928	5.3862	5.1109	4.6369	4.2434	30
31	6.9584	6.5057	6.1083	5.7567	5.4433	5.1623	4.6792	4.2788	31
32	7.0474	6.5834	6.1768	5.8174	5.4976	5.2111	4.7193	4.3123	32
33	7.1323	6.6575	6.2419	5.8752	5.5492	5.2574	4.7572	4.3439	33
34	7.2135	6.7281	6.3040	5.9302	5.5982	5.3014	4.7932	4.3739	34
35	7.2910	6.7956	6.3632	5.9825	5.6448	5.3432	4.8273	4.4023	35
36	7.3652	6.8599	6.4196	6.0323	5.6891	5.3829	4.8597	4.4292	36
37	7.4361	6.9214	6.4734	6.0798	5.7313	5.4207	4.8905	4.4548	37
38	7.5039	6.9801	6.5247	6.1251	5.7715	5.4566	4.9197	4.4790	38
39	7.5688	7.0362	6.5737	6.1682	5.8098	5.4908	4.9475	4.5020	39
40	7.6309	7.0898	6.6205	6.2094	5.8463	5.5234	4.9740	4.5239	40
41	7.6903	7.1411	6.6651	6.2486	5.8812	5.5545	4.9991	4.5447	41
42	7.7471	7.1901	6.7078	6.2861	5.9143	5.5841	5.0231	4.5645	42
43	7.8015	7.2369	6.7486	6.3219	5.9460	5.6123	5.0459	4.5834	43
44	7.8537	7.2818	6.7875	6.3561	5.9762	5.6392	5.0677	4.6013	44
45	7.9036	7.3247	6.8248	6.3888	6.0051	5.6649	5.0884	4.6184	45
46	7.9514	7.3657	6.8604	6.4200	6.0327	5.6894	5.1082	4.6347	46
47	7.9972	7.4050	6.8945	6.4498	6.0590	5.7129	5.1271	4.6502	47
48	8.0411	7.4426	6.9271	6.4783	6.0842	5.7352	5.1451	4.6650	48
49	8.0831	7.4786	6.9583	6.5056	6.1082	5.7566	5.1622	4.6791	49
50	8.1235	7.5131	6.9881	6.5317	6.1312	5.7770	5.1787	4.6926	50

YEARS' PURCHASE

				Rate Per Cent					
Yrs.	11	12	13	14	15	16	18	20	Yrs.
51	8.1621	7.5462	7.0167	6.5566	6.1532	5.7965	5.1943	4.7055	51
52	8.1991	7.5778	7.0440	6.5805	6.1742	5.8152	5.2093	4.7178	52
53	8.2346	7.6081	7.0702	6.6033	6.1943	5.8330	5.2236	4.7295	53
54	8.2687	7.6372	7.0953	6.6252	6.2136	5.8501	5.2373	4.7407	54
55	8.3013	7.6650	7.1193	6.6462	6.2320	5.8664	5.2504	4.7514	55
56	8.3327	7.6917	7.1424	6.6662	6.2496	5.8820	5.2629	4.7617	56
57	8.3627	7.7173	7.1644	6.6854	6.2665	5.8970	5.2749	4.7715	57
58	8.3915	7.7419	7.1856	6.7038	6.2827	5.9113	5.2863	4.7808	58
59	8.4192	7.7654	7.2058	6.7215	6.2982	5.9250	5.2973	4.7898	59
60	8.4457	7.7879	7.2252	6.7384	6.3130	5.9381	5.3078	4.7984	60
61	8.4711	7.8096	7.2439	6.7546	6.3272	5.9507	5.3178	4.8066	61
62	8.4956	7.8303	7.2617	6.7701	6.3408	5.9627	5.3274	4.8144	62
63	8.5190	7.8502	7.2788	6.7850	6.3539	5.9743	5.3366	4.8220	63
64	8.5415	7.8693	7.2952	6.7992	6.3664	5.9853	5.3454	4.8292	64
65	8.5631	7.8877	7.3110	6.8129	6.3783	5.9959	5.3539	4.8360	65
66	8.5838	7.9052	7.3261	6.8260	6.3898	6.0061	5.3620	4.8427	66
67	8.6037	7.9221	7.3406	6.8386	6.4009	6.0158	5.3697	4.8490	67
68	8.6228	7.9383	7.3545	6.8506	6.4114	6.0251	5.3772	4.8550	68
69	8.6411	7.9538	7.3678	6.8622	6.4216	6.0341	5.3843	4.8608	69
70	8.6587	7.9688	7.3806	6.8733	6.4313	6.0427	5.3911	4.8664	70
71	8.6757	7.9831	7.3929	6.8840	6.4406	6.0509	5.3977	4.8717	71
72	8.6919	7.9968	7.4047	6.8942	6.4495	6.0588	5.4040	4.8769	72
73	8.7075	8.0100	7.4160	6.9040	6.4581	6.0663	5.4100	4.8818	73
74	8.7225	8.0227	7.4268	6.9134	6.4664	6.0736	5.4157	4.8865	74
75	8.7368	8.0348	7.4373	6.9224	6.4743	6.0806	5.4213	4.8910	75
76	8.7506	8.0465	7.4473	6.9311	6.4818	6.0873	5.4266	4.8953	76
77	8.7639	8.0577	7.4569	6.9394	6.4891	6.0937	5.4317	4.8995	77
78	8.7766	8.0685	7.4661	6.9474	6.4961	6.0998	5.4366	4.9034	78
79	8.7889	8.0788	7.4750	6.9551	6.5028	6.1057	5.4413	4.9072	79
80	8.8006	8.0888	7.4835	6.9624	6.5092	6.1114	5.4458	4.9109	80
81	8.8119	8.0983	7.4916	6.9695	6.5154	6.1169	5.4501	4.9144	81
82	8.8228	8.1075	7.4995	6.9763	6.5213	6.1221	5.4543	4.9178	82
83	8.8332	8.1163	7.5070	6.9828	6.5270	6.1271	5.4582	4.9210	83
84	8.8432	8.1247	7.5142	6.9890	6.5325	6.1319	5.4621	4.9241	84
85	8.8528	8.1328	7.5212	6.9951	6.5377	6.1365	5.4657	4.9271	85
86	8.8621	8.1406	7.5278	7.0008	6.5428	6.1410	5.4693	4.9300	86
87	8.8710	8.1481	7.5342	7.0064	6.5476	6.1452	5.4726	4.9327	87
88	8.8795	8.1553	7.5404	7.0117	6.5523	6.1493	5.4759	4.9354	88
89	8.8877	8.1623	7.5463	7.0168	6.5567	6.1533	5.4790	4.9379	89
90	8.8956	8.1689	7.5520	7.0217	6.5610	6.1570	5.4820	4.9403	90
91	8.9031	8.1753	7.5574	7.0264	6.5651	6.1607	5.4849	4.9427	91
92	8.9104	8.1814	7.5627	7.0310	6.5691	6.1642	5.4876	4.9449	92
93	8.9174	8.1873	7.5677	7.0353	6.5729	6.1675	5.4903	4.9471	93
94	8.9241	8.1930	7.5726	7.0395	6.5765	6.1707	5.4928	4.9491	94
95	8.9306	8.1984	7.5772	7.0435	6.5800	6.1738	5.4953	4.9511	95
96	8.9368	8.2036	7.5817	7.0474	6.5834	6.1768	5.4976	4.9530	96
97	8.9428	8.2087	7.5860	7.0511	6.5866	6.1796	5.4999	4.9549	97
98	8.9485	8.2135	7.5901	7.0546	6.5898	6.1824	5.5020	4.9566	98
99	8.9540	8.2181	7.5941	7.0581	6.5927	6.1850	5.5041	4.9583	99
100	8.9593	8.2226	7.5979	7.0613	6.5956	6.1875	5.5061	4.9599	100